2

Proceedings of
the International
Congress of
Mathematicians

August 3–11, 1994
Zürich, Switzerland

Birkhäuser Verlag
Basel · Boston · Berlin

Editor:
  S. D. Chatterji
  EPFL
  Département de Mathématiques
  1015 Lausanne
  Switzerland

The logo for ICM 94 was designed by
  Georg Staehelin, Bachweg 6, 8913 Ottenbach, Switzerland.

A CIP catalogue record for this book is available from the
Library of Congress, Washington D.C., USA

Deutsche Bibliothek Cataloging-in-Publication Data

International Congress of Mathematicians <1994, Zürich>:
Proceedings of the International Congress of Mathematicians
1994 : August 3 - 11, 1994, Zürich, Switzerland / [Ed.: S. D.
Chatterji]. - Basel ; Boston ; Berlin : Birkhäuser.
  ISBN 3-7643-5153-5 (Basel…)
  ISBN 0-8176-5153-5 (Boston)
NE: S. D. Chatterji [Hrsg.]

Vol. 2 (1995)

© 1995 Birkhäuser Verlag
  P.O. Box 133
  CH-4010 Basel
  Switzerland

Printed on acid-free paper produced from chlorine-free pulp. TCF ∞
Layout, typesetting by *mathScreen online*, Allschwil

Printed in Germany
  ISBN 3-7643-5153-5
  ISBN 0-8176-5153-5
  9 8 7 6 5 4 3 2 1

Table of Contents

Volume II

Scientific Program

The Work of the Fields Medalists and the Rolf Nevanlinna Prize Winner

Invited One-Hour Addresses at the Plenary Sessions

---

*) The material presented is largely covered by a recent survey under this title by L. Babai
   in: First European Congress of Mathematics, Paris 1992 (A. Joseph et al., eds.), Vol. I,
   Progress in Math. **119**, Birkhäuser Verlag, Basel und Boston 1994, pp. 31–91.

Invited Forty-Five Minute Addresses at the Section Meetings

Section 1. Logic

Section 2. Algebra

Section 3. Number theory

Section 4. Geometry

Section 5. Topology

Section 6. Algebraic geometry

Section 7. Lie groups and representations

Section 11. Partial differential equations

Section 12. Ordinary differential equations and dynamical systems

Section 13. Mathematical physics

Section 14. Combinatorics

Section 15. Mathematical aspects of computer science

Section 16. Numerical analysis and scientific computing

Section 17. Applications of mathematics in the sciences

Section 18. Teaching and popularization of mathematics

Section 19. History of mathematics

Invited Forty-Five Minute Addresses
at the Section Meetings (Continued)

Genuine Representations of the Metaplectic Group and Epsilon Factors

JEFFREY ADAMS[*]

Mathematics Department
University of Maryland
College Park, MD 20742, USA

Introduction

Generalizing results of Waldspurger [14], and motivated in part by the conjectures of Gross-Prasad [5], Kudla has made a series of conjectures about the theta correspondence for the dual pairs $(Sp(2n), O(2n + 1))$ over a local or global field. This paper concerns these local conjectures over $\mathbb{R}$. For a discussion of the case of unitary groups over a p-adic field see [9].

The first main result is that the theta correspondence defines a bijection between the genuine irreducible representations of the metaplectic group $\widetilde{Sp}(2n, \mathbb{R})$, and the irreducible representations of the groups $SO(p, q)$ with $p + q = 2n + 1$ and the parity of q fixed (Theorem 1.4), [2]. In particular this relates the representation theory of the nonlinear group $\widetilde{Sp}(2n, \mathbb{R})$ to that of the linear orthogonal groups.

The second main result relates this lifting to symplectic root numbers, i.e. epsilon factors of symplectic representations of the Weil group of $\mathbb{R}$. A genuine irreducible representation π of $\widetilde{Sp}(2n, \mathbb{R})$ lifts to a unique $SO(p, q)$, and there is a relation between the epsilon factor of this lift, the central character of π, and the Kottwitz invariant of $SO(p, q)$ (Theorem 4.9).

Sections 1 and 2 are joint work with Dan Barbasch [2]. Thanks are due to Steve Kudla for explaining his conjectures and for many useful discussions.

1 Theta correspondence

Let W be a vector space of dimension $2n$ over $\mathbb{R}$ equipped with a nondegenerate symplectic form $<,>$, with isometry group $Sp(W)$. Let V be a vector space of dimension $2n + 1$ over $\mathbb{R}$ with nondegenerate symmetric bilinear form $(,)$ and isometry group $O(V)$. Then $\mathbb{W} = W \otimes V$ comes with the symplectic form $<,> \otimes (,)$, and with the natural embeddings $\iota_V : O(V) \to Sp(\mathbb{W})$ and $\iota_W : Sp(W) \to Sp(\mathbb{W})$, $(Sp(W), O(V))$ is a reductive dual pair in $Sp(\mathbb{W})$ [6].

Let $\widetilde{Sp}(\mathbb{W})$ be a metaplectic group, i.e. a connected two-fold cover of $Sp(\mathbb{W})$. Any two metaplectic groups are isomorphic by a unique map inducing the identity on $Sp(\mathbb{W})$. Unless otherwise noted $\tilde{G}$ will denote the inverse image in $\widetilde{Sp}(\mathbb{W})$ of

---

[*]   Partially supported by NSF Grant #DMS-9401074.

Proceedings of the International Congress
of Mathematicians, Zürich, Switzerland 1994
© Birkhäuser Verlag, Basel, Switzerland 1995

a subgroup G of $Sp(\mathbb{W})$. We say an irreducible representation of $\tilde{G}$ is genuine if it is nontrivial on the kernel of the covering map $p : \tilde{G} \to G$, i.e. if it does not factor to G. The covers of a subgroup arising from different metaplectic groups are canonically isomorphic.

Fix a nontrivial unitary additive character ψ of $\mathbb{R}$, and let $\omega(\psi)$ be the associated oscillator representation of $\widetilde{Sp}(\mathbb{W})$. Restriction of $\omega(\psi)$ to the dual pair defines a bijection between subsets of the irreducible admissible representations of $\widetilde{Sp}(W)$ and $\tilde{O}(V)$ [6, Theorem 1].

Let $\mathbf{K}$ be a maximal compact subgroup of $Sp(\mathbb{W})$, and consider the Fock space $\mathcal{F}(\psi)$ of $\tilde{\mathbf{K}}$-finite vectors in $\omega(\psi)$. For $G = Sp(W)$ or $O(V)$ let $\mathfrak{g}$ be the complexified Lie algebra, K a maximal compact subgroup, and let $\hat{\tilde{G}}_{\text{genuine}}$ be the set of equivalence classes of genuine irreducible $(\mathfrak{g}, \tilde{K})$-modules. The restriction of $\mathcal{F}(\psi)$ to the dual pair defines a bijection between subsets of $\widetilde{Sp}(W)\widehat{}_{\text{genuine}}$ and $\tilde{O}(V)\widehat{}_{\text{genuine}}$ [6, Theorem 2.1]. If $\pi \in \widetilde{Sp}(W)\widehat{}_{\text{genuine}}$ corresponds to $\pi' \in \tilde{O}(V)\widehat{}_{\text{genuine}}$ we write $\pi' = \theta(V, \psi)(\pi)$ and $\pi = \theta(W, \psi)(\pi')$. If π does not occur in the correspondence we say $\theta(V, \psi)(\pi) = 0$, and similarly π'.

We consider correspondences for fixed W as V varies. Let $\mathcal{V}$ be a set of representatives of the isomorphism classes of vector spaces of dimension $2n + 1$ together with a symmetric bilinear form. Because the form is determined up to isomorphism by its signature, $\mathcal{V} = \{V_{p,q} | p + q = 2n + 1\}$ where the signature of $V_{p,q}$ is p, q. The groups $\widetilde{Sp}(W)$ arising from different $\mathbb{W} = W \otimes V$ may be canonically identified via the unique isomorphisms over the identity map of $Sp(W)$. The discriminant $\det(V_{p,q})$ of $V_{p,q}$ is $(-1)^q$.

Let sgn be the one-dimensional representation $\text{sgn}(g) = \text{sgn}(\det(g)) = \det(g)$ of $O(V)$, which by composing with projection $\tilde{O}(V) \to O(V)$ is a nongenuine character of $\tilde{O}(V)$. Tensoring with sgn defines an involution without fixed points on the set of genuine irreducible representations of $\tilde{O}(V)$.

THEOREM 1.1 [2]. *Fix ψ and $\delta = \pm 1$. The theta correspondences $\theta(V, \psi)$ define a bijection between*

$$\widetilde{Sp}(W)\widehat{}_{\text{genuine}}$$

and a subset of

$$\bigcup_{\substack{V \in \mathcal{V} \\ \det(V) = \delta}} \tilde{O}(V)\widehat{}_{\text{genuine}}$$

For every $\pi \in \tilde{O}(V)\widehat{}_{\text{genuine}}$ precisely one of π and $\pi \otimes \text{sgn}$ occurs in the image.

Note: The question of which representations occur in the image is subtle — see Proposition 4.1.

An important role is played by a certain character ξ.

DEFINITION 1.2. *Fix a symplectic vector space W, a metaplectic cover $\widetilde{Sp}(W)$ of $Sp(W)$, a maximal compact subgroup $\tilde{K}$ of $\widetilde{Sp}(W)$, and an additive character ψ. Let $\xi(W, \psi)$ be the character of $\tilde{K}$ acting on the unique $\tilde{K}$-invariant line in*

$\mathcal{F}(\psi)$. We use the same notation to denote this character restricted to the center of $\widetilde{Sp}(W)$.

For $\tilde{O}(V) \subset \widetilde{Sp}(\mathbb{W})$ as above, we use the same notation to denote the restriction of $\xi(\mathbb{W}, \psi)$ from $\widetilde{\mathbf{K}}$ to $\tilde{K} = \widetilde{\mathbf{K}} \cap \tilde{O}(V)$, as well as the unique extension of this to a genuine character of $\tilde{O}(V)$. Finally we define a bijection

$$\tilde{O}(V)\widehat{}_{\text{genuine}} \to O(V)\widehat{}$$
$$\pi \to \overline{\pi} \tag{1.3a}$$

by tensoring with the genuine character ξ^{-1}: for $x \in O(V)$ let

$$\overline{\pi}(x) = \xi(\mathbb{W}, \psi)^{-1}(\tilde{x})\pi(\tilde{x}) \quad (\tilde{x} \in p^{-1}(x)) \tag{1.3b}$$

(independent of the choice of $\tilde{x}$).

If ψ or W is understood we drop them from the notation. Roughly speaking $\xi = \sqrt{\det}$; to be precise under an appropriate isomorphism of K with $U(n)$, ξ^2 factors to the determinant of $U(n)$. Note that the restriction of ξ to the center is independent of the choice of $\tilde{K}$.

We refer to the map $\widetilde{Sp}(W)\widehat{}_{\text{genuine}} \to SO(V)\widehat{}$ given by $\pi \to \overline{\theta(V, \psi)(\pi)}$ restricted to $SO(V)$ as the *modified* theta correspondence.

THEOREM 1.4 [2]. *Fix ψ and $\delta = \pm 1$. The modified theta correspondences define a bijection*

$$\widetilde{Sp}(W)\widehat{}_{\text{genuine}} \longleftrightarrow \bigcup_{\substack{V \in \mathcal{V} \\ \det(V) = \delta}} SO(V)\widehat{}.$$

Recall that the union is over $\{V_{p,q} | p + q = 2n + 1, (-1)^q = \delta\}$. For fixed δ the union is over one group from each isomorphism class of real forms of $SO(2n + 1)$. The correspondence is computed explicitly in [2].

This was conjectured by Kudla, and the case $n = 1$ is given in [14]. A similar result for the dual pairs $(O(2p, 2q), Sp(2n, \mathbb{R}))$ with $p + q = n, n + 1$ is given in [10] (the metaplectic group plays no role here).

The bijection of Theorem 1.4 is one of similarity rather than duality. For example it preserves the property of being a discrete series representation. The same holds for "small" representations, for example the two one-dimensional representations of $SO(n + 1, n)$ correspond to the even halves of the two oscillator representations.

An important problem in the representation theory of reductive groups over a local field is the extension of the local Langlands conjecture [4] to nonlinear groups. Theorem 1.4 suggests that one could define an L-group for genuine representations of $\widetilde{Sp}(2n, \mathbb{R})$ to be the L-group $Sp(2n, \mathbb{C}) \times \mathbb{Z}/2\mathbb{Z}$ of $SO(2n + 1)$. Such a possibility was suggested by Jim Arthur.

For example one could define a genuine L-packet for $\widetilde{Sp}(2n, \mathbb{R})$ to correspond to a (Vogan) L-packet for the groups $SO(p, q)$; then these packets are parametrized by maps of the Weil group into $Sp(2n, \mathbb{C})$. Indeed such a definition may play some role in explaining L-functoriality for dual pairs involving odd orthogonal groups [1], and hence in applications to lifting of automorphic forms.

Some care is necessary with this definition. For example the central character of the representations in such a packet would not be constant (cf. Section 4). More significantly, the Kazhdan-Lusztig character relations are *not* preserved by the bijection. These character relations are fundamental to the proof of Arthur's conjectures for a linear group G [3], in which the representations of G are related to the geometry of a space X. Here X is on the "dual" side; it is a quotient of the space of maps of the Weil group into the L-group of G. Therefore the character relations for genuine representations of $\widetilde{Sp}(2n, \mathbb{R})$ are *not* encoded (at least in the obvious way) by the space X for the groups $SO(p, q)$ $(p + q = 2n + 1)$.

What is lacking is a functorial or geometric explanation of the local Langlands correspondence (for linear groups), in the form of [3] or any other version. Theorem 1.4 might then give an idea of how to extend this to nonlinear groups.

Somewhat similar relations between the representation theory of a nonlinear and a linear group are found in [7] and [13]. In these cases the root systems are self-dual, and the correspondence is between a cover of a group and the group itself.

2 Theta correspondence: Further properties

We collect a few results from [2] that are needed for the application to epsilon factors.

Let $G = Sp(W)$. We choose an isomorphism of K with $U(n)$ so that, with the usual compact Cartan subgroup and positive roots of $U(n)$, $\xi(W, \psi)^2$ (Definition 1.2) corresponds to the weight $(1, \ldots, 1)$. Then $\widetilde{K}^{\wedge}$ is parametrized by weights $(a_1, \ldots, a_n)$ with $a_1 \geq \cdots \geq a_n$ and $a_i \in \mathbb{Z} + \frac{1}{2}$ for all i.

The irreducible representations of the compact group $O(m)$ are parametrized by "weights" $(a_1, \ldots, a_{[\frac{m}{2}]}; \delta)$ with $a_i \in \mathbb{Z}$, $a_1 \geq \cdots \geq a_{[\frac{m}{2}]} \geq 0$, and $\delta = \pm 1$. This parametrization follows [15] and is obtained by restriction from $U(m)$. With $O(m)$ embedded in $U(m)$ in the usual way the representation of $O(m)$ with highest weight $(a_1, \ldots, a_k, 0, \ldots, 0; \delta)$ is the highest weight component of the irreducible representation of $U(m)$ with highest weight

$$(a_1, \ldots, a_k, \overbrace{1, 1, \ldots, 1}^{\frac{1-\delta}{2}(m-2k)}, 0, \ldots, 0).$$

In particular if m is odd the element $-Id$ of the center of $O(m)$ acts by $\delta(-1)^{\sum a_i}$.

Now let $G = O(V)$. If V has signature p, q we may choose an isomorphism, canonical up to inner automorphism, of G with $O(p, q)$ (defined by the form $\mathrm{diag}(I_p, -I_q)$) and of K with $O(p) \times O(q)$.

LEMMA 2.1.

(1) The $\widetilde{K}$-type $\mu = (a_1, \ldots, a_k, 0, \ldots, 0; \delta)(b_1, \ldots, b_\ell, 0, \ldots, 0; \eta)\xi$ is contained in the space of joint harmonics [6] if and only if $k + \frac{1-\delta}{2}(p - 2k) + \ell + \frac{1-\eta}{2}(q - 2\ell) \leq n$.

(2) For π an irreducible representation of $\tilde{O}(V)$, $\theta(W, \psi)(\pi) \neq 0$ if and only if every lowest $\tilde{K}$-type of π (in the sense of Vogan) occurs in the space of joint harmonics. It is enough to check this condition on a single lowest $\tilde{K}$-type μ.

(3) If π is a discrete series representation then the lowest $\tilde{K}$-type of π is of minimal degree (regardless of whether or not $\theta(W, \psi)(\pi) \neq 0$). If $\theta(W, \psi)(\pi)$ is nonzero then it is a discrete series representation.

For induced representations it is useful to have another formulation of condition (2).

Let $(\tilde{M}, \sigma)$ be "Langlands data" for a genuine irreducible representation π of $\tilde{O}(V)$. Thus σ is a discrete series representation of the Levi component $\tilde{M}$ of a cuspidal parabolic subgroup $\tilde{P}$, and π is a constituent quotient of $\mathrm{Ind}_{\tilde{P}}^{\tilde{O}(V)}(\sigma)$. Here M is a Levi factor

$$M \simeq O(V_0) \times GL(1)^a \times GL(2)^b \tag{2.2a}$$

of $O(V)$, for some a, b and a nondegenerate orthogonal subspace V_0 of V and

$$\tilde{M} \simeq \tilde{O}(V_0) \times GL(1)^a \times GL(2)^b \tag{2.2b}$$

is its inverse image in $\tilde{O}(V)$.

Note that M is a member of a dual pair (M, M') where

$$M' \simeq Sp(W_0) \times GL(1)^a \times GL(2)^b \tag{2.3a}$$

is a Levi factor of $Sp(W)$ and $\dim(W_0) + 1 = \dim(V_0)$; $(O(V_0), Sp(W_0))$ is a dual pair in $Sp(\mathbb{W}_0 = V_0 \otimes W_0)$. Let

$$\tilde{M}' \simeq \widetilde{Sp}(W_0) \times GL(1)^a \times GL(2)^b. \tag{2.3b}$$

Let σ_0 be the $\tilde{O}(V_0)$ component of σ, and let $\sigma_0^\dagger = \sigma_0 \xi(\mathbb{W})^{-1} \xi(\mathbb{W}_0)$. Then $\sigma_0^\dagger$ may be identified with a genuine representation of the inverse image of $O(V_0)$ in $\widetilde{Sp}(\mathbb{W})$ and hence is a candidate for a theta-lift to $\widetilde{Sp}(W_0)$.

Write $\sigma = \sigma_0 \otimes \sigma_1 \otimes \cdots \otimes \sigma_{a+b}$ corresponding to the decomposition (2.2b). We normalize the theta-lift for the pair (M, M') so that

$$\theta(M', \psi)(\sigma) = \theta(W_0, \psi)(\sigma_0^\dagger) \otimes \sigma_1^* \otimes \cdots \otimes \sigma_{a+b}^*. \tag{2.4}$$

PROPOSITION 2.5. $\theta(W, \psi)(\pi) \neq 0$ if and only if $\theta(M', \psi)(\sigma) \neq 0$.

Note: If the condition of the proposition holds then $(\overline{M}', \theta(M', \psi)(\sigma)\chi)$ is Langlands data for $\theta(W, \psi)(\pi)$, where $\overline{M}'$ is a certain cover of $\tilde{M}'$ and χ is a certain genuine character on a covering of the GL factors. In some sense the theta correspondence is "functorial".

3 Epsilon factors: Generalities

We establish notation, and some generalities about ϵ-factors over $\mathbb{R}$. This section is independent of the oscillator representation.

For ψ a nontrivial unitary additive character of $\mathbb{R}$ and ϕ a representation of the Weil group $W_{\mathbb{R}}$ of $\mathbb{R}$, we define $\epsilon(\phi) = \epsilon_L(\phi, \psi)$ as in [12, Section 3], cf. [5, Section 9]. This is multiplicative on direct sums. We fix $\psi(x) = e^{2\pi i x}$ and compute $\epsilon(\phi)$ explicitly for ϕ irreducible.

For $\delta = 0, 1$ and $s \in \mathbb{C}$ let $\alpha(\delta, s)(x) = |x|^s \operatorname{sgn}(x)^\delta$, a character of $\mathbb{R}^*$, and also of $W_{\mathbb{R}}$ by the abelianization map $W_{\mathbb{R}} \to \mathbb{R}^*$. For $n \in \mathbb{Z}$ and $s \in \mathbb{C}$ let $\beta(n, s)$ be the two-dimensional representation $\beta(n, s) = \operatorname{Ind}_{\mathbb{C}^*}^{W_{\mathbb{R}}}(|z|^s (z/\bar{z})^{\frac{n}{2}})$ of $W_{\mathbb{R}}$. This is irreducible if and only if $n \neq 0$, and $\beta(n, s) \simeq \beta(-n, s)$. We assume $n \in \mathbb{N}$, and then every irreducible representation of $W_{\mathbb{R}}$ is equivalent to precisely one $\alpha(\delta, s)$ or $\beta(n, s)$. From [12, Section 3] we conclude

$$\epsilon(\alpha(\delta, s)) = i^\delta, \quad \epsilon(\beta(n, s)) = i^{n+1}. \tag{3.1}$$

Now let π be an irreducible representation of $SO(V)$, with $\dim(V) = 2n + 1$. The L-group of $SO(V)$ is isomorphic to $Sp(2n, \mathbb{C}) \times \mathbb{Z}/2\mathbb{Z}$. The L-parameter ϕ of π may be considered as a map $\phi : W_{\mathbb{R}} \to Sp(2n, \mathbb{C})$, which we treat as a representation of $W_{\mathbb{R}}$ via the standard embedding $Sp(2n, \mathbb{C}) \hookrightarrow GL(2n, \mathbb{C})$. Let $\epsilon(\pi) = \epsilon(\phi)$, and for π a representation of $O(V)$ let $\epsilon(\pi) = \epsilon(\pi|_{SO(V)})$.

We identify an infinitesimal character for $SO(V)$ with an element of $\mathfrak{t}^*$ where T is a compact Cartan subgroup. The infinitesimal character of a discrete series representation is of the form $(a_1, \ldots, a_n)$ with $a_i \in \mathbb{Z} + \frac{1}{2}$.

LEMMA 3.2. *Let π be an irreducible representation of $SO(V)$. Then*

(1) *$\epsilon(\pi) = \pm 1$ and is independent of ψ.*

(2) *Suppose π is a discrete series representation of $SO(V)$ with infinitesimal character $(a_1, \ldots, a_n)$. Then $\epsilon(\pi) = \prod_i (-1)^{|a_i| + \frac{1}{2}}$.*

Proof. The first statement is [5, Proposition 9.5]; it follows from [12, Section 3.6] and the fact that ϕ is self-dual and of determinant 1. For (2) the L-parameter of π decomposes as $\beta(2a_1, 0) \oplus \cdots \oplus \beta(2a_n, 0)$, and (2) follows from (3.1). $\qquad\square$

For G a connected reductive algebraic group over a local field F let $\kappa(G) = \pm 1$ be the "Kottwitz invariant" of G [8]. Over $\mathbb{R}$,

$$\kappa(G) = (-1)^{\frac{1}{2}(\dim(G/K) - \dim(G_{qs}/K_{qs}))} \tag{3.3a}$$

where G_{qs} is the quasisplit inner form of G. In particular for $G \simeq SO(p, q)$ with $p + q$ odd

$$\kappa(G) = (-1)^{\frac{1}{8}((p-q)^2 - 1)}. \tag{3.3b}$$

Note that $\kappa(SO(p, q))$ for fixed $p + q$ depends on $p, q \mod (4)$. Let $\kappa(O(V)) = \kappa(SO(V))$.

The irreducible representations of $O(p, q)$ ($p + q$ odd) are obtained by induction from $SO(p, q)$. Such an induced module is the direct sum of two nonisomorphic irreducible summands $\pi \oplus (\pi \otimes \mathrm{sgn})$. It follows that if π is an irreducible discrete series representation of $O(p,q)$ it is determined by a Harish-Chandra parameter for $SO(p,q)$ and one of two possible lowest K-types μ (the other choice being $\mu \otimes \mathrm{sgn}$). A calculation shows the two choices of μ are obtained by taking $\delta = \pm 1$ in

$$\mu = \begin{cases} (x_1,\ldots,x_{[\frac{p}{2}]},1) \otimes (y_1,\ldots,y_\ell,0,\ldots,0;\delta) & p \text{ even}, q \text{ odd} \\ (x_1,\ldots,x_k,0,\ldots,0;\delta) \otimes (y_1,\ldots,y_{[\frac{q}{2}]};1) & p \text{ odd}, q \text{ even}. \end{cases} \tag{3.4}$$

Here $x_{[\frac{p}{2}]} > 0$ (resp. $y_{[\frac{q}{2}]} > 0$) if p (resp. q) is even. For μ of this form write $\delta(\mu) = \delta$.

LEMMA 3.5. *Let π be a discrete series representation of $G = O(p,q)$ ($p+q$ odd) with lowest K-type μ. Then $\delta(\mu)$ and the central character of π are related by:*

$$\delta(\mu) = \epsilon(\pi)\pi(-I)\kappa(G). \tag{3.6}$$

Proof. Assume $p = 2p_0 + 1, q = 2q_0$. Write the Harish-Chandra parameter of π as

$$\lambda = (a_1,\ldots,a_{p_0})(b_1,\ldots,b_{q_0}) \tag{3.7}$$

and μ as in (3.4). The central character of μ is $\mu(-I) = (-1)^{\sum x_i + \sum y_j}\delta$. Now μ (restricted to $SO_e(p,q)$) and λ are related by $\mu = \lambda + \rho_n - \rho_c$ where ρ_n (resp. ρ_c) denotes one-half the sum of the noncompact (resp. compact) roots defined by λ. For an element $\gamma \in \sqrt{-1}\mathfrak{t}_0^*$ let $\Sigma(\gamma)$ be the sum of the coordinates. An easy computation gives $\Sigma(\rho_n) = p_0 q_0 + \frac{1}{2}q_0$ and $\Sigma(\rho_c) = \frac{1}{2}p_0^2 + \frac{1}{2}q_0(q_0 - 1)$, and it follows that $\Sigma(\rho_n - \rho_c) = q_0 - \frac{1}{2}(p_0 - q_0)^2$. Therefore (recall $a_i, b_j \in \mathbb{Z} + \frac{1}{2}$)

$$\begin{aligned} \pi(-I) &= \mu(-I) \\ &= (-1)^{\Sigma(\mu)}\delta(\mu) \\ &= (-1)^{\sum a_i + \sum b_j + q_0 - \frac{1}{2}(p_0 - q_0)^2}\delta(\mu) \\ &= (-1)^{\sum(a_i + \frac{1}{2}) + \sum(b_j + \frac{1}{2}) - \frac{1}{2}(p_0 - q_0) - \frac{1}{2}(p_0 - q_0)^2}\delta(\mu). \end{aligned} \tag{3.8}$$

From this Lemma 3.2(2) gives $\pi(-I)\epsilon(\pi) = (-1)^{-\frac{1}{2}(p_0 - q_0) - \frac{1}{2}(p_0 - q_0)^2}\delta(\mu)$. Finally plugging $p = 2p_0 + 1, q = 2q_0$ into (3.3b) gives $\kappa(G) = (-1)^{-\frac{1}{2}(p_0 - q_0) - \frac{1}{2}(p_0 - q_0)^2}$, and Lemma 3.5 follows. $\qquad\square$

Let $(,)_\mathbb{R}$ be the Hilbert symbol of $\mathbb{R}$. The following identity, which follows by direct calculation, gives a reformulation of Lemma 3.5. Let H be the Hasse invariant and D the discriminant, of an orthogonal space V of dimension $2n + 1$. Then

$$\kappa(O(V)) = H(D, -1)_\mathbb{R}^n(-1, -1)^{n(n+1)/2} \tag{3.9a}$$

$$= HD^n(-1)^{n(n+1)/2} \tag{3.9b}$$

Note: This may be proved by induction on n, which also gives (a) over $\mathbb{Q}_p$.

4 Epsilon factors and the theta correspondence

We return to the setting of Section 1, and the dual pair $(O(V), Sp(W))$.

PROPOSITION 4.1. *Let π be a genuine irreducible representation of $\tilde{O}(V)$. Then $\theta(W, \psi)(\pi) \neq 0$ if and only if*

$$\epsilon(\overline{\pi})\overline{\pi}(-Id)\kappa(O(V)) = 1. \tag{4.2}$$

Recall $\overline{\pi}$ was defined in Definition 1.2.

Proof. Let p, q be the signature of V, and assume $p = 2p_0 + 1, q = 2q_0$ (the other case is similar). We first assume π is a discrete series representation. The (Vogan) lowest $\tilde{K}$-type $\overline{\mu} = \mu \xi^{-1}$ of $\overline{\pi}$ is of the form (3.4). By Lemma 2.1(1) μ occurs in the space of joint harmonics if and only if $k + \frac{1-\delta(\overline{\mu})}{2}(p - 2k) + q_0 \leq n$, and using $p_0 + q_0 = n$ and $k \leq p_0$ this is easily seen to hold if and only if $\delta(\overline{\mu}) = 1$. By Lemma 2.1(3), Theorem 1.1, and [6, Lemma 4.1] $\theta(W, \psi)(\pi) \neq 0$ if and only if $\delta(\overline{\mu}) = 1$, and the result follows upon inserting $\delta(\overline{\mu}) = \epsilon(\overline{\pi})\overline{\pi}(-Id)\kappa(O(V))$ from (3.6).

Now let π be arbitrary, with Langlands data $(\tilde{M}, \sigma)$ as in Section 2. The proof in this case follows from the inductive properties of the three terms on the left-hand side of (4.2). Because the nonvanishing of the theta-lift of π is determined on the $O(V_0)$ factor of M (Proposition 2.5) we need a Lemma saying the GL factors play no role in this condition.

The map $\phi : W_{\mathbb{R}} \to Sp(2n, \mathbb{C})$ corresponding to $\overline{\pi}$ (restricted to $SO(V)$) factors through an embedding of the L-group of M in $Sp(2n, \mathbb{C})$. This embedding takes each $GL(k)$ factor to $GL(k) \times GL(k)$ $(k = 1, 2)$ via the map $g \to (g, {}^tg^{-1})$. The image is contained in a subgroup of $Sp(2n, \mathbb{C})$ isomorphic to

$$Sp(2c, \mathbb{C}) \times GL(1, \mathbb{C})^{2a} \times GL(2, \mathbb{C})^{2b} \tag{4.3}$$

where $\dim(V_0) = 2c + 1$. We group the GL terms together to write $\phi = \phi_0 \times \phi_{GL}$ and $\sigma = \sigma_0 \otimes \sigma_{GL}$.

LEMMA 4.4. $\epsilon(\phi_{GL})\sigma_{GL}(-Id)\kappa(GL) = 1.$

Proof. Because GL is quasisplit $\kappa(GL) = 1$. Because both ϵ and $\sigma(-Id)$ are multiplicative on direct products, the problem reduces to $GL(1)$ and $GL(2)$. For $GL(1)$ suppose σ is the character $\alpha(\delta, s)$ (cf. Section 3). Then $\phi \simeq \alpha(\delta, s) \oplus \alpha(\delta, -s)$ and then $\epsilon(\phi) = i^{2\delta} = (-1)^\delta$. Because $\alpha(\delta, s)(-1) = (-1)^\delta$ this proves the lemma for this case, and the argument is similar for the $GL(2)$ factors.   □

Returning to π, by Proposition 2.5 (using the notation of the proposition)

$$\theta(W, \psi)(\pi) \neq 0 \Leftrightarrow \theta(M', \psi)(\sigma) \neq 0 \tag{4.5a}$$

$$\Leftrightarrow \theta(W_0, \psi)(\sigma_0^\dagger) \neq 0 \tag{4.5b}$$

$$\Leftrightarrow \epsilon(\overline{\sigma}_0^\dagger)\overline{\sigma}_0^\dagger(-Id)\kappa(O(V_0)) = 1 \tag{4.5c}$$

$$\Leftrightarrow \epsilon(\phi_0)\overline{\sigma}_0(-Id)\kappa(O(V_0)) = 1. \tag{4.5d}$$

We proceed to justify these steps. Steps (a) and (b) are Proposition 2.5, and (c) follows from Proposition 4.1 applied to the discrete series representation $\bar{\sigma}_0^\dagger$ if $\tilde{O}(V_0)$. The definition of the operations † and $^-$, together with the definition $\epsilon(\bar{\sigma}_0) = \epsilon(\phi_0)$, gives (d).

Now multiply both sides by $\epsilon(\phi_{GL})\sigma_{GL}(-Id)\kappa(GL) = 1$ to give

$$\Leftrightarrow \epsilon(\phi_0)\epsilon(\phi_{GL})\bar{\sigma}_0(-Id)\sigma_{GL}(-Id)\kappa(O(V_0))\kappa(GL) = 1$$
$$\Leftrightarrow \epsilon(\phi)\bar{\sigma}(-Id)\kappa(M) = 1 \tag{4.6}$$
$$\Leftrightarrow \epsilon(\phi)\bar{\pi}(-Id)\kappa(G) = 1$$

where the last line follows from the fact that $(M, \bar{\sigma})$ is inducing data for $\bar{\pi}$ and $\kappa(M) = \kappa(G)$, which is a basic property of the Kottwitz invariant [8]. This completes the proof. $\qquad\square$

Now suppose $\pi \in \widetilde{Sp}(W)^{\widehat{\ }}_{\text{genuine}}$ and $\pi' = \theta(V, \psi)(\pi) \neq 0$. The centers of $\tilde{O}(V), \widetilde{Sp}(W)$, and $\widetilde{Sp}(\mathbb{W})$ coincide. Therefore there is a relation between the central characters of π and π' which together with Proposition 4.1 gives a relation between $\epsilon(\bar{\pi}')$ and the central character of π.

DEFINITION 4.7. (Waldspurger [14]). For $\pi \in \widetilde{Sp}(W)^{\widehat{\ }}_{\text{genuine}}$ let $\tilde{\epsilon}(\pi, \psi) = \pm 1$ be defined by the relation

$$\xi(W, \psi)^{-1}(x)\pi(x) = \tilde{\epsilon}(\pi, \psi) \cdot Id$$

for all $x \in p^{-1}(-Id)$.

As π and ξ are genuine and the center has order 4 it follows that $\tilde{\epsilon}(\pi, \psi) = \pm 1$ and is independent of the choice of x.

For V a space of dimension m with a symmetric bilinear form, let χ_V be the quadratic character of $\mathbb{R}^*$ given by

$$\chi_V(x) = (x, (-1)^{\frac{m(m-1)}{2}} \det(V))_\mathbb{R}. \tag{4.8}$$

THEOREM 4.9. Suppose $\pi \in \widetilde{Sp}(W)^{\widehat{\ }}_{\text{genuine}}$ has nonzero theta-lift to $\tilde{O}(V)$, and let $\pi' = \theta(V, \psi)(\pi)$. Then

$$\tilde{\epsilon}(\pi, \psi) = \chi_V(-1)^n \epsilon(\bar{\pi}')\kappa(O(V)). \tag{4.10}$$

Proof. Choose $x \in p^{-1}(-Id) \in \widetilde{Sp}(\mathbb{W})$. Then

$$\pi'(x) = \pi'\xi(\mathbb{W}, \psi)^{-1}(x)\xi(\mathbb{W}, \psi)(x)$$
$$= \bar{\pi}'(-Id)\xi(\mathbb{W}, \psi)(x), \tag{4.11a}$$

which by Proposition 4.1 gives

$$\pi'(x) = \epsilon(\bar{\pi}')\kappa(O(V))\xi(\mathbb{W}, \psi)(x) \cdot Id. \tag{4.11b}$$

On the other hand

$$\pi(x) = \tilde{\epsilon}(\pi, \psi)\xi(W, \psi)(x) \cdot Id \tag{4.11c}$$

and setting (b) equal to (c) we need to show

$$\xi(\mathbb{W}, \psi)(x)\xi(W, \psi)^{-1}(x) = \chi_V(-1)^n. \tag{4.12a}$$

It is not hard to see that for $g \in K \simeq U(n)$ and $\iota_W(g) \in \mathbf{K} \simeq U(n(2n+1))$,

$$\det(\iota_W(g)) = \det(g)^{p-q}. \tag{4.12b}$$

Because $\xi^2 = \det$ it follows that

$$\xi(\mathbb{W}, \psi)(x) = \xi(W, \psi)(x)^{p-q} \tag{4.12c}$$

and writing $p - q = p + q - 2q = 2n - 2q + 1$ gives

$$= \det(p(x))^{n-q}\xi(W, \psi)(x)$$
$$= (-1)^{n(n-q)}\xi(W, \psi)(x). \tag{4.12d}$$

Finally (a) follows upon inserting (d) into (a) and the elementary observation $\chi_V(-1) = (-1)^{n-q}$. $\square$

This is a version of "epsilon dichotomy" as conjectured by Kudla (cf. [14] for the case $n = 1$ and [9] for the unitary group case). In the p-adic case there are two quadratic spaces with given discriminant, which are distinguished by their Hasse (equivalently Kottwitz) invariant; epsilon dichotomy refers to the fact that which lift is nonzero is determined by an epsilon factor. In the real case the epsilon factor only determines for which of two families of groups the lift is nonzero; these families are determined by p and $q \mod (4)$.

Theorem 4.9 gives an interpretation of the epsilon factor of a map $\phi : W_{\mathbb{R}} \to Sp(2n, \mathbb{C})$ in terms of the central characters of the lifts to $\widetilde{Sp}(2n, \mathbb{R})$ of the representations in the (Vogan) L-packet of the groups $SO(p, q)$ defined by ϕ. For another interpretation of symplectic root numbers see [11].

As $\kappa(O(p, q))$ (with fixed discriminant) varies with p and q, Theorem 4.9 shows that the central character may fail to be constant on an L-packet for $\widetilde{Sp}(2n, \mathbb{R})$ as defined at the end of Section 1.

References

[1] J. Adams, *L-Functoriality for Dual Pairs*, Astérisque **171–172** (1989), 85–129.

[2] J. Adams and D. Barbasch, *Genuine representations of the metaplectic group*, preprint.

[3] J. Adams, D. Barbasch, and D. Vogan, The Langlands classification and irreducible characters for real reductive groups, Progr. Math. **104**, Birkhäuser, Basel and Boston, 1992.

[4] A. Borel, *Automorphic L-Functions*, in Automorphic Forms, Representations and L-Functions (part 2), A. Borel and W. Casselman, eds., Proc. Sympos. Pure Math. **33**, Amer. Math. Soc., Providence, RI, 1979, 27–61.

[5] B. Gross and D. Prasad, *On the decomposition of a representation of SO_n when restricted to SO_{n-1}*, Canad. J. Math. **44** (1992), 974–1002.

[6] R. Howe, *Transcending classical invariant theory*, J. Amer. Math. Soc. **2** (July, 1989), 535–552.

[7] J.-S. Huang, *The unitary dual of the universal covering group of $GL(n, \mathbb{R})$*, Duke Math. J. **61** (1990), 705–745.

[8] R. Kottwitz, *Sign changes in harmonic analysis on reductive groups*, Trans. Amer. Math. Soc. **278** (1983), 289–297.

[9] S. Kudla, M. Harris, and W. Sweet, *Theta dichotomy for unitary groups*, to appear in J. of AMS.

[10] C. Moeglin, *Correspondance de Howe pour les paires réductives duales, quelques calculs dans le cas Archimedien*, J. Funct. Anal. **85** (1989), 1–85.

[11] D. Prasad and D. Ramakrishnan, *Symplectic root numbers of two-dimensional Galois representations: An interpretation*, preprint.

[12] J. Tate, *Number theoretic background*, in Automorphic Forms, Representations and L-Functions (part 1), A. Borel and W. Casselman, eds., Proc. Symp. Pure Math. **33**, Amer. Math. Soc., Providence, RI, 1979, 275–286.

[13] D. Vogan, *The unitary dual of G_2*, Invent. Math. **116** (1994), 677–791.

[14] J.-L. Waldspurger, *Correspondances de Shimura et quaternions*, Forum Math. **3** (1991), 219–307.

[15] H. Weyl, The Classical Groups, Their Invariants and Representations, Princeton University Press, Princeton, NJ, 1946.

The Irreducible Characters
for Semi-Simple Algebraic Groups and for Quantum Groups

HENNING HAAHR ANDERSEN

Matematisk Institut
Aarhus Universitet
8000 Aarhus C, Denmark

1 The problem

Let G be a reductive algebraic group over an algebraically closed field k of prime characteristic p. The first question that presents itself when we look at finite-dimensional representations of G is the problem of how to determine the irreducible characters. This is the problem we want to discuss in this lecture.

It is well known how a solution of this problem also solves the problem of determining the irreducible modular characters for the finite Chevalley groups $G(\mathbb{F}_q)$ (as well as their twisted versions), see e.g [St]. We shall not comment further on this relation here but shall instead explore connections to analogous problems for semi-simple complex Lie algebras, Kac-Moody algebras, and quantum groups.

An easy reduction allows us first to assume that G is semi-simple and then that G is almost simple. To simplify things we shall make this last assumption from now on.

Choose a maximal torus T of G. Our assumption means that the root system R for G w.r.t. T is indecomposable. Fix a set of positive roots $R_+ \subset R$. All further notation will be introduced as we go along.

2 The classification of irreducible representations

Let $X \simeq \mathbb{Z}^n$ denote the character group of T and let $X^+ \simeq \mathbb{N}^n$ be the subset consisting of the dominant characters (or weights) relative to R_+. Then we have a classification (going back at least to [Ch]) of the finite-dimensional irreducible representations of G via their highest weights. For $\lambda \in X^+$ we let $L(\lambda)$ be the corresponding irreducible representation.

One way to realize the irreducible representations of G is to obtain them as submodules of the G-modules consisting of the global sections of line bundles on the flag manifold for G. Pick a Borel subgroup B in G containing T. Then each $\lambda \in X$ induces a line bundle $\mathcal{L}_\lambda$ on G/B. We set $H^0(\lambda) = \Gamma(G/B, \mathcal{L}_\lambda)$, i.e. $H^0(\lambda) = \{f \colon G \to k \mid f$ is regular and $f(gb) = \lambda(b)^{-1} f(g), g \in G, b \in B\}$. This may also be viewed as the G-module induced by the 1-dimensional B-module λ. If we choose B to correspond to the negative roots $-R_+$ then, $H^0(\lambda) \neq 0$ iff $\lambda \in X^+$. One then checks that for such λ the G-module $H^0(\lambda)$ contains a unique irreducible submodule and this is our $L(\lambda)$.

Proceedings of the International Congress
of Mathematicians, Zürich, Switzerland 1994
© Birkhäuser Verlag, Basel, Switzerland 1995

3 The conjecture

In 1979 G. Lusztig proposed a conjecture [L1] that in terms of an algorithm gives the answer to our problem for $p \geq 2h - 3$, h denoting the Coxeter number. This goes as follows:

If V is an arbitrary finite-dimensional T-module, then $V = \bigoplus_{\lambda \in X} V_\lambda$ where $V_\lambda = \{v \in V \mid tv = \lambda(t)v, t \in T\}$. We set

$$[V] = \sum_\lambda (\dim V_\lambda) e^\lambda \in \mathbb{Z}[X]$$

and call this the (formal) character of V. (Alternatively, we may view $[V]$ as the image of V in the Grothendieck group.)

Elementary character considerations show that for $\lambda \in X^+$ we may write

$$[L(\lambda)] = \sum_{\mu \in X^+} c_{\lambda,\mu}[H^0(\mu)]$$

for some (unique) integers $c_{\lambda,\mu}$. Moreover, $c_{\lambda,\lambda} = 1$ and $c_{\lambda,\mu} = 0$ unless $\mu \leq \lambda$ (in the ordering $\leq$ on X induced by R_+). Now because of Kempf's vanishing theorem (see [Ke] or [A2], [Ha]) the character of $H^0(\mu)$ equals the Euler character of $\mathcal{L}_\mu$ for all $\mu \in X^+$ and is therefore given by Weyl's character formula. Hence determining $[L(\lambda)]$ is equivalent to calculating the integers $(c_{\lambda,\mu})_{\mu \in X^+}$.

As our next step let us point out that we only need to consider a finite number of λ's. We set $X_p = \{\lambda \in X^+ \mid \langle \lambda, \alpha^\vee \rangle < p$ for all simple roots $\alpha\}$ (here $\alpha^\vee$ is the coroot of α). This is called the set of restricted weights.

For any $\lambda \in X^+$ we may write $\lambda = \lambda^0 + p\lambda^1$ for some (unique) $\lambda^0 \in X_p$ and $\lambda^1 \in X_+$. Then Steinberg's tensor product theorem [St] says

$$L(\lambda) \simeq L(\lambda^0) \otimes L(\lambda^1)^{(p)}. \tag{3.1}$$

Here $L(\lambda^1)^{(p)}$ is the Frobenius twist of $L(\lambda^1)$.

It follows that it suffices to determine $[L(\lambda)]$ for λ in the finite set X_p. Equivalently, it suffices to calculate $c_{\lambda,\mu}$ for all $\lambda \in X_p$ and $\mu \leq \lambda$.

Recall that the affine Weyl group W_p is the group generated by the reflections $s_{\alpha,n} : X \to X$, $\alpha \in R_+$, $n \in \mathbb{Z}$, defined by $s_{\alpha,n}.\lambda = s_\alpha.\lambda + np\alpha = \lambda - \langle \lambda + \rho, \alpha^\vee \rangle \alpha + np\alpha$, $\lambda \in X$ with $\rho = \frac{1}{2}\sum_{\alpha \in R_+} \alpha$.

By the linkage principle [A1] we have

$$c_{\lambda,\mu} = 0 \text{ unless } \mu \in W_p.\lambda \tag{3.2}$$

so this reduces further the relevant $c_{\lambda,\mu}$'s to be calculated.

Consider the alcove A_p given by

$$A_p = \{\lambda \in X \mid 0 > \langle \lambda + \rho, \alpha^\vee \rangle > -p, \alpha \in R_+\}.$$

Then the closure of A_p,

$$\bar{A}_p = \{\lambda \in X \mid 0 \geq \langle \lambda + \rho, \alpha^\vee \rangle \geq -p, \alpha \in R_+\},$$

is a fundamental domain for the action of W_p on X, i.e. each $\lambda \in X$ may be written $\lambda = w.\nu$ for some $w \in W_p$ and some unique $\nu \in \bar{A}_p$.

Fix now $\lambda \in X^+$. Let $\nu \in \bar{A}_p$ be the unique element in the W_p-orbit of λ and choose $w = w_\lambda \in W_p$ minimal (in the Chevalley order) such that $\lambda = w.v$ (in the terminology of [J] λ is then in the upper closure of $w.A_p$). We set $c_{w,y} = c_{\lambda,y.\nu}$ for all $y \in W_p$ with $y.\nu \in X^+$. Then we have (cf. [L1])

3.1 CONJECTURE (LUSZTIG 1979). *If* $\langle \lambda + \rho, \alpha^\vee \rangle < p(p - h + 2)$ *for all* $\alpha \in R_+$, *then* $c_{w,y} = (-1)^{l(wy)} P_{y,w}(1)$.

The $P_{y,w}$ appearing in this conjecture is the so-called Kazhdan-Lusztig polynomial associated to the pair (y, w), see [KL1], and l is the length function on W_p.

3.2 REMARKS. **(a)** The assumption in Lusztig's conjecture is only satisfied for all $\lambda \in X_p$ when $p \geq 2h - 2$. There is a variant of the conjecture (see e.g. [Kat]) that gets this bound down to $p \geq h$. To my knowledge there is no conjecture that covers all λ when $p < h$.

(b) One consequence of the conjecture is that the $c_{w,y}$ are independent of p for all relevant w, y. One of the main results in [AJS] says (independently of whether Lusztig's conjecture is valid) that this much is true for all large primes.

(c) For the validity of the conjecture for large p see Section 5 below.

4 Related conjectures

4a. The original Kazhdan-Lusztig conjecture

Let $\mathfrak{g}$ denote the finite-dimensional semi-simple Lie algebra corresponding to G. Then X (identified with the set of integral weights for a Cartan subalgebra in $\mathfrak{g}$) parametrizes the simple $\mathfrak{g}$-modules in the highest weight category $\mathcal{O}$. For $\lambda \in X$ we may realize the corresponding simple $\mathfrak{g}$-module $L(\lambda)$ as being the unique simple quotient of the Verma module $M(\lambda)$ (defined via some Borel subalgebra in $\mathfrak{g}$). With notation analogous to the one used above we write

$$[L(\lambda)] = \sum_{\lambda \in X} c_{\lambda,\mu}[M(\mu)]$$

for certain (unique) $c_{\lambda,\mu} \in \mathbb{Z}$, $c_{\lambda,\lambda} = 1$, and $c_{\lambda,\mu} = 0$ unless $\mu \leq \lambda$. Because the character of $M(\lambda)$ is known, the problem of determining $[L(\lambda)]$ is again equivalent to the calculation of the $c_{\lambda,\mu}$'s.

As before we have a linkage principle that ensures that $c_{\lambda,\mu} = 0$ unless $\mu \in W.\lambda$ where W is the (finite) Weyl group generated by $\{s_\alpha = s_{\alpha,0} \mid \alpha \in R_+\}$. This time we set

$$A = \{\lambda \in X \mid \langle \lambda + \rho, \alpha^\vee \rangle < 0 \text{ for all } \alpha \in R_+\}$$

and we have (see [KL1])

4a.1 CONJECTURE (KAZHDAN-LUSZTIG 1979). *For all* $\lambda \in A$, $y, w \in W$, *we have*

$$[L(w.\lambda)] = \sum_{y \in W} (-1)^{l(yw)} P_{y,w}(1)[M(y.\lambda)].$$

This conjecture was proved in 1981, see [BB] and [BK]. There are, roughly speaking, two ingredients in the proof: a topological part and an algebraic part. The topological part was dealt with by Kazhdan and Lusztig [KL2]. It consists of relating the local intersection cohomology on Schubert varieties in G/B to the Kazhdan-Lusztig polynomials. The algebraic part on the other hand consists of setting up a correspondence between representations of $\mathfrak{g}$ and $\mathcal{D}$-modules on G/B, see [BB] and [BK]. Then the Riemann-Hilbert correspondence makes a bridge to perverse sheaves supported on Schubert varieties.

4a.2 REMARK. There are various generalizations of the Kazhdan-Lusztig conjecture. First we should point out that translation arguments allow us to drop the assumption that λ is regular. Also, it is possible to extend to the case where λ is no longer integral, see e.g. [S].

Now replace $\mathfrak{g}$ by a Kac-Moody algebra. In the symmetrizable case the conjecture and its proof may be carried over to describe those irreducible characters whose highest weights are linked to a dominant weight, see [Ka1], [KT1] or alternatively [C1]. For the negative level case for an affine algebra see Section 4b below.

Let $U_q(\mathfrak{g})$ denote the quantized enveloping algebra over $\mathbb{Q}(q)$ associated with $\mathfrak{g}$ (where $\mathfrak{g}$ now again denotes a finite-dimensional semi-simple Lie algebra). There is a completely analogous conjecture in this case, see [L3]. A proof in this case is obtained by using the Drinfeld equivalence of categories, see [D] or [KL3, III]. In the case where q is specialized to a root of unity, see Section 4c below.

For analogous conjectures in the case of real semi-simple Lie groups, see [V].

4b. The affine Lie algebra negative level case

Let now $\mathfrak{g}$ be an affine Lie algebra with Cartan subalgebra $\mathfrak{h}$. For any $\lambda \in \mathfrak{h}^*$ we have a Verma module $M(\lambda)$ (defined via induction from a Borel subalgebra $\mathfrak{b} \supset \mathfrak{h}$) with a unique simple quotient $L(\lambda)$. Also, in this case the family $\{L(\lambda) \mid \lambda \in \mathfrak{h}^*\}$ constitutes up to isomorphism the set of simple modules in the category $\mathcal{O}$ for $\mathfrak{g}$, see [K].

Denote by $\{\alpha_0, \alpha_1, \ldots, \alpha_n\}$ the set of simple roots and by W the Weyl group generated by the corresponding reflections. Note that W is an affine Weyl group. Choose $\rho \in \mathfrak{h}^*$ such that $\langle \rho, \alpha_i^\vee \rangle = 1$ for all i and set

$$A = \{\lambda \in \mathfrak{h}^* \mid \langle \lambda + \rho, \alpha_i^\vee \rangle \in \mathbb{Z}_{<0}, \ i = 0, 1, \ldots, n\}.$$

Recall from [DGK] that for each $\lambda \in A$, $w \in W$, the Verma module $M(w.\lambda)$ has finite length and its composition factors have the form $L(y.\lambda)$ for certain $y \in W$ with $y \leq w$.

4b.1 CONJECTURE (LUSZTIG 1990). *Let $\lambda \in A$. Then for any $w \in W$ we have*

$$[L(w.\lambda)] = \sum_{y \in W} (-1)^{l(yw)} P_{y,w}(1)[M(y.\lambda)].$$

4b.2 REMARKS. **(i)** Conjecture 4b.1 can be found in [L3], see also [L5]. A proof was announced in 1990 by Casian [C2], but its validity is still controversial. Earlier this year Kashiwara and Tanisaki [KT2] announced a proof along the same lines as the proof of the original Kazhdan-Lusztig conjecture — or rather their generalization, cf. Remark 4a.2, to the symmetrizable Kac-Moody algebra positive level case. To handle the negative level case they have to deal with a category of right $\mathcal{D}$-modules on the infinite-dimensional flag variety with support on finite-dimensional Schubert varieties.

(ii) Let $c \in \mathfrak{g}$ denote the canonical central element. Recall that the level of $\lambda \in \mathfrak{h}^*$ is the complex number $\langle \lambda, c \rangle$ and that c acts on $M(\lambda)$ as multiplication by that number. In particular, the level of ρ is equal to the dual Coxeter number $\hat{h}$ of $\mathfrak{g}$.

Suppose again $\lambda \in A$. Then the level of λ is a negative integer ($\leq -2\hat{h}$) and this level coincides with the level of $w.\lambda$ for all $w \in W$ (as $\langle \alpha_i, c \rangle = 0$ for all i). Hence Conjecture 4b.1 breaks up into a family of conjectures, one for each such negative integer. In order to be able to compare with the quantum case (see below) we now give a reformulation of the conjecture at a fixed negative level:

Let us assume that $\{\alpha_1, \ldots, \alpha_n\}$ are the simple roots for our finite root system R. Then we may identify X with $\{\lambda \in \mathfrak{h}^* \mid \lambda \in \mathbb{R}\alpha_1 \oplus \cdots \oplus \mathbb{R}\alpha_n$ and $\langle \lambda, \alpha_i^\vee \rangle \in \mathbb{Z}, \ i = 1, \ldots, n\}$. Fix $l \in \mathbb{N}$, $l \geq h$, and set (cf. Section 3)

$$A_l = \{\lambda \in X \mid 0 > \langle \lambda + \rho, \alpha^\vee \rangle > -l \text{ for all } \alpha \in R_+\}.$$

Also as in Section 3 let W_l denote the affine Weyl group generated by the reflections $s_{\alpha,n}$, $\alpha \in R_+$, $n \in \mathbb{Z}$, given by $s_{\alpha,n}.\lambda = s_\alpha.\lambda + nl\alpha$, $\lambda \in X$ (then W and W_l may be identified).

Let us for any $\lambda \in X$ denote by $M_l(\lambda)$ the Verma module with highest weight λ on which c acts as multiplication by $-l - \hat{h}$. Its unique simple quotient is then denoted by $L_l(\lambda)$.

The level $-l - \hat{h}$ conjecture then reads

CONJECTURE (LUSZTIG 1990). *Let* $\lambda \in A_l$. *Then for any* $w \in W_l$ *we have*

$$[L_l(w.\lambda)] = \sum_{y \in W_l} (-1)^{l(yw)} P_{y,w}(1)[M_l(y.\lambda)].$$

In the non-simply laced case one should be careful to note that the affine Weyl group coming into play here is the one corresponding to the dual root system, cf. [L5].

(iii) It is shown in [Ku] that the validity of Conjecture 4b.1 for $\lambda = -2\rho$ implies the conjecture for all $\lambda \in \bar{A}$ (the "closure" of A). The above level $-l-h$ conjecture (suitably formulated for $\lambda \in \bar{A}_l$, as in Section 2) therefore holds for all $l \in \mathbb{N}$.

4c. The quantum case at roots of unity

Let $(a_{ij})_{i,j=1,\ldots,n}$ be the Cartan matrix for our root system R. Choose $d_i \in \{1,2,3\}$ minimal such that $(d_i a_{ij})$ is symmetric. Denote by v an indeterminate. If $m \in \mathbb{N}$ we set $[m] = \frac{v^m - v^{-m}}{v - v^{-1}}$ and $[m]! = [m][m-1]\cdots[1]$. If also $t \in \mathbb{N}$ we define $\begin{bmatrix} m \\ t \end{bmatrix} = \prod_{j=1}^{t} \frac{v^{m-j+1} - v^{-m+j-1}}{v^j - v^{-j}}$. For each i we put $v_i = v^{d_i}$ and obtain $[m]_i$, $[m]_i!$, and $\begin{bmatrix} m \\ t \end{bmatrix}_i$ from the above formulas by replacing v by v_i.

The quantum group (or quantized enveloping algebra) associated with (a_{ij}) is the $\mathbb{Q}(v)$-algebra U with generators E_i, F_i, $K_i^{\pm 1}$, $i = 1, \ldots, n$, and relations

$$K_i K_j = K_j K_i, \quad K_i K_i^{-1} = 1 = K_i^{-1} K_i,$$

$$K_i E_j K_i^{-1} = v_i^{a_{ij}} E_j, \quad K_i F_j K_i^{-1} = v_i^{-a_{ij}} F_j$$

$$E_i F_j - F_j E_i = \delta_{ij} \frac{K_i - K_i^{-1}}{v_i - v_i^{-1}}$$

$$\sum_{r+s=1-a_{ij}} (-1)^s \begin{bmatrix} 1 - a_{ij} \\ s \end{bmatrix}_i E_i^r E_j E_i^s = 0$$

$$\sum_{r+s=1-a_{ij}} (-1)^s \begin{bmatrix} 1 - a_{ij} \\ s \end{bmatrix}_i F_i^r F_j F_i^s = 0$$

In addition to the algebra structure on U there is also a comultiplication Δ, a counit ϵ, and an antipode S that make U into a Hopf algebra. We refer to [L4] for these definitions.

Set $\mathcal{A} = \mathbb{Z}[v, v^{-1}]$ and let $U_{\mathcal{A}}$ be Lusztig's $\mathcal{A}$-form in U, i.e. the $\mathcal{A}$-subalgebra generated by $K_i^{\pm 1}$, $E_i^{(m)}$, $F_i^{(m)}$, $i = 1, \ldots, n$, $m \in \mathbb{N}$. Here $E_i^{(m)} = \frac{E_i^m}{[m]!}$ and a similar form holds for $F_i^{(m)}$. Then one checks that $U_{\mathcal{A}}$ is in fact a Hopf subalgebra of U. For any $\mathcal{A}$-algebra we set $U_k = U_{\mathcal{A}} \otimes_{\mathcal{A}} k$ and call this the quantum group over k.

Choose now a nonzero $q \in \mathbb{C}$ and make $\mathbb{C}$ into an $\mathcal{A}$-algebra by specializing v to q. We'll denote the corresponding quantum group over $\mathbb{C}$ by U_q.

The finite-dimensional simple U_q-modules are parametrized by $X^+ \times \{\pm 1\}^n$. For $(\lambda, \epsilon) \in X^+ \times \{\pm 1\}$ we let $L_q(\lambda, \epsilon)$ denote the corresponding simple U_q-module. When $\epsilon = (1, \ldots, 1)$ we write $L_q(\lambda)$ instead of $L_q(\lambda, \epsilon)$. Because the other simple U_q-modules can be obtained (via conjugation by algebra automorphisms of U_q) from the $L_q(\lambda)$'s, we'll limit our discussion to those.

Just as in the modular case (Section 2) we can realize $L_q(\lambda)$ as the socle of an induced module: Let $U_{\mathcal{A}}^0$ (resp. $U_{\mathcal{A}}^-$) be the subalgebra of $U_{\mathcal{A}}$ generated by K_i and

$$\begin{bmatrix} K_i; c \\ m \end{bmatrix} = \prod_{j=1}^{m} \frac{K_i v_i^{c-j+1} - K_i^{-1} v_i^{-c+j-1}}{v_i^j - v_i^{-j}}, \quad c \in \mathbb{Z}, \ m \in \mathbb{N}, \ i = 1, \ldots, n,$$

(resp. by $F_i^{(m)}$, $i = 1, \ldots, n$, $m \in \mathbb{N}$). Set $U_{\mathcal{A}}^{\leq 0} = U_{\mathcal{A}}^- U_{\mathcal{A}}^0$. By U_q^0, U_q^-, and $U_q^{\leq 0}$ we understand the corresponding subalgebras of U_q.

Any $\lambda \in X$ defines a character of $U_{\mathcal{A}}^0$ and $U_{\overline{\mathcal{A}}}^{\leq 0}$ via the formulas

$$\lambda(K_i) = v_i^{\langle \lambda, \alpha_i^\vee \rangle}, \quad \lambda\left(\begin{bmatrix} K_i; c \\ m \end{bmatrix}\right) = \begin{bmatrix} \langle \lambda, \alpha_i^\vee \rangle + c \\ m \end{bmatrix}_i, \quad \lambda(F_i^{(m)}) = 0$$

for all i, m, c. Specializing v to q we get a character of U_q^0 and $U_{\overline{q}}^{\leq 0}$ that we also denote by λ. If now M is a U_q^0-module we set

$$M_\lambda = \{x \in M \mid tx = \lambda(t)x, \ t \in U_q^0\},$$

and if M is a U_q-module we set

$$F(M) = \left\{x \in \bigoplus_{\lambda \in X} M_\lambda \mid E_i^{(s)}x = 0 = F_i^{(s)}x \text{ for all } s \gg 0, \ i = 1, \ldots, n\right\}.$$

Then $F(M)$ is a U_q-submodule of M.

The U_q-module induced by $\lambda \in X$ is (cf. [APW])

$$H_q^0(\lambda) = F\left(\{f \in \mathrm{Hom}_{\mathbb{C}}(U_q, \mathbb{C}) \mid f(bu) = \lambda(b)f(u), \ b \in U_{\overline{q}}^{\leq 0}, \ u \in U_q\}\right).$$

Here $\mathrm{Hom}(U_q, \mathbb{C})$ is a U_q-module via $uf: x \mapsto f(xu)$, $u, x \in U_q$, $f \in \mathrm{Hom}_{\mathbb{C}}(U_q, \mathbb{C})$.

In analogy with the situation in Section 2 we have

$$H_q^0(\lambda) \neq 0 \text{ iff } \lambda \in X^+. \tag{4c.1}$$

We let H_q^i denote the ith right derived functor of induction from $U_{\overline{q}}^{\leq 0}$ to U_q. Then we have the following q-analogue of Kempf's vanishing theorem.

4c.1 THEOREM. *If $\lambda + \rho \in X^+$, then $H_q^i(\lambda) = 0$ for all $i > 0$.*

This result was proved with some restriction on q in [APW] and [AW] by a reduction to the classical case $v = 1$. (See also [PW] for type A.) Recently, Ryom-Hansen [R-H] has established the result in full generality by using some of the nice properties of Kashiwara's crystal base [Ka2].

This result implies (just as in the modular case) that $[H_q^0(\lambda)]$ is given by Weyl's character formula. Hence the problem of determining $[L_q(\lambda)]$ is equivalent to finding the integers $c_{\lambda,\mu}$ for which

$$[L_q(\lambda)] = \sum_{\mu \in X^+} c_{\lambda,\mu}[H_q^0(\mu)].$$

If q is not a root of unity, then $L_q(\lambda) = H_q^0(\lambda)$ for all $\lambda \in X^+$ and the category of finite-dimensional U_q-modules is semi-simple. This follows, for instance, immediately from the q-version of the Borel-Weil-Bott theorem, see [APW].

Assume now for the rest of this subsection that q is a root of unity and let l' denote its order. Set $l = l'$ if l' is odd and $l = \frac{l'}{2}$ if l' is even. We shall assume $l \geq 2\max\{d_i\}$. Then we have a situation very similar to the modular case treated in Section 2. There is a linkage principle [APW], [AW] that says that $c_{\lambda,\mu} = 0$ unless $\mu \in W_l.\lambda$ (Here W_l again denotes the affine Weyl group acting on X as before.) The analogue of Conjecture 3.1 is (see 4b above for the definition of A_l)

4c.2 CONJECTURE (LUSZTIG 1989). *Let $\lambda \in X^+$ and choose $w \in W_l$ minimal such that $w^{-1}.\lambda \in \bar{A}_l$. Then*

$$[L_q(\lambda)] = \sum_{\substack{y \in W_l \\ yw^{-1}.\lambda \in X^+}} (-1)^{l(yw)} P_{y,w}(1)[H_q^0(yw^{-1}.\lambda)].$$

4c.3 REMARKS. **(i)** The difference between this conjecture (which was stated in [L2], [L3], [L5]) and Conjecture 3.1 is that there are no upper bounds on λ. This means that it always (i.e. for all l) gives all finite-dimensional irreducible characters of U_q.

(ii) There is a q-analogue of Steinberg's tensor product theorem, see [L2], [AW], which in the notation from Section 2 says

If $\lambda = \lambda^0 + l\lambda^1$ with $\lambda^0 \in X_l$, $\lambda^1 \in X^+$, then $L_q(\lambda) \simeq L_q(\lambda_0) \otimes L(\lambda')^{[q]}$. (4c.2)

Here $L(\lambda^1)$ is the irreducible representation with highest weight λ^1 of the semi-simple complex Lie algebra corresponding to (a_{ij}) or to ${}^t(a_{ij})$ and $L(\lambda^1)^{[q]}$ is the U_q-module obtained from it via the quantum Frobenius homomorphism, see [L4].

The result reduces the problem of determining $[L_q(\lambda)]$ to the (finite) case $\lambda \in X_l$. It follows from [Kat] that Conjecture 4c.2 is compatible with equation (4c.2).

(iii) Suppose our Cartan matrix (a_{ij}) is symmetric. Then with a very slight bound on l (no bound for type A and at worst $l > 32$ for type E_8) it is shown by Kazhdan and Lusztig [KL3] that the category of finite-dimensional U_q-modules is equivalent to a subcategory of the category $\mathcal{O}_l$ for the affine Kac-Moody algebra corresponding to the extended Cartan matrix associated to (a_{ij}). Here the index l in $\mathcal{O}_l$ means that the modules in the category all have fixed negative level $-l - \hat{h}$ (see 4b). It follows (see Remarks 4b.2 (i) and (iii)) that conjecture 4c.2 holds in this case.

In the case where (a_{ij}) is not symmetric, Lusztig has shown [L5] that a modification of this procedure for proving Conjecture 4c.2 still works provided Conjecture 4b.1 can be extended to certain nonintegral negative level cases.

(iv) Suppose now that k is an arbitrary field that is made into an $\mathcal{A}$-algebra by specializing v to $q \in k \backslash \{0\}$. We'll use the same notation as above.

If $\mathrm{char}(k) = 0$ everything remains unchanged. Also if $\mathrm{char}(k) = p > 0$ and q is not a root of unity, we still have $L_q(\lambda) = H_q^0(\lambda)$ (see [T]).

Suppose on the other hand that $q \in k$ is an lth or a $2l$th root of unity. Then $(l,p) = 1$. We then have an analogue of the Steinberg tensor product theorem completely identical to equation (4c.2) above except that the last factor $L(\lambda^1)$ on the right-hand side is now the modular irreducible module with highest weight λ^1.

In this case the analogous conjecture should read

CONJECTURE. *Suppose $p \geq h$. Let $\lambda \in X^+$ and suppose $\langle \lambda + \rho, \alpha^\vee \rangle < lp$ for all $\alpha \in R_+$. For $w \in W_l$ as before we have*

$$[L_q(\lambda)] = \sum_{\substack{y \in W_l \\ yw^{-1}.\lambda \in X^+}} (-1)^{l(yw)} P_{y,w}(1)[H_q^0(yw^{-1}.\lambda)].$$

Because of the analogue of the Steinberg tensor product theorem just mentioned, this conjecture still gives all irreducible characters for U_q in this case. As evidence for it we note that for fixed l it is easy to deduce it from its characteristic zero counterpart for all $p \gg 0$. Moreover, direct calculations based on the sum formula and the translation principle [T] show that it holds for all p when the Cartan matrix has rank 2 or is type A_3.

5 From the quantum to the modular conjecture

One of the main theorems in [AJS] says that the quantum conjecture (Conjecture 4c.2) implies the modular conjecture (Conjecture 3.1) for large primes. Here we give a brief sketch of how this result is obtained.

Let $\mathfrak{g}$ be the p-Lie algebra of G. Its p-operation is denoted $X \mapsto X^{[p]}$, $X \in \mathfrak{g}$. We have a triangular decomposition of $\mathfrak{g}$, $\mathfrak{g} = \mathfrak{n}^- \oplus \mathfrak{h} \oplus \mathfrak{n}^+$, with $\mathfrak{h}$ being the Lie algebra of T. Then we set $U = U(\mathfrak{g})/I$ where $U(\mathfrak{g})$ is the enveloping algebra of $\mathfrak{g}$ and I is the two-sided ideal in $U(\mathfrak{g})$ generated by $\{X^p - X^{[p]} \mid X \in \mathfrak{n}^- \cup \mathfrak{n}^+\}$. We set $U^0 = U(\mathfrak{h})$ and $U^{\pm} = $ image in U of $U(\mathfrak{n}^{\pm})$ so that $U = U^- U^0 U^+$. The adjoint action of T on $U(\mathfrak{g})$ induces an action of T on U and for $\nu \in X$ we let U_ν denote the ν-weight space in U for this action.

Suppose now that A is a Noetherian commutative ring with a ring homomorphism $\pi : U^0 \to A$. Then we define $\mathcal{C}_A$ to be the category of all $U \otimes A$-modules M such that

$$\text{(i) } M = \bigoplus_{\mu \in X} M_\mu \text{ (as right } A\text{-modules)}$$

for some finitely generated A-modules M_μ with $M_\mu = 0$ except for finitely many $\mu \in X$.

$$\text{(ii) } U_\nu M_\mu \subseteq M_{\nu+\mu}, \ \mu, \nu \in X$$

$$\text{(iii) } sm = m\pi(\tilde{\mu}(s)), \ s \in U^0, \ m \in M_\mu.$$

Here $\tilde{\mu} : U^0 \to U^0$ is the algebra homomorphism that takes H into $H + \mu(H)$, $H \in \mathfrak{h}$.

If $A = k$ and $\pi : U^0 \to k$ is the augmentation map (i.e. $\pi(H) = 0$ for all $H \in \mathfrak{h}$) then it is easy to see that $\mathcal{C}_k$ coincides with the category consisting of all finite dimensional $G_1 T$-modules where G_1 denotes the (first) Frobenius kernel in G.

It is well known that the irreducible $G_1 T$-modules are parametrized by X. We denote the irreducible $G_1 T$-module with highest weight λ by $L_1(\lambda)$. Then by [Cu] we have $L_1(\lambda) = L(\lambda)$ for all $\lambda \in X_p$ and hence to determine $\{[L(\lambda)] \mid \lambda \in X^+\}$ is equivalent to determine $\{[L_1(\lambda)] \mid \lambda \in X_p\}$.

As in the characteristic zero category $\mathcal{O}$ case, the $L_1(\lambda)$'s may be realized as the unique quotients of (baby) Verma modules, which in this context are denoted by $Z_1(\lambda)$, $\lambda \in X$. We say that a module $M \in \mathcal{C}_k$ has a Z-filtration if it may be filtered by submodules whose quotients are isomorphic to $Z_1(\lambda)$'s for appropriate $\lambda \in X$. Among the modules in $\mathcal{C}_k$ that have a Z-filtration we find the projective covers $Q_1(\lambda)$ of the simple modules $L_1(\lambda)$, $\lambda \in X$. Moreover, we have the Brauer-Humphreys reciprocity [H]

For all $\lambda, \mu \in X$ we have $[Q_1(\lambda) : Z_1(\mu)] = [Z_1(\mu) : L_1(\lambda)]$. $\qquad$ (5.1)

Here the first number denotes the number of times $Z_1(\mu)$ occurs in a Z-filtration of $Q_1(\lambda)$, whereas the second number is the composition factor multiplicity of $L_1(\lambda)$ in $Z_1(\lambda)$.

We see that to find $[L_1(\lambda)]$, $\lambda \in X_p$, is equivalent to finding $[Q_1(\lambda)]$, $\lambda \in X_p$, i.e. to determine the numbers $[Q_1(\lambda) : Z_1(\mu)]$, $\lambda \in X_p$, $\mu \in X$.

Let now A be the local ring at the maximal ideal generated by $\{H \mid H \in \mathfrak{h}\}$ in U^0 and let $\pi \colon U^0 \to A$ be the natural embedding. The construction of Verma modules works just as well in $\mathcal{C}_A$ (giving the modules $Z_A(\lambda)$, $\lambda \in X$) and it turns out that the projective modules $Q_1(\lambda)$ lift to projective objects $Q_A(\lambda)$ in $\mathcal{C}_A$, $\lambda \in X$. In the obvious notation we have

$$[Q_A(\lambda) : Z_A(\mu)] = [Q_1(\lambda) : Z_1(\mu)], \ \lambda, \mu \in X. \qquad (5.2)$$

Assume $p \geq h$. It is then well known that $Q_1(\lambda)$ may be obtained as a distinguished summand of the module one gets by applying a sequence of translation functors to the Steinberg module $\mathrm{St}_1 = Z_1((p-1)\rho) = L_1((p-1)\rho) = Q_1((p-1)\rho)$. Hence we are lead to study the endomorphism rings of these translated Steinberg modules. We do this by constructing a fully faithful functor $\mathcal{V} \colon \mathcal{C}_A \to \mathcal{K}_A$ where $\mathcal{K}_A$ is a "combinatorics" category consisting of modules over various localizations of A. Moreover, it is possible to define translation functors on $\mathcal{K}_A$ that correspond to the translation functors on $\mathcal{C}_A$. Some rather elaborate explicit computations verify that this may actually be done independently of k. In fact, we check that if S denotes the symmetric algebra (over $\mathbb{Z}$) of a $\mathbb{Z}$-form of $\mathfrak{h}$ then there is a category $\mathcal{K}_S$ with translation functors such that the objects in $\mathcal{K}_A$ and the corresponding Hom-spaces of interest to us are obtained via the base change $S \to A$.

Suppose $Q \in \mathcal{C}_k$ is a translated Steinberg module (as described above). Then there exists $Q_A \in \mathcal{C}_A$ with $Q_A \otimes k \simeq Q$. Moreover, there exists $\mathcal{Q} \in \mathcal{K}_S$ such that $\mathcal{V}Q_A \simeq \mathcal{Q} \otimes_S A$ and

$$\mathrm{End}_{\mathcal{C}_k}(Q) \simeq \mathrm{End}_{\mathcal{C}_A}(Q_A) \otimes_A k \simeq \mathrm{End}_{\mathcal{K}_A}(\mathcal{V}Q_A) \otimes_A k \simeq \mathrm{End}_{\mathcal{K}_S}(\mathcal{Q}) \otimes_S k \quad (5.3)$$

(for the last step we actually need gradings on $\mathcal{K}_A$ and $\mathcal{K}_S$ that make it possible via dimension counts to establish the desired isomorphism).

Set now $k' = \mathbb{Q}(\zeta)$ where ζ is a primitive lth root of 1 with $l \geq h$. Then U, $\mathcal{C}_k$, $\mathcal{C}_A$ and $\mathcal{K}_A$ have quantum analogues U', $\mathcal{C}_{k'}$, $\mathcal{C}_{A'}$, and $\mathcal{K}_{A'}$. There is a module $Q' \in \mathcal{C}_{k'}$, which corresponds to Q. Setting $\mathcal{E} = \mathrm{End}_{\mathcal{K}_S}(\mathcal{Q}) \otimes_S \mathbb{Z}$ (where $\mathbb{Z}$ is an S-algebra via the augmentation map $S \to \mathbb{Z}$) we get from (5.3)

$$\mathcal{E} \otimes_{\mathbb{Z}} k \simeq \mathrm{End}_{\mathcal{C}_k}(Q). \qquad (5.4)$$

Analogous arguments show

$$\mathcal{E} \otimes_{\mathbb{Z}} k' \simeq \mathrm{End}_{\mathcal{C}'_k}(Q'). \qquad (5.4')$$

It follows that the integers $c_{w,y}$ considered in Section 2 (cf. Remark 3.2b) are independent of k for all fields k with $p = \mathrm{char}(k) \gg 0$. The same is true for the quantum case at roots of unity of order $l \geq h$. In particular, we obtain

5.1 COROLLARY. *Suppose $l = p \gg 0$ and let $\lambda \in X_p$. Then $[L(\lambda)] = [L_q(\lambda)]$.*

5.2 REMARK. This corollary is of course a consequence of Lusztig's conjectures. By the results mentioned in Section 4 it verifies the modular conjecture for $p \gg 0$ (once the nonintegral negative level affine algebra case is proved, see Remark 4c.3 (iii)). This condition on p means that there exists $p_0 \in \mathbb{N}$ depending only on R such that the conjecture holds for all $p > p_0$. Unfortunately, our methods do not give a good estimate for p_0.

References

[A1]   H. H. Andersen, *The strong linkage principle*, J. Reine Angew. Math., **315** (1980), 53–59.

[A2]   H. H. Andersen, *The Frobenius morphism on the cohomology of homogeneous vector bundles on G/B*, Ann. of Math. (2), **112** (1980), 113–121.

[AJS] H. H. Andersen, J. C. Jantzen, and W. Soergel, *Representations of quantum groups at a p-th root of unity and of semisimple groups in characteristic p: Independence of p*, Astérisque, **220** (1994).

[APW] H. H. Andersen, P. Polo, and Wen K., *Representations of quantum algebras*, Invent. Math., **104** (1991), 1–59.

[AW]   H. H. Andersen and Wen K., *Representations of quantum algebras. The mixed case*, J. Reine Angew. Math., **427** (1992), 35–50.

[BB]   A. Beilinson and J. Bernstein, *Localisation de $\mathfrak{g}$-modules*, C. R. Acad. Sci. Paris, **292** (1981), 15–18.

[BK]   J.-L. Brylinski and M. Kashiwara, *Kazhdan-Lusztig conjecture and holonomic system*, Invent. Math., **64** (1981), 387–410.

[C1]   L. Casian, *Formule de multiplicité de Kazhdan-Lusztig dans le case de Kac-Moody*, C.R. Acad. Sci. Paris, **310** (1990), 333–337.

[C2]   L. Casian, *Kazhdan-Lusztig conjecture in the negative level case (Kac-Moody algebras of affine type)*, preprint.

[Ch]   C. Chevalley, *Les systèmes linéaires sur G/B*, Exp. 15 in: Sem. Chevalley (Classification des groupes de Lie algébriques), (1957), Paris.

[Cu]   C. W. Curtis, *Representations of Lie algebras of classical type with application to linear groups*, J. Math. Mech., **9** (1960), 307–326.

[DGK] V. Deodhar, O. Gabber, and V. Kac, *Structure of some categories of infinite-dimensional Lie algebras*, Adv. in Math., **45** (1982), 92–116.

[D]   V. G. Drinfeld, *On quasitriangular quasi-Hopf algebras closely related to $Gal(\bar{Q}/Q)$*, Algebra i Analiz 2, (1990), 149–181.

[Ha]   W. Haboush, *A short proof of the Kempf vanishing theorem*, Invent. Math., **56** (1980), 109–112.

[H]   J. E. Humphreys, Ordinary and modular representations of Chevalley groups, Lecture Notes in Math., **528** (1976), Springer, Berlin and New York.

[J]   J. C. Jantzen, Representations of algebraic groups, Pure and Appl. Math., **131**, (1987) Academic Press, Boston MA.

[K]   V. Kac, Infinite Dimensional Lie Algebras, Third edition, (1990), Cambridge Univ. Press, London and New York.

[Ka1] M. Kashiwara, *Kazhdan-Lusztig conjecture for symmetrizable Kac-Moody Lie algebra*, in: The Grothendieck Festschrift, Vol II, Progr. Math., **87**, (1990) Birkhäuser, Basel and Boston, 407–433.

[Ka2] M. Kashiwara, *Crystal base and Littelmann's refined Demazure character formula*, Duke Math. J., **71** (1993), 839–858.

[KT1]   M. Kashiwara and T. Tanisaki, *Kazhdan-Lusztig conjecture for symmetriza-ble Kac-Moody Lie algebra II*, Progr. Math., **92**, (1990) Birkhäuser, Basel and Boston, 159–195.

[KT2]   M. Kashiwara and T. Tanisaki, *Characters of the negative level highest weight modules for affine Lie algebras*, preprint (1994).

[Kat]   S. Kato, *On the Kazhdan-Lusztig polynomials for affine Weyl groups*, Adv. in Math., **55** (1985), 103–130.

[KL1]   D. Kazhdan and G. Lusztig, *Representations of Coxeter groups and Hecke algebras*, Invent. Math., **53** (1979), 165–184.

[KL2]   D. Kazhdan and G. Lusztig, *Schubert varieties and Poincaré duality*, Proc. Sympos. Pure Math., **36** (1980), 185–203.

[KL3]   D. Kazhdan and G. Lusztig, *Tensor structures arising from affine Lie algebras I-IV*, J. Amer. Math. Soc., **6** (1994), 905–947, 949–1011 and **7** (1994), 335–381, 383–453.

[Ke]   G. Kempf, *Linear systems on homogeneous spaces*, Ann. of Math (2), **103** (1976), 557–591.

[Ku]   S. Kumar, *Toward proof of Lusztig's conjecture concerning negative level representations of affine Lie algebras*, J. Algebra, **164** (1994), 515–527.

[L1]   G. Lusztig, *Some problems in the representation theory of finite Chevalley groups*, Proc. Sympos. Pure Math., **37** (1980), 313–318.

[L2]   G. Lusztig, *Modular representations and quantum groups*, Contemp. Math., **82** (1989), 58–77.

[L3]   G. Lusztig, *On quantum groups*, J. Algebra, **131** (1990), 466–475.

[L4]   G. Lusztig, Introduction to quantum groups, Progr. Math., **110** (1993), Birkhäuser, Basel and Boston.

[L5]   G. Lusztig, *Monodromic systems on affine flag manifolds*, Proc. Roy. Soc. London series A (1994), **445**, 231–246.

[PW]   B. Parshall and Wang J.-P., *Quantum linear groups*, Mem. Amer. Math. Soc., **439** (1991).

[R-H]   S. Ryom-Hansen, *A q-analogue of Kempf's vanishing theorem*, preprint (1994).

[S]   W. Soergel, *Kategorie O, perverse Garben und Moduln über den Koinvarianten zur Weylgruppe*, J. Amer. Math. Soc., **3** (1990), 421–445.

[St]   R. Steinberg, *Representations of algebraic groups*, Nagoya Math. J., **22** (1963), 33–56.

[T]   L. Thams, *Two classical results in the quantum mixed case*, J. Reine Angew. Math., **436** (1993), 129–153.

[V]   D. Vogan, Representations of real reductive Lie groups, Progr. Math., **15** (1981), Birkhäuser, Basel and Boston.

Automorphic Forms on $O_{s+2,2}(\mathbb{R})^+$ and Generalized Kac-Moody Algebras

RICHARD E. BORCHERDS [*]

Department of Mathematics
University of California at Berkeley
CA 94720-3840, USA

ABSTRACT. We discuss how modular forms and automorphic forms can be written as infinite products, and how some of these infinite products appear in the theory of generalized Kac-Moody algebras. This paper is based on my talk at the ICM, and is an exposition of [B5].

1 Product formulas for modular forms

We will start off by listing some apparently random and unrelated facts about modular forms, which will begin to make sense in a page or two. A modular form of level 1 and weight k is a holomorphic function f on the upper half plane $\{\tau \in \mathbb{C} | \Im(\tau) > 0\}$ such that $f((a\tau+b)/(c\tau+d)) = (c\tau+d)^k f(\tau)$ for $\binom{ab}{cd} \in SL_2(\mathbb{Z})$ that is "holomorphic at the cusps". Recall that the ring of modular forms of level 1 is generated by $E_4(\tau) = 1 + 240 \sum_{n>0} \sigma_3(n)q^n$ of weight 4 and $E_6(\tau) = 1 - 504 \sum_{n>0} \sigma_5(n)q^n$ of weight 6, where $q = e^{2\pi i \tau}$ and $\sigma_k(n) = \sum_{d|n} d^k$. There is a well-known product formula for $\Delta(\tau) = (E_4(\tau)^3 - E_6(\tau)^2)/1728$

$$\Delta(\tau) = q \prod_{n>0} (1 - q^n)^{24}$$

due to Jacobi. This suggests that we could try to write other modular forms, for example E_4 or E_6, as infinite products. At first sight this does not seem to be very promising. We can formally expand any power series as an infinite product of the form $q^h \prod_{n>0}(1 - q^n)^{a(n)}$, and if we do this for E_4 we find that

$$
\begin{aligned}
E_4(\tau) &= 1 + 240q + 2160q^2 + 6720q^3 + \cdots \\
&= (1-q)^{-240}(1-q^2)^{26760}(1-q^3)^{-4096240} \cdots
\end{aligned}
\tag{1.1}
$$

but this infinite product does not even converge everywhere because the coefficients are exponentially increasing. In fact, such an infinite product can only converge everywhere if the function it represents has no zeros in the upper half plane, and the only level 1 modular forms with this property are the powers of Δ. On the

---

[*]   Supported by NSF grant DMS-9401186.

Proceedings of the International Congress
of Mathematicians, Zürich, Switzerland 1994
© Birkhäuser Verlag, Basel, Switzerland 1995

other hand there is a vague principle that a function should have a nice product expansion if and only if its zeros and poles are arranged nicely. Well-known examples of this are Euler's product formulas for the gamma function and zeta function, and Jacobi's product formulas for theta functions. (Of course the region of convergence of the infinite product will usually not be the whole region where the function is defined, because it cannot contain any zeros or poles.) The zeros of any modular form are arranged in a reasonably regular way which suggests that some modular forms with zeros might still have nice infinite product expansions.

For a reason that will appear soon we will now look at modular forms of level 4 and weight 1/2 that are holomorphic on the upper half plane but are allowed to have poles at cusps. Kohnen's work [Ko] on the Shimura correspondence suggests that we should look at the subspace A of such forms $f = \sum_{n \in \mathbb{Z}} c(n)q^n$ whose Fourier coefficients $c(n)$ are all integers and vanish unless $n \equiv 0$ or $1 \bmod 4$. It is easy to find the structure of A: it is a 2-dimensional free module over the ring of polynomials $\mathbb{Z}[j(4\tau)]$, where $j(\tau)$ is the elliptic modular function $j(\tau) = E_4(\tau)^3/\Delta(\tau) = q^{-1} + 744 + 196884q + \cdots$. Equivalently, any sequence of numbers $c(n)$ for $n \leq 0$ such that $c(n) = 0$ unless $n \equiv 0$ or $1 \bmod 4$ is the set of coefficients of q^n for $n \leq 0$ of a unique function in A. The space A has a basis consisting of the following two elements:

$$\theta(\tau) = \sum_{n \in \mathbb{Z}} q^{n^2} = 1 + 2q + 2q^4 + \cdots$$

$$\psi(\tau) = F(\tau)\theta(\tau)(\theta(\tau)^4 - 2F(\tau))(\theta(\tau)^4 - 16F(\tau))E_6(4\tau)/\Delta(4\tau) + 60\theta(\tau)$$
$$= q^{-3} + 4 - 240q + 26760q^4 - 85995q^5 + 1707264q^8 - 4096240q^9 + \cdots \tag{1.2}$$

where

$$F(\tau) = \sum_{n>0, n\,\mathrm{odd}} \sigma_1(n)q^n = q + 4q^3 + 6q^5 \cdots.$$

The reader will now understand the reason for all these odd definitions by comparing the coefficients of ψ in (1.2) with the exponents in (1.1). This is a special case of the following theorem.

THEOREM 1.1. ([B5]) *Suppose that B is the space of meromorphic modular forms Φ of integral weight and level 1 for some character of $SL_2(\mathbb{Z})$ such that Φ has integral coefficients, leading coefficient 1, and all the zeros and poles of Φ are at cusps or imaginary quadratic irrationals. Then the map taking $f(\tau) = \sum_{n \in \mathbb{Z}} c(n)q^n \in A$ to*

$$\Phi(\tau) = q^h \prod_{n>0}(1 - q^n)^{c(n^2)}$$

is an isomorphism from A to B. (Here h is a certain rational number in $\frac{1}{12}\mathbb{Z}$ depending linearly on f.) The weight of Φ is $c(0)$, and the multiplicity of the zero of Φ at an imaginary quadratic number of discriminant $D < 0$ is $\sum_{n>0} c(n^2 D)$.

EXAMPLE 1.2. If $f(\tau) = 12\theta(\tau) = 12 + 24q + 24q^4 + \cdots$ then $c(n^2) = 24$ for $n > 0$, so $\Phi(\tau) = q \prod_{n>0}(1 - q^n)^{c(n^2)} = q \prod_{n>0}(1 - q^n)^{24} = \Delta(\tau)$, which is the usual product formula for the Δ function. The fact that $f(\tau)$ is holomorphic corresponds to the fact that $\Delta(\tau)$ has no zeros, and the constant term 12 of $f(\tau)$ is the weight of Δ.

EXAMPLE 1.3. Most of the common modular forms or functions, for example the Eisenstein series E_4, E_6, E_8, E_{10}, and E_{14}, the delta function $\Delta(\tau)$, and the elliptic modular function $j(\tau)$, all belong to the space B and can therefore be written explicitly as infinite products. It is easy to work out the function f corresponding to Φ by using the remarks about weights and multiplicities of zeros at the end of Theorem 1.1; for example, if $\Phi(\tau) = j(\tau) - 1728$, then Φ has a zero of order 2 at every imaginary quadratic irrational of discriminant -4 and has weight 0, so the corresponding function $f \in A$ must be of the form $2q^{-4} + 0q^0 + O(q)$ and must therefore be $2\theta(\tau)(j(4\tau) - 738) - 4\psi(\tau)$.

Theorem 1.1 looks superficially similar to the Shimura correspondence; both correspondences use infinite products to take certain modular forms of half integral weight to modular forms of integral weight. However there are several major differences: the Shimura correspondence uses Euler product expansions, only works for holomorphic modular forms, and is an additive rather than a multiplicative correspondence.

2 Product formulas for $j(\sigma) - j(\tau)$

In this section we describe three different product formulas for $j(\sigma) - j(\tau)$ (where $j(\tau) = \sum_n c(n)q^n = q^{-1} + 744 + \cdots$ is the elliptic modular function).

The simplest one is valid for any σ, τ with large imaginary part (> 1 will do), and is

$$j(\sigma) - j(\tau) = p^{-1} \prod_{m>0, n \in \mathbb{Z}} (1 - p^m q^n)^{c(mn)} \tag{2.1}$$

where $p = e^{2\pi i \sigma}$. This is the denominator formula for the monster Lie algebra; see Example 5.2.

The next product formula was found by Gross and Zagier [GZ]. We let d_1 and d_2 be negative integers that are 0 or 1 mod 4, and for simplicity we suppose that they are both less than -4. Then

$$\prod_{[\tau_1],[\tau_2]} (j(\tau_1) - j(\tau_2)) = \pm \prod_{x \in \mathbb{Z}, n, n' > 0, x^2 + 4nn' = d_1 d_2} n^{\epsilon(n')}$$

where the first product is over representatives of equivalence classes of imaginary quadratic irrationals of discriminants d_1, d_2, and $\epsilon(n') = \pm 1$ is defined in [GZ]. An example of this given by Gross and Zagier is

$$j\left(\frac{1 + i\sqrt{67}}{2}\right) - j\left(\frac{1 + i\sqrt{163}}{2}\right) = -2^{15}3^3 5^3 11^3 + 2^{18}3^3 5^3 23^3 29^3$$

$$= 2^{15}3^7 5^3 7^2 13 \times 139 \times 331.$$

In the first product formula σ and τ were both arbitrary complex numbers with a large imaginary part, and in the second they were both fixed to run over imaginary quadratic irrationals. The third product formula is a sort of cross between these, because we allow τ to be any complex number with a large imaginary part, and make σ run over a set of representatives of imaginary quadratic irrationals of some fixed discriminant d. For simplicity we will assume $d < -4$ and d squarefree. In this case we find that

$$\prod_{[\sigma]}((j(\tau) - j(\sigma))) = q^{-h}\prod_{n \geq 1}(1 - q^n)^{c(n)}$$

where the numbers $c(n)$ are the coefficients of the unique power series in the space A of 1.1 of the form $q^{-d} + O(q)$, and h is the class number of the imaginary quadratic field of $\sqrt{d}$. This follows from Theorem 1.1 because it is easy to see that the product on the left lies in the space B. Conversely, the analogue of this product formula for all values of d together with the Jacobi product formula for the eta function implies Theorem 1.1.

Strangely enough, there seems to be no obvious direct connection between these three product formulas. In particular, assuming any two of them does not seem to be of any help in proving the third one. (Proposition 5.1 of [GZ] almost gives a fourth product formula: it expresses $\log|j(\sigma) - j(\tau)|$ as a limit of an infinite sum.)

3 Automorphic forms for $O_{s+2,2}(\mathbb{R})$

Theorem 1.1 is essentially a specialization of a product formula for automorphic forms on higher dimensional orthogonal groups $O_{s+2,2}(\mathbb{R})^+$. Before giving this generalization we recall the definitions of automorphic forms on orthogonal groups and of rational quadratic divisors.

We will show how to construct the analogue of the upper half plane for these groups. Suppose that L is a Lorentzian lattice of dimension $s + 2$, in other words, a nonsingular lattice of dimension $s + 2$ and signature s. The negative norm vectors in $L \otimes \mathbb{R}$ form two open cones; we choose one of these cones and call it C. We define H to be the subset of vectors $\tau \in L \otimes \mathbb{C}$ such that $\Im(\tau) \in C$, so that if L is one dimensional, then H is isomorphic to the upper half plane. There is an obvious discrete group acting on H generated by the translations $\tau \to \tau + \lambda$ for $\lambda \in L$ and the automorphisms $O_L(\mathbb{Z})^+$ of L that map C into itself. When H is the upper half plane this group is just the group of translations $\tau \to \tau + n$ for $n \in \mathbb{Z}$. In this case we can enlarge the group to $SL_2(\mathbb{Z})$ acting on the upper half plane (by $\binom{ab}{cd}(\tau) = (a\tau + b)/(c\tau + d)$) by adding an extra automorphism $\tau \to -1/\tau$. The analogue of this for unimodular Lorentzian lattices L is the automorphism $\tau \to 2\tau/(\tau, \tau)$.

An automorphic form of weight k on the upper half plane H is a function satisfying the two functional equations

$$f(\tau + n) = f(\tau) \quad (n \in \mathbb{Z})$$
$$f(-1/\tau) = \tau^k f(\tau)$$

(and some conditions about being holomorphic). By analogy with this, if L is an even unimodular Lorentzian lattice, we define an automorphic form on H to be a holomorphic function f on H satisfying the functional equations

$$f(\tau + \lambda) = f(\tau) \quad (\lambda \in L)$$
$$f(w(\tau)) = \pm f(\tau) \quad (w \in O_L(\mathbb{Z})^+)$$
$$f(2\tau/(\tau,\tau)) = \pm((\tau,\tau)/2)^k f(\tau).$$

The group generated by all these transformations is isomorphic to a subgroup of index 2 of the automorphism group of the lattice $M = L \oplus II_{1,1}$, where $II_{1,1}$ is the 2-dimensional even Lorentzian lattice (with inner product matrix $\begin{pmatrix} 0 & 1 \\ 1 & 0 \end{pmatrix}$).

Suppose that $b \in L$ and $a, c \in \mathbb{Z}$, with $(b,b) > 2ac$. The set of vectors $y \in L \otimes \mathbb{C}$ with $a(y,y)/2 + (b,y) + c = 0$ is called a rational quadratic divisor. A rational quadratic divisor in the upper half plane is the same as an imaginary quadratic irrational.

We choose some vector in $-C$ that has a nonzero inner product with all vectors of L, and we write $r > 0$ to mean that $r \in L$ has a positive inner product with this vector.

We have seen in Theorem 1.1 that a modular form with integer coefficients tends to have a nice infinite product expansion if all its zeros are imaginary quadratic irrationals. The next theorem shows that a similar phenomenon occurs for automorphic forms on $O_{s+2,2}(\mathbb{R})^+$, provided we replace imaginary quadratic irrationals by rational quadratic divisors.

THEOREM 3.1. ([B5]) *Suppose that* $f(\tau) = \sum_n c(n)q^n$ *is a meromorphic modular form with all poles at cusps. Suppose also that* f *is of weight* $-s/2$ *for* $SL_2(\mathbb{Z})$ *and has integer coefficients, with* $24|c(0)$ *if* $s = 0$. *There is a unique vector* $\rho \in L$ *such that*

$$\Phi(v) = e^{-2\pi i(\rho,v)} \prod_{r>0} (1 - e^{-2\pi i(r,v)})^{c(-(r,r)/2)}$$

is a meromorphic automorphic form of weight $c(0)/2$ *for* $O_M(\mathbb{Z})^+$. *(Or more precisely, can be analytically continued to a meromorphic automorphic form, because the infinite product does not converge everywhere.) All the zeros and poles of* Φ *lie on rational quadratic divisors, and the multiplicity of the zero of* Φ *at the rational quadratic divisor of the triple* (b, a, c) *(with no common factors) is*

$$\sum_{n>0} c(n^2(ac - (b,b)/2)).$$

We see that just as in Theorem 1.1, the coefficients of negative powers of q in f determine the zeros of Φ, and the constant term of f determines the weight of Φ.

Notice that the zeros of rational quadratic divisors with $a = c = 0$ can be seen as zeros of factors of the infinite product, but the other zeros cannot be seen so easily; they are not zeros of any of the factors of the infinite product and therefore lie outside the region where the infinite product converges.

4 Generalized Kac-Moody algebras

Some of the infinite products giving automorphic forms appear in the theory of generalized Kac-Moody algebras, so in this section we briefly recall some facts about these.

Generalized Kac-Moody algebras are best thought of as infinite-dimensional analogues of finite-dimensional reductive Lie algebras. They can almost be defined as Lie algebras G having the following structure [B4]

(1) G should have a nonsingular invariant bilinear form $(\,,\,)$.

(2) G should have a self-centralizing subalgebra H, called the Cartan subalgebra, such that G is the sum of eigenspaces of H.

(3) The roots of G (i.e., the eigenvalues of H acting on G) should have properties similar to those of the roots of a finite-dimensional reductive Lie algebra. In particular, it should be possible to choose a set of "positive" roots $\alpha > 0$ with good properties, a set of "simple roots", and there should be a "Weyl group" W generated by reflections of real (norm ≥ 0) simple roots. Also G has a "symmetrized Cartan matrix", whose entries are the inner products of the simple roots.

An earlier characterization [B1] identified generalized Kac-Moody algebras as Lie algebras with an "almost positive definite contravariant bilinear form", but the one summarized above is easier to use in practice because it avoids the rather difficult problem of proving positive definiteness.

There is a generalization of the Weyl character formula for the characters of some irreducible highest weight representations of generalized Kac-Moody algebras, and in particular there is a generalization of the denominator formula (coming from the character formula for the trivial representation), which is

$$\sum_{w \in W} \det(w) w(e^\rho S) = e^\rho \prod_{\alpha > 0} (1 - e^\alpha)^{\mathrm{mult}(\alpha)} \tag{4.1}$$

where $\mathrm{mult}(\alpha)$ is the multiplicity of the root α, i.e., the dimension of the corresponding eigenspace. The vector ρ is a "Weyl vector", and S is a correction term depending on the imaginary (norm ≤ 0) simple roots. For finite-dimensional reductive Lie algebras, and more generally for Kac-Moody algebras, there are no imaginary simple roots, so $S = 1$ and we recover the usual Weyl-Kac denominator formula.

The best-known examples of generalized Kac-Moody algebras are the finite-dimensional reductive Lie algebras, the affine Lie algebras, and the Heisenberg Lie algebra (which should be thought of as a sort of degenerate affine Lie algebra). Beyond these there are an enormous number of nonaffine generalized Kac-Moody algebras, which can be constructed by writing down a random symmetrized Cartan matrix, and then writing down some generators and relations corresponding to it. Most of these Lie algebras seem to be of little interest, and it does not usually seem possible to find a clean description of both the root multiplicities and the simple roots. (It is not difficult to find large alternating sums for these numbers by using the denominator formula, but these sums seem too complicated to be of much use; for example, they do not lead to good bounds for the root multiplicities.)

There is a handful of good nonaffine generalized Kac-Moody algebras, for which we can describe both the simple roots and the root multiplicities explicitly. (See the next section for some examples.) These all turn out to have the property that the product in the denominator formula is an automorphic form for an orthogonal group $O_{s+2,2}(\mathbb{R})^+$, where $s + 2$ is the dimension of the Cartan subalgebra. This suggests that this property of the denominator function being an automorphic form can be used to separate out the "interesting" generalized Kac-Moody algebras from the rest. (Something similar happens for the affine Kac-Moody algebras: in this case the denominator function is a Jacobi form [EZ].)

5 Examples

We finish by giving some applications and special cases of Theorem 3.1.

EXAMPLE 5.1.   If we have an automorphic form for the group $O_{s+2,2}(\mathbb{R})^+$ with an infinite product expansion we can restrict it to smaller subspaces to obtain automorphic forms for smaller groups $O_{s-n+2,2}(\mathbb{R})^+$ with infinite product expansions. For example, if we restrict Φ to the multiples τv of a fixed norm $-2N$ vector $v \in L$ (for τ in the upper half plane) we get a modular form of level N. In particular, by specializing the forms in Theorem 3.1 we obtain many ordinary modular forms for $SL_2(\mathbb{R})$ (which is locally isomorphic to $O_{1,2}(\mathbb{R})$) with infinite product expansions, and this can be used to prove Theorem 1.1. Similarly, we can get examples of Hilbert modular forms and genus 2 Siegel modular forms with infinite product expansions by using the fact that the groups $SL_2(\mathbb{R}) \times SL_2(\mathbb{R})$ and $Sp_4(\mathbb{R})$ are locally isomorphic to $O_{2,2}(\mathbb{R})$ and $O_{3,2}(\mathbb{R})$.

EXAMPLE 5.2.   The simplest nontrivial case of Theorem 3.1 is when L is the lattice $II_{1,1}$ and $f(\tau)$ is the elliptic modular function $j(\tau) - 744 = \sum_n c(n)q^n$. In this case Theorem 3.1 says that the infinite product

$$p^{-1} \prod_{m>0, n \in \mathbb{Z}} (1 - p^m q^n)^{c(mn)}$$

is an automorphic function on $H \times H$ (where H is the upper half plane). This product is just the right-hand side of 2.1, and using the fact that it is an automorphic function with known zeros it is easy to identify it as $j(\sigma) - j(\tau)$. This identity 2.1 is the denominator formula 4.1 for the monster Lie algebra, a generalized Kac-Moody algebra acted on by the monster simple group, which is the Lie algebra of physical states of a chiral string on an orbifold of a 26-dimensional torus [B3].

EXAMPLE 5.3.   Suppose we take L to be the 26-dimensional even unimodular lattice $II_{25,1}$, and take $f(\tau)$ to be $1/\Delta(\tau) = \sum_n p_{24}(n+1)q^n = q^{-1} + 24 + 324q^2 + \cdots$. Then by Theorem 3.1 we know that

$$\Phi(v) = e^{-2\pi i(\rho,v)} \prod_{r>0} (1 - e^{-2\pi i(r,v)})^{p_{24}(1-(r,r)/2)} \tag{5.1}$$

is a holomorphic automorphic form of weight $24/2 = 12$. We can identify it explicitly using some facts about singular automorphic forms on $O_{s+2,2}(\mathbb{R})^+$. Any holomorphic automorphic form on $O_{s+2,2}(\mathbb{R})^+$ can be expanded as a power series $\Phi(v) = \sum_{r \in \bar{C}} c(r) e^{-2\pi i(r,v)}$ where the coefficients $c(r)$ are zero unless r lies in the closure $\bar{C}$ of the cone C. If the coefficients $c(r)$ are zero unless r lies on the boundary of C then we say that Φ is singular. It turns out that Φ is singular if and only if its weight is a "singular weight", and for $O_{s+2,2}(\mathbb{R})^+$ the singular weights are 0 and $s/2$. (Moreover any automorphic form of weight less than $s/2$ must be constant of weight 0.) In particular, the form $\Phi(v)$ above has singular weight $12 = 24/2$ so its coefficients $c(r)$ vanish unless $(r,r) = 0$. But for any automorphic form it is easy to find the multiplicities of the coefficients $c(r)$ with $(r,r) = 0$, and if we do this for Φ we find that

$$\Phi(v) = \sum_{w \in W} \det(w) \Delta((v, w(\rho))) \tag{5.2}$$

where ρ is a norm 0 vector and W is the reflection group of the lattice $II_{25,1}$. If we compare 5.1 with 5.2 we obtain the denominator formula for another Lie algebra called the fake monster Lie algebra [B2], which is the Lie algebra of physical states of a chiral string on the torus $\mathbb{R}^{25,1}/II_{25,1}$ [B3].

Incidentally we also get a short proof of the existence of the Leech lattice (a 24-dimensional even unimodular lattice with no roots), because it is not hard to show that if ρ has norm 0 then the lattice $\rho^\perp/\rho$ is extremal (i.e., has no vectors of norm $\leq s/12$), and the fact that $c(\rho) = 1$ is nonzero implies that ρ must have norm 0 because Φ is singular. It is also possible to prove the uniqueness of the Leech lattice and the fact that it has covering radius $\sqrt{2}$ using similar arguments. Unfortunately this argument does not seem to produce examples of extremal lattices in higher dimensions because the forms Φ no longer have singular weight.

The two examples above are particularly simple because the coefficients $c(r)$ vanish for $(r,r) \neq 0$, so it is easy to identify them all. Most of the automorphic forms constructed in Theorem 3.1 do not have this property and seem to be harder to describe explicitly. Moreover, most of them do not seem to be related to generalized Kac-Moody algebras, because all the positive norm roots of a generalized Kac-Moody algebra with root lattice $II_{s+1,1}$ must have norm 2, which means that the function $f(\tau)$ cannot have any terms in q^n for $n \leq -2$.

We conclude with an example of a generalized Kac-Moody algebra related to one of the modular forms in Theorem 1.1.

EXAMPLE 5.4. The product formula

$$E_6(\tau) = 1 - 504 \sum_{n>0} \sigma_5(n) q^n$$

$$= 1 - 504q - 16632q^2 - 122976q^3 - \cdots$$

$$= \prod_{n>0} (1 - q^n)^{c(n^2)}$$

$$= (1-q)^{504}(1-q^2)^{143388}(1-q^3)^{51180024} \cdots$$

where

$$\sum_n c(n^2) q^n$$
$$= \theta(\tau)(j(4\tau) - 1470) - 2\phi(\tau)$$
$$= q^{-4} + 6 + 504q + 143388q^4 + 565760q^5 + 18473000q^8 + 51180024q^9 + O(q^{12})$$

is the denominator formula for a generalized Kac-Moody algebra of rank 1 whose simple roots are all multiples of some root α of norm -2, the simple roots are $n\alpha$ ($\alpha > 0$) with multiplicity $504\sigma_3(n)$, and the multiplicity of the root $n\alpha$ is $c(n^2)$. The positive subalgebra of this generalized Kac-Moody algebra is a free Lie algebra, so we can also state this result by saying that the free graded Lie algebra with $504\sigma_5(n)$ generators of each positive degree n has a degree n piece of dimension $c(n^2)$. There are similar examples corresponding to the infinite products for the Eisenstein series E_{10} and E_{14}. The identity for E_{14} is easy to prove directly because it follows from 2.1 by dividing both sides by $p - q$, setting $p = q$, and using the fact that $j'(\tau) = E_{14}(\tau)/\Delta(\tau)$.

References

[B1] R.E. Borcherds, *Generalized Kac-Moody algebras*, J. Algebra **115** (1988), 501–512.

[B2] R.E. Borcherds, *The monster Lie algebra*, Adv. in Math. **83** no. 1, Sept. 1990.

[B3] R.E. Borcherds, *Sporadic groups and string theory*, in: First european congress of mathematics, Paris, July 6–10, 1992, Vol. **I** (A. Joseph, F. Mignot, F. Murat, B. Prum, R. Rentschler, eds.), Birkhäuser, Basel/Boston, 1994.

[B4] R.E. Borcherds, *A characterization of generalized Kac-Moody algebras*, to appear in J. Algebra, preprint 1993.

[B5] R.E. Borcherds, *Automorphic forms on $O_{n+2,2}(\mathbb{R})$ and infinite products*, Invent. Math. **120** (1995), 161–213.

[EZ] M. Eichler and D. Zagier, The theory of Jacobi forms, Progr. Math. **55**, Birkhäuser, Basel/Boston, 1985.

[GZ] B.H. Gross and D.B. Zagier, *On singular moduli*, J. Reine Angew. Math. **355** (1985), 191–220.

[K]   V.G. Kac, Infinite dimensional Lie algebras, third edition, Cambridge University Press, London and New York, 1990. (The first and second editions (Birkhäuser, Basel, 1983, and C.U.P., 1985) do not contain the material on generalized Kac-Moody algebras.)

[Ko] W. Kohnen, *Modular forms of half integral weight on $\Gamma_0(4)$*, Math. Ann. **248** (1980) 249–266.

Spherical Varieties

Michel Brion

Ecole Normale Supérieure,
46 allée d'Italie
F-69364 Lyon Cedex 7, France

Introduction

Consider a connected reductive algebraic group G over an algebraically closed field k, a Borel subgroup B of G, and a closed subgroup $H \subset G$. The homogeneous space G/H is *spherical* if B acts on it with an open orbit. Examples include flag varieties (H is parabolic in G); more generally, G/H is spherical whenever H contains a maximal unipotent subgroup of G. Another class of examples consists in symmetric spaces; here H is the fixed point set of an involutive automorphism of G. More exotic examples are G_2/SL_3 and the quotient of $SL_2 \times SL_2 \times SL_2$ by its diagonal.

The notion of a spherical homogeneous space has its origin in representation theory (Gel'fand pairs, multiplicity-free spaces). Namely, in characteristic zero, the homogeneous space G/H is spherical if and only if it satisfies the following condition: for any simple, rational G-module M, and for any multiplicative character χ of H, the χ-eigenspace of H in M is zero or a line. If moreover G/H is quasi-affine (e.g. if H is reductive), then this condition can be replaced by the following: the space of H-fixed points in any simple, rational G-module is zero or a line. If G/H is spherical, then its algebra of G-invariant differential operators is commutative; the converse holds when H is reductive.

More generally, define a *spherical variety* as a normal algebraic variety with an action of G and a dense orbit of B. In the case where G is a torus, we recover the definition of a toric variety. The rather abstract notion of a spherical variety leads to a very rich geometry, which is only partially understood. It combines features of flag varieties (e.g. the Bruhat decomposition and the Borel-Weil-Bott theorem) and of symmetric spaces (the little Weyl group and its role in equivariant embeddings; the Harish-Chandra isomorphism). As for toric varieties, the geometry of fans and convex polytopes plays a role, too.

Finally, spherical varieties are a test case for studying actions of reductive groups. Namely, several phenomena, first discovered for spherical varieties, have been generalized to arbitrary varieties with reductive group actions; see work of Knop. However, many results find a simpler and more precise formulation in the case of spherical varieties.

After some preliminaries, we survey several aspects of the theory of spherical varieties where recent progress has occurred, but where basic questions are still not

Proceedings of the International Congress
of Mathematicians, Zürich, Switzerland 1994
© Birkhäuser Verlag, Basel, Switzerland 1995

completely answered: cohomology groups of line bundles, classification of spherical varieties by combinatorial invariants, and orbits of a Borel subgroup. These topics interplay, and they are ordered here in a rather arbitrary way. We refer to Ahiezer's article [A] for an exposition of other topics concerning spherical varieties over $\mathbf{C}$: local structure theorems, orbits of a maximal compact subgroup of G, and relations with symplectic geometry.

1 First properties of spherical varieties

1.1. Complexity and rank of G-varieties

For an algebraic variety X with an action of B, we define its *complexity* $c(X)$ as the minimal codimension of a B-orbit in X. By a classical result of Rosenlicht, $c(X)$ is the transcendence degree of the extension $k(X)^B/k$ where $k(X)$ denotes the function field of X, and $k(X)^B$ its subfield of B-invariants. The set of weights of eigenvectors of B in $k(X)$ is denoted by $\Gamma(X)$. Then $\Gamma(X)$ is a free abelian group of finite rank $r(X)$; this number is called the *rank* of X. A motivation for these notions is the following result, due to Vinberg [Vi] in characteristic zero, and to Knop [Kn6] in general.

THEOREM. *For any G-variety X, and for any closed, B-stable subvariety $Y \subset X$, we have $c(Y) \leq c(X)$ and $r(Y) \leq r(X)$.*

Observe that the spherical varieties are exactly the G-varieties of complexity zero. So the theorem implies readily the following.

COROLLARY. *A G-variety X is spherical if and only if X contains only finitely many B-orbits.*

In particular, any spherical variety contains only finitely many G-orbits, and all of them are spherical. On the other hand, for any nonspherical G-variety X, there exists a G-variety $\tilde{X}$ that is G-birational to X and that contains infinitely many G-orbits (in fact, a family of orbits with $c(X)$ parameters).

The rank of a G-variety is an important invariant; it generalizes the rank of a symmetric space. The G-varieties of rank zero are just unions of flag varieties. There is a very useful classification of homogeneous spaces of rank one (see [A] for the spherical case, and [Po] for the general case). Namely, several theorems on spherical varieties use reduction to rank one.

1.2. Cohomology groups of G-line bundles over spherical varieties

Recall the definition of a G-vector bundle E over a G-variety X. It consists in a vector bundle $p : E \to X$ with a G-action on E, such that p is G-equivariant and that G acts linearly on the fibers of p. Then it is known that the cohomology groups $H^i(X, E)$ are rational G-modules. Assume further that k has characteristic zero; then the groups $H^i(X, E)$ are direct sums of finite-dimensional simple G-modules, with multiplicities. When X is spherical, all multiplicities of the G-module $H^0(X, E)$ are at most equal to the rank of E. This leads to the following (easy) characterization of spherical varieties [Bi], [Br5].

PROPOSITION. *For a normal* quasi-projective *G-variety X, the following conditions are equivalent:*

(i)   *For any G-line bundle $L \to X$, the G-module $H^0(X, L)$ is multiplicity-free.*
(ii) *There exists an ample G-line bundle $L \to X$ such that all multiplicities of the G-modules $H^0(X, L^{\otimes n})$ are bounded independently of the integer $n \geq 1$.*
(iii) *X is spherical.*

The space of sections of a G-line bundle L on a spherical variety X admits a more precise description (it generalizes the well-known correspondence between projective toric varieties with an ample divisor class, and convex polytopes with integral vertices, see [O]). Namely, there exists a dominant weight $\pi(L)$ and a rational convex polytope $\mathcal{C}(L) \subset \Gamma(X) \otimes \mathbf{R}$ such that the set of highest weights of simple submodules of $H^0(X, L)$ is $\pi(L) + (\mathcal{C}(L) \cap \Gamma(X))$. Moreover, we have $\pi(L^{\otimes n}) = n\pi(L)$ and $\mathcal{C}(L^{\otimes n}) = n\mathcal{C}(L)$ for any integer $n \geq 1$. We refer to [Br2] for a description of line bundles over spherical varieties and of their associated polytopes, and to [Br4] for a dual study of curves in spherical varieties, in relation to Mori theory.

   Much less is known about higher cohomology groups; the following qualitative result is proved in [Br5].

THEOREM. *For any spherical variety X, there exists a constant $C(X)$ such that for any G-vector bundle $E \to X$, the multiplicities of all cohomology groups $H^i(X, E)$ are at most $C(X) \operatorname{rank}(E)$.*

In particular, the multiplicities of cohomology groups of line bundles over a fixed spherical variety are uniformly bounded. But a complete description of these groups is only known for toric varieties (see [O]) and for flag varieties (the Borel-Weil-Bott theorem).

1.3. A vanishing theorem for line bundles over spherical varieties
The following vanishing theorem is proved in [BI] by reduction mod p and Frobenius splitting.

THEOREM. *Consider two spherical varieties X and X', a proper G-equivariant morphism $\pi : X \to X'$, and a line bundle $L \to X$ that is generated by its sections over X'. Then $R^i \pi_* L = 0$ for any $i \geq 1$.*

In particular, we have $H^i(X, L) = 0$ for $i \geq 1$, whenever L is a globally generated line bundle over a complete spherical variety X. As another application, consider a spherical variety X and an equivariant desingularization $\pi : \tilde{X} \to X$. Then the theorem can be applied to the trivial line bundle over $\tilde{X}$, and hence *the singularities of spherical varieties are rational.* This property does not hold in general for normal G-varieties with a dense orbit and positive complexity; see [Po].

   The theorem above can be refined by a close study of B-stable curves on spherical varieties, see [Br6]. Namely, to any closed, irreducible, B-stable curve in X, we associate a positive integer $n(C)$, as follows. If C is contained in a projective G-orbit G/Q (with $Q \supset B$) then C is an orbit of a parabolic subgroup $P \supset B$

with P minimal. Denote by α the corresponding simple root, and denote by ρ^Q the half-sum of roots of the unipotent radical of Q. Then $n(C) = 2\langle \rho^Q, \check{\alpha} \rangle$. On the other hand, if C is not contained in any projective orbit, then $n(C) = 1$.

Now, in the vanishing theorem, the assumption that L is generated by its sections over X' can be replaced by: $(L \cdot C) + n(C) > 0$ for any curve C as above, such that $\pi(C)$ is a point. Here $(L \cdot C)$ denotes the degree of the restriction of L to C. In this statement, the numbers $n(C)$ cannot be made smaller in general. For example, when $X = G/B$ and X' is a point, then we obtain the vanishing of $H^i(G/B, L)$ for $i \geq 1$ and L associated to a weight λ such that $\rho + \lambda$ is dominant. This last statement is sharp by the theorem of Borel-Weil-Bott.

2 Classification of spherical varieties

2.1. Embeddings of spherical homogeneous spaces

An *embedding* of a homogeneous space G/H is a normal G-variety with an open G-orbit isomorphic to G/H. The embeddings of a given spherical homogeneous space G/H are classified by combinatorial objects called *colored fans*, which generalize the fans associated with toric varieties. This theory, due to Luna and Vust in characteristic zero, has been simplified and extended to all characteristics by Knop [Kn2]. Here a basic role is played by the set $\mathcal{V}(G/H)$ of G-invariant valuations of the field $k(G/H)$, with rational values. It can be shown that $\mathcal{V}(G/H)$ identifies with a convex polyhedral cone in the $\mathbf{Q}$-vector space $\mathcal{Q}(G/H) := \mathrm{Hom}(\Gamma(G/H), \mathbf{Q})$. In characteristic zero, this cone turns out to be a fundamental domain for some finite reflection group $W(G/H)$ acting on $\mathcal{Q}(G/H)$ (see [Br3]; if G/H is symmetric, then $W(G/H)$ is its little Weyl group). This surprising fact was the starting point for deep investigations of Knop [Kn1,3,4] who defined and studied a little Weyl group $W(X)$ for any G-variety X.

An embedding X of spherical G/H is called *toroidal* if the closure in X of any B-stable divisor in G/H contains no G-orbit. Toroidal embeddings of G/H are classified by fans with support in $\mathcal{V}(G/H)$, i.e. partial subdivisions of $\mathcal{V}(G/H)$ into convex polyhedral cones that contain no line. Smooth, toroidal embeddings are *regular* in the sense of [BDP], i.e. they satisfy the following conditions:

(i)   Each G-orbit closure is smooth, and is the transversal intersection of the smooth orbit closures that contain it.

(ii) The isotropy group of any point x acts on the normal space to the orbit $G \cdot x$ with an open orbit.

Conversely, if a homogeneous space G/H admits a *complete* regular embedding X, then G/H is spherical and X is toroidal. The compactifications of symmetric spaces constructed by DeConcini and Procesi [D] are exactly their smooth, toroidal embeddings; see [Vu].

2.2. The canonical embedding

Consider a spherical homogeneous space G/H, and denote by $N_G(H)$ the normalizer of H in G. Then the quotient group $N_G(H)/H$ is diagonalizable (up to a finite p-group in char. p). Moreover, $N_G(H)/H$ is finite if and only if the valuation cone $\mathcal{V}(G/H)$ contains no line; see [Kn2]. In this case, there is a canonical embedding $\mathbf{X}$

of G/H, namely the toroidal embedding associated with the whole cone $\mathcal{V}(G/H)$. This embedding is projective, and it has at most quotient singularities (this last result holds in characteristic zero). Further, its G-orbit structure is especially simple. Namely, denoting by r the rank of G/H, there are r irreducible, G-stable codimension one orbits $\mathcal{O}_1, \ldots, \mathcal{O}_r$. Moreover, there is an order-reversing bijection between subsets of $\{1, \ldots, r\}$ and orbit closures, given by $I \to \cap_{i \in I} \overline{\mathcal{O}_i}$. In particular, $\mathbf{X}$ contains a unique closed orbit. It turns out that any projective embedding with one closed orbit is dominated by $\mathbf{X}$. On the other hand, an embedding is toroidal if and only if it dominates $\mathbf{X}$.

In characteristic zero, the canonical embedding can be constructed as follows. Denote by $\mathcal{G}$, $\mathcal{H}$ the Lie algebras of G, H, and denote by $\mathbf{L}$ the variety of Lie subalgebras of $\mathcal{G}$. Then $\mathbf{L}$ is a projective G-variety, and the G-orbit closure of $\mathcal{H}$ in $\mathbf{L}$ is a (perhaps nonnormal) projective embedding of $G/N_G(\mathcal{H})$, called the *Demazure embedding*.

THEOREM. *The canonical embedding of G/H is the normalization of the Demazure embedding in $k(G/H)$.*

This result, first discovered by Demazure in some special cases, has been extended by DeConcini and Procesi to symmetric spaces, and to spherical spaces in [Br3]. There it was conjectured that the Demazure embedding is smooth whenever $H = N_G(H)$. A large part of this conjecture has been confirmed by Knop, who proved the following in [Kn5].

THEOREM. *For any spherical homogeneous space G/H with $H = N_G(H)$, the canonical embedding of G/H is smooth.*

2.3. Towards a classification of spherical homogeneous spaces

The problem of classifying spherical spaces by combinatorial invariants is still open; here are some partial results. Besides the well-known classification of symmetric spaces, there is a list of spherical homogeneous spaces G/H with G semisimple and H reductive; see [Kr] for simple G, [Mi] and [Br1] for arbitrary G. This list relies on the description of maximal subgroups of semisimple groups. A more conceptual and fruitful approach has been followed by Luna; it leads e.g. to a complete description of *solvable* spherical subgroups, see [L]. We sketch part of Luna's results.

Assume for simplicity that G is semisimple and adjoint, and that H is equal to its normalizer. By results in 2.1, the valuation cone $\mathcal{V}(G/H) \subset \operatorname{Hom}(\Gamma(G/H), \mathbf{Q})$ can be written uniquely as an intersection of r closed half-spaces, where r denotes the rank of G/H. Let $\gamma_1, \ldots, \gamma_r$ be defining inequations for these subspaces. We normalize the γ_i by demanding that they are in $\Gamma(G/H)$ and that they cannot be divided in this group. It turns out that $\{\gamma_1, \ldots, \gamma_r\} = S(G/H)$ is a basis of a reduced root system $R(G/H)$ with Weyl group $W(G/H)$ (see [Br3], [Kn5]; if G/H is symmetric, then $R(G/H)$ is the reduced root system associated to the restricted roots). The γ_i are called the *spherical roots* of G/H.

By reduction to rank one, the set of all possible spherical roots (for a fixed G) can be described; in particular, this set is finite. For example, the spherical roots for $G = PSL_{n+1}$ are the positive roots and the elements $2\alpha_i$, $\alpha_i + \alpha_j$ $(i < j - 1)$, $\alpha_{i-1} + 2\alpha_i + \alpha_{i+1}$, where $\alpha_1, \ldots, \alpha_n$ are the simple roots.

On the other hand, denote by $\mathcal{D}(G/H)$ the (finite) set of B-orbits of codimension one in G/H. Any $D \in \mathcal{D}(G/H)$ defines a normalized valuation v_D of $k(G/H)$. For any spherical root γ, choose $f_\gamma \in k(G/H)$, which is an eigenvector of B, of weight γ. Then f_γ is uniquely defined up to scalar multiplication. Therefore, the number $v_D(f_\gamma)$ only depends on D and γ. We define a pairing $\rho : \mathcal{D}(G/H) \times S(G/H) \to \mathbf{Z}$ by setting: $\rho(D, \gamma) = -v_D(f_\gamma)$.

A geometric definition of this pairing is as follows. Let $\mathbf{X}$ be the canonical embedding of G/H. Then the G-orbits of codimension one in $\mathbf{X}$ are indexed by the edges of the valuation cone, and hence by $S(G/H)$. For $\gamma \in S(G/H)$ we denote by $\mathcal{O}_\gamma$ the corresponding orbit. By computing the divisor of the rational function f_γ, we obtain the following relations in the Picard group of $\mathbf{X}$:

$$\overline{\mathcal{O}}_\gamma = \sum_{D \in \mathcal{D}(G/H)} \rho(D, \gamma)\overline{D} \ .$$

In the case where G/H is symmetric, the spherical roots are just the simple restricted roots. If moreover the center of the connected component H^0 is finite, then the pairing ρ is given by the Cartan matrix of the restricted root system.

For a spherical homogeneous space G/H with G semisimple adjoint and H equal to its normalizer, the triple $(\mathcal{D}(G/H),\ S(G/H),\ \rho : \mathcal{D}(G/H) \times S(G/H) \to \mathbf{Z})$ should determine G/H uniquely. Furthermore, there should be a characterization of all "admissible" triples by conditions arising from spherical spaces of low rank. This program has been completed by Luna for solvable H (see [L]) and for $G = PGL_{n+1}$ and arbitrary H (work in progress, 1994).

3 Orbits of a Borel subgroup

3.1. B-orbits and cells

It follows from work of Bialynicki-Birula that any complete, smooth spherical variety X has a cellular decomposition. Namely, choose a one-parameter subgroup $\lambda : k^* \to X$ in general position; i.e., the fixed point set X^λ is finite (the existence of λ follows from finiteness of the number of G-orbits in X). Then for any $x \in X^\lambda$, the set

$$X(\lambda, x) = \{z \in X \mid \lim_{t \to 0} \lambda(t)z = x\}$$

is an affine space (the λ-cell of x).

This cellular decomposition is related to the decomposition into B-orbits, as follows. For λ as above, denote by $G(\lambda)$ the set of all $g \in G$ such that $\lambda(t)g\lambda(t)^{-1}$ has a limit when $t \to 0$. Then $G(\lambda)$ is a parabolic subgroup of G. Moreover, we may choose λ in general position, so that $G(\lambda) = B$. In this case, all cells are B-stable. In the case where X is a flag variety, it is known that each cell is a single B-orbit. In the general case, we have the following result [BL] in characteristic zero.

THEOREM. *Let X be a smooth, complete, toroidal variety; choose λ in general position with $G(\lambda) = B$. Then the intersection of each λ-cell with a G-orbit is either empty or a single G-orbit.*

This result extends work of DeConcini and Springer [DS] who obtained formulas for Betti numbers of complete symmetric varieties. Namely, the theorem above suggests a parametrization of B-orbits in X by a set of pairs consisting in a fixed point of a maximal torus of B, and in a cone in the fan of X. This parametrization can be worked out in the symmetric case; the general case is still open.

3.2. Symmetries in the set of B-orbits

Denote by $\mathcal{B}(X)$ the set of B-orbits in the spherical variety X. There is a partial ordering on $\mathcal{B}(X)$ by inclusion of closure; it may be called the *Bruhat order*. If moreover $X = G/H$ is homogeneous, then all closed B-orbits have the same dimension d, namely the dimension of the flag variety of H. Then we define a length function on $\mathcal{B}(X)$ by setting $l(\mathcal{O}) = \dim(\mathcal{O}) - d$. Although no precise description of $\mathcal{B}(X)$ is known in general, this set turns out to have interesting symmetries.

Denote by W the Weyl group of G, identified with the set of double cosets $B\backslash G/B$ via $w \to BwB$. Then W can be turned into a monoid W^*, by defining $w * w'$ as the open double coset in the closure of $BwBw'B$. More concretely, W^* is generated by the set S of simple reflections, with relations $s^2 = s$ for all $s \in S$, and with the braid relations.

Observe that W^* acts on $\mathcal{B}(X)$ by defining $w * \mathcal{O}$ (for $w \in W$ and $\mathcal{O} \in \mathcal{B}(X)$) as the open B-orbit in the closure of $Bw\mathcal{O}$. Moreover, this action has the following properties: (i) $\mathcal{O} \subset w * \mathcal{O}$; (ii) if $\mathcal{O} \subset \mathcal{O}'$ then $w * \mathcal{O} \subset w * \mathcal{O}'$; (iii) l is strictly monotonic for the Bruhat order; and (iv) if $\mathcal{O} \neq s * \mathcal{O}$ then $l(s * \mathcal{O}) = l(\mathcal{O}) + 1$ (properties (iii) and (iv) only make sense for homogeneous X). This observation is implicit in Matsuki's proof [Ma] of the finiteness of $\mathcal{B}(X)$. For homogeneous X, the Bruhat order on $\mathcal{B}(X)$ turns out to be the weakest partial order that satisfies conditions (i)–(iv) above. This is proved in [RS] together with more precise results concerning B-orbits in symmetric spaces.

On the other hand, an action of W on the set $\mathcal{B}(X)$ has been constructed by Knop [Kn6]. This action is not compatible with the Bruhat order, but it has a surprising connection with the little Weyl group $W(X)$. Namely, denoting by X_0 the open B-orbit in X, and by $P(X)$ the stabilizer in G of the set X_0, we have the following result [Kn6].

THEOREM. *For the W-action on $\mathcal{B}(X)$, the isotropy group of X_0 is the semidirect product of $W(X)$ and $W \cap P(X)$.*

In fact, a version of this result holds for any G-variety X, see [Kn6]. In this case, $\mathcal{B}(X)$ is defined as the set of all nonempty, closed, irreducible, and B-stable subvarieties of X. There is still a W^*-action on $\mathcal{B}(X)$; it is used in the proof of the Theorem in Section 1.1. On the other hand, W acts on the subset $\mathcal{B}_0(X)$ consisting in all $Z \in \mathcal{B}(X)$ that contain a family of B-orbits with $c(X)$ parameters.

References

[A]   D. Ahiezer, *Spherical varieties*, preprint, Bochum 1993.

[Bi]   F. Bien, *Orbits, multiplicities, and differential operators*, pp. 199–227, in Representation Theory of Groups and Algebras (J. Adams, ed.), AMS, Providence, RI, 1993.

[BDP] E. Bifet, C. DeConcini, and C. Procesi, *Cohomology of regular embeddings*, Adv. in Math. **82** (1990), 1–34.

[Br1] M. Brion, *Classification des espaces homogènes sphériques*, Compositio Math. **63** (1987), 189–208.

[Br2] M. Brion, *Groupe de Picard et nombres caractéristiques des variétés sphériques*, Duke Math. J. **58** (1989), 397–424.

[Br3] M. Brion, *Vers une généralisation des espaces symétriques*, J. Algebra **134** (1990), 115–143.

[Br4] M. Brion, *Variétés sphériques et théorie de Mori*, Duke Math. J. **72** (1993), 369–404.

[Br5] M. Brion, *Représentations des groupes réductifs dans des espaces de cohomologie*, Math. Ann. **300** (1994), 589–604.

[Br6] M. Brion, *Curves in spherical varieties*, to appear (1995).

[BI] M. Brion and S. P. Inamdar, *Frobenius splitting of spherical varieties*, pp. 207–218, in Algebraic Groups and their Generalizations (W. Haboush, ed.), Amer. Math. Soc., Providence, RI, 1994.

[BL] M. Brion and D. Luna, *Sur la structure locale des variétés sphériques*, Bull. Soc. Math. France **115** (1987), 211–226.

[D] C. DeConcini, *Equivariant embeddings of homogeneous spaces*, Proc. Internat. Congress Math. Berkeley, 1986, 369–377.

[DS] C. DeConcini and T. A. Springer, *Betti numbers of complete symmetric varieties*, pp. 87–108, in Geometry Today (E. Arbarello, ed.), Birkhäuser, Boston, Basel, and Stuttgart 1985.

[Kn1] F. Knop, *Weylgruppe und Momentabbildung*, Invent. Math. **99** (1990), 1–23.

[Kn2] F. Knop, *The Luna-Vust theory of spherical embeddings*, pp. 245–259, in Proc. Hyderabad Conference on Algebraic Groups (S. Ramanan, ed.), Manoj-Prakashan, Madras 1991.

[Kn3] F. Knop, *Über Bewertungen, welche unter einer reduktiven Gruppe invariant sind*, Math. Ann. **295** (1993), 333–363.

[Kn4] F. Knop, *The asymptotic behaviour of invariant collective motion*, Invent. Math. **116** (1994), 309–328.

[Kn5] F. Knop, *Automorphisms, root systems, and compactifications of homogeneous varieties*, to appear in the J. Amer. Math. Soc.

[Kn6] F. Knop, *On the set of orbits for a Borel subgroup*, Comm. Math. Helv. **70** (1995), 285–309.

[Kr] M. Krämer, *Sphärische Untergruppen in kompakten zusammenhängenden Liegruppen*, Compositio Math. **38** (1979), 129–153.

[L] D. Luna, *Sous-groupes sphériques résolubles*, preprint, Grenoble 1993.

[Ma] T. Matsuki, *Orbits on flag manifolds*, Proc. Internat. Congress Math. Kyoto, 1990, 807–814.

[Mi] V. Mikityuk, *On the integrability conditions of invariant hamiltonian systems with homogeneous configuration spaces*, Math. USSR-Sb. **57** (1987), 527–546.

[RS] R. W. Richardson and T. A. Springer, *The Bruhat order on symmetric varieties*, Geom. Dedicata **35** (1990), 389–436.

[O] T. Oda, Convex Bodies and Algebraic Geometry. An Introduction to the Theory of Toric Varieties, Springer, Berlin, Heidelberg, and New York 1988.

[Pa] D. I. Panyushev, *On homogeneous spaces of rank one*, preprint 1994.

[Po] V. L. Popov, *Singularities of closures of orbits*, pp. 133–141, in Israel Math. Conf. Proc., vol. 7, 1993.

[Vi] E. B. Vinberg, *Complexity of action of reductive groups*, Funct. Anal. Appl. **20** (1986), 1–11.

[Vu] T. Vust, *Plongements d'espaces symétriques algébriques: une classification*, Ann. Scuola Norm. Sup. Pisa Cl. Sci. (4) **17** (1990), 165–195.

Rigidity Properties of Group Actions on CAT(0)-Spaces

MARC BURGER

Institut de mathématiques, Université de Lausanne
CH-1015 Lausanne-Dorigny, Switzerland

In this lecture we shall discuss certain aspects of the general rigidity problem of classifying isometric actions of a given group Λ on a CAT(0)-space Y. The CAT(0) property, introduced by Alexandrov [Al], [Wa], generalizes to singular metric spaces the notion of nonpositive curvature. Among such spaces one finds simply connected non-positively curved Riemannian manifolds and Euclidean buildings; in particular, geometric rigidity problems and the linear representation theory of Λ over local fields are put into the same framework. There are various types of additional structures on Λ that lead to different rigidity properties. In this lecture we shall discuss the following three situations.

1. Let Λ be a subgroup of a locally compact group G, satisfying $\Gamma < \Lambda < \mathrm{Com}_G\Gamma$ where Γ is a sufficiently large discrete subgroup of G and $\mathrm{Com}_G\Gamma = \{g \in G : g^{-1}\Gamma g$ and Γ share a subgroup of finite index$\}$ is the commensurator of Γ in G. One expects then that any isometric action of Λ on a CAT(0)-space, which satisfies a suitable geometric irreducibility property, extends continuously to the closure $\overline{\Lambda} \subset G$. Rigidity properties of this type were already established by Margulis in the early 1970s, in the case where Γ is a lattice in a semisimple Lie group G, and led to his arithmeticity criterion [Ma 1], [Ma 4].

In Section 2, we state two recent results in this direction. The first result, due to Margulis, concerns the case where Γ is a cocompact, finitely generated lattice in a locally compact group G, and is based on his theory of generalized harmonic maps. The second, based on ergodic theoretic methods introduced by Margulis in [Ma 3], treats the case where the discrete subgroup $\Gamma < G$ admits a (Γ, G)-boundary (see Section 3 for examples). In Section 5 we apply these results to the study of the commensurator of uniform tree lattices.

2. Let Λ be an irreducible lattice in a group $G = \prod_{\alpha=1}^{n} \mathbb{G}_\alpha(k_\alpha)$, where $\mathbb{G}_\alpha$ is a semisimple algebraic group defined over a local field k_α. In this lecture, a local field is a locally compact nondiscrete field. Thanks to Margulis' work, one has a fairly complete criterion for the existence of a continuous extension to G of an action by isometries of Λ on a CAT(0)-space Y when $\sum_\alpha \mathrm{rank}_{k_\alpha}\mathbb{G}_\alpha \geq 2$, and either the Λ-action comes from a linear representation over a local field [Ma 4, VII. 5,6] or Y is a tree [Ma 2]. The case of irreducible representations leads to his arithmeticity theorem [Ma 4, IX. Theorem A], whereas the case of actions on trees gives a classification of those lattices Λ that are nontrivial amalgams [Ma 2, Theorem 2].

Proceedings of the International Congress
of Mathematicians, Zürich, Switzerland 1994
© Birkhäuser Verlag, Basel, Switzerland 1995

Thus, one would like to understand in general when an isometric action of Λ on Y extends continuously to G. This is understood when Y is a locally compact CAT(-1)-space; see Section 4.

3. Let $\Lambda = \pi_1(M)$ be the fundamental group of a compact manifold M, where M is equipped with a nontrivial action of a semisimple Lie group G of higher rank preserving some geometric structure, for instance an "H-structure" [Zi 2] or a finite measure and a connection [Zi 3], [Sp-Zi]. Following Zimmer's program one expects that the higher rank hypothesis on G implies strong restrictions on the class of CAT(0)-spaces that admit $\pi_1(M)$-actions without fixed points. We illustrate this in Section 4 by a recent result of Adams, generalizing [Sp-Zi, Theorem A].

1 CAT(0)-spaces

A geodesic space is a metric space in which every pair of points x, y is joinable by a geodesic segment, i.e. a continuous curve of length $d(x, y)$. A geodesic triangle is obtained by joining pairwise three points with geodesic segments; it maps sidewise isometrically to a geodesic comparison triangle in the Euclidean plane $\mathbb{E}^2$ or in the hyperbolic plane $\mathbb{H}^2$. A geodesic space is CAT(0) (resp. CAT(-1)) if the comparison maps of its geodesic triangles into $\mathbb{E}^2$ (resp. $\mathbb{H}^2$) are not distance decreasing. Observe that a CAT(-1)-space is CAT(0). Many global geometric properties of Cartan-Hadamard manifolds (see [B-G-S]) generalize to CAT(0)-spaces and we refer the reader to [Gr 2], [Bri-Ha], [B] for general expositions. The following basic properties of the distance function of a CAT(0)-space Y hold:

P1. Convexity of distance: for any geodesic segments $c_1, c_2 : I \to Y$, $I \subset \mathbb{R}$, the function $t \to d(c_1(t), c_2(t))$ is convex;

P2. Uniform convexity of balls: for every $R, a > 0$, there exists $\varepsilon = \varepsilon(R, a) > 0$ such that given any three points $x, y_1, y_2 \in Y$ with $d(x, y_i) \leq R$ and $d(y_1, y_2) \geq a$, the midpoint m on the geodesic segment $[y_1, y_2]$ satisfies $d(x, m) \leq R - \varepsilon$.

A geodesic space Y is UC (uniformly convex) if it satisfies P1 and P2. Whereas P1 implies that Y is uniquely geodesic, P2 often serves as a substitute to local compactness. Finally we mention that a CAT(0)-space Y has a visual boundary $Y(\infty)$; when Y is locally compact and complete, the space $\overline{Y} := Y \sqcup Y(\infty)$ is an Isom(Y)-equivariant compactification of Y.

EXAMPLE (1). L^p-spaces are UC for $1 < p < \infty$. Hilbert spaces are CAT(0); more generally, if S is a finite measure space and X is CAT(0), then $L^2(S, X)$ is CAT(0) [Ko-Sc].

EXAMPLE (2).   Simply connected Riemannian manifolds of sectional curvature $K \leq 0$ (respectively $K \leq -1$) and their convex subsets are CAT(0) (respectively CAT(-1)) [B-G-S].

EXAMPLE (3).   Euclidean buildings are CAT(0) [Bru-Ti], [Bro, VI. 3].

EXAMPLE (4).   For every Coxeter system (W, S), there exists a piecewise Euclidean cell complex Σ which is CAT(0) for the induced length metric and on which W acts properly discontinuously by isometries with compact quotient. Furthermore, Σ admits a piecewise hyperbolic CAT(-1)-structure if and only if W does not contain $\mathbb{Z}^2$. See [Mou] and the expository paper [Da].

EXAMPLE (5). Complexes of groups with a metric of nonpositive curvature are developable and their universal covering is CAT(0) [Ha], [Sp].

EXAMPLE (6). Two dimensional complexes with prescribed link L_v at every 0-cell v and whose 2-cells are regular polygons in $\mathbb{E}^2$ (resp. $\mathbb{H}^2$). Conditions for their existence and for the induced length metric to be CAT(0) (resp. CAT(-1)) are given in [Hag], [Be 1], [Be 2], [B-Br]. First examples of such polyhedra were constructed by Gromov, see [Gr 1, Section 4. C″].

EXAMPLE (7). Metric trees are CAT(-1)-spaces.

2 Rigidity properties of commensurators

2.1. For a finitely generated group Γ and a homomorphism $\pi : \Gamma \to \mathrm{Isom}(Y)$, where Y is a UC space, we introduce the properties:

HP1. For some (and hence every) finite generating set $S \subset \Gamma$, the sublevels of the function $d_S : Y \to \mathbb{R}_+$, $d_S(y) = \max_{\gamma \in S} d(y, \pi(\gamma)y)$, are bounded subsets of Y.

HP2. For every $y_1 \neq y_2$ in Y, there exists $\gamma \in \Gamma$ such that the geodesic segments $[y_1, \pi(\gamma)y_1]$, $[y_2, \pi(\gamma)y_2]$ are not parallel, meaning that if c parametrizes $[y_1, y_2]$, $t \to d(c(t), \pi(\gamma)c(t))$ is not constant.

Observe that if Y is CAT(0), locally compact and complete, HP1 is equivalent to the property that $\pi(\Gamma)$ does not have a fixed point in $Y(\infty)$.

THEOREM 1. [Ma 5] *Let Γ be a finitely generated, cocompact lattice in a locally compact group G, $\Lambda < G$ with $\Gamma < \Lambda < \mathrm{Com}_G\Gamma$ and $\rho : \Lambda \to \mathrm{Isom}(Y)$ a homomorphism into the group of isometries of a complete UC space Y, such that*

(1) Λ *acts c-minimally on Y,*
(2)   *any subgroup of finite index in Γ satisfies HP1 and HP2.*

Then ρ extends continuously to $\overline{\Lambda}$.

REMARKS: (1) A group action by isometries on a geodesic space is *c*-minimal if it admits no nonvoid proper closed convex invariant subspace.

(2) Theorem 1 applies to the case where Y is the Euclidean building associated to a connected k-simple group $\mathbb{H}$ defined over a local field k, the image $\rho(\Lambda) \subset \mathbb{H}(k)$ is Zariski dense in $\mathbb{H}$, and $\rho(\Gamma) \subset \mathbb{H}(k)$ is not relatively compact.

The proof of this theorem relies on uniqueness properties of generalized harmonic maps [Ma 5]. By definition such maps are critical points, in the space of Γ-equivariant measurable maps $\varphi : G \to Y$, of an "energy " functional

$$E(\varphi) := \int_{\Delta(\Gamma)\backslash G \times G} d(\varphi(g_1), \varphi(g_2))^p h(g_1^{-1}g_2)dg_1 dg_2$$

where $p \geq 1$, $\Delta(\Gamma) := \{(\gamma, \gamma) : \gamma \in \Gamma\}$ and h is a suitable positive continuous function. The basic problem for applying this method to discrete subgroups Γ of locally compact groups G is to find a positive continuous function h and an exponent $p > 1$ such that there is at least one equivariant map $\varphi : G \to Y$ of finite energy. When Γ is a finitely generated cocompact lattice in G, $p > 1$, and π satisfies HP1, HP2, there is a unique harmonic map and it is continuous; Theorem 1 follows then from the fact that such a map is automatically Λ-equivariant. These results apply notably when Γ is a cocompact lattice in a connected Lie group, or in $\prod_{\alpha=1}^{n} \mathbb{G}_\alpha(k_\alpha)$ where $\mathbb{G}_\alpha$ is a semisimple connected group defined over a local field k_α, or in the automorphism group $\mathrm{Aut}\,T$ of a uniform tree (see Section 5). Indeed, in all these cases Γ is finitely generated. The above method applies also to irreducible lattices in semisimple connected Lie groups all of whose simple factors are noncompact and not locally isomorphic to $SL(2, \mathbb{R})$. Notice that lattices need not be finitely generated. For example, noncocompact lattices in $G = \mathbb{G}(k)$, where $\mathbb{G}$ is a simple k-group of k-rank one defined over a local field k of positive characteristic, are never finitely generated (see [Lu] for the structure of lattices in $\mathbb{G}(k)$). Also when G is not compactly generated, cocompact lattices are not finitely generated, see 3.3 (b) for an example with dense commensurator.

2.2. Let G be a locally compact group, $H < G$ a closed subgroup, and B a standard Borel space endowed with a Borel action $G \times B \to B$ preserving a σ-finite measure class μ. We call (B, μ) an (H, G)-boundary if it satisfies the properties:

  BP1: the H-action on (B, μ) is amenable;

  BP2: the diagonal H-action on $(B \times B, \mu \times \mu)$ is ergodic.

THEOREM 2. *Let Γ be a discrete subgroup of a locally compact, second countable group G and $\Lambda < G$ with $\Gamma < \Lambda < \mathrm{Com}_G\Gamma$. We assume that there exists (B, μ), which is a (Γ', G)-boundary for any subgroup of finite index $\Gamma' < \Gamma$.*

  A. *Let $\mathbb{H}$ be a connected adjoint k-simple group, where k is a local field, and $\pi : \Lambda \to \mathbb{H}(k)$ a homomorphism such that $\pi(\Lambda)$ is Zariski dense in $\mathbb{H}$ and $\pi(\Gamma) \subset \mathbb{H}(k)$ is not relatively compact. Then π extends continuously to $\overline{\Lambda}$.*

  B. *[Bu-Mo] Let Y be a locally compact, complete $CAT(-1)$-space and $\pi : \Lambda \to \mathrm{Isom}(Y)$ a homomorphism such that $\pi(\Lambda)$ acts c-minimally on Y and $\pi(\Gamma)$ is not elementary. Then π extends continuously to $\overline{\Lambda}$.*

REMARKS: (1) A group of isometries of a $CAT(-1)$-space Y is called elementary, if it has an invariant subset $\Delta \subset \overline{Y}$ consisting of one or two points.

  (2) In the case where Γ is a lattice in the automorphism group of a d-regular tree T_d ($d \geq 3$) and Y is a locally finite tree, Theorem 2B is due to Lubotzky-Mozes-Zimmer [L-M-Z].

  (3) Gao has recently (October 94) proved Theorem 2B in the case where Γ is a divergence group (see Section 3.3) in $G = \mathrm{Isom}(X)$, X is locally compact complete $CAT(-1)$, and Y is separable complete $CAT(-1)$.

  The proof of Theorem 2 relies on uniqueness properties of suitable "boundary" maps. For 2A one shows that, if (B, μ) is a (Γ', G)-boundary for every subgroup of finite index $\Gamma' < \Gamma$, and $\pi : \Gamma \to \mathbb{H}(k)$ is a homomorphism with Zariski dense and unbounded image, there is a proper k-subgroup $\mathbb{L} < \mathbb{H}$ such that for any $\Gamma' < \Gamma$, of finite index, there exists a unique Γ'-equivariant measurable map $\varphi : B \to \mathbb{H}(k)/\mathbb{L}(k)$ (see [Zi 1], [A'C-Bu] for different constructions of boundary maps). In case 2B one shows that, if (B, μ) is a (Γ, G)-boundary and $\pi : \Gamma \to \mathrm{Isom}(Y)$ a homomorphism such that $\pi(\Gamma)$ is not elementary, there is a unique Γ-equivariant measurable map $\varphi : B \to Y(\infty)$ (see [Bu-Mo]).

3 (Γ, G)-boundaries and commensurators

Theorem 2 is of interest when (Γ, G) admits a boundary and $\mathrm{Com}_G\Gamma$ is not discrete. We have the following sources for such groups:

3.1. Arithmetic lattices. Let us consider $G = \prod_{\alpha=1}^{n} \mathbb{G}_\alpha(k_\alpha)$ and $P = \prod \mathbb{P}_\alpha(k_\alpha)$, where $\mathbb{G}_\alpha$ is a connected, simply connected k_α-almost simple, k_α-isotropic group and $\mathbb{P}_\alpha < \mathbb{G}_\alpha$ a minimal k_α-parabolic subgroup of $\mathbb{G}_\alpha$. The homogeneous space $B := G/P$ with its G-invariant measure class is a (Γ, G)-boundary for any lattice $\Gamma < G$. When the lattice Γ is arithmetic, its commensurator is dense in G. By Margulis' arithmeticity theorem this is always the case when $\sum \mathrm{rank}_{k_\alpha} \mathbb{G}_\alpha \geq 2$ and Γ is irreducible.

3.2. Regular tree lattices. Consider $G = \mathrm{Aut}\, T_d$, $d \geq 3$, and $P < G$ the stabilizer of a point in $T_d(\infty)$. Again the homogeneous space $B := G/P$ with its G-invariant measure class is a (Γ, G)-boundary for any lattice $\Gamma < G$. Here BP2 follows from the Howe-Moore property of $\mathrm{Aut}\, T_d$ [Lu-Mo]. We remark that cocompact lattices in $\mathrm{Aut}\, T_d$ have dense commensurators, see Section 5.

3.3. Divergence groups. Let X be a locally compact and complete CAT(-1)-space, $H < \mathrm{Isom}(X)$ a closed subgroup, and

$$\delta_H = \inf\{s > 0 : \ P_x(s) = \int_H e^{-sd(hx,x)}dh < +\infty\}$$

its critical exponent, which does not depend on the choice of $x \in X$. A nonelementary, discrete subgroup $\Gamma < \mathrm{Isom}(X)$ is called a divergence group if $\delta_\Gamma < +\infty$ and $P_x(\delta_\Gamma) = +\infty$. Generalizing the Patterson-Sullivan theory of Kleinian groups (see [Pa], [Su], and [Bo] for compact quotients of CAT(-1)-spaces), one constructs a canonical measure class μ_Γ on $X(\infty)$ that is invariant under $G := \overline{\mathrm{Com}_{\mathrm{Isom}(X)}\Gamma} < \mathrm{Isom}(X)$ and such that $(X(\infty), \mu_\Gamma)$ is a (Γ', G)-boundary for any subgroup Γ' of finite index in Γ. Hence when Γ is a divergence group, Theorem 2 may be applied to any subgroup $\Lambda < \mathrm{Isom}(X)$ with $\Gamma < \Lambda < \mathrm{Com}_{\mathrm{Isom}(X)}\Gamma$. Concerning property BP1, Adams [Ad] has shown that when X is at most of exponential growth, the action on $X(\infty)$ of any closed subgroup of $\mathrm{Isom}(X)$ is universally amenable. We mention the following examples of divergence groups (see [Bu-Mo]):

(a) any lattice in $\mathrm{Isom}(X)$ with $\mathrm{Isom}(X)\backslash X$ compact, is a divergence group;

(b) let $\mathbb{A}$ be a tree of finite groups with edge indexed graph (see [Ba-Ku] for definitions),

$$\overset{b}{\bullet} \!\!\frac{a_1 \ \ 1}{}\!\! \bullet \!\!\frac{a_2 \ \ 1}{}\!\! \bullet \ \ \bullet \cdots$$

where $b \geq 2$ and not all a_i's are 1. Let R be the radius of convergence of the power series $P(x) := \sum_{k=1}^{\infty} (a_k - 1) a_{k-1} \ldots a_1 x^k$, T the universal covering tree, and Γ the fundamental group of $\mathbb{A}$. Then Γ is a divergence group if and only if $P(R) \geq \frac{1}{b-1}$. This happens for example if P is rational. When the edge indexed graphs associated respectively to $\mathrm{Aut}T \backslash T$ and $\mathbb{A}$ coincide (e.g. $b \neq a_i + 1$, $\forall i$) then Γ is cocompact in $\mathrm{Aut}T$ and $\mathrm{Aut}T$ is not compactly generated. One can show that no $(\Gamma, \mathrm{Aut}\ T)$-boundary is a homogeneous space of $\mathrm{Aut}T$. Finally, one can choose the vertex and edge groups of $\mathbb{A}$ in such a way that $\mathrm{Com}_{\mathrm{Aut}\ T}\Gamma$ is dense in $\mathrm{Aut}T$.

(c) Let $\Gamma < \mathrm{Aut}\ T_4$ be the fundamental group of the Cayley graph of the free abelian group on two generators. Then Γ is a divergence group whose commensurator is dense in $\mathrm{Aut}\ T_4$.

4 Lattices in higher rank groups and CAT(-1)-spaces

THEOREM 3. [Ad], [Bu-Mo] *Let Γ be an irreducible lattice in $G = \prod_{\alpha=1}^{n} \mathbb{G}_\alpha(k_\alpha)$, where $\mathbb{G}_\alpha$ is a simply connected, k_α-almost simple, k_α-isotropic group and $\sum_{\alpha=1}^{n} \mathrm{rank}_{k_\alpha} \mathbb{G}_\alpha \geq 2$. Let Y be a locally compact, complete $CAT(-1)$-space and $\pi : \Gamma \to \mathrm{Isom}(Y)$ a homomorphism with nonelementary image. Let $X \subset Y$ be the closed convex hull of the limit set of $\pi(\Gamma)$. Then $\pi : \Gamma \to \mathrm{Isom}(X)$ extends to a continuous homomorphism of G, factoring through a proper homomorphism from a rank one factor of G into $\mathrm{Isom}(X)$.*

When all the almost simple factors of G have rank at least 2, $\pi(\Gamma)$ must be elementary; when $\mathrm{Isom}(Y)$ has finite critical exponent, stabilizers of boundary points are amenable and, as Γ has property T, $\pi(\Gamma)$ is relatively compact and hence fixes a point in Y.

See [Ad, Theorem 11.2] in the case where $G = \mathbb{G}(\mathbb{R})$ $(n = 1)$ and [Bu-Mo] in the general case.

THEOREM 4. [Ad] *Let M be a connected, compact real analytic manifold equipped with a nontrivial real analytic action of a simple Lie group G of real rank ≥ 2. Assume that G preserves a probability measure and a real analytic connection. Then $\pi_1(M)$ cannot act properly discontinuously on a locally compact $CAT(-1)$ space that is at most of exponential growth.*

5 Uniform tree lattices

A uniform tree is a locally finite tree X whose automorphism group contains a discrete subgroup Γ such that $\Gamma \backslash X$ is finite. In particular, such a subgroup Γ is a cocompact lattice in $\mathrm{Aut}\ X$. Uniform trees are characterized by the property that $\mathrm{Aut}\ X \backslash X$ is finite and $\mathrm{Aut}\ X$ is unimodular [Ba-Ku]. It is a remarkable fact that for a uniform tree X, the commensurators of any two cocompact lattices in $\mathrm{Aut}\ X$ are conjugate [Le], [Ba-Ku]. Denoting by $C(X)$ a representative of this conjugacy class of subgroups, we have

THEOREM 5. [Li] *$C(X)$ is dense in $\mathrm{Aut}\ X$.*

This was conjectured by Bass and Kulkarni who proved it for regular trees, see [Ba-Ku].

Theorem 1 and 2B are useful in studying the problem of whether $C(X)$ determines X. The analogous problem of whether Aut X determines X was solved by Bass and Lubotzky, see [Ba-Lu]. Concerning the former, we have a complete answer only in the case of regular trees.

COROLLARY 1. [L-M-Z] *If $\rho : C(T_d) \to C(T_m)$ is an isomorphism, then $d = m$ and ρ is conjugation by an element of Aut T_d.*

In [L-M-Z] it is also shown that $C(T_d)$, $d \geq 3$, is not linear over any field. The proof of this fact used a description of the commensurator of certain cocompact lattices in Aut T_d in terms of "recolorings" of d-regular graphs. This was also used there to give a proof, via length functions, of Theorem 2B for the case where $\Lambda = C(T_d)$, Y is an abitrary tree, and the lattice Γ is cocompact.

Applying Theorem 2 one has

COROLLARY 2. [Bu-Mo] *Let X be a uniform tree all of whose vertices have degree ≥ 3. Assume that Aut X acts c-minimally on X and is not discrete. Then $C(X)$ is not linear over any field.*

Observe that if Aut X is discrete, the group $C(X) =$ Aut X is virtually free and finitely generated, hence linear. In connection with Corollary 2, the subgroup $\text{Aut}^+ X$ generated by all oriented-edge stabilizers plays an important role. It follows from a theorem of Tits [Ti] that if X is uniform and Aut X acts c-minimally, then $\text{Aut}^+ X$ is simple. In the proof of Corollary 2 one shows that if π is a linear representation of $C(X)$, the subgroup Ker $\pi \cap \text{Aut}^+ X$ is nonamenable. It would therefore be interesting to know whether $C(X) \cap \text{Aut}^+ X$ is simple.

Acknowledgments. I thank Norbert A'Campo and Shahar Mozes for sharing their insights with me.

References

[A'C-Bu] N. A'Campo and M. Burger, *Réseaux arithmétiques et commensurateurs d'après G. A. Margulis*, Invent. Math. **116** (1994), 1–25.

[Ad] S. Adams, *Reduction of cocycles with hyperbolic targets*, preprint (1993).

[Al] A. D. Alexandrov, *Über eine Verallgemeinerung der Riemannschen Geometrie*, Schriftenreihe des Forschungsinstituts für Mathematik, Berlin, Heft 1 (1957), 33–84.

[B] W. Ballmann, Lectures on spaces of non-positive curvature, preprint (1994).

[B-Br] W. Ballmann and M. Brin, *Polygonal complexes and combinatorial group theory*, Geom. Dedicata **50** (1994), 165–191.

[B-G-S] W. Ballmann, M. Gromov, and V. Schroeder, Manifolds of nonpositive curvature, Progr. Math. **61**, Birkhäuser, Boston, MA, 1985.

[Ba] H. Bass, *Covering theory for graphs of groups*, J. Pure Appl. Algebra **89** (1993), 3–47.

[Ba-Ku] H. Bass and R. Kulkarni, *Uniform tree lattices*, J. Amer. Math. Soc. **3** (1990), 843–902.

[Ba-Lu]   H. Bass and A. Lubotzky, *Rigidity of group actions on locally finite trees*, preprint.

[Be 1]   N. Benakli, *Polyèdres à géométrie locale donnée*, C.R. Acad. Sci. Paris, t. 313, Série I (1991), 561–564.

[Be 2]   N. Benakli, *Polyèdres hyperboliques à groupe d'automorphisme non discret*, C.R. Acad. Sci. Paris, t. 313, Série I (1991), 667–669.

[Bo]   M. Bourdon, *Actions quasi-convexes d'un groupe hyperbolique, flot géodésique*, Thèse, Univ. de Paris–Sud, 1993.

[Bri-Ha]   M. Bridson and A. Haefliger, Metric spaces of non-positive curvature, in preparation.

[Bro]   K. Brown, Buildings, Springer, Berlin and New York, 1993.

[Bru-Ti]   F. Bruhat and J. Tits, *Groupes réductifs sur un corps local I. Données radicielles valuées*, IHES Publ. Math. **41** (1972), 5–251.

[Bu-Mo]   M. Burger and S. Mozes, *CAT(−1)-spaces, divergence groups and their commensurators*, preprint (1993), revised version (1994).

[Da]   M. W. Davis, *Nonpositive curvature and reflection groups*, Lectures given at the Eleventh Annual Western Workshop in Geometric Topology, Park City, Utah, June 1994.

[Gr 1]   M. Gromov, *Infinite groups as geometric objects*, Proc. Internat. Congress Math., Warszawa (1983), 385–392.

[Gr 2]   M. Gromov, *Hyperbolic groups*, in: Essays in group theory, (S. M. Gersten, ed.), M.S.R.I. Publications **8**, Springer, Berlin and New York (1987), 75–263.

[Ha]   A. Haefliger, *Complexes of group and orbihedra*, in: Group theory from a geometrical viewpoint, (E. Ghys, A. Haefliger, and A. Verjovsky, eds.) ICTP Trieste, World Scientific, Singapore (1991), 504–540.

[Hag]   F. Haglund, *Les polyèdres de Gromov*, C.R. Acad. Sci. Paris t. 313, Série I (1991), 603–606.

[Ko-Sc]   N. J. Korevaar and R. M. Schoen, *Sobolev spaces and harmonic maps for metric space targets*, Comm. Anal. and Geom. 1, **4** (1994), 561–659.

[Le]   F. T. Leighton, *Finite common coverings of graphs*, J. Combin. Theory Ser. B **33** (1982), 231–238.

[Li]   Y.-S. Liu, *Density of commensurability groups of uniform tree lattices*, J. Algebra **165** (1994), 346–359.

[Lu]   A. Lubotzky, *Lattices in rank one Lie groups over local fields*, G.A.F.A. **1** no 4 (1991), 405–431.

[Lu-Mo]   A. Lubotzky and S. Mozes, *Asymptotic properties of unitary representations of tree automorphisms*, in: Harmonic analysis and discrete potential theory (M.A. Picardello, ed.), Plenum Press, New York (1992), 289–298.

[L-M-Z]   A. Lubotzky, S. Mozes, and R. J. Zimmer, *Superrigidity of the commensurability group of tree lattices*, Comm. Math. Helv. **69** (1994), 523–548.

[Ma 1]   G. A. Margulis, *Discrete groups of motions of manifolds of non-positive curvature*, Amer. Math. Soc. Transl. **109** (1977), 33–45.

[Ma 2]   G. A. Margulis, *On the decomposition of discrete subgroups into amalgams*, Selecta Math. Soviet 1 **2** (1981), 197–213.

[Ma 3]   G. A. Margulis, *Arithmeticity of irreducible lattices in semisimple groups of rank greater than one*, Invent. Math. **76** (1984), 93–120.

[Ma 4]   G. A. Margulis, Discrete subgroups of semisimple Lie groups, Ergebn. 3. Folge **17**, Springer (1991).

[Ma 5]   G. A. Margulis, *Superrigidity of commensurability subgroups and generalized harmonic maps*, preprint, preliminary version, June 1994.

[Mou]   G. Moussong, *Hyperbolic Coxeter groups*, Ph.D. thesis, Ohio State Univ. (1988).

[Pa]   S. J. Patterson, *The limit set of a Fuchsian group*, Acta Math. **136** (1976), 241–273.

[Sp]   B. Spieler, *Nonpositively curved orbihedra*, Ph.D. thesis, Ohio State Univ. (1992).

[Sp-Zi]   R. J. Spatzier and R. J. Zimmer, *Fundamental groups of negatively curved manifolds and actions of semisimple groups*, Topology **30** no. 4 (1991), 591–601.

[Su]   D. Sullivan, *Entropy, Hausdorff measures old and new, and limit sets of geometrically finite Kleinian groups*, Acta Math. **153** (1984), 259–277.

[Ti]   J. Tits, *Sur le groupe des automorphismes d'un arbre*, in: Essays on topology and related topics: Mémoire dédié à Georges de Rham (A. Haefliger and R. Narasimhan, eds.), Springer-Verlag, Berlin and New York (1970), 188–211.

[Wa]   A. Wald, *Begründung einer Koordinatenlosen Differentialgeometrie der Flächen*, Ergebnisse eines mathematischen Koloquiums **7** (1935), 2–46.

[Zi 1]   R. J. Zimmer, Ergodic Theory and Semisimple Groups, Birkhäuser, Basel and Boston, MA (1984).

[Zi 2]   R. J. Zimmer, *Actions of semisimple groups and discrete subgroups*, Proc. Internat. Congress Math., Berkeley (1986), 1247–1258.

[Zi 3]   R. J. Zimmer, *Representations of fundamental groups of manifolds with a semisimple transformation group*, J. Amer. Math. Soc. **2** (1989), 201–213.

Smooth Representations of p-adic Groups: The Role of Compact Open Subgroups

COLIN J. BUSHNELL

King's College London
Department of Mathematics
Strand, London WC2R 2LS, United Kingdom

This article concerns the method of investigating the smooth (complex) representations of a reductive p-adic group via the method of restriction to compact open subgroups, and amounts largely to a report on joint work of the author and Kutzko. Let G denote the group of F-points of some connected reductive algebraic group defined over the non-Archimedean local field F. The basic idea of the method is to isolate a family $\mathcal{F}$ of irreducible smooth representations of compact open subgroups of G, and then to describe a given irreducible smooth representation of G in terms of those members of $\mathcal{F}$ that it contains.

This is the approach of [7], where $G = GL(N, F)$ and $\mathcal{F}$ is the family of *simple types*. Simple types are defined directly and explicitly (see Section 3 below). They have the following properties:

(1) an irreducible smooth representation of G contains at most one simple type, up to G-conjugacy;
(2) the irreducible representations containing a fixed simple type can be classified via an isomorphism of Hecke algebras;
(3) an irreducible supercuspidal representation of G contains some simple type, and is induced from a related representation of an open compact mod center subgroup of G.

One can extend (2) to obtain a classification of all the irreducible smooth representations of G; not every such representation contains a simple type, but one reduces to this case via an easy parabolic induction. This classification is obtained without any recourse to either global methods or the Zelevinsky classification [1], [22] (see also [21] for a convenient exposition). Moreover, many properties of the parabolic induction functors central to the Bernstein-Zelevinsky approach can be recovered from the compact open subgroup approach.

When one relates the simple type classification to the Zelevinsky classification, one uncovers what now seems a significant fact: simple types are examples of what we call $\mathfrak{s}$-types, for certain points $\mathfrak{s}$ of the "Bernstein spectrum" $\mathcal{B}(G)$ of G. (Definitions will be given below.) This is the viewpoint of [8]. The notion of $\mathfrak{s}$-type in G can be formulated when G is the group of F-points of any connected reductive group defined over F. One can then ask whether an $\mathfrak{s}$-type exists for every $\mathfrak{s} \in \mathcal{B}(G)$. (The approach of [8] is slightly more general; we have simplified

Proceedings of the International Congress
of Mathematicians, Zürich, Switzerland 1994
© Birkhäuser Verlag, Basel, Switzerland 1995

it here to save space.) If this property holds for G one can take for $\mathcal{F}$ above the set of all $\mathfrak{s}$-types, and one obtains a complete description of the category of all smooth representations of G, along with (slightly weaker) analogues of (1)–(3).

One has the necessary results for $GL(N)$ [8], [9], and many of them for $SL(N)$. They constitute a method whose power we are only beginning to appreciate. For more general groups, little is currently known, but some recent results (see Section 5) seem consistent with this approach.

The axiomatic approach of [8] is so simple that we describe it first, after some preliminaries. We then summarise the existence theorems that lend it substance in the case of linear groups.

Throughout, F denotes a non-Archimedean local field with discrete valuation ring $\mathfrak{o}_F$. We write $\mathfrak{p}_F$ for the maximal ideal of $\mathfrak{o}_F$ and assume that the residue field $\mathsf{k}_F = \mathfrak{o}_F/\mathfrak{p}_F$ is finite.

1 Hecke algebras and compact open subgroups

For this section, G denotes a *locally profinite group*, by which we mean that G is locally compact, totally disconnected, and has a countable base of open sets. We also require that G be *unimodular*. Let dg be some Haar measure on G. The space $\mathcal{H}(G)$ of locally constant, compactly supported functions $\phi : G \to \mathbb{C}$ is then an algebra under convolution,

$$\phi \star \psi(g) = \int_G \phi(x)\psi(g^{-1}x)\,dx, \quad g \in G, \ \phi, \psi \in \mathcal{H}(G).$$

Let K be a compact open subgroup of G, and ρ an irreducible smooth representation of K on some (necessarily finite-dimensional) complex vector space W. This defines a function $e_\rho \in \mathcal{H}(G)$ by

$$e_\rho(g) = \begin{cases} \dfrac{\dim \rho}{\text{meas }K}\, \text{tr}_W(g^{-1}) \text{ if } g \in K, \\ 0 \text{ if } g \in G, \ g \notin K. \end{cases}$$

Here meas denotes measure with respect to dg and tr_W the trace for endomorphisms of W. The function e_ρ is an idempotent element of $\mathcal{H}(G)$, $e_\rho \star e_\rho = e_\rho$, so we get a subalgebra $e_\rho \star \mathcal{H}(G) \star e_\rho$ of $\mathcal{H}(G)$ with unit element e_ρ.

Next, let π be some smooth representation of G on a complex vector space $\mathcal{V}$. As usual, we extend π to give an action of $\mathcal{H}(G)$ on $\mathcal{V}$,

$$\pi(\phi)v = \int_G \phi(g)\pi(g)v\,dg, \quad \phi \in \mathcal{H}(G), \ v \in \mathcal{V}.$$

Let $\mathcal{V}^\rho$ denote the space of ρ-isotypic vectors in $\mathcal{V}$. Thus, $\mathcal{V}^\rho$ is the sum of all irreducible K-subspaces of $\mathcal{V}$ that are equivalent to ρ. We say that $(\pi, \mathcal{V})$ *contains* ρ if $\mathcal{V}^\rho \neq \{0\}$. We have $\mathcal{V}^\rho = \pi(e_\rho)\mathcal{V}$, whence $\mathcal{V}^\rho$ becomes a module over the algebra $e_\rho \star \mathcal{H}(G) \star e_\rho$. This leads to the following elementary observation.

(1.1) *The process* $\mathcal{V} \mapsto \mathcal{V}^\rho$ *induces a bijection between the set of equivalence classes of irreducible smooth representations* $\mathcal{V}$ *of* G *that contain* ρ *and the set of isomorphism classes of simple* $e_\rho \star \mathcal{H}(G) \star e_\rho$-*modules.*

(Here and throughout, if we have a ring R with unit 1_R, we only consider R-modules M with the property $1_R m = m$, $m \in M$.) This straightforward generality gives the first indication that one might try to classify irreducible representations of G via modules over certain Hecke algebras. For this purpose, there is a more convenient algebra. Let $(\rho^\vee, W^\vee)$ denote the contragredient of (ρ, W). Let $\mathcal{H}(G, \rho)$ denote the space of compactly supported functions $\Phi : G \to \mathrm{End}_{\mathbb{C}}(W^\vee)$ such that $\Phi(k_1 g k_2) = \rho^\vee(k_1) \circ \Phi(g) \circ \rho^\vee(k_2)$, $g \in G$, $k_i \in K$. This is an algebra under convolution, called the Hecke algebra of *compactly supported ρ-spherical functions on G.* We then have

(1.2) *There is a canonical isomorphism*

$$\mathcal{H}(G, \rho) \otimes_{\mathbb{C}} \mathrm{End}_{\mathbb{C}}(W) \xrightarrow{\;\approx\;} e_\rho \star \mathcal{H}(G) \star e_\rho$$

of $\mathbb{C}$-algebras with 1.

For this (and other matters above), see [7], Chapter 4. In particular, the two Hecke algebras attached to (K, ρ) are canonically Morita equivalent and have identical module theories. The version $\mathcal{H}(G, \rho)$ is usually more convenient for structural investigations.

Let us recall the classic example in which one uses a method of this kind. Let G denote the set of F-points of some connected reductive algebraic group defined over F. Let K be an Iwahori subgroup of G and let ρ be the trivial character of K. The irreducible smooth (therefore admissible) representations of G that contain ρ, i.e., that have a nontrivial Iwahori-fixed vector, are canonically parametrized by the set of simple modules over the algebra $\mathcal{H}(G, \rho)$. Here, of course, $\mathcal{H}(G, \rho)$ is just the subalgebra of $\mathcal{H}(G)$ consisting of K-bi-invariant functions. One has an explicit description of $\mathcal{H}(G, \rho)$ in terms of generators and relations [13]. When G is additionally of adjoint type, there is a parametrization [14] of the simple modules over $\mathcal{H}(G, \rho)$ in terms of L-group data.

2 The Bernstein decomposition and types

This starts with a précis of parts of [2], in slightly different language. For a more careful translation, see [8]. Again, F denotes a non-Archimedean local field, but in this section G is the group of F-points of a connected reductive algebraic group defined over F. We write $\mathfrak{SR}(G)$ for the category of all smooth (complex) representations of G.

We consider pairs $(L, \boldsymbol{\sigma})$ consisting of (the group of F-points of) a Levi subgroup of G (defined over F) and an irreducible supercuspidal representation $\boldsymbol{\sigma}$ of L. We allow the possibility $L = G$. For such a subgroup L, let $\mathrm{X}_{\mathrm{alg}}(L)$ denote the group of F-rational homomorphisms $L \to GL(1, F)$, and $\mathrm{X}(L)$ the group of homomorphisms $L \to \mathbb{C}^\times$ generated by those of the form $l \mapsto \|\chi(l)\|^s$, where $\chi \in \mathrm{X}_{\mathrm{alg}}(L)$, $s \in \mathbb{C}$, and $\| \;\|$ is the standard absolute value on F. Two such pairs $(L_i, \boldsymbol{\sigma}_i)$ are *equivalent* if there exist $g \in G$ and $\chi \in \mathrm{X}(L_2)$ such that $L_2 = L_1^g = g^{-1} L_1 g$ and $\boldsymbol{\sigma}_2 \equiv \boldsymbol{\sigma}^g \otimes \chi$. We write $\mathcal{B}(G)$ for the set of equivalence classes here, and call it the *Bernstein spectrum* of G.

An irreducible smooth representation $\boldsymbol{\pi}$ of G determines an element $\mathfrak{I}(\boldsymbol{\pi})$ of $\mathcal{B}(G)$ as follows. There exists a parabolic subgroup P of G, with Levi subgroup L say, and an irreducible supercuspidal representation $\boldsymbol{\sigma}$ of L such that $\boldsymbol{\pi}$ is equivalent to a composition factor of the representation of G smoothly induced from the inflation of $\boldsymbol{\sigma}$ to P. We define $\mathfrak{I}(\boldsymbol{\pi})$ to be the equivalence class of $(L, \boldsymbol{\sigma})$ in $\mathcal{B}(G)$.

Given an element $\mathfrak{s} \in \mathcal{B}(G)$, we define a full subcategory $\mathfrak{SR}^{\mathfrak{s}}(G)$ as follows. A smooth representation $\boldsymbol{\pi}$ is an object of $\mathfrak{SR}^{\mathfrak{s}}(G)$ if and only if, for every irreducible subquotient $\boldsymbol{\pi}_0$ of $\boldsymbol{\pi}$, we have $\mathfrak{I}(\boldsymbol{\pi}_0) = \mathfrak{s}$. From [2, (2.10)] we get:

(2.1) *The abelian category* $\mathfrak{SR}(G)$ *is the direct product*

$$\mathfrak{SR}(G) = \prod_{\mathfrak{s} \in \mathcal{B}(G)} \mathfrak{SR}^{\mathfrak{s}}(G).$$

In other words, if $(\pi, \mathcal{V})$ *is a smooth representation of* G, *the space* $\mathcal{V}$ *has a unique maximal* G*-subspace* $\mathcal{V}^{\mathfrak{s}}$ *such that* $\mathcal{V}^{\mathfrak{s}}$ *is an object of* $\mathfrak{SR}^{\mathfrak{s}}(G)$, *and*

$$\mathcal{V} = \coprod_{\mathfrak{s} \in \mathcal{B}(G)} \mathcal{V}^{\mathfrak{s}}.$$

Moreover, for smooth representations $(\pi_i, \mathcal{V}_i)$ *of* G, *we have*

$$\mathrm{Hom}_G(\mathcal{V}_1, \mathcal{V}_2) = \prod_{\mathfrak{s}} \mathrm{Hom}_G(\mathcal{V}_1^{\mathfrak{s}}, \mathcal{V}_2^{\mathfrak{s}}).$$

Now we need some more generalities. Let K be a compact open subgroup of G and ρ be an irreducible smooth representation of K. We use the notation of Section 1; in particular, if $(\pi, \mathcal{V})$ is a smooth representation of G, we write $\mathcal{V}^{\rho}$ for the ρ-isotypic subspace of $\mathcal{V}$. We also write $\mathcal{V}[\rho]$ for the G-subspace of $\mathcal{V}$ generated by $\mathcal{V}^{\rho}$. We let $\mathfrak{SR}_{\rho}(G)$ denote the full subcategory of $\mathfrak{SR}(G)$ whose objects are those $\mathcal{V} \in |\mathfrak{SR}(G)|$ that satisfy $\mathcal{V} = \mathcal{V}[\rho]$. We note in passing that $\mathfrak{SR}_{\rho}(G)$ is not usually very interesting. In particular, it is not always closed under taking subquotients in $\mathfrak{SR}(G)$. However, we do have a functor

$$\mathbf{m}_{\rho} : \mathfrak{SR}_{\rho}(G) \longrightarrow e_{\rho} \star \mathcal{H}(G) \star e_{\rho}\text{-}\mathfrak{Mod}$$

given by $\mathcal{V} \mapsto \mathcal{V}^{\rho}$. Composing with the Morita equivalence implied by (1.2), we get a functor

$$\mathbf{M}_{\rho} : \mathfrak{SR}_{\rho}(G) \longrightarrow \mathcal{H}(G, \rho)\text{-}\mathfrak{Mod}. \tag{2.2}$$

(2.3) DEFINITION. *Let* $\mathfrak{s} \in \mathcal{B}(G)$. *Let* K *be a compact open subgroup of* G *and* ρ *be an irreducible smooth representation of* K. *The pair* (K, ρ) *is called an* $\mathfrak{s}$-*type if it has the following property: an irreducible smooth representation* $(\pi, \mathcal{V})$ *of* G *contains* ρ *if and only if* $\mathfrak{I}(\pi) = \mathfrak{s}$.

We remark that this is slightly less general than the definition in [8]; its principal advantage here is brevity.

(2.4) THEOREM. *Let* $\mathfrak{s} \in \mathcal{B}(G)$ *and let* (K, ρ) *be an* $\mathfrak{s}$-*type. Then, for any smooth representation* $(\pi, \mathcal{V})$ *of* G, *we have* $\mathcal{V}[\rho] = \mathcal{V}^{\mathfrak{s}}$. *Moreover:*

(1) *If* $\mathcal{V}$ *is a smooth representation of* G *such that* $\mathcal{V} = \mathcal{V}[\rho]$, *then every irreducible subquotient of* $\mathcal{V}$ *contains* ρ.

(2) *For any smooth representation* $\mathcal{V}$ *of* G, *the subspace* $\mathcal{V}[\rho]$ *is a direct summand of* $\mathcal{V}$.

(3) *As subcategories of* $\mathfrak{SR}(G)$, *we have*

$$\mathfrak{SR}^{\mathfrak{s}}(G) = \mathfrak{SR}_{\rho}(G).$$

(4) *The functor* $\mathbf{M}_{\rho}$ *of* (2.2) *is an equivalence of categories*

$$\mathbf{M}_{\rho} : \mathfrak{SR}_{\rho}(G) \xrightarrow{\approx} \mathcal{H}(G, \rho)\text{-}\mathfrak{Mod}.$$

This is taken from [8, Section 4]. It demonstrates that the existence of types has powerful repercussions for the structure of $\mathfrak{SR}(G)$. Among other things, it shows that the irreducible representations π with $\mathfrak{I}(\pi) = \mathfrak{s}$ are parametrized by the simple modules over the Hecke algebra of an $\mathfrak{s}$-type.

There are nontrivial examples of types. We shall consider one family below. However, another family has been known for a long time. Let $\mathfrak{s}$ be the class of $(L, \boldsymbol{\sigma})$ in $\mathcal{B}(G)$, where L is a minimal Levi subgroup of G and $\boldsymbol{\sigma}$ is its trivial representation. If K denotes an Iwahori subgroup of G and ρ the trivial representation of G, then (K, ρ) is an $\mathfrak{s}$-type: see [11, Theorem 3.7] for an exposition of this matter (due to Borel [3] and Casselman [12]). In this particular case, (2.4) retrieves, with very little effort, the results of [3] without recourse to the finiteness hypotheses imposed there.

3 Simple types in linear groups

We now construct a family of compact open subgroups of $GL(N, F)$ and a family of irreducible smooth representations of these. This is taken from [7], especially Chapters 3 and 5.

Let V be a finite-dimensional F-vector space and set $A = \mathrm{End}_F(V)$, $G = \mathrm{Aut}_F(V)$. If $\mathfrak{A}$ is a hereditary $\mathfrak{o}_F$-order in A with Jacobson radical $\mathfrak{P}$ (see [7, (1.1)] for a summary or [20] for a complete account), we write $U(\mathfrak{A}) = U^0(\mathfrak{A}) = \mathfrak{A}^{\times}$ and $U^n(\mathfrak{A}) = 1 + \mathfrak{P}^n$, $n \geqslant 1$. We also set $\mathfrak{K}(\mathfrak{A}) = \{x \in G : x^{-1}\mathfrak{A}x = \mathfrak{A}\}$. In fact, $\mathfrak{K}(\mathfrak{A})$ is the G-normalizer of $U(\mathfrak{A})$. Recall that a *stratum in* A is a 4-tuple $[\mathfrak{A}, n, m, b]$, where $\mathfrak{A}$ is a hereditary $\mathfrak{o}_F$-order in A with radical $\mathfrak{P}$, $n > m$ are integers, and $b \in \mathfrak{P}^{-n}$. Two strata $[\mathfrak{A}, n, m, b_i]$ are *equivalent* if $b_1 \equiv b_2 \pmod{\mathfrak{P}^{-m}}$.

A stratum $[\mathfrak{A}, n, m, \beta]$ is called *pure* if the algebra $E = F[\beta]$ is a field, with $E^{\times} \subset \mathfrak{K}(\mathfrak{A})$, and $\beta.\mathfrak{A} = \mathfrak{P}^{-n}$. In this situation, we have an adjoint map $a_{\beta} : A \to A$, $x \mapsto \beta x - x\beta$. We define a quantity $k_0(\beta, \mathfrak{A})$ by

$$k_0(\beta, \mathfrak{A}) = \min\{t \in \mathbb{Z} : \mathfrak{P}^t \cap a_{\beta}(A) \subset a_{\beta}(\mathfrak{A})\}.$$

We then have $k_0(\beta, \mathfrak{A}) = -\infty$ (this is the case $E = F$) or else $k_0(\beta, \mathfrak{A}) \geqslant -n$. If $k_0(\beta, \mathfrak{A}) \leqslant -n$, we call β *minimal over* F.

A pure stratum $[\mathfrak{A}, n, m, \beta]$ is called *simple* if $m < -k_0(\beta, \mathfrak{A})$. The following result (taken from [7, (2.4.1)]) is fundamental:

(3.1) THEOREM. *Let $[\mathfrak{A}, n, m, \beta]$ be a pure stratum. There exists a simple stratum $[\mathfrak{A}, n, m, \gamma]$ equivalent to $[\mathfrak{A}, n, m, \beta]$. The field degree $[F[\gamma] : F]$ divides $[F[\beta] : F]$, and we have equality here if and only if $[\mathfrak{A}, n, m, \beta]$ is simple.*

We take a simple stratum $[\mathfrak{A}, n, 0, \beta]$. To this, we attach two $\mathfrak{o}_F$-orders

$$\mathfrak{H}(\beta, \mathfrak{A}) \subset \mathfrak{J}(\beta, \mathfrak{A})$$

in A as follows. First, set $E = F[\beta]$, $B = \text{End}_E(V)$, and $\mathfrak{B} = \mathfrak{A} \cap B$ (which is a hereditary $\mathfrak{o}_E$-order in B). We write $\mathfrak{H}^n = \mathfrak{H} \cap \mathfrak{P}^n$, $n \geqslant 1$, and similarly for $\mathfrak{J}$. Suppose first that β is *minimal* over F. We set

$$
\begin{aligned}
\mathfrak{H}(\beta, \mathfrak{A}) &= \mathfrak{B} + \mathfrak{P}^{[\frac{n}{2}]+1}, \\
\mathfrak{J}(\beta, \mathfrak{A}) &= \mathfrak{B} + \mathfrak{P}^{[\frac{n+1}{2}]}.
\end{aligned}
\tag{3.2}
$$

Here $[\ ]$ denotes the greatest integer function. Suppose now that β is not minimal over F. Thus, $k_0(\beta, \mathfrak{A}) = -r$, where $0 < r < n$. The stratum $[\mathfrak{A}, n, r, \beta]$ is not simple, but it is equivalent to a simple stratum $[\mathfrak{A}, n, r, \gamma]$ by (3.1). Inductively, we can assume we have defined $\mathfrak{H}(\gamma, \mathfrak{A})$, $\mathfrak{J}(\gamma, \mathfrak{A})$. We then set

$$
\begin{aligned}
\mathfrak{H}(\beta, \mathfrak{A}) &= \mathfrak{B} + \mathfrak{H}^{[\frac{r}{2}]+1}(\gamma, \mathfrak{A}), \\
\mathfrak{J}(\beta, \mathfrak{A}) &= \mathfrak{B} + \mathfrak{J}^{[\frac{r+1}{2}]}(\gamma, \mathfrak{A}).
\end{aligned}
\tag{3.3}
$$

Much effort has to go into proving that $\mathfrak{H}$ and $\mathfrak{J}$ are indeed rings, and that the definitions (3.3) are independent of choices. We write $H^t = 1 + \mathfrak{H}^t$, $J^t = 1 + \mathfrak{J}^t$, $t \geqslant 1$, and $J = \mathfrak{J}^\times$.

The next step is to define a finite family $\mathcal{C}(\mathfrak{A}, \beta)$ ($= \mathcal{C}(\mathfrak{A}, 0, \beta)$ in [7] notation) of abelian characters of the group $H^1(\beta, \mathfrak{A})$. Again the definition is quite straightforward, but much work is required to show it that *is* a definition. For this, we need to choose a continuous character ψ_F of the additive group of F, with conductor $\mathfrak{p}_F$. We set $\psi_A = \psi_F \circ \text{tr}_{A/F}$. For $b \in A$, we write ψ_b for the function $x \mapsto \psi_A(b(x-1))$, $x \in A$. Suppose first that β is *minimal* over F, so that $H^1(\beta, \mathfrak{A}) = U^1(\mathfrak{B})U^{[n/2]+1}(\mathfrak{A})$. The set $\mathcal{C}(\mathfrak{A}, \beta)$ then consists of all characters θ of H^1 such that

$$
\begin{aligned}
\theta \mid U^{[\frac{n}{2}]+1}(\mathfrak{A}) &= \psi_\beta, \\
\theta \mid U^1(\mathfrak{B}) &= \phi \circ \det_B,
\end{aligned}
$$

for some character ϕ of $U^1(\mathfrak{o}_E)$, where $\det_B$ is the determinant mapping $B^\times \to E^\times$. In the general case (3.3), we have $H^1(\beta, \mathfrak{A}) = U^1(\mathfrak{B})H^{[r/2]+1}(\gamma, \mathfrak{A})$. Here $\mathcal{C}(\mathfrak{A}, \beta)$ consists of those characters θ such that

$$
\begin{aligned}
\theta \mid H^{[\frac{r}{2}]+1}(\gamma, \mathfrak{A}) &= \left(\theta_0 \mid H^{[\frac{r}{2}]+1}(\gamma, \mathfrak{A})\right) \cdot \psi_{\beta-\gamma}, \\
\theta \mid U^1(\mathfrak{B}) &= \phi \circ \det_B,
\end{aligned}
$$

for some $\theta_0 \in \mathcal{C}(\mathfrak{A}, \gamma)$ and some character ϕ of $U^1(\mathfrak{o}_E)$.

In the third step, we transfer attention to the group J^1.

(3.4) PROPOSITION. *Let $[\mathfrak{A}, n, 0, \beta]$ be a simple stratum in A as above, and let $\theta \in \mathcal{C}(\mathfrak{A}, \beta)$. There exists a unique irreducible representation η of $J^1(\beta, \mathfrak{A})$ such that $\eta \mid H^1(\beta, \mathfrak{A})$ contains θ. Indeed, the restriction of η to H^1 is a multiple of θ. Further, an element $g \in G$ intertwines the representation η if and only if $g \in JB^\times J$, where $J = J(\beta, \mathfrak{A})$.*

The fourth step is rather more subtle.

(3.5) PROPOSITION. *Let η be as in (3.4). There exists an irreducible representation κ of the group $J(\beta, \mathfrak{A})$ with the following properties:*

(1) *$\kappa \mid J^1(\beta, \mathfrak{A}) = \eta$;*
(2) *every $g \in G$ that intertwines η also intertwines κ.*

Further, these conditions determine κ uniquely up to tensoring with a character of the form $\phi \circ \det_B$, where ϕ is a character of $U(\mathfrak{o}_E)$ that is trivial on $U^1(\mathfrak{o}_E)$.

By definition, we have $J(\beta, \mathfrak{A}) = U(\mathfrak{B})J^1(\beta, \mathfrak{A})$, so $J/J^1 \cong U(\mathfrak{B})/U^1(\mathfrak{B})$. We now assume that $\mathfrak{A}$ is a *principal* order; this has the effect that

$$J(\beta, \mathfrak{A})/J^1(\beta, \mathfrak{A}) \cong GL(f, \mathrm{k}_E)^e, \tag{3.6}$$

for integers e, f such that $ef[E : F] = \dim_F(V)$. We take an irreducible cuspidal representation of $GL(f, \mathrm{k}_E)$, form its e-fold tensor power, and inflate this to a representation σ of $J(\beta, \mathfrak{A})$. We then form the representation $\lambda = \kappa \otimes \sigma$ of $J(\beta, \mathfrak{A})$, with κ as in (3.5). Pairs $(J(\beta, \mathfrak{A}), \lambda)$ constructed in this way are what we call *simple types in G*. (We have omitted a special case here. This corresponds to the situation where $E = F$, $J(\beta, \mathfrak{A}) = U(\mathfrak{A})$, and where η is trivial: see [7, (5.5.10)] for the full definition.)

The next task is to compute the Hecke algebra $\mathcal{H}(G, \lambda)$ for one of these simple types (J, λ). If K is a local field whose residue field has q_K elements, and if 1_m denotes the trivial character of an Iwahori subgroup of $GL(m, K)$, the Hecke algebra $\mathcal{H}(GL(m, K), 1_m)$ has an explicit description in terms of generators and relations: see [7, (5.4)] for the version we have in mind. The main point is that this description depends only on the parameters m and q_K. We therefore write $\mathcal{H}(GL(m, K), 1_m) = \mathcal{H}(m, q_K)$.

(3.7) THEOREM. *Let (J, λ) be a simple type in G as above. We then have*

$$\mathcal{H}(G, \lambda) \cong \mathcal{H}(e, q_E^f),$$

as $\mathbb{C}$-algebras with 1, with e, f as in (3.6).

This isomorphism is canonical up to composition with a rather trivial family of automorphisms of $\mathcal{H}(e, q_E^f)$. Invoking (1.1), the irreducible representations of G that contain λ are parametrized by the simple $\mathcal{H}(e, q_E^f)$-modules. These are known from [14] (which here amounts to a very special case of the Zelevinsky classification).

4 Main theorems for linear groups

Again, we take $G = \operatorname{Aut}_F(V) \cong GL(N, F)$. Suppose we have a simple type (J, λ) in G defined from a simple stratum $[\mathfrak{A}, n, 0, \beta]$ as in Section 3 (the notation of which we continue to use). We call (J, λ) a *maximal* simple type if the parameter e of (3.6) is equal to 1.

(4.1) THEOREM. *Let (J, λ) be a simple type in G as above, and assume it is maximal. Then:*

(1) *The representation λ extends to a representation Λ of $E^\times J(\beta, \mathfrak{A})$.*
(2) *The representation $\pi_0 = \operatorname{Ind}(\Lambda)$ of G smoothly induced by Λ is irreducible and supercuspidal.*
(3) *An irreducible smooth representation π of G contains λ if and only if $\pi \cong \pi_0 \otimes \chi \circ \det$, for some unramified quasicharacter χ of $F^\times$.*
(4) *Let $\mathfrak{s} \in \mathcal{B}(G)$ be the equivalence class of the pair (G, π_0). Then (J, λ) is an $\mathfrak{s}$-type.*

The first three assertions come from [7, Chapter 6] and (4) follows from (3). Note in (3) that the representation π is induced from $\Lambda \otimes \chi \circ \det$. This result has a partial converse.

(4.2) THEOREM. *Let π be an irreducible supercuspidal representation of G, and suppose that π contains some simple type (J, λ). Then (J, λ) is maximal.*

To complete the picture regarding supercuspidal representations, one needs the following:

(4.3) THEOREM. *Let π be an irreducible supercuspidal representation of G. Then π contains some simple type.*

This presently requires a rather elaborate argument, occupying most of Chapter 8 of [7]. There, one invents a new family of "split types"; a representation containing no simple type must contain a split type, and a representation containing a split type cannot be supercuspidal.

We now need to describe the set of representations of G that contain a non-maximal simple type. The following is taken from [7, (8.4.3)], filtered through the viewpoint of [8].

(4.4) THEOREM. *Let (J, λ) be a simple type in $G \cong GL(N, F)$ and define the integer e by (3.6). There exists a maximal simple type (J_0, λ_0) in $GL(N/e, F)$ with the following property:*

View $GL(N/e, F)^e$ as a Levi subgroup of G, and let $\mathfrak{s} \in \mathcal{B}(G)$ denote the equivalence class of the pair $(GL(N/e, F)^e, \pi_0 \otimes \ldots \otimes \pi_0)$, where π_0 is some irreducible (necessarily supercuspidal) representation of $GL(N/e, F)$ containing λ_0. An irreducible smooth representation π then contains λ if and only if $\mathfrak{I}(\pi) = \mathfrak{s}$.

In particular, (J, λ) is an $\mathfrak{s}$-type.

We have so far said nothing concerning uniqueness. Two simple types that occur in the same irreducible representation of G must intertwine in G for elementary reasons, so the uniqueness question is answered by:

(4.5) THEOREM. *Let* (J_i, λ_i), $i = 1, 2$, *be simple types in* $G \cong GL(N, F)$, *and suppose that they intertwine in* G. *There then exists* $g \in G$ *such that* $J_2 = J_1^g$ *and* $\lambda_2 \cong \lambda_1^g$.

In other words, simple types that intertwine are conjugate. This version of the result is [8, (5.6)], but it depends heavily on results from [7].

Remarks:

(1) Underlying the Hecke algebra isomorphism (3.7) is the fact that one can find explicitly the set of $g \in G$ that intertwine a given simple type. The same is true of the intermediate stages of the construction. This is what effectively leads the whole procedure culminating in the simple types.

(2) One can give a systematic, but intrinsically noncanonical, procedure for listing the conjugacy classes of simple types in $GL(N, F)$. See [10] for a discussion of such matters.

(3) The analogues of Theorems (4.1)–(4.5) and (3.7) all hold for $SL(N, F)$; see [8] and [6]. In particular, any irreducible supercuspidal representation of the group $SL(N, F)$ is induced from a compact open subgroup.

(4) The Hecke algebra procedure is also effective for describing the discrete series: see [7, (7.7)] and [6, Section 8].

5 Further results

In the case $G \cong GL(N, F)$, we can combine (4.5) with (2.4) and (3.7) to get a complete description of the factors $\mathfrak{S}\mathfrak{R}^{\mathfrak{s}}(G)$ when $\mathfrak{s} \in \mathcal{B}(G)$ is of the form in (4.5).

For $GL(N, F)$, one can show that there exists an $\mathfrak{s}$-type for *any* $\mathfrak{s} \in \mathcal{B}(G)$. This requires the "semisimple types" of [9], which are constructed directly from simple types. Their Hecke algebras are just what one would expect, namely tensor products of affine Hecke algebras of the form $\mathcal{H}(m, q_F^r)$, where q_F is the cardinality of k_F.

In the general case, where G is just a connected reductive group over F, one can at least ask whether every $\mathfrak{s} \in \mathcal{B}(G)$ gives rise to an $\mathfrak{s}$-type. (There is a slightly looser and more credible version of this idea in [8].) There are no more than hints about this at present. The paper [18] gives a collection of candidates for $\mathfrak{s}$-types for certain $\mathfrak{s}$ in general groups, and [17] suggests further possibilities in classical groups. Historically, simple types arose from refining the fundamental strata of [4], repeatedly using ideas of [15]. Analogues of fundamental strata are available for classical groups in [16], and something very similar for a wide class of groups in [19]. Whether these are susceptible to a similar analysis at present remains to be seen.

In a totally different direction, even the bare précis here shows that simple types in $GL(N)$ are arithmetical, or at least field-theoretical, objects. This leads naturally to the question of whether they reflect any of the "functoriality" properties demanded by the Langlands conjectures. There is some definite evidence to support this, in the context of tamely ramified base-change; see [5].

References

[1]   I. N. Bernstein and A. V. Zelevinsky, *Induced representations of reductive p-adic groups*, Ann. Sci. École Norm. Sup. (4) **10** (1977), 441–472.

[2]   J. Bernstein (rédigé par P. Deligne), *Le "centre" de Bernstein*, Représentations des groupes réductifs sur un corps local, Hermann, Paris, 1984, pp. 1–32.

[3]   A. Borel, *Admissible representations of a semisimple group over a local field with vectors fixed under an Iwahori subgroup*, Invent. Math. **35** (1976), 233–259.

[4]   C. J. Bushnell, *Hereditary orders, Gauss sums and supercuspidal representations of GL_N*, J. Reine Angew. Math. **375/376** (1987), 184–210.

[5]   C. J. Bushnell and G. Henniart, *Local tame lifting for $GL(N)$ I: Simple characters*, preprint (1994).

[6]   C. J. Bushnell and P. C. Kutzko, *The admissible dual of $SL(N)$ II*, Proc. London Math. Soc. (3) **68** (1992), 317–379.

[7]   _____, The admissible dual of $GL(N)$ via compact open subgroups, Princeton University Press, Princeton, NJ, 1993.

[8]   _____, *Smooth representations of p-adic groups: Towards a complete structure theory*, preprint (1994).

[9]   _____, *Semisimple types in $GL(N)$*, preprint (1994).

[10] _____, *Simple types in $GL(N)$: Computing conjugacy classes*, Contemp. Math., to appear.

[11] P. Cartier, *Representations of $\mathfrak{p}$-adic groups: A survey*, Automorphic forms, representations and L-functions (A. Borel and W. Casselman, eds.), Proc. Symposia in Pure Math. XXXIII Part 1, Amer. Math. Soc., Providence, RI, 1979, pp. 111–156.

[12] W. Casselman, *The unramified principal series of $\mathfrak{p}$-adic groups I*, Compositio Math **40** (1980), 387–406.

[13] N. Iwahori and H. Matsumoto, *On some Bruhat decomposition and the structure of the Hecke rings of the $\mathfrak{p}$-adic Chevalley groups*, Publ. Math. IHES **25** (1965), 5–48.

[14] D. Kazhdan and G. Lusztig, *Proof of the Deligne-Langlands conjecture for Hecke algebras*, Invent. Math. **87** (1987), 153–215.

[15] P. C. Kutzko, *Towards a classification of the supercuspidal representations of GL_N*, J. London Math. Soc. (2) **37** (1988), 265–274.

[16] L. E. Morris, *Fundamental G-strata for p-adic classical groups*, Duke Math. J. **64** (1991), 501–553.

[17] _____, *Tamely ramified supercuspidal representations of classical groups II: Representation theory*, Ann. Sci. École Norm. Sup. (4) **25** (1992), 233–274.

[18] _____, *Tamely ramified intertwining algebras*, Invent. Math. **114** (1993), 1–54.

[19] A. Moy and G. Prasad, *Unrefined minimal K-types for p-adic groups*, Invent. Math. **116** (1994), 393–408.

[20] I. Reiner, Maximal Orders, Academic Press, New York, 1975.

[21] F. Rodier, *Représentations de $GL(n,k)$ où k est un corps p-adique*, Astérisque **92–93** (1982), 201–218.

[22] A. V. Zelevinsky, *Induced representations of reductive p-adic groups II: On irreducible representations of $GL(n)$*, Ann. Sci. École Norm. Sup. (4) **13** (1980), 165–210.

Flows on Homogeneous Spaces and Diophantine Approximation

S. G. Dani

Tata Institute of Fundamental Research
Bombay 400 005, India

Dedicated to the memory of Professor K. G. Ramanathan

1. Introduction

Let G be a Lie group and Γ be a lattice in G; that is, Γ is a discrete subgroup such that G/Γ admits a finite Borel measure invariant under the action of G (on the left). On the homogeneous space G/Γ there is a natural class of dynamical systems defined by the actions of subgroups of G. The study of these systems has proved to be of great significance from the point of view of dynamics, ergodic theory, geometry, etc. and has found many interesting applications in number theory. The systems have been explored from various angles; however, I would like to concentrate here on giving an exposition of certain recent developments on the theme of the following classical theorem on orbits of what are called the horocycle flows.

THEOREM 1 (Hedlund, 1936). *Let $G = SL(2, \mathbb{R})$ and Γ be a lattice in G. For $t \in \mathbb{R}$ let $h_t = \begin{pmatrix} 1 & t \\ 0 & 1 \end{pmatrix}$ and let $H = \{h_t \mid t \in \mathbb{R}\}$. Then every H-orbit on G/Γ is either periodic or dense in G/Γ. If G/Γ is compact then all orbits are dense, whereas if G/Γ is noncompact then the set of points with periodic orbits is nonempty and consists of finitely many immersed cylinders, each of which is dense in G/Γ.*

In the compact quotient case the action is "minimal", as there are no proper closed invariant subsets. It may be recalled that by Kronecker's theorem the same holds for the translation flows on the n-dimensional torus, given by actions of $\{(e^{i\alpha_1 t}, \ldots, e^{i\alpha_n t}) \mid t \in \mathbb{R}\}$ where $\alpha_1, \ldots, \alpha_n \in \mathbb{R}$ are linearly independent over $\mathbb{Q}$. Similar minimality results are also known for one-parameter flows (and also affine transformations) on nilmanifolds and more generally for solvmanifolds, thanks to the work of Furstenberg, Parry, Auslander and Brezin etc.; see [D6] for some details and references. In many of these results minimality is concluded from "unique ergodicity", namely uniqueness of the invariant probability measure. The same was proved in the compact quotient case of Theorem 1 by Furstenberg:

Proceedings of the International Congress
of Mathematicians, Zürich, Switzerland 1994
© Birkhäuser Verlag, Basel, Switzerland 1995

THEOREM 2 (Furstenberg, 1972). *Let G, Γ, and H be as in Theorem 1. Suppose further that G/Γ is compact. Then the G-invariant probability measure is the only H-invariant probability measure on G/Γ.*

The result implies in particular that every orbit is "uniformly distributed" with respect to the G-invariant measure; in the general notation, the orbit of $x \in G/\Gamma$ under a one-parameter subgroup $\{u_t\}$ (we implicitly assume t varying in $\mathbb{R}$, in using this notation) is said to be uniformly distributed with respect to a measure μ on G/Γ if for any bounded continuous function φ on G/Γ

$$\frac{1}{T} \int_0^T \varphi(u_t x)\, dt \to \int_{G/\Gamma} \varphi\, d\mu \quad \text{as} \quad T \to \infty.$$

The minimality and unique ergodicity assertions for horocycle flows (in the compact quotient case) were generalized by several authors, including Bowen, Veech, Ellis and Perrizo and also the present author, to what are called horospherical flows, on more general Lie groups (see [D6] for some details).

At this point I should also mention that satisfactory criteria were known by the 1970s for ergodicity of subgroup actions on homogeneous spaces (see [M3] and [D6] for some details). Ergodicity, which means that there is no measurable invariant subset that is nontrivial in the sense that the set as well as its complement are of positive measure, implies in particular that orbits of almost all points are dense. Similarly, by Birkhoff's ergodic theorem it also implies that orbits of almost all points are uniformly distributed with respect to the measure in question. Though the results in this stream pertain mainly to orbits of "almost all" points, many developments in this regard play a crucial role in the understanding of closures of arbitrary orbits in special cases.

Another feature of Theorem 1, of main interest to us in this article, is that for the systems considered, even when there are both dense as well as nondense orbits (as in the noncompact quotient case) the closures of all orbits are "nice" objects geometrically. A search for such a phenomenon in a more general set up got an impetus from an observation of Raghunathan, my teacher, that a well-known conjecture going back to a paper of Oppenheim from 1929, stating that for any nondegenerate indefinite quadratic form in three or more real variables the set of values at integral points is dense in $\mathbb{R}$ unless the form is a scalar multiple of a rational form, would follow if it is shown that all orbits of $SO(2,1)$ on $SL(3,\mathbb{R})/SL(3,\mathbb{Z})$ are either closed or dense. In this context he proposed the following conjecture.

CONJECTURE 1 *Let G be a Lie group and Γ be a lattice in G. Let U be a unipotent subgroup of G (that is, $\operatorname{Ad} u$ is a unipotent linear transformation for all $u \in U$). Then the closure of any U-orbit is a homogeneous set; that is, for any $x \in G/\Gamma$ there exists a closed subgroup F of G such that $\overline{Ux} = Fx$.*

(In [D2] where the conjecture first appeared in print the statement is somewhat weaker, on account of the author's predilection at that time.)

The minimality results mentioned above confirm the conjecture in their respective cases. In a paper in 1986 I verified the conjecture for horospherical flows on not necessarily compact homogeneous spaces of reductive Lie groups; a partial result was obtained earlier in a joint paper with Raghavan and was applied to study orbits of euclidean frames under actions of $SL(n, \mathbb{Z})$ and $Sp(n, \mathbb{Z})$ (see [M3] and [D6] for some details). The result was generalized to all (not necessarily reductive) groups by Starkov (see [D6] for details).

In [D1] and [D2] I obtained a classification of invariant measures of maximal horospherical flows on not necessarily compact homogeneous spaces of reductive Lie groups. The $SL(2, \mathbb{R})$ case of this was later applied to obtain a result on the distribution of the orbits of the horocycle flows as in Theorem 1, in [D3] for the case of $\Gamma = SL(2, \mathbb{Z})$ and in [DS] for any lattice in $SL(2, \mathbb{R})$.

THEOREM 3 (Dani and Smillie, 1984). *Let the notation be as in Theorem 1. Then every nonperiodic orbit of* $\{h_t\}$ *on* G/Γ *is uniformly distributed with respect to the G-invariant probability measure.*

A measure analogue of Raghunathan's conjecture was also formulated in [D2]. In keeping with the set up of Conjecture I, it may be stated as follows.

CONJECTURE 2 *Let G, Γ, and U be as in Conjecture I. Let μ be a U-invariant ergodic probability measure on G/Γ. Then there exists a closed subgroup F of G such that μ is F-invariant and* supp $\mu = Fx$ *for some $x \in G/\Gamma$ (a measure for which this condition holds is called algebraic).*

In a remarkable development both of these conjectures were recently proved by Ratner, through a series of four papers (see [R1], [R2] and the references there). In fact, she proved the results for a larger class of actions; for connected Lie subgroups U her result on invariant measures may be stated as follows.

THEOREM 4 (Ratner, 1991). *Let G be a connected Lie group and Γ be a discrete subgroup of G. Let U be a Lie subgroup that is generated by the unipotent one-parameter subgroups contained in it. Then any finite ergodic U-invariant measure on G/Γ is algebraic.*

(I may mention here that recently Margulis and Tomanov have given another proof of the above theorem in the case of algebraic groups; their proof bears a strong influence of Ratner's original arguments but is substantially different in its approach and methods; see [MT].)

Using Theorem 4 together with a result from [D4] (similar to Theorem 12 below) Ratner proved in [R2] the following result on the distribution of orbits of unipotent one-parameter subgroups, generalizing Theorem 3; the same conclusion was also deduced from Theorem 4 by Nimish Shah [S1] for reductive Lie groups of $\mathbb{R}$-rank 1.

THEOREM 5 (Ratner, 1991). *Let G be a connected Lie group and Γ be a lattice in G. Let $U = \{u_t\}$ be a unipotent one-parameter subgroup of G. Then for any $x \in G/\Gamma$ there exists an algebraic probability measure μ such that the $\{u_t\}$-orbit of x is uniformly distributed with respect to μ.*

From this she deduced Conjecture I and also the following stronger assertion.

THEOREM 6 (Ratner, 1991). *Let G and U be as in Theorem 4. Let Γ be a lattice in G. Then for any $x \in G/\Gamma$, $\overline{Ux}$ is a homogeneous subset with finite invariant measure; that is, there exists a closed subgroup F such that Fx admits a finite F-invariant measure and $\overline{Ux} = Fx$.*

From Raghunathan's observation noted earlier, a case of the theorem with $G = SL(3, \mathbb{R})$ and $\Gamma = SL(3, \mathbb{Z})$ yields Oppenheim's conjecture. The conjecture itself was however proved in the meantime by Margulis (see [M1] and [M2]). The reader is referred to [M3] for an account of the earlier work on the conjecture and to [DM2] and [D5] for some elementary proofs.

Let G, Γ, and $\{u_t\}$ be as in Theorem 5. A point $x \in G/\Gamma$ is said to be *generic* for the $\{u_t\}$-action if there does not exist any proper closed subgroup F containing $\{u_t\}$ such that Fx admits a finite F-invariant measure. By Theorem 5 for any generic point the orbit is uniformly distributed with respect to the G-invariant probability measure on G/Γ. Is the convergence of the averages involved in this (for any fixed bounded continuous function) uniform over compact subsets of the set of generic points? Similarly, what happens if we vary the unipotent one-parameter subgroup? These questions were considered in [DM3] and the results were applied to obtain lower estimates for the number of solutions in large enough sets, for quadratic inequalities as in Oppenheim's conjecture; see Corollary 9 below. One of our results in this direction is the following.

THEOREM 7 (Dani and Margulis, 1992). *Let G be a connected Lie group, Γ be a lattice in G, and m be the G-invariant probability measure on G/Γ. Let $\{u_t^{(i)}\}$ be a sequence of unipotent one-parameter subgroups converging to a unipotent one-parameter subgroup $\{u_t\}$; that is $u_t^{(i)} \to u_t$ for all t, as $i \to \infty$. Let $\{x_i\}$ be a sequence in G/Γ converging to a point $x \in G/\Gamma$. Suppose that x is generic for the action of $\{u_t\}$. Let $\{T_i\}$ be a sequence in $\mathbb{R}^+$, $T_i \to \infty$. Then for any bounded continuous function φ on G/Γ,*

$$\frac{1}{T_i} \int_0^{T_i} \varphi(u_t^{(i)} x_i)\, dt \to \int_{G/\Gamma} \varphi\, dm \quad \text{as } i \to \infty.$$

(I learned later that Burger had pointed out to Ratner in December 1990, before we started the work, that such a strengthening of her theorem can be derived applying her methods; our method is substantially different.)

I will next describe another "uniform version" of uniform distribution that applies to a large class of nongeneric points as well, together with the generic points. The unipotent one-parameter subgroup will also be allowed to vary over compact sets of such subgroups; the class of unipotent one-parameter subgroups of G is considered equipped with the topology of pointwise convergence, when considered as maps from $\mathbb{R}$ to G.

Let G and Γ be as in Theorem 7. Let $\mathcal{H}$ be the class of all proper closed subgroups H of G such that $H \cap \Gamma$ is a lattice in H and $\mathrm{Ad}\,(H \cap \Gamma)$ is Zariski dense in $\mathrm{Ad}\,H$. It turns out that $\mathcal{H}$ is countable (see [DM4]; see also [R1]). For each $H \in \mathcal{H}$ and any subgroup U of G let

$$X(H, U) = \{g \in G \mid Ug \subseteq gH\}.$$

It is easy to see that if $U = \{u_t\}$, $g\Gamma$ is not generic for the $\{u_t\}$-action for any $g \in X(H, U)$, $H \in \mathcal{H}$; conversely, it can be verified that any nongeneric point is contained in the (countable) union $\bigcup_{H \in \mathcal{H}} X(H, U)\Gamma/\Gamma$. The following result deals simultaneously with averages for generic as well as nongeneric points, except for those in certain *compact* subsets from only *finitely* many $X(H, U)\Gamma/\Gamma$.

THEOREM 8 *Let G, Γ, and m be as before. Let $\mathcal{U}$ be a compact set of unipotent one-parameter subgroups of G. Let φ be a bounded continuous function on G/Γ, K be a compact subset of G/Γ, and let $\epsilon > 0$ be given. Then there exist $H_1, \ldots, H_k \in \mathcal{H}$ and a compact subset C of G such that the following holds: for any $U = \{u_t\} \in \mathcal{U}$ and any compact subset F of $K - \cup_{i=1}^k (C \cap X(H_i, U))\Gamma/\Gamma$ there exists a $T_0 \geq 0$ such that for all $x \in F$ and $T > T_0$,*

$$\left| \frac{1}{T} \int_0^T \varphi(u_t x) \, dt - \int_{G/\Gamma} \varphi \, dm \right| < \epsilon.$$

In [DM4] this was proved for singleton $\mathcal{U}$'s; essentially the same proof goes through for the above statement, but the formulation as above was not thought of, the condition having been stated a little differently there. From the result we get the following asymptotic lower estimates for the number of solutions of quadratic inequalities in regions defined by $\{v \in \mathbb{R}^n \mid \nu(v) \leq r\}$, as $r \to \infty$, where ν is a continuous "homogeneous" function on $\mathbb{R}^n$; we call a function ν homogeneous if $\nu(tv) = |t|\nu(v)$ for all $t \in \mathbb{R}$ and $v \in \mathbb{R}^n$. We use the notation $\#$ to indicate cardinality of a set and λ for the Lebesgue measure on $\mathbb{R}^n$.

COROLLARY 9 *Let $n \geq 3$, $1 \leq p < n$, and let $\mathcal{Q}(p, n)$ denote the set of all quadratic forms on $\mathbb{R}^n$ with discriminant ± 1 and signature $(p, n - p)$. Let $\mathcal{K}$ be a compact subset of $\mathcal{Q}(p, n)$ (in the topology of pointwise convergence). Let ν be a continuous homogeneous function on $\mathbb{R}^n$, positive on $\mathbb{R}^n - \{0\}$. Then we have the following:*

(i) for any interval I in $\mathbb{R}$ and $\theta > 0$ there exists a finite subset $\mathcal{E}$ of $\mathcal{K}$ such that each $Q \in \mathcal{E}$ is a scalar multiple of a rational form and for any $Q \in \mathcal{K} - \mathcal{E}$

$$\#\{z \in \mathbb{Z}^n \mid Q(z) \in I, \, \nu(z) \leq r\} \geq (1 - \theta)\lambda(\{v \in \mathbb{R}^n \mid Q(v) \in I, \, \nu(v) \leq r\})$$

for all large r; further, for any compact subset $\mathcal{C}$ of $\mathcal{K} - \mathcal{E}$ there exists $r_0 \geq 0$ such that for all $Q \in \mathcal{C}$ the inequality holds for all $r \geq r_0$.

(ii) if $n \geq 5$, then for $\epsilon > 0$ there exist $c > 0$ and $r_0 \geq 0$ such that for all $Q \in \mathcal{K}$ and $r \geq r_0$

$$\#\{z \in \mathbb{Z}^n \mid |Q(z)| < \epsilon, \, \nu(z) \leq r\} \geq c\lambda(\{v \in \mathbb{R}^n \mid |Q(v)| < \epsilon, \, \nu(v) \leq r\}).$$

The condition in (ii) that $n \geq 5$ is related to Meyer's theorem. It can be verified that the volume terms appearing on the right-hand side of the inequalities are of the order of r^{n-2}. A particular case of interest is of course when ν is the euclidean norm, in which case the regions involved are balls of radius r. I may mention that for a single quadratic form that is not a multiple of a rational form, an estimate as in (i) was obtained by Mozes and myself, with a (possibly small) positive constant in the place $1 - \theta$ as above (unpublished).

Our proofs of Theorems 7 and 8 and other related results (see below) are based on an observation about how polynomial trajectories move "near" algebraic subvarieties in linear spaces; this was also involved in [DM1] and then in [S1]. The observation is used together with Ratner's classification of invariant measures, namely Theorem 4; it may be noted however that we do not assume her uniform distribution theorem (Theorem 5) and our method provides in particular an alternative way (simpler in my view) of deducing it from Theorem 4. The method has also been used recently in [MS], [S2], and [EMS] and has led to several interesting results: Mozes and Shah [MS] show that the set of probability measures on G/Γ that are invariant and ergodic under the action of some unipotent one-parameter subgroup is a closed subset of the space of probability measures. Shah [S2] applies the method to extend Ratner's uniform distribution theorem to polynomial trajectories and multivariable polynomial maps into algebraic groups; using the latter he also generalizes Theorem 5 to actions of higher-dimensional unipotent groups. Eskin, Mozes, and Shah [EMS] consider sequences of the form $\{g_i\mu\}$, where $\{g_i\}$ is a sequence in G and μ is an algebraic probability measure corresponding to some closed orbit of a subgroup H that may not necessarily contain any unipotent element, and give satisfactory conditions for the convergence of $\{g_i\mu\}$ to the G-invariant probability measure. Using the results they make important contributions to the problem of understanding the growth of the number of lattice points on a subvariety of a linear space, within a distance r from the origin, as $r \to \infty$.

2. Distribution of orbit segments

In this section I will sketch the proof of Theorem 7 and also discuss related results; the proofs of Theorem 8 and the other results are similar, though some of them are technically more involved. Let G be a connected Lie group, Γ be a lattice in G, and U be any subgroup of G that is generated by the unipotent one-parameter subgroups contained in it. Let $\mathcal{H}$ and $X(H,U)$, $H \in \mathcal{H}$, be as introduced earlier. By ergodic decomposition, Theorem 4 then implies the following.

COROLLARY 10 *Let G, Γ, and U be as above. Let μ be a finite U-invariant measure such that $\mu(X(H,U)\Gamma/\Gamma) = 0$ for all $H \in \mathcal{H}$. Then μ is G-invariant.*

Another point to be noted is that $X(H,U)$ are essentially "algebraic subsets" of G. Specifically, for each $H \in \mathcal{H}$ there exists a representation $\rho_H : G \to GL(V_H)$, where V_H is a finite-dimensional real vector space, and a $p_H \in V_H$ such that if $\eta_H : G \to V_H$ denotes the orbit map $g \mapsto \rho_H(g)p_H$ and A_H is the Zariski closure of $\eta_H(X(H,U))$ in V_H, then $X(H,U) = \eta^{-1}(A_H) = \{g \in G \mid \eta_H(g) \in A_H\}$; we choose ρ_H to be the hth exterior power of the adjoint representation of G, where $h = \dim H$, and p_H to be a nonzero point in the one-dimensional subspace corresponding to the Lie subalgebra of H; see [DM4] for details.

The question of distribution of orbit segments involved in Theorems 7 and 8 may be formulated as follows. Let $\mathcal{S}$ be the collection of all segments of orbits of unipotent flows, namely subsets σ of the form $\{u_t x \mid 0 \leq t \leq T\}$, where $\{u_t\}$ is a unipotent one-parameter subgroup of G, $x \in G/\Gamma$, and $T > 0$. To each such σ there corresponds canonically a probability measure μ_σ given by the normalized linear measure along the segment; namely $\mu_\sigma(E) = T^{-1}l(\{t \in [0,T] \mid u_t x \in E\})$ for any Borel subset E of G/Γ; l denotes the Lebesgue measure on $\mathbb{R}$. The problem of distribution of orbit segments is essentially the question of understanding the limit points of the collection $\{\mu_\sigma \mid \sigma \in \mathcal{S}\}$, in the space of measures on G/Γ. For convenience we set X to be the one-point compactification of G/Γ and consider the limits in the space of probability measures on X.

Let $\{\sigma_i\}$ be a sequence in $\mathcal{S}$, say $\sigma_i = \{u_t^{(i)} x_i \mid 0 \le t \le T_i\}$, where $\{u_t^{(i)}\}$ is a sequence of unipotent one-parameter subgroups, $\{x_i\}$ is a sequence of points in G/Γ, and $\{T_i\}$ is a sequence of positive real numbers. In considering limits, by rescaling the segments σ_i as above and passing to a subsequence if necessary, we may assume that $u_t^{(i)} \to u_t$ for all $t \in \mathbb{R}$, where $\{u_t\}$ is a unipotent one-parameter subgroup of G. We may also assume that $\{x_i\}$ converges in X and $T_i \to \infty$. We set $U = \{u_t\}$. Our proof of Theorem 7 is achieved via the following

THEOREM 11 *Let the notation be as above. Suppose that $\{x_i\}$ converges in G/Γ, say $x_i \to x \in G/\Gamma$. Let μ be a limit point of $\{\mu_{\sigma_i}\}$. Then $\mu(X(H,U)\Gamma/\Gamma) = 0$ for all $H \in \mathcal{H}$ such that $x \notin X(H,U)\Gamma/\Gamma$.*

The main ingredients in the proof of this are as follows. First we prove that if A is an algebraic subvariety in a vector space V, then for any compact subset C of A and $\epsilon > 0$ there exists a (larger) compact subset D of A such that for any segment the proportion of time spent near C to that spent near D is at most ϵ; specifically, for any neighborhood Φ of D there exists a neighborhood Ψ of C such that for $y \notin \Phi$, any unipotent one-parameter subgroup $\{v_t\}$ of G and $T \ge 0$,

$$l(\{t \in [0,T] \mid v_t y \in \Psi\}) \le \epsilon \, l(\{t \in [0,T] \mid v_t y \in \Phi\}).$$

This depends on certain simple properties of polynomials and the fact that orbits of unipotent one-parameter subgroups in linear spaces are polynomial curves. Now consider V_H and any compact subset C of A_H and let D be the corresponding subset as above. Let $g_i \in G$ be such that $g_i\Gamma = x_i$ for all i. We apply the above to the segments $\{u_t^{(i)} g_i \gamma p_H \mid 0 \le t \le T_i\}$, $\gamma \in \Gamma$. It turns out that there exists a neighborhood Φ of D such that for at most two distinct γ the points $g\gamma p_H$ are contained in Φ, if we restrict to $g \in G$ such that $g\Gamma$ lies in a compact set disjoint from the "self-intersection set" of $X(H,U)\Gamma/\Gamma$, namely the union of its proper subsets of the form $(X(H,U) \cap X(H,U)\alpha)\Gamma/\Gamma$, $\alpha \in \Gamma$. Using this and an inductive argument for the points on the self-intersection set, we can combine the information about the individual segments and conclude that $\mu(\eta_H^{-1}(C)) < \epsilon$. Varying C and ϵ we get that $\mu(\eta_H^{-1}(A)\Gamma/\Gamma) = 0$, as desired.

Observe that Theorem 7 is equivalent to the assertion that if x as in Theorem 11 is not contained in $X(H,U)\Gamma/\Gamma$ for any $H \in \mathcal{H}$, then any limit point μ as in the theorem is nothing but the G-invariant probability measure on G/Γ (viewed as a measure on X with 0 mass on the point at infinity). This would follow from Corollary 10 and Theorem 11 if it is proved that $\mu(\{\infty\}) = 0$, ∞ being the point at infinity. This is achieved using a result from [D4] on the proportion of time spent by the trajectories of unipotent flows in compact subsets of G/Γ; incidentally, this is involved in Ratner's proof as well. The conclusion is transparent from the following variation of the result in question.

THEOREM 12 *Let G be a connected Lie group and Γ be a lattice in G. Let F be a compact subset of G/Γ and let $\epsilon > 0$ be given. Then there exists a compact subset K of G/Γ such that for any unipotent one-parameter subgroup $\{u_t\}$ of G, any $x \in F$ and $T \geq 0$,*

$$l(\{t \in [0,T] \mid u_t x \in K\}) \geq (1-\epsilon)T.$$

When $x \in X(H,U)\Gamma/\Gamma$ for some $H \in \mathcal{H}$ or $x = \infty$, the point at infinity, the limit measure μ may not in general be G-invariant. However applying the method one can get the following.

THEOREM 13 *Let $\{u_t^{(i)}\}$ be a sequence of unipotent one-parameter subgroups such that $u_t^{(i)} \to u_t$ for all t, let $U_i = \{u_t^{(i)}\}$ for all i and $U = \{u_t\}$. Let $\{x_i\}$ be a convergent sequence in G/Γ such that for any compact subset of Φ of G, $\{i \in \mathbb{N} \mid x_i \in (\Phi \cap X(H,U_i))\Gamma/\Gamma\}$ is finite. Let x be the limit of $\{x_i\}$. Then for any $H \in \mathcal{H}$ such that $x \in X(H,U)\Gamma/\Gamma$ there exists a sequence $\{\tau_i\}$ of positive real numbers such that the following holds: if $\sigma_i = \{u_t^{(i)} x_i \mid 0 \leq t \leq T_i\}$, where $\{T_i\}$ is a sequence in $\mathbb{R}^+$, and $\{\mu_{\sigma_i}\}$ converges to μ then*

(i) if $\limsup(T_i/\tau_i) = \infty$ then $\mu(X(H,U)\Gamma/\Gamma) = 0$ and

(ii) if $\limsup(T_i/\tau_i) < \infty$ then there exists a curve $\psi : ([0,1] - D) \to X(H,U)$, where D is a finite subset of $[0,1]$, such that $\eta_H \circ \psi$ extends to a polynomial curve from $[0,1]$ to V_H, $\operatorname{supp}\mu$ meets $\psi(t)N^0(H)\Gamma/\Gamma$ for all $t \in [0,1] - D$ and is contained in their union; here $N^0(H)$ denotes the subgroup of the normalizer of H in G consisting of elements g for which the map $h \mapsto ghg^{-1}$ preserves the Haar measure on H.

An analogue of this result can also be proved for divergent sequences, that is, when $x_i \to \infty$ in X, for ∞ in the place of $X(H,U)\Gamma/\Gamma$. Theorem 13 yields, in particular, an ergodic decomposition for μ. The theorem also readily implies that for $\{u_t^{(i)}\}$ and $\{x_i\}$ as in the hypothesis there exists a sequence $\{\tau_i\}$ of positive real numbers such that the conclusion as in Theorem 7 holds for any sequence $\{T_i\}$ such that $T_i/\tau_i \to \infty$. In the same vein, by a more intricate argument we prove the following.

THEOREM 14 *Let $\{u_t^{(i)}\}$ and $\{x_i\}$ be as in Theorem 13. Then there exists a sequence $\{t_i\}$ in $\mathbb{R}^+$ such that $\{u_{t_i} x_i\}$ has a subsequence converging to a generic point with respect to the limit one-parameter subgroup. Further, $\{t_i\}$ may be chosen from any subset R of $\mathbb{R}^+$ for which there exists an $\alpha > 0$ such that $l(R \cap [0,T]) \geq \alpha T$ for all $T \geq 0$.*

Before concluding this discussion on the dynamics of unipotent flows I would like to mention that analogues of many of these results hold in p-adic and S-arithmetic cases (see [MT]); Ratner has also extended the results in this direction.

3.   Integral solutions of quadratic inequalities

The study of unipotent flows can be applied to various problems in diophantine approximation; see [M3] for a survey of the applications; see also [EMS]. I will however discuss here only the problem of lower estimates for the number of solutions as in Corollary 9. Let $n \geq 3$, $G = SL(n, \mathbb{R})$, and $\Gamma = SL(n, \mathbb{Z})$; the latter is well known to be a lattice in G. One associates to each (not necessarily continuous) function ψ on $\mathbb{R}^n$ vanishing outside a compact subset, a function $\tilde{\psi}$ on G/Γ by setting $\tilde{\psi}(g\Gamma) = \Sigma_{v \in g\mathbb{Z}^n} \psi(v)$; note that as ψ vanishes outside a compact subset, the expression on the right-hand side is actually a finite sum. By a theorem of Siegel, if ψ is integrable on $\mathbb{R}^n$ then $\tilde{\psi}$ is integrable on G/Γ and $\int_{\mathbb{R}^n} \psi \, d\lambda = \int_{G/\Gamma} \tilde{\psi} \, dm$, where λ denotes the Lebesgue measure on $\mathbb{R}^n$ and m is the G-invariant probability measure on G/Γ. Now let Q be a nondegenerate indefinite quadratic form and let $SO(Q)$ be the corresponding special orthogonal group. Let $\{u_t\}$ be a unipotent one-parameter subgroup of $SO(Q)$; the condition $n \geq 3$ ensures that such a one-parameter subgroup exists. Let ψ be the characteristic function of a small open subset B of $\mathbb{R}^n$ such that the values of Q on B form a small subinterval of the given interval I. Let $\kappa > 1$ be a fixed number close to 1 and for $g \in SO(Q)$ and $T > 0$ consider the equality

$$\int_T^{\kappa T} \Sigma_{v \in g\mathbb{Z}^n} \psi(u_t v) \, dt = \int_T^{\kappa T} \tilde{\psi}(u_t g\Gamma) \, dt,$$

which is immediate from the definition of $\tilde{\psi}$. One can see that the left-hand side is bounded by a multiple of the number of integral points in the tubular region $\{g^{-1}u_{-t}v \mid v \in B, T \leq t \leq \kappa T\}$, by a constant depending on B. On the other hand, using Siegel's theorem stated above and Theorem 8 one can see that, except for g in a certain exceptional set, for large T the right-hand side is approximately a constant multiple of the volume of the tubular region as above; further, under certain conditions on B the two constants are almost the same. In these cases the number of integral points in the tubular region as above has a lower bound comparable to its volume. Note also that the range of values of Q over the region is the same as over B. The proof of the first assertion in Corollary 9 is then obtained by showing that, except for a certain finite set of forms determined by the exceptional set in the application of Theorem 8 as above, the regions as in the statement of the Corollary can be closely filled in an essentially nonoverlapping manner by tubular regions for which this observation holds.

The argument for the second assertion is similar, except for the fact that using Meyer's theorem, which asserts that any nondegenerate indefinite quadratic form over $\mathbb{Q}$ in five or more variables has a nontrivial rational zero, we can ensure that unlike in the first case the exceptionalities while applying Theorem 8 do not give rise to any exceptionality with regard to the lower estimates as in the Corollary.

Acknowledgement. I would like to take this opportunity to express my gratitude to S. Raghavan and M. S. Raghunathan who initiated me into the area and to G. A. Margulis with whom I had extensive collaboration. It is also a pleasure to thank Nimish Shah for many helpful discussions.

References

[D1]   S. G. Dani, *Invariant measures of horospherical flows on noncompact homogeneous spaces*, Invent. Math. 47 (1978), 101–138.

[D2]   S. G. Dani, *Invariant measures of horospherical flows on homogeneous spaces*, Invent. Math. 64 (1981), 357–385.

[D3]   S. G. Dani, *On uniformly distributed orbits of certain horocycle flows*, Ergodic Theory Dynamical Systems 2 (1982), 139–158.

[D4]   S. G. Dani, *On orbits of unipotent flows on homogeneous spaces - II*, Ergodic Theory Dynamical Systems 6 (1986), 167–182.

[D5]   S. G. Dani, *A proof of Margulis' theorem on values of quadratic forms, independent of the axiom of choice*, L'enseig. Math. 40 (1994), 49–58.

[D6]   S. G. Dani, *Flows on homogeneous spaces: A review*, Preprint, 1995.

[DM1] S. G. Dani and G. A. Margulis, *On orbits of generic unipotent flows on homogeneous spaces of $SL(3, \mathbb{R})$*, Math. Ann. 286 (1990), 101–128.

[DM2] S. G. Dani and G. A. Margulis, *Values of quadratic forms at integral points: an elementary approach*, L'enseig. Math. 36 (1990), 143–174.

[DM3] S. G. Dani and G. A. Margulis, *On the limit distributions of orbits of unipotent flows and integral solutions of quadratic inequalities*, C. R. Acad. Sci. Paris Serie I, 314 (1992), 698–704.

[DM4] S. G. Dani and G. A. Margulis, *Limit distributions of orbits of unipotent flows and values of quadratic forms*, Adv. in Sov. Math. (AMS Publ.) 16 (1993), 91–137.

[DS]   S. G. Dani and J. Smillie, *Uniform distribution of horocycle orbits for Fuchsian groups*, Duke Math. J. 51 (1984), 185–194.

[EMS] Alex Eskin, Shahar Mozes, and Nimish Shah, *Unipotent flows and counting of lattice points on homogeneous varieties*, Preprint, 1994.

[M1]   G. A. Margulis, *Formes quadratiques indéfinies et flots unipotents sur les espaces homogènes*, C. R. Acad. Sci. Paris Serie I, 304 (1987), 247–253.

[M2]   G. A. Margulis, *Discrete subgroups and ergodic theory*, Number Theory, Trace Formulas and Discrete Subgroups, pp. 377–398, Academic Press, New York and San Diego, CA, 1989.

[M3]   G. A. Margulis, *Dynamical and ergodic properties of subgroup actions on homogeneous spaces with applications to number theory*, Proc. Internat. Congr. Math. (Kyoto, 1990), pp. 193–215, Springer-Verlag, Berlin and New York, 1991.

[MT]   G. A. Margulis and G. M. Tomanov, *Invariant measures for actions of unipotent groups over local fields on homogeneous spaces*, Invent. Math. 116 (1994), 347–392.

[MS]   Shahar Mozes and Nimish Shah, *On the space of ergodic invariant measures of unipotent flows*, Ergodic Theory Dynamical Systems 15 (1995), 149–159.

[R1]   Marina Ratner, *On Raghunathan's measure conjecture*, Ann. of Math. (2) 134 (1991), 545–607.

[R2]   Marina Ratner, *Raghunathan's topological conjecture and distributions of unipotent flows*, Duke Math. J. 63 (1991), 235–280.

[S1]   Nimish Shah, *Uniformly distributed orbits of certain flows on homogeneous spaces*, Math. Ann. 289 (1991), 315–334.

[S2]   Nimish Shah, *Limit distributions of polynomial trajectories on homogeneous spaces*, Duke Math. J., 75 (1994), 711–732.

Singular Automorphic Forms

JIAN-SHU LI[*]

University of Maryland
College Park, MD 20742, USA

In the context of holomorphic forms on tube domains, singular automorphic forms were studied by Maass [Maa], Freitag [Fre], and Resnikoff [Res], among others. A holomorphic form is singular if it is annihilated by certain differential operator(s). It is known that this is the case if and only if all its "nondegenerate" Fourier coefficients vanish, and if and only if it has one of the "singular weights". The equivalence of these properties are far from trivial, and they constitute some of the basic results in the theory.

The representation theoretic treatment of singular automorphic forms was pioneered by Howe [Howb]. By introducing the notion of rank for classical groups over local fields, Howe reduces the study of square integrable singular forms to that of unitary automorphic representations of low rank. The central question is then whether all such singular forms can be generated by theta series. In this note we present (partial) solutions in the form of several inequalities relating multiplicities of automorphic representations.

1. Basic properties of singular forms

Let k be a number field and D a division algebra over k with an involution ι. We assume that the subspace of fixed points of ι is precisely k. Let U be a left vector space over D with a nondegenerate sesquilinear form $(\,,\,)$ such that

$$(x,y)^\iota = \eta(y,x) \qquad (x,y \in U).$$

Here $\eta = \pm 1$ is fixed. Let $G \subseteq GL(U)$ be the Zariski connected component of the identity of the group of isometries of $(\,,\,)$.

Fix a decomposition

$$U = X \oplus U_0 \oplus X^* \tag{1.1}$$

where X and X^* are totally isotropic subspaces of the same dimension, say n, and U_0 is an anisotropic subspace orthogonal to $X \oplus X^*$. The subspaces X and X^* are naturally linear duals of each other under the pairing by $(\,,\,)$. Let $P = MN$ be the maximal parabolic subgroup of G preserving X. Here N is the unipotent radical of P. The Levi factor M may be taken to be the subgroup of G preserving the decomposition (1.1). Thus, $M = GL(X) \cdot G_0$, where G_0 is the subgroup of G leaving X and X^* pointwise fixed.

---

[*]Sloan Fellow. Supported in part by NSF grant No. DMS-9203142.

Proceedings of the International Congress
of Mathematicians, Zürich, Switzerland 1994
© Birkhäuser Verlag, Basel, Switzerland 1995

In what follows, we shall exclude the case $D = k$ and $n = 1$, as there is not much to be said in such case. Let Z be the subgroup of N leaving the subspace U_0 pointwise fixed; this is the center of N. We have a short exact sequence

$$1 \longrightarrow Z \longrightarrow N \longrightarrow \operatorname{Hom}_D(U_0, X) \longrightarrow 1.$$

Let $B(X)$ be the space of all sesquilinear forms on X that are Hermitian (resp. skew-Hermitian) in case $\eta = -1$ (resp. $\eta = 1$). Similarly define $B(X^*)$. These two spaces are naturally linear duals of each other, as we will now describe. Define an involution

$$* : \operatorname{End}_D(U) \longrightarrow \operatorname{End}_D(U), \quad T \longmapsto T^*$$

by the identity

$$(Tu \, , \, v) \; = \; (u \, , \, T^*v) \quad (u \, , \, v \, \in \, U).$$

This involution will preserve $\operatorname{Hom}_D(X \, , \, X^*)$ (which is to be identified with the endomorphisms of U that vanish on $X^* \oplus U_0$ and have their images contained in X^*). Similarly, it preserves $\operatorname{Hom}_D(X^* \, , \, X)$. We have natural isomorphisms

$$B(X) \; \simeq \; \{\, x \, \in \, \operatorname{Hom}_D(X, X^*) \mid x^* \; = \; -\eta x\}$$

$$B(X^*) \; \simeq \; \{\, y \, \in \, \operatorname{Hom}_D(X^*, X) \mid y^* \; = \; -\eta y\}$$

Thus, given $x \in B(X)$ and $y \in B(X^*)$ the composite $x \cdot y$ is an endomorphism of X^*, etc., and the bilinear form

$$< x, y > = tr(x \cdot y) \; = \; tr(y \cdot x)$$

exhibits the duality involved. Here $tr(a)$ denotes the sum of diagonals of $a + a^t$. There is a natural isomorphism

$$B(X^*) \simeq Z, \qquad b \mapsto z(b)$$

where $z(b)$ is given by

$$z(b)(x + u_0 + x^*) = x + u_0 + x^* + b(x^*) \qquad (x \in X, u_0 \in U_0, x^* \in X^*).$$

The example to keep in mind is $D = k$, $\eta = -1$, in which case G is the symplectic group of rank n and $Z = N$.

Let $\mathbb{A}$ be the ring of adeles of k. Fix a nontrivial character ψ of $\mathbb{A}/k$. To each $\beta \in B(X)$ we associate a character ψ_β of $Z \simeq B(X^*)$ by the formula

$$\psi_\beta(z(b)) = \psi(tr(\beta \cdot b)). \tag{1.2}$$

Then the map $\beta \mapsto \psi_\beta$ establishes an isomorphism between (the k-rational points of) $B(X)$ and the Pontryagin dual of $Z(k)\backslash Z(\mathbb{A})$.

Let f be a continuous function on $G(k)\backslash G(\mathbb{A})$. We may expand f in a Fourier series along Z:

$$f(zg) = \sum_{\beta \in B(X)} f_\beta(g)\psi_\beta(z) \qquad (z \in Z(\mathbb{A}), g \in G(\mathbb{A}))$$

where

$$f_\beta(g) = \int_{Z(k)\backslash Z(\mathbb{A})} f(zg)\overline{\psi_\beta(z)}\, dz.$$

DEFINITION 1.1 *Let l be an integer between 0 and n. We say the function f is of rank $\leq l$ if for all $\beta \in B(X)$ with rank $> l$, $f_\beta \equiv 0$. If in addition there is a β of rank l with $f_\beta \neq 0$, then f is of rank l.*

Now let $\pi = \otimes \pi_v$ be an irreducible admissible representation of $G(\mathbb{A})$. For technical reasons, assume also that each π_v is unitary. Then the notion of $Z(k_v)$-rank is defined for each π_v [Howc], [Sca], [Lia]. A convenient characterization is as follows. Let $\psi = \prod_v \psi_v$, where ψ_v is a character of k_v for each v. The Pontryagin dual of $Z(k_v)$ may be identified with $B(X)_v = B(X) \otimes k_v$ by the same formula (1.2), with ψ_v replacing ψ, etc. We denote the character associated to $\tau \in B(X)_v$ by $\psi_{v,\tau}$. Then π_v has rank at most l if for any compactly supported smooth function f on $Z(k_v)$ with the property that $\hat{f}$, the Fourier transform of f, vanishes on the subvariety of $B(X)_v$ of elements of rank $< l$, we have $\pi_v(f) = 0$. Here the Fourier transform of f is defined by the formula

$$\hat{f}(\tau) = \int_{Z(k_v)} f(z) \psi_{v,\tau}(z)\, dz, \qquad \tau \in B(X)_v \,.$$

We say π_v is of rank l if it is of rank $\leq l$ but not of rank $< l$.

Let H be the space of π, and let H^∞ be the subspace of smooth vectors. Assume that H^∞ is realized as a space of smooth functions on $G(k)\backslash G(\mathbb{A})$. We will also assume that for any $\beta \in B(X)$, the linear functional $f \mapsto f_\beta(1)$ is continuous with respect to the smooth topology on H^∞. Then the same is true for any functional $f \mapsto f_\beta(g)$, where g is any fixed element of $G(\mathbb{A})$. In other words the map $f \mapsto f_\beta$ is continuous when $C^\infty(G(\mathbb{A}))$ is given the topology of pointwise convergence.

LEMMA 1.2 *[Howb] The following conditions are equivalent.*

(a) *For any nonzero $f \in H^\infty$, the rank of f is l.*

(b) *For every place v, the representation π_v has rank l.*

(c) *There is at least one place v such that π_v is of rank l.*

Let n_X be the maximal possible rank of elements of $B(X)$. If $D = k$ and $\eta = 1$, $B(X)$ consists of skew-symmetric forms on X and n_X will be the largest even integer not greater than $n = \dim(X)$. In all other cases, we have $n_X = n$. The integer l in the preceding lemma will be called the rank of π. We will say that π is *singular*, or of *low rank*, if $l < n_X$.

Suppose π is singular. Then the $Z(k_v)$-spectrum of π_v is supported on (the closure of) a single equivalence class, say $\mathcal{C}_v$ in $B(X)_v$. We refer to [Howb], [Lia] for this fact as well as the notion of $Z(k)_v$-spectrum. Note that the parabolic $P(k_v)$ acts on $Z(k_v)$ via conjugation and hence dually, on $B(X)_v$. An equivalence class of forms in $B(X)_v$ is nothing but an orbit under $P(k_v)$, or equivalently, under $M(k_v)$. The rank of the orbit $\mathcal{C}_v$ is defined to be that of any of its members.

If $\beta \in B(X)$ then for each place v it defines a form $\beta_v \in B(X)_v$ by extension of scalars. Two global forms β and β' are said to be *locally equivalent* if β_v is equivalent to β'_v for all places v. In the following lemma we do not assume π to be of low rank.

LEMMA 1.3 *Suppose that for each place v the $Z(k_v)$-spectrum of π_v is supported on the closure of a single orbit $\mathcal{C}_v$. Suppose π is automorphic as above. Then the collection of local orbits $\{\mathcal{C}_v\}$ must be coherent. That is, there must be a global form $\beta \in B(X)$ representing the equivalence class $\mathcal{C}_v$ at each place v. In particular, all orbits $\mathcal{C}_v$ must have the same rank, say l. All rank l forms $\beta \in B(X)$ with respect to which π has a nonzero Fourier coefficient must belong to the same local equivalence class, namely the class determined by the collection of local orbits $\{\mathcal{C}_v\}$.*

REMARK 1.4 *Excluding the case $D = a$ quaternion algebra, $\eta = 1$, the Hasse principle implies that a local equivalence class is the same thing as a global equivalence class. In particular, there is exactly one $P(k)$-orbit of β's in $B(X)$ of rank l, with respect to which π has a non-zero Fourier coefficient. In the excluded case $B(X)$ consists of skew-hermitian forms over the quaternion algebra D, and the Hasse principle fails. If s denotes the (even) number of places where D ramifies, then there are exactly 2^{s-2} global equivalence classes within each local equivalence class. See [Sch] for these facts, where the word "isometry" is used in place of "equivalence". In the following we shall refer to this case as "Case Q1":*

$$Case\ Q1: \qquad D \text{ is a quaternion algebra, and } \eta = 1.$$

Let π be as in Lemma 1.3. We fix $\beta \in B(X)$ so that $\beta_v \in \mathcal{C}_v$ for all v. Let R_β be its radical and set $V = X/R_\beta$. Then β induces a nondegenerate form $(\ ,\ )'$ on V. Let G' be its isometry group. Then (G, G') is a reductive dual pair in the sense of [Howa]. To avoid any mention of nonlinear groups, we shall assume that $\dim_D U$ is even in case $D = k, \eta = 1$. For each place v let ω_v be the oscillator representation of the dual pair $G(k_v), G'(k_v)$ associated to the character ψ_v [Howa]. Because of our assumption just made, this will be an ordinary (as opposed to projective) representation of $G(k_v) \times G'(k_v)$. Let σ_v be an irreducible admissible representation of $G'(k_v)$. The *local theta lift* of σ_v, if it exists, is the unique irreducible representation of $G(k_v)$, written $\theta(\sigma_v)$, such that there is a nontrivial intertwining map from ω_v to $\theta(\sigma_v) \otimes \tilde{\sigma}_v$. Here $\tilde{\sigma}_v$ denotes the contragredient of σ_v.

THEOREM 1.5 *Suppose π is unitary, automorphic, and singular. Then for each v there is an irreducible unitary representation σ_v of $G'(k_v)$, and a unitary character χ_v of $G(k_v)$ such that $\pi_v \simeq \chi_v \otimes \theta(\sigma_v)$. Both χ_v and σ_v are unramified for almost all v, and so it makes sense to set $\chi = \otimes \chi_v$, $\sigma = \otimes \sigma_v$. Then χ is automorphic; that is $\chi(\gamma) = 1$ for all $\gamma \in G(k)$.*

2. The first inequality

Let S_∞ be the set of archimedean places of k. We set

$$G_\infty = \prod_{v \in S_\infty} G(k_v).$$

Let $\mathfrak{g}$ be the Lie algebra of G_∞, $U(\mathfrak{g})$ the universal enveloping algebra of $\mathfrak{g}$, and $Z(\mathfrak{g})$ the center of $U(\mathfrak{g})$. Choose a maximal compact subgroup K_∞ of G_∞. Let $\mathcal{A}(G)$

be the space of automorphic forms on G, consisting of functions on $G(k)\backslash G(\mathbb{A})$ of moderate growth that are K_∞ and $Z(\mathfrak{g})$ finite, and are fixed by some open compact subgroup of $G(\mathbb{A}_f)$. Here $\mathbb{A}_f$ denotes the finite adeles of k.

Recall also (from [Cas] or [KRS, p. 496], say) that for each integer n there is defined a Fréchet space $\mathcal{A}_{umg,n}(G)$ consisting of certain smooth functions on $G(k)\backslash G(\mathbb{A})$ of uniform moderate growth. The following can be deduced from work of Casselman [Cas].

LEMMA 2.1 *[KRS] Let π be a smooth irreducible representation of $G(\mathbb{A})$ on a Fréchet space E and let E_{K_∞} be the corresponding subspace of K_∞-finite vectors. Let*

$$A : E_{K_\infty} \longrightarrow \mathcal{A}(G)$$

be a $(\mathfrak{g}, K_\infty) \times G(\mathbb{A}_f)$-intertwining linear map. Then there exists an integer n such that the image of A is contained in $\mathcal{A}_{umg,n}(G)$, and that A extends to a unique continuous $G(\mathbb{A})$-intertwining map from E into $\mathcal{A}_{umg,n}(G)$.

Notation: The space of all intertwining maps described in the above lemma will be denoted $\operatorname{Hom}(\pi, \mathcal{A}(G))$. The dimension of this space is denoted $m(\pi, \mathcal{A}(G))$.

We place ourselves in the setting of the previous section; we shall allow the case $\operatorname{rank}(\beta) = \dim_D V = n_X$. Let $\sigma = \otimes \sigma_v$ be any irreducible admissible unitary representation (automorphic or not) of $G'(\mathbb{A})$. It is known [Lib] that for each v the local theta lift $\theta(\sigma_v)$ exists, and is irreducible unitary. Set $\theta(\sigma) = \otimes_v \theta(\sigma_v)$.

THEOREM 2.2 *Excluding Case Q1, we have*

$$m(\theta(\sigma), \mathcal{A}(G)) \le m(\sigma, \mathcal{A}(G')) . \tag{2.1}$$

We sketch a proof in the case $U_0 = \{0\}$; thus, $N = Z$. Let $X_1 \subseteq X$ be a complement to R_β so that

$$X = X_1 \oplus R_\beta . \tag{2.2}$$

Let $e : X \longrightarrow V$ be the canonical projection. The restriction of e to X_1 will be an isomorphism, which we will denote by e_1. Define a subgroup $G_1 \subseteq GL(X) \subset P$ by

$$G_1 = \{e_1^{-1} \circ h \circ e_1 | h \in G'\} .$$

The group G_1 will act trivially on R_β. It is clear that G_1 preserves ψ_β. Thus, for any $f \in C^\infty(G(k)\backslash G(\mathbb{A}))$ the restriction of f_β to $G_1(\mathbb{A})N(\mathbb{A})$ will be automorphic, i.e. left invariant under $G_1(k)N(k)$.

If $A \in \operatorname{Hom}(\pi, \mathcal{A}(G))$ let A_β be the composition of A with the map $f \mapsto f_\beta|_{G_1N}$, where the last denotes the restriction of f_β to $G_1(k)N(k)\backslash G_1(\mathbb{A})N(\mathbb{A})$. The following lemma implies that A_β corresponds to an intertwining operator from σ to $\mathcal{A}(G')$, and Lemma 1.3 implies that the map $A \mapsto A_\beta$ is injective.

LEMMA 2.3 *Let $\psi_{v,\beta}$ denote the restriction of ψ_β to $Z(k_v)$. Let δ be any smooth representation of $G'(k_v)$ and δ_1 its composition with the isomorphism $G_1 \simeq G'$. Then*

$$\operatorname{Hom}_{G_1(k_v)N(k_v)}(\theta(\sigma_v)^\infty, \delta_1 \otimes \psi_{v,\beta}) \simeq \operatorname{Hom}_{G'(k_v)}(\sigma_v^\infty, \delta) .$$

The long and technical proof of this lemma depends heavily on the construction of the local theta lift in [Lib]. We omit it.

In the Case $Q1$ we may obtain a similar version as follows. Let $\beta = \beta_1, \ldots, \beta_t$ be the rational forms in $B(X)$ that are locally equivalent to β, here $t = 2^{s-2}$ and s is the number of ramified places of the quaternion algebra D. Let G'_j be the isometry group of β_j. For any place v and $1 \le i, j \le t$ we have $G'_i(k_v) \simeq G'_j(k_v)$. Hence $\sigma = \otimes_v \sigma_v$ may be considered a unitary representation of $G'_j(\mathbb{A})$ for any $1 \le j \le t$. To the map $A \in \mathrm{Hom}(\pi, \mathcal{A}(G))$ we may associate the t-tuple $(A_{\beta_1}, \ldots, A_{\beta_t})$. The same argument as before shows that this map is injective. We obtain

THEOREM 2.4 *Suppose D is a quaternion algebra and $\eta = 1$. Then in the above notations we have*

$$m(\theta(\sigma), \mathcal{A}(G)) \le \sum_{j=1}^{t} m(\sigma, \mathcal{A}(G'_j)). \qquad (2.3)$$

3.   Theta series and inner products

We impose a condition on rank $\beta = \dim V$. Set

$$D_0 = \{\xi \in D | \xi^\iota = \eta \xi\}$$

and

$$d = \dim_k D, \; d_0 = \dim_k D_0, \; \varepsilon = \frac{d_0}{d}.$$

We assume

$$2n + \dim_D U_0 = \dim_D U > \dim_D V + 4\varepsilon - 2. \qquad (3.1)$$

Note that because $\dim_D V \le n$, this condition is almost always satisfied.

Set $\omega = \otimes \omega_v$; this is a representation of $G(\mathbb{A}) \times G'(\mathbb{A})$. Let $W = U \otimes_D V$ and endow it with the symplectic form

$$< , >= \mathrm{tr}_{D/k}((,)^\iota \otimes (,)').$$

Here $\mathrm{tr}_{D/k}(d) = d + d^\iota$ for $d \in D$. Let $W = Y \oplus Y^*$ be a direct sum decomposition of W into the maximal totally isotropic subspaces Y and Y^*. We may realize ω on $S(Y(\mathbb{A}))$, the space of Bruhat-Schwartz functions on $Y(\mathbb{A})$. This is the usual Schrödinger model realization of ω [Gel]. Define the theta distribution on this space by

$$\theta(\phi) = \sum_{\xi \in Y(k)} \phi(\xi).$$

As is well known, this distribution is invariant under $G(k) \times G'(k)$. Set

$$\theta_\phi(g) = \theta(\omega(g)\phi), \qquad (\phi \in S(Y(\mathbb{A})), g \in G(\mathbb{A})).$$

Condition (3.1) ensures that θ_ϕ is square integrable on $G(k)\backslash G(\mathbb{A})$. It is easy to see that each θ_ϕ is of rank $\le \dim_D V$. Let Θ be the linear span of all θ_ϕ, $\phi \in S(Y(\mathbb{A}))$, and let $\overline{\Theta}$ be its closure in $L^2(G(k)\backslash G(\mathbb{A}))$. Let Θ_0 be the space of smooth functions in $\overline{\Theta}$ having rank $< \dim_D V$. It is known [Howb] that Θ_0 and its closure $\overline{\Theta}_0$ are invariant under $G(\mathbb{A})$. Thus, we have an invariant orthogonal decomposition

$$\overline{\Theta} = \overline{\Theta}_* \oplus \overline{\Theta}_0$$

where $\overline{\Theta}_*$ is the orthogonal complement of $\overline{\Theta}_0$. Let $\mathcal{P} : \overline{\Theta} \longrightarrow \overline{\Theta}_*$ be the orthogonal projection onto the first summand.

THEOREM 3.1 *For any* $\phi_1, \phi_2 \in S(Y(\mathbb{A}))$ *we have*

$$\int_{G(k)\backslash G(\mathbb{A})} \mathcal{P}(\theta_{\phi_1})(g)\overline{\mathcal{P}(\theta_{\phi_2})(g)}\, dg = \sum_{\gamma \in G'(k)} (\omega(\gamma)\phi_1, \phi_2) \qquad (3.2)$$

where

$$(\omega(g)\phi_1, \phi_2) = \int_{Y(\mathbb{A})} \omega(g)\phi_1(y)\overline{\phi_2(y)}\, dy \qquad (g \in G(\mathbb{A}))$$

is the matrix coefficient associated to ϕ_1, ϕ_2.

It is not difficult to express the right-hand side of (3.2) in terms of (a certain kind of inner product of) the βth Fourier coefficients $(\theta_{\phi_j})_\beta$, $j = 1, 2$. Using this and Mackey Theory, one may deduce

THEOREM 3.2 *Let*

$$L^2(G'(k)\backslash G'(\mathbb{A})) = \int_{\mathcal{M}} \sigma\, d\nu(\sigma)$$

be the direct integral decomposition of $L^2(G'(k)\backslash G'(\mathbb{A}))$ *into irreducible unitary representations of* $G'(\mathbb{A})$. *Then*

$$\overline{\Theta}_* = \int_{\mathcal{M}} \theta(\sigma)\, d\nu(\sigma)$$

is the direct integral decomposition of $\overline{\Theta}_*$ *into irreducible unitary representations of* $G(\mathbb{A})$.

4. The second inequality

Set

$$\mathcal{A}_2(G) = \mathcal{A}(G) \cap L^2(G(k)\backslash G(\mathbb{A}))$$

and let $m(\pi, \mathcal{A}_2(G))$ be the dimension of the space of intertwining operators from (the space of) π to $\mathcal{A}_2(G)$ (cf. Lemma 2.1.). We can now derive the following as an obvious consequence of Theorem 3.2.

THEOREM 4.1 *Let* $\sigma = \otimes \sigma_v$ *be any irreducible unitary representation of* $G'(\mathbb{A})$ *and let* $\theta(\sigma) = \otimes \theta(\sigma_v)$. *Then*

$$m(\theta(\sigma), \mathcal{A}_2(G)) \geq m(\sigma, \mathcal{A}_2(G')). \qquad (4.1)$$

In Case $Q1$ (4.1) is not strong enough. Assume we are in the setting of Theorem 2.4. When $\beta = \beta_1$ is replaced by β_j, $1 \leq j \leq t$, the objects $\theta, \theta_\phi, \Theta$ will be denoted by the same symbols with the superscript j, namely $\theta^j, \theta^j_\phi, \Theta^j$, etc. Similar notations apply to other objects associated to β_j. As was explained in the paragraph before Theorem 2.4, if $\sigma = \otimes_v \sigma_v$ is an irreducible admissible unitary representation of $G'(\mathbb{A})$ then it can be considered one for every $G'_j(\mathbb{A})$, $1 \leq j \leq t$.

For each j we have a closed invariant subspace $\overline{\Theta}^j_*$. We wish to show that there is no linear dependence between these spaces. We do this by showing that if $i \neq j$ then the β_jth Fourier coefficient of any smooth function in $\overline{\Theta}^i_*$ is zero. This is easy to see for elements of Θ^i; the difficult part is to show the same for all smooth functions in $\overline{\Theta}^i_*$. In any event, we arrive at

THEOREM 4.2 *Suppose D is a quaternion algebra and $\eta = 1$. Then*

$$m(\theta(\sigma), \mathcal{A}_2(G)) \geq \sum_{j=1}^{t} m(\sigma, \mathcal{A}_2(G'_j)) \,. \tag{4.2}$$

5. Examples and concluding remarks

If G' is anisotropic over k then $\mathcal{A}_2(G') = \mathcal{A}(G')$. In Case $Q1$, the form $\beta = \beta_1$ is anisotropic over k if and only if all the β_j, $1 \leq j \leq t$, are. Thus, (2.1), (2.3) and (4.1)–(4.2) give

PROPOSITION 5.1 *Suppose β is anisotropic over k. Then*
(a) Excluding Case $Q1$, we have

$$m(\theta(\sigma), \mathcal{A}(G)) = m(\theta(\sigma), \mathcal{A}_2(G)) = m(\sigma, \mathcal{A}(G')) \,. \tag{5.1}$$

(b) Assume D is a quaternion algebra and $\eta = 1$. Then

$$m(\theta(\sigma), \mathcal{A}(G)) = m(\theta(\sigma), \mathcal{A}_2(G)) = \sum_{j=1}^{t} m(\sigma, \mathcal{A}(G'_j)) \,. \tag{5.2}$$

In the case $D = k$ and $\eta = -1$, the second equation in (5.1) was already proven by Howe in [Howb].

The inequalities (2.3) and (4.2) provide many examples of automorphic forms with multiplicity > 1. For example, denoting the trivial one-dimensional representation of $G'(\mathbb{A})$ by $\sigma = 1$, we get

$$m(\theta(1), \mathcal{A}(G)) = m(\theta(1), \mathcal{A}_2(G)) = 2^{s-2} \qquad \text{(Case $Q1$)} \tag{5.3}$$

(recall that $t = 2^{s-2}$ and s is the number of place of k at which D ramifies).

Again suppose we are in Case $Q1$. Consider the special case $\dim_D V = 1$. Then the forms β_j are all represented by multiples of β: we have $\beta_j = t_j \cdot \beta$ where $t_j \in k^\times$ (see [Sch]). It follows that $G'_j = G'$ for $1 \leq j \leq t$. It is easy to see that

$m(\sigma, \mathcal{A}(G')) \leq 1$ for all σ. Let σ be automorphic, that is $m(\sigma, \mathcal{A}(G')) = 1$. Then (2.3) and (4.2) give

$$m(\theta(\sigma), \mathcal{A}(G)) = m(\theta(\sigma), \mathcal{A}_2(G)) = 2^{s-2} \qquad (\text{Case Q1}, \dim_D V = 1). \qquad (5.4)$$

It can be shown in the case $n = 1$, $U_0 = \{0\}$ that $\theta(\sigma)$ is cuspidal if and only if σ is nontrivial.

    The case when $D = k, \eta = -1, \dim V = 2$, and β is split will be referred to as "split binary quadratic". This is the only case where the volume of $G'(k)\backslash G'(\mathbb{A})$ is infinite. Suppose we are not in this case, and that σ is one dimensional. Then clearly

$$m(\sigma, \mathcal{A}(G')) = m(\sigma, \mathcal{A}_2(G')) = 0 \text{ or } 1.$$

Combining this with (2.1) and (4.1) we get

PROPOSITION 5.2 *Suppose that σ is one dimensional. Assume we are not in either the split binary quadratic case or Case Q1. Then*

$$m(\theta(\sigma), \mathcal{A}(G)) = m(\theta(\sigma), \mathcal{A}_2(G)) = m(\sigma, \mathcal{A}(G')) \leq 1. \qquad (5.5)$$

    In the split binary quadratic case we still have (by Theorem 2.2)

$$m(\theta(\sigma), \mathcal{A}(G)) \leq 1$$

and it can be shown that

$$m(\theta(\sigma), \mathcal{A}_2(G)) = 0$$

for *any* σ.

    When $D = k, \eta = -1$, and σ is the trivial character, (5.5) is due to Kudla, Rallis and Soudry [KRS] for $n = 2$, and Kudla and Rallis [KuR] for $n > 2$.

    To end, we remark that the results presented here can be used to compute the dimensions of certain L^2-cohomolgy spaces of some arithmetic manifolds.

References

[Cas]   W. Casselman, *Introduction to the Schwartz space of $\Gamma\backslash G$*, Canad. J. Math. 41 (1989), 285–320.

[Fre]   E. Freitag, Siegelsche Modulfunktionen, Grundlehren der math. Wiss. 54, Springer-Verlag, Berlin-Heidelberg-New York, 1983.

[Gel]   S. Gelbart, Examples of dual reductive pairs, Proc. Sympos. Pure Math. 33, Amer. Math. Soc., Providence, RI, 1979.

[Howa] R. Howe, *θ series and invariant theory*, Automorphic forms, representations, and L-functions, Proc. Sympos. Pure Math. 33, Amer. Math. Soc., Providence, RI, 1979, 275–285.

[Howb] R. Howe, *Automorphic forms of low rank*, Non-Commutative Harmonic Analysis, Lecture Notes in Math. 880, 1980, 211–248.

[Howc] R. Howe, *On a notion of rank for unitary representations of classical groups*, C.I.M.E. Summer School on Harmonic Analysis, Cortona, ed., 1980, 223–331.

[KuR]   S. Kudla and S. Rallis, *A regularized Siegel-Weil formula: The first term identity*, to appear in Ann. of Math. (2).

[KRS] S. Kudla, S. Rallis, and D. Soudry, *On the degree 5 L-function for Sp(2)*, Invent. Math. 107, (1992), 483–541.

[Lia] J-S. Li, *On the classification of irreducible low rank unitary representations of classical groups*, Compositio Math. 71 (1989), 29–48.

[Lib] J-S. Li, *Singular Unitary Representations of Classical Groups*, Invent. Math. 97, (1989), 237–255.

[Maa] H. Maass, Siegel's modular forms and Dirichlet series, Lecture Notes in Math., Springer-Verlag, Heidelberg-New York, 1971.

[Res] H. Resnikoff, *On singular automorphic forms in several complex variables*, Amer. J. Math. 95 (1973), 321–332.

[Sca] R. Scaramuzzi, *A notion of rank for unitary representations of general linear groups*, 1985, Thesis, Yale University.

[Sch] W. Scharlau, Quadratic and Hermitian Forms, Springer-Verlag, Berlin-Heidelberg-New York, 1985.

Gradings on Representation Categories

WOLFGANG SOERGEL

Mathematisches Institut der Universität Freiburg
Albertstr. 23b, 79104 Freiburg, Germany

1. Selfduality of Category $\mathcal{O}$

Let $\mathfrak{g} \supset \mathfrak{b} \supset \mathfrak{h}$ be a complex semisimple Lie algebra, a Borel, and a Cartan. Let $\mathcal{O}$ be the category of all finitely generated $\mathfrak{g}$-modules that are locally finite over $\mathfrak{b}$ and semisimple over $\mathfrak{h}$, see [BGG76]. This category is of interest, as it is a close relative of the category of Harish-Chandra modules for the corresponding simply connected complex algebraic group G, considered as a real Lie group, see [BG80]. For example, for $\mathfrak{g} = \mathfrak{sl}(n, \mathbb{C})$ one takes $G = SL(n, \mathbb{C})$.

It is known [BGG76] that every object of $\mathcal{O}$ has finite length and that there are enough projective and injective objects. Let $U = U(\mathfrak{g})$ be the universal enveloping algebra of $\mathfrak{g}$. For $\lambda \in \mathfrak{h}^*$ consider in $\mathcal{O}$ the Verma module $M(\lambda) = U \otimes_{U(\mathfrak{b})} \mathbb{C}_\lambda$, its unique simple quotient $L(\lambda)$, and an injective hull $I(\lambda)$ of $L(\lambda)$ in $\mathcal{O}$; i.e., an injective object with unique simple submodule $L(\lambda)$. This $I(\lambda)$ is unique only up to nonunique isomorphism. Every simple object of $\mathcal{O}$ is isomorphic to some $L(\lambda)$ for unique $\lambda \in \mathfrak{h}^*$.

Let $Z \subset U$ be the center and $Z^+ = \mathrm{Ann}_Z \mathbb{C}$ the annihilator of the trivial one-dimensional representation of $\mathfrak{g}$. Let L be the direct sum of all simples $L(\lambda)$ with $Z^+ L(\lambda) = 0$. There are but finitely many of those, parametrized by the Weyl group. Let I be the direct sum of their respective injective hulls. We are now ready to state the "selfduality theorem for category $\mathcal{O}$".

THEOREM 1. [Soe90], [BGS91] *There exists an isomorphism of finite-dimensional $\mathbb{C}$-algebras* $\mathrm{End}_\mathfrak{g} I \cong \mathrm{Ext}^\bullet_{\mathcal{O}}(L, L)$. *Furthermore, the graded ring on the right is Koszul.*

REMARKS. 1. Here $\mathrm{Ext}^\bullet$ stands for the direct sum of all Ext^i, made into a ring via the cup-product.

2. The object I is only unique up to nonunique isomorphism, so we cannot expect the isomorphism of the theorem to be canonical. See however [BGS91, 3.8] for more canonicity.

3. Let $(\mathcal{W}, \mathcal{S})$ be the Weyl group and its simple reflections corresponding to $\mathfrak{b} \subset \mathfrak{g}$. Let $\rho \in \mathfrak{h}^*$ be the halfsum of the roots of $\mathfrak{b}$. For $x \in \mathcal{W}$, $\lambda \in \mathfrak{h}^*$, define $x \cdot \lambda = x(\lambda + \rho) - \rho$. Then $L = \oplus_{x \in \mathcal{W}} L(x \cdot 0)$, $I = \oplus_{x \in \mathcal{W}} I(x \cdot 0)$. Let $w_\circ \in \mathcal{W}$ be the longest element of the Weyl group. The isomorphism of the theorem can be chosen in such a way that the projection onto $I(x \cdot 0)$ on the left corresponds to the projection onto $L(x^{-1} w_\circ \cdot 0)$ on the right (which lies in $\mathrm{End}_\mathfrak{g} L = \mathrm{Ext}^0_{\mathcal{O}}(L, L)$).

Proceedings of the International Congress
of Mathematicians, Zürich, Switzerland 1994
© Birkhäuser Verlag, Basel, Switzerland 1995

4. A "Koszul ring" is by definition [Pri70], [BGS91] a positively $\mathbb{Z}$-graded ring $A = \bigoplus_{i \geq 0} A^i$ such that (a) A^0 is semisimple and (b) the left A-module $A^0 = A/A^{>0}$ admits a graded projective resolution $\cdots \to P_2 \to P_1 \to P_0 \twoheadrightarrow A^0$ such that P_i is generated by its elements that are homogeneous of degree i; i.e., $P_i = AP_i^i$. For example, the standard Koszul complex shows that the symmetric algebra over a finite-dimensional vector space is Koszul.

We can reformulate (b) as follows: Let A-gr denote the abelian category of all $\mathbb{Z}$-graded A-modules $M = \bigoplus M^i$. (b)' If $M, N \in A$-gr are concentrated in one degree, say $M = M^m$, $N = N^n$, then $\mathrm{Ext}^i_{A\text{-gr}}(M, N) = 0$ unless $i = n - m$. Thus, Koszulity is an analogon of semisimplicity for $\mathbb{Z}$-graded rings.

5. Set $\mathcal{O}_0 = \{M \in \mathcal{O} \mid (Z^+)^n M = 0 \text{ for } n \gg 0\}$. Because $I \in \mathcal{O}_0$ is an injective generator, the functor

$$\mathrm{Hom}_{\mathfrak{g}}(\,, I) : \mathcal{O}_0 \to \mathrm{End}_{\mathfrak{g}} I\text{-mof}$$

from $\mathcal{O}_0$ to the category of finitely generated $\mathrm{End}_{\mathfrak{g}} I$-modules is an equivalence of categories. The theorem says that $\mathrm{End}_{\mathfrak{g}} I$ admits a $\mathbb{Z}$-grading. Thus, in some sense (which can be made precise) the category $\mathcal{O}_0$ admits a $\mathbb{Z}$-grading and is even "graded semisimple".

The same is true for all other blocks of $\mathcal{O}$. This is proven in [BGS91] for blocks with integral central character and then follows for arbitrary blocks with [Soe90, 2.5].

6. Let us explain what is selfdual about this theorem. If A is any positively $\mathbb{Z}$-graded ring, we may form another positively $\mathbb{Z}$-graded ring $E(A) = \mathrm{Ext}^\bullet_A(A^0, A^0)$. If A is Koszul and A^1 is finitely generated as a left A^0-module, one can prove [Löf86], [BGS91] that $E(A)$ is Koszul too and there is a canonical isomorphism $E(E(A)) = A$. For this reason $E(A)$ is called the Koszul dual of A. But now the Koszul ring A appearing in our theorem is selfdual: Indeed, under the equivalence $\mathcal{O}_0 \cong A$-mof the direct sum L of all simple objects of $\mathcal{O}_0$ corresponds to the direct sum A^0 of all simple objects of A-mof, and hence

$$E(A) = \mathrm{Ext}^\bullet_A(A^0, A^0) \cong \mathrm{Ext}^\bullet_{\mathcal{O}}(L, L) \cong \mathrm{End}_{\mathfrak{g}} I = A.$$

2. Representation of Hecke algebras via Bimodules

To establish the isomorphism of Theorem 1, we will describe both sides combinatorially. This section explains the combinatorics involved, which might be of independent interest. Let $(\mathcal{W}, \mathcal{S})$ be an arbitrary Coxeter system, $\leq$ the Bruhat order on $\mathcal{W}$ and $l : \mathcal{W} \to \mathbb{N}$ the length function, see [Bou81]. The Hecke algebra $\mathcal{H} = \mathcal{H}(\mathcal{W}, \mathcal{S})$ is a free $\mathbb{Z}[t, t^{-1}]$-module $\mathcal{H} = \bigoplus_{x \in \mathcal{W}} \mathbb{Z}[t, t^{-1}] T_x$ with basis T_x indexed by $x \in \mathcal{W}$. Its multiplication is given by the rules $T_x T_y = T_{xy}$ if $l(x) + l(y) = l(xy)$ and $T_s^2 = (t^2 - 1)T_s + t^2 T_e$ for $s \in \mathcal{S}$, where $e \in \mathcal{W}$ denotes the identity.

Suppose now that $\mathcal{S}$ is finite. Consider the geometric representation E of $\mathcal{W}$, its complexification $E_{\mathbb{C}} = E \otimes_{\mathbb{R}} \mathbb{C}$, and the ring $R = R(E_{\mathbb{C}})$ of regular functions on $E_{\mathbb{C}}$ alias the symmetric algebra of $E_{\mathbb{C}}^*$. We equip R with the "even" $\mathbb{Z}$-grading, defined by the prescription $\deg E_{\mathbb{C}}^* = 2$.

Let R-mobf$_{\mathbb{Z}}$-R be the category of all $\mathbb{Z}$-graded R-bimodules that are finitely generated as left modules and as right modules. Let $\langle R$-mobf$_{\mathbb{Z}}$-$R\rangle$ denote its split Grothendieck group; i.e., the free abelian group generated by the objects modulo the relations $\langle A \rangle = \langle A' \rangle + \langle A'' \rangle$ if $A \cong A' \oplus A''$. Certainly this is a ring via $\otimes_R$.

THEOREM 2. [Soe92a] *There is a ring homomorphism $\mathcal{E} : \mathcal{H} \to \langle R$-mobf$_{\mathbb{Z}}$-$R\rangle$ such that $\mathcal{E}(T_s + 1) = \langle R \otimes_{R^s} R \rangle$ for every $s \in \mathcal{S}$ (here R^s are the s-invariants) and $\mathcal{E}(t) = \langle R[-1]\rangle$ (where $R[-1]$ stands for the R-bimodule R with its grading shifted by one in the positive direction).*

Kazhdan and Lusztig [KL80] defined a new basis C_x' of $\mathcal{H}$ over $\mathbb{Z}[t, t^{-1}]$. It can be characterized as follows. Let $i : \mathcal{H} \to \mathcal{H}$ be the involutive algebra homomorphism with $i(t) = t^{-1}, i(T_x) = (T_{x^{-1}})^{-1}$. Then we have

1. $i(C_x') = C_x'$
2. $C_x' = t^{-l(x)} \sum_{x \geq y} P_{x,y}(t) T_y$ with $P_{x,y} \in \mathbb{Z}[t]$, $P_{x,x} = 1$, and $\deg P_{x,y} < l(x) - l(y)$ if $x \neq y$.

Because of property (1) this is often called the selfdual basis. Condition (2) is some kind of minimality condition that ensures uniqueness. The $P_{x,y}$ are called the Kazhdan-Lusztig polynomials. For $s \in \mathcal{S}$ one checks easily $C_s' = t^{-1}(T_s + 1)$.

THEOREM 3. *Suppose $\mathcal{W}$ is a Weyl group. Then there are indecomposable bimodules $B_x \in R$-mobf$_{\mathbb{Z}}$-R such that $\mathcal{E}(C_x') = \langle B_x \rangle$.*

REMARKS.   1. The Krull-Schmidt theorem holds in R-mobf$_{\mathbb{Z}}$-R, hence the B_x are unique up to isomorphism.
2. Here are some examples. We have $B_e = R$, $B_s = R \otimes_{R^s} R[1]$ for $s \in \mathcal{S}$, and $B_{w_\circ} = R \otimes_{R^{\mathcal{W}}} R[l(w_\circ)]$.
3. From the theorem one can deduce the conjectures of Kazhdan and Lusztig concerning composition series of Verma modules in an elementary way [Soe90]. However, the proof of the theorem is not elementary and uses the decomposition theorem of [BBD82]. It would be very interesting to have an elementary proof.
4. It would be very interesting to know whether this theorem holds for an arbitrary Coxeter group, or at least for any finite one.
5. It would be very interesting to know for which finite characteristic analogous results hold. An equivalent problem is to determine the intersection cohomology with coefficients in a finite field of complex Schubert varieties. This would determine some composition factor multiplicities for Weyl modules in the same characteristic.

Let me finish this subsection with some indications on how these theorems are proven. For Theorem 2 one just has to remark that $\mathcal{H}$ is generated over $\mathbb{Z}[t, t^{-1}]$ by the $(T_s + 1)$, $s \in \mathcal{S}$, subject only to the quadratic relations $(T_s + 1)^2 = (T_s +$

$1) + t^2(T_s + 1)$ and relations involving two generators. So we only need to establish isomorphisms

$$R \otimes_{R^s} R \otimes_{R^s} R \cong (R \otimes_{R^s} R) \oplus (R \otimes_{R^s} R[-2])$$

for all $s \in \mathcal{S}$ and isomorphisms between similar objects involving two elements $s, t \in \mathcal{S}$. This is a bit tricky, but not deep.

The only way I know to prove Theorem 3 is to interpret B_x as the equivariant intersection cohomology group of some Schubert variety. Namely choose $G \supset B$ a linear complex algebraic group with a Borel subgroup such that $(\mathcal{W}, \mathcal{S})$ is the corresponding Coxeter system. One knows [Ara86] how to identify the equivariant cohomology ring $H_B^\bullet(G/B)$ with $R \otimes_{R^\mathcal{W}} R$, where R are the regular functions on the Lie algebra of a maximal torus in B. Using this identification we can interpret the equivariant intersection cohomology groups of Schubert varieties as objects of R-mobf$_\mathbb{Z}$-R. One then shows that these objects satisfy the conditions on the B_x required in Theorem 3.

3. Combinatorics of Category $\mathcal{O}$

To prove the "selfduality" Theorem 1, we will show that both sides admit the same combinatorial description. Let us start describing the left-hand side. Choose an exact functor $\mathbb{V} : \mathcal{O}_0 \to \mathbb{C}$-mod that transforms the simple Verma module $M(-2\rho)$ to a one dimensional vector space and annihilates all other simple objects from $\mathcal{O}_0$. Such a functor exists, is unique up to nonunique isomorphism and can be given as $\mathbb{V} = \operatorname{Hom}_\mathfrak{g}(\ , I(-2\rho))$. By functoriality this gives a functor

$$\mathbb{V} : \mathcal{O}_0 \to Z\text{-mod}.$$

PROPOSITION 1. [Soe90] *The functor* $\mathbb{V}$ *is fully faithful on injective objects; i.e., for any two injectives* $I, J \in \mathcal{O}_0$ *it induces an isomorphism*

$$\operatorname{Hom}_\mathfrak{g}(I, J) \to \operatorname{Hom}_Z(\mathbb{V}I, \mathbb{V}J).$$

By the way, an analogous statement holds for every block of $\mathcal{O}$. Now we describe $\mathbb{V}I(x \cdot 0)$. Set $S = S(\mathfrak{h})$ and consider the Harish-Chandra homomorphism $\xi : Z \hookrightarrow S$ normalized by the condition $\xi(z) - z \in U\mathfrak{n}$, where $\mathfrak{n} \subset \mathfrak{b}$ denotes the nilradical. So in particular $\xi^{-1}(S^+) = Z^+$. Via ξ every S-module becomes a Z-module, and in [Soe90] it is proven that $\mathbb{V}I(x^{-1} \cdot 0) \cong B_x/B_x S^+$. From this we deduce

$$\begin{aligned} \operatorname{End}_\mathfrak{g} I &= \operatorname{End}_Z \mathbb{V}I \\ &= \operatorname{End}_Z(\bigoplus_x B_x/B_x S^+) \\ &= \operatorname{End}_S(\bigoplus_x B_x/B_x S^+), \end{aligned}$$

the last equality as $(Z^+)^n I = 0$ for $n \gg 0$ by definition of $\mathcal{O}_0$ and ξ induces isomorphisms $Z/(Z^+)^n \to S/(S^+)^n$.

We next describe combinatorially the right-hand side. Let G be as before, $B \supset N$ connected algebraic subgroups with Lie algebras $\mathfrak{b} \supset \mathfrak{n}$. Let $D^b(G/B)$ denote the bounded derived category with algebraically constructible cohomology of sheaves of complex vector spaces on G/B. Let $D_N^b(G/B) \subset D^b(G/B)$ be the full subcategory of objects that are smooth along N-orbits alias Bruhat cells.

Localization [BB81] along with [Soe86] determines an equivalence of categories (see [Soe89, Prop. 6])

$$D^b(\mathcal{O}_0) \overset{\sim}{\to} D_N^b(G/B)$$

mapping $L(xw_\circ \cdot 0)$ to the intersection cohomology complex $\mathcal{IC}_x = \mathcal{IC}(\overline{Nx^{-1}B/B})$ of the closure of a Bruhat cell. Recall the ring $R = S(\mathfrak{h}^*)$ of regular functions on the Lie algebra of a maximal torus. Via the Borel homomorphism $R \twoheadrightarrow H^\bullet(G/B, \mathbb{C})$ hypercohomology gives us a functor

$$\mathbb{H}^\bullet : D^b(G/B) \longrightarrow R\text{-mod}.$$

PROPOSITION 2. [Soe90] *The functor $\mathbb{H}^\bullet$ induces an isomorphism*

$$\bigoplus_i \operatorname{Hom}_D(\mathcal{IC}_x, \mathcal{IC}_y[i]) \to \operatorname{Hom}_R(\mathbb{H}^\bullet \mathcal{IC}_x, \mathbb{H}^\bullet \mathcal{IC}_y).$$

In addition, we establish the isomorphisms $\mathbb{H}^\bullet \mathcal{IC}_x \cong B_x/B_x R^+$. Let us set $\mathcal{L} = \oplus_x \mathcal{IC}_x$. From the preceding we now deduce

$$\begin{aligned} \operatorname{Ext}_{\mathcal{O}}^\bullet(L, L) &= \bigoplus_i \operatorname{Hom}_D(\mathcal{L}, \mathcal{L}[i]) \\ &= \operatorname{End}_R(\mathbb{H}^\bullet \mathcal{L}) \\ &= \operatorname{End}_R(\bigoplus_x B_x/B_x R^+). \end{aligned}$$

So indeed the left-hand side of the selfduality theorem equals the right-hand side, once we identify S with R; i.e., $\mathfrak{h}$ with $\mathfrak{h}^*$. This already suggests that from the very beginning it would have been more natural to take the Langlands dual Lie algebra on one side of our selfduality isomorphism. This point of view allows a partly conjectural generalization of Theorem 1, as explained in the next section.

4. Real groups

Let X be a complex algebraic variety and H a complex algebraic group acting on X with finitely many orbits. Then we can form a positively graded ring $\operatorname{Ext}_H^\bullet X$, the "geometric extension algebra", as follows. Let $\mathcal{D}_H(X)$ be the equivariant derived category, see [BL92]. Let $\mathcal{L} \in \mathcal{D}_H(X)$ be the direct sum of all simple H-equivariant perverse sheaves on X (i.e. take one from each isomorphism class). Then set $\operatorname{Ext}_H^\bullet X = \bigoplus_i \operatorname{Hom}_{\mathcal{D}_H(X)}(\mathcal{L}, \mathcal{L}[i])$. If $H \subset K$ is a closed subgroup, then the "induction equivalence" gives $\operatorname{Ext}_H^\bullet X = \operatorname{Ext}_K^\bullet(K \times_H X)$.

In our new notation Theorem 1 can be rewritten as

$$\mathcal{O}_0 \cong \operatorname{Ext}_{N^\vee}^\bullet(G^\vee/B^\vee)\text{-mof}$$

where $G^\vee \supset B^\vee \supset N^\vee$ are Langlands dual to $G \supset B \supset N$. Let $\tilde{\mathcal{O}}_0$ be the category of all finite length $\mathfrak{g}$-modules with all their composition factors from $\mathcal{O}_0$. As a variant of the above, one proves [Soe92a]

$$\tilde{\mathcal{O}}_0 \cong \operatorname{Ext}_{B^\vee}^\bullet(G^\vee/B^\vee)\text{-nil}$$

where for a positively $\mathbb{Z}$-graded $\mathbb{C}$-Algebra $A^\bullet$ we denote by A-nil the category of finite-dimensional A-modules annihilated by A^i for $i \gg 0$. Now by [Soe86] the category $\tilde{\mathcal{O}}_0$ is equivalent to the category $\mathcal{H}$ of Harish-Chandra modules for G with trivial central character. On the other hand $\mathrm{Ext}_{B^\vee}^\bullet(G^\vee/B^\vee) = \mathrm{Ext}_{G^\vee}^\bullet(G^\vee \times_{B^\vee} G^\vee/B^\vee)$, and $X = G^\vee \times_{B^\vee} G^\vee/B^\vee$ is just the modified Langlands parameter space associated by [ABV92] to $\mathcal{H}$. So we may write

$$\mathcal{H} \cong \mathrm{Ext}_{G^\vee}^\bullet X\text{-nil},$$

and written in this form it is clear how the theorem should generalize to real and p-adic groups. Some special cases are checked in [Soe92b].

5. Positive characteristic

The Koszulity statement in Theorem 1 seems to have an analogon in positive characteristic. Let R be a root system, p a prime, G the corresponding simply connected semisimple algebraic group over $\bar{\mathbb{F}}_p$, and $\mathfrak{g} = \mathrm{Lie} G$ its Lie algebra. This comes equipped with a formal p-power map $X \mapsto X^{[p]}$, $X \in \mathfrak{g}$. For example, for $R = A_n$ we get $\mathfrak{g} = \mathrm{sl}(n+1, \bar{\mathbb{F}}_p)$ and $X^{[p]} = X^p$ is just the pth power in the matrix ring. Let $U^{[p]} = U(\mathfrak{g})/(X^p - X^{[p]})$ be the restricted enveloping algebra of $\mathfrak{g}$. (Be careful that here X^p stands for the pth power in $U(\mathfrak{g})$!) This is a finite-dimensional $\bar{\mathbb{F}}_p$-algebra.

CONJECTURE 1. [AJS94] *Suppose p is at least the Coxeter number. Then $U^{[p]}$ admits a Koszul grading compatible with its natural $\mathbb{Z}R$-grading.*

REMARK. By a Koszul grading on a ring we mean a $\mathbb{Z}$-grading that makes the ring a Koszul ring. The conjecture says that the category of G_1T-modules should be "graded semisimple". In [AJS94] we construct a $\mathbb{Z}$-grading on $U^{[p]}$ and prove that it gives a Koszul grading on the regular blocks of $U^{[p]}$ if Lusztig's conjecture holds; i.e., if $p \gg 0$ (for R fixed). So the problem is to treat singular blocks and small p.

References

[ABV92] Jeffrey Adams, Dan Barbasch, and David A. Vogan, Jr., The Langlands classification and irreducible characters for real reductive groups, Progr. Math., Birkhäuser, Boston, MA, 1992.

[AJS94] Henning Haahr Andersen, Jens Carsten Jantzen, and Wolfgang Soergel, *Representations of quantum groups at a p-th root of unity and of semisimple groups in characteristic p: Independence of p*, Astérisque **220** (1994), 1–320.

[Ara86] A. Arabia, *Cohomologie T-équivariante de G/B pour un groupe G de Kac-Moody*, C. R. Acad. Sci. Paris Sér. 1 **302** (1986), 631–634.

[BB81] Alexander A. Beilinson and Joseph N. Bernstein, *Localisation de $\mathfrak{g}$-modules*, C. R. Acad. Sci. Paris Sér. 1 **292** (1981), 15–18.

[BBD82] Alexander A. Beilinson, Joseph N. Bernstein, and Pierre Deligne, *Faisceaux pervers*, Astérisque **100** (1982).

[BGS91] Alexander A. Beilinson, Victor Ginsburg, and Wolfgang Soergel, *Koszul duality patterns in representation theory*, preprint, 1991.

[BG80] Joseph N. Bernstein and Sergei I. Gelfand, *Tensor products of finite and infinite representations of semisimple Lie algebras*, Compositio Math. **41** (1980), 245–285.

[BGG76] Joseph N. Bernstein, Israel M. Gelfand, and Sergei I. Gelfand, *Category of $\mathfrak{g}$-modules*, Functional Anal. Appl. **10** (1976), 87–92.

[BL92]   Joseph N. Bernstein and Valery Lunts, *Equivariant sheaves and functors*, Springer Lecture Notes **1578** (1994), (139 pages).

[Bou81] Nicolas Bourbaki, Groupes et algèbres de Lie, vol. 4-6, Masson, Paris, 1981.

[KL80]   David Kazhdan and George Lusztig, *Representations of Coxeter groups and Hecke algebras*, Invent. Math. **53** (1980), 191–213.

[Löf86] Clas Löfwall, *On the subalgebra generated by the one-dimensional elements in the Yoneda Ext-algebra*, Lecture Notes in Math. **1183**, Springer, Berlin and New York, 1986, pp. 291–338.

[Pri70] Stewart B. Priddy, *Koszul resolutions*, Trans. Amer. Math. Soc. **152** (1970), 39–60.

[Soe86] Wolfgang Soergel, *Équivalences de certaines catégories de $\mathfrak{g}$-modules*, C. R. Acad. Sci. Paris Sér. 1 **303** (1986), no. 15, 725–728.

[Soe89] Wolfgang Soergel, $\mathfrak{n}$-*cohomology of simple highest weight modules on walls and purity*, Invent. Math. **98** (1989), 565–580.

[Soe90] Wolfgang Soergel, *Kategorie $\mathcal{O}$, perverse Garben und Moduln über den Koinvarianten zur Weylgruppe*, J. Amer. Math. Soc. **3** (1990), 421–445.

[Soe92a] Wolfgang Soergel, *The combinatorics of Harish-Chandra bimodules*, J. Reine Angew. Math. **429** (1992), 49–74.

[Soe92b] Wolfgang Soergel, *Langlands' philosophy and Koszul duality*, preprint, 1992.

Comparaison d'intégrales orbitales pour des groupes p-adiques

JEAN-LOUP WALDSPURGER

Université Paris 7, CNRS
2 place Jussieu, F-75251 Paris Cedex 05
France

I. Les problèmes

(a) Soient F un corps local non archimédien de caractéristique nulle, à corps résiduel fini, et G un groupe réductif connexe défini sur F. Munissons $G(F)$ d'une mesure de Haar. Plus généralement, pour tout sous-groupe fermé unimodulaire $H \subset G(F)$, munissons H d'une mesure de Haar et $H \backslash G(F)$ de la mesure quotient. Notons $\mathcal{C}_c^\infty(G(F))$ l'espace des fonctions sur $G(F)$, à valeurs complexes, localement constantes et à support compact. Pour $f \in \mathcal{C}_c^\infty(G(F))$ et $x \in G(F)$, posons

$$\phi^G(x, f) = \int_{Z_G(x)^\circ(F) \backslash G(F)} f(y^{-1}xy)dy$$

où $Z_G(x)^\circ$ est la composante neutre du commutant de x dans G. L'intégrale est absolument convergente ([Rao]). Fixons un ensemble de représentations $U^G \subset G(F)$ des classes de conjugaison unipotentes, pour la conjugaison par $G(F)$.

PROPOSITION (Shalika [Sha]). *Pour tout $u \in U^G$, il existe une fonction Γ_u^G sur $G(F)$, dont le germe au voisinage de 1 est uniquement déterminé, de sorte que pour toute $f \in \mathcal{C}_c^\infty(G(F))$, il existe un voisinage V_f de 1 dans $G(F)$ tel que, pour tout $x \in V_f$, on ait l'égalité*

$$\phi^G(x, f) = \sum_{u \in U^G} \Gamma_u^G(x)\phi^G(u, f).$$

Ces fonctions Γ_u^G sont appelées *germes de Shalika*. Le premier problème (et le plus profond) est le :

PROBLÈME A. *Calculer les germes de Shalika Γ_u^G.*

(b) Notons $G(F)_{\text{freg}}$ l'ensemble des éléments fortement réguliers de $G(F)$ (i.e. les $x \in G(F)$ dont le commutant est un tore) et $\bar{F}$ la clôture algébrique de F. Soit T un sous-tore maximal de G (défini sur F). Pour $x \in T(F) \cap G(F)_{\text{freg}}$, l'ensemble $\{y^{-1}xy; y \in (T \backslash G)(F)\}$ est la classe de conjugaison stable de x, i.e. l'ensemble des $x' \in G(F)$ conjugués à x dans $G(\bar{F})$. Le groupe $G(F)$ agit naturellement sur

Proceedings of the International Congress
of Mathematicians, Zürich, Switzerland 1994
© Birkhäuser Verlag, Basel, Switzerland 1995

$(T \backslash G)(F)$. L'ensemble des orbites s'identifie naturellement à un sous-groupe de $H^1(\mathrm{Gal}(\bar{F}/F), T(\bar{F}))$. Notons $k(T/F)$ le groupe dual de ce sous-groupe. Chaque élément de $k(T/F)$ définit une fonction sur $(T \backslash G)(F)$. Munissons $(T \backslash G)(F)$ d'une mesure déduite d'une forme différentielle invariante sur $T \backslash G$. Pour $\kappa \in k(T/F)$, $f \in \mathcal{C}_c^\infty(G(F))$ et $x \in G(F)$, posons

$$\phi^{G,\kappa}(x, f) = \int_{(T \backslash G)(F)} \kappa(y) f(y^{-1} x y) dy.$$

Pour $\kappa = 1$, on écrit $\phi^{G,st}$ au lieu de $\phi^{G,1}$. On conjecture que les intégrales $\phi^{G,\kappa}$ sont égales à des intégrales $\phi^{H,st}$ sur des groupes H plus petits que G, ses "groupes endoscopiques".

Fixons une forme intérieure G^* de G, quasi-déployée sur F, et un isomorphisme "intérieur" $\psi: G \to G^*$. Notons $^L G = \hat{G} \rtimes W_F$ le L-groupe de G ([B] I.2). Soit $(H, \mathcal{H}, s, \xi)$ une donnée endoscopique de G ([LS1] 1.2) : H est un groupe réductif connexe sur F, quasi-déployé, $\mathcal{H}$ est une extension scindée de W_F par $\hat{H}$, s est un élément semi-simple de $\hat{G}$, $\xi: \mathcal{H} \to {}^L G$ est un L-homomorphisme injectif tel que $\xi(\mathcal{H}) \subset Z_{L_G}(s)$, et $\xi(\hat{H}) = Z_{\hat{G}}(s)^\circ$. Il y a une application naturelle

$$\left\{ \begin{array}{c} \text{classes de conjugaison} \\ \text{semi-simples dans } H(\bar{F}) \end{array} \right\} \longrightarrow \left\{ \begin{array}{c} \text{classes de conjugaison} \\ \text{semi-simples dans } G(\bar{F}) \end{array} \right\}.$$

On dit qu'un élément de $H(F)$ est fortement G-régulier si l'image de sa classe est la classe d'un élément fortement régulier de $G(\bar{F})$. Notons $H(F)_{G-\mathrm{freg}}$ l'ensemble de ces éléments. Supposons pour simplifier $\mathcal{H} = {}^L H$ (cf. [LS1] 4.4 pour le cas général). Pour $y \in H(F)_{G-\mathrm{freg}}$ et $x \in G(F)_{\mathrm{freg}}$, on définit un facteur de transfert $\Delta(y, x)$ ([LS1] §3). Il n'est non nul que si les classes sur $\bar{F}$ de y et x se correspondent. Pour $f \in \mathcal{C}_c^\infty(G(F))$ et $y \in H(F)_{G-\mathrm{freg}}$, on pose

$$\phi^{G,H}(y, f) = \sum_x \Delta(y, x) \phi^G(x, f)$$

où l'on somme sur un système de représentants des classes de conjugaison par $G(F)$ dans $G(F)_{\mathrm{freg}}$. Supposons qu'il existe x tel que $\Delta(y, x) \neq 0$. Fixons un tel x, posons $T = Z_G(x)$. On montre qu'il existe $\Delta_0 \in \mathbb{C}^\times$ et $\kappa \in k(T/F)$ tels que

$$\phi^{G,H}(y, f) = \Delta_0 \phi^{G,\kappa}(x, f).$$

PROBLÈME B. *Soit* $f \in \mathcal{C}_c^\infty(G(F))$. *Existe-t-il un "transfert"* $f_H \in \mathcal{C}_c^\infty(H(F))$ *tel que pour tout* $y \in H(F)_{G-\mathrm{freg}}$, *on ait l'égalité* $\phi^{H,st}(y, f_H) = \phi^{G,H}(y, f)$?

Pour $u \in U^G$ et $y \in H(F)_{G-\mathrm{freg}}$, posons

$$\Gamma_u^{G,H}(y) = \sum_x \Delta(y, x) \Gamma_u^G(x),$$

où la somme est comme ci-dessus. Quand $G = H$, on écrit $\Gamma_u^{G,st}$ au lieu de $\Gamma_u^{G,G}$.

PROBLÈME B'. *Soit* $u \in U^G$. *Le germe* $\Gamma_u^{G,H}$ *est-il combinaison linéaire de germes* $\Gamma_v^{H,st}$ *pour* $v \in U^H$?

Langlands et Shelstad on montré qu'une réponse affirmative à cette question pour tout couple (G, H) impliquait l'existence du transfert ([LS3] 2.3).

(c) Soient G et H comme ci-dessus, supposons toujours $\mathcal{H} = {}^L H$, supposons de plus G et H quasi-déployés, déployés sur une extension non ramifiée de F, et qu'il existe une telle extension E/F et un diagramme commutatif

$$
\begin{array}{ccc}
{}^L H & \stackrel{\xi}{\longrightarrow} & {}^L G \\
\downarrow & & \downarrow \\
\hat{H} \rtimes \mathrm{Gal}(E/F) & \stackrel{\xi_E}{\longrightarrow} & \hat{G} \rtimes \mathrm{Gal}(E/F)
\end{array}
$$

Soient K_G un sous-groupe compact maximal hyperspécial de $G(F)$ et $\mathbb{H}(G, K_G)$ l'algèbre des fonctions sur $G(F)$, à valeurs complexes, à support compact et biinvariantes par K_G. Notons $\hat{G}_{ss}$ l'ensemble des éléments semi-simple de $\hat{G}$, $\mathrm{Fr} \in \mathrm{Gal}(E/F)$ l'élément de Frobenius et $(\hat{G}_{ss} \times \{\mathrm{Fr}\})/\hat{G}$ l'ensemble des classes de conjugaison par $\hat{G}$ dans le sous-ensemble $\hat{G}_{ss} \times \{\mathrm{Fr}\} \subset \hat{G} \rtimes \mathrm{Gal}(E/F)$. L'isomorphisme de Satake identifie $\mathbb{H}(G, K_G)$ aux fonctions "polynômes" sur $(\hat{G}_{ss} \times \{\mathrm{Fr}\})/\hat{G}$. Posons des définitions analogues pour le groupe H. Comme ξ_E définit une application

$$
(\hat{H}_{ss} \times \{\mathrm{Fr}\})/\hat{H} \longrightarrow (\hat{G}_{ss} \times \{\mathrm{Fr}\})/\hat{G},
$$

on en déduit une application $b \colon \mathbb{H}(G, K_G) \to \mathbb{H}(H, K_H)$.

PROBLÈME C ("lemme fondamental"). *Existe-t-il $c \neq 0$ tel que pour toute $f \in \mathbb{H}(G, K_G)$ et tout $y \in H(F)_{G-\mathrm{freg}}$, on ait l'égalité*

$$
\phi^{H, st}(y, b(f)) = c\phi^{G, H}(y, f)
$$

(cf. [L1] §3, [H3]) ?

REMARQUES 1. La motivation des problèmes B et C est la stabilisation de la formule des traces de Arthur et Selberg ([L1]). Ce problème s'est posé pour la première fois pour SL(2) dans [Lab-L].

2. On peut généraliser le problème C au cas où l'on se donne un automorphisme θ de G, défini sur F, et un caractère ω de $G(F)$ et où l'on remplace les intégrales $\phi^G(x, f)$ par

$$
\phi_{\theta, \omega}^G(x, f) = \int_{Z_{G, \theta}(x)^\circ(F) \backslash G(F)} \omega(y) f(y^{-1} x \theta(y)) dy
$$

où $Z_{G, \theta}(x) = \{y \in G; y^{-1} x \theta(y) = x\}$. L'intégrale n'est définie que pour les x tels que $\omega|_{Z_{G, \theta}(x)(F)} = 1$. (Cf. [KS].)

II. Premiers résultats sur les problèmes A et B

(a) Le calcul des germes pour les unipotents réguliers (i.e. les $u \in U^G$ tels que $\dim Z_G(u) = \text{rang } G$) et sous-réguliers (i.e. les $u \in U^G$ tels que $\dim Z_G(u) = \text{rang } G + 2$) a été effectué pour $G = \text{GL}(n)$ et $G = \text{SL}(n)$ par Repka ([Re1, Re2, Re3]).

(b) Si T est un sous-tore maximal de G défini sur F, on dit que T est elliptique s'il n'est pas contenu dans un sous-groupe parabolique propre de G défini sur F.

PROPOSITION (Rogawski [Ro]). *Supposons les mesures convenablement normalisées, soit T un sous-tore maximal défini sur F. Alors le germe Γ_1^G, restreint à $T(F) \cap G(F)_{freg}$, est constant au voisinage de l'unité, égal à 1 si T est elliptique, à 0 si T n'est pas elliptique.*

Notons que, si T n'est pas elliptique, les intégrales orbitales $\phi^G(x, f)$, pour $x \in T(F)$, sont égales à des intégrales $\phi^M(x, f_P)$, où M est un sous-groupe de Levi contenant T d'un sous-groupe parabolique propre P de G défini sur F, et f_P une fonction sur $M(F)$ déduite de f. On peut donc se limiter dans nos problèmes au cas où T est elliptique.

(c) Si $G(F)$ est compact modulo son centre, le problème B' se résout positivement pour tout H ([LS2], corollaire 2.5).

III. La variété des étoiles

(a) Soient T un sous-tore maximal de G défini sur F, B un sous-groupe de Borel de G contenant T, non nécessairement défini sur F, Δ l'ensemble des racines simples de T associé à B, W le groupe de Weyl de T (sur $\bar{F}$). Notons $\mathbb{B}$ la variété des sous-groupes de Borel de G. La variété des étoiles S est l'ensemble des $(B(w))_{w \in W} \in \mathbb{B}^W$ tels que pour tous $w \in W$ et $\alpha \in \Delta$, il existe $y \in G(\bar{F})$ tel que $B(w) = y^{-1}By$, $B(s_\alpha w) \subset y^{-1}P_\alpha y$, où s_α est la symétrie associée à α et P_α le sous-groupe parabolique de G contenant B, de rang semi-simple 1, associé à α. Soit X° l'ensemble des $(x, (B(w))_{w \in W}) \in G \times S$ tels que $x \in B(w)$ pour tout $w \in W$ et x est fortement régulier. Soit X la clôture de X° dans $G \times S$. On a des applications

définies ainsi : π est la projection sur le premier facteur; pour $\underline{x} = (x, (B(w))_{w \in W}) \in X$, soit $y \in G(\bar{F})$ tel que $B(1) = y^{-1}By$, alors $\varphi(\underline{x})$ est l'image de yxy^{-1} par la projection naturelle $B \to T$. On peut munir X d'une structure sur F de sorte que π et φ soient définies sur F. Soit $t \in T(F) \cap G(F)_{\text{freg}}$. Alors $\varphi^{-1}(t)(F)$ s'identifie à $(T \backslash G)(F)$. Pour $\kappa \in k(T/F)$ et $f \in \mathcal{C}_c^\infty(G(F))$, on peut interpréter $\phi^{G,\kappa}(t, f)$ comme une intégrale sur $\varphi^{-1}(t)(F)$. Soit $C \subset T$ une courbe définie sur F, passant par l'origine, de tangente à l'origine en position générale. Soient $Y^\circ = \{\underline{x} \in X^\circ; \varphi(\underline{x}) \in C\}$, Y la clôture de Y° dans X. On peut trouver une désingularisation $p: Y_C \to Y$ telle que l'on puisse appliquer la théorie d'Igusa ([I]) à l'application $\varphi \circ p \colon Y_C \to C$. On obtient un développement explicite de $\phi^{G,\kappa}(t,f)$, quand $t \in C(F)$ tend vers 1, en termes de parties principales d'intégrales sur les composantes définies sur F de $(\varphi \circ p)^{-1}(1)$. Les orbites unipotentes interviennent de la façon suivante : pour toute composante E de $(\varphi \circ p)^{-1}(1)$, il existe une (unique) orbite unipotente $\mathcal{O}$ de $G(\bar{F})$ telle que les clôtures de Zariski de $\pi \circ p(E)$ et de $\mathcal{O}$ soient égales.

Cette construction, due à Langlands ([L2]), permet le calcul des germes de Shalika quand on connait une description suffisamment maniable de la désingularisation Y_C.

(b) Le calcul a été mené à bien pour $G = \mathrm{SL}(3)$, $\mathrm{SU}(3)$ ([L2], [LS2]), $\mathrm{GSp}(4)$ ([H1]), $\mathrm{Sp}(4)$ ([H4]), G_2 et $\mathrm{SL}(4)$ ([J1,2]); pour G quelconque et une orbite unipotente sous-régulière ([H2]).

(c) L'utilisation de la variété des étoiles a permis à Shelstad de calculer les germes pour les orbites régulières ([S]). Supposons G quasi-déployé, soient $u \in U^G$ un élément régulier, $T \subset G$ un sous-tore maximal défini sur F, $x \in T(F) \cap G(F)_{\mathrm{freg}}$. On note $\Delta(x)$ la valeur absolue du déterminant de $1 - \mathrm{Ad}(x)$ agissant sur le quotient d'algèbres de Lie $(\mathrm{Lie}\,G)(F)/(\mathrm{Lie}\,T)(F)$. Shelstad définit trois éléments $\mathrm{inv}(x)$, $\mathrm{inv}_T(u)$ et $\mathrm{inv}(T)$ de $H^1(\mathrm{Gal}(\bar{F}/F), T(\bar{F}))$. On a alors l'égalité

$$\Gamma_u^G(x) = \begin{cases} \Delta(x)^{-1/2}, & \text{si } \mathrm{inv}(x) = \mathrm{inv}_T(u)/\mathrm{inv}(T), \\ 0, & \text{sinon,} \end{cases}$$

(il s'agit de l'égalité des germes de ces fonctions au voisinage de 1).

IV. Le cas de $\mathrm{GL}(n)$

Rappelons la définition suivante. Soient F'/F une extension finie, $\mathbb{F}'$ et $\mathbb{F}$ les corps résiduels des anneaux d'entiers de F' et F, $e(F'/F)$ l'indice de ramification, $v_{F'}$ la valuation normalisée de F' et ω_F une uniformisante de F. Soit $x \in F'^{\times}$. On dit que x est F'/F-cuspidal si $\mathrm{pgcd}(v_{F'}(x), e(F'/F)) = 1$ et si la réduction de $x^{e(F'/F)} \omega_F^{-v_{F'}(x)}$ engendre $\mathbb{F}'$ sur $\mathbb{F}$.

Soit T un sous-tore maximal elliptique de $G = \mathrm{GL}(n)$, défini sur F. Identifions $T(F)$ au groupe multiplicatif d'une extension E/F de degré n. Pour tout $u \in U^G$ et tout $x \in E^{\times}$ tel que x soit régulier et $v_E(x-1) > 0$, on définit $\Gamma_u^G(x)$ (cf. plus loin section VI(a)). Supposons E/F modérément ramifiée et $n > 1$. Soit $x \in E^{\times}$, régulier, tel que $v_E(x-1) > 0$. Il existe alors $z \in F^{\times}$, y et $x' \in E^{\times}$ tels que

(i) $x = z(1 + yx')$;
(ii) en posant $F' = F(y)$, y est F'/F-cuspidal;
(iii) $v_F(z-1) > 0$, $v_E(x'-1) > 0$.

Fixons de tels éléments. Soit $G' = Z_G(y)$. Alors $G'(F) \simeq \mathrm{GL}(n', F')$, où $n' = [E : F']$. Pour tous $u \in U^G$, $u' \in U^{G'}$, on définit $p(u, u') \in \mathbb{C}$ et l'on montre que

$$\Gamma_u^G(x) = \sum_{u' \in U^{G'}} p(u, u') \Gamma_{u'}^{G'}(x').$$

Les $p(u, u')$ s'explicitent en fonction des constantes de structure de l'algèbre de Hall des groupes linéaires sur un corps fini. La formule ci-dessus fournit un calcul, par récurrence sur n, des germes pour $\mathrm{GL}(n)$ ([W1,2]).

V. Résultats sur le lemme fondamental

(a) Le problème est résolu pour certains groupes de petit rang : SL(2) ([Lab-L]), U(3) ([BR], [K4]), GSp(4) ([We]).

(b) Pour $G = \mathrm{GL}(n)$, Drinfeld a calculé les intégrales orbitales de certaines fonctions $f \in \mathbb{H}(G, K_G)$ (il n'a pas publié ce calcul, cf. [Lau], theorem 4.6.1).

(c) Soient $G = \mathrm{GL}(n)$, d un diviseur de n, E l'extension non ramifiée de F de degré d, ω un caractère de $F^\times$ dont le noyau est le groupe des normes de $E^\times$. On plonge $H(F) = \mathrm{GL}(n/d, E)$ dans $G(F)$. Pour $x \in H(F)$ et $f \in \mathcal{C}_c^\infty(G(F))$, on pose

$$\phi^{H,G}(x, f) = \int_{Z_G(x)(F)\backslash G(F)} \omega \circ \det(y) f(y^{-1}xy) dy \, .$$

On définit un transfert $b \colon \mathbb{H}(G, K_G) \to \mathbb{H}(H, K_H)$ et un facteur $\Delta^{H,G} \colon H(F)_{G-\mathrm{freg}} \to \mathbb{C}$. Soit p la caractéristique résiduelle de F, supposons $p > n$. On prouve que pour $x \in H(F)_{G-\mathrm{freg}}$ et $f \in \mathbb{H}(G, K_G)$, on a l'égalité

$$\Delta^{H,G}(x)\phi^{H,G}(x, f) = \phi^H(x, b(f)).$$

La preuve se fait par récurrence comme en section IV ([Ka], [W2]).

Le lemme fondamental pour SL(n) s'en déduit (pour $p > n$).

(d) Soient G quelconque vérifiant les hypothèses de section I(c), E une extension non ramifiée de F, Fr le Frobenius de E/F. Le groupe $G(F)$ agit sur lui-même par Fr-conjugaison, i.e. par l'application

$$\begin{aligned} G(E) \times G(E) &\longrightarrow & G(E) \\ (y, x) &\longmapsto & y^{-1}x(\mathrm{Fr}\, y). \end{aligned}$$

Pour $x \in G(E)$, on définit sa classe de Fr-conjugaison, sa classe de Fr-conjugaison stable $\mathrm{Cl}^{\mathrm{Fr},st}(x)$, son centralisateur "tordu" :

$$Z_{G(E),\mathrm{Fr}}(x) = \{y \in G(E); y^{-1}x(\mathrm{Fr}\, y) = x\}.$$

On définit une "norme" N :

$$\left\{ \begin{matrix} \text{classes de Fr-conjugaison} \\ \text{stable dans } G(E) \end{matrix} \right\} \longrightarrow \left\{ \begin{matrix} \text{classes de conjugaison} \\ \text{stable dans } G(F) \end{matrix} \right\}$$

(cf. [K2]). On définit aussi un transfert

$$b \colon \mathbb{H}(G(E), K_{G(E)}) \longrightarrow \mathbb{H}(G(F), K_{G(F)}),$$

avec des notations évidentes. Soient $x \in G(E)$ et $f \in \mathcal{C}_c^\infty(G(E))$. Supposons que $N(\mathrm{Cl}^{\mathrm{Fr},st}(x))$ soit une classe fortement régulière. On pose alors

$$\phi_{\mathrm{Fr}}^{G(E),st}(x, f) = \sum_{x'} \int_{Z_{G(E),\mathrm{Fr}}(x')\backslash G(E)} f(y^{-1}x'(\mathrm{Fr}\, y)) dy$$

où l'on somme sur un système de représentants des classes de Fr-conjugaison dans $\mathrm{Cl}^{\mathrm{Fr},st}(x)$.

On prouve que pour tout $y \in G(F)_{\mathrm{freg}}$ et toute $f \in \mathbb{H}(G(E), K_{G(E)})$, on a les égalités

(i) $\phi^{G(F),st}(y, b(f)) = \phi_{\mathrm{Fr}}^{G(E),st}(x, f)$,

si x est un élément de $G(E)$ tel que $N(\mathrm{Cl}^{\mathrm{Fr},st}(x))$ soit la classe de conjugaison stable de y (alors $\mathrm{Cl}^{\mathrm{Fr},st}(x)$ est uniquement déterminée);

(ii) $\phi^{G(F),st}(y, b(f)) = 0$,

si la classe de conjugaison stable de y n'est pas dans l'image de N ([C]; voir aussi [K1], [AC], [Lab]).

La preuve utilise des arguments globaux (formule des traces). Pour les utiliser, il est essentiel de connaître préalablement le résultat quand f est la fonction caractéristique de $K_{G(E)}$. Ce résultat est dû à Kottwitz ([K2]).

(e) Hales a généralisé la méthode précédente et montré que, dans le cadre de section I(c), le lemme fondamental résultait du même lemme pour la fonction caractéristique de K_G ([H5]).

VI. Résultats reliés

(a) Supposons G classique et quasi-déployé, soit g son algèbre de Lie, que l'on plonge dans une algèbre de matrices. On pose

$$g(F)_{tn} = \{X \in g(F); \lim_{n \to \infty} X^n = 0\}.$$

Soit $K \subset G(F)$ le fixateur d'un sommet de l'immeuble de G, on définit son algèbre de Lie $k \subset g(F)$ et l'on pose

$$K_{tu} = \{x \in K; \lim_{n \to \infty} x^{p^n} = 1\},$$

où p est la caractéristique résiduelle de F. Supposons donné un homéomorphisme $e: k \cap g(F)_{tn} \to K_{tu}$ vérifiant des propriétés proches de celles de l'exponentielle. Pour $u \in U^G$, on définit Γ_u^G sur tout $K_{tu} \cap G(F)_{\mathrm{freg}}$ de sorte que

$$\Gamma_u^G(e(\mu^2 X)) = |\mu|^{d(u)} \Gamma_u^G(e(X))$$

pour tout $X \in k \cap g(F)_{tn}$ tel que $e(X) \in G(F)_{\mathrm{freg}}$ et tout $\mu \in F^\times$ tel que $v_F(\mu) \geq 0$, où $d(u)$ est un certain entier explicite.

PROPOSITION. *Supposons p "assez grand", notons f la fonction caractéristique de K et soit $x \in K_{tu} \cap G(F)_{\mathrm{freg}}$. Alors on a l'égalité*

$$\phi^G(x, f) = \sum_{u \in U^G} \Gamma_u^G(x) \phi^G(u, f).$$

(cf. [W4]).

(b) Dans la situation de section I(c), pour $f \in \mathbb{H}(G, K_G)$ et $x \in G(F)_{\text{freg}}$, on peut exprimer $\phi^G(x, f)$ en termes de la transformée de Satake de f. L'expression fait intervenir des "traces pondérées" de représentations sphériques de $G(F)$ ([W3] dans le cas de GL(n), [A1], §9 et [A2] dans le cas général). Dans certains cas, on peut aussi donner une expression de $\phi^G(u, f)$, pour $u \in U^G$ ([MA]).

(c) Les notions d'intégrales orbitales, de classe de conjugaison stable, de facteur de transfert etc. se descendent aux algèbres de Lie. Fixons sur $g(F)$ une forme bilinéaire $\langle\ ,\ \rangle_g$ symétrique non dégénérée et invariante par l'action adjointe de $G(F)$ et fixons un caractère $\psi: F \to \mathbb{C}^\times$ continu et non trivial. Pour $f \in \mathcal{C}_c^\infty(f(F))$, on définit sa transformée de Fourier $\hat{f} \in \mathcal{C}_c^\infty(g(F))$ par

$$\hat{f}(X) = \int_{g(F)} f(Z)\psi(\langle X, Z \rangle_g) dZ$$

pour tout $X \in g(F)$, où dZ est la mesure autoduale. Pour $X \in g(F)$, on définit une distribution $\hat{\phi}^G(X, \cdot)$ par $\hat{\phi}^G(X, f) = \phi^G(X, \hat{f})$ pour toute $f \in \mathcal{C}_c^\infty(g(F))$. Il existe une fonction $i^G: g(F)_{\text{reg}} \times g(F)_{\text{reg}} \to \mathbb{C}$, localement constante, telle que pour toute $f \in \mathcal{C}_c^\infty(g(F))$, on ait l'égalité

$$\hat{\phi}^G(X, f) = \int_{g(F)} i^G(X, Z) f(Z) \Delta(Z)^{-1} dZ$$

où Δ est l'analogue de la fonction définie en section III(c).

Les distributions $\hat{\phi}^G(X, \cdot)$ doivent être considérées comme des analogues pour l'algèbre de Lie $g(F)$ des caractères de représentations de $G(F)$. Cette interprétation conduit à la conjecture suivante. Dans la situation de section I(b), soit h l'algèbre de Lie de H. On déduit de $\langle\ ,\ \rangle_g$ une forme $\langle\ ,\ \rangle_h$ sur $h(F)$. On définit deux constantes explicites $\gamma_\psi(g)$ et $\gamma_\psi(h)$.

CONJECTURE ([W5]). *Soient $Y \in h(F)_{G-\text{reg}}$ et $Z \in g(F)_{\text{reg}}$. On a l'égalité*

$$\gamma_\psi(h) \sum_{Y', Z'} w(Z')^{-1} \Delta(Z', Z) i^H(Y', Z') = \gamma_\psi(g) \sum_X \Delta(Y, X) i^G(X, Z),$$

où on somme en Y', resp. Z', X, sur un ensemble de représentants des orbites pour l'action adjointe de $H(F)$, resp. $H(F), G(F)$, dans la classe de conjugaison stable de Y, resp. dans $h(F)_{G-\text{reg}}$, dans $g(F)_{\text{reg}}$ et où $w(Z')$ est le nombre d'orbites dans la classe de conjugaison stable de Z'.

Cette conjecture est vérifiée quand Y et Z "tendent vers l'infini". Quand Y et Z "tendent vers 0", elle résout le problème B'.

References

[A1]   J. Arthur, *Towards a local trace formula*, in: Algebraic analysis, geometry, and number theory, Johns Hopkins University Press 1989, 1–24.

[A2]   ———— , *A local trace formula*, Publ. Math. de l'IHES **73** (1991), 5–96.

[AC]   ———— , L. Clozel, Simple algebras, base change and the advanced theory of the trace formula, Annals of Math. Studies **120**, Princeton University Press, Princeton 1989.

[BR]   D. Blasius, J. Rogawski, *Fundamental lemmas for $U(3)$ and related groups*, in: The Zeta functions of Picard modular surfaces, CRM, Montréal, 1992, 395–420.

[B]   A. Borel, *Automorphic L-functions*, in: Automorphic forms, representations and L-functions part 2, Proc. of symp. in pure math. **33**, AMS 1979, 27–61.

[C]   L. Clozel, *The fundamental lemma for stable base change*, Duke Math. J. **61** (1990), 255–302.

[H1]   T. Hales, *Shalika germs on* $\mathrm{GSp}(4)$, in: Orbites unipotentes et représentations II, Astérisque **171–172** (1989), 195–256.

[H2]   ———— , *The subregular germ of orbital integrals*, Memoirs of the AMS **476**, 1992.

[H3]   ———— , *A simple definition of transfer factors for unramified groups*, Contemporary Math. **145** (1993), 109–134.

[H4]   ———— , *The twisted endoscopy of* $\mathrm{GL}(4)$ *and* $\mathrm{GL}(5)$: *transfer of Shalika germs*, prépublication.

[H5]   ———— , *On the fundamental lemma for standard endoscopy: reduction to unit element*, prépublication.

[I]   J. Igusa, Lectures on forms of higher degree, Tata Institute of Fund. Research, Bombay 1978.

[J1]   D. Joyner, *On the transfer of Shalika germs for* G_2, prépublication.

[J2]   ———— , *Langlands-Igusa theory for p-adic* $\mathrm{SL}(4)$, prépublication.

[Ka]   D. Kazhdan, *On liftings*, in: Lie group representations II, Springer LN **1041** (1984), 209–249.

[K1]   R. Kottwitz, *Orbital integrals on* $\mathrm{GL}(3)$, Amer. J. Math. **102** (1980), 327–384.

[K2]   ———— , *Rational conjugacy classes in reductive groups*, Duke Math. J. **49** (1982), 785–806.

[K3]   ———— , *Base change for unit elements of Hecke algebras*, Compositio Math. **60** (1986), 237–250.

[K4]   ———— , *Calculation of some orbital integral*, in: The Zeta functions of Picard modular surfaces, CRM, Montréal, 1992, 349–362.

[KS]   ———— , D. Shelstad, *Twisted endoscopy I: Definitions, norm mappings and transfer factors*.

[Lab]   J.-P. Labesse, *Fonctions élémentaires et lemme fondamental pour le changement de base stable*, Duke Math. J. **61** (1990), 519–530.

[Lab-L] J.-P. Labesse, R. P. Langlands, *L-indistinguishability for* $\mathrm{SL}(2)$, Can. J. Math. **31** (1979), 762–785.

[L1]   R. P. Langlands, *Les débuts d'une formule des traces stable*, Publ. Math. Univ. Paris 7 (1979).

[L2]   ———— , *Orbital integrals on forms of* $\mathrm{SL}(3)$ I, Amer. J. Math. **105** (1983), 465–506.

[LS1]   ———— , D. Shelstad, *On the definition of transfer factors*, Math. Ann. **278** (1987), 219–271.

[LS2]   ———— , ———— , *Orbital integrals on forms of* $\mathrm{SL}(3)$ II, Can. J. Math. **41** (1989), 480–507.

[LS3]   ———— , ———— , *Descent for transfer factors*, in: Grothendieck Festschrift II, Birkhäuser, Basel, Boston, Berlin, 1991, 485–563.

[Lau]   G. Laumon, *Cohomology with compact supports of Drinfeld modular varieties, part I*, prépublication.

[MA]   Magdy Assem, *Some results on unipotent orbital integrals*, Composition Math. **78** (1991), 37–78.

[Rao]   R. Rao, *Orbital integrals in reductive groups*, Ann. of Math. **96** (1972), 505–510.

[Re1]   J. Repka, *Shalika's germs for p-adic* GL(n): *the leading term*, Pac. J. Math. **113** (1984), 165–172.

[Re2]   _____ , *Shalika's germs for p-adic* GL(n): *the subregular term*, Pac. J. Math. **113** (1984), 173–182.

[Re3]   _____ , *Germs associated to regular unipotent classes in p-adic* SL(n), Can. Math. Bull. **28** (1985), 257–266.

[Ro]   J. Rogawski, *An application of the building to orbital integrals*, Compositio Math. **42** (1981), 417–423.

[Sha]   J. Shalika, *A theorem on p-adic semi-simple groups*, Ann. of Math. **95** (1972), 226–242.

[S]   D. Shelstad, *A formula for regular unipotent germs*, in: Orbites unipotentes et représentations II, Astérisque **171–172** (1989), 275–277.

[W1]   J.-L. Waldspurger, *Sur les germes de Shalika pour les groupes linéaires*, Math. Ann. **284** (1989), 199–221.

[W2]   _____ , *Sur les intégrales orbitales tordues pour les groupes linéaires: un lemme fondamental*, Can. J. Math. **43** (1991), 852–896.

[W3]   _____ , *Intégrales orbitales sphériques pour* GL(n) *sur un corps p-adique*, in: Orbites unipotentes et représentations II, Astérisque **171–172** (1989), 279–337.

[W4]   _____ , *Homogénéité de certaines distributions sur les groupes p-adiques*, prépublication.

[W5]   _____ , *Une formule des traces locale pour les algèbres de Lie p-adiques*, prépublication.

[We]   R. Weissauer, *A special case of the fundamental lemma*, prépublication.

L^2-Methods and Effective Results in Algebraic Geometry

JEAN-PIERRE DEMAILLY

Université de Grenoble I, Institut Fourier
URA 188 du CNRS, BP74
F-38402 Saint-Martin d'Hères, France

ABSTRACT. One important problem arising in algebraic geometry is the computation of effective bounds for the degree of embeddings in a projective space of given algebraic varieties. This problem is intimately related to the following question: Given a positive (or ample) line bundle L on a projective manifold X, can one compute explicitly an integer m_0 such that mL is very ample for $m \geq m_0$? It turns out that the answer is much easier to obtain in the case of adjoint line bundles $2(K_X + mL)$, for which universal values of m_0 exist. We indicate here how such bounds can be derived by a combination of powerful analytic methods: theory of positive currents and plurisubharmonic functions (Lelong), L^2 estimates for $\bar{\partial}$ (Andreotti-Vesentini, Hörmander, Bombieri, Skoda), Nadel vanishing theorem, Aubin-Calabi-Yau theorem, and holomorphic Morse inequalities.

1 Basic concepts of hermitian differential geometry

Let X be a complex manifold of dimension n and let F be a C^∞ complex vector bundle of rank r over X. A *connection* D on F is a linear differential operator D acting on spaces $C^\infty(X, \Lambda^{p,q}T_X^\star \otimes F)$ of F-valued differential forms, increasing the degree by 1 and satisfying Leibnitz' rule

$$D(f \wedge u) = df \wedge u + (-1)^{\deg f} f \wedge Du$$

for all forms $f \in C^\infty(X, \Lambda^{a,b}T_X^\star)$, $u \in C^\infty(X, \Lambda^{p,q}T_X^\star \otimes F)$. As usual, we split $D = D' + D''$ into its $(1,0)$ and $(0,1)$ parts, where

$$D' + D'' : C^\infty(X, \Lambda^{p,q}T_X^\star \otimes F) \longrightarrow C^\infty(X, \Lambda^{p+1,q}T_X^\star \otimes F) \oplus C^\infty(X, \Lambda^{p,q+1}T_X^\star \otimes F).$$

With respect to a trivialization $\tau : F_{|\Omega} \xrightarrow{\simeq} \Omega \times \mathbb{C}^r$, a connection D can be written $Du \simeq_\tau du + \Gamma \wedge u$, where $\Gamma = \Gamma' + \Gamma''$ is an arbitrary $(r \times r)$-matrix of 1-forms and d acts componentwise. A standard computation shows that $D^2 u \simeq_\tau \Theta(D) \wedge u$, where $\Theta(D) = d\Gamma + \Gamma \wedge \Gamma$ is a global 2-form on X with values in $\mathrm{Hom}(F, F)$. This form is called the *curvature tensor* of F. In the important case of rank 1 bundles, $\Theta(F) = d\Gamma$ is a d-closed form with complex values; it is well known that the De Rham cohomology class of $\theta(F) := \frac{i}{2\pi}\Theta(F) = \frac{i}{2\pi}D^2$ is the image in De Rham cohomology of the first Chern class $c_1(F) \in H^2(M, \mathbb{Z})$. For any line bundles $F_1, \ldots, F_p$ on X and any compact p-dimensional analytic set Y in X, we set

$$F_1 \cdot \ldots \cdot F_p \cdot Y = \int_Y c_1(F_1) \wedge \ldots \wedge c_1(F_p).$$

Proceedings of the International Congress
of Mathematicians, Zürich, Switzerland 1994
© Birkhäuser Verlag, Basel, Switzerland 1995

If F is equipped with a C^∞ hermitian metric h, the connection D is said to be *compatible with h* if

$$d\langle u, v\rangle_h = \langle Du, v\rangle_h + \langle u, Dv\rangle_h$$

for all smooth sections u, v of F. This is equivalent to the antisymmetry condition $\Gamma^\star = -\Gamma$ (in a unitary frame), i.e. $\Gamma'' = -\Gamma'^\star$. In particular, a compatible connection D is uniquely determined by its $(0,1)$-component D''. If F has a *holomorphic structure*, we precisely have a canonical $(0,1)$-connection $D'' = \overline{\partial}$ obtained by letting $\overline{\partial}$ act componentwise. Hence, there exists a unique $(1,0)$-connection D' that makes $D = D' + \overline{\partial}$ compatible with the hermitian metric. This connection is called the *Chern connection*. Let (e_λ) be a local holomorphic frame of $F_{|\Omega}$ and let $H = (h_{\lambda\mu})$, $h_{\lambda\mu} = \langle e_\lambda, e_\mu\rangle$ be the hermitian matrix representing the metric. Standard computations show that the Chern connection and curvature are given by

$$D' \simeq_\tau \partial + \overline{H}^{-1}\partial\overline{H} \wedge \bullet \, = \overline{H}^{-1}\partial(\overline{H}\bullet), \qquad \Theta(F) = \overline{\partial}(\overline{H}^{-1}\partial\overline{H}).$$

In the special case where F has rank 1, it is convenient to write the unique coefficient $H = h_{11}$ of the hermitian metric in the form $H = e^{-2\varphi}$. The function φ is called the *weight* of the metric in the local coordinate patch Ω. We then find $\Theta(F) = 2\partial\overline{\partial}\varphi$. It is important to observe that this formula still makes sense in the context of distribution theory if φ is just an arbitrary L^1_{loc} function. As we shall see later, the case of logarithmic poles is very important for the applications.

DEFINITION. *A singular hermitian metric on a line bundle F is a metric given in any trivialization* $\tau : F_{|\Omega} \xrightarrow{\simeq} \Omega \times \mathbb{C}$ *by*

$$\|\xi\| = |\tau(\xi)| \, e^{-\varphi(x)}, \qquad x \in \Omega, \ \xi \in F_x,$$

where $\varphi \in L^1_{\mathrm{loc}}(\Omega)$ is an arbitrary function. The associated curvature current is

$$\theta(F) = \frac{\mathrm{i}}{\pi}\partial\overline{\partial}\varphi.$$

The *Lelong-Poincaré* equation states that $\frac{\mathrm{i}}{\pi}\partial\overline{\partial}\log|f| = [D_f]$, where f is a holomorphic or meromorphic function and $[D_f]$ is the current of integration over the divisor of f. More generally, we have

$$\frac{\mathrm{i}}{\pi}\partial\overline{\partial}\log\|\sigma\| = [D_\sigma] - \theta(F)$$

for every section $\sigma \in H^0(X, F)$, as follows from the equality $\|\sigma\| = |f|e^{-\varphi}$, if $f = \tau(\sigma)$. As a consequence, the De Rham cohomology class of $[D_\sigma]$ coincides with the first Chern class $c_1(F)_\mathbb{R} \in H^2_{\mathrm{DR}}(X, \mathbb{R})$.

2 Positivity and ampleness

Let $(z_1, \ldots, z_n)$ be holomorphic coordinates on X and let $(e_\lambda)_{1 \leq \lambda \leq r}$ be an orthonormal frame of F. Let the curvature tensor of F be

$$\Theta(F) = \sum_{1 \leq j, k \leq n, \, 1 \leq \lambda, \mu \leq r} c_{jk\lambda\mu} dz_j \wedge dz_k \otimes e_\lambda^\star \otimes e_\mu.$$

Clearly, this tensor can be identified with a hermitian form on $T_X \otimes F$, namely

$$\widetilde{\Theta}(F)(t) = \sum c_{jk\lambda\mu} t_{j\lambda} \overline{t_{k\mu}}, \qquad t = \sum t_{j\lambda} \frac{\partial}{\partial z_j} \otimes e_\lambda \in T_X \otimes F.$$

DEFINITION. (Kodaira, Nakano, Griffiths) *A holomorphic vector bundle F is*

- *positive in the sense of Nakano if $\widetilde{\Theta}(F)(t) > 0$ for all nonzero tensors $t \in T_X \otimes F$;*
- *positive in the sense of Griffiths if $\widetilde{\Theta}(F)(\xi \otimes v) > 0$ for all nonzero decomposable tensors $\xi \otimes v \in T_X \otimes F$.*

In particular, a holomorphic line bundle F is positive if and only if its weights φ are strictly plurisubharmonic (psh), i.e. if $(\partial^2 \varphi / \partial z_j \bar{\partial} z_k)$ is positive definite.

EXAMPLE 1. Let $D = \sum \alpha_j D_j$ be a divisor with coefficients $\alpha_j \in \mathbb{Z}$ and let $F = \mathcal{O}(D)$ be the associated invertible sheaf of meromorphic functions u such that $\mathrm{div}(u) + D \geq 0$; the corresponding line bundle can be equipped with the singular metric defined by $\|u\| = |u|$. If g_j is a generator of the ideal of D_j on an open set $\Omega \subset X$, then $\tau(u) = u \prod g_j^{\alpha_j}$ defines a trivialization of $\mathcal{O}(D)$ over Ω; thus, our singular metric is associated with the weight $\varphi = \sum \alpha_j \log |g_j|$. By the Lelong-Poincaré equation, we find

$$\frac{\mathrm{i}}{\pi} \Theta(\mathcal{O}(D)) = \frac{\mathrm{i}}{\pi} \partial\bar{\partial}\varphi = [D] \geq 0,$$

where $[D] = \sum \alpha_j [D_j]$ denotes the current of integration over D.

EXAMPLE 2. Assume that $\sigma_1, \ldots, \sigma_N$ are nonzero holomorphic sections of F. Then we can define a natural (possibly singular) hermitian metric on F by

$$\|\xi\|^2 = \frac{|\tau(\xi)|^2}{\sum_{1 \leq j \leq N} |\tau(\sigma_j(x))|^2}$$

with respect to any trivialization τ. The associated weight function is $\varphi(x) = \log \left(\sum |\tau(\sigma_j(x))|^2 \right)^{1/2}$. In this case φ is a psh function; thus, $\mathrm{i}\Theta(F)$ is a closed positive current. Let us denote by Σ the linear system defined by $\sigma_1, \ldots, \sigma_N$ and by $B_\Sigma = \bigcap \sigma_j^{-1}(0)$ its base locus. Let

$$\Phi_\Sigma : X \setminus B_\Sigma \to \mathbb{P}^{N-1}, \qquad x \mapsto (\sigma_1(x) : \sigma_2(x) : \ldots : \sigma_N(x))$$

be the associated map. Then $\theta(F) = \frac{\mathrm{i}}{2\pi} \log(|\sigma_1|^2 + \cdots + |\sigma_N|^2)$ is the pullback over $X \setminus B_\Sigma$ of the Fubini-Study metric ω_{FS} on $\mathbb{P}^{N-1}$.

DEFINITION. *A holomorphic line bundle F over a compact complex manifold X is*

- *very ample, if the map $\Phi_{|F|} : X \to \mathbb{P}^{N-1}$ defined by the complete linear system $|F| = P(H^0(X, F))$ is a regular embedding (this means in particular that $B_{|F|} = \emptyset$);*
- *ample, if mF is very ample for some positive integer m.*

Here we used an additive notation for $\text{Pic}(X) = H^1(X, \mathcal{O}^\star)$, i.e. $mF = F^{\otimes m}$. By Example 2, every ample line bundle F has a smooth hermitian metric with positive definite curvature form; indeed, if $\Phi_{|mF|}$ is an embedding, then we get a positive definite curvature form $\theta(F^{\otimes m}) = \Phi_{|mF|}^\star(\omega_{\text{FS}})$ and we need only extract the mth root of this metric to get the desired smooth metric on F. The converse is also true:

KODAIRA EMBEDDING THEOREM (1954). *A line bundle F is ample if and only if F can be equipped with a smooth hermitian metric of positive curvature.*

In this context, Fujita [Fuj87] has raised the following important conjecture.

CONJECTURE (FUJITA, 1987). *If L is an ample line bundle on a projective n-fold X, then $K_X + (n+1)L$ is globally generated and $K_X + (n+2)L$ is very ample.*

Here $K_X = \Lambda^n T_X^\star$ is the canonical bundle. The example of curves shows that K_X is needed to get a uniform answer (if L is a bundle of degree 1 on a curve, then in general mL does not have any nonzero section unless $m \geq g =$ genus). Also, the example of projective spaces show that Fujita's bounds would be optimal, because $K_{\mathbb{P}^n} = \mathcal{O}(-n-1)$.

Such questions have attracted a lot of attention in recent years. First, the case of surfaces has been completely settled by Reider; in [Rei88] he obtains a very sharp criterion for global generation and very ampleness of line bundles in dimension 2. In higher dimensions, let us mention [Dem90, 93, 94, 95] and the works of Fujita [Fuj87, 94], Kollár [Kol93], Ein-Lazarsfeld [EL92, 93], Lazarsfeld [Laz93], and Siu [Siu93, 94]. Our goal is to describe a few powerful analytic methods that are useful in this context.

3 Bochner technique and vanishing theorems

Let X be a compact complex n-fold equipped with a *Kähler metric*, namely a positive $(1,1)$-form $\omega = i \sum \omega_{jk} dz_j \wedge d\bar{z}_k$ with $d\omega = 0$. Let F be a holomorphic vector bundle on X equipped with a hermitian metric, and let

$$\Delta' = D'D'^\star + D'^\star D', \qquad \Delta'' = D''D''^\star + D''^\star D'',$$

be the complex Laplace operators associated with the Chern connection D. Here the adjoints $D'^\star$, $D''^\star$ are the formal adjoints computed with respect to the L^2 norm $\|u\|^2 = \int_X |u(x)|^2 \, dV_\omega(x)$, where $|u|$ is the pointwise hermitian norm and $dV_\omega = \omega^n/n!$ is the volume form. The fundamental results of Hodge theory imply isomorphisms

$$H^q(X, \Omega_X^p \otimes F) = H_{\bar{\partial}}^{p,q}(X, F) \simeq \mathcal{H}^{p,q}(X, F)$$

between sheaf cohomology groups, Dolbeault $\bar{\partial}$-cohomology groups, and the space $\mathcal{H}^{p,q}$ of harmonic (p,q)-forms $\Delta'' u = 0$. The next fundamental fact is an identity originally used by Bochner to prove vanishing results for Betti numbers. Slightly later, the identity was extended to the complex situation by Kodaira and Nakano.

BOCHNER-KODAIRA-NAKANO FORMULA (1954). *For all* $u = \sum u_{J,K,\lambda} dz_I \wedge d\bar{z}_J \otimes e_\lambda$ *of class* C^∞ *and type* (p,q), *we have*

$$\Delta'' u = \Delta' u + A^{p,q}_{F,\omega} u,$$

where $A^{p,q}_{F,\omega}$ *is the hermitian endomorphism such that* $\langle A^{p,q}_{F,\omega} u, u \rangle =$

$$\sum c_{jk\lambda\mu}\, u_{J,jS,\lambda}\, \overline{u_{J,kS,\mu}} + \sum c_{jk\lambda\mu}\, u_{kR,K,\lambda}\, \overline{u_{jR,K,\mu}} - \sum c_{jj\lambda\mu}\, u_{J,K,\lambda}\, \overline{u_{J,K,\mu}},$$

and the summations are extended to all relevant indices $1 \le j, k \le n$, $1 \le \lambda, \mu \le r$, *and all relevant multiindices* $|J| = p$, $|K| = q$, $|R| = p - 1$, $|S| = q - 1$.

As $\langle \Delta' u, u \rangle = \|D'u\|^2 + \|D'^\star u\|^2 \ge 0$ the Bochner-Kodaira-Nakano formula implies

$$\langle \Delta'' u, u \rangle \ge \int_X \langle A^{p,q}_{F,\omega} u, u \rangle \, dV_\omega.$$

If $A^{p,q}_{F,\omega}$ is positive definite, every (p,q)-harmonic form has to vanish and we conclude that $H^q(X, \Omega^p_X \otimes F) = 0$. In the special case of rank 1 bundles, we can take at each point $x \in X$ simultaneous diagonalizations

$$\omega(x) = i \sum dz_j \wedge d\bar{z}_j, \qquad \Theta(F)(x) = i \sum \gamma_j(x) dz_j \wedge d\bar{z}_j,$$

where $\gamma_1(x) \le \cdots \le \gamma_n(x)$ are the curvature eigenvalues. Then $c_{jj\lambda\mu} = \gamma_j$ and

$$\langle A^{p,q}_{F,\omega} u, u \rangle = \sum_{J,K} \Big(\sum_{j \in K} \gamma_j - \sum_{j \notin J} \gamma_j \Big) |u_{JK}|^2 \ge (\gamma_1 + \cdots + \gamma_q - \gamma_{n-p+1} - \cdots - \gamma_n)|u|^2.$$

Assume now that $i\Theta(F)$ is positive. The choice $\omega = i\Theta(F)$ yields $\gamma_j = 1$ for $j = 1, 2, \ldots, n$ and $\langle A^{p,q}_{F,\omega} u, u \rangle = (p + q - n)|u|^2$. From this, we immediately infer:

AKIZUKI-KODAIRA-NAKANO VANISHING THEOREM (1954). *If* F *is a positive line bundle on a compact complex manifold* X, *then*

$$H^{p,q}(X, F) = H^q(X, \Omega^p_X \otimes F) = 0 \qquad for\ p + q \ge n + 1.$$

The above vanishing result is optimal. Unfortunately, it cannot be extended to semipositive or numerically effective line bundles of bidegrees (p,q) with $p < n$, as shown by a counterexample of Ramanujam [Ram74].

4 Hörmander's L^2 estimates and existence theorems

The basic existence theorem is the following result, which is essentially due to Hörmander [Hö65] and, in a more geometric setting, to Andreotti-Vesentini [AV65].

THEOREM. *Let (X, ω) be a complete Kähler manifold. Let F be a hermitian vector bundle of rank r over X, and assume that $A = A_{F,\omega}^{p,q}$ is positive definite everywhere on $\Lambda^{p,q} T_X^\star \otimes F$, $q \geq 1$. Then for any form $g \in L^2(X, \Lambda^{p,q} T_X^\star \otimes F)$ with*

$$D'' g = 0 \quad and \quad \int_X \langle (A_{F,\omega}^{p,q})^{-1} g, g \rangle \, dV_\omega < +\infty,$$

there exists a $(p, q - 1)$-form f such that $D'' f = g$ and

$$\int_X |f|^2 \, dV_\omega \leq \int_X \langle (A_{F,\omega}^{p,q})^{-1} g, g \rangle \, dV_\omega.$$

The proof can be ultimately reduced to a simple duality argument for unbounded operators on a Hilbert space, based on the a priori inequality

$$\|D'' u\|^2 + \|D''^\star u\|^2 \geq \int_X \langle A_{F,\omega}^{p,q} u, u \rangle \, dV_\omega.$$

The above L^2 existence theorem can be applied in the fairly general context of *weakly pseudoconvex* manifolds (i.e. manifolds possessing a weakly psh exhaustion function), thanks to the fact that every weakly pseudoconvex Kähler manifold (X, ω) carries a *complete* Kähler metric. In particular, the existence theorem can be applied on compact manifolds, pseudoconvex open sets in $\mathbb{C}^n$, Stein manifolds, etc. By regularization arguments, the existence theorem also applies when F is a line bundle and the hermitian metric is a singular metric with positive curvature in the sense of currents. In fact, the solutions obtained with the regularized metrics have weak L^2 limits satisfying the desired estimates. Especially, we get the following more tractable version in the case $p = n$.

COROLLARY 1. *Let (X, ω) be a Kähler weakly pseudoconvex complex manifold of dimension n. Let F be a holomorphic line bundle on X, equipped with a singular metric whose local weights $\varphi \in L_{\text{loc}}^1$ satisfy $i\Theta(F) = 2i\partial\bar{\partial}\varphi \geq \varepsilon\omega$ for some $\varepsilon > 0$. For every $g \in L^2(X, \Lambda^{n,q} T_X^\star \otimes F)$ with $D'' g = 0$, there exists $f \in L^2(X, \Lambda^{p,q-1} T_X^\star \otimes F)$ such that $D'' f = g$ and*

$$\int_X |f|^2 e^{-2\varphi} \, dV_\omega \leq \frac{1}{q\varepsilon} \int_X |g|^2 e^{-2\varphi} \, dV_\omega.$$

This result leads in a natural way to the concept of *multiplier ideal sheaves*, according to Nadel [Nad89]. The basic idea was already implicit in the work of Bombieri [Bom70] and Skoda [Sk72].

MULTIPLIER IDEAL SHEAVES. *Let φ be a psh function on an open subset $\Omega \subset X$. We define $\mathcal{I}(\varphi) \subset \mathcal{O}_X$ to be the sheaf of germs $f \in \mathcal{O}_{\Omega,x}$ such that $|f|^2 e^{-2\varphi}$ is integrable on a small neighborhood V of x with respect to the Lebesgue measure.*

MAIN PROPERTY ([Nad89], [Dem93]). *The ideal sheaf $\mathcal{I}(\varphi) \subset \mathcal{O}_X$ is a coherent analytic sheaf. Its zero variety $V(\mathcal{I}(\varphi))$ is the set of points in a neighborhood of which $e^{-2\varphi}$ is nonintegrable.*

A basic observation is that the zero variety $V(\mathcal{I}(\varphi))$ is closed related to the sublevel sets of Lelong numbers of φ.

DEFINITION. *The Lelong number of a psh function φ at a point $x \in X$ is the limit $\nu(\varphi, x) := \liminf_{z \to x} \varphi(z)/\log|z - x|$. The function φ is said to have a logarithmic pole of coefficient γ if $\gamma = \nu(\varphi, x) > 0$.*

LEMMA ([Sk72]). *Let φ be psh on Ω and let $x \in \Omega$.*

- *If $\nu(\varphi, x) < 1$, then $e^{-2\varphi}$ is integrable near $x \Rightarrow \mathcal{I}(\varphi)_x = \mathcal{O}_{\Omega, x}$.*
- *If $\nu(\varphi, x) \geq n + s$, $s \in \mathbb{N}$, then $e^{-2\varphi} \geq C|z - x|^{-2n - 2s}$ near x and $\mathcal{I}(\varphi)_x \subset m_{\Omega, x}^{s+1}$.*

SIMPLE ALGEBRAIC CASE. Let $\varphi = \sum \alpha_j \log|g_j|$, $\alpha_j \in \mathbb{Q}^+$, be associated with a normal crossing $\mathbb{Q}$-divisor $D = \sum \alpha_j D_j$, $D_j = g_j^{-1}(0)$. An easy computation gives

$$\mathcal{I}(\varphi) = \mathcal{O}\left(-\sum \lfloor \alpha_j \rfloor D_j\right) = \mathcal{O}(-\lfloor D \rfloor),$$

where $\lfloor \alpha_j \rfloor = $ the integral part of α_j. If the assumption on normal crossings is omitted, a desingularization of D has to be used in combination with the following fonctoriality property for direct images.

BASIC FONCTORIALITY PROPERTY. *Let $\mu : X' \to X$ be a modification (i.e. a proper generically $1 : 1$ holomorphic map), and let φ be a psh function on X. Then*

$$\mu_\star\left(\mathcal{O}(K_{X'}) \otimes \mathcal{I}(\varphi \circ \mu)\right) = \mathcal{O}(K_X) \otimes \mathcal{I}(\varphi).$$

Let us now consider the case of general algebraic singularities

$$\varphi \sim \frac{\alpha}{2} \log\left(|f_1|^2 + \cdots + |f_N|^2\right)$$

with $\alpha \in \mathbb{Q}^+$ and f_j holomorphic on an open set $\Omega \subset X$. By Hironaka's theorem, there exists a smooth modification $\mu : \widetilde{X} \to X$ of X such that $\mu^\star(f_1, \ldots, f_N)$ is an invertible sheaf $\mathcal{O}(-D)$ associated with a normal crossing divisor $D = \sum \lambda_j D_j$. Then

$$\mathcal{I}(\varphi) = \mu_\star \mathcal{O}_{\widetilde{X}}\left(\sum (\rho_j - \lfloor \alpha\lambda_j \rfloor)D_j\right),$$

where $R = \sum \rho_j D_j$ is the zero divisor of the jacobian J_μ of the blow-up map. In this context, we get the following important vanishing theorem, which can be seen as a generalization of the Kawamata-Viehweg vanishing theorem (see [Kaw82], [Vie82], [EV86]).

NADEL VANISHING THEOREM ([NAD89], [DEM93]). *Let (X, ω) be a Kähler weakly pseudoconvex manifold, and let F be a holomorphic line bundle over X equipped with a singular hermitian metric of weight φ. Assume that $i\Theta(F) \geq \varepsilon\omega$ for some continuous positive function ε on X. Then*

$$H^q\left(X, \mathcal{O}(K_X + F) \otimes \mathcal{I}(\varphi)\right) = 0 \qquad \text{for all } q \geq 1.$$

Proof. In virtue of Hörmander's L^2 estimates applied on small balls, the $\bar{\partial}$-complex of L^2_{loc} (n, q)-forms is a (fine) resolution of the sheaf $\mathcal{O}(K_X + F) \otimes \mathcal{I}(\varphi)$. The global L^2 cohomology is also zero by the L^2 estimates applied globally on X.□

COROLLARY 2. *Let $x_1, \ldots, x_N$ be isolated points in the zero variety $V(\mathcal{I}(\varphi))$. Then there is a surjective map*

$$H^0(X, K_X + F) \longrightarrow \bigoplus_{1 \leq j \leq N} \left(\mathcal{O}(K_X + L) \otimes \mathcal{O}_X / \mathcal{I}(\varphi)\right)_{x_j}.$$

In particular, if the weight function φ is such that $\nu(\varphi, x) \geq n + s$ at some point $x \in X$ and $\nu(\varphi, y) < 1$ at nearby points, then $H^0(X, K_X + F)$ generates all s-jets at x.

REMARK. It is an easy exercise (left to the reader!) to show that Corollary 2 implies the Kodaira embedding theorem.

5 Numerical criteria for very ample line bundles

The simplest approach to this problem is a recent technique due to [Siu94], which rests merely on Nadel's vanishing theorem and the Riemann-Roch formula. We formulate here a slightly improved version (see also [Dem94, 95]).

THEOREM. *Let L be an ample line bundle on a projective n-fold X. Let $x_j \in X$ and $s_j \in \mathbb{N}$ be given, $1 \leq j \leq N$. For*

$$m \geq m_0 = 2 + \sum_{1 \leq j \leq N} \binom{3n + 2s_j - 1}{n}$$

$H^0(X, 2K_X + mL)$ generates simultaneously jets of order s_j at all points x_j. In particular, $2K_X + mL$ is very ample for $m \geq 2 + \binom{3n+1}{n}$.

Proof. By a result of Fujita, $K_X + mL$ is ample for $m \geq m_0$ (in fact Fujita has shown that $K_X + mL$ is nef for $m \geq m + 1$ and ample for $m \geq n + 2$). The idea is to use a recursion procedure for the construction of psh weights $(\varphi_\nu)_{\nu \geq 1}$ on $K_X + m_0 L$ such that

(α)   the curvature of $K_X + m_0 L$ is positive definite: $i\partial\bar{\partial}\varphi_\nu \geq \varepsilon_\nu \omega$ for some $\varepsilon_\nu > 0$, where ω is the Kähler metric;
(β) $\nu(\varphi_\nu, x_j) \geq n + s_j$ for all j;
(γ) $\mathcal{I}(\varphi_{\nu+1}) \supsetneq \mathcal{I}(\varphi_\nu)$ whenever $\dim V(\mathcal{I}(\varphi_\nu)) > 0$.

Indeed, Nadel's vanishing theorem implies

$$H^q(X, \mathcal{O}(2K_X + mL) \otimes \mathcal{O}/\mathcal{I}(\varphi_\nu)) = 0 \qquad \text{for } m \geq m_0 \text{ and } q \geq 1.$$

Hence, $h^0 = \chi$ is large for some $m \in [m_0, 2m_0 - 1]$, and the existence of a section σ vanishing at order $2(n + s_j)$ at all points x_j follows by the Riemann-Roch formula and an elementary count of dimensions. We then set inductively

$$\varphi_{\nu+1} = \log(e^{\varphi_\nu} + e^{(1-m/2m_0)\psi} |\sigma|^{1/2}),$$

where ψ is a weight for a smooth metric of positive definite curvature on L. Condition (γ) guarantees that the process stops after a finite number of steps. $\square$

One weak point of the above result is that large multiples of L are required. Instead, we would like to find conditions on L implying that $2(K_X + L)$ is very ample. For this, we need a convenient measurement of how large L is.

DEFINITION. *Let L be a numerically effective line bundle, i.e. a line bundle such that $L^p \cdot Y \geq 0$ for all p-dimensional subvarieties Y. For every $S \subset X$, we set*

$$\mu(L, S) = \min_{Y \cap S \neq \emptyset} (L^p \cdot Y)^{1/p},$$

where Y runs over all p-dimensional subvarieties intersecting S. The main properties of this invariant are:

- *Linearity: $\forall k \geq 0, \ \mu(kL, S) = k\,\mu(L, S)$;*
- *Nakai-Moĭshezon criterion: L is ample if and only if $\mu(L, X) > 0$.*

THEOREM ([DEM93]). *Let $s, m \in \mathbb{N}$, $s \geq 1$, $m \geq 2$. If L is ample and satisfies*

$$(m - 1)\,\mu(L, X) \geq 6(n + s)^n - s,$$

then $2K_X + mL$ generates s-jets. Moreover, the result still holds with $6(n + s)^n$ replaced by $12n^n$ if $s = 1$; in particular, $2K_X + 12n^n L$ is always very ample.

Proof. By Corollary 2 of Section 4, the main point is to construct psh weights φ that achieve the desired ideals $\mathcal{I}(\varphi)_{x_j}$ for the jets. This is done by solving a complex Monge-Ampère equation

$$\left(\omega + \frac{\mathrm{i}}{\pi} \partial \bar{\partial} \varphi\right) = f, \qquad \omega = \theta(L),$$

where f is a linear combination of Dirac measures δ_{x_j} and of a uniform density with respect to ω^n. The solution φ does exist by the Aubin-Calabi-Yau theorem, but in general, the poles of φ are not isolated. Hence, the Lelong numbers have to be estimated precisely: this is indeed possible by means of intersection inequalities for positive currents. We refer to [Dem93, 94] for details. □

6 Holomorphic Morse inequalities

The starting point is the following differential geometric asymptotic inequality, in which $X(\leq q, L)$ denotes the set of points $x \in X$ at which $\theta_h(L)(x)$ has at most q negative eigenvalues. The proof is obtained by a careful study of the spectrum of the complex Laplace operator Δ''. See [Dem85, 91] for details.

STRONG MORSE INEQUALITIES ([DEM85]). *Let X be a compact complex n-fold and (L, h) a hermitian line bundle. Then, as $k \to +\infty$,*

$$\sum_{0 \leq j \leq q} (-1)^{q-j} h^j(X, kL) \leq \frac{k^n}{n!} \int_{X(\leq q, L)} (-1)^q \big(\theta_h(L)\big)^n + o(k^n).$$

SPECIAL CASE (ALGEBRAIC VERSION). *Let $L = F - G$, where F and G are numerically effective. Then for all $q = 0, 1, \ldots, n$,*

$$\sum_{0 \leq j \leq q} (-1)^{q-j} h^j(X, kL) \leq \frac{k^n}{n!} \sum_{0 \leq j \leq q} (-1)^{q-j} \binom{n}{j} F^{n-j} \cdot G^j + o(k^n).$$

In particular, for $q = 1$ we get

$$h^0(X, kL) - h^1(X, kL) \geq \frac{k^n}{n!} (F^n - nF^{n-1} \cdot G) - o(k^n).$$

COROLLARY 1. *If F, G are nef and $F^n > 0$, then $k(mF - G)$ has sections as soon as $m > nF^{n-1} \cdot G/F^n$ and $k \gg 0$.*

COROLLARY 2. *If F, G are nef and $F^n > 0$, then $H^0(X, K_X + mF - G) \neq 0$ for some $m \leq nF^{n-1} \cdot G/F^n + n + 1$.*

Proof. Set $m_0 := \lfloor nF^{n-1} \cdot G/F^n \rfloor + 1$. By Corollary 1, $m_0 F - G$ has a psh weight φ with $i\partial\bar{\partial}\varphi \gg 0$; thus, $H^q(X, \mathcal{O}(K_X + mF - G) \otimes \mathcal{I}(\varphi)) = 0$ for $q \geq 1$ and $m \geq m_0$. The Hilbert polynomial is thus equal to

$$h^0(X, \mathcal{O}(K_X + mF - G) \otimes \mathcal{I}(\varphi)) \geq 0,$$

and it must be nonzero for some $m \in [m_0, m_0 + n]$ because there are at most n roots. $\square$

A similar proof yields

COROLLARY 3. *If F, G are nef with $F^n > 0$, and Y is a p-dimensional subvariety, then $H^0(Y, \omega_Y \otimes \mathcal{O}_Y(mF - G)) \neq 0$ for some $m \leq pF^{p-1} \cdot G \cdot Y/F^p \cdot Y + p + 1$, where ω_Y is the L^2 dualizing sheaf of Y.*

A proof by backward induction on $\dim Y$ then yields the following effective version of the big Matsusaka theorem ([Mat72], [KoM83]), improving Siu's result [Siu93].

THEOREM ([SIU93], [DEM94, 95]). *Let F and G be nef line bundles on a projective n-fold X. Assume that F is ample and set $H = \lambda_n(K_X + (n+2)F)$ with $\lambda_2 = 1$ and $\lambda_n = \binom{3n+1}{n} - 2n$ for $n \geq 3$. Then $mF - G$ is very ample for*

$$m \geq (2n)^{(3^{n-1}-1)/2} \frac{(F^{n-1} \cdot (G+H))^{(3^{n-1}+1)/2}(F^{n-1} \cdot H)^{3^{n-2}(n/2-3/4)-1/4}}{(F^n)^{3^{n-2}(n/2-1/4)+1/4}}.$$

In particular mF is very ample for

$$m \geq C_n(F^n)^{3^{n-2}}\left(n + 2 + \frac{F^{n-1} \cdot K_X}{F^n}\right)^{3^{n-2}(n/2+3/4)+1/4}$$

with $C_n = (2n)^{(3^{n-1}-1)/2}(\lambda_n)^{3^{n-2}(n/2+3/4)+1/4}$.

References

[Av65]   A. Andreotti and E. Vesentini, *Carleman estimates for the Laplace-Beltrami equation in complex manifolds*, Publ. Math. I.H.E.S. **25** (1965), 81–130.

[Bom70]   E. Bombieri, *Algebraic values of meromorphic maps*, Invent. Math. **10** (1970), 267–287; Addendum, Invent. Math. **11** (1970), 163–166.

[Dem85]   J.-P. Demailly, *Champs magnétiques et inégalités de Morse pour la d''-cohomologie*, Ann. Inst. Fourier (Grenoble) **35** (1985), 189–229.

[Dem89]   J.-P. Demailly, *Une généralisation du théorème d'annulation de Kawamata-Viehweg*, C. R. Acad. Sci. Paris Série I Math. **309** (1989), 123–126.

[Dem90]   J.-P. Demailly, *Singular hermitian metrics on positive line bundles*, in: Proc. Complex Algebraic Varieties, Bayreuth, April 1990 (K. Hulek, Th. Peternell, M. Schneider and F. Schreyer, eds.), Lecture Notes in Math. **1507**, Springer-Verlag, Berlin and New York, 1992.

[Dem91]   J.-P. Demailly, *Holomorphic Morse inequalities*, Proc. Sympos. Pure Math. **52** (1991), 93–114.

[Dem93]   J.-P. Demailly, *A numerical criterion for very ample line bundles*, J. Differential Geom. **37** (1993), 323–374.

[Dem94]   J.-P. Demailly, L^2 *vanishing theorems for positive line bundles and adjunction theory*, in: Lecture notes of a CIME course, Transcendental Methods in Algebraic Geometry, Cetraro, Italy, July 1994.

[Dem95]   J.-P. Demailly, *Effective bounds for very ample line bundles*, to appear in the special volume of Invent. Math. in honor of R. Remmert (1995).

[EL92]   L. Ein and R. Lazarsfeld, *Seshadri constants on smooth surfaces*, Journées de Géométrie Algébrique d'Orsay, July 1992, Astérisque **282** (1993), 177–186.

[EL93]   L. Ein and R. Lazarsfeld, *Global generation of pluricanonical and adjoint linear series on smooth projective threefolds*, J. Amer. Math. Soc. **6** (1993), 875–903.

[EV86]   H. Esnault and E. Viehweg, *Logarithmic De Rham complexes and vanishing theorems*, Invent. Math. **86** (1986), 161–194.

[Fuj87]   T. Fujita, *On polarized manifolds whose adjoint bundles are not semipositive*, in: Algebraic Geometry, Sendai, 1985 (T. Oda, ed.), Adv. Stud. Pure Math. **10**, North-Holland, Amsterdam and New York, 1987, 167–178.

[Fuj94]   T. Fujita, *Remarks on Ein-Lazarsfeld criterion of spannedness of adjoint bundles of polarized threefold*, preprint, Tokyo Inst. of Technology at Ohokayama, 1994.

[Hö65]   L. Hörmander, L^2*-estimates and existence theorems for the $\bar{\partial}$ operator*, Acta Math. **113** (1965), 89–152.

[Kaw82]   Y. Kawamata, *A generalization of Kodaira-Ramanujam's vanishing theorem*, Math. Ann. **261** (1982), 43–46.

[Kol93]   J. Kollár, *Effective base point freeness*, Math. Ann. **296** (1993), 595–605.

[KoM83]   J. Kollár and T. Matsusaka, *Riemann-Roch type inequalities*, Amer. J. Math. **105** (1983), 229–252.

[Laz93]   R. Lazarsfeld, with the assistance of G. Fernández del Busto, Lectures on linear series, Park City, IAS Math. Series, vol. 3, (1993).

[Mat72]   T. Matsusaka, *Polarized varieties with a given Hilbert polynomial*, Amer. J. Math. **94** (1972), 1027–1077.

[Nad89]   A. M. Nadel, *Multiplier ideal sheaves and Kähler-Einstein metrics of positive scalar curvature*, Proc. Nat. Acad. Sci. U.S.A. **86** (1989), 7299–7300, and Ann. of Math. (2), **132** (1990), 549–596.

[Ram74]   C. P. Ramanujam, *Remarks on the Kodaira vanishing theorem*, J. Indian Math. Soc. (New Ser.) **36** (1972), 41–50; **38** (1974), 121–124.

[Rei88]   I. Reider, *Vector bundles of rank 2 and linear systems on algebraic surfaces*, Ann. of Math. **127** (1988), 309–316.

[Siu93]   Y. T. Siu, *An effective Matsusaka big theorem*, Ann. Inst. Fourier (Grenoble) **43** (1993), 1387–1405.

[Siu94]   Y. T. Siu, *Effective very ampleness*, preprint (1994), to appear in Invent. Math. (1995).

[Sk72]   H. Skoda, *Sous-ensembles analytiques d'ordre fini ou infini dans $\mathbb{C}^n$*, Bull. Soc. Math. France **100** (1972), 353–408.

[Vie82]   E. Viehweg, *Vanishing theorems*, J. Reine Angew. Math. **335** (1982), 1–8.

Meromorphic Functions: Progress and Problems

DAVID DRASIN

Mathematics Department, Purdue University
West Lafayette, IN 47907, USA

Zürich is a special place to workers in meromorphic function theory. Rolf Nevanlinna was Professor both at the ETH and University of Zürich. His address at the 1932 Zürich ICM centered on connections between his new theory of meromorphic functions and the Riemann surface of f^{-1}, a perspective that continues to yield insights. Lars Ahlfors accompanied Nevanlinna to the ETH in 1928, where he developed his fundamental distortion theorem and proved Denjoy's conjecture that an entire function of order ρ has at most 2ρ distinct finite asymptotic values. Zürich has been one of the main venues of the Nevanlinna Colloquia through the years, and the home of Pólya and Pfluger.

Goldberg in [20] and (with Levin and Ostrovskii) [22] has produced thorough surveys whose bibliographies contain $862 + 413$ entries.

1 Introduction

We study (nonconstant) functions f meromorphic in $D(R) = \{|z| < R\}$, $0 < r \le \infty$, in terms of the exhaustion of $D(R)$ by disks $D(r)$, $r < R$. If $a \in \mathbb{C}$, $0 < r < R \le \infty$, set

$$m(r, \infty) = \frac{1}{2\pi} \int_0^{2\pi} \log^+ |f(re^{i\theta})| \, d\theta,$$

$$m(r, a) = \frac{1}{2\pi} \int_0^{2\pi} \log^+ \frac{1}{|f(re^{i\theta}) - a|} \, d\theta \qquad (a \in \mathbb{C}),$$

$$N(r, a) = \int_0^r (n(t, a) - n(0, a))t^{-1} \, dt + n(0, a) \log r,$$

with $n(t, a)$ the number of roots of $f = a$ in $D(t)$ counted with regard to multiplicity; $\overline{N}$ and $\overline{n}$ count multiple roots of $f(z) - a$ once. Define $m(r, f) = m(r, \infty)$, $N(r, f) = N(r, \infty)$, and the (Nevanlinna) characteristic by

$$T(r) = T(r, f) = m(r, f) + N(r, f).$$

The first fundamental theorem asserts that

$$T(r) = T(r, \frac{1}{f - a}) + O(1) \tag{I}$$

Proceedings of the International Congress
of Mathematicians, Zürich, Switzerland 1994
© Birkhäuser Verlag, Basel, Switzerland 1995

for any $a \in \mathbb{C}$ (Jensen's formula). Once (I) is known, we may define all quantities using n and N, and extend them formally to *quasiregular mappings* g [35].

When $R = \infty$ (as we usually assume) Nevanlinna's second fundamental theorem is the inequality

$$\sum_1^q N(r, a_\nu) \geq (q - 2)T(r) + N_1(r) + S(r), \tag{II}$$

where $N_1(r) = N(r, 1/f') + 2N(r, f) - N(r, f')$ measures the total ramification of f in $D(r)$ and the "error term"

$$S(r) \equiv m(r, \frac{f'}{f}) + m(r, \sum_1^q \frac{f'}{f - a_\nu}) + O(1) = o(T(r)) \qquad (r \to \infty, r \notin E), \tag{1}$$

where E has finite measure and is empty when the *order* of f

$$\rho = \limsup_{r \to \infty} T(r)/\log r$$

is finite.

If we set

$$\delta(a) = \liminf_{r \to \infty} \frac{m(r, a)}{T(r)}, \qquad \theta(a) = \liminf_{r \to \infty} \frac{N(r, a) - \overline{N}(r, a)}{T(r)}, \tag{2}$$

so that $\delta(a) \geq 0$, $\theta(a) \geq 0$, $\delta(a) + \theta(a) \leq 1$, (II) is conveniently summarized by

$$\sum_{a \in \overline{\mathbb{C}}} \delta(a) + \theta(a) \leq 2 \tag{II$'$}$$

and, in particular,

$$\sum_{a \in \overline{\mathbb{C}}} \delta(a) \leq 2. \tag{II$''$}$$

These give very precise conditions that a nondegenerate map f defined in $\mathbb{C}$ must satisfy. Further analysis shows that $N(r, a) \sim T(r)$ as $r \to \infty$ for all a outside a set of capacity zero, and we call such a a *normal value* of f.

Inequality (II$'$) is sharp in the sense that any sequence of nonnegative δ_ν, θ_ν (with $\delta_\nu + \theta_\nu \leq 1$) can be associated to any sequence a_ν by some meromorphic function f: f solves the inverse problem. This was obtained by Drasin [5] after a long series of partial results by others (a solution to the restricted problem (II$''$) for entire functions was obtained in [19]). In general (see (4)) f must be of infinite order. We first construct a quasiregular formal solution g to the problem and let ω be a quasiconformal homeomorphism of $\mathbb{C}$ that solves the Beltrami equation

$$\omega_{\bar{z}} = \mu(z)\omega_z \qquad (\mu = g_{\bar{z}}/g_z).$$

where μ is small near ∞, so that $f \equiv g \circ \omega$ is meromorphic in the plane. As ω does not map circles exactly onto circles, it is necessary that $T(r, f)/\log r$ approach infinity quite slowly as $r \to \infty$ to guarantee that f inherit the data $\{\delta_\nu, \theta_\nu\}$ from g. Thus the function f obtained in [5] has 'small' infinite order. No doubt there exist solutions to this inverse problem of arbitrarily rapid growth.

A generalization of (II''), suggested by Nevanlinna, is to consider *small functions* $a(z)$ with $T(r, a) = o(T(r))$, and define $\delta(a)$ as in (2) with $a = a(z)$. Only rather recently has (II'') been established for small functions [32], [41].

Problem (a) Given a countable collection of functions $\{a_\nu(z)\}$ and sequence δ_ν, $0 < \delta_\nu \leq 1$, find a meromorphic function $f(z)$ with $T(r, a_\nu) = o(T(r, f))$ for each ν and $\delta(a_\nu, f) = \delta_\nu$.

(b) Is something more suggestive of (II) valid for small functions:

$$\sum_1^q \bar{N}(r, a_\nu) \geq (q-2)T(f) + o(T(r)) \qquad (r \to \infty, r \notin E)?$$

This would have applications to the uniformization of algebraic curves.

2 Logarithmic derivative, error terms

During the past decade, formal analogies have been observed between value-distribution theory and Diophantine approximation in number theory, and this analogy inspired Osgood's proof of (II'') for small functions. In [42, p. 34] is a "dictionary" between the two subjects. This also intrigued Lang [27], who asked for precise estimates of $S(r)$ in (1), and proposed the bound (3), based on the translation to function theory of his long-standing conjecture on rational approximation of irrational numbers. By an insightful reexamination of R. Nevanlinna's proof of (II) and the analysis of [21], Hinkkanen [25] developed a continuum of estimates for $S(r)$ and $m(r, f'/f)$. The simplest to state is: if $t^{-1}\phi(t) \uparrow$, $\int_1^\infty \phi(t)^{-1}\, dt < \infty$, then

$$S(r) \leq \log \phi(T(r, f)) \qquad (r \notin E), \qquad (3)$$

E of finite measure; examples show that (3) and its variants are very precise.

The best known estimates for $S(r)$ in the multidimensional case have been obtained in [44], based on methods of F. Nevanlinna and Ahlfors. These require that an unbounded term be added to the right side of (3), so they may not be sharp.

Ru and Wong [36] recently used ideas from multidimensional Nevanlinna theory to extend the Thue-Siegel theorem and thus give conditions that limit the number of integral points in the complement of certain hyperplanes in $\mathbb{P}^n$.

By further analyzing f'/f, Fuchs [18] ($\alpha = 1/2$) and Hayman [23] ($\alpha > 1/3$) proved that if $\rho < \infty$, then

$$\sum \delta(a_\nu)^\alpha < \infty. \qquad (4)$$

Weitsman [44], using other methods, obtained (4) when $\alpha = 1/3$. According to [11], Weitsman's theorem together with (II'') give necessary and sufficient conditions that the $\{\delta_\nu\} = \{\delta(a_\nu)\}$ must fulfill for functions of finite order, except in the special situations that $\sum \delta_\nu = 2$ or $\max \delta_\nu = 1$ (almost-entire functions).

It is far more difficult to construct entire functions ($\delta(\infty) = 1$) with $\rho < \infty$ and infinitely many $\delta_\nu > 0$. Arakelyan [1] was the first to do so, and conjectured that for entire functions, (4) should be replaced by $\sum(\log(1/\delta_\nu))^{-1} < \infty$, but this has recently been shown false by Eremenko [13], who in turn asks if (4) holds for all $\alpha > 0$ or if, for all $p > 1$,

$$\sum(\log(1/\delta_\nu))^{-p} < \infty.$$

The only positive result is the theorem of Lewis and Wu [29], which gives (4) with $\alpha = 1/3 - 2^{-260}$. In addition, they prove a conjecture of Littlewood about polynomials, one of whose consequences, stated informally, is that *an entire function assumes most of its values on a small subset of* $\mathbb{C}$. Lewis and Wu used work of Eremenko and Sodin, who had obtained a weak form of the Littlewood conjecture, and one form of this application to entire functions.

3 Extremal functions

If f is rational of degree n, then

$$\sum \delta(a) + \theta(a) = 2 - n^{-1}, \qquad \sum \delta(a) \le 1.$$

What this suggests, and what is true, is that cases of equality in (II$'$) are legion, whereas (II$''$) is usually strict.

THEOREM 1. *If f is meromorphic in the plane with*

$$\sum \delta(a, f) = 2 \qquad \rho < \infty, \tag{5}$$

then each $\delta(a)$ is an integral multiple of ρ^{-1} so that $2\rho - 1$ is a natural number.

(If f is entire, ρ must be a positive integer [34].) F. Nevanlinna and Hille gave examples that show Theorem 1 best possible by considering meromorphic functions $f = w_1/w_2$, where w_1, w_2 are linearly independent solutions to

$$w'' + Pw = 0, \tag{6}$$

P a polynomial; f has order $(n + 2)/2$, where $n = \deg P$. They deduced that the singularities of the Riemann surface of f^{-1} consist of a finite number of logarithmic branch points; even today one way to study Stokes multipliers for solutions to (6) uses the geometry of these surfaces [40, Chapter 8]. In these examples,

$$N_1(r) \equiv 0, \tag{7}$$

and only recently have Bergweiler and Eremenko [4] been able to obtain a function-theoretic proof that (7) characterizes these surfaces: they show that whenever g is meromorphic with $\rho < \infty$, the only possible singularities of the Riemann surface of g^{-1} are algebraic branch points, limits of algebraic branch points, and (by an extension of Ahlfors's theorem) a finite number of logarithmic branch points.

Theorem 1 was conjectured by F. Nevalinna [30] in 1929, and proved by Drasin [6] who performed a *quasiconformal modification* of $f(z^2)$: to introduce a quasiconformal map ω with small dilatation and set $g(z) = \omega \circ f(z^2)$. The role of ω is that 0 and ∞ become (essentially) Picard values of g. This reduces the problem to the (known) case of entire functions, and because of the specific form of ω it is possible to directly read off all information required of f.

Soon after [6] appeared, Eremenko [12] presented an alternate proof of Theorem 1, based on potential theory and convergence properties of δ-subharmonic functions. From this point of view, asymptotic equalities and inequalities become precise relations among the *limit functions* obtained by renormalization and normality considerations. In other papers, some joint with Sodin [16], [17], this viewpoint is systematically pursued and extended to small functions and meromorphic curves; see Section 4(C) for another application. It is very powerful but requires sophisticated potential theory when dealing with the limit functions that arise (however, the translations of these papers are inadequate). As an example of this point of view, their version of (II″) becomes

THEOREM 2. *Let* $u_1, \ldots, u_q$ *be subharmonic in a domain* $\Omega \subset \mathbb{C}$. *If for each* $k \neq \ell$

$$u \equiv \max_j u_j = \max\{u_k, u_\ell\}$$

then $\sum u_j - (q-2)u$ *is subharmonic in* Ω.

Eremenko [14] applies this method to obtain the deepest modern result in the theory: equality is forced in (II′) by a purely geometric hypothesis (compare with (7)). Because functions that satisfy (7) occupy a significant role in several contexts, these insights should have further application (the case $\rho < 1/2$ is in [39]). Eremenko proved

THEOREM 3. *Let* f *be meromorphic in the plane with* $\rho < \infty$ *and suppose that*

$$N_1(r) = o(T(r)) \qquad (r \to \infty). \tag{8}$$

Then $2\rho - 1$ *is a positive integer and* f *satisfies* (5).

Problem. For a given order $\rho < \infty$ find the best upper bound for the left side of (II″) among all entire/meromorphic functions of order ρ.

This is one of the oldest problems in the theory, settled for $\rho < 1$ by [8] and [43] (entire) and by [9] and [27] (meromorphic). I know of no sharp bounds for any values of $\rho > 1$ other than when (5) holds. The conjectured extremals appear in [31, p. 18] and [7]. This problem should be on our list for a long time.

4 Further topics

(A) Picard properties and normal families. A long-standing principle (Bloch) is that properties P that render a function defined in the plane constant may be mated with those that yield normal families for a collection of functions in a domain Ω. Examples of such P are: that f omit three fixed values; that $f \neq 0, f^{(k)} \neq 1$ for some fixed $k \geq 1$; that $f'f^m \neq 1$ for some fixed $m \geq 1$; see [37] and [46] for the history.

A uniform and elegant path between these two settings for a large class of P has been refined in recent years, using renormalizations and compactness arguments. The most recent ingredient is from Pang [33] (the case $h=0$ due to Zalcman): if $\mathcal{F}$ is not normal at $z_0 \in \Omega$ and $-1 < h < 1$ is given, there exist $f_n \in \mathcal{F}$, $z_n \to z_0$, $\rho_n \to 0$, such that

$$g_n(\zeta) \equiv \rho_n^h f_n(z_n + \rho_n \zeta) \to g(\zeta) \tag{9}$$

with g meromorphic in the plane, nonconstant, and of finite order. The other direction goes back to the beginning of the century: if f is meromorphic and nonconstant in the plane, the family $\mathcal{F} = \{f_R(z) = f(z_0 + Rz); R > 0, |z| < 1\}$ cannot be normal in the disk. These principles permit a uniform treatment of these P, and should have further applications.

(B) Complex iteration. Baker [3] used Ahlfors's theory of covering surfaces to prove that the Julia set J (nonnormality set) of iterates of an entire function f is the closure of repelling fixed points (of all orders); for rational maps this is Julia's theorem. These theorems now have a compelling short proof due to Schwick [38], based on (II). In (9), take as $\mathcal{F}$ the family of iterates, $h = 0$, and $z_0 \in J$ such that the system $\{z_0 = f(w_0), w_0 = g(\zeta_0)\}$ has unramified solutions; the desired fix-points arise at once. By (II), there can be at most four exceptional $z_0 \in J$, but as J is perfect these exceptional z_0 may be ignored.

(C) Quasiregular mappings. The limit function technique of Section 3 has been used by Eremenko and Lewis [15] to give a potential-theoretic proof of the

THEOREM OF RICKMAN [35]. *There exists $q_0 = q_0(n, K) < \infty$ such that if $q > q_0$ and $f : \mathbb{R}^n \to \mathbb{R}^n \setminus \{a_1, \ldots, a_q\}$ is K-quasiregular, then f is constant.*

Holopainen and Rickman [26] in turn have used [15] to extend Rickman's theorem to maps $f : \mathbb{R}^n \to M \equiv N \setminus \{a_1, \ldots, a_q\}$, N any oriented compact differentiable n-manifold and quasiconformality defined with respect to a Riemannian metric on M; [15] allows technical matters about path families to be bypassed.

(D) A return to f and f'. Nevanlinna's analysis of $m(r, f'/f)$ yields at once that

$$\limsup_{\substack{r \to \infty \\ r \notin E}} \frac{T(r, f^{(k)})}{T(r, f)} \leq \begin{cases} 1 & (f \text{ entire}) \\ k + 1 & (f \text{ meromorphic}), \end{cases}$$

where $|E| < \infty$. Nevanlinna was optimistic that lower bounds might be as simple, but many counterexamples have been given to this, principally by Toppila. Hayman and Miles [24] combine estimates of $\Re\{zf'/f\}$ with a geometrical analysis of the image of $\{|z| = r\}$ under f and the $\{f^{(k)}\}$ to show that given $K > 1$, there exists a set F_K of positive lower logarithmic density, with

$$\liminf_{\substack{r \to \infty \\ r \in F_K}} \frac{T(r, f^{(k)})}{T(r, f)} \geq \begin{cases} (2eK)^{-1} & (f \text{ entire}) \\ (3eK)^{-1} & (f \text{ meromorphic}), \end{cases}$$

independent of k.

Langley suggests that perhaps $\limsup\limits_{r \to \infty} T(r, f')/T(r, f) = 1$ if f is entire with $\rho < 1/2$; by [28] this would be sharp.

References

[1]   N. U. Arakelyan, *Entire functions of finite order with an infinite set of deficient values*, Dokl. Akad. Nauk. SSSR **170** (1966), 999–1002 (in Russian); Engl. transl.: Soviet Math Dokl. **7** (1966), 1303–1306.

[2]   A. Baernstein, *Proof of Edrei's spread conjecture*, Proc. London Math. Soc **26** (1973), 418–434.

[3]   I. N. Baker, *Repulsive fixpoints of entire functions*, Math. Z. **104** (1968), 252–256.

[4]   W. Bergweiler and A. Eremenko, *On the singularities of the inverse to a meromorphic function of finite order*, Rev. Mat. Iberoamericana, to appear..

[5]   D. Drasin, *The inverse problem of the Nevanlinna theory*, Acta Math. **138** (1977), 83–151.

[6]   ———— , *Proof of a conjecture of F. Nevanlinna concerning functions which have deficiency sum two*, Acta Math. **158** (1987), 1–94.

[7]   D. Drasin and A. Weitsman, *Meromorphic functions with large sums of deficiencies*, Adv. in Math. **15** (1975), 93–126.

[8]   E. A. Edrei, *Locally Tauberian theorems for meromorphic functions of lower order less than one*, Trans. Amer. Math. Soc. **140** (1969), 309–332.

[9]   ———— , *Solution of the deficiency problem for functions of small lower order*, Proc. London Math. Soc. **26** (1973), 435–445.

[10] E. A. Edrei and W. H. J. Fuchs, *Valeurs déficientes et valuers asymptotiques des fonctions méromorphes*, Comm. Math. Helv. **33** (1959), 258–295.

[11] A. E. Eremenko, *The inverse problem of the theory of meromorphic functions of finite order*, Sibirsk. Mat. Zh. **27** (1986), 377–390.

[12] ———— , *A new proof of Drasin's theorem on meromorphic functions of finite order with maximal deficiency sum, I and II*, Teor. Funktsiĭ Funktionals Anal. i Prilozhen. (Kharkov) **51** (1989), 107–116; **52** (1990), 3522–3527; **52** (1990), 3397–3403.

[13] ———— , *A counterexample to the Arakelyan conjecture*, Bull. Amer. Math. Soc. **27** (1992), 159–164.

[14] ———— , *Meromorphic functions with small ramification*, Indiana Univ. Math. J. **42** (1993), 1193–1218.

[15] A. E. Eremenko and J. L. Lewis, *Uniform limits of certain A-harmonic functions with applications to quasiregular mappings*, Ann. Acad. Sci. Fenn. Ser. A I Math. **16** (1991), 361–375.

[16] A. E. Eremenko and M. Sodin, *On meromorphic functions of finite order with maximal deficiency sum*, Teor. Funktsiĭ, Funktionals Anal. i Prilozhen. (Kharkov) **59** (1992), 643–651.

[17] ———— , *Distribution of values of meromorphic functions and meromorphic curves from the point of view of potential theory*, Algebra i Analiz **3** (1991), 131–164 (in Russian); Engl. trans.: St. Petersburg Math. J. **3** (1991), 109–136.

[18] W. H. J. Fuchs, *A theorem on the Nevanlinna deficiencies of meromorphic functions of finite order*, Ann. of Math. (2) **68** (1958), 203–209.

[19] W. H. J. and W. K. Hayman, *An entire function with assigned deficiencies*, Stud. Math. Analysis and Related Topics, Stanford Univ. Press, Stanford, CA, (1962), 117–125.

[20] A. A. Goldberg, *Meromorphic functions*, in v. 10 Serija Math. Analiz, Itogi Nauki i Tekhniki, Moscow (1973) (in Russian); Engl. trans.: J. Soviet Math. **4** (1975), 157–216.

[21] A. A. Goldberg and V. A. Grinshtein, *The logarithmic derivative of a meromorphic function*, Mat. Zametki **19** (1976), 525–530 (in Russian); Engl. transl.: Math. Notes **19** (1976), 320–323.

[22] A. A. Goldberg, B. Ja. Levin, and I. V. Ostrovskii, *Entire and meromorphic functions*, Kompleksaii Analyiz Odna Peremennaja-1, Tom 85; *Serija Sovremennie problemi mathematiki*, Itogi Nauki i Tekhniki, Moscow, 1991.

[23] W. K. Hayman, Meromorphic Functions, Oxford University Press, 1964.

[24] W. K. Hayman and J. Miles, *On the growth of a meromorphic function and its derivatives*, Complex Variables **12** (1989), 245–260.

[25] A. Hinkkanen, *A sharp form of Nevanlinna's second fundamental theorem*, Invent. Math. **108** (1992), 549–574.

[26] I. Holopainen and S. Rickman, *A Picard type theorem for quasiregular mappings of $\mathbb{R}^n$ into n-manifolds with many ends*, Rev. Mat. Iberoamericana **8** (1992), 131–148.

[27] S. Lang, *The error term in Nevanlinna theory*, Duke Math J. **56** (1988), 193–218.

[28] J. Langley, *On the deficiencies of composite entire functions*, Proc. Edinburgh Math. Soc. **36** (1992), 151–164.

[29] J. L. Lewis and J.-M. Wu, *On conjectures of Arakelyan and Littlewood*, J. Analyse Math. **50** (1988), 259–283.

[30] F. Nevanlinna, *Über eine Klasse meromorpher Funktionen*, C.R. 7e Congr. Math. Scand. Oslo (1929), 81–83.

[31] R. Nevanlinna, Le théorème de Picard-Borel et la théorie des fonctions méromorphes, Gauthier-Villars, Paris, 1929.

[32] Ch. Osgood, *Sometimes effective Thue-Siegel-Roth-Nevanlinna bounds, or better*, J. Number Theory **21** (1985), 347–399.

[33] X. Pang, *On normal criterion of meromorphic functions*, Sci. Sinica (5) **33** (1990), 521–527.

[34] A. Pfluger, *Zur Defekt relation ganzer Funktionen endlicher Ordnung*, Comm. Math. Helv. **19** (1946), 91–104.

[35] S. Rickman, Quasiregular Mappings, Ergeb. Math. Grenzgeb., **26** (1993), Springer-Verlag, Berlin and New York.

[36] M. Ru and P-M. Wong, *Integral points of $\mathbb{P}^n \setminus \{2n + 1$ hyperplanes in general position$\}$*, Invent. Math. **106** (1991), 195–216.

[37] J. Schiff, Universitext, Springer-Verlag, Berlin and New York, 1993.

[38] W. Schwick, *Repelling points in the Julia set,*, to appear, Bull. London Math. Soc.

[39] D. Shea, *On the frequency of multiple values of a meromorphic function of small order*, Michigan Math. J. **32** (1985), 109–116.

[40] Y. Sibuya, Global Theory of a Second Order Linear Ordinary Differential Equation with a Polynomial Coefficient, North-Holland, Amsterdam, 1975.

[41] N. Steinmetz, *Eine Verallgemeinerung des zweiten Nevanlinnaschen Hauptsatzes*, J. Reine Angew. Math. **368** (1986), 131–141.

[42] P. Vojta, Diophantine approximations and value distribution theory, Springer-Verlag, Berlin and New York, 1987.

[43] A. Weitsman, *Asymptotic behavior of meromorphic functions with extremal deficiencies*, Trans. Amer. Math. Soc. **140** (1969), 333–352.

[44] _____, *A theorem on Nevanlinna deficiencies*, Acta. Math. **128** (1972), 41–52.

[45] P.-M. Wong, *On the second main theorem in Nevanlinna theory*, Amer. J. Math. **111** (1989), 549–583.

[46] L. Zalcman, *Normal families revisited*, Complex Analysis and Related Topics (J.J.O.O. Weigerinck, ed.), Univ. of Amsterdam, 1993.

Teichmüller Space, Dynamics, Probability

HOWARD MASUR

University of Illinois at Chicago
851 South Morgan
Chicago, IL 60607, USA

Introduction

There are two rather separate sections to this paper. In the first part we indicate how the geometry of Teichmüller space and moduli space can be used to study the dynamics of rational billiards and more generally the dynamics of foliations defined by flat structures or quadratic differentials. In the second part of the paper we study random walks on the mapping class group of a surface and on Teichmüller space and show how the sphere of foliations defined by Thurston can be realized as the boundary of the random walks.

Rational billiards, flat structures, and quadratic differentials

Suppose one has a point mass moving at unit speed in straight lines in a polygon Δ in the plane. At a side the angle of reflection equals the angle of incidence. At a vertex the reflection is undefined. This gives a flow defined on the set of unit tangent vectors to Δ except for those vectors determining orbits that hit a vertex. This flow is called the billiard flow. If the vertex angles are all rational multiples of π, the billiard is called rational.

Billiard flows in domains with smooth boundaries have been studied using standard techniques in dynamical systems. See [S] and the extensive list of references there. These methods have not been applied successfully to rational billiards to answer some of the standard questions in dynamics such as existence and number of periodic orbits, and ergodicity. In this section we describe results on rational billiards that have been obtained using methods in Teichmüller theory.

For each side s of Δ let $r_s \in O(2)$ be the linear part of the reflection in s. That is, r_s is the reflection in the line parallel to s passing through the origin. The rationality assumption on the vertex angles implies that the subgroup Γ of $O(2)$ generated by the r_s is finite. This means that for given initial angle of the billiard ball only finitely many angles appear after all possible reflections.

We may build an invariant surface for the flow. Index copies of Δ by elements of Γ. Then glue copy Δ_{γ_1} to copy Δ_{γ_2} isometrically along the side s, if $\gamma_1 \circ r_s = \gamma_2$. Instead of reflecting the orbit in Δ_{γ_1} in side s, the orbit continues in Δ_{γ_2} in the same direction. The result of these gluings is a surface that locally is isometric to Euclidean space, except at points corresponding to vertices. At these points

Proceedings of the International Congress
of Mathematicians, Zürich, Switzerland 1994
© Birkhäuser Verlag, Basel, Switzerland 1995

a certain number of copies of Δ are glued together to form an angle around the vertex that is $2c\pi$, where c is a positive integer. We call the resulting structure a flat structure with cone angle $2\pi c$ singularities. At the $2\pi c$ singularity the metric can be expressed as

$$ds^2 = dr^2 + (cr d\theta)^2.$$

The billiard flow on Δ with initial angle θ becomes a flow f_t by straight lines on the flat surface.

A classical example is Δ a square. Then Γ is the group of order 4 generated by reflections in the coordinate axes. The flat surface is the flat torus. Four copies of Δ are glued around each vertex so these points do not give rise to singularities. The billiard flow in direction θ gives rise to the linear flow in direction θ on the torus. If the slope is rational, every orbit in that direction is periodic. If the slope is irrational, the Kronecker-Weyl theorem says the flow f_t is minimal, which means every orbit is dense, and furthermore it is *uniquely ergodic*.

DEFINITION. *A flow f_t on a compact space X is uniquely ergodic if for every continuous function h on X,*

$$\lim_{T \to \infty} \frac{1}{T} \int_0^T h(f_t(x))\, dt$$

converges uniformly.

Unique ergodicity is equivalent to saying that X has a unique f_t invariant probability measure μ, and in that case the above limit is

$$\int_X h(x)\, d\mu.$$

In the flat torus case for irrational flows, Lebesgue measure is the unique invariant measure.

Another example of a rational billiard is the

$$(\pi/2, \pi/5, 3\pi/10)$$

right triangle. Here Γ is the dihedral group D_{10}. The flat surface has genus 2 with one cone angle 6π singularity coming from 20 copies of the table glued around the vertex with angle $3\pi/10$.

More generally we define a flat structure with cone angle singularities as the result of gluing a finite number of Euclidean polygons isometrically along sides. We require that every side be glued to exactly one other, and that the total angle around a vertex is $k\pi$, where k is a positive integer. For an example that is not a billiard, glue two copies of a square pairwise along sides. Because two copies of the square are glued around each vertex, the resulting flat surface is a sphere with four cone angle π singularities.

Given a flat structure and a direction θ, the straight lines in that direction give a foliation F_θ. It is a *measured* foliation, in that the local Euclidean structure allows one to define an invariant measure on each transversal. More precisely, we may find local coordinates (x, y) in which the foliation is given by dy. Then the transverse measure is $|dy|$. The foliation is a flow if it is orientable.

We may also describe a flat structure by a meromorphic quadratic differential on a compact Riemann surface with at most simple poles. Away from the vertices, the Euclidean polygons determine local coordinates $z = x + iy$. Because the polygons are glued isometrically, dz^2 is invariant, and so describes a quadratic differential. A cone angle $k\pi$ singularity determines a zero of order $k - 2$. The lines in direction θ are the θ *trajectories* of the quadratic differential. Conversely, suppose $q(z)dz^2$ is a quadratic differential; P is a nonzero point of q, z are local coordinates near P with P corresponding to $z = 0$. Choose a branch of the holomorphic function $q^{1/2}(z)$ near 0 and set

$$w(z) = \int_0^z q^{1/2}(\zeta) \, d\zeta.$$

Then w are new holomorphic natural coordinates and

$$dw^2 = q(z)dz^2.$$

The coordinates w define the local Euclidean structure. We may reconstuct the surface by gluing together rectangles in these coordinates.

A third formulation is via measured foliations. The vertical foliation and the perpendicular horizontal foliation of a quadratic differential define a pair of transverse measured foliations. Conversely, suppose F and G are a pair of transverse measured foliations. We may find local coordinates (x, y) for which F and G are locally given by dx and dy, respectively. Then F and G are the vertical and horizontal foliations, respectively, of a quadratic differential

$$dz^2 = (dx + idy)^2.$$

We are interested in the dynamics of the 1-parameter family of measured foliations F_θ. Unlike the case of the flat torus, in general it is not the case that all leaves of F_θ are either closed or dense. For each θ the surface decomposes into a union of annuli of closed homotopic parallel leaves of F_θ of the same length, and minimal domains; domains in which every leaf of F_θ is dense in that domain ([St], [Z-K], [B-K-M]). Moreover, except for a countable set of θ every leaf of F_θ is dense. The existence of directions with annuli of closed leaves was settled by

THEOREM [M1]. *Let q be any flat structure. For a dense set of directions θ, F_θ has an annulus of closed leaves.*

In the case of the flat torus the growth rate in the number of closed orbits is quadratic, and it is possible to give asymptotic estimates on the growth rate. Veech [V3] was able to find asymptotic estimates for the growth rate for right triangular billiards with one angle π/n. In general we have

THEOREM [M2], [M3]. *For any flat structure q there exist positive constants $c_1 < c_2$ such that the number of parallel families of closed leaves of length $\leq T$ is between $c_1 T^2$ and $c_2 T^2$.*

What is perhaps surprising here is that the quadratic growth rate does not depend on the genus of the surface. We turn now to ergodicity questions. Again, unlike the case of the flat torus, minimality does not imply unique ergodicity. The study of minimal nonergodic phenomena was initiated by Furstenberg [F1] who found an example of a minimal flow on the torus that is not uniquely ergodic. Subsequently, Veech [V1] found an example that can be adapted to rational billiards. Take two copies of the unit circle with a segment of length $1 - \alpha$ marked off on each, counterclockwise from the origin. Rotate a point on a circle counterclockwise by angle θ until it lands in the segment. Then take the same point on the other circle, rotate by θ until landing in the segment, move back to the first circle, and so forth. Clearly, Lebesgue measure on each circle is an invariant measure for this dynamical system. Veech showed that if α is irrational, there are uncountably many θ for which this dynamical system is minimal but not uniquely ergodic. In fact, for each such θ, a set of orbits of positive measure asymptotically spends more than half its time in one circle. The dynamical system is equivalent to a billiard flow in direction θ on a rectangle with sides of length 2 and 1 with a slit of length $1 - \alpha$ centered at the midpoint of one long side. The group of the billiard again has order 4 as in the case of the square, but now the corresponding flat structure is a genus 2 surface with two cone angle 4π singularities corresponding to two copies of the endpoint of the slit. Equivalently, the quadratic differential has two zeros of order 2.

QUESTION. *How common is this minimal nonergodic phenomenon?*

Given a flat structure q let

$$NUE(q) = \{\theta : F_\theta \text{ is minimal but not uniquely ergodic}\}.$$

THEOREM [K-M-S]. *$NUE(q)$ has Lebesgue measure 0.*

This result was sharpended in

THEOREM [M4]. *The Hausdorff dimension of $NUE(q)$ is at most $1/2$.*

The method used to prove the above theorems is to study the action of $SL(2, R)$ on a moduli space of flat structures. Fix a topological surface M of genus g and a finite set $\Sigma = p_1, \dots, p_n \in M$. We say an $(n+1)$-tuple

$$\sigma = (\sigma_1, \dots, \sigma_n; \epsilon),$$

where σ_i are positive integers and $\epsilon = \pm$ is *admissible* if
(1) $\sum_{i=1}^{n}(\sigma_i - 2) = 4g - 4$.
(2) $\epsilon = -$ if σ_i is odd for any i.
(3) $\sigma \neq (\emptyset; -), (1, -1; -), (3, 1; -), (4; -)$.
    We say a flat structure q is *realized* by the data σ, if
(a) q has a cone angle $\sigma_i \pi$ singularity at p_i.

(b) the foliations F_θ are orientable if $\epsilon = +$.

(c) the foliations F_θ are nonorientable if $\epsilon = -$.

Condition (1) is necessary to realize the data σ by the Gauss-Bonnet formula, or equivalently the well-known formula that the sum of the orders of the zeros minus the sum of the orders of the poles of a quadratic differential is $4g - 4$. Because a foliation is orientable in a neighborhood of a singularity if and only if the cone angle is an even multiple of π, clearly (2) is also necessary. It turns out [M-S 2] that (3) is also necessary for σ to be realized, and these necessary conditions are also sufficient.

Now fix an admissible σ. Let $h(M, \Sigma)$ be the group of orientation-preserving homeomorphisms of M that fix Σ, and $h_0(M, \Sigma)$ the subgroup of homeomorphisms homotopic to the identity. Now $h_0(M, \Sigma)$ acts by pullback on the set of flat structures that realize σ. The quotient $\mathcal{Q}(\sigma)$ is a manifold [V2], [M-S 1]. Because a flat structure also gives a complex structure to M we have a map $\mathcal{Q}(\sigma) \to T_g$, where T_g is the Teichmüller space of that genus. The space $\mathcal{Q}(\sigma)$ is called a stratum because for a fixed genus g, the strata of flat structures of that genus with all cone singularities at least 3π fit together to form the bundle of holomorphic quadratic differentials over T_g. There is an $SL(2, R)$ action on each $\mathcal{Q}(\sigma)$. The rotation r_θ

$$\begin{bmatrix} \cos\theta & \sin\theta \\ -\sin\theta & \cos\theta \end{bmatrix}$$

acts by preserving the flat structure q but by rotating directions by angle θ. In complex analytic terms this is the same as multiplying q by $e^{i\theta}$. The diagonal action g_t,

$$\begin{bmatrix} e^t & 0 \\ 0 & e^{-t} \end{bmatrix}$$

is the Teichmüller map with initial quadratic differential q, terminal quadratic differential $g_t(q)$, and maximal dilatation e^{2t}. It deforms the flat structure by contracting vertical lengths by a factor of e^t and by expanding horizontal lengths by e^t. Now let

$$\Gamma(M, \Sigma) = \frac{h(M, \Sigma)}{h_0(M, \Sigma)}.$$

The group Γ acts on each stratum $\mathcal{Q}(\sigma)$ with quotient $\mathcal{MQ}(\sigma)$. The $SL(2, R)$ action commutes with the action of Γ giving an $SL(2, R)$ action on $\mathcal{MQ}(\sigma)$.

We give an indication of how the $SL(2, R)$ action is used to prove Theorem [M1] for flat structures q in the stratum $\mathcal{Q}(6; +)$; in particular for periodic orbits for the

$$(\pi/2, \pi/5, 3\pi/10)$$

billiard. One checks first that there is a dense set of directions θ_0 with a simple closed geodesic in direction θ_0 joining the singularity to itself. Rotate the flat structure using the r_θ action so that the geodesic is vertical. Then apply the diagonal action g_t and let $t \to \infty$. The geodesic shrinks in length. The genus 2 surface degenerates as this curve is pinched, and one sees that any limit flat structure is a flat torus. For t large there is an isometry from the complement of a small neighborhood of the shortened geodesic to the complement of a small convex set in the torus. Find a closed orbit on the flat torus that misses this convex set. The isometry gives a closed orbit on the approximating flat surface $g_t(q)$ and this yields a closed orbit on q in direction θ close to θ_0.

If the flat surface has more singularities, we use the $SL(2, R)$ action to perform two simplifying operations. The first is to coalesce two lower order singularities to a higher order singularity. The second is to squeeze along a simple closed geodesic to reduce the genus. After a combination of these operations, one is reduced as above to a flat torus or possibly a sphere with 4 cone angle π singularities. On these flat structures, we have closed leaves and these are used to find closed leaves on the given flat surface.

The proofs of Theorems [K-M-S] and [M4] on the size of $NUE(q)$ are based on the following ideas. As we saw above, if we apply the diagonal action g_t to a flat structure with a vertical closed geodesic, one or more geodesics become pinched. The Riemann surfaces of the deformed flat structures leave every compact set of the moduli space M_g of Riemann surfaces, the quotient of Teichmüller space by the mapping class group. The next theorem says that minimal nonergodic directions exhibit the same "rational" behavior as directions with a closed geodesic.

THEOREM [M4]. *Suppose q is a flat structure with minimal nonergodic vertical foliation. Then $g_t(q)$ eventually leaves every compact set in M_g.*

Thus, these geodesics are nonrecurrent in M_g. One uses this fact, the $SL(2, R)$ action, and the Deligne-Mumford compactification of M_g in the proofs of Theorems [K-M-S] and [M4].

Our next question concerns lower bounds on the size of $NUE(q)$. Of course for q a flat torus, $NUE(q) = \emptyset$, and the same is true for spheres with 4 cone angle π singularities, because a double cover gives a flat torus. We call strata of these flat structures *exceptional*. Veech [V3] has shown that in every stratum there is a dense set of flat structures for which $NUE(q) = \emptyset$. Concrete examples are provided by billiard tables of right triangles

$$(\pi/2, \pi/n, (n-2)\pi/n).$$

Veech showed that the affine self-maps of the resulting flat structure of the billiard form a lattice in $SL(2, R)$, and in that case minimality implies unique ergodicity. In addition, if θ is a nonminimal direction, then *every* orbit is closed. However, we have

THEOREM [M-S 1]. *For every nonexceptional stratum $\mathcal{Q}(\sigma)$, there exists $\delta = \delta(\sigma) > 0$ such that for almost every $q \in \mathcal{Q}(\sigma)$, $NUE(q)$ has Hausdorff dimension δ.*

Veech's results say that the almost everywhere statement cannot be replaced by everywhere. Because rational billiards form a set of measure 0 inside each stratum, this theorem does not give information about rational billiards. It is an interesting open question to form criteria on a rational billiard to guarantee minimal nonuniquely ergodic directions.

We give an outline of the proof of this theorem to show how Teichmüller theory is used. A metric cylinder is a family of parallel closed leaves. We construct an infinite set of metric cylinders forming a family tree. There will be a single cylinder at the top of this tree. It will have a finite number of offspring. Each offspring will in turn have a finite number of offspring. We construct this tree so that if we follow any infinite sequence of offspring $A_0, A_1 \ldots$, then the corresponding sequence of directions of waist curves $\theta_0, \theta_1, \ldots$ will converge to a limiting direction θ_∞ and the foliation F_{θ_∞} will not be ergodic. The set of limiting directions Λ_q will be a Cantor set of positive Hausdorff dimension.

The tree of cylinders will be constructed inductively. The inductive step is the following: given a cylinder A on the flat surface, we find a collection of cylinders B_i, which are the offspring and are disjoint from A. The number, directions, and lengths of these offspring are chosen to satisfy certain a priori bounds. The bounds are used to show that the limiting directions are not ergodic and the Cantor set of limit directions has positive Hausdorff dimension. The procedure used to find the offspring is much like the procedure described earlier to find closed leaves. We use the $SL(2, R)$ action to shrink the cylinder A and then successively to shrink geodesic segments to limit on a torus or sphere. The disjoint cylinders then arise from closed orbits on the latter flat structures. There are obstructions to carrying out this procedure, such as the ability to limit on a torus or sphere. This is reflected in the almost everywhere statement.

Random walks on Teichmüller space and the mapping class group

In this section we assume M is a closed surface of genus $g \geq 1$ and

$$\Gamma = \text{Diff}^+(M)/\text{Diff}_0(M)$$

is the mapping class group of M. When $g = 1$, T_1, the Teichmüller space of a torus, is hyperbolic space H^2 and Γ is $SL(2, Z)$. In his famous 1976 paper Thurston [T], [F-L-P] introduced the space $\mathcal{PMF}$ of projective measured foliations and proved that it is a natural compactification of Teichmüller space T_g. Moreover, the proper discontinous action of Γ on T_g extends to the natural action on $\mathcal{PMF}$. Thurston used the compactification of T_g by $\mathcal{PMF}$ to give his classification of elements of Γ as periodic, reducible, or pseudo-Anosov, generalizing the classification of elements of $SL(2, Z)$ as finite order, parabolic, and Anosov. Recall also that Γ acts by isometries on T_g with respect to the Teichmüller metric $d_T(\cdot, \cdot)$. In certain contexts a geometric space on which a group acts can be constructed as the boundary of the group. For example, cocompact Fuchsian groups are word or Gromov hyperbolic and the Gromov boundary is the circle at infinity. For geometrically finite Kleinian groups Floyd [Fl] showed that one can put an incomplete metric on the Cayley graph and recover the limit set on the sphere at infinity as the metric completion minus the group. Kerckhoff has asked whether a similar construction of $\mathcal{PMF}$ from Γ is possible. Now Γ is not word hyperbolic because it contains abelian subgroups of rank at least two, and it appears the Floyd boundary reduces to a point. Our approach to this problem is via the theory of the boundary of random walks, introduced by Furstenberg [F2], [F3] to study rigidity of lattices in $SL(n, R)$. We review the relevant definitions restricting ourselves to countable groups.

Let μ be a probability measure on a countable group Γ. This gives a right random walk on Γ;

$$\text{Prob } (\gamma \to \gamma\gamma') = \mu(\gamma').$$

Set the product measure $\mu^\infty = \mu \times \mu \times \cdots$ on the infinite Cartesian product $\Pi_{i=1}^\infty \Gamma$. We have a map

$$(\gamma_1, \gamma_2, \dots) \to (e, \gamma_1, \gamma_1\gamma_2, \dots)$$

from $\Pi_{i=1}^\infty \Gamma$ to the trajectory space Γ^∞ of paths $g_n = \gamma_1 \dots \gamma_n$ through the identity. The image of μ^∞ under this map is the probability measure P^μ on Γ^∞.

Let (B, ν) be a probability space on which Γ acts. Let $\mu * \nu$ be the probability measure on B, which is the image of $\mu \times \nu$ under the action

$$\Gamma \times B \to B.$$

Equivalently, for any measurable $E \subset B$,

$$\mu * \nu(E) = \sum_{\gamma \in \Gamma} \mu(\gamma)\nu(\gamma^{-1}E).$$

The measure ν is said to be *stationary* for μ if

$$\mu * \nu = \nu.$$

DEFINITION. (B, ν) is a *μ-boundary* of Γ if
 (1) ν is a stationary measure for μ.
 (2) $g_n \nu$ converges to a Dirac measure for P^μ a.e. sample paths (g_n).

DEFINITION. (B, ν) is the *Poisson boundary* of (Γ, μ) if it is maximal with respect to (1) and (2); namely, for any (B_1, ν_1) satisfying (1) and (2) there is a Γ equivariant map $\pi : B \to B_1$ such that $\pi\nu = \nu_1$.

There is a connection with harmonic functions. A function $h : \Gamma \to R$ is μ-harmonic if

$$h(\gamma) = \sum_{\gamma' \in \Gamma} \mu(\gamma')h(\gamma\gamma').$$

Condition (1) allows us to introduce the Poisson integral

$$P : L^\infty(B, \nu) \to H^\infty(\Gamma, \mu),$$

where $H^\infty(\Gamma, \mu)$ are the bounded harmonic functions, by the formula

$$P[f](\gamma) = \int_B f(x) \, d(\gamma\nu)(x).$$

A μ-boundary is a Poisson boundary if P is an isomorphism.

THEOREM [K-M]. *Suppose μ is a probability measure on Γ whose support generates Γ as a semigroup. Then there is a unique stationary measure ν_0 on $\mathcal{PMF}$ supported on the uniquely ergodic foliations such that $(\mathcal{PMF}, \nu_0)$ is a μ-boundary of Γ. If in addition, the measure μ has finite first moment with respect to the Teichmüller metric;*

$$\sum_{\gamma \in \Gamma} \mu(\gamma) \, d_T(o, \gamma o) < \infty,$$

where o is some origin in Teichmüller space, then $(\mathcal{PMF}, \nu_0)$ is the Poisson boundary of (Γ, μ).

COROLLARY. *Under the same hypotheses, if μ has finite first moment with respect to the word metric $|\cdot|$ on Γ defined by any finite set of generators;*

$$\sum_{\gamma \in \Gamma} \mu(\gamma) |\gamma| < \infty,$$

then $(\mathcal{PMF}, \nu_0)$ is the Poisson boundary of (Γ, μ).

We sketch some ideas in the proof of Theorem [K-M]. The first point is that stationary measures ν_0 exist for fairly general reasons. Start with any probability measure ν on $\mathcal{PMF}$. Any weak limit of

$$\frac{1}{n}(\nu + \mu * \nu + \mu * (\mu * \nu) + \cdots)$$

is stationary. The next step is to show that ν_0 is nonatomic. One first shows that one can replace μ with a measure supported on all of Γ with the same ν_0 as stationary measure. Then if ν_0 were atomic there would be a point $x \in \mathcal{PMF}$ maximizing $\nu_0(x)$. The definition of stationary gives

$$\nu_0(x) = \sum_{\gamma \in \Gamma} \mu(\gamma)\nu_0(\gamma^{-1}(x)) \leq \sum_{\gamma \in \Gamma} \mu(\gamma)\nu_0(x) = \nu_0(x),$$

which implies $\nu_0(\gamma^{-1}(x)) = \nu_0(x)$ for all γ. Now $\Gamma(x)$ is an infinite set, and so we contradict ν_0 a probability measure.

A basic tool now is the intersection number of measured foliations. For α and β simple closed curves, $i(\alpha, \beta)$ gives the minimum number of intersections for any curves in their homotopy class. The intersection number extends to a continuous function $i(\cdot, \cdot)$ defined on pairs of foliations. Let $\mathcal{MIN}$ be the set of minimal foliations. Define an equivalence relation $\sim$ on $\mathcal{MIN}$ by $F \sim G$ if F and G are topologically equivalent. This is equivalent to the condition

$$i(F, G) = 0.$$

For $H_0 \in \mathcal{MIN}$ let $\widetilde{H}_0$ denote its equivalence class. The equivalence class reduces to H_0 precisely when H_0 is uniquely ergodic. For a foliation F that is not minimal, there is a set of simple closed curves β such that $i(F, \beta) = 0$. There is a graph in the surface consisting of closed critical leaves of F such that the homotopy class of each such β is represented in the graph. We can write $\mathcal{PMF} - \mathcal{MIN}$ as a countable union of sets of foliations that determine the same graph, and an argument similar to the one that proved the measure is nonatomic gives

$$\nu_0(\mathcal{PMF} - \mathcal{MIN}) = 0.$$

Now it is a consequence of the martingale convergence theorem that for P^μ a.e. sample paths (g_n), $g_n\nu_0$ has a limit measure λ and that the limit measure λ is supported on $\mathcal{MIN}$. With the use of the intersection number $i(\cdot,\cdot)$ one then shows that either λ is a Dirac measure or supported on an equivalence class $\widetilde{H}_0$, where $H_0 \in \mathcal{MIN}$. This implies that the space $\mathcal{MIN}/\sim$ with quotient measure $\widetilde{\nu}_0$ is a μ-boundary. We then use Theorem [M4] to conclude that the ν_0 measure of the set of foliations that are not uniquely ergodic is 0. Thus, $(\mathcal{PMF}, \nu_0)$ is itself a μ-boundary.

To show that finite first moment implies that $(\mathcal{PMF}, \nu_0)$ is the Poisson boundary, we apply a theorem of Kaimanovich.

THEOREM [K]. *Let Γ be a countable group and assume μ has finite first moment*

$$\sum_{\gamma \in \Gamma} \mu(\gamma)|\gamma| < \infty$$

with respect to some metric $|\cdot|$ on Γ. Assume μ has finite entropy

$$-\sum_{\gamma \in \Gamma} \mu(\gamma) \log \mu(\gamma) < \infty.$$

Define a reflected measure $\bar{\mu}$ on Γ by $\bar{\mu}(\gamma) = \mu(\gamma^{-1})$ and let $(\bar{B}, \bar{\nu})$ be a $\bar{\mu}$ boundary. Suppose there is a measurable map that assigns to $\nu \times \bar{\nu}$ a.e. pairs $(b, \bar{b})$, a set $A(b, \bar{b}) \subset \Gamma$ such that

$$\lim_{n\to\infty} \frac{1}{n} \log |A(b, \bar{b}) \cap B_n| = 0,$$

where B_n is the ball of radius n centered at the identity. Then (B, ν) is the Poisson boundary of (Γ, μ) and $(\bar{B}, \bar{\nu})$ is the Poisson boundary of $(\Gamma, \bar{\mu})$.

To apply this theorem we show first that finite first moment with respect to the Teichmüller metric implies finite entropy. For this we show that the volume of balls in T_g grows at most exponentially. Now fix a positive number M and an origin $o \in T_g$. For $\nu \times \bar{\nu}$ a.e. pairs

$$(F, \bar{F}) \in \mathcal{PMF} \times \mathcal{PMF}$$

we take the Teichmüller geodesic $l = l_{F,\bar{F}}$ with F and $\bar{F}$ as endpoints. The subset $A(F, \bar{F}) \subset \Gamma$ of group elements $\gamma \in \Gamma$ such that $\gamma(o)$ lies in the M neighborhood of l satisfies the hypotheses of Theorem [K].

As an application of our methods and the methods of Furstenberg, we prove a theorem first proved by Ivanov [I].

THEOREM [K-M]. Γ *is not isomorphic to a lattice in* $SL(n, R)$.

Because Γ contains an element that commutes with every element of a free group of rank 2, it cannot be a lattice in $SL(2, R)$. Suppose Γ were a lattice in $SL(n, R)$ for $n \geq 3$. By [F2] one can construct a probability measure μ on Γ so that the Poisson boundary of (Γ, μ) is the Poisson boundary of $SL(n, R)$, which is a flag space. According to [F2] there exists $\epsilon > 0$ so that there does *not* exist a pair of μ harmonic functions h_1, h_2 defined on Γ satisfying

(i) $0 \leq h_i \leq 1$

(ii) $h_i(e) \geq 1/2 - \epsilon$.

(iii) $\min(h_1(g_n), h_2(g_n)) \to 0$ for any sequence $g_n \to \infty$.

However, just as in the case of a lattice in $SL(2, R)$ we show that we can construct harmonic functions on Γ satisfying (i)–(iii) and therefore Γ cannot be a lattice. (This is how Furstenberg shows that a lattice in $SL(2, R)$ is not isomorphic to a lattice in $SL(n, R)$ for $n \geq 3$.) Applying Theorem [K-M] $(\mathcal{PMF}, \nu_0)$ is a μ-boundary for a unique ν_0. Using the theory of train tracks we can find disjoint closed subsets Q_1 and Q_2 of $\mathcal{PMF}$ and an open set $V \supset \mathcal{PMF} - \mathcal{MIN}$ such that

(a) $\nu_0(Q_i) \geq 1/2 - \epsilon/2$

(b) $\nu_0(V) < \epsilon/2$.

(c) If $F_i \in Q_i, i = 1, 2$, then F_1 and F_2 are not topologically equivalent.

We then define harmonic functions on Γ by the Poisson integral formula

$$h_i(\gamma) = \int_{Q_i - V} d(\gamma \nu_0)(x) = \gamma \nu_0(Q_i - V).$$

By (a) and (b) h_i satisfies (i) and (ii). The fact that (iii) is satisfied comes from a description of all limit measures $\lambda = \lim g_n \nu_0$. One shows that any limit measure (not just a.e. limit measure) $\lambda = \lim g_n \nu_0$ not supported on $\mathcal{MIN}$ is supported on a set of the form

$$E_\alpha = \{F : i(F, \alpha) = 0\},$$

where α is a simple closed curve. If λ is a Dirac measure or supported on an equivalence class $\tilde{H}_0$ then (c) implies $g_n \nu_0(U) \to 1$ on an open set U that is either disjoint from Q_1 or disjoint from Q_2, and so $g_n \nu_0(Q_i) \to 0$ for that set. If λ is supported on E_α, then $g_n \nu_0(V) \to 1$ and so $g_n \nu_0(Q_i - V) \to 0$.

We now give an example of how such measures μ may be constructed. Fursten-berg's construction of measures on lattices used Brownian motion on the relevant symmetric space. The Teichmüller metric is not Riemannian so there is no known useful concept of a Laplacian. Our tool to replace Brownian motion is a geodesic random walk in T_g defined in [M5]. At each point $x \in T_g$ the holomorphic quadratic differentials at x form a Banach space with norm

$$\|q\| = \int_x |q(z)| \, dz^2|.$$

There is a natural Lebesgue measure on the unit ball so that it has measure 1 and this gives a measure on the unit sphere. We fix a number L. We randomly choose a quadratic differential at x. We also randomly choose a number $L \leq t \leq L + 1$

according to Lebesgue distribution. We then move along the Teichmüller geodesic in the direction of the quadratic differential the given distance t. At the new point x' we again choose a random direction and new distance and so forth. This defines a Markov process on T_g. Now T_g is known not to have negative curvature, and in fact the infinitesimal Teichmüller metric is difficult to understand. Nevertheless, we have

THEOREM [M5]. *For L sufficiently large, almost every path in the L geodesic random walk starting at x converges to a point in $\mathcal{PMF}$. The hitting measure ν^x on $\mathcal{PMF}$ for the random walk starting at x is supported on the set of uniquely ergodic foliations, is nonatomic, and positive on open sets. Moreover, $\{\nu^x : x \in T_g\}$ define the same measure class. In addition, there is a Γ invariant $\Omega \subset T_g$ such that Ω/Γ is compact such that for $x \in \Omega$, the expected return time to Ω is uniformly bounded.*

More generally, if θ is a measure on T_g we can define a hitting measure ν_θ on $\mathcal{PMF}$ by

$$\nu_\theta(E) = \int_{T_g} \nu^x(E)\, d\theta(x).$$

An important aspect of the theory is Harnack's inequality, which can be stated as follows. Let $B \subset B'$ be relatively compact open sets. For each $x \in B$ define λ^x to be the first hitting distribution in the complement of B' for Brownian motion starting at x. Then the Radon-Nikodým derivative $d\lambda^x/d\lambda^y$ is uniformly bounded in B. We do not know if a Harnack's inequality holds for the geodesic random walk.

THEOREM [M5]. *There is a modified measure on the sphere of quadratic differentials at each point in T_g such that the L geodesic random walk defined by the modified measure satisfies all the conclusions of the last theorem. In addition, Harnack's inequality is satisfied for the modified random walk.*

For each $x \in T_g$ let π^x denote the probability distribution on T_g for the modified geodesic random walk. It is supported in the domain $\{y : L \leq d_T(x, y) \leq L + 1\}$. A function h on T_g is harmonic if

$$h(x) = \int_{T_g} h(y)\, d\pi^x(y).$$

We let $H(T_g)$ denote the set of bounded harmonic functions. Let ν denote the measure class $\{\nu^x\}$ given by the theorem. There is a Poisson integral operator

$$P : L^\infty(\mathcal{PMF}, \nu) \to H(T_g)$$

defined by

$$P[f](x) = \int_{\mathcal{PMF}} f(y)\, d\nu^x(y).$$

THEOREM [K-M]. *The linear operator P above is an isomorphism so $(\mathcal{PMF}, \nu)$ is the Poisson boundary of the geodesic random walk. Moreover, for each $x \in T_g$, there is a probability measure μ^x on Γ with finite first moment*

$$\sum_{\gamma \in \Gamma} \mu(\gamma) \, d_T(o, \gamma o) < \infty$$

such that $(\mathcal{PMF}, \nu^x)$ is the Poisson boundary of (Γ, μ^x).

References

[B-K-M] C. Boldrighini, C. M. Keane, and F. Marchetti, *Billiards in polygons*, Ann. of Probab. **6** (1978), 532–540.

[F-L-P] A. Fathi, F. Laudenbach, and V. Poenaru, Travaux de Thurston sur les surfaces, Astérisque **66–67** (1979).

[Fl] W. Floyd, *Group completions and limit sets of Kleinian groups*, Invent. Math. **57** (1981), 205–218.

[F1] H. Furstenberg, *Strict ergodicity and transformations of the torus*, Amer J. Math. **88** (1961), 573–601.

[F2] H. Furstenberg, *Poisson boundaries and envelopes of discrete groups*, Bull. Amer. Math. Soc. **73** (1967), 350–356.

[F3] H. Furstenberg, *Random walks and discrete subgroups of Lie groups*, Adv. Probab. Related Topics, vol. 1, Dekker, New York, 1971, pp. 3–63.

[I] N. Ivanov, *Rank of Teichmüller modular groups*, Mat. Zametki **44** (1988), 636–644.

[K] V. Kaimanovich, *The Poisson boundary of hyperbolic groups*, C.R. Acad. Sci. Paris **318** (1994), 59–64.

[K-M] V. Kaimanovich and H. Masur, *The Poisson boundary of the mapping class group and of Teichmüller space*, submitted.

[K-M-S] S. Kerckhoff, H. Masur, and J. Smillie, *Ergodicity of billiard flows and quadratic differentials*, Ann. of Math. (2) **124** (1986), 293–311.

[M1] H. Masur, *Closed trajectories of a quadratic differential with an application to billiards*, Duke Math. J. **53** (1986), 307–313.

[M2] H. Masur, *Lower bounds for the number of saddle connections and closed trajectories of a quadratic differential*, Holomorphic Functions and Moduli, vol. 2, Springer-Verlag, Berlin and New York, 1988.

[M3] H. Masur, *The growth rate of trajectories of a quadratic differential*, Ergodic Theory Dynamical Systems **10** (1990), 151–176.

[M4] H. Masur, *Hausdorff dimension of the set of nonergodic foliations of a quadratic differential*, Duke Math. J. **66** (1992), 387–442.

[M5] H. Masur *Random walk on Teichmüller space and the mapping class group*, to appear in Journal d'Analyse Mathématique.

[M-S 1] H. Masur and J. Smillie, *Hausdorff dimension of sets of nonergodic measured foliations*, Ann. of Math. (2) **134** (1991), 455–543.

[M-S 2] H. Masur and J. Smillie, *Quadratic differentials with prescribed singularities and pseudo-Anosov diffeomorphisms*, Comm. Math. Helv. **68** (1993), 289–307.

[S] Y. Sinai, *Hyperbolic billiards*, Proc. Internat. Congress Math., 1990, vol. 1, Kyoto, pp. 249–260.

[St]   K. Strebel, Quadratic Differentials, Springer-Verlag, Berlin and New York, 1984.

[T]   W. P. Thurston, *On the geometry and dynamics of diffeomorphisms of surfaces*,
         Bull. Amer. Math. Soc. **18** (1988), 417–431.

[V1]   W. Veech, *Strict ergodicity in zero dimensional dynamical systems and the
         Kronecker-Weyl theorem mod 2*, Trans. Amer. Math. Soc. **140** (1969), 1–34.

[V2]   W. Veech, *The Teichmüller geodesic flow*, Ann. of Math. (2) **124** (1986), 441–
         530.

[V3]   W. Veech, *Teichmüller curves in moduli space, Eisenstein series and an appli-
         cation to triangular billiards*, Invent. Math. **97** (1989), 553–583.

[Z-K]   A. Zemylakov and A. Katok, *Topological transitivity of billiards in polygons*,
         Mat. Zametki **18** (1975), 291–300.

Fibering Compact Kähler Manifolds over Projective Algebraic Varieties of General Type

NGAIMING MOK

Department of Mathematics
University of Hong Kong
Pokfulam Road, Hong Kong

Regarding compact Riemann surfaces S, the Uniformization Theorem gives a trichotomy according to the genus of S. Other than the Riemann sphere $\mathbb{P}^1$ (of genus 0) and elliptic curves (of genus 1), S is conformally equivalent to the quotient of the unit disc by a torsion-free cocompact Fuchsian group of Möbius transformations, and as such is equipped with a Hermitian metric of constant negative curvature. For n-dimensional compact complex manifolds X this precise trichotomy in terms of the genus is replaced by the rough classification according to the Kodaira dimension $\kappa(X) = -\infty, 1, 2, \ldots, n$, which is the transcendence degree of the field of meromorphic functions arising from pluricanonical sections, i.e., holomorphic sections of positive powers of the canonical line bundle K_X. When $\kappa(X) = n \geq 1$, X is said to be of general type. They are the analogues of compact Riemann surfaces of genus ≥ 2. In 2 complex dimensions the Enriques-Kodaira classification of compact complex surfaces gives an essentially complete description for X of Kodaira dimension $-\infty, 0, 1$. If X is a compact Kähler surface with $\kappa(X) < 2$ and with infinite fundamental group, then either some finite unramified covering of X is biholomorphic to a compact complex torus, or X is an elliptic surface over a compact Riemann surface S of genus ≥ 1.

In the classification theory of higher-dimensional projective-algebraic and more generally compact Kähler manifolds a central theme is to study holomorphic fibrations. For compact Kähler manifolds X and Y in this article we will say that X can be holomorphically fibered over Y if and only if there exists a surjective holomorphic map $\sigma : X \to Y$ with connected fibers. Our point of departure is the following conjecture on the structure of compact Kähler manifolds X with infinite fundamental groups, which in the case of dimension 2 follows from the Enriques-Kodaira classification of compact complex surfaces.

CONJECTURE. *Let X be a compact Kähler manifold of complex dimension n whose fundamental group $\pi_1(X)$ does not contain an abelian subgroup of finite index and of rank $\leq 2n$. Then for some finite unramified covering X' of X and some modification $\hat{X}$ of X' there exists a surjective holomorphic mapping $\sigma : \hat{X} \to Z$ onto some projective-algebraic manifold Z of general type.*

Proceedings of the International Congress
of Mathematicians, Zürich, Switzerland 1994
© Birkhäuser Verlag, Basel, Switzerland 1995

In the notation of the Conjecture we will say that X' can be meromorphically fibered over Z. The Kähler condition is necessary in the Conjecture. In fact, by a theorem of Taubes [T], given any orientable compact smooth 4-manifold S, there exists some S' with $\pi_1(S') \cong \pi_1(S)$ such that S' admits an anti-selfdual connection. The twistor space Z over S', $\pi_1(Z) \cong \pi_1(S)$, is then a compact complex 3-fold that is non-Kähler and carries no nontrivial meromorphic functions except in very special cases.

1 Holomorphic fibrations arising from the Albanese map

To give credence to the Conjecture and to explain the relevance of the Kähler condition, we remark that it is valid in the special case when $\pi_1(X)$ contains an abelian subgroup of finite index. More generally we have

PROPOSITION 1. *Let X be an n-dimensional compact Kähler manifold whose first Betti number $b_1(X)$ exceeds $2n$. Then for some projective-algebraic variety Z of the general type there exists a surjective holomorphic map $\sigma : X \to Z$.*

Here we say that Z is of the general type if and only if it is birational to a projective-algebraic manifold of the general type. To prove Proposition 1 we consider the Albanese map. As X is Kähler by Hodge Theory the complex dimension of the vector space of closed holomorphic 1-forms ν is at least $n + 1$. Let $\alpha : X \to \mathrm{Alb}(X)$ be the Albanese map and write $S \subset \mathrm{Alb}(X)$ for the image of X under α. Because there exists at least $n + 1$ $\mathbb{C}$-linearly independent ν_i, S cannot be a sub-torus. By a theorem of Ueno (cf. [I]) on subvarieties of abelian varieties, which can be easily adapted for the general Kähler case of compact complex tori, S is a locally trivial holomorphic torus bundle over a projective-algebraic variety Z of the general type.

2 Kähler groups — Summary of relevant results

Following Gromov, the fundamental group of a compact Kähler manifold will be called a Kähler group. The Conjecture may be regarded as one on the representation theory of Kähler groups and is intimately related to factorization theorems. Let X be a compact Kähler manifold and $\Phi : \pi_1(X) \to G$ be a representation into some group G. Replacing X by some finite unramified covering if necessary, we look for meromorphic fibrations of X over some compact Kähler manifold Z such that the representation $\Phi : \pi_1(X) \to G$ arises from some representation $\Psi : \pi_1(Z) \to G$. This type of theorem in which Z can be chosen to have special complex-analytic properties (e.g., being of the general type) will be referred to as a factorization theorem.

In the last fifteen years quite a number of methods have been developed to study Kähler groups. In this article, we will be primarily concerned with the method of harmonic maps. There is first of all the method of harmonic maps into Riemannian manifolds of nonpositive curvature. In [S1, 1980] Siu established the Bochner-Kodaira formula for harmonic maps between compact Kähler manifolds in order to prove strong rigidity for compact quotients N of irreducible bounded symmetric domains of complex dimension ≥ 2. This method was further developed by Siu [S2, 1982] and Zhong [Zh, 1984] to prove holomorphicity or anti-holomorphicity of harmonic maps $f : X \to N$ on compact Kähler manifolds X provided that the rank of f is sufficiently large. Jost-Yau [JY, 1983] resp. Mok [M1, 1985; M2, 1988] used the Bochner-Kodaira formula to prove the strong rigidity of irreducible quotients of bidiscs resp. polydiscs. Sampson [Sa, 1986] generalized Siu's formula to the case when the target manifold is Riemannian. Based on this generalized formula, Carlson-Toledo [CT, 1989] proved factorization theorems for harmonic maps into Riemannian locally symmetric spaces. Mok [M4, 1992] proved factorization theorems for discrete Zariski-dense representations of Kähler groups into semisimple real Lie groups. The existence theory for harmonic maps in this context goes back to Eells-Sampson [ES, 1964] and were generalized by Corlette [C, 1988] and Labourie [L, 1991].) More recently, Gromov-Schoen [GS, 1992] and later Korevaar-Schoen [KS, 1993] and Jost [J, 1993] developed the existence theory for harmonic maps into Bruhat-Tits buildings. In the Kähler case this allows Simpson [Si3, 1991], Katzarkov [K, 1993], Jost-Zuo [JZ, 1993], and Zuo [Z2, 1994] to study Zariski-dense representations of Kähler groups into semisimple groups over local fields and, among other things, to prove factorization theorems. Other methods include the use of Higgs bundles (Simpson [Si1, 1992], Zuo [Z1, 1993]) and the method of L^2-cohomology on universal covers as developed by Gromov [G, 1991].

In this article we will be primarily concerned with the method of harmonic forms and harmonic maps with an emphasis on the use of harmonic maps into Riemannian locally symmetric spaces of the noncompact type. The main objective is to explain a method developed in Mok [M4] using holomorphic foliations and semi-Kähler metrics. This method justifies the Conjecture in the special case when $\pi_1(X)$ admits a Zariski-dense discrete representation into a noncompact semisimple real Lie group. We will also explain how this method can be used to resolve the Conjecture in the special case where $\pi_1(X)$ is of subexponential growth. It will be shown that in this case we have to work with unitary representations on infinite-dimensional Hilbert spaces.

3 The Bochner-Kodaira formula for harmonic maps

Let (X, g) be a compact Kähler manifold and (N, h) be a Riemannian manifold. A smooth map $F : (X, g) \to (N, h)$ is said to be harmonic if it satisfies the Laplace-Beltrami equation

$$\sum_{\alpha, \beta} g^{\alpha\bar\beta} \left[\frac{\partial^2 F^i}{\partial z_\alpha \partial \bar z_\beta} + \sum_{j,k} {}^N\Gamma^i_{jk} \frac{\partial F^j}{\partial z_\alpha} \frac{\partial F^k}{\partial \bar z_\beta} \right] = 0 \, ,$$

where $({}^N\Gamma^i_{jk})$ denotes the Riemann-Christoffel symbols of (N, h). We note that the connection of (X, g) does not enter into the equation. This is due to the Kähler property of (X, g). The terms inside the brackets define coefficients of the complex Hessian $\nabla \bar\partial F$ of F. F is then harmonic if and only if the trace of $\nabla \bar\partial F$ with respect to the Kähler metric g vanishes. We will say that F is pluriharmonic if and only if $\nabla \bar\partial F \equiv 0$. We note that this property does not depend on the choice of the Kähler metric g on X. For harmonic maps on compact Kähler manifolds we have the $\partial\bar\partial$-

Bochner-Kodaira formula of Siu [S1], generalized by Sampson [Sa]. To formulate it we say that a Riemannian manifold (N, h) is of nonpositive sectional curvature in the complexified sense if and only if $R^N(A, B; \bar{B}, \bar{A}) \leq 0$ for the curvature tensor R^N and for any *complexified* tangent vectors A and B. This curvature condition is in particular satisfied by Riemannian locally symmetric manifolds (N, h) of the noncompact type. We have

PROPOSITION 2 (SIU [S1], SAMPSON [SA]). *Let (X, g) be a compact Kähler manifold with fundamental group Γ, (N, h) be a Riemannian manifold of nonpositive sectional curvature in the complexified sense, and $\Phi : \Gamma \to Isom(N, h)$ be a group homomorphism. Let $F : (\tilde{X}, \tilde{g}) \to (N, h)$ be a Φ-equivariant harmonic map on the universal covering space $(\tilde{X}, \tilde{g})$. Then*

$$\int_X \|\nabla \bar{\partial} F\|^2 + H(\partial F \otimes \partial F; \overline{\partial F \otimes \partial F}) = 0 \, ,$$

where $H(\cdot, \bar{\cdot})$ is a positive semidefinite Hermitian form. As a consequence, both terms inside the integrand vanish identically. In particular, F is pluriharmonic.

Here the tensors appearing in the integrand are invariant under Γ and are thus interpreted as tensors defined on the quotient manifold. The first existence theorem for harmonic maps is due to Eells-Sampson [ES]. In the situation of reductive representations into semisimple Lie groups of the noncompact type the existence of a Φ-equivariant map F was due first to Corlette [C]. Labourie then introduced the more general notion of geometrically reductive representations and proved an existence theorem of harmonic maps for such representations [L].

4 Meromorphic foliations and semi-Kähler structures

The first use of Siu's Bochner formula was to prove strong rigidity. Typically, one starts with a smooth homotopy equivalence $f_0 : X \to N$ between two compact Kähler manifolds. Under suitable conditions of nonpositivity of curvature for the target manifold, f_0 is homotopic to a harmonic map, which must then be pluriharmonic according to the Bochner-Kodaira formula. The objective was to exploit the curvature in the Bochner-Kodaira formula to show that actually f is either holomorphic or anti-holomorphic. This is the case when N is a compact quotient of an irreducible bounded symmetric domain of complex dimension ≥ 2. In certain cases, the curvature is not sufficiently negative and the Bochner-Kodaira formula does not yield strong rigidity immediately. This is the case when N is an irreducible compact quotient of the polydisc. In this case one can derive from the formula the existence of associated meromorphic foliations. The existence of holomorphic foliations at generic points was established by Jost-Yau [JY] and the global existence of meromorphic foliations was established by Mok [M2]. Siu [S3] gave a more conceptual proof in terms of the Frobenius condition. Carleson-Toledo then used the method of Siu's proof to establish more generally

PROPOSITION 3 (CARLESON-TOLEDO [CT]). *In the statement of Proposition 2 suppose (N, h) is a Riemannian symmetric space of the noncompact type. Then*

$F^*T_N^{\mathbb{C}}$ can be endowed a holomorphic structure such that ∂F becomes a holomorphic section with values in $\Omega_{\tilde{X}} \otimes F^*T_N^{\mathbb{C}}$. Where ∂F is of constant rank, the distribution $\tilde{x} \to \mathrm{Ker}(\partial F(\tilde{x}))$ defines an integrable holomorphic distribution and thus a holomorphic foliation.

By Φ-equivariance the distribution $\tilde{x} \to \mathrm{Ker}(\partial F(\tilde{x}))$ gives rise to a meromorphic foliation $\mathcal{F}$ on X. This means that there exists a complex-analytic subvariety $V \subset X$ of complex codimension ≥ 2 such that $\mathcal{F}$ defines a holomorphic foliation on $X - V$. On the other hand, there is a Kähler semi-metric on X defined by the pluriharmonic map $F : (\tilde{X}, \tilde{g}) \to (N, h)$, as follows. The Riemannian semi-metric $F^*(h)$, as a symmetric 2-tensor field, decomposes into types in terms of the complex structure of X. The $(1,1)$-part $\tilde{\theta}$ then defines a Hermitian semi-metric, which one can show to be a Kähler semi-metric in the sense that the corresponding $(1,1)$-form $\tilde{\omega}$ is d-closed, as a consequence of the pluriharmonicity of F (cf. Mok [M4]). By Φ-equivariance we obtain a semi-Kähler metric θ with a semi-Kähler form ω. To describe the relationship between the meromorphic foliation $\mathcal{F}$ and the semi-Kähler form ω we introduce the notion of semi-Kähler structures, which we axiomatize as follows

DEFINITION 1. *(Semi-Kähler structures) Let X be a complex manifold. A semi-Kähler structure $(X, \omega, \mathcal{F}, V)$ consists of*
   (a) *a nontrivial closed positive $(1,1)$-current ω on X;*
   (b) *a complex-analytic subvariety $V \subset X$ of codimension ≥ 2, possibly empty, such that ω is smooth on $X - V$;*
   (c) *a holomorphic foliation $\mathcal{F}$ on $X - V$ such that*
   (d) *the closed semipositive $(1,1)$-form ω and $\mathcal{F}$ are compatible on a dense open subset U of $X - V$ in the sense that for any $x \in U$, $\mathrm{Ker}(\omega(x)) = T_x^{1,0}(\mathcal{F})$.*

Returning to our situation of the compact Kähler manifold X, we have

PROPOSITION 4 (MOK [M4, SECTION 4]). *Let (X, g) be a compact Kähler manifold and $F : (\tilde{X}, \tilde{g}) \to (\tilde{N}, \tilde{h})$ be as in the hypothesis of Proposition 2. Let $\mathcal{F}$ be the meromorphic foliation and ω be the semi-Kähler form as defined in the above. Then $(X, \omega, \mathcal{F})$ is a semi-Kähler structure' on X. Furthermore (X, ω) is of nonpositive bisectional curvature in the sense that for any local complex submanifold S on X and any $x \in S$ such that $\omega|_S$ is positive definite at x, $(S, \omega|_S)$ is of nonpositive bisectional curvature.*

Here and henceforth the subvariety $V \subset X$ will sometimes be left out in the notation for a semi-Kähler structure. To give a second example of semi-Kähler structures, again in the context of harmonic maps, consider the situation in Proposition 2 where the d-closed $(1,1)$-form ω is almost everywhere positive-definite on X. Then we have

PROPOSITION 5 (MOK [M4, SECTIONS 5,6]). *We use the same hypothesis as in Proposition 4 and assume furthermore that the Riemannian manifold (N, h) is locally symmetric of the noncompact type and that ω is positive-definite at some point. Then the Ricci form is defined almost everywhere. There exists furthermore a closed positive $(1,1)$-current ρ on X that agrees with the Ricci form wherever the latter is defined. Furthermore, there exists a meromorphic foliation $\mathcal{E}$ on X*

such that $(X, -\rho, \mathcal{E})$ defines a semi-Kähler structure on X. In addition, where $\omega > 0$ and ρ is of maximal rank the leaves of the foliation are totally geodesic with respect to the Kähler form ω.

The proof of Proposition 5 is more elaborate and relies on differentiating the structure equations arising from the Bochner-Kodaira formula for harmonic maps. Because we need to differentiate the curvature formula for (X, ω), the local symmetry of the target manifold (N, h) enters into play, local symmetry being characterized by the fact that the curvature R of (N, h) is parallel.

5 Generating continuous pseudogroups of holomorphic isometries

To make use of semi-Kähler structures on compact Kähler manifolds X our approach is to show that they give rise to meromorphic fibrations or else they exhibit partial local homogeneity, which allows us to study the complex structure of X by using some form of developing maps. To start with we introduce

DEFINITION-PROPOSITION 1. We say that a semi-Kähler structure $(X, \omega, \mathcal{F}, V)$ is factorizable if and only if all leaves L of $\mathcal{F}$ on $X - V$ are closed and the set-theoretic closures $\bar{L}$ on X are compact complex-analytic subvarieties of X. In this case there exists a modification $\rho : \hat{X} \to X$, a compact Kähler manifold Z, and a surjective holomorphic map $\sigma : \hat{X} \to Z$ with generically irreducible fibers such that for all leaves L of $\mathcal{F}$ on $X - V$, $\overline{(\rho^{-1}L)}$ is an irreducible component of a fiber $\sigma^{-1}(z)$, $z \in Z$, and such that $\omega = \sigma^*(\nu)$ for some closed positive $(1,1)$-form ν on Z that is smooth and positive definite almost everywhere.

Given a semi-Kähler structure $(X, \omega, \mathcal{F}, V)$ it is easy to deduce that outside of some singular sets we have a Riemannian foliation. To generate continuous pseudogroups of local holomorphic isometries we study the return maps given by holonomy. This technique is standard in case of Riemannian foliations (without singularities) on compact manifolds. In this case if some leaf is not closed then we can always generate a continuous pseudogroup of holomorphic isometries. In our situation the essential problem is that there are singularities for the meromorphic foliation and for the semi-Kähler form. If there is a leaf on $X - V$ with some limit point on $X - V$ the standard method still applies. For the semi-Kähler structure $(X, \omega, \mathcal{F}, V)$ let p be the complex dimension of S and let ℓ be the complex dimension of leaves. If $\ell > p$ then we are in the range of the Remmert-Stein Extension Theorem, which allows us to conclude that for any closed complex-analytic subvariety S of $X - V$, the topological closure $\bar{S} \subset X$ is a complex-analytic subvariety of X. For $\ell \leq p$ we may have the following difficult situation: every leaf L on $X - V$ is closed and such leaves are always transcendental. To take care of this situation in place of the Remmert-Stein Extension Theorem we have to use the Bishop Extension Theorem, which states that a complex-analytic subvariety $S \subset X - V$ extends complex-analytically across V if (and only if) it is of finite volume with respect to some background Kähler form on X. By measuring leaves we proved

PROPOSITION 6 (MOK [M4, P. 574]). Let $(X, \omega, \mathcal{F}, V)$ be a semi-Kähler structure on a compact complex manifold X. Then either

(a) *the semi-Kähler structure is factorizable, or*

(b) *there exists on $X - V$ a distinguished polydisc $U \cong D \times D'$ such that $(D, \bar{\omega})$ admits a continuous pseudogroup of holomorphic isometries.*

Here D and D' are polydiscs such that $\mathcal{F}$ is holomorphic on U and such that the leaves of $\mathcal{F}|_U$ are of the form $\{a\} \times D'$.

6 A factorization theorem for discrete Zariski-dense representations in semisimple real Lie groups of the noncompact type

In the case of Zariski-dense discrete representations Φ of $\pi_1(X)$ into semisimple Lie groups G of the noncompact type, by exploiting Proposition 6 and analyzing the partial local homogeneity of the semi-Kähler structures arising from the semi-Kähler form and the Ricci form we showed that nonfactorizability of such semi-Kähler structures contradicts with the semisimplicity of G. In order to establish this contradiction, among other things we used some ideas on Hermitian metric rigidity as developed in Mok [M3]. Using the fact that (X, ω) is of nonpositive holomorphic bisectional curvature and the simple Kähler case of the solution to the Grauert-Riemennschneider Conjecture we obtained the following factorization theorem.

THEOREM 1 (MOK [M4]). *Let X be a compact Kähler manifold with fundamental group $\pi_1(X) = \Gamma$ and let G be a semisimple Lie group of the noncompact type. Let $\Phi : \Gamma \to G$ be a discrete Zariski-dense representation. Then there exists a finite unramified covering X' of X and a modification $\hat{X} \to X'$ of X', a nonsingular projective-algebraic variety Z of the general type, a surjective holomorphic map $\sigma : \hat{X} \to Z$, and a representation $\Psi : \pi_1(Z) \to G$ such that $\Phi = \Psi \circ \sigma_*$ on $\Gamma' := \pi(X')$, where $\sigma_* : \pi_1(\hat{X}) \to \pi_1(Z)$ is induced by σ and $\pi_1(\hat{X})$ is canonically identified with $\pi_1(X) = \Gamma'$.*

In the factorization theorem above the fibers can be chosen to be connected, so that $\sigma : \hat{X} \to Z$ realizes $\hat{X}$ as an analytic fiber space over Z. It may happen that the fibers are generically zero dimensional, so that σ is in fact a modification, in which case X is itself of the general type and hence projective-algebraic by Moišezon's Theorem. This is the case in the following result.

THEOREM 2 (MOK [M4]). *Let Ω be a bounded symmetric domain and $\Gamma \subset \mathrm{Aut}(\Omega)$ be a torsion-free discrete group of automorphisms. Let X be a compact Kähler manifold admitting a continuous map $F_o : X \to \Omega/\Gamma$ such that the image of the fundamental class of X is nontrivial. Then X is of the general type and hence projective-algebraic. In particular, compact Kähler manifolds homotopic to complex submanifolds of Ω/Γ are of the general type and projective-algebraic.*

Theorems 1 and 2 may be considered applications of semi-Kähler structures to the study of complex-analytic properties of compact Kähler manifolds. In another direction such structures can also be applied to study the complex structures of universal covering spaces $\tilde{X}$ of compact Kähler manifolds. In Mok [M7] they are used to prove the Steinness of universal covers of certain compact Kähler manifolds, a result related to the Shafarevich Conjecture.

7 Generalization to the nondiscrete case

As explained in Section 2, Theorem 1 can be generalized to include the nondiscrete case by using the theory of harmonic maps into Bruhat-Tits buildings. Let $\Phi : \pi_1(X) = \Gamma \to G$ be a Zariski-dense representation into some semisimple real Lie group of the noncompact type. Either Φ is nonrigid or it is definable over some number field E Galois over $\mathbb{Q}$ (or both). In the latter case by taking all conjugates under the action of $\mathrm{Gal}(E/\mathbb{Q})$ one obtains a representation Ψ of Γ into the group of rational points of some semisimple Lie group defined over $\mathbb{Q}$. Either the denominators of elements of $\Psi(\Gamma)$ are all bounded, in which case Ψ is discrete, or there is some prime p for which there are denominators divisible by any given positive power of p. In this case we say that Ψ is p-unbounded. In the latter case or in the nonrigid case one obtains a Zariski-dense representation into some semisimple Lie group defined over some local field k, which defines a continuous map into some Bruhat-Tits building. The existence theory for harmonic maps into Bruhat-Tits buildings B developed by Gromov-Schoen [GS] (with improvements due to Korevaar-Schoen [KS] and Jost [J]) then leads to a nontrivial harmonic map $f : X \to B$ into such a building B, replacing X by a finite cover if necessary. In [GS] it is proved that the Bochner-Kodaira formula of Siu-Sampson continues to hold despite the singularities, so that f is pluriharmonic outside a proper complex-analytic subvariety. (At a generic point of X the map f can be regarded as a map into some Euclidean space.) As a consequence, outside some complex-analytic subvariety ∂f defines a multivalent vector-valued holomorphic 1-form, which can be lifted to a finite number of globally defined holomorphic 1-forms on some possibly ramified cover X^* of X, called a spectral covering of X. By exploiting the Albanese map on X^* Simpson [Si3] obtained a factorization theorem for harmonic maps into trees. An important ingredient is the Lefschetz Theorem of Simpson [Si2] for holomorphic 1-forms. Jost-Zuo [JZ] and Katzarkov [K] obtained factorization theorems over projective-algebraic varieties. Basing on Jost-Zuo [JZ], Zuo [Z2] proved a factorization theorem over varieties of the general type. The key point is to use spectral coverings and apply the criterion of Kawamata-Viehweg [KV] for the characterization of abelian variety (and of compact complex tori by a simple extension). By this criterion, if a projective-algebraic manifold M admits a ramified covering over an abelian variety, then $\kappa(M) > 0$.

THEOREM 3 (SIMPSON [Si3], JOST-ZUO [JZ], KARTZAKOV [K], AND ZUO [Z2]). *In the notation of Theorem 1 the conclusion holds when the semisimple real Lie group G is replaced by a semisimple Lie group G_k over some local field k and $\Phi(\Gamma)$ is assumed unbounded. As a consequence, Theorem 1 remains valid if we drop the assumption of discreteness on Φ.*

8 Construction of semi-Kähler structures when the fundamental group violates Property (T)

Given a Kähler group, it is difficult to decide whether or not it admits any nontrivial finite-dimensional representations. This limits the scope of application of the factorization theorems explained in the above. We propose that it should be easier to construct semi-Kähler structures of nonpositive holomorphic bisectional curvature. We believe that they exist on any compact Kähler manifold with infinite fundamental group. If one can construct them, the factorization theorems above for finite-dimensional representations should then be applicable for the following reason. Because of seminegativity of holomorphic bisectional curvature imitating the proof of Theorem 1 we expect to have either a factorization theorem over a projective-algebraic manifold of the general type, or else by Proposition 6 there exists a positive-dimensional pseudogroup of local holomorphic isometries. The fundamental group Γ should then act on the Lie algebra of this Lie pseudogroup and as a consequence yield a finite-dimensional representation. One has then to analyze the possiblity that the representation is trivial. When the representation is nontrivial we have then the possibility of applying Ueno's Theorem in the solvable case and applying the factorization theorems above in case the image of the representation is Zariski-dense in a semisimple real Lie group of the noncompact type. Although we are unable to verify all the steps of this strategy in general, there is at least one interesting situation where this strategy can be implemented, which leads in particular to a resolution of the Conjecture in the special case when the fundamental group $\pi_1(X) = \Gamma$ is infinite and of subexponential growth. To start with, one way to construct semi-Kähler structures is to use a nonzero closed holomorphic 1-form ν with values in some locally constant Hilbert bundle (possibly infinite-dimensional). Integrating ν we obtain a holomorphic map $\tilde{F}$ into a Hilbert space H. By pulling back the Euclidean metric we obtain a Γ-equivariant Kähler semi-metric of nonpositive holomorphic bisectional curvature. The associated meromorpohic foliation lifted to $\tilde{X}$ is simply defined by the level sets of $\tilde{F}$. In the case of finite-dimensional unitary representations Φ we have Hodge decomposition and the existence of such a ν amounts to the topological fact that $H^1(\Gamma, \Phi)$ is non-zero. For infinite-dimensional unitary representations this is no longer the case. For compactly generated groups Γ there is a property by Kazhdan, called Property (T), which can be characterized by the fact that $H^1(\Gamma, \Phi) = 0$ for any unitary representation of Γ. For Kähler groups that violate Property (T) of Kazhdan, we can apply the following existence theorem for harmonic forms.

PROPOSITION 7 (MOK [M6]).   *For a compact Riemannian manifold X and a positive integer i suppose there is a unitary representation Φ of $\pi_1(X) = \Gamma$ in some Hilbert space such that $H^i(\Gamma, \Phi) \neq 0$. Then for some possibly nonisomorphic irreducible unitary representation Ψ of Γ there exists on X an E_Ψ-valued harmonic i-form, where E_Ψ stands for the locally constant Hilbert bundle defined by Ψ. As a consequence, for X a compact Kähler manifold whose fundamental group Γ violates Property (T) of Kazhdan, there exists a closed holomorphic 1-form with values in some locally constant Hilbert bundle. In particular, there exists on X a nontrivial semi-Kähler structure of nonpositive holomorphic bisectional curvature.*

The existence part on harmonic forms with local coefficients in the case of $i = 1$ overlaps with a theorem of Korevaar-Schoen (Schoen [Sc]).

9 Resolution of the Conjecture for Kähler groups of subexponential growth

Examples of non-T discrete groups include finitely generated groups of subexponential growth and finitely generated solvable groups. We can now verify the Conjecture for certain non-T Kähler groups. In particular, we have

THEOREM 4 (MOK [M8]). *Suppose X is a complex n-dimensional compact Kähler manifold whose fundamental group Γ is of subexponential growth and does not contain an abelian subgroup of rank $\leq 2n$ and of finite index. Then, some finite covering X' of X can be meromorphically fibered over a projective-algebraic manifold of the general type.*

The only "universal" unitary representations that can be defined for all discrete groups are the left regular representation λ and representations derived from it. When the fundamental group is of subexponential growth, $H^1(\Gamma, \lambda) \neq 0$, so that it is a non-T group. In this case the strategy as explained in Section 8 can be implemented. The easiest case to explain is the case when no subgroup of finite index admits any nontrivial abelian representation. In this case by using the irreducibility of Ψ one can show that the representation of Γ on the finite-dimensional Lie algebra of the pseudogroup of local holomorphic isometries is necessarily nontrivial and must have a Zariski-dense image in some semisimple Lie group of the noncompact type. In general, one can define $\tilde{b}_1(\Gamma)$ to be the maximum of all first Betti numbers of finite unramified coverings of X. From Section 2, Proposition 1 (from Ueno's Theorem) Theorem 4 is obvious if $\tilde{b}_1(\Gamma) > 2n$. Otherwise, we pass to some finite unramified covering X' of X such that $\tilde{b}_1(\Gamma) = b_1(X')$. We consider the Albanese map on X', which must be surjective if the Conjecture were to fail. Then, the difficulty is to construct semi-Kähler structures that are in a certain sense transverse to those arising from closed holomorphic 1-forms. In the case where the fundamental group is of subexponential growth verifying the hypothesis in the theorem the generic smooth fiber of the Albanese fibration is also non-T. We can construct semi-Kähler structures on the fibers and piece them together to obtain a globally defined one transverse to those arising from closed holomorphic 1-forms. For this construction we have to study variations of Hodge structures for harmonic forms with values in locally constant Hilbert bundles.

One application of Theorem 4 is to the study of compact Kähler manifolds with nef anticanonical line bundles. For such manifolds Demailly-Peternell-Schneider [DPS] have proved that their fundamental groups are of subexponential growth. We obtain as a corollary

COROLLARY 1. *Suppose X is a compact Kähler manifold whose anticanonical line bundle is nef and such that the fundamental group Γ is infinite. Then, replacing X by some unramified finite covering space if necessary, either X can be fibered meromorphically over a compact complex torus, or it can be fibered meromorphically over some projective-algebraic manifold of the general type.*

10 Concluding remarks

The strategy as expounded in this article consists of constructing on compact Kähler manifolds with infinite fundamental groups Kähler semi-metrics of "nonpositive curvature". This strategy is also compatible with the use of the Albanese map in Section 1 and the use of harmonic maps into Bruhat-Tits buildings in Section 7. In the case of the Albanese map the use of closed holomorphic 1-forms leads to semi-Kähler structures of nonpositive holomorphic bisectional curvature. In the case of harmonic maps into Bruhat-Tits buildings one uses closed holomorphic 1-forms on spectral coverings. In addition, in the use of the criterion of Kawamata-Viehweg [KV] on ramified coverings of abelian varieties, one can interpret the pullback of the flat metric as defining a Kähler semi-metric of "nonpositive bisectional curvature" where the curvature is supported on the ramification divisor. In the statement of Theorem 4, while the hypothesis that the fundamental group is of subexponential growth remains a strong condition, it gives at least a significant reason to believe in the validity of the Conjecture. It is not unreasonable to believe that compact Kähler manifolds whose fundamental groups are of exponential growth have more to do with "negative curvature" than those with fundamental groups of subexponential gowth. To deal with the Conjecture in the latter case we believe that one should work with nonunitary representations in Hilbert spaces.

References

[CT]   Carleson, J., and Toledo, D., *Harmonic mappings of Kähler manifolds to locally symmetric spaces*, Publ. Math. **69** (1989), 173–201.

[C]   Corlette, K., *Flat G-bundles with canonical metrics*, J. Differential Geom. **28** (1988), 361–382.

[DPS]   Demailly, J.-P.; Peternell, T.; and Schneider M., *Kähler manifolds with semipositive Ricci curvature*, Compositio Math. **89** (1993), 217–240.

[ES]   Eells, J. and Sampson, H., *Harmonic maps of Riemannian manifolds*, Amer. J. Math. **86** (1964), 109–160.

[G]   Gromov, M., *Kähler hyperbolicity and L^2-Hodge Theory*, J. Differential Geom. **33** (1991), 263–292.

[GS]   Gromov, M., and Schoen, R., *Harmonic maps into singular spaces and p-adic superrigidity for lattices in groups of rank one*, Publ. Math. IHES (1992).

[I]   Iitaka, S., Algebraic geometry — An introduction to the birational geometry of algebraic varieties, Graduate Texts in Mathematics, Vol. **76**, Berlin-Heidelberg-New York; Springer, 1982.

[J]   Jost, J., *Equilibrium maps between metric spaces*, preprint.

[JY]   Jost, J., and Yau, S.-T., *Harmonic mappings and Kähler manifolds*, Math. Ann. **262** (1983), 145–166.

[JZ]   Jost, J., and Zuo, K., *Harmonic maps into Tits buildings and factorization of p-adic unbounded and nonrigid representations of π_1 of algebraic varieties*, preprint.

[K]   Katzarkov, L., *Factorization theorems for the representation of the fundamental groups of quasiprojective varieties and some applications*, preprint.

[KV]   Kawamata, Y., and Viehweg, E., *On a characterisation of Abelian varieties in the classification theory of algebraic varieties*, Compositio Math. **41** (1980), 355–360.

[KS]   Korevaar, N. J., and Schoen, R. M., *Sobolev spaces and harmonic maps for metric space targets*, preprint.

[L]   Labourie, F., *Existence d'applications harmoniques tordues à valeurs dans les variétés à courbure négative*, Proc. Amer. Math. Soc. **111** (1991), 878–882.

[M1]   Mok, N., *The holomorphic or anti-holomorphic character of harmonic maps into irreducible quotients of polydiscs*, Math. Ann. **272** (1985), 197–216.

[M2]   Mok, N., *Strong rigidity of irreducible quotients of polydiscs of finite volume*, Math. Ann. **282** (1988), 455–477.

[M3]   Mok, N., Metric rigidity theorems on Hermitian locally symmetric manifolds (Ser. Pure Math., vol. 6), Singapore-New Jersey-London-Hong Kong, World Scientific, Singapore and Teaneck, NJ, 1989.

[M4]   Mok, N., *Factorization of semisimple discrete representations of Kähler groups*, Invent. Math., **110** (1992), 557–614.

[M5]   Mok, N., *Semi-Kähler structures and algebraic dimensions of compact Kähler manifolds*, special volume for Geometry and Global Analysis, Tohôku University, Sendai, Japan, 1993.

[M6]   Mok, N., *Harmonic forms with values in locally constant Hilbert bundles*, to appear in Astérisque, Proceedings of a conference (Orsay, 1993) in honor of J.-P. Kahane.

[M7]   Mok, N., *Steinness of universal coverings of certain compact Kähler manifolds whose fundamental groups are lattices*, preprint.

[M8]   Mok, N., *Fibrations of compact Kähler manifolds with infinite fundamental groups of subexponential growth*, in preparation.

[Sa]   Sampson, J., *Applications of harmonic maps to Kähler geometry*, Contemp. Math. **49** (1986), 125–133.

[Sc]   Schoen, R., Lectures in Geometry and Global Analysis, Sendai, Japan, 1993.

[Si1]   Simpson, C., *Higgs bundles and local systems*, Publ. Math. IHES **75** (1992), 5–95.

[Si2]   Simpson, C., *A Lefschetz theorem for π_o of the integral leaves of a holomorphic one-form*, Compositio Math. **87** (1993), 99–113.

[Si3]   Simpson, C., *Integrality of rigid local systems of rank two of smooth projective varieties*, preprint, 1991.

[S1]   Siu, Y.-T., *The complex analyticity of harmonic maps and the strong rigidity of compact Kähler manifolds*, Ann. of Math. (2) **112** (1980), 73–111.

[S2]   Siu, Y.-T., *Complex-analyticity of harmonic maps, vanishing and Lefschetz theorems*, J. Differential Geom. **17** (1982), 55–138.

[S3]   Siu, Y.-T., *Strong rigidity for Kähler manifolds and the construction of bounded holomorphic functions*, in Discrete Groups in Geometry and Analysis (ed. R. Howe) Proceedings of Conference in March, 1984 in honor of G.D. Mostow, pp. 124–151, Boston-Basel-Stuttgart, Birkhäuser, 1987.

[T]   Taubes, C. H., *The existence of anti-self-dual conformal structures*, preprint.

[Zh]   Zhong, J.-Q., *The degree of strong nondegeneracy of the bisectional curvatures of exceptional bounded symmetric domains*, in Kohn, J. J. et al. (eds) Proc. Intern. Conf. Several Complex Variables, Hangzhou, Boston-Basel-Stuttgart, Birkhäuser, 1984.

[Z1]   Zuo, K., *Factorization of nonrigid Zariski-dense representations of π_1 projective manifolds*, preprint.

[Z2]   Zuo, K., *Kodaira dimension and Chern hyperbolicity of the Shafarevich maps for representations of π_1 of compact Kähler manifolds*, preprint.

Regularity of Fourier Integral Operators

DUONG H. PHONG*

Columbia University
New York, NY 10027, USA

I Introduction

The purpose of this paper is to survey some developments in the study of Radon transforms. These operators and the related oscillatory integrals have long been of interest in harmonic analysis and mathematical physics. Lately, they have emerged as key analytic tools in a wide variety of problems, ranging from partial differential equations to singularity theory and probability. It is not possible for us to describe adequately the progress made in all these areas in this limited space. Instead, we shall focus on the more analytic aspects, and take the opportunity to describe some recent joint work of the author with Elias M. Stein.

II Radon Transforms

Let X and Y be smooth manifolds, and let $\mathcal{C}$ be a smooth submanifold of $X \times Y$. A Radon transform with incidence relation $\mathcal{C}$ is an operator $R : C_0^\infty(Y) \to \mathcal{D}'(X)$ whose kernel is a Dirac measure $\delta_{\mathcal{C}}(x,y)$ supported on $\mathcal{C}$ with smooth density. If $\mathcal{C}_x$ denotes the fiber $\{y \in Y; (x,y) \in \mathcal{C}\}$, we can write more suggestively

$$Rf(x) = \int_{\mathcal{C}_x} f \, d\sigma_x, \tag{1}$$

where $d\sigma_x$ is the measure on $\mathcal{C}_x$ induced by $\delta_{\mathcal{C}}(x,y)$. Because our discussion is local, we always assume enough underlying structure so that (1) is well defined on functions, and that all measures and kernels are compactly supported. Some examples are:

(a) *The Radon transform on d-planes*: Here $Y = \mathbf{R}^n$, X is a subspace of the space $A_{n,d}$ of d-dimensional affine planes in $\mathbf{R}^n$, and $\mathcal{C} = \{(p,y) \in X \times \mathbf{R}^n; y \in p\}$. The classical Radon transform and the X-ray transform correspond, respectively, to $d = n - 1$ and $d = 1$. When $X = A_{n,d}$, f can be recaptured from Rf via the composition formula $R^*R = c_{n,d}(-\Delta)^{-d/2}$. Except in the case of hyperplanes, this problem is however overdetermined because $A_{n,d}$ has dimension $(n-d)(d+1)$. The problem of identifying the n-dimensional submanifolds X that suffice to recapture

---

*)   Work supported by National Science Foundation grant DMS-92-04196.

Proceedings of the International Congress
of Mathematicians, Zürich, Switzerland 1994
© Birkhäuser Verlag, Basel, Switzerland 1995

f (or at least its singularities) has been raised by Gelfand et al. [8] in the complex case, and more recently by Greenleaf and Uhlmann [11] in the real case;

(b) *Spherical means*: For each $t > 0$, let $\mathcal{C}_x = \{y \in \mathbf{R}^3; |x - y| = t\}$, and let $d\sigma_x$ be the surface measure on $\mathcal{C}_x$. The corresponding operator R_t gives the solution at time t of the initial value problem for the wave equation. It is also closely related to restriction phenomena for the Fourier transform;

(c) *Scattering of plane waves by a strictly convex obstacle K in $\mathbf{R}^n$*: Radon transforms $R : \mathbf{R} \times S^{n-1} \to \mathbf{R} \times \partial K$ with incidence relation $\mathcal{C} = \{(s, \omega; t, x); s - t- < x, \omega >= 0\}$ have been shown by Melrose and Taylor [22] to be the key ingredient in the Lax-Phillips scattering operator;

(d) *Convolutions on a group G with a lower-dimensional measure $d\sigma$*: These are evidently Radon transforms with incidence relation $\mathcal{C} = \{(x, y) \in G \times G; x \cdot y^{-1} \in \text{supp } d\sigma\}$. Orbital measures and random walks are well-known examples in representation theory and probability;

(e) *Singular Radon transforms*: Often we need to allow $\mathcal{C}$ and/or the density of the measure $\delta_{\mathcal{C}}(x, y)$ to be singular. Typically, $\mathcal{C}_x$ contains and has conic singularities at x, or $\delta_{\mathcal{C}}$ has fractional singularities along the diagonal $x = y$. Such is the case for operators arising in the Green's function in several complex variables [27], [9], in Guillemin's construction of Zoll-like Lorentz manifolds [14], and in the restricted X-ray transform [11]–[13].

The theory of Fourier integral operators [17] provides a framework for the study of Radon transforms. Henceforth we assume $\dim X = \dim Y$. A Fourier integral operator R is an operator whose kernel $K(x, y)$ can be written locally as

$$K(x, y) = \int_{\mathbf{R}^N} e^{i\phi(x, y, \theta)} a(x, y, \theta)\, d\theta, \tag{2}$$

where the amplitude $a(x, y, \theta)$ is a symbol of class $S^m(X \times Y, \mathbf{R}^N)$, and the phase $\phi(x, y, \theta)$ is a homogeneous function of order 1 in θ, nondegenerate in (x, y, θ) at the critical points of (2). The properties of R depend on the ambient geometry of the wave front set $WF(K)$ of K in $T^*(X) \times T^*(Y)$. The method of stationary phase shows that $WF(K) \subset \Lambda$, where $\Lambda = \{(x, y; d_x\phi, d_y\phi); d_\theta\phi(x, y, \theta) = 0\}$. The variety Λ is smooth and, by construction, Lagrangian. The classic theorem of Hörmander asserts that when the projections π_X and π_Y from $\Lambda \subset T^*(X) \times T^*(Y)$ to each factor have invertible differentials, R^*R and RR^* are pseudo-differential operators of order $2m - \dim X + N$. In particular, R is smoothing of order $m - (\dim X)/2 + N/2$ on Sobolev spaces. When the projections π_X and π_Y drop rank by k, then the order of smoothing can drop by $k/2$.

Because we may view $\mathcal{C}$ locally as $\phi_1(x, y) = \cdots = \phi_\ell(x, y) = 0$, and the Dirac measure $\delta_{\mathcal{C}}(x, y)$ admits an oscillatory integral representation with phase $\theta_1\phi_1 + \cdots + \theta_\ell\phi_\ell$, Radon transforms are Fourier integral operators. Their Lagrangian is just the normal bundle $N^*(\mathcal{C})$ of the incidence relation. When $N^*(\mathcal{C})$ is a local graph, the Radon transform R for submanifolds $\mathcal{C}_x$ of codimension ℓ is then smoothing of order $\frac{1}{2}(n - \ell)$.

The graph condition can be written in terms of a Monge-Ampère determinant

$$\det \begin{pmatrix} 0 & \cdots & 0 & d_x\phi_1 \\ \cdot & & \cdot & \cdot \\ \cdot & & \cdot & \cdot \\ \cdot & & \cdot & \cdot \\ 0 & \cdots & 0 & d_x\phi_\ell \\ d_y\phi_1 & \cdots & d_y\phi_\ell & d^2_{xy}\sum_{j=1}^{\ell}\theta_j\phi_j(x,y) \end{pmatrix} \neq 0 \qquad (3)$$

for any $\theta \in \mathbf{R}^\ell \setminus 0$ [27], [7]. It is satisfied e.g. by the Radon transform on hyperplanes and by convolutions with measures supported on hypersurfaces in $\mathbf{R}^n$ with nonvanishing Gaussian curvature. However, it becomes increasingly restrictive as the codimension increases, and actually can never be satisfied when $\ell > n/2$. When the condition fails, Hörmander's theorem provides bounds based on the rank of π_X and π_Y, but these usually do not reflect the geometry underlying the Radon transform. To illustrate this, we consider the operator $Rf = f \star d\sigma_\gamma$, where γ is a smooth curve in $\mathbf{R}^n$. The natural and generic assumption that γ has nonvanishing torsion implies that $|\hat{d\sigma}_\gamma(\xi)| \leq C|\xi|^{-1/n}$. Thus, R is smoothing of order $1/n$, whereas the order of smoothing guaranteed by the rank alone is just 0. In retrospect, the graph condition is a condition on the second-order derivatives of ϕ_i, and is insensitive to higher orders of contact phenomena.

III Radon Transforms along Curves

This simplest case has seen considerable progress in recent years, and may not be far from a complete understanding. It can be described on some basic models that are motivated as follows [29]–[32]. Consider the classical Radon transform on lines in $\mathbf{R}^2$. Because our discussion is local, we let (y, s) be coordinates in $\mathbf{R}^2$, parametrize a line p by $(x, t) \in \mathbf{R}^2$, where x is the slope of p and t its intercept with the s axis, and introduce a cut-off $\chi \in C_0^\infty(\mathbf{R}^2)$. The Radon transform can then be expressed as

$$Rf(x,t) = \int_{-\infty}^{\infty} f(y, t + S(x,y))\chi(x,y)\, dy = \int_{-\infty}^{\infty} e^{i\lambda t}\left[T(\lambda)\hat{f}(\cdot, \lambda)\right](x)\, d\lambda \qquad (4)$$

where $S(x,y) = xy$, $\hat{f}(y, \lambda)$ is the partial Fourier transform of f with respect to s, and $T(\lambda)$ is an operator-valued multiplier of the form

$$T(\lambda)u(x) = \int_{-\infty}^{\infty} e^{i\lambda S(x,y)}\chi(x,y)u(y)\, dy. \qquad (5)$$

In the present case, $T(\lambda)$ is just a rescaled Fourier transform, and we obtain

$$\|T(\lambda)\| = O(|\lambda|^{-1/2}). \qquad (6)$$

More generally, the graph condition (3) for operators of the form (4) reduces to the nonvanishing of $\partial_x\partial_y S$, and the estimate (6) follows from a well-known argument of Hörmander [16]. To model degeneracies, we consider then operators of the form (4), (5), with a phase $S(x, y)$ satisfying derivative conditions of higher order

$$\partial_x^{m-1}\partial_y S \neq 0, \quad \partial_x \partial_y^{n-1} S \neq 0. \tag{7}$$

THEOREM 1. *Assume (7) on the support of χ. (a) If $S(x, y)$ is an analytic function of (x, y), then the operator $T(\lambda)$ is bounded on $L^2(\mathbf{R})$ with norm*

$$||T(\lambda)|| = O(|\lambda|^{-\frac{m+n-4}{2(mn-m-n)}}). \tag{8}$$

(b) If $S(x, y)$ is merely smooth and $m = n \leq 4$, then

$$||T(\lambda)|| = O(|\lambda|^{-1/n}(\log|\lambda|)^{1/n}). \tag{9}$$

The cases $n = 2, 3$ are special because the variety $\Sigma = \{(x, y); \partial_x \partial_y S(x, y) = 0\}$ is then either empty or a smooth manifold. In these cases, the log terms in (9) can be dropped [16], [26]. More subtle conditions can be formulated for $S(x, y)$ a homogeneous polynomial:

THEOREM 2. *Let $S(x, y)$ be of the form $S(x, y) = \sum_{j=1}^{n-1} a_j x^j y^{n-1-j}$. Then (a) the decay rate $||T(\lambda)|| = O(|\lambda|^{-1/n})$ holds for an arbitrary choice of cut-off function χ if and only if there exist nonzero coefficients a_j and a_k with $j \leq n/2 \leq k$; (b) by contrast, the operator $U(\lambda)$ with a damping factor $|\partial_x \partial_y S(x, y)|^{1/2}$ defined by*

$$U(\lambda)u(x) = \int_{-\infty}^{\infty} e^{i\lambda S(x,y)} |\partial_x \partial_y S(x, y)|^{1/2} \chi(x, y) u(y) dy \tag{10}$$

is bounded on $L^2(\mathbf{R})$ with norm $||U(\lambda)|| = O(|\lambda|^{-1/2})$ for arbitrary coefficients a_j.

The estimates (8)–(10) are based on estimates away from the singular set Σ, so that $\partial_x \partial_y S$ is nonzero, but possibly small. Thus, we divide the (x, y) space into regions $\mathcal{R}_k$ where $|\partial_x \partial_y S| \sim 2^{-k}$, and decompose $T(\lambda)$ accordingly as $T(\lambda) = \sum_k T_k(\lambda)$. Naively, the oscillations of $T_k(\lambda)$ should yield (cf. (6))

$$||T_k(\lambda)|| \leq C(2^{-k}|\lambda|)^{-1/2}. \tag{11}$$

On the other hand, if I_x and I_y denote, respectively, the maxima of the x and y cross-sections of the region $\mathcal{R}_k$, we also have

$$||T_k(\lambda)|| \leq C(I_x I_y)^{1/2}. \tag{12}$$

Combining these two estimates leads to (8)–(10). Surprisingly however, (11) does not seem to hold for arbitrary regions $\mathcal{R}_k$. In practice, we need to decompose $\mathcal{R}_k$ further into simpler shaped boxes, whose contributions are essentially orthogonal. When $S(x, y)$ is analytic, this can be accomplished by parametrizing Σ by Puiseux series [30], [32], whereas in the C^∞ case with $m = n$, it arises via a stopping-time argument [31]. Operators of the form (10) have proved useful in the study of nonlinear dispersive equations [20], [3]. The estimates (8)–(10) lead to the following regularity properties for the Radon transform in 2 dimensions:

THEOREM 3. *Let R be given by (4). Then (a) R is smoothing of order $(m + n - 4)/2(mn - m - n)$ on Sobolev spaces $H_{(s)}(\mathbf{R}^2)$ when $S(x, y)$ is analytic and satisfies (7); (b) when $S(x, y)$ is a homogeneous polynomial satisfying $a_1 a_{n-1} \neq 0$, then R is smoothing from $L^p(\mathbf{R}^2)$ to $L^q(\mathbf{R}^2)$ for $(1/p, 1/q)$ in the intersection of the convex hull of the segment $0 < 1/p = 1/q < 1$ and the point $(2/3, 1/3)$ with the half-plane $(1/p) - (1/q) \leq 1/(n+1)$; (c) when S is smooth and satisfies (7) with $m = n = 3$, then the conclusion of (b) also holds with these values.*

The above bounds have been shown to be sharp by Christ [3]. The general case of Radon transforms along families of curves on a surface is formally close to the above models. In local coordinates (y, s) for $\mathbf{R}^2$, the curve $\mathcal{C}_{(x,t)}$ can be parametrized by $s = t + S(x, y; t)$, and R becomes a vector-valued pseudo-differential operator. Although this variable-coefficient operator poses some new severe difficulties (see e.g. [28]), Seeger [36] has obtained some very general results, in particular that there is a gain of $(m + n - 4)/2(mn - m - n) - \epsilon$, when the Monge-Ampère determinant vanishes of order at most $m - 2$ and $n - 2$ in each set of variables (x, t) and (y, s). For families of curves in a manifold of higher dimension, the problem is more difficult. A simple version based on stratifications is discussed in Section VI.

IV Intermediate-Dimensional Radon Transforms

Unlike Radon transforms along curves or hypersurfaces, the regularity of Radon transforms along intermediate subvarieties is still obscure. Indeed, consider the very simple translation-invariant case $Rf = f \star d\sigma_V$ in $\mathbf{R}^n$, with $d\sigma_V$ a smooth measure supported on a submanifold V of dimension d. As we saw before, nonvanishing torsion when $d = 1$ and nonvanishing Gaussian curvature when $d = n - 1$ are generic and natural geometric conditions that guarantee optimal regularity. However, for intermediate d, no condition has imposed itself as a reasonable measure of how "curved" V is, nor do we know what the best possible situation can be for a given dimension d of the subvariety V. In this section, we shall discuss some approaches to these problems [33]. It turns out that there is a dichotomy in the behavior of the Radon transform with respect to $L^p - L^q$ smoothing on one hand, and Sobolev smoothing on the other hand.

Smoothing in the $L^p - L^q$ sense

From dimensional analysis, it is readily seen that the best smoothing properties we can expect for averaging on a subvariety of dimension d is L^p to L^q, with $(1/p, 1/q)$ in the closed triangle with vertices at $(0, 0)$, $(1, 1)$, and $(n/(2n-d), (n-d)/(2n-d))$ [25]. It is a remarkable recent discovery by Ricci and Travaglini [35] that these sharp $L^p - L^q$ estimates actually hold for a large class of Radon transforms not satisfying the graph condition (3), namely convolutions with orbital measures on generic orbits in compact Lie groups. This suggests looking for less stringent conditions on the Monge-Ampère determinant that can nevertheless guarantee sharp $L^p - L^q$ bounds. The following is a step in this direction:

THEOREM 4. *Let the variety V be the graph of $\mathbf{R}^d \ni t \rightarrow S(t) \in \mathbf{R}^2$, where $S_j(t)$ are quadratic forms, and $d = n - 2$. Assume that, as a function of λ, the Monge-*

Ampère determinant $\det < \lambda, \frac{\partial^2 S(t_0)}{\partial t_j \partial t_k} >$ *vanishes of order* $\leq (n-2)/2$ *on the circle* $|\lambda| = 1$. *Then the sharp* $L^p - L^q$ *smoothing holds in the range indicated above, for measures* $d\sigma_V$ *supported in a small enough neighborhood of* t_0.

The key to this type of behavior is a partial decay estimate for the Fourier transform $\hat{d\sigma_V}$ of the following form

$$|\hat{d\sigma_V}(\xi)| \leq C \prod_{i=1}^{n-d} (1 + | < \xi, \theta_i > |)^{-d/2(n-d)}, \tag{13}$$

where $|\theta_j| = 1$ are suitable directions transverse to the variety V. The analytic tool is a method of stationary phase where however one eigenvalue of the Hessian matrix can be small. There is considerable room for improvement in Theorem 4, but this requires a uniform method of stationary phase with several small eigenvalues that is presently unavailable (cf. Section VI).

Smoothing in the Sobolev sense

To motivate the subsequent approach, we observe that the measure $f \star \sigma_V$ can hardly be expected to be smoothing unless its support generates $\mathbf{R}^n$ locally, in the sense that the mapping $V \times \cdots \times V \ni (x_1, \ldots, x_N) \to x_1 + \cdots + x_N \in \mathbf{R}^n$ be locally surjective for N large enough. In particular, there is no direction $\lambda \in \mathbf{R}^n \setminus 0$ that is orthogonal to all the $T_{x_j}(V)$'s, i.e., is orthogonal to V at N points. This suggests measuring the "curvature" of V by the maximum number μ of points admitting any given direction among its normals. We note that this generalizes the notion of nonvanishing Gaussian curvature for a hypersurface, because that property is simply equivalent to $\mu = 1$ locally. The example of the curve (t, t^N) shows however that the counting of points has to incorporate their multiplicities. The *multiplicity* or *Milnor number* [24], [21] has been introduced before for analytic functions $f : \mathbf{R}^d \to \mathbf{R}$ at an isolated critical point a. It is defined as $\mu = \dim \mathcal{A}(a)/\mathcal{I}[\partial_1 f, \ldots, \partial_d f]$, where $\mathcal{A}(a)$ is the space of germs of analytic functions at a, and $\mathcal{I}[\partial_1 f, \ldots, \partial_d f]$ is the ideal generated by the germs of the partial derivatives of f at a. This leads to

DEFINITION 1. *The analytic submanifold* $V \subset \mathbf{R}^n$ *is said to have nonvanishing* μ-*curvature if for any* $\lambda \in \mathbf{R}^n \setminus 0$, *the function* $\mathbf{R}^d \ni t \to < \lambda, S(t) >$ *has multiplicity at most* μ (*in the above sense*) *at any critical point. Here* $t \to S(t) \in V \subset \mathbf{R}^n$ *is an analytic local parametrization of* V.

THEOREM 5. *Assume* V *is an analytic surface with nonvanishing* μ-*curvature in* $\mathbf{R}^n$. *Then for any* $\epsilon > 0$, *the Radon transform with measure* $d\sigma_V$ *supported on* V *is smoothing of order* $\frac{2}{\mu^{1/2}+1} - \epsilon$ *on Sobolev spaces.*

Theorem 5 is obtained by combining some powerful estimates of Varchenko [38] and Karpushkin [18], [19] for oscillatory integrals, with the bounds provided by Kushnirenko [21] for the multiplicity of a function in terms of an Euler formula for its Newton diagram. We postpone to Section V a more detailed discussion of these developments. It is natural to expect that a similar result holds for analytic subvarieties of any dimension d, with the order of smoothing given by

$$g(\mu, n) = \frac{d}{\mu^{1/d} + 1} - \epsilon. \tag{14}$$

It is intriguing what the best multiplicities $\mu = \mu(n, d)$ can be, given the dimension n of the ambient space, and the dimension d of the subvariety. Evidently, $n/d - 1 \leq \mu(n, d)$, and the gain $g(\mu(n, d), d)$ should be a nonlinear function of d, as it goes from $1/n$ to $(n-1)/2$ when d increases from 1 to $n - 1$. It is easy to construct d-dimensional submanifolds in $\mathbf{R}^n$ with $\mu = (n - d)^d$, but we should be able to do better. For example, the complex curve $z \to (z, z^2, \ldots, z^{n/2})$ viewed as a surface in $\mathbf{R}^n$ achieves $\mu = (\frac{n}{2} - 1)^2$.

So far our discussion has been restricted to Euclidian spaces, but the same notion of μ-curvature can be introduced in any Lie group G. Because the tangent spaces at any point of a variety V can be translated to the origin, we can still say that V has nonvanishing μ-curvature if no cotangent direction at the origin is normal to more than μ of these points. It should be interesting to determine the regularity of convolutions when such a condition is satisfied everywhere, or only outside a lower-dimensional subvariety of V, as is usually the case for conjugacy orbits. Finally, we note that the conditions in Theorems 4 and 5 are in a sense complementary, as they bear, respectively, on the partial derivatives in λ and in t of the Monge-Ampère determinant.

V Sharp Estimates for Oscillatory Integrals

The analysis of Fourier integral operators when the graph condition fails requires estimates for degenerate oscillatory integrals [2]. Two seemingly opposite, but ultimately related, themes dominate our considerations. In the first (cf. (11), (13)), the critical points are nondegenerate, but the estimates have to reflect how close they come to being degenerate. Although the situation for arbitrary dimensions is very complex, we have the following complete answer in one dimension [29]:

THEOREM 6. *Let $P(x)$ be any smooth function in one variable satisfying $|P'(x)| \geq \prod_{i=1}^{d} |x - a_i|$. Assume that the function $((P'(x))^{-1})'$ changes sign only at most N times. Then there is a constant $C_{d,N}$ depending only on d and N such that*

$$\left| \int_a^b e^{i\lambda P(x)} \chi(x) \, dx \right| \leq C_{d,N} \sum_{i=1}^{d} \min_L (|\lambda| \prod_{j \notin L} |a_i - a_j|)^{-1/(|L|+1)}. \tag{15}$$

Here L denotes any cluster of points a_j, with $|L|$ its cardinality.

In the second theme, we consider degenerate critical points, but seek decay rates that are stable under small perturbations. The well-known van der Corput lemma in one dimension is the simplest example of such an estimate. In higher dimensions, the sharpest results to date are due to Varchenko [38] and Karpushkin [18], [19]. We describe them briefly. Let P be an analytic real phase function in $\mathbf{R}^n$, with the origin as a critical point, and let its Newton polyhedron be the convex hull of all the upper quadrants in $\mathbf{R}^n_+$ with vertices at those integers $(k_1, \ldots, k_n)$ whose corresponding monomial $\prod_{i=1}^n x_i^{k_i}$ appears in the Taylor expansion at 0 of P. For each face γ of the Newton polyhedron, let P_γ be the polynomial consisting of the monomials on γ. Then under the generic condition that dP_γ has no zero in $(\mathbf{R} \setminus 0)^n$, Varchenko shows that oscillatory integrals with phase $P(x)$ are bounded by $O(|\lambda|^{-\alpha} (\log |\lambda|)^\beta)$, where α is the inverse of the Newton distance, i.e. the coordinate of the intersection of the line $x_1 = \cdots = x_n$ and a face of the Newton polyhedron, and β is one less than the codimension of the face where the intersection takes place. We have actually encountered α before: in Theorem 1 (a), the decay rate for $\|T(\lambda)\|$ is just $\alpha/2$, when the Newton polyhedron has vertices at $(m-1, 1)$ and $(1, n-1)$. The factor $1/2$ is required by dimensional analysis, as we are dealing here with operators and not just a single integral. The generic condition on P_γ can be effectively removed in dimension 2 by a suitable choice of coordinates. Furthermore, in dimension 2, the theorem of Karpushkin guarantees that Varchenko's estimates are stable, with uniform constants, under deformations $P \to P + Q$, for any Q analytic and sufficiently small. This uniformity is crucial to Sobolev regularity, because the decay rate of a multiplier smoothing of order ϵ has to be $\geq \epsilon$ with constants *uniform* in all directions. Returning to Theorem 5, $d\hat{\sigma}_V$ is given precisely by oscillatory integrals with phase $P(t) = <\lambda/|\lambda|, S(t)>$. The theorems of Varchenko and Karpushkin apply to produce estimates in terms of the Newton distance. By a theorem of Kushnirenko [21], the multiplicity and the Newton diagram are linked by the relation $\mu \geq n! V - (n-1)! \sum_{i=1}^n V_i + (n-2)! \sum_{i<j} V_{ij} - \cdots \pm 1$, where V is the volume of the complement of the Newton polyhedron in the first quadrant, V_i its $(n-1)$-volume on the hyperplane not containing the i-th basis vector, and so on. By maximizing the Newton distance over all Newton diagrams with a fixed multiplicity, we obtain Theorem 5.

In higher dimensions, Varchenko has produced counterexamples to uniform estimates for arbitrary deformations. Nevertheless, uniform estimates can hopefully be formulated for deformations whose multiplicities do not exceed the multiplicity of P. We outline here on simple examples a possible method for such estimates. This method may be more accessible to analysts than Karpushkin's versal families of deformations. It also has the attractive feature of unifying the first theme with the second. First rewrite the $dx_1 \cdots dx_n$ oscillatory integral in projective coordinates $x_1 = x$, $x_i = xu_i$ as an $x^{n-1} dx du_1 \cdots du_{n-1}$ integral. As in Theorem 6, the dx integral can be bounded by

$$\left| \int_0^\infty e^{i\lambda P(x)} \chi(x) x^{n-1} \, dx \right| \leq C_{d,\chi} \sum_{i=1}^d \min_L |a_{i+|L|}|^{n-1} (|\lambda| \prod_{j \notin L} |a_i - a_j|)^{-1/(|L|+1)}$$
$$+ |\lambda|^{-n/(d+1)}, \tag{16}$$

where the $a_i(u)$'s are critical points with respect to dx and have been ordered in increasing order. The difficulty resides with the sum on the right-hand side of (16) when the a_i's are far from 0. However, the actual proof of Theorem 6 shows how to *localize* the effects of these points. The desired estimate for the localized neighborhood of the a_i's follows from the $du_1 \cdots du_{n-1}$ oscillations, unless we are at a point $x = a_i(u)$ that is an intrinsic critical point of the phase function P. In this way we can estimate the contribution to the $dx_1 \cdots dx_n$ integral from the whole $\mathbf{R}^n$, except for some well-defined pockets near each critical point away from the origin. The contributions of the pockets themselves can be obtained by translating the origin to the critical point in their center, and repeating the argument. As an example, we obtain

THEOREM 7. *(a) Let $P(x)$ be a homogeneous polynomial of degree 3 in $\mathbf{R}^2$, and assume that in homogeneous coordinates (x, u), it is of the form $P = x^3 \phi_0(u)$, with $|\phi_0| + |\phi_0'|$ never vanishing. Then for any m, the estimate $O(|\lambda|^{-2/3})$ holds uniformly for oscillatory integrals with phase $P + Q$, where Q is a polynomial of degree $\leq m$ with small coefficients; (b) Let $P(x)$ be a homogeneous polynomial of degree 4 in $\mathbf{R}^3$ that vanishes only at the origin. Then the phase functions $P_\epsilon(x) = P(x) + x_1 x_2^2 - \epsilon x_1^2 x_2 + x_1 x_3^2$ have a critical point tending to 0 as $\epsilon \to 0$, but the estimate $O(|\lambda|^{-3/4})$ still holds uniformly in ϵ for the oscillatory integrals with phase P_ϵ.*

VI Lagrangians with Folds and Cusps

The classification of degeneracies through multiplicities initiated in Section IV for translation invariant operators provides one approach to a theory of degenerate Fourier integral operators. However, the diverse phenomena encountered so far suggest that, as in singularity theory, there may be no best way of describing degeneracies. In fact, Lagrangians with Whitney folds were the first to be studied systematically [22], and we discuss now the stratification viewpoint.

A key example of a Lagrangian with Whitney folds is the normal bundle of the incidence relation (c) in Section II. There the projections π_X and π_Y have one-dimensional kernels along the shadow boundary, i.e., points where the direction ω is tangent to ∂K. Furthermore, the singular variety is transverse to the kernels of π_X and π_Y everywhere. The seminal work of Melrose and Taylor [22] established that such Lagrangians can be conjugated to a canonical one. The corresponding Fourier integral operator is the Airy operator of convolution with $e^{i\lambda x^3}$. This is the model operator of Section III, with $n = 3$ and $a_1 a_2 \neq 0$. It is smoothing of order $1/3$ on Sobolev spaces. Because the order of smoothing would have been $1/2$ for local graphs, we see that Fourier integral operators with Whitney folds lose comparatively $1/6$ derivatives.

Although the projections π_X and π_Y have the same kernel on Λ, their behavior may differ under a stratification. Lagrangians where one projection is a Whitney fold but the other is maximally degenerate have arisen in [11]–[14]. In this case, Greenleaf and Uhlmann have shown that the corresponding Fourier integral operators satisfy Sobolev estimates that lose $1/4$ derivative compared to local graphs. More recently, the assumption that one projection is maximally degenerate has been removed (as anticipated in [11]) first for Radon transforms along curves on surfaces [28], and now in full generality in Greenleaf and Seeger [10].

The methods of Section III provide some insight into these losses of $1/6$ and $1/4$ derivatives. Consider for example the Radon transform (4). For two-sided

Whitney folds, both $\partial_x^2\partial_y S$ and $\partial_x\partial_y^2 S$ do not vanish when $\partial_x\partial_y S = 0$, whereas for one-sided folds, only one of these third derivatives, say $\partial_x^2\partial_y S$, is assumed to be not 0. In both cases, the variety $\{\partial_x\partial_y S = 0\}$ is a smooth manifold, and the (x,y) plane decomposes into simple *strips* where $\partial_x\partial_y S$ does not change sign and has size $\sim 2^{-k}$. If $T_k(\lambda)$ is the corresponding decomposition of the symbol $T(\lambda)$, $\|T_k(\lambda)\| \leq C(2^{-k}|\lambda|)^{-1/2}$ in both cases. However, in the two-sided fold case, the fact that both third derivatives do not vanish implies that the maximum cross-sections I_x and I_y in both directions are of size $\sim 2^{-k}$. Consequently, we also have $\|T_k\| \leq C(I_x I_y)^{1/2} \leq 2^{-k}$, and

$$\|T\| \leq \sum_k \min((2^{-k}\lambda)^{-1/2}, 2^{-k}) = O(|\lambda|^{-1/3}).$$

On the other hand, for one-sided folds we can only assert that $I_x \sim 2^{-k}$ and $I_y \leq 1$, and arrive at the weaker estimate

$$\|T\| \leq \sum_k \min((2^{-k}\lambda)^{-1/2}, 2^{-k/2}) = O(|\lambda|^{-1/4}).$$

The following is a simple example of a further stratification that provides a generalization to families of the notion of nonvanishing torsion for a single curve:

THEOREM 8. *Let* $S(x,y) = (S_1(x,y), \ldots, S_{n-1}(x,y))$ *be an* $(n-1)$ *vector-valued phase function, and consider the family of curves in* $\mathbf{R}^n$ *defined by* $\gamma_{(x,t)} = \{(s,y) \in \mathbf{R}^{n-1} \times \mathbf{R}; s = t + S(x,y)\}$. *Let* $\mathcal{C} = \cup \gamma_{(t,x)}$ *be the corresponding incidence relation, and parametrize its normal bundle* $\Lambda = N^*(\mathcal{C})$ *by* $\Lambda = \{(t, \lambda, x, <\lambda, \partial_x S> ; s, -\lambda, y, <\lambda, \partial_y S>)\}$. *Assume that the two varieties* $\Lambda \cap \{<\lambda, \frac{\partial^m S}{\partial x \partial y^{m-1}}>= 0, 2 \leq m \leq n\}$ *and* $\Lambda \cap \{<\lambda, \frac{\partial^m S}{\partial x^{m-1}\partial y}>= 0, 2 \leq m \leq n\}$ *are empty. Then the Radon transform* R *with incidence relation* $\mathcal{C}$ *and a smooth density is smoothing of order* $\frac{1}{n}$ *(respectively* $\frac{1}{n} - \epsilon$*) on Sobolev spaces for* $S(x,y)$ *analytic (respectively smooth).*

Conditions of the type discussed in Theorem 2(a) are also amenable to an intrinsic formulation in terms of stratifications.

VII Composition of Operators

The regularity of Fourier integral operators whose Lagrangians are local graphs is just one consequence of a precise composition law. In the degenerate case, the wave front calculus [17] shows that a given degeneracy or singularity will in general keep generating new ones upon composition. Nevertheless, there are situations where composition can stabilize, or arrive at known operators. We list here a few examples, in the hope that further investigation can detect some deeper structure:

(a) Repeated convolutions with a measure supported on a lower-dimensional variety that spans are ultimately smoothing [34]. The number of convolutions required to achieve an absolutely continuous density is however not known. It is presumably related to the multiplicity introduced in Section IV, and possibly to the phase transitions discussed by Diaconis for random walks [6];

(b) If R is the Radon transform defined by a family $\mathcal{F}$ of curves, then R^*R is another Radon transform, with the incidence relation given by the cone $\mathcal{C}_x = \cup_{l \ni x} l$ at each point x [11]. The density on $\mathcal{C}_x$ is fractional near x, so that from the microlocal viewpoint, the wave front of R^*R is really the union of two Lagrangians, namely $\Lambda = N^*(\mathcal{C})$ and the diagonal. As shown in [11], Gelfand's cone condition implies that Λ is a flow-out, and the calculus of Antoniono-Guillemin-Melrose-Uhlmann [1], [15], [23] applies;

(c) Consider a Radon transform R of the form (4) in 2 dimensions, satisfying the double-sided Whitney fold conditions. Then

$$|R^*Rf| \leq C|L_{1/2}M_{1/2}f|$$

pointwise, with the operators L_s and M_s defined by

$$(M_s f)(x,t) = \int_{-\infty}^{\infty} f(x, t-v) \frac{dv}{|v|^s}$$

$$(L_s f)(x,t) = \int_{-\infty}^{\infty} f(x-u, t+\Phi(x,y)) \frac{du}{|u|^s}.$$

Here $\Phi(x,y) = S(x, z_c(x,y)) - S(y, z_c(x,y))$, with $z_c(x,y)$ the solution of the critical equation $S'_z(x,z) - S'_z(y,z) = 0$. The $L^p - L^q$ smoothing properties follow by interpolation between $\operatorname{Re} s = 0$ and $\operatorname{Re} s = 1$. Thus, R^*R is controlled in effect by L_s, which for $\operatorname{Re} s = 1$ is a singular Radon transform satisfying the non-infinite order of vanishing of Christ et al. [5] and hence is bounded on L^p for $1 < p < \infty$.

These examples suggest that a cycle of Fourier integral operators closed upon composition must also include singular Radon transforms.

References

[1]   J. Antoniono and G. Uhlmann, *A functional calculus for a class of pseudo-differential operators with singular symbols*, Proc. Symp. Pure Math. **43** (1985), 5–16.

[2]   V. I. Arnold, *Critical points of smooth functions and their normal forms*, Russian Math. Surveys **30** (1975), 1–75.

[3]   M. Christ, *Failure of an endpoint estimate for integrals along curves*, (1993) UCLA preprint.

[4]   M. Christ, A. Nagel, E. M. Stein, and S. Wainger, Singular and maximal Radon transforms, to appear in Ann. of Math. Stud., Princeton Univ. Press, Princeton, NJ.

[5]   M. Cowling, S. Disney, G. Mauceri, and D. Müller, *Damping oscillatory integrals*, Invent. Math. **101** (1990), 237–260.

[6]   P. Diaconis, Group representations in probability and statistics, Lecture Notes-Monograph Series, Inst. Math. Stat. **11** (1988).

[7]   C. Fefferman, *Monge-Ampère equations, the Bergman kernel, and the geometry of pseudo convex domains*, Ann. of Math. **103** (1976), 395–416.

[8]   I. M. Gelfand, M. I. Graev, and Z. Ya. Shapiro, *Integral geometry on k-planes*, Funct. Anal. Appl. **1** (1967), 14–27.

[9] D. Geller, and E. M. Stein, *Estimates for singular convolution operators on the Heisenberg group*, Math. Ann. **267** (1984), 1–15.

[10] A. Greenleaf, and A. Seeger, *Fourier integral operators with fold singularities*, (1993) preprint.

[11] A. Greenleaf, and G. Uhlmann, *Non-local inversion formulas for the X-ray transform*, Duke Math. J. **58** (1989), 205–240.

[12] A. Greenleaf, and G. Uhlmann, *Estimates for singular Radon transforms and pseudo-differential operators with singular symbols*, J. Funct. Anal. **89** (1990), 202–232.

[13] A. Greenleaf, and G. Uhlmann, *Composition of some singular Fourier integral operators and estimates for the X-ray transform*, I, Ann. Inst. Fourier (Grenoble) **40** (1990); II, 1991 preprint.

[14] V. Guillemin, Cosmology in $(2+1)$ dimensions, cyclic models, and deformations of $M_{2,1}$, Ann. of Math. Stud. **121** (1989), Princeton Univ. Press, Princeton, NJ.

[15] V. Guillemin and G. Uhlmann, *Oscillatory integrals with singular symbols*, Duke Math. J. **48** (1981), 251–257.

[16] L. Hörmander, *Oscillatory integrals and multipliers on FL^p*, Arkiv. Mat. **11** (1973), 1–11.

[17] L. Hörmander, The Analysis of Linear Partial Differential Operators I–IV, Springer-Verlag, Berlin and New York (1985).

[18] V. N. Karpushkin, *Uniform estimates of oscillatory integrals with parabolic or hyperbolic phases*, J. Soviet Math. **33** (1986), 1159–1188.

[19] V. N. Karpushkin, *A theorem concerning uniform estimates of oscillatory integrals when the phase is a function of two variables*, J. Soviet Math. **35** (1986), 2809–2826.

[20] C. Kenig, G. Ponce, and L. Vega, *Oscillatory integrals and regularity of dispersive equations*, Indiana Univ. Math. J. **40** (1991), 33–69.

[21] A. G. Kushnirenko, *Polyèdres de Newton et nombres de Milnor*, Invent. Math. **32** (1976), 1–31.

[22] R. Melrose, and M. Taylor, *Near peak scattering and the correct Kirchhoff approximation for a convex obstacle*, Adv. in Math. **55** (1985), 242–315.

[23] R. Melrose, and G. Uhlmann, *Lagrangian intersection and the Cauchy problem*, Comm. Pure Appl. Math. **32** (1979), 483–519.

[24] J. Milnor, Singular points of complex hypersurfaces, Ann. of Math. Stud. **61** (1968), Princeton Univ. Press, Princeton, NJ.

[25] D. Oberlin, *Convolution estimates for some measures on curves*, Proc. Amer. Math. Soc. **99** (1987), 56–60.

[26] Y. Pan, and C. Sogge, *Oscillatory integrals associated to folding canonical relations*, Colloq. Math. **60** (1990), 413–419.

[27] D. H. Phong, and E. M. Stein, *Hilbert integrals, singular integrals, and Radon transforms*, Acta Math. **157** (1986), 99–157; Invent. Math. **86** (1986), 75–113.

[28] D. H. Phong, and E. M. Stein, *Radon transforms and torsion*, Int. Math. Res. Notices **4** (1991), 49–60.

[29] D. H. Phong, and E. M. Stein, *Oscillatory integrals with polynomial phases*, Invent. Math. **110** (1992), 39–62.

[30] D. H. Phong, and E. M. Stein, *Models of degenerate Fourier integral operators and Radon transforms*, Ann. of Math. (2) **140** (1994), 703–722.

[31] D. H. Phong, and E. M. Stein, *On a stopping process for oscillatory integrals*, J. Geom. Anal. **4** (1994), 105–120.

[32] D. H. Phong, and E. M. Stein, *Operator versions of the van der Corput lemma and Fourier integral operators*, Math. Res. Lett. **1** (1994), 27–33.

[33] D. H. Phong, and E. M. Stein, in preparation.

[34] F. Ricci, and E. M. Stein, *Harmonic analysis on nilpotent groups and singular integrals, I. Oscillatory integrals*, J. Funct. Anal. **73** (1987), 179–194; *II. Singular kernels supported on submanifolds*, J. Funct. Anal. **78** (1988), 56–94.

[35] F. Ricci, and G. Travaglini, $L^p - L^q$ *estimates for orbital measures and Radon transform on Lie groups and Lie algebras*, (1994), Torino preprint.

[36] A. Seeger, *Degenerate Fourier integral operators in the plane*, Duke Math. J. **71** (1993), 685–745.

[37] A. Thompson, *Sobolev estimates for singular Radon transforms*, J. Funct. Anal. **112** (1993), 61–96.

[38] A. N. Varchenko, *Newton polyhedra and estimations of oscillatory integrals*, Funct. Anal. Appl. **10** (1976), 175–196.

Finding Structure in Sets with Little Smoothness

STEPHEN W. SEMMES[*]

Department of Mathematics, Rice University
Houston, Texas 77251, USA

There is an interesting tension between the topology of a space and the distribution of its "mass", and this tension is captured roughly by two principles.

First Principle: *Topological information about a space often implies lower bounds on the "size" of the space.*

Second Principle: *Suitable topological conditions on a space (of the type that force lower bounds on the mass, as in the First Principle) in combination with upper bounds on the mass often imply serious restrictions on the geometric complexity of the space.*

The simplest manifestation of the First Principle is the classical fact [HW] that a metric space with topological dimension d must have Hausdorff dimension $\geq d$, and even positive d-dimensional Hausdorff measure. Quantitative lower bounds on the Hausdorff measure can be obtained under quantified topological assumptions (e.g., [Ge], [GP], [Gr2], [V]). (To be honest, the quantitative "topological" conditions in this paper are partially geometric.)

The Second Principle is a complement to the first and is closer to the concerns of this paper. The restrictions predicted by the Second Principle must be spartan, because they are to be derived from simple conditions.

As an example, let K be a compact connected set in $\mathbf{R}^n$. Then $H^1(K) \geq$ diam K, as in the First Principle, where H^s denotes s-dimensional Hausdorff measure (not cohomology!). If also $H^1(K) < \infty$, then K is contained in a rectifiable curve of length $\leq 2H^1(K)$. Thus, K can be parameterized nicely, with bounds, in keeping with the Second Principle. Of course 1-dimensional sets are very special.

Now suppose that $A \subseteq \mathbf{R}^n$ is compact, $A \subseteq B(0,2) \backslash B(0,1)$ (where $B(x,r)$ denotes the ball with center x and radius r), and that A separates 0 from ∞. This implies (as in the First Principle) that $H^{n-1}(A) \geq H^{n-1}(\mathbf{S}^{n-1})$, where $\mathbf{S}^{n-1}$ denotes the unit sphere. Assume also that $H^{n-1}(A) \leq C_0$ for some constant C_0. The Second Principle predicts that A must have some kind of good structure. What does this mean? Is it really correct?

Recall that a compact set can have positive and finite Hausdorff measure without being at all like a nice surface. It could be fractal, like a Cantor set, or a snowflake curve, or a tree-like object. We want to say that this cannot happen

---

*)   Supported by the NSF. Many thanks to those who criticized with alacrity.

Proceedings of the International Congress
of Mathematicians, Zürich, Switzerland 1994
© Birkhäuser Verlag, Basel, Switzerland 1995

when A has the properties above, but the best that we can expect is that A has a substantial part that is well behaved, as A could be the union of a sphere and a fractal. Topological dimension theory provides some information of this type about separating sets like A — e.g., A has a compact subset K that separates 0 from ∞ and is an $(n-1)$-dimensional "Cantor manifold" (so that K cannot be disconnected by a subset of topological dimension $n-3$; see Theorem VI 11 on p. 98 of [HW], and use Brouwer reduction on p. 161 too) — but it does not take advantage of the upper bound on mass very well. Even if A is simple topologically and has no extraneous fractal component it can still have many corners, as is the case for graphs in polar coordinates of generic continuous functions $\alpha : \mathbf{S}^{n-1} \to (1, 2)$ whose distributional gradients are measures with bounded mass. Such functions have some nice structure, though, and are always Lipschitz on some large sets. In general A might enjoy an analogous property.

CONJECTURE 1.   *There exist $\theta > 0$ and $M > 1$, depending only on n and C_0, such that the following is true. If A is as above, then there is a closed subset A_0 of A such that $H^{n-1}(A_0) \geq \theta$ (so that A_0 is a substantial part of A) and A_0 is M-bilipschitz equivalent to a subset of $\mathbf{R}^{n-1}$, which means that there is a mapping $\phi : A_0 \to \mathbf{R}^{n-1}$ such that $M^{-1}|x-y| \leq |\phi(x) - \phi(y)| \leq M|x-y|$ for all $x, y \in A_0$. (That is, ϕ distorts distances only by a bounded factor.)*

When $n = 2$ the conjecture is true. This uses elementary topological arguments and a result of Guy David (see Proposition 8 and Definition 6 on pp. 167–168 of [D1]). When $n > 2$ the conjecture is unsolved. It is known that A must have a Borel subset A_1 with $H^{n-1}(A_1) \geq H^{n-1}(\mathbf{S}^{n-1})$ that is covered by a countable union of C^1 submanifolds; this can be derived from Federer's famous structure theorem (Theorem 3.3.13 on p. 297 of [Fe]) and other facts from geometric measure theory. However, this result says only that A_1 is asymptotically well behaved at its points, without giving any quantitative or large-scale control on the structure of A_1. The point of Conjecture 1 is to make up for these deficiencies.

There is a more restrictive version of Conjecture 1 that has been solved in all dimensions, but before getting to that let us consider another example.

Fix an integer $0 < d < n$, and let D denote the closed unit ball in $\mathbf{R}^d$. Let g denote a topological embedding of D into $\mathbf{R}^n$. For the sake of definiteness assume that $10^{-10n}|x-y|^{10} \leq |g(x) - g(y)| \leq 10^{10n}|x-y|^{\frac{1}{10}}$ for all $x, y \in D$. This gives a bound on the modulus of continuity of g and its inverse, but the specific choice is not important. One can show (as in the First Principle) that there is a universal lower bound on $H^d(g(D))$. Let us assume also that $H^d(g(D)) \leq C_1$ for some constant C_1. This condition prevents $g(D)$ from being a snowflake, and the Second Principle predicts that $g(D)$ must have some good structure. As before, there are examples (like graphs) that show that $g(D)$ can have many corners.

CONJECTURE 2.   *There exist $\eta > 0$ and $N > 0$, depending only on n and C_1, such that if g is as above, then there is a closed subset S of $g(D)$ with $H^d(S) \geq \eta$ such that S is N-bilipschitz equivalent to a subset of $\mathbf{R}^d$.*

Note that g itself might well have no nontrivial bilipschitz piece.

When $d = 1$ this conjecture is true. In this case $g(D)$ is a curve, which we can parameterize by arclength, and one can use the same result from [D1] as before. When $d > 1$ this conjecture is unsolved, although one can again use Federer's structure theorem to show that there is a Borel set $S_0 \subseteq g(D)$ with $H^d(S_0)$ bounded from below and S_0 covered by a sequence of C^1 submanifolds.

These conjectures have known "uniform" versions, in which one replaces the assumptions and conclusions by conditions that are supposed to hold at all scales and locations, with uniform bounds. Here are the precise definitions.

A subset E of $\mathbf{R}^n$ is (Ahlfors) *regular* with dimension d, $0 < d < n$, if it is closed and if there is a constant $C > 1$ such that $C^{-1}r^d \leq H^d(E \cap B(x, r)) \leq Cr^d$ whenever $x \in E$ and $0 < r <$ diam E. The upper mass bound is a uniform version of the sort of mass bound that we would normally want to assume in applying the Second Principle, whereas the lower bound is typically derivable from topological assumptions as in the First Principle. (For simplicity we shall ignore the latter point in the following.) A simple example of a regular set is a d-plane, but there are many fractal sets that are regular. Regular sets can even have noninteger dimension, but we shall deal only with integer dimensions here.

A regular set $E \subset \mathbf{R}^n$ of dimension $d = n - 1$ satisfies *Condition B* (the uniform version of the hypothesis of Conjecture 1) if there is a constant C such that for each $x \in E$ and $0 < r <$ diam E we can find $y_1, y_2 \in B(x, r)$ such that $\mathrm{dist}(y_i, E) \geq C^{-1}r$, $i = 1, 2$, and y_1, y_2 lie in different components of $\mathbf{R}^n \backslash E$. In other words, instead of assuming that E separates 0 from ∞, we require that E separate pairs of points at all scales and locations.

For the uniform version of Conjecture 2, it is better to work with a set E directly rather than embeddings. Here is a suitable condition.

$(*)$ *There is a constant $K > 1$ so that for each $x \in E$ and $r \in (0, K^{-1}$ diam $E)$ there is a (relatively) open subset U of E that is homeomorphic to a d-ball and satisfies $E \cap B(x, r) \subseteq U \subseteq E \cap B(x, Kr)$.*

Actually, this condition is weaker than the hypothesis of Conjecture 1, in that instead of requiring bounds on the moduli of continuity of the relevant maps we make milder assumptions about the way that the topological balls are situated.

Conditions B and $(*)$ are satisfied by planes and compact C^1 manifolds (orientable and of dimension $d = n - 1$ in the case of Condition B), and also by graphs of Lipschitz functions. These conditions capture some features of smoothness without actually requiring smoothness. They prevent cusps and long thin tubes, for instance, but they do permit corners and spirals. Without the assumption of regularity both conditions allow fractal behavior, like snowflakes.

The uniform version of the conclusions of Conjectures 1 and 2 is called uniform rectifiability. A set $E \subset \mathbf{R}^n$ is *uniformly rectifiable* if it is regular with (integer) dimension d and if balls centered on E contain substantial pieces of E that are uniformly bilipschitz equivalent to subsets of $\mathbf{R}^d$, i.e., if there exist constants $\mu > 0$ and $L > 1$ such that for each $x \in E$ and $0 < r <$ diam E there is a closed subset A of $E \cap B(x, r)$ such that $H^d(A) \geq \mu H^d(E \cap B(x, r))$ and A is L-bilipschitz equivalent to a subset of $\mathbf{R}^d$. This precise definition should not be taken too seriously, because it is equivalent to many other conditions. Basically, it means that we can parameterize E, with uniform estimates, if we allow some holes and crossings. It

is equivalent to the existence of a global parameterization with suitable estimates
(but with crossings and spillovers). Practically all competing definitions of uniform
rectifiability are equivalent to the one above [DS1]. Although uniform rectifiability
allows bad behavior, like holes, crossings, and corners, the extent of such behavior
is controlled and reasonably well understood. (See Theorem 5 below.)

THEOREM 3. *Suppose that E is a d-dimensional regular set in $\mathbf{R}^n$. (a) ([D2]) If
$d = n - 1$ and E satisfies Condition B, then E is uniformly rectifiable. (b) ([DS6])
If E satisfies (∗), then E is uniformly rectifiable.*

Theorem 3 provides criteria under which a set can be parameterized with
good estimates. Note the trade-offs in Part (b); we start with homeomorphic parameterizations with no bounds on the way they distort distances, and we get
nonhomeomorphic parameterizations with very good bounds. To put this into perspective, let us recall the striking result of Edwards that there is a finite polyhedron
P homeomorphic to the 5-sphere $\mathbf{S}^5$ such that any homeomorphism $h : P \to \mathbf{S}^5$
maps a certain polygonal circle $C \subseteq P$ to a set with Hausdorff dimension at least 3.
(P is the "join" of a homology sphere and the circle C, and C has the property that
$P\backslash C$ is not simply connected, not even locally near C.) Thus, although P is homeomorphic to $\mathbf{S}^5$, it is not bilipschitz equivalent to $\mathbf{S}^5$ (let alone piecewise-linearly
homeomorphic to $\mathbf{S}^5$), and we cannot even find local bilipschitz coordinates for P
around points in C. See [C2], [Da], [E], [SS]. This implies that it is very difficult
to find nice geometric conditions on a set that ensure the existence of a bilipschitz
parameterization, even locally. (Note that P satisfies (∗), and even stronger versions of it.) Additional examples ([Se7]) further support this conclusion for sets
with dimension ≥ 3, and for quasisymmetric parameterizations too. (The $d = 1$
case is completely different, for simple reasons. The $d = 2$ case is not clear, but
there are positive results for the existence of quasisymmetric parameterizations
when $d = 2$ based on the uniformization theorem, as in Section 5 of [Se5].)

In short, naive expectations about finding homeomorphic parameterizations
with good estimates are wrong, and Theorem 3 offers a compromise.

Uniform rectifiability also has nice properties in terms of analysis.

THEOREM 4. *([DS1]) A d-dimensional regular set E in $\mathbf{R}^n$ is uniformly rectifiable
if and only if the singular integral operator $f(x) \mapsto \int_E k(x - y)f(y)\, dH^d(y)$ is
bounded on $L^2(E)$ whenever $k(x) \in C^\infty(\mathbf{R}^n\backslash\{0\})$ is odd and $|x|^{d+\ell}|\nabla^\ell k(x)|$ is
bounded for all $\ell \geq 0$.*

The "only if" part of Theorem 4 was known before [DS1], using work of
Calderón, Coifman, MacIntosh, Meyer, and David. (See [D1], [D3].) For d-planes
and smooth submanifolds the boundedness of these operators is classical, but the
results for nonsmooth sets are much deeper.

Condition B originally arose in [Se1] (in a stronger form) as a geometric
criterion for the boundedness of singular integral operators on the set E. The
proof in [Se1] was based on analytic methods, and at the time Guy David and
I did not know what could be said about the geometry of E, even though we
believed that it should be pretty good. David then proved Part (a) of Theorem 3,

and I later found a way to use analysis to derive geometric information from (a stronger version of) Condition B in [Se3]. We proved Theorem 4 after that.

It is not known whether a d-dimensional regular set E must be uniformly rectifiable if merely $f(x) \mapsto \int \frac{x-y}{|x-y|^{d+1}} f(y) \, dH^d(y)$ defines a bounded operator on $L^2(E)$. See [DS3], [Ma] for more information. In [Se4], [Se6] there is an operator-theoretic characterization of "almost flatness" for hypersurfaces, but the best parameterizations are not known in this case [Se5]. The role of Sobolev-Poincaré inequalities in this subject is also not clear, but see [Se2] and Section 6 of [DS2].

The method of the proof of Part (a) of Theorem 3 in [D2] is complicated but very general. See [DJ], [DS2] for simpler arguments. In [D2] there is also a higher-codimension version of Part (a) and a criterion for finding large bilipschitz pieces of a Lipschitz map. (See also [J2], the techniques of which are used in [DS2].) The method of [D2] is used again in the proof of Part (b) of Theorem 3 and in [DS3].

These are some of the reasons Guy David and I have been studying uniform rectifiability in recent years, and this paper is a reflection of our joint efforts. A broad goal is to have better technology for dealing with geometric objects with little smoothness, and for managing infinite geometric complexity. These are well-established themes in complex analysis, potential theory, geometric measure theory, and quasiconformal mappings, but they arise in other ways too. For instance, the structure of singular sets of solutions of variational problems is not well understood in general. One often expects singular sets to be smooth almost everywhere, but this is not always clear, and there ought to be better methods for establishing some limited regularity (as in [DS4], [DS5]). The study of compactness properties of families of Riemannian manifolds (à la [Gr1], [Pe]) leads naturally to spaces with little smoothness. There are some very interesting topological finiteness theorems ([Fr2], [GPW], [Pe]), but there ought to be better methods for controlling more than the topology with less than the curvature (as in Theorem 3). Geometric topology generates many questions about quantitative geometric estimates. See [B] for a classical example. Nonsmoothable toplogical 4-manifolds [FQ] are especially intriguing. One of the conclusions of [DoS] is that to do calculus on a general topological 4-manifold one must deal with even "worse" than quasiconformal structures. In other dimensions one can always assume at least a Lipschitz structure [Su], but even then, calculus on topological manifolds is not exactly simple [CST].

Note that the properties of sets discussed here make sense locally, i.e., global effects are not required. The distinction between local and global is a little blurry in the context of scale-invariant estimates. An obstruction for doing something globally often implies an obstruction to doing it locally with a good estimate.

I shall now discuss some more technical aspects of this subject.

Comments about (∗). Sets that are bilipschitz equivalent to $\mathbf{R}^d$ or a compact C^1 submanifold always satisfy (∗), as do finite polyhedra that are topological manifolds (like Edwards' example). Keep in mind that sets that are bilipschitz equivalent to $\mathbf{R}^d$ can still be quite complicated, e.g., they can have infinitely many corners or spirals, even locally, and they can be topologically wild.

Controlled topological conditions like (∗) have appeared before, although "uniform contractibility" conditions are more common. The latter entail requirements like the contractibility of $E \cap B(x, r)$ inside $E \cap B(x, Kr)$. See [Fr1], [Fr2],

[GP], [Pe]. These two types of conditions are closer than they might appear, by a theorem in [Fr1] to the effect that uniform contractibility conditions on a topological manifold E imply conditions like $(*)$ for $E \times \mathbf{R}$. For proving Theorem 3 above one can also work directly with (various) conditions weaker than $(*)$.

Examples of uniformly rectifiable sets. d-planes, compact C^1 manifolds, and finite polyhedra of pure dimension d are all uniformly rectifiable. Some infinite polyhedra are allowed, some are not. The standard construction of a Fox-Artin wild sphere is uniformly rectifiable. If $\{S_j\}$ is a sequence of spheres with radius 2^{-j}, $j \in \mathbf{Z}_+$, and if S_j intersects S_{j+1} for all j, then the closure of $\cup S_j$ is uniformly rectifiable. Bing's "dogbone space" (and other interesting quotients of $\mathbf{R}^3$, as in Section 9 of [Da]) can be realized in a natural way as a uniformly rectifiable set. Uniformly rectifiable sets can have infinitely many spirals, holes, or infinite towers of handles, but not without restriction. (See the discussion of "Carleson sets" below.)

Uniform rectifiability has many stability and invariance properties. An amusing one is that E is uniformly rectifiable if $E \times \mathbf{R}$ is. This type of statement is normally dangerous in the context of bilipschitz equivalence.

The classical theory of rectifiable sets. Fix $d, n \in \mathbf{Z}_+$, $d < n$. A set $E \subseteq \mathbf{R}^n$ is *rectifiable* if there is an $N \subseteq E$ with $H^d(N) = 0$ and a sequence of sets $\{E_j\}$ such that $E \backslash N \subseteq \bigcup E_j$ and each E_j is bilipschitz equivalent to a subset of $\mathbf{R}^d$.

The notion of rectifiability is incredibly stable, in the sense that many other conditions give an equivalent definition. For instance, sequences $\{E_j\}$ of C^1 submanifolds or Lipschitz images of $\mathbf{R}^d$ yield equivalent conditions.

A set A is called *totally unrectifiable* if $H^d(A \cap E) = 0$ for all rectifiable sets E. There are plenty of totally unrectifiable fractal sets A with $0 < H^d(A) < \infty$. The dichotomy between rectifiability and unrectifiability is very clean: given any set A with $H^d(A) < \infty$, there is a rectifiable set E such that $A \backslash E$ is totally unrectifiable.

Rectifiability can be viewed as a property of the asymptotic behavior of a set at almost all of its points. For instance, a set E is rectifiable if and only if there is an approximate tangent d-plane at H^d-almost all points of E. (Roughly speaking, a d-plane P is approximately tangent to E at x if the set of points in E near x but not asymptotically close to P has d-dimensional density zero at x. This definition allows E to contain all points with rational coordinates, even with little surfaces attached, without preventing the existence of approximate tangent planes.) See [Ma] (especially Chapters 16 and 17) and [Pr] for more subtle characterizations of rectifiability in terms of asymptotic behavior at most points.

There is a lot of technology for checking that a set is rectifiable, or that a set has a nontrivial rectifiable part. One of the main tools is the structure theorem of Federer (Theorem 3.3.13 in [Fe]), to the effect that if A is totally unrectifiable, then its projection onto almost every d-plane has measure zero. This is useful for showing that a set must have a large rectifiable piece under suitable topological assumptions (e.g., as in Conjectures 1 and 2 above), as otherwise it would be too scattered. For instance, one can show that a compact set K with topological dimension d and $H^d(K) < \infty$ must have a rectifiable subset E with $H^d(E) > 0$.

See [Fa], [Fe], [Ma], [Pr] for more information (but different terminology).

The theory of rectifiable sets is clean and powerful, but it provides only local asymptotic information, with no control on the behavior of a set at definite scales.

Carleson sets. How smooth are uniformly rectifiable sets, and how can we measure this? They have tangent planes almost everywhere, but we want quantitative information that is meaningful at definite scales. Let E be a d-dimensional regular set in $\mathbf{R}^n$, and let $\epsilon > 0$ be given. We would like to measure the extent to which E looks like a d-plane inside "most" balls B centered on E. Define $\mathcal{G}(\epsilon)$ to be the set of (good) pairs $(x, r) \in E \times \mathbf{R}_+$ for which there is a d-plane $P = P_{x,r}$ such that E looks like P inside $B(x, r)$, i.e., $\mathrm{dist}(y, P) \leq \epsilon r$ for all y in $E \cap B(x, r)$ and $\mathrm{dist}(z, E) \leq \epsilon r$ for all z in $P \cap B(x, r)$. Thus, E is nice if $E \times \mathbf{R}_+ \backslash \mathcal{G}(\epsilon)$ is small.

THEOREM 5. ([DS3]) *A d-dimensional regular set E is uniformly rectifiable if and only if $E \times \mathbf{R}_+ \backslash \mathcal{G}(\epsilon)$ is a Carleson set for every $\epsilon > 0$.*

Carleson sets are relatively sparse subsets of $E \times \mathbf{R}_+$. The precise definition is as follows. Let μ be the restriction of H^d to E, and let λ denote the measure on $E \times \mathbf{R}_+$ obtained by taking the product of μ with the measure $\nu = \frac{dr}{r}$ on $\mathbf{R}_+$. A set $\mathcal{A} \subseteq E \times \mathbf{R}_+$ is a *Carleson set* if there is a constant $C > 0$ such that $\lambda(\mathcal{A} \cap (B(x, r) \times (0, r))) \leq Cr^d$ for all $x \in E$ and $r > 0$.

To understand this condition better set $S(x) = \{t \in (0, 1) : (x, t) \in \mathcal{A}\}$ and $a(x) = \nu(S(x))$, and note that $\nu((0, 1)) = \infty$. If $\mathcal{A}$ is a Carleson set, then $a(x) < \infty$ a.e., and in fact $a(x) \in L^1_{\mathrm{loc}}$. In many cases $a(x) < \infty$ a.e. is about the same as $\inf\{t \in S(x)\} > 0$ a.e., and both correspond to nonquantitative properties like the existence almost everywhere of tangent planes (or differentiability a.e.).

If F is a set of integers, and $\mathcal{A}$ is the set of (x, r) such that $2^j < r < 2^{j+1}$ for some $j \in F$, then $\mathcal{A}$ is a Carleson set if and only if F is finite. If Q is a $(d-1)$-plane, then $\{(x, r) \in E \times \mathbf{R}_+ : \mathrm{dist}(x, Q) < r\}$ is a Carleson set.

In practice we think of $\mathcal{A} \subseteq E \times \mathbf{R}_+$ as the set of balls $B(x, r)$ on which something bad happens, and the Carleson condition imposes scale-invariant bounds on the size of this set. For instance, uniformly rectifiable sets can have infinitely many holes, spirals, or handles, but the extent of such bad behavior is controlled by a Carleson set (via Theorem 5), and simple constructions show that this is sharp.

Similarly, Carleson sets arise in quantitative versions of the fact that Lipschitz functions are differentiable almost everywhere. Let $f : \mathbf{R}^n \to \mathbf{R}$ be a Lipschitz function, and let $\epsilon > 0$ be given. Let $\mathcal{H}(\epsilon)$ be the (good) set of $(x, r) \in \mathbf{R}^n \times \mathbf{R}_+$ such that f can be approximated to within ϵr on $B(x, r)$ by an affine function. For almost all x we have that $(x, r) \in \mathcal{H}(\epsilon)$ for all sufficiently small $r > 0$, as f is differentiable a.e. In fact it is true that $E \times \mathbf{R}_+ \backslash \mathcal{H}(\epsilon)$ is a Carleson set, and there are examples showing that one cannot do much better than this. (However, there is no reasonable version of the "if" part of Theorem 5 for functions.)

See [Ga] for more information about the use of Carleson sets (and their more famous cousins, interpolating sequences and Carleson measures) in analysis, and see [DS4], [DS5] for different applications (to the Mumford-Shah problem).

The "if" part of Theorem 5 is the more substantial of the two. The "only if" part is easier and less surprising (and older — see [DS1]), but it does contain useful information. Theorem 5 has antecedents in traditional geometric measure theory (in terms of "weak tangents"; see [Ma], especially Chapter 16, and p. 40–42 of [DS3]).

Is Theorem 5 the definitive smoothness result for uniformly rectifiable sets? It depends on how the question is interpreted. If we try to measure smoothness in terms of the size of $E \times \mathbf{R}_+ \backslash G(\epsilon)$, then Theorem 5 is exactly the right result. However, uniformly rectifiable sets satisfy better smoothness conditions than the one given in Theorem 5 (which is called the BWGL in [DS3]). There are two basic types of improvements: one can get stronger (quadratic) estimates on the extent to which E is approximated by d-planes, as in Jones' "geometric lemma" [J1], [J3] and its higher-dimensional versions [DS1], or one can control the extent to which the approximating d-planes $P_{x,r}$ spin around, using a geometric version of Carleson's corona construction called the corona decomposition. This last is extremely useful, and I prefer it to good-λ inequalities for analyzing singular integral operators. See [DS1], [DS3], [Se3], and see [Ga] for its analytic antecedent.

Theorem 5 has many variants based on approximating E by other models besides d-planes (e.g., connected sets, convex sets, symmetric sets, unions of d-planes, sets with convex complementary components, Lipschitz graphs, etc., with restrictions on dimensions in some cases), some of which arise in connection with analysis. See [DS3].

This kind of "geometric Littlewood-Paley theory" began in [J1], as far as I can tell. One must give up the linearity of the classical theory, but there is an extra richness in the variety of ways to measure "oscillations" in geometry.

Some deficiencies of current knowledge. The analysis of mass distribution, particularly in connection with topology, is not very well understood. It would be very useful to have good criteria for quantitative rectifiability in terms of projections onto d-planes, just as there are for ordinary rectifiability. Conjectures 1 and 2 would follow from such criteria.

Geometric topology has a lot to say about finding homeomorphic parameterizations of spaces with little smoothness (see, e.g., [C1], [Da], [E]), but not much is known about finding homeomorphisms with good estimates on the extent to which they distort distances. Edwards' strange polyhedral spheres and the examples in [Se7] show that one cannot hope for too much when looking for conditions on a set in $\mathbf{R}^n$ that might ensure the existence of a bilipschitz or quasisymmetric parameterization (or even one whose inverse is Hölder continuous with a not-too-small exponent), at least when the set has dimension $d > 2$. ($d = 2$ is special — see [Se5] — and not fully understood.) The proper reconciliation between topology and estimates is not clear. Carleson sets and corona decompositions should be relevant (e.g., for producing homeomorphisms with Sobolev space estimates under suitable conditions), but they need to be better adapted to topology.

See [B] and the remarks at the end of Section 2 of [FS] for an interesting concrete and classical problem about estimates in geometric topology.

There are many unsolved problems concerning uniform rectifiability and analysis of functions and operators associated to a given set, e.g., in connection with holomorphic or harmonic functions on the complement of the set. In codimensions different from 1 it is often not clear how to formulate nice questions.

Uniform rectifiability is not well understood in terms of intrinsic properties of the set (in which the ambient space $\mathbf{R}^n$ is not used in a serious way, if at all). For instance, one can try to characterize uniform rectifiability in terms of rigidity properties of Lipschitz functions, as in the GWALA (Generalized Weak Approximation of Lipschitz functions by Affine functions) condition in [DS3]. One must be careful, because the Heisenberg group with the usual Carnot metric is very different from Euclidean geometry, even though it has many similar features in terms of analysis in general and Lipschitz functions in particular.

Let me end with a philosophical comment. Many older problems in topology are very appealing geometrically, e.g, how is the structure of a set in $\mathbf{R}^n$ related to the topology of its complement, or to the mappings defined on it (into spheres or Euclidean spaces, say), what does it mean for a space to have topological dimension d, what do such spaces look like, etc. The old-fashioned book [HW] is wonderfully geometric. Traditional topology has dealt effectively with these issues in some ways, but not in terms of geometric estimates, e.g., estimates on the distortion of mappings, on the size of spaces, on geometric complexity, and so forth.

Added in proof: Conjectures 1 and 2 have turned out to be more accessible than I thought or intended. Conjecture 1 has been solved by P. Jones, N. Katz, and A. Vargas, and also by G. David and myself. The two approaches are quite different and both give more information than the minimum requested in the conjecture. G. David and I believe that we can also solve Conjecture 2, but it is more complicated.

There has also been some very surprising progress recently by P. Mattila, M. Melnikov, and J. Verdera [MMV] on the relationship between rectifiability properties of a one-dimensional set and the behavior of the Cauchy integral operator and analytic capacity of the set.

References

[B]   R.H. Bing, *Shrinking without lengthening*, Topology **27** (1988), 487–493.

[C1]   J. Cannon, *The characterization of topological manifolds of dimension n ≥ 5*, in: Proceedings I.C.M. (Helsinki, 1978), 449–454.

[C2]   _____ , *Shrinking cell-like decompositions of manifolds: Codimension 3*, Ann. of Math. (2) **110** (1979), 83–112.

[CST]   A. Connes, D. Sullivan, and N. Teleman, *Quasiconformal mappings, operators on Hilbert space, and local formulae for characteristic classes*, Topology **33** (1994), 663–681.

[Da]   R. Daverman, Decompositions of Manifolds, Academic Press, 1986.

[D1]   G. David, *Opérateurs intégraux singuliers sur certaines courbes du plan complexe*, Ann. Sci. École Norm. Sup. **17** (1984), 157–189.

[D2]   _____ , *Morceaux de graphes lipschitziens et intégrales singulières sur un surface*, Rev. Mat. Iberoamericana **4** (1988), 73–114.

[D3]   _____ , *Wavelets and singular integrals on curves and surfaces*, Lecture Notes in Math. **1465** (1991), Springer-Verlag, Berlin and New York.

[DJ]   G. David and D. Jerison, *Lipschitz approximations to hypersurfaces, harmonic measure, and singular integrals*, Indiana Univ. Math. J. **39** (1990), 831–845.

[DS1]   G. David and S. Semmes, Singular integrals and rectifiable sets in $\mathbf{R}^n$: au-delà des graphes lipschitziens, Astérisque **193** (1991).

[DS2]   _____ , *Quantitative rectifiability and Lipschitz mappings*, Trans. Amer. Math. Soc. **337** (1993), 855–889.

[DS3]   _____ , Analysis of and on uniformly rectifiable sets, Math. Surveys Mono-
         graphs **38** (1993), Amer. Math. Soc., Providence, RI.

[DS4]   _____ , *On the singular sets of minimizers of the Mumford-Shah functional*, to
         appear, J. Math. Pures Appl. (9).

[DS5]   _____ , *Uniform rectifiability and singular sets*, to appear in Ann. I. H. P. Analyse
         Non-Linéaire.

[DS6]   _____ , in preparation.

[DoS]   S. Donaldson and D. Sullivan, *Quasiconformal 4-manifolds*, Acta Math. **163**
         (1989), 181–252.

[E]   R. Edwards, *The topology of manifolds and cell-like maps*, Proceedings I.C.M.
         (Helsinki, 1978), 111–127.

[Fa]   K. Falconer, The Geometry of Fractal Sets, Cambridge Univ. Press, London
         and New York, 1984.

[Fe]   H. Federer, Geometric Measure Theory, Springer-Verlag, Berlin and New York,
         1969.

[Fr1]   S. Ferry, *Counting simple-homotopy types of topological manifolds*, preprint.

[Fr2]   _____ , *Topological finiteness theorems for manifolds in Gromov-Hausdorff
         space*, Duke Math. J. **74** (1994), 95-106.

[FQ]   M. Freedman and F. Quinn, Topology of 4-Manifolds, Princeton Univ. Press,
         Princeton, NJ, 1990.

[FS]   M. Freedman and R. Skora, *Strange actions of groups on spheres*, J. Differential
         Geom. **25** (1987), 75–98.

[Ga]   J. Garnett, Bounded Analytic Functions, Academic Press, New York, 1981.

[Ge]   F. Gehring, *The Hausdorff measure of sets which link in Euclidean space*, in:
         Contributions to Analysis: A Collection of Papers Dedicated to Lipman Bers,
         Academic Press, New York and London, 1974, 159–167.

[GP]   R. Greene and P. Petersen V, *Little topology, big volume*, Duke. Math. J. **67**
         (1992), 273–290.

[GPW]   K. Grove, P. Petersen V, and J.-Y. Wu, *Geometric finiteness theorems via con-
         trolled topology*, Invent. Math. **99** (1990), 205–213; erratum, **104** (1991), 221–222.

[Gr1]   M. Gromov, Structures Métriques pour les Variétés Riemanniennes, (J. La-
         fontaine and P. Pansu, eds.), Cedic/Fernand Nathan, Paris, 1981.

[Gr2]   _____ , *Large Riemannian manifolds, Curvature and Topology of Riemannian
         Manifolds*, Lecture Notes in Math. **1201** (1986), Springer-Verlag, Berlin and
         New York, 108–121.

[HW]   W. Hurewicz and H. Wallman, Dimension Theory, Princeton Univ. Press,
         Princeton, NJ, 1941, 1969.

[J1]   P. Jones, *Square functions, Cauchy integrals, analytic capacity, and harmonic
         measure*, Lecture Notes in Math. **1384** (1989), 24–68, Springer-Verlag, Berlin
         and New York.

[J2]   _____ , *Lipschitz and bi-Lipschitz functions*, Rev. Mat. Iberoamericana **4** (1988),
         115–122.

[J3]   _____ , *Rectifiable sets and the travelling salesman problem*, Invent. Math. **102**
         (1990), 1–15.

[Ma]   P. Mattila, Geometry of sets and measures in Euclidean spaces, Cambridge
         Stud. Adv. Math. **44**, Cambridge Univ. Press., London and New York, 1995.

[MMV]   P. Mattila, M. Melnikov, and J. Verdera, *The Cauchy Integral, Analytic Capacity
         and Uniform Rectifiability*, to appear, Ann. of Math.

[Pe]   P. Petersen V, *Gromov-Hausdorff convergence of metric spaces*, Proc. Sympos.
         Pure Math. **54**, Part 3 (1993), 489–504, Amer. Math. Soc., Providence, RI.

[Pr]   D. Preiss, *Geometry of measures in* $\mathbf{R}^n$: *Distribution, rectifiability, and densities*, Ann.of Math. (2) **125**(1987), 537–643.

[Se1]   S. Semmes, *A criterion for the boundedness of singular integrals on hypersurfaces*, Trans. Amer. Math. Soc. **311** (1989), 501–513.

[Se2]   _____ , *Differentiable function theory on hypersurfaces in* $\mathbf{R}^n$ *(without bounds on their smoothness)*, Indiana Univ. Math. J. **39** (1990), 985–1004.

[Se3]   _____ , *Analysis vs. geometry on a class of rectifiable hypersurfaces in* $\mathbf{R}^n$, Indiana Univ. Math. J. **39** (1990), 1005–1035.

[Se4]   _____ , *Chord-arc surfaces with small constant I*, Adv. in Math. **85** (1991), 198–223.

[Se5]   _____ , *Chord-arc surfaces with small constant II: Good parameterizations*, Adv. in Math. **88** (1991), 170–199.

[Se6]   _____ , *Hypersurfaces in* $\mathbf{R}^n$ *whose unit normal has small BMO norm*, Proc. Amer. Math. Soc. **112** (1991), 403–412.

[Se7]   _____ , *On the nonexistence of bilipschitz parameterizations and geometric problems about* A_∞ *weights*, and *Good metric spaces without good parameterizations*, to appear, La Revista Mat. Iberoamericana.

[SS]   L. Siebenmann and D. Sullivan, *On complexes that are Lipschitz manifolds*, Geometric Topology (J. Cantrell, ed.), Academic Press, New York, 1979, 503–525.

[Su]   D. Sullivan, *Hyperbolic geometry and homeomorphisms*, Geometric Topology, (J. Cantrell, ed.), Academic Press, New York, 1979, 543–555.

[V]   J. Väisälä, *Quasisymmetric embeddings in Euclidean spaces*, Trans. Amer. Math. Soc. **264** (1981), 191–204.

Topological, Geometric and Complex Analytic Properties of Julia Sets

MITSUHIRO SHISHIKURA

The University of Tokyo
Department of Mathematical Sciences
Komaba, Meguro, Tokyo 153, Japan

In this paper, we discuss several aspects of Julia sets, as well as those of the Mandelbrot set. We are interested in topological properties such as connectivity and local connectivity, geometric properties such as Hausdorff dimension and Lebesgue measure, and complex analytic properties such as holomorphic removability.

As one can easily see from the pictures of numerical experiments, there is a huge variety of "shapes" of Julia sets even for polynomials of a simple form $z^2 + c$. And as the parameter c varies, the Julia set can drastically change its shape. In the Mandelbrot set, the blow-ups of different places in the Mandelbrot set can look totally different. Some parts look like the entire Mandelbrot set, and other parts sometimes look like the shape of the corresponding Julia sets. These sets often look very complicated but one can see rich structures inside them. They provide typical examples of "fractals".

Since Douady and Hubbard [DH] started their work on quadratic polynomials, there have been many developments with many new techniques in this field. In this paper, we try to summarize some results from the point of view of the above mentioned properties.

1   Definition of Julia sets and the Mandelbrot set

Let f be a rational map that defines a dynamical system on the Riemann sphere $\bar{\mathbb{C}} = \mathbb{C} \cup \{\infty\}$. We denote by f^n the iteration of f, which is defined inductively by $f^{n+1} = f \circ f^n$. The orbit of f with the initial point $z \in \bar{\mathbb{C}}$ is $\{f^n(z)\}_{n=0}^{\infty}$.

From the point of view of the dynamical systems, we are interested in the orbits of f, invariant sets, and the way these objects change their natures as the parameter of the system varies.

The most important invariant set is the *Julia set*, which is defined in two equivalent ways:

$$J_f = \{ z \in \bar{\mathbb{C}} \mid \{f^n\}_{n=0}^{\infty} \text{ is not normal in any neighborhood of } z \}$$
$$= \overline{\{\text{repelling periodic points of } f\}}.$$

(For "repelling" periodic points, see the definition below.)

Proceedings of the International Congress
of Mathematicians, Zürich, Switzerland 1994
© Birkhäuser Verlag, Basel, Switzerland 1995

If f is a polynomial, then there is another equivalent definition as follows. The *filled-in Julia set* is

$$K_f = \{\, z \in \mathbb{C} \mid \text{the orbit } \{f^n(z)\}_{n=0}^{\infty} \text{ is bounded } \}.$$

Then its boundary coincides with the Julia set: $J_f = \partial K_f$. These sets are known to be compact and perfect.

Let us focus on the family of quadratic polynomials:

$$P_c(z) = z^2 + c,$$

where $c \in \mathbb{C}$ is considered as the parameter of the system. We are particularly interested in this family, because this is the simplest example that is not affine (the dynamics of affine maps is completely understood), and all quadratic polynomials are affine conjugate to P_c with a unique c.

Even for a polynomial of such a simple form, we see that there is a huge variety of Julia sets. This is the main subject of this paper.

Note that 0 is the unique critical point of P_c in $\mathbb{C}$. For simplicity, let us denote:

$$K_c = K_{P_c}, \quad J_c = J_{P_c}.$$

For the connectivity of the Julia sets, we have the following dichotomy:

THEOREM. (Fatou)
If $0 \in K_c$ then both J_c and K_c are connected.
If $0 \notin K_c$ then $J_c = K_c$ (K_c has no interior) and they are totally disconnected.

This suggests the following definition of the *Mandelbrot set*:

$$
\begin{aligned}
M &= \{\, c \in \mathbb{C} \mid J_c \text{ is connected } \} \\
&= \{\, c \in \mathbb{C} \mid 0 \in K_c \,\} \\
&= \{\, c \in \mathbb{C} \mid \{P_c^n(0)\}_{n=0}^{\infty} \text{ is bounded}\}.
\end{aligned}
$$

This set is known to be compact and connected (Douady-Hubbard). Moreover, Douady and Hubbard [DH] gave a combinatorial description of its "shape" in terms of external rays of rational angles.

Let us recall some definitions concerning periodic points.

DEFINITION. A *periodic point* is a point $z_0 \in \bar{\mathbb{C}}$ such that $f^n(z_0) = z_0$ for some integer $n \geq 1$. The smallest n with this property is the period of z_0, and $\lambda = (f^n)'(z_0)$ is the *multiplier* of z_0 if $z_0 \neq \infty$ (for $z_0 = \infty$, it is defined after a change of coordinate).

We call the periodic point z_0 *attracting, indifferent,* or *repelling,* respectively, if $|\lambda| < 1$, $|\lambda| = 1$, or $|\lambda| > 1$. In the indifferent case, it is called *parabolic* if λ is a root of unity, and *irrationally indifferent* otherwise. Furthermore if the map is linearizable at an irrationally indifferent periodic point (i.e. f^n near z_0 is analytically conjugate to $z \mapsto \lambda z$ near 0), then this point is called a *Siegel point*, and if not, a *Cremer point*.

2   Hyperbolicity and stability

DEFINITION. A rational map is called *hyperbolic* if the orbit of every critical point tends to an attracting periodic orbit.

In particular, P_c is hyperbolic either if $c \notin M$ or if P_c has an attracting periodic point (in $\mathbb{C}$).

We have a good understanding of the dynamics of hyperbolic rational maps. For example, they have the following properties:
 – expanding, i.e. for the spherical metric, the derivatives of iterates satisfy

$$\|(f^n)'\| \geq c\lambda^n \ (n \geq 0) \text{ on the Julia set with constants } c > 0 \text{ and } \lambda > 1.$$

 – have a Markov partition.
 – the Julia set has Lebesgue measure zero, Hausdorff dimension strictly less than two, and moreover the Hausdorff measure of the right dimension is positive and finite.
 – if the Julia set is connected, then it is locally connected. (This is the case for hyperbolic P_c with $c \in M$.)
 – are J-stable (see the definition below.)

It follows that if P_c is hyperbolic and $c \in M$ then c is in the interior $\mathrm{int}M$ of M. (It is not difficult to deduce it from the definition.) Thus hyperbolic maps form a very important class of rational maps.

DEFINITION. Given an analytic family of rational maps, a map in this family is called *J-stable* in this family, if a sufficiently close map in the family is topologically conjugate to the first one on the Julia set. If we take the family of all rational maps or all polynomials, we simply say J-stable without referring to the family.

THEOREM. (Mañé-Sad-Sullivan [MSS]) *For any analytic family of rational maps, there is an open dense set of parameters for which the map is J-stable. Moreover, P_c is J-stable if and only if $c \notin \partial M$ (i.e. $c \notin M$ or $c \in \mathrm{int}M$).*

In this sense, one can characterize ∂M as the set of c's such that P_c is **not** J-stable, so it can be considered as the *locus of bifurcation*.

3   Conjectures and some results on M

Various questions on the dynamics of the quadratic family can be stated in terms of the Mandelbrot set. One of the central conjectures in the theory of 1-dimensional complex dynamics is:

CONJECTURE 1. (MLC) *The Mandelbrot set M is locally connected.*

If this is true, then the conformal map from the complement of the unit disk to the complement of M extends to the unit circle continuously. This will complete the combinatorial description of M given by Douady and Hubbard.

For this conjecture, there was some progress by Yoccoz and by Lyubich. To state this we need a definition:

DEFINITION. A quadratic polynomial P_c is called renormalizable, if there exist an integer $k \geq 2$ and simply connected open sets U and V such that $0 \in U \subset \overline{U} \subset V$, $P_c^k : U \to V$ is a proper map of degree 2, and $P_c^n k(0) \in U$ for all $n \geq 0$.

Moreover, if there are infinitely many k's with the other property, then P_c is called *infinitely renormalizable*.

It was shown by Douady and Hubbard that P_c is renormalizable if and only if P_c is in a "small copy" of M in M.

THEOREM 1. (Yoccoz [Y]) *If P_c ($c \in M$) is not infinitely renormalizable then M is locally connected at c.*

Lyubich [L2] has obtained the local connectivity of M at some infinitely renormalizable points.

Another important question is:

CONJECTURE 2. *The set of c's such that P_c is hyperbolic and is open and dense in M.*

It can be easily shown that this set is open. So the density is the question. By Mañé-Sad-Sullivan's theorem, this conjecture is equivalent to

CONJECTURE 2′. *If P_c is J-stable, then it is hyperbolic.*

Using their theory, Douady and Hubbard were able to show:

THEOREM 2. [DH] *Conjecture 1 implies Conjecture 2.*

This shows the importance of Conjecture 1.

As for the geometric aspect of M, we have

THEOREM 3. (Shishikura [S1]) *The boundary of the Mandelbrot set, ∂M, has Hausdorff dimension 2.*

This already shows the complexity of M, which had been observed by many computer experiments. The next question is on the (2-dimensional) Lebesgue measure.

CONJECTURE 3. *The boundary of the Mandelbrot set has Lebesgue measure zero.*

Conjecture 2 and Conjecture 3 (and Mañé-Sad-Sullivan's theorem) will imply:

For almost all c's (with respect to the Lebesgue measure), P_c is hyperbolic.

For Conjecture 3, we have a partial result:

THEOREM 4. (Shishikura [S2]) *The set*

$$\{ z \in \partial M \mid P_c \text{ is not infinitely renormalizable} \}$$

has Lebesgue measure zero.

4   Results on Julia sets

The results on the Mandelbrot set are usually proved through the study of the dynamics in the dynamical plane (z-plane), and such a study will also give several results on the properties of Julia sets. One often needs to classify the maps according to their dynamical behavior, then different techniques are applied to different types of maps.

It is known that P_c has at most one cycle of nonrepelling (i.e. attracting or indifferent) periodic points in $\mathbb{C}$. So first we divide into three cases — (A) $c \notin M$ (i.e. J_c is disconnected), (B) there is a nonrepelling periodic point, or (C) $c \in M$ and all periodic points (in $\mathbb{C}$) are repelling. In fact, we have the following classification into disjoint cases:

A. $c \notin M$, i.e. J_c is disconnected. (In this case, P_c is hyperbolic.)

B. P_c has a nonrepelling periodic point
  B1. P_c has an attracting periodic point. (Hence P_c is hyperbolic.)
  B2. P_c has a parabolic periodic point.
  B3. P_c has an irrationally indifferent periodic point of Siegel type.
  B4. P_c has an irrationally indifferent periodic point of Cremer type.

C. All periodic points of P_c are nonrepelling.
  C1. The critical point 0 is not recurrent and P_c is not infinitely renormalizable.
  C2. The critical point 0 is recurrent and P_c is not infinitely renormalizable.
  C3. P_c is infinitely renormalizable.

The cases A and B1 are hyperbolic cases, and are the easiest ones to understand. The cases B2 and C1 (especially the case where 0 is strictly preperiodic) are next easiest. These are important classes in the Douady-Hubbard's theory on quadratic polynomials. By the work of Yoccoz, we now have a way to study the case C2. He defined two subcases — persistently recurrent and nonpersistently recurrent. The latter is much easier than the former. There is a beautiful result also by Yoccoz on the linearizability of irrationally indifferent periodic points (i.e. on the distinction of B3 and B4). But there are still many subtle open problems, for example, on the topological structure of the Julia sets. Finally, the case C3 leads to a rich structure of Julia sets. Some progress has been made by Sullivan (for real c with bounded type of renormalization), by McMullen (with certain geometric conditions) and by Lyubich (for "high" types of renormalization). This case contains many open problems, and, as we have seen in the previous section, one can give answers to the conjectures by understanding this case.

The cases A and B1 are conjectured to be dense (Conjecture 2). The case C2 (nonpersistently recurrent case) is known to be generic (in the sense of Baire category) in ∂M.

Now let us review several aspects of Julia sets.

Local connectivity of the Julia set

In case A, the Julia set is totally disconnected. So it can never be locally connected.

Let us consider the case $c \in M$. The conformal map from the complement of the unit disk to the complement of the filled-in Julia set conjugates $z \mapsto z^2$ to

P_c. If J_c is locally connected, then by Carathéodory's theorem this map extends continuously to the unit circle. So the dynamics on the Julia set can be presented as a factor of $z \mapsto z^2$ on S^1.

For the cases B1, B2, and C1(strictly preperiodic case), Douady and Hubbard [DH] showed that the Julia set is locally connected. Yoccoz extended it to the case C2 (and all cases of C1).

THEOREM 5. (Yoccoz [Y]) *If all periodic points of P_c are repelling and P_c is not infinitely renormalizable then J_c is locally connected.*

There are some cases in B3 and C3 for which J_c is known to be locally connected (Petersen, Jiang).

On the other hand, there are examples of $c \in M$ such that J_c is not locally connected. Douady and Sullivan showed that this is always the case for B4. And a similar argument applies to some cases of B3, when the critical point is not on the boundary of the Siegel disk (Herman's example). Douady gave a construction of examples of infinitely renormalizable maps of satellite type (i.e. the limit of repeated bifurcations "going out of the cardioid to an attached satellite"). It is important to know the topological structure of the Julia set for these cases. (See the section on holomorphic removability below.)

Lebesgue measure of the Julia set

For the cases A and B1, it was classically known that the Julia set J_c has (2-dimensional) Lebesgue measure zero. This method was extended to the cases B2 and C1 (strictly preperiodic case) by Douady and Hubbard. Using Yoccoz's technique for the local connectivity and McMullen's modulus-area inequality, we can treat the case C2 and have:

THEOREM 6. (Lyubich [L1], Shishikura [S2]) *If all periodic points of P_c are repelling and P_c is not infinitely renormalizable then J_c has Lebesgue measure zero.*

It is a completely open question for the cases B3, B4, and C3.

However there is a striking result announced by Nowicki and van Strien [NvS] that for a large d (even integer) and a certain c, the Julia set has positive measure.

Hausdorff dimension of the Julia set

In hyperbolic cases (A and B1), Ruelle showed (using the idea of Bowen for Kleinian groups) that if δ is the Hausdorff dimension of J_c then the δ-dimensional Hausdorff measure of J_c is positive and finite. Then it follows from the result on the (2-dimensional) Lebesgue measure that the Hausdorff dimension must be strictly less than 2. In fact, one can prove, using a different method, that J_c has Hausdorff dimension less than 2 for cases A, B1, B2, C1. On the other hand, we have:

THEOREM 7. (Shishikura [S1]) *For generic c (in the sense of Baire category) in ∂M, the Julia set J_c has Hausdorff dimension 2.*

As remarked before, for generic $c \in \partial M$, P_c is in the case C2. It is also possible to construct examples in case B4 and C3 such that J_c has Hausdorff dimension 2.

Holomorphic removability

DEFINITION. A closed set X in $\mathbb{C}$ is called *holomorphically removable* if any homeomorphism from an open neighborhood U of X onto an open set of $\mathbb{C}$ that is analytic on $U - X$ is analytic on entire U.

For example, straight lines, circles, and quasi-circles are known to be holomorphically removable. The set (Cantor set) $\times$ (interval)$(\subset \mathbb{R} \times \mathbb{R} \simeq \mathbb{C})$ is not holomorphically removable. It follows from the measurable Riemann mapping theorem that a holomorphically removable set has Lebesgue measure zero.

In the cases A, B1, B2, and C1, Carleson, Jones, and Yoccoz [CJY] proved that the complement of the filled-in Julia set is a "John domain" and that J_c is holomorphically removable. Kahn [K] has proved by developing Yoccoz's method that if all periodic points of P_c are repelling and P_c is not infinitely renormalizable then J_c is holomorphically removable. This in fact implies Theorem 6 by the above remark.

However one can construct examples in cases B4 and C3 (satellite type) such that J_c contains a subset that is C^1-diffeomorphic to (Cantor set)$\times$(interval), hence it is not holomorphically removable.

5   The idea of the proof of Theorems 3 and 7

DEFINITION. A subset X of the Julia set J_c is called a *hyperbolic subset* for P_c if it is closed and forward invariant (i.e. $P_c(X) \subset C$) and if P_c is expanding on X (i.e. $\|(P_c^n)'\| \geq c\lambda^n \ (n \geq 0)$ on X) with some constants $c > 0$ and $\lambda > 1$.
   The *hyperbolic dimension* of P_c is defined by

$$\text{hyp-dim}(P_c) = \sup_{X} \{ \text{ H-dim}(X) \mid X \text{ is a hyperbolic subset of } P_c \},$$

where H-dim(X) is the Hausdorff dimension of the set X.

It follows from the definition and the stability of hyperbolic sets that the function $c \mapsto \text{hyp-dim}(P_c)$ is lower semi-continuous. The Hausdorff dimension of J_c is greater than or equal to the hyperbolic dimension of P_c.

Theorems 3 and 7 can be deduced from the following propositions.

PROPOSITION 8. *For $c \in \partial M$,*

$$H - \dim(\partial M) \geq \text{hyp-dim}(P_c).$$

PROPOSITION 9. *For any $c \in \partial M$, there exists a sequence $\{c_n\}$ in ∂M such that*

$$c_n \to c \quad \text{and} \quad \text{hyp-dim}(P_{c_n}) \to 2 \ (n \to \infty).$$

Theorem 3 follows immediately from these propositions. Theorem 7 follows from Proposition 9 and the lower semi-continuity of hyp-dim(P_c), because $V_n =$

$\{c \in \partial M | \text{hyp-dim}(P_c) > 2 - \frac{1}{n}\}$ is open and dense in ∂M, and for $c \in \cap_n V_n$, we have $\text{H-dim}(J_c) = \text{hyp-dim}(P_c) = 2$.

To prove Proposition 8, we use the "similarity" between M and J_c. For $c_0 \in \partial M$, there is a hyperbolic subset X_{c_0} for P_{c_0} whose Hausdorff dimension is close to the hyperbolic dimension. By the stability of hyperbolic subsets, for c sufficiently close to c_0, there is a hyperbolic subset X_c for P_c such that P_c on X_c is conjugate to P_{c_0} on X_{c_0}. Moreover the conjugacy is bi-Hölder with an exponent close to 1, and X_c depends holomorphically on c. One can show that for $N \geq 0$,

$$\{ c \mid P_c^N(0) \in X_c \} \subset \partial M$$

and that the left-hand side contains a set that resembles X_{c_0}. This is the idea of the proof for Proposition 8.

Proposition 9 is the hardest part. Let us consider the typical case $c = c_0 = \frac{1}{4}$. Then the Julia set $J_{1/4}$ is a Jordan curve and its Hausdorff dimension is *strictly* less than one. From this, one has to *create* a new Julia set whose hyperbolic dimension is close to 2. This can be done by analyzing the bifurcation of parabolic periodic points by means of the theory of Ecalle cylinders, which was developed by Douady, Hubbard, and Lavaurs [DH]. In this case, $z_0 = \frac{1}{2}$ is a parabolic fixed point of $P_{1/4}$, and the interior of $J_{1/4}$ is the parabolic basin, i.e. all orbits in the interior of $J_{1/4}$ tend to $\frac{1}{2}$. If we perturb $P_{1/4}$, then the parabolic fixed point bifurcate into two fixed points and this creates new types of orbits that behave in a more complicated way. By choosing the perturbation carefully, it is possible to create a more complicated Julia set, which contains a hyperbolic subset such that its Hausdorff dimension can be concretely estimated and is shown to be close to 2. This will prove Proposition 9 for $c_0 = \frac{1}{4}$.

For general cases, we can combine this method with Mañé-Sad-Sullivan's result on the density of parabolic points on ∂M.

6   The idea of the proofs of Theorems 1, 4, 5, and 6

The proofs of these theorems are based on Yoccoz's idea of partition of the dynamical and parameter planes, and on his divergence theorem.

Let us suppose that $c \in M$, all periodic points of P_c are repelling, and P_c is not infinitely renormalizable (case C1 or C2). Then Yoccoz constructs a graph consisting of external rays landing at a certain periodic orbit and an equipotential curve. This divides the plane $\mathbb{C}$ into several connected components. A bounded component is called a *piece* of level 0. Then, inductively, a connected component of a piece of level n is called a *piece* of level $(n+1)$. It can be shown that each piece of any level is simply connected and its intersection with the Julia set is connected. Let us denote by $\mathcal{P}_n(x)$ the piece of level n containing x. (Here, for simplicity, we only consider the case where the orbit of x never falls on the periodic orbit used in the construction of the graph.)

It is easy to see that $\mathcal{P}_{n+1}(x) \subset \mathcal{P}_n(x)$, $F_c(\mathcal{P}_n(x)) = \mathcal{P}_{n-1}(P_c(x))$ and the map $P_c : \mathcal{P}_n(x) \to \mathcal{P}_{n-1}(P_c(x))$ is bijective if $0 \notin \mathcal{P}_n(x)$, and is a branched covering of degree 2 if $0 \in \mathcal{P}_n(x)$.

To obtain Theorem 5, we need to show that the diameter of $\mathcal{P}_n(x)$ tends to 0 as $n \to \infty$. (This will prove that $\{\mathcal{P}_n(x) \cap J_c\}_n$ is a fundamental system of neighborhoods that are connected, hence J_c is locally connected at c.)

If 0 is not recurrent, i.e. $P_c^k(0) \notin \mathcal{P}_n(0)$ for some $n \geq 0$ and all $k \geq 1$, then one can obtain "weak hyperbolicity" of the map on the Julia set using the Poincaré metric on the pieces, therefore the diameter of $\mathcal{P}_n(x)$ tends to 0 exponentially.

If 0 is recurrent, then for a suitable choice of the initial graph, there exists a level n_0 such that $\overline{\mathcal{P}_{n+1}(0)} \subset \mathcal{P}_n(0)$, hence $\mathcal{P}_n(0) - \overline{\mathcal{P}_{n+1}(0)}$ is an annulus. Moreover, Yoccoz showed the following:
There is a sequence of levels $\{n_j\}$ such that $n_j \nearrow \infty$ as $j \to \infty$, $P_c^{m_j}(\mathcal{P}_{n_j}(0)) = \mathcal{P}_{n_0}(0)$ and $P_c^{m_j}(\mathcal{P}_{n_j+1}(0)) = \mathcal{P}_{n_0+1}(0)$, where $m_j = n_j - n_0$, with the notation $A_n = \mathcal{P}_n(0) - \overline{\mathcal{P}_{n+1}(0)}$, $P_c^{m_j} : A_{n_j} \to A_{n_0}$ is a covering map between annuli, and (the divergence theorem)

$$\sum_{j=0}^{\infty} \operatorname{modulus}(A_{n_j}) = \infty.$$

(Such a divergence type theorem was first obtained by Branner and Hubbard in the study of cubic polynomials.) From this, one can deduce the local connectivity of J_c at 0, and at other points as well, by the Grötsch principle:

If $\{A_n\}$ is a nested sequence of annuli in a bounded region in $\mathbb{C}$ and if the sum of the moduli of A_n diverges, then the intersection of bounded components of $\mathbb{C} - A_n$ is a point.

In order to obtain Theorem 1, Yoccoz constructs a similar partition for M and shows that there is a nested sequence of annuli $A_{n_j}^M$ surrounding c such that the modulus of $A_{n_j}^M$ is comparable to the modulus of A_n (i.e. the ratio is bounded away from 0 and ∞). Moreover the intersection of the bounded component of A_n^M and M is connected. So this gives the local connectivity of M at c. The local connectivity at c's for which P_c has irrationally indifferent cycles is taken care of by an inequality that was also proved by Yoccoz.

To show Theorems 4 and 6, we use the same partition as Yoccoz, and need to replace the Grötsch principle by McMullen's modulus-area inequality (or by its corollary):

Let $\mathcal{A} = \{A\}$ be a collection of annuli in a bounded region in $\mathbb{C}$. Then the set of points such that the sum of moduli of the annuli in $\mathcal{A}$ that surround this point diverges has Lebesgue measure zero.

In fact, we can make a partition of J_c or the subset of ∂M in Theorem 4, into a countable union of sets, each of which is shown to have measure zero, either by the above result applied to a suitable collection of annuli, or by applying Lebesgue density theorem and Koebe's distortion theorem, etc. For the description of these partitions, we need to use the τ-function, which was invented by Yoccoz to analyze the combinatorial recurrence of the critical point.

References

For general theory of Julia sets and the Mandelbrot set, see [B], [CG], and [DH].

[B]   Beardon, A.F., Iteration of Rational Functions, Springer-Verlag, Berlin/Heidelberg, 1991.

[CG]   Carleson, L., and Gamelin, T., Complex Dynamics, Springer-Verlag, Berlin/Heidelberg, 1993.

[CJY]   Carleson, L., Jones, P., and Yoccoz, J.-C., *Is Julia John?*, preprint 1993.

[DH]   Douady, A., and Hubbard, J.H., *Étude dynamique des polynômes complexes*, Publ. Math. d'Orsay, 1er partie, 84-02; 2me partie, 85-04.

[K]   Kahn, J., Short communication in NATO Advanced Institute on "Real and Complex Dynamics", Hilleroed, 1993.

[L1]   Lyubich, M., *On the Lebesgue measure of the Julia set of a quadratic polynomial*, preprint IMS, SUNY at Stony Brook, 1991/10.

[L2]   ——, *Geometry of quadratic polynomials: Moduli, rigidity and local connectivity*, preprint IMS, SUNY at Stony Brook, 1993/9.

[MSS]   Mañé, R., Sad, P., and Sullivan, D., *On the dynamics of rational maps*, Ann. Sci. École Norm. Sup. (4) **16** (1983), 193–217.

[NvS]   Nowicki, T., and van Strien, S., *Polynomial maps with a Julia set of positive Lebesgue measure: Fibonacci maps*, preprint 1994.

[S1]   Shishikura, M., *The Hausdorff dimension of the boundary of the Mandelbrot set and Julia sets*, to appear in Ann. of Math. (2).

[S2]   ——, in preparation.

[Y]   Yoccoz, J.-C., *On the dynamics of non-infinitely renormalizable polynomials*, in preparation.

Smoothing Estimates for the Wave Equation and Applications

CHRISTOPHER D. SOGGE*

Mathematics Department
University of California
Los Angeles, CA 90024, USA

The purpose of this paper is to go over some recent results in analysis and partial differential equations that are related to regularity properties of solutions of the wave equation

$$\begin{cases} \Box u(t,x) = 0 \\ u(0,x) = f(x), \ \partial_t u(0,x) = g(x). \end{cases} \tag{1}$$

Here, $\Box = \partial^2/\partial t^2 - \Delta$, where Δ is either the Euclidean Laplacian on $\mathbb{R}^n$ or a Laplace-Beltrami operator on a compact n-dimensional manifold M^n (without boundary unless otherwise stated).

Fixed-time $L^p \to L^p$ estimates

In the Euclidean case, the most basic estimate of course is the energy identity:

$$\int_{\mathbb{R}^n} |\nabla_{t,x} u(t,x)|^2 \, dx = \int_{\mathbb{R}^n} \left(|\nabla_x f(x)|^2 + |g(x)|^2 \right) dx, \tag{2}$$

which just follows from integration by parts or a simple application of the Fourier transform. If one is interested in the L^2 norm of the solution, then a related estimate, which also follows directly from Plancherel's theorem, is

$$\|u(t,\cdot)\|_{L^2} \le \|f\|_{L^2} + (1+t^2)^{1/2} \|g\|_{L^2_{-1}}, \tag{3}$$

if L^p_α denotes the usual L^p Sobolev space with norm $\|f\|_{L^p_\alpha} = \|(I - \Delta)^{\alpha/2} f\|_{L^p}$. Both (2) and (3) easily generalize to the setting of manifolds as well.

The fixed-time L^p, $p \ne 2$, behavior of the wave equation is much less favorable, and sharp estimates are harder to obtain. For the Euclidean version, Miyachi [31] and Peral [36] showed that for $1 < p < \infty$

$$\|u(t,\cdot)\|_{L^p} \le C_{p,t} \left(\|f\|_{L^p_{\alpha_p}} + \|g\|_{L^p_{\alpha_p-1}} \right), \quad \alpha_p = (n-1)|\tfrac{1}{2} - \tfrac{1}{p}|. \tag{4}$$

Simple counterexamples show that these estimates are sharp. If one just takes f to be a nontrivial cutoff times either $|x|^{\alpha_p - n/p}$ or $(1 - |x|)^{\alpha_p - 1/p}$, depending on

---

*)   The author was supported in part by the National Science Foundation.

Proceedings of the International Congress
of Mathematicians, Zürich, Switzerland 1994
© Birkhäuser Verlag, Basel, Switzerland 1995

whether p is ≤ 2 or ≥ 2, respectively, then $u(\pm 1, \cdot) \notin L^p(\mathbb{R}^n)$, while $f \in L^p_\alpha$ if $\alpha < \alpha_p$. For small times, Beals [1] extended (4) to variable coefficients.

The proofs of these fixed-time estimates relied on the fact that, in the Euclidean setting, or the small-time manifold setting, the kernel of the solution operator has a very simple form; specifically, that it is a conormal distribution. The techniques of [1], [31], [36] rely on stationary phase and break down when this is not the case. This sort of situation can of course occur for large times on a manifold. Using different techniques related to the plane wave decomposition of the solution of the Euclidean wave equation, Seeger, Stein, and the author [39] showed that (4) holds for all times on a manifold. In fact, a more general result concerning Fourier integral operators holds.

To describe this, let us assume that $a(x, \xi) \in C^\infty$ vanishes for x outside of a compact set and satisfies

$$|\partial_x^\alpha \partial_\xi^\gamma a(x, \xi)| \leq C_{\alpha, \gamma}(1 + |\xi|)^{-|\gamma|}.$$

Assume also that $\varphi(x, \xi)$ is real, homogeneous of degree one in ξ, smooth away from $\xi = 0$, and satisfies

$$\det \partial^2 \varphi / \partial x_j \partial \xi_k \neq 0 \tag{5}$$

on supp a. Then, if $\hat{f}$ denotes the Fourier transform of f, and if we let

$$\mathcal{F}_\alpha f(x) = \int_{\mathbb{R}^n} e^{i\varphi(x, \xi)} a(x, \xi)(1 + |\xi|)^\alpha \hat{f}(\xi) \, d\xi, \tag{6}$$

it was shown in [39] that for $1 < p < \infty$

$$\|\mathcal{F}_\alpha f\|_{L^p(\mathbb{R}^n)} \leq C_p \|f\|_{L^p(\mathbb{R}^n)}, \quad \alpha = -(n-1)|\tfrac{1}{p} - \tfrac{1}{2}|. \tag{7}$$

This contains the fixed-time estimates mentioned above because $u(t, \cdot)$ can always be decomposed into a finite sum of operators of this type acting on the data. The examples showing that (4) is sharp can easily be adapted to this context to show that (7) cannot be improved for conormal operators whose singular support has codimension one, or, more succinctly, for operators with phases satisfying rank $\partial^2 \varphi / \partial \xi_j \partial \xi_k = n - 1$ somewhere.

The main step in the proof of (7) is to show that, even though the limiting L^1 estimate with $\alpha = -(n-1)/2$ is false, dyadic versions of this estimate hold. Specifically, if $\beta \in C_0^\infty((1/2, 2))$, and we set

$$T_\lambda f(x) = \int_{\mathbb{R}^n} e^{i\varphi(x, \xi)} \beta(|\xi|/\lambda) a(x, \xi)(1 + |\xi|)^{-\frac{n-1}{2}} \hat{f}(\xi) \, d\xi,$$

then one has

$$\|T_\lambda f\|_{L^1} \leq C \|f\|_{L^1}, \quad \lambda = 2^k, \ k = 1, 2, \ldots. \tag{6'}$$

To do this, one breaks up the dyadic operator T_λ into "angular pieces," $T_\lambda = \sum_{\nu=1}^{\lambda^{(n-1)/2}} T_\lambda^\nu$, where

$$T_\lambda^\nu f(x) = \int_{\mathbb{R}^n} e^{i\varphi(x, \xi)} \chi_\nu(\xi) \beta(|\xi|/\lambda) a(x, \xi)(1 + |\xi|)^{-\frac{n-1}{2}} \hat{f}(\xi) \, d\xi,$$

with each χ_ν being homogeneous of degree zero, and satisfying $D_\xi^\gamma \chi_\nu(\xi) = O(\lambda^{\frac{|\gamma|}{2}})$, $|\xi| = 1$. Thus, one should think of $\{\chi_\nu\}$ as a smooth partition of unity of $\mathbb{R}^n \backslash 0$, with each term being supported in a cone of aperture $O(\lambda^{-1/2})$, and the derivatives of χ_ν satisfying the natural bounds associated with this decomposition. The reason for this decomposition is that, in the right scale, $\xi \to \varphi(x,\xi)$ behaves like a linear function of ξ. Based on this, it is not hard to show that one can estimate each resulting piece in terms of the size of its symbol:

$$\|T_\lambda^\nu f\|_{L^1} \le C\lambda^{-(n-1)/2} \|f\|_{L^1},$$

which of course yields (6′). This sort of decomposition was first used in the related context of Riesz means by Fefferman [11], as well as by Christ and the author [7] and Córdoba [8]. Also, Smith and the author [42] showed that the estimates (4) also extend to the setting of the wave equation outside of a convex obstacle, and that estimates related to (7) also hold for Fourier-Airy operators.

Space-time $L^p \to L^p$ estimates

Using (4) and (7), one can apply Minkowski's integral inequality to see that $u \in L^p_{\text{loc}}(dtdx)$ if $(f,g) \in L^p_{\alpha_p} \times L^p_{\alpha_p-1}$. However, for many applications it is useful to know whether the regularity assumptions can be weakened, if one considers the space-time rather than spatial regularity properties of u. If $p \le 2$, one can use the counterexample for the sharpness of (4) to see that, for this range of exponents, in general, the local space and space-time regularity properties are the same.

On the other hand, if $n \ge 2$ and $p > 2$ there is an improvement. Specifically, given $p > 2$, there is an $\varepsilon_p > 0$ so that if $S_T = \{(t,x) : 0 \le t \le T\}$

$$\|u(t,x)\|_{L^p(S_T)} \le C_{p,T}(\|f\|_{L^p_{\alpha_q-\varepsilon_p}} + \|g\|_{L^p_{\alpha_p-1-\varepsilon_p}}). \tag{8}$$

Such a "local smoothing estimate" was first obtained in [45] in the Euclidean case when $n = 2$ by the author, using related results from Bourgain [2] and Carbery [5]. Later, a much simpler proof was given by Mockenhaupt, Seeger, and the author [32], and the extension to variable coefficients and higher dimensions was carried out in [33]. The first local smoothing estimate for differential equations seems to go back to Kato [24], who showed that for the K-dV equation there is a local smoothing of order 1 in L^2, due to the dispersive nature of the equation.

As before, (8) generalizes to the setting of Fourier integrals. To be specific, let us consider operators of the form (6), where now, with an abuse of notation, x ranges over $\mathbb{R}^{1+n}$. Then to improve on trivial consequences of (7), one needs certain conditions on the phase. First, one needs the "nondegeneracy condition" that (5) is replaced by the condition that the Hessian appearing there has full rank n everywhere. This amounts to saying that the projection of the canonical relation into $T^*\mathbb{R}^n \backslash 0$ is submersive. In addition, one needs a curvature requirement that is based on the properties of the projection of the canonical relation into the fibers of the bigger cotangent bundle, $T^*\mathbb{R}^{1+n} \backslash 0$. It turns out that the nondegeneracy condition implies that the images of these projections must all be smooth conic hypersurfaces, and they are explicitly given by

$$\Gamma_x = \{\nabla_x \varphi(x,\xi) : \xi \in \mathbb{R}^n \backslash 0\} \subset T_x^* \mathbb{R}^{1+n} \backslash 0 = \mathbb{R}^{1+n} \backslash 0.$$

Our other requirement, "the cone condition," is that at every $\eta \in \Gamma_x$ there are $n-1$ nonvanishing principal curvatures. Together the nondegeneracy condition and the cone condition make up the cinematic curvature condition introduced in [45]. This condition turns out to be the natural homogeneous version of the Carleson-Sjölin condition in [6] (see [33], [46]).

If cinematic curvature holds, it was shown in [33] that (7) can be improved if $n \geq 2$ and $p > 2$, that is, there is an $\varepsilon_p > 0$ so that

$$\|\mathcal{F}_\alpha f\|_{L^p(\mathbb{R}^{1+n})} \leq C\|f\|_{L^p(\mathbb{R}^n)}, \quad \alpha = -(n-1)(\tfrac{1}{2} - \tfrac{1}{p}) + \varepsilon_p. \tag{9}$$

The proof uses the decomposition employed for the fixed-time estimate, along with ideas from the related work of Fefferman [11], Carbery [5], and Córdoba [8] to allow one to exploit the curvature implicit in the cone condition.

The main motivation behind (8) and (9) concerns applications for maximal theorems. For instance, if in $\mathbb{R}^2$, $A_t f(x)$ denotes the average of f over the circle of radius t centered at x,

$$A_t f(x) = \int_{S^1} f(x - t\theta)\, d\theta,$$

then Bourgain [2] showed that

$$\| \sup_{t>0} |A_t f(x)| \|_{L^p(\mathbb{R}^2)} \leq C_p \|f\|_{L^p(\mathbb{R}^2)}, \quad p > 2. \tag{10}$$

Earlier, in effect by using an ingenious square function argument based on bounds that are equivalent to (3), Stein [48] obtained the higher-dimensional version of this result. Stein's theorem, which inspired much of the work described in this paper, says that when $n \geq 3$ the spherical maximal operator is bounded on $L^p(\mathbb{R}^n)$ if $p > n/(n-1)$. He also showed that when $n \geq 2$ no such result can hold if $p \leq n/(n-1)$. Thus, as 2 is the critical Lebesgue exponent for this problem in two dimensions, and as A_t is a Fourier integral operator of order $-1/2$, one cannot use fixed-time estimates like (3) or (4) to obtain (10). Hence, if one wishes to use an argument like Stein's for the two-dimensional setting, harder space-time estimates like (9) are required.

Using these smoothing estimates one can recover (10) in a straightforward manner. In fact, if $\beta \in C_0^\infty((1,2))$, $2 < p < \infty$, and $\varepsilon > 0$, then Sobolev's lemma yields

$$\| \sup_t |\beta(t) A_t f(x)| \|_{L_x^p} \leq C_{\varepsilon,p} \| (I - \partial_t^2)^{\frac{1}{2p} + \frac{\varepsilon}{2}} \beta(t) A_t f(x) \|_{L_{t,x}^p}. \tag{11}$$

However, modulo a trivial error, $f \to (I - \partial_t^2)^{\frac{1}{2p} + \frac{\varepsilon}{2}} \beta(t) A_t f(x)$ is the sum of two operators of the form (6) with $\varphi = x \cdot \xi \pm t|\xi|$ and $\alpha = -\frac{1}{2} + \frac{1}{p} + \varepsilon$. Thus, if $\varepsilon < \varepsilon_p$, (9) implies that the right side of (11) is controlled by the L^p norm of f. This argument thus gives a slightly weaker version of (10), where the supremum in the left is taken over $t \in (1,2)$; however, if one uses Littlewood-Paley theory (see [33], [46]), a simple variation yields the full circular maximal theorem of Bourgain.

This argument of course also applies to certain variable coefficient maximal operators. For instance, if one takes the supremum over small enough radii, one can obtain a variant of (10) where one averages over geodesic circles on two-dimensional Riemannian manifolds (see [45]). Also, it was shown by Iosevich [19] that there is a natural extension of (10) to averages over finite type curves. Also, in [47], the local smoothing estimates (9) were used to show that (10) extends to averages in higher dimensions over hypersurfaces with at least one nonvanishing principal curvature.

Another application was pointed out to us by Wolff. If in the plane

$$A^\delta f(t, x) = \delta^{-1} \int_{|x-y|\in(t,t+\delta)} f(y)\, dy\,,$$

then the above arguments yield for $0 < \delta < 1$

$$\| \sup_{|x|\leq 1} |A^\delta f(t,x)| \|_{L^p([1,2])} \leq C\delta^{\varepsilon-1/p} \|f\|_{L^p(\mathbb{R}^2)}\,, \ \forall \varepsilon < \varepsilon_p\,. \tag{12}$$

It turns out that (9) holds for any $\varepsilon_p < 1/8$ if $p = 4$. From (12) for $p = 4$, one can then deduce, using ideas from Bourgain [3], that a set in the plane containing a translate of a circle of every radius must have Hausdorff dimension $\geq 3/2$. It was shown by Besicovitch (see [9]) that there are such sets of measure zero; however, it is felt that their Hausdorff dimension must always be 2. Showing this is related to a deeper problem concerning the Hausdorff dimension of Besicovitch $(3, 1)$ sets: sets of measure zero in $\mathbb{R}^3$ containing a line in every direction. It is conjectured that such sets must have full Hausdorff dimension. Bourgain [3] showed that they must have dimension $\geq 7/3$, and this result has recently been improved by Wolff [54], who showed that, as in the previous case, the Hausdorff codimension must always be $\leq 1/2$, that is, every Besicovitch $(3, 1)$ set must have dimension $\geq 5/2$.

It was conjectured in [45] that for $p \geq 2n/(n-1)$ there should be local smoothing of order $1/p$, that is, that (9) should hold for all $\varepsilon < 1/p$ for this range of exponents. This conjecture would imply the Bochner-Riesz conjecture, so in higher dimensions it seems presently unattainable. However, in view of the Carleson-Sjölin theorem [6], there might be some hope of verifying the conjecture in $(1 + 2)$-dimensions. If true here, it would imply the conjecture that sets of measure zero in the plane containing a circle of every radius must have full Hausdorff measure. In the radially symmetric case the conjecture for solutions of the Euclidean wave equation was verified in all dimensions by Müller and Seeger [34].

$L^2 \to L^q$ estimates

It is much easier to obtain sharp estimates involving L^2. For instance, using L^2 bounds for Fourier integrals of Hörmander [17] and the Hardy-Littlewood-Sobolev inequality one obtains the sharp estimate

$$\|\mathcal{F}_\alpha f\|_{L^q(\mathbb{R}^n)} \leq C_q \|f\|_{L^2(\mathbb{R}^n)}\,, \quad \alpha = -\tfrac{n}{2} + \tfrac{n}{q}\,, \quad 2 \leq q < \infty\,, \tag{13}$$

if $\mathcal{F}_\alpha$ is as in (7). However, if $\mathcal{F}_\alpha$ sends functions of n-variables to functions of $(1+n)$-variables there is local smoothing of order $1/q$ for a range of exponents if the cinematic curvature condition holds:

$$\|\mathcal{F}_\alpha f\|_{L^q(\mathbb{R}^{1+n})} \le C_q \|f\|_{L^2(\mathbb{R}^n)}, \quad \alpha = -\frac{n}{2} + \frac{n+1}{q}, \quad \frac{2(n+1)}{n-1} \le q < \infty. \tag{14}$$

This result was obtained by Mockenhaupt, Seeger, and the author [33], and it generalizes the important special case of Strichartz [52] for the Euclidean wave equation, which says that if q is as in (14)

$$\|u(t,x)\|_{L^q(\mathbb{R}^{1+n})} \le C_q \left(\|f\|_{\dot{H}^\alpha} + \|g\|_{\dot{H}^{\alpha-1}} \right), \quad \alpha = \frac{n}{2} - \frac{n+1}{q}. \tag{15}$$

Here $\dot{H}^\gamma$ denotes the homogeneous L^2 Sobolev space with norm $\|(-\Delta)^{\gamma/2} f\|_{L^2}$. Earlier partial results go back to Segal [40]. The dual version of (15) is equivalent to the following restriction theorem for the Fourier transform:

$$\int_{\mathbb{R}^n} |\hat{F}(|\xi|, \xi)|^2 \, d\xi / |\xi|^{n-2(n+1)(p-1)/p} \le C_p \|F\|_{L^p(\mathbb{R}^{1+n})}^2, \quad 1 < p \le \frac{2(n+1)}{n+3}.$$

This in turn is related to the earlier L^2 restriction theorem of the Fourier transform for spheres of Stein and Tomas [53]. Inequality (14) contains a local extension of (15) to variable coefficients and the latter was independently obtained by Kapitanski [22]. Strichartz estimates were also obtained by Smith and the author [43] for the wave equation outside of a convex obstacle.

Mixed-norm estimates and minimal regularity for nonlinear equations

Mixed-norm estimates, where different exponents are used for the time and spatial norms, are often useful for applications in analysis and partial differential equations. If we write $(x, y) \in \mathbb{R}^n \times \mathbb{R}^m = \mathbb{R}^{n+m}$, then mixed-norms are given by

$$\|F(x,y)\|_{L_x^q L_y^p(\mathbb{R}^{n+m})} = \left(\int_{\mathbb{R}^n} \left(\int_{\mathbb{R}^m} |F(x,y)|^p \, dy \right)^{q/p} dx \right)^{1/q}.$$

In analysis, typical applications arise from the case where $m = 1$ and $p = 2$, and the estimates involving norms of this types are called square function estimates. We already pointed out that square function estimates for solutions of the Euclidean wave equation (1) lead to Stein's spherical maximal theorem. Related harder estimates that are equivalent to $L_x^4 L_t^2(\mathbb{R}^{2+1})$ estimates for (1) were used by Carbery [5] to obtain estimates for maximal Bochner-Riesz operators in the critical space $L_x^4(\mathbb{R}^2)$.

Square function variants of (15) can also be used to obtain sharp estimates for eigenfunctions. In fact, it was shown in [33] that if one uses L_t^2 instead of L_t^q one can improve (15) and get for $S_T = \{(x,t) : 0 \le t \le T\}$

$$\|u\|_{L_x^q L_t^2(S_T)} \le C_T \left(\|f\|_{\dot{H}^\alpha} + \|g\|_{\dot{H}^{\alpha-1}} \right), \quad \alpha = \frac{n-1}{2} - \frac{n}{q}, \quad \frac{2(n+1)}{n-1} \le q < \infty. \tag{16}$$

Thus, there is a gain of $1/2$ of a derivative over the fixed-time estimates (13). In the Euclidean case, the bounds are independent of T, whereas they may not be on a compact manifold. In [43], Smith and the author showed that these estimates also hold on compact manifolds with concave (i.e., diffractive) boundary. When $q = \infty$ a dyadic version of (16) holds. Using these square function estimates one can obtain "sharp" bounds for eigenfunctions. In fact, if $-\Delta e_\lambda = \lambda^2 e_\lambda$ on either a smooth compact boundaryless manifold or a (relatively) compact manifold with concave boundary, one has

$$\|e_\lambda\|_{L^q(M^n)} \le C(1+\lambda)^{\sigma(q)} \|e_\lambda\|_{L^2(M^n)}, \tag{16$'$}$$

where $\sigma(q) = \frac{n-1}{2} - \frac{n}{q}$ if $\frac{2(n+1)}{n-1} \le q \le \infty$, and $\sigma(q) = \frac{(n-1)(q-2)}{4q}$ if $2 \le q \le \frac{2(n+1)}{n-1}$. The bounds for the first range follow from (16) because if $f = e_\lambda$ and $g = 0$, $u(t,x) = \cos t\lambda \, e_\lambda(x)$, and hence the left side of (16) is comparable to the left side of (16$'$). The bounds for the other range follow via interpolation. Although the bounds in (16$'$) are not sharp in general, a slightly stronger result holds that is always sharp. It says that (16$'$) holds if e_λ is replaced by a function with spectrum in $[\lambda^2, (\lambda+1)^2]$. In the boundaryless case these estimates were first proved by the author [44] using a different, but related approach, based on proving estimates for the standing wave operators $\Delta + \lambda^2$. For the case of manifolds with diffractive boundaries, the estimates are due to Grieser [14] and Smith and the author [43].

It would be very interesting to know to what extent these bounds carry over to the setting of general (relatively) compact manifolds with boundary. Results of Ivrii [20] imply that the L^∞ bounds always hold. However, Grieser [14] showed that one cannot have (16$'$) with $\sigma(q) = \frac{n-1}{2} - \frac{n}{q}$ for the full range $\frac{2(n+1)}{n-1} \le q \le \infty$ if the second fundamental form of the metric has a negative eigenvalue at some point of the boundary. One might expect, though, that in this case they might hold if $q \ge \frac{6n+4}{3n-4}$, which is the largest range allowable by Grieser's counterexample.

Mixed-norm estimates also have important applications for semilinear wave equations. This observation first seems to have been made by Pecher [35], where a variant of the energy estimate (2) was proved, with the right side dominating mixed-norms that scale like the left side of (2). These estimates are in the spirit of the Strichartz estimates (15), and they were used by Pecher to prove scattering theorems for equations like $\Box u = |u|^{\frac{n+2}{n-2}}$ on $\mathbb{R}^{1+n}$ involving data with small energy. Subsequently, these type of mixed-norm estimates, along with (15), were used to prove Grillakis' theorem saying that there are global C^∞ solutions of the critical wave equation $\Box u + u^5 = 0$ in $\mathbb{R}^{1+3}$ with arbitrary C^∞ Cauchy data (see [15], [41], as well as [43] for the extension to the obstacle case).

Mixed-norm estimates and smoothing estimates are also useful for proving existence theorems with minimal regularity for other powers. In [28], Lindblad and the author showed that if $\Box w = F$ in $\mathbb{R}^{1+3}$ and if w has vanishing Cauchy data at $t = 0$, then

$$\|w\|_{L_t^{\frac{2q}{q-2}} L_x^q(\mathbb{R}^{1+3})} \le C_q \|F\|_{L_t^{\frac{q}{q-1}} L_x^{\frac{2q}{q+2}}(\mathbb{R}^{1+3})}, \quad 2 < q < \infty. \tag{17}$$

This inequality is related to earlier Besov space estimates of Ginibre and Velo [12], [13] and Kapitanski [22]. Using it and estimates for the linear Cauchy problem (1) that are related to (15), it was shown in [28] that if $\kappa > 2$ then there are local (weak) solutions of

$$\Box u = |u|^\kappa, \ u(0,x) = f \in \dot{H}^\gamma(\mathbb{R}^3), \ \partial_t u(0,x) = g \in \dot{H}^{\gamma-1}(\mathbb{R}^3), \qquad (18)$$

provided that $\gamma = \max(\frac{3}{2} - \frac{2}{\kappa-1}, 1 - \frac{1}{\kappa-1})$. In the superconformal range, $\kappa \geq 3$, there is also global existence for small data. Scaling arguments show that γ must always be $\geq \frac{3}{2} - \frac{2}{\kappa-1}$, while a counterexample related to an example giving the sharpness of (15) shows that γ must also be $\geq 1 - \frac{1}{\kappa-1}$. The local existence results were also obtained independently by Kapitanski [23].

If one assumes radial symmetry, it turns out that (15) holds for a larger range of exponents, namely, $q > \frac{2n}{n-1}$. For related reasons, the counterexamples in [28] no longer apply, and, consequently, one expects better local and global existence results if f and g in (18) are assumed to be radially symmetric. In fact, recently, Lindblad and the author [29] have established a stronger theorem under these assumptions, using this fact about (15) along with a stronger version of (17) that only holds under the assumptions of radial symmetry, and is proved using ideas from [26] and [34]. Specifically, if $T_\varepsilon = \infty$ for $\kappa > 1 + \sqrt{2}$, $T_\varepsilon = c\varepsilon^{\frac{\kappa(\kappa-1)}{\kappa^2-2\kappa-1}}$ for $2 \leq \kappa < 1 + \sqrt{2}$, and $T_\varepsilon = \exp(c\varepsilon^{-\kappa(\kappa-1)})$ for $\kappa = 1 + \sqrt{2}$, and if $c, \varepsilon > 0$ are sufficiently small there is a weak solution of (18) in $[0, T_\varepsilon) \times \mathbb{R}^3$ provided that the data has $\dot{H}^\gamma \times \dot{H}^{\gamma-1}$ norm $\leq \varepsilon$ with $\gamma = \max(\frac{3}{2} - \frac{2}{\kappa-1}, \frac{1}{2} - \frac{1}{\kappa})$. We already remarked on the necessity of the first condition. The second is also needed because if u_0 is the solution to the linear wave equation with this data, $|u_0|^\kappa$ need not be a distribution if $\gamma < \frac{1}{2} - \frac{1}{\kappa}$. Notice that, for positive powers,

$$\tfrac{1}{2} - \tfrac{1}{\kappa} = \tfrac{3}{2} - \tfrac{2}{\kappa-1} \iff \kappa = 1 + \sqrt{2}.$$

Also, if $\kappa \leq 1 + \sqrt{2}$, it was shown in [21], [27], and [55] that the lifespan bounds in the formula for T_ε are optimal.

This result is of course related to John's [21] existence theorem, which says that if compactly supported data $(f, g) \in C^3 \times C^2$ are fixed, then $\Box u = |u|^\kappa$, $(u(0,x), \partial_t u(0,x)) = \varepsilon(f,g)$ always has a global C^2 solution for small $\varepsilon > 0$ if and only if $\kappa > 1 + \sqrt{2}$. The positive part of John's theorem is proved by an ingenious argument based on iterating in the space with norm

$$\| (1+t)(1+|t-|x||)^{\kappa-2} \sup_{\omega \in S^2} |u(t, |x|\omega)| \|_{L^\infty}.$$

John's argument, following an idea going back to Keller [25], only involves radial estimates because the positivity of the forward fundamental solution for $\Box$ in $\mathbb{R}^{1+3}$ allows one to reduce matters to only proving estimates for radially symmetric functions by first taking supremums over the angular variables $\omega \in S^2$. If one iterates instead in $L^\infty_t L^\kappa_{|x|} L^\infty_\omega$, the arguments used to prove the sharp radial existence theorem mentioned above also allow one to recover John's existence theorem, and the extensions of Lindblad [27] and Zhou [55], which say that there is a solution $u \in C^2([0, T_\varepsilon) \times \mathbb{R}^3)$ if $2 \leq \kappa < 1 + \sqrt{2}$ and $\kappa = 1 + \sqrt{2}$, respectively.

References

[1] M. Beals, L^p *boundedness of Fourier integrals*, Mem. Amer. Math. Soc. **264** (1982).

[2] J. Bourgain, *Averages in the plane over convex curves and maximal operators*, J. Analyse Math. **47** (1986), 69–85.

[3] ———, *Besicovitch type maximal operators and applications to Fourier analysis*, Geom. Functional Anal. **1** (1991), 69–85.

[4] P. Brenner, *On $L_p - L_{p'}$ estimates for the wave equation*, Math. Z. **145** (1975), 251–254.

[5] A. Carbery, *The boundedness of the maximal Bochner-Riesz opearator on $L^4(\mathbb{R}^2)$*, Duke Math. J. (1983), 409–416.

[6] L. Carleson and P. Sjölin, *Oscillatory integrals and a multiplier problem for the disk*, Studia Math. **44** (1972), 287–299.

[7] F. M. Christ and C. D. Sogge, *The weak type L^1 convergence of eigenfunction expansions for pseudo-differential operators*, Invent. Math. **94** (1988), 421–451.

[8] A. Córdoba, *A note on Bochner-Riesz operators*, Duke Math. J. **46** (1979), 505–511.

[9] K. J. Falconer, *The geometry of fractal sets*, Cambridge Univ. Press, Cambridge, 1985.

[10] C. Fefferman, *The multiplier problem for the ball*, Ann. of Math. (2) **94** (1971), 330–336.

[11] ———, *A note on spherical summation multipliers*, Israel J. Math. **15** (1973), 44–52.

[12] J. Ginibre and G. Velo, *Conformal invariance and time decay for nonlinear wave equations, II*, Ann. Inst. H. Poincaré Phys. Théor. **47** (1987), 263–276.

[13] ———, *Scattering theory in the energy space for a class of nonlinear wave equations*, Comm. Math. Phys. **123** (1989), 535–573.

[14] D. Grieser, *L^p bounds for eigenfunctions and spectral projections of the Laplacian near concave boundaries*, thesis, UCLA (1992).

[15] M. G. Grillakis, *Regularity for the wave equation with a critical nonlinearity*, Comm. Pure Appl. Math. **45** (1992), 749–774.

[16] L. Hörmander, *The spectral function of an elliptic operator*, Acta Math. **121** (1968), 193–218.

[17] ———, *Fourier integral operators I*, Acta Math. **127** (1971), 79–183.

[18] ———, *Oscillatory integrals and multipliers on FL^p*, Ark. Mat. **11** (1971), 1–11.

[19] A. Iosevich, *Maximal operators associated to the families of flat curves in the plane*, Duke Math. J. **76** (1994), 633–644.

[20] V. Ivrii, *Precise spectral asymptotics for elliptic operators*, Lecture Notes in Math. 1100, Springer-Verlag, Berlin and New York, 1984.

[21] F. John, *Blow-up of solutions of nonlinear wave equations in three space dimensions*, Manuscripta Math. **28** (1979), 235–265.

[22] L. Kapitanski, *Some generalizations of the Strichartz-Brenner inequality*, Leningrad Math. J. **1** (1990), 693–726.

[23] ———, *Weak and yet weaker solutions of semilinear wave equations*, Brown Univ. preprint, Providence, RI.

[24] T. Kato, *On the Cauchy problem for the (generalized) Korteweg-de Vries equation*, Studies in Applied Math., vol. 8, Academic Press, New York and San Diego, 1983, pp. 93–128.

[25] J. B. Keller, *On solutions of nonlinear wave equations*, Comm. Pure Appl. Math. **10** (1957), 523–530.

[26] S. Klainerman and M. Machedon, *Space-time estimates for null forms and the local existence theorem*, Comm. Pure and Appl. Math. **46** (1993), 1221–1268.

[27] H. Lindblad, *Blow-up for solutions of $\Box u = |u|^p$ with small initial data*, Comm. Partial Differential Equations **15** (1990), 757–821.

[28] H. Lindblad and C. D. Sogge, *On existence and scattering with minimal regularity for semilinear wave equations*, J. Funct. Anal., to appear.

[29] ———, forthcoming.

[30] W. Littman, *The wave operator and L^p-norms*, J. Math. Mech. **12** (1963), 55–68.

[31] A. Miyachi, *On some estimates for the wave operator in L^p and H^p*, J. Fac. Sci. Univ. Tokyo Sect. IA Math. **27** (1980), 331–354.

[32] G. Mockenhaupt, A. Seeger, and C. D. Sogge, *Wave front sets, local smoothing and Bourgain's circular maximal theorem*, Ann. Math. (2) **136** (1992), 207–218.

[33] ———, *Local smoothing of Fourier integrals and Carleson-Sjölin estimates*, J. Amer. Math. Soc. **6** (1993), 65–130.

[34] D. Müller and A. Seeger, *Inequalities for spherically symmetric solutions of the wave equation*, Math. Z., to appear.

[35] H. Pecher, *Nonlinear small data scattering for the wave and Klein-Gordon equations*, Math. Z. **185** (1984), 261–270.

[36] J. Peral, *L^p estimates for the wave equation*, J. Funct. Anal. **36** (1980), 114–145.

[37] D. H. Phong and E. M. Stein, *Hilbert integrals, singular integrals and Radon transforms I*, Acta Math. **157** (1986), 99–157.

[38] A. Seeger and C. D. Sogge, *Bounds for eigenfunctions of differential operators*, Indiana Univ. Math. J. **38** (1989), 669–682.

[39] A. Seeger, C. D. Sogge, and E. M. Stein, *Regularity properties of Fourier integral operators*, Ann. Math. (2) **134** (1991), 231–251.

[40] I. E. Segal, *Space-time decay for solutions of wave equations*, Adv. in Math. **22** (1976), 304–311.

[41] J. Shatah and M. Struwe, *Regularity results for nonlinear wave equations*, Ann. of Math. (2) **138** (1993), 503–518.

[42] H. Smith and C. D. Sogge, *L^p regularity for the wave equation with strictly convex obstacles*, Duke Math. J. **73** (1994), 123–134.

[43] ———, *On the critical semilinear wave equation outside convex obstacles*, Amer. Math. Soc., to appear.

[44] C. D. Sogge, *Concerning the L^p norm of spectral clusters for second order elliptic operators on compact manifolds*, J. Funct. Anal. **77** (1988), 123–134.

[45] ———, *Propagation of singularities and maximal functions in the plane*, Invent. Math. **104** (1991), 349–376.

[46] ———, Fourier integrals in classical analysis, Cambridge Univ. Press, Cambridge, 1993.

[47] ———, *Averages over hypersurfaces with one non-vanishing principal curvature*, Fourier analysis and partial differential equations (J. Garcia-Cuerva, ed.), CRC Press, Boca Raton, FL, 1995, pp. 317–323.

[48] E. M. Stein, *Maximal functions: Spherical means*, Proc. Nat. Acad. Sci. **73** (1976), 2174–2175.

[49] ———, *Oscillatory integrals in Fourier analysis*, Beijing Lectures in Harmonic Analysis, Princeton Univ. Press, Princeton, NJ, 1988, pp. 307–356.

[50] ———, Harmonic analysis real-variable methods, orthogonality, and oscillatory integrals, Princeton Univ. Press, Princeton, NJ, 1993.

[51] R. Strichartz, *A priori estimates for the wave equation and some applications*, J. Funct. Anal. **5** (1970), 218–235.

[52] ———, *Restriction of Fourier transform to quadratic surfaces and decay of solutions to the wave equation*, Duke Math. J. **44** (1977), 705–714.

[53] P. Tomas, *Restriction theorems for the Fourier transform*, Proc. Sympos. Pure Math. **35** (1979), 111–114.

[54] T. Wolff, *An improved bound for Kakeya type maximal functions*, preprint.

[55] Y. Zhou, *Blow up of classical solutions to $\Box u = |u|^{1+\alpha}$ in three space dimensions*, J. Partial Differential Equations Ser. A **5** (1992), 21–32.

A Survey of Möbius Groups

Pekka Tukia

Department of Mathematics
University of Helsinki
Helsinki, Finland

Classical Fuchsian and Kleinian groups are offspring of the theory of functions of one complex variable. Like modern complex analysis they were born in the nineteenth century. They are groups of conformal homeomorphisms of the Riemann sphere identified either with the extended complex plane $\bar{C} = C \cup \{\infty\}$ or with the 2-sphere $S^2 = \{x \in R^3 : |x| = 1\}$, and with discontinuous action on an open nonempty set. Poincaré [Po] had already found out that there is a natural extension of the group action to the upper half-space $H^3 = \{(x_1, x_2, x_3) \in R^3 : x_3 > 0\}$. Poincaré's extension was based on the fact that any conformal homeomorphism of $\bar{C}$ can be represented as a composition of inversions (also called reflections) on spheres, the prototypical inversion being the inversion $x \mapsto x/|x|^2$ of the unit sphere. Obviously, an inversion is extendable to H^3 (or in fact to $\bar{R}^n = R^n \cup \{\infty\}$ with arbitrary dimension n). In this manner it is possible to extend the action of a Kleinian group to H^3.

Not much was made of this extension until very recently. Even the theory of classical Kleinian groups of $\bar{C}$ was a little bit of an oddity and by and large one felt comfortable only with Fuchsian groups, that is with Kleinian groups with an invariant circle S in $\bar{C}$ and with discontinuous action on the complementary domains of S. I do not think that the modern theory of Kleinian groups of $\bar{C}$ began until the 1960s. If one wants to pinpoint the beginning, for me it was Ahlfors' finiteness theorem [Ah2] as well as the Ahlfors conjecture on null measure of the limit set of a finitely generated Kleinian group (now solved for a large class of groups by Bonahon [B]). In the 1980s Thurston revolutionized the area, the major theorem being Thurston's geometrization theorem for Haken 3-manifolds and its implications for Kleinian groups. Here, finally, essential use was made of the fact that Kleinian groups are extendable to H^3.

We currently have a rich theory of Kleinian groups acting on $\bar{C}$ and H^3. As hinted above, it is possible to extend the action of a Kleinian group of $\bar{C}$ to any $\bar{R}^n$. The origin of these groups as Kleinian groups of $\bar{C}$ can be noticed in the fact that $\bar{R}^2 = \bar{C}$ is setwise invariant under the action of the group. If we drop this condition, we obtain groups of $\bar{R}^n$ generated by inversions on $(n-1)$-spheres. As usual, we also regard $(n-1)$-planes as spheres and an inversion on such a plane is just a reflection. We call these groups Möbius groups of $\bar{R}^n$. The groups we study often act discontinuously somewhere; that is, there are points $x \in \bar{R}^n$

Proceedings of the International Congress
of Mathematicians, Zürich, Switzerland 1994
© Birkhäuser Verlag, Basel, Switzerland 1995

having a neighborhood U such that $gU \cap U \neq \emptyset$ for only a finite number of $g \in G$. Failing that, G is at least discrete with respect to some natural topology (all such topologies give the same definition of discreteness).

In contrast to the classical Kleinian groups of $\bar{C}$, the theory of n-dimensional Möbius groups is much less developed. We want to describe here some properties of these groups, with special emphasis on the differences and similarities with the classical theory. Even if the theory is far less extensive than the classical theory, there is more of it that can be covered here. Naturally, I emphasize here the areas that appeal to me and with which I am familiar.

We can start with the following similarity. Although we have here presented n-dimensional Möbius groups as groups generated by inversions in spheres, an equivalent definition would be, as in dimension 2, that they are just groups of conformal homeomorphisms of $\bar{R}^n$, as follows from Liouville's theorem. In the classical case, a Kleinian group was a group of orientation-preserving conformal homeomorphisms; that is, each element in the group is a composition of an even number of inversions. However, in the sequel we also allow orientation-reversing conformal homeomorphisms, sometimes called anticonformal homeomorphisms.

We denote the group of all Möbius transformations of $\bar{R}^n$ by Möb(n). Still one basic similarity with the classical case is that it is possible to extend $g \in$ Möb(n) uniquely to a Möbius transformation of the $(n+1)$-dimensional hyperbolic space $H^{n+1} = \{(x_1, \ldots, x_{n+1}) \in R^{n+1} : x_{n+1} > 0\}$ so that it is a hyperbolic isometry in the hyperbolic metric given by the element of length $|dx|/x_{n+1}$. Thus, in the sequel we will consider Möb(n) also acting on H^{n+1} and on $\bar{H}^{n+1} = H^{n+1} \cup \bar{R}^n$.

In a couple of occasions we use as the model for the hyperbolic space the unit ball $B^{n+1} = \{x \in R^{n+1} : |x| < 1\}$ with boundary the n-sphere S^n. Now the hyperbolic metric is given by the element of length $2|dx|/(1-|x|^2)$. We can always change from H^{n+1} to B^{n+1} by conjugating by some $g \in$ Möb($n+1$).

The existence of Möbius groups

One way to Fuchsian groups is via the uniformization of Riemann surfaces. If we take a Riemann surface of genus at least 2, then the universal cover is the upper half-plane $H^2 = \{z \in C : \text{Im } z > 0\}$ of C and the cover translation group is Fuchsian. Thus, granting some knowledge of Riemann surfaces, we know that there are nontrivial Fuchsian groups of H^2, even such that the quotient H^2/G is a compact surface.

It is of course easy to construct simple Möbius groups. For instance if we take p disjoint $(n-1)$-spheres S_i with disjoint interiors in R^n, then the inversions on S_i generate a discrete group. However, the construction of more complicated Möbius groups for general n is difficult; for instance, to find a Möbius group corresponding to a Fuchsian group obtained from a compact Riemann surface. Such a Möbius group would be a discrete group G such that if G is extended to H^{n+1}, then the quotient H^{n+1}/G is compact. We call such a group *cocompact*.

There are cocompact groups, but their construction is far from trivial. Sullivan [S1] has constructed such groups in general dimension n and they played an important role in the deformation theory of bilipschitz and quasiconformal maps in [S1]. Sullivan's deformation theory later allowed the solution of the problem of extending a quasiconformal homeomorphism of R^n to a quasiconformal homeomorphism of R^{n+1} in [TV]. Thus, the existence of such groups has had some very interesting consequences outside the theory of Möbius groups.

Sullivan's construction is based on the fact that it is possible to represent Möbius transformations of $\bar{R}^n$ as matrices of $O(1, n+1)$; that is, as matrices preserving the bilinear form $x_0^2 + \cdots + x_n^2 - x_{n+1}^2$. Sullivan's construction is algebraic and therefore he prefers to use the group of matrices Γ preserving the bilinear form $x_0^2 + \cdots + x_n^2 - \sqrt{2}x_{n+1}^2$ and with entries of the form $p + q\sqrt{2}$ with p and q integers. The proof of compactness of H^{n+1}/Γ is not trivial.

It would be highly desirable if one could give a geometric description of a cocompact group and of the quotient space. This is possible in low dimensions and these geometric constructions are based on the idea of a fundamental polyhedron. If G is a Kleinian group of H^{n+1}, then a fundamental polyhedron for G is a subset D of H^{n+1} such that the G-transforms of D fill H^{n+1} so that the intersection of two distinct transforms of D is either empty or a common face or subface.

In Poincaré's fundamental polyhedron theorem, one reverses this process and constructs the group G from a would-be hyperbolic fundamental polyhedron $D \subset H^{n+1}$. Thus, one matches faces of D that would be identified by elements of G. Because the G-transforms of D would have to fill H^{n+1} without gaps and overlapping, we obtain some conditions for D. They are simple and natural if $n = 1$. In higher dimensions they become more complicated, although somehow intuitively evident and easily believable and, beginning from Poincaré [Po], there are a number of proofs although most seem to be deficient in rigor, cf. [EP] for a discussion.

This section would not be complete without mention of Thurston's geometrization theorem. Thurston gave conditions for a Haken 3-manifold to be the quotient of a Kleinian group. This result was one of the highlights of mathematics of recent years, and as it has had much coverage we will not discuss it here.

Rigidity

This is perhaps the aspect where the theory of higher-dimensional Möbius groups differs most from the theory of Fuchsian groups and the classical Kleinian groups. It is well known that these groups have nontrivial deformations; i.e., if G is such a group, we can find homeomorphisms f_t, $t \in [0,1]$, with $f_0 = \mathrm{id}$ of $\bar{C}$ conjugating G onto another Kleinian group G_t such that no G_t, $t > 0$, is conjugate by a Möbius transformation to G.

It is often practical to require that each f_t satisfies the regularity condition called quasiconformality (roughly this means that there are uniform bounds for the distortion of an infinitesimal $(n-1)$-sphere), and instead of the nonconjugacy of G_t to G in $\mathrm{Möb}(n)$ to require that for no $t > 0$ the isomorphism $\varphi_t : G \to G_t$ such that $\varphi_t(g) = f_t g f_t^{-1}$ is a conjugation by a Möbius transformation (although it might be that G_t is conjugate to G by some Möbius transformation). This is the way leading to Teichmüller spaces.

Mostow's rigidity theorem asserts definitively that this is not possible for groups $G \subset \mathrm{Möb}(n)$ such that $M_G = H^{n+1}/G$ has finite hyperbolic volume (proved in [M] if M_G is compact). If $\varphi : G \to H$ is any isomorphism between two finite volume Möbius groups of H^{n+1}, $n \geq 2$, then φ is a conjugation in $\text{Möb}(n)$. We will now describe some recent extensions of Mostow's theorem.

In the proof of Mostow's theorem, the first step was to show that if $\varphi : G \to H$ is an isomorphism of finite-volume Möbius groups, then there is a quasiconformal homeomorphism f of the boundary $\bar{R}^n$ of the hyperbolic space inducing φ; that is, $fg = \varphi(g)f$ for $g \in G$. Such a map $f : \bar{R}^n \to \bar{R}^n$ inducing φ is called the boundary map of φ. In the extensions of Mostow's theorem we will consider, we assume the existence of the boundary map (not necessarily quasiconformal nor even a homeomorphism) and the desired conclusion is that the boundary map is in fact a Möbius transformation.

Agard's theorem [Ag1] asserts that a quasiconformal boundary map of an isomorphism $\varphi : G \to H$ of two Möbius groups is a Möbius transformation as soon as G is of the divergence type; that is, the Poincaré series (see (1) below) of G diverges at the exponent $\delta = n$.

In this direction the most powerful theorem seems to be due to Sullivan. His theorem concerns groups with conservative action; i.e., whenever $A \subset R^n$ has positive measure, then $gA \cap A$ has positive measure for infinitely many $g \in G$. In this situation a quasiconformal boundary map is a Möbius transformation, cf. [S2] if $n = 2$ and [S4] for general n.

For nonquasiconformal boundary maps it is possible to show that either f is the restriction of a Möbius transformation or somehow badly behaved. Let f be the boundary map of an isomorphism of groups of the divergence type. If f is a measurable bijection and if f^{-1} is also measurable, then either f is a.e. the restriction of a Möbius transformation or f is singular in the sense that f maps a null set onto a set of full measure, cf. Sullivan [S4] and [T2].

The above theorems have a global character. The following is a more local one. Let $\varphi : G \to H$ be an isomorphism of two Möbius groups that are not finite extensions of abelian groups. If $f : \bar{R}^n \to \bar{R}^n$ is the boundary map of φ, then, unless f is a Möbius transformation, f cannot be differentiable with a nonsingular derivative at any conical limit point of G (cf. the last section for conical limit points), see [T1]. If the group is cocompact, every $x \in \bar{R}^n$ is a conical limit point and hence in this case either f is a Möbius transformation or it cannot be differentiable with a nonsingular derivative at any point $x \in \bar{R}^n$. For $n > 1$, this implies Mostow's theorem. Recently, Ivanov [I] has obtained extensions of these results concerning the situation where the differential of f vanishes but f is differentiable of order $p > 1$ at a point (i.e., f can be approximated by a homogeneous polynomial of order p at the point).

Deformations of conformal structures

Despite Mostow's rigidity theorem, it may be possible to deform a cocompact or finite-volume Möbius group $G \subset \text{Möb}(n)$ if we move to a larger group, for instance to $\text{Möb}(n + 1)$. Geometrically this means the following. Recall that we initially regarded $g \in \text{Möb}(n)$ as a homeomorphism of $\bar{R}^n$, but it was possible to extend g uniquely to a Möbius transformation of H^{n+1} by representing g as a composition of inversions on spheres. Of course, in the same manner we could have extended g to a Möbius transformation of $\bar{R}^{n+1}$ and of H^{n+2}.

Thus, we can regard G as a Möbius group of $\bar{R}^{n+1}$ and of H^{n+2} and now the quotient H^{n+2}/G is no more compact or of finite volume even if this were true for $M_G = H^{n+1}/G$. Hence, Mostow's rigidity theorem does not bind us and it becomes possible to deform G in certain situations, for instance if M_G has a totally geodesic submanifold M_0 of codimension 1. Totally geodesic means that any geodesic of M_G containing two points of M_0 is entirely contained in M_0.

The construction of the deformation is a kind of "bending" around the codimension 1 submanifold M_0. Actual "bending" is done in the universal cover of M_G. It is possible to give a good geometric picture of the deformation if $n = 1$; that is, if G is a Fuchsian group. The visualization of the deformation is easier if we transform the group G by a Möbius transformation so that S^1 becomes the invariant circle. Now the totally geodesic submanifold M_0 is a topological circle of B^2/G.

In addition the upper hemisphere $S_+ = \{(x_1, x_2, x_3) \in S^2 : x_3 > 0\}$ of S^2 is G-invariant and there is a natural G-equivariant identification of the 2-ball B^2 and of S_+. In this identification M_0 is a compact 1-submanifold of the compact surface S_+/G.

Let L_i be the lifts of M_0 to S_+. Then L_i are half-circles orthogonal to S^1. Because M_0 does not have self-intersections, the lines L_i do not intersect. Take one L_i. It divides S_+ into two parts S_i and S_i' forming (with L_i) the hemisphere S_+, and the angle between these two parts is π. Now we bend one of them, say S_i, by a certain amount so that S_i makes an angle α with the old position of S_i. The bending is effected by means of a rotation in the hyperbolic space H^3 around the line L_i and so the new S_i is part of a 2-sphere orthogonal to R^2. The former smooth S_+ now consists of two parts, and the angle between them is $\pi - \alpha$. The earlier circle $\bar{S}_+ \cap R^2$ is now the union of two half-circles making an angle at the points where they intersect.

This of course breaks the symmetry of S_+ and it is no more F-invariant for a nontrivial Möbius group F. However, if we bend consistently at all the lines L_i and if the angle α we chose is not too big, we obtain a nonsmooth topological hemisphere S, invariant under a Möbius group F of $\bar{R}^3$. In addition the action of F and the action of G on S_+ are conjugate: $F = fGf^{-1}$ for a homeomorphism $f : S_+ \to S$. It is possible to extend f to a homeomorphism of $\bar{R}^3$ still conjugating G onto F so that f preserves $\bar{R}^2$. If we regard G and F as groups of $\bar{R}^2$, the construction of F by bending around the half-circles L_i in the hyperbolic three-space is invisible.

In this construction the topologically smooth hemisphere S_+ is transformed to a topological hemisphere S that is smooth except on the bending lines. On the boundary, the smooth circle $S^1 = \partial S_+$ is transformed to a very nonsmooth topological circle $\partial S \subset R^2$, which is a so-called fractal curve with Hausdorff dimension between 1 and 2 and has nowhere tangents. Now ∂S is the invariant (topological) circle of the so-called quasi-Fuchsian group F.

This construction is possible for any discrete Möbius group G of $\bar{R}^n$ such that S^{n-1} is G-invariant and that B^n/G contains a totally geodesic codimension 1 submanifold. It follows that if S_+ is the upper hemisphere of $S^n \setminus S^{n-1}$, then also S_+/G contains a totally geodesic submanifold M_0 such that the lifts of M_0

are $(n-1)$-dimensional half-spheres orthogonal to $\bar{R}^n$. As above, it is possible to "bend" around lifts of M_0 corresponding to a not too large angle $\alpha \in [0, \alpha_0]$. In the construction $\bar{R}^n$ remains invariant, but S^{n-1} is transformed to a very nonsmooth n-sphere. The conformal structure of components of $(\bar{R}^n \setminus S^{n-1})/G$ is changed and so we can change the conformal structure, although we could not change the hyperbolic structure.

Thus, there is a deformation theory for higher-dimensional Möbius groups. The example we have described is the so-called Mickey Mouse example of Thurston [Th, 8.7.3]. It seems that Apanasov [Ap1] was the first to construct a nontrivial deformation of $G \subset \text{Möb}(n)$ in $\text{Möb}(n+1)$. Apanasov has also found some ways to deform Möbius groups other than "bending" [Ap2]. See also Kourouniotis [K] as well as Johnson and Millson [JM].

Dynamic and ergodic properties

In this section, we change from H^{n+1} to the ball B^{n+1} and assume that G is a discrete Möbius group on B^{n+1} and on its boundary S^n.

The *limit set* $L(G)$ of G is the set of accumulation points of an orbit Gz, $z \in B^{n+1}$ (it is independent of z). Because G is discrete, all the accumulation points of Gz are on the boundary and hence $L(G) \subset S^n$. If $L(G)$ contains more than two points, G is nonelementary and in this situation we will now consider the dynamic behavior of G on $L(G)$. In contrast to earlier topics, here the qualitative dynamic behavior of G on $L(G)$ is much the same for the classical groups ($n = 1$ or $n = 2$) and for the higher-dimensional Möbius groups.

In this connection, the *exponent of convergence* δ_G of G is an important constant. It mirrors the growth of the orbital counting function; that is, the number $N(r)$ of points in a given orbit Gz, $z \in B^{n+1}$, whose hyperbolic distance from a given point is less than r. The exponent of convergence is defined by means of the Poincaré series of G, which can be given as the sum

$$P_\delta(z) = \sum_{g \in G} e^{-\delta d(z, g(z))}. \tag{1}$$

Here $z \in B^{n+1}$ is an arbitrary basepoint whose choice does not affect the convergence or divergence of (1). The number δ is the exponent of the Poincaré series and there is a certain critical value δ_G, the exponent of convergence of G, such that (1) converges for $\delta > \delta_G$ and diverges for $\delta < \delta_G$; for $\delta = \delta_G$ the series may converge or diverge. It is a standard result that $\delta_G \leq n$ [Ah1] and the Poincaré series converges for $\delta = n$ if the action of G is discontinuous somewhere on S^n.

An alternative way to define the Poincaré series, more reminiscent of the classical definition, would be to use the series

$$\sum_{g \in G} |g'(z)|^\delta. \tag{2}$$

Here $|g'|$ is the operator norm of the derivative. For groups of B^{n+1} (but not of H^{n+1}), the convergence of (1) is equivalent to that of (2).

In the study of the dynamic behavior of G on $L(G)$, the Patterson-Sullivan measure μ is of decisive importance. It was first defined by Patterson [Pa] for $n = 1$. Sullivan [S3], [S6] extended the definition for $n > 1$ and proved the biggest part of many of the following results. The measure μ is a probability measure supported by the limit set and satisfies the transformation rule

$$\mu(gX) = \int_X |g'|^\delta \, d\mu \qquad (3)$$

for $g \in G$ and a Borel subset X of $L(G)$, and where δ is the exponent of convergence δ_G. The Patterson-Sullivan measure can be simply constructed by taking a basepoint $z \in B^{n+1}$, an exponent $\delta > \delta_G$, putting a point mass of weight $|g'(z)|^\delta$ at $g(z)$, and normalizing so that the total mass is 1. Call this measure μ_δ, let $\delta \to \delta_G$, and take the weak limit. In this manner we obtain a probability measure supported by $L(G)$ and satisfying (3) (this follows easily from the form (2) of the Poincaré series), provided that the Poincaré series diverges at the exponent of convergence. If it converges, then the construction can be modified.

We call a finite measure satisfying (3) a *conformal measure*; the number δ is the dimension of the measure. For instance, the n-dimensional Hausdorff measure on S^n is a familiar example of a conformal measure. As another example, if G is of compact type; that is, $(\bar{B}^{n+1} \setminus L(G))/G$ is compact, then the δ_G-dimensional Hausdorff measure is a nontrivial conformal measure on $L(G)$, cf. [S3].

If the Poincaré series diverges at the exponent of convergence, we have a good picture of the dynamic behavior. Now the measure μ is ergodic, not only with respect to the action of G on $L(G)$, but also the action of $(x, y) \mapsto (g(x), g(y))$ of G on $L(G) \times L(G)$ is ergodic with respect to $\mu \times \mu$. Here we mean by ergodicity that every G-invariant subset has either full or null measure.

This situation can also be characterized geometrically. We say that $x \in L(G)$ is a *conical limit point* if, given a point $z \in B^{n+1}$, there are $g_i \in G$ such that $g_i(z) \to x$ and $g_i(z)$ are at bounded hyperbolic distance from the line segment joining 0 and x. If the Poincaré series diverges at the exponent of convergence, the conical limit points have full μ-measure; otherwise, the μ-measure is 0.

These conditions can be reversed and the following three conditions are equivalent for a nontrivial conformal measure μ of dimension δ on $L(G)$:

   1. The Poincaré series diverges at the exponent δ.
   2. The conical limit points have full μ-measure.
   3. The action of G on $L(G) \times L(G)$ is ergodic with respect to $\mu \times \mu$.
   If one of these conditions is true, then $\delta = \delta_G$, and μ is uniquely determined up to multiplication by a constant.

The proof of the fact that convergence at the exponent of convergence implies null measure for conical limit points is fairly easy [N, Theorem 4.4.1], but the other direction is much more complicated. There are two main lines of proof, one of them due to Sullivan [S2] and the other to Thurston (given, e.g., in [Ah1]). These have been generalized for general conformal measures in [N] and [T3], respectively, cf. also [H].

The equivalence of conditions 2 and 3 above is due to the fact that the ergodicity of G on $L(G) \times L(G)$ is equivalent to the ergodicity of the geodesic flow on the hyperbolic convex hull H_G of $L(G)$ (i.e., $H_G \subset B^{n+1}$ is the smallest closed and hyperbolically convex set whose closure contains $L(G)$; if $L(G) = S^n$, then H_G is just B^{n+1}), and this is equivalent to the full measure of conical limit points. This result goes back to Hopf with contributions from many mathematicians, cf. the discussion in [S2]. In the form stated above it can be found in [N]. Now the uniqueness and the fact that $\delta = \delta_G$ follow from [S3, Theorem 21] as there can be no atoms.

We found that the ergodicity on the product $L(G) \times L(G)$ could be characterized by means of the conical limit points. A similar characterization is possible for the conservativity of the action of G. We say that G acts *conservatively* on X with respect to μ if, whenever $A \subset X$ is measurable and $\mu(A) > 0$, then $\mu(gA \cap A) > 0$ for infinitely many $g \in G$. If G is a group of measurable maps of a measure space A, then the following is valid in a fairly general situation. We can divide A into two measurable and disjoint G-invariant parts, $A = A' \cup A''$, so the action is conservative in A' and that the action of G on A'' has a measurable fundamental set F; that is, F contains exactly one point from each orbit Gz, $z \in A''$ (cf. [Ag2] if $G \subset \text{Möb}(2)$ and μ is atomless).

If μ is the Hausdorff n-measure on S^n, then Sullivan [S2] has characterized the conservative part of the action of a discrete Möbius group in S^n as the horospherical limit set. A horosphere at $x \in S^n$ is an n-sphere $S \subset B^{n+1} \cup \{x\}$ such that ∂S is tangent to S^n at x. A point $x \in S^n$ is in the horospheric limit set of G, if inside every horosphere at x there is an infinite number of points from any orbit Gz, $z \in B^{n+1}$. The same characterization for the conservativity of the action is valid for arbitrary conformal measures, at least if we change the definition of the horospheric limit set $\mathcal{H}(G)$ so that $x \in \mathcal{H}(G)$ if and only if, given $z \in B^{n+1}$, there is a horosphere S at x such that inside S there is an infinite number of points of Gz (cf. [T4]).

The property that the action of G is ergodic on $L(G)$ is an intermediate property between conservative action on $L(G)$ and ergodic action on $L(G) \times L(G)$. It is curious that there does not seem to be a geometric description of this situation, like the horospheric and conical limit point sets in the two other cases.

The strongest form of ergodicity, ergodicity of the action on $L(G) \times L(G)$, is frequently met because some common categories of groups have this property. If $(\bar{B}^{n+1} \setminus L(G))/G$ is compact, then every $x \in L(G)$ is a conical limit point and hence the product action is ergodic. More generally, the product action is ergodic if G is geometrically finite [S6]; i.e., the action of G on B^{n+1} has a finite-sided fundamental polyhedron.

The weaker types of ergodic action are also met, although they are not so obvious. Let λ be the Hausdorff 1-measure. There are Fuchsian groups on B^2 such that the action is conservative but not ergodic on S^1 with respect to λ, as well as groups with ergodic action on S^1 but not on $S^1 \times S^1$, see [T2, 4F] (note that the inclusion relation of O_{HB} and O_G was stated in the wrong way in [T2]). The construction of the group with conservative but nonergodic action on S^1 is based on the following observations. Let G be a Fuchsian group on B^2 such that G is the universal cover translation group of $\bar{C} \setminus X$ where X is a closed and bounded subset of the real line. By [Pm, Ex. 4] the action of G is conservative on S^1 if and only if $\lambda(X) = 0$. If the capacity of X is positive, the argument of [T2, 4F] implies that the action is nonergodic on S^1.

Some recent results allow the addition of some groups to the category of ergodic but not product-ergodic groups. Suppose that G is a finitely generated, totally degenerate Kleinian group of $\bar{C}$; that is, $\bar{C} \setminus L(G)$ is simply connected. Then Bonahon's results [B] imply that the areal measure of $L(G)$ vanishes and we can infer from Bishop-Jones [BJ, Theorem 13.2] that $\delta_G = 2$. Because $L(G) \neq S^n$, the Poincaré series P_δ converges for $\delta = 2$ and so P_δ converges for $\delta = \delta_G$. Hence the action of G on $L(G) \times L(G)$ is not ergodic with respect to the product of any conformal measure on $L(G)$. On the other hand, Sullivan [S5] has constructed some such groups with conformal measure μ of dimension 2 on $L(G)$ such that the action is ergodic on $L(G)$ with respect to μ.

The groups constructed by Sullivan are obtained from a Fuchsian group F by taking to the limit so that one of the complementary domains of $L(F)$ shrinks away. On the other hand, it is possible to take the limit in such a way that both complementary domains of $L(F)$ shrink away so that S^2 is the limit set of the limit group. Such a group is geometrically tame by Bonahon [B], and Thurston [Th, 8.11] implies that the geodesic flow is ergodic and hence the action of G is ergodic on $S^2 \times S^2$ with respect to the areal measure.

References

[Ah1] L. V. Ahlfors, *Möbius transformations in several dimensions*, Lecture notes at the University of Minnesota, 1981.

[Ah2] —, *Finitely generated Kleinian groups*, Amer. J. Math. **86** (1964), 413–429.

[Ag1] S. Agard, *A geometric proof of Mostow's rigidity theorem for groups of divergence type*, Acta Math. **151** (1983), 231–252.

[Ag2] —, *Mostow rigidity on line: A survey*, in: Holomorphic Functions and Moduli II (D. Drasin et al., eds.), MSRI Publications **11**, 1988, 1–12.

[Ap1] B. Apanasov, *Non-triviality of Teichmüller space for Kleinian groups in space*, in: Riemann Surfaces and Related Topics (I. Kra and B. Maskit, eds.), Ann. of Math. Stud. **97**, Princeton University Press, Princeton, 1981, 21–31.

[Ap2] —, *Deformations of conformal structures on hyperbolic manifolds*, J. Differential Geometry **35** (1992), 1–20.

[B] F. Bonahon, *Bouts des variétés hyperboliques de dimension 3*, Ann. of Math. (2) **124** (1986), 71–158.

[BJ] C. J. Bishop and P. W. Jones, *Hausdorff dimension and Kleinian groups*, to appear.

[EP] D. B. A. Epstein and C. Petronio, *An exposition of Poincaré's polyhedron theorem*, to appear.

[H] S. Hong, *Patterson-Sullivan measure and groups of divergence type*, Bull. Korean Math. Soc. **30** (1993), 223–228.

[I] N. V. Ivanov, *Action of Möbius transformations on homeomorphisms: Stability and rigidity*, to appear.

[JM] D. Johnson and J. J. Millson, *Deformation spaces associated to compact hyperbolic manifolds*, in: Discrete Groups in Geometry and Analysis (R. Howe, ed.), Birkhäuser, Basel and Boston, 1987, 48–106.

[K] C. Kourouniotis, *Deformations of hyperbolic structures*, Math. Proc. Cambridge Philos. Soc. **98** (1985), 247–261.

[M]   G. D. Mostow, *Quasiconformal self-mappings in n-space and the rigidity of hyperbolic space forms*, Inst. Hautes Études Sci. Publ. Math. **34** (1968), 53–104.

[N]   P. J. Nicholls, The ergodic theory of discrete groups, London Math. Soc. Lecture Note Ser. **143**, 1989.

[Pa]   S. J. Patterson, *The limit set of a Fuchsian group*, Acta Math. **136** (1976), 241–273.

[Po]   H. Poincaré, *Mémoire sur les groupes Kleinéens*, Acta Math. **3** (1883), 49–92.

[Pm]   Ch. Pommerenke, *On Fuchsian groups of the accessible type*, Ann. Acad. Sci. Fenn. Ser. A I Math. **7** (1982), 249–258.

[S1]   D. Sullivan, *Hyperbolic geometry and homeomorphisms*, in: Geometric Topology, Proc. of the 1977 Georgia Topology Conference (J. C. Cantrell, ed.), Academic Press 1979, 543–555.

[S2]   —, *On the ergodic theory at infinity of an arbitrary discrete group of hyperbolic motions*, in: Riemann Surfaces and Related Topics (I. Kra and B. Maskit, eds.), Ann. of Math. Stud. **97**, Princeton University Press, Princeton, 1981, 465–496.

[S3]   —, *The density at infinity of a discrete group of hyperbolic motions*, Inst. Hautes Études Sci. Publ. Math. **50** (1979), 171–202.

[S4]   —, *Discrete conformal groups and measurable dynamics*, Bull. Amer. Math. Soc. N.S. **6** (1982), 57–73.

[S5]   —, *Growth of positive harmonic functions and Kleinian group limit sets of geometrically finite Kleinian groups of zero planar measure and Hausdorff dimension two*, in: Geometry Symposium Utrecht 1980 (E. Looijenga, D. Siersma, and F. Takens, eds.), Lecture Notes in Math. **894**, Springer-Verlag, Berlin and New York, 1981, 127–144.

[S6]   —, *Entropy, Hausdorff measures old and new, and limit sets of geometrically finite Kleinian groups*, Acta Math. **153** (1984), 259–277.

[Th]   W. P. Thurston, Geometry and topology of three-manifolds, Lecture Notes at the Princeton University, 1980.

[T1]   P. Tukia, *Differentiability and and rigidity of Möbius groups*, Invent. Math. **82** (1985), 557–578.

[T2]   —, *A rigidity theorem for Möbius groups*, Invent. Math. **97** (1989), 405–431.

[T3]   —, *The Poincaré series and the conformal measure of conical and Myrberg limit points*, J. Analyse Math. **62** (1994), 241–259.

[T4]   —, *Conservative action and the horospheric limit set*, to appear.

[TV]   P. Tukia and J. Väisälä, *Quasiconformal extension from dimension n to n + 1*, Ann. of Math. (2) **115** (1982), 331–348.

Geometric and Dynamical Aspects of Real Submanifolds of Complex Space

SIDNEY M. WEBSTER

Department of Mathematics, University of Chicago,
Chicago, Illinois 60637, USA

Introduction

This work is founded on an analogy between complex analysis and classical mechanics, which at first glance may not seem too meaningful. Our purpose is to show, however, that not only is it useful as a formal guide, but that the interplay is substantial at the level of mathematical proof.

The solvable, or integrable, problems of mechanics are well known: the pendulum, Kepler's problem, certain tops, geodesics on a tri-axial ellipsoid, ... They have been studied extensively in recent times, as well as classically. They shed a great deal of light on the general theory. What are the *explicitly* solvable problems of complex analysis?

More precisely, consider the Riemann mapping problem. For which domains in the complex plane is this problem explicitly solvable? Perhaps the best example from the point of view here is the interior E of an ellipse. In 1869 H. A. Schwarz gave an explicit formula for the conformal map of E onto a disc, using trigonometric and elliptic functions. (In 1868 E. Mathieu had shown that the eigenvalue problem for the laplacian was explicitly solvable; i.e. separable, for E. The billiard problem for E was shown to be integrable in 1927 by G. D. Birkhoff, although the essential reason was known to the classical geometers.) The mapping problem is also solvable for domains bounded by hyperbolas, parabolas, lemniscates, and cuspidal cubics.

Recently in [25], a large family of other domains for which the Riemann mapping problem is explicitly solvable has been found. It includes domains bounded by real branches of certain nonsingular cubics, binodal quartics, and bicuspidal quartics. These curves are termed [18] circular, bicircular, and cartesian, respectively. Although they have been extensively studied in the past, apparently not much has been written on the mapping problem for the domains that they bound. What is perhaps most interesting is that *all* the examples mentioned here have a common explanation: the bounding curves admit *double valued reflection* [25].

What happens in $\mathbf{C}^n$ for $n \geq 2$? Of course, there is no Riemann map, as was shown by Poincaré. Instead, for a bounded smooth strongly pseudoconvex domain we may consider the problem of finding a number of intrinsic objects: the Bergman kernel, the Szegö kernel relative to an appropriate surface measure, the Carathéodory and Kobayashi metrics, the Moser normal form, the chains of Cartan-Chern-Moser, ... None of these are explicitly known, except for those few

Proceedings of the International Congress
of Mathematicians, Zürich, Switzerland 1994
© Birkhäuser Verlag, Basel, Switzerland 1995

domains with abundant symmetries. Are some of these problems solvable for ellipsoidal domains? (Just recently, we have found that a relevant dynamical problem constitutes an integrable system for ellipsoids.)

For $n \geq 2$ one can also consider real submanifolds of higher codimension. That the case of a real n-manifold M^n in $\mathbf{C}^n$ has special significance was realized by Bishop [3]. Generically, M is either totally real or has complex tangents along a subset of codimension two. If M is also real analytic, it is locally the fixed point set of an anti-holomorphic involution, or reflection. Near nondegenerate complex tangents with nonvanishing Bishop invariant γ, this reflection becomes double valued. Some problems considered for M are to find: analytic discs bounding on M, the hull of holomorphy of M, a biholomorphic flattening (i.e. transformation into a real hyperplane) of M, and the normal form of [14] near suitable complex tangents. Considerable progress on the last two of these problems has been made recently in the Ph.D. thesis of Gong [7].

1. Double valued reflection. Complexification

A multiple valued reflection on a complex manifold is an anti-holomorphic, involutive correspondence assigning to each point z a complex subvariety Q_z,

$$z \mapsto Q_z, \ z \in Q_w \Leftrightarrow w \in Q_z. \tag{1.1}$$

In the double valued case Q_z is zero dimensional and has at most, and in general, two points. Thus we have

$$Q_{z_0} = \{z_1, z_1'\}, \ Q_{z_1} = \{z_0, z_2\}, \ Q_{z_2} = \{z_1, z_3\}, \ \ldots;$$

generating a sequence

$$z_0 \mapsto z_1 \mapsto z_2 \mapsto z_3 \mapsto \ldots . \tag{1.2}$$

A similar sequence is generated if we choose z_1'. The central problem is to understand the dynamics of this process.

Consider an analytic real submanifold of codimension l,

$$M = \{z \in \mathbf{C}^n | R(z, \bar{z}) = 0, \ R = \bar{R} = (r^1, \ldots, r^l)\}. \tag{1.3}$$

Its complexification $\mathcal{M}$ gives rise to a family of complex subvarieties of $\mathbf{C}^n$,

$$\mathcal{M} = \{R(z, \zeta) = 0\}, \ Q_w \cong \mathcal{M} \cap \{\zeta = \bar{w}\}. \tag{1.4}$$

This process plays a key role in many areas of complex analysis:

- differential invariants and normal form(Segre, Cartan, [4], [6]);

- algebraicity of holomorphic maps [20], [21], [26] and of analytic sets [19];

- biholomorphic classification of ellipsoids [20], [21], [26];

- single valued reflection principle (Lewy, Pinchuk, [22], [17], [5], [1], [24]);

- boundary regularity of biholomorphic maps [15].

We believe that the usefulness of complexification is far from being exhausted. It can shed light on nearly every aspect of complex analysis relating to real submanifolds. We have only given a smattering of relevant references.

The most natural (single valued) reflection fixing M is that on its complexification

$$\rho : \mathcal{M} \to \mathcal{M}, \ \rho(z, \zeta) = (\overline{\zeta}, \overline{z}). \tag{1.5}$$

Reflection about M in $\mathbf{C}^n$ is double valued precisely when the two mappings

$$\pi_i : \mathcal{M} \to \mathbf{C}^n, \ \pi_1(z, \zeta) = z, \pi_2(z, \zeta) = \zeta, \tag{1.6}$$

are two-fold branched coverings. They then have covering involutions

$$\tau_i : \mathcal{M} \to \mathcal{M}, \ \pi_i \circ \tau_i = \pi_i, \ \tau_i^2 = I, \ i = 1, 2, \ \tau_2 = \rho\tau_1\rho. \tag{1.7}$$

The process (1.2) is described in a single valued way by the dynamics of the map

$$\sigma = \tau_1\tau_2. \tag{1.8}$$

It is *reversible*; i.e. conjugate to its inverse by an involution.

2. Real algebraic curves in the complex plane

In this case $M \subset \mathbf{C}$, R is a single real polynomial, and $\mathcal{M} \subset \mathbf{C}^2$ is a complex algebraic curve. If $\mathcal{M}$ admits two distinct two-fold branched coverings π_1, π_2 onto $\mathbf{P}_1$, then a classical result says that it is either a rational curve or an elliptic curve. In both cases it is easy to classify the possible data $(\mathcal{M}, \tau_1, \rho)$ [25].

If $\mathcal{M}$ has genus zero, one gets fractional linear involutions, which are classified [14] as hyperbolic, elliptic, or parabolic. These yield the ellipse, hyperbola, and parabola, respectively. For the ellipse, for example, π_1 is the "Zhoukowski map". If f is the Schwarz-Riemann map of the ellipse onto the right half plane, then $f \circ \pi_1 \circ exp$ is readily given by a Weierstrass sigma quotient relative to a suitable rectangular lattice [25].

In the genus one case, $\mathcal{M} = \mathbf{C}/\Lambda$ must admit a reality structure; i.e. the modular function must be real. Then a suitable anti-holomorphic involution ρ and a suitable holomorphic involution τ_1 on $\mathcal{M}$ must be determined. A function π_1 invariant by τ_1 is given in terms of the Weierstrass $\wp$-function of Λ. This leads to a real quartic curve $M \subset \mathbf{C}$. For a particular case [25] we gave the Riemann map for the simply connected domain bounded by a component of M in the form $\hat{\phi} \circ \wp^{-1}$. Here $\hat{\phi}$ is a Weierstrass sigma quotient relative to another lattice $\hat{\Lambda}$ determined by τ_1.

3. Surfaces in $\mathbf{C}^2$

Somewhat ironically, double valued reflection was first systematically studied in the higher dimensional case. In [14] we gave a holomorphic normal form for analytic real surfaces $M^2 \subset \mathbf{C}^2$ having complex tangent at 0 with nonexceptional Bishop invariant γ. This was derived from a holomorphic normal form for the involution pair τ_1, τ_2 on the complex surface $\mathcal{M}$. The fixed point sets of these maps are complex curves meeting at 0. For an elliptic complex complex tangent ($0 < \gamma < 1/2$) 0 is a hyperbolic fixed point of σ, and the normal form converges, giving, in particular, a holomorphic flattening of M near 0.

If $1/2 < \gamma < \infty$, the surface is hyperbolic, but the map σ has 0 as an elliptic fixed point, which entails the usual possibly complicated dynamics. For a countable dense set of γ in this range (the exceptional ones) the normal form does not exist. But even for nonexceptional $\gamma > 1/2$, there are surfaces that cannot be flattened [14], so that the transformation into normal form must therefore diverge. Holomorphic flattening of M is equivalent to the existence of an integral, or invariant function, for the reversible map σ.

If M is already holomorphically flat, does its normal form then converge? This reduces to the following question. Suppose that a reversible transformation with elliptic fixed point at 0 in the plane has a nontrivial integral. Does its transformation into normal form converge? There are many parallels between reversible and area preserving maps. Birkhoff [2] showed that the answer is affirmative in the area preserving case. Surprisingly, the answer is no for reversible maps, as was shown by Gong [10]. In [11] he gives a further study, showing that the level curves of the integral can be transverse to the Birkhoff curves in the reversible case, whereas the two families of curves must coincide in the area preserving case. (Birkhoff curves are curves of points that are mapped radially by some iterate of the map.) This has a dynamical significance beyond its application to complex analysis.

Gong [8] also makes a study of surfaces under unimodular transformation of $\mathbf{C}^2$. Generally, there are more invariants, even at totally real points. For this case he makes use of Vey's unimodular Morse lemma. He also derives a unimodular normal form at nondegenerate complex tangents, sorting out the rather complicated jumble of invariants. In the case $\gamma = 0$ he proves convergence of the normalizing map, provided that all the invariants vanish, by adapting a KAM argument of Moser [13].

In [23] we showed that real Lagrangian surfaces ($\mathrm{Re}(dz_1 \wedge dz_2) = 0$ on M) are are all formally equivalent, near generic complex tangents, via unimodular holomorphic transformation. In [24] a similar result was given in holomorphic contact geometry and applied to normalize real hypersurfaces with generic Levi-form degeneracies. However, Gong [7] showed, using a remark of Moser on the linearized problem, that the normalizing transformation generally diverges. The theory reduces to that of a parabolic pair of involutions. Such involution pairs also arise in the theory of glancing hypersurfaces [12] in the standard symplectic space $\mathbf{R}^{2n}$. The corresponding normal forms also diverge, in general. This was shown by Oshima [16] for $n \geq 3$. Gong [9], by adapting the above methods, has extended this to $n \geq 2$.

References

[1] M. S. Baouendi and L. P. Rothschild, *A general reflection principle in C^2*, J. Funct. Anal. 99(1991)409–442.

[2] G. D. Birkhoff, *Surface transformations and their dynamical applications*, Acta Math. 43(1920)1–119.

[3] E. Bishop, *Differentiable manifolds in complex Euclidean space*, Duke Math. J. 32(1965)1–22.

[4] S. S. Chern and J. K. Moser, *Real hypersurfaces in complex manifolds*, Acta Math. 133(1974)219–271.

[5] K. Diederich and S. M. Webster, *A reflection principle for degenerate real hypersurfaces*, Duke Math. J. 47(1980)835–843.

[6] J. J. Faran, *Segre families and real hypersurfaces*, Invent. Math. 60 (1980) 135–172.

[7] X. Gong, *Real analytic submanifolds under unimodular transformation*, Thesis, University of Chicago, August, 1994.

[8] ———, *Normal forms of real surfaces under unimodular transformations near elliptic complex tangents*, Duke Math. J. 74(1994)145–157.

[9] ———, *Divergence for the normalizations of real analytic glancing hypersurfaces*, Comm. Partial Differential Equations 19(1994)643–654.

[10] ———, *On the convergence of normalizations of real analytic surfaces near hyperbolic complex tangents*, to appear.

[11] ———, *Fixed points of elliptic reversible transformations with integrals*, to appear.

[12] R. B. Melrose, *Equivalence of glancing hypersurfaces*, Invent. Math. 37 (1976) 165–191.

[13] J. K. Moser, *Analytic surfaces in C^2 and their local hulls of holomorphy*, Ann. Acad. Sci. Fenn. Ser. A I Math. Dissertationes 10(1985)397–410.

[14] J. K. Moser and S. M. Webster, *Normal forms for real surfaces in C^2 near complex tangents and hyperbolic surface transformations*, Acta Math., 150 (1983) 255–296.

[15] L. Nirenberg, S. Webster, and P. Yang, *Local boundary regularity of holomorphic mappings*, Comm. Pure Appl. Math. 33(1980)305–338.

[16] T. Oshima, *On analytic equivalence of glancing hypersurfaces*, Sci. Papers College Gen. Ed. Univ. Tokyo 28(1978)51–57.

[17] S. I. Pinchuk, *The reflection principle for manifolds of codimension two*, Soviet Math. Dokl. 34(1987)129–132. (from Dokl. Akad. Nauk. SSSR 289(1986)).

[18] G. Salmon, *A treatise on the higher plane curves*, Dublin, 1879.

[19] A. Sukhov, *On algebraicity of complex analytic sets*, Math. USSR-Sb. 74 (1993) 419–426 (from Mat. Sbornik 182 (1991)).

[20] S. M. Webster, *On the mapping problem for algebraic real hypersurfaces*, Invent. Math. 43(1977)53–68.

[21] ———, *Some birational invariants for algebraic real hypersurfaces*, Duke Math. J. 45(1978)39–46.

[22] ———, *On the reflection principle in several complex variables*, Proc. Amer. Math. Soc. 71(1978)26–28.

[23] ———, *Holomorphic symplectic normalization of a real function*, Ann. Scuola Norm. Sup. Pisa Cl. Sci. (4) 19(1992) 69–86.

[24] ———, *Holomorphic contact geometry of a real hypersurface*, to appear in Gunning-Kohn Conf. Proc.

[25] ———, *Double valued reflection in the complex plane*, to appear.

[26] ———, *Non-degenerate analytic real hypersurfaces*, to appear.

The Classification Problem for Amenable C*-Algebras

GEORGE A. ELLIOTT

Matematisk Institut   and   Department of Mathematics
Københavns Universitet   University of Toronto
DK-2100 København Ø   Toronto, Ontario
Denmark   Canada M5S 1A1

1. For thirty-five years, one of the most interesting and rewarding classes of operator algebras to study has been the approximately finite-dimensional C*-algebras of Glimm and Bratteli ([39], [7]).

Recall that a separable C*-algebra A is said to be approximately finite-dimensional (or AF) if it is generated by an increasing sequence

$$A_1 \subseteq A_2 \subseteq \cdots \subseteq A$$

of finite-dimensional sub-C*-algebras: $A = (\cup A_n)^-$.

It should be recalled that a finite-dimensional C*-algebra is isomorphic to a finite direct sum of matrix algebras over the complex numbers.

One of the striking properties of AF algebras is that they can be classified. A classification in terms of the multiplicity data involved in a given increasing sequence of finite-dimensional subalgebras was obtained in [39] and [7]. A classification in terms of an invariant — the ordered group K_0 — was obtained in [21] and [22]. (Of course, the ordered K_0-group can be computed from the multiplicity data. It can in fact be seen directly to contain the same asymptotic information concerning this data that appears in the classification of [39] and [7].)

The classification works equally well for the AF algebra — the norm closure of the union of the increasing sequence — and for the union itself.

2. For thirty years, the only known AF algebras were those actually given in terms of an increasing sequence of finite-dimensional C*-algebras.

An example for which this was not the case was given by Blackadar in [3]. Blackadar's construction still involved giving an increasing sequence of C*-algebras, of a rather special form, but these were no longer finite dimensional. The algebras Blackadar used were direct sums of matrix algebras over $C(\mathbb{T})$, the algebra of continuous functions on the circle.

The AF algebra constructed by Blackadar was not itself new. It was in fact one of the most familiar such algebras — one of the infinite tensor products of matrix algebras over the complex numbers considered by Glimm (the case of the algebra of 2×2 matrices repeated infinitely often).

Exploiting the symmetry of his construction, though, Blackadar was able to construct a finite-order automorphism of this C*-algebra that was essentially new

Proceedings of the International Congress
of Mathematicians, Zürich, Switzerland 1994
© Birkhäuser Verlag, Basel, Switzerland 1995

— it could not arise through any construction involving only finite-dimensional C*-algebras. More precisely, it could not leave invariant any increasing sequence of finite-dimensional subalgebras with dense union — or even leave invariant the union of such a sequence. This was because the fixed point subalgebra could be shown — by K-theoretical considerations — not to be an AF algebra. (The fixed point subalgebra could be easily calculated, as the closure of the increasing union of the fixed point subalgebras of the algebras in Blackadar's construction, because the automorphism was of finite order.)

One way of putting this, perhaps, is as follows. With Blackadar's construction, the theory of AF algebras changed from being basically algebraic (with a little spectral theory thrown in — which was almost optional) to being essentially topological in nature. In view of the circles appearing in the construction, one could even say that an AF algebra was revealed as a topological object itself, in a real sense. It would no longer be sufficient to view an AF algebra as just a topological completion of a locally finite-dimensional algebra.

3. Topological constructions of AF algebras similar to Blackadar's were soon given by other authors ([48], [8], [37], [11], [13], and [65]). In addition to revealing more and more structure in various AF algebras, this work had other consequences.

Perhaps most surprisingly, it led in a natural way to the question of classifying C*-algebras constructed in terms of increasing sequences, but which did not happen to be AF algebras. To begin with, if some algebras were not AF, why weren't they? In certain constructions, involving a choice of embedding at each stage, the non-vanishing of the K_1-group of the resulting algebra appeared to be the only obstruction to its being AF. (Other obstructions were observed later, but these were also K-theoretical in nature.)

It was just a small step from this to the idea that the complete K-theoretical data, i.e., the (pre-) ordered group K_0, the group K_1, and the space of traces on the algebra (at least in the simple, stable case), should determine the algebra, within a certain class (indeed, that this data might be all there was to see).

This idea has now been borne out in a number of investigations, beginning with [24]. Some of these will be summarized below. So successful has it been, in fact, that one now expects these invariants to determine isomorphism within the class of all stable, non type I, separable, amenable, simple C*-algebras.

4. Recall that a C*-algebra is amenable (equivalently, nuclear — see [17], [42]) if, and only if, its bidual is an amenable von Neumann algebra (see [16], [15]). Amenability itself, in either setting, is defined in terms of a fixed point property (innerness of certain derivations). It is equivalent to amenability of the unitary group — with the weak topology in the C*-algebra setting and the weak* topology in the von Neumann algebra setting ([44], [55]).

Accordingly, on very general grounds, a classification of separable amenable C*-algebras might be hoped for in analogy with the classification of amenable von Neumann algebras with separable pre-dual due to Connes, Haagerup, Krieger, and Takesaki.

5. Until recently, perhaps the main reason for considering the class of amenable C*-algebras as the target for classification was that on account of the Choi-Effros lifting theorem (together with Voiculescu's theorem — which does not use amenability), two homomorphisms from a separable amenable C*-algebra into the Calkin algebra are unitarily equivalent if, and only if, they give rise to the same Kasparov KK-element. This uniqueness theorem resembles very much results that have already proved useful for classifying special classes of amenable C*-algebras.

6. Recently, Kirchberg has shown that any separable, amenable, unital, simple C*-algebra that contains a sequence of copies of the Cuntz algebra $\mathcal{O}_2$ approximately commuting with each element of the algebra must be isomorphic to $\mathcal{O}_2$ ([47]). This very strong isomorphism theorem — a characterization of $\mathcal{O}_2$ — although not formulated in terms of K-theoretical invariants, comes very close to being so.

The desired K-theoretical form of the characterization would be that a nonzero, separable, amenable, unital, simple C*-algebra with the same K-groups as $\mathcal{O}_2$, namely, zero, and with no traces, is isomorphic to $\mathcal{O}_2$. One would hope to deduce the existence of a sequence of embeddings of $\mathcal{O}_2$ as above from the K-theoretical hypotheses.

7. A more explicit description of the invariant under consideration for stable simple C*-algebras is as follows.

(i) The K_0-group, with its natural pre-order structure. (The positive cone, K_0^+, consists of the elements arising from projections in the algebra or, in the nonstable case, in matrix algebras over the algebra.)

(ii) The K_1-group.

(iii) The space T^+ of densely defined, lower semicontinuous, positive traces, with its natural structure of topological convex cone. (This structure is most easily introduced by identifying the traces with the positive tracial linear functionals on the Pedersen ideal — the smallest dense two-sided ideal — and considering the pointwise convex operations and the topology of pointwise convergence.)

(iv) The natural pairing of the cone of traces with the K_0-group. (Any positive tracial functional on the Pedersen ideal can be restricted to a hereditary sub-C*-algebra contained in the Pedersen ideal — on which it must be bounded — and then extended to the algebra with unit adjoined and to matrix algebras over this algebra. In this way, by Brown's stabilization theorem, one gets a functional on K_0 of the original algebra, which can be shown to be independent of the choice of hereditary sub-C*-algebra.)

The nonstable case would involve additional information — but in the unital case presumably just the K_0-class of the unit. The nonsimple case would involve even more information — one should keep track of the ideals and also the associated K-theory and KK-theory data. We shall not consider these cases here.

8. As well as the question of the completeness of the invariant (now established for a fairly large class of algebras), there is also the question of its range. What properties characterize the objects arising as above from separable, amenable, stable, simple C*-algebras?

There are certain properties that the invariant is known to have, in the separable, amenable, stable, simple case.

(i) The K_0-group is countable (and abelian), and the pre-order structure is simple (if $g > 0$ and h is any element, then $-ng \leq h \leq ng$ for some $n = 1, 2, \dots$).

(ii) The K_1-group is countable (and abelian).

(iii) The cone of traces is nonzero whenever $K_0 \neq K_0^+$ and, when nonzero, has a compact base that is a Choquet simplex.

(iv) The pairing of the cone of traces with K_0 gives rise to all positive functionals on K_0, unless $K_0^+ = 0$.

It is an interesting question whether these properties characterize what arises as the invariant in the stable, simple case.

In particular, it is not known whether K_0^+ must be weakly unperforated. (If $g \in K_0$ and $ng \in K_0^+ \setminus 0$ for some $n = 2, 3, \ldots$, then must $g \in K_0^+$?)

9. If G is any simple pre-ordered abelian group (i.e., $G^+ + G^+ \subseteq G^+$ and for any $g \in G^+ \setminus 0$ and $h \in G$ there exists $n = 1, 2, \ldots$ such that $-ng \leq h \leq ng$), then each of the two subgroups $G^+ \cap -G^+$ and $G^+ - G^+$ is equal either to 0 or to G. It follows that there are three cases:

(1) $G^+ = 0$;

(2) $G^+ \cap -G^+ = 0$ and $G^+ - G^+ = G$ (i.e., G is an ordered group);

(3) $G^+ = G$.

The intersection of any two of these cases is the case $G = 0$.

10. If A is any amenable simple C*-algebra, then the three cases enumerated above for the simple pre-ordered group $K_0 A$ may be combined in a natural way with the two main cases for $T^+ A$, as follows.

Case (1): $K_0^+ = 0$; $T^+ \neq 0$.

Case (2): $K_0^+ \cap -K_0^+ = 0$, $K_0^+ - K_0^+ = K_0 \neq 0$; $T^+ \neq 0$.

Case (3): $K_0^+ = K_0$; $T^+ = 0$.

These cases are now clearly disjoint. In fact, they are also exhaustive. To see this, assume that A does not belong either to Case (1) or to Case (2), and let us verify that it belongs to Case (3). First, let us show that $T^+ = 0$. By the "zero-one law" of Section 9, it is enough to consider the case that $K_0^+ \cap -K_0^+ \neq 0$. If $\tau \in T^+$ then τ is zero on $K_0^+ \cap -K_0^+$, and as this is not zero, in particular τ is zero on a nonzero projection. Because A is simple, $\tau = 0$.

Second, let us show that $K_0^+ = K_0$. By what we have just proved, it is enough to show that this follows from the property $T^+ = 0$. If $K_0^+ \neq K_0$, then $K_0 \neq 0$, and by Section 9 either $K_0^+ = 0$ or K_0 is an ordered group. In the case that K_0 is a (nonzero) ordered group, by [6] A has a nonzero quasi-trace, which by [43] is a trace. (In dealing with this case, we may assume for convenience that A is unital.) In the case that $K_0^+ = 0$, but $K_0 \neq 0$, A is stably projectionless. (Otherwise, on passing to the stabilization of A, i.e., to $A \otimes \mathcal{K}$, where $\mathcal{K}$ is the C*-algebra of compact operators on a separable infinite-dimensional Hilbert space, as A is simple, by Brown's stabilization theorem, A is isomorphic to $B \otimes \mathcal{K}$ for some unital C*-algebra B — any nonzero unital hereditary sub-C*-algebra of A. Hence, $K_0 A = (K_0 A)^+ - (K_0 A)^+$, in contradiction with $K_0 A \neq 0$, $(K_0 A)^+ = 0$.) On the other hand, from $T^+ = 0$ it follows by [5] (on using [43] again) that $A \otimes \mathcal{K}$ has an infinite projection — in particular, a nonzero projection. This contradiction shows that the hypotheses $T^+ = 0$ and $K_0^+ \neq K_0$ are incompatible, as desired.

This argument also establishes statements 8 (iii) and 8 (iv). Statements 8 (i) and 8 (ii) follow from the fact that close projections or unitaries are homotopic, together with the use of Brown's stabilization theorem as above.

11. In Case (1), the invariant consists of the countable abelian groups K_0 and K_1, together with the cone T^+, in duality with K_0.

Examples of Case (1), which consists of the stably projectionless simple amenable C*-algebras, were first constructed by Blackadar in [1].

Additional examples are constructed in [36], [63], and [27].

12. In Case (2), the invariant consists of the countable ordered abelian group K_0, the countable abelian group K_1, and the simplicial cone T^+, in duality with K_0. The pairing of T^+ with K_0 is positive on K_0^+, and all positive functionals on K_0 arise from this pairing ([6], [43]).

This case consists of the stably finite unital simple amenable C*-algebras, and the C*-algebras stably isomorphic to these.

Constructions in [2], [4], [36], [63], and [27] realize all possibilities for the invariant in which K_0 is weakly unperforated. (The algebra $\mathcal{K}$ need not be used.)

Whether the ordered group K_0 must be weakly unperforated (i.e., whether $ng > 0$ for some $n = 2, 3, \ldots$ always implies $g \geq 0$) is an interesting question.

13. In Case (3), the invariant consists of the countable abelian groups K_0 and K_1.

Examples exhausting all pairs of such groups are now known. (The algebra 0 need not be used.) (See [57], [59], and [35].)

The first such examples are of course the Cuntz algebras, for which the invariant was computed in [18].

14. In Case (1), no isomorphism results have yet been obtained.

15. In Case (2), the isomorphism theorem of [22] for AF algebras was succeeded (twenty years later) by the following result ([24], [25], [26]).

Let A and B be stable separable amenable simple C*-algebras in Case (2). Assume that each of A and B is the closure of the union of an increasing sequence of sub-C*-algebras isomorphic to finite direct sums of matrix algebras over $C(\mathbb{T})$ (cf. above). Suppose that the invariant for A is isomorphic to the invariant for B. More explicitly, suppose that there are isomorphisms, of ordered groups, $\varphi_0 : K_0A \to K_0B$, of groups, $\varphi_1 : K_1A \to K_1B$, and of topological convex cones, $\varphi_T : T^+B \to T^+A$, such that φ_0 and φ_T respect the pairing of T^+ with K_0, i.e.,

$$< \tau, \varphi_0 g > \; = \; < \varphi_T \tau, g >, \quad g \in K_0A, \tau \in T^+B.$$

It follows that A and B are isomorphic. Furthermore, there exists an isomorphism $\varphi \colon A \to B$ giving rise to a given triple of isomorphisms $(\varphi_0, \varphi_1, \varphi_T)$ as above.

More recently, this isomorphism theorem has been generalized to the case that the circle $\mathbb{T}$ is replaced by an arbitrary compact metrizable space of finite dimension; the space may vary, but the dimension must (so far) be assumed to be bounded. This result was proved by Gong, Li, and the author in [31]. As well as using the result of [26] (the case of circles), this theorem uses results in [60], [50], [29], [33], [32], [51], [33], [20], [40], [49], [19], and [41].

All examples obtained in this way can in fact be obtained using spaces of dimension three or less.

The algebras obtained by such a construction are still rather special; in particular, the ordered K_0-group has the Riesz decomposition property and is weakly unperforated. If the algebra $\mathcal{K}$ is excluded from consideration, then the group K_0 is noncyclic. Such ordered groups were considered in [23], and using the decomposition result for such ordered groups proved in [23] (due to Effros, Handelman, and Shen in the torsion-free case), it was shown in [30] that every simple countable ordered abelian group with these properties arises as ordered K_0 in this class of algebras; at the same time, every countable abelian group arises as K_1.

The question of what can arise as the tracial cone in this construction, and as the pairing of this with K_0, has been answered by Villadsen in [64] (building on results of Thomsen in [61] and [62]). Villadsen showed that, in the case that only circles are used, the only special properties of the invariant, in addition to those described for Case (2) in Section 8, and in addition to the properties of K_0 described above, are that, first, K_0 and K_1 are torsion free, and, second, every extreme ray of the tracial cone gives rise to an extreme ray of the cone of positive functionals on K_0.

Combining Villadsen's methods with the result mentioned before, one sees that the range of the invariant for the class of algebras described above (with spaces of dimension three in place of circles) is the same as in the case considered by Villadsen (circles) except with torsion allowed in K_0 and K_1. In other words, beyond the specifications for Case (2) in Section 8, K_0 is weakly unperforated, noncyclic, and has the Riesz property, and extreme rays of T^+ yield extreme rays in the cone of positive functionals on K_0.

When more general examples are considered (for instance, based on subhomogeneous building blocks), it will be necessary to exclude the algebra $\mathcal{K}$. As pointed out in Section 12, all possibilities for the invariant in Case (2) allowed in Section 8, with K_0 weakly unperforated, are realizable — without using $\mathcal{K}$.

16. In Case (3), no isomorphism results were known until very recently.

Three years ago, in a tour de force of mathematical physics, Bratteli, Kishimoto, Rørdam, and Størmer showed in [14], using the anticommutation relations of quantum field theory, that the shift on the infinite tensor product of 2×2 matrix algebras, M_{2^∞} (cf. Section 2), has the noncommutative Rokhlin property of Voiculescu. Using this, they were able to conclude that the tensor product of the Cuntz algebra $\mathcal{O}_2$ with M_{2^∞} is isomorphic to $\mathcal{O}_2$.

Using the known connection between the Rokhlin property and stability (first appearing in the work of Connes, and studied in the C*-algebra context by Herman and Ocneanu in [45]), a number of authors — most notably, Rørdam — pushed rapidly forward to classify a very large natural class of algebras in Case (3), so large that it could very well consist of everything — i.e., all stable, separable, amenable, nonzero, simple C*-algebras in Case (3). (See [57], [9], [53], [59], and [35].)

More precisely, the class of algebras in Case (3) for which the invariant considered — the K_0- and K_1-groups — has been shown to be complete exhausts the invariant and has certain natural properties. First of all, the class can be described in a rather abstract way (introduced in [59]): within the class of stable, separable, amenable, nonzero, simple C*-algebras in Case (3), consider those that are purely infinite, i.e., in which every nonzero hereditary sub-C*-algebra contains a projection that is infinite in the sense of Murray and von Neumann — equivalent to a proper subprojection. (This property may be automatic.) Consider those C*-algebras in this class such that for every other algebra in the class, any KK-element into it is realized by a nonzero homomorphism, and a unique one up to approximate unitary equivalence (to within an arbitrarily small tolerance on each finite subset). Within this subclass, the K_0-group and K_1-group determine an algebra up to isomorphism. This subclass contains algebras with arbitrary K_0- and K_1-groups.

Second, the preceding class is closed under the operation of passing to the closure of an increasing sequence $A_1 \subseteq A_2 \subseteq \cdots$ with A_i belonging to the class. In fact, each A_i may be allowed to be a direct sum of members of the class, provided that the closure of the union is assumed to be simple, which is no longer automatic.

Although no stable, separable, amenable, nonzero, simple C*-algebra in Case (3) is known not to belong to the class, and many well-known ones are known to belong — for instance, the Cuntz algebra $\mathcal{O}_n$ with n finite and even [57], or $n = \infty$ [54], and the tensor products $\mathcal{O}_n \otimes \mathcal{O}_m$ with both n and m finite and even [9] — already the algebra $\mathcal{O}_3$ is not known to belong.

17. The isomorphism theorems described in Sections 15 and 16 have been applied in a number of different ways.

The AF algebra case of the isomorphism theorem was used by Blackadar in his constructions of stably projectionless C*-algebras in [1] and of what he called unital projectionless C*-algebras in [2].

This case of the result was also used in the thermodynamical phase diagram construction of [10] (the construction of a C*-algebraic dynamical system with a prescribed bundle of simplices as the bundle of KMS states at various inverse temperatures — and with prescribed ground and ceiling state spaces at inverse temperatures $\pm\infty$ — as well as a generalization of this to include more thermodynamical variables).

What might perhaps be called the $A\mathbb{T}$ case of the theorem (inductive limits of sequences of direct sums of matrix algebras over $C(\mathbb{T})$ — instead of $C(\mathrm{pt})$) has been used in computing the automorphism group of the irrational rotation C*-algebra A_θ for each irrational number θ between 0 and 1. Using the fact that this simple C*-algebra is $A\mathbb{T}$ (established in [28]), it was shown in [34] that the automorphism group of A_θ is an extension of a topologically simple group by $\mathrm{GL}(2, \mathbb{Z})$ — the latter group arising as the image under the action of the automorphism group on $K_1 A_\theta = \mathbb{Z}^2$. Although the question of the triviality of this extension was left open, the realization of the full automorphism group of $K_1 A_\theta$ by automorphisms of the algebra answered a well-known question. (Earlier, only the automorphisms of $\mathbb{Z}^2$ with determinant $+1$ had been realized in this way.)

Before it was known that the irrational rotation C*-algebras are $A\mathbb{T}$, it had been proved by Putnam in [56] that the simple C*-algebras arising (as crossed products) from actions of $\mathbb{Z}$ on the Cantor set are $A\mathbb{T}$. Therefore, these algebras come under the purview of the isomorphism theorem. Because these algebras all have the same K_1-group (namely, $\mathbb{Z}$), and their traces are separated by K_0, the invariant reduces just to the ordered group K_0 (together with the class of the unit of the algebra). In [46], using a construction based on a Bratteli diagram for the ordered group (a representation of it as an inductive limit of finite direct sums of copies of $\mathbb{Z}$ — which always exists in the $\mathbb{AT}$ case), it was shown that every simple unital $\mathbb{AT}$ algebra with the property that the traces are separated by K_0 arises as above from an action of $\mathbb{Z}$ on the Cantor set. Thus, the simple C*-algebras arising from actions of $\mathbb{Z}$ on the Cantor set can be both characterized and classified. The classification, viewed at the level of the $\mathbb{Z}$-action, was shown in [38] to amount to a refinement of orbit equivalence — called strong orbit equivalence. Ordinary orbit equivalence was shown also to be determined by the K-theoretical invariant: quite remarkably, as shown in [38], the K_0-group of the crossed product modulo the elements zero on traces (together with the class of the unit) is a complete invariant for orbit equivalence.

It is easy to see that the C*-algebras arising (as crossed products) from tensor product actions of $\mathbb{Z}$ on M_{2^∞} (cf. above) are $\mathbb{AT}$. In [12], it was shown that for a simple C*-algebra arising in this way (the most likely outcome), there are only two possibilities for the K-theoretical invariant. K_0 and K_1 are always the same, and either there is a unique tracial state, or the extreme tracial states form a circle. Hence by the isomorphism theorem (or, rather, the unital variant of it), precisely two simple C*-algebras arise in this way.

Recently, using methods also used in the isomorphism theorem of Section 16, Lin showed in [52] that two almost commuting self-adjoint matrices are close to commuting ones, thus solving a well-known problem. This result turns out to have important implications for the classification question.

There have been fewer applications so far of the isomorphism theorem in Case (3), described in Section 16, but as Rørdam pointed out in [57], it was not even known before that $\mathcal{O}_8 \otimes M_3$ is isomorphic to $\mathcal{O}_8$.

An interesting subclass of Case (3) consists of the so-called Cuntz-Krieger algebras. Because these include all $\mathcal{O}_n$ with n finite (and not just n even), as mentioned above this class is not known to be contained in the class for which the isomorphism theorem has been established. In spite of this, Rørdam showed in [58] that the isomorphism theorem could be applied to two particular Cuntz-Krieger algebras, which were known (by earlier work of Cuntz) to be critical for deciding the classification question. As a consequence, two simple C*-algebras in the Cuntz-Krieger class are isomorphic if and only if they have the same K-groups (together with the class of the unit in K_0). In particular, $\mathcal{O}_5 \otimes M_3 \cong \mathcal{O}_5$.

18. In addition to the isomorphism theorem of Sections 15 and 16, one has the following homomorphism theorem (proved by similar methods).

Let A and B be two stable, separable, amenable, simple, non type I C*-algebras belonging to either of the classes considered in Sections 15 and 16. Then any homomorphism between the invariants (in the appropriate sense) arises from a homomorphism between the algebras. (Presumably, this result also holds more generally.)

Hence by [28], one recovers the Pimsner-Voiculescu embedding of the irrational rotation C*-algebra in an AF algebra with the same ordered K_0-group. (This approach shows that the embedding is unique — up to approximate unitary equivalence.)

References

[1] B. Blackadar, *A simple C*-algebra with no nontrivial projections*, Proc. Amer. Math. Soc. **78** (1980), 504–508.

[2] B. Blackadar, *A simple unital projectionless C*-algebra*, J. Operator Theory **5** (1981), 63–71.

[3] B. Blackadar, *Symmetries of the CAR algebra*, Ann. of Math. (2) **131** (1990), 589–623.

[4] B. Blackadar, O. Bratteli, G. A. Elliott, and A. Kumjian, *Reduction of real rank in inductive limits of C*-algebras*, Math. Ann. **292** (1992), 111–126.

[5] B. Blackadar and J. Cuntz, *The structure of stable algebraically simple C*-algebras*, Amer. J. Math. **104** (1982), 813–822.

[6] B. Blackadar and M. Rørdam, *Extending states on preordered semigroups and the existence of quasitraces on C*-algebras*, J. Algebra **152** (1992), 240–247.

[7] O. Bratteli, *Inductive limits of finite dimensional C*-algebras*, Trans. Amer. Math. Soc. **171** (1972), 195–234.

[8] O. Bratteli, G. A. Elliott, D. E. Evans, and A. Kishimoto, *Actions of finite groups on AF algebras obtained by folding the interval*, K-Theory, to appear.

[9] O. Bratteli, G. A. Elliott, D. E. Evans, and A. Kishimoto, *On the classification of C*-algebras of real rank zero, III: The infinite case*, preprint.

[10] O. Bratteli, G. A. Elliott, and A. Kishimoto, *The temperature state space of a C*-dynamical system, II*, Ann. of Math. (2) **123** (1986), 205–263.

[11] O. Bratteli, D. E. Evans, and A. Kishimoto, *Crossed products of totally disconnected spaces by $\mathbb{Z}_2 * \mathbb{Z}_2$*, Ergodic Theory Dynamical Systems **13** (1994), 445–484.

[12] O. Bratteli, D. E. Evans, and A. Kishimoto, *The Rokhlin property for quasi-free automorphisms of the Fermion algebra*, Proc. London Math. Soc., to appear.

[13] O. Bratteli and A. Kishimoto, *Non-commutative spheres, III : Irrational rotations*, Comm. Math. Phys. **147** (1992), 605–624.

[14] O. Bratteli, A. Kishimoto, M. Rørdam, and E. Størmer, *The crossed product of a UHF-algebra by a shift*, Ergodic Theory Dynamical Systems **13** (1993), 615–626.

[15] M.-D. Choi and E. G. Effros, *Separable nuclear C*-algebras and injectivity*, Duke Math. J. **43** (1976), 309–322.

[16] A. Connes, *Classification of injective factors*, Ann. of Math. (2) **104** (1976), 73–115.

[17] A. Connes, *On the cohomology of operator algebras*, J. Funct. Anal. **28** (1978), 248–253.

[18] J. Cuntz, *K-theory for certain C*-algebras*, Ann. of Math. (2) **113** (1981), 181–197.

[19] M. Dadarlat, *Approximately unitarily equivalent morphisms and inductive limit C*-algebras*, preprint.

[20] M. Dadarlat, *Reduction to dimension three of local spectra of real rank zero C*-algebras*, preprint.

[21] J. Dixmier, *On some C*-algebras considered by Glimm*, J. Funct. Anal. **1** (1967), 182–203.

[22] G. A. Elliott, *On the classification of inductive limits of sequences of semisimple finite-dimensional algebras*, J. Algebra **38** (1976), 29–44.

[23] G. A. Elliott, *Dimension groups with torsion*, Internat. J. Math. **1** (1990), 361–380.

[24] G. A. Elliott, *On the classification of C*-algebras of real rank zero*, J. Reine Angew. Math. **443** (1993), 179–219.

[25] G. A. Elliott, *A classification of certain simple C*-algebras*, Quantum and Non-Commutative Analysis (editors, H. Araki et al.), Kluwer, Dordrecht, 1993, pages 373–385.

[26] G. A. Elliott, *A classification of certain simple C*-algebras, II*, Ann. of Math. (2), to appear.

[27] G. A. Elliott, *An invariant for simple C*-algebras*, in preparation.

[28] G. A. Elliott and D. E. Evans, *The structure of the irrational rotation C*-algebra*, Ann. of Math. (2) **138** (1993), 477–501.

[29] G. A. Elliott and G. Gong, *On inductive limits of matrix algebras over the two-torus*, preprint.

[30] G. A. Elliott and G. Gong, *On the classification of C*-algebras of real rank zero, II*, preprint.

[31] G. A. Elliott, G. Gong, and L. Li, *On simple inductive limits of matrix algebras over higher dimensional spaces, II*, in preparation.

[32] G. A. Elliott, G. Gong, H. Lin, and C. Pasnicu, *Homomorphisms, homotopies, and approximations by circle algebras*, C. R. Math. Rep. Acad. Sci. Canada **16** (1994), 45–50.

[33] G. A. Elliott, G. Gong, H. Lin, and C. Pasnicu, *Abelian C*-subalgebras of C*-algebras of real rank zero and inductive limit C*-algebras*, preprint.

[34] G. A. Elliott and M. Rørdam, *The automorphism group of the irrational rotation C*-algebra*, Comm. Math. Phys. **155** (1993), 3–26.

[35] G. A. Elliott and M. Rørdam, *Classification of certain infinite simple C*-algebras, II*, preprint.

[36] G. A. Elliott and K. Thomsen, *The state space of the K_0-group of a simple separable C*-algebra*, Geom. and Functional Anal. **4** (1994), 522–538.

[37] D. E. Evans and A. Kishimoto, *Compact group actions on UHF algebras obtained by folding the interval*, J. Funct. Anal. **18** (1991), 346–360.

[38] T. Giordano, I. F. Putnam, and C. Skau, *Topological orbit equivalence and C*-crossed products*, preprint.

[39] J. G. Glimm, *On a certain class of operator algebras*, Trans. Amer. Math. Soc. **95** (1960), 318–340.

[40] G. Gong, *On inductive limits of matrix algebras over higher dimensional spaces, Part II*, preprint.

[41] G. Gong, *On simple inductive limits of matrix algebras over higher dimensional spaces*, preprint.

[42] U. Haagerup, *All nuclear C*-algebras are amenable*, Invent. Math. **74** (1983), 305–319.

[43] U. Haagerup, *Every quasi-trace on an exact C*-algebra is a trace*, preprint.[1]

[44] P. de la Harpe, *Moyennabilité du groupe unitaire et propriété P de Schwartz des algèbres de von Neumann*, Algèbres d'opérateurs (Séminaire, Les Plans-sur-Bex, Suisse 1978) (editor, P. de la Harpe), Lecture Notes in Math. **725**, Springer-Verlag, Berlin and New York, 1979, pages 220–227.

[45] R. H. Herman and A. Ocneanu, *Stability for integer actions on UHF C*-algebras*, J. Funct. Anal. **59** (1984), 132–144.

[46] R. H. Herman, I. F. Putnam, and C. Skau, *Ordered Bratteli diagrams, dimension groups and topological dynamical systems*, Internat. J. Math. **3** (1992), 827–864.

[47] E. Kirchberg, Lecture at ICM Satellite Meeting, Geneva, 1994.

[48] A. Kumjian, *An involutive automorphism of the Bunce-Deddens algebra*, C. R. Math. Rep. Acad. Sci. Canada **10** (1988), 217–218.

---

[1]It should be noted that the result of [43] is stated only in the unital case. E. Kirchberg has informed the author that for simple C*-algebras he has been able to extend the result to the nonunital case.

[49] L. Li, Ph.D. thesis, University of Toronto, 1994.

[50] H. Lin, *Approximation by normal elements with finite spectra in C*-algebras of real rank zero*, preprint.

[51] H. Lin, *Homomorphisms from C(X) into C*-algebras*, preprint.

[52] H. Lin, *Almost commuting self-adjoint matrices and applications*, preprint.

[53] H. Lin and N. C. Phillips, *Classification of direct limits of even Cuntz-circle algebras*, preprint.

[54] H. Lin and N. C. Phillips, *Approximate unitary equivalence of homomorphisms from $\mathcal{O}_\infty$*, preprint.

[55] A. L. T. Paterson, *Nuclear C*-algebras have amenable unitary groups*, Proc. Amer. Math. Soc. **124** (1992), 719–721.

[56] I. F. Putnam, *On the topological stable rank of certain transformation group C*-algebras*, Ergodic Theory Dynamical Systems **10** (1990), 187–207.

[57] M. Rørdam, *Classification of inductive limits of Cuntz algebras*, J. Reine Angew. Math. **440** (1993), 175–200.

[58] M. Rørdam, *Classification of Cuntz-Krieger algebras*, K-Theory, to appear.

[59] M. Rørdam, *Classification of certain infinite simple C*-algebras*, J. Funct. Anal., to appear.

[60] H. Su, Ph.D. thesis, University of Toronto, 1992.

[61] K. Thomsen, *Inductive limits of interval algebras: The tracial state space*, Amer. J. Math. **116** (1994), 605–620.

[62] K. Thomsen, *On the range of the Elliott invariant*, J. Funct. Anal., to appear.

[63] K. Thomsen, *On the ordered K_0-group of a simple C*-algebra*, preprint.

[64] J. Villadsen, *The range of the Elliott invariant*, J. Reine Angew. Math., to appear.

[65] S. G. Walters, *Inductive limit automorphisms of the irrational rotation C*-algebra*, Comm. Math. Phys., to appear.

Note added in proof on December 1, 1994:

Building on his results announced in [47] — basically, that for any separable, amenable, purely infinite, simple C*-algebra A, the tensor product with $\mathcal{O}_\infty$ is isomorphic to A, and the tensor product with $\mathcal{O}_2 \otimes \mathcal{K}$ is isomorphic to $\mathcal{O}_2 \otimes \mathcal{K}$ — Kirchberg has now almost solved the classification problem in Case (3). More precisely, the problem is solved for the class of separable amenable simple algebras in Case (3) that are purely infinite (cf. Section 16) and that satisfy the so-called universal coefficient theorem in KK. (In other words, these algebras are classified by their K_0- and K_1-groups — together with the K_0-class of the unit in the unital case.) The remaining problem, therefore, is to show that these two properties always hold.

The same result has also been obtained by Phillips — using the results of Kirchberg announced in [47] (stated above).

Note added in proof on July 14, 1995:

Recently, Villadsen has constructed a simple amenable C*-algebra in Case (2) with perforated positive cone K_0^+. (Cf. Sections 8, 12 above.)

Recent Results in the Theory of Infinite-Dimensional Banach Spaces

W. T. Gowers

Department of Mathematics, University College London,
Gower Street, London WC1E 6BT, England

Many of the best-known questions about separable infinite-dimensional Banach spaces are of at least one of the following forms.

(1) If X is an arbitrary space, must it have a "nice" subspace?
(2) If X is an arbitrary space, are there nonobvious examples of operators on X?
(3) If certain operators are known to be defined on a space X, does this imply anything nonobvious about the structure of X?

Many such questions have been solved in the last three years, and surprising connections between them have been discovered. The purpose of this paper is to explain these developments. Unless otherwise stated, all spaces and subspaces mentioned will be infinite-dimensional separable Banach spaces.

We begin by discussing questions of the first kind above. It is clear straight away that not every space has a Hilbert space, the nicest space of all, as a subspace; the space ℓ_1 is but one of many obvious counterexamples. However, if one asks whether every space contains c_0 or ℓ_p for some $1 \leq p < \infty$, then one already has a very simple question to which the answer is not at all obvious. In fact, this question was not answered until the early 1970s, when Tsirelson [T] used a clever inductive procedure to define a counterexample. The proof that his example does not contain c_0 or ℓ_p is surprisingly short (this is even more true of the dual of his space as presented by Figiel and Johnson [FJ]), but the ideas he introduced have been at the heart of the recent progress.

There were two further weakenings of the notion of "nice" that left questions not answered by Tsirelson's example. For the first, recall that a *Schauder basis* (we will often say simply *basis*) of a Banach space X is a sequence $(x_n)_{n=1}^{\infty}$ such that every vector in X has a unique expression as a norm-convergent sum of the form $\sum_{n=1}^{\infty} a_n x_n$. A simple result proved in the early 1930s by Mazur (see [LT2]) is that every Banach space has a subspace with a basis. Whether every separable Banach space had a basis was a famous open problem, answered negatively by Enflo [En] in 1973.

The definition of a Schauder basis is unlike that of an algebraic basis for a vector space in that the order of the x_n is important. A basis $(x_n)_{n=1}^{\infty}$ with the property that $(x_{\pi(n)})_{n=1}^{\infty}$ is a basis for every permutation π of the positive integers is called an *unconditional* basis. It was shown by Mazur that $(x_n)_{n=1}^{\infty}$ is

Proceedings of the International Congress
of Mathematicians, Zürich, Switzerland 1994
© Birkhäuser Verlag, Basel, Switzerland 1995

an unconditional basis if and only if there is a constant C such that, for every sequence $(a_n)_{n=1}^{\infty}$ of scalars and any sequence $(\epsilon_n)_{n=1}^{\infty}$ with each ϵ_n of modulus one, we have the inequality

$$\left\|\sum_{n=1}^{\infty} \epsilon_n a_n x_n\right\| \leq C\left\|\sum_{n=1}^{\infty} a_n x_n\right\|.$$

Such a basis is called C-unconditional. Notice that this implies that, for every subset $A \subset \mathbb{N}$, the projection $\sum_{n \in \mathbb{N}} a_n x_n \mapsto \sum_{n \in A} a_n x_n$ has norm at most C. Thus, if X has an unconditional basis, then there are many nontrivial projections on X. A sequence that is an unconditional basis for its closed linear span is called an *unconditional basic sequence*.

We may now ask whether every space has a subspace with an unconditional basis, or, equivalently, contains an unconditional basic sequence (the unconditional basic sequence problem). This question was first asked as soon as unconditional bases were defined in the 1940s and appears in print in [BP2]. The lack of a proof after three or four decades led many people to begin to suspect that the answer was no, but there still seemed to be some chance of a positive answer to a yet weaker question: Does every space contain c_0, ℓ_1, or a reflexive subspace? The reason this is a weaker question is that a result of James [J1] states that a space with an unconditional basis not containing c_0 or ℓ_1 must itself be reflexive, so a positive answer would be implied by a positive answer to the first question.

In the summer of 1991, Maurey and I independently discovered spaces *not* containing unconditional basic sequences. There is not room to explain the constructions here, but we can make a few remarks. First, we introduce some notation. If X is a Banach space with a given basis $(e_n)_{n=1}^{\infty}$, then the *support* of a vector $x = \sum_{n=1}^{\infty} a_n e_n \in X$ is just the set of n for which a_n is nonzero. Given $x, y \in X$, we write $x < y$ to mean that every element of the support of x is less than every element of the support of y. A sequence $x_1 < x_2 < \cdots$ of nonzero vectors is called a *block basis*, and a subspace generated by a block basis is called a *block subspace*. A simple but very useful result of Bessaga and Pelczynski [BP1] states that every subspace Y of a space X with a basis has a further subspace Z isomorphic to a block subspace of X. This reduces many problems about subspaces of Banach spaces to ones about block subspaces.

The unconditional basic sequence problem is no exception. It is straightforward to show that X contains no C-unconditional basic sequence if and only if every block subspace $Y \subset X$ contains a sequence of vectors $x_1 < \cdots < x_n$ such that

$$\left\|\sum_{i=1}^{n} x_i\right\| > C\left\|\sum_{i=1}^{n} (-1)^n x_i\right\|.$$

(This is not quite true if X has complex scalars. In that case, sequences satisfying the above inequality exist in every block subspace provided X contains no $2C$-unconditional basic sequence.) Therefore X contains no unconditional basic sequence at all if and only if such a finite sequence can be found in every block subspace for every C. Notice that the condition on the supports of $x_1, \ldots, x_n$ is important, because otherwise a constant sequence would do.

Of great importance to us in constructing such a space was noticing that, for any *fixed* C, a space which had just been constructed by Schlumprecht [S1] could be renormed so that every block subspace did contain such a "C-conditional" sequence. However, Schlumprecht's space has a 1-unconditional basis, so any renorming has a C-unconditional basis for *some* C. To find a space with no unconditional basic sequence at all, one somehow had to use the ideas from the renorming of Schlumprecht's space but produce a nonisomorphic space. This led to considerable conceptual and technical difficulties. Two other points are worth making. The first is that Schlumprecht's space was, like Tsirelson's space and indeed our spaces, constructed inductively (at first glance, the definition appears to be circular). The advantage of his space over Tsirelson's was that it had stronger properties related to the so-called distortion of Banach spaces. Indeed, his space played an important part in his solution with Odell [OS1] of a famous problem known as the distortion problem. (In one formulation this asks: Must every space isomorphic to a Hilbert space have a subspace almost isometric to a Hilbert space?) Again, there is not room to explain this connection (some indication can be found in [OS2]), but there is one and it is important. The second point is that the spaces that Maurey and I found were not in fact distinct, indicating that, however complicated, our approach was a natural one. The result, with some extensions that will be described in a moment, appears in a joint paper [GM1].

The first indication that the space X_1 we constructed was also relevant to questions about operators was an observation of Johnson. He pointed out to us that we could alter our argument(s) and show that every continuous projection on X_1 had finite rank or corank, and moreover that every subspace of X_1 had the same property. This is equivalent to saying that no subspace Y of X_1 can be written as a topological direct sum $W \oplus Z$ with W and Z infinite dimensional. A space with this property he called *hereditarily indecomposable*. Another equivalent form of the property, which brings out its strangeness, is that, for any two infinite-dimensional subspaces Y and Z of X_1 and any $\epsilon > 0$, there exist unit vectors $y \in Y$ and $z \in Z$ such that $\|y - z\| < \epsilon$. In a certain sense, the angle between any two infinite-dimensional subspaces is zero.

Thus, our space gave a strong answer to a question of Lindenstrauss [L2]: Can every space be decomposed as a topological direct sum of two infinite-dimensional subspaces? (Such a space is simply called *decomposable*.) It showed that, for an arbitrary space X, one could not in general expect $L(X)$, the space of operators on X, to contain interesting examples of projections.

Another important class of operators is isomorphisms onto proper subspaces, and the next step was the discovery that these also did not have to exist in general. I constructed a variant X_U of the space X_1 above, which had an unconditional basis [G1]. However, it was not isomorphic to any proper subspace, and therefore solved the so-called hyperplane problem, which had its origins in Banach's book [B]. This was the question of whether an arbitrary space is isomorphic to its closed subspaces of codimension one. (Note that if Y and Z are distinct such subspaces, then $Y \simeq (Y \cap Z) \oplus \mathbb{C} \simeq Z$, so any two subspaces of the same finite codimension are isomorphic.) Equivalently, X_U is not isomorphic to $X_U \oplus \mathbb{C}$. (Actually, as Maurey recently pointed out to me, the precise question asked by Banach was whether every space X is isomorphic to a *subspace* of its hyperplanes, but this is also answered negatively by X_U.)

It was a little strange that an unconditional basis was needed to make the above construction work, although it did mean that a question of the third kind was answered. The situation became clearer with the following result [GM1], which underlines the importance of hereditary indecomposability. Recall that an operator $T : X \to Y$ is *Fredholm* if the dimensions $\alpha(T)$ and $\beta(T)$ of the kernel of T and Y/TX, respectively, are finite, and that the *index* of such an operator is defined to be $\alpha(T) - \beta(T)$. A *strictly singular* operator $S : X \to Y$ is one for which there is no subspace Z of X such that the restriction of S to Z is an isomorphism. Equivalently, for every $Z \subset X$ and every $\epsilon > 0$ there exists $z \in Z$ with $\|Sz\| < \epsilon \|z\|$. The strictly singular operators share many of the smallness properties of the compact operators (indeed, for several spaces X, the strictly singular, and compact operators in $L(X)$ coincide). For example, if T is Fredholm and S is strictly singular then $T + S$ is Fredholm with the same index as T.

THEOREM 1. [GM1] *Let X be a hereditarily indecomposable space with complex scalars. Then every operator $T \in L(X)$ can be written in the form $\lambda I + S$, where $\lambda \in \mathbb{C}$ and S is strictly singular. In particular, every operator on X is either strictly singular or Fredholm with index zero.*

It is easy to see that an isomorphism onto a proper subspace cannot be strictly singular or Fredholm with index zero, so the space X_1 discussed earlier is another counterexample to Banach's question. (In fact the space X_1 can be real or complex. In the case of real scalars, we have a direct proof that it satisfies the conclusions of Theorem 1. In general, a real hereditarily indecomposable space is isomorphic to no proper subspace.) Thus, in a certain sense, $L(X_1)$ is trivial and the answer to the second question with which we started is a simple no. On the other hand, it is not known whether there exists a space on which every operator is a *compact* perturbation of a multiple of the identity.

We have left behind the question of whether an arbitrary space must contain c_0, ℓ_1, or a reflexive subspace. For this problem, unlike the unconditional basic sequence problem, it is not enough to find a space such that every (block) subspace contains a *finite* sequence of some kind. The problem is in this sense more genuinely infinite dimensional, and for this reason it was not obvious whether the techniques used for constructing the hereditarily indecomposable space X could be extended to give a counterexample. In the end, however, it turned out to be possible [G2], and the resulting space X_{CLR} had a slightly stronger property. No subspace $Y \subset X_{CLR}$ contains ℓ_1 or has a separable dual. Note that it is not obvious that there even exists a space with a nonseparable dual not containing ℓ_1. This was a question asked by Banach and answered independently by examples of James [J2] and Lindenstrauss and Stegall [LS].

It would seem, then, that there is no reasonable sense of the word "nice" for which one can (truthfully) say that every space has a nice subspace. However, this conclusion, we shall see later, is premature.

Let us concentrate on questions of the third kind. We have already seen that the existence of plenty of projections on a space does not guarantee that there are nontrivial isomorphisms to subspaces. What about the reverse? If we are given a space that is isomorphic to a proper subspace, must there be nontrivial projections? There are many other questions of this general kind, and to answer some of them Maurey and I generalized the results of our first paper. We proved [GM2] that, under certain conditions on an algebra $\mathcal{A}$, one can construct a space $X = X(\mathcal{A})$ such that $\mathcal{A}$ embeds in an obvious way into $L(X)$, and every operator on X is a small perturbation of (the image of) an element of $\mathcal{A}$. Sometimes "small" in this statement simply means "strictly singular", and sometimes it means a slight weakening that is nevertheless strong enough for applications.

To illustrate, let us consider what can be said about a space if it has a basis and the shift with respect to that basis is an isometry. From Theorem 1 we know straight away that the space is not *hereditarily* indecomposable, but this does not imply that there are nontrivial projections defined on the whole space. We also know that, writing S for the shift and L for the left shift, every operator of the form $\sum_{n=0}^{\infty} a_n S^n + \sum_{n=1}^{\infty} b_n L^n$ is continuous if $\sum_{n=0}^{\infty} |a_n| + \sum_{n=1}^{\infty} |b_n| < \infty$. In other words, the algebra $\mathcal{A}$ of convolutions by absolutely summable sequences on $\mathbb{Z}$ embeds into the algebra of operators on the space. (This is not quite a homomorphism because $LS \neq SL$, but finite-rank perturbations do not matter to us.) Our theorem gives a space X_S for which the shift is an isometry, such that every operator in $L(X_S)$ is a strictly singular perturbation of an element of $\mathcal{A}$. It is straightforward to deduce from this that every projection on X_S has finite rank or corank. Thus, if Y is a subspace of X_S and there is a continuous projection onto Y (such a subspace is called *complemented*), then Y is finite codimensional. (Recall that subspaces are infinite dimensional unless it is otherwise stated.) Because all subspaces of the same finite codimension are isomorphic, the existence of the shift guarantees that every complemented subspace of X_S is isomorphic to X_S. A space with this property is called *prime*. Before this example, the only known examples were c_0 and ℓ_p for $1 \leq p \leq \infty$. Apart from ℓ_∞, these were shown to be prime by Pelczynski [P]. Lindenstrauss [L1] proved it for ℓ_∞, the only known nonseparable example.

The general philosophy here is that, often, the existence of certain operators on a space X implies little more than that the algebra $\mathcal{A}$ generated by those operators embeds into $L(X)$, which is obvious anyway. Two more examples are as follows. For every positive integer $n \geq 2$ there exists a space X_n such that two finite-codimensional subspaces of X_n are isomorphic if and only if they have the same codimension modulo n, and there exists a space Z_n such that the product spaces Z_n^r and Z_n^s are isomorphic if and only if r and s are equal modulo $n - 1$. For example, there is a space isomorphic to its subspaces of codimension two but not to its hyperplanes, and there is a space isomorphic to its cube but not to its square. For the second class of examples, one obtains algebras $\mathcal{A}_n$ which resemble the C*-algebras $\mathcal{O}_n$, which were analyzed using K-theory by Cuntz [C]. Our proof that the algebras $\mathcal{A}_n$ do not contain isomorphisms between Z_n^r and Z_n^s when r and s are not equal modulo $n - 1$ was a modification of his argument. Notice that if $X \simeq X^3$ but $X \not\simeq X^2$, then we have an example of two nonisomorphic spaces X and Y such that either embeds into the other as a complemented subspace. This gives a negative answer to the so-called Schroeder-Bernstein problem for Banach spaces, originally solved in [G3]. We have another example if we take the space X_2 above, which is isomorphic to $X_2 \oplus \mathbb{C}^2$ but not to $X_2 \oplus \mathbb{C}$.

The theorem proved in [GM2] gives fairly general circumstances under which the third question from the beginning of this paper has a negative answer. However, there are certain very strong assumptions one can make about a space where our theorem has nothing to say. For example, a famous result of Lindenstrauss and Tzafriri [LT1], solving a problem known as the complemented subspaces problem, states that if every closed subspace of a Banach space is complemented, then the space must be isomorphic to a Hilbert space. If we consider isomorphisms instead of projections, we get the following question of Banach [B]: If X is a space isomorphic to all its (closed infinite-dimensional) subspaces, must X be isomorphic to a Hilbert space? A space isomorphic to all its subspaces is nowadays called *homogeneous*. Are there examples other than ℓ_2?

Szankowski [Sz] generalized Enflo's solution to the basis problem, by showing that, unless a space is very close to a Hilbert space in a certain technical sense, then it has a subspace without a basis (or even the approximation property). Because we have Mazur's result that every space has a subspace with a basis, it follows that a homogeneous space must be close to a Hilbert space. On the other hand, Johnson [Jo1] showed that a variant of Tsirelson's space (the 2-*convexified* version) has the property that every quotient of every subspace has a basis, but the space does not contain ℓ_2.

Until recently, the most powerful result in the positive direction was also due to Johnson [Jo2]. He showed that if X and X^* are homogeneous and have what is known as the *GL-property*, a property related to, but much weaker than, having an unconditional basis, then X is isomorphic to ℓ_2. Then, early in 1993, Komorowski and Tomczak-Jaegermann proved the following result.

THEOREM 2. [KT] *Let X be a Banach space of cotype q for some $q < \infty$. Then either X contains ℓ_2 or X contains a subspace without an unconditional basis.*

It does not matter too much here what it means to be of cotype q. Suffice it to say that, by Szankowski's result, a homogeneous space is too close to ℓ_2 to fail the condition of Theorem 2. It follows that a homogeneous space not isomorphic to (and therefore not containing) ℓ_2 must have a subspace without an unconditional basis. But because the space is homogeneous, it does not even contain an unconditional basic sequence. Another way of saying this is that a homogeneous space with an unconditional basis must be ℓ_2. Theorem 2 therefore uses a stronger assumption on the space than Johnson, but the assumption about the dual space is no longer necessary. This theorem of Komorowski and Tomczak-Jaegermann provided a remarkable link between Banach's homogeneous spaces problem and the unconditional basic sequence problem. Moreover, it was potentially very useful, because the examples that had been constructed of spaces not containing unconditional basic sequences had, as we have seen in some cases, nowhere near enough operators to be homogeneous — quite the reverse!

Let us consider the prime space X_S mentioned earlier. It is an easy consequence of the results of [GM2] that it contains no unconditional basic sequence.

We also know that it is not hereditarily indecomposable. However, it can be shown that X_S has a hereditarily indecomposable subspace. The same is true of the spaces X_n and Z_n. Could this be a general phenomenon? The answer is yes.

THEOREM 3. [G4] *Every Banach space X has a subspace Y that either has an unconditional basis or is hereditarily indecomposable.*

The solution to Banach's question is now easy. We have seen that a homogeneous space X not isomorphic to ℓ_2 contains no unconditional basic sequence. Therefore, by Theorem 3, it contains a hereditarily indecomposable subspace. Hence, as X is homogeneous, X is itself hereditarily indecomposable. But this, in the light of Theorem 1, is a very strong contradiction.

Before we move on, it is worth pointing out that there is no connection between the proofs of Theorem 2 and Theorem 3. It was a historical accident that they were proved in the order that they were. Similarly, Theorem 3 is completely independent of the actual existence of hereditarily indecomposable spaces, except that nobody thought to look at the notion of hereditary indecomposability until an example of a space exhibiting it was produced.

We now return to the word "nice". The reason one would like to find "nice" subspaces is that one can say more about them than about general spaces. But if we take this as a vague definition of niceness, then Theorem 3 *does* give us a nice subspace. For, in a sense, we know everything there is to know about a space if we know that it is hereditarily indecomposable. If, on the other hand, it has an unconditional basis, we also have a lot of information about it. This is not just a quibble about words either, because we have given an example of a problem where this kind of niceness gave exactly the sort of control that was needed. From this point of view one would say that the questions that were traditionally asked about nice subspaces were of the wrong kind (although more examples of Banach spaces were needed to make this clear). It is more fruitful to look for theorems such as the last two, where one obtains subspaces with one of two or more very different properties.

Indeed, such theorems already exist, Rosenthal's ℓ_1-theorem [R1] being the most famous example (see also [R2]). It is interesting that infinite Ramsey theory, which can be used to prove Rosenthal's theorem [F], was also important in the proof of Theorem 3. We now briefly outline a very general and essentially combinatorial result, of which Theorem 3 is an easy consequence. We shall need a small amount of notation.

Given a space X with a fixed basis, let $\Sigma = \Sigma(X)$ denote the set of all finite sequences $x_1, \ldots, x_n$ of vectors of norm at most 1, such that $x_1 < \cdots < x_n$. (The meaning of "$<$" was given earlier.) Given a subset $\sigma \subset \Sigma$, let us say that it is *large* if every block subspace Y (still infinite dimensional) contains some sequence in σ.

Every set $\sigma \subset \Sigma$ defines a two-player game as follows. The first player, S, chooses a block subspace X_1 of X, then the second player, P, chooses a point $x_1 \in X_1$ of finite support and norm at most 1. Then S chooses X_2 and P chooses $x_2 \in X_2$ of norm at most 1 such that $x_1 < x_2$. They continue in this way, alternately picking subspaces and points in them. The game is won by P if at some stage the sequence $(x_1, \ldots, x_n)$ is in σ. If it goes on forever without this happening, then S wins.

Clearly, if P is to have any chance of winning, then σ must at least be large, because otherwise S could repeatedly choose the same subspace containing no sequence in σ. In fact, saying that P has a winning strategy for the set σ is a much stronger statement than saying that σ is large, and that is the point of the next theorem. If $\Delta = (\delta_1, \delta_2, \dots)$ is a sequence of positive reals and $\sigma \subset \Sigma$, we will write σ_Δ for the set of sequences $(x_1, \dots, x_n) \in \Sigma$ "within Δ of σ"; that is, such that there exists $(y_1, \dots, y_n) \in \sigma$ with $\|x_i - y_i\| \leq \delta_i$ for $1 \leq i \leq n$.

THEOREM 4. [G4] *Let X be a Banach space with a basis and let σ be a large subset of $\Sigma(X)$. Then, for every positive sequence $\Delta = (\delta_1, \delta_2, \dots)$ there exists a block subspace $Y \subset X$ such that, if S and P play the above game in the space Y, then P has a winning strategy for obtaining sequences in σ_Δ.*

Ignoring perturbations, this says that if σ is large, then P has a winning strategy in some subspace. The proof of the theorem (and of the main result of [G5], which we shall mention in a moment) is related to arguments due to Galvin and Prikry [GP] and Ellentuck [E] concerning infinite versions of Ramsey's theorem.

Now one of the remarks earlier was that if X contains no unconditional basic sequence, then for every C the set σ of sequences $x_1 < \dots < x_n$ with the property that $\|\sum_{i=1}^n x_i\| > C \|\sum_{i=1}^n (-1)^i x_i\|$ is large. Theorem 4 allows us to drop to a subspace Y in which P has a winning strategy for finding such sequences (say with C replaced by $C/2$). Suppose S plays a strategy that simply involves alternating between two subspaces W and Z of Y. Then P's strategy guarantees the existence of one of these sequences with the odd-numbered x_i's in W and the even-numbered ones in Z. Letting w and z be the sums of the odd-numbered and even-numbered vectors, respectively, we have the inequality $\|z + w\| > C \|z - w\|$. If C is large, it follows easily that $z/\|z\|$ and $w/\|w\|$ are close.

We have not quite managed to find *arbitrarily* close unit vectors in Z and W, because we made a fixed choice of C at the beginning of the argument above. However, one can repeat it for a sequence of C_n tending to infinity and obtain a nested sequence $Y_1 \supset Y_2 \supset \dots$ of subspaces such that for each n the argument works for C_n in the subspace Y_n. It is then easy to check that a "diagonal" subspace generated by a block basis $y_1, y_2, \dots$ with $y_n \in Y_n$ is hereditarily indecomposable. Thus, Theorem 3 follows from Theorem 4. An adaptation of this proof that is in some ways more direct was found by Maurey and appears in [M].

It is noticeable that, from the point of view of the general theory of Banach spaces, certain examples of spaces appear to be more natural than others. The property that distinguishes one of these natural examples is that every subspace has a further subspace that is not interestingly different from the whole space. For example, every block subspace of Tsirelson's space mentioned earlier turns out to be a "Tsirelson-type" space (this statement can be made quite precise), Schlumprecht's space has the very strong property that every subspace has a further subspace not only isomorphic to the whole space but complemented inside it (see [S2]), and every subspace of the hereditarily indecomposable space X_1 has a further subspace that, although definitely not isomorphic to X_1, can be described in an almost identical way. Of the classical spaces, only c_0 and the ℓ_p-spaces are natural in this sense, because every classical space contains one of them. The spaces

X_S, X_n and Z_n (for $n \geq 2$) are good examples of unnatural spaces because they lose what little structure they have when one passes to an appropriate subspace (recall that they have hereditarily indecomposable subspaces).

It is possible that a precise definition of "natural" will emerge, but even without it, one can attempt to classify the natural spaces. Theorem 3 shows that they either have an unconditional basis or are hereditarily indecomposable. In [G5], a generalization is presented of Theorem 4 to analytic sets of infinite sequences (in a certain sensible topology), and more information is given about natural spaces with an unconditional basis. Loosely speaking, there is a theorem that states that such a space either has many isomorphisms between its subspaces or has none that do not follow trivially from the existence of the unconditional basis. The space X_U mentioned earlier is an example where the second possibility holds.

From another point of view, the nonclassical "natural" spaces mentioned above are extremely unnatural. They all have Tsirelson-type inductive definitions, as opposed to formulae for their norms. Given a vector with support of size n, it does not even seem to be possible to calculate its norm in one of these spaces in a time polynomial in n. It is tempting to ask whether there is a meta-theorem that states that a norm that is in some sense directly defined must give a space that has some ℓ_p-space or c_0 as a subspace. At the moment there is not even a precise conjecture along these lines, but it could be that there is, waiting to be discovered, a theory of "easily described" Banach spaces and linear operators very different indeed from the theory of general spaces as outlined in this paper.

References

[B]   S. Banach, Théorie des Opérations Linéaires, Warsaw, 1932.

[BP1]   C. Bessaga and A. Pelczynski, *On bases and unconditional convergence of series in Banach spaces*, Studia Math. **17** (1958), 151–164.

[BP2]   C. Bessaga and A. Pelczynski, *A generalization of results of R. C. James concerning absolute bases in Banach spaces*, Studia Math. **17** (1958), 165–174.

[C]   J. Cuntz, *K-theory for certain C^*-algebras*, Ann. of Math. **113** (1981), 181–197.

[E]   E. Ellentuck, *A new proof that analytic sets are Ramsey*, J. Symbolic Logic **39** (1974), 163–165.

[En]   P. Enflo, *A counterexample to the approximation property in Banach spaces*, Acta Math. **130** (1973), 309–317.

[F]   J. Farahat, *Espaces de Banach contenant ℓ_1 d'après H. P. Rosenthal*, Séminaire Maurey-Schwartz, École Polytechnique, 1973–4.

[FJ]   T. Figiel and W. B. Johnson, *A uniformly convex Banach space which contains no ℓ_p*, Compositio Math. **29** (1974), 179–190.

[GP]   F. Galvin and K. Prikry, *Borel sets and Ramsey's theorem*, J. Symbolic Logic **38** (1973), 193–198.

[G1]   W. T. Gowers, *A solution to Banach's hyperplane problem*, Bull. London Math. Soc. **26** (1994), 523–530.

[G2]   W. T. Gowers, *A Banach space not containing c_0, ℓ_1 or a reflexive subspace*, Trans. Amer. Math. Soc. **344** (1994), 407–420.

[G3]   W. T. Gowers, *A solution to the Schroeder-Bernstein problem for Banach spaces*, submitted.

[G4]   W. T. Gowers, *A new dichotomy for Banach spaces*, preprint.

[G5] W. T. Gowers, *Analytic sets and games in Banach spaces*, preprint IHES M/94/42.

[GM1] W. T. Gowers and B. Maurey, *The unconditional basic sequence problem*, J. Amer. Math. Soc. **6** (1993), 851–874.

[GM2] W. T. Gowers and B. Maurey, *Banach spaces with small spaces of operators*, preprint IHES M/94/44.

[J1] R. C. James, *Bases and reflexivity of Banach spaces*, Ann. of Math. (2) **52** (1950), 518–527.

[J2] R. C. James, *A separable somewhat reflexive Banach space with non-separable dual*, Bull. Amer. Math. Soc. (New Ser.) **80** (1974), 738–743.

[Jo1] W. B. Johnson, *Banach spaces all of whose subspaces have the approximation property*, in: Special Topics of Applied Mathematics, Proceedings GMD, Bonn 1979, North-Holland, Amsterdam 1980, 15–26.

[Jo2] W. B. Johnson, *Homogeneous Banach Spaces*, in: Geometric Aspects of Functional Analysis, Israel Seminar 1986–7, Lecture Notes in Math., Springer-Verlag, Berlin and New York, 1988, 201–203.

[KT] R. Komorowski and N. Tomczak-Jaegermann, *Banach spaces without local unconditional structure*, Israel J. Math. **89** (1995), 205–226.

[L1] J. Lindenstrauss, *On complemented subspaces of m*, Israel J. Math. **5** (1967), 153–156.

[L2] J. Lindenstrauss, *Some aspects of the theory of Banach spaces*, Adv. in Math. **5** (1970), 159–180.

[LS] J. Lindenstrauss and C. Stegall, *Examples of separable spaces which do not contain ℓ_1 and whose duals are non-separable*, Studia Math. **54** (1975), 81–105.

[LT1] J. Lindenstrauss and L. Tzafriri, *On the complemented subspaces problem*, Israel J. Math. **9** (1971), 263–269.

[LT2] J. Lindenstrauss and L. Tzafriri, Classical Banach Spaces I: Sequence Spaces, Springer-Verlag, Berlin and New York 1977.

[M] B. Maurey, *Quelques progrès dans la compréhension de la dimension infinie*, Journée Annuelle 1974, Société Mathématique de France.

[OS1] E. Odell and T. Schlumprecht, *The distortion problem*, Acta Math., (to appear).

[OS2] E. Odell and T. Schlumprecht, *Distortion and stabilized structure in Banach spaces; new geometric phenomena for Banach and Hilbert spaces*, this volume.

[P] A. Pelczynski, *Projections in certain Banach spaces*, Studia Math. **19** (1960), 209–228.

[R1] H. Rosenthal, *A characterization of Banach spaces containing ℓ_1*, Proc. Nat. Acad. Sci. U.S.A. **71** (1974), 2411–2413.

[R2] H. Rosenthal, *A subsequence principle characterizing Banach spaces containing c_0*, Bull. Amer. Math. Soc., to appear.

[S1] T. Schlumprecht, *An arbitrarily distortable Banach space*, Israel J. Math. **76** (1991), 81–95.

[S2] T. Schlumprecht, *A complementably minimal Banach space not containing c_0 or ℓ_p*, Seminar Notes in Functional Analysis and PDEs, LSU 1991–2, 169–181.

[Sz] A. Szankowski, *Subspaces without the approximation property*, Israel J. Math. **30** (1978), 123–129.

[T] B. S. Tsirelson, *Not every Banach space contains ℓ_p or c_0*, Functional Anal. Appl. **8** (1974), 139–141.

Exact C*-Algebras, Tensor Products, and the Classification of Purely Infinite Algebras

EBERHARD KIRCHBERG

Humboldt-Universität zu Berlin, Mathematisches Institut
Unter den Linden 6
D-10099 Berlin, Germany

1. Our survey (and the reference list) does not reflect the history of tensor products of operator algebras. Here we make use of tensor product functors on the category of C*-algebras as a unifying principle. An application of our theory to the classification problem of Elliott [13] can be found at the end of this paper. Throughout the paper *algebra* means C*-algebra and *vN-algebra* means von Neumann algebra. $\mathcal{L}(H)$ is the algebra of bounded operators on a Hilbert space of infinite dimension.

2.Tpf's. We call a bifunctor $(A, B) \Rightarrow A \otimes^\alpha B$ a *tensor product functor* (tpf) if it is obtained by completing of the algebraic tensor product $A \odot B$ of *-algebras in a functional way with respect to suitable C*-norms $\| \cdot \|_\alpha$.

We impose on $\otimes^\alpha$ mild nondegeneracy conditions such as $A \otimes^\alpha B \subseteq A \otimes^\alpha B^{**}$ and $A \otimes^\alpha D \subseteq A \otimes^\alpha B$ if D is a hereditary C*-subalgebra of B and similar conditions on the "first variable" A instead of the "second variable" B. Here B^{**} is the second conjugate vN-algebra of B, and for C*-subalgebras $E \subseteq A$, $F \subseteq B$ the notation $E \otimes^\beta F \subseteq A \otimes^\alpha B$ means that the algebraic inclusion $E \odot F \subset A \odot B$ extends to a C*-algebra monomorphism if we complete w.r.t. $\| \cdot \|_\beta$ and $\| \cdot \|_\alpha$ respectively. It is useful to relax the definition of tpf's and to consider also *partial* tensor product functors (ptpf's) $B \Rightarrow A \otimes^\alpha B$, where A is a fixed C*-algebra and where we impose the nondegeneracy conditions on $A \otimes^\alpha (\cdot)$. For example, if $A = N$ is a vN-algebra then one gets a ptpf $B \Rightarrow N \otimes^{\mathrm{nor}} B$ by considering the l.u.b. of all C*-norms defined by *-representations of $N \odot B$ that are normal on N.

The nondegeneracy conditions allow one to extend the functor uniquely to completely positive contractions, "decomposable" maps (cf. def. after Cor. 4.5 in [23]), C*-triple systems, Hilbert C*-modules, C*-systems, and C*-spaces (as e.g. $B/(L + R)$ for closed left and right ideals L and R of B [23]). In general a tpf does not extend to complete contractions, operator systems, or operator spaces [12], [4]. By our requirements on ptpf's, $A \otimes^\alpha C \subseteq A \otimes^\alpha B$ if there is a conditional expectation from B^{**} onto C^{**} ($\cong \sigma(B^{**}, B^*)$-closure of C in B^{**}). Thus, the study of a ptpf (resp. tpf) can be reduced to separable B (resp. separable A and B), cf. [20, Lemma 3.4, and Prop. 3.1(ii)]. The functoriality is essential: the C*-norm on $N \odot N$, which is defined by the identity correspondence (cf. [6], Ch. V, App. B) of the vN-algebra $N = VN(F_2)$ (generated by the regular representation of the free group on two generators F_2) cannot come from a ptpf.

Proceedings of the International Congress
of Mathematicians, Zürich, Switzerland 1994
© Birkhäuser Verlag, Basel, Switzerland 1995

We call a ptpf $A \otimes^\alpha (\cdot)$ *short exact* (w.r.t. the second variable) if

$$0 \to A \otimes^\alpha B \to A \otimes^\alpha C \to A \otimes^\alpha D \to 0$$

is short exact whenever $0 \to B \to C \to D \to 0$ is short exact. The partial tpf $A \otimes^\alpha (\cdot)$ is *injective* if $A \otimes^\alpha C \subseteq A \otimes^\alpha B$ for every C*-subalgebra C of B. Note that every ptpf is projective: $A \otimes^\alpha B$ maps onto $A \otimes^\alpha (B/J)$ for every ideal J of B.

3. Examples. The maximal (or universal) tpf $\otimes^{\max}$ is given by the maximal C*-norm on $A \odot B$. $\otimes^{\max}$ is short exact in both variables. The spatial (or minimal) C*-algebra tensor product functor $\otimes = \otimes^{\min}$ is given by $A \odot B \subset \mathcal{L}(H \otimes K)$ if $A \subset \mathcal{L}(H)$ and $B \subset \mathcal{L}(K)$. $\otimes$ is injective. The epimorphisms from $A \otimes^{\max} B$ onto $A \otimes^\alpha B$ and from $A \otimes^\alpha B$ onto $A \otimes B$ define functor transformations for every tpf $\otimes^\alpha$. There exist other injective tpf's, because $\mathcal{L}(H) \otimes^{\max} \mathcal{L}(H) \neq \mathcal{L}(H) \otimes^{\min} \mathcal{L}(H)$, [17]. One can define in a functorial way from a given C*-norm $\|\cdot\|_\delta$ on $\mathcal{L}(H) \odot \mathcal{L}(H)$ a uniform operator space tensor norm $\alpha = (\alpha_n)_{n \geq 1}$ on the category of operator spaces in the sense of [4, Def. 5.9] such that the completions of $M_n(A \odot B)$ with respect to α_n coincide with the canonical C*-algebra matrix norms of the completion of $A \odot B$ with respect to the C*-norm $\|\cdot\|_\delta$ (restricted to $A \odot B$) if A and B are C*-subalgebras of $\mathcal{L}(H)$. Thus, the conjecture at the end of [4] is wrong, and by [20] there exists an algebra A with unique C*-norm on $A^{op} \odot A$ such that A is not approximately injective in the sense of [10]. Note that A has the WEP of [32] and is not amenable. Therefore A is not locally reflexive (see below).

The question of whether there is only one short exact tpf (in both variables) is equivalent to the question of whether there is only one C*-norm on $C^*(F_2) \odot C^*(F_2)$, where $C^*(F_2)$ means the full group C*-algebra of the free group F_2 on two generators. The latter question turns out to be equivalent to the question of whether every II_1-factor with separable predual is a subfactor of the ultrapower of the hyperfinite II_1-factor, or to the question of whether the predual of every vN-algebra is finitely representable in the trace class operators [20, Prop. 8.1].

We now define tpf's, that are injective w.r.t. the first variable and short exact w.r.t. the second variable.

(i) Assume that $(\cdot) \otimes^\alpha B$ is a partial tpf (w.r.t. the first variable). Let $A \otimes^{\alpha, il} B$ be the closure of $A \odot B$ in $\mathcal{L}(K) \otimes^\alpha B$, where K is a Hilbert space and A is a C*-subalgebra of $\mathcal{L}(K)$. Then $A \Rightarrow A \otimes^{\alpha, il} B$ is an injective ptpf (w.r.t. the first variable). If moreover $\otimes^\alpha$ is short exact with respect to the second variable, then $\otimes^{\alpha, il}$ is a tpf that is injective w.r.t. the first variable and short exact w.r.t. the second variable.

(ii) Assume that $A \otimes^\alpha (\cdot)$ is a partial tpf. We denote by $A \otimes^{\alpha, pr} B$ the completion of $A \odot B$ w.r.t. the maximal C*-norm among the norms induced by the canonical embeddings of $A \odot B$ into $(A \otimes^\alpha C)/(A \otimes^\alpha J)$. Then $B \Rightarrow A \otimes^{\alpha, pr} B$ is a short exact partial tpf, and $A \otimes^{\alpha, pr} B$ is a tpf if $\otimes^\alpha$ is a tpf. If $\otimes^\alpha$ is a tpf that is injective w.r.t. the first variable then $\otimes^{\alpha, pr}$ is injective w.r.t. the first variable and short exact w.r.t. the second variable.

THEOREM 1. ([21, Th. 1.1]) *If B has the lifting property then $N \otimes^{\mathrm{nor}} B = N \otimes^{\max} B$.*

It follows that there is a unique C*-norm on $\mathcal{L}(H) \odot C^*(F_\infty)$. This is equivalent to the uniqueness of the tpf which is injective with respect to the first variable and short exact with respect to the second variable. In particular $A \otimes^{\beta,il} B = A \otimes^{\alpha,pr} B$ for every tpf $\otimes^\alpha$ that is injective with respect to the first variable and every tpf $\otimes^\beta$ that is short exact w.r.t. the second variable, e.g. $\otimes^{\max,il} = \otimes^{\min,pr}$.

4. Growth conditions. Let $\| \cdot \|_\alpha \geq \| \cdot \|_\beta$ be C*-norms on $A \odot B$. Assume that B is generated by $X = X^*$, a self-adjoint linear subspace of B with $1 \in X$. We denote by X^n the span of $\{y_1 y_2 \ldots y_n \colon y_i \in X\}$ and by $\Theta(n)$ the norm of the operator given by $A \otimes^\beta B \supseteq X^n \odot B \to X^n \odot B \subseteq A \otimes^\alpha B$.

Then $\Theta(2n) \geq \Theta(n)^2$ and $\dim(X^n) \geq \Theta(n) \geq 1$. Thus $\| \cdot \|_\alpha = \| \cdot \|_\beta$ if and only if $\Theta(n) \equiv 1$ if and only if $\lim \Theta(n)^{1/n} = 1$. The latter is always the case if the filtration of B defined by X has subexponential growth, i.e. $\lim \dim(X^n)^{1/n} = 1$. In particular *a unital C*-algebra B is nuclear if B has a dense self-adjoint filtration $B_n \subseteq B$, $1 \in B_n^* = B_n$, $B_n B_m \subset B_{n+m}$ of subexponential growth* $\lim \dim(B_n)^{1/n} = 1$ [27]. If G is a discrete group and $C_r^*(G)$ is nuclear, then G is amenable [32]. It gives a proof of the following well-known result: if G is discrete, finitely generated, and has subexponential growth, then G is amenable.

5. Exactness. Let $A \otimes^\alpha (\cdot)$ be a ptpf or $\otimes^\alpha$ a tpf. A is called $\otimes^\alpha$-*exact* if $B \Rightarrow A \otimes^\alpha B$ is an exact functor (i.e. is short exact and injective w.r.t. the second variable). A is *exact* if A is $\otimes^{\min}$-exact. In [2] and [10] the following properties have been studied: A has property (C') (resp. $(C''), (C)$) if for every C*-algebra B the canonical inclusion $A^{**} \odot B^{**} \subseteq (A \otimes B)^{**}$ induces on $A \odot B^{**}$ (resp. on $A^{**} \odot B$, on $A^{**} \odot B^{**}$) the minimal C*-norm. Obviously (C) implies (C') and (C''). Conversely (C'') and (C') together imply (C). It is easy to see that (C') implies exactness of A. Every nuclear C*-algebra satisfies property (C), and (C) passes to quotient algebras and C*-subalgebras.

Property (C'') is equivalent to the (operator space) *local reflexivity* of A, i.e. for every finite-dimensional subspace $X \subseteq \mathcal{L}(H)$ and every complete contraction $T \colon X \to A^{**}$ there exists a net of complete contractions $T_\sigma \colon X \to A$ such that $f(T(y)) = \lim f(T_\sigma(y))$ for every $y \in X$ and $f \in A^*$. Here it suffices to consider self-adjoint unital X and completely positive unital T.

Local reflexivity passes to quotient algebras and C*-subalgebras. An extension $0 \to A \to E \to B \to 0$ of locally reflexive C*-algebras A and B is locally reflexive if and only if the epimorphism $\pi \colon E \to B$ is locally liftable in the sense that for every finite-dimensional operator space $X \subset B$ there exists a completely contractive map $\sigma \colon X \to E$ with $\pi\sigma = id_X$. This implies (by the 3×3-Lemma) that *an extension E of exact C*-algebras A, B is exact if and only if E is locally reflexive.*

6. From exactness to property (C). We consider functors $B \Rightarrow \mathcal{P}(B)$ from the category of C*-algebras into the sets of completely positive (= c.p.) contractions $CPC(B, N)$ from an algebra B into a given vN-algebra N. We assume that

(i) $\mathcal{P}(B)$ is a point-weakly closed convex set of c.p. contractions from B into N such that $Vh \in \mathcal{P}(C)$ if $V \in \mathcal{P}(B)$ and $h \colon C \to B$ is a c.p. contraction,

(ii) $\varphi(\cdot)1 \in \mathcal{P}(B)$ for every state φ on B, and $\mathcal{P}(B)$ contains $\sum_{i,j} n_i^* V(b_i^*(\cdot)b_j)n_j$ if $V \in \mathcal{P}(B)$, $b_1, \dots, b_k \in B$ with $\|\sum b_i b_i^*\| \le 1$ and $n_1, \dots, n_k \in N$ with $\sum n_i^* n_i \le 1$,

(iii) if $W \in CPC(B,N)$ and if there exist $\gamma \in \mathbb{R}_+$ and $V \in \mathcal{P}(B)$ such that $\gamma V - W$ is c.p., then $W \in \mathcal{P}(B)$.

A functor satisfying (i)–(iii) is called a *cp-functor*. A cp-functor $B \Rightarrow \mathcal{P}(B)$ is *short exact* if $\{\, V \in \mathcal{P}(B),\ J \text{ is a closed ideal of } B \text{ and } V(J) = 0 \,\}$ together imply the existence of $W \in \mathcal{P}(B/J)$ such that $V = W\pi_J$, where $\pi_J \colon B \to B/J$ is the quotient map. The cp-functor $B \Rightarrow \mathcal{P}(B)$ is called *injective* if $V \in \mathcal{P}(B)$ and $B \subseteq A$ imply the existence of $W \in \mathcal{P}(A)$ with $V = W|B$. $B \Rightarrow \mathcal{P}(B)$ is *exact* if it is short exact and injective. An example of an injective cp-functor is given by $B \Rightarrow \mathcal{P}_{\mathrm{nuc}}(B)$, where $\mathcal{P}_{\mathrm{nuc}}(B)$ is the set of weakly nuclear c.p. contractions from B into N. It is an open question whether or not $\mathcal{P}_{\mathrm{nuc}}(B)$ is exact (for every vN-algebra N).

A c.p. contraction $V \colon B \to C$ (resp. $V \colon B \to N$) is *nuclear* (resp. *weakly nuclear*) if V is the point norm limit (resp. the point $\sigma(N, N_*)$-limit) of maps $S_\sigma T_\sigma$, where T_σ is a c.p. contraction from B into a matrix algebra M_{k_σ} and S_σ is a c.p. contraction from M_{k_σ} into C (resp. N). By a separation argument this is equivalent to saying that $V \otimes^{\max} id \colon B \otimes^{\max} E \to C \otimes^{\max} E$ (resp. $\pi(V \otimes^{\max} id) \colon B \otimes^{\max} E \to N \otimes^{\mathrm{nor}} E$, where π is the canonical epimorphism $N \otimes^{\max} E \to N \otimes^{\mathrm{nor}} E$) factorizes through $B \otimes E$ for every algebra E.

For a short exact cp-functor $B \Rightarrow \mathcal{P}(B)$ there exist, for every $V \in \mathcal{P}(B)$, an algebra Q, a *-homomorphism $h \colon B \to Q$, and a $W \in \mathcal{P}(Q)$ such that $V = Wh$ and W has the following maximality property: if $k \colon Q \to C$ is a *-homomorphism and $U \in \mathcal{P}(C)$ with $Uk = W$, then $U(C_1) = W(Q_1)$, where Q_1, C_1 mean the closed unit balls. This comes from the fact that short exact cp-functors respect inductive limits. The Kaplansky density theorem is equivalent to $A^{**} = A_\tau / \ker(\tau - \lim)$, where A_τ means the C*-algebra of $\tau(A^{**}, A^*)$-convergent bounded nets in A indexed by a suitable directed set, e.g. the set of the $\sigma(A^*, A^{**})$-compact subsets of A^*. It implies that we can choose Q as a vN-algebra and W as a normal c.p. contraction.

If the cp-functor $B \Rightarrow \mathcal{P}(B)$ is exact (i.e. moreover injective) and N is a properly infinite vN-algebra, then $h \colon B \to Q$ and $W \in \mathcal{P}(Q)$ can be found such that Q is an injective vN-algebra and W is normal.

Let $B \Rightarrow A \otimes^\alpha B$ be a ptpf, $M \subseteq \mathcal{L}(H)$ a vN-algebra of infinite multiplicity (i.e. M' is properly infinite), and $T \colon A \to M$ a c.p. contraction such that $T(A_1)$ is weakly dense in M_1. Then $B \Rightarrow \mathcal{P}_{T,\alpha}(B) = \{V \in CPC(B, M') \colon T \odot V \colon A \odot B \to \mathcal{L}(H) \text{ is } \|\cdot\|_\alpha\text{-continuous on } A \odot B\}$ defines a cp-functor (into M'). $\mathcal{P}_{T,\alpha}$ is short exact (resp. injective, exact) if $A \otimes^\alpha (\cdot)$ is short exact (resp. injective, exact). Note that $A \otimes^\alpha (\cdot)$ is exact iff $A \otimes^\alpha (\cdot)$ is injective and $A \otimes^{\alpha,pr} (\cdot) = A \otimes^\alpha (\cdot)$. In particular (by Theorem 1) $A \otimes B = A \otimes^{\min,pr} B = A \otimes^{\max,il} B \subseteq \mathcal{L}(H) \otimes^{\max} B$ iff A is exact and $A \subseteq \mathcal{L}(H)$, i.e. A is *nuclearly embeddable* in the sense that the inclusion map $A \hookrightarrow \mathcal{L}(H)$ is nuclear. Together this gives Theorem 2:

THEOREM 2. *If $A \otimes^\alpha (\cdot)$ is a ptpf and A is $\otimes^\alpha$-exact, then for every algebra B and for every commuting pair $d_1 \colon A \to \mathcal{L}(K)$, $d_2 \colon B \to \mathcal{L}(K)$ of *-representations* such that $d_1 \odot d_2 \colon A \odot B \to \mathcal{L}(K)$ is $\| \cdot \|_\alpha$-continuous there exists an injective W^*-algebra Q, a normal c.p. contraction $W \colon Q \to d_1(A)'$, and a *-homomorphism $h \colon B \to Q$ such that $d_2 = W \circ h$. Thus, A satisfies properties (C), (C'), and (C''), A is exact, and $A \otimes^\alpha B = A \otimes B \subseteq \mathcal{L}(H) \otimes^{\max} B$ for every C^*-algebra B.

In conjunction with [2] we get that A is exact iff A satisfies (C) iff A satisfies (C') iff A is nuclearly embeddable, and every exact C^*-algebra is locally reflexive.

It is unknown whether local reflexivity implies exactness. $\otimes^{\max}$-exact algebras are the nuclear (= amenable) algebras. From Theorem 2 we see that C^*-subalgebras and quotients of A are exact if A is exact, that exactness is preserved under inductive limits, cross products by (strongly Voiculescu-) amenable (quantum deformed) Kac algebra (e.g. by coactions of locally compact groups or by actions of amenable groups). If A and B are exact then $A \otimes B$ and $A \oplus B$ are exact. Amalgamated reduced free products along finite-dimensional C^*-subalgebras of A and B are again exact if A and B are exact (one can modify the proof of [21, Th. 7.2] with the help of Corollary 7 such that it proves the general result).

By a continuous bundle of C^*-algebras $(A, C_0(\Omega) \subseteq M(A))$ over a locally compact space Ω we mean the C^*-algebra A of continuous sections (vanishing at infinity) of a continuous field $(A_x)_{x \in \Omega}$ of C^*-algebras. In [28] the following results have been obtained: an algebra B is exact iff $(A \otimes B, C_0(\Omega) \otimes 1 \subseteq M(A \otimes B))$ is a continuous bundle of C^*-algebras for every continuous bundle of C^*-algebras $(A, C_0(\Omega) \subseteq M(A))$. A continuous bundle of C^*-algebras $(A, C_0(\Omega) \subseteq M(A))$ is exact iff every fibre A_x is exact and, for every algebra B, $(A \otimes B, C_0(\Omega) \otimes 1 \subseteq M(A \otimes B))$ is a continuous bundle of C^*-algebras. $(A \otimes^{\max} B, C_0(\Omega) \otimes 1 \subseteq M(A \otimes^{\max} B))$ is a continuous bundle for every continuous bundle $(A, C_0(\Omega) \subseteq M(A))$ if and only if B is nuclear. Similar results exist for cross-products of continuous bundles by l.c. groups [30]. It is unknown if exact C^*-algebras with the lifting property are nuclear. This would be true if there were only one short exact tpf. Because exact C^*-algbras are locally reflexive, an exact C^*-algebra is nuclear iff it has the WEP of Lance [32], [10].

7. Exact locally compact groups.

Let G be a l.c. (= locally compact) group, A an algebra, and $\alpha \colon G \to \mathrm{Aut}(A)$ a strongly continuous group homomorphism (i.e. $g \in G \mapsto \alpha(g)a$ is a continuous map from G in A for every $a \in A$). Then $(A, \alpha) = (G, \alpha, A)$ is a $(G$-$)$ covariant system. They form a category $CS(G)$. The morphisms from (A, α) to (B, β) are the *-homomorphisms $\varphi \colon A \to B$ with $\beta(g)\varphi(a) = \varphi(\alpha(g)a)$.

The full (= universal) crossed product $(A, \alpha) \Rightarrow C^*(G, \alpha, A) = A \rtimes_\alpha G$ (= C^*- hull of the convolution algebra $L_1(G, \alpha, A)$) is a short exact functor from $CS(G)$ into the category of C^*-algebras. The following are equivalent [30]:

(i) $(A, \alpha) \Rightarrow C^*(G, \alpha, A)$ is injective (i.e. is exact), (ii) G is amenable, and (iii) if G acts fibre-wise on a continuous bundle of C^*-algebras $(A, C_0(X) \subseteq M(A))$ then $(A \rtimes_\alpha G, C_0(X) \subseteq M(A) \subseteq M(A \rtimes_\alpha G))$ is a continuous bundle of C^*-algebras.

The equivalences (i)–(iii) remain valid for those subclasses of the q.d. (= quantum deformed) Kac algebras for which the weak and the strong Voiculescu amenability are equivalent. A q.d. Kac algebra (M, Φ) is weakly Voiculescu amenable if there exists an invariant mean on M. (M, Φ) is strongly Voiculescu *amenable* if the trivial representation is weakly contained in the regular representation. The regular representation d_r is defined as the zero of the semigroup of quasi-equivalence classes of *unitary* (in general non-self-adjoint) representations of the convolution algebra (M_*, Φ_*) of (M, Φ). Strong and weak amenability are the same if the regular representation weakly contains a central state. For discrete (undeformed) Kac algebras all amenability definitions are equivalent [36].

A more interesting functor is $(A, \alpha) \Rightarrow A \rtimes_{\alpha,r} G = C_r^*(G, \alpha, A) =$ image of $C^*(G, \alpha, A)$ by the (regular) representation into the multiplier algebra of $A \otimes \mathcal{K}(L_2(G))$. The functor $(A, \alpha) \Rightarrow C_r^*(G, \alpha, A)$ is injective. We say that a l.c. group G is *exact* if $(A, \alpha) \Rightarrow C_r^*(G, \alpha, A)$ is an exact functor. $C_r^*(G, \alpha, A)$ *is an exact C*-algebra if A is exact and G is exact*, because $C_r^*(G, \alpha, A) \otimes B = C_r^*(G, \alpha \otimes id, A \otimes B)$.

In particular $C_r^*(G)$ is exact as a C*-algebra if G is exact as a l.c. group. *If G is discrete, then G is an exact group if and only if $C_r^*(G)$ is an exact C*-algebra* [30]. A l.c. group G is exact if G is amenable. We don't know if $SL_2(\mathbb{R})$ and $SL_3(\mathbb{R})$ are exact.

The statements remain valid for q.d. Kac algebras with similar definitions, as long as we consider a subclass where weak amenability is equivalent to the strong amenability.

If G is a closed discrete subgroup of a connected Lie group H and A is a cocompact amenable closed subgroup of H then $C_r^*(G)$ is a unital C*-subalgebra of $C_r^*(G, C(H/A))$, which is nuclear by [34]. Thus, *discrete closed subgroups of Lie groups have exact reduced group C*-algebras* (Connes). The action of a (discrete) hyperbolic group G on its Gromov boundary ∂G is amenable, cf. [1]. It follows that $C_r^*(G, C(\partial G))$ is nuclear and that $C_r^*(G)$ *is exact for hyperbolic groups* G. It is unknown if $C_r^*(G)$ is exact for every "bolic" discrete group G (in the sense of Skandalis).

The functor $(A, \alpha) \Rightarrow C_r^*(G, \alpha, A)$ is the composition of $(A, \alpha) \Rightarrow (C_r^*(G, \alpha, A), \hat\alpha)$ with the functor that forgets the action $\hat\alpha$, where the action $\hat\alpha$ means the action of the dual Kac algebra $\hat G = (VN(G), \text{Kronecker product})$ (or the dual "coaction" of G). The stable cocycle equivalence $(A, \alpha) \cong (C_r^*(\hat G, \hat\alpha, C_r^*(G, \alpha, A)), \hat{\hat\alpha})$ of Takai-Takesaki is functorial. Therefore, $(A, \alpha) \Rightarrow (C_r^*(G, \alpha, A), \hat\alpha)$ is an exact functor. Thus, for discrete groups G we have that G is exact iff $C_r^*(G)$ is exact iff the functor $(A, \hat\alpha) \Rightarrow A$ (which forgets a coaction $\hat\alpha$ of G on A) is exact. The latter is a saturation condition for invariant hereditary C*-subalgebras under (globally) saturated actions of the compact Kac algebras $\hat G$.

8. Nonexact algebras and weak exactness. Let G be a l.c. group. We denote by λ and ϱ the left and right regular representations of $C^*(G)$ on $L_2(G)$ respectively. Here $C^*(G)$ means the full (= universal) group C*-algebra of G. G has the *factorization property* (F) if $\lambda \odot \varrho \colon C^*(G) \odot C^*(G) \to \mathcal{L}(L_2(G))$ is $\|\cdot\|_{\min}$-continuous. G has property (F) if G is discrete and has an injective group homomorphism into a l.c. group H with property (F) [20]. In particular maximally almost periodic discrete groups have property (F). By a lemma of Selberg, a finitely generated group G is maximally almost periodic if and only if G is residually finite (profinite). In [22] it is shown that for a class of discrete groups G containing the groups with Kazhdan's property (T) (and e.g. the finitely generated free groups) we have that property (F) for G implies that G is residually finite. In the case of property (T) groups a simplified proof has been given by A. Valette [42, p. 37]. In [19] we stated the following generalization of results of [39], [40], and [32]:

Let G be a l.c. group with property (F) and J the kernel of the regular representation $C^*(G) \to C_r^*(G) \subseteq \mathcal{L}(L_2(G))$. If the regular representation weakly contains a central state, then

$$0 \to C^*(G) \otimes J \to C^*(G) \otimes C^*(G) \to C^*(G) \otimes C_r^*(C) \to 0$$

is exact if and only if G is amenable.

A proof is suggested in [19] along the lines of [39], [40], and [32], and is carried out in [20, Sec. 7]. The result holds for q.d. Kac algebras too. In particular $C^*(G)$ is not locally reflexive and therefore not exact for discrete nonamenable maximally almost periodic G, e.g. for discrete subgroups G of noncompact simple Lie groups that have finite covolume. Is $C^*(G)$ nonexact for a nonamenable Burnside group G?

A sequence $\Phi = (\varphi_1, \varphi_2, \dots)$ of irreducible states on an algebra B is *free* if the corresponding c.p. contraction $V_\Phi(b): = (\varphi_1(b), \varphi_2(b), \dots)$ from B into l_∞ contains the open unit ball of $c_0 (\subseteq l_\infty)$ in the image of the open unit ball of B. We call B *weakly sub-Rickart* if V_Φ maps the closed unit ball of B onto the closed unit ball of l_∞ for every free sequence Φ of irreducible states on B. If B is weakly sub-Rickart and is not a subalgebra of a matrix algebra over an abelian algebra then B contains a C*-subalgebra C that has $\mathcal{L}(l_2)$ as a quotient algebra. Thus, *locally reflexive weakly sub-Rickart algebras are subhomogeneous.* If B has the property that for $a, b \in B_+$ with $ab = 0$ there exists $c \in B_+$ with $ac = 0$, $cb = b$, then B is weakly sub-Rickart. Therefore Corona algebras $M(A)/A$ and AW^*-algebras are examples of weakly sub-Rickart algebras. In particular, a vN-algebra M is exact (as C*-algebra) if and only if M is a finite sum of matrix algebras over abelian C*-algebras.

We say that a vN-algebra M is *weakly exact* if the cp-functor $B \Rightarrow \mathcal{P}_{id,\min}(B)$ into $CPC(B, M' \overline{\otimes} \mathcal{L}(H))$ for $id \colon M \to M$ is exact. Equivalently, $B \Rightarrow \mathcal{P}_{nuc}(B) \subseteq CPC(B, M)$ is exact. If A is exact and $T \colon A \to M$ is a c.p. contraction such that $T(A_1)$ is weakly dense M_1, then M is weakly exact. If $N \subseteq M$ is a vN-subalgebra of M and there exists a net $V_\sigma \colon M \to N$ of normal unital c.p. maps such that $V_\sigma | N$ tends point-weakly to id_N, then N is exact if M is exact. $M \overline{\otimes} \mathcal{L}(H)$ is weakly exact iff M is weakly exact iff M' is weakly exact. A vN-algebra M on a separable Hilbert space is weakly exact if and only if the factors M_γ arising in the (μ-measurable) central decomposition of M are weakly exact for μ-almost every γ. Thus, the open question of whether every vN-algebra is weakly exact can be limited to II_1-factors with separable preduals. It is not known if the ultrapower of the hyperfinite II_1-factor (and hence every separable subfactor) is weakly exact.

B is exact if and only if B is locally reflexive and B^{**} is weakly exact (consider $\mathcal{P}_{\varepsilon,\min}(\cdot)$ for $\varepsilon \colon B \hookrightarrow B^{**}$ and use Theorem 2, or see [21]).

9. Approximation properties.
Let $A \subseteq M$ be a C*-subalgebra of a vN-algebra M. Assume that there exists a net of finite rank linear maps $V_\sigma \colon A \to M$ such that for every $b \in A \otimes \mathcal{K}$ and $\omega \in (M \overline{\otimes} \mathcal{L}(H))_*$, we have that $\omega(V_\sigma \otimes id(b))$ tends to $\omega(b)$. Then V_σ tends to the inclusion map $\varepsilon \colon A \hookrightarrow M$ in the $\sigma(B(A, M), A \hat{\otimes}^{op} M_*)$-topology, where $A \hat{\otimes}^{op} M_*$ means the maximal (projective) operator space tensor

product in the sense of [4], cf. [15, Lemma 1.6, Remark 18]. If $a \in A \odot \mathcal{L}(H)$ and $f \in (M \otimes \mathcal{L}(H))^*$ such that f is partially $\sigma(M, M_*)$-continuous, then $\varphi_{a,f}(V) := f(V \otimes id(a))$ satisfies $\|\varphi_{a,f}\|_\wedge \subseteq \|f\|\|a\|_{\min}$ and $\varphi_{a,f}$ is $\sigma(CB(A, M), A \hat{\otimes}^{op} M_*)$-continuous, because $f|M \otimes X$ is $\sigma(M \otimes X, (M \overline{\otimes} \mathcal{L}(H))_*)$-continuous, where $X = \{\psi \otimes id(a): \psi \in A^*\}$. It follows that $\mathcal{P}_{\varepsilon,\min}(\cdot)$ is an exact cp-functor if $A = \varepsilon(A)$ is weakly dense in M. In the cases $A = M$ and $M = A^{**}$ we get (cf. [15, Th. 2.2], [11], [25]):

PROPOSITION 3. (i) *If M is a vN-algebra with the weak slice map property S_σ (i.e. with w^*OAP of [15]), then M is weakly exact.* (ii) *If A is locally reflexive and has the OAP (i.e. has the slice map property for the compact operators $\mathcal{K}$ [31]), then A has the general slice map property (i.e. has the strong OAP). In particular, A is exact.*

The exactness of closed operator subspaces of $\mathcal{K}$ implies that extensions of C*-algebras with OAP have the OAP [25]. On the other hand, if A is separable and unital and does not have the local lifting property (e.g. if $A = C_r^*(SL_2(\mathbb{Z}))$) then $SA = C_0(\mathbb{R}) \otimes A$ has a quasi-diagonal extension E by the compact operators such that $0 \to \mathcal{K} \to E \to C_0(\mathbb{R}) \otimes A \to 0$ is not semisplit (i.e. does not have a c.p. lift) [20]. In particular E is not locally reflexive. Thus, E is not exact and does not have the strong OAP. The example shows that exactness cannot be characterized by properties of the second conjugate vN-algebra. In [15] it is shown for discrete G that $C_r^*(G)$ has the strong OAP iff $C_r^*(G)$ has the OAP iff $VN(G)$ has the w*OAP. An algebra B is exact iff $B \otimes X = \{d \in B \otimes Y: f \otimes id(d) \in X, \forall f \in B^*\}$ for every pair $X \subset Y$ of operator spaces such that there is a completely bounded projection from Y^{**} onto X^{**} (use property (C')). Conjectures: $C^*(SL_2(\mathbb{Z}))$, $C_r^*(SL_3(\mathbb{Z}))$, and some C*-subalgebra of M_{2^∞} do not have the Grothendieck AP (and thus do not have the OAP).

10. Embeddings of exact C*-algebras. Let $A \subset \mathcal{L}(H)$ be a separable exact C*-algebra. Because A is nuclearly embeddable there exists a sequence of unital c.p. maps $V_n: M_{k_n} \to M_{k_{n+1}}$ from matrix algebras into matrix algebras such that $\Delta(A) \subseteq X := \operatorname{indlim}(V_n) \subset l_\infty(\mathcal{L}(H))/c_0(\mathcal{L}(H))$, where $\Delta(a) = (a, \dots) + c_0(\mathcal{L}(H))$ is the diagonal embedding. X is an example of a nuclear operator system. By an operator system Y we mean a unital self-adjoint closed subspace of $\mathcal{L}(K)$ for some Hilbert space K (with the matrix norms and matrix order unit structure coming from $\mathcal{L}(K)$, see [9]). There is a general theory of operator spaces (see e.g. [12]). Most of the definitions (e.g. c.p., exact, nuclear, ...) carry over to operator systems. Stinespring dilations of the V_n's lead to Theorem 4:

THEOREM 4. [24] *For separable operator systems X, the following properties (i)–(iv) on X are equivalent: (i) X is nuclear. (ii) The second conjugate operator system X^{**} is injective. (iii) There are matrix algebras $B_n = M_{k_n}$ and u.c.p. maps $V_n: B_n \to B_{n+1}$ such that $X \cong \operatorname{indlim}(V_n: B_n \to B_{n+1})$ in the category of operator systems. (iv) There exists a closed left ideal L in the CAR-algebra M_{2^∞} such that $X \cong M_{2^\infty}/(L^* + L)$, where $L^* = \{a^*: a \in L\}$ and the right-hand side means the operator space quotient.*

Here $\cong$ means unital completely isometric isomorphism. A classification of the positions in M_{2^∞} of the essential hereditary C*-subalgebras of M_{2^∞} would imply a classification of the separable unital nuclear C*-algebras. There exist separable nuclear operator systems that have no completely isometric unital embedding into an exact algebra [29]. If L is a closed left ideal in a C*-algebra A, π denotes the quotient map $A \to A/(L^* + L)$ and a, b are in the open unit ball of A then there exists c in the open unit ball of A with $\|a - c\| \leq f(\|\pi(a) - \pi(b)\|)$ and $\pi(c) = \pi(b)$, where $f(t) = t + (2t)^{1/2}$, [23, Cor. 3.2]. It follows that π maps the closed unit ball of A onto the closed unit ball of $A/(L^* + L)$. Thus, Theorem 4, Choi's inequality [5], and [10] together imply Corollary 5:

COROLLARY 5 ([24, Cor. 1.5]). *If A is a unital separable exact algebra then there exists a semisplit extension $0 \to D \to E \to A \to 0$ of A by a hereditary C*-subalgebra D of the CAR-algebra M_{2^∞} such that E is a unital C*-subalgebra of the CAR-algebra.*

There exists a unitary U in M_{2^∞} such that A is isomorphic to a quotient of the relative commutant of U in M_{2^∞} by an AF-ideal [21]. By $\mathcal{O}_2$ we mean the Cuntz algebra on two generators [7]. By the inclusion $M_{2^\infty} \subset \mathcal{O}_2$ (or Glimm's theorem) there is a semisplit extension $0 \to \mathcal{O}_2 \otimes \mathcal{K} \to F \to A \to 0$ such that F is a unital C*-subalgebra of $\mathcal{O}_2$. An algebra B has been called *purely infinite* (= pi) if B is simple and nonzero, and every nonzero hereditary C*-subalgebra of B contains an infinite projection ([8, remark on p. 186]). $\mathcal{O}_2$ is a purely infinite algebra, [7]. An application of a lemma of Glimm on pi algebras gives a Voiculescu type generalization of the Weyl-von Neumann theorem. $\mathcal{K}_B$ and $M(\mathcal{K}_B)$ mean $B \otimes \mathcal{K}$ and the multiplier algebra of $B \otimes \mathcal{K}$.

THEOREM 6 [26]. *Assume that B is a C*-algebra, C is a separable unital C*-subalgebra of $M(\mathcal{K}_B)$, A is a purely infinite C*-subalgebra of $\mathcal{K}_B$ such that $CA \subset A$ and $A \cap C$ contains a strictly positive element of $\mathcal{K}_B$, and that $V: C \to M(\mathcal{K}_B)$ is a unital completely positive map with $V(C \cap \mathcal{K}_B) = \{0\}$. If $b^*V(.)b$ is nuclear for every $b \in \mathcal{K}_B$ then there exists a sequence of isometries $s_n \in M(\mathcal{K}_B)$ such that $V(y) - s_n^* y s_n \in \mathcal{K}_B$ and $\lim \|V(y) - s_n^* y s_n\| = 0$ for every $y \in C$.*

It follows that there is a sequence u_n of unitary operators in $M(\mathcal{K}_B)$ with $(y \oplus V(y)) - u_n^* y u_n \in \mathcal{K}_B$ and $\lim \| (y \oplus V(y)) - u_n^* y u_n \| = 0$ for $y \in C$ if V is moreover a *-homomorphism. The extension group $Ext^{-1}(A, \mathcal{O}_2) \cong KK(C_o(\mathbb{R}) \otimes A, \mathcal{O}_2)$ is zero because the identity id of O_2 is homotopic to $2id$ and the Kasparov functor KK is homotopy invariant [8], [18].

COROLLARY 7 [26]. *There exists a unital *-isomorphism h_1 from A onto a unital C*-subalgebra of $\mathcal{O}_2$ if A is a separable unital exact algebra.*

h_1 is unique up to unitary homotopy (see below). For a separable unital nuclear algebra A we get moreover that the unital c.p. lift of A in E and $h_1(A)$ in Corollary 5 and Corollary 7 are ranges of completely contractive projections on M_{2^∞} and $\mathcal{O}_2$ respectively.

11. Classification of pi-sun algebras. Assume that A is unital separable and exact, $h_1: A \to \mathcal{O}_2$ and $h_2: \mathcal{O}_2 \otimes \mathcal{O}_2 \to \mathcal{O}_2$ are unital *-monomorphisms (from Corollary

7), and $h_0(a) := h_2(h_1(a) \otimes 1)$. Further let B be a unital algebra that contains a (fixed) unital copy of $\mathcal{O}_2$. Then $h \oplus (h_0 \otimes id_{\mathcal{K}}) \colon \mathcal{K}_A \to \mathcal{K}_B$ makes sense and is a *-monomorphism that is nuclear if h is nuclear. We say that $k \colon \mathcal{K}_A \to \mathcal{K}_B$ is *unitary homotopic* to h if there exists a strongly continuous map $t \in \mathbb{R}_+ \mapsto u(t)$ into the unitaries of $M(\mathcal{K}_B)$ such that $\lim_{t \to \infty} u(t)^* h(a) u(t) = k(a)$ for every $a \in \mathcal{K}_A$. Every nuclear *-homomorphism $h \colon \mathcal{K}_A \to \mathcal{K}_B$ canonically defines an extension $[h - 0] \colon 0 \to S\mathcal{K}_B \to E \to \mathcal{K}_A \to 0$ in $Ext_{nuc}(\mathcal{K}_A, S\mathcal{K}_B) \cong KK_{nuc}(A, B)$, cf. [37]. The map $\varphi \colon h \in Hom_{nuc}(\mathcal{K}_A, \mathcal{K}_B) \mapsto [h - 0] \in KK_{nuc}(A, B)$ is a semigroup morphism, where we take sums $h \oplus k$ on the unitary equivalence classes of nuclear homomorphisms h and k from $\mathcal{K}_A$ into $\mathcal{K}_B$.

THEOREM 8 [26]. $\varphi \colon h \mapsto [h-0]$ *is a semigroup epimorphism from* $Hom_{nuc}(\mathcal{K}_A, \mathcal{K}_B)$ *onto* $KK(A, B)$, *and* $[h - 0] = [k - 0]$ *in* $KK(A, B)$ *if and only if* $h \oplus (h_0 \otimes id_{\mathcal{K}})$ *and* $k \oplus (h_0 \otimes id_{\mathcal{K}})$ *are unitary homotopic.*

Since KK is homotopy invariant (and "scaling" invariant), it suffices to prove the related result for nuclear *-monomorphisms

$$h \colon \mathcal{K}_A \to C_b(\mathbb{R}_+, \mathcal{K}_B)/C_0(\mathbb{R}_+, \mathcal{K}_B)$$

and unitary equivalence.

We say that A is *pi-sun* if A is a purely infinite separable unital nuclear C*-algebra. In the stable isomorphism class of a pi-sun algebra A there is up to isomorphism exactly one algebra A^{st} that contains a unital copy of $\mathcal{O}_2$ [8]. A modification of a lemma of Elliott and Theorem 8 lead to Theorem 9:

THEOREM 9 [26]. *If* A *and* B *are pi-sun algebras with* $A = A^{st}$ *and* $B = B^{st}$, *then for every KK-equivalence* $z \in KK(A, B)$ *there exists a unital *-isomorphism* h *from* A *onto* B *such that* $z = [h - 0]$.

It follows that *homotopic pi-sun algebras are isomorphic* and that $A \cong B$ if $K_*(A) \cong K_*(B)$ and A and B are pi-sun algebras in the bootstrap class (i.e. if A and B satisfy the UCT of KK-theory, cf. [3]).

Using Corollary 7, Theorem 8, and Theorem 9 one can show that every separable nuclear algebra is KK-equivalent to a pi-sun algebra. We get that Elliott's conjecture [13] in case $T^+ = 0$ is equivalent to a positive answer on questions (Q1) and (Q2):

(Q1) Is every simple nuclear stably infinite algebra purely infinite?

(Q2) Is every separable nuclear algebra in the bootstrap class?

Question (Q2) is equivalent to the simplest looking case of Elliott's conjecture:

(Q2′) Is a pi-sun algebra A with $K_*(A) = 0$ isomorphic to $\mathcal{O}_2$?

In November 95 (when we found that Theorem 6 and Corollary 7 yield proofs of Theorem 8 and Theorem 9) we were informed by G. Elliott, that N. C. Phillips (Eugene, Oregon) had announced a programme of a proof of Theorem 9 using E-theory and the particular isomorphisms $A \otimes \mathcal{O}_2 \cong \mathcal{O}_2$ and $A \otimes \mathcal{O}_\infty \cong A$ for pi-sun A (consequences of Theorem 9). We showed the latter isomorphisms together with a proof of Cor.7 in our lecture at ICM Satellite Meeting in Geneva (1994).

Another application of exactness can be found in [14] where Haagerup gave a proof that (2-)quasitraces on unital exact algebras are traces.

The paper has profited from discussions with R. Archbold, D. Bisch, B. Blackadar, J. Cuntz, E. Effros, G. Elliott, U. Haagerup, G. Kasparov, G. Pisier, M. Rørdam, J.Ruan, D. Voiculescu, and S. Wassermann.

Added in proof: Recently G. Pisier found a simpler proof of Theorem 1.

References

[1] S. Adams, *Boundary amenability for word hyperbolic groups and an application to smooth dynamics of simple groups*, Topology **33** (1994), 765–783.

[2] R. J. Archbold and C. J. K. Batty, *C^*-tensor norms and slice maps*, J. London Math. Soc. **22** (1980), 127–138.

[3] B. Blackadar, K-Theory for operator algebras, Springer, New York, Berlin, and Heidelberg, 1986.

[4] D. P. Blecher and V. I. Paulsen, *Tensor products of operator spaces*, J. Funct. Anal. **99** (1991), 262–291.

[5] M.-D. Choi, *A Schwarz inequality for positive linear maps on C^*-algebras*, Illinois J. Math. **18** (1974), 565–574.

[6] A. Connes, Noncommutative Geometry, Academic Press, New York and San Diego, 1994.

[7] J. Cuntz, *Simple C^*-algebras generated by isometries*, Comm. Math. Phys. **57** (1977), 173–185.

[8] J. Cuntz, *K-theory for certain C^*-algebras*, Ann. of Math. (2) **113** (1981), 181–197.

[9] E. G. Effros, *Aspects of noncommutative order*, Lecture Notes in Math. **650**: C^*-algebras and applications to physics, Springer, Berlin, Heidelberg, and New York, 1978.

[10] E. G. Effros and U. Haagerup, *Lifting problems and local reflexivity for C^*-algebras*, Duke Math. J. **52** (1985), 103–128.

[11] E. G. Effros and Z.-J. Ruan, *On approximation properties for operator spaces*, Internat. J. Math. (IJM) **1** (1990), 163–187.

[12] E. G. Effros and Z.-J. Ruan, *Recent developments in operator spaces*, in Current topics in operator algebras, Proceedings of the Satellite Conference of ICM-90, ed. H.Araki et al., World Scientific, Singapore and Teaneck, NJ (1991), 146–164.

[13] G. A. Elliott, *The classification problem for amenable C^*-algebras*, in Proceedings of the International Conference of Mathematicians, Zürich 1994, Birkhäuser, Basel 1995.

[14] U. Haagerup, *Quasitraces on exact C^*-algebras are traces*, notes.

[15] U. Haagerup and J. Kraus, *Approximation properties for group C^*-algebras and group von Neumann algebras*, Trans. Amer. Math. Soc. **344** (1994), 667–699.

[16] T. Huruya and S.-H. Kye, *Fubini products of C^*-algebras and applications to C^*-exactness*, Publ. Res. Inst. Math. Sci., Kyoto Univ. **24** (1988), 765–773.

[17] M. Junge and J. Pisier, *Bilinear forms on exact operator spaces and $B(H) \otimes B(H)$*, preprint, to appear in Geom. Functional Anal.

[18] G. G. Kasparov, *The operator K-functor and extensions of C^*-algebras*, Math. USSR-Izv. **16** (1981), 513–572.

[19] E. Kirchberg, *Positive maps and C^*-nuclear algebras*, in Proc. Inter. Conference on Operator Algebras, Ideals and Their Applications in Theoretical Physics, Leipzig, September 12–20, 1977, pp. 255–257, Teubner Texte, Leipzig, 1978.

[20] E. Kirchberg, *On non-semisplit extensions, tensor products and exactness of group C*-algebras*, Invent. Math. **112** (1993), 449–489.

[21] E. Kirchberg, *Commutants of unitaries in UHF algebras and functorial properties of exactness*, J. Reine Angew. Math. **452** (1994), 39–77.

[22] E. Kirchberg, *Discrete groups with Kazhdan's property T and factorization property are residually finite*, Math. Ann. **299** (1994), 551–563.

[23] E. Kirchberg, *On restricted perturbations in inverse images and a description of normalizer algebras in C*-algebras*, J. Funct. Anal. **129** (1995), 1–34.

[24] E. Kirchberg, *On subalgebras of the CAR-algebra*, J. Funct. Anal. **129** (1995), 35–63.

[25] E. Kirchberg, *On the matricial approximation property*, preprint.

[26] E. Kirchberg, *The classification of purely infinite C*-algebras using Kasparov's theory*, preprint.

[27] E. Kirchberg and G. Vaillant, *On C*-algebras having linear, polynomial and subexponential growth*, Invent. Math. **108** (1992), 635–652.

[28] E. Kirchberg and S. Wassermann, *Operations on continuous bundles of C*-algebras*, preprint, to appear in Math. Ann.

[29] E. Kirchberg and S. Wassermann, *C*-algebras generated by operator systems*, preprint.

[30] E. Kirchberg and S. Wassermann, *Exact groups and continuous bundles of C*-algebras*, preprint.

[31] J. Kraus, *The slice map problem and approximation properties*, J. Funct. Anal. **102** (1991), 116–155.

[32] E. C. Lance, *On nuclear C*-algebras*, J. Funct. Anal. **12** (1973), 157–176.

[33] G. Pisier, *Exact operator spaces*, preprint, to appear in Colloque sur les Algèbres d'Opérateurs (Orléans 1992), Astérisque (Soc. Math. France).

[34] M. A. Rieffel, *Strong Morita equivalence of certain transformation group C*-algebras*, Math. Ann. **222** (1976), 7–22.

[35] M. Rørdam, *Classification of inductive limits of Cuntz algebras*, J. Reine Angew. Math. **440** (1993), 175–200.

[36] Z.-J. Ruan, *Amenability of Hopf von Neumann algebras and Kac algebras*, preprint.

[37] G. Skandalis, *Une notion de nucléarité en K-théorie (d' aprés J.Cuntz)*, K-Theory **5** (1988), 1–7.

[38] D. Voiculescu, *Around quasidiagonal operators*, Integral Equations Operator Theory **17** (1993), 137–149.

[39] S. Wassermann, *On tensor products of certain group C*-algebras*, J. Funct. Anal. **23** (1976), 239–254.

[40] S. Wassermann, *Tensor products of free-group C*-algebras*, Bull. London Math. Soc. **22** (1990), 375–380.

[41] S. Wassermann, *C*-algebras associated with groups with Kazhdan's property T*, Ann. of Math. (2) **134** (1991), 423–431.

[42] S. Wassermann, Exact C*-algebras and related topics, Lecture Notes Series, no. 19, Global Analysis Research Center, Seoul National University, 1994.

Distortion and Stabilized Structure in Banach Spaces; New Geometric Phenomena for Banach and Hilbert Spaces

E. ODELL[1] AND TH. SCHLUMPRECHT[2]

[1]Department of Mathematics
The University of Texas at Austin
Austin, TX 78712, USA

[2]Department of Mathematics
Texas A&M University
College Station, TX 77843, USA

Many of the fundamental research problems in the geometry of normed linear spaces can be loosely phrased as: Given a Banach space X and a class of Banach spaces $\mathcal{Y}$ does X contain a subspace $Y \in \mathcal{Y}$? As a Banach space X is determined by its unit ball $B_X \equiv \{x \in X : \|x\| \leq 1\}$ the problem can be rephrased in terms of the geometry of convex sets: Can a given unit ball B_X be sliced with a subspace to obtain a set in some given class of unit balls? A result of this type is the famous theorem of Dvoretzky (see also [L], [M6], [M4], [MS], [FLM]).

THEOREM 1 [D]. *For every $\varepsilon > 0$ and integer k there exists an integer $n = n(k, \varepsilon)$ such that if X is an n-dimensional normed space then X contains a k-dimensional subspace E with $d(\ell_2^k, E) < 1 + \varepsilon$.*

Before proceeding, we define the terms of Theorem 1. For $1 \leq p \leq \infty$, $\ell_p^k = (\mathbb{R}^k, \|\cdot\|_p)$ where $\|(a_i)_1^k\|_p \equiv (\sum_{i=1}^k |a_i|^p)^{1/p}$ if $p < \infty$ and $\|(a_i)_1^k\|_\infty \equiv \max_{i \leq n} |a_i|$. ($\ell_p$ and c_0 are defined analogously.) Thus, ℓ_2^k is a k-dimensional Euclidean space. Banach spaces X and Y are *isomorphic* if there exists an *isomorphism* — a bounded linear invertible operator — from X onto Y. We define the Banach-Mazur multiplicative distance between isomorphic spaces X and Y by

$$d(X, Y) = \inf\{\|T\| \, \|T^{-1}\| : T \text{ is an isomorphism from } X \text{ onto } Y\} \, ;$$

to get a true metric one takes $\log d(X, Y)$. Geometrically $d(X, Y) < K$ means that the two unit balls B_X and B_Y can be placed in the same linear space Y by an affine transformation $T : X \to Y$ so that $B_Y \subseteq TB_X \subseteq KB_Y$. Distance close to 1 means that the convex bodies TB_X and B_Y are geometrically close.

---

[1]Partially supported by NSF and TARP 235.
[2]Research supported by NSF.

Proceedings of the International Congress
of Mathematicians, Zürich, Switzerland 1994
© Birkhäuser Verlag, Basel, Switzerland 1995

The Distortion Problem and Tsirelson's Space

Theorem 1 was the starting point of what has come to be called the local theory of Banach spaces. This theory, whose development has exploded over the last quarter century, has proved to be both rich and deep (see, e.g., [P1], [P2], [Pe], [T-J1], [MS], [M5]). In this paper we shall discuss primarily infinite-dimensional geometry: Given a certain infinite-dimensional unit ball B_X and a certain class C of infinite-dimensional unit balls does some infinite-dimensional slice of B_X belong to C? Infinite-dimensional geometry was overshadowed in the 1980s by the progress in local theory. The outstanding problems seemed intractable. Then in the 1990s most of the famous problems were quickly solved. The new development that made this possible was a deeper understanding of Tsirelson's construction (more about this later) as exemplified by the construction of the Banach space S [S1], [S2]. One might initially view these new Banach spaces as merely a class of pathological examples, but this is not the case. As we shall see, the geometry of S tells us much about the geometry of the Hilbert space; we obtain information about Hilbert space that indeed may well be otherwise truly intractable.

Dvoretzky's theorem says that if the dimension of X is sufficiently large then one can slice B_X with a k-dimensional subspace and obtain (up to ε) an ellipsoid. Put another way, one can find a Euclidean norm $|\cdot|$ on a given $X = (\mathbb{R}^n, \|\cdot\|)$ and a k-dimensional subspace E so that

$$\big| |x| - 1 \big| < \varepsilon \text{ if } x \in S_E = \{x \in E : \|x\| = 1\} \ .$$

In the late 1960s Milman extended Dvoretzky's theorem in connection with the study of uniformly continuous real valued functions f defined on S_X. Let X be infinite-dimensional. Given $\varepsilon > 0$ and any integer k, does there exist a k-dimensional subspace $E \subseteq X$ so that

$$\operatorname{osc}(f, S_E) \equiv \sup\{|f(x) - f(y)| : x, y \in S_E\} < \varepsilon \ ?$$

We say $a \in \gamma(f)$, the spectrum of f, if for any $\varepsilon > 0$ and k there exists $E \subseteq X$ of dimension k with

$$|f(x) - a| < \varepsilon \text{ for all } x \in S_E \ .$$

THEOREM 2 [M1], [M4]. *For any uniformly continuous f on S_X, the spectrum $\gamma(f)$ is nonempty.*

Milman defined the notion of an infinite-dimensional spectrum $\gamma_\infty(f)$ ($a \in \gamma_\infty(f)$ if for $\varepsilon > 0$ there exists an infinite-dimensional $Y \subseteq X$ so that $|f(y) - a| < \varepsilon$ for all $y \in S_Y$) and asked whether $\gamma_\infty(f) \neq \emptyset$. In particular, one can ask the question where f is an *equivalent norm* $|||\cdot|||$ on X (i.e., Id : $(X, \|\cdot\|) \to (X, |||\cdot|||)$ is an isomorphism) and this brings us to the topic of distortion. Henceforth, $X, Y, Z, \ldots$ will refer to separable infinite-dimensional real Banach spaces. $Y \subseteq X$ means that Y is a closed linear subspace of X.

DEFINITION 3. (a) *Let $\lambda > 1$. $(X, \|\cdot\|)$ is λ-distortable if there exists an equivalent norm $|||\cdot|||$ on X so that for all $Y \subseteq X$*

$$\sup\left\{\frac{|||x|||}{|||y|||} : x, y \in S_Y\right\} > \lambda \ .$$

(b) Let $f : S_X \to \mathbb{R}$ be uniformly continuous. f is *oscillation stable* on X if for all $\varepsilon > 0$ and all $Y \subseteq X$ there exists $Z \subseteq Y$ with $\mathrm{osc}(f, S_Z) < \varepsilon$.

Thus, X does not contain a distortable subspace Y iff every equivalent norm on X is oscillation stable. James [J] proved that ℓ_1 and c_0 are not distortable.

PROBLEMS. Which Banach spaces are distortable? For which Banach spaces X are all uniformly continuous functions $f : S_X \to \mathbb{R}$ oscillation stable?

Milman connected these questions with what was an outstanding open problem at that time: Does every X contain an isomorph of ℓ_p for some $1 \le p < \infty$ or c_0? A wonderful characterization of spaces containing ℓ_1 was given by Rosenthal [R1], who has recently also characterized spaces containing c_0 [R3], [R4]. Krivine and Maurey [KM], inspired by a theorem of Aldous [A], proved that the answer is yes for a large class of X's (the *stable* Banach spaces). Milman [M2], [M3] proved that if X is not distortable then X contains c_0 or ℓ_p for some $1 \le p < \infty$. Tsirelson's famous example ([T], see also [FJ]) of a space T not containing c_0 or ℓ_p $(1 \le p < \infty)$ showed that there are distortable spaces.

Tsirelson's space was the first truly nonclassical Banach space. Classical Banach spaces (ℓ_p, L_p, $C(K)$, H_p, ...) have their norms defined explicitly by a given formula. To describe T (we follow the description given in [FJ]) we need some terminology.

Let c_{00} be the linear space of finitely supported real valued functions on $\mathbb{N}$. If $x \in c_{00}$ and $E \subseteq \mathbb{N}$, $Ex \in c_{00}$ is defined by $Ex(i) = x(i)$ if $i \in E$ and 0 otherwise. For $E, F \subseteq \mathbb{N}$, $E < F$ means that $\max E < \min F$. A sequence of subsets of $\mathbb{N}$, $(E_i)_{i=1}^n$, is *admissible* if $\{n\} \le E_1 < \cdots < E_n$. T is the completion of c_{00} under the norm

$$\|x\| = \max \left(\|x\|_\infty, \sup \tfrac{1}{2} \sum_{i=1}^n \|E_i x\| \right)$$

where the "sup" is taken over all admissible collections.

Thus, the norm in T is not given explicitly but is rather the solution of an equation (of course it must be verified that such a norm exists). A detailed study of Tsirelson's space appears in [CS]. The wonderful things that could be done with such implicit norm descriptions were not fully realized until the last five years.

Consider the problem: Is Hilbert space, ℓ_2, distortable? This can be shown to be equivalent to finding sets $A, B \subseteq S_{\ell_2}$ so that both sets are *asymptotic* (A is asymptotic if $A \cap S_Y \ne \emptyset$ for all $Y \subseteq \ell_2$) and $D(A, B) \equiv \inf\{\|a - b\| : a \in A, b \in B\} > 0$. Indeed (roughly) one can use sets of the form $\overline{co}(A \cup -A)$ to construct the unit ball of a distorting norm. How does one find such sets? Every $x \in S_{\ell_2}$ looks like every other element; x could be the first element of an orthonormal basis for ℓ_2. This sort of problem was hampering the solution of many of the outstanding problems of the sort "does every X contain a nice Y?" If not, how could one distinguish the "bad vectors" in every subspace? In order to overcome similar problems in local theory, probability has been very useful. For example, in regard to Dvoretzky's theorem one can actually show that probabilistically most k-dimensional slices of X are (up to ε) ellipsoids. Infinite-dimensional geometry lacks this tool. Although the method of infinite combinatorics (e.g., Ramsey theory) has had some success it has been unable to penetrate many problems. It is worth noting here that Gowers [G4] has discovered a wonderful new combinatorial principle.

Tsirelson's space T gave a clue as to how to proceed. It could be seen how to explicitly distort T; one could identify two types of different vectors present in all infinite-dimensional subspaces of T and see that T was $2 - \varepsilon$ distortable for all $\varepsilon > 0$. However more was needed. Tomczak-Jaegermann [T-J2] proved, for example, that if T is *arbitrarily distortable* (λ-distortable for all $\lambda > 1$) then ℓ_2 is arbitrarily distortable. This gave the first possible connection between the infinite-dimensional geometry of a nonclassical space and that of ℓ_2. It is still open as to whether T is more than 2-distortable.

In [S1] the second named author gave the first example of an arbitrarily distortable space S. This was the first of a new generation of Tsirelson type spaces. Its construction led to the solution of the unconditional basic sequence problem by Gowers and Maurey [GM1]. This second generation example led to many others and many open problems were solved (see e.g., [G2], [G3], [G5], [OS3], [KT-J], [GM2], [AD]). The space S is defined by an implicit Tsirelson type norm. For $x \in c_{00}$ set

$$\|x\| = \max \left(\|x\|_\infty, \sup_{(E_i)_1^n \in \mathcal{F}_n} \frac{1}{\phi(n)} \sum_{i=1}^n \|E_i x\| \right)$$

where $\phi(n) = \log_2(n+1)$ and $\mathcal{F}_n = \{(E_i)_1^n : E_i \subseteq \mathbb{N} , E_1 < \cdots < E_n\}$. S is the completion of c_{00} under this norm. The norms that arbitrarily distort S are

$$\|x\|_n = \max \left(\|x\|_\infty , \sup_{(E_i)_1^n \in \mathcal{F}_n} \frac{1}{\phi(n)} \sum_{i=1}^n \|E_i x\| \right) .$$

In fact one has that S is more than arbitrarily distortable. The following definition is due to Gowers and Maurey [GM1].

DEFINITION 4. (A_n, A_n^*) is an A.B.S. (*asymptotic biorthogonal sequence*) for X if for all n, $A_n \subseteq S_X$, $A_n^* \subseteq B_{X^*}$, A_n is asymptotic in X and for some sequence $\varepsilon_n \downarrow 0$:

  i) for all n and $x \in A_n$ there exists $x^* \in A_n^*$ with $x^*(x) > 1 - \varepsilon_n$ and
  ii) if $n \neq m$, $x \in A_n$ and $x^* \in A_m^*$, then $|x^*(x)| < \varepsilon_{\min(n,m)}$.

We shall say that X is *biorthogonally distortable* if X admits an A.B.S.

It turns out that the norms $\| \cdot \|_n$ can be used to describe an explicit A.B.S. in S ([GM1], see also [S1], [S2]). Remarkably, one can transfer the A.B.S. of S to ℓ_2. This transference cannot be linear and involves knowing detailed information about the A.B.S. in S. We ultimately have the following theorems from [OS2] (see also [OS1]).

THEOREM 5. Let $1 < p < \infty$. Then ℓ_p is biorthogonally distortable and hence is arbitrarily distortable.

Combined with the work of Milman and James cited above this gives

THEOREM 6. *If X is not distortable then for all $Y \subseteq X$ either ℓ_1 or c_0 embeds isomorphically into Y.*

Gowers [G1] proved that every uniformly continuous $f : S_{c_0} \to \mathbb{R}$ is oscillation stable. The fact that ℓ_2 is distortable yields (via the Mazur map discussed below) that ℓ_1 fails this property.

THEOREM 7. *If every uniformly continuous $f : S_X \to \mathbb{R}$ is oscillation stable then c_0 embeds isomorphically into every $Y \subseteq X$.*

Before proceeding to describe the transfer mechanism alluded to above we need some terminology. Let $(x_i) \subseteq X \setminus \{0\}$ and $K \geq 1$. (x_i) is *K-basic* in X if for all $n < m$ and $(a_i)_1^m \subseteq \mathbb{R}$, $\|\sum_1^n a_i x_i\| \leq K\|\sum_1^m a_i x_i\|$. If in addition the closure of $\langle x_i \rangle$, the linear span of (x_i), equals X then (x_i) is called a *basis* for X. (x_i) is *monotone* basic if $K = 1$. (x_i) is *K-unconditional* if for all $(a_i)_1^m \subseteq \mathbb{R}$ and $(\varepsilon_i)_1^m \in \{-1, 1\}^m$, $\|\sum_1^m \varepsilon_i a_i x_i\| \leq K\|\sum_1^m a_i x_i\|$. If X has a 1-unconditional basis (e_i) then X may be viewed as the completion of c_{00} under some norm $\|\cdot\|$ which makes X into a lattice: $\| |x| \| = \|x\|$ for all $x \in X$ where $|\sum a_i e_i| \equiv \sum |a_i| x_i$. A *block basis* (y_n) of a basic sequence (x_n) is a nonzero sequence given by $y_n = \sum_{i \in F_n} a_i x_i$ for some sequence $(a_i) \subseteq \mathbb{R}$ and $F_1 < F_2 < \cdots$.

Extending an argument in [GM1], which is in turn an extension of an argument in [MR], one can prove

THEOREM 8 [OS2]. *Let X have a basis (e_i) and assume that X is biorthogonally distortable. Let $n \in \mathbb{N}$ and $\varepsilon > 0$. Then there exists an equivalent norm $||| \cdot |||$ on X with the property that if $(y_i)_1^n$ is any finite monotone basis and $(w_i)_1^\infty$ is any block basis of (e_i) then there exists a block basis $(z_i)_1^n$ of (w_i) that satisfies: for all $(a_i)_1^n \subseteq \mathbb{R}$,*

$$\left\| \sum_1^n a_i y_i \right\| \leq \left\| \left\| \sum_1^n a_i z_i \right\| \right\| \leq (1 + \varepsilon) \left\| \sum_1^n a_i y_i \right\|.$$

For example, given n, ℓ_2 can be renormed so as to contain in every $Y \subseteq \ell_2$, n-dimensional subspaces whose unit balls are (up to ε) n-cubes $(B_{\ell_\infty^n})$.

The Entropy Map and Uniform Homeomorphisms

Let $1 < p < \infty$. The Mazur map $M_p : S_{\ell_1} \to S_{\ell_p}$ given by $M_p(a_i)_1^\infty = (\text{sign } a_i |a_i|^{1/p})_1^\infty$ is a *uniform homeomorphism* (i.e., a uniformly continuous bijection with uniformly continuous inverse) [Ma], [Ri]. For $1 < p < \infty$, ℓ_p is distortable iff S_{ℓ_p} contains asymptotic sets A, B with $D(A, B) > 0$. Because M_p also preserves block bases it is easily seen that ℓ_p (for any $1 < p < \infty$) is distortable iff S_{ℓ_1} contains a pair of separated asymptotic sets. We shall use a generalized Mazur map to transfer the A.B.S. from S to ℓ_p going through S_{ℓ_1}. In order to define this map, which was considered earlier by Lozanovskii [Lo] and Gillespie [Gi], we need some further terminology.

Let (e_i) be a normalized 1-unconditional basis for X. Let $S_X^+ = \{x \in S_X : x = |x|\}$. If h is a finitely supported vector in $S_{\ell_1}^+$ and $y = \sum y_i e_i \in X^+ \equiv \{x \in X : x = |x|\}$ we define the *entropy* $E(h, y) = \sum h_i \log(y_i)$ (where $0 \log 0 \equiv 0$). It is easy to show that there exists a unique $x \in S_X^+$ having the same support as h

that maximizes $E(h, y)$ on S_X^+. Define $F_X(h)$ to be that x. If X is reasonably nice (e.g., uniformly convex and uniformly smooth) the *entropy map* F_X extends to a uniform homeomorphism between S_{ℓ_1} and S_X. In fact, $F_{\ell_p} = M_p$. More generally, one can prove the following. X is said to *contain* ℓ_∞^n's *uniformly* if there exists $C < \infty$ and $E_n \subseteq X$ with $d(E_n, \ell_\infty^n) \leq C$ for all n.

THEOREM 9 [OS2]. *Let X be a Banach space with an unconditional basis. Then S_X and S_{ℓ_1} are uniformly homeomorphic iff X does not contain ℓ_∞^n's uniformly.*

The "necessity" is due to Enflo [E]. This theorem has been extended to more general lattices independently by Chaatit [C] and Daher [Da]. For other results on uniform homeomorphisms between Banach spaces, see the fine survey paper by Benyamini [B].

The problem that presents itself in trying to use the map F_X to transfer a known pair of separated asymptotic sets to S_{ℓ_1} is that, unlike the Mazur map, F_X need not preserve block bases and thus one cannot conclude that $F_X^{-1}(A)$ is asymptotic in S_{ℓ_1} if A is asymptotic in S_X. We originally achieved the existence of a pair of separated asymptotic sets in S_{ℓ_1} by an indirect argument. Maurey [Mau1] then gave us an elegant argument showing us how our ideas could be used in conjunction with F_{S^*} to show that ℓ_2 admits an A.B.S. We define a sequence of sets $B_k \subseteq S_{\ell_1}$ as follows. Let (A_k, A_k^*) be the specific A.B.S. constructed in S using $\|\cdot\|_n$ (see [OS2]). Let

$$B_k = \left\{ \frac{x_k^* \circ x_k}{\|x_k^* \circ x_k\|_{\ell_1}} : x_k \in A_k, \ x_k^* \in A_k^*, \text{ and } |x_k^*|(|x_k|) \geq 1 - \varepsilon_k \right\}.$$

Here $x^* \circ x$ denotes the sequence obtained by pointwise multiplication. One obtains ultimately the following theorem. A set C of sequences is a *lattice* set if $x \in C$ iff $|x| \in C$. C is *spreading* if $x = (x_i) \in C$ iff $y = (0, 0, \ldots, 0, x_1, 0, 0, x_2, \cdots) \in C$ no matter how the 0's are placed.

THEOREM 10 [OS2]. *The sets $C_k \equiv M_2(B_k)$ are asymptotic, lattice, and spreading in ℓ_2. Moreover, for some sequence $\varepsilon_k \downarrow 0$, $\langle |x_k|, |x_\ell| \rangle < \varepsilon_{\min(k,\ell)}$ if $k \neq \ell$, $x_k \in C_k$, and $x_\ell \in C_\ell$.*

Here $\langle \cdot, \cdot \rangle$ denotes the inner product on ℓ_2. Thus, the sets C_k are nearly mutually orthogonal. Maurey [Mau3] has shown that one can also produce the sets C_k in such a way that each C_k is *symmetric* ($x = (x_i) \in C_k$ iff $x_\pi = (x_{\pi(i)}) \in C_k$ for all permutations π of $\mathbb{N}$). One can in certain circumstances show that the map F_X behaves much like the Mazur map. The following result applies for example to $X = S^{(2)}$, the convexification of S.

THEOREM 11. *Let X be a space with a normalized 1-unconditional basis (e_i). Assume that there exists $C > 0$ so that for all n and all block bases $(x_i)_1^n$ of $(e_i)_n^\infty$,*

$$(C \log n)^{-1} \left(\sum_1^n \|x_i\|^2 \right)^{1/2} \leq \left\| \sum_1^n x_i \right\| \leq \left(\sum_1^n \|x_i\|^2 \right)^{1/2}.$$

Let (y_i) be any block basis of the unit vector basis for ℓ_1, $\varepsilon > 0$, $m \in \mathbb{N}$. There exists a normalized block basis $(z_i)_1^m$ of (y_i) with $(F_{X^*} z_i)_1^m$ $(1 + \varepsilon)$-equivalent to the unit vector basis of ℓ_2^m and having the property that if $(\alpha_i)_1^m \in S_{\ell_1}^m$ then

$$\left\| F_{X^*} \left(\sum_1^n \alpha_i z_i \right) - \sum_1^m \operatorname{sign} \alpha_i |\alpha_i|^{1/2} F_{X^*} z_i \right\| < \varepsilon .$$

Questions concerning the relationships between the notions of being distortable, arbitrarily distortable, and biorthogonally distortable remain open. No example of a distortable space of bounded distortion is known, although T is a prime candidate.

Further Developments

A number of further developments in distortion theory have been obtained. Maurey [Mau2] has proved that every super-reflexive space with an unconditional basis contains an arbitrarily distortable subspace. Tomczak-Jaegermann has proved [T-J3] that the Schatten classes C_p of operators on a Hilbert space are arbitrarily distortable for $1 < p < \infty$. Milman and Tomczak-Jaegermann [MiT-J] have shown that if X is of bounded distortion then X contains an *asymptotic ℓ_p* space; i.e., there exist C, $1 \leq p \leq \infty$ and a basic sequence (x_i) in X so that if $n \in \mathbb{N}$ and $(y_i)_1^n$ is a normalized block basis of $(x_i)_n^\infty$, then $(y_i)_1^n$ is C-equivalent to the unit vector basis of ℓ_p^n. Argyros and Deliyanni [AD] have produced an asymptotic ℓ_1 space that is arbitrarily distortable. They have also constructed such a space that does not contain an unconditional basic sequence. Casazza, Kalton, Kutzarova, and Mastylo have proven [CKKM] that for any C, ℓ_2 can be renormed so as to not contain any C-unconditional basic sequence and yet still satisfy certain modulus of convexity and smoothness conditions.

Maurey and Tomczak-Jaegermann [MT-J] have proved that if $X_\theta = (X_0, \ell_2)_\theta$ is a complex interpolation space with $0 < \theta < 1$ and if X_0 is an asymptotic ℓ_p space $(1 \leq p \leq \infty)$, then X_θ is biorthogonally distortable. They have also shown the following. For every $D > 1$ there exists an equivalent symmetric norm $|\cdot|$ on ℓ_2 such that for any finite number of unitary operators (or even into isomorphisms) $T_1, \ldots, T_N$, the norm $|||x||| \equiv \sum_{i=1}^N |T_i x|$ is a D-distortion of ℓ_2. By definition, distorted norms cannot be "corrected" by passing to a subspace. The above shows that one cannot always correct a norm on ℓ_2 by applying a finite number of unitary operators. Both types of "corrections" are possible in the finite-dimensional setting. Dvoretzsky's theorem gives the subspace correction, and the unitary operator correction is due to Bourgain, Lindenstrauss, and Milman [BLM].

Krivine's Theorem and Spreading Models

The successes of local theory as obtained in Theorems 1 and 2 cannot be passed on to infinite-dimensional geometry. However, there are structural notions that combine elements of both the finite- and infinite-dimensional theories. The notion of a spreading model is due to Brunel and Sucheston ([BS]; see also [BL], [O]). A normalized basic sequence (e_i) is a *spreading model* of a normalized basic sequence

(x_i) in some Banach space X if for all $n \in \mathbb{N}$, $n < i_1 < \cdots < i_n$, and $(a_i)_1^n \subseteq [-1,1]^n$,

$$(1) \qquad \left| \left\| \sum_{j=1}^n a_j x_{i_j} \right\| - \left\| \sum_{j=1}^n a_j e_j \right\| \right| < \frac{1}{n} .$$

If (x_i) is weakly null then (e_i) is unconditional. From Ramsey theory and Rosenthal's theorem [R1] one obtains that every X admits an unconditional spreading model. It was recently shown [OS3] that not every X has a spreading model containing c_0 or ℓ_p for some $1 \le p < \infty$.

Krivine [K] proved the following beautiful local theorem (see also [Le], [R2], [MS]).

THEOREM 12. *Let $C \ge 1$, $n \in \mathbb{N}$ and $\varepsilon > 0$. There exists $m = m(C, n, \varepsilon) \in \mathbb{N}$ so that if $(x_i)_1^m$ is a C-basic sequence then there exist $1 \le p \le \infty$ and a block basis $(y_i)_1^n$ of $(x_i)_1^m$ so that $(y_i)_1^n$ is $1 + \varepsilon$-equivalent to the unit vector basis of ℓ_p^n.*

In [OS3] an example is constructed of a basic sequence (x_i) with the property that for all $1 \le p \le \infty$, ℓ_p is block finitely represented in all block bases of (x_i). Krivine's theorem gives rise to the following stabilization principle (see [ORS]).

THEOREM 13. *For all $C > 0$, $\varepsilon > 0$, and $n \in \mathbb{N}$ there exists $m = m(C, n, \varepsilon)$ so that if $(x_i)_1^m$ is C-basic and if $f : S_{\langle x_i \rangle_1^n} \to \mathbb{R}$ is C-Lipschitz then there exists a block basis $(y_i)_1^m$ of $(x_i)_1^n$ so that $\mathrm{osc}(f \mid S_{\langle y_i \rangle_1^m}) < \varepsilon$.*

This result, in turn, gives rise to an extended notion of spreading model [ORS]. Given (E_n), a sequence of finite-dimensional subspaces of X with $\dim E_n \to \infty$, there exist integers $k_n \uparrow \infty$, subspaces G_n of E_{k_n} with $\dim G_n \to \infty$ and a spreading model (e_i) so that (1) is valid whenever $x_{i_j} \in S_{G_{i_j}}$.

More recently Maurey, Milman and Tomczak-Jaegermann [MMT-J] have defined another asymptotic infinite-dimensional concept that evidently yields a richer theory than that of spreading models.

References

[A]   D. J. Aldous, *Subspaces of L^1 via random measures*, Trans. Amer. Math. Soc. **267** (1981), 445–453.

[AD]   S. Argyros and I. Deliyanni, *Examples of asymptotically ℓ^1 Banach spaces*, preprint.

[BL]   B. Beauzamy and J.-T. Lapresté, Modèles étalés des espaces de Banach, Travaux en Cours, Hermann, Paris, 1984.

[B]   Y. Benyamini, *The uniform classification of Banach spaces*, Longhorn Notes 1984–85, The University of Texas at Austin, 1985.

[BLM]   J. Bourgain, J. Lindenstrauss, and V. D. Milman, *Minkowski sums and symmetrizations*, GAFA-Seminar 86–87, Lecture Notes in Math. **1376**, Springer-Verlag, Berlin and New York, 1989, pp. 278–287.

[BS]   A. Brunel and L. Sucheston, *B-convex Banach spaces*, Math. Systems Theory **7** (1974), 294–299.

[CKKM]   P. G. Casazza, N. J. Kalton, Denka Kutzarova, and M. Mastylo, *Complex interpolation and complementably minimal spaces*, preprint.

[CS]   P. G. Casazza and T. J. Shura, Tsirelson's space, Lecture Notes in Math., vol. 1363, Springer-Verlag, Berlin and New York, 1989.

[C]   F. Chaatit, *Uniform homeomorphisms between unit spheres of Banach lattices*, Pacific J. Math. to appear.

[Da]   M. Daher, Homéomorphismes uniformes entre les sphères unites des éspaces d'interpolation, thesis, Université Paris 7.

[D]   A. Dvoretzky, *Some results on convex bodies and Banach spaces*, Proc. Sympos. Linear Spaces, Jerusalem, 1961, pp. 123–160.

[E]   P. Enflo, *On a problem of Smirnov*, Ark. Mat. **8** (1969), 107–109.

[FJ]   T. Figiel and W. B. Johnson, *A uniformly convex Banach space which contains no ℓ_p*, Compositio Math. **29** (1974), 179–190.

[FLM]   T. Figiel, J. Lindenstrauss, and V. D. Milman, *The dimension of almost spherical sections of convex bodies*, Acta. Math. **139** (1977), 53–94.

[Gi]   T. A. Gillespie, *Factorization in Banach function spaces*, Indag. Math. **43** (1981), 287–300.

[G1]   W. T. Gowers, *Lipschitz functions on classical spaces*, European J. Combin. **13** (1992), 141–151.

[G2]   _____, *A solution to the Banach hyperplane problem*, Bull. London Math. Soc., to appear.

[G3]   _____, *A space not containing c_0, ℓ_1 or a reflexive subspace*, Trans. Amer. Math. Soc., to appear.

[G4]   _____, *A new dichotomy for Banach spaces*, preprint.

[G5]   _____, *Recent results in the theory of infinite-dimensional Banach spaces*, these proceedings.

[GM1]   W. T. Gowers and B. Maurey, *The unconditional basic sequence problem*, J. Amer. Math. Soc. **6** (1993), 851–874.

[GM2]   _____, *Banach spaces with small spaces of operators*, preprint.

[J]   R. C. James, *Uniformly nonsquare Banach spaces*, Ann. of Math. (2) **80** (1964), 542–550.

[KT-J]   R. Komorowski and N. Tomczak-Jaegermann, *Banach spaces without local unconditional structure*, Israel J. Math., to appear.

[K]   J. L. Krivine, *Sous espaces de dimension finie des espaces de Banach réticulés*, Ann. of Math. (2) **104** (1976), 1–29.

[KM]   J. L. Krivine and B. Maurey, *Espaces de Banach stables*, Israel J. Math. **39** (1981), 273–295.

[Le]   H. Lemberg, *Nouvelle démonstration d'un théorème de J.L. Krivine sur la finie représentation de ℓ_p dans un espaces de Banach*, Israel J. Math. **39** (1981), 341–348.

[L]   J. Lindenstrauss, *Almost spherical sections; their existence and their applications*, Jahresber. Deutsch. Math.-Verein (1992), 39–61.

[Lo]   G. Ya. Lozanovskii, *On some Banach lattices*, Siberian Math. J. **10** (1969), 584–599.

[Mau1]   B. Maurey, private communication (1992).

[Mau2]   _____ , *A remark about distortion*, Oper. Theory: Adv. Appl. **77** (1995), 131–142.

[Mau3]   _____ , *Symmetric distortion in ℓ_2*, Oper. Theory: Adv. Appl. **77** (1995), 143–147 to appear.

[MMT-J]   B. Maurey, V. D. Milman, and N. Tomczak-Jaegermann, *Asymptotic infinite-dimensional theory of Banach spaces*, Oper. Theory: Adv. Appl. **77** (1995), 149–175.

[MR]   B. Maurey and H. Rosenthal, *Normalized weakly null sequences with no unconditional subsequences*, Studia Math. **61** (1971), 77–98.

[MT-J]   B. Maurey and N. Tomczak-Jaegermann, private communication (1994).

[Ma]   S. Mazur, *Une remarque sur l'homéomorphisme des champs fonctionnels*, Studia Math. **1** (1930), 83–85.

[M1]   V. D. Milman, *The infinite dimensional geometry of the unit sphere of a Banach space*, Soviet Math. Dokl. **8** (1967), 1440–1444, (trans. from Russian).

[M2]   _____ , *The spectrum of bounded continuous functions which are given on the unit sphere of a B-space*, Funktsional Anal. i Prilozhen **3** (1969), 67–79, M.R. 40 4740.

[M3]   _____ , *Geometric theory of Banach spaces II, geometry of the unit sphere*, Russian Math. Surveys **26** (1971), 79–163, (trans. from Russian).

[M4]   _____ , *A new proof of the theorem of A. Dvoretsky on sections of convex bodies*, Functional Anal. Appl. **5** (1971), 28–37.

[M5]   _____ , *The concentration phenomenon and linear structure of finite-dimensional normed spaces*, Proc. Internat. Congress Math., Berkeley, CA (1986), 961–975.

[M6]   _____ , *Dvoretsky's theorem — thirty years later*, Geom. Functional Anal. **4** (1992), 455–479.

[MS]   V. D. Milman and G. Schechtman, *Asymptotic theory of finite dimensional normed spaces*, Lecture Notes in Math., vol. 1200, Springer-Verlag, Berlin and New York, 1986, p. 156.

[MiT-J]   V. Milman and N. Tomczak-Jaegermann, *Asymptotic ℓ_p spaces and bounded distortion* (Bor-Luh Lin and W. B. Johnson, eds.),, Contemp. Math. **144** (1993), 173–195.

[O]   E. Odell, *Applications of Ramsey theorems to Banach space theory*, Notes in Banach spaces (H. E. Lacey, ed.), Univ. Texas Press, Austin, TX, pp. 379–404.

[ORS]   E. Odell, H. Rosenthal, and Th. Schlumprecht, *On weakly null FDD's in Banach spaces*, Israel J. Math. **84** (1993), 333–351.

[OS1]   E. Odell and Th. Schlumprecht, *The distortion of Hilbert space*, Geom. Functional Anal. **3** (1993), 201–207.

[OS2]   _____ , *The distortion problem*, Acta Math. **173** (1994), 259–281.

[OS3]   ———, *On the richness of the set of p's in Krivine's theorem*, Oper. Theory: Adv. Appl. **77** (1995), 177–198.

[Pe]   A. Pełczyński, *Structural theory of Banach spaces and its interplay with analysis and probability*, Proc. Internat. Congress Math. (1983), 237–269.

[P1]   G. Pisier, *Finite rank projections on Banach spaces and a conjecture of Grothendieck*, Proc. Internat. Congress Math. (1983), 1027–1039.

[P2]   ———, *The volume of convex bodies and Banach space geometry*, Cambridge Tracts in Math., vol. 94, Cambridge Univ. Press, Cambridge and New York, 1989.

[Ri]   M. Ribe, *Existence of separable uniformly homeomorphic non isomorphic Banach spaces*, Israel J. Math. **48** (1984), 139–147.

[R1]   H. Rosenthal, *A characterization of Banach spaces containing ℓ_1*, Proc. Nat. Acad. Sci. U.S.A. **71** (1974), 2411–2413.

[R2]   ———, *On a theorem of Krivine concerning block finite representability of ℓ_p in general Banach spaces*, J. Funct. Anal. **28** (1978), 197–225.

[R3]   ———, *A subsequence principle characterizing Banach spaces containing c_0*, Bull. Amer. Math. Soc. **30** (1994), 227–233.

[R4]   ———, *A characterization of Banach spaces containing c_0*, J. Amer. Math. Soc., to appear.

[S1]   Th. Schlumprecht, *An arbitrarily distortable Banach space*, Israel J. Math. **76** (1991), 81–95.

[S2]   ———, *A complementably minimal Banach space not containing c_0 or ℓ_p*, Seminar Notes in Functional Analysis and PDE's, LSU, 1991–92, pp. 169–181.

[T-J1]   N. Tomczak-Jaegermann, *Banach-Mazur distances and finite dimensional operator ideals*, Pitman Monographs **38** (1989).

[T-J2]   ———, private communication (1991).

[T-J3]   ———, *Distortions on Schatten classes C_p*, preprint.

[T]   B. S. Tsirelson, *Not every Banach space contains ℓ_p or c_0*, Functional Anal. Appl. **8** (1974), 138–141.

Operator Algebras and Conformal Field Theory

ANTONY J. WASSERMANN

Department of Pure Mathematics and Mathematical Statistics
University of Cambridge
16 Mill Lane, Cambridge CB2 1SB, United Kingdom

1 Introduction

We report on a programme to understand unitary conformal field theory (CFT) from the point of view of operator algebras. The earlier stages of this research were carried out with Jones, following his suggestion that there might be a deeper "subfactor" explanation of the coincidence between certain braid group representations that had turned up in subfactors, statistical mechanics, and conformal field theory. (Most of our joint work appears in Section 10.) The classical *additive* theory of operator algebras, due to Murray and von Neumann, provides a framework for studying unitary Lie group representations, although in specific examples almost all the hard work involves a quite separate analysis of intertwining operators and differential equations. Analogously, the more recent *multiplicative* theory provides a powerful tool for studying the unitary representations of certain infinite-dimensional groups, such as loop groups or Diff S^1. It must again be complemented by a detailed analysis of certain intertwining operators, the primary fields, and their associated differential equations. The multiplicative theory of von Neumann algebras has appeared in three separate but related guises: first in the algebraic approach to quantum field theory (QFT) of Doplicher, Haag and Roberts; then in Connes' theory of bimodules or correspondences and their tensor products; and last (but not least) in Jones' theory of subfactors. Our results so far include:

(1) Several new constructions of subfactors.
(2) Nontrivial algebraic QFT's in $1 + 1$ dimensions with finitely many sectors and noninteger statistical (or quantum) dimension ("algebraic CFT").
(3) A definition of quantum invariant theory without using quantum groups at roots of unity.
(4) A computable and manifestly unitary definition of fusion for positive energy representations ("Connes fusion") making them into a tensor category.
(5) Analytic properties of primary fields ("constructive CFT").

Proceedings of the International Congress
of Mathematicians, Zürich, Switzerland 1994
© Birkhäuser Verlag, Basel, Switzerland 1995

2 Classical Invariant Theory

The irreducible unitary (finite-dimensional) representations of $G = SU(N)$ can be studied in two distinct approaches. These provide a simple but important prototype for developing the theory of positive energy loop group representations and primary fields.

Borel-Weil Approach. This constructs all irreducible representations uniformly in a Lie algebraic way via highest weight theory. The representations are described as quotients of Verma modules, that is in terms of lowering and raising operators. This approach gives an important *uniqueness* result — such a representation is uniquely determined by its highest weight — but has the disadvantage that it is not manifestly unitary.

Hermann Weyl Approach. This starts from a special representation, $V = \mathbb{C}^N$ or ΛV, and realizes all others in the tensor powers $V^{\otimes \ell}$ or $(\Lambda V)^{\otimes \ell}$. The key to understanding the decomposition of $V^{\otimes \ell}$ is *Schur-Weyl duality*: $\mathrm{End}_G V^{\otimes \ell}$ is the image of $\mathbb{C}S_\ell$, where the symmetric group S_ℓ acts by permuting the tensor factors in $V^{\otimes \ell}$. This sets up a one-one correspondence between the irreducible representations of G and the symmetric groups and gives a manifestly unitary construction of the irreducible representations of G (on multiplicity spaces of S_ℓ). The irreducible unitary representation V_f with character χ_f and signature $f : f_1 \geq \cdots \geq f_N (= 0)$ is generated by the vector $e_f = e_1^{\otimes (f_1 - f_2)} \otimes (e_1 \wedge e_2)^{\otimes (f_2 - f_3)} \otimes \cdots$ in $(\Lambda V)^{\otimes \ell}$. The signature can be written in the usual way as a Young diagram and we then have the tensor product rule $V_f \otimes V_\square = \oplus V_g$, where g runs over all diagrams that can be obtained by adding one box to f.

Thus, the Borel-Weil Lie algebraic approach leads to *uniqueness* results, whereas the Hermann Weyl approach leads to *existence* results and an explicit *construction*, giving analytic unitary properties.

3 Fermions and Quantization

Let H be a complex Hilbert space. Bounded operators $a(f)$ for $f \in H$ are said to satisfy the *canonical anticommutation relations* (CAR) if $[a(f), a(g)]_+ = 0$, $[a(f), a(g)^*]_+ = (f, g) \cdot I$, where $f \mapsto a(f)$ is $\mathbb{C}$-linear and $[x, y]_+ = xy + yx$. The *complex wave representation* π of the CAR on fermionic Fock space $\mathcal{F} = \Lambda H$ is given by $a(f)\omega = f \wedge \omega$. It is irreducible. Now the equations $c(f) = a(f) + a(f)^*$, $a(f) = \frac{1}{2}(c(f) - ic(if))$ give a correspondence with *real* linear maps $f \mapsto c(f)$ such that $c(f) = c(f)^*$ and $[c(f), c(g)]_+ = 2\mathrm{Re}\,(f, g) \cdot I$. Any projection P in H defines a new complex structure on H, by taking multiplication by i as i on PH and $-i$ on $P^\perp H$. So through c, this gives a new irreducible representation π_P of the CAR on Fock space $\mathcal{F}_P$. By considering approximations by finite-dimensional systems, Segal showed that $\pi_P \cong \pi_Q$ iff $P - Q$ is Hilbert-Schmidt. This leads to the following *quantization criterion*. Any $u \in U(H)$ gives a Bogoliubov automorphism of the CAR, $\alpha_u : a(f) \mapsto a(uf)$. The automorphism α_u is said to be implemented in $\mathcal{F}_P$ if $a(uf) = Ua(f)U^*$ for some unitary $U \in U(\mathcal{F}_P)$, unique up to a phase. The quantization criterion states that α_u is implemented in $\mathcal{F}_P$ iff $[u, P]$ is Hilbert-Schmidt. Thus, we get a homomorphism from the subgroup of implementable unitaries into $\mathcal{PU}(\mathcal{F}_P)$, the *basic* projective representation. As a special case, there are *canonical quantizations*: any unitary u with $uPu^* = P$ is canonically implemented in Fock space; and if $uPu^* = I - P$, then u is canonically implemented by a conjugate-linear isometry in Fock space.

4 Positive Energy Representations

Let $G = SU(N)$ and define the loop group $LG = C^\infty(S^1, G)$, the smooth maps of the circle into G. The diffeomorphism group of the circle Diff S^1 is naturally a subgroup of Aut LG with the action given by reparametrization. In particular the group of rotations Rot $S^1 \cong U(1)$ acts on LG. We look for projective representations $\pi : LG \to PU(H)$ that are both *irreducible* and have *positive energy*. This means that π should extend to $LG \rtimes$ Rot S^1 so that $H = \oplus_{n \geq 0} H(n)$, where the $H(n)$'s are eigenspaces for the action of Rot S^1, i.e. $r_\theta \xi = e^{in\theta}\xi$ for $\xi \in H(n)$, and dim $H(n) < \infty$ with $H(0) \neq 0$. Because the constant loops G commute with Rot S^1, the $H(n)$'s are automatically G-modules.

Uniqueness. An irreducible positive energy representation π on H is uniquely determined by its *level* $\ell \geq 1$, a positive integer specifying the central extension or 2-cocycle of LG, and its *lowest energy space* $H(0)$, an irreducible representation of G. Only finitely many irreducible representations of G occur at level ℓ: their signatures must satisfy the quantization condition $f_1 - f_N \leq \ell$ and form a set $\mathbf{Y}_\ell$.

Existence/Analytic Properties. Let $H = L^2(S^1) \otimes V$ and let P be the projection onto the Hardy space $H^2(S^1) \otimes V$ of functions with vanishing negative Fourier coefficients (or equivalently boundary values of functions holomorphic in the unit disc). The semidirect product $LG \rtimes \text{Diff}^+ S^1$ acts unitarily on H and satisfies the quantization criterion for P, so gives a projective representation of $LG \rtimes \text{Diff}^+ S^1$ in $\mathcal{F}_P$. The irreducible summands of $\mathcal{F}_P^{\otimes \ell}$ give all the level ℓ representations of LG and this construction shows that any positive energy representation extends to $LG \rtimes \text{Diff}^+ S^1$ ("invariance under reparametrization").

If H is a positive energy representation of level ℓ, the C^∞ vectors H^∞ for Rot S^1 are acted on continuously by $LG \rtimes$ Rot S^1 (or more generally $LG \rtimes \text{Diff } S^1$) and its Lie algebra. This can be seen in a variety of ways, using representations of the Heisenberg group or the infinitesimal version of the fermionic construction. If $\mathfrak{g} = \text{Lie}(G)$, then $\text{Lie}(LG) = L\mathfrak{g} = C^\infty(S^1, \mathfrak{g})$. Its complexification is spanned by the functions $e^{in\theta}x$ with $x \in \mathfrak{g}$. Let $x(n)$ be the corresponding unbounded operators on H^∞ (or H^0, the subspace of finite energy vectors) and let d be the self-adjoint generator for Rot S^1 (so that $r_\theta = e^{i\theta d}$). Then $[x(n), y(m)] = [x, y](n + m) + \ell n \, \delta_{n+m,0} \, \text{tr}(xy) \cdot I$ and $[d, x(n)] = -nx(n)$.

5 Formal Conformal Field Theory

Purely as motivation, we sketch the standard approach to CFT based on formal quantum fields.

State-Field Correspondence. For fixed level ℓ, there should be a one-one correspondence between states $v \in H = \oplus_{f \in \mathbf{Y}_\ell} H_f^0$ and field operators $\phi(v, z) = \sum_n \phi(v, n)z^{-n-h_{ij}}$, where $\phi_{ij}(v, n) : H_j^0 \to H_i^0$ and z is a formal parameter. The field "creates the state from the vacuum", i.e. $\phi(v, 0)\Omega = v$. Fields are first defined for vectors $x \in H_0(1) \cong \mathfrak{g}$ by $x(z) = \sum \pi(x(n))z^{-n-1}$. For $a \in H_0$, fields $\phi(a, z)$ are uniquely determined by $\phi(a, 0)\Omega = a$, rotation invariance $[d, \phi(v, n)] = -n\phi(v, n)$ and the gauge condition $x(z)\phi(a, w) \sim \phi(a, w)x(z)$. (This notation means that, on taking matrix coefficients, one side is the analytic continuation of the other, with the domains of definition given by decreasing moduli of arguments.) The fields $\phi(a, z)$ for $a \in H_0^0$ form a *vertex algebra*, Borcherds' analogue of a commutative ring. Commutativity and associativity are replaced by $\phi(a, z)\phi(b, w) \sim \phi(b, w)\phi(a, z)$ and $\phi(a, z)\phi(b, w) \sim \phi(\phi(a, z - w)b, w)$. The operator product expansion (OPE) is obtained by expanding the right-hand side of this last equation as a power series in $(z - w)$: the resulting coefficients are the fields arising from the fusion of $\phi(a, z)$ and $\phi(b, w)$. In this sense the $x(z)$'s generate the vertex algebra. The H_i's become modules over the vertex algebra and the gauge condition defines general fields as intertwiners. The fields corresponding to vectors in $H_i(0)$ are called *primary fields*. Other secondary fields are obtained by successive fusion with $x(z)$'s.

Braiding-Fusion Duality. If neither a nor b lies in H_0, the commutativity and associativity relations must be replaced by braiding and fusion relations:

$$\phi_{ij}^p(a, z)\phi_{jk}^q(b, w) \sim \sum_h \alpha_h \phi_{ih}^q(b, w)\phi_{hk}^p(a, z) \text{ (where } a \in H_p, b \in H_q).$$
$$\phi_{ij}^p(a, z)\phi_{jk}^q(b, w) \sim \sum_h \beta_h \phi_{ik}^h(\phi_{hq}^p(a, z - w)b, w).$$

These are first proved as identities between lowest energy matrix coefficients of primary fields and follow in general by fusion. The matrix coefficients give a vector-valued function $f(\zeta)$ of one variable $\zeta = z/w$. It satisfies the Knizhnik-Zamolodchikov ODE $f'(z) = z^{-1}Pf(z) + (1 - z)^{-1}Qf(z)$, with P, Q constant matrices. α_h and β_h are entries in the matrices connecting the solutions at 0 with the solutions at ∞ and 1 respectively. The evident algebraic relations between these two matrices constitute "braiding-fusion duality".

6 Construction of Primary Fields

Let H_i, H_j be positive energy representations of level ℓ and let W be an irreducible representation of G. A *primary field* of charge W is a continuous linear map $\phi : H_i^\infty \otimes C^\infty(S^1, W) \to H_j^\infty$ that commutes with the action of $LG \rtimes \mathrm{Rot}\, S^1$. This makes sense because H_i and H_j are projective representations with the same cocycle, whereas $C^\infty(S^1, W)$ is an ordinary representation, with LG acting by pointwise multiplication and $\mathrm{Rot}\, S^1$ by rotation. Any $f \in C^\infty(S^1, W)$ determines a "smeared field" $\phi(f) : H_i^\infty \to H_j^\infty$, which must satisfy the covariance relation $\phi(g \cdot f) = \pi_j(g)\phi(f)\pi_i(g)^*$ for $g \in LG \rtimes \mathrm{Rot}\, S^1$.

Uniqueness. A primary field ϕ is uniquely determined by its initial term $H_i(0) \otimes W \to H_j(0)$, which commutes with G. The charge W must have signature f satisfying $f_1 - f_N \leq \ell$. Moreover the initial term must satisfy an algebraic quantization condition with respect to $SU(2) \subset SU(N)$: $(*)$ when cut down to irreducible summands of $SU(2)$, the resulting intertwiners $V_p \otimes V_q \to V_r$ can only be non-zero if $p + q + r \leq \ell$ where the spins p, q, r are half integers $\leq \ell/2$.

Existence/Construction. Primary fields for the vector representation of G come from compressing fermions $P_j(a(f) \otimes I \otimes \cdots \otimes I)P_i$, where P_i, P_j are projections onto H_i, H_j summands of $\mathcal{F}^{\otimes \ell}$. More generally, primary fields arise from (antisymmetric) external tensor products of fermions, parallelling the explicit construction of highest weight vectors in $(\Lambda V)^{\otimes \ell}$. For $v \in V$, define $v_m(\theta) = e^{im\theta} v$ in $C^\infty(S^1, V)$ and $a(v, m) = a(v_m)$. Introduce the formal Laurent series $a(v, z) = \sum_m a(v, m) z^{-m}$. At level one, the primary field for $\Lambda^k V$ corresponds to compressions of the formal Laurent series $\phi(e_1 \wedge \cdots \wedge e_k, z) = a(e_1, z) a(e_2, z) \cdots a(e_k, z)$ (essentially an external tensor product as the e_i's are orthogonal). At level ℓ, the primary fields of signature f arise as formal Laurent series $\phi(w, z)$, uniquely specified by $\phi(e_f, z) = P_j(\phi(e_1, z)^{\otimes(f_1 - f_2)} \otimes \phi(e_1 \wedge e_2, z)^{\otimes(f_2 - f_3)} \otimes \cdots) P_i$ and G-covariance. All possible primary fields arise in this way because an intertwiner satisfies $(*)$ iff it appears as a component of the map $\Lambda \otimes \Lambda \to \Lambda$, $\alpha \otimes \beta \mapsto \alpha \wedge \beta$, where Λ is the exterior algebra $(\Lambda V)^{\otimes \ell}$.

This fermionic construction of the primary fields makes manifest their continuity properties on H_i^∞. In particular the primary fields for the vector representation or its dual must satisfy the same kind of L^2 bounds as fermions, $\|\phi(f)\| \leq A\|f\|_2$, underlining Haag's philosophy that QFT can and should be understood in terms of (algebras of) *bounded* operators. Here there is no choice.

7 The K-Z ODE and Braiding of Primary Fields

When $f, g \in C^\infty(S^1) \otimes V$ have disjoint support, the corresponding smeared fermi fields satisfy the anticommutative exchange rule $a(f) a(g)^* = -a(g)^* a(f)$. Similarly, if a and b are test functions supported in the upper and lower semicircle of S^1, there are *braiding relations*

$$\phi_{f_0}^f(a) \phi_{0\square}^{\overline{\square}}(b) = \sum \lambda_g \, \phi_{fg}^{\overline{\square}}(e^{\mu_g} \cdot b) \phi_{g\square}^f(e^{\nu_g} \cdot a), \tag{1}$$

where the constants λ_g, μ_g, ν_g are to be determined and $e^\tau(\theta) = e^{i\tau\theta}$. The λ_g's arise as the entries of the matrix connecting the solutions at 0 and ∞ of a matrix-valued ODE as follows. Fusion of the $x(z)$'s shows that, up to an additive constant, d is given by $L_0 \equiv (N + \ell)^{-1}[\sum_i \frac{1}{2} x_i(0)^* x_i(0) + \sum_{n>0,i} x_i(n)^* x_i(n)]$ (the *Segal-Sugawara* formula), where (x_i) is an orthonormal basis of $\mathfrak{g}$. Let $f(z) = \sum \langle \phi(v_2, n) \phi(v_3, -n) v_4, v_1^* \rangle z^n$, the reduced 4-point function with values in $(V_f \otimes V_f^* \otimes V_\square \otimes V_\square^*)^G$. The two expressions, when d and L_0 are inserted, between the two field operators can be simplified using the commutation relations with primary fields. After the change of variable $z \mapsto (1 - z)^{-1}$, they lead to the Knizhnik-Zamolodchikov ODE $f'(z) = z^{-1} P f(z) + (1 - z)^{-1}(P - Q) f(z)$, where P and Q are self-adjoint $(N \times N)$-matrices with P having distinct eigenvalues, Q proportional to a rank one projection, and P, Q in general position. There is then an essentially unique choice of (non-orthogonal) basis so that

$$P = \begin{pmatrix} 0 & 1 & & 0 \\ 0 & 0 & 1 & \\ & & & 0 \\ & & & 1 \\ a_1 & & & a_N \end{pmatrix}, \qquad Q = \begin{pmatrix} 0 & 0 & & 0 \\ & & & \\ & & & \\ b_1 & & & b_N \end{pmatrix}.$$

This is the matrix-valued ODE for the generalized hypergeometric equation. Entries of the transport matrices relating solutions at 0 and 1 are calculated by an extension of the classical method of Gauss and two tricks: the unitary of the transport matrix when the constant $N + \ell$ is made imaginary; and Karamata's Tauberian theorem.

The inner product of both sides of (1) with lowest energy vectors can be expressed through integrals involving a, b, and the branches of $f(z)$ at 0 and ∞ (viewed as vector-valued distributions on S^1). The transport matrix between the branches therefore gives the braiding coefficients (and phase corrections) for the inner products with lowest energy vectors. Using lowering and raising operators, they also work for inner products with arbitrary finite energy vectors and hence, by continuity, with all smooth vectors.

8 Von Neumann Algebras

It is perhaps most natural to define von Neumann algebras as "the symmetry algebras of unitary groups". Thus, if H is a complex Hilbert space, von Neumann algebras $M \subseteq B(H)$ are of the form $M = G'$, where G is a subgroup of the unitary group $U(H)$ and the *commutant* or symmetry algebra of $\mathcal{S} \subseteq B(H)$ is $\mathcal{S}' = \{T : Tx = xT$ for all $x \in \mathcal{S}\}$. If $\mathcal{S}^* = \mathcal{S}$, then $\mathcal{S}''$ coincides with the von Neumann algebra generated by $\mathcal{S}$ (i.e. the smallest von Neumann algebra containing $\mathcal{S}$). It is also the strong or weak operator closure of the unital *-algebra generated by $\mathcal{S}$.

If M is a von Neumann algebra, its *center* $Z(M) = M \cap M'$ is an Abelian von Neumann algebra, so of the form $L^\infty(X, \mu)$ for some measure space (X, μ). If X is atomic, then M is canonically a direct sum of *factors*, von Neumann algebras with trivial center, with one factor for each point of X. In general M has an essentially unique direct integral decomposition $M = \int_X^\oplus M_x \, d\mu(x)$, where each M_x is a factor, so the study of von Neumann algebras reduces to that of factors.

Any von Neumann algebra is generated by its projections. Because $M = M'' = (M')'$, these projections correspond to invariant subspaces or submodules for the von Neumann algebra M'. Unitary equivalence of M'-modules translates into a notion of equivalence of projections ("Murray-von Neumann equivalence"). If in addition M is a factor, then simple set-theoretic type arguments show that M falls into one of three types: (I) M has minimal projections; (II) M has projections not equivalent to any proper subprojection; (III) every nonzero projection in M is equivalent to a proper subprojection (so that they are all equivalent).

The type I factors have the form $B(K)$ for some Hilbert space K. In the type II case, Murray and von Neumann defined a countably additive dimension function on projections with range $[0, 1]$ or $[0, \infty]$, with two projections equivalent iff they have the same dimension. This leads to the notion of "continuous dimension" for any M-module. The two possibilities for the range give a further subdivision into type II_1 and type II_∞ factors. Any type II_∞ factor is the von Neumann algebra of infinite matrices with values in some type II_1 factor. For type II_1 factors, Murray and von Neumann proved that the dimension function can be linearized to give a *trace* tr on M, i.e. a state with $\text{tr}(ab) = \text{tr}(ba)$. Conversely, any factor admitting such a tracial state must be a type II_1 factor.

9 Modular Theory

Modular theory has its roots implicitly in QFT (Haag-Araki duality for bosons) and explicitly in statistical physics (the lattice models of Haag-Hugenholtz-Winninck). Independently Tomita proposed a general theory for any von Neumann algebra, developed in detail by Takesaki. For *hyperfinite* von Neumann algebras (those approximable by an increasing sequence of finite-dimensional algebras), Hugenholtz and Wierenga gave a more elementary approach based on the lattice model proof.

Tomita-Takesaki Theory. Let $M \subset B(H)$ be a von Neumann algebra and let $\Omega \in H$ (the "vacuum vector") be a unit vector such that $M\Omega$ and $M'\Omega$ are dense in H. It is then possible to define an operator $S = S_M : M\Omega \to M\Omega$, $a\Omega \mapsto a^*\Omega$. S is conjugate-linear, densely defined, and closeable with closure $\overline{S} = S_{M'}^*$. Let $\overline{S} = J\Delta^{1/2}$ be the polar decomposition of $\overline{S}$, so that J is a conjugate-linear isometry with $J^2 = I$ and Δ is a positive unbounded operator not having 0 as an eigenvalue. Then $JMJ = M'$ and $\Delta^{it}M\Delta^{-it} = M$. Thus, $x \mapsto Jx^*J$ gives an isomorphism between M^{op} (M with multiplication reversed) and M' and $\sigma_t(x) = \Delta^{it}x\Delta^{-it}$ gives a one-parameter group of automorphisms of M and M'.

Connes' (2×2)-Matrix Trick. Connes' fundamental observation was that the image of σ_t in the outer automorphism group of M is independent of the choice of the state Ω, and thus can be used to provide further intrinsic invariants of M.

"Trivial" Example (von Neumann). Let A be a unital *-algebra and tr a tracial state on A. Let $L^2(A, \mathrm{tr})$ be the Hilbert space completion of A for the inner product $\mathrm{tr}(b^*a)$. If λ and ρ denote the actions of A on $L^2(A, \mathrm{tr})$ by right and left multiplication, then $\lambda(A)'' = \rho(A)'$ and $\Delta = I$. In particular, if $A = \mathbb{C}[\Gamma]$ where Γ is a discrete countable group and tr is the Plancherel trace $\mathrm{tr}(\gamma) = \delta_{\gamma,1}$, then $L^2(A, \mathrm{tr}) = \ell^2(\Gamma)$ and λ and ρ become the usual left and right regular representations. If Γ has infinite (non-identity) conjugacy classes, e.g. if $\Gamma = S_\infty$, then $\lambda(\Gamma)''$ is a factor with a trace, so a type II_1 factor.

Easy Consequences. Connes' (2×2)-matrix trick shows that if M is a type I or II factor, then the modular group σ_t must be inner. Hence, if the fixed point algebra M^σ equals $\mathbb{C}$, i.e. σ_t is *ergodic*, then M is a type III factor (in fact, III_1).

Classification of Type III Factors. Connes' "essential spectrum" is defined as $S(M) = \bigcap \mathrm{Sp}(\Delta_\Omega)$, where Ω ranges over vectors cyclic for M and M'. Then $\Gamma = S(M) \cap \mathbb{R}_+^*$ is a closed subgroup of $\mathbb{R}_+^*$, so the type III factors can be subdivided further into: type III_0 when $\Gamma = \{1\}$; type III_λ when $\Gamma = \lambda^{\mathbb{Z}}$ with $\lambda \in (0, 1)$; and III_1 when $\Gamma = \mathbb{R}_+^*$. In the type III_0 case, a further invariant is the "flow of weights", an ergodic flow on a Lebesgue space (the action $\hat{\sigma}_\tau = \mathrm{id} \otimes \mathrm{Ad}\, m(e^{i\tau \cdot})$ of $\mathbb{R}$ on the center of $(M \otimes B(L^2(\mathbb{R})))^{\sigma \otimes \mathrm{Ad}\lambda}$). Thanks to the work of von Neumann, Connes, and Haagerup (completed in 1985), any hyperfinite factor is uniquely determined by its type (and flow of weights). In particular, the hyperfinite type II_1 and III_1 factors are unique.

Takesaki Devissage. If $N \subset M$ is a von Neumann subalgebra, normalized by Δ^{it}, then Δ and J restrict to the corresponding operators for N on the closure of $N\Omega$. This result allows one to pass from the modular operators for a theory to those of a subtheory. Thus, if M is a hyperfinite type III_1 factor with σ_t ergodic, so too is N.

10 Haag Duality and Local Loop Groups

Geometric Modular Theory for Fermions on S^1. Let I be an open interval of S^1 and let I^c be the complementary open interval. Let $\mathrm{Cliff}(I)$ be the *-algebra generated by $a(f)$ with $f \in L^2(I) \otimes V$. Then Haag-Araki duality holds: $\mathrm{Cliff}(I)'' = \mathrm{Cliff}(I^c)'$ (graded commutant). This follows directly from the more important fact that the modular operators are *geometric*. Taking I and I^c to be the upper and lower semicircles, this means that J is the canonical quantization of the flip $z \mapsto \bar{z}$, sending $f(z)$ to $\bar{z}f(\bar{z})$. Δ^{it} is the canonical quantization of the Möbius flow fixing the endpoints of I. This is proved directly by "reduction to one-particle states": S is the canonical quantization of an operator s. The polar decomposition of s gives that of S and was computed directly with Jones by two methods: by an analytic continuation argument à la Bisognano-Wichmann; or by considering representations of the algebra generated by two projections. The local algebra $\mathrm{Cliff}(I)''$ is manifestly hyperfinite. Moreover $\mathrm{Cliff}(I)''$ is a type III$_1$ factor by the ergodicity of σ_t, because Δ^{it} is the direct sum of the trivial representation and copies of the regular representation of $\mathbb{R}$.

Loop Group Subfactors (Jones-Wassermann). Let $L_I G$ be the local loop group consisting of loops concentrated in I, i.e. loops equal to 1 off I, and let π_i be an irreducible positive energy representation of level ℓ. Haag-Araki duality and the fermionic construction of π_i imply that operators in $\pi(L_I G)$ and $\pi(L_{I^c} G)$, defined up to a phase, actually commute ("locality"). Thus, we get the canonical inclusion:

$$\pi_i(L_I G)'' \subseteq \pi_i(L_{I^c} G)'. \tag{2}$$

Consequences of Takesaki Devissage. Because the modular operators for the fermionic free field theory are geometric and the loop group representations are constructed as subtheories, Takesaki devissage can be applied to the geometric inclusion of local algebras on $\mathcal{F}_P^{\otimes \ell}$, $\pi^{\otimes \ell}(L_I G)'' \subset (\mathrm{Cliff}(I)^{\otimes \ell})''$. It has the following consequences:

Haag Duality in the Vacuum Sector. If π_0 is the vacuum representation at level ℓ (so that the lowest energy subspace, generated by the vacuum vector, gives the trivial representation of G), then $\pi_0(L_I G)'' = \pi_0(L_{I^c} G)'$. Moreover an argument of Reeh-Schlieder shows that the vacuum vector is cyclic for $\pi_0(L_I G)''$, and hence $\pi_0(L_I G)'$. The corresponding modular operators are geometric. So in general the inclusion (2) *measures the failure of Haag duality.*

Local Equivalence. $\pi_0|_{L_I G} \cong \pi_i|_{L_I G}$, so that all positive energy representations at level ℓ become unitarily equivalent when restricted to the local loop groups. (Note that smeared vector primary fields give explicit bounded intertwiners.)

Type of Local Algebras. $\pi_i(L_I G)''$ is isomorphic to the hyperfinite type III$_1$ factor. Hyperfiniteness can also be deduced more directly, independently of the Connes-Haagerup classification, by the *factorization property*. This property, inherited from fermions, means that the representations π_0 and $\pi_0 \otimes \pi_0$ of $L_{I \cup J} G = L_I G \times L_J G$ are unitarily equivalent if I and J are nontouching disjoint intervals. So if $I_n \uparrow I$, there is a type I factor B_n lying between $\pi_0(L_{I_n} G)''$ and $\pi(L_{I_{n+1}} G)''$. $B_n \uparrow \pi_0(L_I G)''$ forces hyperfiniteness (the "Dick trick").

Generalized Haag Duality. Let $\pi = \bigoplus \pi_i$ on $H = \bigoplus H_i$, the direct sum of all the level ℓ representations and let ϕ be the primary field for the vector representation. Then $\pi(L_I G)'' = \langle \phi(f), \phi(f)^* : f \in C_c^\infty(I) \otimes V \rangle' \cap (\bigoplus B(H_i))$.

Von Neumann Density. Let I_1 and I_2 be *touching* intervals obtained by removing a point from the interval I. Then $\pi(L_{I_1} G)'' \vee \pi(L_{I_2} G)'' = \pi(L_I G)''$ ("irrelevance of points"). Jones and I first deduced this from a stronger result: the pullback of the quotient strong operator topology on LG under the map $LG \to PU(\mathcal{F}_P)$ makes $L_{I_1} G \times L_{I_2} G$ *dense* in $L_I G$. Von Neumann density also follows by taking commutants in generalized Haag duality and noting that, because of its L^2 bounds, ϕ "does not see points".

Irreducibility. If $L^{\pm 1} G = L_I G \times L_{I^c} G$ is the subgroup of LG consisting of loops trivial to all orders at ± 1, then irreducible positive energy representations of LG stay irreducible and inequivalent when restricted to $L^{\pm 1} G$.

11 Connes Fusion and Braiding

Connes defined an associative tensor operation ("Connes fusion") on bimodules over (type III) von Neumann algebras. Let $X = {}_A X_B$ be an (A, B)-bimodule and $Y = {}_B Y_C$ a (B, C)-bimodule. Let (H_0, Ω) be a "trivial" (B, B)-bimodule defined by modular theory. Let $\mathcal{X} = \mathrm{Hom}_{B^{\mathrm{op}}}(H_0, X)$, $\mathcal{Y} = \mathrm{Hom}_B(H_0, Y)$ and define $X \boxtimes Y$ as the Hilbert space completion of $\mathcal{X} \otimes \mathcal{Y}$ with inner product $\langle x_1 \otimes y_1, x_2 \otimes y_2 \rangle = (x_2^* x_1 y_2^* y_1 \Omega, \Omega)$. It is naturally an (A, C)-bimodule. If ${}_A X_B$ and ${}_B Y_A$ are irreducible, Y is called *conjugate* to X iff $X \boxtimes Y$ and $Y \boxtimes X$ both contain the trivial bimodule at least once. Y is then unique up to isomorphism and the trivial bimodule appears exactly once. Any homomorphism $\rho : A \to B$ defines an (A, B)-bimodule, because ρ makes H_0 an A-module. Connes fusion corresponds to composition of homomorphisms. Because all modules over a type III factor are equivalent, every bimodule arises this way. Many properties of Connes fusion can be proved in the homomorphism picture.

Definition of Fusion (State-Field Correspondence). For representations of LG, the bimodule point of view comes through restricting to $L_I G \times L_{I^c} G$ and Connes fusion can be defined without explicit reference to von Neumann algebras. Let X, Y be positive energy representations of LG at level ℓ. Replace states $\xi \in X, \eta \in Y$ by intertwiners $x \in \mathcal{X} = \mathrm{Hom}_{L_{I^c} G}(H_0, X)$, $y \in \mathcal{Y} = \mathrm{Hom}_{L_I G}(H_0, Y)$. The "fields" x, y create the states $\xi = x\Omega$, $\eta = y\Omega$ from the vacuum. The inner product on $\mathcal{X} \otimes \mathcal{Y}$ is given by the *four-point formula* $\langle x_1 \otimes y_1, x_2 \otimes y_2 \rangle = \langle x_2^* x_1 y_2^* y_1 \rangle$ (vacuum expectation). The Hilbert space completion $X \boxtimes Y$ naturally supports a projective representation of $L_I G \times L_{I^c} G$.

Braiding Properties of Bounded Intertwiners. By hermiticity, the braiding relations (1) for smeared primary fields can be written symbolically as

$$a_{f0} b_{\square 0}^* = \sum \lambda_g b_{gf}^* a_{g\square}, \qquad a_{g\square} b_{\square 0} = \varepsilon_g b_{gf} a_{f0}, \tag{3}$$

where $|\varepsilon_g| = 1$ in the second Abelian relation. We call $a = a_{f0}$ and $b = b_{\square 0}$ the *principal parts*. Letting $A_1 = a_{f0}$, $A_2 = \bigoplus |\lambda_g|^{1/2} a_{g\square}$, $B_1 = b_{\square 0}$, $B_2 = \bigoplus \varepsilon_g |\lambda_g|^{1/2} b_{gf}$, the braiding relations (3) take the form $A_1 B_1^* = B_2^* A_2$, $A_2 B_1 =$

B_2A_1: these equations are unchanged if the A_i's or B_j's are replaced by their phases. The a_{ij}'s become bounded after such a "phase correction". Each set of intertwiners (c_{ij}) can be modified in three steps so that (3) still holds but with a and b *unitary*:(i) replace c_{ij} by $\sum 2^{-n}\pi_i(g_n)c_{ij}\pi_j(u_n)$, with (g_n) a dense subgroup of L_IG and u_n partial isometries in $\pi_0(L_IG)''$ with $u_iu_j^* = \delta_{ij}I$, $\sum u_i^*u_i = I$; (ii) make a phase correction on (c_{ij}) so that the principal part c satisfies $cc^* = I$; (iii) replace c_{ij} by $c_{ij}\pi_j(u)$ where u is a partial isometry in $\pi_0(L_IG)''$ with $u^*u = I$, $uu^* = c^*c$. If now $x : H_0 \to H_f$ and $y : H_0 \to H_\square$ are arbitrary intertwiners, their *nonprincipal* parts are defined by $x_{ij} = a_{ij}\pi_j(a_{f0}^*x)$ and $y_{pq} = b_{pq}\pi_q(b_{\square 0}^*y)$. They satisfy the analogues of (3).

Computation (Braiding-Fusion Duality). To prove the fusion rules $H_f \boxtimes H_\square \cong \oplus H_g$, where now $g \in \mathbf{Y}_\ell$, it suffices to define an explicit isometry U of $H_f \boxtimes H_\square$ into $\oplus H_g$, which is an intertwiner for $L^{\pm 1}G = L_IG \times L_{I^c}G$; for by Schur's lemma and irreducibility for $L^{\pm 1}G$, U must be unitary making $H_f \boxtimes H_\square$ a positive energy representation. By the braiding relation for intertwiners,

$$\|x \otimes y\|^2 = \langle x_{f0}^*x_{f0}y_{\square 0}^*y_{\square 0}\rangle = \sum \lambda_g \langle x_{f0}^*y_{gf}^*x_{g\square}y_{\square 0}\rangle = \sum |\lambda_g| \langle y_{\square 0}^*x_{g\square}^*x_{g\square}y_{\square 0}\rangle.$$

The coefficients have to be positive, as the equation can be interpreted as writing a vector state as a linear combination of inequivalent pure states. Thus, only the non-vanishing of the λ_g's is important. Now define $U(x \otimes y) = \oplus |\lambda_g|^{1/2}x_{g\square}y_{\square 0}\Omega$.

Braiding. The braiding map $b : X \boxtimes Y \to Y \boxtimes X$ is the unitary given by $b(x \otimes y) = e^{-\pi iL_0} \cdot (e^{i\pi L_0}ye^{-i\pi L_0} \otimes e^{i\pi L_0}xe^{-i\pi L_0})$. Under the "concrete" isomorphism U on $H_\square \boxtimes H_\square$, $UbU^*(\oplus|\lambda_g|^{1/2}x_{g\square}y_{\square 0}\Omega) = \oplus|\lambda_g|^{1/2}y_{g\square}x_{\square 0}\Omega = \oplus|\lambda_g|^{1/2}\mu_g x_{g\square}y_{\square 0}\Omega$, so that $UbU^* = \mu_gI$ on H_g. In general $H_1 \boxtimes \cdots \boxtimes H_n$ can also be defined and computed using a $2n$-point function, after having divided S^1 into n intervals. The b's have a very simple concrete form, especially on $H_\square^{\boxtimes n}$ where only vector primary fields are invoked. This realization makes manifest their braiding and cabling properties.

Closure under Fusion and Conjugation. By associativity and induction: each irreducible positive energy representation H_i appears in some $H_\square^{\boxtimes n}$; the H_i's are closed under Connes fusion; each H_i has a (unique) conjugate $\overline{H_i}$.

General Fusion Rules (Faltings' Trick). The fusion coefficients N_{ij}^k are given by $H_i \boxtimes H_j = \oplus N_{ij}^k H_k$. Braiding shows that $H_i \boxtimes H_j \cong H_j \boxtimes H_i$. Let $\mathcal{R}$ be the representation ring of formal sums $\sum m_i H_i$. $\mathcal{R}$ is commutative with an identity and an involution. Thus, the complexification $\mathcal{R}_\mathbb{C}$ is a finite-dimensional *-algebra with a nondegenerate positive trace $\mathrm{tr}(\sum c_i H_i) = c_0$. So $\mathcal{R}_\mathbb{C} \cong \mathbb{C}^M$, where $M = |\mathbf{Y}_\ell|$. The fusion rules for $H_{\lambda^k V}$ are deduced by combining the method used for $H_\square$ with properties of $\mathcal{R}$. From these fusion rules, the characters of $\mathcal{R}_\mathbb{C}$ are given by $[H_f] \mapsto \mathrm{ch}(H_f, h) = \chi_f(D(h))$ where $h \in \mathbf{Y}_\ell$ and $D(h) \in SU(N)$ is the diagonal matrix with $D(h)_{kk} = \exp(2\pi i(h_k + N - k - H)/(N + \ell))$ where $H = (\sum h_k + N - k)/N$. Thus, the N_{ij}^k's can be computed using the multiplication rules for the basis $\mathrm{ch}(H_f, \cdot)$ of $C(\mathbf{Y}_\ell)$. They agree with the Verlinde formulas in Kac's book.

Summary. The positive energy representations $H_0, \ldots, H_M$ at a fixed level ℓ become a braided ribbon C* tensor category.

12 Subfactors

Let $N \subset M$ be an inclusion of type II_1 factors in $B(H)$, so that H becomes an (M, N^{op})-bimodule. They act on $L^2(M, tr)$. Let $e = e_1$ be the projection onto $L^2(N)$ and $M_1 = \langle M, e_1 \rangle''$, the *Jones basic construction*.

Definition. N is of *finite index* in M iff M_1 is a type II_1 factor. Its *(Jones) index* is given by $[M : N] = tr(e_1)^{-1} = \dim_N L^2(M)$. So M is finite dimensional as an N-module. There is an equivalent probabilistic definition due to Pimsner-Popa. The projection $e : L^2(M) \to L^2(N)$ restricts to a "conditional expectation" $E : M \to N$ satisfying $E(x) \geq \lambda x$ for $x \geq 0$ where $\lambda = [M : N]^{-1}$. The index yields the best possible value of $\lambda > 0$.

Higher Relative Commutants (Subfactor Invariants). It turns out that $[M_1 : M] = [M : N]$, so the basic construction can be iterated to get a tower:

$$N \subset M \subset^{e_1} M_1 \subset^{e_2} M_2 \subset^{e_3} \cdots$$

The higher relative commutants are $A_n = M' \cap M_n$, $B_n = M_1' \cap M_n$. They are finite-dimensional von Neumann algebras, so direct sums of matrix algebras. The inclusions $B_n \subset A_n$ increase to an inclusion of type II_1 factors $B \subset A$. The inclusion $N \subset M$ is said to have *finite depth* if the centers of A_n and B_n have uniformly bounded dimension. The inclusion is *irreducible* iff $N' \cap M = \mathbb{C}$.

Bimodule Picture. $L^2(M)$ is a bimodule over (M, M), (M, N), (N, M), and (N, N). The algebras A_n and B_n encode the decomposition and branching rules for the bimodules $L^2(M)^{\boxtimes m}$, fused over N.

Popa's Finite Depth Classification Theorem. If the inclusion of hyperfinite type II_1 factors $N \subset M$ has finite depth and is irreducible, then $N \subset M \cong B^{op} \subset A^{op}$. A version of the same theorem also holds in the hyperfinite type III_1 case, provided that the Pimsner-Popa inequality is taken as the definition of finite index and the inclusion $B^{op} \subset A^{op}$ is replaced by its tensor product with the hyperfinite type III_1 factor.

13 Quantum Invariant Theory Subfactors

Classical Invariant Theory Subfactors. If V is a representation of G, we get an inclusion of type II_1 factors $(\cup_m \mathbb{C} \otimes \text{End}_G V^{\otimes m})'' \subset (\cup_m \text{End}_G V^{\otimes m+1})''$ with Jones index $\dim(V)^2$. When $G = SU(N)$ and $V = \mathbb{C}^N$, the right-hand side is generated by $S_\infty = \cup S_n$ and the left-hand side is obtained by applying the shift endomorphism $\rho(s_i) = s_{i+1}$ where $s_i = (i, i+1)$. The higher relative commutants are given by:

$$
\begin{array}{ccccccc}
A_n : & \text{End}_G V & \subset & \text{End}_G V \otimes \overline{V} & \subset & \text{End}_G V \otimes \overline{V} \otimes V & \subset \\
 & \cup & & \cup & & \cup & \\
B_n : & \mathbb{C} & \subset & \text{End}_G \overline{V} & \subset & \text{End}_G \overline{V} \otimes V & \subset
\end{array}
\tag{4}
$$

Braid Group Subfactors (Jones-Wenzl). Let tr be a positive definite trace on the infinite braid group $B_\infty = \cup B_n$, generated by $g_1, g_2, g_3, \ldots$ with relations $g_i g_{i+1} g_i = g_{i+1} g_i g_{i+1}$ and $g_i g_j = g_j g_i$ if $|i - j| \leq 2$. Suppose that tr has the *Jones-Markov* property $tr(a g_n^{\pm 1}) = \mu tr(a)$ for $a \in B_n = \langle g_1, \ldots, g_{n-1} \rangle$. Form

$L^2(\mathbb{C}B_\infty, \mathrm{tr})$ and let π be the left unitary action of B_∞. Assume in addition that the algebras $\pi(\mathbb{C}B_n)$ are finite-dimensional and that the dimensions of their centers are uniformly bounded ("finite depth"). The braid group subfactor is given by the inclusion $\pi(g_2, g_3, \ldots)'' \subset \pi(g_1, g_2, \ldots)''$ and as for the symmetric group arises from a shift $\rho(g_i) = g_{i+1}$. It has index $|\mu|^{-2}$. More generally, Wenzl considered the irreducible parts of the inclusions $\pi(g_{m+1}, g_{m+2}, \ldots)'' \subset \pi(g_1, g_2, \ldots)''$, obtained by reducing by minimal projections in the relative commutant. The first examples arose by taking $g_i = ae_i + b$ with the e_i's Jones projections and a, b constants. Most other examples arose from the solutions of the quantum Yang-Baxter equations associated with quantum groups at roots of unity and restricted solid-on-solid models in statistical mechanics. Unfortunately, the positivity of the trace here only followed after a very detailed analysis of $\pi(\mathbb{C}B_n)$ using q-algebraic combinatorics.

Quantum Invariant Theory Subfactors. There is a more direct construction of the braid group subfactors. It is more conceptual, manifestly unitary, and allows a direct computation of the higher relative commutants. The data $(G, V, \otimes)$ is replaced by $(LG, H, \boxtimes)$: $(\cup_m \mathbb{C} \otimes \mathrm{End}_{LG} H^{\boxtimes m})'' \subset (\cup_m \mathrm{End}_{LG} H^{\boxtimes m+1})''$. If H corresponds to the vector representation, the right-hand side is generated by B_∞ and the left-hand side is obtained from the shift $\rho(g_i) = g_{i+1}$. The Jones index equals the square of the *quantum dimension* of H. This is given by $d(H_f) = \mathrm{ch}(H_f, 0)$ and is the unique positive character of $\mathcal{R}$. Thanks to "Wenzl's lemma", the higher relative commutants are obtained by replacing $(G, V, \otimes)$ by $(LG, H, \boxtimes)$ in (4).

14 Doplicher-Haag-Roberts Formalism

Algebraic QFT gives a translation from the bimodule to the homomorphism point of view. For fixed H_i, let $I \subset\subset J$ and $I^c \subset\subset K$ and take unitary intertwiners $U : H_0 \to H_i$ for $L_J G$ and $V : H_0 \to H_i$ for $L_K G$. Set $M = \pi_0(L_I G)''$. Then $\rho_i(x) = V^* U x U^* V$ defines a *DHR endomorphism* of M and the loop group inclusion $\pi_i(L_I G)'' \subset \pi_i(L_{I^c} G)'$ is isomorphic to the inclusion $\rho_i(M) \subset M$. The endomorphism ρ_i is *localized* in $I_1 = S^1 \backslash \overline{K} \subset\subset I$, in the sense that it fixes loop group elements supported in $I \backslash I_1$. Let T be a diffeomorphism, supported in I, with $T(I_1)$ disjoint from I_1 in a clockwise sense. Define the statistics operator by $g = u^* \rho_i(u)$, where $u = T^* U \rho_i(T)$. Then $g \rho_i(g) g = \rho_i(g) g \rho_i(g)$ and g lies in $\rho_i^2(M)' \cap M$. Hence $g_k = \rho_i^{k-1}(g)$ gives a unitary representation of B_∞. Under the bimodule-endomorphism correspondence, the results on Connes fusion imply: $\rho_i^{k+1}(M)' \cap M \cong \mathrm{End}_{LG} H_i^{\boxtimes k+1}$, with $g_1, \ldots, g_k$ identified with the Connes braiding; the Jones index of the loop group subfactor is $d(H_i)^2$; and the higher relative commutants for the loop group subfactor agree with those of the corresponding quantum invariant theory subfactor.

15 The Main Result on Subfactors

Because the higher relative commutants agree, Popa's finite depth classification theorem implies:

Theorem (Jones-Wassermann Conjecture). *The loop group inclusion of hyperfinite type III_1 factors $\pi_i(L_I G)'' \subset \pi_i(L_{I^c} G)'$ is isomorphic to the tensor product of the*

*hyperfinite type III_1 factor with the quantum invariant theory inclusion of type II_1
factors $N_0 = (\cup \mathbb{C} \otimes \mathrm{End}_{LG} H_i^{\boxtimes m})'' \subset (\cup \mathrm{End}_{LG} H_i^{\boxtimes m+1})'' = M_0$.*

This result may be sharpened using the inclusion $M \hookrightarrow M_2(M)$, $x \mapsto x \oplus \rho(x)$.

THEOREM. *There is an automorphism α of M and a unitary $u \in M$ such that
$\alpha \rho = \mathrm{Ad}\, u\, \rho \alpha$ and, if $\rho_1 = \alpha \rho$ and $M_1 = (\cup \rho_1^m(M)' \cap M)''$, then $M = M_1 \overline{\otimes} M^{\rho_1}$,
$N = \rho_1(M_1) \otimes M^{\rho_1}$. M^{ρ_1} is isomorphic to the hyperfinite type III_1 factor and
the inclusion $N_1 \subset M_1$ is isomorphic to $N_0 \subset M_0$ by an isomorphism preserving
endomorphisms and braid group operators.*

16 Future Directions

WZW Models. Other constructions of subfactors can be obtained by taking other
compact simple groups G, not necessarily simply connected. The theory for the B,
C, D series seems to follow from the (3×3)-matrix ODE of Fateev-Dotsenko.

Minimal Models. The theory has been developed along similar lines by Loke for
discrete series representations of Diff S^1 for central charge $c < 1$.

Conformal Inclusions. A subgroup H of G gives a conformal inclusion if the level
one representations of LG remain finitely reducible when restricted to H. The
basic construction M_1 for the inclusion $N = \pi_0(L_I H)'' \subset \pi_0(L_I G)'' = M$ can be
identified with $\pi_0(L_{I^c} H)'$, so $N \subset M_1$ is a loop group inclusion. So $N \subset M$ has
finite index and depth. For example the conformal inclusion $SU(2) \subset SO(5)$ gives
the Jones subfactor of index $3 + \sqrt{3}$.

Disjoint Intervals. If the circle is divided up into $2n$ disjoint intervals and I is the
union of n alternate intervals, the inclusion $\pi_i(L_I G)'' \subset \pi_i(L_{I^c} G)'$ has finite index
and probably finite depth. It is related to higher genus CFT.

Fusion. Connes fusion can be viewed as glueing together two circles along a common semicircle. This picture seems to be a unitary boundary value of Graeme
Segal's holomorphic proposal for fusion, based on a disc with two smaller discs removed. When the discs shrink to points on the Riemann sphere, Segal's definition
should degenerate to the algebraic geometric fusion of Kazhdan-Lusztig et al.

References

[1] R. Borcherds, *Vertex algebras, Kac-Moody algebras and the Monster*, Proc. Nat. Acad. Sci. U.S.A. **83** (1986), 3068–3071.

[2] A. Connes, Non-commutative Geometry, Chapter V, Appendix B, Academic Press, San Diego, CA, 1994.

[3] S. Doplicher, R. Haag, and J. Roberts, *Local observables and particle statistics I, II*, Comm. Math. Phys. **23** (1971) 119–230 and **35** (1974) 49–85.

[4] K. Fredenhagen, K.-H. Rehren, and B. Schroer, *Superselection sectors with braid group statistics and exchange algebras*, Comm. Math. Phys. **125** (1989), 201–226.

[5] P. Goddard and D. Olive, *Kac-Moody and Virasoro algebras*, Adv. Ser. Math. Phys. Vol. 3, World Scientific, Singapore, 1988.

[6] R. Haag, Local quantum physics, Springer-Verlag, Berlin and New York, 1992.

[7] V. Jones, *Index for subfactors*, Invent. Math. **72** (1983), 1–25.

[8] V. Jones, *von Neumann algebras in mathematics and physics*, Proc. Internat. Congress Math. Kyoto 1990, 121–138.

[9] V. Kac, Infinite dimensional Lie Algebras, 3rd edition, Cambridge Univ. Press, 1990.

[10] V. Kazhdan and G. Lusztig, *Tensor structures arising from affine Lie algebras IV*, J. Amer. Math. Soc. **7** (1994), 383–453.

[11] V. Knizhnik and A. Zamolodchikov, *Current algebra and Wess-Zumino models in two dimensions*, Nuclear Phys. B **247** (1984), 83–103.

[12] T. Nakanishi and A. Tsuchiya, *Level-rank duality of WZW models in conformal field theory*, Comm. Math. Phys. **144** (1992), 351–372.

[13] S. Popa, *Classification of subfactors: The reduction to commuting squares*, Invent. Math. **101** (1990), 19–43.

[14] S. Popa, *Classification of subfactors and of their endomorphisms*, Lecture notes from C.B.M.S. conference, to appear.

[15] A. Pressley and G. Segal, Loop Groups, Oxford Univ. Press, London and New York, 1986.

[16] G. Segal, *Notes on conformal field theory*, unpublished manuscript.

[17] A. Tsuchiya and Y. Kanie, *Vertex operators in conformal field theory on $\mathbb{P}^1$ and monodromy representations of braid group*, Adv. Stud. Pure Math. **16** (1988), 297–372.

[18] H. Wenzl, *Hecke algebras of type A_n and subfactors*, Invent. Math. **92** (1988), 249–383.

[19] H. Wenzl, *Quantum groups and subfactors of type B, C and D*, Comm. Math. Phys. **133** (1990), 383–432.

Brownian Motion, Heat Kernels, and Harmonic Functions

RICHARD F. BASS*

Department of Mathematics, University of Washington
Seattle, WA 98195, USA

ABSTRACT. Although the boundary behavior of harmonic functions is an old subject (Fatou's theorem was proved in 1906), interesting results are still being obtained today. In this article we discuss some recent results concerning harmonic functions, heat kernels, and related topics that have been obtained using Brownian motion. In the following sections we will discuss the heat kernels for the Neumann Laplacian, the boundary Harnack principle, the Martin boundary, conditional lifetimes, and the conditional gauge theorem.

1 Heat kernels and reflecting Brownian motion

Let $p^D(t, x, y)$ denote the Neumann heat kernel for a domain D. This is the fundamental solution to the heat equation $\partial u / \partial t = (1/2)\Delta u$ with Neumann boundary conditions. (Having Neumann boundary conditions means that the normal derivative of u is 0 on the boundary of D.) $p^D(t, x, y)$ is also, of course, the transition density of reflecting Brownian motion in D.

What can one say about $p^D(t, x, y)$ as a function of the domain D? The heat kernel with Dirichlet boundary conditions (that is, $u = 0$ on the boundary of D) is easily seen to decrease as the domain D becomes smaller. One might expect that for Neumann boundary conditions, $p^{D_1}(t, x, y)$ should be greater than $p^{D_2}(t, x, y)$ if $D_1 \subseteq D_2$. A small room should warm up faster than a large room. A little thought shows that one must assume D_1 and D_2 are convex. With this assumption, the above monotonicity holds if D_2 is a ball centered at x [13], if a sphere about x separates ∂D_1 and ∂D_2 [19], if $\bar{D}_1 \subseteq D_2$ and t is sufficiently small [12], or if t is sufficiently large [13].

It turns out, however, that this domain monotonicity need not always hold. Fix $t > 0$. Let $\varepsilon > 0$ and set

$$D_2 = \{re^{i\theta} : r > 0, 0 < \theta < 3\pi/4\}, \qquad D_1 = D_2 + (-\varepsilon, \varepsilon),$$

$$x_\varepsilon = (-\varepsilon, \varepsilon), \qquad y_\varepsilon = (1, \varepsilon).$$

THEOREM 1.1. *If ε is sufficiently small,*

$$p^{D_2}(t, x_\varepsilon, y_\varepsilon) > p^{D_1}(t, x_\varepsilon, y_\varepsilon).$$

*) The research described here was partially supported by grants from the National Science Foundation.

Proceedings of the International Congress
of Mathematicians, Zürich, Switzerland 1994
© Birkhäuser Verlag, Basel, Switzerland 1995

It is easy to modify this example to come up with D_1 and D_2 that are bounded, smooth, and strictly convex, and with $x, y \in D_1 \subseteq \bar{D}_1 \subseteq D_2$ such that $p^{D_2}(t, x, y) > p^{D_1}(t, x, y)$.

To prove Theorem 1.1 is not difficult. The method used in [8] obtains an explicit expression for $p^{D_2}(t, 0, \cdot)$ from symmetry considerations and then uses translation invariance to get $p^{D_1}(t, x_\varepsilon, y_\varepsilon)$. To estimate $p^{D_2}(t, x_\varepsilon, \cdot)$, we show that it can be written as Uf for a function f, where U denotes the potential with respect to the Green function for reflecting Brownian motion in D_2. Conformal mapping then reduces this to an estimate on reflecting Brownian motion in the upper half plane.

The domains D_1 and D_2 are examples of Lipschitz domains. In two dimensions, conformal mapping is a useful tool, but good estimates for heat kernels with Neumann boundary conditions can be obtained in Lipschitz domains even when the dimension is three or higher, and some upper bounds are available even in Hölder domains [11]. A Hölder domain is one whose boundary is locally the graph of a Hölder continuous function.

2 Boundary Harnack principle

Harnack's inequality says that the values of a positive harmonic function are comparable in the interior of a domain D. Not much can be said, in general, near the boundary. However, if D is a Lipschitz domain, say, and two harmonic functions u and v both vanish on the same portion of the boundary, then u and v tend to 0 near that portion of that boundary at the same rate.

This is called the boundary Harnack principle, and is a very useful tool. For example, it can be used to identify the Martin boundary in Lipschitz domains and to prove Fatou theorems for Lipschitz domains (see [3], [4]). Using probabilistic techniques, it was proved in [2] and [5] that the boundary Harnack principle holds in Hölder domains of order α for $\alpha > 0$. It also holds in twisted Hölder domains of order α provided that $\alpha > 1/2$, where twisted Hölder domains are generalizations of Hölder domains in much the same way that John domains are generalizations of Lipschitz domains.

Let D^c denote the complement of D. As a sample of the kind of theorem that is proved in [2] and [5], we have

THEOREM 2.1. *Suppose D is a Hölder domain of order α, $\alpha > 0$. Let K be a compact set, and V an open set containing K. Let $x_0 \in D$. If u and v are two positive harmonic functions in D such that $u(x_0) = v(x_0) = 1$, u and v vanish at the points of $\partial D \cap V$ that are regular for D^c, and u and v are bounded in a neighborhood of $\partial D \cap V$, then*

$$u(x)/v(x) \le c, \qquad x \in K \cap D,$$

where c is a constant that depends on D, K, V, and x_0, but not u or v.

If D happens to be the domain above the graph of a Hölder continuous function Γ, and

$$Q = \big\{(x_1, \dots, x_d) : \Gamma(x_1, \dots, x_{d-1}) < x_d < a + \Gamma(x_1, \dots, x_{d-1}),$$
$$|(x_1, \dots, x_{d-1})| \le R\big\}$$

for positive numbers a and R, it is not too hard to get estimates for the probability that Brownian motion exits Q through the upper boundary of Q and the probability that Brownian motion exits Q through the sides of Q. The key to the proof of Theorem 2.1 is to combine these estimates to show that Brownian motion conditioned to stay in D is not too likely to exit a box such as Q by creeping along the boundary of D, but rather must go at least a certain amount into the interior of D.

If one replaces Brownian motion by certain diffusions corresponding to divergence form operators, one gets the analogue of Theorem 2.1 with harmonic replaced by L-harmonic, where an L-harmonic function h is one with $Lh = 0$. Here

$$Lh(x) = \sum_{i,j=1}^{d} \frac{\partial}{\partial x_i}\left(a_{ij}(x)\frac{\partial h}{\partial x_j}\right)(x),$$

and the matrix a_{ij} is assumed to be bounded, uniformly positive definite, and measurable (no smoothness is required). If instead of divergence form operators such as the above, one has nondivergence operators like

$$Lh(x) = \sum_{i,j=1}^{d} a_{ij}(x)\frac{\partial^2 h}{\partial x_i \partial x_j}(x)$$

with the same assumptions on a_{ij}, it turns out that the boundary Harnack principle holds for positive L-harmonic functions in Hölder domains of order α when $\alpha > 1/2$. It need not hold for domains when $\alpha < 1/2$, unless the domain satisfies an additional regularity condition. See [9].

3 Martin boundary

Let $x_0 \in D$ and let $g_D(x, y)$ be the Green function for the domain D with pole at y. The (minimal) Martin boundary $\partial_M D$ in a domain is an ideal boundary such that every positive harmonic function h in a domain D can be written

$$h(x) = \int_{\partial_M D} M(x, y)\mu(dy)$$

for some measure μ supported on $\partial_M D$ in one and only one way, where $M(x, y)$, the Martin kernel, is the appropriate extension of $g_D(x, y)/g_D(x_0, y)$. For Lipschitz domains the Martin boundary can be identified with the Euclidean boundary (first proved by Hunt and Wheeden [18]), and it is not hard to show this fact from scaling properties of Lipschitz domains and the boundary Harnack principle. Because Theorem 2.1 says that the boundary Harnack principle holds for Hölder domains, one might wonder whether the Martin boundary must equal the Euclidean boundary in all Hölder domains. The answer is no. However, we have the following theorem from [7].

THEOREM 3.1. *Let*

$$\gamma(x) = bx\frac{\log\log(1/x)}{\log\log\log(1/x)}.$$

Provided b is sufficiently small, the Martin boundary of D may be identified with the Euclidean boundary if D is a C^γ domain.

A C^γ domain is one in which the boundary can be represented locally as the graph of a function whose modulus of continuity is no worse than γ.

This is suggestive of the laws of the iterated logarithm beloved by probabilists, but $x \log \log(1/x)$ is not the right borderline function. In fact for every $b > 0$ there is a C^γ domain with $\gamma(x) = bx \log \log(1/x)$ for which the Martin boundary is different than the Euclidean boundary.

The proof of Theorem 3.1 involves a careful estimate of the constant c in Theorem 2.1 for C^γ domains.

4 Conditional lifetimes

Around 1983 K.L. Chung posed the conditional lifetime problem. Suppose $\mathbb{E}_z^x$ represents the expectation of Brownian motion in a domain D conditioned to exit D at the point z. Let τ_D represent the exit time of D for Brownian motion. The conditional lifetime problem is to determine for which domains $\mathbb{E}_z^x \tau_D$ is finite.

Cranston and McConnell [16] showed that in two dimensions, the expected conditional lifetime is bounded by a constant that depends only on the area of D and not on x or z. In three dimensions the lifetime can be infinite even for bounded domains [16], but must be finite in bounded Lipschitz domains [14]. Various refinements have been obtained, and in the last few years, some extensive generalizations have been proved ([1], [6], [17]). For example, we have

THEOREM 4.1. *Suppose $p > d - 1$ and D is a domain of the form*

$$D = \big\{(x_1, \ldots, x_d) : |(x_1, \ldots, x_{d-1})| < 1, \ 0 > x_d > -f(x_1, \ldots, x_{d-1})\big\},$$

where f is a nonnegative function that is in $L^p(\mathbb{R}^{d-1})$. Then $\mathbb{E}_z^x \tau_D < c$, where c is a finite constant depending only on f and not x or z.

Many refinements of Theorem 4.1 are possible. Brownian motion can be replaced by diffusions corresponding to operators L in either divergence or nondivergence form. Twisted Hölder domains of order α are possible (it turns out that the critical α becomes $1/3$). A domain need only have its boundary be given locally by the graph of an L^p function.

Various proofs of theorems such as Theorem 4.1 have been given. In [17], the Girsanov transformation is the principal tool; in [1] heat kernel estimates play an important part. The proof in [6] is based on the observation that the amount of time Brownian motion spends in a strip of width r is proportional to r^2. Combining this estimate with the techniques used in [14] and [16] give Theorem 4.1 and its relatives.

5 Conditional gauge

In solving the Dirichlet problem using probability theory, it is possible to replace $(1/2)\Delta u(x)$ by $(1/2)\Delta u(x) + q(x)u(x)$, the Schrödinger operator, if one uses the Feynman-Kac formula. Expectations of expressions such as $\exp(\int_0^{\tau_D} q(X_s)\,ds)$ then play a role. To study local behavior of the solution to the Dirichlet problem, such as the Fatou theorems, Martin boundary, harmonic measure, etc., one uses conditioned Brownian motion. It should be no surprise, then, that to study the local behavior of solutions to equations involving the Schrödinger operator, it is necessary to look at $\mathbb{E}_z^x \exp(\int_0^{\tau_D} q(X_s)\,ds)$. Here X_s is Brownian motion, τ_D is the exit time from D, and $\mathbb{E}_z^x$ represents the expectation of Brownian motion started at x and conditioned to exit D at z.

The above expectation is called the conditional gauge. It need not be finite even when q is constant and D is an interval in one dimension. However, when it is finite, many of the local properties of harmonic functions carry over to functions that are harmonic with respect to the operator $(1/2)\Delta + q$.

The conditional gauge theorems are results that say that under certain assumptions on q and D, if the conditional gauge is finite for one pair (x, z) with $x \neq z$, then it is finite for all pairs (x, z), and the values of the conditional gauge are comparable. For example, if $d \geq 3$, D is a Lipschitz domain in $\mathbb{R}^d$, and q is in the Kato class, then the conditional gauge theorem holds [15]. For q to be in the Kato class means that

$$\limsup_{\varepsilon \to 0} \sup_{x \in D} \int_{D \cap B(x, \varepsilon)} |q(y)|\,|x - y|^{2-d}\,dy = 0,$$

where $B(x, \varepsilon)$ is the ball of radius ε about x. When $d = 2$, $|x - y|^{2-d}$ in the definition is replaced by $\log(1/|x - y|)$.

Various partial results have been proved for the case $d = 2$, but by analogy to the conditional lifetime results, one would expect that the conditional gauge theorem ought to be true in bounded domains in the plane, with no further assumptions on the domain necessary. This turns out to be correct [10].

The key to the conditional gauge theorem is to get a sufficiently good estimate on the Green function for conditioned Brownian motion in D, namely on the ratio

$$\frac{g_D(x, y)g_D(y, z)}{g_D(x, z)},$$

where g_D is the Green function for ordinary Brownian motion in D. In [10] we obtain the bound

THEOREM 5.1. *If D is a bounded domain in the plane, then*

$$\frac{g_D(x, y)g_D(y, z)}{g_D(x, z)} \leq c\big[1 + \log^+(1/|x - y|) + \log^+(1/|y - z|)\big],$$

where c depends only on the diameter of the domain and $\log^+ w = \max(0, \log w)$.

Actually, certain unbounded domains are also allowed, such as domains contained in a strip or domains that have finite area.

If x is a distance 1 from a point y, then the probability that Brownian motion starting at x makes a loop around y before exiting $B(y, 2)$ or hitting $B(y, 1/2)$ is comparable to the probability of hitting $B(y, 1/2)$ before exiting $B(y, 2)$. The key to Theorem 5.1 is to show that the same is true even if in addition we kill Brownian motion on hitting a set K, provided the capacity of K is sufficiently small.

References

[1] R. Bañuelos, *Intrinsic ultracontractivity and eigenvalue estimates for Schrödinger operators*, J. Funct. Anal. **100** (1991), 181–206.

[2] R. Bañuelos, R. F. Bass, and K. Burdzy, *Hölder domains and the boundary Harnack principle*, Duke Math. J. **64** (1991), 195–200.

[3] R. F. Bass, Probabilistic Techniques in Analysis, Springer-Verlag, Heidelberg, 1995.

[4] R. F. Bass and K. Burdzy, *A probabilistic proof of the boundary Harnack principle*, Seminar on Stochastic Processes 1989, 1–16, Birkhäuser, Boston, 1990.

[5] _____ *A boundary Harnack principle for twisted Hölder domains*, Ann. of Math. (2) **134** (1991), 253–276.

[6] _____ *Lifetimes of conditioned diffusions*, Probab. Theory Related Fields **91** (1992), 405–444.

[7] _____ *The Martin boundary in non-Lipschitz domains*, Trans. Amer. Math. Soc. **337** (1993), 361–378.

[8] _____ *On domain monotonicity of the Neumann heat kernel*, J. Funct. Anal. **116** (1993), 215–224.

[9] _____ *The boundary Harnack principle for non-divergence form operators*, J. London Math. Soc. (2), **50** (1994), 157–169.

[10] _____ *Conditioned Brownian motion in planar domains*, Probab. Theory Related Fields **101** (1995), 479–493.

[11] R. F. Bass and P. Hsu, *Some potential theory for reflecting Brownian motion in Hölder and Lipschitz domains*, Ann. Probab. **19** (1991), 486–508.

[12] R. A. Carmona and W. Zheng, *Reflecting Brownian motions and comparison theorems for Neumann heat kernels*, J. Funct. Anal. **123** (1994), 109–128.

[13] I. Chavel, *Heat diffusion in insulated convex domains*, J. London Math. Soc. (2) **34** (1986), 473–478.

[14] M. Cranston, *Lifetime of conditioned Brownian motion in Lipschitz domains*, Z. Wahrsch. **70** (1985), 335–340.

[15] M. Cranston, E. Fabes, and Z. Zhao, *Conditional gauge and potential theory for the Schrödinger operator*, Trans. Amer. Math. Soc. **307** (1988), 171–194.

[16] M. Cranston and T. R. McConnell, *The lifetime of conditioned Brownian motion*, Z. Wahrsch. **65** (1983), 1–11.

[17] B. Davis, *Intrinsic ultracontractivity and the Dirichlet Laplacian*, J. Funct. Anal. **100** (1991), 163–180.

[18] R. A. Hunt and R. L. Wheeden, *Positive harmonic functions on Lipschitz domains*, Trans. Amer. Math. Soc. **132** (1970), 507–527.

[19] W. S. Kendall, *Coupled Brownian motions and partial domain monotonicity for the Neumann heat kernel*, J. Funct. Anal. **86** (1989), 226–236.

Interaction and Hierarchy in Measure-Valued Processes

DONALD DAWSON

Department of Mathematics and Statistics
Carleton University
Ottawa, Canada K1S 5B6

1 Introduction

Measure-valued Markov processes include both stochastic particle systems and an important class of stochastic partial differential equations. Examples of these arise in a variety of contexts such as mathematical genetics and reaction diffusion equations in random media. This paper will survey results on the long-time behavior of the class of *critical measure-valued processes*, that is, processes satisfying a mean conservation law. In the long-time limit processes of this type can develop an equilibrium structure or experience nonequilibrium cluster formation. The major objectives of this field are to give a precise description of these phenomena for special classes of critical systems and to develop an approach to classify such systems into broad "universality" classes that share the same large scale behaviors.

We consider a multitype population with *space of types* Θ that is spatially distributed in an *environmental space* E. Let $C(E)$, $\mathcal{B}_+(E)$, and $\mathcal{M}(E)$, denote the continuous functions, nonnegative Borel measurable functions, and Radon measures, respectively, on the topological space E. The system at time $t \in \mathbb{R}_+$ is described by a measure $X(t)$ on $E \times \Theta$. The $\mathcal{M}(E \times \Theta)$-valued Markov process $X(t)$ is said to be *critical* if there exists a conservative Markov semigroup $\{T_t : t \geq 0\}$ with generator $(\mathfrak{G}, D(\mathfrak{G}))$ on $C(E)$ such that

$$E\left(\int_E \int_\Theta \phi(x)\psi(y)X_t(dx,dy)\right) = \int_E \int_\Theta T_t\phi(x)\psi(y)X_0(dx,dy).$$

In other words, for $\phi \in D(\mathfrak{G})$, and $\psi \in C(\Theta)$

$$M_t(\phi \otimes \psi) := \int_E \int_\Theta \phi(x)\psi(y)X_t(dx,dy) - \int_0^t \int_E \int_\Theta \mathfrak{G}\phi(x)\psi(y)X_s(dx,dy)\,ds$$

is a martingale. To complete the description of the process it is necessary to specify the structure of these martingales. Sections 2–4 are devoted to different martingale structures and the long-time behavior of the resulting processes. In the final section we return to the question of universality classes.

Proceedings of the International Congress
of Mathematicians, Zürich, Switzerland 1994
© Birkhäuser Verlag, Basel, Switzerland 1995

2 Finite-type Systems

The simplest setting in which to describe these phenomena is the system of stochastic differential equations for a monotype population

$$dx_i(t) = \sum_{j \in S} q_{i-j}[x_j(t) - x_i(t)] \, dt + \sqrt{2g(x_i(t))} \, dw_i(t) \qquad (2.1)$$

where $x_i(0) \geq 0$, $i \in S$, $E = S$ is a countable abelian group, $\{q_{i-j}\}_{i,j \in S}$ is the Q-matrix of a symmetric random walk on S, and $\{w_i\}$ are independent Brownian motions.

Three important classes are:

(1) *interacting continuous state branching processes*: state space $(\mathbb{R}_+)^S$ and $g(x) = x$,

(2) *interacting Fisher-Wright diffusions*: state space $[0,1]^S$ and $g(x) = x(1-x)$

(3) *nonstationary Anderson parabolic model*: state space $(\mathbb{R}_+)^S$ and $g(x) = x^2$.
(This class will not be discussed below but has been studied in [CM].)

There is in fact a close relationship between the first two classes. To illustrate this we consider the two-type system of stochastic differential equations

$$dx_i(t) = \sum_{j} q_{i-j}[x_j(t) - x_i(t)]dt + \sqrt{x_i(t)}dw_{1,i}(t)$$

$$dy_i(t) = \sum_{j} q_{i-j}[y_j(t) - y_i(t)]dt + \sqrt{y_i(t)}dw_{2,i}(t)$$

where $\{w_{1,i}\}$ and $\{w_{2,i}\}$ are independent. Letting $r_i(t) = x_i(t) + y_i(t)$ and $z_i(t) := \frac{x_i(t)}{x_i(t)+y_i(t)}$ we get

$$dr_i(t) = \sum_{j} q_{i-j}[r_j(t) - r_i(t)]dt + \sqrt{r_i(t)}d\tilde{w}_{1,i}(t)$$

$$dz_i(t) = \frac{1}{r_i(t)} \sum_{j} q_{i-j}r_j(t)[z_j(t) - z_i(t)]dt + \sqrt{\frac{z_i(t)(1 - z_i(t))}{r_i(t)}}d\tilde{w}_{2,i}(t) \,.$$

The collection $\{r_i(t)\}$ is an interacting system of critical continuous state branching (CCSB) processes. In the biological setting it is more realistic to assume that there is a locally finite carrying capacity and that at low densities the population is supercritical, which leads to the *stochastic logistic system*

$$dr_i(t) = \sum_{j} q_{i-j}[r_j(t) - r_i(t)]dt + (c_1 r_i(t) - c_2 r_i^2(t))dt + \sqrt{r_i(t)}d\tilde{w}_{1,i}(t) \qquad (2.2)$$

with $c_1, c_2 \geq 0$.

The system (2.2) is not critical. However, if $c_1 = c_2 = c$, $r_i(0) \equiv 1$ in (2.2), and we take the limit $c \to \infty$, we obtain the system conditioned to have $r_i(t) = 1 \, \forall i$ and $t \geq 0$. Under this conditioning the collection $\{z_i(t)\}$ forms an interacting system of Fisher-Wright diffusions, which is critical. This means that the two species model constrained to have constant total population leads to the Fisher-Wright system of population genetics (cf. [EM] for the infinite-type analogue that gives rise to interacting Fleming-Viot processes).

We begin the discussion of the long-time behavior of these systems of stochastic differential equations with the following result.

THEOREM 2.1 [CG]. *Assume that*

$$g : [0, 1] \to [0, 1] \text{ is Lipschitz and } g > 0 \text{ on } (0, 1). \tag{2.3}$$

Assume that the initial measure μ is shift ergodic and has density $\theta = \int x_0 \, d\mu$.

(a) *If $\{q_i\}$ is transient, then the law $\mathcal{L}(x(t))$ converges weakly as $t \to \infty$ to the stationary measure ν_θ, which is translation invariant mixing and has density θ.*

(b) *If $\{q_i\}$ is recurrent, then $\mathcal{L}(x(t)) \Rightarrow (1 - \theta)\delta_0 + \theta\delta_1$ as $t \to \infty$.*

Thus, the *global equilibrium* (a) versus *local clustering* (b) dichotomy holds for the entire *universality class* governed by (2.1) and (2.3).

The persistence-extinction phase transition behavior for the spde model in $\mathbb{R}^1$ corresponding to the logistic system (2.2) has been obtained in [MT] (in particular, for fixed $c_2 > 0$, extinction for small c_1 and persistence for large c_1). The next section will be devoted to this for a large class of CCSB systems.

3 Persistence and Extinction in Critical Branching Systems

An important property of continuous state branching systems is that their laws are infinitely divisible, which opens up the possibility of using the many tools of infinitely divisible processes and measures. In this section we survey results on the long-time behavior of spatially homogeneous monotype branching systems in $\mathbb{R}^d$. In addition these processes arise as limits of branching particle systems, and the analogous questions for the latter have been intensively studied in recent years (cf. [GW]). The class of processes we consider are characterized by two indices, one associated to the mean semigroup T_t, and the other to the continuous branching mechanism. To be precise, we take T_t to be the semigroup of a symmetric stable process of index α or a d-dimensional Brownian motion ($\alpha = 2$) with generator Δ_α, and the branching mechanism to be $(1+\beta)$ CCSB. In fact to each of these can be associated a natural universality class (where the associated mechanisms are in the domains of attraction of these), but we restrict ourselves to the canonical objects. Because these processes are infinitely divisible, they can be characterized by their Laplace functionals. In particular the (α, d, β)-superprocess has transition Laplace functional

$$E\left[\exp\left(-\int \phi(x)X_t(dx)\right)\Big|X(0) = \mu\right] = \exp\left(-\int u(t, x)\mu(dx)\right)$$

for every $\phi \in \mathcal{B}_+(\mathbb{R}^d)$. The function $u(t, x)$ is the solution of the initial value problem

$$\frac{\partial v}{\partial t} = \Delta_\alpha v - \gamma v^{1+\beta}, \quad v(0, x) = \phi(x) \tag{3.1}$$

with $0 < \beta \leq 1$, $0 < \alpha \leq 2$. These processes can be defined with state space $\mathcal{M}_p(\mathbb{R}^d) = \{\mu : \int((1 + |x|)^2)^{-p}\mu(dx) < \infty\}$, with $p \in (d/2, (d + \alpha)/2)$. Note that Lebesgue measure λ belongs to this class and is invariant for the symmetric stable and Brownian semigroups. The following persistence-extinction dichotomy describes the long time behavior of these processes.

THEOREM 3.1 [D1], [K], [DP], [GW]. *The (α, d, β) process X_t with initial measure $X_0 = \theta\lambda$, $\theta > 0$, converges in distribution as $t \to \infty$ to an equilibrium random field X_∞ with law ν_θ. There exists a critical dimension $d_c = \frac{\alpha}{\beta}$ such that*

(1) *if $d \leq d_c$, then $X_\infty = 0$ with probability one (local extinction)*
(2) *if $d > d_c$, then $E[X_\infty] = \theta\lambda$ (persistence).*

In fact in this setting local extinction is equivalent to $\|u(t)\|_1 \to 0$ as $t \to \infty$ and this reduces to an analytical estimate.

This leaves open the possibility that there could be a locally finite invariant measure with infinite mean measure in low dimensions. However Bramson, Cox, and Greven [BCG] proved that δ_0 is the only invariant measure in dimensions $d = 1, 2$ if $\beta = 1$. They also recently proved in dimensions $d \geq 3$ that $\{\nu_\theta, \ \theta \in [0, \infty)\}$ coincides with the set of all extremal invariant measures.

In order to describe the nature of the cluster formation consider the space-time-mass rescaling

$$X_t^K(A) := K^{-\frac{1}{\beta}}X_{Kt}(K^{1/d\beta}A).$$

THEOREM 3.2 [DF]. *Let $d < \alpha/\beta$. Then X_K converges in distribution to the pure atomic process $\{X_t^0\}_{t\geq 0}$ in which X_t^0 is Poisson with intensity $(\gamma\beta t)^{-1/\beta}X(0)$ and the mass of each atom evolves according to a continuous state branching.*

Roughly speaking, "at time K there are clumps of mass of order $K^{1/\beta}$ with interclump distance of order $K^{1/\beta}$". To describe ergodic behavior consider the *occupation time process* $Y_t := \int_0^t X_s \, ds$.

THEOREM 3.3 [I], [FG]. *For $d > \frac{\alpha}{\beta}$ with probability 1, $\lim_{t\to\infty} t^{-1}Y_t = \lambda$ (in the vague topology). For $d = \frac{\alpha}{\beta}$, $\lim_{t\to\infty} t^{-1}Y_t = \xi\lambda$ where ξ is a nondegenerate infinitely divisible random variable with mean one.*

Theorems 3.1 and 3.3 imply that in the critical dimension $d = \frac{\alpha}{\beta}$ the mass of each bounded open set goes to zero in probability but nevertheless is visited by clumps of mass at arbitrarily large times.

Let $\phi \geq 0$ be Hölder continuous with compact support and $\int \phi(x) \, dx = 1$. Then Iscoe and Lee [IL] and Lee [L] have obtained the following large deviation result for the $(2, d, 1)$-superprocess:

$$- \lim_{t\to\infty} a(t, d)^{-1} \log P\left\{\frac{1}{t} \int \phi(x)Y_t(dx) > c\right\} > 0$$

when c is greater than but close to 1 where $a(t, 3) = t^{1/2}$, $a(t, 4) = t/\log t$, and $a(t, d) = t$ for $d \geq 5$. Deuschel and Wang [DW] recently obtained the corresponding strong large deviation principle in $d = 3$ and a weak large deviation principle in $d = 4$. This initially surprising large deviation behavior of the occupation time measure in dimensions $d = 3, 4$ is reminiscent of that of independent random walks or Brownian motions in dimensions $d = 1, 2$ and suggests some type of null recurrence phenomenon, which we will make explicit in Theorem 3.5.

An important tool in developing an understanding of these long-time phenomena is the *historical process* introduced in [DP], [DYN], and [LG]. The historical process is an enriched version of the basic measure-valued process that carries with it information on the genealogical and migrational history of the population. For simplicity we describe it only in the case $\alpha = 2$, $\beta = 1$. In order to describe the equilibrium measure we consider the process with time parameter in $(-\infty, \infty)$. Let $\mathcal{C}_{s,t} = C([s, t), \mathbb{R}^d)$, $\mathcal{C} = \mathcal{C}_{-\infty, \infty}$, and $\mathcal{M}(\mathcal{C}_{s,t})$ denote the space of measures on $\mathcal{C}_{s,t}$. Given the Brownian motion $\{w(t) : t \in \mathbb{R}\}$ with unnormalized entrance law $\{\mu_s\}$ we associate the $\mathcal{C}$-valued process $\hat{w}$ defined by $\hat{w}(t) := \{w(s \wedge t)\}_{s \in (-\infty, \infty)}$ and the associated nonhomogeneous semigroup $S_{s,t}f := E(f(\hat{w}(t))|\hat{w}(s))$, $t \geq s$. The *historical process* is an $M(\mathcal{C})$-valued time inhomogeneous Markov process with transition Laplace functional, $\phi \in \mathcal{B}_+(\mathcal{C})$

$$E\left(\exp(-\int \phi(y) H_t(dy))\,\middle|\, H_s\right) = \exp\left(-\int V_{s,t}\phi(y) H_s(dy)\right)$$

where

$$V_{s,t}\phi(y) = S_{s,t}\phi(y) - \gamma \int_s^t S_{s,r}((V_{r,t}\phi)^2)\, dr.$$

For $s < u < t$ let $r_{s,u}H_t$ denote the restriction of H_t to $\mathcal{C}_{s,u}$. Then H_t can be represented as a *Poisson random field of clan measures* in $\mathcal{M}(\mathcal{C}_{s,t})$ with Poisson intensity $\frac{1}{\gamma(t-u)}H_a$ and typical clan measure Ξ_t, which can be interpreted as the *descendent population* from an individual alive at time u with history y' and $r_{s,u}\Xi_t = \Xi_t(\mathcal{C})\delta_{y'}$. In fact this clan system can be identified with a time inhomogeneous binary branching historical particle system starting from one particle with history y at time s and branching rate $\frac{1}{\gamma(t-u)}$, $s \leq u < t$ (cf. [DP]).

For $t > s$ let H_t^I denote a realization of the historical process H_t conditioned to stay alive forever starting from a finite initial measure $H_s = \eta_s \in \mathcal{M}(\mathcal{C}_{-\infty,s})$. Then ([RR], [EP], [E])

$$P_{s,\eta_s}(H_t^I \in A) = \langle \eta_s, 1 \rangle^{-1} P_{s,\eta_s}(\mathbf{1}_A(H_t)\langle H_t, 1\rangle)$$

where $\langle \mu, f \rangle := \int f\, d\mu$. The corresponding *normalized Campbell measure* of H_t is then

$$Q_{s,\eta_s}(H_t^I \in A, \xi_t \in B) := \langle \eta_s, 1 \rangle^{-1} P_{s,\eta_s}[\mathbf{1}_A(H_t) H_t(B)].$$

Under Q_{s,η_s}, $\{\xi_t\}_{t \geq s}$ is a Brownian motion with initial law $\langle \eta_s, 1 \rangle^{-1}\eta_s(dy)$ and the conditional immortal clan law, Q_s^ξ of Ξ_t^I, the clan associated to ξ, corresponds to the measure-valued process obtained if ξ "throws off historical process excursions at constant rate" 2γ. More precisely,

$$Q_s^\xi[\exp^{-\langle \Xi_t^I, f \rangle}] = \exp\left(-\int_s^t 2\gamma V_{u,t}f(\xi_u)\, du\right).$$

Let μ_λ^t denote the law of Brownian motion stopped at time t with the unique entrance law having Lebesgue marginals. Let

$$R_{s,t}(A) := \int Q_s^\xi(A)\mu_\lambda^t(d\xi)\,.$$

Then $\lim_{s\to-\infty} R_{s,t} = R_{-\infty,t}$ exists as a locally finite measure if and only if $d \geq 3$. Moreover, $R_{-\infty,t}$ is supported by the set of *infinite clan measures* at time t, namely measures on $\mathcal{C}_{-\infty,t}$ such that any two paths in the support have a common ancestral trajectory. A typical infinite clan containing an individual located at x at time t is constructed by running a Brownian trajectory ξ backwards to time $-\infty$ and then collecting the mass at time t corresponding to $Q_{-\infty}^\xi$.

THEOREM 3.4 [DP]. (a) *Stationarity. Assume that H_s has the infinite law $R_{-\infty,s}$ and $t > s$. Then H_t has law $R_{-\infty,t}$.*
(b) *Equilibrium clan decomposition. The equilibrium random field X_∞ (cf. Theorem 3.1) is an infinitely divisible random measure with canonical measure $\hat{R}(B) = R_{-\infty,0}(\{\pi_0\mu \in B\})$ where $\pi_0\mu(A) := \mu(\{\xi : \xi(0) \in A\})$, $A \in \mathcal{B}(\mathbb{R}^d)$, $\mu \in \mathcal{M}(\mathcal{C})$.*

This leads to a description of the equilibrium dynamics as the motion of a countable collection of infinite clan measures. The recurrence phenomenon alluded to above in dimensions $d = 3, 4$ is made explicit as follows.

THEOREM 3.5 [SW]. *Consider the immortal clan process $\{\Xi_t^I\}$ in dimensions $d \geq 3$. Then*

  (1) Ξ_t^I *populates each ball $B \subset \mathbb{R}^d$ at arbitrarily early and late times if $d = 3, 4$.*
  (2) *If $d \geq 5$, then Ξ_t^I populates a fixed ball over only a finite time horizon.*

The immortal clans also exhibit an important scaling relation. If Ξ_0^I denotes such a clan measure, then $k^{-2}\pi_0\Xi_0^I(kA) \overset{d}{=} \pi_0\Xi_0^I(A)$. Also $\hat{R}(\{\mu : \pi_0\mu(B(0,r)) > 0\}) = cr^{d-2}$. The last two properties also form the key to understanding the long-time behavior of *two level super-Brownian motion*. Wu [W] and Gorostiza, Hochberg, and Wakolbinger [GHW] have established that $d = 4$ is the critical dimension for the stability-clustering dichotomy in this case.

4 Equilibria and Clustering in Stepping Stone Models

In this section we consider the analogue of the interacting system of Fisher-Wright diffusions for the infinite-type case $\Theta = [0,1]$ again constrained to have constant total mass one at each site. The state space is $(\mathcal{M}_1([0,1]))^S$ and the process is denoted by $\{x_\xi\}_{\xi \in S}$.

The mathematical characterization of this system of *interacting Fleming-Viot (FV) processes* is given as the unique solution of a martingale problem with generator L given by (4.1) below. In turn this martingale problem establishes a connection between the probability law of this process and that of a simpler process, called the *dual process*. The dual process $\pi(t)$ is a system of interacting partition elements that has the form $\pi(t) = (\tilde{\pi}_k^n(t), \tilde{\pi}_k(t))$ with

$$\tilde{\pi}_k^n : \{1, \ldots, n\} \to \{1, \ldots, k\} \text{ and } \tilde{\pi}_k : \{1, \ldots, k\} \to S\,.$$

This means that $\tilde{\pi}_k^n$ is a partition of $\{1, \ldots, n\}$ and $\tilde{\pi}_k$ assigns locations to each element of the partition. The process evolves as follows:

(1) the partition elements perform continuous time symmetric random walks on S with rates $q_{\xi - \xi'}$

(2) each pair of partition elements during the period they both reside at an element of S coalesces to the partition element equal to the union of the two partition elements at rate d_0.

In order to relate this process to the interacting FV system we require the family of "test functions"

$$F((f, \pi), ((\mu_\xi)_{\xi \in S})) =$$

$$\int_{[0,1]} \cdots \int_{[0,1]} f(u_{\pi_k^n(1)}, \ldots, u_{\pi_k^n(n)}) \mu_{\tilde{\pi}_k(1)}(du_1) \ldots \mu_{\tilde{\pi}_k(k)}(du_k)$$

where $f \in C([0,1]^n)$.

The generator L of the interacting Fleming-Viot system has the usual form for a diffusion, namely a second order differential operator but in this case involving the derivative $\frac{\delta F}{\delta \mu_\xi}(u) := \frac{d}{d\varepsilon} F(\mu + \varepsilon \cdot \delta_u)|_{\varepsilon=0}$. It is given by

$$LF((\mu_\xi)) = \sum_{\xi, \xi' \in S} q_{\xi - \xi'} \int_{[0,1]} \frac{\delta F}{\delta \mu_\xi}(u)(\mu_{\xi'}(du) - \mu_\xi(du))$$

$$\hspace{8cm} (4.1)$$

$$+ \frac{d_0}{2} \sum_{\xi \in S} \int_{[0,1]} \int_{[0,1]} \frac{\delta^2 F}{\delta \mu_\xi \delta \mu_\xi}(u, v)[\mu_\xi(du)\delta_u(dv) - \mu_\xi(du)\mu_\xi(dv)].$$

The first term corresponds to spatial migration and the second to continuous sampling. In the case $S = \{1\}$, the process is the *Fleming-Viot diffusion* with no mutation.

For functions F that belong to the special class of test functions defined above it can be shown that

$$LF((f, \pi), ((\mu_\xi)_{\xi \in S})) = KF((f, \pi), ((\mu_\xi)_{\xi \in S}))$$

where K is the generator of the partition-valued dual $\pi(t)$ with $\pi(0) = (\tilde{\pi}_n^n, \tilde{\pi}_n)$. On the left-hand side the operator L acts on the variables (μ_ξ) and on the right-hand side the operator K acts on the variable π.

It then follows (cf. [D2, Section 5.5]) that

$$E_{(\mu_\xi)_{\xi \in S}}(F((f, \pi(0)), X(t))) = E_{\pi(0)}(F((f, \pi(t)), X(0)))$$

where $\pi(0) = (\tilde{\pi}_n^n(0), \tilde{\pi}_n(0))$ and $X(0) = ((\mu_\xi)_{\xi \in S})$.

THEOREM 4.1. *If $\{q_\xi\}$ is transient, then the process X has a one parameter family of nontrivial ergodic invariant measures $(\{\nu_\mu\}_{\{\mu \in M_1(E)\}})$ such that ν_μ has single* *site mean measure μ. If $\{q_\xi\}$ is recurrent, then the extreme invariant measures are $\delta_{\delta_a}, a \in [0,1]$, that is, there is global fixation.*

Refer to [S] for the case of finitely many alleles, and [DGV] for the measure-valued case. The proofs are based on the coalescing random walk structure of the dual process.

The study of long-time behavior in more detail has been carried out in two cases. The first involves simple random walk in the case $S = \mathbb{Z}^d$. In the second case, $S = \Omega_N$, the *hierarchical group* introduced in [SF] and defined by

$$\Omega_N = ((\xi_1, \xi_2, \dots) : \xi_i \in \{0, \dots, N-1\}, \xi_i = 0, \text{ a.a. } i).$$

The *hierarchical distance* is a metric on Ω_N defined by $d(\xi, \xi') = \max\{i : \xi_i \neq \xi_i'\}$ and the transition rates for the random walk are given by

$$q_{\xi - \xi'}^N = \sum_{k=1}^{\infty} \frac{c_{k-1}}{N^{2k-1}} 1\{d(\xi', \xi) \leq k\}. \tag{4.2}$$

Here c_k / N^k represents the rate of jumping $k+1$ levels in the hierarchy. If $c_k = r^k$, then the random walk is transient if $r > 1$ and recurrent if $r \leq 1$. The subset of Ω_N given by $\{\xi' : d(\xi, \xi') \leq k\}$ is called the *k-block* containing ξ and the *k-block average* is

$$X_{\xi,k}^N(t) := N^{-k} \sum_{\xi':d(\xi,\xi')\leq k} x_{\xi'}^N(t) \text{ and } X_{\xi,0}^N := x_\xi^N \quad \forall \xi \in \Omega_N. \tag{4.3}$$

The mechanism of diffusive clustering was first discovered by Cox and Griffeath [CGR] in the case of the voter model on $\mathbb{Z}^d$ in the critical dimension $d = 2$. The analogue for the system of equations of the form (2.1) on Ω_N in the case $c_k \equiv 1$ was recently established by Fleischmann and Greven [FGR] and in the hierarchical mean field limit in [DG]. For the interacting Fleming-Viot system on Ω_N with $c_k \equiv 1$ and mean measure θ this becomes

$$\{X_{0,[\alpha t]}^N(N^t)\}_{\alpha \geq 0} \xrightarrow[t \to \infty]{fdd} \{Z(0 \vee \log \frac{1}{\alpha})\}_{\alpha \geq 0}$$

where Z is a Fleming-Viot process with no mutation and initial measure θ.

5 Equilibria and Universality in the Hierarchical Mean Field Limit

By letting N tend to ∞ in (4.3) the time scales for the block averages of different sizes completely separate making possible a rigorous multiple time scale analysis. The first step involves the consideration of the *k-block averages* in the *natural time scale* T_k^N, $X_{\xi,k}^N(T_k^N)$ where $\frac{T_k^N}{N^k} \to \infty$, $\frac{T_k^N}{N^{k+1}} \to 0$. In the *hierarchical mean field limit* (HMFL),

$$\{X_{\xi,j}^N(T_k^N)\}_{\{j=k+1,k,\dots,1,0\}} \xrightarrow[N\to\infty]{} \{\Xi_j^k\}_{\{j=k+1,k,\dots,1,0\}}$$

where $\{\Xi_j^k\}$ is a reverse time inhomogeneous $M_1[0,1]$-valued Markov chain called the *interaction Markov chain* with $\Xi_{k+1}^k = \theta$ and transition kernel $\Gamma_{\theta_{j+1}}^j(d\theta_j) := P(\Xi_j^k \in d\theta_j | \Xi_{j+1}^k = \theta_{j+1})$ (which depends on $c_0, \ldots, c_j$ but not k). For $k > j$, we have the $(k - j + 1)$ step transition on $M_1[0,1]$

$$\mu_\theta^{k,j}(d\theta_j) = \int_{M_1[0,1]} \cdots \int_{M_1[0,1]} \Gamma_\theta^k(d\theta_k)\Gamma_{\theta_k}^{k-1}(d\theta_{k-1}) \ldots \Gamma_{\theta_{j+1}}^j(d\theta_j).$$

A hierarchical mean field *global equilibrium* $\{\Xi_j^\infty\}$ is defined by an entrance law for this Markov chain which describes the macroscopic behavior in exponential (or superpolynomial) time scales.

GLOBAL EQUILIBRIUM AND SPATIAL ERGODIC THEOREM [DGV].

(1) *There is a one parameter family* $\{\mu_\theta^{\infty,j}\}$ *with* $\theta \in M_1[0,1]$ *of nontrivial extremal entrance laws for the interaction chain (global equilibria) if and only if the transience condition for the* $\{q_\xi\}$ *is satisfied. Otherwise the only extremal entrance laws are* $\{\delta_{\delta_a}, a \in [0,1]\}$.

(2) *Under the global equilibrium with mean measure* θ, Ξ_j^∞ *is pure atomic and* $\lim_{j\to\infty} \Xi_j^\infty = \theta$ *a.s.*

(3) *In exponential time scales* $T^N = E^N$, *there is convergence to equilibrium:*

$$\{X_{\xi,j}^N(E^N)\}_{\{j=\ldots,2,1,0\}} \underset{N\to\infty}{\Longrightarrow} \{\Xi_j^k\}_{\{j\in=\ldots,2,1,0\}}.$$

In the representation $\Xi_j^\infty = \sum_k m_j(k)\delta_{x_k}$, $m_j(k)$ denotes the mass of the immortal type (clan) $x_k \in [0,1]$ in the j-block containing a fixed ξ. The asymptotic behavior of $\{m_j(k)\}$ as $j \to \infty$ has been studied in [DGV].

We have seen above in both measure-valued CCSB systems and interacting FV systems the formation of isolated clusters in low dimensions, diffusive clustering in critical dimensions, and global equilibria in high dimensions. As mentioned in Section 1 some phenomena can be established for much broader classes. In fact we expect that there are large universality classes that share the same large scale behavior. There are significant difficulties in carrying out this research program for systems that cannot be studied directly with Laplace functionals or dual processes, but some progress has recently been made. The latter is based on renormalization group ideas from statistical physics. We will illustrate this by returning to the system (2.1), this time on the hierarchical group, and then considering the hierarchical mean field limit with $q_{\xi-\xi'}^N$ as in (4.2) with $c_k \equiv 1$. Then for each k, the real-valued processes $X_{0,k}^N(N^k t)$ converge as $N \to \infty$ to a diffusion with generator

$$g_k(x)\frac{\partial^2}{\partial x^2} + (\theta - x)\frac{\partial}{\partial x}$$

$g_0 = g$, and

$$g_{(k+1)}(\theta) = \frac{\int_0^1 \exp\left[-\int_\theta^x \frac{y-\theta}{g_k(y)}dy\right] dx}{\int_0^1 g_k^{-1}(x) \exp\left[-\int_\theta^x \frac{y-\theta}{g_k(y)}dy\right] dx}.$$

THEOREM [DG], [BCGH]. *Let g satisfy the conditions of Theorem 2.1. Then*

$$\lim_{k \to \infty} k g_k(\theta) = \theta(1 - \theta) \quad \text{uniformly on } [0, 1].$$

If in addition $\liminf_{x \downarrow 0} x^{-2} g(x) > 0$ *and* $\liminf_{x \uparrow 1} (1 - x)^{-2} g(x) > 0$, *then there exist* $0 < c_g < C_g < \infty$ *such that for all sufficiently large k*

$$\frac{c_g}{k} < \sup_{\theta \in (0,1)} \left[\frac{|k g_k(\theta) - \theta(1 - \theta)|}{\theta(1 - \theta)} \right] < \frac{C_g}{k}.$$

This result gives a precise meaning to the *Fisher-Wright universality class* in this setting and perhaps can serve as a prototype for the study of other universality questions. For example, an infinite-type analogue has recently been obtained in [DGV].

References

[BCGH] J.-B. Baillon, Ph. Clément, A. Greven, and F. den Hollander, *On the attracting orbit of a nonlinear transformation arising from renormalization of hierarchically interacting diffusions*, Canad. J. Math. (1994).

[BCG]   M. Bramson, J. T. Cox, and A. Greven, *Ergodicity of a critical spatial branching process in low dimensions*, Ann. Probab. **21** (1993), 1946–1957.

[CM]   R. A. Carmona and S. A. Molchanov, *Parabolic Anderson problem and intermittency*, Mem. Amer. Math. Soc. **518** (1994).

[CG]   J. T. Cox and A. Greven, *Ergodic theorems for infinite systems of interacting diffusions*, Ann. Probab. (1994).

[CGR]   J. T. Cox and D. Griffeath, *Diffusive clustering in the two dimensional voter model*, Ann. Probab. **14** (1986), 347–370.

[D1]   D. A. Dawson, *The critical measure diffusion*, Z. Wahr. Verw. Geb. **40** (1977), 125–145.

[D2]   D. A. Dawson, *Measure-valued Markov processes*, Ecole d'Été de Probabilités de Saint Flour XXI, Lecture Notes in Math., vol. 1541, P.L. Hennequin, ed., Springer-Verlag, Berlin and New York, 1993, pp. 1–260.

[DF]   D. A. Dawson and K. Fleischmann, *Strong clumping of critical space-time branching models in subcritical dimension*, Stoch. Proc . Appl. **30** (1988), 193–208.

[DG]   D. A. Dawson and A. Greven, *Hierarchical models of interacting diffusions: Multiple time scale phenomena, phase transition and pattern of cluster formation*, Probab. Theory Related Fields **96** (1993), 435–473.

[DP]   D. A. Dawson and E. A. Perkins, *Historical Processes*, Mem. Amer. Math. Soc. **454** (1991), 1–179.

[DGV]   D. A. Dawson, J. Vaillancourt, and A. Greven, *Equilibria and quasiequilibria for infinite collections of interacting Fleming-Viot processes*, Trans. Amer. Math. Soc., to appear.

[DW]   J.-D. Deuschel and K. Wang, *Large deviations for occupation time functionals of branching Brownian particles and super-Brownian motion*, preprint (1994).

[DYN]   E. B. Dynkin, *Branching particle systems and superprocesses*, Ann. Probab. **19** (1991), 1157–1194.

[EM]   A. Etheridge and P. March, *A note of superprocesses*, Probab. Theory Related Fields **89** (1991), 141–147.

[E]   S. N. Evans, *Two representations of a conditioned superprocess*, J. Royal Soc. Edinburgh **123A** (1993), 959–971.

[EP]   S. N. Evans and E. A. Perkins, *Measure-valued Markov branching processes conditioned on non-extinction*, Israel J. Math. **71** (1990), 329–337.

[FG]   K. Fleischmann and J. Gärtner, *Occupation time processes at a critical point*, Math. Nachr. **125** (1986), 275–290.

[FGR]   K. Fleischmann and A. Greven, *Diffusive clustering in an infinite system of hierarchically interacting diffusions*, Probab. Theory Related Fields **98** (1994), 517–566.

[GHW]   L. G. Gorostiza, K. J. Hochberg, and A. Wakolbinger, *Persistence of a critical super-2 process*, preprint (1993).

[GW]   L. G. Gorostiza and A. Wakolbinger, *Long time behavior of critical branching particle systems and applications*, Measure-valued processes stochastic partial differential equations, D. A. Dawson, ed., CRM Proceedings and Lecture Notes, vol. 5, Amer. Math. Soc., Providence, RI, 1994, pp. 119–138.

[I]   I. Iscoe, *Ergodic theory and local occupation time for measure-valued Brownian motion*, Stochastics **18** (1986), 197–243.

[IL]   I. Iscoe and T.-Y. Lee, *Large deviations for occupation times of measure-valued branching Brownian motions*, Stochastics and Stochastics Reports **45** (1993), 177–209.

[K]   O. Kallenberg, *Stability of critical cluster fields*, Math. Nachr. **77** (1977), 7–43.

[L]   T.-Y. Lee, *Some limit theorems for super-Brownian motion and semilinear differential equations*, Ann. Probab. **21** (1993), 979–995.

[LG]   J.-F. Le Gall, *Brownian excursions trees and measure-valued branching processes*, Ann. Probab. **19** (1991), 1399–1439.

[MT]   C. Mueller and R. Tribe, *A stochastic pde arising as the limit of a long range contact process and its phase transition*, preprint (1994).

[RR]   S. Roelly-Coppoletta and A. Rouault, *Processus de Dawson-Watanabe conditionné par les futurs lointains*, C.R. Acad. Sci. Paris vol. **309** Série I (1989), 867–872.

[SF]   S. Sawyer and J. Felsenstein, *Isolation by distance in a hierarchically clustered population*, J. Appl. Probab. **20** (1983), 1–10.

[S]   T. Shiga, *Ergodic theorems and exponential decay of sample paths for certain interacting diffusion systems*, Osaka J. Math. **29** (1992), 789–807.

[SW]   A. Stöckl and A. Wakolbinger, *On clan recurrence and transience in time stationary branching Brownian particles systems*, Measure-valued processes stochastic partial differential equations, D. A. Dawson, ed., CRM Proceedings and Lecture Notes, vol. 5, Amer. Math. Soc., Providence, RI, 1994, pp. 213–220.

[W]   Y. Wu, *Asymptotic behavior of the two level measure branching process*, Ann. Probab. **22** (1994).

Abstract Statistical Estimation and Modern Harmonic Analysis

DAVID L. DONOHO

Department of Statistics, Stanford University
Berkeley, CA 94708, USA

1. Nonparametric Estimation

Suppose that t_i are equispaced points in the unit interval $t_i = i/n$, and we observe

$$y_i = f(t_i) + \sigma z_i, \quad i = 1, \ldots, n, \tag{1}$$

where the z_i are i.i.d. $N(0,1)$. Our goal is to recover the object f from these noisy observations. In order to do so, we must know something about f (otherwise we have n observations and $2n$ unknowns). In the branch of statistics called *nonparametric regression*, it is traditional to assume quantitative smoothness information about f, often of the form $f \in \mathcal{F}$, where $\mathcal{F}$ is a ball in a functional class, for example an L^2-Sobolev ball $\{f : \|f^{(m)}\|_{L^2} \leq C\}$. Performance is then measured by considering the minimax risk

$$M(n, \mathcal{F}) = \min_{\hat{f}(\cdot)} \max_{f \in \mathcal{F}} E\|\hat{f}(\mathbf{y}^{(n)}) - f\|_{L^2(T)}^2. \tag{2}$$

There is a considerable body of literature in the field of mathematical statistics to evaluate the minimax risk and to describe optimal and near-optimal estimators $\hat{f}$ under various assumptions — the articles [2], [22], [30] are good starting points. This literature focuses on the question: What is the best way to recover f if all we know is that f has certain smoothness properties?

In the author's view, modern harmonic analysis makes it possible to recast the problems of this literature in a more modular form, separating out key results into two components, one falling in the domain of harmonic analysis and one in the domain of statistical decision theory. This separation makes it possible for statisticians to avoid re-inventing the wheel; for solving the harmonic analysis part of their question they can fully exploit recent advances from modern harmonic analysis, rather than doing an elementary and possibly inadequate job from scratch. This separation also makes it possible for statisticians to do what they do best, and which no-one else will do for them: namely, solve problems of statistical decision theory. In the author's opinion, this separation will also suggest new questions in statistics; for example, if mathematical statisticians are not making use of all the tools of modern harmonic analysis, why not? Is this a sign that there are new problems they could be attacking and are not yet doing so? And so on.

Proceedings of the International Congress
of Mathematicians, Zürich, Switzerland 1994
© Birkhäuser Verlag, Basel, Switzerland 1995

2. Abstract Statistical Estimation

To connect the concrete problem (2) with deeper mathematical questions, we discuss a continuum model: the *White Noise Experiment* [23]. In a standard example of this, the object we would like to recover, f, is an unknown function on the index set T (e.g. $[0,1]$) and we have observations

$$Y^\epsilon(dt) = f(t)dt + \epsilon W(dt), \qquad t \in T. \tag{3}$$

Here W is a standard Wiener process on T, and ϵ a formal noise level parameter.

A typical minimax problem in this abstract model is of the form

$$M^*(\epsilon, \mathcal{F}) = \min_{\hat{f}(\cdot)} \max_{f \in \mathcal{F}} E\|\hat{f}(Y^\epsilon) - f\|^2_{L^2(T)}. \tag{4}$$

Here f is known only to lie in $\mathcal{F}$, a class of functions, the same type of class as in (2).

The model (3) and the model (1) are closely related. Indeed, when $\mathcal{F}$ is the same in both models, and supposing that $\mathcal{F}$ is nice enough, for example a subset of Hölder$(1/2 + \eta, C)$ for some $\eta > 0$, we generally have [3]

$$M(n, \mathcal{F}) \sim M^*(\sigma/\sqrt{n}, \mathcal{F}), \quad n \to \infty,$$

which says in some sense that problem (4) at noise level $\epsilon_n = \sigma/\sqrt{n}$ is essentially the same as the problem (2). More importantly, if the minimax risk $M^*(\epsilon_n, \mathcal{F})$ in the white noise model is attained by the use of a simple estimator $\hat{f}^{(\epsilon)}$, then generally one has an induced estimator attaining the minimax risk in the sampled-data model. So by studying the continuum model one learns enough to "solve" the finite-sample model.

2.1. The Abstract Normal Mean

The white noise model is equivalent to a type of infinite-dimensional normal mean problem. To see this, suppose we are given an orthogonal basis $(\phi_i)_{i=0}^\infty$, a C.O.N.S. for $L^2(T)$. We define $\theta_i = \langle f, \phi_i \rangle$, and $y_i = \int_T \phi_i(t) Y^\epsilon(dt)$, and $z_i = \int_T \phi_i(t) W(dt)$. Using this we can rewrite (3) as

$$y_i = \theta_i + \epsilon \cdot z_i, \qquad i = 1, 2, 3, \dots, \tag{5}$$

where $\theta = (\theta_i)$ is a vector in sequence space, (z_i) is a Gaussian white noise (i.i.d. $N(0,1)$), and ϵ is a formal noise level parameter, taking the same value as in the continuum model. Moreover, the orthogonality of the ϕ_i gives the isometry $\|\hat{f} - f\|^2_{L^2(T)} = \|\hat{\theta} - \theta\|^2_{\ell^2}$, where $\hat{f} \sim \sum_i \hat{\theta}_i \phi_i$. Hence, letting $\Theta = \Theta(\mathcal{F})$ denote the collection of sequences arising as coefficients of members of $\mathcal{F}$, (4) corresponds exactly to

$$M^*(\epsilon, \Theta) = \min_{\hat{\theta}(\cdot)} \max_{\theta \in \Theta} E\|\hat{\theta}(y) - \theta\|^2_2. \tag{6}$$

In fact, in some cases Θ happens to have a nice description. For example, if T is the circle $[0, 2\pi)$, and $\mathcal{F} = \{f : \|f^{(m)}\|_{L^2} \le C\}$, then if we choose ϕ_i as the usual real sine and cosine basis of $L^2[0, 2\pi)$, Θ turns out to be an infinite-dimensional ellipsoid $\{\sum_i a_i \theta_i^2 \le C^2\}$, for appropriate a_i. In short, the minimax problem (4) is transformed, quite literally by *harmonic* analysis, into the problem of estimating a normal mean when that mean is known to lie in an ellipsoid in ℓ^2.

2.2.   The Bounded Normal Mean   For certain special sets Θ — those with a nice geometry — we know how to obtain minimax and nearly minimax estimators, thanks to the efforts of researchers who have studied the bounded normal mean problem. Interesting cases that have been studied include the following:

- ℓ^p-balls: $\Theta = \Theta_p(C) = \{\theta : \|\theta\|_{\ell^p_d} \leq C\}$.
- Hyper-Rectangles: $\Theta = \Theta_\infty(\tau) = \{\theta : |\theta_i| \leq \tau_i\}$.
- Ellipsoids: $\Theta = \Theta_2(a) = \{\theta : \sum_i |\theta_i|^2 a_i \leq C\}$.
- ℓ^p-bodies: $\Theta = \Theta_p(a) = \{\theta : \sum_i |\theta_i|^p a_i \leq C\}$.
- Hyper-Crosses: $\Theta = \Theta_0(k,d) = \{\theta : \#\{i : |\theta_i| \neq 0\} \leq k\}$.
- Cartesian Products of the above.

In addition to studies of the minimax risk over such sets Θ, there have also been efforts to study the minimax linear risk

$$M_L^*(\epsilon, \Theta) = \min_{\hat{\theta}(\cdot) \text{ linear}} \max_{\Theta} E\|\hat{\theta}(\mathbf{y}) - \theta\|_2^2 \qquad (7)$$

and to compare this with the minimax risk.

In these cases, there are a variety of "soft" results characterizing minimax estimators and identifying nearly minimax estimators. For example, minimax estimators are typically "shrinkers", satisfying $\|\nabla \cdot \hat{\theta}(\mathbf{y})\|_2 < 1$. There are also quantitative results identifying behaviors of minimax risk in an asymptotic sense, as $\epsilon \to 0$. Two specific cases seem to occur.

Case (a):   Linear estimators are nearly minimax. This means that for an appropriate diagonal affine transformation $\hat{\theta} = A\mathbf{y} + b$ we have a risk close to the minimax risk, and

$$M_L^*(\epsilon, \Theta) < C \cdot M^*(\epsilon, \Theta),$$

with C small (for example, with ellipsoids $C \leq$ the Ibragimov-Khas'minskii constant $\mu^* \approx 5/4$, uniformly over all $\epsilon > 0$ [14]).

Case (b):   Linear estimators perform badly — C cannot be effectively controlled independently of ϵ and the dimension d. In those cases, coordinatewise *thresholding* estimators behave well; i.e. estimators of the form

$$\hat{\theta}_i = (|y_i| - \lambda_i)_+ \cdot \text{sgn}(y_i) \qquad (8)$$

work well [14], [8], [9].

(Note for Banach Spacers: the division into cases where linear estimators work and don't work has to do with the geometry of the set Θ, and interacts with the notion of 2-convex and 2-concave sets appearing in Lindenstrauss and Tsafriri [14].)

2.3.   Lifting   These results suggest that one can estimate f in the white noise model (3) by following a commutative diagram:

1. Transform the white noise observations into the abstract observations (5);
2. Apply a minimax or nearly-minimax estimator in the abstract setting;
3. Transform the result back to the original setting.

The composition of these three steps defines an estimator in the white noise model. Moreover, due to the orthogonality of the transform, the performance in the white noise model is the same as the performance in the abstract model. Thus, if $\Theta = \Theta(\mathcal{F})$ is the induced body of coefficient sequences of a class $\mathcal{F}$, and one can find a minimax linear estimator for Θ in the abstract model, the induced estimator $\hat{f}$ is also minimax linear in the white noise model.

We might say that methods of estimating a multivariate normal mean "lift" up to methods of estimating a function from white noise observations.

The only possible catch in the above program comes in the following issue: given a functional class $\mathcal{F}$, we need to find an orthogonal sequence that transforms $\mathcal{F}$ into a set Θ for which minimax estimators are understood. So far, minimax estimators are understood essentially only for very special sets of the type listed earlier; these are special in part because they possess a tremendous degree of symmetry — in particular, orthosymmetry.

3. Examples

So far, we have claimed that if we have a minimax estimation problem $(Y^\epsilon, \mathcal{F})$ and also a complete orthonormal system that transforms a functional ball $\mathcal{F}$ into a geometrically very special set, such as an ellipsoid, then multivariate normal decision theory comes into play, and furnishes us with minimax or near-minimax estimates in the original function estimation problem by "lifting".

One suspects at this point that such situations are quite rare. Fortunately they include almost all the cases of $\mathcal{F}$ that statisticians had previously been studying in connection with models (1) and (3), and many new situations they have only just begun to study.

3.1. Pinsker's Theorem The simplest and most elegant example has already been alluded to, where T is the circle $[0, 2\pi)$, and $\mathcal{F} = \{f : \|f^{(m)}\|_{L^2} \le C\}$. Choose ϕ_i as the usual real sine and cosine basis of $L^2[0, 2\pi)$, Θ turns out to be an infinite-dimensional ellipsoid $\{\sum_i a_i \theta_i^2 \le C^2\}$, for appropriate a_i. The solution of this bounded normal mean problem was found by Pinsker [29], who found the minimax linear shrinkage estimator and showed that this estimator was asymptotically minimax as $\epsilon \to 0$. The "lifting" of this normal mean estimator to the nonparametric regression model is due to Nussbaum [28]. The "lifting" to spectral density estimation and density estimation contexts is due to Efroimovich and Pinsker [15], [16].

3.2. Wavelets As the reader might suspect, outside the case of L^2 smoothness spaces, there is no orthogonal basis that exactly transforms balls $\mathcal{F}$ into simple objects like ellipsoids. However, a key fact about wavelet bases is that, after renorming of standard smoothness spaces, they map coefficient bodies in such spaces into nice geometrical objects [26], [19].

Suppose that T is the circle $[0, 2\pi)$, and $\mathcal{F}$ is either an L^p-Sobolev ball $\{f : \|f^{(m)}\|_{L^p} \le C\}$, $1 < p < \infty$, or, more generally, a Besov ball or a Triebel ball (compare [27], [20]). Choose ϕ_i as the usual periodized Meyer wavelet basis [26],

[27]. Then we can equivalently re-norm the Hölder, Sobolev, Besov, or Triebel space so that the induced body of wavelet coefficients $\Theta = \Theta(\mathcal{F})$ turns out to have a nice shape. For example, a Hölder class maps into a hyper-rectangle, an L^2-Sobolev class maps into an ellipsoid; a ball from the Bump algebra $B_{1,1}^1$ transforms to an ℓ^1-body $\{\sum_i a_i|\theta_i| \leq C\}$, for appropriate a_i. The general Besov body transforms into a set $\|\theta\|_{b_{p,q}^s} \leq C$, where

$$\|\theta\|_{b_{p,q}^s} = \left(\sum_j 2^{jsq} \left(\sum_{k=0}^{2^j-1} |\theta_{2^j+k}|^p \right)^{q/p} \right)^{1/q}$$

and the general Triebel body transforms into a set $\|\theta\|_{f_{p,q}^s} \leq C$

$$\|\theta\|_{f_{p,q}^s} = \left(\int_0^1 \left(\sum_{j,k} 2^{jsq} |\theta_{2^j+k}|^q 1_{[k/2^j,(k+1)/2^j)}(t) \right)^{p/q} dt \right)^{1/p}.$$

Donoho and Johnstone [11] carefully studied the normal mean estimation problems arising from the bodies generated by the wavelet transform and found that in every case, simple thresholding was nearly-minimax, that is minimax to within constant multiples independent of $\epsilon > 0$. "Lifting" this result to the nonparametric regression problem says that for every classical smoothness condition in the Hölder, Sobolev, Triebel, or Besov scale, simple thresholding of wavelet coefficients with appropriate levels is minimax to within constant factors.

3.3. Unconditional Basis The Wavelet and Fourier examples are special cases of a more general picture. They are examples where the corresponding functional classes $\mathcal{F}$ admit orthogonal unconditional bases [27], [20], [21]. Such classes transform isometrically, under the appropriate orthogonal basis into solid, orthosymmetric sets. Nearly-minimax estimation can be treated in such cases by "lifting" simple thresholding estimators using the orthogonal basis.

In detail, suppose we have a functional space F that has an unconditional orthogonal basis, i.e. for a certain special basis (ϕ_i), the norm obeys

$$\left\| \sum_i \pm_i \theta_i \phi_i \right\|_F \leq C \cdot \left\| \sum_i \theta_i \phi_i \right\|_F,$$

for all sequences of signs $\pm_i$. Then the space can be given an equivalent quasi-norm in which functional balls $\mathcal{F} = \{\|f\|_F \leq C\}$ map into coefficient bodies $\Theta = \Theta(\mathcal{F})$ that are solid and orthosymmetric. Here by solid and orthosymmetric, we mean that if $\theta \in \Theta$ and $\xi = (s_i\theta_i)_i$ with $|s_i| \leq 1$ for every i, then $\xi \in \Theta$ as well.

Suppose we have data (3) with $f \in \mathcal{F}$. Then, as before, this is equivalent to data in the abstract model (5) with a priori information $\theta \in \Theta$, with Θ a solid orthosymmetric body. To solve this, suppose, as in [14], [8], [9], [13], that Θ has tail n-widths $d_m^* = \sup\{\sum_{i>m} \theta_i^2 : \theta \in \Theta\}$ that decay as fast as some power of the index, $d_m^* \leq \text{Const} \cdot m^{-\beta}$, for some $\beta > 0$. Construct the estimator $\hat{\theta}^{(\epsilon,\beta)}$, as follows. Set $m(\epsilon) = \lfloor \epsilon^{-(2.001/\beta)} \rfloor$ so that $d_{m(\epsilon)}^* = o(\epsilon^2)$ and set $\lambda(\epsilon) = \sqrt{2\log(m(\epsilon))}$. The estimator sets all coordinates $i > m$ of $\hat{\theta}_i^{(\epsilon,\beta)} = 0$. And, in the first m coordinates, it sets $\hat{\theta}_i^{(\epsilon,\beta)} = \eta_\lambda(y_i)$, where η_λ is the soft threshold nonlinearity (8).

THEOREM 1

$$\sup_{\theta \in \Theta} E\|\hat{\theta}^{\epsilon,\beta} - \theta\|_2^2 \leq O(\log(\epsilon^{-1})) \cdot M^*(\epsilon,\Theta), \qquad \epsilon \to 0.$$

In words, simple thresholding is minimax to within logarithmic factors.

Remark 1. Because the estimator $\hat{\theta}^{\epsilon,\beta}$ depends only very weakly on Θ (through the exponent β), a single estimator can be nearly minimax over a very wide range of solid orthosymmetric sets. "Lifting" this insight via wavelet bases, one can posit smoothness information of the form $f \in \mathcal{F}(s,C)$, where s is a smoothness parameter and C is a scaling parameter, with both s and C unspecified, and have a single estimator nearly minimax over a wide range of s and C. See for example [9], [13].

Remark 2. It is easy to give examples of spaces outside the usual range of smoothness spaces where unconditional structure is present. Modulation spaces in Time-Frequency analysis [17] admit orthogonal unconditional bases [6], [18] based on so-called Wilson bases with Gabor-like elements. In the multivariate setting, the Mixed Smoothness spaces studied by Vladimir Temlyakov are anisotropic and fall outside the usual range of isotropic Besov, Triebel, Sobolev, etc., spaces; for those spaces unconditional bases can be constructed from tensor products of 1-d wavelets having different widths in the different directions. In all such cases, noise removal in the continuum model can be effected by "lifting" the technique of simple thresholding using the orthogonal basis.

At this most general level, the thresholding technique is no longer necessarily a "smoother" — it is more of a "de-noiser", as Coifman likes to say.

4. Adaptation

Much of the activity in modern harmonic analysis has moved away from the fixed-orthogonal-basis concept. Rather than assuming that one fixed orthogonal basis — Fourier, wavelets, or something else — will provide the answer to our questions, one constructs a special expansion *adapted to the problem at hand*. In some cases this might be a special orthogonal expansion; in other cases it might be a non-orthogonal expansion — an "atomic decomposition". Such ideas have been used for example, in decomposing pseudo-differential operators, in study of boundedness of general operators, and even (Coifman tells me) in Fefferman's proof of Carleson's theorem on the almost-everywhere convergence of Fourier series.

A leader in applying this type of thinking to signal processing has been Coifman, who along with Meyer developed libraries of special libraries of orthogonal bases — wavelet packets and cosine packets — based on the tiling of the discrete-time time-frequency plane by assorted Heisenberg tiles, and who worked with Wickerhauser to create algorithms to rapidly select best-adapted orthogonal bases.

Such ideas may be of use in statistics as well; if we have a noisy oscillatory signal and want to remove noise, the issue is one of de-noising rather than smoothing, and the idea of finding a best-adapted time-frequency basis for the purposes of de-noising is an attractive possibility. A conceptual problem, however, in dealing with such possibilities, is the fact that construction of special bases or other expansions adapted to *noisy data* seems intrinsically perilous. One fears that the selected basis might be heavily influenced by the presence of noise. This suggests some new questions for statistics: (a) how to select an orthogonal basis in the presence of noise; and (b) how to develop a theory of what is the *best* performance possible.

A concrete result in this direction has recently been obtained by the author and Johnstone [12]. Suppose we have observations $y_i = s_i + z_i$, $i = 1, \ldots, n$, where (s_i) is signal and (z_i) is i.i.d. Gaussian white noise. Suppose we have available a library $\mathcal{L}$ of orthogonal bases, such as the wavelet packet bases or the cosine packet bases of Coifman and Meyer. We wish to select, adaptively based on the noisy data (y_i), a basis in which best to recover the signal ("de-noising"). Let M_n be the total number of distinct vectors occurring among all bases in the library and let $t_n = \sqrt{2 \log(M_n)}$. (For wavelet packets, $M_n = n \log_2(n)$.)

Let $y[\mathcal{B}]$ denote the original data y transformed into the Basis $\mathcal{B}$. Choose $\lambda > 8$ and set $\Lambda_n = (\lambda \cdot (1 + t_n))^2$. Define the entropy functional

$$\mathcal{E}_\lambda(y, \mathcal{B}) = \sum_i \min(y_i^2[\mathcal{B}], \Lambda_n^2).$$

Let $\hat{\mathcal{B}}$ be the best orthogonal basis according to this entropy:

$$\hat{\mathcal{B}} = \arg\min_{\mathcal{B} \in \mathcal{L}} \mathcal{E}_\lambda(y, \mathcal{B}).$$

Define the hard-threshold nonlinearity $\eta_t(y) = y 1_{\{|y| > t\}}$. In the empirical best basis, apply hard-thresholding with threshold $t = \sqrt{\Lambda_n}$:

$$\hat{s}_i^*[\hat{\mathcal{B}}] = \eta_{\sqrt{\Lambda_n}}(y_i[\hat{\mathcal{B}}]).$$

THEOREM 2 *With probability exceeding* $\pi_n = 1 - e/M_n$,

$$\|\hat{s}^* - s\|_2^2 \leq (1 - 8/\lambda)^{-1} \cdot \Lambda_n \cdot \min_{\mathcal{B} \in \mathcal{L}} E\|\hat{s}_\mathcal{B} - s\|_2^2.$$

Here the minimum is over all ideal procedures working in all bases of the library, i.e. in basis $\mathcal{B}$, $\hat{s}_\mathcal{B}$ is just $y_i[\mathcal{B}] 1_{\{|s_i[\mathcal{B}]| > 1\}}$.

In short, the basis-adaptive estimator achieves a loss within a logarithmic factor of the ideal risk that would be achievable if one had available an oracle that would supply perfect information about the ideal basis in which to de-noise, and also about which coordinates were large or small.

5. Speculation

The way is clear for mathematical statisticians to study new estimation problems — problems outside the realm of traditional "smoothing", problems where estimation involves "de-noising" rather than smoothing. The key is to understand classes $\mathcal{F}$ for which adaptive bases or adaptive atomic decompositions are the best way to proceed. In the fixed basis case we had the concept of unconditional basis; but for the more general setting we need some new concept. The author and Johnstone have been working on problems of edge-preserving de-noising and de-noising of chirps, and have been attempting to find such a substitute in those cases; hopefully harmonic analysts will find something even better.

6. Acknowledgments

The picture I sketched here relies on the work of many others, including my mentors, Peter Bickel and Lucien Le Cam, on the pioneering work of Ildar Ibragimov and Rafail Khas'minskii in the white noise model, and of many researchers in the former Soviet Union who developed the white noise model — Bentkus, Golubev, Lepskii, Korostelev, Kuks and many others. Finally, I must thank my collaborators, Iain Johnstone, Gerard Kerkyacharian, Richard Liu, Brenda MacGibbon, and Dominique Picard.

There are many applications of wavelets in statistics; this article presents only one point of view. The survey article [13] and some of the articles by my coworkers [24], [25] may give useful pointers. [13] also includes references to material about solutions of minimax estimation problems that *don't* use "wavelets ideas" — for example, the important work of Nemirovskii, Polyak, and Tsybakov, of van de Geer, of Birgé and Massart, and of Mammen. Owing to space limitations I cannot say more.

Finally, Raphy Coifman, Albert Cohen, Ingrid Daubechies, Ron DeVore, Brad Lucier, and Yves Meyer have all done a lot to stimulate my thinking through pleasant informal conversations.

References

[1] P. J. Bickel, *Minimax estimation of a normal mean when the parameter space is restricted*, Ann. Statist., **9** (1981), 1301–1309.

[2] J. Bretagnolle and C. Carol-Huber, *Estimation des densités: risque minimax*, Z. Wahrscheinlichkeitstheorie und Verw. Gebiete, **47** (1979), 119–137.

[3] L.D. Brown and M.G. Low, *Asymptotic equivalence of nonparametric regression and white noise* (1990), to appear, Ann. Statist.

[4] R. R. Coifman, Y. Meyer, and M. V. Wickerhauser, *Wavelet analysis and signal processing*, in Wavelets and Their Applications, M. B. Ruskai et al. (eds.), Jones and Bartlett, Boston, MA, 1992, 153–178.

[5] R. R. Coifman and M. V. Wickerhauser, *Entropy-based algorithms for best-basis selection*, IEEE Trans. Inform. Theory, **38** (1992), 713–718.

[6] I. Daubechies, S. Jaffard, and J. L. Journé, *A simple Wilson orthonormal basis with exponential decay*, SIAM J. Math. Anal., **22** (1991), 554–572.

[7] R. A. DeVore, B. Jawerth, and V. Popov, *Compression of Wavelet Decompositions*, Amer. J. Math., **114** (1991), 737–785.

[8] D. L. Donoho, *Unconditional bases are optimal bases for data compression and for statistical estimation*, Appl. Comput. Harmonic Analy., **1** (1993), 100–115.

[9] D. L. Donoho, *De-noising via soft-thresholding*, (1992), to appear, IEEE Trans. Inform. Theory, May 1995.

[10] D. L. Donoho and I. M. Johnstone, *Minimax risk over ℓ_p-balls for ℓ^q-error*, Probab. Theory Related Fields **99** (1994), 277–303.

[11] D. L. Donoho and I. M. Johnstone, *Minimax estimation via wavelet shrinkage*, (1994), to appear, Ann. Statist.

[12] D. L. Donoho and I. M. Johnstone, *Ideal de-noising in a basis chosen from a library of orthonormal bases*, C. R. Acad. Sci. Paris A, **319** (1994), 1317–1322.

[13] D. L. Donoho, I. M. Johnstone, G. Kerkyacharian, and D. Picard, *Wavelet Shrinkage: Asymptopia?* J. Roy. Statist. Soc. **B 57** (1995), 301–369.

[14] D. L. Donoho, R. C. Liu, and K. B. MacGibbon, *Minimax risk over hyperrectangles, and implications*, Ann. Statist., **18** (1990), 1416–1437.

[15] S. Y. Efroimovich and M. S. Pinsker, *Estimation of square-integrable [spectral] density based on a sequence of observations*, Problems Inform. Transmission, (1982), 182–196.

[16] S. Y. Efroimovich and M. S. Pinsker, *Estimation of square-integrable probability density of a random variable*, Problems Inform. Transmission, (1983), 175–189.

[17] H. G. Feichtinger, *Atomic characterizations of modulation spaces through Gabor-type representations*, Rocky Mountain J. Math., **19** (1989), 113–126.

[18] H. G. Feichtinger, K. Gröchenig, and D. Walnut, *Wilson bases and modulation spaces*, Math. Nachr., **155** (1992), 7–17.

[19] M. Frazier and B. Jawerth, *A discrete transform and decomposition of distribution spaces*, J. Functional Anal., **93** (1990), 34–170.

[20] M. Frazier, B. Jawerth, and G. Weiss, *Littlewood-Paley Theory and the study of function spaces*, NSF-CBMS Regional Conf. Ser. in Mathematics, **79** (1991), American Math. Soc., Providence, RI.

[21] K. Gröchenig, *Unconditional bases in translation- and dilation-invariant function spaces on R^n*, in Constructive Theory of Functions Conference Varna 1987, B. Sendov et al., eds., Bulgarian Acad. Sci., 174–183.

[22] I. A. Ibragimov and R. Z. Has'minskii, *Bounds for the risk of nonparametric regression estimates*, Theory Probab. Appl., **27** (1982), 84–99.

[23] I. A. Ibragimov and R. Z. Has'minskii, *On nonparametric estimation of the value of a linear functional in a Gaussian white noise*, Teor. Veroyatnost. i Primenen., **29** (1984), 19–32.

[24] I. M. Johnstone, G. Kerkyacharian, and D. Picard, *Estimation d'une densité de probabilité par méthode d'ondelettes*, C. R. Acad. Sci. Paris (A) **315**(1992), 211–216.

[25] G. Kerkyacharian and D. Picard, *Density estimation in Besov spaces*, Statist. Prob. Lett. **13** (1992), 15–24.

[26] P. G. Lemarié and Y. Meyer, *Ondelettes et bases Hilbertiennes*, Rev. Math. Iberoamericana, **2** (1986), 1–18.

[27] Y. Meyer, Ondelettes, Hermann, Paris, 1990.

[28] M. Nussbaum, *Spline smoothing and asymptotic efficiency in L_2*, Ann. Statist., **13** (1985), 984–997.

[29] M. S. Pinsker, *Optimal filtering of square integrable signals in Gaussian white noise*, Problems Inform. Transmission (1980), 120–133.

[30] C. Stone, *Optimal global rates of convergence for nonparametric estimators*, Ann. Statist., **10** (1982), 1040–1053.

Quasi-Regular Dirichlet Forms and Applications

Zhi-Ming Ma[*]

Probability Laboratory, Institute of Applied Mathematics
Academia Sinica
P.O. Box 2734, Beijing 100080, People's Republic of China

1   Introduction

Since the celebrated result of Fukushima on the connection between regular Dirichlet forms and Hunt processes in 1971, the theory of Dirichlet forms has been rapidly developed and has brought a wide range of applications in various related areas of mathematics and physics (see e.g. the three new books [BH 91], [MR 92], [FOT 94] and references therein). In this survey paper I shall mainly discuss the development of quasi-regular Dirichlet forms and their applications.

Roughly speaking, quasi-regular Dirichlet forms on general state spaces are those Dirichlet forms that are associated with right continuous strong Markov processes. Recently an analytic characterization of quasi-regular Dirichlet forms has been found [AM 91c], [AM92], [AMR92a,b,c] [MR 92], [AMR 93a,b], which has completed the solution of a long-standing open problem of this area. The characterization condition has been proved to be checkable in quite general situations [RS 93], and the framework of quasi-regular Dirichlet forms has been shown to be especially useful in dealing with very singular or infinite-dimensional problems. Applications are e.g. in the study of singular Schrödinger operators [AM 91a,b], loop or path spaces over Riemannian manifolds [ALR 93], [DR 92], infinite-dimensional stochastic differential equations [AR 91], Quantum field theory [AR 90], large deviation theory [Mu 93], non-symmetric Ornstein-Uhlenbeck processes [Sch 93], measure valued processes [ORS 93], and Markov uniqueness for infinite dimensional operators [ARZ 93].

It was also proved that a Dirichlet form is quasi-regular if and only if it is quasi-homeomorphic to a regular Dirichlet form on a locally compact separable metric space [AMR 92c], [MR 92], [CMR 93]. Hence most of the results known for regular Dirichlet forms can be transferred to the quasi-regular case. This transfer method has been used e.g. in the study of absolute continuity of symmetric diffusions [Fi 94], transformation of local Dirichlet forms by supermartingale multiplicative functionals [Ta 94], and measures charging no exceptional sets and corresponding additive functionals [Kuw 94].

Concerning the history, it should be mentioned that the analytic part of the theory of Dirichlet forms goes back to the pioneering papers of Beurling and Deny

---

[*]   Supported in part by NNSFC

[BeDe 58, 59]), whereas the more recent probabilistic part was initiated by the fundamental work of Fukushima [Fu 71a,b, 76, 80] and Silverstein [Si 74] combining regular Dirichlet forms and Hunt processes on locally compact separable metric spaces. At the same time the above-mentioned two authors already touched the study of the connection between nonregular Dirichlet forms and Markov processes by a method of regular representation [Fu 71b], [Si 74]. The case of local Dirichlet forms in infinite-dimensional space, leading to associated diffusion processes, was studied originally by Albeverio and Høegh-Krohn in a rigged Hilbert space setting [AH 75], [AH 77a]. More recently, many people have made contributions in the research direction towards extending the theory to also cover the nonregular case (e.g. [Dy 82], [Le 83], [FiGe], Fi 89]), especially the infinite-dimensional state spaces case (e.g.[Kus 82], [AR 90]). See [MR 92] for more detailed historical information in this connection.

It should be mentioned that recently the area of Dirichlet forms is also very active in other research directions. In what follows for each activity I shall mention only one or two related papers with the hope that the interested reader may find further information from the references therein. In addition to the three new books mentioned at the beginning of this paper, other activities have concerned further extensions of the framework of Dirichlet forms, which includes the study of semi-Dirichlet forms [MOR 93], [AMR 94], positivity preserving forms [MR 93], time dependent (parabolic) Dirichlet forms [O 93], [Sta 94], and noncommutative Dirichlet forms ([Li 94], originated by [AH 77b]), as well as more detailed studies of additive functionals and smooth measures [Fu 94], [FuLe 91]. Further problems that have been discussed are uniqueness problems [ARZ 93], maximum Markovian extensions [Ta 92] and essential self-adjointness of Dirichlet operators [AKoR 93], Dirichlet forms and diffusions on fractals [FuSh 92], [KZ 92], application of Dirichlet forms to pseudo differential operators [Ja 92], [JaHo 94], application of Dirichlet forms to heat kernel estimates and geometry [Stu 94], application of Dirichlet forms to the study of nonlinear differential equations [CWZ 93], [Z 94], application to multiparameter processes [Hi 94], [So 94], and application to the study of Feynman-Kac semigroups [ABM 91], etc. It is fair to say that the area of Dirichlet forms has been extremely active in recent years and it seems that this will even increase in the future.

The remainder of this paper is organized as follows. The analytic characterization of quasi-regular Dirichlet forms will be described in Section 2. Two examples of applications of quasi-regular Dirichlet forms (Schrödinger operator with singular potentials, construction of diffusions on loop spaces) will be discussed in Section 3.

2   Quasi-regular Dirichlet Forms and Markov Processes

For the reader's convenience we begin this section with the definition of Dirichlet forms. For simplicity throughout this section let E be a metrizable Lusin space, i.e., a Borel subset of a Polish space. But we remark that all results represented in this section can be extended to general topological spaces E in an appropriate way (cf. Remark 2.7 below and [MR 92]). Let m be a σ-finite measure on $(E, \mathcal{B})$. Denote by (,) the inner product of the space $L^2(E; m)$. For a (real-valued) bilinear form $\mathcal{E}$ with domain $D(\mathcal{E})$, which is a linear subspace of $L^2(E; m)$, the *symmetric* *part $\tilde{\mathcal{E}}$ of $\mathcal{E}$ is defined by*

$$\tilde{\mathcal{E}}(u,v) = \frac{1}{2}[\mathcal{E}(u,v) + \mathcal{E}(v,u)]; \quad u,v \in D(\mathcal{E}).$$

DEFINITION 2.1 A bilinear form $(\mathcal{E}, D(\mathcal{E}))$ with $D(\mathcal{E})$ dense in $L^2(E;m)$ is called a *coercive closed form* if:

(i) Its symmetric part $(\tilde{\mathcal{E}}, D(\mathcal{E}))$ is positive definite and closed on $L^2(E;m)$, i.e., $D(\mathcal{E})$ equipped with the inner product $\tilde{\mathcal{E}}_\alpha := \tilde{\mathcal{E}} + \alpha(,)$, $\alpha > 0$, is a Hilbert space for some (hence all) $\alpha > 0$.

(ii) (*Sector Condition*) There exists a constant $K > 0$ (called the *continuity constant*) such that

$$|\mathcal{E}_1(u,v)| \le K\mathcal{E}_1(u,u)^{1/2}\mathcal{E}_1(v,v)^{1/2} \qquad \text{for all} \quad u,v \in D(\mathcal{E}). \qquad (2.1)$$

$(\mathcal{E}, D(\mathcal{E}))$ is called a *Dirichlet form* on $L^2(E;m)$ if in addition:

(iii) (*Dirichlet property*) For every $u \in D(\mathcal{E})$, $u^\# := u^+ \wedge 1 \in D(\mathcal{E})$ and $\mathcal{E}(u \pm u^\#, u \mp u^\#) \ge 0$.

Note that (ii) is equivalent to saying that $\mathcal{E}$ is a continuous functional on $D(\mathcal{E}) \times D(\mathcal{E})$ with respect to the product topology induced by the norm $\tilde{\mathcal{E}}_1^{1/2}$. Let $(L, D(L))$ be the *generator* of a coercive form $(\mathcal{E}, D(\mathcal{E}))$ on $L^2(E;m)$, i.e., the unique closed linear operator on $L^2(E;m)$ such that $D(L) \subset D(\mathcal{E})$ and $\mathcal{E}(u,v) = (-Lu, v)$ for all $u \in D(L)$, $v \in D(\mathcal{E})$. Let $(T_t)_{t>0}$ be the strongly continuous contraction semigroup on $L^2(E;m)$ generated by L. We say that $(\mathcal{E}, D(\mathcal{E}))$ has the *semi-Dirichlet* property if in 2.1 (iii) it holds only that $\mathcal{E}(u + u^\#, u - u^\#) \ge 0$. Then the semi-Dirichlet property for $(\mathcal{E}, D(\mathcal{E}))$ is equivalent to the *sub-Markov property* for the semigroup, i.e., if $f \in L^2(E;m)$ with $0 \le f \le 1$ then $0 \le T_t f \le 1$ for all $t > 0$. Hence, if $\hat{L}, (\hat{T}_t)_{t>0}$ denote the corresponding dual (i.e., adjoint objects in $L^2(E;m)$), then the Dirichlet property of $(\mathcal{E}, D(\mathcal{E}))$ is equivalent with the sub-Markov property for both $(T_t)_{t>0}$ and $(\hat{T}_t)_{t>0}$.

DEFINITION 2.2

(i) An increasing sequence $(F_k)_{k \in \mathbb{N}}$ of closed subsets of E is called an $\mathcal{E}$-*nest* if

$$\cup_{k \in \mathbb{N}} D(\mathcal{E})_{F_k} \text{ is } \tilde{\mathcal{E}}_1^{1/2}\text{-dense in } D(\mathcal{E}),$$

where $D(\mathcal{E})_{F_k} := \{u \in D(\mathcal{E}) | u = 0 \ m\text{-a.e. on } F_k^c\}$ and $F_k^c := E \setminus F_k, k \in \mathbb{N}$.

(ii) A set $N \subset E$ is called $\mathcal{E}$-*exceptional* if $N \subset \cap_{k \in \mathbb{N}} F_k^c$ for some $\mathcal{E}$-nest $(F_k)_{k \in \mathbb{N}}$. A property holds $\mathcal{E}$-quasi-everywhere (abbreviated $\mathcal{E}$-q.e.) if it holds outside some $\mathcal{E}$-exceptional set.

(iii) An $\mathcal{E}$-q.e. defined function u on E is called $\mathcal{E}$-*quasi-continuous* if there exists an $\mathcal{E}$ nest $(F_k)_{k \in \mathbb{N}}$ such that $u|_{F_k}$ is continuous for all $k \in \mathbb{N}$. (In this case we write $u \in C(\{F_k\})$.)

DEFINITION 2.3 A Dirichlet form $(\mathcal{E}, D(\mathcal{E}))$ is called *quasi-regular* if:

(i) There exists an $\mathcal{E}$-nest consisting of compact subsets of E.

(ii) There exists an $\tilde{\mathcal{E}}_1^{1/2}$-dense subset of $D(\mathcal{E})$ whose elements have $\mathcal{E}$-quasi-continuous m-versions.

(iii) There exist $u_n \in D(\mathcal{E})$, $n \in \mathbb{N}$, having $\mathcal{E}$-quasi-continuous m-versions $\tilde{u}_n$, $n \in \mathbb{N}$, and an $\mathcal{E}$-exceptional set $N \subset E$ such that $\{\tilde{u}_n | n \in \mathbb{N}\}$ separates the points of $E \setminus N$.

Next we establish the correspondence between quasi-regular Dirichlet forms and Markov processes. In what follows we add an isolated point Δ to E. Set $E_\Delta := E \cup \{\Delta\}$ with Borel σ-field $\mathcal{B}(E_\Delta)$ and extend m to $\mathcal{B}(E_\Delta)$ by setting $m(\{\Delta\}) = 0$. Let $\mathbf{M} = (\Omega, \mathcal{F}, (\mathcal{F}_t), (X_t), (P_z)_{z \in E_\Delta})$ be a normal strong Markov process with state space E, with life time ζ, cemetery Δ, and shift operators θ_t, $t \geq 0$. In this paper $\mathbf{M}$ is called a *right process* if it is Borel right in the sense of [Sh 88 (20.1)], (i.e., $t \mapsto X_t(\omega)$ is right continuous on $[0, \infty[$ for P_z-a.e. $\omega \in \Omega$ and all $z \in E$, and its transition function is Borel).

Now let $(\mathcal{E}, D(\mathcal{E}))$ be a Dirichlet form on $L^2(E; m)$ and $(T_t)_{t>0}, (\hat{T}_t)_{t>0}$ the corresponding sub-Markovian strongly continuous contraction semigroups on $L^2(E; m)$. We say a right process $\mathbf{M}$ with state space E is associated (resp. coassociated) with $(\mathcal{E}, D(\mathcal{E}))$ if

$$P_t f := E.[f(X_t)] \text{ is an } m\text{-version of } T_t f(\text{resp. } \hat{T}_t f) \text{ for all } \quad f : E \to \mathbb{R},$$
$$\text{such that } f \text{ is } \mathcal{B}(E)\text{-measurable, } m\text{-square integrable, and all } t > 0, \tag{2.2}$$

where as usual we use the same symbol f for the L^2-class determined by the function f and $E_z[\ ]$ denotes expectation w.r.t. P_z. Note that in this case the *transition semigroup* $(P_t)_{t>0}$ of $\mathbf{M}$ "respects m-classes of functions". We say that a pair $(\mathbf{M}, \hat{\mathbf{M}})$ of right processes with state space E is *associated with* $(\mathcal{E}, D(\mathcal{E}))$ if $\mathbf{M}$ is associated and $\hat{\mathbf{M}}$ is coassociated with $(\mathcal{E}, D(\mathcal{E}))$.

Note in general that for $(T_t)_{t>0}$ as above there does not exist a reasonable Markov process satisfying (2.2).

The following result shows that the quasi-regularity condition is exactly an analytic characterization of those Dirichlet forms that are associated with right processes.

THEOREM 2.4. *A Dirichlet form $(\mathcal{E}, D(\mathcal{E}))$ on $L^2(E; m)$ is quasi-regular if and only if there exists a pair $(\mathbf{M}, \hat{\mathbf{M}})$ of right processes associated with $(\mathcal{E}, D(\mathcal{E}))$. In this case $(\mathbf{M}, \hat{\mathbf{M}})$ is always properly associated with $(\mathcal{E}, (D(\mathcal{E}))$ in the sense that $P_t f$ (resp. $\hat{P}_t f$) is an $\mathcal{E}$-quasi-continuous m-version of $T_t f$ (resp. $\hat{T}_t f$) for all $f : E \to R, \mathcal{B}(E)$-measurable, m-square integrable, and all $t > 0$.* (2.3)

Let $\mathbf{M} = (\Omega, \mathcal{F}, (\mathcal{F}_t), (X_t), (P_z)_{z \in E_\Delta})$ be a right process with state space E and resolvent $(R_\alpha)_{\alpha > 0}$, i.e.,

$$R_\alpha f(z) := \int_0^\infty e^{-\alpha t} E_z[f(X_t)] \, dt, \quad \alpha > 0, \quad z \in E, \tag{2.4}$$

for $f \in \mathcal{B}_b(E)$ (i.e., $f : E \to \mathbb{R}, f$ $\mathcal{B}(E)$-measurable, bounded). We call **M** *m-sectorial* if for one (and hence all) $\alpha > 0$ there exists a constant $K_\alpha > 0$ such that for all $f, g \in \mathcal{B}_b(E)$, m-square integrable,

$$|(R_\alpha f, g)| \leq K_\alpha (R_\alpha f, f)^{1/2} (R_\alpha g, g)^{1/2} \tag{2.5}$$

(cf. property (2.1)). Suppose now that m is αR_α-supermedian for all $\alpha > 0$ (i.e., $\int \alpha R_\alpha f \, dm \leq \int f \, dm$ for all f $\mathcal{B}(E)$-measurable, $f \geq 0$). Then each R_α "lifts" to a positive definite bounded linear operator G_α on $L^2(E; m)$ which satisfies the sector condition. It can be seen (cf. [MR 92, Chap. II, Sect. 5] and also [MR 92 Chap. IV, Sect. 2]) that $(G_\alpha)_{\alpha>0}$ is a sub-Markovian strongly continuous contraction resolvent. Hence (cf. e.g. [MR 92, Chap. I]) there exists a corresponding Dirichlet form $(\mathcal{E}, D(\mathcal{E}))$ on $L^2(E; m)$ such that **M** is associated with $(\mathcal{E}, D(\mathcal{E}))$ in the sense of (2.2). By 2.4, $(\mathcal{E}, D(\mathcal{E}))$ is quasi-regular and **M** is properly associated with $(\mathcal{E}, D(\mathcal{E}))$. We say that two right processes **M** and **M'** are *m-equivalent* if they share a common invariant set S such that their transition functions coincide on S and $m(E \setminus S) = 0$. We say that two pairs of right processes $(\mathbf{M}, \hat{\mathbf{M}})$ and $(\mathbf{M'}, \hat{\mathbf{M}'})$ are *m-equivalent* if **M** is m-equivalent to **M'** and $\hat{\mathbf{M}}$ is m-equivalent to $\hat{\mathbf{M}'}$. It was proved in [MR 92, Chap. IV. 6.4] that if $(\mathbf{M}, \hat{\mathbf{M}})$ and $(\mathbf{M'}, \hat{\mathbf{M}'})$ are two pairs of m-sectorial right processes that are properly associated with the same quasi-regular Dirichlet form $(\mathcal{E}, D(\mathcal{E}))$, then $(\mathbf{M}, \hat{\mathbf{M}})$ and $(\mathbf{M'} \, \hat{\mathbf{M}'})$ are m-equivalent. Therefore from the above discussion we have

THEOREM 2.5. *The relation (2.3) establishes a one-to-one correspondence between all the m-equivalence classes of pairs of m-sectorial right processes* $(\mathbf{M}, \hat{\mathbf{M}})$ *on E and all the quasi-regular Dirichlet forms on* $L^2(E; m)$.

REMARK 2.6 Definitions 2.2 and 2.3, and Theorems 2.4 and 2.5 are taken from [AMR 93a], the results of which have been announced in [AMR 92b]. The notion of quasi-regular Dirichlet forms appeared first in [AMR 92a]. A different version of the conditions (i)–(iii) in Definition 2.3 was first formulated for symmetric Dirichlet forms in [AM 91c, 92]. A systematical study of quasi-regular Dirichlet forms with many examples is contained in [MR 92].

REMARK 2.7

(i)   If we consider m-tight special standard processes instead of right processes, then Theorem 2.4 and Theorem 2.5 extend to the case where E is merely a Hausdorff topological space such that its Borel σ-algebra coincides with its Baire σ-algebra (c.f. [MR 92]). Morover, there is a one-to-one correspondence between all the quasi-regular semi-Dirichlet forms and all the equivalence classes of sectorial m-tight special standard processes; see [MOR 93] for details.

(ii) In addition to the above basic correspondence stated in Theorem 2.5 (and Remark 2.7(i)), the following further correspondences between quasi-regular (semi-)Dirichlet forms and m-sectorial Markov processes have been found:

- One-to-one correspondence between strictly quasi-regular semi-Dirichlet forms and equivalence classes of Hunt processes ([AMR 94], see also [MR 92] for the Dirichlet form case)   (2.6)

- One-to-one correspondence between quasi-regular Dirichlet forms having the local property and equivalence classes of diffusions ([AMR 93b], [MR 92], extending previous correponding work in [Fu 80])   (2.7)

3   Applications

3.1 Schrödinger operators with singular potentials.

As a simple application of quasi-regular Dirichlet forms consider the Schrödinger operator $-L = \dfrac{\Delta}{2} + V$ on $L^2(\mathbb{R}^d; dx)$, $d \geq 2$. Here the potential V is given by

$$V(x) = \sum_{i=1}^{\infty} c_i |x - x_i|^{-\alpha_i} \qquad (3.1.1)$$

where $\{x_i\}_{i \in \mathbb{N}}$ is the totality of points in $\mathbb{R}^d$ with rational coefficients, $\{c_i\}_{i \in \mathbb{N}}$ a sequence of strictly positive numbers, and $\alpha_i \geq d$ for all $i \in \mathbb{N}$. Note that with the above data the potential function V is singular in every neighborhood of each point $x \in \mathbb{R}^d$. In order to obtain the operator L we consider the classical Dirichlet form $(\mathcal{E}, D(\mathcal{E}))$ on $L^d(\mathbb{R}^d; m)$, where m is the Lebesgue measure and $(\mathcal{E}, D(\mathcal{E}))$ is defined by

$$D(\mathcal{E}) = H^{1,2}(R^d) := \{u \in L^2(R^d; m)| \int |\nabla u|^2 m(dx) < \infty\}$$

$$\mathcal{E}(u, v) = \int \nabla u \cdot \nabla v \; m(dx); \qquad u, v \in D(\mathcal{E}). \qquad (3.1.2)$$

Here and henceforth the derivatives are taken in the distributional sense unless otherwise stated. Let $\mu(dx) = V(x)dx$ and consider the perturbation $(\mathcal{E}^\mu, D(\mathcal{E}^\mu))$, which is defined by

$$D(\mathcal{E}^\mu) = \{u \in D(\mathcal{E})| \int \tilde{u}^2 \mu(dx) < \infty\}$$

$$\mathcal{E}^\mu(u, v) = \mathcal{E}(u, v) + \int \tilde{u}\tilde{v}\mu(dx). \qquad (3.1.3)$$

Here $\tilde{u}$ stands for a quasi-continuous m-version of u. It was proved in [AM 91a] that if μ is a smooth measure in the sense of [Fu 80] then $(\mathcal{E}^\mu, D(\mathcal{E}^\mu))$ is again a Dirichlet form. Moreover, for any given sequence of real numbers $\{\alpha_i\}_{i \in \mathbb{N}}$, one can always find a sequence of strictly positive numbers $\{c_i\}_{i \in \mathbb{N}}$ such that $\mu(dx) = V(x)dx$ with V defined by (3.1.1) is a smooth measure, and hence the perturbed form $(\mathcal{E}^\mu, D(\mathcal{E}^\mu))$ is a Dirichlet form. Let L with domain $D(L)$ be the generator of $(\mathcal{E}^\mu, D(\mathcal{E}^\mu))$, then $-L = -\dfrac{\Delta}{2} + V$ in the distributional sense and its domain $D(L)$ is a dense subset of $L^2(\mathbb{R}^d; m)$. Note that in this case there is no continuous function (except for the zero function) in $D(\mathcal{E}^\mu)$. Hence $(\mathcal{E}^\mu, D(\mathcal{E}^\mu))$ is by no means a regular Dirichlet form in the sense of [Fu 80]. Nevertheless one can prove that $(\mathcal{E}^\mu, D(\mathcal{E}^\mu))$ is always a quasi-regular Dirichlet form and hence it is associated with a diffusion process on $I\!\!R^d$. For a detailed discussion see [AM 91a,b] and references therein. In [AM 91a] we also obtained an improvement of the Kato-Lax-Milgram-Nelson theorem, and necessary and sufficient conditions for a perturbed form (by signed smooth measures) to be bounded below in a general contex. It turns out that the perturbed form is bounded below if and only if the associated Feynman-Kac functional is a strongly continuous semigroup on $L^2(E;m)$. See also [Ma 90], [ABM 91] for the application of Dirichlet forms to Feyman-Kac semigroups.

3.2 Construction of diffusions on pinned loop spaces. Let (M, g, ∇, o) be given, where M is a d-dimensional compact Riemannian manifold without boundary, g is a Riemannian metric on M, ∇ is a g-compatible covariant derivative, and $o \in M$ is a fixed base point. It will always be assumed that the covariant derivative ∇ is *torsion skew symmetric*, i.e., if T is the torsion tensor of ∇, then $g\langle T\langle X, Y\rangle, Z\rangle \equiv 0$ for all vector fields X, Y, and Z on M. We denote by $\mathcal{L}(M)$ the set of continuous paths $\sigma : [0,1] \mapsto M$ such that $\sigma(0) = \sigma(1) = o$ and we equip $\mathcal{L}(M)$ with the topology of uniform convergence. The pinned Wiener measure concentrated on $\mathcal{L}(M)$ will be denoted by ν. The corresponding real L^2-space is denoted by $L^2(\nu)$. Recall that the coordinate maps $\sum_s : W(M) \to M$ given by $\sum_s(\sigma) = \sigma(s)$ are M-valued semi-martingales relative to the measures ν. Therefore it is possible to define a stochastic parallel translation operator $H_s(\sigma) : T_o M \to T_{\sigma(s)} M$ for ν almost every path σ.

Let H be the reproducing kernel Hibert space consisting of functions $h : [0,1] \to T_o M$ such that $h(0) = h(1) = 0$, h is absolutely continuous, and $(h, h)_H := \int_0^1 |h'(s)|^2 \, ds < \infty$, where $|v|^2 := g_0\langle v, v\rangle$ for $v \in T_o M$. A function $F : \mathcal{L}(M) \mapsto I\!\!R$ is said to be a *smooth cylinder function* if F can be represented as $F(\sigma) = f(\sigma(s_1), \ldots, \sigma(s_n))$, where $f : M^n \mapsto I\!\!R$ is a smooth function and $0 \leq s_1 \leq s_2 \cdots \leq s_n \leq 1$. Let $\mathcal{F}C^\infty$ denote the set of all smooth cylinder functions. Note that $\mathcal{F}C^\infty$ is dense in $L^2(\nu)$ because it separates the points of $\mathcal{L}(M)$. Given $h \in H \cap C^1$ and a smooth cylinder function $F(\sigma)$ as above, the h-derivative of F is

$$\partial_h F(\sigma) := \sum_{i=1}^n g_{\sigma(s_i)}\langle \nabla_i f(\overrightarrow{\sigma}), H_{s_i}(\sigma) h(s_i)\rangle, \tag{3.2.1}$$

where $\overrightarrow{\sigma} := (\sigma(s_1), \ldots, \sigma(s_n))$, $\nabla_i f(\overrightarrow{\sigma}) \in T_{\sigma(s_i)} M$ is the gradient of the function f relative to the ith variable while the remaining variables are held fixed.

Let

$$G(s,t) := \begin{cases} s(1-t) & \text{if } s \leq t \\ t(1-s) & \text{if } s \geq t \end{cases}, \quad \text{which is the Green's function}$$

for the operator $-\dfrac{d^2}{ds^2}$ with Dirichlet boundary conditions at both $s = 0$ and $s = 1$. Then one can check that G is a reproducing kernel for H. For F as in (3.2.1) we set

$$DF(\sigma)(s) = \sum_{i=1}^n G(s, s_i) H_{s_i}(\sigma)^{-1} \nabla_i f(\overrightarrow{\sigma}), \tag{3.2.2}$$

where $F(\sigma) = f(\vec{\sigma})$ as specified before (3.2.1). Then

$$\partial_h F(\sigma) = (DF(\sigma), h)_H \qquad \text{for all} \quad h \in H \cap C^1. \tag{3.2.3}$$

It was proved in [DR 92] that up to ν-equivalence DF is the unique function from $\mathcal{L}(M)$ to H that satisfies (3.2.3). We call DF the *gradient* of F. We now define

$$\mathcal{E}(F, K) := \int_{\mathcal{L}(M)} (DF, DK)_H \nu(d\sigma) \quad \text{for} \quad F, K \in \mathcal{F}C^\infty. \tag{3.2.4}$$

One can prove that $(\mathcal{E}, \mathcal{F}C^\infty)$ is closable on $L^2(\nu)$ and its closure $(\mathcal{E}, D(\mathcal{E}))$ is a symmetric Dirichlet form on $L^2(\nu)$ having the local property. Moreover, one can check that $(\mathcal{E}, D(\mathcal{E}))$ is quasi-regular and hence by Theorem 2.4 and Remark 2.7(ii) there exists a diffusion process $I\!\!M = (\Omega, \mathcal{F}, (\mathcal{F}_t)_{t\geq 0}, (X_t)_{t\geq 0}, (P_z)_{z \in E_\Delta})$ associated with $(\mathcal{E}, D(\mathcal{E}))$, i.e., for all $f \in L^2(\nu)$ and all $t > 0$,

$$P_t f := E.[f(X_t)] \text{is an } \mathcal{E}\text{-quasi-continuous } \nu\text{-version of } T_t f$$
$$\text{where } T_t := e^{tL} \text{ and } L \text{ is the generator of } (\mathcal{E}, D(\mathcal{E})). \tag{3.2.5}$$

Note that L is an infinite-dimensional Ornstein-Uhlenbeck operator over the non-linear space $\mathcal{L}(M)$, and L is of great interest in both mathematics and physics. Similarly one can construct diffusions on the path space over M. For a detailed discussion of this subsection see [DR 92]. See also [ALR 93] for the construction of diffusions on free loop spaces over $I\!\!R^d$.

References

[ABM 91] S. Albeverio, Blanchard and Z-M. Ma, *Feynman-Kac semigroups in terms of signed smooth measures*, in Internat. Ser. Numer. Math. **102**, 1–31, Birkhäuser-Verlag, Basel/Boston 1991.

[AH 75] S. Albeverio and R. Høegh-Krohn, *Quasi-invariant measure, symmetric diffusion processes and quantum fields*, in Les Méthodes mathématiques de la théorie quantique des champs, Colloques Internationaux du C.R.N.S. **248**, Marseille, 223–27 juin 1975, C.N.R.S., 1976.

[AH 77a] S. Albeverio and R. Høegh-Krohn, *Dirichlet forms and diffusion processes on rigged Hilbert spaces*, Z. Wahrsch. Verw. Geb. **40** (1977), 1–57.

[AH 77b] S. Albeverio and R. Høegh-Krohn, *Dirichlet forms and Markov semigroups on C^*-algebras*, Comm. Math. Phys. **56** (1977), 173–187.

[AKoR 93] S. Albeverio, Yu. Kondratiev, and M. Röckner, *Dirichlet operators via stochastic analysis*, preprint 1993, to appear in J. Funct. Anal.

[ALR 93] S. Albeverio, R. Leándre, and M. Röckner, *Construction of a rotational invariant diffusion on the free loop space*, C. R. Acad. Sci. Paris, t. 316, Série I (1993), 287–292.

[AM 91a] S. Albeverio and Z-M. Ma, *Perturbation of Dirichlet forms — lower semi-boundedness, closability and form cores*, J. Funct. Anal. **99** (1991), 332–356.

[AM 91b] S. Albeverio and Z-M. Ma, *Diffusion processes associated with singular Dirichlet forms*, in Stochastic Analysis and Applications (A.B. Cruzeiro et al., eds.), Birkhäuser-Verlag, Basel/Boston 1991.

[AM 91c] S. Albeverio and Z-M. Ma, *Necessary and sufficient condition for the existence of m-perfect processes associated with Dirichlet forms*, in Lecture Notes in Math. **1485**, 374–406, Springer-Verlag, Berlin/Heidelberg, 1991.

[AM 92]   S. Albeverio and Z-M. Ma, *A general correspondence between Dirichlet forms and right processes*, Bull. Amer. Math. Soc., (New series) **26** (1992), 245–252.

[AMR 92a] S. Albeverio, Z-M. Ma, and M. Röckner, *A Beurling-Deny type structure theorem for Dirichlet forms on general state space*, in Ideas and Methods in Mathematical Analysis, Stochastics, and Applications, Cambridge University Press, Cambridge, 1992.

[AMR 92b] S. Albeverio, Z-M. Ma, and M. Röckner, *Non-symmetric Dirichlet forms and Markov processes on general state spaces*, C.R. Acad. Sci. Paris Série I **314** (1992), 77–82.

[AMR 92c] S. Albeverio, Z-M. Ma, and M. Röckner, *Regularization of Dirichlet spaces and applications*, C.R. Acad. Sci. Paris Série I **314** (1992), 859–864.

[AMR 93a] S. Albeverio, Z-M. Ma, and M. Röckner, *Quasi-regular Dirichlet forms and Markov processes*, J. Funct. Anal. **111** (1993), 118–154.

[AMR 93b] S. Albeverio, Z-M. Ma, and M. Röckner, *Local property of Dirichlet forms and diffusions on general state spaces*, Math. Ann. **296** (1993), 677–686.

[AMR 94] S. Albeverio, Z-M. Ma, and M. Röckner, *Characterization of (non-symmetric) semi-Dirichlet forms associated with Hunt processes*, preprint, 1994.

[AR 90]   S. Albeverio and M. Röckner, *Classical Dirichlet forms on topological vector spaces — closability and a Cameron-Martin formula*, J. Funct. Anal. **88** (1990), 395–436.

[AR 91]   S. Albeverio and M. Röckner, *Stochastic differential equations in infinite dimensions, solutions via Dirichlet forms*, Probab. Theory Related Fields **89** (1991), 347–386.

[ARZ 93] A. Albeverio and M. Röckner, T-S. Zhang, *Markov uniqueness for a class of infinite dimensional Dirichlet operators*, in Stochastic Processes and Optimal Control (H.J. Engelbert et al., eds.), Stochastic Monographs **7**, 1–26, Gordon & Breach, New York, 1993.

[BeDe 58] A. Beurling and J. Deny, *Espaces de Dirichlet*, Acta Math. **99** (1958), 203–224.

[BeDe 59] A. Beurling and J. Deny, *Dirichlet spaces*, Proc. Nat. Acad. Sci. U.S.A. **45** (1959), 208–215.

[BH 91]   N. Bouleau and F. Hirsch, Dirichlet forms and analysis on Wiener space, de Gruyter, Berlin/New York 1991.

[CMR 93] Z. Chen, Z.-M. Ma, and M. Röckner, *Quasi-homeomorphism of Dirichlet forms*, preprint 1993, to appear in Nagoya Math. J.

[CWZ 93] Z. Q. Chen, R. J. Williams, and Z. Zhao, *On the existence of positive solution of semilinear elliptic equations with Dirichlet boundary conditions*, preprint 1993, to appear in Math. Ann.

[DR 92]   B. Driver and M. Röckner, *Construction of diffusions on path and loop spaces of compact Riemannian manifolds*, C.R. Acad. Sci. Paris Série I (1992), 603–608.

[Dy 82]   E. B. Dynkin, *Green's and Dirichlet spaces associated with fine Markov processes*, J. Funct. Anal. **47** (1982), 381–418.

[Fi 89]   P. Fitzsimmons, *Markov processes and nonsymmetric Dirichlet forms without regularity*, J. Funct. Anal. **85** (1989), 287–306.

[Fi 94]   P. J. Fitzsimmons, *Absolute continuity of symmetric diffusions*, preprint 1994.

[FiGe]   P. Fitzsimmons and R. Getoor, *On the potential theory of symmetric Markov processes*, Math. Ann. **281** (1988), 495–512.

[Fu 71a]   M. Fukushima, *Dirichlet spaces and string Markov processes*, Trans. Amer. Math. Soc. **162** (1971), 185–224.

[Fu 71b]   M. Fukushima, *Regular representation of Dirichlet forms*, Trans. Amer. Math. Soc. **155** (1971), 455–473.

[Fu 76]   M. Fukushima, *Potential theory of symmetric Markov processes and its applications*, in Lecture Notes in Math. **550**, Springer-Verlag, Berlin and New York, 1976.

[Fu 80]   M. Fukushima, *Dirichlet forms and Markov processes*, North Holland, Amsterdam/Oxford/New York, 1980.

[Fu 94]   M. Fukushima, *On a decomposition of additive functionals in the strict sense for a symmetric Markov processes*, preprint 1994, to appear in Proc. ICDFSP, (Z.-M.Ma et al., eds.), Walter de Gruyter, Berlin and Hawthorne, NY.

[FuLe 91]   M. Fukushima and Y. LeJan, *On quasi-supports of smooth measures and closability of pre-Dirichlet forms*, Osaka J. Math. **28** (1991), 837–845.

[FOT 94]   M. Fukushima, Y. Oshima, and M. Takeda, Dirichlet Forms and Symmetric Markov Processes, Walter de Gruyter, Berlin 1994.

[FuSh 92]   M. Fukushima and T. Shima, *On a spectral analysis for the Sierpinshi gasket*, Potential Analysis **1** (1992), 1–35.

[Hi 94]   F. Hirsch, *Potential theory related to some multiparameter processes*, preprint 1994.

[Kus 82]   S. Kusuoka, *Dirichlet forms and diffusion processes on Banach spaces*, J. Fac. Sci. Univ. Tokyo, Sect. **IA 29** (1982), 387–400.

[KZ 92]   S. Kusuoka and X.-Y. Zhou, *Dirichlet forms on fractals, Poincaré constant and resistance*, Probab. Theory Related Fields **93** (1992), 169–196.

[Kuw 94]   K. Kuwae, *Permanent sets of measures charging no exceptional sets and the Feynman-Kac formula*, preprint 1994, to appear in Forum Math.

[Ja 92]   N. Jacob, *Feller semigroups, Dirichlet forms and pseudo differential operators*, Forum Math. **4** (1992), 433–446.

[JaHo 94]   N. Jacob and W. Hoh, *On the Dirichlet problems for pseudo differential operators generating Feller semigroups*, preprint 1994.

[Le 83]   Y. LeJan, *Quasi-continuous functions and Hunt processes*, J. Math. Soc. Japan **35** (1983), 37–42.

[Li 94]   M. Lindsay, *Non-commutative Dirichlet forms*, preprint 1994, to appear in Proc. ICDFSP (Z.-M. Ma et al., eds.), Walter de Gruyter, Berlin and Hawthorne, NY.

[Ma 90]   Z.-M. Ma, *Some new results concerning Dirichlet forms, Feynman-Kac semigroups and Schrödinger equations*, in Probability Theory in China (Contemp. Math.), Amer. Math. Soc., Providence, RI, 1990.

[MOR 93]   Z.-M. Ma, L. Overbeck, and M. Röckner, *Markov processes associated with semi-Dirichlet forms*, SFB 256, No. 281, 1993, to appear in Osaka. J. Math.

[MR 92]   Z.-M. Ma and M. Röckner, An introduction to the theory of (non-symmetric) Dirichlet forms, Springer-Verlag, Berlin and New York, 1992.

[MR 93]   Z.-M. Ma and M. Röckner, *Markov processes associated with positivity preserving coercive forms*, preprint 1993.

[Mu 93]   S. Mück, *Large deviations w.r.t. quasi-every starting point for symmetric right processes on general state spaces*, preprint 1993.

[O 93]   Y. Oshima, *Time dependent Dirichlet forms and its applications to a transformation of space-time Markov processes*, preprint 1993, to appear in Proc. ICDFSP, (Z.-M. Ma et al., eds.), Walter de Gruyter, Berlin and Hawthorne, NY.

[ORS 93]   L. Overbeck, M. Röckner, and B. Schmuland, *An analytic approach to Fleming-Viot processes with interactive selection*, preprint 1993.

[RS 93]   M. Röckner and B. Schmuland, *Quasi-regular Dirichlet forms, examples and counterexamples*, preprint 1993, to appear in Canad. J. Math.

[Sch 93]   B. Schmuland, *Non-symmetric Ornstein-Uhlenbeck processes in Banach space via Dirichlet forms*, Canad. J. Math. **45** (6) (1993), 1324–1338.

[Sh88]   M. Sharpe, General Theory of Markov Processes, Academic Press, INC. 1988.

[ShI 92]   I. Shigekawa and S. Taniguchi, *Dirichlet forms on separable metric spaces*, in Probability Theory and Mathematical Statistics (A.N. Shiryaev et al., eds.), World Scientific, Singapore, 1992.

[Si 74]   M. L. Silverstein, Symmetric Markov processes, Lecture Notes in Math. **426**, Springer-Verlag, Berlin/Heidelberg/New York, 1974.

[So 94]   S. Q. Song, *Construction d'un processus à deux paramètres à partir d'un semi groupe à un paramètre*, preprint 1994.

[Sta 94]   W. Stannat, *Generalized Dirichlet forms and associated Markov processes*, preprint 1994.

[Stu 94]   K-T. Sturm, *Analysis on local Dirichlet spaces, I–III*, preprint 1994.

[Ta 92]   M. Takeda, *The maximum Markovian self-adjoint extensions of generalized Schrödinger operators*, J. Math. Soc. Japan **44** (1992), 113–130.

[Ta 94]   M. Takeda, *Transformations of local Dirichlet forms by supermartingale multiplicative functionals*, preprint, 1994 to appear in Proc. ICDFSP (Z.-M. Ma et al., eds.), Walter de Gruyter, Berlin and Hawthorne, NY.

[Z 94]   W. Zheng, *Conditional propagation of chaos and a class of quasi-linear PDE*, preprint 1994.

A Surface View of First-Passage Percolation

CHARLES M. NEWMAN[*]

Courant Institute of Mathematical Sciences
New York University
251 Mercer Street
New York, NY 10012, USA

ABSTRACT. Let $\tilde{B}(t)$ be the set of sites reached from the origin by time t in standard first-passage percolation on $\mathbf{Z}^d$, and let B_0 (roughly $\lim \tilde{B}(t)/t$) be its deterministic asymptotic shape. We relate the $t \to \infty$ microstructure of the surface of $\tilde{B}(t)$ to spanning trees of time-minimizing paths and their transverse deviations and to curvature properties of B_0. The most complete results are restricted to $d = 2$.

1 The Growing Shape of First-Passage Percolation

In standard first-passage percolation [HW], [K1], one begins with i.i.d. nonnegative random variables $\tau(e)$ on some $(\Omega, \mathcal{F}, P)$, indexed by the nearest neighbor edges e of $\mathbf{Z}^d$. The passage time $T(r)$ for a finite (site self avoiding) path r consisting of edges $e_1, \ldots, e_n$ is simply $\sum_i \tau(e_i)$ and the passage time $T(u, v)$ between two sites $u, v \in \mathbf{Z}^d$ is the inf of $T(r)$ over all paths r from u to v. From any site x, one may consider the stochastically growing region $\tilde{B}^x(t) = \{y \in \mathbf{Z}^d : T(x, y) \leq t\}$.

The surface of $\tilde{B}^x(t)$ (for fixed x and increasing t) is one of a variety of physically interesting models of growing interfaces [KS]. In this note we report on some initial progress in understanding the microscopic structure of this surface when $t \to \infty$. The results, which have partly been obtained in collaboration with Cristina Licea and Marcelo Piza, are presented here in a preliminary form with no attempt at stating optimal hypotheses and with only sketches of most proofs. Complete proofs and improved hypotheses will be presented in future papers.

Our hypotheses on the common distribution of the $\tau(e)$'s are:

A. $\tau(e)$ is a continuous random variable.
B. $E(e^{\alpha \tau(e)}) < \infty$ for some $\alpha > 0$.
C. $P(\tau(e) > u) > 0$ for every $u < \infty$.

Hypotheses A and B are much more than sufficient for the celebrated shape theorem [R], [CD], [K1] which, roughly speaking, implies that for fixed x, $t^{-1}\tilde{B}^x(t)$ converges a.s. to B_0, a nonrandom, compact, convex subset of $\mathbf{R}^d$ (symmetric about the origin) with nonempty interior. Hypothesis A is also sufficient (and necessary) to ensure that for every u, v there is a.s. a unique time-minimizing path (which we denote $M(u, v)$) between u and v — i.e., such that $T(M(u, v)) = T(u, v)$.

---

[*]   Research supported in part by NSF Grant DMS-9209053.

Proceedings of the International Congress
of Mathematicians, Zürich, Switzerland 1994
© Birkhäuser Verlag, Basel, Switzerland 1995

Closely related to the analysis we present here are some recent results on fluctuations of $\tilde{B}^x(t)$ as $t \to \infty$. One major result, due to Kesten and Alexander [K2], [A] is, roughly speaking, that the deviation of $\tilde{B}^x(t)$ from tB_0 is $O(t^{1/2} \log t)$. Another version of their result (more suited to our present purposes) may be expressed in terms of the norm $g(y)$ (on $\mathbf{R}^d$) associated with the shape $B_0 : g(y) = \inf\{\lambda > 0 : y/\lambda \in B_0\}$. Note that the norm $g(y)$ and the Euclidean norm $\|y\|$ are bounded by multiples of each other. The shape theorem is essentially the statement that as $\|y\| \to \infty$, $T(0,y)/g(y) \to 1$; Kesten and Alexander show that for any $\epsilon > 0$,

$$P(|T(0,y) - g(y)| \geq \lambda) \leq C_1 \exp(-C_2\lambda/\|y\|^{1/2}), \qquad (1.1)$$

for $\|y\|^{1/2+\epsilon} \leq \lambda \leq \|y\|^{3/2-\epsilon}$.

In [NP], the Kesten and Alexander results, which concern longitudinal deviations of the surface of $\tilde{B}^x(t)$, are used to bound the transverse deviations of the finite time-minimizing paths, $M(x,y)$. The result of [NP], roughly speaking, is that for any $\epsilon > 0$, $M(x,y)$ stays within distance $\|x - y\|^{3/4+\epsilon}$ of the straight line joining x and y when $\|y\| \to \infty$ while $y/\|y\| \to \hat{y}$; here, the unit vector $\hat{y}$ is required to be a "direction of curvature" for B_0. In Section 3 of this paper (see the first two propositions there), using (1.1), we extend these transverse bounds to semi-infinite time-minimizing paths. This extension, which is crucial to our analysis, is based on an assumption of "uniform curvature" for B_0, which we discuss below.

To study the microstructure of the $t \to \infty$ limit of the surface of $\tilde{B}(t)$ (i.e., $\tilde{B}^x(t)$ with $x = 0$), one natural approach is to pick a sequence y_n of sites in $\mathbf{Z}^d$ with $\|y_n\| \to \infty$ and $y_n/\|y_n\| \to \hat{y}$ and then ask whether $\tilde{B}(t)$, at the time it reaches y_n, when viewed from y_n, has a limit *in distribution* as $n \to \infty$: $\lim_n(\tilde{B}(T(0,y_n)) - y_n)$. We remark that if B_0 has a unique tangent plane at $\bar{y} = \hat{y}/g(\hat{y})$ (which is *not* part of our uniform curvature assumption), then this limit should have a boundary "surface" (necessarily passing through the origin) that is asymptotically parallel to that tangent plane.

The first step in our analysis is to replace $\tilde{B}(T(0,y_n)) - y_n$ by the equidistributed $\tilde{B}^{-y_n}(T(-y_n,0))$, which we denote by $\tilde{B}[-y_n]$. The advantage of this replacement is that now there is a chance for an *almost sure* limit. Indeed, to have such a limit, it suffices if for each site x, $T(-y_n,x) - T(-y_n,0)$ converges a.s. to some $H(x)$ (nonzero for $x \neq 0$) because then the a.s. limit of $\tilde{B}[-y_n]$ is just $\{x : H(x) \leq 0\}$. Thus, we are led to the natural question of whether $T(u,x_n) - T(v,x_n)$ has an a.s. limit (nonzero for $u \neq v$) as x_n tends to infinity in some direction $\hat{x}$. To state a theorem, we need one more definition.

We say that B_0 (or its corresponding norm g) is *uniformly curved* if for some $C > 0$ and any $z = \alpha z_1 + (1 - \alpha)z_2$ with $g(z_1) = g(z_2) = 1$ and $\alpha \in [0,1]$,

$$1 - g(z) \geq C[\min(g(z - z_1), g(z - z_2))]^2 . \qquad (1.2)$$

We remark that this will be the case if for some $\rho < \infty$ and any point z' on the surface of B_0 there is a ball of radius at most ρ with z' on its surface that contains the entire interior of B_0. Unfortunately, there is in general very little information known about the shape of B_0, and thus, at present, no proof (for any distribution of the $\tau(e)$'s) that B_0 is uniformly curved. We also remark that weakened versions of (1.2) where the square in the R.H.S. is replaced by any finite power are sufficient to yield the conclusions of the next theorem, but we do not know of a proof of this weakened hypothesis either.

THEOREM 1.1. *Set $d = 2$. Assume Hypotheses A, B, and C on the distribution of the $\tau(e)$'s and also assume that B_0 is uniformly curved. Let ν be any continuous distribution (i.e., probability measure with no atoms) on the unit sphere of $\mathbf{R}^2$; then for ν-almost every $\hat{x}$, the following is true almost surely: for every u, v in $\mathbf{Z}^2$, there is an $H^{\hat{x}}(u, v)$ (nonzero for $u \neq v$) such that*

$$\lim_{\substack{\|x\| \to \infty \\ x/\|x\| \to \hat{x}}} [T(u, x) - T(v, x)] = H^{\hat{x}}(u, v) . \tag{1.3}$$

The proof of Theorem 1.1 is given at the end of the next section after several preliminary results concerning spanning trees of time-minimizing paths.

2 Spanning Trees

For each x in $\mathbf{Z}^d$, denote by $R(x)$ the union over all $y \in \mathbf{Z}^d$ of the time-minimizing paths $M(x, y)$. Slightly abusing notation, we regard $R(x)$ both as a set of edges and as the graph with that edge set and vertex set $\mathbf{Z}^d$. It is easy to see that for each x, the graph $R(x)$ is a tree, spanning all of $\mathbf{Z}^d$. Thus, there must be at least one semi-infinite path in $R(x)$ (starting from x). The next three theorems concern these semi-infinite paths; together they yield Theorem 1.1 (as we explain below) by giving an explicit construction of $H^{\hat{x}}(u, v)$ in terms of another spanning tree constructed out of (some of) the semi-infinite paths from the $R(x)$'s. Sketches of the proofs of the three theorems are given in Section 3. The first theorem, valid for any d, is the one related to transverse deviations of semi-infinite minimizing paths. The second and third theorems are currently restricted to $d = 2$.

If a semi-infinite path r consisting of the edges $(x_0, x_1), (x_1, x_2), \ldots$ has the property that $x_n/\|x_n\| \to \hat{x}$, we say that r has direction $\hat{x}$. Let D denote the event that every semi-infinite path in every $R(x)$ has a direction. For each unit vector $\hat{x}$, denote by $D_E(\hat{x})$ (resp., $D_U(\hat{x})$) the event that for every x in $\mathbf{Z}^d$ there exists at least one (resp., at most one) semi-infinite path (starting from x) in $R(x)$ with direction $\hat{x}$.

THEOREM 2.1. *Assume Hypotheses A and B and also that B_0 is uniformly curved. Then $P(D) = 1$ and $P(D_E(\hat{x})$ occurs for every $\hat{x}) = 1$.*

THEOREM 2.2. *Set $d = 2$ and assume Hypothesis A. Let ν be any continuous distribution on the unit sphere of $\mathbf{R}^2$; then for ν-almost every $\hat{x}$, $P(D_U(\hat{x})) = 1$.*

THEOREM 2.3. *Set $d = 2$. Assume Hypotheses A and C and also that $\hat{x}$ is a deterministic direction with $P(D_U(\hat{x})) = 1$. Then there is zero probability that there exist semi-infinite paths r_u and r_v (starting from some u and v) in $R(u)$ and $R(v)$ that both have direction $\hat{x}$ and that are site-disjoint.*

Proof of Theorem 1.1. By Theorems 2.1 and 2.2, $P(D) = 1$ and we may restrict attention to a deterministic $\hat{x}$ with $P(D_E(\hat{x})) = 1 = P(D_U(\hat{x}))$. On $D \cap D_E(\hat{x}) \cap D_U(\hat{x})$, we denote by s_u the unique semi-infinite path (starting from u) in $R(u)$ with direction $\hat{x}$; a key observation is then that s_u is the limit of $M(u, x_n)$ for *any* sequence x_n such that $\|x_n\| \to \infty$ and $x_n/\|x_n\| \to \hat{x}$. It follows that if s_u and s_v are not site-disjoint, then $(i)s_u \cap s_v = s_w$ for some $w = W(u,v)$ (which may equal u or v), (ii) the edge set of $s_u \cup s_v$ is the disjoint union of $M(u,w), M(v,w)$ and s_w, and (iii) $T(u,x) - T(v,x) \to T(u,w) - T(v,w)$ as $\|x\| \to \infty$, $x/\|x\| \to \hat{x}$. Consider now the graph $S[\hat{x}]$, with site set $\mathbf{Z}^2$ and edge set, the union of s_u over all $u \in \mathbf{Z}^2$. It follows from (i) and (ii) that $S[\hat{x}]$ has no loops and is thus a forest (a union of trees). By Theorem 2.3 it is a single tree (a.s.), which clearly spans all of $\mathbf{Z}^2$. From (iii), we obtain (1.3) with

$$H^{\hat{x}}(u,v) = T(u, W(u,v)) - T(v, W(u,v)) \ . \tag{2.1}$$

Note that replacing the minus by a plus in the R.H.S. of (2.1) would yield the natural distance between u and v in the spanning tree $S[\hat{x}]$. $H^{\hat{x}}(u,v)$ is nonzero for $u \neq v$ because Hypothesis A implies that $P(T(u,y) = T(v,y)) = 0$ for every y. $\square$

3 Sketches of Proofs

We begin with some notation and a definition needed for Theorem 2.1. Denote by $\theta(x,y)$ the angle (in $[0, \pi]$) between nonzero x and y in $\mathbf{R}^d$ and by $\mathcal{C}(x, \epsilon)$ the cone of y's in $\mathbf{R}^d$ with $\theta(y,x) \leq \epsilon$. If R is an infinite (nearest neighbor) tree on $\mathbf{Z}^d$ containing the origin 0, and $x \in R$, we denote by $R^{\text{out}}[x]$ the set of sites v in R such that the path in R between v and 0 touches x. For h a positive function on $(0, \infty)$ (generally decreasing to zero), we define R to be *h-straight* if for all but finitely many x in R,

$$R^{\text{out}}[x] \subseteq \mathcal{C}(x, h(\|x\|)) \ . \tag{3.1}$$

Theorem 2.1 is an immediate consequence of the next two propositions.

PROPOSITION 3.1. *Suppose R is a spanning tree in $\mathbf{Z}^d$ that is h-straight, where $h(L) \to 0$ as $L \to \infty$. Then every semi-infinite path in R (starting from 0) has a direction. Furthermore, for every unit vector $\hat{x}$, there is at least one semi-infinite path in R (starting from 0) with direction $\hat{x}$.*

Proof. Let $x_0, x_1 \ldots$ be the sequence of sites in a semi-infinite path in R. Then by (3.1), we have for large m that $\theta(x_n, x_m) \leq h(\|x_m\|)$ for $n \geq m$. It follows that $x_n/\|x_n\|$ converges. To prove the second conclusion of the proposition, note that simply because R is spanning, one can inductively find, for any given $\hat{x}$, a semi-infinite path $(y_0, y_1, \ldots)$ in R such that for each j, $R^{\text{out}}[y_j]$ contains a sequence (depending on j) $x_1, x_2, \ldots$ with $x_n/\|x_n\| \to \hat{x}$. But the already proved first conclusion shows that $y_j/\|y_j\|$ tends to some $\hat{y}$ and then (3.1) implies that $\hat{x} \cdot \hat{y} = 1$ so $\hat{y} = \hat{x}$. $\square$

PROPOSITION 3.2. *Assume Hypotheses A and B and also that B_0 is uniformly curved. Then for any $\epsilon > 0$, the spanning tree $R(0)$ is a.s. h-straight with $h(L) = L^{-(1/4-\epsilon)}$.*

Proof. We will show that for any $0 < \delta < 1/4$, $R = R(0)$ is h-straight with $h(L) = C_\delta L^{-\delta}$. Our strategy is to use the inequality (1.1) of Kesten and Alexander [K2], [A] to bound the probability of the event $G(x, x')$, that the time-minimizing path $M(0, x)$ passes through x', or equivalently that $x \in R^{\text{out}}[x']$, or equivalently that $T(0, x') + T(x', x) \leq T(0, x)$. Let $T_c(u, v) = T(u, v) - g(v - u)$ and $\Delta(x, x') = g(x') + g(x - x') - g(x) \geq 0$; then

$$P(G(x, x')) = P(T_c(0, x') + T_c(x', x) - T_c(0, x) \leq -\Delta(x, x')) \qquad (3.2)$$

$$\leq \sum_{i=1}^{3} (P|T_c(0, w_i)| \geq \Delta(x, x')/3),$$

where $w_1 = x'$, $w_2 = x - x'$, and $w_3 = x$. $\qquad\qquad\square$

Let $A_{x'}$ denote the set of sites y in the cone $\mathcal{C}(x', (g(x'))^{-\delta})$ with $g(y)/g(x')$ in $[1 - (g(x'))^{-2\delta}, 2]$ and denote by $\partial A_{x'}$ its boundary (i.e., the sites not in $A_{x'}$ that are nearest neighbors of some site in $A_{x'}$). Define $\partial_i A_{x'}$ (resp., $\partial_o A_{x'}$), the inside (resp., outside) boundary, as those boundary sites in the just-mentioned cone, with $g(y)/g(x')$ below $1 - (g(x'))^{-2\delta}$ (resp., above 2) and define $\partial_s A_{x'}$, the side boundary as those boundary sites not in the cone. Define $G(x')$ to be the event that $R^{\text{out}}[x']$ touches $\partial_i A_{x'} \cup \partial_s A_{x'}$. We claim first that a.s. $G(x')$ occurs for only finitely many x' and second that hence R is h-straight with $h(L) = C_\delta L^{-\delta}$.

To justify the first claim we bound $P(G(x'))$ by the sum over x in $\partial_i A_{x'} \cup \partial_s A_{x'}$ of $P(G(x, x'))$, and then use (3.2). To apply (1.1), we note the following. For x in the inside boundary,

$$\Delta(x, x') \geq g(x') - g(x) \geq (g(x'))^{1-2\delta} \geq C_3 \|x'\|^{1-2\delta} .$$

For x in the side boundary, one can use the uniform curvature condition (1.2) with $z_1 = x'/g(x')$, $z_2 = (x - x')/g(x - x')$, and $z = x/[g(x') + g(x - x')]$ to again bound $\Delta(x, x')$ from below by a multiple of $\|x'\|^{1-2\delta}$. For either case, each w_i appearing in (3.2) has $\|w_i\|$ bounded between multiples of $\|x'\|$ and $\|x'\|^{1-2\delta}$. Because $1 - 2\delta > 1/2$, (1.1) can be applied with exponent proportional to $\|x'\|^{1-2\delta}/\|w_i\|^{\frac{1}{2}} \geq \|x'\|^{\frac{1}{2}-2\delta}$. The overall bound is thus

$$P(G(x')) \leq C_4 \|x'\|^d \exp(-C_5 \|x'\|^{\frac{1}{2}-2\delta}) . \qquad (3.3)$$

Because $\delta < 1/4$, the Borel-Cantelli lemma now yields the first claim.

The justification of the second claim is now deterministic. On an a.s. event, we have that for all large $g(x')$, $R^{\text{out}}[x']$ is contained in the union of $A_{x'}$ and the union of $R^{\text{out}}[x]$ for x in $\partial_o A_{x'}$. For every $x \in \mathcal{C}(x', \epsilon_1)$, the cone $\mathcal{C}(x, \epsilon_2)$ is contained in $\mathcal{C}(x', \epsilon_1 + \epsilon_2)$; thus, it follows by induction that for $m = 1, 2, \ldots$

$$R^{\text{out}}[x'] \subseteq \mathcal{C}(x', \epsilon_m(x')) \cup \left(\bigcup_{x \in \partial^m(x')} R^{\text{out}}[x] \right), \qquad (3.4)$$

where $\epsilon_m(x') = \sum_{j=0}^{m-1}(2^j g(x'))^{-\delta}$ and $\partial^m(x')$ is the set of x in $\mathcal{C}(x', \epsilon_m(x'))$ with $g(x) \geq 2^m g(x')$. Now the intersection of $R^{\text{out}}[x]$ with any finite subset of $\mathbf{Z}^d$ is eventually empty as $\|x\| \to \infty$ and so we may let $m \to \infty$ in (3.4) to conclude that $R^{\text{out}}[x']$ is contained in $\mathcal{C}(x', (1 - 2^{-\delta})^{-1}(g(x'))^{-\delta})$ for large $g(x')$, which implies the second claim.

Proof of Theorem 2.2. If $e = \{u, v\}$ is an edge in $R = R(0)$ with $v \in R^{\text{out}}[u]$ and $R^{\text{out}}[v]$ infinite, we may *for $d = 2$* inductively define an *infinite* path $r^+(e)$ in $R^{\text{out}}[u]$ starting with e so that each succesive step makes as counterclockwise a turn as possible (among steps leading to infinite paths). Suppose $\hat{x}$ is such that there exist r_1 and r_2, two distinct infinite paths in R with direction $\hat{x}$. Let r denote the one that is located asymptotically clockwise to the other. $r_1 \cap r_2$ is $M(0, u)$ for some u; let $e = \{u, v\}$ denote the first edge of r after u. It follows that $r^+(e)$ has direction $\hat{x}$. Because a similar argument works for any $R(x)$ and there are only countably many e's and x's, it follows that (for P-a.e. fixed ω) $I(\omega, \hat{x})$, the indicator of the complement of $D_U(\hat{x})$, vanishes except for countably many $\hat{x}$'s. Integrating I with respect to the product of P and ν and applying Fubini's Theorem completes the proof. $\square$

Proof of Theorem 2.3. As in the earlier proof of Theorem 1.1, let s_u denote the (unique, if it exists) semi-infinite path (from u) in $R(u)$ with direction $\hat{x}$ and let $S = S[\hat{x}]$ be the union over all u of s_u. Here S is either empty, a single tree, or a forest of two or more trees. To rule out the third case, we mimic the proof structure of [BK].

Part 1: Let N be the number of (infinite) trees in S. We show that $P(N \geq 2) > 0$ implies $P(N = \infty) > 0$. This uses ergodicity *and* the fact that $d = 2$.

Part 2: Assume (w.l.o.g.) that $\theta(\hat{x}, (1, 0)) < \pi/2$. We show that $P(N \geq 3) > 0$ implies $P(F_k) > 0$ for some k where F_k is the event that some tree in S touches $Q_k = \{(0, 0), (0, 1), \dots, (0, k)\}$ but no other site with x-coordinate ≤ 0. This is a "local perturbation" argument in which one begins with three trees and then "chops off" the middle one (using Hypothesis C) by greatly increasing τ_e for each of the $k + 1$ edges into Q_k from the left.

Part 3: We show that $P(F_k) > 0$ is impossible. This uses a deterministic lack-of-space argument, as in [BK]. Consider a regular array of nonintersecting translates of Q_k in an $(L \times L)$ square and the corresponding translates of F_k. $P(F_k) > 0$ implies a positive probability that cL^2 of these events occur simultaneously and hence that there are cL^2 disjoint (infinite) trees all exiting a boundary of size $c'L$, which is impossible for large L. $\square$

Note added in proof: For a complete proof of Theorem 2.3 (in fact, of an improved version which does not require Hypothesis C), see C. Licea and C.M. Newman, *Geodesics in two-dimensional first-passage percolation*, Ann. Probab., in press.

References

[A]   K. Alexander, *Approximation of subadditive functions and convergence rates in limiting-shape results*, Ann. Probab., in press.

[BK]   R. Burton and M. Keane, *Density and uniqueness in percolation*, Comm. Math. Phys., **121** (1989), 501–505.

[CD]   J. T. Cox and R. Durrett, *Some limit theorems for percolation processes with necessary and sufficient conditions*, Ann. Probab., **9** (1981) 583–603.

[HW] J. M. Hammersley and D. J. A. Welsh, *First-passage percolation, subadditive processes, stochastic networks and generalized renewal theory*, in Bernoulli, Bayes, Laplace Anniversary Volume (J. Neyman and L. Lecam, eds.), Springer-Verlag, Berlin and New York (1965), 61–110.

[K1]   H. Kesten, *Aspects of first-passage percolation*, in Lecture Notes in Math., **1180**, Springer-Verlag, Berlin and New York (1986), 125–264.

[K2]   H. Kesten, *On the speed of convergence in first-passage percolation*, Ann. Appl. Probab., **3** (1993), 296–338.

[KS]   J. Krug and H. Spohn, *Kinetic roughening of growing surfaces*, in Solids Far from Equilibrium: Growth, Morphology and Defects (C. Godrèche, ed.), Cambridge Univ. Press, London and New York (1991), 479–582.

[NP]   C. M. Newman and M. S. T. Piza, *Divergence of shape fluctuations in two dimensions*, Ann. Probab., **23** (1995), in press.

[R]   D. Richardson, *Random growth in a tesselation*, Proc. Cambridge Philos. Soc., **74** (1973), 515–528.

Quantum Stochastic Calculus

K.R. PARTHASARATHY

Indian Statistical Institute, New Delhi 110016, India

To Alberto Frigerio

ABSTRACT. The basic integrator processes of quantum stochastic calculus, namely, creation, conservation, and annihilation, are introduced in the Hilbert space of square integrable Brownian functionals. Stochastic integrals with respect to these processes and a quantum Itô's formula are described. As an application two examples of quantum stochastic differential equations are discussed. A continuous time version of Stinespring's theorem on completely positive maps in C^*-algebras is exploited to formulate the notion of a quantum Markov process and indicate how classical Markov processes are woven into the fabric of the Schödinger-Heisenberg dynamics of quantum theory.

1. Introduction

Let $\{X(t), t \in I\}$ be a commuting family of self-adjoint operators in a complex separable Hilbert space $\mathcal{H}$, I being an interval on the line. For any fixed unit vector u in $\mathcal{H}$ consider the functions:

$$\varphi^u_{t_1, t_2, \ldots, t_n}(x_1, x_2, \ldots, x_n) = \langle u, e^{i \sum_j x_j X(t_j)} u \rangle \tag{1.1}$$

where $(x_1, x_2, \ldots, x_n) \in \mathbb{R}^n$, $t_j \in I$, and $< \cdot, \cdot >$ denotes the scalar product, which is linear in the second variable. Then (1.1) is a consistent family of characteristic functions of finite-dimensional probability distributions. By Kolmogorov's theorem, $\{X(t)\}$ together with u determine a real-valued stochastic process with a law P whose n-dimensional distributions have Fourier transforms given by (1.1). We say that the family $\{X(t)\}$ of *observables executes a stochastic process with the law P in the vector state u.* This at once suggests the possibility of construction of models of stochastic processes by a differential analysis of expressions of the form $dX(t) = X(t + dt) - X(t)$ in terms of some basic and universal operator-valued functions of a time variable. In this context we are reminded of the Schrödinger-Heisenberg dynamics in quantum theory where $X(t) = e^{itH} X_0 e^{-itH}$, H and X_0 are self-adjoint operators, and $dX(t) = i(HX(t) - X(t)H)dt$ but $\{X(t)\}$ is seldom commutative. Is it possible to introduce some universal noise differentials and realize some of the well-known classical processes? Starting from standard Brownian motion we explore this problem and present a few illustrations.

Proceedings of the International Congress
of Mathematicians, Zürich, Switzerland 1994
© Birkhäuser Verlag, Basel, Switzerland 1995

2. The basic noise processes of quantum stochastic calculus

We first express the standard Brownian motion (SBM) on $\mathbb{R}$ as a family $\{Q(t), t \geq 0\}$ of self-adjoint operators together with a vector state. To this end, denote by $(\Omega, \mathcal{F}, P)$ the probability space of SBM so that a sample point ω in Ω is a Brownian trajectory B. For any $0 \leq s < t < \infty$ denote by $\mathcal{F}(s,t) \subset \mathcal{F}$ the smallest σ-algebra generated by the random variables $\{B(b) - B(a), s \leq a < b \leq t\}$ and write $\mathcal{H}(s,t) = L^2(\Omega, \mathcal{F}(s,t), P)$. The independent increments property of B implies that for any partition $0 = t_0 < t_1 < \cdots < t_n < \cdots$ of $\mathbb{R}_+ = [0, \infty)$, the Hilbert space $\mathcal{H} = L^2(P)$ can be expressed as a countable tensor product: $\mathcal{H} = \mathcal{H}(t_0, t_1) \otimes \mathcal{H}(t_1, t_2) \otimes \cdots$ with respect to the stabilizing sequence $\{1\}$ of unit vectors 1, the constant function identically equal to 1 on Ω. In other words, $\mathcal{H}$ can be visualized as a *continuous tensor product* of Hilbert spaces. A family $\{X(t), t \geq 0\}$ of (not necessarily bounded) operators on $\mathcal{H}$ is called an *adapted process* (with respect to B) if, roughly speaking, for each $t, X(t) = X_0(t) \otimes I(t, \infty)$ where $X_0(t)$ is an operator in $\mathcal{H}(0, t)$ and $I(t, \infty)$ is the identity operator in $\mathcal{H}(t, \infty)$. Care is needed regarding domains of unbounded operators and for details we refer to [20]. Denote by $I(s,t)$ the identity operator in $\mathcal{H}(s,t)$. Define the commuting family $\{Q(t), t \geq 0\}$ of self-adjoint multiplication operators in $\mathcal{H}$ by

$$[Q(t)f](B) = B(t)f(B) \tag{2.1}$$

on the domain $\{f | \int(1 + B(t)^2)|f(B)|^2 P(dB) < \infty\}$. Then $\{Q(t)\}$ is an adapted process that executes SBM in the vector state 1.

For any $\varphi \in L^2(\mathbb{R}_+)$ introduce the *exponential* random variable $e(\varphi)$ in $\mathcal{H}$ defined by

$$e(\varphi)(B) = \exp\left(\int_0^\infty \varphi \, dB - \frac{1}{2}\int_0^\infty \varphi(s)^2 \, ds\right) \tag{2.2}$$

and its normalized form

$$e_0(\varphi) = (\exp -\frac{1}{2}||\varphi||^2)e(\varphi). \tag{2.3}$$

Then $e_0(\varphi)$ is a unit vector, $\langle e(\varphi_1), e(\varphi_2)\rangle = \exp\langle\varphi_1, \varphi_2\rangle$ for any φ_1, φ_2 in $L^2(\mathbb{R}_+)$ and $\{e(\varphi), \varphi \in L^2(\mathbb{R}_+)\}$ is a total and linearly independent set in $\mathcal{H}$. Each $e(\varphi)$ has the important factorizability property: $e(\varphi) = e(\varphi 1_{[0,t]}) \otimes e(\varphi 1_{(t,\infty)})$ for all $t > 0$.

For any $x \in \mathbb{R}, t \geq 0$ there exist unique unitary operators $U_x(t), V_x(t)$ satisfying

$$U_x(t)e(\varphi) = e(e^{ix1_{[0,t]}}\varphi), \tag{2.4}$$

$$V_x(t)e(\varphi) = [\exp(-\frac{x^2}{2}t + x\int_0^t \varphi(s)\,ds)]e(\varphi - x1_{[0,t]}) \tag{2.5}$$

for all $\varphi \in L^2(\mathbb{R}_+), 1_{[a,b]}$ denoting the indicator of $[a,b]$. This is easily established by showing that $U_x(t)$ and $V_x(t)$ preserve scalar products on the total set of exponential random variables. The unitarity of $V_x(t)$ is also the Cameron-Martin theorem. Each of the families $\{U_x(t), x \in \mathbb{R}, t \geq 0\}, \{V_x(t), x \in \mathbb{R}, t \geq 0\}$ is commutative, strongly continuous in x and $U_x(t)U_y(t) = U_{x+y}(t)$, $V_x(t)V_y(t) = V_{x+y}(t)$. Hence, by Stone's theorem, there exist unique self-adjoint operators $\Lambda(t), P(t)$ satisfying

$$U_x(t) = e^{ix\Lambda(t)}, V_x(t) = e^{ixP(t)} \tag{2.6}$$

and each of the families $\{\Lambda(t)\}$ and $\{P(t)\}$ is a commutative adapted process. If $M(t)$ is any one of the operators $Q(t), \Lambda(t), P(t)$ defined by (2.1) and (2.6) then for $s < t$ the "increment" $M(t) - M(s)$ factorizes as $I(0, s) \otimes M_0(s, t) \otimes I(t, \infty)$ where $M_0(s, t)$ is a self-adjoint operator in $\mathcal{H}(s, t)$. Furthermore, M satisfies the *martingale property:* $\langle e(\varphi), (M(t) - M(s))e(\psi)\rangle = 0$ whenever φ and ψ are supported in $[0, s]$ and $0 \le s < t < \infty$. With due care paid to unbounded operators the following theorems hold:

THEOREM 2.1 *For any $\varphi \in L^2(\mathbb{R}_+)$, in the vector state $e_0(\varphi)$ defined by (2.3), $\{Q(t) - 2\int_0^t Re\, \varphi(s)\, ds\}$ and $\{P(t) - 2\int_0^t Im\, \varphi(s)\, ds\}$ execute SBM whereas $\{\Lambda(t)\}$ executes a Poisson process with intensity measure λ on $\mathbb{R}$, given by $d\lambda = |\varphi(t)|^2 dt$.*

THEOREM 2.2 *The operators $\{Q(t), \Lambda(t), P(t)\}$ obey the following commutation relations:*

$$[\Lambda(t), Q(s)] = -iP(s \wedge t), [\Lambda(t), P(s)] = iQ(s \wedge t),$$

$$[Q(t), P(s)] = 2i\, s \wedge t,$$

$$e^{i\theta\Lambda(\infty)}Q(t)e^{-i\theta\Lambda(\infty)} = (\cos\,\theta)Q(t) + (\sin\,\theta)P(t), \theta \in \mathbb{R},$$

where $s \wedge t$ denotes the minimum of s and t.

Because there exists a rich theory of stochastic integration with respect to Brownian motion, and the Poisson process (and, more generally, local semimartingales), Theorem 2.1 raises the very natural question of whether there could be a fruitful theory of integration with respect to the adapted operator processes Q, Λ, and P. In order to facilitate computations in such an investigation it is covenient to introduce the operators

$$A(t) = \frac{1}{2}(Q(t) + iP(t)), A^\dagger(t) = \frac{1}{2}(Q(t) - iP(t)). \tag{2.7}$$

Then we have the eigen relation $A(t)e(\varphi) = (\int_0^t \varphi(s)ds)e(\varphi)$ and the adjoint relation $\langle e(\varphi), A(t)e(\psi)\rangle = \langle A^\dagger(t)e(\phi), e(\psi)\rangle$ for all $\varphi, \psi \in L^2(\mathbb{R})$. Borrowing from the terminology of free field theory we call $A^\dagger = \{A^\dagger(t)\}$, $\Lambda = \{\Lambda(t)\}$, and $A = \{A(t)\}$ the *creation, conservation,* and *annihilation* processes respectively. The commutation relations in Theorem 2.2 assume the form

$$[A^\dagger(t), A^\dagger(s)] = [\Lambda(t), \Lambda(s)] = [A(t), A(s)] = 0,$$

$$[\Lambda(t), A(s)] = -A(t \wedge s), [\Lambda(t), A^\dagger(s)] = A^\dagger(t \wedge s),$$

$$[A(t), A^\dagger(s)] = s \wedge t, e^{i\theta\Lambda(\infty)}A(t)e^{-i\theta\Lambda(\infty)} = e^{-i\theta}A(t).$$

$A^\dagger$, Λ and A are the basic noise processes with respect to which quantum stochastic integrals will be defined.

REMARKS. Theorem 2.2 is nothing but the canonical commutation relations (CCRs) of a free boson field expressed in terms of SBM. This connection between SBM and CCR was first observed by Segal [24]. The pair $(Q(t), P(t))$ as a quantum Wiener process was first introduced and investigated by Cockroft and Hudson [8]. It follows from Theorems 2.1 and 2.2 that, for any fixed angle θ, the adapted process $\{(\cos \theta)Q(t) + (\sin \theta)P(t), t \geq 0\}$ executes SBM in the vector state 1. A slightly more delicate result is the fact that for any $\lambda > 0$, the adapted process $\{\Lambda(t) + \sqrt{\lambda} Q(t) + \lambda t\}$ executes a Poisson process with intensity parameter λ in the vector state 1.

3. Integration with respect to the basic noise processes and quantum Ito's formula

In order to describe the integrands of our calculus we need an initial Hilbert space $\mathcal{H}_0$, called the *system* Hilbert space along with the *noise* or *bath* Hilbert space $\mathcal{H}$ where the creation, conservation, and annihilation processes are defined. The operators of interest will be in $\tilde{\mathcal{H}} = \mathcal{H}_0 \otimes \mathcal{H}$ where the system and noise interact. If we write $\tilde{\mathcal{H}}(0,t) = \mathcal{H}_0 \otimes \mathcal{H}(0,t)$ then $\tilde{\mathcal{H}} = \tilde{\mathcal{H}}(0,t) \otimes \mathcal{H}(t,\infty)$ for every $t \geq 0$. If $X = \{X(t)\}$ is a family of operators in $\tilde{\mathcal{H}}$ such that, roughly speaking, $X(t) = X_0(t) \otimes I(t,\infty)$ where $X_0(t)$ is an operator in $\tilde{\mathcal{H}}(0,t)$ for each t then we say that X is an adapted process. (A little more precisely, we demand that for a rich class of vectors $f \in \mathcal{H}_0$, $\varphi \in L^2(\mathbb{R}_+)$, the identity $X(t)f \otimes e(\varphi) = \{X_0(t)f \otimes e(1_{[0,t]}\varphi)\} \otimes (1_{[t,\infty)}\varphi)$ should hold.) If M is any of $A^\dagger, \Lambda, A$ we look upon $M(t)$ as the operator $I_{\mathcal{H}_0} \otimes M(t)$ in $\tilde{\mathcal{H}}$. For any adapted process X it is important to note that for $s < t$, $X(s)$ is operating in $\tilde{\mathcal{H}}(0,s)$, $M(t) - M(s)$ is operating in $\mathcal{H}(s,t)$, and hence $X(s)$ and $M(t) - M(s)$ commute with each other in a wide sense. An adapted process X is said to be *simple* if there exists a partition $0 = t_0 < t_1 < \cdots < t_n < \cdots\cdots$. of $\mathbb{R}_+$ such that

$$X = X(0)1_{[0,t_1]} + \sum_{j=1}^{\infty} X(t_j)1_{(t_j,t_{j+1}]}.$$

For any four simple adapted processes E, F, G, H, simplicity being with respect to the same partition given by $\{t_j\}$, we write

$$\int_0^t (E\, dA^\dagger + F\, d\Lambda + G\, dA + H\, ds)$$

$$= \sum_{j=1}^{\infty} \{E(t_{j-1})(A^\dagger(t_j \wedge t) - A^\dagger(t_{j-1} \wedge t)) + F(t_{j-1})(\Lambda(t_j \wedge t) - \Lambda(t_{j-1} \wedge t))$$

$$+ G(t_{j-1})(A(t_j \wedge t) - A(t_{j-1} \wedge t)) + H(t_{j-1})(t_j \wedge t - t_{j-1} \wedge t)\}.$$

Then the right-hand side is another adapted process. Such an integration can be completed to a fairly rich class of quadruples (E, F, G, H) of adapted processes. If an adapted process X has the form

$$X(t) = X(0) + \int_0^t (E\, dA^\dagger + F\, d\Lambda + G\, dA + H\, ds)$$

we write

$$dX = E\,dA^\dagger + F\,d\Lambda + G\,dA + H\,dt$$

and say that X has initial value $X(0)$. If dM denotes any one of $dA^\dagger, d\Lambda, dA, dt$ then for any adapted process E, $EdM = (dM)E$. If $dX_i = E_i dA^\dagger + F_i d\Lambda + G_i dA + H_i dt$, $i = 1, 2$ then one has a *quantum Ito's formula* for the differential $d(X_1 X_2)$ of the product process $X_1 X_2 = \{X_1(t) X_2(t)\}$:

$$
\begin{aligned}
d(X_1 X_2) \quad &= \quad (E_1 X_2 dA^\dagger + F_1 X_2 d\Lambda + G_1 X_2 dA + H_1 X_2 dt) \\
&\quad + (X_1 E_2 dA^\dagger + X_1 F_2 d\Lambda + X_1 G_2 dA + X_1 H_2 dt) \\
&\quad + (F_1 E_2 dA^\dagger + F_1 F_2 d\Lambda + G_1 F_2 dA + G_1 E_2 dt),
\end{aligned}
$$

which can be abbreviated as

$$d(X_1 X_2) = X_1 dX_2 + (dX_1)X_2 + dX_1 dX_2 \tag{3.1}$$

where the Itô correction term $dX_1 dX_2$ is computed by bilinearity and the multiplication table

	$dA^\dagger$	$d\Lambda$	dA	dt
dA	dt	dA	0	0
$d\Lambda$	$dA^\dagger$	$d\Lambda$	0	0
$dA^\dagger$	0	0	0	0
dt	0	0	0	0

$$\tag{3.2}$$

If (dM_1, dM_2) is any ordered pair from the set $\{dA^\dagger, d\Lambda, dA, dt\}$ then $dM_1 dM_2 = 0$ whenever the order of creation, conservation, and annihilation is not violated. This is an enriched Wick ordering with $d\Lambda$ included.

Because $Q(t) = A(t) + A^\dagger(t)$ the quantum Itô's formula implies $(dQ)^2 = dt$, which is the classical Itô's formula for SBM. If $N_\lambda(t) = \Lambda(t) + \sqrt{\lambda}Q(t) + \lambda t$ then $(dN_\lambda)^2 = dN_\lambda$, which is the classical formula for the Poisson process. Itô's formula as described by (3.1) and (3.2) is derived entirely from the CCR, which encapsulates the Heisenberg uncertainty principle. Thus, the Itô correction in the case of Brownian motion as well as the Poisson process can be attributed to the uncertainty principle.

The quantum Itô's formula suggests the possibility of expressing and analyzing models of stochastic processes in terms of quantum stochastic differential equations (q.s.d.e's). We illustrate this by two examples.

EXAMPLE 3.1 Let $\mathcal{H}_0 = \mathbb{C}$ so that $\tilde{\mathcal{H}} = \mathcal{H}$. Consider the equation

$$dX = (c - 1)X d\Lambda + dA^\dagger + dA, X(0) = xI \tag{3.3}$$

where X is an unknown adapted process and c is a real constant satisfying $-1 \leq c < 1$. It has an explicit solution $\{X(c, x, t), t \geq 0\}$ in terms of stochastic integrals satisfying the following: (i) $\|X(c, x, t)\| \leq |x| + \sqrt{2t(1 - c)^{-1}}$; (ii) for each fixed c, x it is a commutative adapted process; (iii) in the vector state 1, it is a martingale as well as a Markov process obeying the classical Ito's formula in terms of paths:

$$df(X(t)) = (L_c f)(X(t-))dX(t) + M_c f(X(t))dt$$

where $(L_c f)(y) = [(c-1)y]^{-1}(f(cy) - f(y))$, $(M_c f)(y) = [(c-1)y]^{-2}(f(cy) - f(y) - (c-1)yf'(y))$. In the literature these are known as Azéma martingales, and their properties have been studied in detail by Emery [9], Chebotarev and Fagnola [7], and the author [19], [21].

EXAMPLE 3.2 Let L_i, $1 \le i \le 4$, be bounded operators in the initial or system Hilbert space $\mathcal{H}_0$. Viewing each L_i as a constant adapted process $\{L_i(t)\}$ where $L_i(t) = L_i \otimes I_{\mathcal{H}}$ in $\tilde{\mathcal{H}} = \mathcal{H}_0 \otimes \mathcal{H}$, consider the exponential type q.s.d.e.:

$$dU = (L_1 dA^\dagger + L_2 d\Lambda + L_3 dA + L_4 dt)U, \quad U(0) = I \tag{3.4}$$

where U is an unknown adapted process. By following a Picard type iterative scheme and using the quantum Itô's formula it is possible to show that (3.4) admits a unique unitary operator-valued adapted process $U = \{U(t)\}$ as a solution if and only if the quadruple (L_1, L_2, L_3, L_4) has the form

$$L_1 = L, \quad L_2 = S - I, \quad L_3 = -L^* S, \quad L_4 = -iH - \frac{1}{2}L^* L \tag{3.5}$$

where S is unitary and H is self-adjoint. When $L = 0$, $S = I$, (3.4) reduces to the Schrödinger equation $dU = -iHU\,dt$. Thus, (3.4) may be interpreted as a *Schrödinger equation in the presence of noise*. This suggests that a detailed study of (3.4) would be fruitful when the operators L_i are not necessarily bounded. Just to give a flavor of this we present two simple examples. Consider the two equations

$$dU = (L_i dA^\dagger - L_i^* dA - \frac{1}{2}L_i^* L_i dt)U, \quad U(0) = I, \quad i = 1, 2 \tag{3.6}$$

where

$$L_1 = \begin{pmatrix} 0 & 0 & 0 & \cdots \\ \lambda_0 & 0 & 0 & \cdots \\ 0 & \lambda_1 & 0 & \cdots \\ . & . & . & \cdots \\ . & . & . & \cdots \end{pmatrix}, \quad L_2 = \begin{pmatrix} 0 & \mu_1 & 0 & 0 & \cdots \\ 0 & 0 & \mu_2 & 0 & \cdots \\ 0 & 0 & 0 & & \cdots \\ . & . & . & & \cdots \\ . & . & . & & \cdots \end{pmatrix} \tag{3.7}$$

are operators in $\ell^2(\{0, 1, 2, \ldots\})$, $\lambda_0, \lambda_1, \ldots, \mu_1, \mu_2, \ldots$ being complex scalars. Then a unique unitary solution for (3.6) in the first case exists if and only if $\sum_j |\lambda_j|^{-2} = \infty$. This should remind the reader of Feller's criterion for a pure birth process with birth rates $\{|\lambda_j|^2\}$ to be nonexplosive. In fact there is a very close connection between the existence of unitary solutions of (3.4) and the conservativity of an associated minimal Markov semigroup in the sense of Feller [4]. In the second case there always exists a unitary solution. This corresponds to the fact that pure death processes are always conservative. It is interesting to note that the birth rates $\{|\lambda_j|^2\}$ and death rates $\{|\mu_j|^2\}$ are being replaced by their corresponding complex amplitudes $\{\lambda_j\}$ and $\{\mu_j\}$ in our description. This is in tune with the spirit of quantum theory. If $e_0, e_1, \ldots$ is the canonical basis of $\ell^2(\{0, 1, 2, \ldots\})$ and e_j labels the state that there are j particles in the system then L_1 in (3.7) is a creation operator in the sense that $L_1 e_j = \lambda_j e_{j+1}$ whereas L_2 in (3.7) is an annihilation operator in the sense that $L_2 e_j = \mu_j e_{j-1}$. When the system's creation operator teams up with the differential of the creation process in (3.6) we get a pure birth process whereas in the opposite case we get a pure death process in the noisy Schrödinger picture.

REMARKS. There are now three accounts of quantum stochastic calculus available in the form of books and lecture notes. [20] is a fairly self-contained and leisurely account of the subject whereas the lecture notes by Meyer [17] and Biane [5] contain more up-to-date, brisk, and lively accounts based on Maassen's kernel calculus [15] and are also aimed at an expert audience of probabilists. The quantum Itô's formula and its application to noisy Schrödinger equations with bounded coefficients first appeared in [13] by Hudson and the author. A heuristic and preliminary form of Itô's formula without the conservation process appeared earlier in [14], [12]. The discussion on birth and death processes through noisy Schrödinger equations has been taken from Fagnola [11]. The topic of noisy Schrödinger equations with unbounded operator coefficients has recently witnessed considerable progress from the hands of several authors including Mohari, Chebotarev, Fagnola, Sinha, Bhat, etc. The bibliographical details may be found in [17].

4. The case of several degrees of freedom

Let $\underline{B} = (B_1, B_2, \ldots, B_n)$ denote the sample trajectory of a standard n-dimensional Brownian motion described by the probability measure P. For any $\underline{\varphi} = (\varphi_1, \varphi_2, \ldots, \varphi_n)$ in the n-fold direct sum $L^2(\mathbb{R}_+) \oplus \cdots \oplus L^2(\mathbb{R}_+)$ define the exponential random variable $e(\underline{\varphi})$ in $\mathcal{H} = L^2(P)$ by

$$e(\underline{\varphi})(\underline{B}) = \exp\left(\int_0^\infty \underline{\varphi} \cdot d\underline{B} - \frac{1}{2} \int_0^\infty \sum_j \varphi_j^2(s) \, ds \right).$$

The analogues of the adapted processes $Q(t)$ and $P(t)$ of Section 2 are now defined by

$$[Q_j(t)f](\underline{B}) = B_j(t)f(\underline{B}), \quad f \in \mathcal{H},$$

$$[e^{ixP_j(t)}e(\underline{\varphi})](\underline{B}) = \exp\left(-\frac{x^2}{2}t + x \int_0^t \varphi_j(s) \, ds\right)e(\underline{\varphi} - x 1_{[0,t]}e_j)$$

for all $e(\underline{\varphi})$, where $x \in \mathbb{R}$ and e_j is the jth element in the canonical basis of $\mathbb{C}^n$. To define the analogue of $\Lambda(t)$ we exploit the natural representation of the group $U(n)$ in $\mathcal{H}$. For any $(n \times n)$ hermitian matrix H define $\Lambda_H(t)$ in $\mathcal{H}$ as the Stone generator of the one parameter unitary group $e^{ix\Lambda_H(t)}$ satisfying

$$e^{ix\Lambda_H(t)}e(\underline{\varphi}) = e(e^{ix 1_{[0,t]}H}\underline{\varphi}), \quad x \in \mathbb{R}$$

for all $e(\underline{\varphi})$. For any $(n \times n)$-matrix K define

$$\Lambda_K(t) = \Lambda_{\frac{K+K^*}{2}}(t) + i\Lambda_{\frac{K-K^*}{2i}}(t).$$

If $K = E_{ij}$ is the elementary matrix with 1 in the ijth position and 0 elsewhere write

$$\Lambda_i^j(t) = \Lambda_{E_{ij}}(t), \quad t \geq 0, \ 1 \leq i, j \leq n$$

and put

$$\Lambda_0^j(t) = \frac{Q_j(t) + iP_j(t)}{2}, \quad \Lambda_j^0(t) = \frac{Q_j(t) - iP_j(t)}{2}, \quad \Lambda_0^0(t) = tI.$$

Then the canonical commutation relations (CCRs) can be expressed as

$$[\Lambda_j^i(s), \Lambda_\ell^k(t)] = \delta_\ell^i \Lambda_j^k(s \wedge t) - \delta_j^k \Lambda_\ell^i(s \wedge t),$$

$$[\Lambda_j^i(s), \Lambda_0^k(t)] = -\delta_j^k \Lambda_0^i(s \wedge t), [\Lambda_j^i(s), \Lambda_k^0(t)] = \delta_k^i \Lambda_j^0(s \wedge t)$$

for $1 \leq i, j, k, \ell \leq n$. All the processes $\{\Lambda_j^i(t), 0 \leq i, j \leq n\}$ are adapted to the Brownian motion. $\Lambda_j^i(t)$ and $\Lambda_i^j(t)$ are adjoint to each other on the linear manifold generated by exponential random variables. For a rich class of adapted processes $\{L_j^i(t)\}$, $0 \leq i, j \leq n$ in $\mathcal{H}_0 \otimes \mathcal{H} = \tilde{\mathcal{H}}$ it is possible to define stochastic integrals of the form $\int_0^t L_j^i(s) \, d\Lambda_i^j(s)$ where repeated indices indicate summation over them. The quantum Itô's formula is expressed by

$$d\Lambda_j^i d\Lambda_\ell^k = \hat{\delta}_\ell^i d\Lambda_j^k, \quad 0 \leq i, j, k, \ell \leq n,$$

where $\hat{\delta}_\ell^i$ is the modified Kronecker delta given by $\hat{\delta}_\ell^i = \delta_\ell^i$ if $(i, \ell) \neq (0, 0)$ and $\hat{\delta}_\ell^i = 0$ otherwise.

As an application of the quantum Itô's formula consider the exponential q.s.d.e.

$$dU = (L_j^i d\Lambda_i^j)U, \quad U(0) = I \tag{4.1}$$

where L_j^i, $0 \leq i, j \leq n$, are bounded operators in $\mathcal{H}_0$, viewed also as constant adapted processes. Then the following theorem holds:

THEOREM 4.1 *There exists a unique unitary operator-valued solution $\{U(t)\}$ for (4.1) if and only if the following conditions are fulfilled.*
(a)

$$L_j^i = \begin{cases} S_j^i - \delta_j^i & \text{if } i \geq 1, j \geq 1, \\ L_i & \text{if } i \geq 1, j = 0, \\ -\sum_k L_k^* S_j^k & \text{if } i = 0, j \geq 1, \\ -iH - \frac{1}{2}\sum_k L_k^* L_k & \text{if } i = j = 0; \end{cases}$$

(b) $((S_j^i))_{1 \leq i, j \leq n}$ is a unitary operator in the n-fold direct sum $\mathcal{H}_0 \oplus \cdots \oplus \mathcal{H}_0$;
(c) H is a bounded self-adjoint operator in $\mathcal{H}_0$.

COROLLARY 4.2 *For any bounded operator X in $\mathcal{H}_0$ define $j_t^0(X) = U(t)^*(X \otimes I_\mathcal{H})U(t)$ where $U(t)$ is determined by Theorem 4.1. Then j_t^0 is a * unital homomorphism from $\mathcal{B}(\mathcal{H}_0)$ into $\mathcal{B}(\mathcal{H}_0 \otimes \mathcal{H})$ satisfying*

$$dj_t^0(X) = j_t^0(\theta_j^i(X))d\Lambda_i^j(t)$$

where $\mathcal{B}(\mathcal{K})$ denotes the $*$ algebra of all bounded operators in any Hilbert space $\mathcal{K}$ and $\theta^i_j : \mathcal{B}(\mathcal{H}_0) \to \mathcal{B}(\mathcal{H}_0)$ are maps defined by

$$
\theta^i_j(X) = \begin{cases}
(S^k_i)^* X S^k_j - \delta^i_j X & \text{if } i \geq 1, j \geq 1, \\
(S^k_i)^* [X, L_k] & \text{if } i \geq 1, j = 0, \\
[L^*_k, X] S^k_j & \text{if } i = 0, j \geq 1, \\
i[H, X] - \frac{1}{2} \sum_k (L^*_k L_k X + X L^*_k L_k - 2 L^*_k X L_k) & \text{if } i = 0, j = 0
\end{cases}
$$

with the convention that repeated index implies summation.

   If $\mathcal{A}_0 \subset \mathcal{B}(\mathcal{H}_0)$ is an abelian $*$ algebra and the maps θ^i_j leave $\mathcal{A}_0$ invariant then the family $\{j^0_t(X), X \in \mathcal{A}_0, t \geq 0\}$ is commutative.

REMARKS. Theorem 4.1 and Corollary 4.2 are proved in [13]. Mohari and Sinha [18] have proved an important and useful generalization of this result when the number of degrees of freedom is infinite. For details see [19]. Accardi, Frigerio, and Lu have studied several physical models leading to q.s.d.e.'s of the noisy Schrödinger type. A number of references on this topic may be found in [2]. Several interesting connections between quantum groups and q.s.d.e.'s have been discussed by Schürmann [23].

5.   Quantum Markov processes

A discrete time Markov process in classical probability theory is determined by a family $(X_i, \mathcal{F}_i)$, $i = 0, 1, 2, \ldots$ of measurable spaces, an initial distribution μ_0 on $(X_0, \mathcal{F}_0)$, and transition probabilities $P_i(x_i, dx_{i+1})$ from $(X_i, \mathcal{F}_i)$ to $(X_{i+1}, \mathcal{F}_{i+1})$ for each i. Then there exists a unique probability measure P_μ on the infinite product space $(\Omega, \mathcal{F}) = \bigotimes_{i=0}^{\infty} (X_i, \mathcal{F}_i)$ such that the projection of P_μ on $\bigotimes_{i=0}^{n} (X_i, \mathcal{F}_i)$ is given by

$$
P^n_\mu(E_0 \times E_1 \times \cdots \times E_n)
$$
$$
= \int_{E_0 \times E_1 \times \cdots \times E_n} \mu(dx_0) P_0(x_0, dx_1) P_1(x_1, dx_2) \cdots P_{n-1}(x_{n-1}, dx_n)
$$

for all $E_i \in \mathcal{F}_i$, $0 \leq i \leq n$, $n = 0, 1, 2, \ldots$. The probability space $(\Omega, \mathcal{F}, P_\mu)$ describes the Markov process with initial distribution μ and transition probability $P_i(\cdot, \cdot)$ for transition from a state at time i to a new state at time $i+1$. Denote by $\mathcal{A}_i$ the commutative $*$ algebra of all complex-valued bounded random variables on $(X_i, \mathcal{F}_i)$. Introduce the positive unital operators $T(i, i+1) : \mathcal{A}_{i+1} \to \mathcal{A}_i$ defined by

$$
(T(i, i+1)g)(x_i) = \int g(x_{i+1}) P_i(x_i, dx_{i+1}).
$$

For any $i \leq k$ define $T(i, k) : \mathcal{A}_k \to \mathcal{A}_i$ by $T(i, k) = T(i, i+1) T(i+1, i+2) \cdots T(k-1, k)$ with the convention $T(i, i) = $ identity. The family $\{T(i, k), i \leq k\}$ of transition operators obeys the Chapman-Kolmogorov equations: $T(i, k) T(k, \ell) = T(i, \ell)$ for $i \leq k \leq \ell$. Let $\mathcal{H}$ be the Hilbert space $L^2(P_\mu)$ and $F(i)$ the Hilbert space projection on the subspace of functions depending only on the first $i + 1$ coordinates

$(x_0, x_1, \ldots, x_i)$ of $\omega = (x_0, x_1, x_2, \ldots)$ in Ω. Then $\{F(i)\}$ is an increasing sequence of projections in $\mathcal{H}$. For any $g \in \mathcal{A}_i$ define the operator $j_i(g)$ in $\mathcal{H}$ by

$$[j_i(g)\varphi](\omega) = g(x_i)(F(i)\varphi)(\omega), \quad \omega = (x_0, x_1, \ldots), \quad \varphi \in \mathcal{H}. \tag{5.1}$$

Then j_i is a $*$ homomorphism from $\mathcal{A}_i$ into the $*$ algebra $\mathcal{B}(\mathcal{H})$ of all bounded operators in $\mathcal{H}$. The Markov property of the stochastic process $(\Omega, \mathcal{F}, P_\mu)$ is encapsulated in the operator relations

$$j_k(1) = F(k), F(i)j_k(g)F(i) = j_i(T(i,k)g), \quad g \in \mathcal{A}_k, \quad i \le k, \tag{5.2}$$

$$\langle u, j_0(g_0)j_1(g_1) \ldots j_n(g_n)v \rangle = \int (\bar{u}vg_0)(x_0)g_1(x_1) \ldots g_n(x_n)P_\mu(d\omega) \tag{5.3}$$

for all u, v in the range of $F(0)$, $g_i \in \mathcal{A}_i$, $i = 0, 1, 2, \ldots, n, n = 0, 1, 2, \ldots$. We may call the triple $(\mathcal{H}, F, \{j_k\}, k = 0, 1, 2, \ldots)$ consisting of the Hilbert space $\mathcal{H}$, the filtration of projections $F(k)$ increasing in k, and the family $\{j_k\}$ of $*$ (but nonunital) homomorphisms, a Markov process with transition operators $\{T(i,j), i \le j\}$. Such a description carries forward to the continuous time case without any extra effort.

We now quantize the picture given above. To this end consider a C^*-algebra $\mathcal{A}_t$ of bounded operators on a Hilbert space $\mathcal{K}_t$ for every $t \ge 0$. The time index t here may be discrete or continuous. It is useful to imagine any hermitian element $X \in \mathcal{A}_t$ as a real-valued bounded observable concerning a system at time t. For $0 \le s \le t < \infty$, let $T(s,t) : \mathcal{A}_t \to \mathcal{A}_s$ be a linear, unital, and completely positive map satisfying the following: (i) $T(s,s)$ is the identity map on $\mathcal{A}_s$; (ii) $T(r,t) = T(r,s)T(s,t)$ for all $0 \le r \le s \le t < \infty$. We say that $\{T(s,t)\}$ is a family of *transition operators*. Complete positivity of $T(s,t)$ is equivalent to the property that for any $(n \times n)$-matrix $((Y_{ij}))$ of elements in $\mathcal{A}_t$ the matrix $((T(s,t)(Y_{ij})))_{1 \le i,j \le n}$ is a positive operator in the n-fold direct sum $\mathcal{K}_s \oplus \cdots \oplus \mathcal{K}_s$ whenever $((Y_{ij}))$ is a positive operator in the n-fold direct sum $\mathcal{K}_t \oplus \cdots \oplus \mathcal{K}_t$. With this notation, the following generalization of Stinespring's theorem [25] holds:

THEOREM 5.1 *Let $\mathcal{A}_t$ be a unital C^*-algebra of operators in a Hilbert space $\mathcal{K}_t$ for every $t \ge 0$ and let $T(s,t) : \mathcal{A}_t \to \mathcal{A}_s$, $s \le t$ be a family of transition operators. Then there exists a Hilbert space $\mathcal{H}$, an increasing family $\{F(t)\}$ of projections in $\mathcal{H}$, a family of contractive $*$ homomorphisms $j_t : \mathcal{A}_t \to \mathcal{B}(\mathcal{H})$, $t \ge 0$ and a unitary isomorphism V from $\mathcal{K}_0$ onto the range of $F(0)$ satisfying the following: (i) $j_t(I_t) = F(t), I_t$ being the identity in $\mathcal{K}_t$; (ii) for any $0 \le s \le t < \infty$, $X \in \mathcal{A}_t$, $F(s)j_t(X)F(s) = j_s(T(s,t)(X))$; (iii) the set $\{j_{t_1}(X_1) \cdots j_{t_n}(X_n)Vu, t_1 > \cdots > t_n = 0, X_i \in \mathcal{A}_{t_i}, u \in \mathcal{K}_0\}$ is total in $\mathcal{H}$; (iv) $j_0(X)V = VX$ for all $X \in \mathcal{A}_0$; (v) for any $u, v \in \mathcal{K}_0$, $t_1 > t_2 > \cdots > t_n = 0$, $X_i, Y_i \in \mathcal{A}_{t_i}$,*

$$\langle j_{t_1}(X_1) \ldots j_{t_n}(X_n)Vu, j_{t_1}(Y_1) \cdots j_{t_n}(Y_n)Vv \rangle$$
$$= \langle u, X_n^*(\cdots T(t_3, t_2)(X_2^* T(t_2, t_1)(X_1^* Y_1)Y_2) \cdots)Y_n v \rangle.$$

The quadruple $(\mathcal{H}, F, \{j_t\}, V)$ satisfying properties (i) - (iv) is unique up to a unitary isomorphism.

Our Hilbert space-theoretic description of a classical Markov process in terms of (5.2) and (5.3) together with Theorem 5.1 motivate the following definition:

suppose $\mathcal{A}_t, \mathcal{K}_t$ and $T(s,t)$, $s \leq t$ are as in Theorem 5.1. Then any quadruple $(\mathcal{H}, F, \{j_t\}, V)$ consisting of a Hilbert space $\mathcal{H}$, an increasing family of projections $\{F(t)\}$ in $\mathcal{H}$, contractive $*$ homomorphisms j_t from $\mathcal{A}_t$ into $\mathcal{B}(\mathcal{H})$, and a unitary isomorphism V from $\mathcal{K}_0$ onto the range of $F(0)$ is called a *conservative Markov flow* with transition operators $T(\cdot, \cdot)$ if properties (i), (ii), and (iv) of Theorem 5.1 are fulfilled. The unique such flow satisfying, in addition, property (iii) is said to be *minimal*. It should be interesting to obtain a differential description of the minimal flow in the case of continuous time. For such a description, are the basic integrators $\{\Lambda_j^i(t)\}$ with $0 \leq i, j < \infty$ sufficient?

EXAMPLE 5.2 In the notation of Section 4 consider the case when $\mathcal{H}_0$ is an n-dimensional Hilbert spae with an orthonormal basis $\{e_j\}$, $1 \leq j \leq n$. For any $u, v \in \mathcal{H}_0$, use the Dirac notation to define the operator $|u><v|$ satisfying $|u><v|w = \langle v, w \rangle u$. Consider the q.s.d.e. (4.1) where L_j^i are as in condition (a) of Theorem 4.1 with $S_j^i = |e_j><e_i|$, $L_i = |\ell_i><e_i|$, $1 \leq i, j \leq n$. Then the maps θ_j^i of Corollary 4.2 leave the diagonal algebra $\mathcal{A}_0$ of operators in the basis $\{e_k\}$ invariant and hence the family $\{j_t^0(X), X \in \mathcal{A}_0, t \geq 0\}$ is commutative. Define $j_t(X) = j_t^0(X)F(t)$ where $F(t)$ is the projection on the subspace $\tilde{\mathcal{H}}(0, t) = \mathcal{H}_0 \otimes \mathcal{H}(0, t)$ in $\tilde{\mathcal{H}}$. Let $Vu = u \otimes 1$ for all $u \in \mathcal{H}_0$. Then the quadruple $(\tilde{\mathcal{H}}, F, \{j_t\}, V)$ yields a Markov flow with a differential description given by Corollary 4.2. In fact it describes the classical Markov chain with generating matrix $((\omega_{ij}))$ given by

$$\omega_{ij} = \begin{cases} |\langle \ell_i, e_j \rangle|^2 & \text{if } i \neq j, \\ -\sum_{r:r \neq i} |\langle \ell_i, e_r \rangle|^2 & \text{if } i = j. \end{cases}$$

Thus, the intensity ω_{ij} of transition probability from the state i to the state j is hidden in our quantum description in the corresponding amplitude $\langle \ell_i, e_j \rangle$. It is possible that the flow thus described by q.s.d.e. is not minimal. It should be interesting to know whether minimality can be achieved by a q.s.d.e.

REMARKS. That the study of stochastic processes can be reduced to studying the dynamics of changing representations of a $*$ unital algebra is a fundamental idea proposed by Accardi, Frigerio, and Lewis in [1]. Theorem 5.1 has been taken from Bhat and the author [3]. Meyer [16] pointed out the possibility of realizing Markov chains through q.s.d.e. More general Markov processes were realized by the author and Sinha [22] using the idea of an Evans-Hudson flow. The subject is undergoing many interesting developments for which we refer to [16], [17], [4].

References

[1] Accardi, L.; Frigerio, A.; and Lewis, J. T., *Quantum stochastic processes*, Publ. Res. Inst. Math. Sci., Kyoto Univ. 18, 97–133 (1982).

[2] Accardi, L., and Lu, Y. G., From Markovian approximation to a new type of quantum stochastic calculus, Quantum Probability and Related Topics VII (ed. Accardi, L.) World Scientific, Singapore (1992).

[3] Bhat, B. V. R., and Parthasarathy, K. R., *Kolmogorov's existence theorem for Markov processes in C^* algebras*, Proc. Indian Acad. Sci. Math. Sci. 104, 253–262 (1994).

[4] Bhat, B. V. R., and Parthasarathy, K. R., *Markov dilations of nonconservative quantum dynamical semigroups and a quantum boundary theory*, to appear in Ann. Inst. H. Poincaré.

[5] Biane, Ph., *Calcul stochastique non-commutatif*, Cours présenté à l'école d'été de probabilités de Saint Flour, Août 1993.

[6] Chebotarev, A. M., and Fagnola, F., *Sufficient conditions for conservativity of quantum dynamical semigroups*, J. Funct. Anal. 118, 131–153 (1993).

[7] Chebotarev, A. M., and Fagnola, F., *On quantum extensions of the Azéma martingale semigroup*, to appear in Sém. Prob.

[8] Cockroft, A. M., and Hudson, R. L., *Quantum mechanical Wiener processes*, J. Multivariate Anal. 7, 107–124 (1977).

[9] Emery, M., *On the Azéma martingales*, Sém. Prob. XXIII, Lecture Notes in Math., Springer, Berlin and New York, 1372, 66–87 (1989).

[10] Evans, M. P., and Hudson, R. L., *Multidimensional diffusions*, Lecture Notes in Math., Springer, Berlin and New York, 1303, 69–88 (1988).

[11] Fagnola, F., *Pure birth and death processes as quantum flows in Fock space*, Sankhyā Ser. A, 53, 288–297 (1991).

[12] Hudson, R. L.; Krandikar, R.L.; and Parthasarathy, K. R., *Towards a theory of noncommutative semimartingales adapted to Brownian motion and a quantum Ito's formula*, Theory and Applications of Random Fields, Bangalore 1982, Lecture Notes in Control and Inform. Sciences 49, 96–110 (1983) Springer, Berlin and New York.

[13] Hudson, R. L., and Parthasarathy, K. R., *Quantum Ito's formula and stochastic evolutions*, Comm. Math. Phys. 93, 301–323 (1984).

[14] Hudson, R. L., and Streater, R. F., *Ito's formula is the chain rule with Wick ordering*, Phys. Lett. A 86, 277–279 (1981).

[15] Maassen, H., Quantum Markov processes on Fock space described by integral kernels, Quantum Prob. and Appl. II (ed. Accardi, L., and Waldenfels, von W.) Lecture Notes in Math., Springer, Berlin and New York, 1136, 361–374 (1985).

[16] Meyer, P. A., *Chains de Markov finies et representations chaotique*, Strasbourg preprint (1989).

[17] Meyer, P. A., Quantum Probability for Probabilists, Lecture Notes in Math., Springer, Berlin and New York, 1538 (1993).

[18] Mohari, A., and Sinha, K. B., *Quantum stochastic flows with infinite degrees of freedom and countable state Markov processes*, Sankhyā Ser. A 52, 43–57 (1990).

[19] Parthasarathy, K. R., *Azéma martingales and quantum stochastic calculus*, Proc. R.C. Bose Symp. on Prob. and Stat., Wiley Eastern, New Delhi, 551–569 (1990).

[20] Parthasarathy, K. R., An Introduction to Quantum Stochastic Calculus, Monographs in Mathematics, Birkhäuser Verlag, Basel (1992).

[21] Parthasarathy, K. R., *Azéma martingales with drift*, Indian Statistical Institute, New Delhi, preprint (1994).

[22] Parthasarathy, K. R., and Sinha, K. B., Markov chains as Evans-Hudson diffusions in Fock space, Sem. Prob. XXIV, Lecture Notes in Math., Springer, Berlin and New York, 1426, 362–369 (1989).

[23] Schürmann, M., White Noise on Bialgebras, Lecture Notes in Math., Springer, Berlin and New York, 1544 (1993).

[24] Segal, I. E., *Tensor algebras over Hilbert spaces*, Trans. Amer. Math. Soc. 81, 106–134 (1956).

[25] Stinespring, W. F., *Positive functions on C^* algebras*, Proc. Amer. Math. Soc. 6, 211–216 (1955).

Measure-Valued Branching Diffusions and Interactions

Edwin A. Perkins

Department of Mathematics
University of British Columbia
Vancouver, BC V6T 1Z2, Canada

1 Introduction

Dawson-Watanabe superprocesses (or measure-valued branching diffusions) provide a stochastic model for a population undergoing random critical (or near critical) reproduction and spatial migration. They arise in a variety of contexts including population genetics, stochastic partial differential equations (PDEs) and interacting particle systems. I will describe some of these connections and then present some sample path properties of super-Brownian motion. An excellent survey of the field may be found in [D2]. Generality will be sacrificed for the sake of accessibility.

The independence properties of these processes make them mathematically tractable. Recently there has been interest in the introduction of new techniques to study more complex models in which particles interact with each other ([DM], [DK], [P5]). The final section will present one approach to a class of these processes.

2 Dawson-Watanabe Superprocesses and Their Historical Processes

An approximating branching particle system is constructed from the following ingredients:

(i) a Hunt process ξ taking values in a Polish space E (for spatial migration)
(ii) an offspring law ν on $\mathbb{Z}_+$ with mean one and variance one
(iii) a constant branching rate $\gamma > 0$.

Fix a finite nonzero measure m on E ($m \in M_F(E)$), let $N \in \mathbb{N}$, and start a collection of i.i.d. particles $\{\xi_0^i : i \leq Nm(E)\}$ with law $m(\cdot)/m(E)$. Particles then follow independent copies of ξ on $[i(N\gamma)^{-1}, (i+1)(N\gamma)^{-1}]$, $i \in \mathbb{Z}_+$, and at $t = i(N\gamma)^{-1}$ each particle dies and is replaced by a random number of offspring distributed according to ν. All the individual motions and family sizes are independent of each other. A continuous branching model in which particles branch at rate γ would lead to the same limiting law. This gives a tree $\{\xi^\alpha : \alpha \in I\}$ of branching ξ-processes. The current state of this randomly branching and migrating population is given by the random measure, which assigns to any $A \subset E$ the value

$$X_t^N(A) = N^{-1} \times \text{(number of particles in } A \text{ at time } t).$$

Proceedings of the International Congress
of Mathematicians, Zürich, Switzerland 1994
© Birkhäuser Verlag, Basel, Switzerland 1995

As $N \to \infty$, $\mathbb{P}(X^N \in \cdot)$ converges weakly to a law $\mathbb{P}_m$ on $\Omega_X = C(\mathbb{R}_+, M_F(E))$ (see [W], [D1], [DP, Theorem 7.13]). $\mathbb{P}_m$ is the law of the Dawson-Watanabe ξ-superprocess (D.W. process) and depends only on m, γ, and the choice of ξ. Under these laws the canonical process $X_t(\omega) = \omega_t$ is an $M_F(E)$-valued diffusion (e.g. [F1]).

Traditionally ξ was taken to be a Feller process taking values in a locally compact space. Dynkin initiated a sequence of generalizations including branching according to an additive functional of ξ and allowing ξ to be a time-inhomogeneous right process (e.g. [Dy2]). Note that $\xi^t = \xi(\cdot \wedge t)$ is a time-inhomogeneous Hunt process taking values in the Polish space $D(\mathbb{R}_+, E) = D(E)$ of right-continuous E-valued paths with left limits, equipped with the Skorokhod topology. Working, as above, with this inhomogeneous Hunt process in place of ξ, we define

$$H_t^N(A) = N^{-1} \sum_{\alpha \in I} 1(\xi^\alpha(\cdot \wedge t) \in A, \ \xi^\alpha \text{ alive at time } t), \quad A \subset D(E) \equiv D.$$

As $N \to \infty$, $\mathbb{P}(H^N \in \cdot)$ converges weakly to a law $\mathbb{Q}_m$ on

$$\Omega_H = \{H \in C(\mathbb{R}_+, M_F(D)) : \ y = y^t \ H_t - \text{a.a. } y \ \forall t \geq 0\}$$

(see [DP, Theorem 7.15]). $\mathbb{Q}_m$ (and more generally $\mathbb{Q}_{\tau,m}$ for $\tau \geq 0$ and $m \in M_F(D)$ with $y = y^\tau \ m - \text{a.e.}$) is the law of the ξ-historical process $H_t(\omega) = \omega_t$ starting at $(0, m)$ (respectively (τ, m)). $(H, \mathbb{Q}_{\tau,m})$ is a time-inhomogeneous $M_F(D)$-valued diffusion. Therefore the ξ-historical process is just the D.W. process associated with the Markov process ξ^t and, assuming ξ is not too degenerate, H_t will record the family trees of the individuals in the population at time t. Clearly the associated D.W. process is given by $X_t(A) = H_t(y : y_t \in A)$, but in general it is not possible to recover H from X (see Theorem 5.4).

The historical process was introduced independently by Dawson and Perkins to study the small scale and large scale behavior of D.W. processes, by Dynkin to construct the "exit measures" from a set (see Section 3), and by Le Gall in the development of the path-valued process (see [L1]).

To prove convergence in the above limit theorems, one needs a convenient characterization of the limiting law. Let A be the suitably defined infinitesimal generator of ξ (e.g. the weak generator defined in [F1] or the strong generator if ξ is Feller) on its domain $D(A)$. Let $\mu(\phi)$ denote the integral of ϕ with respect to μ and $\mathcal{F}_t^X = \sigma(X_s : s \leq t)$. The following is a special case of a result in [F2].

THEOREM 2.1. $\mathbb{P}_m$ is the unique law on Ω_X such that

$$\forall \phi \in D(A) \qquad X_t(\phi) = m(\phi) + \int_0^t X_s(A\phi) \, ds + M_t(\phi)$$

$$M_t(\phi) \text{ is a continuous } \mathcal{F}_t^X \text{-local martingale with } M_0 = 0 \qquad \text{(MP)}$$

$$\langle M(\phi) \rangle_t = \int_0^t X_s(\gamma \phi^2) \, ds.$$

It is relatively easy to see that each limit point of $\mathbb{P}(X^N \in \cdot)$ satisfies (MP). Uniqueness in (MP) will be outlined in the next section.

3 Some Areas of Application

(a) A Non-linear PDE. If $\phi : E \to \mathbb{R}_+$ is bounded and measurable, let $U_t = U_t\phi$ denote the unique mild solution of

$$\frac{\partial U_t}{\partial t} = A(U_t) - \frac{\gamma}{2}(U_t)^2, \qquad U_0 = \phi. \tag{NLE}$$

Using (MP) and some Itô calculus, one can show that $N_s = \exp(-X_s(U_{t-s}))$ ($s \le t$) is a martingale (although there are some details to check – see [F2]) and therefore

$$\mathbb{P}_m(\exp(-X_t(\phi))) = \exp(-m(U_t\phi)). \tag{LT}$$

As only (MP) was used to derive (LT), this gives uniqueness in (MP) because (LT) identifies the law of X_t and a standard result then implies uniqueness of the law of X. Note also that the multiplicative dependence on m in (LT) is a consequence of the strong independence between individuals in the population.

Clearly (LT) allows one to translate estimates on asymptotic results on $U_t\phi$ (or solutions to related equations) into probability estimates on X. Iscoe's work contains many nice examples of this technique (e.g. [I]).

Dynkin ([Dy1]) extended this connection to boundary value problems. To illustrate his approach we consider only super-Brownian motion. Let $D \subset \mathbb{R}^d$ be a bounded open set with a regular boundary for the classical Dirichlet problem. Modify the branching particle systems described earlier by starting N particles at $x \in D$ and stopping the particles and their branching mechanisms when each exits D at time τ_D. Let X_D^N denote the random measure, which assigns mass N^{-1} to each exit location.

THEOREM 3.1. ([Dy1], [Dy2]) *(a)* $X_D^N \overset{w}{\to} X_D$ *as* $N \to \infty$.
(b) If $g : \partial D \to \mathbb{R}_+$ *is continuous then* $\mathbb{P}_{\delta_x}(\exp(-X_D(g))) = \exp(-u(x))$ *where*
u is the unique solution of

$$\triangle u = \gamma u^2 \text{ on } D, \quad \lim_{\substack{y \to x \\ y \in D}} u(y) = g(x) \quad \forall x \in \partial D. \tag{BVP}$$

This probabilistic representation of solutions to (BVP) is reminiscent of the classical representation of solutions to the Dirichlet problem in terms of Brownian motion. The exit measures X_D may be defined directly from historical Brownian motion H by ($t_i^n = i2^{-n}$)

$$X_D(\phi) = \mathbb{P} - \lim_{n \to \infty} \sum_{i=1}^{\infty} \int \phi(y(t_i^n)) 1(t_{i-1}^n \le \tau_D(y) < t_i^n)\, H_{t_i^n}(dy).$$

Theorem 3.1 has led to several results for super-Brownian motion and solutions of (BVP) and related equations. See [L2] for an example of the latter. Dynkin's characterization of polar sets for super-Brownian motion is a good example of the former. Let $S(\mu)$ denote the closed support of a measure μ and let $R = \lim_{\delta \to 0+} \text{cl}(\cup_{t \ge \delta} S(X_t))$ denote the range of X. If $g_d(x) = |x|^{-d}$ for $d \ge 1$ and

$g_0(x) = 1 + \log^+ \frac{1}{x}$ recall that Cap $(g_d)(K) = 0$ iff $\int_K \int_K g_d(x-y)\, d\nu(x)\, d\nu(y) < \infty$ implies $\nu(K) = 0$. If $d \leq 3$ then R hits points (see [DIP]). If $d \geq 4$ then (see [Dy1]) for each analytic set K in $\mathbb{R}^d$ and each nonzero $m \in M_F(\mathbb{R}^d)$

$$K \cap R = \emptyset \; \mathbb{P}_m - \text{a.s.} \Leftrightarrow \text{Cap}(g_{d-4})(K) = 0$$

The necessity of the capacity condition was first established in [P3] by a direct probabilistic argument. Dynkin's proof of the more delicate converse proceeds in two steps (take K compact without loss of generality):

(1) $v(x) = -\log \mathbb{P}_{\delta_x}(R \cap K = \emptyset)$ is the maximal solution of $\triangle v = \gamma v^2$ on K^c

(2) By [BP] Cap $(g_{d-4})(K) = 0 \Leftrightarrow$ solutions of $\triangle v = \gamma v^2$ on K^c are bounded.

It follows that $\inf_{x \in K^c} \mathbb{P}_{\delta_x}(R \cap K = \emptyset) \geq p > 0$ and a stopping argument then shows $p = 1$ from which the polarity of K follows.

(b) Fleming-Viot Processes. There are several close connections (e.g. [KS], [EM], [P4]) between D.W. processes and the Fleming-Viot processes used in population genetics to model selection, mutation, and random sampling in the distribution of genotypes in a population. E is now a space of genotypes and ξ models the mutation of an offspring from the type of its parent. In the approximating particle systems, the population size N now is fixed and hence we must modify the branching mechanism at $t = i(N\gamma)^{-1}$. At these times particle $j(j \leq N)$ is replaced by k_j offspring where $(k_1, \ldots, k_N)$ has a multinomial distribution with parameters $(N; N^{-1}, \ldots, N^{-1})$. This models an essentially infinite number of potential offspring for each parent (e.g. fish eggs) that, due to limited resources, is then culled down to the constant population size. In the absence of selective advantages, each parent is equally likely to produce an offspring that reaches maturity and the above multinomial distribution arises. If $V_t^N(A)$ is the proportion of types in A at time t, then the Fleming-Viot process V (with law $\bar{\bar{\mathbb{P}}}_m$) is a probability-valued process that is the weak limit of V^N as $N \to \infty$.

A trivial calculation shows that if $(k_i, i \leq N)$ are i.i.d. Poisson (1) random variables then $P((k_1, \ldots, k_N) \in \cdot | \sum_1^N k_i = N)$ is multinomial $(N; N^{-1}, \ldots, N^{-1})$. Now take ν to be Poisson (1) in the definition of X^N above (and $m(E) = 1$) to see that $\mathbb{P}((X_s^N, s \leq N) \in \cdot | X_s^N(E) = 1 \; \forall s \leq N) = \mathbb{P}((V_s^N, s \leq N) \in \cdot)$. Letting $N \to \infty$ suggests (but does not prove) the following result from [EM].

THEOREM 3.2. $\mathbb{P}_m(X \in \cdot | \forall t \leq \epsilon^{-1}, |X_t(E) - 1| < \epsilon) \overset{w}{\to} \bar{\bar{\mathbb{P}}}_m(V \in \cdot)$ as $\epsilon \downarrow 0$.

More generally one can identify the regular conditional distribution of $X/X(E)$ given $X.(E) = f(\cdot)$ as a Fleming-Viot process with sampling rate $f(\cdot)^{-1}$ ([P4]).

(c) Stochastic Partial Differential Equations.

THEOREM 3.3. ([KS], [Re2]) *If X is super-Brownian motion in one spatial dimension, then $\mathbb{P}_m$ a.s. $X_t(dx) = u(t,x)dx \; \forall t > 0$ where $u : (0, \infty) \times \mathbb{R} \to \mathbb{R}_+$ is the jointly continuous unique (in law) solution of*

$$\frac{\partial u}{\partial t}(t,x) = \frac{\triangle}{2}u(t,x) + \sqrt{\gamma u(t,x)}\dot{W}, \tag{3.1}$$

where $\dot{W}$ is a space-time white noise.

(3.1) has several interesting features. If $\sqrt{\gamma u}$ is replaced by u^α for $\alpha \in (0,1)$ but $\alpha \neq 1/2$, uniqueness in law remains unresolved. Solutions to these equations may be interpreted as the density of a measure-valued branching process with a density dependent branching rate. An associated historical process was used in [MP] to derive qualitative properties of the solutions. An associated historical process was also used in recent work of Mueller and Tribe to study the wave fronts of solutions to a class of stochastic PDEs with Wright-Fisher noise.

(3.1) is also a rare instance when a nonlinear stochastic PDE driven by white noise can be solved in higher dimensions. In [Re1] (3.1) was solved as an infinitesimal difference equation. Although for $d > 1$ the resulting solution was infinite on an infinitesimal set, when integrated out, it produced super-Brownian motion.

(d) Aldous' Continuum Random Tree. Aldous (see [A1]) constructed a random probability on $\mathbb{R}^d$ (integrated super-Brownian excursion or ISE) by embedding the compact continuum random tree into $\mathbb{R}^d$ by running d-dimensional Brownian motions along the edges of the tree. In terms of super-Brownian motion, ISE is

$$\lim_{\epsilon \downarrow 0} \mathbb{P}_{\epsilon \delta_0} \left(\int_0^\infty X_t \, dt \in \cdot \mid \int_0^\infty X_t(\mathbb{R}^d) \, dt = 1 \right),$$

i.e., the law of $\int_0^\infty X_t \, dt$ under the canonical measure associated with the infinitely divisible law $\mathbb{P}_{\delta_0}$ conditioned on its total mass being one. (The description in terms of Le Gall's path-valued process is perhaps the simplest.) Aldous conjectured that ISE is the rescaled limit as $n \to \infty$ of uniform measure on n-vertex lattice trees animals for $d > 8$ (super-Brownian motion has double points iff $d < 8$ ([P3], [DIP])).

4 Properties of Super-Brownian Motion

Throughout this section X_t and H_t denote super-Brownian motion and its associated historical process, respectively (i.e., $X_t \in M_F(\mathbb{R}^d)$, $H_t \in M_F(C(\mathbb{R}_+, \mathbb{R}^d))$). Recently there have been many precise results on the qualitative behavior of X. What kind of measure is X_t? How fast can mass propagate? What can be said about the topological features of its closed support $S(X_t)$?

Dawson and Hochberg showed that for $d \geq 2$ and t fixed, X_t is a.s. a singular measure supported by a Borel set of dimension two ([DH]). Here is a refinement.

First, some notation.

$$\phi_d(r) = \begin{cases} r^2 \log \log 1/r & d \geq 3 \\ r^2 (\log 1/r)(\log \log \log 1/r) & d = 2 \end{cases}$$

$\phi - m(A)$ is the Hausdorff ϕ-measure of A.

THEOREM 4.1. (a) If $d \geq 2$, $\exists C_d > 0$ such that for all $m \in M_F(\mathbb{R}^d)$ and all $t > 0$

$$X_t(A) = C_d \phi_d - m(A \cap S(X_t)) \quad \forall \text{ Borel set } A \quad \mathbb{P}_m - a.s.$$

(b) If $d \geq 3$ $\exists 0 < C'_d \leq C''_d < \infty$ such that for all $m \in M_F(\mathbb{R}^d)$

$$C'_d \phi_d - m(A \cap S(X_t)) \leq X_t(A) \leq C''_d \phi_d - m(A \cap S(X_t))$$
$$\forall \text{ Borel } A, t > 0 \ \mathbb{P}_m - a.s.$$

(c) If $d = 2$ $X_t \perp dx$ $\forall t > 0$ $\mathbb{P}_m - a.s.$ $\forall m \in M_F(\mathbb{R}^d)$.

It is possible to prove (b) and (c) using a cluster decomposition of the Palm measure associated with the historical process. The original proofs in [P1], [P2] implicitly used the same ideas in a nonstandard model for the branching particle systems. (a) for $d \geq 3$ is obtained from (b) in [DP] through a $0-1$ law for the Palm of the canonical measure associated with H_t. The result for $d = 2$ is proved in [LP] and relies on Le Gall's path-valued process to introduce time dynamics into the study of $S(X_t)$. Note that all of these arguments work with an "enriched" model for X from which ancestral relationships can be derived.

Theorem 4.1 shows that X_t spreads its mass over $S(X_t)$ in a uniform manner. Although the equality of C_d' and C_d'' in (b) remains open, for $d \geq 3$ it is possible to recover X_t from $S(X_t)$ for all $t > 0$ a.s. by considering a Lebesgue measure on the ϵ-sausage of $S(X_t)$ and letting $\epsilon \downarrow 0$ (see [P7]). These results reduce the study of X to the study of its support process $S(X.)$.

THEOREM 4.2. ([DIP], [DP]) $\forall c > 2 \ \exists \delta(c, \omega) > 0 \ \mathbb{Q}_m$ - a.s. such that $\forall t > 0$

$$S(H_t) \subset \{y : |y_r - y_s| \leq c((r-s)\log^+ 1/(r-s))^{1/2} \text{ whenever}$$
$$|r - s| < \delta\}.$$

The result is false for $c < 2$.

The result implies that $S(X_t)$ propagates no faster than $(2 + \epsilon)(s \log^+ 1/s)^{1/2}$. Note that this is faster than a single Brownian path for which $c = \sqrt{2}$ is critical by Levy's modulus. There are diffusions ξ for which the associated D.W. process X propagates instantaneously ([DP, Sec. 8]).

It is easy to use Theorem 4.2 to strengthen the original cluster argument of Dawson and Hochberg and give a simple proof of $\dim S(X_t) \leq 2$ (which of course is immediate from Theorem 4.1). An elementary argument, using Kolmogorov's result that the survival probability of a critical Galton-Watson process after n generations is asymptotically cn^{-1}, shows that

$$S(H_t(y^{t-\epsilon} \in \cdot)) = \{w_1, \ldots, w_{M(\epsilon)}\}$$

where conditional on $\mathcal{F}_{t-\epsilon}^H$, $M(\epsilon)$ is Poisson $(2H_{t-\epsilon}(1)\epsilon^{-1})$. Note that $M(\epsilon)$ is the number of ancestors at $t-\epsilon$ of the entire population at t. Decompose X_t as $\sum_{i=1}^{M(\epsilon)} X_t^i$ where $X_t^i(\cdot) = H_t(y_t \in \cdot : y^{t-\epsilon} = w_i)$ so that

$$S(X_t) = \cup_{i=1}^{M(\epsilon)} S(X_t^i). \tag{4.1}$$

Theorem 4.2 implies $\operatorname{diam}(S(X_t^i)) \leq (4 + \eta)(\epsilon \log 1/\epsilon)^{1/2}$ (for $\epsilon < \delta(\omega, 2 + \eta/2)$) and a trivial Borel-Cantelli argument shows that (4.1) implies $\dim S(X_t) \leq 2$.

As $S(X_t^i)$ and $S(X_t^j)(i \neq j)$ are conditionally independent sets of zero $x^2 - m$ (Theorem 4.1) they will be disjoint if $d \geq 4$ (e.g. see [DIP]). It then follows easily from (4.1) and the above bound on $\operatorname{diam}(S(X_t^i))$ that $S(X_t)$ is a.s. totally disconnected if $d \geq 4$. A weaker result is given in [T] for $d = 3$, but the total disconnectedness of $S(X_t)$ remains unresolved for $d = 2$ or 3.

5   A Class of Interactive Branching Measure-Valued Diffusions

Let $b(t, K, y) \in \mathbb{R}^d, \sigma(t, K, y) \in \mathbb{R}^{d \times d}$, and $\gamma(t, K, y) \in \mathbb{R}_+$ be predictable mappings on $\mathbb{R}_+ \times \Omega_H \times C(\mathbb{R}_+, \mathbb{R}^d)$. Our goal is to construct and characterize an interactive historical process $K_t(dy)$ where a "particle" y^t in population K_t is subject to a drift $b(t, K, y)$, diffusion matrix $\sigma(t, K, y)$, and branching rate $\gamma(t, K, y)$ (continuous time branching). This is carried out in [P5] and [P6].

EXAMPLE 5.1. $b(t, K, y) = \int \tilde{b}(y_t' - y_t) K_t(dy'), \ \sigma(t, K, y) = \int \tilde{\sigma}(y_t' - y_t) K_t(dy')$. Each particle y_t' in the population could, for example, exert an attractive drift $\tilde{b}(y_t' - y_t) K_t(dy')$ on every other particle y_t. If $\tilde{\sigma}(x) = p_\epsilon(x) I_{d \times d}$, particles will diffuse at a rate proportional to the approximate density of nearby particles. $\tilde{b}$ and $\tilde{\sigma}$ must be bounded Lipschitz functions for our theory to apply.

EXAMPLE 5.2. (A variant of Adler's goats) $b = \int_0^t \int \nabla p_\epsilon(y_s' - y_t) K_s(dy') e^{-\alpha(t-s)} ds$, $\gamma = \exp(-\int_0^t \int p_\epsilon(y_s' - y_t) K_s(dy') e^{-\alpha(t-s)} ds)$. These goat-like particles tend to drift away from regions where the population has recently grazed and depleted the resources. They also reproduce at a lower rate in such regions.

To show that the martingale problem for K is well posed, we will use analogues of the key tools in finite-dimensional diffusion theory: Itô's stochastic differential equations (SDEs), the Girsanov change of measure techniques and the Stroock-Varadhan martingale problem.

(a) Tree-indexed Stochastic Integrals and SDEs. Let H be an historical Brownian motion on $(\Omega, \mathcal{F}, \mathcal{F}_t, \mathbb{P})$ with $\gamma = 1$ and $H_0 = m \neq 0$ ($H_t \in M_F(C)$ where $C = C(\mathbb{R}_+, \mathbb{R}^d)$). Let $(\hat{\Omega}, \hat{\mathcal{F}}_t) = (\Omega \times C, \mathcal{F}_t \times \mathcal{C}_t)$ ($\mathcal{C}_t = \sigma(y_s, s \leq t)$). If T is a bounded $(\mathcal{F}_t)$-stopping time, the associated Campbell measure on $\hat{\Omega}$ (whose elements will be denoted (ω, y)) is given by

$$\hat{\mathbb{P}}_T(A \times B) = \mathbb{P}(1_A(\omega) H_T(B)) \, \mathbb{P}(H_0(1))^{-1}.$$

Under $\hat{\mathbb{P}}_T$, y is an $(\hat{\mathcal{F}}_t)$-Brownian motion stopped at T and therefore $\int_0^t \sigma(s, \omega, y) \, dy(s)$ may be defined for the class $D(I)$ of $(\hat{\mathcal{F}}_t)$-predictable functions $\sigma(s, \omega, y) \in \mathbb{R}^{n \times d}$ such that $\int_0^t ||\sigma(s, \omega, y)||^2 \, ds < \infty \ H_t$-a.a. $y \ \forall t > 0$ a.s.

THEOREM 5.1.   (a) If $\sigma \in D(I)$ there is an $\mathbb{R}^n$-valued $(\hat{\mathcal{F}}_t)$-predictable process $I(\sigma, t, \omega, y)$ such that for all T as above,

$$I(\sigma, t \wedge T, \omega, y) = \int_0^t \sigma(s, \omega, y) \, dy(s) \quad \forall t \geq 0 \ \hat{\mathbb{P}}_T - a.s.$$

(b) If $\tilde{I}(\sigma, t, \omega, y)$ is another such process, then

$$\tilde{I}(\sigma, s, \omega, y) = I(\sigma, s, \omega, y) \quad \forall s \leq t \ H_t - a.a. \ y \ \forall t \geq 0 \ \mathbb{P} - a.s.$$

Therefore $I(\sigma, t, \omega, y) \equiv \int_0^t \sigma(s, \omega, y) \, dy(s)$ is the stochastic integral of σ along the branch y up to time t in the Brownian tree ω. (b) is a trivial application of the predictable section theorem.

Consider the stochastic differential equation that we denote by $(HSE)_{\gamma, \sigma, b}$

(i) $\quad Y_s(\omega, y) = y_0 + \int_0^s \sigma(r, K, Y) \, dy(r) + \int_0^s b(r, K, Y) \, dr \quad \forall s \leq t \;\; H_t - \text{a.a. } y$

$$\forall t \geq 0 \;\; \mathbb{P} - \text{a.s}$$

(ii) $\quad K_t(\phi) = \int \phi(Y(\,\cdot\,\wedge t)(\omega, y))\gamma(t, K, Y)(\omega, y)H_t(dy) \quad \forall t \geq 0, \phi \in b\mathcal{C} \;\; \mathbb{P} - \text{a.s.}$

By (i), $Y(\omega, y)$ solves an Itô equation along the branch y with drift $b(s, K, \cdot\,)$ and diffusion matrix $\sigma(s, K, \cdot\,)$. Recalling that H_t is essentially uniform measure on those trajectories that are alive at time t, one sees that for $\gamma = 1$, K_t is uniform measure over the Y^t's and the solution K of (HSE) should be the desired process. In general, the transformation of the branching rate γ into a mass factor in (HSE) will require some further adjustments.

THEOREM 5.2. *Assume (γ, σ, b) are Lipschitz continuous, γ is bounded, and (H_0, K_0) satisfy (HSE) (ii) at $t = 0$.*
   (a) There is a pathwise unique solution (K, Y) to $(HSE)_{\gamma, \sigma, b}$.
   (b) If K_0 is deterministic, K is a predictable function of H and its law depends only on (K_0, γ, σ, b).

The proof is a contraction mapping argument. A form of Itô's lemma for the tree-indexed integrals is used to iterate the maps (which arise from (HSE)) in the appropriate Banach space. The Lipschitz condition (L) on (σ, b) is the natural one using the Vaserstein metric on $M_F(C)$ and the associated sup-norm on $\Omega_H \times C$. The above result is false if this same condition is imposed on γ and hence the Lipschitz hypothesis on γ is more restrictive (see (C_3) in [P6, Sec. 5]). This hypothesis is satisfied, for example, if $\gamma(t, K, y) = \int_{-\infty}^t \phi_\epsilon(s)\tilde{\gamma}(s^+, K, y)ds$ where $\phi_\epsilon \geq 0$ is smooth and integrable, and $\tilde{\gamma}$ satisfies (L), or if $\gamma(t, K, y) = \tilde{\gamma}(t, y_t)$ and $\tilde{\gamma}$ is sufficiently smooth.

Although the solution K of $(HSE)_{\gamma, \sigma, b}$ will branch at rate $\gamma(t, K, y)$, the introduction of γ as a mass factor also leads to additional zeroth and first order terms in the generator and hence this is not the desired process. To use Dawson's Girsanov theorem to make the appropriate changes we need a martingale problem for K.

(b) A Historical Martingale Problem. Let $K_t(\omega) = \omega_t$ on Ω_H and set $D_0 = \{\bar{\psi}(y) = \psi(y_{t_1}, \ldots, y_{t_n}) : t_i \geq 0, \psi \in C_0^\infty(\mathbb{R}^{nd})\}$. For $\bar{\psi}$ as above and $\bar{y}(t) = (y(t \wedge t_1), \ldots, y(t \wedge t_n))$, $\nabla\bar{\psi}(t, y)$ is the vector in $\mathbb{R}^d$ with jth component $\sum_{i=0}^{n-1} 1(t < t_{i+1})\psi_{id+j}(\bar{y}(t))$ and

$$\bar{\psi}_{ij}(t, y) = \sum_{k=0}^{n-1}\sum_{l=0}^{n-1} 1(t < t_{k+1} \wedge t_{l+1})\psi_{kd+i \; ld+j}(\bar{y}(t)), \quad 1 \leq i, j \leq d.$$

Let $a(t, K, y) \in S^d$ (symmetric $(d \times d)$ matrices) and $g(t, K, y) \in \mathbb{R}$ be predictable maps on $\mathbb{R}_+ \times \Omega_H \times C$, and for $\bar{\psi} \in D_0$ define

$$A_K \bar{\psi}(t, y) = \frac{1}{2} \sum_{i=1}^{d} \sum_{j=1}^{d} a_{ij}(t, K, y) \bar{\psi}_{ij}(t, y) + b(t, K, y) \cdot \nabla \bar{\psi}(t, y) + g(t, K, y) \bar{\psi}(y^t).$$

The last term arises by giving certain individuals y in a population K a selective advantage ($g > 0$) or disadvantage ($g < 0$) in branching. For $\tau \geq 0$ and $m \in \Omega_H^\tau = \{K \in \Omega_H : K_{\cdot \wedge \tau} = K.\}$ consider the martingale problem (denoted by $(\mathrm{HMP})_{\gamma,a,b,g}^{\tau,m}$)

$$\forall \bar{\psi} \in D_0 \quad K_t(\bar{\psi}) = K_\tau(\bar{\psi}) + M_t(\bar{\psi}) + \int_\tau^t \int A_K \bar{\psi}(s, y) K_s(dy) \, ds$$

$M_t(\bar{\psi})$ is a continuous $(\mathcal{F}_t^K)$-local martingale, $K^\tau = m$, and

$$< M(\bar{\psi}) >_t = \int_\tau^t \int \gamma(s, K, y) \bar{\psi}(y^s)^2 K_s(dy) \, ds.$$

If $\gamma(s, K, y) \equiv \gamma$ is constant, $a = I_{d \times d}$, and $b = g = 0$, then Itô's lemma shows that $A_K \bar{\psi}(t, y) = \sum_{i=1}^{d} \frac{1}{2} \bar{\psi}_{ii}(t, y)$ is the generator of $t \to B^t$ (B a Brownian motion). The martingale problem (MP) for D.W. processes therefore suggests that $\mathbb{Q}_{\tau,m}$ (the law of historical Brownian motion) is the unique solution of $(\mathrm{HMP})_{\gamma,I,0,0}^{\tau,m}$ and this indeed is the case ([P6]). It is easy to see that the processes described at the outset will satisfy $(\mathrm{HMP})_{\gamma,\sigma\sigma^*,b,0}^{\tau,m}$.

THEOREM 5.3. *Assume that (γ, b, σ) are as in Theorem 5.2, g is bounded, $\gamma^{-1}(t, K, y) \leq L(t, K_t^*(1))$, and $a(t, K, y) = \sigma\sigma^*(t, K, y)$ is strictly positive definite for $K_t - a.a.$ y and for all $t \geq 0$.*
(a) *There is a unique law $\hat{\mathbb{Q}}_{\tau,m}$ on Ω_H satisfying $(HMP)_{\gamma,a,b,g}^{\tau,m}$ that depends continuously on (τ, m).*
(b) *If $(\gamma, a, b, g)(t, K, y) = (\tilde{\gamma}, \tilde{a}, \tilde{b}, \tilde{g})(t, K_t, y)$, then $\hat{\mathbb{Q}}_{\tau,m} = \tilde{\mathbb{Q}}_{\tau,m(\tau)}$ on $\sigma(K_s, s \geq \tau)$ and $(K_t, \tilde{\mathbb{Q}}_{\tau,m_0})$ is a time-inhomogeneous strong Markov process.*

The proof of (a) proceeds in two steps:
(1) $(\mathrm{HSE})_{\gamma,\sigma,b}^{\tau,m} \Leftrightarrow (\mathrm{HMP})_{\gamma,\sigma\sigma^*,b+\tilde{b}(\sigma\sigma^*,\gamma),\tilde{g}(\gamma,b)}^{\tau,m}$ for appropriate $\tilde{b}, \tilde{g}$. ($\Rightarrow$) is a simple exercise in stochastic calculus for tree-indexed integrals. ($\Leftarrow$) requires the nondegeneracy conditions on (γ, σ) to invert (HSE) and define a historical Brownian motion in terms of K such that K solves $(\mathrm{HSE})_{\gamma,\sigma,b}^{\tau,m}$. Theorem 5.2 now shows that $(\mathrm{HMP})_{\gamma,\sigma\sigma^*,b+\tilde{b},\tilde{g}}^{\tau,m}$ is well posed.
(2) By redefining b and using Dawson's Girsanov theorem ([D2, Ch. 10]) to reset $\tilde{g}$ to g, one sees that $(HMP)_{\gamma,a,b,g}^{\tau,m}$ is well posed.

REMARKS. (a) The existence of solutions to $M_F(\mathbb{R}^d)$-valued martingale problems similar to (HMP) was known by tightness arguments (see [M] or [MR]). It is easy to project down solutions K to (HMP) via $X_t(\,\cdot\,) = K_t(y_t \in \,\cdot\,)$ to obtain Feller solutions to these Markovian martingale problems, which are the weak limits of the natural interactive branching particle systems.

(b) Dawson and March [DM] have adapted the parametrix method of Stroock and Varadhan to prove that a class of interactive Fleming-Viot martingale problems is well posed. This approach seems to be complementary both in the techniques used and in the areas of application.

(c) The stochastic equation approach appears to be somewhat robust. Donnelly and Kurtz [DK] have incorporated it with their "look-down" processes to study Markovian spatial interactions in a general class of population models that includes both Fleming-Viot and Dawson-Watanabe processes.

The historical process has been the key tool in the study of super-Brownian motion X, its connections with nonlinear boundary value problems, and the construction of interactive models. Although in general this information is lost when dealing with X itself, it is interesting to note that this is not the case for sufficiently high dimensions.

THEOREM 5.4. ([BaP]) *Let $\mathbb{Q}_m$ denote the law of historical Brownian motion H in d dimensions and let $X_t(\,\cdot\,) = H_t(y_t \in \cdot\,)$ be the associated super-Brownian motion. Let $\mathcal{F}_t^H$ (resp., $\mathcal{F}_t^X$) be the σ-field generated by $(H_s, s \leq t)$ (resp., $(X_s, s \leq t)$) and the $\mathbb{Q}_m$-null sets. If $d \geq 5$, $\mathcal{F}_t^X = \mathcal{F}_t^H$ $\forall t$ and if $d \leq 3$ $\mathcal{F}_t^X \neq \mathcal{F}_t^H$ $\forall t > 0$.*

References

[A1]   D. Aldous, *Tree-based models for random distribution of mass*, J. Statist. Phys. **73** (1993), 625–641.

[BP]   P. Baras, and M. Pierre, *Singularités éliminables pour des équations semi-linéaires*, Ann. Inst. Fourier **34** (1984), 185–206.

[BaP] M. T. Barlow, and E. A. Perkins, *On the filtration of historical Brownian motion*, Ann. Probab. **22** (1994), 1273–1294.

[D1]   D. A. Dawson, *Stochastic evolution equations and related measure-valued processes*, J. Multivariate Anal. **5** (1975), 1–52.

[D2]   D. A. Dawson, *Measure-valued Markov processes*, in: Ecole d'Eté de Probabilités de Saint Flour XXI - 1991 (P.L. Hennequin, ed.), Lect. Notes in Math. **1541**, Springer-Verlag Berlin, Heidelberg, and New York, 1993, 2–260.

[DH]   D. A. Dawson, and K. J. Hochberg, *The carrying dimension of a stochastic measure diffusion*, Ann. Probab. **7** (1979), 693–703.

[DIP] D. A. Dawson, I. Iscoe, and E. A. Perkins, *Super-Brownian motion: Path properties and hitting probabilities*, Probab. Theory Related Fields **83** (1989), 135–205.

[DM]   D. A. Dawson, and P. March, *Resolvent estimates for Fleming-Viot operators and uniqueness of solutions to related martingale problems*, to appear in J. Functional Analysis (1995).

[DP]   D. A. Dawson, and E. A. Perkins, Historical Processes, Mem. Amer. Math. Soc. **454** (1991).

[DK]   P. E. Donnelly, and T. G. Kurtz, *Particle representations for measure-valued population models*, preprint, 1994.

[Dy1] E. B. Dynkin, *A probabilistic approach to one class of nonlinear differential equations*, Probab. Theory Related Fields **89** (1991), 89–115.

[Dy2] E. B. Dynkin, *Superprocesses and partial differential equations*, Ann. Probab. **21** (1993), 1185–1262.

[EM]   A. Etheridge, and P. March, *A note on superprocesses*, Probab. Theory Related Fields **89** (1991), 141–147.

[F1]   P. J. Fitzsimmons, *Construction and regularity of measure-valued branching processes*, Israel J. Math. **64** (1988), 337–361.

[F2]   P. J. Fitzsimmons, *On the martingale problem for measure-valued Markov branching processes*, in: Seminar on Stochastic Processes 1991 (E. Cinlar, K.L. Chung, and M.J. Sharpe, eds.), Birkhäuser, Boston, Basel, and Berlin, 1992, 39–51.

[I]   I. Iscoe, *On the supports of measure-valued critical branching Brownian motion*, Ann. Probab. **16** (1988), 200–221.

[KS]   N. Konno, and T. Shiga, *Stochastic differential equations for some measure-valued diffusions*, Probab. Theory Related Fields **79** (1988), 201–225.

[L1]   J.-F. Le Gall, *A class of path-valued Markov processes and its application to superprocesses*, Probab. Theory Related Fields **95** (1993), 25–46.

[L2]   J.-F. Le Gall, *The Brownian snake and solutions of $\triangle u = u^2$ in a domain*, to appear in Probab. Theory Related Fields (1995).

[LP]   J.-F. Le Gall, and E. A. Perkins, *The Hausdorff measure of the support of two-dimensional super-Brownian motion*, to appear in Ann. Probab. (1995).

[MR]   S. Meleard and S. Roelly, *Interacting branching measure processes. Some bounds for the support*, Stochastics and Stochastic Reports **44** (1993), 103–122.

[M]   M. Metivier, *Weak convergence of measure-valued processes using Sobolev-imbedding techniques*, in: Proc. Trento 1985 SPDE and Applications (G. Da Prato, and L. Tubaro, eds.), Lect. Notes Math. **1236**, 172–183, Springer-Verlag, Berlin, Heidelberg, and New York 1987.

[MP]   C. Mueller, and E. A. Perkins, *The compact support property for solutions to the heat equation with noise*, Probab. Theory Related Fields **93** (1992), 325–358.

[P1]   E. A. Perkins, *A space-time property of a class of measure-valued branching diffusions*, Trans. Amer. Math. Soc. **305** (1988), 743–795.

[P2]   E. A. Perkins, *The Hausdorff measure of the closed support of super-Brownian motion*, Ann. Inst. H. Poincaré **25** (1989), 205–224.

[P3]   E. A. Perkins, *Polar sets and multiple points for super-Brownian motion*, Ann. Probab. **18** (1990), 453–491.

[P4]   E. A. Perkins, *Conditional Dawson-Watanabe processes and Fleming-Viot processes*, in: Seminar on Stochastic Processes 1991 (E. Cinlar, K. L. Chung, and M. J. Sharpe, eds.), Birkhäuser, Boston, Basel, and Berlin, 1992, pp. 143–156.

[P5]   E. A. Perkins, *Measure-valued branching diffusions with spatial interactions*, Probab. Theory Related Fields **94** (1992), 189–245.

[P6]   E. A. Perkins, *On the martingale problem for interactive measure-valued branching diffusions*, Mem. Amer. Math. Soc. **549** (1995).

[P7]   E. A. Perkins, *The strong Markov property of the support of super-Brownian motion*, in: The Dynkin Festschrift (M. Freidlin, ed.), Progress in Math., Birkhäuser, Boston, Basel, and Berlin, 1994, 307–326.

[Re1]   M. Reimers, *Hyperfinite methods applied to critical branching diffusion*, Probab. Theory Related Fields **81** (1989), 11–27.

[Re2]   M. Reimers, *One-dimensional stochastic partial differential equations and the branching measure diffusion*, Probab. Theory Related Fields **8** (1989), 319–340.

[T]   R. Tribe, *The connected components of the closed support of super-Brownian motion*, Probab. Theory Related Fields **84** (1991), 75–87.

[W]   S. Watanabe, *A limit theorem of branching processes and continuous state branching*, J. Math. Kyoto Univ. **8** (1968), 141–167.

Diffusion Processes in Random Environments

HIROSHI TANAKA

Department of Mathematics
Faculty of Science and Technology
Keio University
Yokohama 223, Japan

1 Introduction

Problems concerning limiting behavior of random processes in random environments have been discussed mostly in the framework of random walks (e.g., see [1], [4], [6], [15], [17], [23], [24]). Most of the problems, naturally, can also be treated in the framework of diffusion processes. We give here a survey of some recent results concerning diffusion processes in random environments, mainly of one dimension, with emphasis on the following two examples of problems. The use of methods and results in theory of diffusion processes makes our argument transparent.

(I) Localization by random centering (depending only on the environment) of a diffusion in a one-dimensional Brownian environment.
(II) Limit theorems for a diffusion in a one-dimensional Brownian environment with drift.

We also give a brief survey concerning

(III) results for a diffusion in a multidimensional Brownian environment.

2 A diffusion in a one-dimensional Brownian environment (with drift)

Let P be the Wiener measure on $\mathbf{W} = C(\mathbf{R}) \cap \{W : W(0) = 0\}$. The processes $\{W(t), t \geq 0, P\}$ and $\{W(-t), t \geq 0, P\}$ are thus independent Brownian motions. Let $\Omega = C[0, \infty)$ and denote by $\omega(t)$ the value of a function $\omega(\in \Omega)$ at time t. For fixed W and a given constant κ we consider a probability measure P_W on Ω such that $\{\omega(t), t \geq 0, P_W\}$ is a diffusion process with generator

$$\mathcal{L}_W = \frac{1}{2} e^{W(x,\kappa)} \frac{d}{dx} \left(e^{-W(x,\kappa)} \frac{d}{dx} \right)$$

and starting at 0, where $W(x, \kappa) = W(x) - \frac{1}{2}\kappa x$. It is well known that a version of $\mathbf{X}_W = \{\omega(t), t \geq 0, P_W\}$ can be obtained from a Brownian motion through a scale change and a time change. We can regard W and $\{\omega(t), t \geq 0\}$ as defined on the probability space $(\mathbf{W} \times \Omega, \mathcal{P})$ where $\mathcal{P}(dW\,d\omega) = P(dW)P_W(d\omega)$. The process $\mathbf{X} = \{\omega(t), t \geq, \mathcal{P}\}$ is then called a *diffusion in a Brownian environment (with drift*

Proceedings of the International Congress
of Mathematicians, Zürich, Switzerland 1994
© Birkhäuser Verlag, Basel, Switzerland 1995

if $\kappa \neq 0$). Symbolically one may write $d\omega(t) = dB(t) - \frac{1}{2}W'(\omega(t), \kappa)dt$ where $B(t)$ is a Brownian motion independent of $W(\cdot)$; however, this stochastic differential equation has no rigorous meaning.

When $\kappa = 0$, $\mathbf{X}$ is a diffusion model of Sinai's random walk in a random environment [23]. In this case Schumacher [22] and Brox [3] showed that $\mathbf{X}$ exhibits the same asymptotic behavior as Sinai's random walk, namely, that the limit distribution of $(\log t)^{-2}\omega(t)$ as $t \to \infty$ exists. Kesten [14] obtained the explicit form of the limit distribution. Golosov [7] also obtained a similar explicit form for a reflecting random walk model. Some generalizations of these results were done in [10] and [25]. The problem (I) stated in the introduction is to elaborate the result of [22] and [3] by taking account of a random centering that depends only on the environment W. This will be discussed in the next section. It is to be noted that a similar localization result was already obtained by Golosov [6] for reflecting random walks on $\mathbf{Z}^+$.

The problem (II) is concerned with the case $\kappa \neq 0$ and may be regarded as a diffusion analogue of what was discussed by Kesten-Kozlov-Spitzer [15], Solomon [24], and Afanas'ev [1]. Here we are mainly interested in limit theorems concerning the first passage time $T_x = \inf\{t > 0 : \omega(t) = x\}$ as $x \to \infty$. As will be seen in Section 4, the result varies with κ and naturally is compatible with those of [15] and [1].

3 Localization by random centering in the case $\kappa = 0$

The argument of [3] relies on the notion of a valley introduced in [23]; in order to state only the result, however, it is adequate to start simply with the definition of the "bottom"(denoted by b_λ) of a suitable valley around the origin. Given a Brownian environment $W = \{W(x), x \in \mathbf{R}\}$, let us define $b_\lambda = b_\lambda(W)$ following [14] for each $\lambda > 0$. Setting

$$W^\#(x) = W(x) - \min_{[x \wedge 0, x \vee 0]} W,$$

$$d_\lambda^+ = \min\{x > 0 : W^\#(x) = \lambda\}, \quad V_\lambda^+ = \min_{[0, d_\lambda^+]} W,$$

$$d_\lambda^- = \max\{x < 0 : W^\#(x) = \lambda\}, \quad V_\lambda^- = \min_{[d_\lambda^-, 0]} W,$$

we first determine b_λ^+ and b_λ^- by $W(b_\lambda^+) = V_\lambda^+$ and $W(b_\lambda^-) = V_\lambda^-$, respectively (such $b_\lambda^\pm$ are uniquely determined with P-measure 1 for each fixed $\lambda > 0$), and then define $b_\lambda = b_\lambda(W)$ by

$$b_\lambda(W) = \begin{cases} b_\lambda^+ & \text{if } M_\lambda^+ \vee (V_\lambda^+ + \lambda) < M_\lambda^- \vee (V_\lambda^- + \lambda), \\ b_\lambda^- & \text{if } M_\lambda^+ \vee (V_\lambda^+ + \lambda) > M_\lambda^- \vee (V_\lambda^- + \lambda), \end{cases}$$

where $M_\lambda^+ = \max\{W(x) : 0 \leq x \leq b_\lambda^+\}$ and $M_\lambda^- = \max\{W(x) : b_\lambda^- \leq x \leq 0\}$. When $\lambda = 1$ we write $b = b(W)$ suppressing the suffix 1. We also define $W_\lambda(\in \mathbf{W})$ for each $\lambda > 0$ and $W \in \mathbf{W}$ by $W_\lambda(x) = \lambda^{-1}W(\lambda^2 x), x \in \mathbf{R}$. Then $\{W_\lambda, P\}$ is equivalent in law to $\{W, P\}$ and hence the distribution of $b(W_\lambda)$ is independent of $\lambda > 0$.

Let $\mathbf{X} = \{\omega(t), t \geq 0, \mathcal{P}\}$ be a diffusion in a Brownian environment ($\kappa = 0$) starting at 0. According to Schumacher [22] and Brox [3]

$$\lambda^{-2}\omega(e^\lambda) - b(W_\lambda) \to 0 \tag{3.1}$$

in probability with respect to $\mathcal{P}$ as $\lambda \to \infty$.

Localization by random centering arises from the following question: Under what scaling does the left-hand side of (3.1) admit a nondegenerate limit distribution? The answer is simply that $\omega(e^\lambda) - \lambda^2 b(W_\lambda)$ does. To state the result more precisely we need to introduce another probability measure Q on $\mathbf{W}$, defined in such a way that $\{W(x), x \geq 0, Q\}$ and $\{W(-x), x \geq 0, Q\}$ are independent Bessel processes of index 3 starting at 0. Let μ_W be the probability measure in $\mathbf{R}$ of the form $const. \exp\{-W(x)\}dx$; it is well defined for almost all W with respect to Q because $\exp(-W) \in L^1(\mathbf{R})$, Q-a.s. For an integer $k \geq 1$ we set $\mu_W^k = \mu_W \otimes \cdots \otimes \mu_W$ (the k-fold product) and $\mu^k = \int \mu_W^k Q(dW)$.

THEOREM 1 ([26], [28]). *For any $t_1, \ldots, t_k$ with $0 < t_1 < \cdots < t_k$ the joint distribution of $\omega(e^\lambda t_j) - b_\lambda(W), 1 \leq j \leq k$, with respect to $\mathcal{P}$ converges to μ^k as $\lambda \to \infty$.*

This theorem was proved in [26] for $k = 1$. The case $k \geq 1$ was proved in [28] by making use of Ogura's theorem stated below. Suppose we are given a sequence of diffusion operators

$$L_n = \frac{d}{m_n(dx)}\frac{d}{dS_n(x)}, \quad n \geq 1,$$

and denote by $X_n^x(t)$ the diffusion process with generator L_n starting at x. We assume that the following conditions (i), (ii), and (iii) are satisfied.

(i) $S_n(0) = 0$ and $S_n(x)$ tends to ∞ or $-\infty$ accordingly as $x \to \infty$ or $x \to -\infty$; for each x, $S_n(x) \to 0$ as $n \to \infty$.

(ii) The measure m_n converges vaguely to some nontrivial finite measure m as $n \to \infty$.

(iii) The measure $\widetilde{m_n} = m_n \circ S_n^{-1}$ converges vaguely to $c\delta_0$ as $n \to \infty$, where S_n^{-1} is the inverse function of S_n, $c = m(\mathbf{R}) > 0$, and δ_0 is the δ-measure at 0.

OGURA'S THEOREM ([21]; see also [26]). *For any $\varepsilon \in (0, 1)$ and an integer $k \geq 1$ we set*

$$T_{k,\varepsilon} = \{(t_1, \ldots, t_k) \in \mathbf{R}^k : \varepsilon \leq t_1 < t_k \leq 1/\varepsilon, t_j - t_{j-1} \geq \varepsilon \ (1 \leq \forall j \leq k)\} \tag{3.2}$$

and consider a sequence $\{x_n\}$ satisfying

$$|S_n(x_n)| \leq 1/\varepsilon, \quad n \geq 1. \tag{3.3}$$

Then for any continuous functions f_j in $\mathbf{R}$ with compact supports, $1 \leq j \leq k$,

$$E\left\{\prod_{j=1}^k f_j(X_n^{x_n}(t_j))\right\} \to \prod_{j=1}^k \int f_j \, dm_0$$

as $n \to \infty$ uniformly in $\{x_n\}$ satisfying the condition (3.3) and in $(t_1, \ldots, t_k) \in T_{k,\varepsilon}$, where m_0 is the probability measure $c^{-1}m$.

It is known (see [3], Lemma 1.3) that, for fixed W, the process $\{\omega(\lambda^4 t), t \geq 0, P_W\}$ is equivalent in law to $\{\lambda^2 \omega(t), t \geq 0, P_{\lambda W_\lambda}\}$. This combined with the fact that $b_\lambda(W) = \lambda^2 b(W_\lambda)$ implies that the process $\{\omega(e^\lambda t) - b_\lambda(W), t \geq 0, P_W\}$ is equivalent in law to $\{\lambda^2(\omega(\lambda^{-4} e^\lambda t) - b(W_\lambda)), t \geq 0, P_{\lambda W_\lambda}\}$; in addition, W_λ and W are identical in law. Therefore, for the proof of Theorem 1 it is enough to show

$$\int E_{\lambda W} \left\{ \prod_{j=1}^{k} f_j(\lambda^2(\omega(\lambda^{-4} e^\lambda t_j) - b(W))) \right\} P(dW)$$

$$\longrightarrow \int \left\{ \prod_{j=1}^{k} \int f_j \, d\mu_W \right\} Q(dW), \quad \lambda \to \infty. \tag{3.4}$$

For fixed W the generator of the diffusion process $\{\lambda^2(\omega(\lambda^{-4} e^\lambda t) - b(W)), t \geq 0, P_{\lambda W}\}$ is given by $\{d/m_\lambda^W(dx)\}\{d/dS_\lambda^W(x)\}$, where

$$S_\lambda^W(x) = 2e^{-\lambda} \int_0^x \exp\{\lambda(W(\lambda^{-2} y + b) - W(b))\} \, dy,$$

$$m_\lambda^W(dx) = \exp\{-\lambda(W(\lambda^{-2} x + b) - W(b))\} dx.$$

LEMMA 1 ([28]). (i) $S_\lambda^W(x)$ tends to 0 as $\lambda \to \infty$ with P-measure 1.
(ii) If we regard m_λ^W and $\widetilde{m}_\lambda^W = m_\lambda^W \circ (S_\lambda^W)^{-1}$ as random variables taking values in the space of Radon measures in $\mathbf{R}$ equipped with the topology of vague convergence, then the joint distribution (under P) of m_λ^W and $\widetilde{m}_\lambda^W$ converges to the joint distribution (under Q) of $\exp\{-W(x)\} dx$ and $c_W \delta_0$ as $\lambda \to \infty$ where $c_W = \int \exp\{-W(x)\} \, dx$.

Making use of Lemma 1 and Ogura's theorem we can prove (3.4) and hence Theorem 1. For details see [28].

A similar localization problem was discussed in [11] when $\{W(x)\}$ is a step process arising from a random walk that is assumed to converge in law, under a suitable scaling, to a strictly stable process.

4 Limit theorems in the case $\kappa \neq 0$

Let $\mathbf{X} = \{\omega(t), t \geq 0, \mathcal{P}\}$ denote the diffusion in a Brownian environment with drift ($\kappa \neq 0$), and set $T_x = \inf\{t > 0 : \omega(t) = x\}, \overline{\omega}(t) = \max\{\omega(s) : 0 \leq s \leq t\}$ and $\underline{\omega}(t) = \inf\{\omega(s) : s \geq t\}$.

THEOREM 2 ([13]). (i) If $\kappa > 1$, then

$$\lim_{x \to \infty} T_x/x = \frac{4}{\kappa - 1}, \quad \mathcal{P} - a.s.,$$

$$\lim_{t \to \infty} \omega(t)/t = \frac{\kappa - 1}{4}, \quad \mathcal{P} - a.s.$$

(ii) If $\kappa = 1$, then $(x \log x)^{-1} T_x$ converges to 4 in probability (w.r.t. $\mathcal{P}$) as $x \to \infty$ and each of

$$t^{-1}(\log t)\overline{\omega}(t), \quad t^{-1}(\log t)\omega(t) \quad \text{and} \quad t^{-1}(\log t)\underline{\omega}(t)$$

converges to $1/4$ in probability (w.r.t. $\mathcal{P}$) as $t \to \infty$.

(iii) *If* $0 < \kappa < 1$, *then*

$$\lim_{x \to \infty} \mathcal{P}\{x^{-1/\kappa} T_x \le t\} = F_\kappa(t), \quad t > 0,$$

$$\lim_{t \to \infty} \mathcal{P}\{t^{-\kappa} \overline{\omega}(t) \le x\} = \lim_{t \to \infty} \mathcal{P}\{t^{-\kappa} \omega(t) \le x\}$$

$$= \lim_{t \to \infty} \mathcal{P}\{t^{-\kappa} \underline{\omega}(t) \le x\} = 1 - F_\kappa(x^{-1/\kappa}), \quad x > 0,$$

where F_κ is the distribution function of a one-sided stable distribution with Laplace transform $\exp(-c\lambda^\kappa)$.

REMARK. *The constant c in the Laplace transform $\exp(-c\lambda^\kappa)$ is given by*

$$c = \left\{ 2^{1-\kappa} \Gamma(\kappa) \int_0^\infty \phi(x)^{-2} dx \right\}^{-1},$$

where $\phi(x)$ is the solution of $\frac{d}{dM(x)} \cdot \frac{d\phi}{dx} = 2\phi, \phi(0) = 1, \phi'(0) = 0; M(x)$ is given by $M(x) = 2\gamma(\rho^{-1}(x))$, where $\gamma(x) = \int_0^x z^{-\kappa} e^{-4z} dz$ and $\rho^{-1}(x)$ is the inverse function of $\rho(y) = \int_0^y z^{\kappa-1} e^{4z} \, dz$.

Theorem 2 is a diffusion analogue of (a part of) the results for random walks due to Kesten et al. [15]. We do not give a detailed proof here but we remark that our method of the proof, in particular, of (ii) and (iii) is different from that of [14] and is based on the following lemma due to Kotani.

KOTANI'S LEMMA (1988, unpublished; see [13]). *Let $\lambda > 0$. Then for $t \ge 0$*

$$E_W \left\{ e^{-\lambda T_t} \right\} = \exp\left\{ -\int_0^t U_\lambda(s) \, ds \right\}, \quad \mathcal{P} - a.s.,$$

where $U_\lambda(t)$ is the unique stationary positive solution of

$$dU_\lambda(t) = U_\lambda(t) dW(t) + \left\{ 2\lambda + \frac{1-\kappa}{2} U_\lambda(t) - U_\lambda(t)^2 \right\} dt.$$

By virtue of Kotani's lemma, for the proof of (iii) it is enough to show that, with $\lambda = \xi x^{-1/\kappa}$,

$$\lim_{x \to \infty} \mathcal{E} \left\{ \exp(-\xi T_x / x^{1/\kappa}) \right\}$$

$$= \lim_{\lambda \downarrow 0} E \left\{ \exp\left(-\int_0^{\lambda^{-\kappa} \xi^\kappa} U_\lambda(s) \, ds \right) \right\} = \exp(-c\xi^\kappa),$$

and the key point in proving the last equality is the use of Kasahara's continuity theorem [9] concerning Krein's correspondence (e.g. see [16]). A full proof is given in [13].

The following theorem is a diffusion analogue of the result of Afanas'ev [1].

THEOREM 3 ([12]). (i) If $-2 < \kappa < 0$, then

$$P\{T_x < \infty\} \sim const.x^{-3/2}\exp(-\kappa^2 x/8), \quad x \to \infty,$$

where

$$const. = 2^{\frac{5}{2}+\kappa}\Gamma(-\kappa)^{-1}\int_0^\infty\int_0^\infty\int_0^\infty\int_0^\infty z(a+z)^{-1}a^{-\kappa-1}e^{-a/2}y^{-\kappa}e^{-\lambda z}u\sinh u\,da\,dy\,dz\,du,$$

$(\lambda = 2^{-1}(1+y^2) + y\cosh u)$.

(ii) If $\kappa = -2$, then

$$P\{T_x < \infty\} \sim (2/\pi)^{1/2}x^{-1/2}\exp(-x/2), \quad x \to \infty.$$

(iii) If $\kappa < -2$, then

$$P\{T_x < \infty\} \sim \frac{-\kappa-2}{-\kappa-1}\cdot\exp\{(\kappa+1)x/2\}, \quad x \to \infty.$$

The proof of (i) relies on an explicit representation of the distribution of a certain Brownian functional due to Yor ([29], see the formula (6.e)).

5 A diffusion in a multidimensional Brownian environment

One generalization of the model discussed in Section **3** to a multidimensional case is to take a Lévy's Brownian motion with a multidimensional time as an environment. Let $\{W(x), x \in \mathbf{R}^d, P\}$ be a Lévy's Brownian motion with a d-dimensional time that is supposed to be an environment. For a frozen Brownian environment W let $\mathbf{X}_W = \{\omega(t), t \geq 0, P_W\}$ be a diffusion process with generator $2^{-1}(\Delta - \nabla W \cdot \nabla)$ starting at 0. Existence of such a diffusion is guaranteed by the result of Nash ([20]). As in a one-dimensional case we call $\mathbf{X} = \{\omega(t), t \geq 0, \mathcal{P}\}$ a diffusion in a d-dimensional Brownian environment, where $\mathcal{P}(dW\,d\omega) = P(dW)P_W(d\omega)$. A similar diffusion model appeared in a heuristic argument of [18]. Durrett [4] obtained rigorous results on recurrence and localization for random walks on $\mathbf{Z}^d$ described by a certain random potential having asymptotic self-similarity and stationary increments. The diffusion $\mathbf{X}$ may be regarded as the continuous time analogue of what was discussed in Example 2 ($\beta = 1$) of [4]. Recently Mathieu [19] considered the diffusion $\mathbf{X}$ itself and discussed its long time asymptotic behavior.

THEOREM 4 ([27]; see also [4] for random walks). $\mathbf{X}_W$ is recurrent for almost all Brownian environments W for any dimension d.

This theorem can easily be proved by making use of Ichihara's recurrence test ([8], see Theorem A) concerning symmetric diffusions. We can also use Fukushima's recurrence criterion [5] in terms of the associated Dirichlet form. $\mathbf{X}_{|W|}$ is also recurrent, P-a.s.; however, $\mathbf{X}_{-|W|}$ is transient, P-a.s., for any $d \geq 2$ as can be proved by using Ichihara's transience test ([8], see Theorem B). From the argument of [27] it is also easy to see that Theorem 4 remains valid when $\{W(x)\}$ is replaced by any continuous random field $\{V(x)\}$ in $\mathbf{R}^d$ satisfying the following conditions (i), (ii), and (iii).

(i) Self-similarity: there exists $\alpha > 0$ such that the law of $\{\lambda^{-1}V(\lambda^\alpha x)\}$ equals that of $\{V(x)\}$, denoted by P, for each $\lambda > 0$.

(ii) $\{T_t, t \in \mathbf{R}\}$ is ergodic, where T_t is a P-preserving transformation from $C(\mathbf{R}^d)$ onto itself defined by $(T_t V)(x) = e^{-t/\alpha}V(e^t x), x \in \mathbf{R}^d$.

(iii) $\min\{V(x) : |x| = 1\} > 0$ with positive probability.

The argument of [19] entails the following theorem.

THEOREM 5 ([19]; see also [4] for random walks). *Localization takes place for* $\mathbf{X}$ *in the sense that*

$$\lim_{N\to\infty} \overline{\lim_{\lambda\to\infty}} \mathcal{P}\{\lambda^{-2}\max(|\omega(t)| : 0 \le t \le e^\lambda) \ge N\} = 0.$$

It seems that there is no proof of the existence of the limiting distribution of $\lambda^{-2}\omega(e^\lambda)$ as $\lambda \to \infty$. It is to be noted, however, that Mathieu [19] gave the existence proof together with an explicit representation of the limiting distribution of $\lambda^{-2}\omega(e^\lambda)$ in terms of the local time of $|W|$ at level 0 *when W is replaced by* $|W|$.

The above results on recurrence and localization rely heavily on the (asymptotic) self-similarity of W as well as the symmetry of $\mathbf{X}_W$. Without these conditions the situation will change much. In the case of random walks there is a profound work by Bricmont and Kupiainen [2].

References

[1] V. I. Afanas'ev, *On a maximum of a transient random walk in random environment*, Theor. Probab. Appl. **35** (1990), 205–215.

[2] J. Bricmont and A. Kupiainen, *Random walks in asymmetric random environments*, Comm. Math. Phys. **142** (1991), 345–420.

[3] T. Brox, *A one-dimensional diffusion process in Wiener medium*, Ann. Probab. **14** (1986), 1206–1218.

[4] R. Durrett, *Multidimensional random walks in random environments with subclassical limiting behavior*, Comm. Math. Phys. **104** (1986), 87–102.

[5] M. Fukushima, *On recurrence criteria in the Dirichlet space theory*, in Local Times to Global Geometry, Control and Physics (ed. Elworthy), Research Notes in Math., Longman Scientific & Technical, Essex **150** (1986), 100–110.

[6] A. O. Golosov, *Localization of random walks in one-dimensional random environments*, Comm. Math. Phys. **92** (1984), 491–506.

[7] A. O. Golosov, *On limiting distributions for a random walk in a critical one-dimensional random environment*, Russian Math. Surveys **41** (1986), 199–200.

[8] K. Ichihara, *Some global properties of symmetric diffusion processes*, Publ. Res. Inst. Math. Sci., Kyoto Univ. **14** (1978), 441–486.

[9] Y. Kasahara, *Spectral theory of generalized second order differential operators and its applications to Markov processes*, Japan. J. Math. (New Ser.) **1** (1975), 67–84.

[10] K. Kawazu, Y. Tamura, and H. Tanaka, *One-dimensional diffusions and random walks in random environments*, Lecture Notes in Math., Springer-Verlag, Berlin and New York, **1299** (1988), 170–184.

[11] K. Kawazu, Y. Tamura and H. Tanaka, *Localization of diffusion processes in one-dimensional random environment*, J. Math. Soc. Japan **44** (1992), 515–550.

[12] K. Kawazu and H. Tanaka, *On the maximum of a diffusion process in a drifted Brownian environment*, Lecture Notes in Math. (Séminaire de Probabilités XXVII), Springer-Verlag, Berlin and New York, **1557** (1993), 78–85.

[13] K. Kawazu and H. Tanaka, *A diffusion process in a Brownian environment with drift*, to appear in J. Math. Soc. Japan.

[14] H. Kesten, *The limit distribution of Sinai's random walk on random environment*, Phys. A **138** (1986), 299–309.

[15] H. Kesten, M. V. Kozlov and F. Spitzer, *A limit law for random walk in a random environment*, Compositio Math. **30** (1975), 145–168.

[16] S. Kotani and S. Watanabe, *Krein's spectral theory of strings and generalized diffusion processes*, Lecture Notes in Math., Springer-Verlag, Berlin and New York, **923** (1982), 235–259.

[17] A. V. Letchikov, *Localization of one-dimensional random walks in random environments*, Soviet Sci. Rev. Sect. C: Math. Phys. Rev. **8** (1989), 173–220.

[18] E. Marinari, G. Parisi, D. Ruelle, and P. Windey, *On the interpretation of $1/f$ noise*, Comm. Math. Phys. **89** (1983), 1–12.

[19] P. Mathieu, *Zero white noise limit through Dirichlet forms, with application to diffusions in a random medium*, Probab. Theory Related Fields **99** (1994), 549–580.

[20] J. Nash, *Continuity of solutions of parabolic and elliptic equations*, Amer. J. Math. **80** (1958), 931–953.

[21] Y. Ogura, *One-dimensional bi-generalized diffusion processes*, J. Math. Soc. Japan **41** (1989), 213–242.

[22] S. Schumacher, *Diffusions with random coefficients*, Contemp. Math. **41** (1985), 351–356.

[23] Ya. G. Sinai, *The limiting behavior of a one-dimensional random walk in a random medium*, Theor. Probab. Appl. **27** (1982), 256–268.

[24] F. Solomon, *Random walks in a random environment*, Ann. Probab. **3** (1975), 1–31.

[25] H. Tanaka, *Limit distribution for 1-dimensional diffusion in a reflected Brownian medium*, Lecture Notes in Math. (Séminaire de Probabilités XXI), Springer-Verlag, Berlin and New York, **1247** (1987), 246–261.

[26] H. Tanaka, *Limit theorem for one-dimensional diffusion process in Brownian environment*, Lecture Notes in Math., Springer-Verlag, Berlin and New York, **1322** (1988), 156–172.

[27] H. Tanaka, *Recurrence of a diffusion process in a multidimensional Brownian environment*, Proc. Japan Acad. Ser. A Math. Sci. **69** (1993), 377–381.

[28] H. Tanaka, *Localization of a diffusion process in a one-dimensional Brownian environment*, Comm. Pure Appl. Math. **47** (1994), 755–766.

[29] M. Yor, *On some exponential functionals of Brownian motion*, Adv. Appl. Probab. **24** (1992), 509–531.

Analytical and Numerical Aspects of Fluid Interfaces

J. Thomas Beale

Mathematics Department, Duke University,
Durham, NC 27708, USA

1. Introduction

Water waves are familiar in everyday experience and illustrate the rich variety of phenomena observed in wave motion. The exact equations are difficult to deal with directly because of the free boundary and the inherent nonlinearity. However, approximate treatments, especially linear theory and shallow water theory, as well as numerical computations, have led to the understanding of many important aspects. We concentrate here on qualitative properties of the equations of motion, linearized about an arbitrary solution, and the design of convergent numerical methods of a special type, called boundary integral methods.

The boundary integral formulation of the equations of water waves leads to a natural approach for computing time-dependent motions. The moving interface is tracked directly, and only quantities on the interface need be computed. However, numerical instabilities are difficult to avoid, because of the unusual form of the equations, the nonlinearity, and the lack of dissipation. We summarize here the results of a study of the motion of two-dimensional water waves in this formulation. This work is done in conjunction with Hou and Lowengrub; details are given in [5], [6]. We analyze the behavior of the equations linearized about an arbitrary, time-dependent solution, which may be far from equilibrium. A structure for these linearized equations emerges that is analogous to the familiar case at equilibrium. The well-posedness of the linearized equations is evident from this structure, in contrast to other cases of fluid interfaces that are ill-posed. When the equations are discretized, the issue of numerical stability is closely related to the linear well-posedness in the continuous case. By maintaining a similar structure for the discretized equations, we find that computational methods can be designed that are numerically stable, even for fully nonlinear motion, and converge to the exact solution. The methods treated here are closely related to those of Baker, Meiron, and Orszag [2], [3]. Calculations of breaking waves in [3], [6], and in earlier works illustrate the capability of these methods.

The approach used here is a familiar one for analyzing approximations to nonlinear partial differential equations. In general, for an evolution equation of the form $u_t = F(u)$, an infinitesimal variation $\dot{u}(t)$ from an exact solution $u(t)$ will satisfy $\dot{u}_t = dF(u)\dot{u}$, where $dF(u)$ is the Fréchet derivative of the mapping F on appropriate spaces. To decide whether this linearized equation is well-posed in the sense of Hadamard, we only need to keep account of the most important terms in $dF(u)\dot{u}$; other terms, which might affect the stability of the underlying solution

Proceedings of the International Congress
of Mathematicians, Zürich, Switzerland 1994
© Birkhäuser Verlag, Basel, Switzerland 1995

u, can be treated generically. Now suppose that we introduce a spatial discretization in the unknown u and the mapping F, replacing the original evolution by $u_t^h = F^{(h)}(u^h)$, where h indicates the spatial scale. Then the error $u^{\mathrm{err}} = u^h - u$ between the discrete solution u^h and the exact solution u will satisfy an equation of the form $u_t^{\mathrm{err}} = dF^{(h)}(u)u^{\mathrm{err}} + r$, where r includes the consistency error of u in the discrete scheme and nonlinear terms in u^{err}. To control the error we need growth estimates for u^{err} much like the well-posedness estimates for $\dot{u}$ in the continuous case. However, the discrete operator $dF^{(h)}$ could easily introduce numerical instabilities, i.e., rapid growth in high wave numbers, even if the continuous problem is well-posed. Thus analysis has a role in selecting numerical schemes that do not encounter numerical instabilities.

2.   Water Waves

In the simplest model of water waves, the fluid is taken to be incompressible and inviscid, with constant density, which we set to 1. Thus the fluid flow is governed by the usual Euler equations with the force of gravity. The fluid is bounded above by a free surface or interface that moves with the fluid. The pressure at the interface matches the pressure above, provided surface tension is neglected; we ignore the motion of the air above, so that the pressure on the surface is zero. (The effect of surface tension will be described below.) We customarily assume that the fluid motion is irrotational. The fluid velocity should then be of the form $\nabla\phi$, the gradient of a scalar potential ϕ. For irrotational flow the momentum equation can be integrated to give Bernoulli's equation

$$\phi_t + \tfrac{1}{2}|\nabla\phi|^2 + p + gy = 0 \tag{1}$$

in the fluid domain, assuming the motion is at rest at infinity. Moreover, the velocity has divergence zero, so that

$$\Delta\phi = 0. \tag{2}$$

An important consequence is that the potential ϕ in the fluid domain, and thus the velocity field, are determined by the value of ϕ on the interface. (For simplicity we assume the fluid is of infinite depth; the L^2 norm of $\nabla\phi$ should be finite.) It is natural to describe the state of the system by giving the location of the interface and the value of ϕ restricted to the interface. The evolution of these two variables is determined by the velocity field and by (1), with the pressure set to zero; (2) acts as a side condition, needed to find $\nabla\phi$ on the surface.

From the description above we can quickly obtain the familiar linear model of water waves near equilibrium; it will serve as a guide for the more general case. We consider motions slightly perturbed from fluid at rest in the lower half-plane $\{y < 0\}$. If the interface is $y = \eta(x;t)$, we match its vertical velocity to that of the fluid and use (1) to obtain the linearized equations of motion

$$\eta_t = \phi_y, \qquad \phi_t = -g\eta. \tag{3}$$

Here $\phi = \phi(x, y; t)$, and the equations above are evaluated at $y = 0$. It is a simple matter to determine ϕ_y on $y = 0$ from $\phi(\cdot, 0)$, using (2) in $\{y < 0\}$ and a Fourier transform in x,

$$\hat{\phi}_y(k, 0) = |k| \hat{\phi}(k, 0) \,. \tag{4}$$

It follows that the general solution for η is a superposition of special solutions

$$\eta(x, t) = e^{ikx} e^{\pm i\omega t}, \qquad \omega = \sqrt{gk} \,. \tag{5}$$

This is the familiar dispersion relation for linear waves in deep water without surface tension. We note, for later reference, that η_t in (3) is given by the nonlocal operator (4) applied to $\phi(\cdot, 0)$. We can think of this operator as HD, where $D = \partial/\partial x$ and H is the Hilbert transform; it also appears in the general case below. Many important phenomena of water waves have been explained using this model. For instance, Kelvin was led to the method of stationary phase by his study of water waves; he used it to derive the pattern of a wake behind a ship in the linear model of deep water.

If the direction of gravity is reversed above, we have exponential growth like $\exp(\sqrt{gk}\, t)$ in (5). The evolution equations are then ill-posed, because of this unbounded growth in high wave numbers. Such effects, often due to a heavy fluid over a vacuum or light fluid, are called Rayleigh-Taylor instabilities (e.g., see [13]). A different kind of ill-posedness, called Kelvin-Helmholtz instabilities, occurs in a vortex sheet, i.e., an interface between two layers of the same fluid with different tangential velocities. In that case the growth is like $\exp(kt)$. A survey of numerical methods for the ill-posed problem of vortex sheet motion is given in [10].

3. The Boundary Integral Formulation

We now describe the boundary integral formulation of the exact equations of water waves. For details, see [2], [3], [6], [7]. Quantities on the moving interface will be treated as functions of a Lagrangian variable α; i.e., α is a material coordinate. For two-dimensional flow, it is convenient to use complex notation. We write the interface as $z(\alpha; t)$, with either $z - \alpha \to 0$ as $|\alpha| \to \infty$, or $z - \alpha$ periodic with period 2π; $\partial z/\partial t$ will be given by the fluid velocity. The state of the system is specified by the curve $z(\alpha; t)$ and the velocity potential $\phi(\alpha; t)$ on the interface. According to (2), the extension of ϕ to the fluid domain is harmonic, and it follows that the velocity, in the complex form $w = u - iv$, is analytic in z. We write the velocity w at a point z below the surface as a Cauchy integral

$$w = \frac{1}{2\pi i} \int \frac{1}{z - z(\alpha')} \gamma(\alpha')\, d\alpha'$$

with γ to be found in terms of ϕ. (This is equivalent to writing ϕ as a double layer potential.) The limiting velocity at the free surface is

$$w(\alpha) = \frac{1}{2} \frac{\gamma(\alpha)}{z_\alpha(\alpha)} + \frac{1}{2\pi i} \int \frac{\gamma(\alpha')}{z(\alpha) - z(\alpha')}\, d\alpha' \,. \tag{6}$$

We can determine γ from the condition $D_\alpha \phi = \phi_x x_\alpha + \phi_y y_\alpha = \mathrm{Re}(w z_\alpha)$. Using (6) we obtain

$$\phi_\alpha(\alpha) = \frac{1}{2} \gamma(\alpha) + \mathrm{Re}\left(\frac{z_\alpha(\alpha)}{2\pi i} \int \frac{\gamma(\alpha')}{z(\alpha) - z(\alpha')}\, d\alpha' \right), \tag{7}$$

an integral equation of the second kind, which can be solved for γ in either of the two cases of interest. The kernel is an adjoint double layer potential. Finally, given z, ϕ at time t, solving (7) for γ, and finding w from (6), we have the evolution equations

$$\frac{\partial z}{\partial t} = w^*, \qquad \frac{\partial \phi}{\partial t} = \tfrac{1}{2}|w|^2 - gy, \tag{8}$$

as functions of $(\alpha; t)$. Here $z = x + iy$ and w^* is the complex conjugate of w. This form of Bernoulli's equation differs from (1) because of the change to Lagrangian variables.

Computational methods have been based on this or related formulations for some time. Longuet-Higgins and Cokelet [11] used a method of this type to calculate plunging breakers. An approach like the above for an interface between two different fluids was suggested long ago by Birkhoff. Such a method was developed successfully by Baker, Meiron, and Orszag [3]. Further work has been extensive; more references are given in [6]. However, numerical instabilities have been observed even in careful studies. Linear analysis has led to an understanding of their sources [4], [8], [12].

4. The Linearized Equations

We will not attempt to deal with the full evolution equations (6)–(8) directly; instead we treat the equations linearized about an arbitrary solution. We will obtain linear equations for a variation $\dot{z}, \dot{\phi}$ from an exact solution z, ϕ of (6)–(8), with coefficients depending on z, ϕ. In doing so, we will keep explicitly only the terms that affect well-posedness, treating the rest more crudely. Thus, e.g., in the equation for $\partial \dot{z}/\partial t$, terms proportional to $D_\alpha \dot{z}$ must be kept explicitly, but not terms like $\dot{z}$ itself, even though they might cause exponential growth.

We begin by varying the integral term w_0 in (6). We have

$$\dot{w}_0(\alpha) = \frac{1}{2\pi i} \int \frac{\dot{\gamma}(\alpha')}{z(\alpha) - z(\alpha')} \, d\alpha' - \frac{1}{2\pi i} \int \frac{\dot{z}(\alpha) - \dot{z}(\alpha')}{(z(\alpha) - z(\alpha'))^2} \gamma(\alpha') \, d\alpha'. \tag{9}$$

We can expect that the highest order dependence in $\dot{w}_0$ on $\dot{\gamma}, \dot{z}$ comes from α' near α. Thus, with $z(\alpha) - z(\alpha') \approx z_\alpha(\alpha)(\alpha - \alpha')$, the important part of the first term should be $(2iz_\alpha)^{-1} H\dot{\gamma}$, where H is the Hilbert transform, defined by

$$(Hf)(\alpha) = \frac{1}{\pi} \int_{-\infty}^{\infty} \frac{f(\alpha')}{\alpha - \alpha'} \, d\alpha'. \tag{10}$$

Similarly, we expect to approximate the second term by a simpler quantity, which can be identified as $\gamma(2iz_\alpha^2)^{-1} HD_\alpha \dot{z}$. More precisely, we can show that

$$\dot{w}_0 = \frac{1}{2iz_\alpha} H\dot{\gamma} - \frac{\gamma}{2iz_\alpha^2} HD_\alpha \dot{z} + A_{-\infty}(\dot{\gamma}) + A_0(\dot{z}). \tag{11}$$

Here and below, we use the notation A_{-r} for a linear operator bounded from H^j to H^{j+r}, H^j being the Sobolev space of functions with j derivatives in L^2, for j as allowed by the smoothness of the underlying solution. Similarly, $A_{-\infty}$ denotes an operator with arbitrary smoothing. It is a useful fact that the commutator of H with a multiplication operator is a smoothing operator. Taking into account the first term in w, and using this fact, we have

$$\dot{w} = \frac{1}{2z_\alpha}(I - iH)\left(\dot{\gamma} - \frac{\gamma}{z_\alpha}\dot{z}_\alpha\right) + A_{-\infty}(\dot{\gamma}) + A_0(\dot{z}). \tag{12}$$

By varying (7) we can relate $\dot{\gamma}$ in (12) to $\dot{\phi}, \dot{z}$. After solving for $\dot{\gamma}$ and some further manipulation we can re-express (12) as

$$\dot{w} = z_\alpha^{-1}(I - iH)D_\alpha\dot{F} + A_{-\infty}(\dot{\phi}) + A_0(\dot{z}), \tag{13}$$

where

$$\dot{F} = \dot{\phi} - \text{Re}(w\dot{z}) = \dot{\phi} - u\dot{x} - v\dot{y}. \tag{14}$$

This expression for $\dot{w}$ gives us evolution equations for $\dot{z}$. Let $\dot{z}^N, \dot{z}^T$ be the normal and tangential components of $\dot{z}$, with respect to the underlying curve $z(\alpha)$, N being the outward normal, and let $\dot{\delta} = \dot{z}^T + H\dot{z}^N$. Using $\dot{z}_t = \dot{w}^*$, we find that

$$\dot{z}_t^N = \sigma H D_\alpha\dot{F} + A_{-\infty}(\dot{\phi}) + A_0(\dot{z}) \tag{15}$$
$$\dot{\delta}_t = A_{-\infty}(\dot{\phi}) + A_0(\dot{z}) \tag{16}$$

with $\sigma = |z_\alpha|^{-1}$. In this form it appears that only the normal component of $\dot{z}$ is important. Also, the form of (13) suggests that we use the modified varied potential $\dot{F}$, rather than $\dot{\phi}$, in our linearized evolution equations. Accordingly, we differentiate $\dot{F}$ in t and vary Bernoulli's equation to find $\dot{\phi}_t$ to obtain, after some cancellation, $\dot{F}_t = -g\dot{y} - u_t\dot{x} - v_t\dot{y}$. Comparison with the Euler equations in Lagrangian form shows that the vector $(u_t, v_t + g)$ is in fact $-\nabla p$, a normal vector at the surface, because $p = 0$ there. Thus, we can rewrite the above as

$$\dot{F}_t = -c\dot{z}^N, \qquad c = c(\alpha, t) = (u_t, v_t + g) \cdot N. \tag{17}$$

Equations (15)–(17) are the evolution equations for the variation, with $\dot{z}^N, \dot{\delta}, \dot{F}$ in place of the original variables $\dot{z}, \dot{\phi}$.

Several qualitative conclusions are evident from the linearized evolution equations (15)–(17). The Fourier symbol of H is $-i\,\text{sgn}\,k$, so that the operator $\Lambda = HD$ in (15) has symbol $|k|$, and is positive and symmetric. The same operator appeared in the equations linearized at equilibrium (3), (4), except that now it occurs with respect to the α-variable. In fact, in the special case at equilibrium, with $z \equiv \alpha$ and $w \equiv 0$, we have $\sigma = 1$, $c = g$, and equations (15), (17) reduce to the earlier system (3). In general, we see from (14), (15) that $\partial\dot{z}^N/\partial t$ contains a term like $\Lambda\dot{z}^N$, a term that would produce Kelvin-Helmholtz instabilities, were it not for a compensating term in the $\dot{F}$ equation. Indeed, if we multiply (15) by $(c/\sigma)\dot{z}^N$ and (17) by $\Lambda\dot{F}$, the two principal terms are the same except for opposite signs. As a consequence, we can show that the energy form

$$\int\left((\dot{z}^N)^2 + \dot{\delta}^2 + \dot{F}(\Lambda + 1)\dot{F}\right) d\alpha$$

grows at most exponentially, provided $c(\alpha, t)$ has a positive lower bound. Similar estimates hold for higher Sobolev norms. The positivity of c means that, in the underlying flow, the fluid is not accelerating downward, normal to itself, as fast as the normal component of the acceleration of gravity. This condition can be viewed as a natural generalization of the criterion of Taylor for horizontal interfaces. If it is violated, we should expect the interface to develop Rayleigh-Taylor instabilities, and the linearized equations to be ill-posed, as in the equilibrium case. The following theorem is proved in [5]. Related results are given in [1].

THEOREM 4.1 *Assume that a smooth solution of the exact water wave equations (6)–(8) exists for some time interval, and assume that*

$$(u_t, v_t + g) \cdot N > c_0 > 0 \tag{18}$$

for some c_0. Then the linearized equations (15)–(17) are well-posed.

5. The Numerical Method

The equations of motion of water waves, written in the boundary integral formulation (6)–(8), are naturally suited for computation, but discretizations must be chosen for the singular integrals and the derivatives. These choices affect critically the accuracy and stability of the numerical method. We consider the simplest case in which the interface is periodic, i.e., $z(\alpha; t) - \alpha$ is periodic with period 2π. Similarly, γ is periodic, and also ϕ, provided the velocity at infinity is zero. In the integrals of (6), (7) we now integrate over one period and replace the Cauchy kernel by the sum over periodic images, so that our equations become

$$z_t^* = \frac{\gamma(\alpha)}{2z_\alpha(\alpha)} + \frac{1}{4\pi i} \int_{-\pi}^{\pi} \cot\left(\frac{z(\alpha) - z(\alpha')}{2}\right) \gamma(\alpha') \, d\alpha' \tag{19}$$

$$\phi_\alpha(\alpha) = \frac{\gamma(\alpha)}{2} + \mathrm{Re}\left(\frac{z_\alpha(\alpha)}{4\pi i} \int_{-\pi}^{\pi} \cot\left(\frac{z(\alpha) - z(\alpha')}{2}\right) \gamma(\alpha') \, d\alpha'\right) \tag{20}$$

$$\phi_t = \tfrac{1}{2}|w|^2 - gy \,. \tag{21}$$

In discretizing these equations, we can expect the error between the computed and exact solutions to satisfy, approximately, a discretized form of the linearized equations (15)–(17). However, the discretization could easily destroy the balance of terms observed before and lead to rapid growth in the high wave numbers. To produce a numerical scheme, we replace the α-interval with N equally spaced points, $\alpha_j = jh$, where $h = 2\pi/N$. We compute $z_j(t), \phi_j(t), \gamma_j(t)$, where $z_j(t)$ approximates $z(\alpha_j; t)$ etc. Our first issue is the discretization of the singular integral. We use a choice that has been found to be practical as well as simple, a sum over alternating points, omitting the point with the singularity:

$$\int_{-\pi}^{\pi} \cot\left(\frac{z_j - z(\alpha')}{2}\right) \gamma(\alpha') \, d\alpha' \approx \sum_{k-j \text{ odd}} \cot\left(\frac{z_j - z_k}{2}\right) \gamma_k \cdot 2h \,.$$

This "alternate quadrature" is high order accurate and has the advantage of avoiding a special contribution at the singular point; see [6] for references.

It would be reasonable to expect that, with this quadrature rule and a difference operator for the α-derivative, we would obtain a usable numerical scheme from (19)–(21). However, as can be seen from the analysis below, the resulting scheme would be numerically unstable at equilibrium; see [7]. To discuss this further, we use the discrete Fourier transform. For a discrete function $\{f_j\}$ on the periodic interval, the discrete transform and its inverse are

$$\hat{f}_k = \frac{1}{N} \sum_{j \in I} f_j e^{-ik\alpha_j}, \qquad f_j = \sum_{k \in I} \hat{f}_k e^{ik\alpha_j}.$$

Here I is the set of integers $\{-N/2 + 1 \le k \le N/2\}$, assuming N even. We will write a discrete derivative operator in the form

$$D_h^{(\rho)} f_j = \sum_k \rho(kh) ik \hat{f}_k e^{ik\alpha_j},$$

for some ρ with $\rho(0) = 1$, $\rho(\pi) = 0$. The standard difference operators have this form; the order of accuracy is the order to which $\rho(\xi) \to 1$ as $\xi \to 0$. Of course, if the derivative is spectral, i.e., applied in the Fourier transform, ρ can be chosen with infinite order. Once a derivative operator, or equivalently a function ρ, is chosen, we also use a regularization based on ρ; for arbitrary periodic f_j we define $f_j^{(\rho)}$ by multiplying $\hat{f}_k$ by $\rho(kh)$. Thus, e.g., $D_h^{(\rho)} f_j = D_h^{(1)} f_j^\rho$. Similarly, we define $z_j^{(\rho)}$ by applying ρ to the transform of $z_j - \alpha_j$.

The numerical scheme we use for (19)–(21) is obtained by using the alternate quadrature for the integrals, replacing the α-derivative with an operator $D_h^{(\rho^r)}$, and replacing z_j inside the integrals with the regularized form $z_j^{(\rho)}$. The scheme so obtained is

$$\frac{dz_j^*}{dt} = \frac{\gamma_j}{2 D_h^{(\rho)} z_j} + \frac{1}{4\pi i} \sum_{k-j \text{ odd}} \cot\left(\frac{z_j^{(\rho)} - z_k^{(\rho)}}{2}\right) \gamma_k \cdot 2h \equiv w_j \qquad (22)$$

$$D_h^{(\rho)} \phi_j = \frac{\gamma_j}{2} + \mathrm{Re}\left(\frac{D_h^{(\rho)} z_j}{4\pi i} \sum_{k-j \text{ odd}} \cot\left(\frac{z_j^{(\rho)} - z_k^{(\rho)}}{2}\right) \gamma_k \cdot 2h\right) \qquad (23)$$

$$\frac{d\phi_j}{dt} = \tfrac{1}{2}|w_j|^2 - gy_j. \qquad (24)$$

For this version we are able to show full numerical stability and convergence. The reason for the regularization of z_j in the singular integrals will become evident in the discussion of stability below. We now state the convergence result; see [6].

THEOREM 5.1 *Suppose an exact solution of the water wave equations (19)–(21) is smooth and satisfies the acceleration condition (18) for some time interval. Suppose the numerical scheme (22)–(24) is initialized with this solution, and solved with a choice of $D_h^{(\rho)}$ that is rth order accurate, $r \ge 4$. Then the computed interface $\{z_j\}$ is close to the exact interface z to order h^r in discrete L^2 norm at each time,*

$$|z_j(t) - z(\alpha_j; t)|_{\ell^2} \le C h^r.$$

Similarly, ϕ_j is accurate to order h^r and γ_j to order h^{r-1} in discrete L^2 norm.

To outline the proof, we write equations for the errors $\dot{z}_j(t) \equiv z_j(t) - z(\alpha_j; t)$, etc., and attempt to estimate their growth as before in the continuous case. The estimates are not as straightforward in the discrete case, but several observations will illustrate the essential points. If we compare the sum in (22) for w_j with the corresponding sum for the exact w, the terms linear in $\dot{z}_j, \dot{\gamma}_j$ are

$$\frac{1}{2\pi i} \sum_{k-j \text{ odd}} \frac{\dot{\gamma}_k}{z(\alpha_j)^{(\rho)} - z(\alpha_k)^{(\rho)}} 2h \; - \; \frac{1}{2\pi i} \sum_{k-j \text{ odd}} \frac{\gamma(\alpha_k)(\dot{z}_j^{(\rho)} - \dot{z}_k^{(\rho)})}{\left(z(\alpha_j)^{(\rho)} - z(\alpha_k)^{(\rho)}\right)^2} 2h$$

in analogy with (9). (We have expanded the periodic sum, with k now unbounded.) The most important contribution to the first term is $(2iz_\alpha)^{-1} H_h \dot{\gamma}$, where H_h is the discrete Hilbert transform with alternate quadrature. It is helpful that this discrete Hilbert transform has the same Fourier symbol as before, i.e., $(H_h f)\hat{}_k = -i \operatorname{sgn} k \, \hat{f}_k$, $\; 0 < k < N/2$. In a similar way we can find the symbol of the important part of the second term and identify that part as

$$-\frac{\gamma}{2iz_\alpha^2} H_h D_h^{(1)} \dot{z}_j^{(\rho)} \; = \; -\frac{\gamma}{2iz_\alpha^2} H_h D_h^{(\rho)} \dot{z}_j \; .$$

Thus, the regularization of z_j in the singular integral leads to a term involving $D_h^{(\rho)} \dot{z}_j$, and our two terms are analogous to those in (11). Without $z_j^{(\rho)}$, we have a mismatch leading to the instability referred to above. Moreover, when we add the first term in $\dot{w}$ to these, we can combine all the terms as we did before in (12), obtaining

$$\dot{w} \approx \frac{1}{2z_\alpha}(I - iH_h)\left(\dot{\gamma} - \frac{\gamma}{z_\alpha}D_h^{(\rho)}\dot{z}\right) \; .$$

It might appear that we could simply choose $\rho \equiv 1$ and maintain the structure just observed. However, there would be trouble from the terms we have omitted from our discussion. Just as in spectral methods, "aliasing" errors can arise from products, and it seems necessary to have some cut-off in the derivative operator in the high modes. In addition, there are aliasing errors in discrete integral operators with smooth kernels. Surprisingly, with the alternate quadrature, such an operator may not gain derivatives at all. However, we can show that if K is a smooth function of (α, α'), then at least

$$\sum_{k-j \text{ odd}} K(\alpha_j, \alpha_k) f_k^{(\rho)} 2h \; = \; A_{-1}(f_j) \; .$$

This fact is needed in handling sums that occur above with a smooth kernel multiplying $\dot{\gamma}$. When we assess the error in (23) and solve for $\dot{\gamma}$, the ρ-smoothing enters through the derivative of ϕ as well as the smoothing of z in the sum, and using the above fact, we find, in analogy with (13), that

$$\dot{w} = z_\alpha^{-1}(I - iH_h)D_h^{(\rho)}\dot{F} \; + \; A_0(\dot{z}) + A_0(\dot{\phi}) + O(h^r) \; .$$

The remainder of the convergence proof is then parallel to the earlier treatment of the continuous linearized equations. Several improvements and generalizations are discussed in [6]. We have treated only the "semi-discrete" case, leaving the time discretization as a separate issue; cf. [8]. The related work [4] uses Fourier analysis in a similar way to identify numerical instabilities near equilibrium.

To test this numerical scheme, we have calculated in detail a wave that turns over and starts to break [6]. The time discretization was the Adams-Bashforth method. No instabilities were observed until well after the wave became vertical. Once it overturns, condition (18) is violated, and we expect Rayleigh-Taylor instabilities. In that regime we filter the high modes to prevent rapid growth of numerical errors, as in other calculations of ill-posed interfaces [10]. Once the wave begins breaking, physical effects neglected in this model should become important. Earlier calculations of breaking waves include [11], [2], [8].

6.   The Effect of Surface Tension

Surface tension is a cohesive force that tends to stabilize the high wave numbers. With surface tension included in the physical model, the pressure at the interface has a discontinuity proportional to the (mean) curvature. Thus, the earlier form of Bernoulli's equation in (8), (21) is replaced by

$$\phi_t = \tfrac{1}{2}|w|^2 - gy + \tau\kappa, \tag{25}$$

where κ is the curvature and τ the coefficient of surface tension. In contrast to the earlier case, the linearized equations are well-posed without qualification [5]:

THEOREM 6.1 *For any smooth solution of the water wave equations with surface tension, the linearized equations are well-posed.*

To see the reason, we return to the earlier linearization procedure. We again have (15), (16) for $\dot{z}^N, \dot{\delta}$. However, surface tension is now the dominant term in Bernoulli's equation, and neglecting terms that were essential before, we find

$$\dot{F}_t = \tau\dot{\kappa} + A_0(\dot{z}) = \tau\sigma\left(\sigma\dot{z}_\alpha^N\right)_\alpha + A_0(\dot{z}). \tag{26}$$

Again we obtain an energy estimate, but with a different balance between the two important terms in the $\dot{z}^N$- and $\dot{F}$-equations. The direction of gravity does not enter, because it affects only a lower order term, so the motion is well-posed even if water is above the surface.

The numerical scheme already described can be adapted to include surface tension, but further care is needed with the high wave numbers. Assume the cut-off function ρ is chosen so that $\rho'(\pi) = 0$ as well as $\rho(\pi) = 0$. Our choice of expression for the curvature is related to a product rule for discrete derivatives. For f a smooth function of α and any $u \in \ell^2$, we have

$$D_h^{(\rho)}[f(\alpha_j)u_j] = f(\alpha_j)D_h^{(\rho)}u_j + f_\alpha(\alpha_j)u_j^q + hA_0(u_j),$$

where $(u^q)\hat{}_k = q(kh)\hat{u}_k$ and $q = d(\xi\rho(\xi))/d\xi$. We discretize the curvature as

$$\kappa_j = \frac{D_h x_j^q D_h^2 y_j - D_h y_j^q D_h^2 x_j}{((D_h x_j^q)^2 + (D_h y_j^q)^2)^{3/2}} \tag{27}$$

where D_h is $D_h^{(\rho)}$. This choice is made so that the variation $\dot{\kappa}_j$ is a discrete form of (26); see [6], [7]. The numerical scheme is the same as in (22)–(24) except that $\tau\kappa_j$ is added to the discrete Bernoulli's equation (24). Convergence is proved in [6]:

THEOREM 6.2 *For a smooth solution of the water wave equations with surface tension, (19), (20), (25), the modified numerical scheme, with κ_j given by (27) and ρ as above, converges in the same sense as before.*

For the more general case of two different fluids separated by an interface with surface tension, a similar but more complicated numerical scheme, as in [3], can be shown to converge; these results will be reported elsewhere.

For many realistic situations, the coefficient of surface tension is small, and difficulties arise with time discretization due to stiffness. An explicit method would require a very small time step once large curvature develops. Hou, Lowengrub, and Shelley [9] have found a way to overcome this problem. They identify the highest order contribution to the surface tension term and treat it implicitly, thereby allowing a larger time step. We expect that this approach can be used with the spatial discretization discussed above.

References

[1] V. K. Andreev, Stability of Unsteady Motions of a Fluid with a Free Boundary, VO "Nauka", Novosibirsk, 1992 (in Russian).

[2] G. Baker, *Generalized vortex methods for free-surface flows*, Waves on Fluid Interfaces, R. Meyer, ed., Univ. of Wisconsin Press, Madison, WI 1983, 53–81.

[3] G. Baker, D. Meiron, and S. Orszag, *Generalized vortex methods for free-surface flow problems*, J. Fluid Mech., **123** (1982), 477–501.

[4] G. Baker and A. Nachbin, *Stable methods for vortex sheet motion with surface tension*, preprint, 1992.

[5] J. T. Beale, T. Y. Hou, and J. S. Lowengrub, *Growth rates for the linearized motion of fluid interfaces away from equilibrium*, Comm. Pure Appl. Math., **46** (1993), 1269–1301.

[6] J. T. Beale, T. Y. Hou, and J. S. Lowengrub, *Convergence of a boundary integral method for water waves*, to appear in SIAM J. Numer. Anal.

[7] J. T. Beale, T. Y. Hou, J. S. Lowengrub, and M. J. Shelley, *Spatial and temporal stability issues for interfacial flows with surface tension*, Math. Comput. Modelling, **20** (1994), 1–27.

[8] J. W. Dold, *An efficient surface-integral algorithm applied to unsteady gravity waves*, J. Comput. Phys., **103** (1992), 90–115.

[9] T. Y. Hou, J. Lowengrub, and M. Shelley, *Removing the stiffness from interfacial flows with surface tension*, J. Comput. Phys., **114** (1994), 312–338.

[10] R. Krasny, *Computing vortex sheet motion*, Proc. Internat. Congress Math. Kyoto 1990, (1990), 1573–83.

[11] M. S. Longuet-Higgins and E. D. Cokelet, *The deformation of steep surface waves on water, I. A numerical method of computation*, Proc. Roy. Soc. London Ser. A, **350** (1976), 1–26.

[12] A. J. Roberts, *A stable and accurate numerical method to calculate the motion of a sharp interface between fluids*, IMA J. Appl. Math., **31** (1983), 13–35.

[13] D. L. Sharp, *An overview of Rayleigh-Taylor instability*, Phys. D, **12** (1984), 3–18.

Morse Theory in Differential Equations

KUNG-CHING CHANG

Department of Mathematics, Peking University
Beijing 100871, People's Republic of China

1 Introduction

In the study of closed geodesics, Marston Morse developed his theory on the calculus of variations in the large. The Morse inequalities, which link on one hand, the numbers of critical points in various types of a function, and on the other hand, the topological invariants of the underlying manifold, play an important role in Morse theory. Naturally, they provide an estimate for the number of critical points of a function by using the topology of the manifold. Hopefully, this topological method would deal with the existence and the multiplicity of solutions of certain nonlinear differential equations. However, in this theory, the manifold is compact, and the functions are assumed to be C^2 and to have only nondegenerate critical points; all of these restrict the applications seriously. In contrast, Leray-Schauder degree theory has become a very useful topological method. In 1946, at the bicentennial conferences of Princeton University, there was much discussion of their contrast. M. Shiffman hoped that the two methods could be brought closer together "so that they may alter and improve each other, and also so that each may fill out the gaps in the scope of the other" [Pr]. Since then, great efforts have been made to extend the Morse theory. We only mention a few names of the pioneers as follows: R. Bott, E. Rothe, R. S. Palais, S. Smale, D. Gromoll, W. Meyer, A. Marino, and G. Prodi.

The minimax principle, another important topological method in dealing with the existence of critical points, is a twin of Morse theory. As was mentioned in his book [Mo], Morse confessed that the second Morse inequality "is essentially Birkhoff's minimax principle, although not stated by Birkhoff in precisely this form", and that " whose minimax principle was the original stimulus of the present investigation". In this direction, G. Birkhoff, L. Ljusternik, L. Schnirelmann, M. A. Krasnoselski, R. S. Palais, A. Ambrosetti, and P. H. Rabinowitz have made great contributions.

Melting together the classical Morse theory, the minimax principle, and the L-S degree theory for potential operators into a unified framework, the extended Morse theory becomes a powerful tool in nonlinear analysis.

Proceedings of the International Congress
of Mathematicians, Zürich, Switzerland 1994
© Birkhäuser Verlag, Basel, Switzerland 1995

2 Basic Results

Let M be a C^2 complete Banach-Finsler manifold, and $f \in C^1(M, \mathbb{R}^1)$. The critical set of f is denoted by K, and $\forall\, c \in \mathbb{R}^1$, $K_c = K \cap f^{-1}(c)$. The level set $f_c = \{x \in M \,|\, f(x) \leq c\}$. According to Brézis and Nirenberg, f is said to satisfy the $(PS)_c$ condition if any sequence $x_j \in M$, along which $f(x_j) \to c$ and $f'(x_j) \to \theta$, is subconvergent.

Morse theory consists of two parts: the local and global theories.

2.1 Local theory

For an isolated critical point, one attaches a series of topological invariants, which are called the critical groups, in describing the local behavior of f near p:

$$C_*(f, p) = H_*(f_c \cap U, (f_c \setminus \{p\}) \cap U; G), \tag{2.1}$$

where $* = 0, 1, 2, \ldots$, $c = f(p)$, U is an isolated neighborhood of p, G is an Abelian coefficient group, and $H_*(\cdot, \cdot; G)$ stands for the singular relative homology groups with coefficients in G.

Comparing with the Leray-Schauder index of f' at p, where f' is assumed to be a compact perturbation of identity, and M is a Hilbert space, we have

$$\mathrm{ind}_{LS}(f', p) = \sum_{q=0}^{\infty} (-1)^q \,\mathrm{rank}\, C_q(f, p). \tag{2.2}$$

Thus, the critical groups provide more information than the L–S index. They can be computed or estimated by the Morse index and the nullity of p; e.g., for isolated p, it is a local minimizer, if and only if rank $C_q(f, p) = \delta_{q0}$. Furthermore, they enjoy the homotopy invariance as the L–S index does (see [Ch4]).

2.2 Global theory

Let a, b be regular values of f, and let f satisfy $(PS)_c$ $\forall\, c \in [a, b]$. The following two principles provide the existence of critical points:

I.   If the pair (f_b, f_a) is nontrivial, i.e., f_a is not a deformation retract of f_b, then $K \cap f^{-1}(a, b) \neq \varnothing$ (Noncritical Interval Theorem).

In particular, if $H_*(f_b, f_a; G)$ is nontrivial, say $0 \neq [\tau] \in H_q(f_b, f_a; G)$, then

$$c = \inf_{\tau \in [\tau]} \, \sup_{x \in |\tau|} \, f(x) \tag{2.3}$$

is a critical value, where $|\tau|$ is the support of τ.

This principle includes the Mountain Pass Lemma as well as its high dimensional counterparts (homological link, either finite of infinite dimensional) as special cases [Ch7, 6]. For instance, if c is a Mountain Pass value, then f_{c+0} is path connected, but f_{c-0} is not.

II.   For two nontrivial classes $[\tau_1], [\tau_2] \in H_*(f_b, f_a; G)$, $[\tau_1]$ is called subordinate to $[\tau_2]$, denoted by $[\tau_1] < [\tau_2]$, if $\exists\, w \in H^*(f_b, G)$ with dim $w > 0$ such that $[\tau_1] = [\tau_2] \cap w$.

Assume that $[\tau_1] < [\tau_2] < \cdots < [\tau_l]$ are subordinate classes in $H_*(f_b, f_a; G)$. Let

$$c_i = \inf_{\tau \in [\tau_i]} \sup_{x \in |\tau|} f(x) \quad , \qquad i = 1, 2, \ldots, l.$$

If $c = c_1 = c_2 = \cdots = c_l$, then $Cat\ (K_c) \geq l$.

This principle provides multiplicity results (see [Ch6, 8], [CL1], and [Li]), and is based on a deformation theorem.

Morse inequalities, which link up these two theories, read as follows

$$M(f, a, b; t) = P(M, a, b; t) + (1 + t)\ Q(t), \tag{2.4}$$

where $M(f, a, b; t) = \sum_{q=0}^{\infty} M_q(a, b)\ t^q$, $P(M, a, b; t) = \sum_{q=0}^{\infty} \mathrm{rank}\ H_q(f_b, f_a)\ t^q$, $Q(t)$ is a formal series with nonnegative coefficients, and

$$M_q(a, b) = \sum_{j=1}^{l} \mathrm{rank}\ C_q(f, p_j). \tag{2.5}$$

In the above we have assumed that $f \in C^1(M, \mathbb{R}^1)$ satisfies $(PS)_c\ \forall\ c \in [a, b]$, and that $K \cap f^{-1}(a, b) = \{p_1, \ldots, p_l\}$ [Ch4].

A direct consequence of Morse inequalities is that, if the critical value c is obtained in (2.3), then $\exists\ p \in K_c$ satisfying $C_q(f, p) \neq 0$. Morse inequalities are also used in proving the existence and the multiplicity of critical points.

G-invariant Morse theory is also extended to this generality; i.e., instead of the isolated critical points, we investigate isolated critical orbits, see [Hi], [Wa1], [Ch7, 6], [CP], [BCP].

If we are dealing with a functional whose critical points may not be isolated, the critical groups can also be extended to define on isolated critical sets. Indeed, this is a special case of the Conley's homotopy index, which is defined on isolated invariant sets for a semi-flow. For pseudo-gradient flow, the Conley index becomes extremely simple [Ch4], [Da], and the corresponding Morse inequalities were extended by Conley and Zehnder [CZ], see also [Be].

Can we reduce the differentiability of the functional? The first step towards this extension was to consider a nondifferentiable, but locally Lipschitzian functional for which the generalized gradient $\partial f(p)$, in the sense of Clarke, is set valued. One extended the notions of critical point p, by assuming $\theta \in \partial f(p)$, and of the $(PS)_c$ condition, by using $\|\partial f(p)\| = \mathrm{Min}\{\|w\| | w \in \partial f(p)\}$ as a replacement of $|f'(p)|$ in the definition. Indeed, the deformation theorems hold [Ch1]. The extensions to C^1-perturbations of convex l.s.c. functions [Sz1] as well as to l.s.c. functions with φ-monotone subdifferential of order 2 [GMT] were explored afterwards. Recently, by introducing a notion, called the weak slope $\|d f(p)\|$ of a contionuous function f defined on a metric space, which replaces $\|\partial f(p)\|$ for locally Lipschitzian functions, De Giovanni and his colleagues successfully extended deformation theorems to continuous functions [CDM]. Remarkably, this unifies all the above-mentioned diverse extensions.

3 Applications and extensions

Principle I is frequently used; it is exactly the topological foundation in many
PDE literatures, although it might not be formulated apparently in this form. We
mention just a few results of recent years.

It is known that the Yang Mills functional does not satisfy the (PS) condition.
However, under a suitable symmetric assumption, when we restrict the functional
on the symmetric subspace, it does; then the minimizer exists in each component.
This gives a new proof of the result of Sadun and Segert on the existence of
nonminimal YM fields on S^4, see [Pa].

The existence of infinitely many closed geodesics on compact, connected,
simply connected Riemannian manifolds is based on the fact that the integral
homology of the loop space does not vanish for infinitely many dimensions. The
same idea was used to study those on certain noncompact Riemannian manifolds
[BG], geodesics in static space-time [BFG], and the periodic orbits of a Hamiltonian
system of N-body type (strong force) [FH].

As a replacement of the Ljusternik-Schnirelmann multiplicity theorem, Principle II provides estimates of critical groups. Applications can be found in [Ch1],
[CJ].

In Principle I, the (PS) condition is only used in proving the "Noncritical Interval Theorem". There are cases in which the deformation theorem holds but not
the (PS) condition; e.g., the harmonic maps. Eells and Sampson first noticed that
the heat flows for harmonic maps can be used as deformations. In particular, for
2-dimensional harmonic maps, with or without boundary conditions, not only the
"Noncritical Interval Theorem" and the Morse inequalities, but also the "Morse
handlebody theorem" hold below the energy level of the first bubbling. Some multiple solution results obtained previously by Brézis-Coron-Jost and Benci-Coron
are considered as consequences [Ch5].

In some cases, the (PS) condition fails, but the $(PS)_c$ sequence is subcon-
vergent if along which the Morse indices are under control. Combining with a
perturbation argument, the Morse theory is applied, see [Gh].

The above-mentioned Morse theory has been applied to study multiple solutions for second order elliptic boundary value problems of the following type:

$$\begin{cases} -\triangle u = \phi(x, u) & \text{in } \Omega, \\ u \in H_0^1(\Omega), \end{cases} \tag{3.1}$$

where Ω is a bounded domain in $\mathbb{R}^n$, $n \geq 3$, with smooth boundary, and $\phi \in
C^1(\bar{\Omega} \times \mathbb{R}^1, \mathbb{R}^1)$ satisfying the growth condition

$$|\phi'_\tau(x, \tau)| \leq C(1 + |\tau|^{\frac{n+2}{n-2}}). \tag{3.2}$$

There are many examples showing how the Morse theory has been applied,
see [Ch6, 8], [Wa2].

Because the maximum principle holds for second order elliptic equations,
an ordered Banach space degree theory has been developed, which enriches the

study of multiple solutions for this kind of problem. In particular, it admits the sub-and super-solution method and the cutoff technique. In contrast, in applying the Morse theory to these problems, we would rather take C_0^1 than H_0^1 as the coordinate working space.

Two problems arise: (1) One may assume that the corresponding functional

$$f(u) = \int_\Omega \frac{1}{2} |\nabla u|^2 - \Phi(x, u) \tag{3.3}$$

(where Φ is the primitive of ϕ with resp. to u) satisfies $(PS)_c$ on H_0^1, but it can not be transferred to C_0^1. (2) How to compute the critical groups in C_0^1? For (1), it was proved that C_0^1 is invariant under the gradient flow, see [Ch3]. Although $(PS)_c$ does not hold on C_0^1, the deformation property is still valid. As for (2), we have

$$C_*(f, u) = C_*(f|_{C_0^1}, u), \tag{3.4}$$

see [Ch9]. Under a weaker assumption, Brézis and Nirenberg proved this conclusion for minimizers [BN].

So far, all our discussions are only concerned with manifold M without boundary. However, in the early 1930s, Morse and van Schaack studied functions with boundary conditions defined on bounded smooth domains in $\mathbb{R}^n$. This result has been applied to PDE in [Ch2]. Infinite-dimensional Morse theory under general boundary conditions has been investigated. Namely, let M be a C^2 Hilbert Riemannian manifold with a C^2 boundary, which is a codimension 1 submanifold. Let $n(x)$ be the unit outward normal vector field defined on $\Sigma = \partial M$. For a function $f \in C^1(M, \mathbb{R}^1)$ without critical point on Σ, let $\Sigma_- = \{x \in \Sigma | (f'(x), n(x)) \le 0\}$ and $\hat{f} = f|_{\Sigma_-}$. The following theorem holds ([CL2], [Ro]):

THEOREM. *If f and $\hat{f}$ have only isolated critical points, and both satisfy $(PS)_c$ $\forall c \in [a, b]$, in which a, b are regular values for f and $\hat{f}$, then*

$$M(f, a, b; t) + M(\hat{f}, a, b; t) = P(M, a, b; t) + (1 + t) Q(t), \tag{3.5}$$

where M, P, and Q have the same meaning as in (2.4).

The proof depends on a modified deformation theorem for manifolds with boundaries. Because the negative gradient flow directs inward on $\Sigma \backslash \Sigma_-$, only f on Σ_- has contribution. A similar consideration has been used by Majer [Ma] and has been applied to study the N-body type problem (strong force), in which a boundary defined by a level set of another related functional is introduced to avoid the lack of (PS) condition [MT].

Using the above considerations, the modified Morse inequalities (3.5) can be extended to manifolds with corners.

When M is replaced by a locally convex subset S of itself (the local convexity depends on the special atlas), we call p the critical point of f on S, if it satisfies the variational inequality $< f'(p), v >\ \ge 0\ \forall\ v\ \in\ T_p(S)$, the tangent cone of S at p. Again, if one uses the notation $\|x^*\|_p = \sup\{< x^*, v > |v \in T_p(S), \|v\| \le 1\}$, for $x^* \in T_p(M)^*$, and replaces $f'(x_n) \to \theta$ by $\| - f'(x_n)\|_{x_n} \to 0$ in the

$(PS)_c$ condition; then the Morse theory is extendible. Applications to the Plateau problem for minimal surfaces were studied by [St], [CE], [JS].

In dealing with strongly indefinite functionals; i.e., the Morse indices for critical points being infinite, for example, the variational problems arising from the periodic solutions of Hamiltonian systems or the nonlinear wave equations, etc., all the critical groups are trivial. One has to redefine the critical groups in keeping the information obtained from the neighborhood of a critical point. Analogous to the above Gromoll-Meyer theory, an infinite-dimensional cohomology theory is introduced to modify the definition. Obviously, a Galerkin approximation scheme is needed, and the modified critical groups are defined to be the direct limits of sequences of critical groups on the approximate subspaces, see [Sz2]. However, to a periodic orbit $B(t)$ of Hamiltonian systems, the Maslov index $i(B)$ plays the same role as the Morse index of a nondegenerate critical point does in finite-dimensional Morse theory. A systematic study of the Maslov index has been made by Zehnder, Amann, Conley, Long, Salamon, etc. Following the same idea, for a degenerate periodic orbit $B(t)$, a pair $(i(B), n(B))$ is defined to be the Maslov index, see [Lo]. Again the Morse theory is applied. We have the following result for asymptotically linear Hamiltonian systems:

THEOREM. ([CLL], [Fe]) *Let* $H \in C^2([0,1] \times \mathbb{R}^{2n}, \mathbb{R})$ *be 1-periodic in* t. *We assume that there exist* $2n \times 2n$ *symmetric matrix functions* $B_0(t)$ *and* $B_\infty(t)$ *that are continuous and 1-periodic, satisfying*

$$\left| H_x(t,x) - B_0(t)x \right| = 0 \left(|x| \right) \qquad as \quad |x| \to 0 , \tag{3.6}$$

$$h(t,x) \equiv H(t,x) - \frac{1}{2} \left(B_\infty(t)x, x \right) \to 0 , \tag{3.7}$$

and

$$\left| h_x(t,x) \right| \to 0 , \tag{3.8}$$

as $|x| \to \infty$ *uniformly in* $t \in [0,1]$.

If further,

$$\left| H_{xx}(t,x) \right| \leq C \left(1 + |x|^s \right) \tag{3.9}$$

for some $C > 0$, $s > 1$, *and for all* $(t,x) \in [0,1] \times \mathbb{R}^{2n}$, *then the Hamiltonian system*

$$-J \frac{dx}{dt} = H_x(t,x) \tag{3.10}$$

possesses a nontrivial solution when one of the following three cases occurs:

(1) $\int_0^1 H(t,\theta)\, dt = 0$,

(2) $\int_0^1 H(t,\theta)\, dt > 0$ *and* $i(B_\infty) \notin [i(B_0), i(B_0) + n(B_0)]$,

(3) $\int_0^1 H(t,\theta)\, dt < 0$ *and* $i(B_\infty) + n(B_\infty) \notin [i(B_0), i(B_0) + n(B_0)]$.

As a consequence, (3.10) possesses a nontrivial solution if

$$[i(B_0), i(B_0) + n(B_0)] \cap [i(B_\infty), i(B_\infty) + n(B_\infty)] = \varnothing .$$

For degenerate $B_\infty(t)$, it is called strong resonance if (3.6) and (3.7) hold. In the case where $B_\infty(t)$ is nondegenerate, one may replace them by the following:

$$|H_x(t, x) - B_\infty(t) x| = 0(|x|) \qquad \text{as} \qquad |x| \to \infty \quad \text{uniformly in } t. \qquad (3.11)$$

This theorem covers all previously known results [AZ], [CZ], [LZ], [LL], [Sz2], [Ch7] as special cases.

4 Critical points at infinity

The (PS) condition, or, more generally, the compactness condition, which implies the deformation theorem, is basic point in Morse theory. However, in many interesting PDE problems arising in geometry and physics, the lack of compactness occurs and then the Morse theory is not directly applicable. It is a great challenge to extend Morse theory to these problems.

4.1 Compactifying the underlying manifold

In some problems, we may compactify the manifold M by adding infinity points, and extend the functional onto the new space such that the deformation lemma and then Morse theory hold for the enlarged space. The critical groups are used to distinguish the genuine critical points from the critical points at infinity. We present here an example: the Nirenberg's problem showing how the method works.

Given a function K on S^2, one asks if there exists a metric g that is pointwise conformal to the canonical metric g_0 on S^2 such that K is the Gaussian curvature of (S^2, g). This is equivalent to solving the following PDE:

$$-\triangle_{g_0} u + 1 = K\, e^{2u} \qquad \text{on} \quad S^2, \qquad (4.1)$$

which is the Euler-Lagrange equation of functional

$$J(u) = \fint (|\nabla u|^2 + 2u) - \log \fint K\, e^{2u} \qquad (4.2)$$

defined on the manifold

$$M = \{u \in H^1(S^2)|\ \fint e^{2u} = 1\}, \qquad (4.3)$$

where $\fint$ denotes the average over S^2. And when $\fint K\, e^{2u} \leq 0$, we set $J(u) = +\infty$. It is easily seen that the (PS) condition fails. However, noticing that M is conformal invariant, one splits the manifold into a product: $M = M_0 \times \overset{\circ}{B}{}^3$, where

$$M_0 = \{u \in M|\ \fint x_i\, e^{2u} = 0, i = 1, 2, 3\}, \qquad (4.4)$$

and $\overset{\circ}{B}{}^3 \cong S^2 \times [1, \infty)/S^2 \times \{1\}$ is the moduli space of the conformal group; i.e., $\forall\, a = (Q, t) \in S^2 \times [1, \infty), \exists\, a\ 1 - 1$ correspondence to the conformal transform ϕ_a on S^2. For each u, let a be the unique solution of the system:

$$\fint x_i\, e^{2u_{\phi_a}} = 0, \qquad i = 1, 2, 3, \qquad (4.5)$$

where $u_{\phi_a} = u \circ \phi_a + \frac{1}{2} \log \det(\phi_a')$.

Then $u \mapsto (u_{\phi_a}, a)$ is a diffeomorphism.

The "compactified" manifold $\overline{M}$ is defined to be $M_0 \times \overline{B^3}$, and the extended functional reads as

$$
\tilde{J}(u) = \begin{cases} J(u) & \text{if } u = (w, s(t)\,Q) \in M_0 \times \overset{\circ}{B^3} \\ f\,|(\nabla w|^2 + 2w) - \log\ K(Q) & \text{if } u = (w, Q) \in M_0 \times S^2, \end{cases} \tag{4.6}
$$

where s is a diffeomorphism: $[1, \infty) \to [0, 1)$, with $1 - t^{-2} \ln t$ as asymptotics.

Now, $\overline{M}$ is a Banach manifold with boundary, on which the (PS) condition for $\tilde{J}$ is gained. Although $\tilde{J}$ may fail to be C^1 at ∂M, one can deform $\tilde{J}$ into a C^1-functional preserving the same critical set, and then the Morse theory under the general boundary condition is applied. Assume

$$
K \in C^2, \text{Max } K > 0, \text{and } \triangle K(x) \neq 0 \text{ if } \nabla K(x) = 0 \text{ and } K(x) > 0 \tag{4.7}
$$

Let a, b be regular values of both K and e^{-J} such that only
finite critical points of both K and e^{-J} have values in (a, b) $\tag{4.8}$

and let

$$
\Omega = \{x \in S^2 | \ K(x) > 0, \triangle K(x) < 0\},
$$
$$
Cr_0(a, b) = \{x \in \Omega | \ K(x) \in (a, b), \text{and } x \text{ is a loc. max.}\},
$$
$$
Cr_1(a, b) = \{x \in \Omega | \ K(x) \in (a, b), \text{and } x \text{ is a saddle point}\}.
$$

THEOREM. *Under the assumptions (4.7) and (4.8), we have*

$$
\sum (M_q + \mu_q - \beta_q)\,t^q = (1 + t)\,Q(t), \tag{4.9}
$$

where M_q is the qth Morse type number of J,

$$
\beta_q = H_q(J_{-\log a}, J_{-\log b}), \tag{4.10}
$$

and

$$
\mu_q = \begin{cases} 0 & q \geq 2 \\ \#\ Cr_q(a, b) & q = 0, 1. \end{cases} \tag{4.11}
$$

This theorem improves previously known results, see [CD], [CY1, 2], and covers some new ones. In this connection, the critical groups are used to distinguish critical points from infinity. By perturbation, one may avoid the finiteness of critical points in the assumption. As a consequence: (4.1) is solvable if (4.7) and deg $(\Omega, \nabla K, \theta) \neq 1$.

The same method has been applied to study Nirenberg's problem on the hemisphere S^2_+, preserving ∂S^2_+ as a geodesic (see [LiL]) and the prescribing geodesic curvature problem. Given a function k on the unit circle $S^1 = \partial D$, one asks: Is there a flat metric g that is pointwise conformal to the Euclidean metric such that k is the geodesic curvature with resp. to g? The later, in turn, is reduced to the following nonlinear boundary value problem: find $u \in C^{\infty}(\overline{D})$, satisfying

$$\begin{cases} \triangle u = 0 & \text{in} \quad D \\ \dfrac{\partial u}{\partial n} + 1 = k\ e^u & \text{on} \quad S^1 . \end{cases} \qquad (4.12)$$

Similar conclusions drawn from Morse inequalities have been obtained by [CL4].

The method has also been applied to study the strong resonance problem [CL1], in particular, the Theorem ([CLL], [Fe]) in Section 3.

4.2 Bahri's theory

We cannot finish our talk without referring to the profound contribution due to Bahri, who has also been invited to be a speaker at the Congress. My abilities fall short of my desires to introduce his far-reaching work. I shall satisfy myself with giving a list of his main results in this direction, see [Ba1].

○ *Critical Sobolev exponent nonlinear elliptic equation [BB]*

Let (M^n, g) be a compact, n-dimensional, Riemannian manifold without boundary. Let $q \in L^{\infty}(M)$, and let $\triangle_g$ be the Laplace-Beltrami operator. Assume that $-\triangle_g + q$ is coercive. Then the equation

$$\begin{cases} (-\triangle_g + q)\, u = u^{\frac{n+2}{n-2}} & \text{on } M, \\ u > 0 & \text{on } M, \end{cases} \qquad (4.13)$$

has a solution, if one of the following conditions is satisfied:

(1) $3 \le n \le 5$,
(2) $n = 6, 7$, and $H_3(M, \mathbb{Z}_2) \ne 0$,
(3) n is arbitrary, and $H_1(M, \mathbb{Z}_2)$ or $H_2(M_1, \mathbb{Z}_2)$ is nonzero.

For the Dirichlet problem on bounded domains, see [BC2].

○ *Prescribing scalar curvature problem on S^n*

This is an analogous problem to Nirenberg's problem. One solves the following equation:

$$\begin{cases} L_{g_0}\, u = K(x)\, u^{\frac{n+2}{n-2}} & \text{on } S^n, \\ u > 0 & \text{on } S^n, \end{cases} \qquad (4.14)$$

where L_{g_0} is the conformal Laplacian with respect to the standard metric g_0. Assume that $K \in C^2$ is positive and has only nondegenerate critical points $p_1, \ldots, p_l$ such that $\triangle K(p_j) \ne 0$, $j = 1, \ldots, l$. Let $k_j = \text{ind}(p_j)\ \forall j$.

For $n = 3$, (4.14) is solvable if

$$\sum_{\triangle K(p_j) < 0} (-1)^{k_i} \ne -1.$$

Under additional conditions, this holds for $n = 4$, see [BC1] and [Zh].
For $n \ge 7$, a recent result was announced in [Ba3].

For arbitrary n, under different assumptions, by using a priori estimates, see Yanyan Li [LY].

○ *N-body type problem, [BR], [Ri].*

○ *Another proof of the Yamabe problem, [Ba2].*

All these results are based on the combined use of the Morse theory, of a kind of compactification of critical points at infinity, and of a penetrating analysis of the interactions of bubbles.

References

[AZ]   Amann, H., and Zehnder, E., *Nontrivial solutions for a class of nonresonance problems and applications to nonlinear differential equations*, Ann. Scuola Norm Sup. Pisa Cl. Sci. (4) **7** (1980), 539–630.

[Ba1]   Bahri, A., Critical point at infinity in some variational problems, Pitman Res. Notes Math. Longman, London, **182** (1989).

[Ba2]   ——, *Another proof of the Yamabe conjecture for locally conformally flat manifolds*, Nonlinear Anal. TMA **20** (1993), 1261–1278.

[Ba3]   ——, *The scalar cuvature problem on spheres of dimension larger or equal than 7*, preprint (1994).

[BB]   Bahri, A., and Brezis, H., *Elliptic differential equations involving the Sobolev critical exponent on manifolds*, preprint (1994).

[BC1]   Bahri, A., and Coron, J. M., *On a nonlinear elliptic equation involving the critical Sobolev exponent: the effect of the topology of the domain*, Comm. Pure Appl. Math. **41** (1988), 253–294.

[BC2]   ——, *The scalar-curvature problem on the standard three-dimensional sphere*, J. Funct. Anal. **95** (1991), 106–172.

[BR]   Bahri, A., and Rabinowitz, P. H., *Periodic solutions of Hamiltonian systems of three-body type*, Analyse Non Linéaire **6** (1991), 561–649.

[BCP]   Bartch, T.; Clapp, M.; and Puppe, D., *A mountain pass theorem for actions of compact Lie groups*, J. Reine Angew. Math. **419** (1991).

[Be]   Benci, B., *A new approach to the Morse-Conley theory and some applications*, Ann. Mat. Pura Appl. (IV) **158** (1991), 231–305.

[BFG]   Benci, V.; Fortunato, D.; and Giannoni, F., *On the existence of multiple geodesics in static space-time*, Analyse Non Linéaire **8** (1991), 79–102.

[BG]   Benci, V., and Giannoni, F., *On the existence of closed geodesics on noncompact Riemannian manifolds*, Duke Math. J. **68** (1992), 195–215.

[BN]   Brézis, H., and Nirenberg, L., H^1 *versus* C^1 *local minimizers*, preprint (1993).

[CY1]   Chang, A. S. Y., and Yang, P., *Prescribing Gaussian curvature on* S^2, Acta Math. **159** (1987), 214–259.

[CY2]   ——, *Conformal deformation of metric on* S^2, JDG **23** (1988), 259–296.

[Ch1]   Chang, K. C., *Variational methods for non-differentiable functionals and their applications to PDE*, J. Math. Anal. Appl. **80** (1981), 102–129.

[Ch2]   ——, *Solutions of asymptotically linear operator equations via Morse theory*, Comm. Pure Appl. Math. **34** (1981), 693–712.

[Ch3]   ——, *A variant mountain pass lemma*, Sci. Sinica Ser. A, **26** (1983), 1241–1255.

[Ch4]   ——, Infinite dimensional Morse theory and its applications, Univ. de Montreal **97**, 1985.

[Ch5]   ——, *Morse theory for harmonic maps, Variational methods*, (Ed., H. Berestycki), PNLDE **4**, Birkhäuser, Basel and Boston, (1990), 431–446.

[Ch6]   ———, *Critical groups, Morse theory and applications to semilinear elliptic BVP_s*, Chinese Math. into the 21st Century, (Ed., Wu Wen-tsun and Cheng Min-de), Peking Univ. Press. (1991), 41–65.

[Ch7]   ———, *On the homology method in the critical point theory*, Partial Diff. Eqs. and Related subjects, (Ed., M. Miranda), Pitman Res. Notes Math. **269** (1992), 59–77.

[Ch8]   ———, *Infinite dimensional Morse theory and multiple solution problems*, PNLDE **6**, Birkhäuser, Basel and Boston, 1993.

[Ch9]   ———, *H^1 versus C^1 isolated critical points*, Peking Univ., preprint (1993).

[CE]   Chang, K. C., and Eells, J., *Unstable minimal surface coboundaries*, Acta Math. Sinica **2**, (1986), 233–247.

[CJ]   Chang, K. C., and Jiang, M. Y., *The Lagrange intersection for $(\mathbb{C}P^n, \mathbb{R}P^n)$*, Manuscripta Math. **68** (1990), 89–100.

[CL1]   Chang K. C., and Liu, J. Q., *A strong resonance problem*, Chinese Ann. Math. **11** B.2 (1990), 191–210.

[CL2]   ———, *Morse theory under general boundary conditions*, J. Systems Sci Math. Sci. **4** (1991), 78–83.

[CL3]   ———, *On Nirenberg's problem*, Internat. J. Math. **4** (1993), 35–58.

[CL4]   ———, *A prescribing geodesic curvature problem*, ICTP preprint (1993).

[CLL]   Chang, K. C.; Liu, J. Q.; and Liu, M. J., *Nontrivial periodic solutions for strong resonance Hamiltonian Systems*, Peking Univ. preprint (1993).

[CD]   Chen, W. X., and Ding W. Y., *Scalar curvature on S^2*, TAMS, **303** (1987), 365–382.

[CP]   Clapp, M., and Puppe, D., *Critical point theory with symmetries*, J. Reine Angew. Math. **418** (1991), 1–29.

[CZ]   Conley, C. C., and Zehnder, E., *Morse type index theory for flows and periodic solutions for Hamiltonian equations*, Comm. Pure Appl. Math. **37** (1984), 207–253.

[CDM]   Corvellec, J. N.; De Giovanni, M.; and Marzocchi, M., *Deformation properties for continuous functionals and critical point theory*, Univ. di Pisa, preprint (1992).

[Da]   Dancer, E. N., *Degenerate critical points, homotopy indices and Morse inequalities*, J. Reine Angew. Math. **350** (1984), 1–22.

[Fe]   Fe, G. H., Ph.D. thesis, Nanjing Univ. (1994).

[FH]   Fadell, E., and Hsseini, S., *Infinite cup length in free loop spaces with an application to a problem of the N-body type*, Analyse Non Linéaire **9** (1992), 305–320.

[Gh]   Ghoussonb, N., *Duality and perturbation methods in critical point theory*, Cambridge Tracts in Math. **107**, Cambridge Univ. Press, Cambridge and New York, 1993.

[GMT]   De Giorgi, E.; Marino, A.; and Tosques, M., *Problemi di evoluzione in spazi metrici e curve di massima pendenza*, Atti Accad. Naz. Lincei Rend. Cl. Sci. Fis. Mat. Natur. (8) **68** (1980), 180–187.

[Hi]   Hingston, N., *Equivariant Morse theory and closed geodesics*, JDG **19** (1984), 85–116.

[JS]   Jost, J., and Struwe, M., *Morse-Conley theory for minimal surfaces of varying topological type*, Invent. Math. **102**, (1990), 465–499.

[Li]   Liu, J. Q., *A Morse index of a saddle point*.

[LiL]   Liu, J. Q., and Li, P. L., *Nirenberg's problem on the 2-dimensional hemisphere*, Internat. J. of Math. (1994),

[LL]   Li, S. J., and Liu, J. Q., *Morse theory and asymptotically linear Hamiltonian systems*, JDE, **78** (1989), 53–73.

[Ly]   Li, Y., *Prescribing scalar curvature on S^n and related problems*, Part 1 and Part 2, preprint (1994).

[Lo]   Long, Y. M., *Maslov index, degenerate critical points and asymptotically linear Hamiltonian systems*, Science in China, Ser. A **33** (1990), 1409–1419.

[LZ]   Long, Y. M., and Zehnder, E., *Morse theory for forced oscillations of asymptotically linear Hamiltonian systems*, Stochastic Process, Physics, and Geometry, World Sci. Press, (1990), 528–563.

[Ma]   Majer, P., *Variational methods on manifolds with boundary*, SISSA preprint (1991).

[MT]   Majer, P., and Terracini, S., *Periodic solutions to some problems of n-body type*, Arch. Rational Mech. Anal. **124** (1993), 381–404.

[Mo]   Morse, M., *The calculus of variations in the large*, Amer. Math. Soc. Colloq. Publ. no. **18**, Providence, RI, 1934.

[Pa]   Parker, T. H., *A Morse theory for equivariant Yang-Mills*, Duke Math. J. **66** (1992), 337–356.

[Pr]   Problems of Mathematics, Princeton University Bicentennial Conferences, Princeton, NJ 1947.

[Ri]   Riahi, H., *Periodic orbits of N-body type problems*, Ph.D. thesis, Rutgers Univ. (1993).

[Ro]   Rothe, E., *Critical point theory in Hilbert space under general boundary conditions*, J. Math. Anal. Appl. **2** (1965), 357–409.

[SZ]   Salamon, D., and Zehnder, E., *Morse theory for periodic solutions of Hamiltonian systems and the Morse index*, Comm. Pure Appl. Math. **45** (1992), 1303–1360.

[St]   Struwe, M., *Plateau's problem and the calculus of variations*, Princeton, Tokyo, 1989.

[Sz1]   Szulkin, A., *Minimax principles for lower semi continuous functions and applications to nonlinear BVP_s*, Analyse Non Linéaire **3** (1986), 77–109.

[Sz2]   ——, *Cohomology and Morse theory for strong indefinite functionals*, Math. Z. **209** (1992), 375–418.

[Wa1]   Wang, Z. Q., *Equivariant Morse theory for isolated critical orbits and its applications to nonlinear problems*, Lecture Notes in Math. **1306**, Springer-Verlag, Berlin and New York (1988), 202–221.

[Wa2]   ——, *On a superlinear elliptic equation*, Analyse Non Linéaire **8** (1991), 43–58.

[Zh]   Zhang, D., Ph.D. thesis, Stanford Univ. (1990).

Analyse microlocale et mécanique des fluides en dimension deux

Jean-Yves Chemin

Laboratoire d'Analyse Numérique, Université Paris 6
BP 187, 2 place Jussieu, F-75230 Paris Cedex 05, France

Introduction

Dans ce texte, nous avons cherché à donner un aperçu sur quelques résultats récents démontrés sur l'équation d'Euler incompressible en dimension deux. Nous avons voulu mettre l'accent sur l'importance que peuvent prendre l'aspect géométrique et les idées venues de l'analyse microlocale dans la description précise de la régularité des solutions du système d'Euler.

1. Présentation du système d'Euler

Nous allons commencer par rappeler quelques propriétés de base sur le système d'Euler relatif à un fluide parfait incompressible.

$$(E) \begin{cases} \partial_t v + \operatorname{div} v \otimes v &= -\nabla p \\ \operatorname{div} v &= 0 \\ v_{|t=0} &= v_0, \end{cases}$$

Le tourbillon (ou vorticity en anglais), est, par définition, le rotationnel du champ de vecteurs v. C'est la quantité clef pour comprendre le système d'Euler.

CONVENTION. *Lorsque la dimension est 2, on identifie les matrices antisymétriques avec les réels. Ainsi, on note $\omega(v) = \partial_1 v^2 - \partial_2 v^1$ le tourbillon de v. En dimension supérieure, on note $\Omega(v) = (\Omega_j^i(v))_{1 \le i,j \le d}$ avec $\Omega_j^i(v) = \partial_j v^i - \partial_i v^j$. On omet de noter explicitement (v) en l'absence de toute ambiguïté.*

L'importance du tourbillon dans l'étude des solutions du système d'Euler incompressible (E) vient du fait qu'il détermine le champ des vitesses. C'est la loi de Biot et Savart qui affirme que

$$v^i(x) = c_d \sum_k \int_{\mathbf{R}^d} \frac{x^k - y^k}{|x - y|^d} \Omega_k^i(y) dy. \tag{1}$$

La pression est une inconnue du système. En supposant le champ des vitesses suffisamment régulier, il vient, en dérivant l'équation sur le champ de vecteurs v,

$$-\Delta p = \sum_{j,k=1}^{d} \partial_j \partial_k (v^j v^k). \tag{2}$$

Proceedings of the International Congress
of Mathematicians, Zürich, Switzerland 1994

Lorsque r est un nombre réel, nous désignerons par C^r l'espace de Hölder d'indice r. On a alors le théorème d'existence locale en temps suivant.

THÉORÈME 1.1 *Soient r et a deux réels strictement supérieurs à 1 et v_0 un champ de vecteurs de divergence nulle appartenant à l'espace C^r. Supposons que ∇v_0 appartient à L^a. Il existe alors un unique temps $T^\star$ maximal et une unique solution (v,p) de (E) tels que $v \in L^\infty_{loc}([0,T^\star[;C^r)$ et $(\nabla v, \nabla p)$ appartienne à l'espace $L^\infty_{loc}([0,T^\star[;L^a)$. De plus, on a*

$$T^\star < +\infty \Rightarrow \int_0^{T^\star} \|\Omega(t)\|_{L^\infty}\,dt = +\infty.$$

A propos de ce théorème, il faut citer les travaux de L. Lichtenstein(voir [14]). La nécessité d'étudier l'évolution du tourbillon apparaît clairement. En différentiant le système (E), il vient

$$\partial_t \Omega + v \cdot \nabla \Omega + \Omega \cdot \nabla v = 0 \quad \text{avec} \quad (\Omega \cdot \nabla v)^i_j = \sum_{k=1}^d \Omega^k_j \partial_k v^i - \Omega^k_i \partial_k v^j. \quad (3)$$

Lorsque la dimension d vaut 2, il résulte de (3) que

$$\partial_t \omega + v \cdot \nabla \omega = 0 \quad \text{et donc} \quad \|\omega\|_{L^a} = \|\omega_0\|_{L^a}. \quad (4)$$

On déduit alors du théorème 1.1 le corollaire suivant.

COROLLAIRE 1.1 *On suppose que $d = 2$. Soient r et a deux réels strictement supérieurs à 1 et v_0 un champ de vecteurs de divergence nulle de classe C^r. Supposons que ∇v_0 appartient à L^a. Il existe alors une unique solution (v,p) de (E) telle que $v \in L^\infty_{loc}(\mathbf{R}^+;C^r)$ et $(\nabla v, \nabla p) \in L^\infty_{loc}(\mathbf{R}^+;L^a)$.*

2. Le théorème de Yudovich

Le bon cadre lorsque les solutions sont peu régulières est celui des perturbations d'énergie cinétique finie de solutions stationnaires particulières.

DÉFINITION 2.1 *On appelle champ de vecteurs stationnaire et on note σ, tout champ de vecteurs de la forme*

$$\sigma = \left(-\frac{x^2}{r^2} \int_0^r \rho g(\rho)d\rho, \frac{x^1}{r^2} \int_0^r \rho g(\rho)d\rho\right) \quad \text{où} \quad g \in C^\infty_0(\mathbf{R} \setminus \{0\}).$$

On vérifie facilement que le champ de vecteurs σ est une solution stationnaire de l'équation d'Euler. Nous pouvons maintenant présenter la définition suivante.

DÉFINITION 2.2 *Soit m un réel, on désigne par E_m l'ensemble des champs de vecteurs de divergence nulle v du plan tels qu'il existe un champ de vecteurs stationnaire σ vérifiant*

$$\int_{\mathbf{R}^2} \omega(\sigma) = m \quad et \quad v - \sigma \in L^2.$$

On démontre facilement que, si $(1+|x|)d\omega(x)$ est une mesure bornée et si ω appartient à l'espace de Sobolev $H^{-1}(\mathbf{R}^2)$, alors le champ de vecteurs associé appartient à $E_{\int d\omega}$. Énonçons le théorème de Yudovich, démontré en 1964 dans [20].

THÉORÈME 2.1 *Soient m un réel et v_0 un élément de E_m. Supposons en outre que ω_0 appartienne à $L^\infty \cap L^a$ avec $1 < a < +\infty$. Il existe alors une unique solution (v, p) de (E) appartenant à l'espace $C(\mathbf{R}; E_m) \times L^\infty_{loc}(\mathbf{R}; L^2)$ et telle que le tourbillon ω du champ de vecteurs v appartienne à $L^\infty(\mathbf{R}^3) \cap L^\infty(\mathbf{R}; L^a(\mathbf{R}^2))$.*

De plus, ce champ de vecteurs v possède un flot. Plus précisément, il existe une unique application ψ continue de $\mathbf{R} \times \mathbf{R}^2$ dans $\mathbf{R}^2$ telle que

$$\psi(t, x) = x + \int_0^t v(s, \psi(s, x))ds.$$

En outre, il existe une constante C telle que $\psi(t) - \mathrm{Id} \in C^{\exp(-Ct\|\omega_0\|_{L^\infty \cap L^a})}$.

L'existence et l'unicité des solutions vont résulter du lemme ci-après.

LEMME 2.1 *Soient (v_1, p_1) et (v_2, p_2) deux solutions du système d'Euler incompressible (E). Il existe une constante C ne dépendant que de $\|v_i(0) - \sigma\|_{L^2}$ et de $\|\omega_i(0)\|_{L^\infty \cap L^a}$ telle que*

$$\|v_1(0) - v_2(0)\|^2_{L^2} \leq e^{-a(\exp(Ct)-1)}$$
$$\Rightarrow \|v_1(t) - v_2(t)\|^2_{L^2} \leq \|v_1(0) - v_2(0)\|^{2\exp(-Ct)}_{L^2} e^{a(1-\exp(-Ct))}.$$

Il s'agit d'estimer la fonction $I(t) \overset{\text{déf}}{=} \|(v_1 - v_2)(t)\|^2_{L^2}$. Par intégration par parties, on obtient que $I'(t) \leq 2\|v_1(t) - v_2(t)\|^{\frac{2}{b}}_{L^\infty} I(t)^{1-\frac{1}{b}} \|\nabla v_2(t)\|_{L^b}$. D'après la conservation de la norme L^∞ du tourbillon et de la norme L^1, le champ de vecteurs v est borné sur $\mathbf{R}^3$. Il en résulte que

$$I'(t) \leq 2MI(t)^{1-\frac{1}{b}}\|\nabla v_2(t)\|_{L^b}. \tag{5}$$

Il est bien connu que les multiplicateurs de Fourier envoient L^a dans lui-même. Nous utiliserons le résultat précis suivant, démontré par exemple dans [18].

$$\|\nabla v\|_{L^b} \leq C\frac{b^2}{b-1}\|\Omega(v)\|_{L^b}. \tag{6}$$

Il en résulte que $\|\nabla v_2(t)\|_{L^b} \leq Ca\|\omega_2(0)\|_{L^\infty \cap L^a}$. Posons $J_\eta(t) = \eta + I(t)$. Toutes les inégalités écrites dans la suite ne seront valables que sous l'hypothèse que $\eta + I(t) \leq 1$. Il résulte de l'inégalité (5) que $J'_\eta(t) \leq CJ_\eta(t)^{1-\frac{1}{b}}$. En prenant $b = a - \log J_{\epsilon,\eta}(t)$, on obtient $J'_\eta(t) \leq e\alpha(t)(a - \log J_\eta(t))J_\eta(t)$. Après intégration, il vient, en passant par deux fois à l'exponentielle, $J_\eta(t) \leq e^{a(1-e^{-Ct})}J_\eta(0)^{\exp -Ct}$. En faisant tendre η vers 0 et en passant à la limite, on conclut la démonstration du lemme.

La preuve du résultat sur le flot ψ se fait par des calculs très analogues à ceux qui nous ont servi à démontrer ce lemme.

Soit ω_0 la fonction sur le plan $\mathbf{R}^2$ nulle en dehors de $[-1,1] \times [-1,1]$, impaire en les deux variables x_1 et x_2 et valant 2π sur $[0,1] \times [0,1]$. On considère le champ de vecteurs v_0 dont ω_0 est le tourbillon. Dans [3], H. Bahouri et l'auteur démontrent que le flot de la solution de Yudovich associée n'appartient pas, à l'instant t, à $C^{\exp -t}$; le théorème de Yudovich est donc optimal.

3. Structures géométriques stables

Le but de ce paragraphe est d'exposer des résultats de persistance de structures géométriques dans les fluides parfaits incompressibles. La motivation initiale de ces questions est le problème des poches de tourbillon. Supposons que le tourbillon soit, à l'instant initial, la fonction caractéristique d'un ouvert borné D_0 dont le bord est de classe de Hölder $C^{1+\epsilon}$. D'après le théorème 2.1 et la relation (4), il existe un unique champ de vecteurs solution des équations d'Euler sur $\mathbf{R} \times \mathbf{R}^2$, dont le tourbillon est, à l'instant t, la fonction caractéristique d'un ouvert borné D_t dont la topologie reste inchangée. Deux questions très naturelles se posent alors : le bord de l'ouvert reste-t-il régulier à temps petit? si oui, que se passe-t-il pour les temps grands?

Considèrons γ_0 un plongement du cercle $\mathbf{S}^1$ de classe $C^{1+\epsilon}$ dont l'image est le bord de l'ouvert D_0. Le champ de vecteurs solution est alors complètement déterminé par le bord de l'ouvert. Cherchons alors si la fonction γ définie par $\partial_t \gamma(t,s) = v(t, \gamma(t,s))$ est le paramétrage d'une courbe lisse. D'après la loi de Biot-Savart, la Formule de Green entraîne que

$$\partial_t \gamma(t,s) = \frac{1}{2\pi} \int_0^{2\pi} \log |\gamma(t,s) - \gamma(t,\sigma)| \, \partial_\sigma \gamma(t,\sigma) d\sigma. \qquad (7)$$

On a le théorème suivant.

THÉORÈME 3.1 *Soient ϵ appartenant à l'intervalle $]0,1[$ et γ_0 une fonction de l'espace $C^{1+\epsilon}(\mathbf{S}^1; \mathbf{R}^2)$ injective et dont la différentielle ne s'annule pas. Il existe alors une unique solution $\gamma(t,s)$ de l'équation (7) appartenant à l'espace $L_{loc}^\infty(\mathbf{R};$ $C^{1+\epsilon}(\mathbf{S}^1; \mathbf{R}^2))$ et qui est, pour tout temps, un plongement du cercle.*

Nous allons démontrer un théorème général de persistance des structures géométriques non singulières pour le système d'Euler incompressible qui contiendra bien sûr le théorème ci-dessus. Le concept important sera celui de régularité tangentielle par rapport à une famille substantielle X de champs de vecteurs de classe C^ϵ. Ce concept, introduit par J.-M. Bony dans l'étude de la propagation des singularités microlocales des équations semi-linéaires (voir [5]), fût ensuite introduit dans le cadre des équations aux dérivées partielles quasi-linéaires par S. Alinhac (voir [1]) et par l'auteur (voir [7]).

Il nous faut trouver une condition suffisante pour que, si une fonction u est bornée, les fonctions $\partial_i \partial_j \Delta^{-1} u$ le soient aussi. Ceci nous amène à introduire les définitions suivantes. Dans toute la suite, on désignera par ϵ un réel de $]0,1[$.

DÉFINITION 3.1 *Soit $X = (X_\lambda)_{\lambda \in \Lambda}$ une famille de champs de vecteurs de classe C^ϵ ainsi que leur divergence. Une telle famille est dite substantielle si et seulement si l'on a $I(X) = \inf_{x \notin \Sigma} \sup_{\lambda \in \Lambda} |X_\lambda(x)| > 0$.*

Définissons maintenant la notion de régularité tangentielle par rapport à une famille substantielle de champs de vecteurs.

$$N_\epsilon(X) = \frac{1}{\epsilon} \sup_{\lambda \in \Lambda} \frac{\|X_\lambda\|_\epsilon + \|\operatorname{div} X_\lambda\|_\epsilon}{I(X)},$$

$$\|u\|_{\epsilon,X} = N_\epsilon(X)\|u\|_{L^\infty} + \sup_{\lambda \in \Lambda} \frac{\|X_\lambda(x,D)u\|_{\epsilon-1}}{I(X)}.$$

DÉFINITION 3.2 *Soit X une famille substantielle de champs de vecteurs de classe C^ϵ ainsi que leur divergence (on suppose $\Sigma = \emptyset$). On désigne par $C^\epsilon(\Sigma, X)$ l'ensemble des distributions u appartenant à L^∞ telles que, pour tout λ, on ait*

$$X_\lambda(x,D)u \stackrel{\text{déf}}{=} \operatorname{div}(uX_\lambda) - u\operatorname{div} X_\lambda \in C^{\epsilon-1}.$$

Introduit tout d'abord dans [8] pour étudier le système d'Euler, ce type de régularité a permis de démontrer dans [9] et [10] le théorème de persistance ci-dessous.

THÉORÈME 3.2 *Soient ϵ un réel de l'intervalle $]0,1[$, a un réel supérieur à 1 et $X_0 = (X_{0,\lambda})_{\lambda \in \Lambda}$ une famille substantielle de classe C^ϵ sur le plan. On considère un champ de vecteurs v_0 sur $\mathbf{R}^2$ appartenant à $C_\star^1$ et dont le gradient est dans L^a. Si ω_0 appartient à $C^\epsilon(X_0)$, alors, il existe une unique solution v de (E) telle que $v \in L_{loc}^\infty(\mathbf{R}; Lip)$ et $\nabla v \in L^a$. De plus, si ψ désigne le flot de v, alors, pour tout λ, $X_{0,\lambda}(x,D)\psi \in L_{loc}^\infty(\mathbf{R}; C^\epsilon)$. Enfin, si $X_{t,\lambda} = \psi(t)^\star X_{0,\lambda}$, alors, la famille $X_t = (X_{t,\lambda})_{\lambda \in \Lambda}$ est substantielle et l'on a $N_\epsilon(X_t) \in L_{loc}^\infty(\mathbf{R})$ et $\|\omega(t)\|_{\epsilon,X_t} \in L_{loc}^\infty(\mathbf{R})$.*

Postérieurement, A. Bertozzi et P. Constantin ont redémontré dans [4] le cas particulier du théorème 3.1. De plus, P. Serfati a donné dans [17] une nouvelle démonstration du théorème 3.2. Une version en dimension trois et à temps petit de ce théorème a été démontré par P. Gamblin et X. Saint-Raymond dans [13]. Enfin, P. Serfati démontre dans [16] que l'équation (7) est, pour les petites perturbations du cercle, une équation différentielle ordinaire.

Vérifier que ce théorème entraîne bien le théorème 3.1 est un exercice que nous laissons au lecteur. Régularisons la donnée initiale. Le corollaire 1.1 d'existence globale de solutions régulières affirme l'existence d'une solution globale v_n du système (E). Le point important de la présente preuve consiste à démontrer une estimation a priori sur la norme Lipschitz d'une solution régulière du système (E), puis de passer à la limite. Pour cela, on utilise l'estimation stationnaire suivante.

THÉORÈME 3.3 *Il existe une constante C telle que, pour tout ϵ de l'intervalle $]0,1[$, et pour tout a supérieur ou égal à 1, on ait la propriété suivante :*

Soient X une quelconque famille substantielle de champs de vecteurs de classe C^ϵ et ω une fonction appartenant à $C^\epsilon(X) \cap L^a \cap L^\infty$. Si v est un champ de vecteurs de gradient $L^{a+\epsilon}$ et de tourbillon ω, alors le gradient de v est borné et l'on a

$$\|\nabla v\|_{L^\infty} \leq Ca\|\omega\|_{L^a} + \frac{C}{\epsilon}\|\omega\|_{L^\infty} \log\left(e + \frac{\|\omega\|_{\epsilon,X}}{\|\omega\|_{L^\infty}}\right).$$

Il est un cas particulier où la démonstration de ce théorème est particulièrement simple. Supposons que la famille substantielle X soit réduite au seul champ de vecteurs ∂_1 et accessoirement, que le support de la transformée de Fourier de ω ne rencontre pas l'origine. Il est évident que $\|X\|_\epsilon = I(X) = 1$. Classiquement, on a

$$\|f\|_{L^\infty} \leq \frac{C}{\epsilon}\|f\|_0 \log\left(e + \frac{\|f\|_\epsilon}{\|f\|_0}\right). \tag{8}$$

Ainsi, pour j valant 1 ou 2, il vient

$$\|\partial_1\partial_j\Delta^{-1}\omega\|_{L^\infty} \leq \frac{C}{\epsilon}\|\partial_1\partial_j\Delta^{-1}\omega\|_0 \log\left(e + \frac{\|\partial_1\partial_j\Delta^{-1}\omega\|_\epsilon}{\|\partial_1\partial_j\Delta^{-1}\omega\|_0}\right).$$

Comme les multiplicateurs de Fourier opèrent dans les espaces de Hölder et comme $\partial_2^2 = \Delta - \partial_1^2$, on a

$$\|\partial_i\partial_j\Delta^{-1}\omega\|_{L^\infty} \leq \|\omega\|_{L^\infty} + \frac{C}{\epsilon}\|\omega\|_0 \log\left(e + \frac{\|\partial_1\omega\|_{\epsilon-1}}{\|\omega\|_0}\right).$$

D'où le théorème dans ce cas très particulier. Pour le cas général, qui utilise le calcul paradifférentiel de Bony (voir [6]), nous renvoyons le lecteur à [9] et [10].

Il s'agit de contrôler la norme Lipschitz du champ de vecteurs solution. Pour cela, nous allons démontrer que, pour tout temps t, on a

$$\|\nabla v(t)\|_{L^\infty} \leq \widetilde{N}(X_0,\epsilon,\omega_0) \exp\frac{Ct\|\omega_0\|_{L^\infty}}{\epsilon^2} \quad \text{avec}$$

$$\widetilde{N}(X_0,\epsilon,\omega_0) = Ca\|\omega_0\|_{L^a} + \frac{C}{\epsilon}\|\omega_0\|_{L^\infty} \log\frac{\|\omega_0\|_{\epsilon,X_0}}{\|\omega_0\|_{L^\infty}}. \tag{9}$$

On transporte les données géométriques, c'est-à-dire la famille substantielle par le flot du champ de vecteurs v et l'on applique l'inégalité du théorème 3.3 à chaque instant. Définissons donc la famille $X_t = (X_{t,\lambda})_\lambda \in \Lambda$ par

$$X_{t,\lambda}(x) = \psi_\star(t)X_{0,\lambda}(x) = \left(X_{0,\lambda}(x,D)\psi(t)\right)(\psi^{-1}(t,x)). \tag{10}$$

Le point décisif de la démonstration est la majoration de $\|\omega(t)\|_{\epsilon,X_t}$. En utilisant le calcul paradifférentiel et les relations qui suivent,

$$\begin{cases} \partial_t X_{0,\lambda}(x,D)\psi(t,x) &= \nabla v(t,\psi(t,x))X_{0,\lambda}(x,D)\psi(t,x) \\ X_{0,\lambda}(x,D)\psi(0,x) &= X_{0,\lambda}(x), \end{cases}$$

$$\partial_t X_{t,\lambda} + v \cdot \nabla X_{t,\lambda} = X_{t,\lambda}(x,D)v,$$

$$\partial_t \operatorname{div} X_{t,\lambda} + v \cdot \nabla \operatorname{div} X_{t,\lambda} = X_{t,\lambda}(x,D)\operatorname{div} v,$$

on démontre que

$$\frac{\|\omega(t)\|_{\epsilon,X_t}}{\|\omega(t)\|_{L^\infty}} \leq C \frac{\|\omega_0\|_{\epsilon,X_0}}{\|\omega_0\|_{L^\infty}} \exp\left(\frac{C}{\epsilon}\int_0^t \|\nabla v(\tau)\|_{L^\infty} d\tau\right).$$

Le théorème 3.3 permet d'en déduire que

$$\|\nabla v(t)\|_{L^\infty} \leq Ca\|\omega_0\|_{L^a} + \frac{C}{\epsilon}\|\omega_0\|_{L^\infty} \log\frac{\|\omega_0\|_{\epsilon,X_0}}{\|\omega_0\|_{L^\infty}} + \frac{C}{\epsilon^2}\|\omega_0\|_{L^\infty}\int_0^t \|\nabla v(\tau)\|_{L^\infty} d\tau.$$

Le lemme de Gronwall assure alors l'estimation (9).

4. Le problème des nappes de tourbillon

Il s'agit de trouver une solution à l'équation d'Euler lorsque le tourbillon est la mesure de longueur d'une courbe de classe C^1. Il est très facile de vérifier que ces données intiales rentrent dans le cadre du théorème ci-dessous.

THÉORÈME 4.1 *Soient m un réel et v_0 un élément de E_m. Supposons que ω_0, son tourbillon à l'instant initial, soit une mesure bornée de partie singulière positive. Il existe alors un couple (v,p) solution du système d'Euler (E) tel que $(v,p) \in L_{loc}^\infty(\mathbf{R}; E_m) \times L_{loc}^\infty(\mathbf{R}; \mathcal{F}^{-1}(L^2 + L^\infty))$.*

De plus, à chaque instant t, le tourbillon ω_t est une mesure bornée de partie singulière positive et de masse totale inférieure ou égale à celle de ω_0.

Ce théorème a été démontré par J.-M. Delort en 1990 dans [11]. Signalons que, dans [15], A. Majda redémontre ce théorème dans le cas où ω_0 est positif. Le premier ingrédient de la démonstration est le théorème suivant, qui se démontre en étudiant attentivement la pression.

THÉORÈME 4.2 *Soient m un réel et v un champ de vecteurs de divergence nulle appartenant à l'espace $L_{loc}^\infty(\mathbf{R}; E_m)$. Les deux conditions suivantes sont équivalentes.*
(i) Il existe $p \in L_{loc}^\infty(\mathbf{R}; \mathcal{S}'(\mathbf{R}^2))$ telle que $\mathcal{F}_x p \in L_{loc}^\infty(\mathbf{R}; L^2 + L^\infty)$ et telle que (v,p) soit solution du système (E).
(ii) Le champ de vecteurs satisfait

$$\partial_t v + A(D)\begin{pmatrix}(v^1)^2 - (v^2)^2 \\ v^1 v^2\end{pmatrix} = 0 \tag{11}$$

où $A(D)$ est défini par $A(\xi) = i\begin{pmatrix}\xi_1\xi_2^2|\xi|^{-2} & \xi_2(\xi_2^2 - \xi_1^2)|\xi|^{-2} \\ -\xi_1^2\xi_2|\xi|^{-2} & \xi_1(\xi_1^2 - \xi_2^2)|\xi|^{-2}\end{pmatrix}.$

Cette formulation faible, spécifique à la dimension deux, a été clairement dégagée par J.-M. Delort dans [11] et est implicitement contenue dans [12] et dans [2].

Après régularisation de la donnée initiale par convolution avec une approximation de l'identité positive, le théorème 1.1 assure l'existence de solutions globales associées aux données régularisées. Notons $(v_n)_{n\in\mathbf{N}}$ la suite des solutions associées. D'après l'équation (11), il suffit de passer à la limite sur les termes $v^1 v^2$ et $(v^1)^2 - (v^2)^2$. Remarquons qu'un changement d'axe permet de se ramener à un seul cas, par exemple celui du terme $v^1 v^2$. Le tourbillon ω_n appartient à $L^\infty \cap L^1$.

D'après la loi de Biot-Savart, on a, pour toute fonction g indéfiniment différentiable à support compact sur $\mathbf{R} \times \mathbf{R}^2$,

$$
\begin{aligned}
\Delta(g, v_n) &= \int_{\mathbf{R}^5} G(t, x, y) d\mu_n(t, x, y) \quad \text{avec} \\
G(t, x, y) &= -\frac{1}{4\pi^2} \int \frac{z_2 - y_2}{|z - y|^2} \frac{z_1 - x_1}{|z - x|^2} g(t, z) dz \quad \text{et} \\
d\mu_n(t, x, y) &= \omega_n(t, x) \omega_n(t, y) dt dx dy.
\end{aligned}
\tag{12}
$$

D'après la conservation du tourbillon (4), la suite $(\mu_n)_{n \in \mathbf{N}}$ est une suite bornée de mesures bornées. On peut supposer que la suite $(\mu_n)_{n \in \mathbf{N}}$ converge faiblement vers une mesure bornée μ. (Nous omettons systématiquement de noter les extractions.)

Il est très facile de démontrer que la fonction G est continue en dehors de la diagonale et nulle à l'infini. La forme particulière de la non-linéarité entraîne que la fonction G est bornée. Nous avons besoin des deux lemmes d'intégration suivants:

LEMME 4.1 *Soit X un espace métrique localement compact σ-compact. On considère une suite bornée de mesures bornées $(\mu_n)_{n \in \mathbf{N}}$ convergeant faiblement vers μ. Si la suite $(|\mu_n|)_{n \in \mathbf{N}}$ converge faiblement vers ν, alors, pour toute fonction borélienne bornée, nulle à l'infini et continue en dehors d'un fermé N, ν-négligeable, on a*

$$
\lim_{n \to \infty} \int f d\mu_n = \int f d\mu.
$$

LEMME 4.2 *Soit μ une mesure bornée sur $\mathbf{R}^2$ appartenant à $H^{-1}(\mathbf{R}^2)$. La mesure μ est alors diffuse.*

D'après (4), on a $\omega_n(t, x) = f_n^+(t, x) - f_n^-(t, x) + \omega_n^s(t, x)$
$$
\begin{aligned}
\text{avec} \quad & f_n^{\pm}(t, x) = (\rho_n \star f_0^{\pm})(\psi_n^{-1}(t, x)) \\
\text{et} \quad & \omega_n^s(t, x) = (\rho_n \star \omega_0^s)(\psi_n^{-1}(t, x)).
\end{aligned}
\tag{13}
$$

On peut supposer que les suite $(\omega_n^s)_{n \in \mathbf{N}}$ et $(f_n^{\pm})_{n \in \mathbf{N}}$ convergent faiblement vers des mesure notées ω^s et $\mu^{\pm}$. Les familles $(f_n^{\pm})_{n \in \mathbf{N}}$ sont uniformément intégrables, donc faiblement compactes dans L^1. En utilisant la relation (12), il vient

$$
< v_n^1 v_n^2, g > = \int_{\mathbf{R}^5} G(t, x, y) d\mu_n(t, x, y).
$$

Le lemme 4.2 assure que les hypothèses du lemme 4.1 sont satisfaites, d'où le théorème 4.1.

References

[1] S. Alinhac, *Interaction d'ondes simples pour des équations complètement non linéaires*, Annales Scientifiques de l'Ecole Normale Supérieure, **21**, 1988, pages 91–133.

[2] S. Alinhac, *Un problème de concentration évanescente pour les flots non stationnaires et incompressibles en dimension deux*, Communication in Mathematical Physics, **127**, 1990, pages 585–596.

[3] H. Bahouri et J.-Y. Chemin, *Equations de transport relatives à des champs de vecteurs non-lipschitziens et mécanique des fluides*, Prépublication de l'Ecole Polytechnique n⁰1059 1993, à paraître dans Archiv for Rationnal Mechanics and Analysis.

[4] A. Bertozzi et P. Constantin, *Global regularity for vortex patches*, Communication in Mathematical Physics, **152**(1), 1993, pages 19–28.

[5] J.-M. Bony, *Propagation des singularités pour des équations aux dérivées partielles non linéaires*, Séminaire EDP de l'Ecole Polytechnique, 1979–1980.

[6] J.-M. Bony, *Calcul symbolique et propagation des singularités pour les équations aux dérivées partielles non linéaires*, Annales de l'Ecole Normale Supérieure, **14**, 1981, pages 209–246.

[7] J.-Y. Chemin, *Calcul paradifférentiel précisé et application à des équations aux dérivées partielles non linéaires*, Duke Mathematical Journal, **56**(1), 1988, pages 431–469.

[8] J.-Y. Chemin, *Sur le mouvement des particules d'un fluide parfait, incompressible, bidimensionnel*, Inventiones Mathematicae, **103**, 1991, pages 599–629.

[9] J.-Y. Chemin, *Persistance des structures géométriques liées aux poches de tourbillon*, Séminaire Equations aux Dérivées Partielles de l'Ecole Polytechnique, 1990–1991.

[10] J.-Y. Chemin, *Persistance de structures géométriques dans les fluides incompressibles bidimensionnels*, Annales de l'Ecole Normale Supérieure, **26**(4), 1993, pages 1–26.

[11] J.-M. Delort, *Existence de nappes de tourbillon en dimension deux*, Journal of the American Mathematical Society, **4**(3), 1991, pages 553–586.

[12] R. Di Perna et A. Majda, *Concentrations in regularizations for 2-D incompressible flows*, Communication in Pure and Applied Mathematics, **40**, 1987, pages 301–345.

[13] P. Gamblin et X. Saint-Raymond, *On three-dimensional vortex patches*, Prépublication de l'Université d'Orsay, 1993.

[14] L. Lichtenstein, *Über einige Existenzprobleme der Hydrodynamik homogener unzusammendrückbarer, reibungsloser Flüßigkeiten und die Helmholtzschen Wirbelsätze*, Mathematische Zeitschrift, **23**, 1925, pages 89–154 ; **26**, 1927, pages 196–323 ; **28**, 1928, pages 387–415 et **32**, 1930, pages 608–725.

[15] A. Majda, *Remarks on weak solutions for vortex sheets with a distinguished sign*, Indiana University Mathematics Journal, **42**(3), 1993, pages 921–939.

[16] P. Serfati, *Etude mathématique de flammes infiniment minces en combustion. Résultats de structure et de régularité pour l'équation d'Euler incompressible*, Thèse de l'Université de Paris VI, 1992.

[17] P. Serfati, *Une preuve directe d'existence globale des vortex patches 2D*, Notes aux Comptes-Rendus de l'Académine des Sciences de Paris, **318**, Série I, 1993, pages 515–518.

[18] A. Torchinsky, Real variable methods in harmonic analysis, Pure and Applied Mathematics, **123**, Academic Press, New York.

[19] W. Wolibner, *Un théorème d'existence du mouvement plan d'un fluide parfait, homogène, incompressible, pendant un temps infiniment long*, Mathematische Zeitschrift, **37**, 1933, pages 698–726.

[20] V. Yudovich, *Non stationnary flows of an ideal incompressible fluid*, Zh. Vych. Math. **3**, 1963, pages 1032–1066.

Some Mathematical Problems of Fluid Mechanics

PETER CONSTANTIN

Department of Mathematics
University of Chicago
Chicago, IL 60637, USA

ABSTRACT. In this paper I will describe results concerning incompressible fluids at high Reynolds numbers — everyday fluid turbulence. I will address two broad subjects: integrals of quantities associated with the flow, and formation of large gradients.

1. Bulk Dissipation

Torque in Taylor-Couette flow, drag in flow past an obstacle, and heat transfer in Rayleigh-Bénard convection are examples of bulk dissipation quantities. They are measured in physical experiments [2], [18] and represent perhaps the simplest, most important and most reliable ways to measure turbulent flows in physical systems driven at the boundary. These quantities depend on the key parameters (Reynolds number, Rayleigh number) in a reproducible manner: they obey empirical laws. I will describe a method [12], [6], [4] to estimate rigorously such quantities directly from the equations of motion.

I will use a simplified Taylor-Couette setting to illustrate this method. Let us consider the incompressible Navier-Stokes equations

$$\partial_t u + u \cdot \nabla u + \nabla p = \Delta u \tag{1}$$

$$\nabla \cdot u = 0$$

in a 2-D strip, with boundary conditions

$$u(x + \ell, y, t) = u(x, y, t)$$

$$u(x, 0, t) = 0; \qquad u(x, 1, t) = \mathbf{Re}\, \hat{x}$$

The Reynolds number $\mathbf{Re}$ is a large positive number, $\hat{x}$ is the direction of the x axis, and $\ell > 0$ represents the aspect ratio. Consider

$$< \|u\|^2 >= \lim_{T \to \infty} \sup \frac{1}{T} \int_0^T \|u(t)\|^2 \, dt,$$

where $\|u\|^2 = D(u)$ is the bulk dissipation:

$$\|u\|^2 = \int_0^\ell \int_0^1 |\nabla u|^2 \, dx \, dy.$$

Proceedings of the International Congress
of Mathematicians, Zürich, Switzerland 1994
© Birkhäuser Verlag, Basel, Switzerland 1995

The problem is to estimate $< \|u\|^2 >$ on solutions as a function of $\mathbf{Re}$ as $\mathbf{Re} \to \infty$.

The inhomogeneous boundary conditions together with the finiteness of the kinetic energy and of the dissipation determine an affine set $\mathcal{U}$ in function space. We consider any time independent function $b \in \mathcal{U}$ ("the background") and associate to it a corresponding linear operator $\mathcal{L}_b$. In the Taylor-Couette case this operator is computed as follows:

$$\mathcal{L}_b = A + 2\mathcal{S}_b \,,$$

where

$$\mathcal{S}_b v = \mathbf{P}\left(\frac{((\nabla b) + (\nabla b)^*)}{2} v \right),$$

$\mathbf{P}$ is the Leray projector on divergence-free vectors in L^2, and A is the Stokes operator $\mathbf{P}(-\Delta)$. Let $\lambda(b)$ denote the bottom of the spectrum of $\mathcal{L}_b$ in L^2.

THEOREM 1 *Assume that $b \in \mathcal{U}$ satisfies $\mathbf{P}(b \cdot \nabla b) = 0$ and $\lambda(b) \geq 0$. Then every solution $u(t)$ of (1) in $\mathcal{U}$ satisfies*

$$< \|u(t)\|^2 > \leq \|b\|^2.$$

The interpretation of this result is the following. Take any steady solution of the inviscid (Euler) equation $\mathbf{P}(b \cdot \nabla b) = 0$ with the correct boundary conditions ($b \in \mathcal{U}$) and compute its quadratic form stability as if it were a steady solution of the Navier-Stokes equation with half the given viscosity ($\frac{1}{2}$ in our nondimensional setting). If the solution is stable ($\lambda(b) \geq 0$) then its bulk dissipation is an upper bound for the long time average bulk dissipation of *any* solution of the viscous problem.

Examples of such functions are obtained by choosing flat shear flow backgrounds with sharp boundary layers. Choosing the size of the boundary layer of the order of $\mathbf{Re}^{-1}$ ensures that $\lambda(b) > 0$. Using Theorem 1 one can prove that

$$< \|u(t)\|^2 > \leq C\mathbf{Re}^3.$$

Moreover, C can be estimated explicitly and the results agree with the physical experiment [18], [12], [6], [4]. In order to improve the estimate of the prefactor C one is naturally lead to variational problems with spectral side conditions. The set of shear backgrounds

$$\mathcal{X} = \left\{ b \in \mathcal{U}; b = \begin{pmatrix} \mathbf{Re}\psi(y) \\ 0 \end{pmatrix}, \psi(0) = 0, \psi(1) = 1 \right\}$$

poses already nontrivial problems. Note that for $b \in \mathcal{X}$, $\mathbf{P}(b \cdot \nabla b) = 0$. In $\mathcal{X}$ consider the subset

$$\mathcal{C}_{\mathcal{X}} = \{ b \in \mathcal{X}; \lambda(b) > 0 \} \,.$$

This is a convex set. The minimization problem is to compute

$$\inf_{b \in \mathcal{C}_{\mathcal{X}}} \|b\|^2.$$

This problem leads to new, nonlinear Orr-Sommerfeld-like equations [6]; their study is in progress. A simpler problem is obtained by strengthening the constraints. Instead of the set $\mathcal{C}_\mathcal{X}$, which consists of $b \in \mathcal{X}$ such that $(\mathcal{L}_b v, v)_{L^2} > 0$ for all *divergence-free* $v \neq 0$, we consider a smaller set $\tilde{\mathcal{C}}_\mathcal{X}$ where we require the shear background to be stable under a larger class of perturbations, dropping the condition that they be divergence-free vectors. Of course, this leaves fewer backgrounds and hence yields a larger infimum. If $b \in \mathcal{X}$ then this more stringent stability condition is equivalent to:

$$\int_0^1 \left[(v'(y))^2 - \mathbf{Re}\psi'(y)(v(y))^2 \right] dy > 0$$

for all $v \in H_0^1\left((0,1)\right), v \neq 0$. Thus, the problem of finding

$$\inf_{b \in \tilde{\mathcal{C}}_\mathcal{X}} \|b\|^2$$

has the same structure as the problem we started with: we seek

$$\inf_{u \in \mathcal{C}_\mathcal{U}} D(u)$$

where

$$D(u) = \int_0^1 |u'(y)|^2 \, dy,$$

the affine set $\mathcal{U}$ is

$$\mathcal{U} = \{ u \in H^1((0,1)); \; u(0) = 0, u(1) = \mathbf{Re} \},$$

the operator associated to a background u is

$$\mathcal{L}_u v = -v'' - u'v \, ,$$

and the spectral condition is

$$\mathcal{C}_\mathcal{U} = \{ u \in \mathcal{U}; (\mathcal{L}_u v, v)_{L^2} > 0 \, \text{for all } v \in V, v \neq 0 \}.$$

$(V = H_0^1(0,1))$. The equations associated to the infimum are

$$\mathcal{L}_u g = 0, \quad (g,g)_{L^2} = 1$$

and

$$\left(\frac{\delta D}{\delta u}(u) \right) w = m \left(\frac{\delta \lambda}{\delta u}(u) \right) w \, ,$$

which in this simple case is

$$-u'' = m(g^2)'.$$

Using the boundary conditions $u \in \mathcal{U}$ and substituting in the equation for the ground state g we obtain the steady cubic Schrödinger equation:

$$-g'' + \left(m(g^2 - 1) - \mathbf{Re} \right) g = 0$$

with the homogeneous boundary conditions and normalization

$$g(0) = g(1) = 0, \quad \int_0^1 g^2(y)\, dy = 1.$$

This problem has an explicit solution at each Reynolds number in terms of Jacobi elliptic functions $\mathrm{sn}(u|p)$ [6]. The asymptotic behavior of this expression is

$$\lim_{\mathbf{Re} \to \infty} \mathbf{Re}^{-3} \left(\inf_{u \in \mathcal{C}_U} D(u) \right) = \frac{1}{12}.$$

This yields a better agreement with the experiment, as expected.

2. Scaling Exponents

The mathematical framework for the statistical study of turbulence is the theory of invariant measures for the Navier-Stokes equations [14], [13], [19]. An important theory proposing the existence of a universal behavior was set forth by Kolmogorov [16]. Based on dimensional analysis, this theory's predictions have been verified in numerous experiments to a surprisingly large degree. The main assertions of this theory are: "the dissipation rate of a turbulent flow

$$\epsilon = \nu < |\nabla u(x,t)|^2 >$$

is constant and independent of Reynolds number" and "there exists a range of lengths where the energy transfer is local and universal and depends only on ϵ." The Reynolds number is

$$Re = \frac{UL}{\nu},$$

where U is a typical velocity difference across a typical distance L and ν is the kinematic viscosity.

From dimensional analysis it follows that the length below in which viscous dissipation effects are dominant is given by

$$\ell_K := \left(\frac{\nu^3}{\epsilon} \right)^{\frac{1}{4}}.$$

The interval $[\ell_K, L]$ is called the inertial range. The variation $< |u(x+y) - u(x)| >$ of velocity across a distance $r = |y|$ can also be determined from dimensional analysis if r belongs to the inertial range: the only velocity one can form with r and ϵ is

$$< |u(x+y) - u(x)| > = (\epsilon |y|)^{\frac{1}{3}}.$$

This is the Kolmogorov-Obukhov law. It is one of the most important predictions of this theory. The relation

$$\epsilon = \frac{U^3}{L}$$

follows. Note that $\epsilon > 0$ is bounded independently of viscosity. The experimental and theoretical evidence seems to indicate the possibility of small corrections to the

Kolmogorov-Obukhov law [17]. More precisely, the equal time generalized structure functions

$$< |u(x+y,t) - u(x,t)|^m >^{\frac{1}{m}}$$

scale with distance like powers

$$< |u(x+y,t) - u(x,t)|^m >^{\frac{1}{m}} \sim U \left(\frac{|y|}{L} \right)^{\zeta_m},$$

where the value of the exponents ζ_m is close to $\frac{1}{3}$ but depends on m. The ζ_m must be nonincreasing in m because of the Hölder inequality.

In this section I will describe some results [8], [5] regarding the scaling of velocity structure functions for the Navier-Stokes equations. I start with the assumptions. We consider ensembles of solutions of the Navier-Stokes equations in the whole 3-D space. We assume that there exist uniform bounds for the velocities in the ensembles

$$\sup_{x,t} |u(x,t)| \le U \quad (*).$$

This assumption implies regularity of the solutions. We will consider driving body forces B that are bounded uniformly,

$$\sup_{x,t} |B(x,t)| \le B \quad (**).$$

The forces are deterministic. These are the standing assumptions. We define the averaging procedure M_ρ by

$$M_\rho(f(x,t)) = \text{AV} \sup_{x_0 \in R^3} \lim_{T \to \infty} \sup \frac{1}{T} \frac{3}{(4\pi\rho^3)} \int_0^T \int_{B_\rho(x_0)} f(x,t) \, dx \, dt.$$

AV means ensemble average and $\sup_{x_0}$ is a supremum over all Euclidean balls B_ρ with center x_0 and radius ρ.

We set

$$\epsilon_{(\rho)} = \nu M_{\frac{\rho}{2}} \left(|\nabla u(x,t)|^2 \right),$$

and

$$s_m^{(\rho)}(y) = [M_\rho \left(|u(x+y,t) - u(x,t)| \right)^m]^{\frac{1}{m}}.$$

We denote

$$\mathbf{Re} = \frac{U\rho}{\nu}.$$

We have an additional mild assumption:

$$\int_{|y| \le \rho} s_1^{(\rho)}(y) \frac{dy}{|y|^3} \le cU \quad (***).$$

There are no assumptions of homogeneity or isotropy. If

$$cU \left(\frac{|y|}{\rho} \right)^{\zeta_m} \le s_m^{(\rho)}(y) \le CU \left(\frac{|y|}{\rho} \right)^{\zeta_m}$$

holds on an interval

$$\mathbf{Re}^{-\frac{1}{1+\zeta_m}} \leq \frac{|y|}{\rho} \leq 1 ,$$

then we say that we have m-scaling. (This definition implies that the local Reynolds number $\frac{|y| s_m^{(\rho)}(y)}{\nu}$ ranges between c and $C\mathbf{Re}$ in the scaling region.) Here are the main results.

THEOREM 2 *Assume* $(*), (**), (***)$. *Then*

$$\epsilon_{(\rho)} \leq C \left(\frac{U^3}{\rho} + BU + \nu \frac{U^2}{\rho^2} \right) .$$

Note that the dissipation is bounded uniformly as $\nu \to 0$. We show that structure functions for the pressure are bounded in terms of structure functions for the velocity. Once the pressure is controlled, then the result follows via local energy inequalities. We relate the second structure function to the first.

THEOREM 3 *Assume* $(*), (**), (***)$. *Then*

$$s_2^{(\rho)}(y) \leq cU \frac{|y|}{\rho} \mathbf{Re}^{\frac{1}{2}} .$$

In particular, if 1 and 2-scaling occur, then

$$\zeta_1 \geq \zeta_2 \geq \frac{1}{3}.$$

Corrections to the Kolmogorov-Obukhov exponent of $\zeta = \frac{1}{3}$ are referred to by the name of intermittency. They are believed to be connected to the existence of statistically significant large variations in the velocity gradients, over small regions in physical space. We have some mathematical evidence for this connection.

THEOREM 4 *Assume* $(*), (**), (***)$, *and 4-scaling. If*

$$\nu M_{\frac{\rho}{4}} \left(|\nabla u(x+y,t) - \nabla u(x,t)|^2 \right) \geq c\epsilon_{(\frac{\rho}{2})} > 0$$

for y satisfying

$$\frac{|y|}{\rho} \leq C(\mathbf{Re})^{-\beta} ,$$

then

$$\zeta_4 \leq \frac{1}{4\beta}.$$

It is widely believed that $\zeta_3 = \frac{1}{3}$. The evidence is both numerical and theoretical. The traditional theoretical arguments are based on the assumption of homogeneity and isotropy [1]. A conjecture of Onsager corresponds to the statement that $\zeta_3 > \frac{1}{3}$ implies $\epsilon = 0$; a mathematical formulation and a proof of this conjecture [11] offer additional arguments in support of $\zeta_3 = \frac{1}{3}$. It follows that if ζ_4 makes sense, then it must be equal or less then $\frac{1}{3}$. If $\beta = \frac{3}{4}$, then it follows that $\zeta_4 = \frac{1}{3}$. If $\beta > \frac{3}{4}$, however, then the preceding result gives sufficient and testable conditions for intermittency corrections.

3. Stretching of Vortex Lines

The problem of formation of singularities or near singularities in solutions of the 3-D Euler equations is of great importance for the understanding of fluid turbulence. One is interested in the qualitative nature of the singular or near singular flows, the effect viscosity has on them, and the relevance these singular events have on the large scale energetic features of the flow.

It is well known [15] that, for the 3-D incompressible Euler equations, the time integral of the maximum of the vorticity magnitude must diverge if finite time blow up occurs. The vorticity is the antisymmetric part of the spatial gradient of the velocity. It is also easy to prove that the singularities cannot occur without small scales developing [5] in the vorticity. By that I mean that large spatial gradients in the vorticity magnitude must develop, at a fast enough rate.

In this section I want to emphasize the role played by the direction field associated to the vorticity. In 2-D this is a field of parallel lines; the 2-D equations have no finite time singularities. I want to present evidence suggesting that singularities in the direction field of vorticity are perhaps necessary for 3-D finite time blow up. I also would like to address the effect of viscosity.

The evidence I would like to present is two-fold: analytical and numerical. The 3-D incompressible Euler equations are

$$D_t \omega = \omega \cdot \nabla u , \tag{2}$$

where $u = u(x, t)$ is the velocity field assumed to be divergence free, $\nabla \cdot u = 0$,

$$D_t = \partial_t + u(x, t) \cdot \nabla$$

denotes the material derivative (derivative along fluid particles), and ω is the vorticity:

$$\omega = \nabla \times u.$$

We refer to the right-hand side in (2) as the stretching term. It vanishes in the corresponding 2-D equation.

The 3-D incompressible Euler equations (2) are equivalent to the requirement that the vector field

$$\Omega = \omega(x, t) \cdot \nabla$$

commute with the material derivative:

$$[D_t, \Omega] = 0.$$

The integral lines of Ω are material, i.e., they are carried by the flow. Their length element is $|\omega|$. They stretch according to

$$D_t |\omega| = \alpha |\omega|. \tag{3}$$

The stretching rate alpha is given by

$$\alpha(x) = (\nabla u(x)) \, \xi(x) \cdot \xi(x) \tag{4}$$

and the direction field ξ by

$$\xi(x) = \frac{\omega(x)}{|\omega(x)|}. \tag{5}$$

The region $\{x : |\omega(x)| > 0\}$ is material. Both α and ξ are defined in it. The stretching rate α has a remarkable integral representation

$$\alpha(x) = \text{P. V.} \int D\left(\hat{y}, \xi(x+y), \xi(x)\right)|\omega(x+y)|\frac{dy}{|y|^n}. \tag{6}$$

Here n is the dimension of space (3 for Euler equations) and the geometric factor D vanishes not only in the spherical average, but also if the vectors $\xi(x+y)$ and $\xi(x)$ are parallel or antiparallel. More precisely, if $\cos\phi = \xi(x) \cdot \xi(x+y)$, then $|D| \leq |\sin\phi|$.

Based on this property we conjecture that if the direction field ξ is smooth in regions of high vorticity then blow up does not occur.

In order to test this conjecture we consider active scalars — 2-D models of the 3-D Euler equations [4], [5], [9], [10] — and investigate them analytically and numerically. These equations obey the same commutation relation

$$[D_t, \Omega] = 0$$

that determines the 3-D Euler equation; the difference is in the constitutive laws which relate the coefficients ω of Ω to the velocity u. Specifically, the active scalars are functions $\theta(x,t)$ that obey

$$(\partial_t + u \cdot \nabla)\theta = 0 \tag{7}$$

with

$$u(x,t) = \int a(x-y)\omega(y,t)\,dy\,, \tag{8}$$

where the function a is given and is smooth away from the origin and

$$\omega = \nabla^\perp\theta. \tag{9}$$

In (9) $\nabla^\perp$ denotes the gradient rotated counterclockwise by 90 degrees. The 2-D Euler equation corresponds to $a(x) = \frac{1}{2\pi}\log(|x|)$. The natural analogue of the 3-D Euler equation constitutive law is $a(x) = \frac{1}{|x|}$. This defines the quasi-geostrophic active scalar equation (QGASE). The QGASE is physically significant in its own right: it is a model for atmospheric temperature in a geostrophic approximation. Its theoretical resemblance to 3-D Euler equations is remarkable: *all* the statements about Euler equations made above are valid for the QGASE. The vortex lines correspond to iso-theta lines. The stretching equation, the Kato-Beale-Majda criterion, the integral representation of the stretching rate α, its depletion if the direction field ξ is smooth, all apply equally well to the QGASE as to the 3-D Euler equation.

The conjecture relating smoothness of ξ to absence of blow up can be proved [10]. The numerical evidence of [10] supports strongly two statements. First sharp fronts do form in finite time (sharp means large gradients of θ, i.e., large $|\omega|$; based on the present computations, blow up cannot be predicted). Second there is a marked difference in the rate of development of these sharp fronts caused by the nature of ξ. There exist initial data for which ξ develops only antiparallel ($\sin\phi = 0$) singularities. The formation of fronts is then depleted. For other initial data, a saddle point in θ provides a Lipschitz singularity in the ξ direction field. This is the source of much more intense gradient growth.

Now I will address briefly the role of viscosity. In the 3-D incompressible Navier-Stokes equations there exist suitable weak solutions that satisfy

$$< |\omega(x,t)||\nabla\xi(x,t)|^2 > \leq \frac{\Gamma}{\nu},$$

where $\nu > 0$ is the viscosity, Γ is given in terms of the initial data and $< \cdots >$ is an appropriate space and time average [3], [7]. Consequently, typical regions of high vorticity have Lipschitz ξ.

Moreover, if one assumes

Assumption (A)

There exist constants $\Omega > 0$ and $\rho > 0$ such that

$$|P^\perp_{\xi(x,t)}(\xi(x+y,t))| \leq \frac{|y|}{\rho}$$

holds if both $|\omega(x,t)| > \Omega$ and $|\omega(x+y,t)| > \Omega$, and $0 \leq t \leq T$, ($P^\perp_{\xi(x)}\xi(x+y)$ is the projection of $\xi(x+y)$ orthogonal to $\xi(x)$), then [7]:

THEOREM 5 *Under Assumption (A) the solution of the initial value problem for the Navier–Stokes equation is smooth (C^∞) on the time interval $[0,T]$.*

The interpretation of this result is that Lipschitz (and antiparallel Lipschitz) singularities are smoothed out by the viscosity.

References

[1] G. Batchelor, The Theory of Homogeneous Turbulence, Cambridge Univ. Press, Cambridge, 1960.

[2] B. Castaing, G. Gunaratne, F. Heslot, L. P. Kadanoff, A. Libchaber, S. Thomae, X.-Z. Wu, S. Zaleski, and G. Zanetti, *Scaling of hard thermal turbulence in Rayleigh–Benard connection*, J. Fluid Mech. **204**, 1 (1989).

[3] P. Constantin, *Navier-Stokes equations and area of interfaces*, Comm. Math. Phys. **129**, 241–266 (1994).

[4] P. Constantin, *Geometric statistics in turbulence*, SIAM Rev., **36**, 73–98 (1994).

[5] P. Constantin, Geometric and analytic studies in turbulence, Lecture Notes in Appl. Math. **100**, Springer-Verlag, Berlin and New York (1994).

[6] P. Constantin and C. R. Doering, *Variational bounds on energy dissipation in incompressible flows, I Shear flow*, Phys. Rev. E **49**, 4087–4099 (1994).

[7] P. Constantin and Ch. Fefferman, *Direction of vorticity and the problem of global regularity for the Navier-Stokes equations*, Indiana Univ. Math. J. **42**, 775 (1993).

[8] P. Constantin and Ch. Fefferman, *Scaling exponents in fluid turbulence: some analytic results*, Nonlinearity **7**, 41 (1994).

[9] P. Constantin, A. Majda, and E. Tabak, *Singular front formation in a model for quasi-geostrophic flow*, Phys. Fluids **6**, 9 (1994).

[10] P. Constantin, A. Majda, and E. Tabak, *Formation of strong fronts in the 2-D quasigeostrophic thermal active scalar*, Nonlinearity, **7** 1495–1533 (1994).

[11] P. Constantin and W. E. E. Titi, *Onsager's conjecture on the energy conservation for solutions of Euler's equation*, Comm. Math. Phys., **165**, 207–209 (1994).

[12] Ch. R. Doering and P. Constantin, *Energy dissipation in shear driven turbulence*, Phys. Rev. Lett. **69**, 1648–1651, (1992).

[13] C. Foias, *Statistical study of Navier-Stokes equations I and II*, Rend. Sem. Mat. Univ. Padova **48**, 219–348 (1972) and **49**, 9–123 (1973).

[14] E. Hopf, *Statistical hydrodynamics and functional calculus*, J. Rat. Mech. Anal.**1**, 87–123 (1952).

[15] T. Kato, J.T. Beale, and A. Majda, *Remarks on the breakdown of smooth solutions for the 3-D Euler equations*, Comm. Math. Phys. **94**, 61–66 (1984).

[16] A. N. Kolmogorov, *Local structure of turbulence in an incompressible fluid at very high Reynolds numbers*, Dokl. Akad. Nauk. SSSR **30**, 299–303 (1941).

[17] A.Y.S. Kuo and S. Corrsin, *Experiments on internal intermittency and fine-structure distribution functions in fully turbulent fluid*, J. Fluid Mech. **50**, 285 (1971).

[18] D. P. Lathrop, J. Fineberg, and H. L. Swinney, *Turbulent flow between concentric rotating cylinders at large Reynolds number*, Phys. Rev. Lett. **68**, 1515–1518 (1992).

[19] M. J. Vishik and A. V. Fursikov, *Solutions statistiques homogènes des systèmes différentiels paraboliques et du système de Navier-Stokes*, Ann. Scuola Norm. Sup. Pisa Cl. Sci. **4**, 531–576 (1977); and *Homogeneous in x space-time statistical solutions of the Navier-Stokes equations and individual solutions with infinite energy*, Dokl. Akad. Nauk. SSR **239**, 1025–1032 (1978).

Hyperbolic Systems of Conservation Laws

Constantine M. Dafermos

Division of Applied Mathematics
Brown University
Providence, RI 02912, USA

0. Introduction

Conservation laws are first order systems of quasilinear partial differential equations in divergence form; they express the balance laws of continuum physics for media with "elastic" response, in which internal dissipation is neglected. The absence of internal dissipation is manifested in the emergence of solutions with jump discontinuities across manifolds of codimension one, representing, in the applications, phase boundaries or propagating shock waves. The presence of discontinuities makes the analysis hard; the redeeming feature is that solutions are endowed with rich geometric structure. Indeed, the most interesting results in the area have a combined analytic-geometric flavor.

The paper will survey certain aspects of the theory of hyperbolic systems of conservation laws. Any attempt to be comprehensive would fail because of space limitations. Major developments in the future will likely come from the exploration of systems in several space dimensions, which presently is terra incognita. Accordingly, the author has opted to emphasize the multidimensional setting, at the expense of the one-space dimensional case, where past and present achievements mostly lie. Many important recent results in one-space dimension will only be briefly mentioned, whereas others will not be referenced at all. Similarly, the bibliography is far from exhaustive. Finally, the exciting research program that addresses the connection between systems of conservation laws and the kinetic equations will not be discussed here as it will be surveyed in this volume by Perthame [P].

1. Conservation Laws and Continuum Physics

A system of n *conservation laws* has the general form

$$\operatorname{div} G = 0, \tag{1.1}$$

where G is defined on an open subset $\mathcal{B}$ of $\mathbb{R}^m$ and takes values in the space of $(n \times m)$ matrices. The motivation for the terminology is that, under mild regularity conditions on G, (1.1) is equivalent to

$$\oint_{\partial \Omega} GN dS = 0 \tag{1.2}$$

Proceedings of the International Congress
of Mathematicians, Zürich, Switzerland 1994
© Birkhäuser Verlag, Basel, Switzerland 1995

for all $\Omega \subset \mathcal{B}$ of finite perimeter, with boundary $\partial\Omega$; N denotes the unit normal on $\partial\Omega$.

The conservation property is robust: assume $\hat{X} = \hat{X}(X)$ is any bilipschitz homeomorphism of $\mathcal{B}$ to $\hat{\mathcal{B}} \subset \mathbb{R}^m$, such that the Jacobian matrix $J = \partial\hat{X}/\partial X$ satisfies $\det J \geq a > 0$, a.e. on $\mathcal{B}$. Then an L^∞ field G satisfies on $\mathcal{B}$ the conservation law (1.1), in the sense of distributions, if and only if the L^∞ field

$$\hat{G}(\hat{X}) := [\det J(X(\hat{X}))]^{-1} J(X(\hat{X})) G(X(\hat{X})) \tag{1.3}$$

satisfies on $\hat{\mathcal{B}}$ the conservation law

$$\operatorname{div} \hat{G} = 0. \tag{1.4}$$

The development of the theory of conservation laws over the past two centuries has been motivated to a great extent by problems arising in continuum physics. Continuum theories are demarcated by a prescribed set of conservation laws. Thus, continuum mechanics is identified by conservation of mass and momentum; continuum thermomechanics by conservation of mass, momentum, and energy; etc. [TN]. In the context of statics, $\mathbb{R}^m$ is physical space, of dimension one, two, or three, whereas in the context of dynamics $\mathbb{R}^m$ is space-time, of dimension two, three, or four. The conservation laws of continuum physics may be cast in referential (Lagrangian) or spatial (Eulerian) form. Passing from Lagrangian to Eulerian formulation involves a change of variable of the form (1.3) and is therefore allowable even when the fields are merely in L^∞ [Da5].

In continuum physics, the set of conservation laws is complemented with *constitutive relations* that specify the nature of the medium. Here we focus attention on media with "elastic" response in which G is determined by a *state vector* U in $\mathbb{R}^n$ through a smooth function

$$G = F(U). \tag{1.5}$$

Thus, (1.1) and (1.5) yield a first order quasilinear system of n equations from which the n-vector field U is to be determined.

The umbilical cord that joins the theory of systems of conservation laws with continuum physics is still vital for the proper development of the subject and should not be severed.

2. Jump Discontinuities and Oscillations

The issues to be discussed here are pertinent to statics as well as to dynamics. The crucial factor is whether the system (1.1), (1.5) may admit solutions U on $\mathbb{R}^m$ with jump discontinuities of the following form: for some unit vector N in $\mathbb{R}^m$ and two distinct vectors U_-, U_+ in $\mathbb{R}^n$, $U = U_-$ on the half-space $X^T N < 0$ and $U = U_+$ on the half-space $X^T N > 0$. Such a function satisfies (1.1), (1.5) in the sense of distributions if and only if the *Rankine-Hugoniot jump conditions* hold:

$$[F(U_+) - F(U_-)]N = 0. \tag{2.1}$$

It is clear that (2.1) cannot hold if for every U in $\mathbb{R}^n$ and unit vector N in $\mathbb{R}^m$ the $(n \times n)$ matrix $DF(U)N$ is nonsingular. This condition, characterizing "elliptic" systems, typically prevails in stable static problems. Ellipticity fails in statics, and there are discontinuous solutions of the form described above, when the model allows for *phase transitions*, so that the plane $X^T N = 0$ is realized as a *phase boundary*. In dynamics, solutions with jump discontinuities, considered above, are common and have been studied extensively over the past 150 years, beginning with the pioneering papers of Stokes [St] and Riemann [Ri]. In that context the plane $X^T N = 0$ is realized as a propagating *shock wave*.

When (2.1) holds, one may construct solutions U of (1.1), (1.5) by partitioning $\mathbb{R}^m$ into slabs by means of any finite family of parallel, codimension-one planes normal to N and then assigning, in alternating order, the values U_- and U_+ on the slabs contained between adjacent planes. This construction may yield highly oscillatory solutions, a manifestation of instability. Indeed, by judiciously selecting the family of parallel planes of jump discontinuity, it is easy to construct a sequence $\{U_j\}$ of solutions converging, as $j \to \infty$, in L^∞ weak*, to an L^∞ function U that does not satisfy (1.1), (1.5).

In situations involving phase transitions, highly oscillatory configurations that may model material microstructure [BJ] are resisted by capillarity, which induces penalization each time a phase boundary forms [CGS] (this active area of research lies beyond the scope of the present exposition). On the other hand, in elastodynamics rapid oscillations are quenched by the effects of internal dissipation (such as Newtonian viscosity), which has been neglected when postulating the constitutive relation (1.5) but whose legacy survives in so-called *entropy inequalities*

$$\operatorname{div} h \le 0 \qquad (2.2)$$

with

$$h = q(U), \qquad (2.3)$$

assumed to be satisfied by *admissible solutions* of (1.1), (1.5). The motivation for the name "entropy" is that, in the applications to continuum physics, (2.2) is related, directly or indirectly, to the Clausius-Duhem inequality, which expresses the second law of thermodynamics [TN].

In the tradition of continuum physics, it is required that every classical (i.e., Lipschitz continuous) solution of (1.1), (1.5) be admissible; i.e., satisfy automatically (2.2), (2.3). This will be the case if and only if there is a smooth function P from $\mathbb{R}^n$ to $\mathbb{R}^n$ such that

$$P(U)^T DF(U) = Dq(U). \qquad (2.4)$$

Thus, P must satisfy the integrability condition

$$DP(U)^T DF(U) = DF(U)^T DP(U). \qquad (2.5)$$

Of course, we are only interested in the nontrivial case $P \not\equiv$ constant. Note that (2.5) holds, with $P(U) = U$, when the system (1.1), (1.5) is symmetric; i.e., $DF(U)^T = DF(U)$. Conversely, when (2.5) holds and P is a diffeomorphism of $\mathbb{R}^n$ to $\mathbb{R}^n$, the change $V = P(U)$ in the state vector renders the resulting system of conservation laws symmetric [FL].

Once the constitutive relation (1.5) has been laid down, the functions $q(U)$ that may be candidates for (2.3) are to be determined through (2.4), after integrating (2.5). Now (2.5) imposes $\frac{1}{2}n(n-1)m$ conditions on the n components of P and so it constitutes an overdetermined system, unless either $n = 1$ and m arbitrary or $n = 2$ and $m = 1$. Even so, the conservation laws of continuum physics are always equipped with a natural choice of $q(U)$, simply because when constitutive relations are laid down, $F(U)$ and $q(U)$ are simultaneously determined in such a fashion that (2.4) holds.

A solution involving two constant states U_- and U_+, experiencing a jump discontinuity across a plane $X^T N = 0$, as discussed earlier in this section, will satisfy the entropy admissibility criterion (2.2), (2.3) if and only if

$$[q(U_+) - q(U_-)]N \leq 0. \tag{2.6}$$

Note that when (2.6) holds as a strict inequality the oscillatory solutions with the laminated structure, described above, are ruled inadmissible. The central question is whether entropy inequalities (2.2), (2.3) induce stability, in general. An affirmative answer would hinge on whether the following assertion is true. Suppose $\{U_j\}$, bounded in L^∞, is any sequence of approximate admissible solutions of (1.1), (1.5); i.e., as $j \to \infty$, $\{\operatorname{div} F(U_j)\}$ converges to zero and $\{\operatorname{div} q(U_j)\}$ converges to a nonpositive distribution. Then the L^∞ weak* limit U of $\{U_j\}$ is a solution of (1.1), (1.5). The veracity of this statement, which was conjectured by Tartar [T] and elaborated by DiPerna [Di6], has been established, thus far, only for very special systems, as we shall see in Section 4.

Another important issue pertains to the geometric structure of the set of points of discontinuity of weak solutions of (1.1), (1.5). Suppose U is an L^∞ solution of (1.1), (1.5) on $\mathcal{B}$, which is of class BV; i.e., the distributional partial derivatives of U are Radon measures on $\mathcal{B}$ [EG]. Then $\mathcal{B}$ is decomposed into the union of three, pairwise disjoint subsets $\mathcal{C}, \mathcal{J}, \mathcal{I}$ with the following properties: $\mathcal{I}$ has $(m-1)$-dimensional Hausdorff measure zero; $\mathcal{J} \cup \mathcal{I}$ is the countable union of Lipschitz manifolds of codimension one; U is Lebesgue approximately continuous on $\mathcal{C}$; with every point $\overline{X}$ of $\mathcal{J}$ there is associated a unit vector N in $\mathbb{R}^m$ such that U attains distinct Lebesgue approximate limits U_- and U_+ at $\overline{X}$ relative to the half-spaces $(X - \overline{X})^T N < 0$ and $(X - \overline{X})^T N > 0$, respectively; furthermore, U_-, U_+, and N are interrelated by the Rankine-Hugoniot condition (2.1).

3. Hyperbolicity

Henceforth, we focus attention on dynamics so $\mathbb{R}^m$ is space-time with typical point $X = (x, t)$, where t, the time, takes values in $\mathbb{R}$ and x, the position in space, takes values in $\mathbb{R}^k (k = m - 1)$. For simplicity, we assume that in (1.5) the last column of the $(n \times m)$ matrix $F(U)$ is just U, in which case (1.1), (1.5) become

$$\partial_t U + \sum_{\alpha=1}^{k} \partial_\alpha F_\alpha(U) = 0, \tag{3.1}$$

with $\partial_t = \partial/\partial t$ and $\partial_\alpha = \partial/\partial x_\alpha$, $\alpha = 1, \ldots, k$. The $F_\alpha(U)$ take values in $\mathbb{R}^n$. Similarly, upon labeling the last column of $q(U)$ in (2.3) as $\eta(U)$, we rewrite (2.2),

(2.3) as

$$\partial_t \eta(U) + \sum_{\alpha=1}^{k} \partial_\alpha q_\alpha(U) \leq 0. \tag{3.2}$$

Following Lax [L2], it has become standard to call $\eta(U)$ *entropy*, even though it never coincides with the physical entropy (in certain cases $-\eta$ is the physical entropy). $q_\alpha(U), \alpha = 1, \ldots, k$, is the associated *entropy flux*. From (2.4) we deduce $P(U)^T = D\eta(U)$ so (2.5) now reduces to

$$D^2\eta(U)DF_\alpha(U) = DF_\alpha(U)^T D^2\eta(U) , \quad \alpha = 1, \ldots, k. \tag{3.3}$$

The system (3.1) enjoys maximal freedom to propagate waves when it is *hyperbolic*: For each U in $\mathbb{R}^n$ and any unit vector ν in $\mathbb{R}^k$, the $(n \times n)$ matrix

$$\Lambda(U; \nu) = \sum_{\alpha=1}^{k} \nu_\alpha DF_\alpha(U) \tag{3.4}$$

has real eigenvalues $\lambda_1(U; \nu) \leq \cdots \leq \lambda_n(U; \nu)$ and n linearly independent eigenvectors $R_1(U; \nu), \ldots, R_n(U; \nu)$.

Henceforth we shall be assuming that our system (3.1) is hyperbolic and seek a solution $U(x, t)$ on the half-space $\{(x, t) : x \in \mathbb{R}^k, t \in \mathbb{R}^+\}$, with prescribed initial conditions:

$$U(x, 0) = U_0(x) , \quad x \in \mathbb{R}^k. \tag{3.5}$$

The problem (3.1), (3.5) has been solved for the single equation $n = 1$ [K]: When $U_0 \in L^\infty(\mathbb{R}^k) \cap L^1(\mathbb{R}^k)$, there is a unique admissible solution that satisfies (3.2) for any convex entropy function $\eta(U)$ and is stable in L^1:

$$\|U(\cdot, t) - \overline{U}(\cdot, t)\|_{L^1(\mathbb{R}^k)} \leq \|U_0(\cdot) - \overline{U}_0(\cdot)\|_{L^1(\mathbb{R}^k)} , \quad 0 \leq t < \infty. \tag{3.6}$$

In particular, if U_0 is in BV so is U. New, unexpected regularity properties of these solutions have recently been uncovered [LPT], so even this, supposedly well-plowed, corner of the field continues to yield interesting results.

The remaining sections focus on systems, $n \geq 2$, for which the theory is still in the stage of development.

4. Systems in One-Space Dimension

Hyperbolic systems of conservation laws in one space variable:

$$\partial_t U + \partial_x F(U) = 0 \tag{4.1}$$

have been studied intensively. The direction of research over the past 40 years was set by the pioneering paper of Lax [L1]. An enormous amount of information has been amassed; see the monograph [Sm]. Here I paint a highly impressionist picture, the goal being to contrast the single-space from the multispace dimensional case.

With the exception of certain systems with very special structure, analytical results have been derived only on solutions taking values in a small neighborhood of a constant state. Over the years, the system of one-dimensional (isentropic or nonisentropic) gas dynamics has served as the prototypical example and this has steered the research effort towards *strictly hyperbolic* systems, where the eigenvalues of $DF(U)$ are strictly separated: $\lambda_1(U) < \cdots < \lambda_n(U)$.

Shocks have to satisfy admissibility conditions that are motivated by (and are compatible with) entropy inequalities (2.6). The issue of admissibility of shocks of moderate strength in strictly hyperbolic systems is now well understood [Li]. By contrast, the case where strict hyperbolicity fails even at a single state involves phenomena that are presently poorly understood. Research in that direction [SS] lies beyond the scope of this exposition.

An important property of systems (4.1) is that they admit solutions in the form of *wave fans*: $U(x,t) = V(x/t)$. An *i*-wave fan, $i = 1,\ldots,n$, in a small neighborhood of a constant state $\overline{U}$, is a solution of this type with $V(\xi)$ taking all its variation in a small interval about $\xi_i = \lambda_i(\overline{U})$. The simplest examples of an *i*-wave fan are the admissible *i-shock*, in which V is a step function with a single jump near ξ_i, and the *i-rarefaction wave*, in which V is Lipschitz continuous. In particular, when the system is *genuinely nonlinear* [L1], any admissible *i*-wave fan is either an admissible *i*-shock or an *i*-rarefaction wave. For general, strictly hyperbolic systems, however, an *i*-wave fan may contain both *i*-shocks and *i*-rarefaction waves, with the property that any *i*-shock adjacent to an *i*-rarefaction wave is a *contact discontinuity* on the side of the interface [Li].

In the *Riemann problem* for (4.1), a jump discontinuity between two nearby constant states U_- and U_+ is resolved into a wave fan, which is a superposition of *n* *i*-wave fans, one from each characteristic family [L1], [Li]. The solution of the Riemann problem describes the local structure of any BV solution of (4.1) [Di1] and also provides the testing ground for admissibility criteria for solutions [Da3].

The solution of the Riemann problem has been used as a building block for constructing solutions of (4.1) under initial data with small total variation. The most effective construction is due to Glimm [G]: the initial data are approximated by a step function and a local approximate solution is then constructed by resolving the jump discontinuities into wave fans, as in the solution of the Riemann problem. Before waves originating at different jump points begin to interact, the clock is stopped, the approximate solution is again approximated by a step function, and the process is repeated for another time step, and so on. In order to secure the consistency of the resulting algorithm, the approximating by step functions has to be done in a special way, namely the random choice method of Glimm or the equidistribution choice method of Liu [Li]. The study of wave interactions yields a priori estimates on the family of approximate solutions that induce compactness and allow us to pass a.e. to the limit. The resulting solution U inherits bounds

$$TV_x U(\cdot,t) \le cTV_x U_0(\cdot) \ , \quad 0 < t < \infty, \tag{4.2}$$

$$\int_{-\infty}^{\infty} |U(x,t) - U(x,\tau)| \, dx \le c|t - \tau|TV_x U_0(\cdot) \ , \quad 0 < \tau < t < \infty, \tag{4.3}$$

which guarantee, in particular, that it is a function of class BV. Stronger conclusions obtain [GL] for systems of two conservation laws, $n = 2$, because in that case the state may be represented in *Riemann invariant* coordinates, in terms of which the two characteristic families decouple to higher order. The recent study of wave interactions by Young [Y] reveals that some of the properties formerly believed to be peculiar to systems of just two conservation laws have analogs in larger systems as well; but also hints that new, potentially destabilizing wave resonance phenomena may occur in solutions of such systems.

An alternative method of constructing solutions by solving Riemann problems is *front tracking* [Br], [Da1], [Di2], [Ris]: In the solution of the Riemann problem, rarefaction waves are approximated by fans of (inadmissible) rarefaction shocks of very small amplitude. Thus, all wave interactions become shock interactions that may be resolved by solving new Riemann problems. The difficulty is, of course, that, as the number of shocks proliferates, the time between successive shock interactions may become very short and the construction may grind to a stop. To prevent this from happening, the algorithms involve some procedure for eliminating shocks of negligible amplitude. Bressan and Colombo have demonstrated [BC] that if $U(x,t)$ and $\overline{U}(x,t)$ are any two solutions of systems of two conservation laws (4.1) with initial data $U_0(x)$ and $\overline{U}_0(x)$, respectively, constructed by either the random choice method or the front tracking algorithm, then

$$\|U(\cdot,t) - \overline{U}(\cdot,t)\|_{L^1(\mathbb{R})} \leq c\|U_0(\cdot) - \overline{U}_0(\cdot)\|_{L^1(\mathbb{R})}, \quad 0 < t < \infty. \qquad (4.4)$$

There are strong indications that (4.4) should hold for any pair of admissible solutions, irrespective of the method of construction, at least when $n = 2$, and possibly even when $n > 2$. Such a result would settle in a definitive manner the issue of uniqueness of solutions. Interesting uniqueness theorems are currently known [Di3], [LX] but apply only for solutions endowed with regularity not necessarily present in arbitrary BV solutions. Properties of general solutions, regardless of how they were constructed, may be derived by the method of generalized characteristics [Da4].

Over the past decade, considerable effort has been exerted to construct solutions of (4.1) in L^∞ by the method of compensated compactness [T]. The idea of this approach has already been explained in Section 2: a sequence $\{U_j\}$ of approximate solutions, compatible with entropy inequalities, is constructed. The expectation is that the entropy inequalities will quench rapid oscillations and hence $\{U_j\}$ will converge a.e. to a solution of (4.1). A great number of entropy inequalities has to be employed in the process, and this has limited so far the applicability of the method to systems endowed with a rich family of entropies, and most notably systems of two conservation laws, $n = 2$. The first result in this direction [Di4] dealt with strictly hyperbolic, genuinely nonlinear systems of two conservation laws: By employing the family of Lax entropies [L2], it is shown that oscillations are quenched by the effect of genuine nonlinearity. The system of (one-dimensional) isentropic gas dynamics with equation of state $p \propto \rho^\gamma$, in which strict hyperbolicity only fails at the vacuum state $\rho = 0$, has similarly been treated [C1], [Di5], [LPS] for the full range $\gamma > 1$ of the adiabatic exponent. The method has also been successfully employed [CK] for genuinely nonlinear systems of two conservation laws in which strict hyperbolicity fails at a single state (*umbilic point* with hyperbolic degeneracy). On the other hand, the case of strictly hyperbolic systems that are not genuinely nonlinear is still imperfectly explored. In the extreme situation of linearly degenerate characteristic fields, rapid oscillations may propagate [C2], [S].

5. Stability

We now return to several space dimensions and consider solutions $U(x,t)$ of (3.1), (3.5) in $L^\infty(\mathbb{R}^k \times \mathbb{R}^+)$. It is easy to see that (3.1) induces some (very weak) continuity in the time direction [Da7]:

$$t \mapsto U(\cdot, t) \text{ continuous in } L^\infty \text{ weak}^*. \qquad (5.1)$$

Assume (3.1) is endowed with a uniformly convex entropy function $\eta(U)$. As noted in Section 2, (3.1) is then symmetrizable. When U_0 lies in a Sobolev space of sufficiently high order, the initial-value problem (3.1), (3.5) admits a classical solution on some time interval. The stability of shock fronts has also been investigated. See the monograph [M]. When (3.2) is interpreted, in the usual fashion, as

$$\int_0^\infty \int_{\mathbb{R}^k} [\eta(U)\partial_t \varphi + \sum_{\alpha=1}^k q_\alpha(U)\partial_\alpha \varphi] \, dx \, dt + \int_{\mathbb{R}^k} \eta(U_0(x))\varphi(x,0) \, dx \geq 0 \qquad (5.2)$$

for all nonnegative test functions φ with compact support in $\mathbb{R}^k \times \mathbb{R}^+$, then any L^∞ solution compatible with it and normalized by (5.1) satisfies

$$\int_{\mathbb{R}^k} \eta(U(x,t)) \, dx \leq \int_{\mathbb{R}^k} \eta(U_0(x)) \, dx \ , \quad 0 < t < \infty. \qquad (5.3)$$

It follows from (5.3) that the map $t \mapsto U(\cdot, t)$ is actually strongly continuous at $t = 0$. Because the value $t = 0$ has no special status, it is conceivable that the interpretation of the entropy inequality (3.2) should be strengthened to yield $\int_{\mathbb{R}^k} \eta(U(x,t)) \, dx$ nonincreasing on $\mathbb{R}^+$ and thus render $t \mapsto U(\cdot, t)$ strongly continuous for every value of t.

Always assuming $\eta(U)$ is uniformly convex, (5.3) implies

$$\|U(\cdot, t)\|_{L^2(\mathbb{R}^k)} \leq c_2 \|U_0(\cdot)\|_{L^2(\mathbb{R}^k)} \ , \quad 0 < t < \infty. \qquad (5.4)$$

In fact, a more general result has been established [Da2], [Di3]: if $\overline{U}(x,t)$ is any Lipschitz solution of (3.1) with initial data $\overline{U}_0(x)$ and $U(x,t)$ is any L^∞ solution of (3.1), (3.5) satisfying (5.2), then

$$\|U(\cdot, t) - \overline{U}(\cdot, t)\|_{L^2(\mathbb{R}^k)} \leq ce^{\alpha t} \|U_0(\cdot) - \overline{U}_0(\cdot)\|_{L^2(\mathbb{R}^k)} \ , \quad 0 < t < \infty. \qquad (5.5)$$

In particular, whenever they exist, Lipschitz solutions are unique within the broader class of L^∞ solutions that satisfy an entropy inequality with a uniformly convex entropy function. The reader should be aware, however, that just one entropy inequality is generally insufficient to single out a unique solution with discontinuities, and thus more discriminating admissibility criteria are needed for that purpose. The question of uniqueness of L^∞ solutions is open.

Estimates (5.4), (5.5) may leave the impression that for systems of conservation laws in several space dimensions L^2-stability prevails. On the other hand, (3.6) and (4.4) seem to indicate that L^1-stability is the natural one. The question then arises of whether estimates

$$\|U(\cdot, t)\|_{L^p(\mathbb{R}^k)} \le c_p \|U_0(\cdot)\|_{L^p(\mathbb{R}^k)} , \quad 0 < t < \infty, \tag{5.6}$$

are valid for any $p \ne 2$. When (3.1) is linear, it is known [B] that (5.6) holds for any $1 \le p \le \infty$ if and only if the Jacobian matrices of the $F_\alpha(U)$ commute:

$$DF_\alpha DF_\beta = DF_\beta DF_\alpha , \quad \alpha, \beta = 1, \ldots, k. \tag{5.7}$$

Rauch [R] notes that if (5.6) holds for solutions of a quasilinear system (3.1) then it must also hold for solutions of the system resulting by linearization of (3.1) about any constant state. Thus, (5.7) is a necessary condition for (5.6) in the quasilinear case as well. For systems of two conservation laws, $n = 2$, and solutions taking values in a small neighborhood of the state $U = 0$, it has been shown [Da6] that (5.7) is also sufficient for (5.6) to hold with $p \in [1, 2]$ and, in certain cases, even for $p = \infty$. The proof is based on the observation that when (5.7) is satisfied the matrix $\Lambda(U; \nu)$ in (3.4) has n linearly independent eigenvectors $R_1(U), \ldots, R_n(U)$, independent of ν. In that case (3.3) reduces to

$$R_i(U)^T D^2 \eta(U) R_j(U) = 0 , \quad i, j = 1, \ldots, n; \ i \ne j. \tag{5.8}$$

In particular, if $n = 2$ (5.8) yields one linear, second order hyperbolic equation for determining the entropy function $\eta(U)$. By solving appropriate Goursat problems, one may construct, on some neighborhood of the origin, solutions $\eta(U)$ of (5.8) that are convex functions with growth $\eta(U) \sim |U|^p$ at $U = 0$. Then (5.6) follows from (5.3).

In view of the above, systems with the property (5.7) may serve as a training ground for exploring conservation laws in several space dimensions. This class, however, is very special and does not include the systems arising in continuum physics. The failure of (5.6), with $p = 1$, is disturbing, because it implies [R] that estimates like (4.2), (4.3) will fail as well and hence solutions will not generally lie in the space BV. Recall from Section 2 that in BV solutions, discontinuities organize as propagating shock waves, a very desirable feature. It would be of considerable interest to know whether such structure is present even in L^∞ solutions, without the BV property. The investigation should begin with $k = n = 1$, because the answer is unknown even in that very simple case. The geometric structure of solutions at codimension higher than one is extremely intricate [CH].

A different line of investigation would aim towards understanding why (5.6) fails when $p \ne 2$. A familiar phenomenon that contributes to this effect is focussing and defocussing of waves. If that is the principal cause for the failure of (5.6), some kind of *renormalization* of solutions may conceivably restore the BV property. This is mere speculation, however, as the author is unaware of any concrete supporting evidence.

In the above considerations, it was important that $\eta(U)$ be convex. In continuum physics, convexity of the entropy function is a natural assumption for certain models (e.g., elastic fluids) while in other cases (e.g., elastic solids) it would be in violation of the laws of physics [TN]. We must weaken the assumption of convexity so that compatibility with physics is preserved while retaining implications on stability, like (5.4) and (5.5). As a minimum requirement we should ask:

$$t \mapsto \eta(U(\cdot, t)) \text{ lower semicontinuous in } L^\infty \text{ weak}^*. \tag{5.9}$$

Indeed, (5.9) together with (5.1) yield the desirable (5.3) for any L^∞ solution satisfying (5.2).

What assumptions on $\eta(U)$ (short of convexity) would guarantee (5.9)? In continuum physics (elastodynamics, electrodynamics, etc.) the fluxes in the conservation laws (3.1) often satisfy a condition

$$A_\alpha F_\beta(U) + A_\beta F_\alpha(U) = 0 \,, \quad \alpha, \beta = 1, \ldots, k, \tag{5.10}$$

for some family of $(\ell \times n)$ matrices A_α, $\alpha = 1, \ldots, k$. This in turn implies that any solution U of (3.1) satisfies the so-called *involution* [Bo], [Da2]

$$\sum_{\alpha=1}^{k} A_\alpha \partial_\alpha U = 0 \tag{5.11}$$

given that the initial data U_0 do so. In that case (5.9) will hold provided $\eta(U)$ is *A-quasiconvex* in the sense of Dacorogna [D], namely

$$\int_{\mathcal{C}} \eta(\overline{U}) \, dx \le \int_{\mathcal{C}} \eta(U) \, dx \tag{5.12}$$

for every hypercube $\mathcal{C}$ in $\mathbb{R}^k$, each fixed $\overline{U}$ in $\mathbb{R}^n$, and any field U in L^∞ that satisfies (5.11) and whose mean value over $\mathcal{C}$ is $\overline{U}$.

With an involution (5.11) is associated a *wave cone*:

$$\mathcal{K} = \bigcup_\nu \ker \sum_{\alpha=1}^{k} \nu_\alpha A_\alpha, \tag{5.13}$$

where the union extends over all unit vectors ν in $\mathbb{R}^k$. It is easy to see that any nondegenerate characteristic direction of (3.1) as well as the amplitude of any nondegenerate shock must lie on $\mathcal{K}$. Furthermore, if $\eta(U)$ is A-quasiconvex, then it must be convex in the direction of $\mathcal{K}$ [D]. All these conditions are physically motivated in the applications to continuum physics.

In [Da7] it is shown that when $\eta(U)$ is A-quasiconvex the analog

$$\|U(\cdot, t) - \overline{U}(\cdot, t)\|_{L^2(\mathbb{R}^k)} \le c(t) \|U_0(\cdot) - \overline{U}_0(\cdot)\|_{L^2(\mathbb{R}^k)} \,, \quad 0 < t < \infty \tag{5.14}$$

of the stability estimate (5.5) holds for any Lipschitz solution $\overline{U}(x, t)$ and any L^∞ solution $U(x, t)$ of (3.1), taking values in a small neighborhood of the origin and satisfying (5.2), provided that the corresponding initial data $\overline{U}_0(x)$ and $U_0(x)$ have compact support and satisfy the involution (5.11).

References

[BJ] J. M. Ball and R. D. James, *Fine phase mixtures as minimizers of energy*, Arch. Rational Mech. Anal. **100** (1987), 15–52.

[Bo] G. Boillat, *Involutions des systèmes conservatifs*, C.R. Acad. Sci. Paris **307** (1988), 891–894.

[B] P. Brenner, *The Cauchy problem for symmetric hyperbolic systems in L^p*, Math. Scand. **19** (1966), 27–37.

[Br] A. Bressan, *Global solutions to systems of conservation laws by wave-front tracking*, J. Math. Anal. Appl. **170** (1992), 414–432.

[BC] A. Bressan and R. M. Colombo, *The semigroup generated by 2×2 conservation laws*, to appear.

[CGS] J. Carr, M. E. Gurtin, and M. Slemrod, *Structured phase transitions on a finite interval*, Arch. Rational Mech. Anal. **86** (1984), 317–351.

[CH] T. Chang and L. Hsiao, The Riemann Problem and Interactions of Waves in Gas Dynamics. Longman, Essex, 1989.

[C1] G.-Q. Chen, *The compensated compactness method and the system of isentropic gas dynamics*, preprint MSRI-00527-91, Berkeley 1990.

[C2] G.-Q. Chen, *The method of quasidecoupling for discontinuous solutions to conservation laws*, Arch. Rational Mech. Anal. **121** (1992), 131–185.

[CK] G.-Q. Chen and P.T. Kan, *Hyperbolic conservation laws with umbilic degeneracy*, Arch. Rational Mech. Anal. **130** (1995), 231–276.

[D] B. Dacorogna, Weak Continuity and Weak Lower Semicontinuity of Nonlinear Functionals, Springer-Verlag, Berlin and New York, 1982.

[Da1] C. M. Dafermos, *Polygonal approximations of solutions of the initial value problem for a conservation law*, J. Math. Anal. Appl. **38** (1972), 33–41.

[Da2] C. M. Dafermos, *Quasilinear hyperbolic systems with involutions*, Arch. Rational Mech. Anal. **94** (1986), 373–389.

[Da3] C. M. Dafermos, *Admissible wave fans in nonlinear hyperbolic systems*, Arch. Rational Mech. Anal. **106** (1989), 243–260.

[Da4] C. M. Dafermos, *Generalized characteristics in hyperbolic systems of conservation laws*, Arch. Rational Mech. Anal. **107** (1989), 127–155.

[Da5] C. M. Dafermos, *Equivalence of referential and spatial field equations in continuum physics*, Notes Numer. Fluid Mech. **43** (1993), 179–183.

[Da6] C. M. Dafermos, *L^p Stability for systems of conservation laws in several space dimensions*, SIAM J. Math. Anal., to appear.

[Da7] C. M. Dafermos, *Entropy for hyperbolic systems of conservation laws in several space dimensions*, to appear.

[Di1] R. J. DiPerna, *Singularities of solutions of nonlinear hyperbolic systems of conservation laws*, Arch. Rational Mech. Anal. **60** (1975), 75–100.

[Di2] R. J. DiPerna, *Global existence of solutions to nonlinear hyperbolic systems of conservation laws*, J. Differential Equations **20** (1976), 187–212.

[Di3] R. J. DiPerna, *Uniqueness of solutions to hyperbolic conservation laws*, Indiana Univ. Math. J. **28** (1979), 137–188.

[Di4] R. J. DiPerna, *Convergence of approximate solutions to conservation laws*, Arch. Rational Mech. Anal. **82** (1983), 27–70.

[Di5] R. J. DiPerna, *Convergence of the viscosity method for isentropic gas dynamics*, Comm. Math. Phys. **91** (1983), 1–30.

[Di6] R. J. DiPerna, *Compensated compactness and general systems of conservation laws*, Trans. Amer. Math. Soc. **292** (1985), 383–420.

[EG] L. C. Evans and R. F. Gariepy, Measure Theory and Fine Properties of Functions, CRC Press, Ann Arbor, 1992.

[FL] K. O. Friedrichs and P. D. Lax, *Systems of conservation equations with a convex extension*, Proc. Nat. Acad. Sci. U.S.A. **68** (1971), 1686–1688.

[G] J. Glimm, *Solutions in the large for nonlinear hyperbolic systems of equations*, Comm. Pure Appl. Math. **18** (1965), 697–715.

[GL] J. Glimm and P. D. Lax, *Decay of solutions of systems of nonlinear hyperbolic conservation laws*, Mem. Amer. Math. Soc. **101** (1970).

[K] S. N. Kruzkov, *First order quasilinear equations in several independent variables*, Math. USSR-Sbornik **10** (1970), 217–243.

[L1] P. D. Lax, *Hyperbolic systems of conservation laws*, Comm. Pure Appl. Math. **10** (1957), 537–566.

[L2] P. D. Lax, *Shock waves and entropy*, Contributions to Functional Analysis (E. A. Zarantonelo, ed.), 603–634. Academic Press, New York, 1971.

[LX] P. LeFloch and Z.-P. Xin, *Uniqueness via the adjoint problems for systems of conservation laws*, Comm. Pure Appl. Math. **46** (1993), 1499–1533.

[LPS] P. L. Lions, B. Perthame, and P. Souganidis, *Compensated compactness applied to isentropic gas dynamics*, to appear.

[LPT] P. L. Lions, B. Perthame, and E. Tadmor, *A kinetic formulation of multidimensional scalar conservation laws and related equations*, J. Amer. Math. Soc. **7** (1994), 169–191.

[Li] T. P. Liu, *Admissible solutions of hyperbolic conservation laws*, Mem. Amer. Math. Soc. **240** (1981).

[M] A. Majda, Compressible Fluid Flow and Systems of Conservation Laws in Several Space Variables. Springer-Verlag, Berlin and New York, 1984.

[P] B. Perthame, *Kinetic equations and hyperbolic systems of conservation laws*, this volume.

[R] J. Rauch, *BV estimates fail for most quasilinear hyperbolic systems in dimensions greater than one*, Comm. Math. Phys. **106** (1986), 481–484.

[Ri] B. Riemann, *Über die Fortpflanzung ebener Luftwellen von endlicher Schwingungsweite*, Gött. Abh. Math. Cl. **8** (1860), 43–65.

[Ris] N. N. Risebro, *A front-tracking alternative to the random choice method*, Proc. Amer. Math. Soc. **117** (1993), 1125–1139.

[SS] D. Schaeffer and M. Shearer, *Riemann problems for nonstrictly hyperbolic 2×2 systems of conservation laws*, Trans. Amer. Math. Soc. **304** (1987), 267–306.

[S] D. Serre, *Oscillations non linéaires des systèmes hyperboliques: Methodes et résultats qualitatifs*, Analyse Non Linéaire **8** (1991), 351–417.

[Sm] J. Smoller, Shock Waves and Reaction-Diffusion Equations. Springer-Verlag, Berlin and New York, 1982.

[St] G. G. Stokes, *On a difficulty in the theory of sound*, Phil. Mag. **33** (1848), 349–356.

[T] L. Tartar, *Compensated compactness and applications to partial differential equations*, Nonlinear Analysis and Mechanics, vol. IV (R.J. Knops, ed.), 136–212, Pitman, London, 1979.

[TN] C. A. Truesdell and W. Noll, *The nonlinear field theories of mechanics*. Handbuch der Physik III/3. Springer-Verlag, Berlin and New York, 1965.

[Y] R. Young, *On elementary interactions for hyperbolic conservation laws*, SIAM Rev., to appear.

Eigenfunctions and Harmonic Functions in Convex and Concave Domains

DAVID JERISON*

Department of Mathematics
Massachusetts Institute of Technology
Cambridge, MA 02139, USA

Let Ω be a bounded, convex, open subset of $\mathbf{R}^N$. This paper concerns the behavior of positive harmonic functions that vanish on $\partial\Omega$. We will consider both the case in which the function is defined in Ω and the case in which the function is defined in the complement of Ω. We will also discuss eigenfunctions defined in Ω. The theme is to study how the shape of Ω influences the size of solutions to these basic elliptic equations.

The simplest measure of the shape of Ω is its eccentricity, which we define as

$$\operatorname{ecc} \Omega = \frac{\text{diameter}\,\Omega}{\text{inradius}\,\Omega}$$

Our estimates will fall into two categories, global and local estimates. The global estimates are those estimates on the size of solutions that are uniform as the eccentricity tends to infinity. The local estimates (Section 5) are those estimates that are valid when the eccentricity is bounded above. Local estimates are valuable only because they are valid uniformly up to the boundary. One main focus will be on the normal derivative of the solution at the boundary and its interaction with the Gauss curvature of the boundary.

1 The first nodal line

It is well known that the Dirichlet problem for Ω has eigenvalues $0 < \lambda_1 < \lambda_2 \leq \lambda_3 \leq \ldots$ Define u as the second eigenfunction

$$\Delta u = -\lambda_2 u \text{ in } \Omega \text{ and } u = 0 \text{ on } \partial\Omega$$

The set on which u vanishes is known as the first nodal set:

$$\Lambda = \{z \in \Omega : u(z) = 0\}$$

Λ divides Ω into two components $\Omega_+ = \{z \in \Omega : u(z) > 0\}$ and $\Omega_- = \{z \in \Omega : u(z) < 0\}$.

1991 *Mathematics Subject Classification.* Primary 35J25, 35B65; secondary 35J05.
Key words and phrases. Eigenfunctions, Green's function, convex domains, harmonic measure, capacity.
*) The author was partially supported by NSF grant DMS-9401355.

Proceedings of the International Congress
of Mathematicians, Zürich, Switzerland 1994
© Birkhäuser Verlag, Basel, Switzerland 1995

There are several results showing that the first nodal set meets the boundary. (See [P], [L], [J2], [J4], and especially [M].) These results motivate the more quantitative estimates of the first two eigenfunctions that follow.

THEOREM 1.1 [J5]. *There is an absolute constant C such that if Ω is a convex domain in $\mathbf{R}^2$, then*

$$\text{diameter } \Lambda \leq C \text{ inradius } \Omega$$

This theorem has no content unless the eccentricity is large. The first and most basic idea in the proof is that the shape of the eigenfunction is governed not by the eccentricity, but by a quantity that we call the "length scale" L associated to Ω. The number L is, by definition, the length of the rectangle R contained in Ω with lowest Dirichlet eigenvalue $\lambda_1(R)$. This rectangle can have a very different diameter from Ω. If Ω is a rectangle, then R is the same rectangle. But if Ω is a right triangle with legs of length 1 and N, with $N > 1$, then L is comparable to $N^{1/3}$. In general, $N^{1/3} \leq L \leq N$; the example of a trapezoid shows that any intermediate value is possible. It is not hard to prove, using differential inequalities of Carleman type, that the first and second eigenfunctions of Ω have a large percentage of their mass supported in R, with an exponential decay outside R. The methods of Theorem 1.1 give a precise location for the nodal line. In particular, the nodal line is near the middle of R, not Ω.

The second idea in the proof is a competition for energy between the two halves $\Omega_\pm$ of Ω. Rotate so that the projection of Ω onto the y-axis is the shortest of all projections, and define $I(x) = \{(x,y) : (x,y) \in \Omega\}$. The goal is to show that $I(x)$ is close to Λ for some x, so we consider the two domains obtained by cutting Ω in half at $I(x)$. When x is near the middle of the rectangle R, a shift of $I(x)$ to the left or the right by a unit distance increases/decreases the eigenvalues of the two halves by the same amount as in the rectangle R. Thus, Ω *resembles the rectangle R in the perturbation of its eigenvalues.* We will return to this idea in Section 4.

In addition to eigenvalue estimates, the proof of Theorem 1 uses several barrier arguments and the generalized maximum principle. Moreover, the proof provides a detailed approximation to the first and second eigenfunctions in terms of the solution to an ordinary differential equation that is naturally associated with the domain.

There is a unique segment $I = I(x^0)$ that divides Ω into two regions with the same first Dirichlet eigenvalue. With inradius $\Omega = 1$, Theorem 1 is equivalent to

THEOREM 1.1'. *There is an absolute constant C such that* dist $(\Lambda, I) \leq C$

Theorem 1.1' has recently been sharpened to:

THEOREM 1.2 [GJ]. dist $(\Lambda, I) \to 0$ *as* $N \to \infty$.

It seems plausible that the methods of Theorem 1.2 will ultimately lead to the estimate (with inradius $\Omega = 1$)

$$\text{dist } (\Lambda, I) \leq C(s + e^{-cL})$$

where s is the largest difference between slopes of the boundary curve at the top and bottom of Ω within a distance $\log L$ of Λ. This rate is best possible for triangles, rectangles, and their perturbations.

2 Green's function and harmonic measure

Green's function for Ω with pole at the origin is the function G satisfying

$$\Delta G = \delta \text{ in } \Omega \quad G = 0 \text{ on } \partial\Omega$$

Harmonic measure (with pole at 0) is the measure $d\omega$ on $\partial\Omega$ such that

$$u(0) = \int_{\partial\Omega} f \, d\omega$$

whenever u is continuous in $\bar{\Omega}$ and satisfies $\Delta u = 0$ in Ω and $u = f$ on $\partial\Omega$. Green's formula implies that

$$d\omega = (\partial G/\partial n) d\sigma$$

where $d\sigma$ is the surface measure on $\partial\Omega$.

The Schwarz-Christoffel formula [A] for the conformal mapping of the upper half-plane to a polygon is

$$\Phi'(z) = Ae^{i\theta} \prod_{k=1}^{m} (z - x_k)^{-\beta_k} \quad \Phi(i) = 0 \qquad (2.1)$$

where $\text{Im } z \geq 0$, $A > 0$, and θ, β_k, and x_k are real numbers. One of the difficulties with this formula is that it need not represent a mapping that is globally one-to-one. However, in the convex case, we assume that $\beta_k > 0$ and $\sum \beta_k = 2$ and (for convenience) x_k is an increasing sequence. Then Φ maps each segment $I_k = [x_{k-1}, x_k]$ onto a side $\Phi(I_k)$ of a convex polygon with exterior angle $\pi\beta_k$ at the vertex $\Phi(x_k)$, and the mapping is globally one-to-one from the half-plane to the polygon. (We use the convention that $I_1 = [x_m, \infty] \cup (-\infty, x_1]$.) The factors A and $e^{i\theta}$ correspond to dilation and rotation. The choice of $\Phi(i) = 0$ corresponds to a translation.

A much more serious shortcoming of this formula as a tool in conformal mapping is that the given data are not the locations of the vertices of the polygon. Rather, they are the angles of the polygon and the numbers x_k, which are points on the real axis. As we shall see, this formula solves a kind of inverse problem for harmonic measure. Notice that the sequence of values x_k contains the same information as the sequence of values

$$\frac{1}{\pi} \int_{I_k} \frac{dx}{1 + x^2} = \omega(\Phi(I_k))$$

The left-hand side is the Poisson integral or harmonic measure of I_k in the upper half-plane, with pole at i. But because the conformal mapping Φ preserves harmonic measure, this equals the harmonic measure $\omega(\Phi(I_k))$. Now the problem solved can be rephrased as: *Given the normals to the sides of a polygon and the harmonic measure of each side, find the polygon.*

To generalize this inverse problem consider the Gauss mapping $g : \partial\Omega \to S^n$, $n = N - 1$, which takes a point $x \in \partial\Omega$ to the outer unit normal at x. (This mapping is defined almost everywhere with respect to the surface measure $d\sigma$. Hence, it is defined almost everywhere with respect to harmonic measure.) We define the measure $d\mu = g_*(d\omega)$ on the sphere by

$$\mu(E) = \omega(g^{-1}(E)) \text{ for any Borel subset } E \subset S^n \qquad (2.2)$$

THEOREM 2.3 [J3], [J1]. *If $d\mu$ is a positive measure on S^n of total mass 1, then there exists a convex domain Ω containing the origin such that $g_*(d\omega) = d\mu$.*

Note that there is no requirement that Ω be bounded. For example, if $d\mu$ is a delta mass at $e \in S^n$, then Ω is a half-space perpendicular to e. There is a direct relationship between the size of $(\partial G/\partial n)$ and the rate at which G vanishes, that is, its size at the boundary. Thus this inverse problem is saying something about what rates are permissible. However, as we will see in Section 5, this theorem says much less about $\partial G/\partial n$ by itself than about the relationship between $\partial G/\partial n$ and Gauss curvature.

3 The Minkowski problem and variants

The inverse problem for harmonic measure resembles the classical Minkowski problem, which asks when one can construct a convex polyhedron given the normals to the faces and their areas. The generalized problem is to solve

$$g_*(d\sigma) = d\nu \tag{3.1}$$

In particular, if $d\nu = (1/K)d\xi$, where $d\xi$ is the uniform measure on the sphere, then K is the Gauss curvature of $\partial\Omega$.

However, (3.1) has very different global compatibility conditions than those of (2.2). The projection body function of Ω is the function defined for $e \in S^n$ by

$$P(e) = P_\Omega^{\mathrm{vol}}(e) = \int_{S^n} (e \cdot \xi)_+ \, d\nu(\xi) \tag{3.2}$$

This is the n-volume of the projection of Ω onto the hyperplane perpendicular to e. It is easy to see that $P(e) = P(-e)$, and this can be written

$$\int_{S^n} (e \cdot \xi) \, d\nu(\xi) = 0 \text{ for all } e \in S^n \tag{3.3}$$

In other words, we have N linear necessary conditions for the existence of Ω. We will assume that Ω is bounded, so that the total surface area is finite. It follows that

$$\min P > 0 \text{ and } \max P < \infty \tag{3.4}$$

To explain how to find Ω solving (3.1), we consider the Minkowski support function

$$h_\Omega(\xi) = \sup_{x \in \Omega} x \cdot \xi \tag{3.5}$$

The Minkowski sum is

$$A + B = \{a + b : a \in A \quad b \in B\}$$

With ν defined by (3.1), we have

$$\mathrm{vol}\,\Omega = \frac{1}{N} \int_{S^n} h_\Omega(\xi) \, d\nu(\xi), \tag{3.6}$$

$$\frac{d}{dt}\mathrm{vol}\,(\Omega + tA)\big|_{t=0} = \int_{S^n} h_A(\xi) \, d\nu(x) \tag{3.7}$$

These formulas underlie the following theorem of Minkowski.

THEOREM 3.8. *If ν is a positive measure on S^n satisfying (3.3) and (3.4), then the minimizers of*

$$\inf\{\int_{S^n} h_\Omega \, d\nu : \mathrm{vol}\,\Omega \geq 1\}$$

satisfy $g_*(d\sigma) = cd\nu$ where c is a suitable constant.

The constant c is a Lagrange multiplier. The minimizing domains are unique up to translation and have volume 1. One can dilate to solve (3.1).

The necessary conditions (3.3) and (3.4) are quite different from the total mass 1 constraint for harmonic measure. Moreover, there is no variational formula for the harmonic measure problem. However, there are two analogous variational problems involving canonical functions on the interior and exterior of Ω.

Let u_1 denote the first Dirichlet eigenfunction of Ω, normalized to have $L^2(\Omega)$ norm 1. For $N \geq 3$, define the equilibrium potential U for Ω as the function satisfying

$$\Delta U = 0 \text{ in } \mathbf{R}^N \backslash \Omega, \quad U = 1 \text{ on } \partial\Omega, \quad \text{and} \quad U(x) \to 0 \text{ as } x \to \infty$$

It is well known that

$$U(x) = \gamma A_N |x|^{2-N} + O(|x|^{1-N}) \text{ as } x \to \infty$$

where A_N is chosen so that $A_N|x|^{2-N}$ is the fundamental solution and γ is the capacity of Ω, $\mathrm{cap}\,\Omega$. In particular, $U - 1$ is a multiple of Green's function for the complement of Ω with pole at infinity.

Let ν_1 and ν_2 be positive measures satisfying the same constraints as ν. If in Theorem 3.8, $\mathrm{vol}\,\Omega$ is replaced by $\lambda_1(\Omega)$, and ν by ν_1, there exists a minimizer solving

$$g_*((\partial u_1/\partial n)^2 d\sigma) = cd\nu_1 \tag{3.9}$$

If in Theorem 3.8 $\mathrm{vol}\,\Omega$ is replaced by $\mathrm{cap}\,(\Omega)$ and ν by ν_2, the minimizer solves

$$g_*((\partial U/\partial n)^2 d\sigma) = cd\nu_2 \tag{3.10}$$

The formulas analogous to (3.6) and (3.7) are classical variational formulas due to Poincaré and Hadamard [Po], [GS], [PS]:

$$\lambda_1(\Omega) = \frac{1}{2}\int_{S^n} h_\Omega(\xi) \, d\nu_1(\xi) \text{ where } d\nu_1 = g_*((\partial u_1/\partial n)^2 d\sigma) \tag{3.11}$$

$$\mathrm{cap}\,\Omega = \frac{1}{N-2}\int_{S^n} h_\Omega(\xi) \, d\nu_2(x) \text{ where } d\nu_2 = g_*((\partial U/\partial n)^2 d\sigma) \tag{3.12}$$

$$\frac{d}{dt}\lambda_1(\Omega + tA)\big|_{t=0} = -\int_{\partial\Omega} h_A(\xi) \, d\nu_1(\xi) \tag{3.13}$$

$$\frac{d}{dt}\mathrm{cap}\,(\Omega + tA)\big|_{t=0} = \int_{\partial\Omega} h_A(\xi) \, d\nu_2(\xi) \tag{3.14}$$

4 Bounds on eccentricity

The fundamental *a priori* bound in the Minkowski problem is

$$\operatorname{ecc} \Omega \leq C_N \max_{S^n} P_\Omega^{\mathrm{vol}} / \min_{S^n} P_\Omega^{\mathrm{vol}} \tag{4.1}$$

This is the starting point of the regularity theory of [CY]. Similarly, consider

$$P_\Omega^{\mathrm{eig}}(e) = \int_{S^n} (e \cdot \xi)_+ \, d\nu_1(\xi) \tag{4.2}$$

$$P_\Omega^{\mathrm{cap}}(e) = \int_{S^n} (e \cdot \xi)_+ \, d\nu_2(\xi) \tag{4.3}$$

A key estimate for the analogous theory of the first eigenvalue is that the ratio $\max P_\Omega^{\mathrm{eig}} / \min P_\Omega^{\mathrm{eig}}$ controls $\operatorname{ecc} \Omega$; this is also true for the capacitary projection body function [J6].

To motivate some conjectures, we present an easy proof of (4.1) based on an important lemma of John.

LEMMA 4.4. *There is a dimensional constant C_N such that for every open, bounded, convex domain Ω there is an ellipsoid $\mathcal{E}$ such that*

$$\mathcal{E} \subset \Omega \subset C_N \mathcal{E}$$

where multiplication by C_N represents a dilation from the center of mass of $\mathcal{E}$.

A proof of Lemma 4.4 due to Cordoba and Gallegos can be found in [dG, p. 133].

Suppose that the ellipsoid has semiaxes $e_1, \ldots, e_N$ of lengths $a_1 \leq a_2 \leq \cdots \leq a_N$. Then Lemma 4.4 implies

$$P(e_k) \approx a_1 a_2 \cdots \hat{a}_k \cdots a_N$$

(Here $\approx$ means comparable up to a dimensional constant factor.) Therefore,

$$\frac{\max P}{\min P} \approx \frac{a_2 \cdots a_N}{a_1 \cdots a_{N-1}} = \frac{a_N}{a_1} \approx \operatorname{ecc} \Omega$$

Lemma 4.4 implies

$$P_\Omega^{\mathrm{vol}} \approx P_\mathcal{E}^{\mathrm{vol}}$$

This suggests the conjecture

$$P_\Omega^{\mathrm{cap}} \approx P_\mathcal{E}^{\mathrm{cap}} \tag{4.5}$$

The order of magnitude of P^{vol} gives exactly enough information to recover the order of magnitude of the quadratic form defining $\mathcal{E}$. It might be worthwhile to compare P^{cap} to the virtual mass and polarization associated to Ω [SS].

CONJECTURE 4.6.
$$P^{\text{eig}}_\Omega \approx P^{\text{eig}}_R$$
where R is the parallelepiped of lowest first eigenvalue contained in Ω.

This conjecture is true in $\mathbf{R}^2$. But little is known in higher dimensions [J4].

There is a direct relationship between Conjecture 4.6 and Section 1. Let S denote the unit segment with endpoints 0 and e. Then

$$P^{\text{eig}}(e) = -\frac{d}{dt}\lambda_1(\Omega + tS)\big|_{t=0}$$

Thus, the projection body function measures the rate of change of the eigenvalue with respect to a lengthening of Ω in the direction of e. For an $(L \times 1)$-rectangle R,

$$P^{\text{eig}}_R(e_1) \approx 1, \quad P^{\text{eig}}_R(e_2) \approx 1/L^3 \tag{4.7}$$

If R is the optimal rectangle of Ω, then (4.7) also holds with Ω in place of R. This implies Conjecture 4.6 for $\mathbf{R}^2$. The second assertion in (4.7), and the only nontrivial one, is an infinitesimal variant of the property mentioned in Section 1 on the effect of moving $I(x)$ by a unit distance.

5 Regularity and the Monge-Ampère equation

Let ϕ be the convex function for which the graph $(x, \phi(x))$ corresponds locally to $\partial\Omega$. The equation

$$g_*(d\sigma) = d\nu = (1/K)d\xi \tag{5.1}$$

can be written locally as a Monge-Ampère equation

$$\det \nabla^2\phi(x) = F(\nabla\phi(x)) \tag{5.1'}$$

for a suitable function F depending on K. Similarly, the equations

$$g_*((\partial G/\partial n)d\sigma) = d\mu = \rho d\xi \tag{5.2}$$
$$g_*((\partial u_1/\partial n)^2 d\sigma) = d\nu_1 = \rho d\xi \tag{5.3}$$
$$g_*((\partial U/\partial n)^2 d\sigma) = d\nu_2 = \rho d\xi \tag{5.4}$$

can be written in the form

$$\det \nabla^2\phi(x) = F(\nabla\phi(x))(\partial G/\partial n)(x, \phi(x)) \tag{5.2'}$$
$$\det \nabla^2\phi(x) = F(\nabla\phi(x))(\partial u_1/\partial n)^2(x, \phi(x)) \tag{5.3'}$$
$$\det \nabla^2\phi(x) = F(\nabla\phi(x))(\partial U/\partial n)^2(x, \phi(x)) \tag{5.4'}$$

The regularity theorem for Gauss curvature is due to Nirenberg, Pogorelov, Calabi, Cheng, and Yau.

THEOREM 5.5 [CY]. *If $K > 0$, $K \in C^\infty(S^n)$, then every Ω that solves (5.1) is C^∞.*

THEOREM 5.6 [J6], [J3], [J1]. *If $\rho > 0$, $\rho \in C^\infty(S^n)$, then every Ω that solves (5.2), (5.3), or (5.4) is C^∞.*

The extra factors present serious new difficulties over equation (5.1). In equations (5.2) and (5.3) the normal derivative can vanish. In the case of (5.4), the normal derivative can be infinite. But recently Caffarelli [C1], [C2], [C3], [C4] has simplified and sharpened the regularity theory of the Monge-Ampère equation by developing a notion of scaling that takes care of problems like the ones above. The scaling comes from consideration of the convex sets

$$F = \{x \in \mathbf{R}^n : \phi(x) < a \cdot x + b = \ell(x)\} \tag{5.7}$$

The connection with the Monge-Ampère equation is as follows. First, the function $\phi(x) - \ell(x)$ has the same Hessian as ϕ. Second, this new function has the boundary value 0 on ∂F. Third, a linear transformation of the x variable just changes the right-hand side of the Monge-Ampère equation by a constant multiplicative factor. The point is that even though the set F in $\mathbf{R}^n$ can have completely uncontrolled eccentricity, the John lemma in $\mathbf{R}^n$ implies that one can always make a linear change of variable to transform F into a renormalized set F' with diameter and inradius both comparable to 1. Finally, the shape of these level sets F is closely related to the size of the first and second derivatives of ϕ.

Caffarelli's theory says that to obtain an *a priori* estimate, one must discover *scale-invariant* estimates on the right-hand side. The set (5.7) corresponds to a set on $\partial\Omega$ of the form

$$E = \partial\Omega \cap H \text{ where } H \text{ is any half-space} \tag{5.8}$$

We will call these sets slices. The key estimate for (5.2) is that if the ecc Ω is bounded above, then for all slices E,

$$\max_E(\partial G/\partial n) \leq C \frac{1}{\sigma(E)} \int_{\frac{1}{2}E} (\partial G/\partial n) \, d\sigma \tag{5.9}$$

where $(1/2)E$ is defined as the graph over $(1/2)F$, and the dilation of F uses as its origin the center of mass of F. In particular, this implies a scale-invariant doubling condition on slices, $\omega(E) \leq C\omega(\frac{1}{2}E)$. The doubling condition is familiar in the context of Lipschitz domains [HW], [JK] for balls. The novelty here is that the slices do not have uniformly bounded eccentricity. (Here we equate the eccentricity of E with the eccentricity in $\mathbf{R}^n$ of F.) The uniformity is that C depends only on the eccentricity of Ω in $\mathbf{R}^N$, not E. In a general convex domain, the eccentricity of a slice is not under control. For example, consider the case of a slice parallel to an edge of a polyhedron.

The eigenfuncion u_1 vanishes at the same rate as G at the boundary, so (5.9) also implies a doubling property for the measure $(\partial u_1/\partial n)^2 d\sigma$. (The second power is no problem because we even have control of the maximum.) The exterior problem is different. The doubling condition is false. But a weaker condition holds, namely, there exists $\epsilon > 0$ and C such that for all slices E,

$$\int_E \delta^{1-\epsilon}(\partial U/\partial n)^2 \, d\sigma \leq C \int_{(1/2)E} (\partial U/\partial n)^2 \, d\sigma \tag{5.10}$$

where δ is a scale-invariant distance to the boundary of E. (More precisely, δ is the distance to the boundary in the renormalized set F'.) This is the weakest estimate of its kind that still yields to the scale-invariant Caffarelli technique.

Estimates (5.9) and (5.10) are optimal local estimates for the rate of vanishing of G, u_1, and $1 - U$ at the boundary, and they reflect the relationship with Gauss curvature. For a relationship between G and mean curvature and scalar curvature, see the optimal estimates on third derivatives of G in [FJ].

6 Uniqueness

The solution to the harmonic measure problem (Theorem 2.3) is almost certainly unique up to dilation. This can be proved using (2.1) in dimension 2, and in the case of smooth data there is a proof in [J3]. McMullen and Thurston [MT] have given a maximum principle argument for uniqueness up to dilation that applies at least to polyhedra and most smooth domains.

Consider convex combinations of two convex bodies

$$\Omega_t = (1 - t)\Omega_0 + t\Omega_1 \quad 0 \le t \le 1$$

The Brunn-Minkowski inequality says that

$$\text{vol}\,(\Omega_t)^{1/N}$$

is a concave function of t. It is linear if and only if Ω_0 and Ω_1 are a translate and dilate of each other. This uniqueness implies uniqueness up to translation in the Minkowski problem (3.1). The analogous property that

$$\text{cap}\,(\Omega_t)^{1/(N-2)}$$

is a concave function of t was proved by Borell [B]. The fact that this function is linear if and only if Ω_0 is a translate and dilate of Ω_1 is proved in [CJL]. This implies uniqueness up to translation in the capacity problem (5.4) when $N \ge 4$. Note, however that equation (5.4) has homogeneity of degree $N - 3$. When $N = 3$, (5.4) is dilation invariant; there is only one multiple of $d\nu_2$ for which there is a solution, and that solution is unique up to translation and dilation.

Finally, the analogue of the Brunn-Minkowski inequality for eigenvalues follows from [BL], [BL1]. However, the case of equality has not yet been analyzed.

References

[A]   L. Ahlfors, Complex Analysis, 2nd ed., McGraw-Hill, New York, 1966.
[B]   C. Borell, *Capacitary inequalities of the Brunn-Minkowski type*, Math. Ann. **263** (1983), 179–184.
[BL]   H. J. Brascamp and E. H. Lieb, *Some inequalities for Gaussian measures and the long-range order of the one-dimensional plasma*, Functional Integration and Its Applications (A.M. Arthurs, ed.), Clarendon Press, Oxford, 1975.

[BL1] _____, *On extensions of the Brunn-Minkowski and Prékopa Leindler theorems, including inequalities for log concave functions, and with an application to the diffusion equation*, J. Funct. Anal. **22** (1976), 366–389.

[C1] L. A. Caffarelli, *Interior a priori estimates for solutions of fully non-linear equations*, Ann. of Math. (2) **131** (1989), 189–213.

[C2] _____, *A localization property of viscosity solutions to the Monge-Ampère equation and their strict convexity*, Ann. of Math. (2) **131** (1990), 129–134.

[C3] _____, *Interior $W^{2,p}$ estimates for solutions of the Monge-Ampère equation*, Ann. of Math. (2) **131** (1990), 135–150.

[C4] _____, *Some regularity properties of solutions to the Monge-Ampère equation*, Comm. Pure Appl. Math. **44** (1991), 965–969.

[CJL] L. A. Caffarelli, D. Jerison, and E. Lieb, *On the case of equality in the Brunn-Minkowski inequality for capacity*, in preparation.

[CY] S.-Y. Cheng and S.-T. Yau, *On the regularity of the solution of the n-dimensional Minkowski problem*, Comm. Pure Appl. Math. **29** (1976), 495–516.

[dG] M. deGuzman, *Differentiation of integrals in $\mathbf{R}^n$*, (Lecture Notes in Math. vol. **481**), Springer-Verlag, Berlin, Heidelberg, and New York, 1975.

[FJ] S. J. Fromm and D. Jerison, *Third derivative estimates for Dirichlet's problem in convex domains*, Duke Math. J. **73** (1994), 257–268.

[GS] P. R. Garabedian and M. Schiffer, *Convexity of domain functionals*, J. Analyse Math. **2** (1953), 281–368.

[GJ] D. Grieser and D. Jerison, work in progress.

[HW] R. R. Hunt and R. L. Wheeden, *On the boundary values of harmonic functions*, Trans. Amer. Math. Soc. **132** (1968), 307–322.

[J1] D. Jerison, *Harmonic measure in convex domains*, Bulletin AMS **21** (1989), 255–260.

[J2] _____, *The first nodal line of a convex planar domain*, Internat. Math. Research Notes **1** (1991), 1–5 (in Duke Math. J. vol. 62).

[J3] _____, *Prescribing harmonic measure on convex domains*, Invent. Math. **105** (1991), 375–400.

[J4] _____, *The first nodal set of a convex domain*, Essays in Fourier Analysis in honor of E.M. Stein (C.F. Fefferman, ed.), Princeton Univ. Press, Princeton, NJ, 1993.

[J5] _____, *The diameter of the first nodal line of a convex planar domain*, Ann. of Math. (2), to appear.

[J6] _____, *The Minkowski problem for electrostatic capacity*, preprint.

[JK] D. Jerison and C. E. Kenig, *Boundary value problems in Lipschitz domains*, Studies in P.D.E., MAA Stud. Math. (W. Littman, ed.), vol. 23, 1982, pp. 1–68.

[L] C. S. Lin, *On the second eigenfunction of the Laplacian in $\mathbf{R}^2$*, Comm. Math. Phys. **111** (1987), 161–166.

[MT] C. McMullen and W. Thurston, personal communication.

[M] A. Melas, *On the nodal line of the second eigenfunction of the Laplacian in $\mathbf{R}^2$*, J. Differential Geom. **35** (1992), 255–263.

[P] L. E. Payne, *On two conjectures in the fixed membrane eigenvalue problem*, J. Appl. Math. Phys. (ZAMP) **24** (1973), 721–729.

[Po] H. Poincaré, Figures d'équilibre d'une masse fluide, Paris, 1902.

[PS] G. Pólya and M. Schiffer, *Convexity of functionals by transplantation*, J. Analyse Math. (1953), 245–345.

[SS] M. Schiffer and G. Szegö, *Virtual mass and polarization*, Trans. Amer. Math. Soc. **67** (1949), 130–205.

Kinetic Equations and Hyperbolic Systems of Conservation Laws

Benoît Perthame

Laboratoire d'Analyse Numérique
Université Pierre et Marie Curie
and
Institut Universitaire de France
4 place Jussieu
F-75252 Paris Cedex 05, France

ABSTRACT. We present several recent results concerning the global existence of solutions to kinetic equations, which are, generally, semi-linear or quasi-linear hyperbolic partial differential equations of transport type. The most famous of them are certainly the Boltzmann or Vlasov-Poisson equations. We describe more precisely some general tools that can be used for their analysis: compactness results and dispersive effects. Then we give a new point of view on their fluid limits. This allows us to recover some nonlinear hyperbolic systems of conservation laws by a singular perturbation according to the "mean free path".

I Introduction

Kinetic equations are, usually nonlinear, partial differential equations based on a first order transport operator. The most famous models are neutrons transport, Boltzmann equation for gases with binary collisions, and Vlasov-Poisson and Vlasov-Maxwell equations for self-interacting plasmas. But many other equations are of interest for physical kinetics, statistical and fluid mechanics, biology, metallurgy, or semi-conductors modelling. The common main structure to these kinetic models is that of a simple transport operator describing the evolution of a microscopic density $f(t, x, \xi)$ of particles

$$\frac{\partial}{\partial t} f(t, x, \xi) + \xi \cdot \nabla_x f(t, x, \xi) = Q(t, x, \xi). \tag{I.1}$$

Here $t \geq 0$ represents time, and $x \in \mathbb{R}^n$ represents the position of particles with velocity $\xi \in \mathbb{R}^n$. The notation Q will denote a given function or several possible operators acting on f and depending upon the physics of the model. We will give other examples below, but let us present first the Vlasov-Poisson model. It consists in taking, in (I.1)

$$Q = -E(t, x) \cdot \nabla_\xi f(t, x, \xi), \tag{I.2}$$

$$E(t, x) = \pm \nabla_x U(t, x), \quad \varepsilon_0 \Delta U = \rho(t, x) - \rho_0, \tag{I.3}$$

Proceedings of the International Congress
of Mathematicians, Zürich, Switzerland 1994
© Birkhäuser Verlag, Basel, Switzerland 1995

where $\varepsilon_0 > 0$ is the given dielectric constant, ρ_0 is a given nonnegative constant (an electric charge in the applications), and

$$\rho(t, x) = \int_{\mathbb{R}^n} f(t, x, \xi) \, d\xi \qquad (\text{I.4})$$

is the macroscopic charge induced by the density f itself. The sign $\pm$ in equation (I.3) depends on the model, charged plasma as here, or gravity forces (in astrophysics for instance).

Recently, many progresses have been made in the mathematical understanding of these models. We describe some of them in Section II, as well as several general mathematical tools that can be used to analyze the transport equations. Our first example is the *averaging compactness*, which shows that averages in ξ of the solution to equation (I.1) win some regularity in t and x. We also present some *dispersive estimates* that provide decay in the variables ξ and t.

Other classical problems concerning kinetic equations range in singular limits. Rescaling the equations according to a "mean free path", it is possible to recover the so-called fluid or macroscopic limits. This is the case for example of the systems of Euler or Navier-Stokes equations for incompressible or compressible flows that can be derived from the Boltzmann equation. In the last section of this paper, we will describe recent results in that direction. We will show that the relation with nonlinear hyperbolic systems of conservation laws is broader than expected; several systems can actually be written *exactly* as kinetic equations.

There are many other mathematical problems related to kinetic equations that we cannot present here. Let us mention some examples, restricting ourselves to recent results using partial differential equations methods: the derivation of kinetic models from quantum mechanics (Schrödinger equations) see Lions and Paul [18], Lions and Perthame [20]; or from statistical physics using PDE methods: Pulvirenti *et al.* [30], Perthame and Pulvirenti [27], numerical methods, inverse and scattering problems, models for specific applications (Markowitch *et al.* [24] or [26] for instance), and qualitative properties such as large time asymptotics for which many progresses have been realized recently.

II Analysis of kinetic equations

The analysis of nonlinear kinetic equations relies on several tools: derivation of a priori estimates, functional analysis, and specific manipulations for each model. Here, we will restrict our presentation to some very general estimates and compactness results. But first, we would like to begin with some recent examples of global existence results for large initial data. This means that we completely skip the questions of the existence of global small solutions with uniqueness. A review for this can be found in [13].

Concerning global existence of weak solutions to kinetic equations, let us first mention the result of Di Perna and Lions proving the existence of *renormalized* solutions to the Cauchy problem for the Boltzmann equation for general initial datum. The Boltzmann model corresponds, in the general equation (I.1), to an operator

$$Q = \int_{\mathbb{R}^n \times S^{n-1}} [f(\xi')f(\xi_\star') - f(\xi)f(\xi_\star)]B(|\xi - \xi_\star|, |(\xi - \xi_\star) \cdot w|) \, d\xi_\star \, dw. \quad \text{(II.1)}$$

In (II.1) we have skipped, for the ease of notation, the dependency of f upon t and x. The function $B(\cdot, \cdot)$ is a given positive kernel. Finally, the post-collisional velocities $(\xi', \xi_\star')$ are obtained from the pre-collisional velocities $(\xi, \xi_\star)$ by the relations

$$\xi' = \xi - (\xi - \xi_\star, w)w, \qquad \xi_\star' = \xi_\star + (\xi - \xi_\star, w)w, \quad \text{(II.2)}$$

which are merely a parametrization, by the unit vector w, of the manifold of velocities satisfying the conservation of momentum and energy:

$$\xi + \xi_\star = \xi' - \xi_\star', \; |\xi|^2 + |\xi_\star|^2 = |\xi'|^2 - |\xi_\star'|^2.$$

The difficulty in solving this model is that the only known bounds for solutions of (I.1), (II.2) are

$$\int_{\mathbb{R}^{2N}} f(t, x, \xi)(|x|^2 + |\xi^2| + |\ln f|) \, dx \, d\xi \leq C(T) \text{ for all } t \leq T, \quad \text{(II.3)}$$

whenever the initial data $f(t = 0, x, \xi)$ satisfies them.

Because f is a priori nonnegative this estimate is merely an L^1 estimate, which is not enough to give a meaning to the quadratic operator Q in (II.1). The idea of renormalization is simply to transform (I.1) into

$$\frac{\partial \beta(f)}{\partial t} + \xi \cdot \nabla_x \beta(f) = \beta'(f)Q. \quad \text{(II.4)}$$

Of course this equation is equivalent to (II.1) for smooth solutions and Q, but for the Boltzmann equation it is not. The advantage of the renormalized equation is that it is now meaningful in $L^1(\mathbb{R}^{2N}_{x,\xi})$ whenever the function $\beta'(f)$ decays at least as $\frac{1}{f}$ at infinity. Then, Di Perna and Lions, in [11], show that, indeed, $\beta'(f)Q$ is well defined in L^1, and they overcome the difficulty to build a solution (this requires deep functional analysis, a simpler and more general version being presented in Lions [17]).

Among the other models for which global existence of weak solutions has been obtained recently, let us mention the Vlasov-Maxwell equations by Di Perna and Lions [12], and the Bhatnagar-Gross and Kruck (BGK) model by the author [25]. Strong solutions, with uniqueness and regularity, could also be achieved in a number of cases: three-dimensional Vlasov-Poisson equations by Pfaffelmoser [29], Batt and Rein [4], Lions and Perthame [19], the Vlasov-Poisson-Fokker-Planck equation by Bouchut [5], and the BGK model by Pulvirenti and the author [27]. All these results are based on precise estimates related to the special structure of the model. We do not wish to give details on these results but on some general lemmas that are often used in the above papers; they explain that, despite the solutions to the transport equation (I.1) can be handled explicitly by the method of characteristics, they satisfy some very strong properties that cannot be seen through this method.

LEMMA 1. *The solutions to the transport equation* (I.1) *satisfy, for any test function* $\varphi \in \mathcal{D}(\mathbb{R}^n)$,

(i) $\quad \| \int f(t,x,\xi)\varphi(\xi) \, d\xi \|_{H^{1/2}(\mathbb{R}^{n+1})} \le C(n, \operatorname{supp}\varphi) \|f\|_{L^2(\mathbb{R}^{n+1})}^{1/2} \|Q\|_{L^2(\mathbb{R}^{2n+1})}^{1/2},$

(ii) $\quad$ *if* $\quad Q = (I - \Delta_\xi)^{m/2} g$, $g \in L^p(\mathbb{R}^{2n} \times \mathbb{R})$, *and* $f \in L^p(\mathbb{R}^{2n} \times \mathbb{R})$, *then*

$$\int_{\mathbb{R}^n} f(t,x,\xi)\varphi(\xi) \, d\xi \in B_q^{s,p}(\mathbb{R}^n \times \mathbb{R})$$

with $s = \frac{1}{(m+1)\max(p,p')}$, $q = \max(p,2)$.

Lemma 1 is known as an "averaging lemma". It explains how the hyperbolic equation (I.1), which propagates singularities in the phase space (x,ξ), can also regularize the data (Q here, but it also regularizes the initial data). The first result in that direction was given by Golse *et al.* [15] and extended in [14] to the case (i). The case (ii) was successively proved in Di Perna, Lions, and Meyer in [11] and [12]. It is a keystone in proving the existence of weak solutions in [11], [10] or the asymptotic limit of the radiative transfer equations (a nonlinear version of the neutrons transport equations) in Bardos *et al.* [3].

The second general result we wish to mention is due to the author [25], and a more direct proof is given in [20].

LEMMA 2. *Let f satisfy* (I.1) *and denote* $f(t=0,x,\xi) = f^o(x,\xi)$. *Then*

$$\int_{-\infty}^{+\infty} \int_{\mathbb{R}^{2n}} \frac{|\xi|}{1 + |x|^{1+\alpha}} |f(t,x,\xi)| \, dx \, d\xi \, dt \le C(N,\alpha) \int_{\mathbb{R}^{2n}} |f^o(x,\xi)| \, dx \, d\xi. \quad \text{(II.5)}$$

Although Lemma 1 shows that extra-integrability in the Fourier variable is won, Lemma 2 yields extra-integrability in the velocity variable ξ (the price to pay is a localization in the variable x). It is also possible to prove a strong decay in the time variable.

LEMMA 3. *With the notations of Lemma 1 and*

$$\rho(t,x) = \int_{\mathbb{R}^n} f(t,x,\xi) \, d\xi \quad \text{(II.6)}$$

we have

$$|\rho(t,x)| \le \frac{1}{t^n} \int_{\mathbb{R}^n} \sup_{\xi \in \mathbb{R}^n} |f_o(x,\xi)| \, dx.$$

This result was widely used in the context of the Vlasov-Poisson equation (and also for related models) by Bardos and Degond [1].

III Fluid equations

The kinetic equations can be parametrized, thus raising singular perturbation problems. The most classical one (still unresolved) is the compressible gas dynamics limits of the Boltzmann equation, which can be formally obtained through a Hilbert or Chapman-Enskog expansion (see Cercignani's book [7] for instance).

Several singular limits of kinetic equations could be justified recently. Let us mention for instance, still limiting our examples to global results, the porous medium limit of radiative transfer equations by Bardos *et al.* [3] and the different incompressible limits of the Boltzmann equations by De Masi, Esposito, and Lebowitz [9], and Bardos, Golse, and Levermore [2], which can be obtained through a very interesting scaling of both the initial data and the equation. Concerning the compressible limit of the Boltzmann equation, we would like to mention also the recent extension of the Chapman-Enskog and Grad hierarchies developed in Levermore [16]. The Vlasov-Poisson equation can also be scaled. A first example is through the dielectric constant in the Poisson equation, so as to generate singular limits, as in Brenier and Grenier [6] who relate their results to incompressible gas dynamics. A second example is the "variational inequality" obtained by Degond and Raviart [8] (see also the paper by Degond in [26]) to describe special monokinetic boundary conditions for the Vlasov-Poisson equation.

In this section, we could like to concentrate on a new point of view on the relations between hyperbolic systems of nonlinear conservation laws and kinetic equations. It explains that the two formalisms are closer than expected and some conservation laws can be formulated exactly as kinetic equations. Here we give two examples of kinetic formulations: the scalar case and the system of isentropic gas dynamics.

The first example of kinetic formulation is given by Lions, Perthame, and Tadmor [22] and concerns scalar conservation laws,

$$\partial_t u + \sum_{i=1}^{n} \partial_{x_i} A_i(u) = 0, \ t \geq 0, x \in \mathbb{R}^n. \tag{III.1}$$

Here the given functions, called fluxes $A_i(.)$ are smooth. The main difficulty with nonlinear hyperbolic equations such as (III.1) is that, even for smooth initial data, the solution cannot remain smooth and discontinuities, called shocks, appear in finite time. Associated to these discontinuities, uniqueness is lost and it is well known, from the theories of Lax and Kruzkov for instance, that entropy inequalities have to be added in order to describe the right physical discontinuities of (III.1). These entropy inequalities are

$$\partial_t S(u) + \sum_{i=1}^{n} \partial_{x_i} H_i(u) \leq 0 \quad \text{in } \mathcal{D}'(\mathbb{R}_+ \times \mathbb{R}^n), \tag{III.2}$$

for any convex "entropy" $S(.)$, and denoting

$$a_i(.) = A_i'(.), \quad H_i'(.) = S'(.)a_i(.). \tag{III.3}$$

Of course, (III.2) stands as an equality for smooth solutions of the scalar conservation law (III.1). In [22], it is proved that $u(t,x)$ satisfies the equations (III.1), (III.2) if and only if

$$\partial_t \chi(u,\xi) + a(\xi) \cdot \partial_x \chi(u,\xi) = \partial_\xi m, \tag{III.4}$$

for some nonpositive, bounded measure $m(t,x,\xi)$ and

$$\chi(u,\xi) = \begin{cases} +1, & \text{if } \leq \xi \leq u, \\ -1, & \text{if } u \leq \xi \leq 0, \\ 0, & \text{otherwise.} \end{cases} \tag{III.5}$$

Notice that the measure m describes the entropy dissipation in shocks and thus vanishes in smoothness regions of $u(t,x)$.

This formulation appears as a singular limit, as ε tends to 0, of the following Boltzmann type of model

$$\partial_t f(t,x,\xi) + a(\xi) \cdot \partial_x f(t,x,\xi) = \frac{\chi(u(t,x),\xi) - f(t,x,\xi)}{\varepsilon}, \tag{III.6}$$

where the nonlinearity enters through the definition of the macroscopic density

$$u(t,x) = \int_{\mathbb{R}} f(t,x,\xi) \, d\xi.$$

This approach allows us to give a new existence proof for entropy solutions to (III.1), and is also related to numerical schemes for scalar conservation laws. It also gives a "statistical physics" approach as presented in Perthame and Pulvirenti [28]. Let us finally give another example of application of this formulation. Under some nondegeneracy assumptions on the fluxes $a(.)$, regularizing effects in Sobolev spaces have been obtained in [22], (and in [23] for the isentropic gas dynamics presented below for $\gamma = 3$). The proof relies on the averaging compactness that we have presented in Lemma 1.

The second example of kinetic formulation that we wish to present here is the isentropic gas dynamics developed by the same authors in [23]. This system is

$$\begin{cases} \partial_t \, \rho + \partial_x \, \rho u = 0, \ t \geq 0, x \in \mathbb{R}, \\ \partial_t \, \rho u + \partial_x (\rho u^2 + \kappa \rho^\gamma) = 0. \end{cases} \tag{III.7}$$

Here $\gamma > 1$ is a given parameter. We choose for simplicity $\kappa = \theta^2/\gamma$, $\theta = (\gamma - 1)/2$. The entropy inequalities are now

$$\partial_t \eta + \partial_x H \leq 0 \quad \text{in } \mathcal{D}'(\mathbb{R}_+ \times \mathbb{R}), \tag{III.8}$$

for any convex "entropy" $\eta(\rho, \rho u)$, satisfying

$$\eta_{\rho\rho} + \gamma \kappa \rho^{\gamma-3} \eta_{uu} = 0, \tag{III.9}$$

which is a necessary and sufficient condition for the existence of an entropy flux $H(\rho, \rho u)$ such that (III.8) holds as an equality for smooth solutions of (III.7).

In [23], it is proved that $(\rho,\ \rho u)$ is a solution to the system (III.7). (III.8) is exactly equivalent to writing

$$\partial_t \chi(\rho, \xi - u) + \partial_x [\theta\xi + (1 - \theta)u]\ \chi(\rho, \xi - u) = \partial_{\xi\xi} m, \qquad \text{(III.10)}$$

for some nonnegative, bounded measure $m(t, x, \xi)$ and now

$$\chi(\rho, w) = (\rho^{\gamma-1} - w^2)_+^\lambda, \quad \lambda = \frac{3 - \gamma}{2(\gamma - 1)}.$$

Again, the measure m vanishes in smoothness regions of the solution (ρ, u).

The formulation (III.10) of the conservation laws (III.1) is also close to classical kinetic equations because the new, kinetic parameter ξ has been added. The main difference with the kinetic formulation (III.1) is that the advection velocity $\theta\xi + (1 - \theta)u$ is obtained combining the kinetic velocity ξ and the macroscopic velocity $u(t, x)$. This difference makes it difficult to apply the general tools of kinetic theory as described in Section II, and in particular the averaging lemma. Nevertheless, some purely "kinetic" applications of the formulation (III.10) can be given. New estimates for (III.7) can be obtained using the moments lemma (Lemma 2 above). Also, a weak stability result in L^∞, and the global existence of weak entropy solutions, can be achieved for all $\gamma > 1$ (see Lions, Perthame, and Souganidis [21]) thus generalizing the known results in that direction.

References

[1]   C. Bardos and P. Degond, *Global existence for the Vlasov-Poisson equation in 3 space variables with small initial data*, Ann. Inst. H. Poincaré, Anal. Non Linéaire **2** (1985), 101–118.

[2]   C. Bardos, F. Golse, and D. Levermore, *Fluid dynamic limits of kinetic equations, I. Formal Derivations*, J. Statist. Phys. **63** (1) (1991), 323–344; *II. Convergence proofs for the Boltzmann equation*, to appear.

[3]   C. Bardos, F. Golse, B. Perthame, and R. Sentis, *The nonaccretive radiative transfer equations. Existence of solutions and Rosseland approximation*, J. Funct. Anal. **77** (1988), 434–460.

[4]   J. Batt and G. Rein, *Global classical solutions of the periodic Vlasov-Poisson system in three dimensions*, C.R. Acad. Sci. Paris, Série I **313** (1991), 411–416.

[5]   F. Bouchut, *Existence and uniqueness of a global regular solution for the Vlasov-Poisson-Fokker-Planck system in three dimensions*, J. Funct. Anal. **111** (1993), 239–258.

[6]   Y. Brenier and E. Grenier, *Limite singulière du système de Vlasov-Poisson dans le régime de quasi-neutralité: le cas indépendant du temps*, C. R. Acad. Sci. Série I **318** (1994), 121–127.

[7]   C. Cercignani, The Boltzmann Equation and its Applications, Springer-Verlag, Berlin and New York, 1989.

[8]   P. Degond and P.A. Raviart, *An asymptotic analysis of the one-dimensional Vlasov-Poisson system, the Child-Langmuir law*, Asymptotic Anal. **4** (1991), 187–214.

[9]   A. De Masi, R. Esposito, and J.L. Lebowitz, *Incompressible Navier-Stokes and Euler limits of the Boltzmann equation*, Comm. Pure Appl. Math. **42** (1989), 1189–1214.

[10] R.J. Di Perna and P.L. Lions, *On the Cauchy problem for the Boltzmann equation: Global existence and weak stability results*, Ann. of Math. (2) **130** (1989), 321–366.

[11] R.J. Di Perna and P.L. Lions, *Global weak solutions of Vlasov-Maxwell systems*, Comm. Pure Appl. Math. **42** (1989), 729–757.

[12] R.J. Di Perna, P.L. Lions, and Y. Meyer, L^p *regularity of velocity averages*, Anal. Inst. H. Poincaré Ann. Non Linéaire **8** (3–4) (1991), 271–287.

[13] R. Glassey, *Lecture on the Cauchy problem in transport theory*, preprint.

[14] F. Golse, P.L. Lions, B. Perthame, and R. Sentis, *Regularity of the moments of the solution of a transport equation*, J. Funct. Anal. **76** (1988), 110–125.

[15] F. Golse, B. Perthame, and R. Sentis, *Un résultat de compacité pour les équations du transport*, C. R. Acad. Sci. Paris, Série I **301** (1985), 341–344.

[16] D. Levermore, *Moment closure hierarchies for kinetic theories*, preprint.

[17] P.L. Lions, *Compactness in Boltzmann's equation via Fourier integral operators and applications*, preprint CEREMADE n° 9301, 9305, 9316 (1993), to appear in R.I.M.S. Kyoto.

[18] P.L. Lions and Th. Paul, *Sur les mesures de Wigner*, Rev. Mat. Iberoamericana **9** (3) (1993), 553–618.

[19] P.L. Lions and B. Perthame, *Propagation of moments and regularity for the 3-dimensional Vlasov-Poisson system*, Invent. Math. **105** (1991), 415–430.

[20] P.L. Lions and B. Perthame, *Lemmes de moments, de moyenne et de dispersion*, C. R. Acad. Sci. Paris Série I **314** (1992), 801–806.

[21] P.L. Lions, B. Perthame, and P.E. Souganidis, *Existence of entropy solutions to isentropic gas dynamics system*, to appear in C.P.A.M.

[22] P.L. Lions, B. Perthame, and E. Tadmor, *A kinetic formulation of multidimensional scalar conservation laws and related equations*, J. of Amer. Math. Soc. **7** (1) (1994), 169–191.

[23] P.L. Lions, B. Perthame, and E. Tadmor, *Kinetic formulation of isentropic gas dynamics and p-systems*, to appear in Comm. Math. Phys. (1994).

[24] P.A. Markowitch, C.A. Ringhofer, and C. Schmeiser, Semiconductors equations, Springer-Verlag, Berlin and New York (1989).

[25] B. Perthame, *Global existence of solutions to the BGK model of Boltzmann equations*, J. Differential Equations **82** (1) (1989), 191–205.

[26] B. Perthame, ed., Advances in Kinetic Theory and Computing, selected papers, World Scientific, Singapore (1995).

[27] B. Perthame and M. Pulvirenti, *Weighted L^p bounds and uniqueness for the Boltzmann-BGK model*, Arch. Rational Mech. Anal. **125** (3) (1993), 289–295.

[28] B. Perthame and M. Pulvirenti, *On some large systems of random particles which approximate scalar conservation laws*, Asymptotic Anal. **10** (3) (1995), 263–278.

[29] K. Pfaffelmoser, *Global classical solutions of the Vlasov-Poisson system in three dimensions for general initial data*, J. Differential Equations **95** (1992), 281–303.

[30] M. Pulvirenti, W. Wagner, and B. Zavelani-Rossi, *Convergence of particle systems for the Boltzmann equation*, preprint IAAS n° 49, Berlin (1993).

The Cauchy Problem for Harmonic Maps on Minkowski Space

JALAL SHATAH[*]

Courant Institute of Mathematical Sciences
251 Mercer Street
New York, NY 10012, USA

Introduction

In this article we shall be reporting on recent progress in the study of harmonic maps from Minkowski space (M, η) into a Riemannian manifold (N, g). These maps (also called wave maps or sigma models) are solutions to the wave equation with partial derivatives replaced by covariant derivatives. These equations are naturally nonlinear because the image lives on a manifold instead of a vector space, as is the case for the linear wave equation. A useful way to describe the problem would be when the target manifold N is a hypersurface in $\mathbb{R}^{k+1}$. In this case if $u \in N \subset \mathbb{R}^{k+1}$ and $n(u)$ is the unit normal to N at u, then the equations are given by:

$$\eta^{\alpha\beta} D_\alpha \partial_\beta u = \partial_0^2 u - \Delta u + (\partial_0 u \cdot \partial_0 n(u) - \partial_i u \cdot \partial_i n(u)) \, n(u) = 0 \ .$$

For a general manifold N the equations can be written in local coordinates:

$$\partial^\mu \partial_\mu U^a + {}^N\Gamma^a_{bc}(U)\partial_\mu U^b \partial^\mu U^c = 0, \tag{1}$$

where $\eta = \operatorname{diag}(-1, 1, \ldots, 1)$, and we sum over repeated indices. These equations are relevant to various theories in physics (see Misner [12]), and moreover they serve as a good model for studying various properties of solutions to nonlinear wave equations such as existence, uniqueness, global regularity, and development of singularities. We are also interested in how the geometry or the topology of N influences solutions of the Cauchy problem for the map $U : M \to N$:

$$\begin{cases} \partial^\mu \partial_\mu U^a + {}^N\Gamma^a_{bc}(U)\partial_\mu U^b \partial^\mu U^c = 0 \\ U(0, x) = U_0 \\ \partial_t U(0, x) = U_1. \end{cases} \tag{2}$$

The basic identity for solutions of the above equation is conservation of energy:

$$E(U) = \frac{1}{2} \int g_{ab}(\partial_0 U^a \partial_0 U^b + \partial_i U^a \partial_i U^b) \, dx,$$

[*]Research supported in part by the National Science Foundation grant DMS-8857773.

Proceedings of the International Congress
of Mathematicians, Zürich, Switzerland 1994
© Birkhäuser Verlag, Basel, Switzerland 1995

which is a dimensionless quantity in two space dimensions (this can be seen by scaling $U_\lambda(x,t) \overset{\text{def}}{=} U(\lambda x, \lambda t)$, which keeps the equation invariant). Therefore, with respect to the energy norm, these equations are characterized as subcritical in one space dimension, critical in two space dimensions, and supercritical in three or more space dimensions. These characterizations indicate that it may be possible to prove regularity in one or two space dimensions using finite energy. For higher dimensions new estimates would be needed to prove regularity. Most of our discussion here is concerned with solutions that have co-rotational symmetry, and these solutions exist when the target manifold N is a rotationally symmetric product manifold defined by:

$$N = [0, \phi^*) \times_G \mathbb{S}^{\kappa-1} \ ,$$

where $\phi^* \in \mathbb{R}^+ \cup \{+\infty\}$ and $G : \mathbb{R} \to \mathbb{R}$ is smooth and odd, $G(0) = 0$, $G'(0) = 1$. On N we have the "polar" coordinates $(\phi, \chi) \in [0, \phi^*) \times \mathbb{S}^{\kappa-1}$. In these coordinates the metric of N takes the form:

$$d\phi^2 + G^2(\phi)d\chi^2 \ ,$$

where $d\chi^2$ is the standard metric of $\mathbb{S}^{\kappa-1} \hookrightarrow \mathbb{R}^\kappa$. Using spatial polar coordinates on M $(t, r, \omega) \in \mathbb{R} \times \mathbb{R}^+ \times \mathbb{S}^{n-1}$ co-rotational maps are defined by:

$$U^j(t, x) = \phi(t, r) \cdot \chi_0^j(\omega) \ ,$$

where $(U^1, \ldots, U^k)$ are normal coordinates on N, $\chi_0^j(\omega)$ is a *harmonic polynomial map of degree* $\ell > 0$, and ϕ is a solution to:

$$
\begin{cases}
\phi_{tt} - \phi_{rr} - \dfrac{n-1}{r}\phi_r + \dfrac{k}{r^2}f(\phi) = 0, \\
\phi(0, r) = \phi_0, \\
\phi_t(0, r) = \phi_1,
\end{cases}
\tag{3}
$$

with $f(\phi) \overset{\text{def}}{=} G(\phi)G'(\phi)$, and $k \overset{\text{def}}{=} \ell(\ell + n - 2)$.

1. Local Existence and Uniqueness

Equation (2) is a Cauchy problem for a system of semilinear wave equations, and thus the local in time existence of classical solutions is standard provided the initial data is regular enough. If we view (2) as a general nonlinear system $\Box U^a = F^a(U, DU)$, then scaling shows that the Cauchy problem lacks estimates for initial data $U(0, \cdot) \in H^s(\mathbb{R}^n, N)$ and $U_t(0, \cdot) \in H^{s-1}(\mathbb{R}^n, N)$, and $s \leq \frac{n}{2}$, where H^s is the familiar Sobolev space of functions $f : \mathbb{R}^n \to N \hookrightarrow \mathbb{R}^K$ whose s derivative is in L^2. Using energy estimates alone, local existence and uniqueness requires $s > \frac{n}{2} + 1$ (e.g. see Choquet-Bruhat [2]) . This differentiability requirement was lowered to $s > \frac{n+1}{2}$ by Ponce and Sideris [14] using Strichartz type estimates.

However the above equations are not arbitrary but have a special structure. Using this fact Klainerman and Machedon [10] showed that the differentiability requirement can be reduced to $s = \frac{n+1}{2}$ (see also Beals and Bezard [1] and Struwe [21]). Recently Klainerman and Machedon proved the following theorem, which is based on null forms estimates for the wave equation in three space dimensions:

THEOREM 1 *In three dimensions the Cauchy problem is well posed for $s > \frac{3}{2}$.*

In the co-rotational problem the differentiability requirement for the initial data can be reduced to $s = \frac{n}{2}$, and this is optimal in the sense that we have nonuniqueness for $s < \frac{n}{2}$ (as will be shown later). The following theorem is due to Tahvildar-Zadeh and the author [18]:

THEOREM 2 *There exists a $T^* > 0$ such that for $n \geq 2$ the Cauchy problem (2) with co-rotational initial data (U_0, U_1),*

$$U_0 \in H^{n/2}_{loc}(\mathbb{R}^n, N), \qquad U_1 \in H^{(n-2)/2}_{loc}(\mathbb{R}^n, N),$$

has a unique solution U such that for every $z_0 \in [0, T^*) \times \mathbb{R}^n$,

$$U \ \in \ L^\infty([0, t_0), H^{\frac{n}{2}}_{loc}(\mathbb{R}^n, N)) \cap L^q([0, t_0), \dot{B}^{\sigma,q}_{loc}(\mathbb{R}^n, N)),$$

$$\partial_t U \ \in \ L^\infty([0, t_0), H^{\frac{n}{2}-1}_{loc}(\mathbb{R}^n, N)).$$

where $\sigma = \frac{n-1}{2}$ and $q = \frac{2(n+3)}{n+1}$.

Here $\dot{B}^{s,p}(\Omega)$ denotes the homogeneous Besov space on a domain Ω (see [16]). Sketch of the proof: the proof is based on Strichartz estimates that were used by Struwe and the author [16] to show regularity of critical semilinear wave equations. Suppose that ϕ is a solution of (3) with $\ell = 1$. Define a function $v : M \to \mathbb{R}$ by setting $\phi = rv$, then v satisfies:

$$\begin{cases} v_{tt} - v_{rr} - \dfrac{m-1}{r} v_r = v^3 Z(rv), \\ v(0, r) = v_0 = \phi_0/r, \\ v_t(0, r) = v_1 = \phi_1/r, \end{cases} \qquad (4)$$

with $m = n+2$. If we ignore Z, we are left with a wave equation in m space dimension and cubic nonlinearity. Using Strichartz estimates and energy inequalities we can easily obtain that the Cauchy problem (2) has unique local solutions in the stated space.

2. Regularity

For small initial data there are several global existence results that are due to Choquet-Bruhat [2], Klainerman [9], Kovalyov [11], and Sideris [19]. However in this article we are mainly interested in initial data without size restrictions. All results stated herein are for initial data of arbitrary size.

In one space dimension global regularity results for these maps were obtained by Gu [7] and later by Ginibre and Velo [4], where they used the special structure of the equations. In higher dimensions the problem is more difficult and indeed solutions may not stay regular; this will be illustrated later in the case of three or more spatial dimensions. In two space dimensions the regularity results are possible because the problem is critical with respect to the energy norm. The first results on semilinear wave equations with critical exponents were obtained by Struwe [20]; later this was generalized by Grillakis [5].

Let us consider the co-rotational problem in two space dimensions. As observed before, such a map satisfies the Ansatz:

$$\phi = \phi(t, r), \qquad \chi = \chi(\omega) ,$$

where $\chi = \ell\omega$, with $\ell \in \mathbb{N}$. The equation for ϕ is:

$$-\phi_{tt} + \phi_{rr} + \frac{1}{r}\phi_r - \frac{\ell^2}{r^2}f(\phi) = 0, \qquad f \stackrel{\text{def}}{=} GG'. \tag{5}$$

In an unpublished work Christodoulou and the author (1988) proved regularity of solutions to equation (3) for the case when N is the hyperbolic plane and $\ell = 1$. Tahvildar-Zadeh and the author [17] proved regularity of solutions provided N is geodesically convex ($G' > 0$). This condition was weakened by Grillakis [6] to obtain the following theorem:

THEOREM 3 *Let the function G in the metric of N satisfy the condition:*

$$G(s) + sG'(s) > 0 \qquad for \ s > 0. \tag{6}$$

Then the Cauchy problem with smooth data for an equivariant wave map from M into N has smooth solutions defined for all time.

A different proof for the above theorem was given by Tahvildar-Zadeh and the author [18], based on their proof for the geodesically convex case, *and*, a modified version of the Morawetz identity [13]. The proof is carried out in two main steps. The first step uses a gravitational collapse argument due to Christodoulou. The premise of this argument is to evolve the equation in r instead of t. This allows us to obtain estimates everywhere except on a small cone, whose vertex is at the origin. The estimates on this small cone are obtained using the slightly modified Morawetz identity. The second step of the proof uses a modification of Struwe and the author's [16] application of the localized Strichartz estimates to these problems. Details of the proof can be found in [18].

Another class of solutions that have global regularity is that of spherically symmetric maps. In this case the nonlinear terms in equation (2) remain as products of derivatives of U. For this reason the equations are more difficult to analyze. In two space dimensions this problem was considered by Christodoulou and Tahvildar-Zadeh [3] under the assumptions that the target manifold N is geodesically convex and is well behaved at infinity. They proved the following theorem:

THEOREM 4 *In two space dimensions the Cauchy problem (2) with smooth spherically symmetric inital data has a smooth solution defined for all time.*

3. Development of Singularities

Because equations (2) are invariant under scaling $U_\lambda(x, t) \stackrel{\text{def}}{=} U(\lambda x, \lambda t)$ then singularities can develop by simply forcing the solution to concentrate at a point. This suggests looking for solutions that are self-similar:

$$U(x,t) = \theta(x/t). \tag{7}$$

These solutions are constant on rays emanating from the origin, thus possibly leading to a derivative singularity at the origin. Substituting this Ansatz into equation (2) we obtain the hamonic map equation from the hyperbolic space $\mathbb{H}^n$ into the target manifold N. This is to be expected because if we put the coordinates:

$$\sigma = \sqrt{t^2 - |x|^2}, \qquad \rho = \frac{|x|}{t} \ , \tag{8}$$

on the Minkowski space $M = \mathbb{R}^{n+1}$, then $\mathbb{H}^n$ is the hypersurface given by $\sigma = 1$. Therefore solutions of (1) that are independent of σ are harmonic maps from $\mathbb{H}^n$ into N. If such maps exist and are regular (including regularity at infinity), we can solve the Cauchy problem (2) using these maps as initial data at $t = -1$. In this situation the solution inside the light cone will be given by the function $U(x,t) = \theta(x/-t)$, which is singular at the origin.

In the co-rotational case this construction can be carried out by substituting $\phi(r,t) = \theta(r/t)$ in equation (3). The equation for θ will be:

$$\theta'' + \left(\frac{n-1}{\rho} + \frac{(n-3)\rho}{1-\rho^2}\right)\theta' - \frac{k}{\rho^2(1-\rho^2)}f(\theta) = 0. \tag{9}$$

First we consider the case $n = 3$. In this case a necessary condition for the solution θ to be smooth is $f(\theta(1)) = 0$, or equivalently $G'(\theta(1)) = 0$ (i.e. N is not geodesically convex). This condition also turns out to be sufficient under a non-degeneracy assumption as indicated in the following theorem of Tahvildar-Zadeh and the author [18]:

THEOREM 5 *Let M be the $3 + 1$-dimensional Minkowski space, and N a rotationally symmetric 3-manifold with Riemannian metric $ds^2 = d\phi^2 + G^2(\phi)d\chi^2$, $\chi \in S^2$, and $G(0) = 0$, $G'(0) = 1$. Let θ_* be the smallest positive zero of G', and assume that $G''(\theta_*) \neq 0$. Then there is a class of smooth initial data such that the corresponding Cauchy problem for a co-rotational map from M into N has a solution that blows up in finite time.*

Sketch of the proof: as stated earlier, the idea is to construct a smooth solution to equation (9). This can be done by setting a variational problem to minimize the functional:

$$\mathcal{E}[\psi] = \frac{1}{2}\int_0^1 \psi'^2 + k\frac{G^2(\psi) - G^2(\theta_*)}{\rho^2(1-\rho^2)} \ \rho^2 d\rho \ ,$$

over the space $X = \{u \in H^1(B_1), u(1) = \theta_*\}$. Note that on X, $\mathcal{E}$ is bounded from below. By standard arguments we can show that $\mathcal{E}$ achieves its minimum at some bounded function θ that is monotone. Using equation (9), the boundary condition, and the fact that functions in X are Hölder continuous away from the origin, we obtain that θ is a smooth function away from the origin. To show that θ is regular everywhere, observe that θ maps any compact set $\Omega \subset\subset \mathbb{H}^3$ into a strictly convex neighborhood of N. By the regularity results for the elliptic harmonic map (e.g. [8]), θ is also regular in the interior.

For $n > 3$ the condition for the solution θ of (9) to be regular becomes weaker. For example, in five space dimensions Cazenave, Tahvildar-Zadeh, and the author (1993) proved that if N has positive curvature then θ is regular, and therefore solutions to the Cauchy problem (3) develop singularities in finite time. In higher dimensions the conditions on N become analytical with no obvious geometric interpretation. In two space dimensions there are no nontrivial self-similar solutions [21].

4. Weak Solutions

The weakest solutions that we can define for equation (1) are finite energy solutions. However the existence of such solutions is not known for the general problem because the nonlinear term is not weakly compact for solutions in H^1. For a special class of target manifolds, namely symmetric spaces, we can show existence of finite energy solutions because in these cases equation (1) is equivalent to a k-dimensional system of conserved currents generated by the isometries of N [15]. We can also show existence of finite energy solutions for the co-rotational case [18]. However these solutions are not unique in three or more space dimensions. This will be illustrated for the case where the target manifold is $\mathbb{S}^3$. In this case the self-similar solution can be found explicitly [22]:

$$\theta = 2\tan^{-1}\rho. \tag{10}$$

Direct calculations show that $\phi(r,t) = \theta(\frac{r}{t})$ is a weak solution for all time. Moreover we can construct another weak solution in the following way:

$$\xi(r,t) \stackrel{\text{def}}{=} \begin{cases} \theta(\frac{r}{t}) & r \geq t \\ \frac{\pi}{2} & r \leq t. \end{cases}$$

Note that for $t < 0$, $\xi = \theta$ is a smooth function. Therefore we have a smooth solution for $t < 0$ that develops a singularity at $t = 0$, which can be continued as a weak solution for $t > 0$ in at least two different ways. In general when the target manifold is not geodesically convex we can only show that there are two different solutions for the same finite energy initial data (see [18]). Finally, these nonunique solutions are in H^s for $s < \frac{n}{2}$, thus proving that Theorem 2 is optimal.

References

[1] M. Beals and M. Bezard, *Low regularity solutions for field equations*, preprint, 1992.

[2] Y. Choquet-Bruhat, *Global existence for nonlinear σ-models*, Rend. Sem. Mat. Univ. Politec. Torino, Fascicolo Speciale, pages 65–86, 1988.

[3] D. Christodoulou and A. Tahvildar-Zadeh, *On the regularity of spherically symmetric wave maps*, Comm. Pure Appl. Math, 46:1041–1091, 1993.

[4] J. Ginibre and G. Velo, *The Cauchy problem for the $O(n)$, $\mathbb{C}P(n-1)$ and $G\mathbb{C}(n,p)$ models*, Ann. Physics, 142:393–415, 1982.

[5] M. Grillakis, *Regularity and asymptotic behavior of the wave equation with a critical nonlinearity*, Ann. of Math. (2), 132:485–509, 1990.

[6] M. Grillakis, *Classical solutions for the equivariant wave map in 1+2-dimensions*, to appear in Indiana Univ. Math. J.

[7] C. H. Gu, *On the Cauchy problem for harmonic maps defined on two-dimensional Minkowski space*, Comm. Pure Appl. Math., 33:727–737, 1980.

[8] J. Jost, Harmonic Mappings Between Riemannian Manifolds, Aust. Nat. Univ. Press, Canberra, 1983.

[9] S. Klainerman, *The null condition and global existence to nonlinear wave equations*, Lectures in Appl. Math., 23:293–326, 1986.

[10] S. Klainerman and M. Machedon, *Space-time estimates for null forms and the local existence theorem*, Comm. Pure Appl. Math, 46:1221–1268, 1991.

[11] M. Kovalyov, *Long-time behaviour of solutions of a system of nonlinear wave equations*, Comm. Partial Differential Equations, 12:471–501, 1987.

[12] C. W. Misner, *Harmonic maps as models for physical theories*, Phys. Rev. D., 18(12):4510–4524, 1978.

[13] C. Morawetz, *Time decay for the nonlinear Klein-Gordon equation*, Proc. Royal Soc., A 306:291–296, 1968.

[14] G. Ponce and T. Sideris, *Local regularity of nonlinear wave equations in three space dimensions*, Comm. Partial Differential Equations, to appear, 1992.

[15] J. Shatah, *Weak solutions and development of singularities in the $SU(2)$ σ-model*, Comm. Pure. Appl. Math., 41:459–469, 1988.

[16] J. Shatah and M. Struwe, *Regularity results for nonlinear wave equations*, Ann. of Math. (2), 138:503–518, 1993.

[17] J. Shatah and A. Tahvildar-Zadeh, *Regularity of harmonic maps from Minkowski space into rotationally symmetric manifolds*, Comm. Pure Appl. Math., 45:947–971, 1992.

[18] J. Shatah and A. Tahvildar-Zadeh, *On the Cauchy problem for equivariant wave maps*, Comm. Pure Appl. Math., 47:719–753, 1994.

[19] T. Sideris, *Global existence of harmonic maps in Minkowski space*, Comm. Pure Appl. Math., 42:1–13, 1989.

[20] M. Struwe, *Globally regular solutions to the u^5 Klein-Gordon equation*, Ann. Scuola Norm. Sup. Pisa Cl. Sci. (4), 15:495–513, 1988.

[21] M. Struwe, Geometric evolution problems, Park City Geom. Ser. of the Amer. Math. Soc., Providence, RI, 1992.

[22] N. Turok and D. Spergel, *Global texture and the microwave background*, Phys. Rev. Lett., 64(23):2736–2739, 1990.

Interface Dynamics in Phase Transitions

PANAGIOTIS E. SOUGANIDIS

Department of Mathematics
University of Wisconsin-Madison
Madison, WI 53706, USA

0 Introduction

The problems discussed in this paper have the following common theme. An, in time t, evolving, in time t, order parameter that, depending on the specific context, describes the different phases of a material or the total (averaged) magnetization of a stochastic system or the temperature of a reacting-diffusing system, etc., approaches, typically as $t \to \infty$, the equilibrium states of certain systems. Depending on the values of a threshold parameter, such systems either have a unique or more than one equilibrium state. The existence of multiple equilibrium states can be associated to phase transitions.

When more than one equilibrium state exists, the evolving order parameter develops, for $t \gg 1$, interfaces, which are the boundaries of the regions where it converges to the different equilibria. The problem is then to justify the appearance of these interfaces and to understand in a qualitative way their dynamics, geometry, regularity, etc..

In addition to the general situation described above, interfaces (fronts, surfaces) in $\mathbb{R}^N$ evolving with normal velocity

$$V = v(Dn, n, x, t), \tag{0.1}$$

where n and Dn are the exterior normal vector to the surface and its gradient, respectively, arise also in geometry, in image processing, in the theory of turbulent flame propagation and combustion, etc..

Typical examples of interface dynamics appearing in the aforementioned areas are, among others, the general anisotropic motion

$$V = -\mathrm{tr}[\theta(n, x, t)Dn] + c(n, x, t), \tag{0.2}$$

a special case of which is the motion by mean curvature

$$V = -\mathrm{tr}\ Dn, \tag{0.3}$$

the motion by Gaussian curvature

$$V = \kappa_1 \cdots \kappa_{N-1},$$

where $\kappa_1, \ldots, \kappa_{N-1}$ are the principal curvatures of Γ_t, etc.

Proceedings of the International Congress
of Mathematicians, Zürich, Switzerland 1994
© Birkhäuser Verlag, Basel, Switzerland 1995

The main mathematical characteristic of such evolutions is the development of singularities in finite time, independently of the smoothness of the initial surface. A great deal of work has been done during the last few years to interpret the evolution past the singularities and to study and validate the different models mentioned earlier.

The outcome of this work has been the development of a weak notion of evolving fronts called generalized front propagation. The generalized evolution $\{\Gamma_t\}_{t\geq0}$ with normal velocity (0.1) starting with a given surface $\Gamma_0 \subset \mathbb{R}^N$ is defined for all $t \geq 0$, although it may become extinct in finite time. Moreover, it agrees with the classical differential-geometric flow, as long as the latter exists. The generalized motion may, on the other hand, develop singularities, change topological type, and exhibit various other pathologies.

In spite of these peculiarities, the generalized motion $\{\Gamma_t\}_{t\geq0}$ has been proven to be the right way to extend the classical motion past singularities. Some of the most definitive results in this direction are about the fact that the generalized evolution (0.2) governs the asymptotic behavior of solutions of semilinear reaction-diffusion equations and systems. Such equations are also often used in continuum mechanics to describe the time evolution of an order parameter determining the phases of a material (phase field theory).

Another recent striking application of the generalized front propagation is the fact that it governs the macroscopic behavior, for large times and in the context of grain coarsening, of a number of stochastic interacting particle systems like the stochastic Ising model with long-range interactions and general spin flip dynamics. Such systems are standard Gibbsian models used in statistical mechanics to describe phase transitions. It turns out that the generalized front propagation not only describes the limiting behavior of such systems but also provides a theoretical justification, from the microscopic point of view, of several phenomenological sharp interface models in phase transitions.

The paper is organized as follows. Section 1 is devoted to the description of the generalized front propagation. Section 2 describes a simple model from the phase field theory and explains its relationship with the evolving fronts. Section 3 discusses the asymptotics of the stochastic models. Finally, Section 4 discusses a mathematical result that provides general criteria for asymptotic problems in order for them to yield in the limit generalized front propagation.

Due to the limitations on length, it will not be possible to discuss in this paper the way moving fronts relate to image processing and to turbulent flame propagation. Instead, I refer to Alvarez, Guichard, Lions, and Morel [AGLM] for the former and to Majda and Souganidis [MS] for the latter. Because of the same constraint, the theorems presented here will be stated without most of the hypotheses, which, although natural, take some space to state. Instead I will refer to specific references for the exact statements. Finally, the length limitation will definitely inhibit the number of given references.

1 The generalized front propagation

One of the most successful approaches for understanding the generalized evolution past the singularities, which is known as the level set approach, consists of identifying the moving front as the level set (for definiteness the zero level set) of the solution of a fully nonlinear partial differential equation of the form

$$u_t = F(D^2 u, Du, x, t) \quad \text{in} \quad \mathbb{R}^N \times (0, \infty) \tag{1.1}$$

with F related to v in (0.1).

The level set approach was initially suggested for numerical computations by Osher and Sethian [OS] — see also Barles [Ba] for a first-order model for flame propagation. Later this approach was developed by Evans and Spruck [ES] for motion by mean curvature and, independently, by Chen, Giga, and Goto [CGG], who considered more general geometric motions. The results of [ES] and [CGG] were extended by, among others, Goto [G], Gurtin, Soner and Souganidis [GSS], Ishii and Souganidis [IS], etc.. In all these works the analysis is based on the theory of viscosity solutions to fully nonlinear first- and second-order parabolic equations, which were introduced by Crandall and Lions [CL] and Lions [L].

A more intrinsic alternative characterization of the weak evolution is known as the distance function approach. This approach, which was introduced by Soner [So1] and later extended in more general situations by Barles, Soner, and Souganidis [BSS], is based upon checking whether the signed distance function to the propagating front satisfies in some sense (1.1) (see [So1], [BSS] for the exact definition). For the special case of the mean curvature, the distance function criterion reduces to requiring that the signed distance d to the front satisfies, in the viscosity sense, the inequalities

$$d_t - \Delta d \begin{cases} \geq 0 & \text{in } \{d > 0\}, \\[2mm] \leq 0 & \text{in } \{d < 0\}. \end{cases} \tag{1.2}$$

Notice that (1.2) is sharp because it has to hold even in the case of the classical motion. The point here, however, is that in general d is only Lipschitz continuous in x and semicontinuous in t.

Either approach allows for the existence and uniqueness, under some conditions, of moving fronts $\{\Gamma_t\}_{t \geq 0}$. There are, of course, a number of very interesting and rather important questions related to such evolutions, like the equivalence of the several different approaches, the creation of interior and the regularity of the evolving sets. Such issues will not be discussed here. Instead I refer to [ES], [BSS], and to Ilmanen [I].

Concluding this section, I describe a new formulation of the generalized evolution of fronts, which is developed in Barles and Souganidis [BS]. This new formulation, which turns out to be equivalent to the aforementioned ones, is related to the notion of "barriers" introduced by DeGiorgi [D] and provides a more intuitive and definitely a more geometric way to understand the weak propagation.

Although this description will appear to be cumbersome, it is, on the contrary, rather natural, because it uses, in some sense, smooth fronts as "test sets" in the definition of the motion, in the same way that smooth functions are used to test the definition of solutions of (1.1). It also provides a powerful tool to study asymptotic problems of the type referred to in the introduction with very strong anisotropies, because it reduces everything to the study of smooth evolutions.

In order to describe this new formulation, it is more convenient to think of the front Γ_t as the boundary of an open set $\Omega_t \subset \mathbb{R}^N$ and to define the evolution of Ω_t as follows:

A family $(\Omega_t)_{t \in [0,T]}$ of open subsets of $\mathbb{R}^N$ is said to propagate with normal velocity V if and only if, for any $(x,t) \in \mathbb{R}^N \times (0,T)$ and for any open subsets $\mathcal{O}_t$ and $\tilde{\mathcal{O}}_t$ of $\mathbb{R}^N$ with smooth boundary such that, for some $r > 0$,

$$\mathcal{O}_t \subset\subset \Omega_t \cap B_r(x) \quad \text{and} \quad \tilde{\mathcal{O}}_t \subset\subset (\mathbb{R}^N \backslash \overline{\Omega}_t) \cap B_r(x),$$

there exists $h_0 > 0$ depending only on r and the C^3-norm of $\partial\mathcal{O}_t$ and $\partial\tilde{\mathcal{O}}_t$ such that, for any $h \in (0, h_0)$ and $\alpha > 0$ small enough if $\{\mathcal{O}_{t+h}\}_{h \in [0,h_0]}$ and $\{\tilde{\mathcal{O}}_{t+h}\}_{h \in [0,h_0]}$ are the smooth evolutions of $\mathcal{O}_t$ and $\tilde{\mathcal{O}}_t$ with normal velocity $V + \alpha$ and $V - \alpha$ respectively, then, for all $h \in (0, h_0)$,

$$\mathcal{O}_{t+h} \subset \Omega_{t+h} \quad \text{and} \quad \tilde{\mathcal{O}}_{t+h} \subset \mathbb{R}^N \backslash \overline{\Omega}_{t+h}.$$

It should be noted that all the above have their own limitations, because they only apply to front motions, for which one can prove a comparison or (avoidance) principle. It remains an open problem to find an appropriate way to describe other evolutions, like, for example, the Hele-Shaw motion, past their singularities, as well as evolutions related to multiphases (e.g., triple junctions).

2 Phase field theory — Asymptotics of reaction-diffusion equations

Reaction-diffusion equations are used in the theory of phase transitions to describe the evolution of an order parameter that identifies different phases of a material. For example, the equation

$$u_t - \Delta u + W'(u) = 0, \tag{2.1}$$

where W is a double-well potential was proposed by Allen and Cahn [AC] in this context to describe phase transitions in polycrystalline materials. Such equations are also related to the stochastic Ginzburg-Landau model, an equation for first-order phase transitions. Reaction-diffusion equations arise as the mean field equations for a class of stochastic Ising models with local interactions and fast stirring. Finally such equations with multiple scales and perhaps random coefficients also appear as models in the theory of turbulent flame propagation and combustion.

Typically, as $t \to \infty$, solutions of (2.1) develop interfaces, which can be thought of as the boundaries of the regions where they converge, in the limit $t \to \infty$, to the different equilibria of $W(u)$. It is, of course, of interest to understand the propagation of these interfaces, as this provides a considerable insight for the dynamics of the phase transitions modeled by (2.1).

A rather natural way to understand the interfacial dynamics, as $t \to \infty$, is to appropriately scale space and time so as to reproduce in finite time the long time behavior and to keep the interface in bounded space regions. The first natural scaling is $(x,t) \to (\lambda^{-1}x, \lambda^{-1}t)$. It leads to asymptotic problems of the type

$$u_t^\lambda - \lambda \Delta u^\lambda + \lambda^{-1} W'(u^\lambda) = 0. \tag{2.2}$$

Another possible scaling is $(x, t) \to (\lambda^{-1}x, \lambda^{-2}t)$, which yields

$$u_t^\lambda - \Delta u^\lambda + \lambda^{-2} W'(u^\lambda) = 0. \tag{2.3}$$

In the rest of this section, to give a flavor of the type of results one can obtain here, I will concentrate on the case of a potential W with only two wells at ± 1 of equal depth, i.e. $W(-1) = W(1)$ and a local maximum at 0. In this case, the appropriate asymptotic problem to analyze is (2.3). This problem had been the object of considerable study ranging from formal analysis to local in time results; see, for example, Rubinstein, Sternberg, and Keller [RSK], DeMottoni and Schatzman [DS], Chen [C], etc.. The first global in time result based on the motion of the generalized front evolution was obtained by Evans, Soner, and Souganidis [ESS], who proved the following result. For the exact assumptions as well as generalizations and references predating [ESS], I refer to [ESS] and [BSS].

THEOREM [ESS]. *Let u^λ be the solution of (2.3) starting at $t = 0$ at u_0. If $(\Omega_t)_{t>0}$ is the generalized mean curvature evolution starting at $\Omega_0 = \{u_0 > 0\}$, then, as $\lambda \to 0$, $u^\lambda \to 1$ in Ω_t and $u^\lambda \to -1$ in $\mathbb{R}^N \backslash \bar{\Omega}_t$.*

Below and in order to give an idea of the type of arguments involved in proving such a result, I sketch a "possible" proof, without paying much attention to all the technical details, referring instead to [ESS] and [BSS]. To this end, let $q : \mathbb{R} \to \mathbb{R}$ be the unique standing wave solving

$$\ddot{q} = W'(q) \text{ in } \mathbb{R}, \quad \dot{q} > 0 \text{ and } q(\pm\infty) = \pm 1 \text{ and } q(0) = 0.$$

For the purpose of this exposition it is enough to assume that

$$u^\lambda = q\left(\frac{d_0}{\lambda}\right) \qquad \text{on } \mathbb{R}^N \times \{0\},$$

d_0 being the signed distance function from $\Gamma_0 = \partial\Omega_0$, i.e. $d_0(x) > 0$ if $x \in \Omega_0$ and $d_0(x) < 0$ if $x \in \mathbb{R}^N \backslash \bar{\Omega}_0$. Writing

$$u^\lambda = q\left(\frac{Z^\lambda}{\lambda}\right),$$

one easily finds that Z^λ satisfies

$$\begin{cases} Z_t^\lambda - \Delta Z^\lambda - \lambda^{-1} Q(\lambda^{-1} Z^\lambda)(|DZ^\lambda|^2 - 1) = 0 \text{ in } \mathbb{R}^N \times (0, \infty), \\ Z^\lambda = d_0 \text{ on } \mathbb{R}^N \times \{0\}, \end{cases}$$

with $Q(\xi) = \ddot{q}\dot{q}^{-1}(\xi)$.

Since $|DZ^\lambda| \leq 1$ in $\mathbb{R}^N \times \{0\}$, a simple maximum principle type argument yields that $|DZ^\lambda| \leq 1$ in $\mathbb{R}^N \times (0, \infty)$. But then the fact that

$$\xi \text{ and } \dot{q} > 0$$

give

$$Z_t^\lambda - \Delta Z^\lambda \begin{cases} \geqq 0 & \text{in } \{Z^\lambda > 0\}, \\[2mm] \leqq 0 & \text{in } \{Z^\lambda < 0\}. \end{cases}$$

Assuming for the moment that $Z^\lambda \to Z$ locally uniformly as $\lambda \to 0$, it is easy to see that

$$Z_t - \Delta Z \begin{cases} \geqq 0 & \text{in } \{Z > 0\}, \\[2mm] \leqq 0 & \text{in } \{Z < 0\}. \end{cases}$$

Finally, going back to the equation satisfied by Z^λ, it appears to be possible to prove that $|DZ| = 1$ in $\mathbb{R}^N \times (0, \infty)$ and hence that

$$Z(x,t) = d(x, \{y : Z(y,t) = 0\}).$$

In view of (1.2), the last two facts together with $q(\pm\infty) = \pm 1$ yield the result, provided all the above can be made precise. This can be done, but it requires using the machinery associated with viscosity solutions, which allows for passage to the limit under very weak hypotheses.

The "proof" sketched above appears in [BSS], which studies asymptotic problems like (2.2) and (2.3) in more general situations. The same paper is also a very good source of references regarding the history of the problem. It should be noted that the special case of the Allen-Cahn equation (2.3) can also be studied using the Brakke notion of mean curvature, which characterizes the front as a varifold. This was done by Ilmanen [I] and later refined by Soner [So2].

3 Macroscopic limits of partical systems — Stochastic Ising models

Stochastic Ising models are the canonical Gibbsian models used in statistical mechanics to describe phase transitions. Describing in detail such models is beyond the scope of this paper. Instead below, abusing if needed the mathematical rigor at some points, I present a brief summary of these models and refer to DeMasi and Presutti [DP] and Spohn [Sp] for the complete theory.

To this end one considers the lattice $\mathbb{Z}^N$, the spin $\sigma(x) = \pm 1$ at $x \in \mathbb{Z}^N$, the configuration (sample) space $\Sigma = \{-1, 1\}^{\mathbb{Z}^N}$ and the Gibbs (equilibrium) measures μ^β on Σ which depend on the inverse temperature $\beta > 0$ and the Hamiltonian (energy) function H, which is given by

$$H(\sigma) = -\sum_{x \neq y} J(x,y)\sigma(x)\sigma(y) - h\sum_x \sigma(x);$$

here $J \geqq 0$ is the interaction potential and h is the external magnetization field. The assumption that $J \geqq 0$ means that one deals with ferromagnetic Ising systems.

It turns out (see [DP], [Sp]) that for any $\beta > 0$, as long as $h \neq 0$, there exists a unique Gibbs measure. On the other hand, if $h = 0$ there exists β_c such that for $\beta < \beta_c$ still there exists a unique Gibbs measure, but for $\beta > \beta_c$ there exist at least two probability measures $\mu_\pm$ on Σ such that any linear combination

$\alpha\mu_- + (1 - \alpha)\mu_+$ ($\alpha \in [0, 1]$) is also a Gibbs measure. In this case one says that there is a phase transition.

Studying the phase transitions from the dynamic point of view for $\beta > \beta_c$ amounts to introducing some dynamics, i.e. a Markov process on Σ which has the Gibbs measure as invariant measures and to analyzing the way this process evolves any initial distribution (measure) to the equilibria Gibbs measures. Convenient quantities (order parameters) to analyze in this context are the moments of the evolving measures, the first one (moment) being the total magnetization m, which will, of course, develop an interface for large times. The shape and evolution of this interface is of great interest theoretically and in the applications.

A very general example of a dynamics that has the Gibbs measures as invariant measures is the spin-flip dynamics, which, loosely speaking, is a sequence of flips σ^x, where

$$\sigma^x(y) = \begin{cases} \sigma(y) & \text{if } y \neq x, \\ -\sigma(x) & \text{if } y = x, \end{cases}$$

with rate

$$c(x, \sigma) = \Psi(\Delta_x H),$$

for an appropriate Ψ, where $\Delta_x H$ is the energy difference due to a spin flip at x. The only restriction on Ψ, which is related to the requirement that the Gibbs measures are invariant for the dynamics, is that it satisfies the balance law

$$\Psi(r) = \Psi(-r)e^{-r}.$$

In view of the previous discussion, one is interested in the behavior of the system as $t \to \infty$. Another classical limit, known as the Lebowitz-Penrose limit, is to study the behavior also as the interaction range tends to infinity. In this limit, known in the physics literature as grain coarsening, there is a law of large numbers effect that dampens the oscillations and causes the whole collection to evolve deterministically. An important question is whether these limits commute and, if not, whether there is a particular scaling or scalings for which one can study both.

These issues are addressed by Katsoulakis and Souganidis [KS3] for the general dynamics described above, with long range interactions, with rate

$$c_\gamma(x, \sigma) = \Psi(\Delta_x H_\gamma),$$

where γ^{-1} is the interaction range,

$$H_\gamma(x) = -\sum_{y \neq x} J_\gamma(x, y)\sigma(x)\sigma(y)$$

and

$$J_\gamma(x, y) = \gamma^N J(\gamma(x - y));$$

the nonnegative interaction potential J is assumed to satisfy some integrability conditions and to be symmetric, i.e., $J(z) = J(-z)$ but not isotropic.

The associated mean field equation for the averaged magnetization, i.e., the equation obtained as $\gamma \to 0$ but with t kept fixed (mesoscopic limit), was shown by DeMasi, Orlandi, Presutti, and Triolo [DOPT] and Katsoulakis and Souganidis [KS3] to be

$$m_t + \Phi(\beta(J * m))[m - \tanh \beta J * m] = 0 \quad \text{in } \mathbb{R}^N \times (0, \infty),$$

where Φ is given by

$$\Phi(r) = \Psi(-2r)(1 + e^{-2r})^{-1}.$$

After the appropriate scaling, namely $(x, t) \to (\lambda^{-1}x, \lambda^{-2}t)$, the mean field equation becomes the asymptotic problem

$$m_t^\lambda + \lambda^{-2}\Phi(\beta(J^\lambda * m^\lambda))[m^\lambda - \tanh \beta J^\lambda * m^\lambda] = 0 \quad \text{in } \mathbb{R}^N \times (0, \infty),$$

where $J^\lambda(x) = \lambda^{-N}J(\lambda^{-1}x)$. The asymptotic behavior of m^λ was shown in [KS3] to be governed by a generalized front Γ_t moving with the anisotropic normal velocity

$$V = -\mathrm{tr}(\theta(n)Dn). \tag{3.1}$$

The matrix $\theta(n)$, which appears because of a nonlinear averaging effect taking place at the limit $\lambda \to 0$, is not described by any microscopic considerations but rather is given by the following explicit Green-Kubo-type formula

$$\theta(n) = \frac{\beta}{2} \left(\int \frac{\dot{q}^2(\xi, n)}{\Phi(\beta \int J(y)q(\xi + y \cdot n, n)dy)(1 - q^2(\xi, n))} \, d\xi \right)^{-1} \times$$

$$\left[\iint J(y)\dot{q}(\xi, n)[\dot{q}(\xi + u \cdot n, n)(y \otimes y) + D_n q(\xi + y \cdot n, n) \otimes y + \right.$$

$$\left. y \otimes D_n q(\xi + y \cdot n, n)]dy d\xi \right].$$

Here $q(\cdot, n)$ is the standing wave associated with the mean field equation at the direction n, and $\dot{q}$ and $D_n q$ denote derivatives with respect to ξ and n respectively. Notice that because J is not assumed to be isotropic, these standing waves depend in a nontrivial way on their direction.

The two terms in the product defining θ have a clear geometric meaning, since the one inside the brackets is related to the surface tension of the interface and the other is the mobility (see [Sp]).

The macroscopic behavior of the particle system is described by the following theorem.

THEOREM [KS3]. *There exists $\rho^* > 0$ such that under some assumptions on the initial distribution, for any $\beta > \beta_c$ there exists $m_\beta > 0$ such that for any scaling $\lambda(\gamma)$ decaying to 0 slower than γ^{ρ^*},*

$$\lim_{\gamma \to 0} \sup_{x \in M_n} |E_\mu \prod_{i=1}^n \sigma_{t\lambda^{-2}(\gamma)}(x_i) - m_\beta^n \prod_{x_i \in \mathcal{N}_t^\gamma} (-1)| = 0,$$

where $\mathcal{N}_t^\gamma = \{x \in \mathbb{Z}^N \text{ s.t. } \gamma\lambda(\gamma)x \in \mathbb{R}^N \backslash \overline{\Omega}_t\}$, $(\Omega_t)_{t \in [0,T]}$ moving with normal velocity (3.1) and $M_n = \{\underline{x} \in (\mathbb{Z}^N)^n : x_1 \neq \cdots \neq x_n\}$.

This theorem, which is global in time, can be thought of as providing, from the microscopic point of view, a theoretical justification for the sharp interface models used in continuum mechanics, as well as for the numerical simulations performed in the physics community to study moving interfaces. The matrix $\theta(n)$ also settles the issue about the exact form and relation between the mobility and the surface tension of the interface.

The long time behavior of stochastic Ising models with isotropic potentials and Glauber dynamics as well as for nearest neighbor interaction and fast stirring was studied by Katsoulakis and Souganidis in [KS1] and [KS2] — the short time analysis in this context was carried out, respectively, by [DOPT] and Bonaventura [Bo]. Finally, Jerrard [J] studied the asymptotics of a local version of the mean field equation with isotropic potential.

4 A general theory

Here I present a rather general theory that allows to prove rigorously the appearance of interfaces and to identify their dynamics for a large class of asymptotic problems, which satisfy some general principles. The main point is that if one can prove the appearance of moving interfaces, as long as they remain smooth, then the appearance of moving interfaces is valid also for the generalized evolution with the same dynamics, provided that these dynamics satisfy a "comparison principle".

To this end let $(u_\lambda)_{\lambda>0}$ be a uniformly bounded family of functions in, say, $\mathbb{R}^N \times (0,T)$. The desired result is that, as $\lambda \to 0$,

$$
u_\lambda(x,t) \to
\begin{cases}
a & \text{if } (x,t) \in \mathcal{O} = \bigcup_{t \in (0,T)} \Omega_t \times \{t\}, \\[2mm]
b & \text{if } (x,t) \in \mathbb{R}^N \backslash \overline{\mathcal{O}} = \bigcup_{t \in (0,T)} (\mathbb{R}^N \backslash \overline{\Omega}_t) \times \{t\},
\end{cases}
$$

for some $a, b \in \mathbb{R}$ with $a < b$, with the family $(\Omega_t)_{t \in [0,T]}$ propagating with some normal velocity V.

The key assumptions, in a simplified form, on the family $(u_\lambda)_{\lambda>0}$ are:

(H1) *Causality*: For any $\lambda > 0$ and any $t \geq 0$ and $h > 0$, there exists a family of maps $S^\lambda_{t,t+h} : L^\infty(\mathbb{R}^N) \to L^\infty(\mathbb{R}^N)$ such that

$$
u_\lambda(\cdot, t+h) = S^\lambda_{t,t+h} u_\lambda(\cdot, t) \text{ in } \mathbb{R}^N.
$$

(H2) *Monotonicity*: For any $u, v \in L^\infty(\mathbb{R}^N)$, $\lambda > 0$, $t \geq 0$ and $h > 0$,

$$
\text{if } u \leq v \text{ in } \mathbb{R}^N, \text{ then } \quad S^\lambda_{t,t+h} u \leq S^\lambda_{t,t+h} v \text{ in } \mathbb{R}^N.
$$

(H3) *Existence of equilibria*: For all $\lambda > 0$, there exists $a_\lambda, b_\lambda \in \mathbb{R}$ such $a_\lambda < b_\lambda$ and $S^\lambda_{t,t+h} a_\lambda = a_\lambda$, $S^\lambda_{t,t+h} b_\lambda = b_\lambda$ for all $t \geq 0$ and $h > 0$. Moreover,

$$
a_\lambda \leq u_\lambda(\cdot, 0) \leq b_\lambda \quad \text{in } \mathbb{R}^N
$$

and there exists $a, b \in \mathbb{R}$ such that $a < b$ and, as $\lambda \to 0$,

$$a_\lambda \to a \quad \text{and} \quad b_\lambda \to b.$$

(H4) *Consistency*: For any $(x,t) \in \mathbb{R}^N \times (0,T)$ and for any open subset $\mathcal{O}$ of $\mathbb{R}^N$ with smooth boundary such that $\mathcal{O} \subset B_r(x)$ for some $r > 0$, there exists $\delta > 0$ and $h_0 > 0$, depending only on r and the C^3-norm of $\partial\mathcal{O}$ such that, as $\lambda \to 0$,

$$S^\lambda_{t,t+h}[(b_\lambda - \delta)\mathbf{1}_{\overline{\mathcal{O}}} + a_\lambda \mathbf{1}_{\mathbb{R}^N \setminus \overline{\mathcal{O}}}] \to b \quad \text{in } \mathcal{O}_{t+h},$$

$(\mathcal{O}_{t+h})_{h \in (0,h_0)}$ been the smooth evolution of $\mathcal{O}_t = \mathcal{O}$ by the law $V + \alpha$ for all small $\alpha > 0$.

A similar statement should hold for $S^\lambda_{t,t+h}(b_\lambda \mathbf{1}_{\mathcal{O}} + (a_\lambda + \delta)\mathbf{1}_{\mathbb{R}^N \setminus \overline{\mathcal{O}}})$ outside a family of smooth sets evolving with normal velocity $V - \alpha$.

The result obtained in [BS], again stated in a somehow simplified form, is:

THEOREM [BS]. *Assume that (H1)–(H4) hold and set, for $t > 0$,*

$$\overset{*}{\lim\sup}\, u_\lambda(x,t) = \limsup_{\substack{(y,s) \to (x,t) \\ \lambda \to 0}} u_\lambda(y,s); \quad \liminf_{*} u_\lambda(x,t) = \liminf_{\substack{(y,s) \to (x,t) \\ \lambda \to 0}} u_\lambda(y,s).$$

If $\Omega^1_0 = \mathbb{R}^N \setminus \overline{\Omega}^2_0$ and if the evolution of the family $(\Omega_t)_{t \in [0,T]}$ with normal velocity V does not create interior, then $\Omega^1_t = \Omega_t$ and $\Omega^2_t = \mathbb{R}^N \setminus \overline{\Omega}^1_t$, i.e. $u_\lambda \to b$ in Ω_t and $u_\lambda \to a$ in $\mathbb{R}^N \setminus \overline{\Omega}_t$.

In most of the examples (H1) and (H3) are given by the problem and (H2) follows from some maximum principle-type argument. The only difficulty lies in checking (H4) which amounts to proving a result similar to the one to be proved but only for smooth data, for compact smooth fronts and for small time, as small as needed. In other words checking (H4) amounts to showing that the formal asymptotics, that one usually derives in such problems, can be justified under all the needed regularity assumptions. This is not an easy task, but nevertheless considerably easier than working for the general problem with no regularity available.

Finally, to make the statement of the theorem a bit shorter, I omitted an assumption, which amounts to having that Ω^1_t and Ω^2_t are non-empty for small $t > 0$. In the examples this can be shown by some short-time analysis, which, although messy to write, is considerably easier than checking (H4).

I conclude remarking that this theorem is the basis for proving the result of the previous section as well as to study the asymptotics of reaction-diffusion equations with anisotropies and oscillatory coefficients.

References

[AC]   S. M. Allen and J. W. Cahn, *A macroscopic theory for antiphase boundary motion and its application to antiphase domain coarsening*, Acta Metal. **27** (1979), 1085–1095.

[AGLM]   L. Alvarez, F. Guichard, P.-L. Lions, and J.-M. Morel, *Axioms and fundamental equations of image processing*, Arch. Rational Mech. Anal. **123** (1992), 199–257.

[Ba]   G. Barles, *Remark on a flame propagation model*, Rapport INRIA No. 464.

[BS]   G. Barles and P. E. Souganidis, *A new approach to front propagation problems: Theory and Application*, preprint.

[BSS]   G. Barles, H. M. Soner, and P. E. Souganidis, *Front propagation and phase field theory*, SIAM J. Control Optim. **31** (1993), 439–469.

[Bo]   L. Bonaventura, *Mouvement par courbure dans un system de spin avec interaction*, preprint.

[C]   X. Chen, *Generation and propagation of interfaces in reaction diffusion systems*, J. Differential Equations **96** (1992), 116–141.

[CGG]   Y.-G. Chen, Y. Giga, and S. Goto, *Uniqueness and existence of viscosity solutions of generalized mean curvature flow equations*, J. Differential Geom. **33** (1991), 749–786.

[CL]   M. G. Crandall and P.-L. Lions, *Viscosity solutions of Hamilton-Jacobi equations*, Trans. Amer. Math. Soc. **277** (1983), 1–42.

[D]   E. DeGiorgi, *Boundaries, barriers, motion of manifolds*, Proceedings of Capri Workshop, 1990.

[DP]   A. DeMasi and E. Presutti, Mathematical methods for hydrodynamic limits, Lecture Notes in Math. **1501**, Springer-Verlag, Berlin New York, 1991.

[DOPT]   A. DeMasi, E. Orlandi, E. Presutti, and L. Triolo, *Glauber evolution with Kač potentials: I. Mesoscopic and macroscopic limits, interface dynamics*, Nonlinearity **7** (1994), 633–696.

[DS]   P. DeMottoni and M. Schatzman, *Geometric evolution of developed interfaces*, Trans. Amer. Math. Soc., in press.

[ESS]   L. C. Evans, H. M. Soner, and P. E. Souganidis, *Phase transitions and generalized motion by mean curvature*, Comm. Pure Appl. Math. **XLV** (1992), 1097–1123.

[ES]   L. C. Evans and J. Spruck, *Motion of level sets by mean curvature I*, J. Differential Geom. **33** (1991), 635–681, II, Trans. Amer. Math. Soc. **330** (1992), 635–681, III, J. Geom. Anal. **2** (1992), 121–150.

[G]   S. Goto, *Generalized motion of hypersurfaces with superlinear growth speed in curvature*, Differential Integral Equations **7** (1994), 323–343.

[GSS]   M. Gurtin, H. M. Soner, and P. E. Souganidis, *Anisotropic motion of an interface relaxed by the formation of infinitesimal wrinkles*, J. Differential Equations, **119** (1995), 54–108.

[I]   T. Ilmanen, *Convergence of the Allen-Cahn equation to Brakke's motion by mean curvature*, J. Differential Geom. **38** (1993), 417–461.

[IS]   H. Ishii and P. E. Souganidis, *Generalized motion of noncompact hypersurfaces with velocity having arbitrary growth on the curvature tensor*, Tôhoko Math. J. **47** (1995), 227–250.

[J]   R. Jerrard, *Fully nonlinear phase transitions and generalized mean curvature motion*, Comm. Partial Differential Equations **20** (1995), 223–265.

[KS1]   M. Katsoulakis and P. E. Souganidis, *Generalized motion by mean curvature as a macroscopic limit of stochastic Ising models with long range interactions and Glauber dynamics*, Comm. Math. Phys. **169** (1995), 61–97.

[KS2]   M. Katsoulakis and P. E. Souganidis, *Interacting particle systems and generalized front propagation*, Arch. Rational Mech. Anal. **127** (1994), 133–157.

[KS3]   M. Katsoulakis and P. E. Souganidis, *Generalized front propagation and macroscopic limits of Stochastic Ising models with long range anisotropic spin flip dynamics*, preprint.

[L]   P.-L. Lions, *Optimal control of diffusion processes. Part 2: Viscosity solutions and uniqueness*, Comm. Partial Differential Equations **8** (1983), 1229–1276.

[MS]   A. Majda and P. E. Souganidis, *Large scale front dynamics for turbulent reaction-diffusion equations with separated velocity scales*, Nonlinearity **7** (1994), 1–30.

[OS]   S. Osher and J. Sethian, *Fronts moving with curvature dependent speed: Algorithms based on Hamilton-Jacobi equations*, J. Comput. Phys. **79** (1988), 12–49.

[RSK]   J. Rubinstein, P. Sternberg, and J. B. Keller, *Fast reaction, slow diffusion and curve shortening*, SIAM J. Appl. Math. **49** (1989), 116–133.

[So1]   H. M. Soner, *Motion of a set by the curvature of its boundary*, J. Differential Equations **101** (1993), 313–372.

[So2]   H. M. Soner, *Ginzburg-Landau equation and motion by mean curvature I: Convergence, II: Development of the initial interface*, J. Geom. Anal; to appear.

[Sp]   H. Spohn, Large Scale Dynamics of Interacting Particles, Springer-Verlag, Berlin and New York, 1991.

Fully Nonlinear Elliptic Equations and Applications to Geometry

JOEL SPRUCK

Johns Hopkins University
Baltimore, Maryland 21218, USA

In this paper we will describe some recent advances in the theory of fully nonlinear elliptic equations that are motivated by some basic geometric problems. For example, one can ask, when does a smooth Jordan curve in R^3 bound a surface of positive constant Gauss curvature? The theme of this talk is roughly that such geometric problems often suggest the proper formulation of purely analytic partial differential equation (PDE) results. As an example, in 1984 [2] it was shown that the classical Monge-Ampère boundary value problem

$$\begin{cases} \det u_{ij} = \psi(x) & \text{in} \quad \Omega \\ u = \phi & \text{on} \quad \partial\Omega, \end{cases} \tag{$*$}$$

where ϕ, ψ, Ω smooth, $\psi_0 = \inf_\Omega \psi > 0$, and Ω *strictly convex*, always has a (unique) strictly convex solution $u \in C^\infty(\bar{\Omega})$.

It was also shown that the Dirichlet problem for surfaces of constant positive Gauss curvature K_0

$$\begin{cases} \det u_{ij} = K_0(1 + |\nabla u|^2)^{\frac{n+2}{2}} & \text{in} \quad \Omega \\ u = \phi & \text{on} \quad \partial\Omega, \end{cases} \tag{$**$}$$

where ϕ, Ω smooth and Ω *strictly convex*, has a (unique) strictly convex solution $u \in C^\infty(\bar{\Omega})$ for K_0 sufficiently small depending on the boundary data.

From the analytic point of view, these results are essentially best possible. However, from the point of view of geometry, one wants to solve $(*)$ or $(**)$ in *domains Ω of arbitrary geometry*, as we will explain later. Of course, this is not always possible but it is important to understand the obstructions to solvability. For example, Guan and I [11] proved that for general Ω, if there is a strictly convex strict subsolution of $(*)$ or $(**)$ for given boundary data ϕ, then $(*)$ or $(**)$ has a (unique) smooth solution.

We now proceed to describe how fully nonlinear elliptic equations arise from a surface described by a relation among its principal curvatures.

We call a hypersurface $S \subset R^{n+1}$, whose principal curvatures $\kappa = (\kappa_1, \ldots, \kappa_n)$ satisfy a relation of the form

$$f(\kappa) = \psi(x) > 0,$$

Proceedings of the International Congress
of Mathematicians, Zürich, Switzerland 1994
© Birkhäuser Verlag, Basel, Switzerland 1995

where ψ is a given smooth function of position and f is smooth and symmetric, a Weingarten surface. The most familiar examples arise by choosing $\psi \equiv 1$ and $f(\kappa) = \sigma_r(\kappa)$, $1 \leq r \leq n$, where $\sigma_r(\kappa)$ denotes the rth elementary symmetric function. Thus, up to a constant, σ_1 is the mean curvature, σ_2 is the scalar curvature, and σ_n is the Gauss-Kronecker curvature.

What does it mean that S is elliptic? One way to answer this question is to locally represent S as a graph $x_{n+1} = u(x)$ with $x \in \Omega \subset R^n$. Then the first and second fundamental forms of S are given by $g_{ij} = \delta_{ij} - \frac{u_i u_j}{W^2}$, $b_{ij} = \frac{u_{ij}}{W}$, where $W^2 = 1 + |Du|^2$. The κ_i are determined by the relation

$$\det(b_{ij} - \lambda g_{ij}) = 0 \,,$$

or equivalently, the κ_i satisfy

$$\det(a_{ij} - \lambda \delta_{ij}) = 0 \,,$$

where $a_{ij} = (g^{ij})^{\frac{1}{2}}(b_{ij})(g^{ij})^{\frac{1}{2}}$, (g^{ij}) is the inverse matrix to (g_{ij}), and $(g^{ij})^{\frac{1}{2}}$ is its positive square root (see [4, Lemma 1.1]).

We do not write down a_{ij} here but note only that a_{ij} is linear in $D^2 u$. Therefore, the function u satisfies a fully nonlinear equation of the form

$$f(\kappa) = F(a_{ij}) \equiv G(D^2 u, Du) = \psi \,,$$

and we say S is elliptic if G is elliptic, i.e. $\left(\frac{\partial G}{\partial u_{ij}}\right)$ is positive definite. If we choose coordinates so that

$$(a_{ij}) = (\kappa_1, \ldots, \kappa_n)_{\mathrm{diag}} \,,$$

then (see [4])

$$L = \frac{\partial G}{\partial u_{ij}} \partial_i \partial_j = \Sigma f_{\kappa_i} \partial_i^2$$

and L is elliptic if $f_{\kappa_i} > 0 \; \forall i$.

Another equivalent but more geometric way to see that S is elliptic is to compute the Jacobi operator $\mathcal{L}$ associated to a normal variation gN. Such a variation gives rise to a one-parameter family S_t with $S_0 = S$, where by definition

$$\mathcal{L}g = \frac{d}{dt} f(\kappa_t)\Big|_{t=0} \,.$$

In curvature coordinates, one can compute

$$\mathcal{L} = \sum_i \kappa_i^2 f_{\kappa_i} \,, \quad L = \sum_i f_{\kappa_i} \nabla_i^2$$

so again $\mathcal{L}$ is elliptic if $f_{\kappa_i} > 0 \; \forall i$. When $\psi(x) \equiv \psi_0$, a positive constant, one also can derive the following important identities:

$$\mathcal{L}\nu = 0 \,, \quad X = \sum_i \kappa_i f_{\kappa_i} \nu \,,$$

where ν is the unit normal and X is the position vector. These formulas give an elegant generalization of the well-known case of mean curvature ($f = \sum_i \kappa_i$) and are important for the analytic study of these Weingarten surfaces.

The function $f(\lambda)$ is naturally defined in an open convex cone with vertex at 0 (the ellipticity cone), containing the positive cone

$$\Gamma^+ = \{\lambda \in \mathbb{R}^n : \lambda_i > 0 \ \forall i\} .$$

We call S admissible if at every point $P \in S$, $\kappa(P) \in \Gamma$. For example, when $f(\lambda) = \sigma_r^{\frac{1}{r}}(\lambda)$, $1 \leq r \leq n$, Γ_r is the component of $\{\lambda \in \mathbb{R}^n : f(\lambda) > 0\}$ containing $(1, \ldots, 1)$. Then (see [3]) $f_{\lambda_i} > 0 \ \forall i$ in Γ_r, $\Gamma_1 = [\lambda : [\lambda_i > 0\}$, $\Gamma_n = \Gamma^+$, and $\Gamma_k \subseteq \Gamma_{k-1}$. Moreover, f is *concave* in Γ_r. This concavity is particularly important because it is equivalent to the concavity of $F(a_{ij}) = f(\kappa)$. In particular, our fully nonlinear equation $G(D^2u, Du) = \psi$ is concave on D^2u and this is precisely the class of fully nonlinear elliptic equations for which we have a good regularity theory.

We now concentrate on the Gauss curvature and formulate some basic geometric problems. Let $\Gamma = (\Gamma_1, \ldots, \Gamma_m) \subset R^{n+1}$ be a smooth disjoint collection of closed codimension 2 embedded submanifolds (for example, curves in R^3). The basic question is the existence of a hypersurface of positive constant Gauss curvature K (we will call such a surface a K-hypersurface) spanning Γ. Of course, the range of allowing K depends on Γ. An elementary necessary condition ($n = 2$) is that each Γ_i does not contain inflection points. However, even for a single Γ there are topological obstructions for Γ to bound an immersed S with $K(S) > 0$, see [14], [7]. We make the

CONJECTURE. *Suppose Γ bounds a strictly locally convex hypersurface S with $K(S) > K_0 > 0$. Then Γ bounds a K_0-hypersurface.*

This problem is difficult and open in this generality. Let's look at some appealing special cases.

EXAMPLE 1. Let $\Gamma = \{C_1, C_2\}$ with C_1, C_2 strictly convex in parallel planes. It is evident that Γ bounds a strictly convex annulus and for $K > 0$ sufficiently small we expect at least one solution.

EXAMPLE 2 (extreme Γ). Let S be a strictly convex compact (without boundary) hypersurface and let $\Gamma = \partial D$ for D a finitely connected subdomain of S with smooth boundary. Set $K_0 = \inf_{P \in S} K(P)$. For $0 < K < K_0$, we expect a K-hypersurface solution.

Both examples were recently settled by Guan and myself [11] by finding the solution as a radial graph $X = \rho(x)x$, $x \in \Omega \subset S^n$, where $\rho > 0$ satisfies a Monge-Ampère type equation with appropriate boundary values, and Ω is a smooth domain in S^{n-1} of arbitrary geometry obtained by projecting appropriately the given Γ. Thus, we arrive at the following formulation of our boundary value problem on S^n:

Let Ω be a smooth domain on $S^n \subset R^{n+1}$. We seek a smooth strictly locally convex hypersurface, which can be represented as

$$X(x) = \rho(x)x, \quad \rho > 0, \quad x \in \Omega, \tag{1}$$

with Gauss curvature

$$K[X(x)] = \psi(x), \quad x \in \Omega, \tag{2}$$

and boundary values

$$X(x) = \phi(x)x \quad \text{on} \ \partial\Omega, \tag{3}$$

where $\psi \in C^\infty(\overline{\Omega})$, $\phi \in C^\infty(\partial\Omega)$, $\psi, \phi > 0$.

We assume that

$$\Omega \text{ does not contain any hemisphere,} \tag{4}$$

and that there exists a smooth strictly locally convex radial graph $\bar{X}(x) = \bar{\rho}(x)x$ over Ω satisfying

$$\begin{aligned} K[\bar{X}(x)] &\geq \psi(x) + \delta_0 & \text{in} \ \Omega, \\ \bar{\rho} &= \phi & \text{on} \ \partial\Omega, \end{aligned} \tag{5}$$

for some $\delta_0 > 0$.

The main result of [11] may be stated as follows.

THEOREM 1 *Under conditions (4) and (5), there exists a smooth strictly locally convex radial graph $X(x) = \rho(x)x$ with $\rho \leq \bar{\rho}$ that satisfies (2) and (3). Moreover any such solution satisfies the a priori estimate*

$$\|\rho\|_{C^{2,\alpha}(\overline{\Omega})} \leq C, \quad \rho \geq c_0 > 0, \quad \kappa_i \geq \epsilon_0, \quad i = 1, \dots, n, \tag{6}$$

where the κ_i are the principal curvatures of X and C, c_0, ϵ_0 are uniform constants.

In general, solutions to (2) and (3) are not unique and we find the unique solution that is closest to the subsolution.

COROLLARY. *Let M be a strictly convex hypersurface in R^{n+1}, and let $\Gamma = (\Gamma_1, \dots, \Gamma_m)$ be strictly extreme; that is, $\Gamma = \partial D$ for a smooth subdomain $D \subset M$. Then for $0 < K < K(M)$, Γ bounds an embedded K-hypersurface that is contained inside M.*

Proof. In order to apply Theorem 1 we need only observe that if we choose our origin of coordinates strictly inside the convex hull of $M - D$, then D radially projects onto a domain Ω satisfying (4) and (5). □

In the same way we can also prove

COROLLARY. *Let M be a globally strictly convex hypersurface with boundary in R^{n+1}. Then for $0 < K < K(M)$, ∂M bounds an embedded K-hypersurface that is contained "inside" M.*

It is also of interest to consider a polyhedral version of Example 1.

EXAMPLE 3. Let Σ be the boundary of a convex polyhedron in R^{n+1} and let $\Gamma = (\Gamma_1, \dots, \Gamma_m) \subset \Sigma$ be a collection of smooth strictly convex codimension 2 surfaces such that each face of Σ contains at most one Γ_i strictly in its interior. For $K > 0$ sufficiently small find a K-hypersurface spanning Γ.

COROLLARY. *For K sufficiently small, there is an embedded K-hypersurface that solves Example 3. In particular, any two strictly convex curves in parallel planes bound some K-surface.*

Proof. We observe that there is no topological obstruction to the construction of a strictly convex hypersurface spanning Γ (without loss of generality we may suppose $m \geq 2$). For example, it is easy to construct a piecewise smooth strictly convex (in a generalized sense) hypersurface S spanning Γ that is smooth in a neighborhood of Γ. The interior of S can then be smoothed to obtain a strictly globally convex smooth hypersurface M spanning Γ. Again choosing the origin in the convex hull of Γ allows us to apply Theorem 1 for $0 < K < K(M)$. $\qquad\square$

One of the surprising and important features of Theorem 1 with respect to the theory of Monge-Ampère equations is that there are no geometric assumptions on $\partial\Omega$. That is, all of the analytic and geometric obstructions to proving existence are now embodied in the assumption of the existence of a strict subsolution of (2) and (3). This has led us to a careful reexamination of the local regularity estimates of [2], [12]. In particular, it is worthwhile to state a global result from [11] that contains an optimal smooth existence result for graphs over domains in R^n of prescribed Gauss curvature. This result is also important for the geometric problems in H^{n+1} I will discuss next.

The classical boundary value problem of Monge-Ampère type in R^n may be formulated as follows. Find a strictly convex $u \in C^\infty(\bar{\Omega})$ admissible solution satisfying

$$\begin{cases} \det u_{ij} &= \psi(x, u, Du) \quad \text{on} \quad \Omega \\ u &= \phi \qquad\qquad \text{on} \quad \partial\Omega, \end{cases} \tag{7}$$

where Ω is smooth, ϕ, ψ smooth, $\psi_0 = \inf_\Omega \psi > 0$. The case $\psi(x, u, p) = K(x, u)$ $(1 + p^2)^{\frac{n+2}{2}}$, $K(x, u) \geq K_0 > 0$ corresponds to prescribed Gauss curvature, but it is useful for applications to allow general ψ.

For Ω, ϕ arbitrary, (7) may not be solvable but the point of view here is to find a suitable condition that removes the obstruction for given boundary values ϕ.

THEOREM 2 *Assume*

    (i) *There is a smooth strictly convex $\underline{u}$ satisfying*

$$\begin{aligned} \det \underline{u}_{ij} &\geq \psi(x, \underline{u}, D\underline{u}) + \delta_0 \quad in \quad \Omega \\ \underline{u} &= \phi \qquad\qquad\qquad on \quad \partial\Omega \end{aligned}$$

    (ii) $\psi^{\frac{1}{n}}(\cdot, \cdot, p)$ *is convex in p.*

Then there is a smooth admissible solution u of (7); $u \geq \underline{u}$. (If $\psi_u \leq 0$, the solution is unique.) Moreover, $\|u\|_{C^{2+\alpha}(\bar{\Omega})} \leq C$ for controlled $\alpha \in (0,1), C > 0$.

The condition (ii) is technical, but is satisfied for prescribed Gauss curvature $\psi = K(x, u)(1+p^2)^{\frac{n+2}{2}}$. The theorem essentially says that if we have a subsolution for the problem we can deform it to a solution. This is a very powerful tool for geometric problems because a subsolution is often evident (as in Examples 1, 2,

3 above). We give another illustration of this by next considering a fundamental problem for Gauss curvature in hyperbolic space (joint work with Rosenberg [15]).

The basic problem we consider is to find a complete (embedded) strictly convex K-hypersurface S in H^{n+1} with given asymptotic boundary $\Gamma \in \partial_\infty H^{n+1}$. Recall that the Gauss curvature $K = K_{\text{ext}} - 1$, where K_{ext} is the extrinsic, i.e. the determinant of the second fundamental form. Thus, S is convex for $K > -1$.

Our approach is to construct S as the limit of "K-graphs" over a fixed compact domain Ω in a horosphere. For this it is convenient to use the half-space model

$$H^{n+1} = \{(x, x_{n+1}) \in R^{n+1} : x \in R^n , \quad x_{n+1} > 0\}$$

with metric

$$ds^2 = \frac{1}{x_{n+1}{}^2} \sum_{i=1}^{n} dx_i^2 .$$

We introduce the oriented distance from (x, x_{n+1}) to the horosphere $P_1 = \{x_{n+1} = 1\}$, given by $y = \ln x_{n+1}$, and look for S as a graph $y = f(x) x \in \Omega \subset P_1$ with respect to the vertical geodesics. On Ω we use the Euclidean metric.

LEMMA.   *The equation for K is*

$$K + 1 = \frac{\det(f_{ij} + 2f_i f_j + e^{-2f} \delta_{ij})}{e^{-2nf}(1 + e^{2f}|\nabla f|^2)^{\frac{n+2}{2}}} .$$

DEFINITION (admissible solution). We say f is "hyperbolic strictly locally convex" if $\{f_{ij} + 2f_i f_j + e^{-2f} \delta_{ij}\} > 0$ in Ω.

The basic boundary value problem for f then is

$$\begin{cases} \det(f_{ij} + 2f_i f_j + e^{-2f} \delta_{ij}) &= \psi(x, f, \nabla f) \quad \text{in} \quad \Omega \\ f &= \phi \qquad\qquad \text{on} \quad 2\Omega \end{cases} \tag{8}$$

with f, ψ, Ω smooth, $\psi_o = \inf_\Omega \psi > 0$. The case $\psi = (K + 1)e^{-2nf}(1 + e^{2f}|\nabla f|^2)$, $K + 1 = K(x, f) + 1 \geq \varepsilon_0 > 0$ corresponds to prescribed Gauss curvature $K = K(x, f)$.

Then we have exactly the same theorem as for the classical case [15].

THEOREM 3 *Assume*

(i) There is a strict admissible subsolution $\underline{u}$ of $(**)$ for the given boundary values ϕ.

(ii) $\psi^{\frac{1}{n}}(\cdot, \cdot, p)$ is convex.

Then there is a smooth admissible solution $f \in C^\infty(\Omega)$ to (8). Moreover, $\|f\|_{C^{2+\alpha}(\bar\Omega)} \leq C$ for controlled $\alpha \in (0, 1)$, $C > 0$.

COROLLARY.   Let $K \in (-1, 0)$, $\Omega \subset P_c = \{x_{n+1} = c\}$ smooth. Then there exists a smooth K-hypersurface S with $\partial S = \partial \Omega$ that can be chosen as a graph in horosphere coordinates.

Proof. We consider Ω as sitting in P_1 and apply the theorem with $f = \ln c$, $\underline{u} \equiv \ln c$, that is the horosphere P_c satisfies $K \equiv 0$ so $\Omega \subset P_c$ is a strict admissible subsolution. $\qquad\square$

It is important for the following discussion to note that there is a remarkable simplification of $(**)$ by the change of variable $u = e^{2f}$, $\varphi = e^{2\phi}$, $\underline{u} = e^{2\underline{f}}$. Then f admissible is equivalent to $\{u_{ij} + 2\delta_{ij}\} > 0$ and

$$\begin{cases} \det(u_{ij} + 2\delta_{ij}) = 2^n u^n \psi(x, \tfrac{1}{2}\ln u, \tfrac{1}{2}\tfrac{\nabla u}{u}) = \tilde{\psi}(x, u, \nabla u) & \text{in} \quad \Omega \\ \qquad\qquad\qquad\qquad u = \varphi & \text{on} \quad \partial\Omega \ . \end{cases}$$

For the case of prescribed Gauss curvature, $\tilde{\psi} = (K+1)(1 + \frac{|\nabla u|^2}{4u})^{\frac{n+2}{2}}$.

Given a prescribed $\Gamma = \partial\Omega \subset P_\infty = \{x_{n+1} = 0\}$ we vertically translate Ω to $\Omega_c \subset P_c$ and take $\Gamma_c = \partial\Omega_c$. Applying the Corollary, we find for $K \in (-1,0)$ a family of smooth admissible solutions $u(x,c)$ to

$$\begin{cases} \det(u_{ij} + 2\delta_{ij}) = 2^n (K+1)(1 + \frac{|\nabla u|^2}{4n})^{\frac{n+2}{2}} & \text{on} \quad \Omega \\ \qquad\qquad\qquad\qquad u = c^2 & \text{on} \quad \partial\Omega \ . \end{cases} \qquad (9)$$

Our goal is to pass to the limit as $c \to 0$. Of course, as we are "going to infinity" in H^{n+1} there is a degeneracy on the right-hand side of (9). Using the rich geometry of hyperbolic space, we are able to construct appropriate comparison functions and prove the basic

PROPOSITION. $\|u(x;c)\|_{C^2(\Omega)} \leq C$ *with C independent of c as $c \to 0$.*

By elliptic regularity [1] this implies

COROLLARY. $\|u(x,c)\|_{C^{2+\alpha}(\Omega')} \leq C(\Omega')$ *independent of c for $\Omega' \subset\subset \Omega$.*

Now we can pass to the limit as $c \to 0$ and obtain

THEOREM 4 *Let $\Gamma = \partial\Omega \subset \partial_\infty H^{n+1}$ be smooth. Then for $K \in (-1,0)$, $\Gamma = \partial_\infty S$ for S an embedded K-hypersurface of H^{n+1}. Moreover, S can be represented as a graph $X_{n+1} = \sqrt{u(x)}$ over Ω with $u \in C^\infty(\Omega) \cap C^{1,1}(\bar\Omega), u > 0$ on Ω. $u = 0$ on $\partial\Omega$.*

REMARK. For Γ Jordan the theorem holds with $u \in C^\infty(\Omega) \cap C^{0,1}(\bar\Omega)$.

Finally, it is a remarkable property of H^3 that all of the K-surfaces we construct are canonically unique for a single Jordan curve boundary.

DEFINITION. A Jordan curve Γ on P_∞ is the asymptotic homological boundary of a surface S in H^3 if for $c > 0$ small, $S \cap P_c$ contains a connected component Γ_c such that $\Gamma_c \to \Gamma$ as $c \to 0$, and Γ_c is homologous to zero on S. We write $\Gamma = \partial_\infty S$.

THEOREM 5 *Let Ω be a bounded simply connected domain on P_c, respectively P_∞, with Jordan boundary Γ. Then there are exactly embedded K-surfaces S in H^3 with $\partial S = \Gamma$, respectively $\partial_\infty S = \Gamma$. Each surface is a graph over one of the two components of $P_\infty - \Gamma$. Moreover, any immersed K-surface with $\partial S = \Gamma$, respectively, $\partial_\infty S = \Gamma$, is embedded and thus one of the two graph solutions.*

The main fact used in the proof is that the Jacobi operator is of the form $\mathcal{L} = L + HK$, with $H > 0$, $K < 0$, so that $\mathcal{L}$ has no kernel. By our compactness results, we can then deform in P_c or P_∞ and foliate space by K-graphs. This allows us to use the maximum principle via "sweeping arguments."

In conclusion we just remark that the ideas presented here are not very special to the equation of Gauss curvature. For example, Guan [8], [10], [9] has shown that they hold for large classes of (nongeometric) fully nonlinear elliptic equations with suitable invariance properties. However, it remains a formidable task to solve the general geometric problems we have described for Gauss curvature and to extend even the partial results to other functions of curvature.

References

[1] L. A. Caffarelli, J. J. Kohn, L. Nirenberg, and J. Spruck, *The Dirichlet problem for nonlinear second-order elliptic equations, II. Complex Monge-Ampère and uniformly elliptic equations*, Comm. Pure Appl. Math. **38** (1985), 209–252.

[2] L. A. Caffarelli, L. Nirenberg, and J. Spruck, *The Dirichlet problem for nonlinear second-order elliptic equations, I. Monge-Ampère equations*, Comm. Pure Appl. Math. **37** (1984), 369–402.

[3] L. A. Caffarelli, L. Nirenberg, and J. Spruck, *The Dirichlet problem for nonlinear second-order elliptic equations, III. Functions of the eigenvalues of the Hessian*, Acta Math. **155** (1985), 211–301.

[4] L. A. Caffarelli, L. Nirenberg, and J. Spruck, *Nonlinear second-order elliptic equations, IV. Starshaped compact Weingarten hypersurfaces*, in Current Topics in P.D.E., Y. Ohya, K. Kasahara, and N. Shikmakura, eds., Kinokuniya Co., Tokyo, 1986, pp. 1–26.

[5] L. A. Caffarelli, L. Nirenberg, and J. Spruck, *Nonlinear second-order elliptic equations, V. The Dirichlet problem for Weingarten hypersurfaces*, Comm. Pure Appl. Math. **41** (1988), 47–70.

[6] L. C. Evans, *Classical solutions of fully nonlinear, convex, second order elliptic equations*, Comm. Pure Appl. Math. **35** (1982), 333–363.

[7] H. Gluck and L.-H. Pan, *Knot theory in the presence of curvature*, preprint.

[8] B. Guan, *Boundary value problems for surface of prescribed Gauss curvature and Monge-Ampère equations*, Ph. D. thesis, Univ. of Massachusetts at Amherst, 1992.

[9] B. Guan, *The Dirichlet problem for a class of fully nonlinear elliptic equations*, Comm. Partial Differential Equations **19** (1994), 399–415.

[10] B. Guan, *Existence and regularity of hypersurfaces of prescribed Gauss curvature and boundary*, preprint.

[11] B. Guan and J. Spruck, *Boundary value problems on S^n for surfaces of constant Gauss curvature*, Ann. of Math. (2) **138** (1993), 601–624.

[12] D. Hoffman, H. Rosenberg, and J. Spruck, *Boundary value problems for surfaces of constant Gauss curvature*, Comm. Pure Appl. Math. **45** (1992), 1051–1062.

[13] N. V. Krylov, *Boundedly inhomogeneous elliptic and parabolic equations in a domain*, Izvestia Math. **47** (1983), 95–108.

[14] H. Rosenberg, *Hypersurfaces of constant curvature in space forms*, Bull. Sci. Math. (2) **117** (1993), 211–239.

[15] H. Rosenberg and J. Spruck, *On the existence of convex hypersurfaces of constant Gauss curvature in hyperbolic space*, to appear in J. Differential Geometry.

[16] N.S. Trudinger, *Fully nonlinear, uniformly elliptic equations under natural structure conditions*, Trans. Amer. Math. Soc. **278** (1983), 751–769.

Lower-Semicontinuity of Variational Integrals and Compensated Compactness

VLADIMÍR ŠVERÁK

Department of Mathematics, University of Minnesota
Minneapolis, MN 55455, USA

I   Lower-Semicontinuity and Quasiconvexity

We consider variational integrals

$$I(u) = \int_\Omega f(Du(x)) \, dx$$

defined for (sufficiently regular) functions $u : \Omega \to \mathbf{R}^m$. Here Ω is a bounded open subset of $\mathbf{R}^n$, $Du(x)$ denotes the gradient matrix of u at x, and $f : M^{m \times n} \to \mathbf{R}$ is given, $M^{m \times n}$ denoting the space of real $(m \times n)$-matrices. We are interested in the case $m, n \geq 2$.

Natural questions regarding I are those of existence, and then other properties (such as regularity) of minimizers of I in appropriate classes of functions. Questions regarding the existence of minimizers have been studied in Morrey's 1952 paper [Mo1], where the following notion was introduced: we say that f is *quasiconvex* if for any matrix $A \in M^{m \times n}$ and any smooth function $\varphi : \Omega \to \mathbf{R}^m$ compactly supported in Ω the inequality $\int_\Omega f(A + D\varphi) \, dx \geq \int_\Omega f(A) \, dx$ holds. The class of quasiconvex functions is independent of Ω. (See [Mo1], [Mo2].)

The basic result obtained in [Mo1] is that, under certain technical assumptions, quasiconvexity of f is necessary and sufficient for the weak sequential lower-semicontinuity of I on appropriate Sobolev spaces. Optimal results in this direction can be found in [AF1]. (Once we know that I is weakly sequentially lower-semicontinuous, we can use the direct method of the calculus of variations to obtain existence results. Of course, some technical assumptions are needed.)

In 1984 Evans proved that quasiconvexity of f gives also, when slightly strengthened, partial regularity of minimizers of I, see [Ev] and also [AF2].

In the next section we will see that quasiconvex functions also appear naturally in PDE problems that are not directly related to minimizers of variational integrals.

It is not difficult to verify that for $n = 1$ or $m = 1$ quasiconvexity reduces to convexity. On the other hand, for $n \geq 2$ and $m \geq 2$ there always exist nonconvex quasiconvex functions. (A typical example in the case $m = n$ is $f(X) = \det X$.) In fact, it turns out that it may be very difficult to decide whether or not a given function is quasiconvex. For specific examples see [AD], [DM], [Sv1]. In this

Proceedings of the International Congress
of Mathematicians, Zürich, Switzerland 1994
© Birkhäuser Verlag, Basel, Switzerland 1995

connection, the following simpler notions have been introduced, see [Ba1], [Da], [Mo2]:

- f is *rank-one convex* if for each matrix $A \in M^{m \times n}$ and each rank-one matrix $B \in M^{m \times n}$ the function $t \to f(A + tB)$ is convex. (For C^2–functions rank-one convexity is the same as the so-called *Legendre-Hadamard condition*, see [Ba1], [Mo2].)

- f is *polyconvex* if $f(X) =$ convex function of subdeterminants of the matrix X. (For example, $f : M^{2 \times 2} \to \mathbf{R}$ is polyconvex if there exists a convex function $G : M^{2 \times 2} \times \mathbf{R} \to \mathbf{R}$ such that $f(X) = G(X, \det X)$ for each $X \in M^{2 \times 2}$.) The important role of the subdeterminants stems from the fact that they are (the only) *null Lagrangians*, see [Ba1], [Mo2].

It is well known that rank-one convexity (RC) is a necessary condition for quasiconvexity (QC) and that polyconvexity (PC) is a sufficient condition for quasiconvexity. In other words, PC$\Rightarrow$QC$\Rightarrow$RC. We remark that in principle it should be relatively easy to decide whether or not a given function is rank-one convex or polyconvex (although actual computations can be lengthy and tedious). It is therefore of great interest to know whether or not there are further relations between the three notions of convexity introduced above. A classical result in this direction is that a quadratic function f is quasiconvex if and only if it is rank-one convex. This can be proved by using the Fourier transformation (and, in fact, using the Fourier transformation seems to be the only way to prove it).

It turns out that there are quasiconvex functions that are not polyconvex, see [Te], [Se], [Ba3], [AD], [Sv2].

For a long time it was an open problem whether or not RC$\Rightarrow$QC. It turns out that for $n \geq 2$, $m \geq 3$ this fails; see [Sv4] where an example is given that shows that for $n \geq 2$, $m \geq 3$ there exists a quartic polynomial that is rank-one convex but not quasiconvex. The case $n \geq 2$, $m = 2$ remains open. We remark that even in the case $n = m = 2$ the implication RC$\Rightarrow$QC would have far-reaching consequences.

II   Quasiconvexity and Compensated Compactness

Let $K \subset M^{m \times n}$ be a closed set. The purpose of this section is to show how quasiconvexity is related to compactness properties of sets of (approximate) solutions of the system

$$(1) \qquad\qquad\qquad\qquad Du \in K.$$

Our approach to this question is based on ideas from the theory of compensated compactness initiated by Tartar and Murat [Ta1], [Ta2], [Mu1], [Mu2], and also on ideas from [BJ1], [DP], and [KP].

We denote by $\mathcal{M}$ the set of all boundedly supported probability measures on $M^{m \times n}$. For $\nu \in \mathcal{M}$ we denote by $\bar\nu$ its center of mass, i.e. $\bar\nu = \int_{M^{m \times n}} X \, d\nu(X)$. We use the usual notation $\langle \nu, f \rangle = \int_{M^{m \times n}} f(X) \, d\nu(X)$. We define

$$\mathcal{M}^{qc}(K) = \{\nu \in \mathcal{M}, \ \text{supp}\,\nu \subset K \text{ and } \langle \nu, f \rangle \geq f(\bar\nu) \text{ for each quasiconvex } f\}.$$

We also define $\mathcal{M}^{rc}(K)$ and $\mathcal{M}^{pc}(K)$ in a similar way, replacing the class of quasiconvex functions in the definition of $\mathcal{M}^{qc}(K)$ by the class of rank-one convex functions and polyconvex functions respectively. It is easy to see that
$$\mathcal{M}^{pc}(K) = \{\nu \in \mathcal{M},\ \text{supp}\,\nu \subset K \text{ and } \langle \nu, f \rangle = f(\bar{\nu}) \text{ for each subdeterminant } f\}.$$
Because PC$\Rightarrow$QC$\Rightarrow$RC, we have $\mathcal{M}^{rc}(K) \subset \mathcal{M}^{qc}(K) \subset \mathcal{M}^{pc}(K)$.

We say that a set of measures is *trivial* if it contains only Dirac masses.

One of the reasons the sets introduced above are of interest is the following result, which follows directly from ideas developed in [Ta1], [Ta2], [BJ1], and [KP]. (The result is well known to experts, and the author does not claim any originality here.)

PROPOSITION. *The following conditions are equivalent:*
(i) For each sequence of functions $u_j : \Omega \to \mathbf{R}^m$ satisfying $|Du_j| \leq c$ (for some $c > 0$) and $\int_\Omega \text{dist}(Du_j, K) \to 0$, the sequence Du_j is compact in $L^1(\Omega)$.
(ii) $\mathcal{M}^{qc}(K)$ is trivial (i.e. the only measures in $\mathcal{M}^{qc}(K)$ are Dirac masses).

The proof of (ii)$\Rightarrow$(i) is based on considering the Young measures of (subsequences of) the sequence Du_j, an idea due to Tartar [Ta1], [Ta2]. See also [BJ1], [BJ2], [KP]. The proof of (i)$\Rightarrow$(ii) follows from results in [KP]. (Heuristically one can understand the implication (i)$\Rightarrow$(ii) in the following way. We note that if (1) has a nontrivial Lipschitz solution u such that $u(x) = Ax$ at the boundary of Ω for some $A \in M^{m \times n}$, then we can easily construct nontrivial periodic solutions of (1) in $\mathbf{R}^n$ and obtain noncompact sequences of solutions by using the invariance of (1) under the scaling $u(x) \to \epsilon u(\frac{x}{\epsilon})$. The main point now is that every measure ν in $\mathcal{M}^{qc}(K)$ can be viewed as an "almost solution" of (1) with the affine boundary condition $A = \bar{\nu}$.)

REMARK. It is known that if $\mathcal{M}^{rc}(K)$ is sufficiently nontrivial, then (1) can be expected to admit surprisingly wild exact solutions, which can be constructed by using an adaptation of Gromov's "convex integration", see [Gr, p. 218] and [MS]. It is not clear whether under some reasonable assumptions on K one could replace $\mathcal{M}^{rc}(K)$ by $\mathcal{M}^{qc}(K)$ in these results.

EXAMPLES.

1. An obvious necessary condition for $\mathcal{M}^{rc}(K)$ to be trivial (and hence also for $\mathcal{M}^{qc}(K)$ to be trivial) is that $\text{rank}(X - Y) \geq 2$ for each $X, Y \in K$, $X \neq Y$.

2. A sufficient condition for $\mathcal{M}^{qc}(K)$ to be trivial is that there exists a rank-one convex quadratic function f such that $f(X - Y) < 0$ for each $X, Y \in K$, $X \neq Y$. (To prove this we recall that each rank-one convex quadratic function is quasiconvex, see Section I. We can use this to infer that for each $\nu \in \mathcal{M}^{qc}(K)$ we have

$$0 \geq \int_{K \times K} f(X - Y)\, d\nu(X)\, d\nu(Y) = \int_{K \times K} (f(X) - Df(X)Y + f(Y))\, d\nu(X)\, d\nu(Y)$$
$$\geq f(\bar{\nu}) - Df(\bar{\nu})\bar{\nu} + f(\bar{\nu}) = 0.$$

Hence $\nu \times \nu$ is supported on the diagonal of $K \times K$ and therefore ν must be a Dirac mass.

3. Let $m = n = 2$ and let $K \subset M^{2 \times 2}$. Then $\mathcal{M}^{pc}(K)$ is trivial if and only if either $\det(X - Y) > 0$ for each $X, Y \in K$, $X \neq Y$, or $\det(X - Y) < 0$ for each $X, Y \in K$, $X \neq Y$. Hence if $K \subset M^{2 \times 2}$ is connected, then a necessary and sufficient condition for any of the sets $\mathcal{M}^{rc}(K), \mathcal{M}^{qc}(K), \mathcal{M}^{pc}(K)$ to be trivial is that $\det(X - Y) \neq 0$ for each $X, Y \in K$. See [Sv7] for details.

4. Let $K = \{A_1, A_2, A_3\} \subset M^{m \times n}$. Then $\mathcal{M}^{qc}(K)$ is trivial if and only if $\mathrm{rank}(X - Y) \geq 2$ for each $X, Y \in K$, $X \neq Y$. See [Sv3], [Sv5]. Note that in this case we can have a situation where $\mathcal{M}^{qc}(K)$ is trivial and $\mathcal{M}^{pc}(K)$ is nontrivial.

5. Let $m = n = 2$. Let $A = \begin{pmatrix} 2 & 0 \\ 0 & 1 \end{pmatrix}$, $B = \begin{pmatrix} -1 & 0 \\ 0 & 2 \end{pmatrix}$, and $K = \{A, B, -A, -B\}$. (This set K, which was first considered by Tartar, plays an important role in many examples.) Then $\mathrm{rank}(X - Y) \geq 2$ for each $X, Y \in K$, $X \neq Y$, but $\mathcal{M}^{rc}(K)$ is non-trivial. Using quasiconvex functions from [Sv5], one can prove that for this set K we have $\mathcal{M}^{qc}(K) = \mathcal{M}^{rc}(K)$. On the other hand $\mathcal{M}^{pc}(K) \neq \mathcal{M}^{rc}(K)$. (The computation of $\mathcal{M}^{rc}(K)$ and $\mathcal{M}^{pc}(K)$ in this case is an easy but instructive exercise.)

6. Let F be a quasiconvex (or polyconvex) function on $M^{l \times 2}$ and let us consider the functional $I(u) = \int_\Omega F(Du)$ and its Euler-Lagrange equation

$$(2) \qquad \qquad \mathrm{div}\, DF(Du) = 0.$$

This can be rewritten as $DF(Du) = -DvJ$, where $J = \begin{pmatrix} 0 & -1 \\ 1 & 0 \end{pmatrix}$ and u, v are unknown functions from $\Omega \subset \mathbf{R}^2$ to $\mathbf{R}^l$. Letting

$$K = \left\{ \begin{pmatrix} X \\ DF(X)J \end{pmatrix}, X \in M^{l \times 2} \right\} \subset M^{2l \times 2}$$

we see that (2) can be rewritten as $DU \in K$, where $U = \begin{pmatrix} u \\ v \end{pmatrix}$ is a function from Ω to $\mathbf{R}^{2l}$. One can now ask under which conditions on F the set $\mathcal{M}^{qc}(K)$ is trivial. This seems to be open. (Not much is known about $\mathcal{M}^{rc}(K)$ and $\mathcal{M}^{pc}(K)$ either. It is easy to see that if F is strictly rank-one convex, then $\mathrm{rank}(A - B) \geq 2$ for each $A, B \in K$, $A \neq B$. The opposite implication has been studied in [Ba2].) Of course, if $F = F_0 + L$ for some strictly convex F_0 and a null Lagrangian L, then $\mathcal{M}^{qc}(K)$ (and, in fact, $\mathcal{M}^{pc}(K)$) is trivial.

The triviality of $\mathcal{M}^{qc}(K)$ for K constructed from (2) should be equivalent, modulo technicalities, to compactness of interesting classes of approximate weak solutions of (2). It should also be related to partial regularity of weak solutions of (2). On the other hand, if $\mathcal{M}^{rc}(K)$ is sufficiently nontrivial, then one can hope to construct very wild exact solutions of (2) by using an adaptation of the method of convex integration from [Gr]; see the remark following the previous Proposition. It can happen that in the situation considered in this example the set $\mathcal{M}^{qc}(K)$ is nontrivial even under the assumption that F is strictly polyconvex. In fact, there is an example [Sv8] of a strictly polyconvex quartic polynomial F on $M^{6 \times 2}$ such that (2) admits a smooth, nontrivial periodic solution $u : \mathbf{R}^2 \to \mathbf{R}^6$.

7. Let us consider an $(l \times 2)$-hyperbolic system

$$(3) \qquad \qquad u_t + f(u)_x = 0$$

together with the entropy condition

$$\eta(u)_t + q(u)_x \leq 0,$$

where $u : (0,T) \times (a,b) \to \mathbf{R}^l$ is the unknown function, $f : \mathbf{R}^l \to \mathbf{R}^l$ is a given function satisfying suitable (hyperbolicity) assumptions, and (η, q) is a suitable entropy pair (see [Ta1], [Ta2], [DP] for details). In connection with questions regarding compactness properties of weak solutions of this system, one can consider the set

$$K = \left\{ \begin{pmatrix} f(u), & -u \\ q(u), & -\eta(u) \end{pmatrix} , u \in \mathbf{R}^l \right\} \subset M^{(l+1)\times 2}.$$

It is not difficult to see that the question of whether or not $\mathcal{M}^{qc}(K)$ is trivial is equivalent, modulo technicalities, to questions regarding compactness of certain classes of approximate entropic solutions of (3). It is not known under which conditions the set $\mathcal{M}^{qc}(K)$ (resp. $\mathcal{M}^{rc}(K)$, $\mathcal{M}^{pc}(K)$) is trivial. (Of course, we have the obvious necessary condition that rank$(A - B) \geq 2$ for each $A, B \in K$, $A \neq B$. This assumption is equivalent to the condition that the system under consideration does not admit shock waves that preserve the entropy η.)

It is clear that a number of problems in PDE can be formulated as questions about $\mathcal{M}^{qc}(K)$ for suitable sets K. One of the main motivations for studying $\mathcal{M}^{qc}(K)$ for "nonstandard" sets K has been the paper [BJ1], where a model for certain phase transformations is proposed in which the sets $\mathcal{M}^{qc}(K)$ play an important role. See also [BJ2], [Ko], [Sv6].

References

[AF1]   E. Acerbi and N. Fusco, *Semicontinuity problems in the calculus of variations*, Arch. Rational Mech. Anal. **86** (1986), 125–145.

[AF2]   E. Acerbi and N. Fusco, *A regularity theorem for minimizers of quasiconvex integrals*, Arch. Rational Mech. Anal. **99** (1987), 261–281.

[AD]   J. J. Alibert and B. Dacorogna, *An example of a quasiconvex function not polyconvex in dimension two*, Arch. Rational Mech. Anal. **117** (1992), 155-166.

[Ba1]   J. M. Ball, *Convexity conditions and existence theorems in nonlinear elasticity*, Arch. Rational Mech. Anal. **63** (1978), 337–403.

[Ba2]   J. M. Ball, *Strict convexity, strong ellipticity, and regularity in the calculus of variations*, Math. Proc. Cambridge Philos. Soc. **87** (1980), 501–513.

[Ba3]   J. M. Ball, *Remarks on the paper 'Basic calculus of variations'*, Pacific J. Math., **116**, No.1 (1985).

[BJ1]   J. M. Ball and R. D. James, *Fine phase mixtures as minimizers of energy*, Arch. Rational Mech. Anal. **100** (1987), 13–52.

[BJ2]   J. M. Ball and R. D. James, *Proposed experimental tests of a theory of fine microstructures and the two-well problem*, preprint.

[Da]   B. Dacorogna, Direct Methods in the Calculus of Variations, Springer-Verlag, Berlin/Heidelberg 1989.

[DM]   B. Dacorogna and P. Marcellini, *A counterexample in the vectorial calculus of variations*, in Material Instabilities in Continuum Mechanics (J. M. Ball ed.), Oxford Sci. Publ., Oxford Univ. Press, New York, 1988, 77–83.

[DP]   J. P. DiPerna, *Compensated compactness and general systems of conservations laws*, Trans. Amer. Math. Soc. **292** (1985), 383–420.

[Ev]   L. C. Evans, *Quasi-convexity and partial regularity in the calculus of variations*, Arch. Rational Mech. Anal. **95** (1986), 227–252.

[Gr]   M. Gromov, Partial Differential Relations, Springer-Verlag, Berlin/Heidelberg 1986.

[KP]   D. Kinderlehrer and P. Pedregal, *Characterization of Young measures generated by gradients*, Arch. Rational Mech. Anal. **115** (1991), 329–367.

[Ko]   R. V. Kohn, *The relaxation of a double-well energy*, Cont. Mech. Thermodyn. **3** (1991), 192–236.

[Mo1]   Ch. B. Morrey, *Quasi-convexity and the lower semicontinuity of multiple integrals*, Pacific J. Math. **2** (1952), 25–53.

[Mo2]   Ch. B. Morrey, Multiple Integrals in the Calculus of Variations, Springer-Verlag, Berlin/Heidelberg, 1966.

[MS]   S. Müller and V. Šverák, in preparation.

[Mu1]   F. Murat, *Compacité par compensation*, Ann. Scuola Norm. Sup. Pisa **5** (1978), 489–507.

[Mu2]   F. Murat, *Compacité par compensation: condition necessaire et suffisante de continuité faible sous une hypotheses de rang constant*, Ann. Scuola Norm. Sup. Pisa Cl. Sci. (4) **8** (1981), 69–102.

[Se]   D. Serre, *Formes quadratiques et calcul des variations*, J. Math. Pures Appl. (9), **62** (1983), 177–196.

[Sv1]   V. Šverák, *Examples of rank-one convex functions*, Proc. Roy. Soc. Edinburgh Sect A, **114 A** (1990), 237–242.

[Sv2]   V. Šverák, *Quasiconvex functions with subquadratic growth*, Proc. Roy. Soc. Lond. Ser. A **433** (1991), 723–725.

[Sv3]   V. Šverák, *On regularity for the Monge-Ampère equation without convexity assumptions*, preprint.

[Sv4]   V. Šverák, *Rank-one convexity does not imply quasiconvexity*, Proc. Roy. Soc. Edinburgh Sect. A **120** (1992), 185–189.

[Sv5]   V. Šverák, *New examples of quasiconvex functions*, Arch. Rational Mech. Anal. **119** (1992), 293–300.

[Sv6]   V. Šverák, *On the problem of two wells*, in Microstructure and phase transition (D. Kinderlehrer, R. D. James, M. Luskin, and J. L. Ericksen, eds.), IMA Vol. Math. Appl. **54**, Springer-Verlag, Berlin/Heidelberg, 1993.

[Sv7]   V. Šverák, *On Tartar's conjecture*, Ann. Inst. H. Poincaré, Anal. Non Linéaire **10**, no. 4 (1993), 405–412.

[Sv8]   V. Šverák, to appear.

[Ta1]   L. Tartar, *Compensated compactness and applications to partial differential equations*, in Nonlinear Analysis and Mechanics: Heriot-Watt Symposium IV, Pitman Research Notes in Mathematics **39** (1979), 136–212.

[Ta2]   L. Tartar, *The compensated compactness method applied to systems of conservations laws*, in Systems of Nonlinear Partial Differential Equations (J. M. Ball ed.), NATO ASI Series **C 111**, Reidel, Dordrecht 1982.

[Te]   F. J. Terpstra, *Die Darstellung der biquadratischen Formen als Summen von Quadraten mit Anwendung auf die Variationsrechnung*, Math. Ann. **116** (1938), 166–180.

The Riemann-Hilbert Problem and Fuchsian Differential Equations on the Riemann Sphere

A.A. BOLIBRUCH

Steklov Mathematical Institute
Vavilov str. 42
Moscow 117966
Russia

1. Introduction

(1) The Riemann-Hilbert problem concerns a certain class of linear ordinary differential equations (ODEs) in the complex domain. Let the system

$$\frac{dy}{dx} = B(x)y \qquad (1)$$

with unknown vector function $y = (y^1, \ldots, y^p)^t$ (t means transposition) have singularities $a_1, \ldots, a_n$; that is, $B(x)$ is holomorphic in $S := \bar{\mathbb{C}} \setminus \{a_1, \ldots, a_n\}$ (where $\bar{\mathbb{C}}$ is the Riemann sphere). The system is called *Fuchsian at a_i* (and a_i is a *Fuchsian singularity* of the system) if $B(x)$ has a pole there of order at most one. The system is *Fuchsian* if it is Fuchsian at all a_i. Let all $a_i \neq \infty$. Then

$$B(x) = \sum_{i=1}^{n} \frac{1}{x - a_i} B_i, \quad \sum_{i=1}^{n} B_i = 0. \qquad (2)$$

Consider a loop g, starting at a point $x_0 \in S$ and lying in S. Under analytic continuation along this loop the germ $\hat{Y}(x)$ at x_0 of a fundamental matrix $Y(x)$ to (1) is transformed to $Y'(x) = \hat{Y}(x)G^{-1}$, $G \in GL(p, \mathbb{C})$. The correspondence $g \mapsto G$ generates a linear representation

$$\chi : \pi_1(S, x_0) \longrightarrow GL(p, \mathbb{C}), \qquad (3)$$

which is called *a monodromy representation* of system (1) or simply a *monodromy*.

The monodromy for the pth order linear ODE

$$y^{(p)} + q_1(x)y^{(p-1)} + \cdots + q_p(x)y = 0 \qquad (4)$$

is just the same as for the pth order system, describing the behavior of the vectors $\left(y, \frac{dy}{dx}, \ldots, \frac{d^{p-1}y}{dx^{p-1}}\right)^t$, where y satisfies the equation. Equation (4) is called *Fuchsian at a point a* if its coefficients $q_1(x), \ldots, q_p(x)$ are holomorphic in some punctured neighborhood of this point and $q_i(x) = r_i(x)/(x - a)^i$, $i = 1, \ldots, p$, where $r_1(x), \ldots, r_p(x)$ are functions holomorphic at a.

Proceedings of the International Congress
of Mathematicians, Zürich, Switzerland 1994
© Birkhäuser Verlag, Basel, Switzerland 1995

(2) At the end of the 1850s, Riemann was the first to mention the problem of the reconstruction of a Fuchsian equation from its monodromy representation in a note. In 1990 Hilbert included it on his list of "Mathematical Problems" under the 21st number. It was formulated as follows [Hi]:

Prove that there always exists a linear differential equation of Fuchsian type with given singular points and a given monodromy group.

A tradition has been established in mathematical literature that the problem for Fuchsian systems is usually called the Riemann-Hilbert problem. (Note here that it was known at that time that the problem for a Fuchsian scalar ODE had a negative solution. This follows from the fact that a Fuchsian equation of pth order with singularities $a_1, \ldots, a_n$ contains fewer parameters than the set of classes of conjugate representations (3). This goes back to Poincaré [Poi], who calculated the difference between these two numbers of parameters. So in general it is impossible to construct a Fuchsian equation without an appearance of additional singularities. The number of such additional singularities is presented in Section 3.)

For a number of years people thought that the Riemann-Hilbert problem was completely solved by Plemelj [Pl] in 1908. Only recently was it realized that there was a gap in his proof (for the first time this was observed by Kohn [Koh] and Arnold, and Il'yashenko [AI]). It turned out that Plemelj obtained a positive answer to a problem similar to the Riemann-Hilbert problem, but concerning so-called regular systems instead of Fuchsian ones. Here is the definition of them.

Let (1) be a system with singularities $a_1, \ldots, a_n$. It is called *regular at a_i* (and a_i is *a regular singularity* for this system) if any of its solutions has at most polynomial (in $1/|x - a_i|$) growth at a_i as x tends to a_i, remaining inside some sector with the vertex at a_i (without going around this point).

The system is called *regular* if it is regular at all a_i. Any Fuchsian system is regular (see [Ha]), but a regular system need not be Fuchsian (Plemelj was able to find systems in a broader class than that required by Hilbert). Note here that for a scalar ODE the notions of Fuchsianity and regularity coincide.

Plemelj used the theory of singular integral equations to construct a regular system with prescribed singular points and the monodromy. Then he transformed the constructed system to the system with the same singularities and monodromy, which was Fuchsian at all points except one. Let $g_1, \ldots, g_n$ be loops at x_0 such that g_i "goes around" a_i without "going around" any $a_j \neq a_i$. Denote the homotopic classes of the loops by the same letters. It follows from Plemelj's paper, [Pl] that

If at least one of the matrices $\chi(g_1), \ldots, \chi(g_n)$ is semisimple (diagonalizable), then the answer to the Riemann-Hilbert problem is positive.

In the 1920s Lappo-Danilevskii [LD] proved that

If all $\chi(g_i)$ are sufficiently close to the identity matrix I, then the answer is positive too.

In 1979 Dekkers [Dek] showed that

In the case $p = 2$ the answer is positive (independently of n).[1]

---

[1] Lappo-Danilevskii and Dekkers did not pretend to solve the Riemann-Hilbert problem (at that time there was the opinion that this problem was solved by Plemelj), but the results formulated above follow immediately from their results.

In 1957 Röhrl [R] published another approach to the same problem using some arguments from the theory of Riemann surfaces and the algebraic geometry. The developments of this approach will be discussed in Section 3.

In 1989 a negative solution to the Riemann-Hilbert problem was found in [Bo1], [Bo2]. It turned out that there exist representations (3) that cannot be representations of any Fuchsian systems. This result is explained in Section 2. Here we also discuss some sufficient conditions for representation (3) still to be realized as the monodromy representation of some Fuchsian system.

The methods, developed in the process of solution of the Riemann-Hilbert problem, can also be applied to problems of somewhat different types; e.g., to the problem of Birkhoff standard form, which is discussed in Section 4.

2. The Riemann-Hilbert problem

The first counterexample to the Riemann-Hilbert problem concerns the case $p = 3$, $n = 4$. Consider the system (1) with

$$B(x) = \frac{1}{x^2} \begin{pmatrix} 0 & 1 & 0 \\ 0 & x & 0 \\ 0 & 0 & -x \end{pmatrix} + \frac{1}{6(x+1)} \begin{pmatrix} 0 & 6 & 0 \\ 0 & -1 & 1 \\ 0 & -1 & 1 \end{pmatrix} \tag{5}$$

$$+ \frac{1}{2(x-1)} \begin{pmatrix} 0 & 0 & 2 \\ 0 & -1 & -1 \\ 0 & 1 & 1 \end{pmatrix} + \frac{1}{3\left(x - \frac{1}{2}\right)} \begin{pmatrix} 0 & -3 & -3 \\ 0 & -1 & 1 \\ 0 & -1 & 1 \end{pmatrix}.$$

It is singular at $a_0 = 0$, $a_1 = -1$, $a_2 = 1$, and $a_3 = \frac{1}{2}$ and the point ∞ is its point of holomorphy. The points a_1, a_2, a_3 are Fuchsian singularities, however a_0 is not Fuchsian, but a pole of order 2. Thus, the system is not Fuchsian, but it is possible to show that it is regular [Bo2].

Our system has some monodromy. Denote the monodromy matrices $\chi(g_i)$ of the system by G_i, $i = 0, 1, 2, 3$.

THEOREM 1 *There exists no a Fuchsian system with the same singularities and monodromy.*

The idea of the proof is as follows. Because $B(x) = \begin{pmatrix} 0 & * \\ 0 & B'(x) \end{pmatrix}$, where

$$B'(x) = \frac{1}{x} \begin{pmatrix} 1 & 0 \\ 0 & -1 \end{pmatrix} + \frac{1}{6(x+1)} \begin{pmatrix} -1 & 1 \\ -1 & 1 \end{pmatrix} \tag{6}$$

$$+ \frac{1}{2(x-1)} \begin{pmatrix} -1 & -1 \\ 1 & 1 \end{pmatrix} + \frac{1}{3(x - \frac{1}{2})} \begin{pmatrix} -1 & 1 \\ -1 & 1 \end{pmatrix},$$

system (1), (5) has Fuchsian quotient system (1), (6) with the monodromy matrices G'_i and $G_i = \begin{pmatrix} 1 & * \\ 0 & G'_i \end{pmatrix}$. From the form (5) of the matrix $B(x)$ it follows that each matrix G_i can be transformed to Jordan normal form consisting of one block with the eigenvalue 1. (It is easy to check for G_1, G_2, G_3, but it is more subtle work for G_0.)

It can be deduced from the form of G_i mentioned above that the existence of a Fuchsian system with the monodromy matrices $G_0, \ldots, G_3$ implies the existence of a Fuchsian system with the monodromy matrices $G_0', \ldots, G_3'$ with some hard restrictions on the asymptotics of its solutions at the singular points. More precisely, such a system would have equal (up to logarithms) orders of asymptotics for all its solutions at all singular points. But it is possible to show that such a system does not exist (see [Bo2], [AB], [Bo6]). Thus, the monodromy representation of system (1), (5) cannot be realized as the monodromy representation of any Fuchsian system. This means that

The Riemann-Hilbert problem has in general a negative solution.

Exponents of asymptotics of Fuchsian system (1), (2) coincide with eigenvalues of its coefficient matrices B_i. This follows from the fact that a fundamental matrix $Y(x)$ of a Fuchsian system in a neighborhood of a singular point a_i can be presented as follows (see [G], [Le]):

$$Y(x) = U_i(x)(x - a_i)^{\Lambda_i} (x - a_i)^{E_i} S_i, \tag{7}$$

where U_i is holomorphically invertible at a_i, $\Lambda_i = \mathrm{diag}(\lambda_i^1, \ldots, \lambda_i^p)$, $\lambda_i^j \in \mathbb{Z}$, $\lambda_i^1 \geq \cdots \geq \lambda_i^p$, $E_i = \frac{1}{2\pi i} \log \chi(g_i)$ is upper-triangular with eigenvalues ρ_i^j, such that $0 \leq \mathrm{Re}\rho_i^j < 1$, $S_i \in \mathrm{GL}(p, \mathbb{C})$.

The numbers λ_i are called *valuations of the system* and the numbers $\beta_i^j = \lambda_i^j + \rho_i^j$ are called *exponents*. It follows from (7) that $y_j = u_j(x)(x - a_i)^{\beta_i^j}[1 + o(|\log^p(x - a_i)|)]$ for the j-column of the matrix $Y S_i^{-1}$.

The exponents of system (1), (6) are as follows: $\beta_0^1 = 1$, $\beta_0^2 = -1$, $\beta_i^j = 0$, $i = 1, 2, 3$, $j = 1, 2$. (One can see that the orders of asymptotics of the solutions to this system at zero do not coincide.)

For a Fuchsian system (1), (2) of two equations consider the number $\gamma(B) = \sum_{i=1}^p (\lambda_i^1 - \lambda_i^2)$. The number $\gamma(\chi) = \min_B \gamma(B)$, where the minimum is taken over all Fuchsian systems with the given monodromy, is called *the Fuchsian weight of the representation* χ.

It is clear that $\gamma(\chi) = 0$ if and only if there exists a Fuchsian system with the monodromy χ whose valuations are equal at each singular point. (For system (1), (6) the condition $\gamma(\chi') = 0$ is equivalent to the existence of a Fuchsian system with the same monodromy and such that all orders of asymptotics of solutions to the system at all singular points are equal.)

The following theorem describes all representations (3) of dimension three that cannot be realized by any Fuchsian system [Bo2].

THEOREM 2 *Representation* χ *of dimension* $p = 3$ *cannot be realized as the monodromy representation of any Fuchsian system, if and only if the following three conditions hold:*

(i) the representation χ *is reducible;*

(ii) each matrix $\chi(g_i)$ *can be reduced to a Jordan normal form, consisting of only one block;*

(iii) the corresponding two-dimensional subrepresentation or quotient representation χ' *has nonzero Fuchsian weight.*

(The calculation of the Fuchsian weight of the monodromy of system (1), (6) is the most subtle part of the proof of Theorem 1.)

It turns out that all counterexamples to the Riemann-Hilbert problem in dimension $p = 3$ are unstable in the following sense. If one slightly perturbs the singular points $a_1, \ldots, a_n$ without changing the monodromy matrices $\chi(g_1), \ldots, \chi(g_n)$, then the answer to the Riemann-Hilbert problem can become positive (see [Bo2], [AB]). The first stable counterexample appears in the case $p = 4, n = 3$ ([Bo4]).

All counterexamples concern reducible representation. The following statement explains the cause.

THEOREM 3 *For any irreducible representation (3) the Riemann-Hilbert problem has a positive solution.*

Proof. Consider a regular system with the given monodromy, Fuchsian at points $a_2, \ldots, a_n$ (such a system always exists, see [Pl]). Present a fundamental matrix $Y(x)$ of the system at a_1 as follows:

$$Y(x) = V(x)(x - a_1)^A (x - a_1)^{E_1}, \tag{8}$$

where $V(x)$ is meromorphic at a_1, E_1 is the same as in (7), $A = \mathrm{diag}(b_1, \ldots, b_n)$,

$$b_i - b_{i+1} > d, \; i = 1, \ldots, p - 1, \; d > 0. \tag{9}$$

Due to Sauvage's lemma (see [Ha]) there exists a matrix Γ_1 meromorphic on all the Riemann sphere, holomorphically invertible outside of a_1 and such that

$$\Gamma_1(x)V(x) = (x - a_1)^C U'(x), \tag{10}$$

where $C = \mathrm{diag}(c_1, \ldots, c_p)$, $c_i \in \mathbb{Z}$, $c_1 \geq \cdots \geq c_p$, and $U'(x)$ is holomorphically invertible at a_1.

It is proved in [Bo3], [Bo4] that if (3) is irreducible, then the numbers c_i from (10) satisfy the following inequalities:

$$c_1 - c_p \leq \sum_{i=1}^{p}(c_1 - c_i) \leq \frac{(p-2)n(n-1)}{2}. \tag{11}$$

Due to some technical lemma (see [Bo4] or [S], where this lemma is called Kimura's lemma) there exists a matrix Γ_2 meromorphic on all the Riemann sphere, holomorphically invertible outside of a_1, and such that

$$\Gamma_2(x)(x - a_1)^C U'(x) = U(x)(x - a_1)^D, \tag{12}$$

where D is a diagonal matrix, obtained by some permutation of the diagonal elements of the matrix C, and $U(x)$ is holomorphically invertible at a_1.

Take $d > (p(p - 2)n(n - 1))/2$ in (9) and consider the system with the fundamental matrix $Y'(x) = \Gamma_2 \Gamma_1 Y(x)$. Because of holomorphy of the matrices Γ_i and Γ_i^{-1} outside of a_1 the new system is still Fuchsian at $a_2, \ldots, a_n$. At the point a_1 we have from (10), (12):

$$Y'(x) = U(x)(x - a_1)^{D+A}(x - a_1)^{E_1}. \tag{13}$$

It follows from (9), (11), and the choice of d that $D + A$ is the diagonal matrix whose diagonal elements form a nonincreasing sequence. Because E_1 is upper-triangular, we obtain that the matrix $(x - a_1)^{D+A} E_1 (x - a_1)^{-D-A}$ is holomorphic. Straightforward calculation shows that the coefficient matrix $B'(x) = (dY'/dx) \cdot (Y'(x))^{-1}$ of the constructed system has a pole of order one at a_1, therefore the system is Fuchsian throughout all the Riemann sphere. The theorem is proved. (The independent proof of the theorem in presented in [Ko].) β

Denote by χ_i the local representation, determined by the monodromy matrix $\chi(g_i)$. The following statement gives the sufficient conditions for representation (3) to be realized by some Fuchsian system (see [Bo5], [Bo6], [AB]).

THEOREM 4 *Let representation (3) be reducible and χ^1, χ^2 be its subrepresentation and quotient representation. Suppose χ^1 is irreducible and χ^2 can be realized by some Fuchsian system. If for some i the local representation χ_i is a direct sum of χ_i', χ_i'', where χ_i' is a subrepresentation of χ^1, then the representation (3) also can be realized as the monodromy representation of some Fuchsian system.*

A negative solution of the Riemann-Hilbert problem means that, as distinct from a local situation, the class of Fuchsian systems and the class of systems with regular singular points are not meromorphically equivalent globally throughout the Riemann sphere. The following statements concern regular systems that cannot be transformed to Fuchsian ones (see [AB], [Bo6]).

THEOREM 5 *Any system (1) with regular singular points is a subsystem (quotient system) of some system with the same singular points meromorphically equivalent to the Fuchsian system (1), (2).*

PROPOSITION 1 *For arbitrary representation (3) there exists a regular system (1) Fuchsian off the point a_1 such that the order of pole of the system at a_1 does not exceed the number $(n-2)p(p-2)/2 + pn + 1$.*

THEOREM 6 *For sufficiently large p and n the codimension of non-Fuchsian representations in the moduli space of all representations is equal to $(p-1)(2n-1)$.*

(Questions concerning a stratification of representations and their codimension in $(\mathrm{GL}(p, \mathbb{C}))^{n-1}$ are also considered in [Ko].)

It is well known that in a neighborhood of a singular point each Fuchsian scalar equation with the help of some meromorphic transformation can be transformed to a Fuchsian system (see [Ha]). The following statement shows it is true to fact globally too [Bo3].

THEOREM 7 *For any Fuchsian equation (4) on the Riemann sphere there always exists a Fuchsian system with the same singular points and the same monodromy.*

3. Vector bundles associated with monodromy

Representation (3) determines a vector bundle F on $\bar{\mathbb{C}} \setminus \{a_1, \ldots, a_n\}$ with a holomorphic connection ∇.

Consider all extensions of F on the whole $\bar{\mathbb{C}}$ that provide for ∇ at most logarithmic singularities at $a_1, \ldots, a_n$. Locally any such extension at a_i is determined by matrices Λ_i, S_i from (7). Denote by F^λ the extension of F on the whole $\bar{\mathbb{C}}$ with the help of the matrices Λ_i, S_i from (7) (see [Bo2], [Bo4]). If all $\Lambda_i = 0$, then the corresponding extension F^0 is called *the canonical extension* (cf. [Del]).

If for some λ the bundle F^λ is holomorphically trivial, then the connection ∇ determines some Fuchsian system on $\bar{\mathbb{C}}$ with the given singularities and monodromy (and conversely).

Consider the decomposition of F^λ into the direct sum of line bundles

$$F^\lambda \cong \mathcal{O}(-c_1^\lambda) \oplus \cdots \oplus \mathcal{O}(-c_p^\lambda), \qquad (14)$$

where $c_1^\lambda \geq \ldots \geq c_p^\lambda$. The set $c_1^\lambda, \ldots, c_p^\lambda$ is called *the splitting type* of F^λ. It completely determines the holomorphic type of the bundle.

Now we introduce the following concepts. The number

$$\gamma(\lambda) = \sum_{i=1}^p (c_1^\lambda - c_i^\lambda)$$

is called *the weight of F^λ*, and the number $\gamma_m(\chi) = \sup_\lambda \gamma(\lambda)$ is called *the maximal Fuchsian weight of χ*.

The important property of the family $\{F^\lambda\}$, constructed by irreducible representation (3), is its *property of finiteness*: $\gamma_m(\chi) \leq ((p-2)n(n-1))/2$, which is another, geometric form of inequality (11) (see [Bo4]). Exactly this property plays the crucial role in proving Theorem 3. Note here that statement (12) establishes the connection between the splitting types of vector bundles F^λ and asymptotics of solutions to corresponding Fuchsian systems.

THEOREM 8 *The weight of the canonical extension F^0 for a two-dimensional representation χ coincides with its Fuchsian weight: $\gamma(0) = \gamma(\chi)$.*

Now Theorem 2 can be formulated for arbitrary p as follows (see [Bo4], [AB]).

THEOREM 9 *Let (3) be a reducible representation with subrepresentation χ', and let each monodromy matrix $\chi(g_i), i = 1, \ldots, n$, can be reduced to a Jordan normal form, consisting of only one block. Then the representation χ can be realized by a Fuchsian system if and only if $\gamma_\chi(0) = 0$, or equivalently $\gamma_{\chi'}(0) = \gamma_{\chi/\chi'}(0) = 0$, where $\gamma_\chi(0)$ is the weight of the canonical extension of the vector bundle, constructed by the representation χ.*

In terms of Fuchsian weight it is possible to express the number of additional so-called "apparent" singularities, arising under an attempt to construct a Fuchsian scalar differential equation with prescribed singularities and monodromy.

THEOREM 10 *The minimal possible number m_0 of additional apparent singularities of a Fuchsian equation with a given irreducible monodromy (3) is equal to the following one:*

$$m^0 = \frac{(n-2)p(p-1)}{2} - \gamma_m(\chi).$$

(Here m_0 is taken with multiplicities, which coincide with orders of additional zeroes of a Wronskian of the corresponding "minimal" Fuchsian equation.)

4. Birkhoff standard form

Consider a linear system of differential equations

$$x\frac{dy}{dx} = C(x)y, \quad C(x) = x^r \sum_{n=0}^{\infty} C_n x^{-n}, \quad C_0 \neq 0, \quad r \geq 0, \qquad (15)$$

where $C(x)$ is a matrix of size (p, p) and the power series converges in some neighborhood of ∞.

Under a transformation $z = \Gamma(x)y$ system (15) is transformed to the system

$$x\frac{dz}{dx} = B(x)z, \quad B(x) = x\frac{d\Gamma}{dx}\Gamma^{-1} + \Gamma C(x)\Gamma^{-1}. \qquad (16)$$

If $\Gamma(x)$ is holomorphically invertible in some neighborhood of ∞, then such a transformation is called *analytic.*

If the matrix $B(x)$ in (16) is a polynomial in x of degree r, then (16) is called *a Birkhoff standard form for (15).*

Birkhoff [Bi] claimed that each system (15) can be analytically transformed to a Birkhoff standard form, but Gantmacher [G] presented a counterexample to this statement. It turned out that Birkhoff's proof was valid only for the case when a monodromy matrix of system (15) was diagonalizable.

Let us call system (15) *reducible* (or *generic*) if there exists a holomorphically invertible in some neighborhood of ∞ matrix $\Gamma(x)$ such that under the transformation with help of this matrix, system (15) is transformed to system (16) with a lower diagonal block matrix $B(x) = \begin{pmatrix} B' & 0 \\ * & B'' \end{pmatrix}$.

For $p = 2$ Jurkat, Lutz, and Peyerimhoff [JLP], and for $p = 3$ Balser [Ba] proved that *each irreducible system (15) can be analytically transformed to a Birkhoff standard form.* In [Bo7] the analogous result is proved for arbitrary p.

THEOREM 11 *Each irreducible system (15) can be analytically transformed to a Birkhoff standard form.*

The idea of the proof is as follows. Consider a fundamental matrix $Y(x)$ to the system (15) with the following factorization $Y(x) = M(x)x^E$, where $M(x)$ is a single-valued matrix function with nonvanishing $\det M(x)$ in some punctured neighborhood O of ∞, and E is the same as in (7). The matrix $M(x)$ defines some vector bundle on $\bar{\mathbb{C}}$ with coordinate neighborhoods $O, \mathbb{C}$. Denote this bundle by F

and consider all its admissible extensions F^A. Each such extension is determined by the transition function $M(x)x^{-A}$, where A is the same as in (9).

Because of (14) there exist a holomorphically invertible in some neighborhood of ∞ matrix $T(x)$ and holomorphically invertible in the complex plane matrix $U'(x)$ such that

$$T(x)M(x)x^{-A} = x^C U'(x), \tag{17}$$

where C is the splitting type of F^A.

It turns out that for an irreducible system (15) the family $\{F^A\}$ has the finiteness property; i.e., $c_i - c_{i+1} \le r$, $i = 1, \ldots, p - 1$. Take $d = pr$ in (9) and consider for the matrix $x^C U'(x)$ the matrix Γ_2 from (12). Under analytic at ∞ transformation $Y'(x) = \Gamma_2 T(x)Y(x)$ our original system (15) is transformed to system (16), whose fundamental matrix in complex plane has the form

$$Y'(x) = U(x)x^{D+A}x^E.$$

In a way similar to that in Theorem 3 we obtain that diagonal elements of the matrix $D + A$ are in nonincreasing order and the matrix $L = x^{D+A}Ex^{-D-A}$ is the entire matrix function. Therefore, $B(x) = x\frac{dY'}{dx}(Y')^{-1} = x\frac{dU}{dx}U^{-1} + U(D + A + L)U^{-1}$ is an entire matrix function too. Because $B(x)$ has pole of order r at ∞, this completes the proof.

References

[AB] D. V. Anosov and A. A. Bolibruch, The Riemann-Hilbert Problem, Aspects of Mathematics, Vieweg, Braunschweig/Wiesbaden, 1994.

[AI] V. I. Arnold and Yu. S. Il'yashenko, *Ordinary differential equations*, Dynamical Systems 1, D. V. Anosov and V. I. Arnold (eds.), Encyclopaedia of the Mathematical Sciences, 1, Springer, Berlin and New York, 1988.

[Ba] W. Balser, *Analytic transformation to Birkhoff standard form in dimension three*, Funkcial. Ekvac., **33:1** (1990), 59–67.

[Bi] G. D. Birkhoff, Collected mathematical papers. Volume 1, Dover Publ., Inc., New York, 1968.

[Bo1] A. A. Bolibruch, *The Riemann-Hilbert problem on the complex projective line* (in Russian), Mat. Zametki, **46:3** (1989), 118–120.

[Bo2] A. A. Bolibrukh, *The Riemann-Hilbert problem*, Russian Math. Surveys, **45:2** (1990), 1–47.

[Bo3] A. A. Bolibrukh, *Construction of a Fuchsian equation from a monodromy representation*, Math. Notes Acad. Sci. USSR **48:5** (1990), 1090–1099.

[Bo4] A. A. Bolibruch, *Fuchsian systems with reducible monodromy and the Riemann-Hilbert Problem*, Lecture Notes in Math., **1520**, Springer, Berlin and New York, 1992, 139–155.

[Bo5] A. A. Bolibrukh, *On sufficient conditions for the positive solvability of the Riemann-Hilbert Problem*, Math. Notes Acad. Sci. USSR, **51:2** (1992), 110–117.

[Bo6] A. A. Bolibruch, *Hilbert's twenty-first problem for Fuchsian linear systems*, in: Developments in Mathematics: The Moscow School, V. Arnold, and M. Monastyrsky, (eds.),Chapman & Hall, London and Madras, 1993, 54–99.

[Bo7] A. A. Bolibruch, *On analytic transformation to Birkhoff standard form* (in Russian), Proc. Russian Acad. Sci., **334:5** (1994), 553–555.

[Dek] W. Dekkers, *The matrix of a connection having regular singularities on a vector bundle of rank 2 on $P^1(\mathbb{C})$*, Lecture Notes in Math., **712**, Springer, Berlin and New York, 1979, 33–43.

[Del] P. Deligne, Equations différentielles a points singuliers réguliers, Lecture Notes in Math., **163**, Springer, Berlin and New York, 1970.

[G]   F. R. Gantmacher, Theory of Matrices. Volume II, Chelsea, New York, 1959.

[Ha]   P. Hartman, Ordinary Differential Equations, John Wiley, New York, 1964.

[Hi]   D. Hilbert, Mathematische Probleme, Nachr. Ges. Wiss., Göttingen, 1900, 253–297.

[JLP] W. B. Jurkat, D. A. Lutz, and A. Peyerimhoff, *Birkhoff invariants and effective calculations for meromorphic differential equations*, Part I, J. Math. Anal. Appl., **53**(1976), 438–470.

[Koh] A. Kohn Treibich, Un resultat de Plemelj, Progr. Math., **37**, Birkhäuser, Boston, MA, 1983.

[Ko]   V. P. Kostov, *Fuchsian systems on CP^1 and the Riemann-Hilbert Problem*, C. R. Acad. Sci. Paris, **315**, Série I (1992), 143–148.

[LD]   I. Lappo-Danilevskii, Mémoires sur la théorie des systèmes des equations différentielles linéaires, Chelsea, New York, 1953.

[Le]   A. H. M. Levelt, *Hypergeometric functions*, Nederl. Akad. Wetensch. Proc. Ser. A **64** (1961).

[Pl]   J. Plemelj, Problems in the Sense of Riemann and Klein, Interscience, New York, 1964.

[Poi] H. Poincaré, *Sur les groupes des équations linéaires*, Acta Math., **5** (1884), 201–312.

[R]   H. Röhrl, *Das Riemann-Hilbertsche Problem der Theorie der linearen Differentialgleichungen*, Math. Ann., **133** (1957), 1–25.

[S]   Y. Sibuya, *Linear differential equations in the complex domain: Problems of analytic continuation*, Translations of math. monographs, AMS, Providence, RI, 1990.

Invariant Sets of Hamiltonian Systems and Variational Methods

SERGEY V. BOLOTIN*

Department of Mathematics and Mechanics
Moscow State University
Vorobyevy gory
Moscow 119899, Russia

1 Introduction

We study the problem on the existence of homoclinic trajectories to Mather minimizing invariant sets (multidimensional generalization of Aubry-Mather sets) of positive definite time-periodic Hamiltonian systems [19]. These sets are supports of invariant probability measures in the phase space minimizing the average action $\int L \, d\mu$ for a suitable calibration of the Lagrangian L. For natural systems with $L = \|v\|^2/2 - V(x)$, the minimizing set is $\Gamma = \{V = h\}$, $h = \max V$, and for time-periodic systems with reversible L the minimizing sets consist of brake orbits of minimal action. For natural Hamiltonian systems, the existence of homoclinics to Γ was proved in [3] using the Maupertuis-Jacobi functional $\int \sqrt{h - V(x)} \, \|dx\|$, and for reversible time-periodic systems in [4] using Hamilton's functional (see also [5], [16]). For nonreversible systems (for example, natural systems with gyroscopic forces), in general there are no Mather sets of simple structure. We extend the above existence results to arbitrary minimizing sets replacing homoclinic trajectories by semihomoclinic ones in Birkhoff's sense [2]. A similar problem was studied in [20].

  A main source of examples of minimizing sets is in perturbations theory. Let the Poincaré map of a time-periodic Hamiltonian system have a Lagrangian manifold M^m fibrated to n-dimensional invariant tori. If the frequency $\omega \in \mathbf{R}^n$ is Diophantine and some positive definiteness conditions are satisfied, then at least one torus $\Gamma \subset M$ survives a C^∞-small perturbation of the Hamiltonian and becomes minimizing and whiskered [1]. If ω is nearly resonant, or the perturbation is only C^2-small, then the torus is replaced by minimizing sets near Γ. Integrable Hamiltonian systems have families of resonant Lagrangian tori M_I, $I \in \mathbf{R}^n$, fibrated to n-dimensional invariant tori. Homoclinics to whiskered tori Γ_I of the perturbed system form a basis for Arnold's diffusion: exponentially slow drift of the integral I [1], [9]. Gaps in $\mathbf{R}^n\{I\}$ corresponding to non-Diophantine $\omega(I)$ are represented by Mather sets Γ_I having semihomoclinic trajectories.

*)   Work supported by the Russian Foundation of Basic Research and ISF.

Proceedings of the International Congress
of Mathematicians, Zürich, Switzerland 1994
© Birkhäuser Verlag, Basel, Switzerland 1995

Recently many authors have studied homoclinics to equilibria of Hamiltonian systems by different variational methods. See, for example, [10]–[12], [15], [22] and the papers cited there. Homoclinics to invariant tori were studied in [6], and to invariant sets in [7]. This paper is partly based on [7], [8].

2 Minimizing invariant sets

We consider a time-periodic Hamiltonian system with compact configuration manifold M^m and Hamiltonian $H \in C^2(P)$ on the phase space $P = T^*M \times \mathbf{T}$, $\mathbf{T} = \mathbf{R}/\mathbf{Z}$. Following Mather [19], we assume that the Hessian of the Hamiltonian $H(x, y, t)$ in momentum $y \in T_x^*M$ is positive definite, H is superlinear in momentum: $H(x, y, t)/\|y\| \to \infty$ as $\|y\| \to \infty$, and the phase flow $g^t : P \to P$ of the Hamiltonian vector field ξ is complete.

These assumptions hold for classical Hamiltonian systems. Main results can be generalized to systems with $H(x, y, t) = F(x, y, \phi_t(\cdot))$ where ϕ_t is a flow on a compact metric space (for example, almost periodic systems).

We identify P and $TM \times \mathbf{T}$ by using the Legendre transform $z = (x, y, t) \in P \to (x, v, t) \in TM \times \mathbf{T}$, $v = H_y \in T_xM$. Let $L \in C^2(TM \times \mathbf{T})$ be the Lagrangian $L(x, v, t) = \langle y, v \rangle - H(x, y, t)$ and $\mathfrak{M}$ the set of g^t-invariant Borel probability measures μ with compact support in P. The *average action* [19] of a measure $\mu \in \mathfrak{M}$ is $\mathcal{A}(\mu) = \int L \, d\mu$. The rotation vector $\rho(\mu) \in H_1(M, \mathbf{R})$ is defined by the equation $\langle c, \rho(\mu) \rangle = \int \langle \omega(x), v \rangle \, d\mu$ for any cohomology class $c = [\omega] \in H^1(M, \mathbf{R})$ represented by a closed 1-form ω on M. Mather [19] proved that for any $\rho \in H_1(M, \mathbf{R})$ the functional $\mathcal{A}$ has a minimum on $\mathfrak{M}_\rho = \{\mu \in \mathfrak{M} : \rho(\mu) = \rho\}$, and for any minimum point $\mu \in \mathfrak{M}_\rho$ there exists $c \in H^1(M, \mathbf{R})$ such that μ is a minimum point of the functional $\mathcal{A}_c(\mu) = \mathcal{A}(\mu) - \langle c, \rho(\mu) \rangle$ on $\mathfrak{M}$.

DEFINITION 2.1. Minimum points of $\mathcal{A}_c$ are called *c-minimal measures*, and the closure of union of their supports the *c-minimizing set* Γ_c.

Ergodic properties of minimal measures are studied in [17], [18]. The set Γ_c is a Lipschitz graph [19], i.e. a graph of a Lipschitz section $y = p(x, t) \in T_x^*M$ on a compact set $\Sigma_c = \pi(\Gamma_c) \subset M \times \mathbf{T}$, where $\pi : P \to M \times \mathbf{T}$ is the projection. For a Lipschitz graph $\Gamma \subset P$, the projection of the Poincaré-Cartan 1-form $\lambda = \langle y, dx \rangle - H \, dt$ is a Lipschitz 1-form on $\Sigma = \pi(\Gamma) \subset M \times \mathbf{T}$:

$$\Lambda(x, t) = \langle p(x, t), dx \rangle - H(x, p(x, t), t) \, dt, \qquad (x, t) \in \Sigma.$$

DEFINITION 2.2. The set Γ is *strongly isotropic* if $\Lambda = \Omega|_\Sigma$, where Ω is a locally exact Lipschitz 1-form on $M \times \mathbf{T}$.

Thus, there is $W \in C^{1+\mathrm{Lip}}(M \times \mathbf{T})$, a closed 1-form ω on M, and $a \in \mathbf{R}$ such that

$$p(x, t) = W_x + \omega(x), \quad f(x, t) = -W_t - H(x, p(x, t), t) = a \qquad (2.1)$$

for $(x, t) \in \Sigma$. If an invariant set Γ is strongly isotropic and Σ is the set of minimum points for f, then Γ is a c-minimizing set with $c = [\omega]$. We call such sets *strictly minimizing*. If μ is a c-minimal measure and $\Gamma = \mathrm{supp}(\mu)$, then Ω can be taken from the cohomology class $(c, a) \in H^1(M \times \mathbf{T}, \mathbf{R})$, $a = \min \mathcal{A}_c$. In particular,

$L|_\Gamma = \langle \lambda, \xi \rangle = -\langle \omega(x), v \rangle - a - \dot{W}$ is a full derivative. Mañé [18] proved that for ergodic μ there is $W \in \mathrm{Lip}(\Sigma)$ satisfying the last condition. For isotropic Γ, performing the canonical transformation

$$(x, y, t, H) \to (x, y - \omega(x) - W_x, t, H + W_t + a), \qquad (2.2)$$

we may assume that $L|_\Gamma \equiv 0$ and $\min \mathcal{A} = 0$. Probably support of any minimal invariant measure is strongly isotropic, but this is proven only for regular measures.

DEFINITION 2.3. A minimal measure μ is *regular* if $\Sigma = \pi(\Gamma)$ is a C^1 submanifold in $M \times \mathbf{T}$, and the measure $\nu = \pi_*(\mu)$ on Σ is nonsingular.

THEOREM 2.1. *If μ is regular, then Γ is strongly isotropic.*

COROLLARY 2.1. *If μ is regular, the action variable $J = [\lambda|_\Gamma] \in H^1(\Gamma, \mathbf{R})$ is correctly defined by $\langle J, [\gamma] \rangle = \oint_\gamma \lambda$ for any closed curve $\gamma \subset \Gamma$, and $J = \pi^*(c, a)$, where the map $\pi^* \colon H^1(M \times \mathbf{T}, \mathbf{R}) \to H^1(\Gamma, \mathbf{R})$ is induced by the projection.*

If Γ is an invariant torus with invariant Lebesgue measure, then Γ is isotropic and J the usual action variable. If the torus is ergodic, minimality is not needed [14].

DEFINITION 2.4. The set Γ is *weakly isotropic* if Λ belongs to the closure of the set $\{\Omega|_\Sigma : d\Omega = 0,\ [\Omega] = (c, a)\}$ of closed Lipschitz forms on Σ in the weak $L^2(\nu)$-topology.

THEOREM 2.2. *If μ is a minimal measure, then $\Gamma = \mathrm{supp}(\mu)$ is weakly isotropic.*

It is easy to see that if μ is regular and Γ weakly isotropic, Γ is strongly isotropic. Thus, Theorem 2.2 implies Theorem 2.1. Performing the transformation (2.2) with $W = 0$, or a calibration $L_c = L - \langle \omega, v \rangle - a$ of the Lagrangian, we may assume that $(c, a) = 0$ and thus $\mathcal{A}(\mu) = \min \mathcal{A} = 0$. Then Definition 2.4 is equivalent to $\int \langle \eta, \Lambda \rangle \, d\nu = 0$ for all vector fields η on $M \times \mathbf{T}$ such that $\int \langle \eta, df \rangle \, d\nu = 0$ for all $f \in C^\infty(M \times \mathbf{T})$. Theorem 2.2 follows from

PROPOSITION 2.1. *Let μ be a minimal measure. Then $\int \langle \lambda, \eta \rangle \, d\mu = 0$ for every μ-preserving Lipschitz vector field η on P.*

Proof. For a vector field ζ on P and $z = (x, y, t) \in P$ let $(\zeta_x, \zeta_t) \in T_x M \times \mathbf{R}$ be the projection of $\zeta(z)$ to $M \times \mathbf{T}$. Birkhoff's ergodic theorem implies that for any μ-preserving Lipschitz vector field ζ on P such that $\zeta_t > 0$,

$$\int_\Gamma L(x, \zeta_x/\zeta_t, t)\zeta_t \, d\mu(z) \geq 0, \qquad z = (x, y, t), \qquad (2.3)$$

and for the Hamiltonian vector field ξ the equality holds. We set $\zeta = \xi + \varepsilon\eta$ and differentiate (2.3) by ε at $\varepsilon = 0$. Because $\xi_t = 1$ and $\xi_x = v$, we obtain:

$$0 = \frac{\partial}{\partial \varepsilon}\Big|_{\varepsilon=0} \int_\Gamma L\left(x, \frac{v + \varepsilon\eta_x}{1 + \varepsilon\eta_t}, t\right)(1 + \varepsilon\eta_t) \, d\mu(z)$$

$$= \int_\Gamma (\langle y, \eta_x \rangle - H(z)\eta_t) \, d\mu = \int_\Gamma \langle \lambda, \eta \rangle \, d\mu.$$

$\square$

3 Almost asymptotic and almost homoclinic trajectories

Let μ be a minimal measure, $\Gamma = \operatorname{supp}(\mu)$, and $\Sigma = \pi(\Gamma)$.

THEOREM 3.1. *Let $f\colon \Gamma \to M$ be a continuous map. For any $\varepsilon, \delta > 0$ there exists $T \leq C_1/(\varepsilon^2\delta)$ such that for $z \in \Gamma$ with μ-probability $1 - \varepsilon$ there exists a trajectory $\sigma(t) = (q(t), p(t), t) \in P$ such that:*

(1) $q(t(z)) = f(z)$, $q(t(z)+T) = x(t(z)+T)$ *and* $\|p(t(z)+T)-y(t(z)+T)\| \leq \delta$;
(2) *Hamilton's action* $I_{t(z)}^{t(z)+T}(\sigma) = \int_{t(z)}^{t(z)+T} L(\sigma(t))\,dt \leq C_2/\varepsilon$.

Here $t(z) \bmod \mathbf{Z}$ is the $\mathbf{T}$-component of z, and $(x(t), y(t), t) = \hat{z}(t) = g^{t-t(z)}(z)$ is the trajectory of $z \in \Gamma$. The constants $C_{1,2} > 0$ are independent of ε and δ.

COROLLARY 3.1. *For μ-almost all points $z \in \Gamma$ (for all if μ is ergodic) and any $\delta > 0$ there is a δ-asymptotic trajectory to $\hat{z}(t)$ satisfying (1) of Theorem 3.1.*

In particular, every trajectory in Γ is unstable. If $L|_\Gamma$ is a full derivative (for example, μ is ergodic), then Corollary 3.1 holds for all points of Γ.

DEFINITION 3.1. *A trajectory $\sigma(t) = (q(t), p(t), t) \in P$, $a \leq t \leq b$, is δ-homoclinic to a trajectory $(x(t), y(t), t) \in P$ if $q(a) = x(a)$, $q(b) = x(b)$ and $\|p(a) - y(a)\| \leq \delta$, $\|p(b) - y(b)\| \leq \delta$. The homotopy class $[\sigma]$ of a δ-homoclinic trajectory is the class of the curve $q|_{[a,b]} \circ \left(x|_{[a,b]}\right)^{-1}$ in the set Π of free homotopy classes of loops in M.*

THEOREM 3.2. *For any $f \in \Pi$ there exist $C_{1,2} > 0$ such that for any $\varepsilon, \delta > 0$ and $z \in \Gamma$ with μ-probability $1 - \varepsilon$ the trajectory $\hat{z}(t)$ has a δ-homoclinic trajectory $\sigma\colon [t(z), t(z) + T] \to P$ such that $[\sigma] = f$, $T \leq C_1/(\delta^2\varepsilon)$, and $I_{t(z)}^{t(z)+T}(\sigma) \leq C_2/\varepsilon$.*

COROLLARY 3.2. *For μ-almost all trajectories in Γ (all if μ is ergodic) and any $\delta > 0$ there exists a δ-homoclinic from a given homotopy class.*

Proof of Theorem 3.2. We can assume $\mathcal{A}(\mu) = \min \mathcal{A} = 0$. Let $\widetilde{M}$ be the universal covering of M. Nontrivial $f \in \Pi$ is represented by a translation $f\colon \widetilde{M} \to \widetilde{M}$. We lift H and L to translation invariant functions on $\widetilde{P} = T^*\widetilde{M} \times \mathbf{R}$. For given $z \in \Gamma$ denote by $\widetilde{z}(t) = (\widetilde{x}(t), \widetilde{y}(t), t) \in \widetilde{P}$ any trajectory covering $\hat{z}(t) \in \Gamma$. Take $T > 0$ and let $\gamma_z\colon [t(z), t(z) + T] \to \widetilde{M}$ be a minimizer of Hamilton's action $I_{t(z)}^{t(z)+T}(\gamma)$ on the set of curves $\gamma\colon [t(z), t(z) + T] \to \widetilde{M}$ joining the points $\widetilde{x}(t(z))$ and $f(\widetilde{x}(t(z) + T))$.

The function $S(z, T) = I_{t(z)}^{t(z)+T}(\gamma_z)$ does not depend on the choice of the covering trajectory $\widetilde{z}(t) \in \widetilde{P}$. For any $h > 0$, set $F(T) = \int S(z, T)\,d\mu(z)$ and $F_h(T) = F(T) + hT$. Then F is a continuous nonincreasing function on $\mathbf{R}^+$ and, because $\inf \operatorname{dist}(x, f(x)) > 0$ and L is superlinear in velocity, $F(T) \to \infty$ as $T \to 0$. Because $\min \mathcal{A} = 0$, F is bounded from below as $T \to \infty$. We take for T a minimum point of F_h and denote by $\sigma(t) = (q(t), p(t), t) \in P$, $t(z) \leq t \leq t(z) + T$, the trajectory such that $\gamma_z(t) \in \widetilde{M}$ covers $q(t) \in M$.

Let $\alpha_{z,\varepsilon}\colon [t(z) + \varepsilon, t(z) + T] \to \widetilde{M}$ and $\beta_{z,\varepsilon}\colon [t(z), t(z) + T + \varepsilon] \to \widetilde{M}$ be smooth variations of γ_z joining the points $\widetilde{x}(t(z) + \varepsilon)$, $f(\widetilde{x}(t(z) + T))$ and $\widetilde{x}(t(z))$,

$f(\widetilde{x}(t(z) + T + \varepsilon))$, respectively. Differentiating $I_{t(z)}^{t(z)+T+\varepsilon}(\alpha_{z,\varepsilon})$ and $I_{t(z)+\varepsilon}^{t(z)+T}(\beta_{z,\varepsilon})$ by ε at $\varepsilon = 0$ and taking into account that μ is g^T-invariant, L is f-invariant, and $\int L \, d\mu = 0$, we obtain

$$
\int (\langle p(t(z)), v(z) \rangle - H(\sigma(t(z))) - L(z)) \, d\mu(z)
$$

$$
= \int (\langle p(t(z) + T), v(g^T(z)) \rangle - H(\sigma(t(z) + T)) - L(g^T(z))) \, d\mu(z) = h.
$$

For small $h > 0$, Young's and Chebyshev's inequalities yield Theorem 3.2. $\qquad\square$

For a given homotopy class $f \in \Pi$, the almost homoclinic trajectory $\sigma(t)$ may coincide with $\hat{z}(t)$. To avoid this, we need to factorize Π by $\pi_1(\Gamma)$, but, because Γ may be fractal, this group is not well defined. Thus, we pass to homology groups. Let Γ be connected and $G = \widehat{H}_1(M \times \mathbf{T}, \Sigma, \mathbf{Z})$ be the Čech homology group.

DEFINITION 3.2. The *homology class* $h(\sigma) \in G$ of a δ-homoclinic trajectory $\sigma \subset P$ is the homology class of the curve $\gamma = \pi(\sigma) \subset M \times \mathbf{T}$ with $\partial\gamma \subset \Sigma$.

COROLLARY 3.3. *For any $g \in G$ there is $C > 0$ such that for any $\varepsilon, \delta > 0$:*

(1) *for $z \in \Gamma$ with probability $1 - \varepsilon$ (for all z if μ is ergodic) the trajectory $\hat{z}(t) \in \Gamma$ has a δ-homoclinic $\sigma \colon [t(z), t(z) + T] \to P$ such that $h(\sigma) = g$;*
(2) *Hamilton's action of σ is bounded by C/ε;*
(3) *if U is a neighborhood of Σ in $M \times \mathbf{T}$ such that g has nonzero image in $H_1(M \times \mathbf{T}, U, \mathbf{Z})$, then $\pi(\sigma(t)) \notin U$ for some $t \in (t(z), t(z) + T)$.*

DEFINITION 3.3 (Birkhoff [2]). *An orbit $z(t) \in P$ is* semiasymptotic *to an invariant set Γ as $t \to \infty$ if it is stable in Lagrange's sense, $z(t) \notin \Gamma$, and every recurrent trajectory in the ω-limit set of $z(t)$ is contained in Γ.* The definitions of an orbit semiasymptotic to Γ as $t \to -\infty$ is similar. We denote the unions of semiasymptotic and asymptotic orbits by $W^{\pm}(\Gamma) \supset W^{s,u}(\Gamma)$, respectively.

There are minimal measures with no semiasymptotic trajectories to their supports. For a natural system, every point in the minimizing set $\Gamma = \{V = h\}$ supports a minimal measure, but asymptotic orbits may exist only for points in $\partial\Gamma$.

THEOREM 3.3 [7]. *Let $\Gamma \subset P$ be a c-minimizing invariant set. Then $\pi(W^{\pm}(\Gamma)) = M \times \mathbf{T}$. If Γ is strictly minimizing, then $\pi(W^{s,u}(\Gamma)) = M \times \mathbf{T}$.*

If $\pi(\Gamma) \neq M \times \mathbf{T}$, this implies instability of Γ, but not of individual trajectories in Γ. For natural systems, Theorem 3.3 was established in [16].

THEOREM 3.4. *Let Γ be a c-minimizing set and Λ a connected component of Γ such that $G = \widehat{H}_1(M \times \mathbf{T}, \pi(\Lambda), \mathbf{Z}) \neq 0$. Then:*

(1) *there exists an orbit semiasymptotic to Λ as $t \to -\infty$ and to Γ as $t \to \infty$;*
(2) *if Γ is strictly minimizing, the number of action minimizing homoclinics to Γ is at least $2(\operatorname{rank} \widehat{H}_1(M, \pi(\Gamma), \mathbf{Z}) + \operatorname{rank} \widehat{H}_0(\Gamma, \mathbf{Z}))$.*

Theorems 3.3 and 3.4 follow from Theorem 3.1 and Corollary 3.3 in the limit $\delta \to 0$.

For the existence of heteroclinic orbits connecting different minimizing sets Γ_c and $\Gamma_{c+\Delta c}$, additional assumptions are needed (for integrable systems, there are no such orbits). In [20], Mather suggested a generalization of Peierls's barrier in the Aubry-Mather theory to the multidimensional case. Let $h_c(x, y) = \inf_\gamma I_c(\gamma)$, where $\gamma : [0, a] \to M$, $a \in \mathbf{Z}$, connects x and y, and I_c is a modified action for the calibrated Lagrangian L_c. The barrier function on M is $B_c(x) = \inf\{h_c(y, x) + h_c(x, z) - h_c(y, z), \ y, z \in \Sigma_c^0\}$, $\Sigma_c^0 = \Sigma_c \cap \{t = 0\}$. Then connection is possible for directions $\Delta c \in (i_* \widehat{H}_1(\Sigma_c^0, \mathbf{R}))^\perp \subset H^1(M, \mathbf{R})$.

This barrier does not provide a possible direction for the case of Arnold's diffusion. However, if $G \neq 0$, one can construct a modified barrier closely related to the Poincaré-Melnikov-Arnold integral [1].

4 Hyperbolic minimizing sets

If the minimizing set Γ is connected and M simply connected, some regularity assumptions are needed for proving the existence of semihomoclinic orbits. The reason is that homoclinics cannot be obtained by minimizing the action functional and so we need a manifold structure on the function space. The following definition is inspired by the notion of a whiskered (or hyperbolic) torus [1, 13].

DEFINITION 4.1. An invariant manifold $\Gamma \subset P$ is called *whiskered* if:

(1) $W^{s,u}(\Gamma) \subset P$ are $(m + 1)$-dimensional submanifolds in P;
(2) for any point $z \in \Gamma$ the tangent spaces to $W^{s,u}$ are direct sums $T_z W^{s,u} = T_z\Gamma \oplus E_z^{s,u}$, where the vector bundles $E^{s,u}$ are g^t-invariant;
(3) there exist $\alpha > \beta > 0$ such that

$$\|dg^{\pm t}(z)\zeta\| \leq Ce^{\mp \alpha t}\|\zeta\| \quad \text{for all} \quad \zeta \in E_z^{s,u}, \quad t \geq 0,$$

$$\|dg^t(z)\zeta\| \leq Ce^{\beta t}\|\zeta\| \quad \text{for all} \quad \zeta \in T_z\Gamma, \quad t \in \mathbf{R}.$$

THEOREM 4.1. *Suppose that a minimizing set* $\Gamma = \Gamma_c$ *is a whiskered manifold and it is isotropic (for example, carries a regular minimal measure). Then:*

(1) Γ *has a semihomoclinic orbit. Moreover, the sets* $W^u \cap W^+ \setminus \Gamma$ *and* $W^- \cap W^s \setminus \Gamma$ *are nonempty. Thus,* $W^s \cap \overline{W}^u \setminus \Gamma \neq \emptyset$ *and* $W^u \cap \overline{W}^s \setminus \Gamma \neq \emptyset$;
(2) *if* Γ *is strictly minimizing, then* $W^u \cap W^s \setminus \Gamma \neq \emptyset$.

The actions of homoclinics in (2) are bounded by an a priori constant. If their number is finite, there is an infinite number of multibump homoclinics. For transversal homoclinics, this was discovered by Poincaré and Birkhoff. See [10]–[12], [22], and the papers cited there for variational treatment of homoclinics to equilibria. A similar method applies in the present case.

COROLLARY 4.1 [7]. *If* Γ *is a minimizing whiskered invariant torus, then assertions of Theorem 4.1 hold.*

For Γ a contractible periodic orbit, this was proven in [5] (see also [16]). Homoclinics to invariant tori can be studied by the same method [6]–[8].

COROLLARY 4.2 [4]. *Let the system be reversible. If the minimum of the function V on M, $V(x) = I_0^{1/2}(g^t(x,0,0))$, is strict and nondegenerate, then the corresponding brake orbit has a homoclinic trajectory. If the minimum is only strict but $\pi_1(M) \neq 0$, there exists a set of generators of $\pi_1(M)$ containing homoclinic orbits.*

Locally a whiskered isotropic minimizing manifold Γ is *strictly minimizing*: it is given by (2.1), and $\Sigma = \pi(\Gamma)$ is a nondegenerate critical manifold of zero index for f. We perform the canonical transformation (2.2). In the new variables, $\Gamma \subset \{y = 0\}$ is a nondegenerate critical manifold of zero index for L and $L|_\Gamma \equiv 0$.

PROPOSITION 4.1. *There exist $C > 0$ and a neighborhood U of Σ such that:*

(1) *for any $\delta > 0$, Γ has a δ-homoclinic $\sigma \colon [a,b] \to P$ with $I_a^b(\sigma) \leq C$;*

(2) *$\pi(\sigma(t)) \notin U$ for some $t \in (a,b)$;*

(3) *let $\alpha = \inf\{t > a : \pi(\sigma(t)) \in \partial U\}$, $\beta = \sup\{t < b : \pi(\sigma(t)) \in \partial U\}$. Then $\operatorname{dist}(\sigma(\alpha), W^u(\Gamma)) \to 0$, $\operatorname{dist}(\sigma(\beta), W^s(\Gamma)) \to 0$, and $\alpha - a \to \infty$, $b - \beta \to \infty$, as $\delta \to 0$.*

This implies (1) of Theorem 4.1. Indeed, $\sigma(\alpha) \in P$ is uniformly bounded as $\delta \to 0$. Take a sequence $\delta \to 0$ such that $\sigma(\alpha) \to z \in W^u(\Gamma)$. By (3), the trajectory $\hat{z}(t)$ is asymptotic to Γ as $t \to -\infty$, and it is semiasymptotic as $t \to \infty$.

PROPOSITION 4.2. *There exists $C > 0$ such that for any $h > 0$ there is a trajectory $\sigma(t) = (q(t), p(t), t)$, $a \leq t \leq b$, such that:*

(1) *$q(a) \in \Sigma_a$, $q(b) \in \Sigma_b$, $p(a) \perp T_{q(a)}\Sigma_a$, $p(b) \perp T_{q(b)}\Sigma_b$;*

(2) *$H(\sigma(a)) - \langle p(a), v(q(a), 0, a)\rangle = H(\sigma(b)) - \langle p(b), v(q(b), 0, b)\rangle = h$;*

(3) *$I_a^b(\sigma) + h(b-a) \leq C$.*

Here $\Sigma_t = \{x : (x,t) \in \Sigma\} \subset M$. Proposition 4.1 can be deduced from Proposition 4.2 by estimating the behavior of the solution $\sigma(t)$ of the Hamiltonian equations near a whiskered manifold Γ [8].

Proof of Proposition 4.2. Define the function space Ω as a submanifold in the Hilbert manifold $W^{1,2}([0,1], M) \times \mathbf{T} \times \mathbf{R}^+$:

$$\Omega = \{\omega = (\gamma, a, T) : \gamma(0) \in \Sigma_a, \ \gamma(1) \in \Sigma_{a+T}\}.$$

To each point $\omega \in \Omega$ there corresponds a curve $q \colon [a,b] \to M$, $q(t) = \gamma((t-a)/T)$, where $a \bmod \mathbf{Z}$ and $b = a + T$. If $L \in C^2(TM \times \mathbf{T})$ is quadratic in v for large $\|v\|$ (the general case is reduced to this one), the action functional

$$S(\omega) = I_a^b(q) = T \int_0^1 L(\gamma(s), \gamma'(s)/T, a + sT) \, ds$$

is locally $C^{1+\mathrm{Lip}}$ on Ω. For any $h > 0$, set $S_h = S + hT$. A point $\omega \in \Omega^C = \{S_h \leq C\}$ is a critical point of S_h iff the corresponding trajectory $\sigma(t) = (q(t), p(t), t)$, $a \leq t \leq b$, satisfies Proposition 4.1. Thus, it is sufficient to prove that S_h has a critical point. Projections of trajectories in Γ form a manifold

$$\Lambda = \{\omega = (\gamma_x, a, T) \in \Omega : \gamma_x(s) = \pi_M(g^{sT}(x, 0, a)), \ x \in \Sigma_a\}$$

of trivial critical points of S, and $S|_\Lambda \equiv 0$. It is easy to see that S has no other critical points. On the other hand, S_h has no critical points in Λ.

There exists a compact set $K \subset \Omega$ noncontractible into Λ. Let $C > \max_K S$. Then $K \subset \Omega^C$ for small $h > 0$. The problem in proving that S_h has a critical point in Ω^C is loss of the Palais-Smale condition as $T \to 0$ and $T \to \infty$. For $h > 0$, the second problem is eliminated by the assumption $\min \mathcal{A} = 0$. Hence, S_h satisfies the Palais-Smale condition for $T \geq d > 0$. Because $\mathrm{ind}_\Gamma L = 0$, for small $d > 0$ it is possible to construct a pseudogradient flow retracting $\Omega^C \cap \{T \leq d\}$ onto $\Lambda \cap \{T \leq d\}$. The proof is completed by combining the gradient and pseudogradient flows [8]. $\square$

If there is a family $\{\Gamma_c\}$ of whiskered strictly minimizing manifolds, and the number of homoclinics in Ω^C is finite, one can use the methods of [10]–[12], [22] to construct connecting orbits.

5 Perturbations of invariant Lagrangian manifolds

Under certain twist conditions, exact symplectic maps are Poincaré maps of positive definite Hamiltonian systems [21]. Let N^{2m} be a symplectic manifold and $g \colon N \to N$ a symplectic map having a compact invariant Lagrangian manifold $M^m \subset N$. Let $g|_M$ be quasiperiodic with frequency vector $\omega \in \mathbf{T}^n$, $n < m$. Thus, $g|_M = f_\omega$, where $f \colon \mathbf{T}^n \times M \to M$ is a free group action of $\mathbf{T}^n$ on M. We may assume that ω is nonresonant: $\langle \omega, k \rangle \notin \mathbf{Z}$ for all $k \in \mathbf{Z}^n \setminus \{0\}$. If this is not so, then $g|_M$ is quasiperiodic with frequency vector in $\mathbf{T}^{n-r}$, where r is multiplicity of the resonance.

Because M is Lagrangian, its neighborhood U in N can be identified with a neighborhood of M in $T^*M\{x, y\}$. For any $x \in M$, the Hessian of the generating function S on U (defined by $dS = g^*\langle y, dx \rangle - \langle y, dx \rangle$) gives a quadratic function K on T^*M. Let $\overline{K}$ be the average over the $\mathbf{T}^n$-action on T^*M:

$$\overline{K} = \int_{\mathbf{T}^n} K \circ f_\theta^* \, d\theta.$$

The quadratic form $\overline{K}$ on T^*M is invariantly defined. The twist condition is that M is *positively definite*: $\overline{K}(x, y)$ is positive definite in y for all $x \in M$. Let $\| \cdot \|$ be the corresponding Riemannian metric on M.

EXAMPLE 5.1. INTEGRABLE MAP. Let $N = \mathbf{T}^m \times \mathbf{R}^m\{x, y\}$ with standard symplectic structure and $g(x, y) = (x + \Omega(y), y)$. Let $M = \mathbf{T}^m \times \{y\}$ be a resonant invariant torus: $A = \{k \in \mathbf{Z}^n : \langle \Omega(y), k \rangle \in \mathbf{Z}\}$ is an Abelian group of rank $m - n$. Then $g|_M$ is quasiperiodic with frequency $\omega \in \mathbf{T}^n$ and M is positively definite iff the symmetric matrix Ω_y is positive definite. Invariant tori M_I with resonance group A are parametrized by the action variable $I \in D \subset \mathbf{R}^n$ (momentum of the $\mathbf{T}^n$-action on $\mathbf{T}^m$ producing the foliation $\langle x, k \rangle = \mathrm{const}$, $k \in A$).

Consider a C^1-small exact symplectic perturbation $g_\varepsilon \in C^2(N \times [0, \varepsilon_0], N)$ of the map g. Let S_ε be the generating function. We set $V = -\partial S_\varepsilon / \partial \varepsilon|_{\varepsilon=0}$ and define the Poincaré function $\overline{V}$ as the average of $V|_M$ over the $\mathbf{T}^n$-action. Suppose that maximum h of $\overline{V}$ on $M/\mathbf{T}^n$ is strict and set $\Gamma_0 = \{x \in M : \overline{V}(x) = h\}$. Let $i \colon \Gamma_0 \to M$ be the inclusion and $E = i^* H^1(M, \mathbf{R}) \subset \mathbf{R}^n = H^1(\Gamma_0, \mathbf{R})$.

THEOREM 5.1 [7]. *For any $C > 0$ and all sufficiently small $\varepsilon > 0$ there exists $\delta \to +0$ as $\varepsilon \to 0$ such that:*

(1) *for any $I \in E$ such that $|I| \leq C\varepsilon$, the map g_ε has a compact invariant set*

$$\Gamma = \Gamma_I \subset \left\{ (x,y) : x \in B, \; \|y\|^2 \leq 2\varepsilon \max_B (h - \overline{V}) + o(\varepsilon \delta^2) \right\},$$

where $B = \{x \in M : \mathrm{dist}(x, \Gamma_0) \leq \delta\}$;

(2) *for any $x \in M$ there exists $y \in T_x^* M$ such that the orbit $g_\varepsilon^k(x,y)$ is semi-asymptotic to Γ in $W_\varepsilon = \left\{ (x,y) : \|y\|^2 \leq 2\varepsilon \max_M (h - \overline{V}) + o(\varepsilon) \right\}$;*

(3) *if $G = H_1(M, \Gamma_0, \mathbf{Z}) \neq 0$, then Γ has not less than $2 \, \mathrm{rank} \, G$ semihomoclinic orbits in W.*

If Γ_I is an invariant torus close to Γ_0, then I is its action variable. The sets Γ_I are c-minimizing with $I = i^* c$ for an appropriate Hamiltonian system. If $E = \mathbf{R}^n$, for any C and small ε there exist at least $n + 1$ points $I_k \in \mathbf{R}^n$, $|I_k| \leq C\varepsilon$, such that the sets Γ_{I_k} support invariant minimal measures with strictly extremal rotation vectors $O(\sqrt{\varepsilon})$-close to $i_* \omega$, and so they are strictly ergodic [17]. Theorem 5.1 can be reduced to Theorem 3.4 by using Delaunay's method [8]. Mather [19] proved the existence of $\Gamma(I)$ for $m = n$ and Diophantine ω.

COROLLARY 5.1. *For an exact perturbation of a completely integrable positive definite symplectic map and given resonance group of rank $m - n$, to any $I \in D \subset \mathbf{R}^n$ there corresponds an invariant set Γ_I in an $o(\sqrt{\varepsilon})$-neighborhood of M_I and at least $2(m - n)$ semihomoclinic orbits.*

Now let $g_\varepsilon \in C^\infty(N \times [0, \varepsilon_0], N)$, $\omega \in \mathbf{T}^n$ Diophantine, and maximum of $\overline{V}$ on $M / \mathbf{T}^n$ nondegenerate. Then one of the sets Γ_I is a whiskered torus.

THEOREM 5.2 [7]. *For sufficiently small $\varepsilon > 0$:*

(1) *g_ε has an n-dimensional whiskered torus Γ_ε with rotation vector ω;*
(2) *Γ_ε smoothly depends on ε and $W^{s,u}(\Gamma_\varepsilon)$ on $\sqrt{\varepsilon}$;*
(3) *if i^* is surjective, then $\pi(W^{s,u}) = M$, where π is the projection, and;*
(4) *Γ_ε has a homoclinic orbit in the $O(\sqrt{\varepsilon})$-neighborhood W_ε of M;*
(5) *if $G \neq 0$, then Γ_ε has at least $2 \, \mathrm{rank} \, G$ minimizing homoclinics.*

Statements (1) and (2) are a KAM-type result proved under slightly different assumptions in [13], [23]. The rest can be deduced from Theorem 4.1 [8]. If homoclinics in (4) and (5) are isolated, there are an infinite number of multibump homoclinics. For a perturbation of an integrable map, $2(m-n)$ homoclinics in (5), if isolated, form a basis for Arnold's diffusion of the action $I \in D$ [1], [9]. In Arnold's example, all sets Γ_I are whiskered tori. Recently Bessi treated this example by variational methods. In general, there are gaps in D corresponding to minimizing sets Γ_I. Establishing Arnold's diffusion requires computation of the barrier function for Γ_I. When the Poincaré-Melnikov-Arnold integral method works [1], the barrier can be estimated. However, in Arnold's diffusion case we are stuck with a problem of exponentially small intersection angles, which presents great difficulties even for a standard map. For a simpler analogous problem, diffusion was established in [9].

References

[1] V. I. Arnold, *Instability of dynamical systems with many degrees of freedom*, Dokl. Akad. Nauk SSSR **5** (1964), 581–585 (Russian).

[2] G. D. Birkhoff, Dynamical systems, Amer. Math. Soc. Colloq. Publ. IX., New York, 1927.

[3] S. V. Bolotin, *Libration motions of natural dynamical systems*, Vestnik Moskov. Univ. Ser. I Mat. Mekh. **6** (1978), 72–77 (Russian).

[4] _____ , Libration motions of reversible Hamiltonian systems, Dissertation, Moscow State University, 1981.

[5] _____ , *The existence of homoclinic motions*, Vestnik Moskov. Univ. Ser. I Mat. Mekh. **6** (1983), 98–103 (Russian).

[6] _____ , *Homoclinic orbits to invariant tori in the perturbation theory of Hamiltonian systems*, Prikl. Mat. Mekh. **54** (1990), 497–501 (Russian).

[7] _____ , *Homoclinic orbits to minimal tori of Lagrangian systems*, Vestnik Moskov. Univ. Ser. I Mat. Mekh. **6** (1992), 34–41 (Russian).

[8] _____ , Homoclinic orbits to invariant tori of Hamiltonian systems, preprint (to be published in Amer. Math. Soc. Advances of Soviet Mathematics), Moscow, 1992.

[9] L. Chierchia and G. Galavotti, Drift and diffusion in phase space, CARR Reports in Mathematical Physics, no. 15, Italy, 1992.

[10] V. Coti Zelati, I. Ekeland, and E. Seré, *A variational approach to homoclinic orbits in Hamiltonian systems*, Math. Ann. **288** (1990), 133–160.

[11] V. Coti Zelati and P. H. Rabinowitz, *Homoclinic orbits for second order Hamiltonian systems possessing superquadratic potentials*, J. Amer. Math. Soc. **4** (1991), 693–727.

[12] F. Giannoni and P. H. Rabinowitz, *On the multiplicity of homoclinic orbits on Riemannian manifolds for a class of second order Hamiltonian systems*, Nonlinear Differential Equations and Applications **1** (1993), 1–49.

[13] S. M. Graff, *On the conservation of hyperbolic tori for Hamiltonian systems*, J. Differential Equations **15** (1974), 1–69.

[14] M. R. Herman, *Existence et non existence de tores invariants par des diffeomorphismes symplectiques*, preprint (1988).

[15] H. Hofer and K. Wysocki, *First order elliptic systems and the existence of homoclinic orbits in Hamiltonian systems*, Math. Ann. **288** (1990), 133–160.

[16] V. V. Kozlov, *Calculus of variations in large and classical mechanics*, Uspekhi. Mat. Nauk **40** (1985), 33–60 (Russian).

[17] R. Mañé, *On the minimizing measures of Lagrangian dynamical systems*, Nonlinearity **5** (1992), 623–638.

[18] _____ , *Generic properties and problems of minimizing measures of Lagrangian systems* (1993), preprint.

[19] J. Mather, *Action minimizing invariant measures for positive definite Lagrangian systems*, Math. Z. **227** (1991), 169–207.

[20] _____ , *Variational construction of connecting orbits*, Ann. Inst. Fourier (Grenoble) **43** (1993), 1349–1386.

[21] J. Moser, *Monotone twist mappings and the calculus of variations*, Ergodic Theory Dynamical Systems **6** (1986), 401–413.

[22] E. Séré, *Existence of infinitely many homoclinics in Hamiltonian systems*, Math. Z. **209** (1992), 27–42.

[23] D. V. Treshchev, *The fracture mechanism of resonant tori of Hamiltonian systems*, Mat. Sb. (New Ser.) **180** (1989), 1325–1346 (Russian).

Rotation Vectors for Surface Diffeomorphisms

JOHN FRANKS

Department of Mathematics
Northwestern University
Evanston, IL 60208-2730, USA

ABSTRACT. We consider the concepts of rotation number and rotation vector for area preserving diffeomorphisms of surfaces and their applications. In the case that the surface is an annulus A the rotation number for a point $x \in A$ represents an average rate at which the iterates of x rotate around the annulus. More generally the rotation vector takes values in the one-dimensional homology of the surface and represents the average "homological motion" of an orbit.

There are two main results. The first is that if 0 is in the interior of the convex hull of the recurrent rotation vectors for an area preserving diffeomorphism f isotopic to the identity, then f has a fixed point of positive index. The second result asserts that if f has a vanishing mean rotation vector, then f has a fixed point of positive index.

Applications include the result that an area preserving diffeomorphism of A that has at least one periodic point must in fact have infinitely many interior periodic points. This is a key step in the proof of the theorem that every smooth Riemannian metric on S^2 has infinitely many distinct closed geodesics. Another application is a new proof of the Arnold conjecture for area preserving diffeomorphisms of closed oriented surfaces.

In this article we consider area preserving diffeomorphisms of compact surfaces. We are concerned with "rotation vectors" defined in terms of homology that can be associated to the points of the surface. Two main results are announced. The first is that if f is an area preserving diffeomorphism of a compact surface that is homotopic to the identity and 0 is in the interior of the convex hull of the rotation vectors of f, then f has a fixed point of positive index. The second result concerns the mean rotation vector of f. It asserts that if this vector vanishes then f has a fixed point of positive index. An expanded version of these results including detailed proofs will appear in [4].

A special case of this second result — when the surface in question is a surface of genus zero — was proved in [3]. This result provided a key ingredient in the proof that any Riemannian metric on S^2 has infinitely many closed geodesics.

There are several applications of these results including a new proof of the Arnold conjecture for area preserving diffeomorphisms of compact surfaces.

Proceedings of the International Congress
of Mathematicians, Zürich, Switzerland 1994
© Birkhäuser Verlag, Basel, Switzerland 1995

1 Homological Rotation Vectors

The idea of rotation number for homeomorphisms of the circle or annulus goes back at least to Poincaré. The idea of an analogous concept with values in homology can be traced back at least to Schwartzman [6].

Homological rotation vectors have subsequently been considered by many authors. In particular for surfaces of genus zero results similar to those below were given in [3]. We also consider the "mean rotation vector" for an invariant measure μ that can be expressed as the integral of the rotation vectors of points with respect to μ (see (1.3) below).

We are interested in investigating the existence of periodic orbits for area preserving diffeomorphisms on a surface. We will consider a compact surface M and focus on the case where the diffeomorphism $f : M \to M$ is isotopic to the identity, and M has negative Euler characteristic. We begin by formulating a definition of "homological rotation vector" for such a diffeomorphism $f : M \to M$.

We fix a metric on M of constant negative curvature. We assume that each boundary component is a geodesic. Even more, we assume that one can form M by taking a convex geodesic polygon in hyperbolic space and making identifications of some of the edges.

Pick a base point b_0 in the interior of the polygon whose sides are identified to form M. We want to define a function γ that assigns to each $x \in M$ a geodesic segment γ_x in M from b_0 to x, in such a way that the correspondence $x \to \gamma_x$ is measurable. We do this using the fact that M is a convex geodesic polygon in hyperbolic space with some edges identified, and with b_0 in its interior. We then let γ_x be the unique geodesic segment from b_0 to x if x is in the interior of this polygon. For each pair of edges that are identified we pick one and choose γ_x be the unique geodesic segment from b_0 to x that, when lifted back to the polygon, ends on the chosen edge.

Let $f_t(x)$ be a homotopy from $f_0 = id : \widetilde{M} \to \widetilde{M}$ to $f_1 = f$. Because the Euler characteristic of M is negative f_t is unique up to homotopy. This means that if g_t is another homotopy with $g_0 = id$ and $g_1 = f$, then there is a homotopy from f_t to g_t, i.e. a map $H : M \times [0,1] \times [0,1] \to M$ such that $H(x,t,0) = f_t(x)$ and $H(x,t,1) = g_t(x)$.

For any point $x \in M$ we want to construct a path in M from x to $f^n(x)$ and then form a loop with the segments γ_x and $\gamma_{f^n(x)}$. To do this we observe that if $\pi : \widetilde{M} \to M$ is the universal covering space of M, there is a canonical lift of f to a diffeomorphism $F : \widetilde{M} \to \widetilde{M}$; namely, F is that lift obtained by lifting the homotopy f_t from the identity to f to form a homotopy on $\widetilde{M}$ starting at the identity on $\widetilde{M}$. The other end of this homotopy is then defined to be F. The uniqueness of f_t up to homotopy implies that F does not depend on the choice of homotopy from the identity to f. F is the unique lift whose extension to the ideal points at infinity of $\widetilde{M}$ has all those points as fixed points.

Consider the path $\alpha(n,x)$ from x to $f^n(x)$ in M that is given by

$$\alpha(n,x)(t) = f_t^n(x).$$

Again the homotopy class of this path relative to its endpoints is independent of the choice of the homotopy f_t because of the uniqueness (up to homotopy) of this homotopy.

For each $x \in M$ let $h_n(x, f)$ be the closed loop based at b_0 formed by the concatenation of γ_x, the path $\alpha(n, x)$ in M from x to $f^n(x)$ and $\gamma_{f^n(x)}$ traversed backwards. If the diffeomorphism f is clear from the context, we will abbreviate $h_n(x, f)$ to $h_n(x)$.

Note that if $*$ denotes concatenation of based loops then $h_n(x) * h_m(f^n(x))$ is homotopic to $h_{n+m}(x)$. We will denote by $[h_n(x)]$ the homology class in $H_1(M, \mathbb{R})$ of the loop $h_n(x)$. Note that $[h_{n+m}(x)] = [h_n(x)] + [h_m(f^n(x))]$. We can now formulate the definition of homology rotation vector.

(1.1) DEFINITION. *Let M be a compact surface that has negative Euler characteristic and may have nonempty boundary. Suppose $f : M \to M$ is a diffeomorphism that is isotopic to the identity map. The homological rotation vector of $x \in M$ is an element of $H_1(M, \mathbb{R})$ denoted $\mathcal{R}(x, f)$, and is defined as*

$$\mathcal{R}(x, f) = \lim_{n \to \infty} \frac{[h_n(x)]}{n}$$

if this limit exists.

Let μ be a smooth f-invariant measure on M. The homology classes $[h_1(x)] \in H_1(M, \mathbb{R})$ depend measurably on x. In fact there is a closed set of measure zero in M (consisting of the "edges" of the polygon and their inverse images under f) on the complement of which the function $[h_1(x)]$ is locally constant.

(1.2) LEMMA. *If $f : M \to M$ is as in (1.1) and f preserves the smooth measure μ then the function $[h_1(x)]$ defined on M with values in $H_1(M, \mathbb{R})$ is bounded and μ measurable, hence integrable.*

The Birkhoff ergodic theorem asserts that $\mathcal{R}(x, f)$ is a μ-measurable function of x and that

$$\int \mathcal{R}(x, f) \, d\mu = \int [h_1(x, f)] \, d\mu.$$

(1.3) DEFINITION. *Let M be a compact surface with negative Euler characteristic. Suppose $f : M \to M$ is a diffeomorphism of the surface M that is isotopic to the identity map and preserves a measure μ. The mean rotation vector of f is an element of $H_1(M, \mathbb{R})$ denoted $\mathcal{R}_\mu(f)$, and is defined as*

$$\mathcal{R}_\mu(f) = \int \mathcal{R}(x, f) \, d\mu.$$

A key property of the mean rotation vector is that it is a homomorphism from the group of μ-invariant diffeomorphisms isotopic to the identity to $H_1(M, \mathbb{R})$. The proof of this fact is really just the change of variables formula for integration.

(1.4) PROPOSITION. *Suppose f and g are diffeomorphisms of the compact surface M that are isotopic to the identity and preserve a smooth measure μ. Then*

$$\mathcal{R}_\mu(f \circ g) = \mathcal{R}_\mu(f) + \mathcal{R}_\mu(g).$$

The Poincaré recurrence theorem implies that the set of recurrent points in M has full measure. Thus, the set of points for which $\mathcal{R}(x, f)$ exists and that is recurrent under f has full measure. Moreover, any rotation vector is essentially a convex combination of rotation vectors of a set of recurrent points as the following proposition shows.

(1.5) PROPOSITION. *Let x be a point for which the limit $\mathcal{R}(x, f) \in H_1(M, \mathbb{R})$ exists and let $\varepsilon > 0$ be arbitrary. Then there are points y_i, $0 \leq i \leq n$, in the set of f recurrent points with the property that $\mathcal{R}(y_i, f)$ exists and there are rationals c_i such that*

$$\left\| \mathcal{R}(x, f) - \sum_{i=0}^{n} c_i \mathcal{R}(y_i, f) \right\| < \varepsilon.$$

2 Handel's Fixed Point Theorem

Ultimately our aim is to prove the existence of fixed points. The tool we use for this purpose is a beautiful result of Handel on fixed points of homeomorphisms of the disk. In our application the disk in question will be the compactification of the universal covering space of a surface. We proceed with a statement of Handel's result.

Suppose that $h : D \to D$ is an orientation preserving homeomorphism of the disk $D = \mathbb{D}^2$. We are interested in the orbits of points whose alpha and omega limit sets each consist of single (but distinct) points in ∂D. Given two such points $x, y \in D$ we will say that their orbits $orb(x)$ and $orb(y)$ are *linked* provided the points $\alpha(y)$ and $\omega(y)$ separate the points $\alpha(x), \omega(x)$ in ∂D, or equivalently if the straight line segments from $\alpha(x)$ to $\omega(x)$ and from $\alpha(y)$ to $\omega(y)$ intersect in a single interior point of D.

We say that orbits $orb(x_i)$, $1 \leq i \leq n$, form an *oriented cycle of links* if there is an oriented polygon in the interior of D whose i^{th} side is a segment of the line segment from $\alpha(x_i)$ to $\omega(x_i)$ and the orientation of this side is consistent with the orientation of this line segment from $\alpha(x_i)$ to $\omega(x_i)$.

(2.1) THEOREM [5]. *Suppose h is a homeomorphism of $\mathbb{D}^2$. If there are points $x_i \in \mathbb{D}^2$ whose orbits form an oriented cycle of links, then h has a fixed point in the interior of $\mathbb{D}^2$.*

It seems likely that a slightly stronger conclusion to this theorem is valid, namely that there exists an interior fixed point of positive index. This stronger conclusion is important for most of our applications. Although this conclusion is not known to hold in general, it does hold in the special setting where we wish to apply it. This result is also due to Handel.

Suppose $f : M \to M$ is a diffeomorphism of a compact surface with negative Euler characteristic and f is isotopic to the identity. It is possible and convenient to provide M with a metric of constant negative curvature with each boundary component being a geodesic. Let $\widetilde{M}$ denote the universal covering space of M and let $C(\widetilde{M})$ denote the circle or Cantor set of "points at infinity" of $\widetilde{M}$, which can be added to $\widetilde{M}$ to compactify it. We think of $\widetilde{M}$ as a convex subset of the hyperbolic disk and $C(\widetilde{M})$ as a Cantor set in the boundary circle of the hyperbolic disk or the entire boundary if M has no boundary components.

If $F : \widetilde{M} \to \widetilde{M}$ is the lift of f obtained from lifting the homotopy of f to the identity, then F extends to $C(\widetilde{M})$ by setting $F(z) = z$ for all $z \in C(\widetilde{M})$. If $x \in \widetilde{M}$ is not a periodic point of F, but $\pi(x) \in M$ is periodic, then it is not difficult to see that $\alpha(x, F)$ and $\omega(x, F)$ are each a single point in $C(\widetilde{M})$ and these points are distinct.

Indeed, one can find these points as follows: if $\pi(x)$ has period n, let γ be the loop from $\pi(x)$ to itself traced by $\pi(x)$ under the homotopy of f^n to the identity that is the nth iterate of the homotopy from f to the identity. The curve γ cannot be null homotopic because that would imply that x is periodic under F. Let γ_0 be the closed geodesic in M freely homotopic to γ and given the same orientation as γ. If Γ is the lift of γ that starts at x, then there is a lift Γ_0 of γ_0 that is a uniformly bounded distance from Γ. The curve Γ_0 is then a geodesic in the hyperbolic plane that has its ends at $\alpha(x, F)$ and $\omega(x, F)$ in $C(\widetilde{M})$.

The setting in which we will apply Handel's fixed point theorem is this: the disk D we consider is $\widetilde{M} \cup C(\widetilde{M})$, and the points x_i whose orbits form the oriented cycle of links will all have the property that $\pi(x_i)$ is a periodic point of f.

(2.2) PROPOSITION (HANDEL). *Suppose $f : M \to M$ is a diffeomorphism homotopic to the identity where M is a surface of finite type with negative Euler characteristic. Let F be the canonical lift of f to the universal covering space $\widetilde{M}$. Suppose that there are points $x_1, x_2, \ldots, x_k \in \widetilde{M}$ whose orbits form an oriented cycle of links for F and such that each $\pi(x_i)$ is a periodic point of f in the interior of M. Then F has a fixed point in the interior of $\widetilde{M}$. If the fixed points of F project to a finite set in M, then F has a fixed point of positive index.*

3 The Convex Hull of the Rotation Set

We can now state the first of our two main results.

(3.1) THEOREM. *Let M be a compact orientable surface with negative Euler characteristic. Suppose $f : M \to M$ is a diffeomorphism of the surface M that is isotopic to the identity map and preserves a smooth probability measure μ and has at most finitely many interior fixed points. If 0 is in the interior of the convex hull of the rotation set $\mathcal{R}(f) = \cup \mathcal{R}(x, f)$, then f has an interior fixed point of positive index. In fact, the canonical lift of f to its universal cover $F : \widetilde{M} \to \widetilde{M}$ has an interior fixed point that is of positive index.*

The idea of the proof is to use the hypothesis about 0 being in the interior of the convex hull of the rotation set to construct a finite collection of "periodic

ε-chains" for f with special properties. An element of this collection is a sequence $x_1, x_2, \ldots, x_n = x_1$ with the property that for a small C^0 perturbation of f it will actually be a periodic orbit. In fact, the perturbation can be constructed simultaneously for all the ε-chains in the collection. Moreover, this can be done with a perturbation sufficiently small that the perturbed diffeomorphism g has precisely the same fixed points as f and agrees with f on a neighborhood of this fixed point set.

There is one additional property of this construction. For each of the g periodic orbits made from periodic ε-chains for f we can form a loop from x_1 to itself traced by x_1 under the homotopy of g^n to the identity that is the nth iterate of the homotopy from g to the identity (n is the period of the point x_1). In this way we form a closed loop γ_j for g periodic orbit. The final property required of this construction is that there are positive integers a_j such that

$$\sum a_j [\gamma_j] = 0,$$

where $[\gamma_j] \in H_1(M)$ is the integral homology class determined by γ_j.

To complete the proof one shows that this last condition implies that if $G : \widetilde{M} \to \widetilde{M}$ is the canonical lift of g to the universal covering space of M, then there are lifts of the g periodic orbits on M to orbits of G that satisfy the hypothesis of Handel's result (2.2). It follows that G (and hence g) has a fixed point of positive index and as f and g agree on a neighborhood of their common fixed point set, the same is true of f.

The proof of the following result is similar to that of (3.1). The hypothesis that the mean rotation vector is zero is adequate to show that 0 is in the interior of the convex hull of rotation vectors of diffeomorphisms in an arbitrarily small C^0 neighborhood of f, and this is adequate to carry out the proof much as before.

(3.2) THEOREM. *Suppose M is an oriented compact surface with negative Euler characteristic and $f : M \to M$ is a diffeomorphism isotopic to the identity that preserves the smooth measure μ and has $\mathcal{R}_\mu(f) = 0$. Then there is an interior fixed point of f that has positive index and that lifts to a fixed point of F the canonical lift of f to the universal covering space $\widetilde{M}$.*

4 Applications

A heuristic guiding the study of area preserving diffeomorphisms, especially exact symplectic diffeomorphisms, is that they should have properties like the time one map of the flow associated with an autonomous Hamiltonian vector field. In particular we might hope to find estimates on the number and nature of fixed points based on the analogous results for critical points of smooth functions (the Hamiltonian function). The following result, sometimes called the Arnold conjecture, was proved by Floer [2] and Sikorav [7].

(4.1) THEOREM. *Suppose M is a compact oriented surface with nonpositive Euler characteristic and without boundary and suppose $f : M \to M$ is a diffeomorphism isotopic to the identity that preserves the smooth measure μ and has $\mathcal{R}_\mu(f) = 0$.*

Then f has at least three distinct fixed points. If f has finitely many fixed points, then at least two of them have positive index.

Sketch of Proof. Clearly we need only consider the case when f has finitely many fixed points. We sketch the case when the Euler characteristic of M is negative. By Theorem (3.1) above there is a fixed point z of positive index. The fact that z comes from the fixed point of the canonical lift of f to the universal cover $\widetilde{M}$ implies that f is isotopic to the identity relative to z.

Let N denote M with the fixed point z blown up. Then there is a natural identification of $H_1(N, \mathbb{R})$ with $H_1(M, \mathbb{R})$ induced by the obvious map from N to M. Clearly, if $f_0 : N \to N$ is the blown up version of f, then for any $x \in int(N)$ for which $\mathcal{R}(x, f)$ is defined we have $\mathcal{R}(x, f) = \mathcal{R}(x, f_0)$. Thus, $\mathcal{R}_\mu(f_0) = 0$ and f_0 satisfies the hypothesis of (3.1). It follows that f_0 has an interior fixed point of positive index and hence f has two fixed points of positive index. Because the Euler characteristic of M is less than or equal to zero, there must also be a fixed point of f of negative index, making at least three fixed points. $\square$

Another way in which the heuristic mentioned above guides us is by suggesting that if a diffeomorphism of a surface has some number of fixed points that are like minima for a Hamiltonian function, then there must another fixed point that is like a maximum. In some circumstances this is the case. Suppose we are given an area preserving diffeomorphism $f : M \to M$ and a finite set of fixed points P such that f is homotopic to the identity on $M \setminus P$. If the Euler characteristic of $M \setminus P$ is negative and we "blow up" the points of P, replacing each p with a boundary component C_p, then the homological rotation number is defined for each point in the boundary of the resulting surface.

In particular if this rotation vector is not zero, we have a well-defined notion of "which way" each boundary component is rotating compared with the orientation that boundary component inherits from M. We will say that a fixed point $p \in P$ is of *minimum type* (respectively *maximum type*) provided $\mathcal{R}(z, f)$ is a positive (respectively negative) multiple of $[C_p]$ for $z \in C_p$, where the orientation of C_p is that induced from an orientation of M.

(4.2) THEOREM. *Suppose $f : S^2 \to S^2$ is an area preserving diffeomorphism that has a finite fixed point set consisting of at least three points. If $P = \{p_1, p_2, \dots, p_m\}$, $m > 1$, is a set of fixed points each of which is of minimum type, then f has two additional fixed points not in P, whose indices have opposite sign. If all the fixed points of f are generic, there is one of maximum type.*

As a final application of the results in Section 3 we cite the following result from [3].

(4.3) THEOREM. *Suppose $f : A \to A$ is an area preserving diffeomorphism of the closed annulus that is isotopic to the identity. If f has at least one periodic point (which may be on the boundary), then f must have infinitely many interior periodic points.*

We note that the result published in [3] claims this theorem for homeomorphisms (as opposed to diffeomorphisms) and for the open annulus. However, the proof given there is not valid in this generality.

The idea of the proof of this result as a consequence of (3.2) is not difficult. This theorem is true for f if and only if it is true for some iterate of f. Thus in a proof by contradiction we may replace f by an iterate and assume that f has finitely many interior periodic points, all fixed.

Blowing up all these fixed points we obtain a diffeomorphism $f_0 : M_0 \to M_0$ of a compact surface M_0 of genus zero and negative Euler characteristic. An argument using Thurston's classification theorem for surface homeomorphisms and the Poincaré-Birkhoff theorem shows that if f_0 is not isotopic to the identity, then it must have infinitely many periodic points contradicting our assumption. Assuming that f_0 is isotopic to the identity we consider the mean rotation vector $\mathcal{R}_\mu(f_0)$. If this is nonzero, then an argument appealing again to the Poincaré-Birkhoff theorem shows that f_0 has infinitely many periodic points. On the other hand if $\mathcal{R}_\mu(f_0) = 0$, then Theorem (3.2) above implies the existence of an additional interior fixed point. So in either case we contradict the assumption that the finitely many periodic points of M have all been blown up.

This theorem supplies one important ingredient in the proof of the following result (see [1]).

(4.4) THEOREM. *Suppose S^2 is equipped with an arbitrary smooth Riemannian metric. Then there are infinitely many closed geodesics for this metric that are distinct as point sets.*

References

[1] Victor Bangert, *On the existence of closed geodesics on two-spheres*, Internat. J. Math. **4** (1993), 1–10.

[2] A. Floer, *Proof of the Arnold Conjecture for surfaces and generalizations to certain Kähler manifolds*, Duke Math. J. **51** (1986), 1–32.

[3] J. Franks, *Geodesics on S^2 and periodic points of annulus homeomorphisms*, Invent. Math. **108** (1992), 403–418.

[4] J. Franks, *Rotation vectors and fixed points of area preserving surface diffeomorphisms*, to appear.

[5] Michael Handel, *A fixed point theorem for planar homeomorphisms*, preprint.

[6] S. Schwartzman, *Asymptotic cycles*, Ann. of Math. (2) **66** (1957), 270–284.

[7] J.-C. Sikorav, *Points fixes d'une application symplectique homologue à l'identité*, J. Differential Geom. **22** (1985), 49–79.

A Priori Estimates and Regularity of Nonlinear Waves

Manoussos G. Grillakis[*]

Department of Mathematics, University of Maryland
College Park, MD 20742, USA

In this paper I would like to explain certain nonlinear hyperbolic systems of equations and the questions that are raised when one tries to understand the behavior of their solutions. In each case I will consider the simplest possible problem in order to demonstrate the main ideas.

Consider a map

$$z(x): \quad \mathbb{R} \times \mathbb{R}^2 \quad \longmapsto \quad D \subset \mathbb{C}, \tag{1}$$

where D is a subset of the complex plane with a conformal metric on it given by $g(z, \bar{z})dzd\bar{z}$, where g is the conformal factor. One can also think of D as a Riemann surface. The underlying space is $\mathbb{R} \times \mathbb{R}^2$ and is equipped with a Minkowski metric $\eta^{\mu\nu} = \text{diag}(-1, 1, 1)$. I will use the notation $\partial^\mu = \eta^{\mu\nu}\partial_\nu$ with $\mu, \nu = 0, 1, 2$ to raise indices and summation over repeated indices is implied. Later I will identify $x^0 \equiv t$ with the time variable and write $x = (x^1, x^2)$ for the space variables when the situation calls for it. Now consider the Lagrangian

$$\mathfrak{L}(z, \bar{z}) = \frac{1}{2}\int_{\mathbb{R} \times \mathbb{R}^2} \{g(z, \bar{z})\partial_\mu z\partial^\mu \bar{z}\}\, dx. \tag{2}$$

A formal calculation of the derivative $\frac{\delta\mathfrak{L}(z,\bar{z})}{\delta\bar{z}} = 0$ gives a system of equations

$$\Box z - (\frac{\partial}{\partial z}\log(g(z, \bar{z})))\partial_\mu z\partial^\mu z = 0, \tag{3}$$

where $\Box = -\partial_\mu\partial^\mu = \partial_t^2 - \triangle_x$ is the D'Alembertian with respect to the metric $\eta^{\mu\nu}$. Because the Lagrangian in (2) is translation invariant equations (3) preserve an integral called energy, i.e.

$$E(t) = \int_{\mathbb{R}^2} \frac{1}{2}g(z, \bar{z})[|\nabla z|^2 + |z_t|^2]dx \tag{4}$$

is independent of time, $E(t) = E(0)$, for all t. This is a fundamental property of equations (3).

---

[*]   Research supported by a PYI and a Sloan Fellowship.

Proceedings of the International Congress
of Mathematicians, Zürich, Switzerland 1994
© Birkhäuser Verlag, Basel, Switzerland 1995

Performing an analytic change of variables $z = f(w)$ changes the conformal factor

$$g(z, \bar{z}) \longmapsto g(f(w), \overline{f}(w))|f'(w)|^2 \tag{5}$$

but does not change the nature of the equations. The relevant quantity is the curvature of the target surface given by

$$\frac{\partial}{\partial \bar{z}} \frac{\partial}{\partial z} \log(g) = -\frac{1}{2}\mathcal{K}g \tag{6}$$

where $\mathcal{K} = 1, 0, -1$ is the curvature and gives respectively a sphere, a flat plane, and a hyperbolic plane. The nature of the nonlinear terms in equations (3) should depend only on the curvature; in particular $\mathcal{K} = 0$ gives the linear wave equation. It is natural to try to understand how the curvature affects the nonlinear terms. For this purpose let me rewrite the metric in local coordinates as follows:

$$ds^2 = d\rho^2 + f^2(\rho)d\theta^2, \tag{7}$$

where $d\rho$ is the radial infinitesimal distance and $d\theta$ is the angular infinitesimal increase; $f(\rho) = \sin \rho, \rho, \sinh \rho$; correspond to the sphere, the flat plane, and the hyperbolic plane respectively. The Lagrangian in these coordinates is

$$\mathcal{L}(\rho, \theta) = \frac{1}{2} \int_{\mathbb{R} \times \mathbb{R}^2} [\partial_\mu \rho \partial^\mu \rho + f^2(\rho) \partial_\mu \theta \partial^\mu \theta] \, dx. \tag{8}$$

The rigid motions of the target manifold generate three divergence-free vector fields, namely

$$\begin{cases} \omega_\mu^1 = f^2(\rho)\partial_\mu\theta, \\ \omega_\mu^2 = \cos\theta\partial_\mu\rho - f(\rho)f'(\rho)\sin\theta\partial_\mu\theta, \\ \omega_\mu^3 = \sin\theta\partial_\mu\rho + f(\rho)f'(\rho)\cos\theta\partial_\mu\theta, \end{cases} \tag{9}$$

so that

$$\partial_\mu \omega^{\alpha,\mu} = 0, \qquad \alpha = 1, 2, 3. \tag{9a}$$

Now call

$$z = \rho e^{i\theta}, \qquad \omega_\mu = \omega_\mu^2 + i\omega_\mu^3,$$

then

$$\omega_\mu = \partial_\mu z + ig(|z|)z\omega_\mu^1, \tag{10}$$

where

$$g(\rho) = \frac{f(\rho)f'(\rho) - \rho}{\rho f^2(\rho)}. \tag{10a}$$

From (10) we get the system of equations for z

$$\Box z - i\partial_\mu(g(|z|)z\omega^{1,\mu}) = 0, \tag{11a}$$

$$\partial_\mu \omega^{1,\mu} = 0. \tag{11b}$$

The energy is now given by the conserved integral

$$\frac{1}{2}\int_{\mathbb{R}^2}\{|\nabla z|^2 + |z_t|^2 + (f^2(\rho) - \rho^2)(\theta_t^2 + |\nabla\theta|^2)\}\,dx.$$

Notice that $f^2(\rho) - \rho^2 \geq 0$ if $f(\rho) = \sinh\rho$, i.e. $\mathcal{K} = -1$, whereas $f^2(\rho) - \rho^2 \leq 0$ if $f(\rho) = \sin\rho$, i.e. $\mathcal{K} = 1$. If we set $z = u^1 + iu^2$, equations (11) have the general form

$$\Box u^\alpha + A^\alpha_{\beta\gamma}(u)l(u^\beta, u^\gamma) = 0, \qquad \alpha, \beta, \gamma = 1, 2\,, \tag{12}$$

where $A^\alpha_{\beta\gamma}(u)$ are given functions of $z = u^1 + iu^2$ and

$$l(u^\beta, u^\gamma) = \partial_\mu u^\beta \partial^\mu u^\gamma. \tag{12a}$$

Equations (11) are of critical type; this means that the nonlinear terms in the equation are as strong as the linear estimates available. This poses certain interesting questions about the possible linear estimates that we will discuss later.

There are certain reasonable conjectures that can made about the equations (3) or (11).

CONJECTURE 1 If $\mathcal{K} = 1$, i.e. the surface is the sphere, then solutions blow up if the initial energy $E(0)$ is sufficiently large.

CONJECTURE 2 If $\mathcal{K} = -1$, i.e. the target surface is the hyperbolic plane, then solutions are regular provided that the initial data are smooth.

CONJECTURE 3 If $\mathcal{K} = \pm 1$, i.e. there is no restriction on the target surface, then solutions are regular provided that the initial energy is sufficiently small and the initial data smooth.

There are certain special cases that have already been answered. Write $z = \rho e^{i\theta}$. Assuming radial symmetry, i.e. $z(t, r)$ with $r = |x|$, then Conjecture 2 is true. This is a remarkable result due to Christodoulou and Tahvildar-Zadeh [3]. Assuming corotational symmetry, i.e. $\rho(t, r)$ and $\theta = k\varphi$ with $(t, x) = (t, r, \varphi)$ and $k \in \mathbb{Z}$, then Conjecture 2 is true, see [7], [17], [16]. Conjecture 1 is open even in the simple corotational case.

Despite some success there is still no general method of attack for equations (12), however there is a related critical equation namely

$$\Box u + u^5 = 0, \qquad (t, x) \in \mathbb{R} \times \mathbb{R}^2 \tag{I}$$

for which regularity was proved recently, see [22], [5], [6], [16], [11] that can help to motivate a method of attack. The proof of (I) is based on two ingredients. First one shows, using energy estimates, that the nonlinear part of the energy, namely

$$\int u^6\,dx,$$

does not concentrate in an appropriate sense, then this information is used in certain space-time estimates due to Strichartz and Littman with further improvements and generalizations, see [4], [13], [9], [21], [10], to prove that $u \in L^8(\mathbb{R}^4)$.

This is the crucial step, further regularity can be shown by a simple bootstrap argument.

The space-time estimate in 2+1 dimensions can be stated as follows. Consider the equation

$$\begin{cases} \Box u = f, & (t,x) \in \mathbb{R} \times \mathbb{R}^2, \\ u(0,x) = u_0, \quad u_t(0,x) = u_1, & x \in \mathbb{R}, \end{cases} \tag{13}$$

then

$$\|D^{\frac{1}{2}} u\|_{L^6(\mathbb{R}^3)} \leq \int \|f(t,\cdot)\|_{L^2}\, dt + \|\nabla u_0\|_{L^2} + \|u_1\|_{L^2}. \tag{13a}$$

There is a tradeoff in the above estimate where only half a derivative is estimated in L^6 over the space-time. This should be compared with the standard energy estimate, which is

$$\sup_{0 \leq t \leq T} \|Du(t,\cdot)\|_{L^2(\mathbb{R}^2)} \leq \int_0^T \|f(t,\cdot)\|_{L^2}\, dt + \|\nabla u_0\|_{L^2} + \|u_1\|_{L^2}. \tag{14}$$

To complement the above, Harmse [9] proved that the estimate

$$\|Du\|_{L^p} \leq C\|\Box u\|_{L^p}$$

cannot be true if $p \neq 2$. However the energy estimate can be improved in some sense. Part of what follows has been determined in collaboration with Fang. Consider the equation

$$\begin{cases} (\Box + m^2)u = 0, & (t,x) \in \mathbb{R} \times \mathbb{R}^2, \\ u(0,x) = 0, \quad u_t(0,x) = f, & x \in \mathbb{R}, \end{cases}$$

and the quantity

$$\epsilon(t) = \left(\int_{\mathbb{R}} |\nabla u^2|^2 dx \right)^{\frac{1}{2}}, \tag{15}$$

then

$$\|\epsilon\|_{bmo} \leq C\|f\|_{L^2}^2, \tag{16}$$

where bmo is an appropriate local version of BMO. The appearance of the mass term in the equation is only for technical reasons. The problem with estimates like (13a) is that they cannot handle equations (12). Consider the caricature equation

$$\Box u = l(u,u). \tag{17}$$

The right-hand side of inequality (13a) is of the form

$$\int \|l(u,u)\|_{L^2(\mathbb{R}^2)}\, dt\,,$$

which cannot be balanced with the left-hand side of the inequality. However inequality (13a) is not optimal; in particular the right-hand side of (13a) can be improved. Take the Fourier transform in space-time variables for the equation

$$\Box u = f ,$$

which gives

$$(|\xi|^2 - \tau^2)\hat{u} = \hat{f}.$$

This calculation is only formal but it motivates the correct idea. First notice that

$$|\xi|^2 - \tau^2 = (|\xi| + |\tau|)(|\xi| - |\tau|).$$

A more careful calculation shows that at the expense of making the estimates local in time we can replace

$$|\xi| - |\tau| \sim ||\xi| - |\tau|| + 1.$$

Next, let me call

$$D = |\xi| + |\tau|, \quad \Lambda = ||\xi| - |\tau|| + 1, \quad Q = D\Lambda, \tag{18}$$

then formally we have the estimate

$$\|D^{\frac{1}{2}}Q^{\frac{1}{2}}\hat{u}\|_{L^2} \leq \|\frac{\hat{f}}{\Lambda^{\frac{1}{2}}}\|_{L^2}.$$

Unfortunately the above estimate is only formal. A direct but lengthy calculation shows that (13) can be estimated by

$$\|D^{\frac{1}{2}}Q^{\lambda}\hat{u}\|_{L^2(\mathbb{R}^3)} \leq \|\frac{D^{\frac{1}{2}}\hat{f}}{Q^{1-\lambda}}\|_{L^2(\mathbb{R}^3)} + \|u_0\|_{H^{1+\frac{\epsilon}{2}}} + \|u_1\|_{H^{\frac{\epsilon}{2}}}, \tag{19}$$

where

$$\lambda = \frac{1+\epsilon}{2} \quad \text{and} \quad \epsilon > 0.$$

Estimate (19) has the disadvantage that $\lambda > \frac{1}{2}$ but the right-hand side is better because

$$\frac{D^{\frac{1}{2}}}{Q^{1-\lambda}} = \frac{D^{\frac{\epsilon}{2}}}{\Lambda^{\frac{1-\epsilon}{2}}}.$$

The idea now is to try to estimate the solutions in the norm

$$N_\epsilon(u) = \|D^{\frac{1}{2}}\Lambda^{\lambda}\hat{u}\|_{L^2(\mathbb{R}^3)} \quad ; \quad \lambda = \frac{1+\epsilon}{2} \tag{20}$$

and this will be possible if we manage to balance the right-hand side with $N_\epsilon(u)$. For simplicity consider only the caricature equation (17). We would like to show that

$$\|\frac{D^{\frac{1}{2}}}{Q^{1-\lambda}}\hat{l}(u,u)\|_{L^2} \leq N_\epsilon^2(u). \tag{20a}$$

The crucial observation here, due to Bourgain [1], is that

$$l(u, u) = \Box u^2 - 2u\Box u,$$

hence, if I denote by $\hat{\Box} = |\xi|^2 - \tau^2$,

$$\hat{l}(u, u) = \hat{\Box}\hat{u} * \hat{u} - 2\hat{u} * \hat{\Box}\hat{u}. \tag{21}$$

Observe that

$$\frac{D^{\frac{1}{2}}\hat{\Box}}{Q^{1-\lambda}} \sim D^{\frac{1}{2}}Q^\lambda,$$

hence the first term in (21) will give the estimate

$$\|D^{\frac{1}{2}}Q^\lambda(\hat{u} * \hat{u})\|_{L^2} \leq \|D^{\frac{1}{2}}Q^\lambda u\|_{L^2}^2. \tag{22}$$

Estimate (22) is not optimal, if we call

$$M_{\epsilon,\delta}(u) = \|D^{\lambda+\frac{1}{2}}\Lambda^{\frac{1-\delta}{2}}\hat{u}\|_{L^2} \quad ; \qquad \delta < \epsilon,$$

then a better estimate holds, namely

$$\|D^{\frac{1}{2}}Q^\lambda(\hat{u} * \hat{u})\|_{L^2} \leq N_\epsilon(u)M_{\epsilon,\delta}(u). \tag{22a}$$

There is a good reason why estimate (22) is correct, one can show easily that

$$\|u\|_{C^\epsilon} \leq N_\epsilon(u),$$

hence u is bounded, which seems to be the crucial information. In other words, if $u \in C^\epsilon$ then $u^2 \in C^\epsilon$. On the other hand, the $N_0(u)$ norm only implies that u is in every L^p space and at best can be in BMO, but then u^2 is not necessarily in BMO. To handle the second term in (21) consider an arbitrary function $h(t, x)$ and use a duality argument

$$\left\langle \hat{h}, \frac{D^{\frac{1}{2}}}{Q^{1-\lambda}}(\hat{u} * \hat{\Box}\hat{u}) \right\rangle = \left\langle D^{\frac{1}{2}}Q^\lambda\hat{u}, \frac{\hat{\Box}}{D^{\frac{1}{2}}Q^\lambda}(\hat{u} * \frac{D^{\frac{1}{2}}}{Q^{1-\lambda}}\hat{h}) \right\rangle.$$

Because

$$\frac{\hat{\Box}}{D^{\frac{1}{2}}Q^\lambda} \sim \frac{Q^{1-\lambda}}{D^{\frac{1}{2}}},$$

it is enough to show that

$$\|\frac{Q^{1-\lambda}}{D^{\frac{1}{2}}}(\hat{u} * \frac{D^{\frac{1}{2}}}{Q^{1-\lambda}}\hat{h})\|_{L^2} \leq N_\epsilon(u)\|\hat{h}\|_{L^2}. \tag{23}$$

This inequality is subtler than (22) but still not optimal, it also has the advantage that it can be used to handle equations of the form

$$\Box u = g(u)l(u, u).$$

In any case, combining the estimates in (22) and (23) gives an estimate of the form

$$N_\epsilon(u) \leq N_\epsilon^2(u) + \|u_0\|_{H^{1+\frac{\epsilon}{2}}} + \|u_1\|_{H^{\frac{\epsilon}{2}}},$$

hence if the term $\|u_0\|_{H^{1+\frac{\epsilon}{2}}} + \|u_1\|_{H^{\frac{\epsilon}{2}}}$ is small then $N_\epsilon(u) < 1$. This is the crucial step; further regularity can be obtained by differentiating equation (17) and applying the same estimates. Call $v = \partial u$ any derivative, then

$$\Box v = l(v, u)$$

and estimate (19) gives

$$N_\epsilon(v) \leq \|\frac{D^{\frac{1}{2}}}{Q^{1-\lambda}} \hat{l}(u, v)\|_{L^2} + C. \tag{24}$$

In a similar manner we want the estimate

$$\|\frac{D^{\frac{1}{2}}}{Q^{1-\lambda}} \hat{l}(u, u)\|_{L^2} \leq N_\epsilon(u) N_\epsilon(v),$$

which however is the same estimate as (20a). Now inequality (24) becomes

$$N_\epsilon(v) \leq N_\epsilon(u) N_\epsilon(v) + C,$$

which will give $N_\epsilon(v)$ bounded because $N_\epsilon(u) < 1$. Notice that the estimates are not optimal and there is room for some improvement; however it seems impossible to have the estimates that we want if $\epsilon = 0$, i.e. $\lambda = \frac{1}{2}$. So far we have ignored the special structure of the equations (11). It is possible that the estimates are correct if we know a priori that u is continuous which seems to be the critical barrier to overcome. Estimates similar to (22), (23) have been recently obtained by Klainerman and Machedon in $3 + 1$ dimensions [12].

References

[1] Bourgain, J., *Fourier transform restriction phenomena for certain lattic subsets and applications to nonlinear evolution equations*, GAFA **3** (1993), 209–262.

[2] Brenner, P., *On $L^p - L^q$ estimates for the wave equation*, Math. Z. **145** (1975), 251–254.

[3] Christodoulou, D. and Tahvildar-Zadeh, T., *On the regularity of spherically symmetric wave maps*, CPAM **XLVI** (1993), 1041–1091.

[4] Ginibre, J. and Velo, G., *The global Cauchy problem for the nonlineear Klein-Gordon equations*, Math. Z. **189** (1985), 487–505.

[5] Grillakis, M. G., *Regularity and asymptotic behaviour of the wave equation with a critical nonlinearity*, Ann. of Math. (2) **132** no. 1 (1990), 485–509.

[5] Grillakis, M. G., *Regularity for the wave equation with a critical nonlinearity*, CPAM **XLV** (1992), 749–774.

[7] Grillakis, M. G., *Classical Solutions for the equivariant wave map in $1 + 2$ dimensions*, preprint.

[8] Grillakis, M. G. and Fang, Y., *Apriori estimates for the wave equations in $2 + 1$ dimensions*, preprint.

[9] Harmse, Jorgen, *On Lebesque space estimates for the wave equation*, Indiana Univ. Math. J. **39** (1990), 229–248.

[10] Kapitanski, L. V., *Some generalizations of the Strichartz-Brenner inequality*, Math. J. **1** (1990), 693–726.

[11] Kapitanski, L., *The Cauchy problem for semilinear wave equations*, preprint.

[12] Klainerman, S. and Machedon, M., *Smoothing estimates for null forms and applications*, preprint.

[13] Littman, W., *Multipliers in L^p and interpolation*, Bull. Amer. Math. Soc. **83** (1956), 482–492.

[14] Pecher, H., *L^p-Abschätzungen und klassische Lösungen für nichtlineare Wellengleichungen I*, Math. Z. **150** (1976), 159–183.

[15] Pecher, H., *L^p-Abschätzungen und klassische Lösungen für nichtlineare Wellengleichungen II*, Manuscripta Math. **20** (1977), 227–244.

[16] Shatah, J. and Struwe, M., *Regularity results for nonlinear wave equations*, Ann. of Math. (2) **183** (1993), 503–518.

[17] Shatah, J. and Tahvildar-Zadeh, T., *Regularity of harmonic maps from the Minkowski space into rotationally symmetric manifolds*, CPAM **45** (1992), 947–971.

[18] Stein, E. M., *Interpolation of linear operators*, Trans. Amer. Math. Soc. **83** (1956), 482–492.

[19] Stein, E. M., *Oscillatory integrals in Fourier Analysis*, Beijing Lectures in Harmonic Analysis, Princeton Univ. Press, Princeton, NJ, 1986, 307–355.

[20] Strauss, W., Nonlinear Wave Equations, CONM AMS **73N**, 1989.

[21] Strichartz, S. R., *A priori Estimates for the wave equation and some applications*, J. Funct. Anal. **5** (1970), 218–235.

[22] Struwe, M., *Globally regular solution to the u^5 Klein-Gordon equation*, Ann. Scuola Norm. Sup. Pisa Cl. Sci. (4) **15** (1988), 495–513.

[23] Struwe, M., *Semilinear wave equations*, Bull. Amer. Math. Soc. **261** (1992), 53–85.

[24] Thomas, P., *A restriction theorem for the Fourier transform*, Bull. Amer. Math. Soc. **81** (1975), 477–478.

Applications of Dynamics to Compact Manifolds of Negative Curvature

FRANÇOIS LEDRAPPIER

CNRS, Centre de Mathématiques   Laboratoire de Probabilités
École Polytechnique   and   Université Paris VI
F-91128 Palaiseau, France   F-75230 Paris, France

ABSTRACT. Consider closed Riemannian manifolds with negative sectional curvature. There are three natural dynamics associated with the Riemannian structure: the geodesic flow on the unit tangent bundle, the dynamics of the invariant foliations of the geodesic flow, and the Brownian motion on the universal cover of the manifold. These dynamics define global asymptotic objects such as growth rates or measures at infinity. For locally symmetric negatively curved spaces, these objects are easy to compute and to describe. In this paper, we survey some of their properties and relations in the general case.

1 Measures at infinity

Let (M, g) be a closed Riemannian manifold with negative sectional curvature and let $\pi : (\widetilde{M}, \tilde{g}) \to (M, g)$ be the universal cover of M, endowed with the canonically lifted metric $\tilde{g}$. The space $(\widetilde{M}, \tilde{g})$ is a simply connected Riemannian manifold with negative curvature; in particular, the space $(\widetilde{M}, \tilde{g})$ is a Hadamard manifold and the geometric boundary $\partial \widetilde{M}$ is defined as the space of ends of geodesics (see e.g. [BGS]). The geometric boundary $\partial \widetilde{M}$ is homeomorphic to a sphere. For any x in $\widetilde{M}$ write τ_x for the homeomorphism between the unit sphere $S_x \widetilde{M}$ in the tangent space at x and $\partial \widetilde{M}$ defined by associating to a unit vector v in $S_x \widetilde{M}$ the end $\tau_x(v)$ of the geodesic σ_v starting at v. In this section are defined natural families of finite positive measures on the boundary indexed by $x, x \in \widetilde{M}$.

(a) **Lebesgue visibility measures.** Let λ_x denote the image measure under τ_x of the Lebesgue measure on the unit sphere $S_x \widetilde{M}$. It follows from [A], [ASi] that for x and y in $\widetilde{M}$, the measures λ_x and λ_y have the same negligible sets and that the density $\frac{d\lambda_y}{d\lambda_x}$ admits a (Hölder) continuous version on $\partial \widetilde{M}$ (the metric on $\partial \widetilde{M}$ will be recalled below). Write λ for the common measure class of the $\lambda_x, x \in \widetilde{M}$.

(b) **Harmonic measures.** Let Δ be the Laplace-Beltrami operator on C^2-functions on $\widetilde{M}$, $\Delta = \text{div grad}$. A function u on $\widetilde{M}$ is called harmonic if $\Delta u = 0$. The Dirichlet problem is solvable on $\widetilde{M} \cup \partial \widetilde{M}$ ([An], [S]): let f be a continuous function on $\partial \widetilde{M}$; there is a unique harmonic function u_f on $\widetilde{M}$ such that for all ξ in $\partial \widetilde{M}$,

Proceedings of the International Congress
of Mathematicians, Zürich, Switzerland 1994
© Birkhäuser Verlag, Basel, Switzerland 1995

$\lim_{z \to \xi} u_f(z) = f(\xi)$. For any x in $\widetilde{M}$, the mapping $f \to u_f(x)$ defines a probability measure ω_x on $\partial \widetilde{M}$. The measure ω_x is called the harmonic measure of the point x. For x and y in $\widetilde{M}$, the measures ω_x and ω_y have the same negligible sets and the density $\frac{d\omega_y}{d\omega_x}$ admits a Hölder continuous version on $\partial \widetilde{M}$ called the Poisson kernel and denoted $k(x, y, \cdot)$ ([ASn], [Aa]). Write ω for the common measure class of the $\omega_x, x \in \widetilde{M}$.

(c) Margulis-Patterson measures. For two points (ξ, η) in $\partial \widetilde{M}$, and x in $\widetilde{M}$, define the Gromov product $(\xi, \eta)_x$ by:

$$(\xi, \eta)_x = \lim_{\substack{y \to \xi \\ z \to \eta}} \frac{1}{2}\left(d(x, y) + d(x, z) - d(y, z)\right)$$

(see e.g. [GH]). Set $d_x(\xi, \eta) = \exp{-(\xi, \eta)_x}$ and define balls, spherical Hausdorff measures, and spherical Hausdorff dimension as if d_x was a distance on $\partial \widetilde{M}$ (in fact, there is $\alpha > 0$ so that d_x^α is a distance on $\partial \widetilde{M}$). Let H be the spherical Hausdorff dimension of $\partial \widetilde{M}$. The spherical H-Hausdorff measure ν_x is positive and finite and the measure ν_x is called the Margulis-Patterson measure of the point x. For x, y in $\widetilde{M}$, the measures ν_x and ν_y have the same negligible sets.

Recall that for x in $\widetilde{M}$, ξ in $\partial \widetilde{M}$, the Busemann function $b_{x,\xi}$ is a function on $\widetilde{M}$ defined by

$$b_{x,\xi}(y) = \lim_{t \to \infty} d(y, \sigma_v(t)) - t ,$$

where σ_v is the geodesic in $\widetilde{M}$ starting at $v = \tau_x^{-1}\xi$. Then the density $\frac{d\nu_y}{d\nu_x}$ is given by

$$\frac{d\nu_y}{d\nu_x}(\xi) = \exp{-Hb_{x,\xi}(y)} .$$

Write ν for the common measure class of the $\nu_x, x \in \widetilde{M}$. The construction of this measure is essentially given in [M2]. The presentation and the properties given here are derived from [H1], [Ka3], and [L3].

(d) General properties. Let γ be an isometry of $\widetilde{M}$. Then the action of γ extends to $\partial \widetilde{M}$ and to measures on $\partial \widetilde{M}$. By naturality for $\mu = \lambda, \omega$, or ν:

$$\mu_{\gamma x} = \gamma \mu_x .$$

For $\mu = \lambda, \omega$, or ν, define a positive Radon measure $\tilde{\mu}$ on $S\widetilde{M}$ by setting

$$\int df\tilde{\mu} = \int \left(\int_{\partial \widetilde{M}} f(\tau_x^{-1}\xi) \, d\mu_x(\xi)\right) d\text{vol}(x)$$

for any continuous function f on $S\widetilde{M}$ with compact support. The measure $\tilde{\mu}$ is invariant under the action of γ, and therefore defines a finite positive measure $\bar{\mu}$ on the quotient space SM.

In the case when $\widetilde{M}$ is a symmetric space of negative curvature, there is a compact group K_x of isometries of $\widetilde{M}$ that fixes x and acts transitively on $\partial\widetilde{M}$. Let m_x be the unique K_x-invariant probability measure on $\partial\widetilde{M}$. It follows from the above invariance relation that if (M, g) is locally symmetric there are constants a, b such that for all x in $\widetilde{M}$

$$a\lambda_x = \omega_x = b\nu_x = m_x \ .$$

Conversely, assume that there is a constant a, b, or c such that one of the following equalities of measures holds for all x in $\widetilde{M}$:

$$a\lambda_x = \omega_x, b\nu_x = \omega_x \text{ or } c\lambda_x = \nu_x \ .$$

Then the space (M, g) is locally symmetric. In order to prove this result, set for x in $\widetilde{M}$ and ξ in $\partial\widetilde{M}$:

$$B(x, \xi) = \Delta_y \ b_{x,\xi} \ (y)_{|y=x} \ ,$$

and observe that either hypothesis implies that B is constant ([L3], [Y]). A key result is that the function B is constant if and only if the space (M, g) is locally symmetric. This is immediate in dimension 2 and can be checked directly in dimension 3 (see e.g. [Kn]). In higher dimensions, the proof combines results from [FL], [BFL], and [BCG]. This result is used in the other characterizations of locally symmetric spaces that are given below.

2 Geodesic flow

The geodesic flow $(\theta_t)_{t\in\mathbf{R}}$ is a one-parameter group of diffeomorphisms of the unit tangent bundle SM, defined as follows: for v in SM, write $\{\sigma_v(t), t \in \mathbf{R}\}$ for the unit-speed geodesic starting at v. Then for any real t, $\theta_t v$ is the speed vector of the geodesic σ at $\sigma_v(t)$. A flot $(\theta_t)_{t\in\mathbf{R}}$ is called Anosov if there exist a metric $|| \ ||$ on TSM, numbers $C > 0$ and $\chi < 1$, and a Whitney decomposition of TSM as $E^{ss} \oplus E^{uu} \oplus \mathbf{R}X$, where X is the vector field generating the flow and for v in $E^{ss}, t > 0, ||D\theta_t v|| \leq C\chi^t ||v||$, for v in $E^{uu}, t > 0, ||D\theta_{-t} v|| \leq C\chi^t ||v||$.

Because of negative curvature, the geodesic flow is Anosov ([A]).

(a) Topological entropy. The number H is the topological entropy of the geodesic flow ([B]). There is a function c on M such that, uniformly on $\widetilde{M}$,

$$\lim_{R\to\infty} \ \exp(-HR) \ \text{vol } B(x, R) = c(\pi x),$$

where $B(x, R)$ is the ball of radius R about x in $(\widetilde{M}, \tilde{g})$ and vol $B(x, R)$ its Riemannian volume ([M1]). Because $c(\pi x)$ is proportional to $\nu_x(\partial\widetilde{M})$, the function c is C^∞. The function c is in general not constant ([Kn]).

(b) Metric entropy. The measure $\bar{\lambda}$ is the Liouville measure; it is finite and invariant under the geodesic flow. Write $h_{\bar{\lambda}}$ for the Kolmogorov-Sinai entropy of the system

$$\left(SM, \frac{\bar{\lambda}}{\bar{\lambda}(SM)}, \theta_1 \right) \; ; \; h_{\bar{\lambda}} = \frac{\int B \, d\bar{\lambda}}{\bar{\lambda}(SM)} \qquad \text{(see[ASi]) .}$$

From [LY] it follows that $h_{\bar{\lambda}}$ is the Hausdorff dimension of the λ measure class on $\partial \widetilde{M}$, i.e.

$$h_{\bar{\lambda}} = \inf \left\{ \text{Hausdorff dimension } (A) : A \subset \partial \widetilde{M}, \lambda(A) > 0 \right\} .$$

From the variational principle ([BR]) it follows that $h_{\bar{\lambda}} \leq H$ with equality if and only if the measure classes λ and ν coincide. In dimension 2, $h_{\bar{\lambda}} = H$ if and only if the curvature is constant ([K1]). In higher dimensions, the entropy rigidity problem is whether $h_{\lambda} = H$ if and only if the space (M, g) is locally symmetric.

(c) Regularity of the stable direction. In general, the distribution $E^s = E^{ss} \oplus \mathbf{R}X$ is only Hölder continuous. If the distribution is C^2, then $h_{\bar{\lambda}} = H$ ([H5]). If the distribution is C^∞, then the space (M, g) is locally symmetric (this follows again from [BFL] and [BCG]). The properties are more precise in the case of surfaces: the distribution E^s in C^1 ([Ho]), even $C^{1+\Lambda_*}$ ([HK]). If the distribution is $C^{1+o(s \cdot |\log s|)}$, then it is C^∞ ([HK]) and the curvature is constant ([Gh]). This discussion is a particular case of the analogous discussion for general Anosov flows (see [Gh], [BFL], [H4], and [F]).

3 Brownian motion on $\widetilde{M}$

Recall that Δ is the Laplace-Beltrami operator on $\widetilde{M}$ and write $p(t, x, y)$ for the fundamental solution of the equation $\frac{\partial u}{\partial t} = \Delta u$. The properties below reflect asymptotic properties of the Brownian motion on $\widetilde{M}$.

(a) Growth rates. There is a positive number ℓ such that, for all x in $\widetilde{M}$,

$$\ell = \lim_{t \to \infty} \frac{1}{t} \int_{\widetilde{M}} d(x, y) p(t, x, y) \, \text{dvol} \, (y)$$

(see [Gu]) and ℓ is given by $\ell = \frac{\int B d\bar{\omega}}{\bar{\omega}(SM)}$ ([Ka1]). There is another positive number h such that for all x in $\widetilde{M}$,

$$h = \lim_{t \to \infty} -\frac{1}{t} \int_{\widetilde{M}} p(t, x, y) \log p(t, x, y) \, \text{dvol} \, (y)$$

([Ka1]). Finally denote δ the spectral gap of Δ:

$$\delta = \inf_{f \in C_K^2(\widetilde{M})} \frac{-\int f \Delta f \, \text{dvol}}{\int f^2 \, \text{dvol}} .$$

(b) Relations. The following inequalities hold:

(1) $h \leq \ell H$ with equality if and only if the measure classes ω and ν coincide ([L2]),

(2) $\ell^2 \leq h$ with equality if and only if the space (M, g) is locally symmetric ([Ka1]), and

(3) $4\delta \leq h$ with equality if and only if the space (M, g) is locally symmetric ([L4]).

Observe that other sharp inequalities can be directly derived from the above three:

$$\ell \leq H, \ h \leq H^2, \ \text{or} \ 4\delta \leq H^2 \ .$$

Let m be the only θ-invariant probability measure on SM such that, if $\tilde{m}$ denotes the isometry-invariant extension of m to $\widetilde{SM}$, the projection of $\tilde{m}$ by $\tau = \{\tau_x, x \in \widetilde{M}\}$ on $\partial \widetilde{M}$ belongs to the ω-class ([L2], see also [H2], [Ka3]). Then h/ℓ is the Kolmogorov-Sinai entropy of the system $(SM, m; \theta_1)$ and also the Hausdorff dimension of the measure class ω. The first relation follows from the variational principle for the geodesic flow. The proof of the other two relations is based on an integral formula satisfied by the measure $\bar{\omega}$.

(c) Measure rigidity. The question again arises as to whether measure classes at infinity can coincide with ω only when the space (M, g) is locally symmetric. In dimension 2, the curvature is constant if and only if the measure classes ω and λ coincide ([K2], [L1]) or if and only if the measure classes ω and ν coincide ([L3], [H3]). Observe that this problem makes sense for other objects such as graphs. There are examples of finite graphs that are neither homogeneous nor bipartite, but such that some pair of natural measures at infinity has the same negligible sets [Ls].

4 Invariant foliations

Recall that the distribution E^{ss} is continuous in TSM and that it admits integral manifolds W^{ss} defined by

$$W^{ss}(v) = \{w : \lim_{t \to +\infty} d(\theta_t v, \theta_t w) = 0\}$$

(see [A]).

The W^{ss} form a continuous foliation with smooth leaves and there is a natural metric on the leaves, lifted from the metric g on M through the canonical projection. Let Δ^{ss} be the Laplace-Beltrami operator along the leaves W^{ss}. Then, for any continuous function f on SM, which is C^2 along the W^{ss} leaves, $\int \Delta^{ss} f \, d\bar{\nu} = 0$.

The measure $\bar{\nu}$ is — up to multiplication by a constant factor — the unique measure with that property (the proof of this uses results from [Ka2] and [BM]). The measure $\bar{\nu}/\bar{\nu}(SM)$ can also be seen as the limit of averages on large spheres in $\widetilde{SM}$ ([Kn]). In particular $H = \frac{\int B d\bar{\nu}}{\bar{\nu}(SM)}$.

(a) **Stable foliation.** The manifolds W^s given by $W^s(v) = \bigcup_{t \in \mathbf{R}} \theta_t W^{ss}(v)$ form a continuous foliation, with smooth leaves and with $TW^s = E^s$. Consider again the metric on the leaves lifted from the metric g on M, and let Δ^s be the corresponding Laplace-Beltrami operator. The measure $\bar{\omega}$ is — up to multiplication by a constant factor — the unique measure on SM satisfying $\int \Delta^s f \, d\bar{\omega} = 0$ for any continuous f, which is C^2 along the W^s leaves ([G]).

For a continuous function f on SM write $\tilde{f}$ for the continuous function on $\widetilde{M} \times \partial \widetilde{M}$ given by

$$\tilde{f}(x, \xi) = f \cdot \pi(x, \tau_x^{-1} \xi) .$$

Then for $t > 0$, there is a function $Q_t f$ on SM such that:

$$\widetilde{Q_t f}\,(x, \xi) = \int p(t, x, y) \; \tilde{f}(y, \xi) \; d\mathrm{vol}\,(y) .$$

The operator Q_t is the leafwise heat operator $Q_t = \exp t \, \Delta^s$.

There is a Hölder norm $|\;|$ on functions on SM with the following property: there are $C > 0$ and $\chi < 1$ such that for all $t > 0$ any function f on SM:

$$\left| Q_t \, f - \frac{\int f d\bar{\omega}}{\mathrm{vol}\, M} \right| < c \, \chi^t \, |f|$$

([L5]).

From this follow asymptotic properties of the Brownian motion on $\widetilde{M}$ and a decomposition theorem for closed regular leafwise 1-forms ([L6]). As a consequence define for $s \in \mathbf{R}$ the function $\varphi(s)$ by

$$\varphi(s) = \lim_{t \to \infty} \frac{1}{t} \; \log \; \max_{(x, \xi)} \int p(t, x, y) \; k^s(x, y, \xi) \; d\mathrm{vol}\,(y) .$$

The function φ is convex and analytic in a neighborhood of 0. The space (M, g) is locally symmetric if and only if $\varphi(s) = as(s - 1)$ for some constant a (in fact a is then the common value of ℓ^2, H^2, h, or 4δ) or if and only if we have

$$2 \, \varphi'(0) + \varphi''(0) = 0 .$$

References

[A]   D.V. Anosov, *Geodesic flow on closed Riemannian manifolds with negative curvature*, Proc. Steklov Inst. Math. **90** (1967).

[Aa]   A. Ancona, *Negatively curved manifolds, elliptic operators and the Martin boundary*, Ann. of Math. (2) **125** (1987), 495–536.

[An]   M.T. Anderson, *The Dirichlet problem at infinity for manifold of negative curvature*, J. Differential Geom. **18** (1983), 701–721.

[ASi]   D.V. Anosov and Ya. Sinai, *Some smooth ergodic systems*, Russian Math. Surveys **22:5** (1967), 103–167.

[ASn]   M.T. Anderson and R. Schoen, *Positive harmonic functions on complete manifolds of negative curvature*, Ann. of Math. (2) **121** (1985), 429–461.

[B]   R. Bowen, *Symbolic dynamics for hyperbolic flows*, Amer. J. Math. **95** (1973), 429–460.

[BCG]   G. Besson, G. Courtois, and S. Gallot, *Entropies et rigidités des espaces localement symétriques de courbure strictement négative*, to appear in GAFA.

[BFL]   Y. Benoist, P. Foulon, and F. Labourie, *Flots d'Anosov à distributions stable et instable différentiables*, J. Amer. Math. Soc. **5** (1992), 33–74.

[BGS]   W. Ballmann, M. Gromov, and V. Schroeder, Manifolds of non-positive curvature, Progr. Math. **61**, Birkhäuser, Basel, Boston, 1985.

[BM]   R. Bowen and B. Marcus, *Unique ergodicity for horocycle foliations*, Israel J. Math. **26** (1977), 43–67.

[BR]   R. Bowen and D. Ruelle, *The ergodic theory of Axiom-A flows*, Invent. Math. **29** (1975), 181–202.

[F]   P. Foulon, *Rigidité entropique des flots d'Anosov en dimension 3*, preprint (1994).

[FL]   P. Foulon and F. Labourie, *Sur les variétés compactes asymptotiquement harmoniques*, Invent. Math. **109** (1992), 97–111.

[G]   L. Garnett, *Foliations, the ergodic theorem and Brownian Motion*, J. Funct. Anal. **51** (1983), 285–311.

[Gh]   E. Ghys, *Flots d'Anosov dont les feuilletages stables sont différentiables*, Ann. Sci. École Norm. Sup. (4) **20** (1987), 251–270.

[GH]   E. Ghys and P. de la Harpe (éds.), Sur les groupes hyperboliques d'après M. Gromov, Progr. Math. **83**, Birkhäuser, Basel, Boston, 1990.

[Gu]   Y. Guivarc'h, *Sur la loi des grands nombres et le rayon spectral d'une marche aléatoire*, Astérisque **74** (1980), 47–98.

[H1]   U. Hamenstädt, *A new description of the Bowen-Margulis measure*, Ergodic Theory Dynamical Systems **9** (1989), 455–464.

[H2]   U. Hamenstädt, *An explicit description of the harmonic measure*, Math. Z. **205** (1990), 287–299.

[H3]   U. Hamenstädt, *Time-preserving conjugacies of geodesic flows*, Ergodic Theory Dynamical Systems **12** (1992), 67–74.

[H4]   U. Hamenstädt, *Regularity of time-preserving conjugacies for contact Anosov flows with C^1 Anosov splitting*, Ergodic Theory Dynamical Systems **13** (1993), 65–72.

[H5]   U. Hamenstädt, *Invariant two-forms for geodesic flows*, preprint (1993).

[HK]   S. Hurder and A. Katok, *Differentiability, rigidity and Godbillon Vey classes for Anosov flows*, Publ. IHES **72** (1990), 5–61.

[Ho]   E. Hopf, *Statistik der geodätischen Linien in Mannigfaltigkeiten negativer Krümmung*, Ber. Verh. Sächs. Akad. Wiss. Leipzig **91** (1939), 261–304.

[K1]   A. Katok, *Entropy and closed geodesics*, Ergodic Theory Dynamical Systems **2** (1982), 339–365.

[K2]   A. Katok, *Four applications of conformal equivalence to geometry and dynamics*, Ergodic Theory Dynamical Systems **8**** (1988), 115–140.

[Ka1]   V.A. Kaimanovich, *Brownian Motion and harmonic functions on covering manifolds. An entropy approach*, Soviet Math. Dokl. **33** (1986), 812–816.

[Ka2]   V.A. Kaimanovich, *Brownian Motion on foliations: Entropy, invariant measures, mixing*, Funct. Anal. Appl. **22** (1988).

[Ka3]   V.A. Kaimanovich, *Invariant measures of the geodesic flow and measures at infinity on negatively curved manifolds*, Ann. Inst. H. Poincaré Phys. Théor. **53** (1990), 361–393.

[Kn]   G. Knieper, *Spherical means on compact Riemannian manifolds of negative curvature*, to appear in J. Diff. Geom. Appl.

[L1]   F. Ledrappier, *Propriété de Poisson et courbure négative*, C.R. Acad. Sci. Paris **305** (1987), 191–194.

[L2]   F. Ledrappier, *Ergodic properties of Brownian Motion on covers of compact negatively-curved manifolds*, Bol. Soc. Brasil. Mat. **19** (1988), 115–140.

[L3]   F. Ledrappier, *Harmonic measures and Bowen-Margulis measures*, Israel J. Math. **71** (1990), 275–287.

[L4]   F. Ledrappier, *A heat kernel characterization of asymptotic harmonicity*, Proc. Amer. Math. Soc. **118** (1993), 1001–1004.

[L5]   F. Ledrappier, *Central limit theorem in negative curvature*, to appear in Ann. Probab. (1995).

[L6]   F. Ledrappier, *Harmonic 1-forms on the stable foliation*, Bol. Soc. Bras. Mat. **25** (1994), 121–138.

[Ls]   R. Lyons, *Equivalence of boundary measures on cocompact trees*, preprint (1993).

[LY]   F. Ledrappier and L-S Young, *The metric entropy of diffeomorphisms*, Ann. of Math. (2) **122** (1985), 509–574.

[M1]   G.A. Margulis, *Applications of ergodic theory to the investigation of manifolds of negative curvature*, Funct. Anal. Appl. **3** (1969), 335–336.

[M2]   G.A. Margulis, *Certain measures associated with U-flows on compact manifolds*, Funct. Anal. Appl. **4** (1970), 55–67.

[S]   D. Sullivan, *The Dirichlet problem at infinity for a negatively curved manifold*, J. Diff. Geom. **18** (1983), 723–732.

[Y]   C. Yue, *On the Sullivan conjecture*, Random & Comp. Dynamics **1** (1992), 131–142.

On the Borderline of Real and Complex Dynamics

MIKHAIL LYUBICH

Mathematics Department and Institute for Math. Sciences
State Univ. of New York, Stony Brook, NY 11794, USA

1 Introduction

1.1. Overview. We will describe recent developments in several intimately related problems of complex and real one-dimensional dynamics: rigidity of polynomials and local connectivity of the Mandelbrot set, measure of Julia sets, and attractors of quasi-quadratic maps. A combinatorial basis for this study is provided by the Yoccoz puzzle. The main problem is to understand the geometry of the puzzle. Our main geometric result is that in the quadratic case its principal moduli grow linearly. Renormalization ideas are strongly involved in the discussion. The interplay between real and complex dynamics enlightens both. In the end we will briefly discuss a new geometric object which can be associated to a rational function, a hyperbolic orbifold 3-lamination.

1.2. Polynomial dynamics: Definitions and notation. For the reader's convenience and to fix the notations we will give here the definitions of some basic objects in holomporphic dynamics.

Let $P(z) = z^d + a_1 z^{d-1} + \ldots + a_d$ be a monic polynomial of degree $d \geq 2$, and P^n its n-fold iterate. The basin of ∞ is the set of points escaping to ∞: $D(\infty) = \{z : P^n z \to \infty\}$. The *filled Julia set* is its complement: $K(P) = \mathbf{C} \backslash D(\infty)$. The Julia set $J(P)$ is the common boundary of $D(\infty)$ and $K(P)$.

The Julia set (and the filled Julia set) is connected if and only if none of the critical points escapes to ∞. In this case there is a unique conformal map $R : D(\infty) \to \{z : |z| > 1\}$, normalized by $R(z) \sim z$ as $z \to \infty$. Note that $R \circ P \circ R^{-1} : z \mapsto z^d$. The *external rays* and *equipotentials* of P are defined as the R-preimages of the straight rays $\{re^{i\theta} : 1 < r < \infty\}$ and round circles $\{re^{i\theta} : 0 \leq \theta \leq 2\pi\}$.

A polynomial P is called *hyperbolic* if the orbits of all critical points converge to attracting cycles in $\mathbf{C}$ or ∞. It is called *postcritically finite* if the orbits of all critical points are finite.

Two polynomials P_1 and P_2 are called *topologically (conformally, quasi-conformally) conjugate* if there is a homeomorphism (conformal isomorphism, quasi-conformal map correspondingly) $h : \mathbf{C} \to \mathbf{C}$ such that $P_1 = h^{-1} \circ P_2 \circ h$.

2 Rigidity Conjecture

2.1. Our main object will be the quadratic family $P_c : z \mapsto z^2 + c$, $c \in \mathbf{C}$. A key problem in the modern holomorphic dynamics is to classify quadratics up to topological conjugacy. Here is the main conjecture.

Proceedings of the International Congress
of Mathematicians, Zürich, Switzerland 1994
© Birkhäuser Verlag, Basel, Switzerland 1995

QUADRATIC RIGIDITY CONJECTURE. *Any nonhyperbolic quadratic polynomial P_b is not conjugate to any other quadratic polynomial P_c.*

Let us look closer at the meaning of this conjecture from the point of view of the bifurcation diagram in the parameter plane. The *Mandelbrot set M* is defined as the set of c for which the Julia set $J(P_c)$ is connected. A component H of int M is called *hyperbolic* if it is filled with hyperbolic quadratics. The hyperbolic component contains one special point c_H, its *center*, where the critical point is periodic. The Rigidity Conjecture would assert that the Mandelbrot set splits into the following topological classes: hyperbolic components of int M punctured at their centers, and single points.

A possible nonhyperbolic component Q of int M is called "queer". It is still true that all polynomials within Q are topologically conjugate (see [L0], [MSS]). So the Rigidity Conjecture would imply absence of queer components, and hence density of hyperbolic maps in the quadratic family. This conjecture is a special case of the so-called Fatou Conjecture (see [F], p. 73, and discussion in [McM3]).

QUADRATIC FATOU CONJECTURE. *Hyperbolic polynomials are dense in the quadratic family.*

There is another famous conjecture due to Douady and Hubbard:

MLC CONJECTURE. *The Mandelbrot set is locally connected.*

It turns out to be true, though not at all obvious, that this conjecture is stronger than both of the above (compare [DH1]). So we have the following implications:

$$\text{MLC Conjecture} \Rightarrow \text{Rigidity Conjecture} \Rightarrow \text{Fatou Conjecture.}$$

Though the Rigidity Conjecture is formally weaker than MLC, so far progress has been made simultanuously in both by means of the same ideas and methods (at least, outside the boundaries of the hyperbolic components). On the other hand, there is an ergodic approach to the Fatou Conjecture that may settle it before the other two (see [MSS], [McM2]).

2.2. Copies of the Mandelbrot set and Douady-Hubbard renormalization. The Mandelbrot set contains many "little copies of itself" canonically homeomorphic to the whole set M but different from M (see [DH1], [D], [M1]). Each copy arises from a hyperbolic component H of M. A map P_c (and the corresponding parameter value) is called *Douady-Hubbard (DH) renormalizable* if c belongs to a little copy of the Mandelbrot set (we will sometimes say just "renormalizable" if it cannot be confused with a generalized notion from 3.6). If there are two nested Mandelbrot copies containing c, then c is *twice renormalizable*, etc. In particular, we can classify the quadratics as *finitely* or *infinitely DH renormalizable*.

A copy M' is called *maximal* if it is not contained in any other copy. Let $\mathcal{M}$ denote the set of all maximal copies of M, and let $\sigma : \cup_{M' \in \mathcal{M}} M' \to M$ be the map whose restriction onto any copy $M' \in \mathcal{M}$ is the canonical homeomorpism onto M. To any infinitely renormalizable $c \in M$ we can associate its *DH combinatorial type*. This is a sequence $\tau(c) = [M_1, M_2, \ldots]$ of maximal copies $M_n \in \mathcal{M}$, defined by $\sigma^n(c) \in M_n$. Moreover, any such sequence is realized for some parameter value.

In 3.4 we will make clear the dynamical meaning of DH renormalization.

2.3. Rigidity theorems. Let us start with a rigidity result that marked the beginning of a new stage in the field (and resolved a problem of monotonicity of topological entropy in the quadratic family, see [MT, Section 13], [DH3]). We will specify it for the quadratic family.

THEOREM 2.1 (THURSTON). *Any postcritically finite quadratic is rigid.*

The next breakthrough was made in the work of Branner and Hubbard on cubic maps with one escaping critical point [BH], and Yoccoz's work on quadratics (see the discussion in [H], [L3] and [M2]). Again we will state only the quadratic result.

THEOREM 2.2 (YOCCOZ). *Any quadratic polynomial P_c that is at most finitely renormalizable and has no attracting periodic points in the finite plane is rigid. Moreover, MLC holds at c.*

(Note that quadratics with attracting periodic point are hyperbolic.) The following result of the author settles many infinitely renormalizable cases:

THEOREM 2.3 [L4]. *There is a family $\mathcal{S} \subset \mathcal{M}$ of maximal Mandelbrot copies such that if $\tau(c) = [M_1, M_2, \ldots]$ with $M_n \in \mathcal{S}$ then P_c is rigid. Moreover, MLC holds at c.*

This family $\mathcal{S}$ is specified by a property of *sufficiently high* combinatorial height (see Section 3 for the definition). This condition becomes especially efficient on the real line, since it can be complemented by the following rigidity result of Sullivan (see [S] and [MvS]).

THEOREM 2.4 (SULLIVAN). *Let $c \in \mathbf{R}$ be an infinitely renormalizable parameter value, $\tau(c) = [M_0, M_1, \ldots]$ with all M_n selected from a finite family of maximal copies of M. Then P_c is rigid on the real line.*

Combining the methods of Theorem 2.3 and Theorem 2.4 we obtain the following.

COROLLARY 2.5. *Any nonhyperbolic real quadratic polynomial is rigid on the real line.*

It follows that hyperbolic maps are dense on the real line. This result had been earlier announced by Swiatek [Sw], who approached it from the real point of view. A related rigidity result was also proven by McMullen (see [McM2]). The latter one asserts that any real non-hyperbolic quadratic polynomial is quasi-conformally rigid (that is, the quasi-conformal class of such a map consists of this map only).

Let us note that the main content of Theorem 2.3 is the so-called complex *a priori* bounds, which yield much more than rigidity. We will discuss these issues later on.

3 Combinatorial framework

3.1. DH polynomial-like maps. Let U' and U be two topological disks with $\operatorname{cl} U' \subset U$, and $f : U' \to U$ be a holomorphic branched covering map. Such a map is called

DH polynomial-like map. It is called *DH quadratic-like* if $\deg f = 2$. In this case we always put the critical point of f at the origin 0.

One can naturally define the filled Julia set of a polynomial-like map as the set of nonescaping points: $K(f) = \{z : f^n z \in U' : n = 0, 1, \ldots\}$. The Julia set is defined as $J(f) = \partial K(f)$. These sets are connected if and only if all critical points are nonescaping, that is, belong to $K(f)$.

Actually one should view a polynomial-like map as a germ near its filled Julia set, so that the choice of the U' and U is not canonical. Given a polynomial-like map $f : U' \to U$, we can consider a *fundamental annulus* $A = U' \backslash U$. Let $\mathrm{mod}(f) = \sup \mathrm{mod}\, A$, where A runs over all fundamental annuli of f. The control of moduli of appropriate polynomial-like maps is a key issue of the renormalization theory (see [S], [McM2]).

If there is a quasi-conformal conjugacy h between two polynomial-like maps f and g (near the filled Julia sets), with $\bar{\partial} h = 0$ almost everywhere on the filled Julia set $K(f)$, then f and g are called *hybrid* equivalent. A *hybrid class* $\mathcal{H}(f)$ is the space of maps hybrid equivalent to f modulo conformal equivalence. According to Sullivan, $\mathcal{H}(f)$ should be viewed as an infinitely dimensional Teichmüller space. In contrast with the classical Teichmüller theory this space has a preferred point: *Any hybrid class of polynomial-like maps with connected Julia set contains a unique polynomial of the same degree* (Straightening Theorem [DH2]). In particular, the hybrid classes of quadratic-like maps $z \mapsto z^2 + c$ are labeled by the points $c = c(f) \in M$ of the Mandelbrot set.

Given a DH polynomial-like map with connected Julia set, we can define external rays and equipotentials near the filled Julia set by choosing some hybrid conjugacy to a polynomial (of course these curves are not uniquely defined).

3.2. Limbs. Let P_c be a quadratic polynomial with both fixed points being repelling, and let α_c be the dividing fixed point, so that $J(P_c) \backslash \{\alpha_c\}$ is disconnected. There are finitely many external rays $R_i(\alpha_c)$ landing at α_c, which are cyclically permuted by P_c with combinatorial rotation number $\rho(\alpha_c) = q_c/p_c$ (see [H]).

This rotation number can be easily read off from the position of c at the Mandelbrot set. Let b be the parabolic bifurcation point on the main cardioid of M where $P_b'(\alpha_b) = e^{2\pi i q/p}$. The connected component of $M \backslash \{b\}$, which does not contain the origin, is called the *primary limb* L_b of M with *root* at b. It turns out that if $c \in L_b$ then $\rho(\alpha_c) = q/p$. Similarly, given a hyperbolic component H attached to the main cardioid, we can consider *secondary limbs* L_b attached to H at bifurcation points $b \in \partial H$.

We refer to a *truncated limb* if we remove from it a neighborhood of its root.

3.3. Yoccoz puzzle. Let $J(P_c)$ be connected with both fixed points being repelling. Let E be an equipotential of P_c. The rays $R_i(\alpha)$ landing at α cut the domain bounded by E into p closed topological disks $Y_i^{(0)}$, $i = 0, \ldots, p-1$, called *puzzle pieces of zero depth*.

Let us define *puzzle pieces* $Y_i^{(n)}$ of *depth* n as the connected components of $f^{-n} Y_k^{(0)}$. They form a finite tiling of the neighborhood of $K(f)$ bounded by $f^{-n} E$. For every depth there is one puzzle piece containing the critical point. It is called *critical* and is labeled as $Y^{(n)} \equiv Y_0^{(n)}$.

The Yoccoz puzzle provides us with the Markov family of puzzle pieces to play with. There are several different ways to do this: by means of the Branner-Hubbard tableaux [BH], or by means of the Yoccoz τ-function (unpublished), or by means of the principal nest and generalized renormalization ([LM], [L1]–[L4]), as will be described below.

3.4. Principal nest. Let us say that a map $f = P_c$ is *immediately DH renormalizable* if the orbit of 0 does not escape $Y^{(1)}$ under iterates of f^p. In this case c belongs to a copy of M attached to the main cardioid.

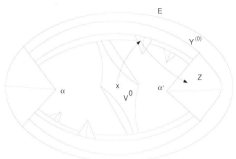

Figure 1. Puzzle.

If P_c is not immediately renormalizable, then there is a t such that $f^{tp}0 \in Z$, where Z is a noncritical puzzle piece of depth 1 attached to $\alpha' = -\alpha$ (see Figure 1). Then let us construct *the (short) principal nest* $V^1 \supset V^2 \supset \ldots$ of puzzle pieces in the following way.

Let t be the first moment when $f^{tp}0 \in Z$. Then let $V^0 \ni 0$ be the pull-back of Z along the orbit $\{f^n0\}_0^{tp}$, that is, the critical puzzle piece such that $f^{tp}V^0 = Z$. Further, let us define V^n as the pull back of V^{n-1} corresponding to the first return of the critical point 0 back to int V^n. Then we have a double branched covering $g_n = f^{l(n)} : V^n \to V^{n-1}$, where $l(n)$ is the corresponding return time.

Let us call a return to level $n - 1$ *central* if $g_n0 \in V^n$. If we have several consecutive central returns, we refer to a *central cascade* of puzzle pieces. Let $\chi(f)$ denote the number of noncentral levels in the principle nest. We call it the *height* of f. In other words, $\chi(f)$ is the number of *different* quadratic-like maps in the sequence g_n (recall that we think of quadratic-like maps as germs, see Section 3.1). The height is finite if and only if f is renormalizable.

Renormalizable maps can be easily recognized in terms of the principal nest. Namely, a map f is *DH* renormalizable if and only if it is either immediately renormalizable, or there is a level N such that the critical point 0 does not escape V^N under iterates of g_N.

In the immediately renormalizable case a little enlargement $U' \supset Y^{(1)}$ provides us with a quadratic-like map $f^p : U' \to U$ with non-escaping critical point (compare [DH1], [M2]). When f is renormalizable, but not immediately, then $g_N : V^N \to V^{N-1}$ is a quadratic-like map with non-escaping critical point. In both cases the corresponding quadratic-like map up to conformal equivalence is called the *DH renormalization Rf* of f. Note that the hybrid class of RP_c is labeled by $\sigma(c) \in M$, where σ is the map from Section 2.2.

Now we can study Rf by the same means as f: cut the Julia set by external rays, consider the Yoccoz puzzle and the principle nest, etc. If Rf is also renormalizable, we will repeat the procedure, and so on.

In such a way we obtain a sequence of renormalizations $R^m f$ and the corresponding *principal nests* of puzzle pieces:

$$\ldots \supset Y^{(m,1)} \supset V^{m,0} \supset V^{m,1} \supset \ldots \supset V^{m,t(m)} \supset V^{m,t(m)+1} \supset \ldots.$$

We truncate this nest at a DH renormalizable level. In the immediately renormalizable case this nest is reduced to one piece $Y^{(m,1)}$. In the finitely renormalizable case the nest corresponding to the last renormalization is infinite.

Let $A^{m,n} = V^{m,n-1} \backslash V^{m,n}$ denote the principal nest of annuli.

3.5. Generalized polynomial-like maps. Let $\{U_i\}$ be a finite or countable family of topological discs with disjoint interiors compactly contained in a topological disk U. We call a map $g : \cup U_i \to U$ a *(generalized) polynomial-like map* if $g : U_i \to U$ is a branched covering of finite degree which is univalent on all but finitely many U_i. The DH polynomial-like maps correspond to the case of a single disk U_0. All concepts introduced before for DH polynomial-like maps can be readily extended to the generalized situation: The filled Julia set $K(g)$ is the set of all non-escaping points, the Julia set $J(g)$ is its boundary, etc. Let us say that a polynomial-like map g is of finite type if its domain consists of finitely many disks U_i.

GENERALIZED STRAIGHTENING THEOREM. *Any polynomial-like map of finite type is hybrid equivalent to a polynomial with the same number of nonescaping critical points.*

Let us call a (generalized) polynomial-like map a *(generalized) quadratic-like map* if it has a single (and nondegenerate) critical point. In such a case we will always assume that 0 is the critical point, and label the discs U_i in such a way that $U_0 \ni 0$.

3.6. Generalized renormalization. Philosophically the dynamical renormalization is the first return map to an appropriate piece of the space considered up to conjugacy. Let us make this precise in our quadratic-like setting.

Let $f : \cup U_i \to U$ be a generalized quadratic-like map, and $V \ni 0$ be a topological disk satisfying the following property: $f^n(\partial V) \cap V = \emptyset$, $n = 1, 2, \ldots$ Then the first return map $g : \cup V_i \to V$ has the following structure: it is defined on a union of disjoint topological disks V_i compactly contained in V, and univalently maps all of them except the critical one onto V. The critical disk V_0 (if it exists) is two-to-one mapped onto V. Moreover, if the orbit of 0 under iterates of f infinitely many times visits V then it is nonescaping under iterates of g.

Let us restrict g on the union of domains V_i visited by the critical orbit. We call this map considered up to affine rescaling the *generalized renormalization* $T_V f$ of f on the domain V.

Let us define the n-fold generalized renormalization of a DH quadratic-like map f, $T^n f \equiv g_n : \cup V_i^n \to V^{n-1}$, as the generalized renormalization of f on the piece V^{n-1} of the principal nest. This sequence of renormalizations is our key to understanding geometry of the map.

3.7. Special families of Mandelbrot copies. Note that the height function $\chi(P_c)$ is constant over any copy M' of the Mandelbrot set. So we can use the notation $\chi(M')$. Let $\mathcal{S} \subset \mathcal{M}$ be a family of maximal copies of the Mandelbrot set. Let us call it *special* if it satisfies the following property: for any truncated secondary limb L there is a height χ_L such that $\mathcal{S}$ contains all copies $M' \subset L$ of the Mandelbrot set with $\chi(M') \geq \chi_L$.

Let f be an infinitely DH renormalizable quadratic-like map with $\tau(f) = [M_1, M_2, \ldots]$. Let us say that it is of $\mathcal{S}$-type if $M_n \in \mathcal{S}$, $n = 0, 1, \ldots$

4 Geometry of the puzzle

The main geometric problem is to gain control of sizes and shapes of puzzle pieces. To this end we need to bound the moduli of the annuli $A^n = V^{n-1} \backslash V^n$ in the principal nest (we skip the first index m when we work within a fixed renormalization level). The following lemma allows us to begin.

LEMMA 4.1 (INITIAL MODULUS). *Let P_c be a quadratic polynomial with c ranging over a truncated secondary limb L^{tr}. Then $\mathrm{mod}(A^1) \geq \nu > 0$ with ν depending only on L^{tr}.*

The rough reason is that configuration of external rays of P_c has bounded geometry when c ranges over a truncated limb. The next theorem is our main geometric result [L4]:

THEOREM 4.2 (MODULI GROWTH). *Let $n(k)$ count the noncentral levels in the short principle nest. Then $\mathrm{mod}\, A_{n(k)+1} \geq Ck$, where C depends only on $\mathrm{mod}\, A_1$.*

The proof is based upon combinatorial and geometric analysis of the cascade of generalized renormalizations. The above two results yield the complex a priori bounds for maps of special type:

THEOREM 4.3 (A PRIORI BOUNDS). *There is a special family $\mathcal{S}$ of Mandelbrot copies with the following property. If P is an infinitely renormalizable quadratic polynomial of $\mathcal{S}$-type, then $\mathrm{mod}(R^n P) \geq \mu(\mathcal{S}) > 0$.*

Proof. Let us fix a big $Q > 0$. Let f be a quadratic-like map with $\tau(f) = [M_0, M_1, \ldots]$, where the Mandelbrot copy M_n belongs to a truncated limb L_n. Assume $\mathrm{mod}\, f > Q$. By Lemma 4.1 the modulus of the first annulus in the long principal nest is definite: $\mathrm{mod}\, A^{0,1} \geq C(Q)\nu(L_0) > 0$. If the height of M_n is sufficiently big (depending on L_n), then by Theorem 4.2 the modulus of the last annulus $A^{0,t(0)+1}$ of the short nest will be at least Q. Hence $\mathrm{mod}(Rf) \geq Q$, and we can repeat the argument.

5 Rigidity and pullback argument

Let us call two infinitely renormalizable polynomials *combinatorially equivalent* if they have the same type. The topological classes are clearly contained in the combinatorial ones. So the Rigidity Conjecture would follow if we knew that the combinatorial classes are single points (which is actually equivalent to MLC for infinitely renormalizable quadratics). A well-known approach to this problem is based upon the following remark: if all polynomials within the combinatorial class $\mathcal{C}_b$ of P_b are quasi-conformally conjugate, then this class is reduced to a single point $\{b\}$. Indeed, combinatorial classes are clearly closed. On the other hand, quasi-conformal classes are open unless they are single points (by varying of Beltrami differentials). Hence $\mathcal{C}_b$ must be a single point.

Let us have a finite family $\mathcal{L}$ of truncated secondary limbs L_i. Denote by $\mathcal{S}(\mathcal{L}, \chi) \subset \mathcal{M}$ all Mandelbrot copies contained in $\cup L_i$ whose height is at least χ. The following result implies the Rigidity Theorem 2.3.

THEOREM 5.1. *There is a χ depending on $\mathcal{L}$ such that any two quadratic-like maps f and $\tilde{f}$ of the same $\mathcal{S}(\mathcal{L}, \chi)$-type are quasi-conformally conjugate.*

The method we use for the proof is called "the pullback argument". The idea is to start with a qc map respecting some dynamical data, and then pull it back so that it will respect some new data on each step. In the end it becomes (with some luck) a qc conjugacy. This method originated in the works of Thurston (see [DH3]), McMullen (see [McM1], Prop. 8.1) and Sullivan (see [MvS]) (perhaps, it can actually be tracked further down, in the setting of Kleinian groups). Then it was developed in several other works, for more complicated combinatorics (see Kahn [K] and Swiatek [Sw]). In particular, using this method, Jeremy Kahn gave a new proof of the Yoccoz Rigidity Theorem 2.2.

Our way is to pull back through the cascade of generalized renormalizations. The geometric bounds of the previous section are the crucial ingredients of the argument. The linear growth of moduli (Theorem 4.2) keeps the dilatation of pullbacks bounded until the next DH renormalization level, while complex a priori bounds (Theorem 4.3) allow us to penetrate through the next level.

Another method to prove Theorem 2.3 is to transfer the geometric results of Section 4 into the parameter plane (in preparation).

6 Real dynamics

6.1. Scaling factors. A C^3-map $f : [-1, 1] \to [-1, 1]$ is called *quasi-quadratic* if it has a negative Schwarzian derivative and a single nondegenerate critical point. Assume that this map has a fixed point α with negative multiplier (otherwise it is dynamically trivial), and let α' be the dynamically symmetric point: $f\alpha = f\alpha'$. Let $I_0 = [\alpha, \alpha']$. Assume also that the critical point is recurrent. Then we can consider the first return of the critical point to I^0, and pull I^0 along the corresponding orbit. This gives us an interval $I^1 \subset I^0$. Now we can consider the first return to I^1 and the corresponding pullback, and so on. In such a way we construct the real counterpart of the principal nest: $I^0 \supset I^1 \supset \dots$. Moreover, all combinatorial notions such as central returns, DH renormalization, generalized renormalization, etc. are readily transferred to the real case (in the case of a real quadratic polynomial P_c, $c \in \mathbf{R}$ they are just the "real traces" of the corresponding complex notions).

Let $|J|$ denote the length of an interval J. Let us define the scaling factors λ_n as the ratios $|I^n|/|I^{n-1}|$. The real counterpart of Theorem 4.2 is the following result:

THEOREM 6.1 (GEOMETRY DECAYS). *Let $n(k)$ count noncentral returns in the principal nest. Then $\lambda_{n(k)+1} \leq Cq^k$, where $C > 0$ and $q < 1$ depend only on the initial geometry of f.*

In the quadratic-like case this result follows from Theorem 4.2. In the DH nonrenormalizable case it was proven in [L2] by passing to limits of generalized renormalizations and proving that they are generalized quadratic-like maps. In general we prove Theorem 6.1 by extending f to the complex plane so that it is asymptotically conformal near the real line, and using the "quadratic-like" technique with exponentially small errors [L5].

The first application of this geometric result was to the problem of attractors (see the next section). Other applications are to come.

6.2. Fibonacci maps. These are important examples satisfying some extremal combinatorial properties. Though they are nonrenormalizable in the usual sense, they can be treated as infinitely renormalizable in the generalized sense described above. The domains of these renormalizations $T^n f = g_n : V_0^n \cup V_1^n \to V_0^{n-1}$ consist of two puzzle pieces on all levels.

The geometric properties of quasi-quadratic Fibonacci maps from the renormalization point of view were studied in [LM], where the following asymptotic formula for the scaling factors was proven:

$$\lambda_n \sim a(\frac{1}{2})^{n/3}, \tag{6.1}$$

and hence the geometry of the postcritical Cantor set $\omega(0)$ is exponentially decaying.

Let $\mathcal{F}_d$ denote the class of S-unimodal Fibonacci maps with critical point 0 of type x^d. In contrast with (6.1), the postcritical set of an $f \in \mathcal{F}_d$ has bounded geometry for $d > 2$: the scaling factors stay away from 1 and 0. In this case the renormalization approach and complex bounds of [LM], [L2] combined with Sullivan's [S] or McMullen's [McM4] arguments lead to the following result:

THEOREM 6.3. *Let d be an even integer. Then there is only one real Fibonacci polynomial $x \mapsto x^d + c$ of degree d (rigidity). If $d > 2$ then the generalized renormalizations $T^n f$ of any Fibonacci map $f \in \mathcal{F}_d$ converge to a cycle of period two independent of the initial map f.*

Keller and Nowicki [KN] have studied the real geometry of the higher degree Fibonacci maps, and obtained the following bounds:

$$C_1/d \leq \lambda_n(f) \leq C_2/d, \tag{6.2}$$

where the constants are universal in degree. So λ_n^d is bounded from both sides uniformly on d. This indicates a possibility of a new universality phenomenon near the critical value:

PROBLEM. *Study the asymptotical shape of the puzzle pieces fV^n as $d \to \infty$.*

Note that in the quadratic case the Fibonacci renormalization theory implies that the shape of high level puzzle pieces near the critical point imitates the filled Julia set of $z \mapsto z^2 - 1$ [L5] (see Figure 2).

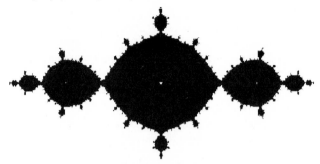

Figure 2. A puzzle piece for the degree 2 Fibonacci map.

7 Attractors and measure of the Julia set

7.1. Quasi-quadratic case. Our first application of geometric Theorem 6.1 was the following result, which resolved Milnor's problem on attractors [M3] for quasi-quadratic maps:

THEOREM 7.1 [L2]. *Let* $f : [-1,1] \to [-1,1]$ *be a DH nonrenormalizable quasi-quadratic map without attracting fixed points. Then* $\omega(x) = [f0, f^2 0]$ *for Lebesgue almost all* $x \in [-1,1]$.

A theoretical alternative would be a "wild" Cantor attractor $A = \omega(0)$ attracting almost all $x \in [-1,1]$ (see [L3] for a survey on this problem, and Section 7.2 below).

The complex counterpart of Theorem 7.1 is the following theorem by the author [L1] and M. Shishikura (unpublished). The cubic case with one escaping critical point had been earlier treated by McMullen (see [BH]).

THEOREM 7.2. *If* P_c *is at most a finitely renormalizable quadratic polynomial without irrational neutral periodic points then* mes $J(P_c) = 0$.

7.2. Higher degree Fibonacci maps. When the work [LM] on quasi-quadratic Fibonacci maps was done, the author suggested an approach to the problem of attractors and the measure problem for higher degree Fibonacci maps. It was based upon consideration of a random walk on the principal nest of annuli with transition maps corresponding to the generalized renormalization. Drift to the left for this random walk corresponds to existence of a "wild" measure-theoretic attractor in the real setting, and positive measure of the Julia set in the complex setting.

Together with F. Tangerman, the author carried out a computer experiment based on this random walk approach to figure out if there are "wild attractors" in higher degrees. The experiment gave the positive answer already for degree 6. Recently this method has been carried out rigorously:

THEOREM 7.3 [BKNS]. *If* d *is sufficiently big, then any Fibonacci map* $f \in \mathcal{F}_d$ *has a Cantor attractor:* $\omega(x) = \omega(0)$ *for Lebesgue almost all* x *(even though this map is topologically transitive on* $[f0, f^2 0]$*).*

A similar experiment in the complex plane carried out jointly with S. Sutherland has recently shown positive measure of the Julia set for the degree 32 Fibonacci map. Nowicki and van Strien have proven this as a rigorous result:

THEOREM 7.4 [SN]. *The Fibonacci polynomial $z \mapsto z^d + c$ of sufficiently high degree has a Julia set of positive measure.*

The experiment and the proof are based upon the same random walk idea and the Fibonacci renormalization theory (Theorem 6.3). The amazing new geometric ingredient of [SN] is a rigorous analysis of the sizes and shapes of the puzzle pieces of the principal nest (compare Figure 3).

Figure 3. Pricipal nest for the degree 6 Fibonacci map.

PROBLEM. *Prove that there is a critical exponent δ such that Fibonacci maps of power higher than δ have a Cantor attractor, whereas those of smaller degree do not. Prove the same result for the measure of the Julia set (with a different exponent).*

There is a remarkable connection between the problem of measure of the Julia set and the rigidity problem. Namely: *If the Julia set $J(f)$ has zero measure then f is quasi-conformally rigid* [MSS]. Theorems 6.3 and 7.4 show, however, that these problems are not equivalent.

8 Rigidity and hyperbolic orbifold 3-laminations

A great insight in the theory of Kleinian groups comes from the third dimension (see [Th]), which allows one to relate hyperbolic geometry to the action at infinity. This gives powerful tools for the rigidity problems and the Ahlfors measure problem. So far nothing like this has appeared in holomorphic dynamics (though 3D analogy plays an important role in the work of McMullen, see [McM4]). Yair Minsky and the author have recently made an attempt to fill in this gap [LMin]. It was inspired by Sullivan's work on Riemann surface laminations which play the role of Riemann surfaces associated to Kleinian groups (see [MvS], [S]).

Let me briefly outline our construction. Let f be a rational function. Consider the natural extension $\hat{f} : \mathcal{N}_f \to \mathcal{N}_f$, where $\mathcal{N}_f$ is the space of backward orbits of

f. This space contains a regular part $\mathcal{R}_f$ that is decomposed into a union of *leaves* with a natural conformal structure. All these leaves are either hyperbolic or parabolic planes.

TYPE PROBLEM. *Are there hyperbolic leaves except Siegel disks and Herman rings?*

Parabolic leaves are conformally equivalent to $\mathbf{C}$, and hence bear an intrinsic affine structure preserved by dynamics. Unfortunately, this structure is not necessarily continuous in the transversal direction. To make it continuous one should strengthen the topology of $\mathcal{R}_f$. To get a reasonable object one then should complete $\mathcal{R}_f$. This procedure adds some singular leaves, with orbifold affine structure. An object that we build in such a way can be called an *affine orbifold lamination*.

The next step is to attach hyperbolic orbifold 3-leaves to these affine 2-leaves. This gives us a *hyperbolic orbifold 3-lamination* $\mathcal{H}_f$. The map $\hat{f}$ can be extended to this space as hyperbolic isometries on the leaves, and it acts properly discontinuously on $\mathcal{H}_f$. The final step is to take the quotient $\mathcal{H}_f/\hat{f}$. This hyperbolic orbifold 3-lamination is our candidate for a role similar to that which hyperbolic 3-manifolds play in the theory of Kleinian groups.

Having such an object in hand, we can define its convex core. A map f is called *convex cocompact* if this convex hull is compact.

THEOREM 8.1. *Let f be a postcritically finite rational function. Then the hyperbolic orbifold 3-lamination $\mathcal{H}_f/f$ is convex cocompact.*

This leads to a three-dimensional proof of the Thurston Rigidity Theorem 2.1, which follows the same lines as the proof of the Mostow rigidity theorem, with the substitution of "lamination" for "manifold".

Acknowledgement. I thank John Milnor and Curt McMullen for many useful comments on the manuscript. I also thank Scott Sutherland and Brian Yarrington for making the computer pictures.

References

[BH]   B. Branner and J. H. Hubbard, *The iteration of cubic polynomials, Part II*, Acta Math. **169** (1992), 229–325.

[BKNS] H. Bruin, G. Keller, T. Nowicki, and S. van Strien, *Absorbing Cantor sets in dynamical systems: Fibonacci maps*, preprint IMS Stony Brook, 1994/1.

[D]   A. Douady, *Chirurgie sur les applications holomorphes*, Proc. Internat. Congress Math. Berkeley, **1** (1986), 724–738.

[DH1] A. Douady and J. H. Hubbard, *Étude dynamique des polynômes complexes*, Publ. Math. Orsay, 84–02 and 85–04.

[DH2] A. Douady and J. H. Hubbard, *On the dynamics of polynomial-like maps*, Ann. Sci. École Norm. Sup. (4), **18** (1985), 287–343.

[DH3] A. Douady and J. H. Hubbard, *A proof of Thurston's topological characterization of rational functions*, Acta Math. **171** (1993), 263–297.

[F]   P. Fatou, *Sur les équations fonctionnelles*, Bull. Soc. Math. France, **48** (1990), 33–94.

[H]   J. H. Hubbard, *Local connectivity of Julia sets and bifurcation loci: Three theorems of J.-C. Yoccoz*, Topological Methods in Modern Math., A Symposium in Honor of John Milnor's 60th Birthday, Publish or Perish, Houston, TX, 1993.

[K]   J. Kahn, *Holomorphic removability of Julia sets*, manuscript in preparation.

[KN]   G. Keller and T. Nowicki, *Fibonacci maps re(al)visited*, Ergod. Th. & Dynam. Syst., **15** (1995), 99–120.

[L0]   M. Lyubich, *Some typical properties of the dynamics of rational maps*, Russian Math. Surveys **38** (1983), 154–155.

[L1]   M. Lyubich, *On the Lebesgue measure of the Julia set of a quadratic polynomial*, preprint IMS at Stony Brook, 1991/10.

[L2]   M. Lyubich, *Combinatorics, geometry and attractors of quasi-quadratic maps*, Ann. Math. **140** (1994), 347–404.

[L3]   M. Lyubich, *Milnor's attractors, persistent recurrence and renormalization*, Topological Methods in Modern Mathematics, A Symposium in Honor of John Milnor's 60th Birthday, Publish or Perish, Houston, TX, 1993.

[L4]   M. Lyubich, *Geometry of quadratic polynomials: Moduli, rigidity and local connectivity*, preprint IMS Stony Brook, 1993/9.

[L5]   M. Lyubich, *Teichmüller space of Fibonacci maps*, preprint IMS Stony Brook, 1993/12.

[LM]   M. Lyubich and J.Milnor, *The unimodal Fibonacci map*, preprint IMS Stony Brook, 1991/15; J. Amer. Math. Soc., **6** (1993), 425–457.

[LMin] M. Lyubich and Y.Minsky, *Laminations in holomorphic dynamics*, preprint IMS at Stony Brook, 1994/20.

[MSS]   R. Mañé, P. Sad, and D. Sullivan, *On the dynamics of rational maps*, Ann. Sci. École Norm. Sup. (4), **16** (1983), 193–217.

[McM1] C. McMullen, *Families of rational maps*, Ann. Math., **125** (1987), 467–493.

[McM2] C. McMullen, Complex dynamics and renormalization, Annals of Math. Studies, **135**, Princeton Univ. Press, 1994.

[McM3] C. McMullen, *Frontiers of holomorphic dynamics*, Bulletin AMS, **331** (1994), 155–172.

[McM4] C. McMullen, *Renormalization and 3-manifolds which fiber over the circle*, preprint, 1994.

[MvS]   W. de Melo and S. van Strien, *One dimensional dynamics*, Springer-Verlag, Berlin and New York, 1993.

[M1]   J. Milnor, *Self-similarity and hairiness in the Mandelbrot set*, Computers in geometry and topology, Lecture Notes in Pure and Appl. Math., **114** (1989), 211–257.

[M2]   J. Milnor, *Local connectivity of Julia sets: Expository lectures*, preprint IMS Stony Brook, 1992/11.

[M3]   J. Milnor, *On the concept of attractor*, Comm. Math. Phys, **99** (1985), 177–195, and **102** (1985), 517–519.

[MT]   J. Milnor and W.Thurston, *On iterated maps of the interval*, Dynamical Systems, Proc. U. Md., 1986–87, ed. J. Alexander, Lecture Notes in Math., **1342** (1988), 465–563.

[SN]   S. van Strien and T. Nowicki, *Polynomial maps with a Julia set of positive Lebesgue measure: Fibonacci maps*, preprint IMS at Stony Brook 1994/3.

[Sw]   G. Swiatek, *Hyperbolicity is dense in the real quadratic family*, preprint IMS at Stony Brook 1992/10.

[S]   D. Sullivan, *Bounds, quadratic differentials, and renormalization conjectures*, Amer. Math. Soc. Centennial Publications. **2**: Mathematics into Twenty-first Century (1992).

[Th]   W. Thurston, *The geometry and topology of three-manifolds*, preprint, 1979.

Ergodic Variational Methods:
New Techniques and New Problems

RICARDO MAÑÉ †

Former address: I.M.P.A.
                   E. Dona Castorina 110, J. Botânico
                   22460 Rio de Janeiro, Brazil

Introduction

Our subject will be *minimizing measures* of Lagrangian dynamical systems. This is a class of invariant probabilities μ of the flow generated by the Euler-Lagrange equation associated to a periodic Lagrangian on a closed manifold, selected by the property of minimizing the μ-average of the Lagrangian among all the invariant probabilities with a given asymptotic cycle (in the sense of Schwartzmann [S]). This concept was introduced by Mather [Ma 2] in a successful attempt to produce an analog of the Aubry-Mather theory for systems with more than one degree of freedom. His results on minimizing measures, when applied to periodic Lagrangians on the circle, recover the main results of the Aubry-Mather theory of twist maps. The main concepts and results of Mather's work will be recalled below.

Our objective is to show, through properties proved in [M 2], that the theory of minimizing measures becomes substantially sharper when restricted to *generic* Lagrangians; where genericity will be understood in a natural way to be defined in the next section. On the other hand, while producing new and stronger results, the generic viewpoint raises many interesting questions that seem beyond the range of the techniques employed in [M 2]. Some of them will be posed below; chosen to exhibit the most visible aspects of the incompleteness of our knowledge on which the ultimate sharpening possibilities of the generic approach really are.

I   Minimizing Measures

Let M be a closed Riemannian manifold. By a Lagrangian on M we shall mean a C^∞ function $L: TM \times \mathbf{R} \to \mathbf{R}$, periodic (say of period 1) in the time variable, such that the second derivative of its restriction to the fibers $T_x M$ is uniformly positive definite. In fact the weaker hypothesis required in [Ma 2] would suffice. Such a Lagrangian generates a flow $\varphi_t: TM \times S^1 \hookleftarrow$ (where $S^1 = \mathbf{R}/\mathbf{Z}$) defined by:

$$\varphi_t(x_o, v_o, t_o) = (x(t + t_o), \dot{x}(t + t_o), (t + t_o) \mathrm{mod}\, \mathbf{Z})$$

where $x: \mathbf{R} \to M$ is the solution of the Euler-Lagrange equation associated to L, with $x(t_o) = x_o$, $\dot{x}(t_o) = v_o$.

Proceedings of the International Congress
of Mathematicians, Zürich, Switzerland 1994
© Birkhäuser Verlag, Basel, Switzerland 1995

Let $\mathcal{M}^*(L)$ be the set of φ-invariant probabilities with compact support. To each $\mu \in \mathcal{M}^*(L)$ we associate its action $S_L(\mu)$ defined by:

$$S_L(\mu) = \int L \, d\mu$$

and its homology (or asymptotic cycle; [S], [Ma 2]) $\rho(\mu)$ defined by satisfying:

$$\langle [\omega], \rho(\mu) \rangle = \int \omega \, d\mu$$

for every closed 1-form ω on M. It always exists, is unique, and $\rho: \mathcal{M}^*(L) \to H_1(M, \mathbf{R})$ is surjective [Ma 2].

DEFINITION A *minimizing measure* of L is a $\mu \in \mathcal{M}^*(L)$ such that:

$$S_L(\mu) = \min\{S_L(\nu) \,/\, \nu \in \mathcal{M}^*(L), \rho(\nu) = \rho(\mu)\}.$$

For every $\gamma \in H_1(M, \mathbf{R})$, the set $\mathcal{M}_\gamma(L)$ of $\mu \in \mathcal{M}^*(L)$ with $\rho(\mu) = \gamma$ is *nonempty* [Ma 2].

The *action function* of L, $\beta_L: H_1(M, \mathbf{R}) \to \mathbf{R}$, is defined by:

$$\beta_L(\gamma) = \min\{S_L(\nu) \,/\, \nu \in \mathcal{M}^*(L), \rho(\nu) = \gamma\}.$$

This is a *convex* and *superlinear* function [Ma 2], and there exist interesting connections between the convex theoretical properties of β_L at a point γ and the dynamical properties of measures in $\mathcal{M}_\gamma(L)$. For instance, if γ is an *extremal* point of β_L (i.e. $(\gamma, \beta_L(\gamma))$ is an extremal point of the epigraph of β_L) then $\mathcal{M}_\gamma(L)$ contains ergodic elements, and if γ is a *strictly extremal* point of β_L (i.e. if there exists a hyperplane in $H_1(M, \mathbf{R}) \times \mathbf{R}$ touching the graph of β_L at $(\gamma, \beta_L(\gamma))$) then $\mathcal{M}_\gamma(L)$ contains measures supported in minimal sets [M 2], [Ma 2]. For properties concerning the vertex and smoothness of β_L, see [Ma 1] and [D].

Now we can introduce and exemplify the generic viewpoint. Let $C^\infty(M \times S^1)$ be the space of C^∞ functions on $M \times S^1$ endowed with the C^∞ topology. We say that a property holds for *generic* Lagrangians (or generically) if given any Lagrangian L there exists a residual subset $\mathcal{A} \subset C^\infty(M \times S^1)$ such that the property holds for every Lagrangian of the form $L_o + \psi$, with $\psi \in \mathcal{A}$.

THEOREM I.1. [M 2]. *Generically, the set* $\mathrm{Ext}(L)$ *of extremal points of* β_L *contains a residual subset* $\mathcal{A} \subset \mathrm{Ext}(L)$ *such that if* $\gamma \in \mathcal{A}$, $M_\gamma(L)$ *consists of a single measure that moreover is uniquely ergodic and its support is the limit in the Hausdorff metric of a sequence of periodic orbits.*

Observe that $\mathrm{Ext}(L)$ is always a Baire set, i.e. its residual subsets are dense.

PROBLEM I For a generic L, does the property of I.1 hold for every extremal point?

PROBLEM II For a generic L, does $\mathrm{Ext}(L)$ contain a dense subset $\mathcal{A}_o \subset \mathrm{Ext}(L)$ such that $\gamma \in \mathcal{A}_o$ implies that $\mathcal{M}_\gamma(L)$ contains a measure supported in a periodic orbit?

THEOREM I.2. *If L is a generic Lagrangian, for every ergodic minimizing measure μ there exists a sequence $\{\mu_n\}$ of uniquely ergodic minimizing measures such that $\mu_n \to \mu$ weakly and $\mathrm{supp}(\mu_n) \to \mathrm{supp}(\mu)$ in the Hausdorff metric.*

PROBLEM III   For generic Lagrangians, is every ergodic minimizing measure uniquely ergodic?

II   The Cohomological Approach and the Coboundary Property

The dual approach to minimizing measures is based on the fact that if ω is a closed 1-form on M then the Lagrangians L and $L - \omega$ generate the same flow. Then $\mathcal{M}^*(L) = \mathcal{M}^*(L - \omega)$ and:

$$S_{L-\omega}(\mu) = S_L(\mu) - \langle \omega, \rho(\mu) \rangle$$

for all $\mu \in \mathcal{M}^*(L)$. This also shows that $S_{L-\omega}(\mu)$ depends only on $[\omega] \in H^1(M, \mathbf{R})$. Define $\mathcal{M}^\omega(L)$ as the set of $\mu \in \mathcal{M}^*(L)$ such that:

$$S_{L-\omega}(\mu) = \min\{S_{L-\omega}(\nu) \,/\, \nu \in \mathcal{M}^*(L)\}.$$

Then $\mathcal{M}^\omega(L)$ depends only on $[\omega]$. To relate the sets $\mathcal{M}^\omega(L)$ with the sets $\mathcal{M}_\gamma(L)$ we define $c(\omega)$ as the maximum of the $c \in \mathbf{R}$ such that $\beta_L(x) \geq \langle [\omega], x \rangle + c$ for all $x \in H_1(M, \mathbf{R})$, and $K(\omega)$ by

$$K(\omega) = \{x \in H_1(M, \mathbf{R}) \,/\, \beta_L(x) = \langle \omega, x \rangle + c(\omega)\}.$$

Then it can be checked that:

$$\mathcal{M}^\omega(L) = \bigcup\{\mathcal{M}_\gamma(L) \,/\, \gamma \in K(\omega)\},$$
$$\bigcup_\omega \mathcal{M}^\omega(L) = \bigcup_\gamma \mathcal{M}_\gamma(L).$$

Therefore all the minimizing measures are found through the sets $\mathcal{M}^\omega(L)$. Denote by $\Lambda^\omega(L) \subset TM \times S^1$ the closure of the union of the supports of the measures in $\mathcal{M}^\omega(L)$. Among the main results of Mather in [Ma 2] are the *compacity* of $\Lambda^\omega(L)$ and the *graph property*, i.e. that if $\pi: TM \times S^1 \to M$ denotes the canonical projection then $\pi/\Lambda^\omega(L)$ is injective and its inverse is Lipschitz. The Aubry-Mather theory follows from this remarkable property and a theorem of Moser [Mo] describing twist maps as time one maps of flows of periodic Lagrangians on the circle.

The "converse" of the definition of $\Lambda^\omega(L)$ is true:

THEOREM II.1. *If the support of $\mu \in \mathcal{M}^*(L)$ is contained in $\Lambda^\omega(L)$, then $\mu \in \mathcal{M}^\omega(L)$.*

From this it follows easily that:

COROLLARY II.2. *Every $\mathcal{M}^\omega(L)$ contains a measure supported in a minimal set. If $\mathcal{M}^\omega(L)$ contains a single measure μ, then μ is uniquely ergodic.*

The theorems stated in Section I rely on the following property, which has its own intrinsic interest.

THEOREM II.3. (Coboundary Property [M 2]). *If $\mu \in \mathcal{M}^\omega(L)$ is ergodic, then $(L - \omega - c(\omega))/\operatorname{supp}(\mu)$ is a Lipschitz coboundary, i.e. there exists a Lipschitz function $V: \operatorname{supp}(\mu) \to \mathbf{R}$ such that:*

$$(L - \omega - c(\omega))/\operatorname{supp}(\mu) = \frac{dV}{d\varphi}$$

where

$$\frac{dV}{d\varphi}(\theta) := \lim_{t \to 0} \frac{1}{t}(V(\varphi_t(\theta)) - V(\theta)).$$

The following result shows that for generic Lagrangians, the set $\mathcal{M}^\omega(L)$ consists, for most $[\omega] \in H^1(M, \mathbf{R})$, of a single measure, which, by Corollary II.2, has to be uniquely ergodic.

THEOREM II.4. [M 2]. *(a) Given a closed 1-form ω on M, the set $\mathcal{M}^\omega(L)$ consists, for a generic L, of a single measure.*

(b) For a generic Lagrangian L, there exists a residual subset $\mathcal{A} \subset H^1(M, \mathbf{R})$ such that $[\omega] \in \mathcal{A}$ implies that $\mathcal{M}^\omega(L)$ consists of a single measure.

PROBLEM IV Is it true that for a generic Lagrangian L there exists a dense set $\mathcal{A} \subset H^1(M, \mathbf{R})$ such that $[\omega] \in \mathcal{A}$ implies that $\mathcal{M}^\omega(L)$ consists of a single measure supported in a hyperbolic periodic orbit?

It follows from the results in [D] that the set of classes $[\omega]$ with this property is open for generic Lagrangians.

When $M = S^1$ the answer is positive. This follows from Mather's result on the differentiability of the action function, when $M = S^1$, at irrational values of $H_1(M, S^1)$, plus the generic properties of the vertex of β_L (where also another proof of the differentiability is given). Observe that when $M = S^1$, the answer to Problems I, II, and III is affirmative for every *Lagrangian*.

The proof of Theorem II.3 uses a reformulation of the space on which the variational principle is applied. We introduce a space $\mathcal{M}$ of probabilities on $TM \times S^1$, so large that $\mathcal{M} \supset \mathcal{M}^*(L)$ for *every* L, and such that for every L and ω there exists $\mu \in \mathcal{M}$ satisfying:

$$(*) \qquad \int (L - \omega)\, d\mu = \min\left\{ \int (L - \omega)\, d\nu \,/\, \nu \in \mathcal{M} \right\}.$$

Moreover, it is proved that this property *implies* the L-invariance of μ. Then $\mu \in \mathcal{M}^\omega(L)$. The existence of μ's satisfying (*) follows from the fact that $\mathcal{M}$ is naturally embedded as a convex subset of a locally convex space and sets of the form

$$\left\{ \nu \,/\, \int (L - \omega)\, d\nu \leq C \right\}$$

are compact for all L, ω, and C. This translates the analysis of $\mathcal{M}^\omega(L)$ into questions on strictly extremal points (also called exposed points) of compact convex sets in locally convex spaces. Some care has to be taken with the fact that compact convex subsets of locally convex spaces (even Fréchet spaces) may have no strictly extremal points (as opposed to the case of Banach spaces).

References

[D]   J. Delgado, *Vertex and differentiability of the action function of Lagrangians*, Thesis, IMPA 1993.

[M 1]   R. Mañé, *Minimizing measures of Lagrangian systems*, Nonlinearity **5** (1992), 623–638.

[M 2]   R. Mañé, *Generic properties and problems of minimizing measures*, preprint I.M.P.A. 1993.

[Ma 1] J. Mather, *Differentiability of the minimum average action as a function of the rotation numbers*, Bol. Soc. Brasil. Mat. **21** (1990), 57–70.

[Ma 2] J. Mather, *Action Minimizing measures for positive definite Lagrangian systems*, Math. Z. **207** (1993), 169–207.

[Mo]   J. Moser, *Monotone twist mappings and the calculus of variations*, Ergodic Theory and Dynamical Systems **6** (1986), 401–413.

[S]   S. Schwartzmann, *Asymptotic cycles*, Ann. of Math. (2) **66** (1957), 27–284.

Homoclinic Bifurcations and Persistence of Nonuniformly Hyperbolic Attractors

MARCELO VIANA[*]

Marcelo Viana, IMPA, Est. D. Castorina 110, Jardim Botânico,
22460 Rio de Janeiro, Brazil

1. Introduction

Let $\varphi\colon M \longrightarrow M$ be a general smooth transformation on a riemannian manifold. A main object of study in dynamics is the asymptotic behavior of the orbits $\varphi^n(z) = \varphi \circ \cdots \circ \varphi(z)$, $z \in M$, as time n goes to infinity. Typical forms of behavior — occurring for "many" $z \in M$ — are, of course, of particular relevance and this leads us to the notion of attractor. By an *attractor* we mean a (compact) φ-invariant set $\Lambda \subset M$ that is dynamically indivisible and whose basin — the set of points $z \in M$ for which $\varphi^n(z) \to \Lambda$ as $n \to +\infty$ — is a large set. Dynamical indivisibility can be expressed by the existence of a dense orbit in Λ (if Λ supports a "natural" φ-invariant measure, one may also require that φ be ergodic with respect to such a measure). As for the basin, it must have positive Lebesgue volume or, even, nonempty interior; in all the cases we will consider here the basin actually contains a full neighborhood of the attractor.

In addition, we want to focus on forms of asymptotic behavior that are typical also from the point of view of the dynamical system: we call an attractor *persistent* if it occurs for a large set of maps near φ. "Large" is to be understood in this context in a measure-theoretical sense: positive Lebesgue measure set of parameter values in every generic family of transformations containing φ. On the other hand, stronger forms of persistence — e.g. with "large set" meaning a full neighborhood of φ — hold in some important situations to be described below.

In the simplest case, Λ reduces to a single periodic orbit of φ. Although the presence of a large or, even more so, an infinite number of these periodic attractors — possibly with high periods and strongly intertwined basins — may render the behavior of individual orbits rather unpredictable, rich asymptotic dynamics comes more often associated with the presence of nonperiodic attractors (having, in many cases, an intricate geometric structure). Indeed, there is a large amount of numerical evidence for the occurrence of such nontrivial attractors in a wide range of situations in dynamics, from mathematical models of complex natural phenomena to even the simplest abstract nonlinear systems. A striking feature of many of these systems is the phenomenon of *exponential sensitivity with respect to initial conditions*: typical (pairs of) orbits of nearby points move away from

---

[*]This work was partially supported by a J. S. Guggenheim Foundation Fellowship.

Proceedings of the International Congress
of Mathematicians, Zürich, Switzerland 1994
© Birkhäuser Verlag, Basel, Switzerland 1995

each other exponentially fast as they approach the attractor. Note the profound consequences: measurement imprecisions and round-off errors tend to be amplified under iteration and so, in practice, the long-term behavior of trajectories in the basin of the attractor is unpredictable (or "chaotic").

A conceptual framework for the understanding of such *chaotic dynamics* is currently under active development. Two main general problems in this context are to describe the (dynamical, geometric, ergodic) structure of chaotic attractors, and, to identify the mechanisms responsible for their formation and persistence. A fairly complete solution to these problems is known in the special case of uniformly hyperbolic (or Axiom A) attractors, see e.g. [Sh2], [Bo], and this is a basic ingredient here. On the other hand, uniform hyperbolicity *per se* is seldom observed in dynamical systems arising from actual phenomena in the experimental sciences, where sensitivity with respect to initial conditions is quite more often related to nonuniformly hyperbolic behavior. This last notion can be defined as follows. We say that φ has *Lyapunov exponents* $\lambda_1, \ldots, \lambda_l$ at $z \in M$ if the tangent space may be split $T_z M = E_1 \oplus \cdots \oplus E_l$ in such a way that

$$\lim_{n \to +\infty} \frac{1}{n} \log \|D\varphi^n(z)v\| = \lambda_j \quad \text{for every } v \in E_j \backslash \{0\} \text{ and } 1 \leq j \leq l.$$

By Oseledec's theorem such a splitting exists at almost every point, relative to any finite φ-invariant measure. Then we call the system *nonuniformly hyperbolic* [Pe], if $\lambda_j \neq 0$ for all j and for almost all points (with respect to the relevant measure under consideration); see [Pe]. Note that occurrence of some positive Lyapunov exponent corresponds precisely to (infinitesimal) exponential sensitivity around the trajectory of z. Also, in the situations to be considered here, existence of positive Lyapunov exponents is the key ingredient for nonuniform hyperbolicity; the fact that all the remaining exponents are strictly negative then follows from elementary considerations.

The dynamics of nonuniformly hyperbolic attractors is, in general, rather unstable under perturbations of the system and this means that more subtle mechanisms of dynamical persistence occur in this general context than in the Axiom A case (where persistence comes along with structural stability and is, ultimately, an instance of transversality theory). The comprehension of such mechanisms is then directly related to the general study of bifurcations of dynamical systems. This is, in fact, the departing point of the program towards a theory of sensitive dynamics recently proposed by Palis and underlying Section 2. below. The basic strategy is to focus on a convenient set of well-defined bifurcation processes — this set should be dense among all (non-Axiom A) systems exhibiting interesting dynamical behavior — and to determine which are the persistent forms of dynamics in generic parametrized families unfolding such bifurcations (once more, persistence is meant in the sense of positive Lebesgue measure of parameter values). See e.g. [PT] for precise formulations and an extended discussion.

A central role is played here by the processes of *homoclinic bifurcation* — that is, creation and/or destruction of transverse intersections between the stable and the unstable manifolds of a same hyperbolic saddle, see Figure 1 — which, by themselves, encompass all presently known forms of interesting behavior in this setting of discrete dynamical systems. The study of homoclinic bifurcations and of their interplay with other main processes of dynamical modification provides a most promising scenario for the understanding of complicated asymptotic behavior, especially in low dimensions, and in Section 2. we discuss some of the results already substantiating this scenario.

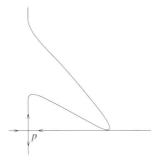

Figure 1: A homoclinic tangency

On the other hand, several of these results actually extend to manifolds of arbitrary dimension and this is an area of considerable ongoing progress. A very interesting topic is the construction and analysis of the properties of *multidimensional* nonuniformly hyperbolic attractors. By multidimensionality we mean existence of several directions of stretching, i.e. several positive Lyapunov exponents (this also implies that the attractor has topological dimension larger than 1). A discussion of recent developments and open problems on this topic occupies most of Section 3.

2. Bifurcations and attractors

Jakobson's theorem [Ja] provided the first rigorous situations of persistence of chaotic dynamics in a strictly nonuniformly hyperbolic setting: *for a positive measure set of values of $a \in (1,2)$ the quadratic real map $q_a(x) = 1 - ax^2$ admits an invariant probability measure μ_a which is absolutely continuous with respect to the Lebesgue measure. Moreover, μ_a is ergodic and has positive Lyapunov exponent:*

$$\lim_{n \to +\infty} \frac{1}{n} \log |Dq_a^n| = \int \log |Dq_a| \, d\mu_a > 0, \quad \mu_a - almost \ everywhere.$$

On the other hand, Benedicks-Carleson [BC] proved that complicated behavior is also abundant in another important nonlinear model, the Hénon family of diffeomorphisms of the plane $H_{a,b}(x,y) = (1 - ax^2 + by, x)$: *for a positive measure set of parameter values $H_{a,b}$ exhibits a compact invariant set $\Lambda_{a,b} \subset \mathbb{R}^2$ (the closure of the unstable manifold of a fixed saddle-point) satisfying*

(i) *The basin $W^s(\Lambda_{a,b}) = \{z \in \mathbb{R}^2 : H_{a,b}^n(z) \to \Lambda_{a,b} \text{ as } n \to +\infty\}$ contains a neighborhood of $\Lambda_{a,b}$;*

(ii) *There exists $\hat{z} \in \Lambda_{a,b}$ whose orbit $\{H_{a,b}^n(\hat{z}) : n \geq 0\}$ is dense in $\Lambda_{a,b}$.*

Moreover, *this dense orbit may be taken exhibiting a positive Lyapunov exponent:*

(iii) $\left\| DH_{a,b}^n(\hat{z})u \right\| \geq c\sigma^n$ for some $c > 0$, $\sigma > 1$, and $u \in \mathbb{R}^2$ and all $n \geq 0$;

(iv) $\left\| DH_{a,b}^n(\hat{z})v \right\| \to 0$ as $|n| \to \infty$ for some $v \in \mathbb{R}^2$, $v \neq 0$ (and so $\Lambda_{a,b}$ is not uniformly hyperbolic).

A stronger formulation of the sensitivity property (iii) is contained in the construction by [BY] of an SBR-measure $\mu_{a,b}$ supported on the "strange" attractor $\Lambda_{a,b}$: $H_{a,b}$ has a positive Lyapunov exponent $\mu_{a,b}$-almost everywhere (and at every point in a positive Lebesgue volume subset of the basin). An alternative construction of these SBR-measures also giving new information on the geometry of the attractor is being provided in [JN].

Let us outline the mechanism yielding positive Lyapunov exponents in these two situations. A common feature of these and other important models is the combination of fairly hyperbolic behavior, in most of the dynamical space, with the presence of *critical regions* where hyperbolicity breaks down. In the case of q_a the critical region is just the vicinity of the critical point $x = 0$, where the map is strongly contracting. For Hénon maps, criticality corresponds to the "folding" occurring near $x = 0$, which obstructs the existence of invariant cone fields. Then the proofs of the previous results require a delicate control on the recurrence of the critical region, in order to prevent nonhyperbolic effects from accumulating too strongly. In the 1-dimensional case, for instance, one must impose a convenient lower bound on $|q_a^n(0)|$ for each $n > 0$. This translates into a sequence of conditions on the parameter, which are part of the definition of the positive measure set in the statement. The argument is rather more complex in the Hénon case but it still follows the same basic strategy of *control of the recurrence through exclusion of parameter values*. The dynamical persistence displayed by the maps one gets after these exclusions is all the more remarkable in view of their instability: although an arbitrarily small perturbation of the parameter may destroy the chaotic attractor (e.g. creating periodic attractors, see [Ur]), it is a likely event (positive probability) that the attractor will actually remain after the perturbation.

Starting from these models, we now discuss a number of results and open problems leading to a quantitative and qualitative description of the occurrence of attractors in the general setting of homoclinic bifurcations. Let us begin by defining this setting in a more precise way than we did before. We consider generic smooth families of diffeomorphisms $\varphi_\mu : M \longrightarrow M$, $\mu \in \mathbb{R}$, such that φ_0 exhibits some nontransverse intersection between the stable and the unstable manifolds of a hyperbolic saddle-point p; recall Figure 1. In this section we take M to be a surface. Genericity means that this *homoclinic tangency* is nondegenerate — quadratic — and unfolds generically with the parameter: the two invariant manifolds move with respect to each other with nonzero relative speed, near the tangency. We also suppose $|\det D\varphi_0(p)| \neq 1$ and in what follows we consider $|\det D\varphi_0(p)| < 1$ (in the opposite case just replace φ_μ by φ_μ^{-1}). Then, see e.g. [TY], return-maps to a neighborhood of the tangency contain small perturbations of the family of singular maps $(x, y) \mapsto (1 - ax^2, 0)$. Combining this fact with an extension of the methods

in [BC] one can prove that *Hénon-like attractors* — i.e. satisfying (i)–(iv) above
— occur in a persistent way whenever a homoclinic tangency is unfolded:

THEOREM 1 [MV]   *There exists a positive Lebesgue measure set of values of μ,
accumulating at $\mu = 0$, for which φ_μ has Hénon-like attractors close to (in a
const $|\mu|$-neighborhood of) the orbit of tangency.*

This should also be compared with the well-known theorem of Newhouse on
abundance of periodic attractors:

THEOREM 2 [Ne]   *There exist intervals $I \subset \mathbb{R}$ accumulating at $\mu = 0$ and residual
(Baire second category) subsets $B \subset I$ such that for every $\mu \in B$ the diffeomorphism φ_μ has infinitely many periodic attractors close to the orbit of tangency.*

These two contrasting forms of asymptotic behavior are, actually, strongly
interspersed: the values of μ one gets in both the proofs of these results are accu-
mulated by other parameters corresponding to new homoclinic tangencies [Ur].

PROBLEM 1: (Palis) Can any diffeomorphism exhibiting a Hénon-like attractor,
resp. infinitely many periodic attractors, be approximated by another one having
a homoclinic tangency?

PROBLEM 2: Can Newhouse's phenomenon occur for a set S of parameter values
with positive Lebesgue measure?

The answer to Problem 2 is usually conjectured to be negative but it is as
yet unknown. Note that the sets B constructed in the proof of Theorem 2 have
zero measure [TY]. An interesting related question is formulated replacing above
"positive Lebesgue measure" by "positive Lebesgue density at $\mu = 0$", that is

$$\lim_{\varepsilon \to 0} \frac{m(S \cap [-\varepsilon, \varepsilon])}{2\varepsilon} > 0, \qquad m = \text{Lebesgue measure.}$$

Ongoing progress seems to indicate that the answer to this last question is negative,
even if one replaces S by the set of parameter values corresponding to existence
of *some* periodic attractor near the tangency. A similar problem can be posed for
nonuniformly hyperbolic attractors as in Theorem 1:

PROBLEM 3: Can Hénon-like attractors occur for a set of parameter values having
positive density at $\mu = 0$?

Although this last problem remains open in the context of homoclinic tangencies, it admits a complete, positive answer in a closely related setting of bifurcations: the unfolding of critical saddle-node cycles. By a *saddle-node k-cycle*,
$k \geq 1$, of a diffeomorphism φ we mean a finite set of periodic points $p_1, p_2, \ldots, p_k$
such that

- p_1 is a saddle-node (eigenvalues 1 and λ, with $|\lambda| < 1$) and p_i is a hyperbolic
  saddle for each $2 \leq i \leq k$;

- $W^u(p_{i-1})$ and $W^s(p_i)$ have points of transverse intersection, for all $2 \leq i \leq k$,
  and $W^u(p_k)$ intersects the interior of $W^s(p_1)$.

Following [NPT], we call the saddle-node cycle *critical* if $W^u(p_k)$ has a non-transverse intersection with some leaf F of the strong stable foliation of $W^s(p_1)$. Figure 2 describes such a cycle in the case $k = 1$ (in this case we actually require $W^u(p_1)$ to be contained in the interior of $W^s(p_1)$).

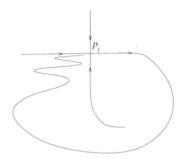

Figure 2: A critical saddle-node cycle

Now we consider the unfolding of such cycles by generic families of diffeomorphims $\varphi_\mu \colon M \longrightarrow M$, $\mu \in \mathbb{R}$. More precisely, we suppose that φ_0 has some critical saddle-node cycle satisfying a few mild assumptions: the saddle-node is nondegenerate and unfolds generically with the parameter μ, and the criticality — i.e. the nontransverse intersection between $W^u(p_k)$ and F — is quadratic. A theorem of [NPT] asserts that such families always go through homoclinic tangencies, at parameter values arbitrarily close to zero. A converse is also true (Mora): critical saddle-node cycles are formed whenever a homoclinic tangency is unfolded. On the other hand, the present setting is special in that Hénon-like attractors always occur for a *positive fraction* of the parameter values near the one corresponding to the cycle. This is the only bifurcation mechanism known to exhibit such a strong accumulation by chaotic attractors.

THEOREM 3 [DRV] *Let $(\varphi_\mu)_\mu$ be a generic family of diffeomorphisms unfolding a critical saddle-node cycle as above. Then the set of parameter values for which φ_μ exhibits Hénon-like attractors has positive Lebesgue density at $\mu = 0$.*

The proof of Theorem 3 is based on a combination of Theorem 1 with a careful analysis of the distribution of homoclinic tangencies in parameter space, cf. previous remarks. This construction yields Hénon-like attractors that are related to orbits of homoclinic tangency and so have a semi-local nature. Although this is unavoidable in the generality of the statement above, attractors of a much more global type can be found in some relevant cases, by using a more direct approach. We mention the case of 1-cycles, recall Figure 2. If φ_0 has a critical 1-cycle then it is not difficult to find a compact domain R containing $W^u(p_1)$ and such that $\varphi_0(R) \subset \text{interior}(R)$. Then, for a sizable portion of the parameter values near zero the asymptotic behavior of all the points in the domain R (which depends only on the bifurcating diffeomorphism φ_0) is driven by a *unique, global,* nonuniformly hyperbolic attractor:

THEOREM 4 [DRV] *For an open class of families $(\varphi_\mu)_\mu$ unfolding a critical saddle-node 1-cycle, there is a set of values of μ with positive Lebesgue density at $\mu = 0$ for which $\Lambda_\mu = \bigcap_{n\geq 0} \varphi_\mu^n(R)$ is a Hénon-like attractor.*

3. Multidimensional expansion

The unfolding of homoclinic tangencies or saddle-node cycles in higher dimensions leads, more often, to the formation of periodic points with positive unstable index (some expanding eigenvalue) and/or of "strange saddles", see [Ro]. In order to have attractors one makes an assumption of (local) *sectional dissipativeness*: the product of any two of the eigenvalues associated to the saddle p exhibiting the tangency, resp. to the saddle-node p_1 involved in the cycle, has norm less than 1. On the other hand, under this assumption Theorems 1–4 do generalize to manifolds of arbitrary dimension, see [PV], [V1]. In particular, persistent Hénon-like attractors may occur in any ambient manifold.

Now, the attractors one finds in such a sectionally dissipative setting are special in that they exhibit at most one direction of stretching (one single positive Lyapunov exponent). This is also related to the fact that the Hénon-like attractors in the previous paragraph always have topological dimension 1. Our goal in this section is to present a construction of persistent nonuniformly hyperbolic attractors with multidimensional character: *typical orbits in their basin exhibit several stretching directions*. In more precise terms, at Lebesgue almost every point z in the basin there is a splitting $T_z M = E^+ \oplus E^-$ such that

$$\liminf \frac{1}{n} \log \left\| D\varphi^n(z)v^+ \right\| > 0 > \limsup \frac{1}{n} \log \left\| D\varphi^n(z)v^- \right\| \text{ for } v^\pm \in E^\pm \backslash \{0\}$$

and $\dim E^+ > 1$. Previously known examples have been restricted to rather structured situations, such as Axiom A diffeomorphisms or the persistently transitive examples in [Sh1] or [Ma]. In these last examples, obstruction to uniform hyperbolicity comes from the presence of saddles with different stable indices but the dynamics is actually fairly uniform (in particular, they admit everywhere-defined continuous invariant cone fields).

These examples of *multidimensional attractors* we now describe are the first ones in the presence of *critical* behavior (in the sense of Section 2.) In fact, the basic idea here is to couple nonuniform models such as Hénon maps, with convenient uniformly hyperbolic systems. On the other hand, the attractors we obtain in this way are considerably more robust than the low-dimensional Hénon-like ones: *they persist in a whole open set of diffeomorphisms*. Let us sketch this construction in a simple situation, details being provided in [V2]. We start by considering diffeomorphisms of the form

$$\varphi \colon \mathrm{T}_3 \times \mathbb{R}^2 \longrightarrow \mathrm{T}_3 \times \mathbb{R}^2, \quad \varphi(\Theta, x, y) = (g(\Theta), f(\Theta, x, y))$$

where g is a solenoid map on the solid torus $\mathrm{T}_3 = \mathrm{S}^1 \times \mathrm{B}^2$, see e.g. [Sh2], and $f(\Theta, x, y) = (a(\Theta) - x^2 + by, -bx)$. Here b is a small positive number, a is some nondegenerate function (e.g. a Morse function) with $1 < a(\Theta) < 2$, and we take the solenoid to be sufficiently expanding along the S^1-direction. Then, for an appropriate choice of these objects, φ is contained in an open set of diffeomorphisms exhibiting a multidimensional nonuniformly hyperbolic attractor:

THEOREM 5 [V2] *There is a compact domain $K \subset \mathbb{R}^2$ such that for every diffeomorphism $\psi \colon T_3 \times \mathbb{R}^2 \longrightarrow T_3 \times \mathbb{R}^2$ sufficiently close to φ (in the C^3-sense), $\psi(T_3 \times K) \subset \mathrm{interior}(T_3 \times K)$, and ψ has two stretching directions at Lebesgue almost every point of $T_3 \times K$.*

A crucial fact distinguishing these examples from the quadratic models in Section 2. is that their critical regions are too large for the same kind of recurrence control as we described there to be possible. In order to motivate this remark we observe that in the (singular) limit $b = 0$ the critical set of φ coincides with $\{\det D\varphi = 0\}$, a codimension-1 submanifold, and, therefore, is bound to have robust intersections with (some of) its iterates. In other words, close *returns* of the critical region back to itself cannot be avoided by any sort of parameter exclusions, which means that we are forced to deal with the accumulation of contracting/nonhyperbolic effects associated to such returns. This is done through a statistical type of argument, which we can (very roughly) sketch as follows. Given $z \in T_3 \times \mathbb{R}^2$, the nonhyperbolic effect introduced at each time $\nu \geq 1$ for which $\varphi^\nu(z)$ is close to the critical region is estimated in terms of an appropriate integrable function $\Delta_\nu(z)$. The definition of $\Delta_\nu(z)$ in the actual situation of Theorem 5 — with $b > 0$ — is fairly complicated and we just mention that in the (much simpler) limit case $b = 0$ one may take $\Delta_\nu(z) = -\log|x_\nu|$, where x_ν is the x-coordinate of $\varphi^\nu(z)$. Then we derive two crucial stochastic properties of these Δ_ν:

(1)   The expected (i.e. average) value of Δ_ν is small for each $\nu \geq 1$;
(2)   the probability distributions of Δ_μ and Δ_ν are (fairly) independent from each other if $|\mu - \nu|$ is large enough.

This allows us to use probabilistic arguments (of large deviations type) to conclude that, for most trajectories, the overall nonhyperbolic effect corresponding to iterates near the critical region is smaller than (i.e. dominated by) the hyperbolic contribution coming from the iterates taking place away from that region.

The proof of (2) above is based on the fast decay of correlations exhibited by uniformly hyperbolic systems such as solenoids and, in fact, this seems to be the key property of the map g for what concerns our construction (in its present form the proof makes use of a few other properties of solenoid maps, in an apparently less important way). This suggests that a similar type of argument should apply if the solenoid is replaced in the construction above by more general (not necessarily uniformly hyperbolic) maps having such fast mixing character. As a first step in this direction we pose

PROBLEM 4: Prove that $\varphi(x, y) = (g(x), a(x) - y^2)$ has two positive Lyapunov exponents for a large set of choices of $a(x)$, where g is some convenient — possibly multimodal — smooth transformation of the real line exhibiting chaotic behaviur in the sense of Jakobson's theorem.

Finally, in the view of the discussion in the Introduction, one should try to relate the present topic with the general study of bifurcations of higher-dimensional smooth systems, in the spirit of Section 2. Again, a first step may be

PROBLEM 5: Describe generic bifurcation mechanisms leading to the formation of multidimensional nonuniformly hyperbolic attractors.

Acknowledgement: The author is grateful to the hospitality of the CIMAT-Guanajuato, UCLA, and Princeton University during the preparation of this work.

References

[BC] M. Benedicks and L. Carleson, *The dynamics of the Hénon map*, Ann. of Math. 133 (1991), 73–169.

[BY] M. Benedicks and L.-S. Young, *SBR-measures for certain Hénon maps*, Invent. Math. 112-3 (1993), 541–576.

[Bo] R. Bowen, Equilibrium states and the ergodic theory of Anosov diffeomorphisms, Lect. Notes in Math. 470 (1975), Springer-Verlag, Berlin and New York.

[DRV] L. J. Diaz, J. Rocha, and M. Viana, *Strange attractors in saddle-node cycles: prevalence and globality*, preprint IMPA and to appear.

[Ja] M. Jakobson, *Absolutely continuous invariant measures for one-parameter families of one-dimensional maps*, Comm. Math. Phys. 81 (1981), 39–88.

[JN] M. Jakobson and S. Newhouse, *Strange attractors in strongly dissipative surface diffeomorphisms*, in preparation.

[Ma] R. Mañé, *Contribution to the stability conjecture*, Topology 17 (1978), 383–396.

[MV] L. Mora and M. Viana, *Abundance of strange attractors*, Acta Math. 171 (1993), 1–71.

[Ne] S. Newhouse, *The abundance of wild hyperbolic sets and nonsmooth stable sets for diffeomorphisms*, Publ. Math. IHES 50 (1979), 101–151.

[NPT] S. Newhouse, J. Palis, and F. Takens, *Bifurcations and stability of families of diffeomorphisms*, Publ. Math. IHES 57 (1983), 7–71.

[PT] J. Palis and F. Takens, Hyperbolicity and sensitive chaotic dynamics, Cambridge University Press, London and New York, 1993.

[PV] J. Palis and M. Viana, *High dimension diffeomorphisms displaying infinitely many periodic attractors*, Ann. of Math. 140 (1994), 205–250.

[Pe] Ya. Pesin, *Characteristic Lyapounov exponents and smooth ergodic theory*, Russian Math. Surveys 32 (4) (1977), 55–114.

[Ro] N. Romero, *Persistence of homoclinic tangencies in higher dimensions*, thesis IMPA 1992 and to appear, Ergodic Theory Dynamical Systems.

[Sh1] M. Shub, Topologically transitive diffeomorphisms on T^4, Lecture Notes in Math. 206 (1971), 39, Springer-Verlag, Berlin and New York.

[Sh2] M. Shub, Global Stability of Dynamical Systems, Springer-Verlag, Berlin and New York, 1987.

[TY] L. Tedeschini-Lalli and J. A. Yorke, *How often do simple dynamical processes have infinitely many coexisting sinks?*, Comm. Math. Phys. 106 (1986), 635–657.

[Ur] R. Ures, *On the approximation of Hénon-like strange attractors by homoclinic tangencies*, thesis IMPA 1993 and to appear, Ergodic Theory Dynamical Systems.

[V1] M. Viana, *Strange attractors in higher dimensions*, Bull. Braz. Math. Soc. 24 (1993), 13–62.

[V2] M. Viana, *Multidimensional nonhyperbolic attractors*, preprint IMPA and to appear.

Ergodic Theory of Attractors

LAI-SANG YOUNG*

Department of Mathematics
University of California
Los Angeles, CA 90024-1555, USA

We begin with an overview of this article. Consider a dynamical system generated by a diffeomorphism f with an attractor Λ. We assume $f|\Lambda$ is sufficiently complex that it is impossible to have exact knowledge of every orbit. The ergodic theory approach, which we will take, attempts to describe the system in terms of the average or statistical properties of its "typical" orbits.

If Λ is an Axiom A attractor, then it follows from the work of Sinai, Ruelle, and Bowen ([S],[R1],[BR]) in the 1970s that orbits starting from almost all initial conditions have a common asymptotic distribution. "Almost all" here refers to a full Lebesgue measure set in the basin of attraction of Λ. We will call this invariant measure the *SRB measure* of (f, Λ).

In the late 1970s and early 1980s the idea of a nonuniformly hyperbolic system was developed and the notion of an SRB measure was extended to this more general context. In Section 1 of this article I will define SRB measures and describe some of their ergodic and geometric properties, including their entropy and dimension.

Although one could formally define SRB measures and study them abstractly, the question of how prevalent they are outside of the Axiom A category is not well understood. The first nonuniform (dissipative) examples for which SRB measures were constructed are the Hénon attractors. In Section 2, I will discuss briefly the analysis by Benedicks and Carleson [BC1], [BC2] of certain parameter values of the Hénon maps, and the subsequent work of Benedicks and myself [BY1] on the construction of SRB measures for these parameters.

In Section 3, I would like to present a recent work, also joint with Benedicks [BY2], in which we study stochastic processes of the form $\{\varphi \circ f^i\}_{i=0,1,2,...}$ where f is a "good" Hénon map, the underlying measure is SRB, and φ is a Hölder continuous observable. We prove for these random variables the exponential decay of correlations and a central limit theorem.

Although the results in Sections 2 and 3 are stated only for the Hénon family, our methods of proof are not particularly model-specific. I will conclude with some remarks on the types of situations to which these methods may apply.

*)   This research was partially supported by The National Science Foundation.

Proceedings of the International Congress
of Mathematicians, Zürich, Switzerland 1994
© Birkhäuser Verlag, Basel, Switzerland 1995

1 Some ergodic and geometric properties of SRB measures

Let f be a C^2 diffeomorphism of a finite-dimensional manifold M and let $\Lambda \subset M$ be a compact f-invariant set. We call Λ an attractor if there is a set $U \subset M$ with positive Riemannian measure such that for all $x \in U$, $f^n x \to \Lambda$ as $n \to \infty$.

Given an f-invariant Borel probability measure μ, let $\lambda_1 > \lambda_2 > \cdots > \lambda_r$ denote the distinct Lyapunov exponents of (f, μ) and let E_i be the corresponding subspaces in the tangent space of each point. Stable and unstable manifolds are defined a.e. on sets with negative and positive Lyapunov exponents. They are denoted by W^s and W^u respectively.

Let (f, μ) be such that $\lambda_1 > 0$ a.e., and let η be a measurable partition on M. Let $W^u(x)$ and $\eta(x)$ denote respectively the unstable manifold and element of η containing x. We say that η is *subordinate to* W^u if for a.e. x, $\eta(x) \subset W^u(x)$ and contains an open neighborhood of x in $W^u(x)$. For a given η, let $\{\mu_x^\eta\}$ denote a canonical family of conditional probabilities of μ with respect to η. We will use m_x^η to denote the Riemannian measure induced on $\eta(x)$ as a subset of the immersed submanifold $W^u(x)$.

DEFINITION 1. *Let (f, μ) be as above. We say that μ has absolutely continuous conditional measures on W^u if for every measurable partition η subordinate to W^u, μ_x^η is absolutely continuous with respect to m_x^η for a.e. x.*

This definition has its origins in [S] and [R1]; in its present form it first appeared in [LS].

For Axiom A attractors the invariant measure we called *SRB* in the introduction has several equivalent definitions, one of which is that it has absolutely continuous conditional measures on W^u. My feelings are that as a working definition, this property is the most useful and the most straightforward to generalize. I therefore take the liberty to introduce the following definition:

DEFINITION 2. *Let f and Λ be as in the beginning of this section. An f-invariant Borel probability measure μ on Λ is called an SRB measure if $\lambda_1 > 0$ a.e. and μ has absolutely continuous conditional measures on W^u.*

The physical significance of this property is that the set of points whose future trajectories are generic with respect to an SRB measure forms a positive Lebesgue measure set. This is because we can "integrate out" from the attractor along W^s using the absolute continuity of the stable foliation. More precisely:

THEOREM 1 [P] [PS]. *Let μ be an ergodic SRB measure of f and assume that $\lambda_i \neq 0 \ \forall i$. Then there is a set $\tilde{U} \subset M$ with positive Lebesgue measure such that if φ is a continuous function defined on a neighborhood of Λ then*

$$\frac{1}{n} \sum_{i=0}^{n-1} \varphi(f^i x) \to \int \varphi \, d\mu \quad \text{for every } x \in \tilde{U}.$$

In general, entropy and Lyapunov exponents are different invariants, although both measure the complexity of a dynamical system. With respect to its SRB measure, however, the entropy of a map is equal to the sum of its positive Lyapunov exponents. Indeed, SRB measures are precisely the extreme points in the following variational principle:

THEOREM 2 [P], [R2], [LS], [L1], [LY1]. *Let μ be an f-invariant Borel probability measure. Then*

$$h_\mu(f) \le \int \sum_{\lambda_i > 0} \lambda_i \cdot \dim E_i \, d\mu \;\; ;$$

and equality holds if and only if μ is SRB.

For arbitrary invariant measures, the difference between entropy and the sum of positive Lyapunov exponents can be understood in terms of the dimension of the measure. It is shown in [LY2] that if μ is ergodic, then corresponding to each $\lambda_i \ne 0$ there is a number δ_i with

$$0 \le \delta_i \le \dim E_i \quad \text{such that} \quad h_\mu(f) = \sum_{\lambda_i > 0} \lambda_i \cdot \delta_i \; = - \sum_{\lambda_i < 0} \lambda_i \cdot \delta_i \; .$$

The number δ_i has the geometric interpretation of being the dimension of μ "in the direction of E_i"; it is equal to h_i/λ_i where h_i is the entropy "in the direction of E_i". (See [LY2] for precise definitions.)

These ideas have led to the following result on the dimension of SRB measures. For a finite measure μ, we write $\dim(\mu) = \alpha$ if for $\mu-$ a.e. x,

$$\lim_{r \to 0} \frac{\log \mu B(x,r)}{\log r} = \alpha$$

where $B(x,r)$ is the ball of radius r about x.

THEOREM 3 [L2], [LY2]. *Let μ be an SRB measure. We assume that (f,μ) is ergodic, and that $\lambda_i \ne 0 \; \forall i$. Then*

$$\dim(\mu) = \sum_{\text{all } i} \delta_i$$

where the δ_i's are as above. In particular, $\delta_i = \dim E_i$ for all i with $\lambda_i > 0$.

It is not known at this time whether this notion of dimension is well defined for arbitrary invariant measures. For a special case, see e.g. [Y1].

2 SRB measures for Hénon maps

As we mentioned in the introduction, it follows from the work of Sinai, Ruelle, and Bowen that every Axiom A attractor admits an SRB measure. It is natural to wonder to what extent this is true without the hypothesis of Axiom A. Mathematically very little has been proven, although the existence of SRB measures in general situations is often taken for granted in numerical experiments and by the physical scientist.

To the best of my knowledge, the first dissipative, genuinely nonuniformly hyperbolic attractor for which SRB measures were constructed were the Hénon attractors. (By "dissipative" I mean not volume preserving: if a volume preserving diffeomorphism has a positive Lyapunov exponent a.e. then its volume measure satisfies the condition in Definition 2.) The Hénon maps are a 2-parameter family of maps $T_{a,b} : \mathbb{R}^2 \to \mathbb{R}^2$ defined by

$$T_{a,b} : \begin{pmatrix} x \\ y \end{pmatrix} \mapsto \begin{pmatrix} 1 - ax^2 + y \\ bx \end{pmatrix}.$$

It is not hard to see that there is an open region in parameter space for which $T_{a,b}$ has an attractor; and that for (a, b) in the region, there is a continuous family of invariant cones on $\{|x| > \delta\}$ (δ depending on parameters) but that the attractor is not Axiom A.

In [BC2] Benedicks and Carleson proved that for b sufficiently small there is a positive measure set of a's for which $T = T_{a,b}$ has a positive Lyapunov exponent on a dense subset of Λ. In addition to proving this result they devised a machinery for analyzing DT^n, the derivatives of the iterates of T, for certain orbits with controlled behavior. Without getting into the specifics of their machinery, let me try to explain the essence of these ideas.

Some of the ideas go back to 1 dimension, so let me first explain how expanding properties are proved for the quadratic family $f_a : x \to 1 - ax^2$, $x \in [-1, 1]$, $a \in [0, 2]$. Jacobson [J] proved in 1981 that for a positive measure set of parameters a, f_a admits an invariant measure absolutely continuous with respect to Lebesgue and has a positive Lyapunov exponent a.e. Roughly speaking, the "good" parameters are those for which the derivatives along the critical orbit have exponential growth. Away from the critical point 0, we could think of the map as essentially expanding, and for x near 0, the orbit of x stays near that of 0 for some period of time, giving $(f^n)'x \sim 2ax \cdot (f^{n-1})'(f0) \sim 2ax \cdot \lambda^{n-1}$ for some $\lambda > 1$. These ideas have been used by various authors studying 1-dimensional maps (see e.g. [CE] and [BC1] as well as [J]).

An obstacle to proving hyperbolicity in dimensions greater than 1 is the switching of expanding and contracting directions. For a (2×2) matrix A that is not an isometry, let $s(A)$ denote the direction that will be contracted the most by A. Suppose that for $m, n > 0$ we have proved hyperbolicity for the stretches from $T^{-m}x$ to x and from x to $T^n x$. In order to extend this hyperbolic behavior all the way from $T^{-m}x$ to $T^n x$ we must control $\angle(s(DT_x^{-m}), s(DT_x^n))$, the angle between $s(DT_x^{-m})$ and $s(DT_x^n)$.

In some sense then, the set of points x where $\angle(s(DT_x^{-m}), s(DT_x^n)) \to 0$ as $m, n \to \infty$ plays the role of the critical point in 1 dimension. An essential difference, however, is that the exact location of this "critical set" cannot be known ahead of time. To identify points with the property above, one must prove the hyperbolicity of DT_x^{-m} and DT_x^n for arbitrarily large m and n, but the behavior of these derivatives in turn depends on how the orbit of x interacts with the critical set. This almost seems like circular reasoning, but can in fact be achieved through inductive arguments. In dimensions greater than 1, the inductive character of the analysis is both more prominent and more essential than in 1 dimension.

What Benedicks and Carleson did in [BC2] was to identify and control — for a positive measure set of parameters — a critical set $\mathcal{C}$ as described above. We stress that their inductive procedure goes through only on a positive measure set of parameters. Furthermore they showed that for certain orbits approaching this set, the loss of hyperbolicity is $\sim \text{dist}(f^n x, \mathcal{C})$, and that subsequent recovery is guaranteed.

Building on this machinery, Benedicks and I constructed SRB measures for Hénon maps corresponding to these "good" parameters.

THEOREM 4 [BC2, [BY1]. *Let $\{T_{a,b}\}$ be the Hénon family. Then for each suffi- ciently small b, there is a positive measure set Δ_b such that for each $a \in \Delta_b$, $T = T_{a,b}$ admits an SRB measure μ. This SRB measure is unique; its support is all of Λ; and (T, μ) is isomorphic to a Bernoulli shift.*

As a corollary to this theorem and to Theorem 1, we have a positive Lebesgue measure set in $\mathbb{R}^2$ consisting of points the statistics of whose future trajectories are governed by μ. If for instance one is to pick a point in this set and to plot its first N iterates for some sufficiently large N, then the resulting picture is essentially that of μ. It follows from our proof of Theorem 4 that this set of generic points fills up a large part of the basin of Λ; we believe (but have not yet proved) that it in fact fills up the entire basin up to a set of measure zero.

In [BC2] the analysis is focused mostly on the "bad set" $\mathcal{C}$. Part of the proof of Theorem 4 consists of adapting and globalizing these ideas to unstable manifolds. We then prove the existence of μ by pushing forward Lebesgue measure m on a piece of unstable leaf γ. A key observation is that it is only necessary to consider a positive percentage of these pushed-forward measures. Roughly speaking we show that for a positive density set of integers n, there are subsets γ_n of γ with $m(\gamma_n)$ bounded away from 0 such that for each n,

(i) $|DT^n|_{\gamma_n}| \geq c\lambda^n$ for some $\lambda > 1$;
(ii) $T^n(\gamma_n)$ is the union of (many) roughly parallel curves of a fixed length.

An SRB measure is then extracted from the Cesàro averages of $T_*^n(m|_{\gamma_n})$.

Although the result above is stated only for the Hénon family, it holds for families with similar qualitative properties, such as those that appear in certain homoclinic bifurcations [MV].

We close this discussion by remarking that one cannot expect *all* attractors — or even all attractors with the general appearance of the Hénon attractors — to admit SRB measures. Periodic sinks are easily created near homoclinic tangencies [N], and the presence of sinks substantially complicates the dynamical picture. Nonhyperbolic periodic points are also not conducive to the existence of invariant measures with smooth conditional measures on unstable manifolds [HY]. The question of existence of SRB measures in general is not one that is likely to be resolved in the near future.

It seems, though, that the time has come to attempt the following type of questions: Given a "typical" or "generic" 1-parameter family of dynamical systems that are hyperbolic on large parts of their phase spaces without being uniformly hyperbolic everywhere, is it reasonable to expect that a positive measure set of them will admit SRB measures? (This is the "attractor" or "dissipative" version; one could also formulate similar questions for the positivity of Lyapunov exponents for conservative systems.) I will come back with some brief remarks on this in Section 4.

3 Decay of correlations for Hénon maps

Independent identically distributed random variables are "chaotic" in the sense that it is impossible to predict the future from knowledge of the past, yet their distributions obey very simple limit laws. One might wonder if the same is true for processes coming from chaotic dynamical systems. For example, if f has an attractor Λ and μ is its SRB measure, what can be said about the random variables $\{\varphi \circ f^i\}_{i=0,1,2,\dots}$ where φ is a reasonable function on Λ?

I would like to report on some recent results in this direction.

THEOREM 5 [BY2]. *Let $\{T_{a,b}\}$ be the Hénon family, and let $T = T_{a,b}$ be any one of the maps in Theorem 4 shown to admit an SRB measure μ. Let $\mathcal{H}_\beta$ denote the set of Hölder continuous functions on Λ with exponent β. Then there exists $\tau < 1$ such that for all $\varphi, \psi \in \mathcal{H}_\beta$, there is a constant $C = C(\varphi, \psi)$ such that*

$$\left| \int \varphi \cdot (\psi \circ T^n) \, d\mu - \int \varphi \, d\mu \cdot \int \psi \, d\mu \right| \leq C\tau^n \quad \forall n \geq 1.$$

The main ideas of our proofs are as follows. Given that "horseshoes" are building blocks of uniformly hyperbolic systems, the following seems to be a natural generalization to the nonuniform setting: let Δ_0 be a rectangular lattice obtained by intersecting local stable and unstable manifolds. Suppose that Δ_0 intersects unstable manifolds in positive Lebesgue measure sets, and that it is the disjoint union of a countable number of "s-subrectangles" $\Delta_{0,i}$, each one of which is mapped under some power of T, say under T^{R_i}, hyperbolically onto a "u-subrectangle" of Δ_0. (A subset $X \subset \Delta_0$ is called an "s-subrectangle" of Δ_0 if for every local stable leaf γ used to define Δ_0, either $x \cap \gamma = \phi$ or $x \cap \gamma = \Delta_0$.) Let $R(x) = R_i$ for $x \in \Delta_{0,i}$. Then we may regard the dynamics of T as something like the discrete time version of a special flow built under the return time function R over a uniformly hyperbolic "horseshoe" with infinitely many branches.

For the "good" Hénon maps in Theorem 4, we show that sets with the properties of Δ_0 above are easily constructed. Furthermore, because of the rapid recovery after each visit to the critical set $\mathcal{C}$, the return time function R has the property that $\mu\{R > n\} < C\theta^n$ for some $\theta < 1$. This enables us to show that there is a gap in the spectrum of the Perron-Frobenius operator, proving exponential decay of correlations. (A similar tower construction is used in [Y2].)

Using this spectral property of the Perron-Frobenius operator we obtain also the central limit theorem for $\{\varphi \circ T^i\}_{i=0,1,2,\dots}$:

THEOREM 6 [BY2]. *Let (T, μ) be as in Theorem 5, and let $\varphi \in \mathcal{H}_\beta$ be a function with $\int \varphi d\mu = 0$ and $\varphi \neq \psi \circ T - \psi$. Then*

$$\frac{1}{\sqrt{n}} \sum_{i=0}^{n-1} \varphi \circ T^i \xrightarrow{\text{distribution}} \mathcal{N}(0, \sigma)$$

where $\mathcal{N}(0, \sigma)$ is the normal distribution and $\sigma > 0$ is given by

$$\sigma = \lim_{n \to \infty} \left[\frac{1}{n} \int \left(\sum_{i=0}^{n-1} \varphi \circ T^i \right)^2 d\mu \right]^{1/2}.$$

4 Final remarks

The proofs of Theorems 4, 5, and 6 involve technical estimates specific to the Hénon maps, but I would like to point out that the ideas behind them are not model-specific and may be quite general.

Very roughly speaking, the existence and mixing properties of SRB measures seem to be related to the rates at which arbitrarily small pieces of unstable manifolds grow to a fixed size (which is more than just the existence of a positive Lyapunov exponent pointwise). To formulate something more precisely, one could look for a set with the properties of Δ_0 in the last section, and study the return time function R. If R is integrable with respect to Lebesgue measure on W^u-leaves, then an SRB measure exists; and if, in addition to that, R has an exponentially decaying tail estimate as in Section 3, then the system has the exponential mixing property provided all powers of the map are ergodic.

In general, I doubt that it is possible to determine the nature of R from the overall appearance of a dynamical system. If, however, there is a recognizable "bad set" — in the sense that away from this set the map is uniformly hyperbolic (with no discontinuities), and when an orbit gets near it there is a quantifiable loss in hyperbolicity followed by a "recovery period" — then, as observed in [BY2], there are often natural candidates for Δ_0, and the character of the return time function R is directly related to the rate of recovery after each encounter with the "bad set". In particular, if the recovery is "exponential" (meaning it takes $\sim \log \frac{1}{\delta}$ iterates to recover fully from a loss $\sim \delta$) then R has an exponentially decaying tail estimate.

Obvious examples that fit into this "bad set-recovery" scenario are large classes of piecewise uniformly hyperbolic maps, including certain billiards, where the "bad set" is the set of singularity curves (see also the recent preprint [Li]), and quadratic maps of the interval whose critical orbits carry positive Lyapunov exponents (see Section 3). It is less obvious a priori that the Hénon maps fit into this category; indeed the various notions there have to be interpreted with a bit more care. The rate of recovery is exponential in these examples, but not in e.g. [HY].

It is certainly not the case that all nonuniformly hyperbolic systems have recognizable "bad sets", nor am I suggesting a generic theorem that can be applied to all "bad set-recovery" type scenarios. I wish only to point out that many of the known nonuniform examples belong in this category, and I hope that the methods discussed here will shed some light on the ergodic properties of systems with these characteristics.

References

[BC1]   Benedicks, M., and Carleson, L., *On iterations of* $1 - ax^2$ *on* $(-1, 1)$, Ann. of Math. (2) **122** (1985) 1–25.

[BC2]   _____, *The dynamics of the Hénon map*, Ann. of Math. (2) **133** (1991) 73–169.

[BY1]   Benedicks, M., and Young, L.-S., *SBR measures for certain Hénon maps*, Invent. Math. **112** (1993) 541–576.

[BY2]   _____, *Decay of correlations for certain Hénon maps*, in preparation.

[BR]   Bowen, R., and Ruelle, D., *The ergodic theory of Axiom A flows*, Invent. Math. **29** (1975) 181–202.

[CE]   Collet, P., and Eckmann, J.-P., *On the abundance of aperiodic behavior for maps on the interval*, Comm. Math. Phys. **73** (1980), 115–160.

[HY]   Hu, H., and Young, L.-S., *Nonexistence of SBR measures for some systems that are "almost Anosov"*, Erg. Th. & Dynam. Sys. **15** (1995), 67–76.

[J]   Jacobson, M., *Absolutely continuous invariant measures for one-parameter families of one-dimensional maps*, Comm. Math. Phys. **81** (1981) 39–88.

[L1]   Ledrappier, F., *Propriétés ergodiques des mesures de Sinai*, Publ. Math. IHES **59** (1984) 163–188.

[L2]   _____, *Dimension of invariant measures*, Teubner-Texte Math. **94**, Teubner, Leipzig (1987) 116–124.

[LS]   Ledrappier, F., and Strelcyn, J.-M., *A proof of the estimation from below in Pesin entropy formula*, Erg. Th. & Dynam. Sys. **2** (1982) 203–219.

[LY1]   Ledrappier, F., and Young, L.-S., *The metric entropy of diffeomorphisms, Part I: Characterization of measures classifying Pesin's entropy formula*, Ann. of Math. (2) **122** (1985) 509–574.

[LY2]   _____, *The metric entropy of diffeomorphisms, Part II: Relations between entropy, exponents and dimension*, Ann. of Math. (2) **122** (1985) 540–574.

[Li]   Liverani, C., *Decay of correlations*, preprint.

[MV]   Mora, L., and Viana, M., *Abundance of strange attractors*, to appear in Acta Math.

[N]   Newhouse, S., *Lectures on dynamical systems*, Progr. Math. **8**, Birkhäuser, Basel and Boston (1980) 1–114.

[P]   Pesin, Ya. B., *Characteristic Lyapunov exponents and smooth ergodic theory*, Russian Math. Surveys **32** (1977) 55–114.

[PS]   Pugh, C., and Shub, M., *Ergodic attractors*, Trans. Amer. Math. Soc., **312** no. 1 (1989) 1–54.

[R1]   Ruelle, D., *A measure associated with Axiom A attractors*, Amer. J. Math. **98** (1976) 619–654.

[R2]   _____, *An inequality of the entropy of differentiable maps*, Bol. Soc. Brasil Mat. **9** (1978) 83–87.

[S]   Sinai, Ya. G., *Gibbs measures in ergodic theory*, Russian Math. Surveys **27**, no. 4 (1972) 21–69.

[Y1]   Young, L.-S., *Dimension, entropy and Lyapunov exponents*, Erg. Th. & Dynam. Sys. **2** (1982) 109–129.

[Y2]   _____, *Decay of correlations for certain quadratic maps*, Comm. Math. Phys. **146** (1992) 123–138.

Noncommutative Geometry and Quantum Hall Effect

JEAN BELLISSARD

Université Paul Sabatier, URA 505
CNRS and Laboratoire de Physique Quantique
118, Route de Narbonne, 31062-Toulouse Cedex, France

ABSTRACT. A mathematical framework based on noncommutative geometry is proposed to describe the Integer Quantum Hall Effect (IQHE). It takes localization effects into account. It permits us to prove rigorously that the Hall conductivity is quantized and that plateaus occur when the Fermi energy varies in a region of localized states.

1. Introduction

In 1880, Hall [14] undertook the classical experiment that led to the so-called Hall effect. A century later, von Klitzing and his co-workers [17] showed that the Hall conductivity was quantized at very low temperatures as an integer multiple of the universal constant e^2/h. Here e is the electron charge and h is Planck's constant. This is the Integer Quantum Hall Effect (IQHE). This discovery led to a new accurate measurement of the fine structure constant and a new definition of the standard of resistance [21].

On the other hand, during the 1970s, A. Connes [8],[10] extended most of the tools of differential geometry to noncommutative C^*-algebras, thus creating a new branch of mathematics called *noncommutative geometry*. The main new result obtained in this field was the definition of cyclic cohomology and the proof of an index theorem for elliptic operators on a foliated manifold. He recently extended this theory to what is now called *quantum calculus* [11].

After the works by Laughlin [19] and especially by Kohmoto, den Nijs, Nightingale, and Thouless [23] (called TKN_2 below), it became clear that the quantization of the Hall conductance at low temperature had a geometric origin. The universality of this effect had then an explanation. Moreover, as proposed by Prange [20], [16], Thouless [22] and Halperin [15], the Hall conductance plateaus, appearing while changing the magnetic field or the charge-carrier density, are due to localization. Neither the original Laughlin paper nor the TKN_2 one, however, could give a description of both properties in the same model. Developing a mathematical framework able to reconcile topological and localization properties at once was a challenging problem. Attempts were made by Avron et al. [1] who exhibited quantization but were not able to prove that these quantum numbers were insensitive to disorder. In 1986, Kunz [18] went further and managed to prove this for disorder small enough to avoid filling the gaps between Landau levels.

Proceedings of the International Congress
of Mathematicians, Zürich, Switzerland 1994
© Birkhäuser Verlag, Basel, Switzerland 1995

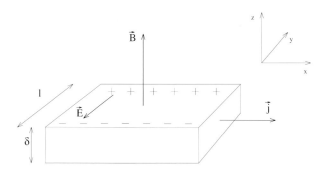

Figure 1: *The classical Hall effect: The sample is a thin metallic plate of width δ. The magnetic field is uniform and perpendicular to the plate. The current density $\vec{j}$ parallel to the x axis is stationary. The magnetic field pushes the charges as indicated creating the electric field $\vec{\mathcal{E}}$ along the y direction. The Hall voltage is measured between opposite sides along the y axis.*

But in [3], [5], [4], we proposed to use noncommutative geometry to extend the TKN_2 argument to the case of arbitrary magnetic field and disordered crystal. It turned out that the condition under which plateaus occur was precisely the finiteness of the localization length near the Fermi level. This work was rephrased later on by Avron et al. [2] in terms of charge transport and relative index, filling the remaining gap between experimental observations, theoretical intuition, and mathematical frame.

It is our aim in this talk to describe the main steps of this construction. The reader interested by details of the physical phenomena or of the mathematical proofs is kindly invited to look into the recent work [7].

2. IQHE: experiments and theories

Let us consider a very flat conductor, considered as two-dimensional, placed in a constant uniform magnetic field B in the z direction perpendicular to the plane Oxy of the plate (see Figure 1). If we force a constant current $\vec{j}$ in the x direction, the electron fluid will be submitted to the Lorentz force perpendicular to the current and the magnetic field creating an electric field $\vec{\mathcal{E}}$ along the y axis. In a stationary state, writing that the total force acting on the charge vanishes leads to the relation $\vec{j} = \boldsymbol{\sigma}\vec{\mathcal{E}}$ with a (2×2) antidiagonal antisymmetric matrix with matrix element $\pm\sigma_H$ given by

$$\sigma_H = \nu\frac{e^2}{h} , \qquad \nu = \frac{nh}{eB} ,$$

where n is the two-dimensional density of charge carriers, h is Planck's constant, e is the electron charge, and ν is called the *filling factor*. We remark that the sign of σ_H depends upon the sign of the carrier charge. In particular, the orientation of the Hall field will change when passing from electrons to holes. This observation is commonly used nowadays to determine which kind of particles carry the current.

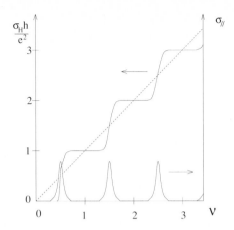

Figure 2: *Schematic representation of the experimental observations in the IQHE. The Hall conductivity σ_H is drawn in units of e^2/h versus filling factor ν. The dashed line shows the Hall conductivity of the Landau Hamiltonian without disorder. The direct conductivity $\sigma_{//}$ is shown in arbitrary units.*

The quantity $R_H = h/e^2$ is called the Hall resistance. It is a universal constant with value $R_H = 25812.80\Omega$. R_H can be measured directly with an accuracy better than 10^{-8} in QHE experiments. Since January 1990, this has been the new standard of resistance at the National Bureau of Standards [21].

Lowering the temperature below $1K$ leads to the observation of plateaus for integer values of the Hall conductance (see Figure 2). The accuracy of the Hall conductance on the plateaus is better than 10^{-8}. For values of the filling factor corresponding to the plateaus, the direct conductivity $\sigma_{//}$, namely the conductivity along the current density axis, vanishes: the sample becomes insulating. To summarize:

(i) At very low temperatures, in the limit of large sample size, and provided the system can be considered as two-dimensional, Hall plateaus appear at integer values of the Hall conductance in units of the inverse Hall resistance.

(ii) On plateaus the sample is an insulator. This is due to disorder in the sample, which produces the localization of charge carriers wave functions.

(iii) For the Hall plateaus with large index (namely indices ≥ 2) one can ignore the Coulomb interaction between charge carriers without too much error.

3.   The Kubo-Chern formula

Because we can ignore Coulomb interactions between particles, the fermion fluid made of the charge carriers is entirely described by the one-particle theory. The quantum motion can be derived from the data of a self-adjoint operator called the Hamiltonian of the system. A typical example of a one-particle Hamiltonian involved in the QHE for spinless particles is given by

$$H_\omega = \frac{(\vec{P} + e\vec{A})^2}{2m_*} + V_\omega(\vec{x}) , \tag{1}$$

where $\vec{P}$ is the 2D momentum operator and m_* is the effective mass of the particle, $\vec{A} = (A_1, A_2)$ is the vector potential given by the magnetic field, and $V_\omega(\vec{x})$ describes the potential created by disorder in the plate. Here ω, which denotes the configuration of disorder, can be seen as a point in a compact metrizable Hausdorff space Ω on which the translation group $\mathbf{R}^2$ acts by homeomorphisms. Then the covariance condition $V_\omega(\vec{x} - \vec{a}) = V_{T^{\vec{a}}\omega}(\vec{x})$ expresses the fact that moving the sample and changing the reference axis backward are equivalent.

Such a model is typical but may be replaced by others, such as lattice approximants, or the particle with spin model. In any case, the one-particle Hamiltonian describing the fermion fluid satisfies the following general properties:

(i) The translation group $\mathcal{G}$ acting on the sample is $\mathbf{R}^2$ or $\mathbf{Z}^2$. It acts by homeomorphisms $T^a, a \in \mathcal{G}$, on the space Ω of the disorder configurations. It also acts by unitary projective representation $T(a), a \in \mathcal{G}$, on the one-particle Hilbert space $\mathcal{H}$.

(ii) The one-particle Hamiltonian is a norm resolvent strongly continuous family $(H_\omega)_{\omega \in \Omega}$ of self-adjoint operators on $\mathcal{H}$, bounded from below and satisfying the covariance condition $T(a)H_\omega T(a)^{-1} = H_{T^a\omega}$.

Such a Hamiltonian actually generalizes the case of a periodic operator, namely the case for which there is a sufficiently large discrete subgroup of $\mathcal{G}$ leaving the Hamiltonian invariant. In this latter case, the Bloch theorem permits us to describe the quantum motion in term of quasi-momenta k belonging to the so-called *Brillouin zone*, which is a manifold diffeomorphic to a torus. Both magnetic field and disorder break this translation symmetry in a non-trivial way, so that the notion of Brillouin zone becomes meaningless in the classical sense. Actually it is still possible to describe such a manifold in terms of noncommutative geometry by replacing the algebra of continuous functions over the Brillouin zone by a noncommutative C^*-algebra. In our case this C^*-algebra is nothing but the one generated by the bounded functions of the H_ω's, $\omega \in \Omega$. It turns out that it is a closed subalgebra of the twist crossed product $\mathcal{A} = C^*(\Omega, \mathcal{G}, B) = \mathcal{C}(\Omega) \times_B \mathcal{G}$ [6] where the product is twisted by a module defined by the magnetic field. A differential and integral calculus exists on such an algebra, making it a noncommutative differential manifold that we have proposed to call the *noncommutative Brillouin zone*. More precisely in our two-dimensional situation, we can define two derivations ∂_i ($i = 1, 2$) by using the position operators $X_i, i = 1, 2$, as

$$(\partial_i A)_\omega = \imath[X_i, A_\omega] , \qquad A \in \mathcal{A} .$$

Thus, $\mathcal{C}^1$ elements of $\mathcal{A}$ are well defined.

The integral depends upon the choice of a $\mathcal{G}$-invariant ergodic probability $\mathbf{P}$ on Ω. It is then given by the *trace per unit area* $\mathcal{T}_P$, namely

$$\mathcal{T}_P(A) = \lim_{\Lambda \uparrow \mathcal{G}} \frac{1}{|\Lambda|} \mathrm{Tr}_\Lambda(A_\omega) , \qquad A \in \mathcal{A} , \text{ for } \mathbf{P}\text{-almost all } \omega\text{'s}, \qquad (2)$$

where Λ denotes a sequence of squares in $\mathcal{G}$ centered at the origin and covering $\mathcal{G}$ and Tr_Λ is the restriction to Λ of the usual trace.

Because of the Fermi statistics obeyed by the charge carriers (electrons or holes), two different particles of the fluid must occupy different quantum eigenstates of the Hamiltonian H_ω. In the limit of zero temperature they occupy the levels of lowest energy, namely all eigenstates with energy lower than some maximal one E_F called the *Fermi level*. We will denote by $P_{F,\omega}$ the corresponding eigenprojection of the Hamiltonian.

Standard results in transport theory permit us to compute the conductivity in terms of the linear response of the fermion fluid under the influence of an external field. This is the famous *Green-Kubo* formula. In the QHE-limit, namely in the limit of (i) zero temperature, (ii) infinite sample size, (iii) negligible collision processes, and (iv) vanishingly small electric fields, the direct conductivity either vanishes or is infinite, whereas the transverse conductivity, when defined, is given by

$$\sigma_H = \frac{e^2}{h} \mathbf{Ch}(P_F) = \frac{e^2}{h} 2\iota\pi \ \mathcal{T}_P(P_F[\partial_1 P_F, \partial_2 P_F]) . \qquad (3)$$

It turns out that $\mathbf{Ch}$ is nothing but the noncommutative analog of a Chern character. Thus, Kubo's formula gives rise to a Chern character in the QHE limit. This is why we propose to call eq. (3) the *Kubo-Chern formula*, associating Japan with China.

The main properties of the noncommutative Chern character are the following.

(i) *Homotopy invariance*: Given two equivalent $\mathcal{C}^1$ projections P and Q in $\mathcal{A}$, such that there is $U \in \mathcal{C}^1(\mathcal{A})$ with $P = U^*U$ and $Q = UU^*$, then $\mathbf{Ch}(P) = \mathbf{Ch}(Q)$. This is actually what happens if P and Q are homotopic in $\mathcal{C}^1(\mathcal{A})$.

(ii) *Additivity*: Given two $\mathcal{C}^1$ orthogonal projections P and Q in $\mathcal{A}$, such that $PQ = QP = 0$, then $\mathbf{Ch}(P \oplus Q) = \mathbf{Ch}(P) + \mathbf{Ch}(Q)$.

In particular, the homotopy invariance shows that $\mathbf{Ch}(P_F)$ is a topological quantum number, if $P_F \in \mathcal{C}^1(\mathcal{A})$. One of the main results of noncommutative geometry is that this Chern character is an integer. Thus, thanks to eq. (3) we get the Hall conductance quantization. Unfortunately, $P_F \notin \mathcal{C}^1(\mathcal{A})$ in general, unless the Fermi level belongs to a spectral gap. We will see, however, in Section 5. below, that this Chern character is still well defined and quantized precisely whenever the Fermi level lies in a region of localized states. Moreover, changing the value of the filling factor produces the moving of the Fermi level, which does not change the Chern character as long as the localization length stays bounded.

4.   The four traces way

In this section we use four different traces that are technically needed to express the complete results of this theory. The first one is the usual trace on matrices or on trace-class operators. The second one, introduced in Section 3. above, is the *trace per unit volume*. The third one is the *graded trace* or *supertrace* introduced in this section below. This is the first technical tool proposed by Connes [8] to define the cyclic cohomology and constitutes the first important step in proving quantization of the Hall conductance [4]. The last one is the *Dixmier trace* defined by Dixmier in 1964 [12] and of which the importance for quantum differential calculus was emphasized by Connes [9], [10], [11]. It will be used in connection with Anderson's localization.

Let $\mathcal{H}$ be the physical one-particle Hilbert space of Section 3.. We then build the new Hilbert space $\hat{\mathcal{H}} = \mathcal{H}_+ \oplus \mathcal{H}_-$ with $\mathcal{H}_\pm = \mathcal{H}$. The grading operator G and the "Hilbert transform" F are defined as follows:

$$G = \begin{pmatrix} +1 & 0 \\ 0 & -1 \end{pmatrix} , \qquad\qquad F = \begin{pmatrix} 0 & \frac{X}{|X|} \\ \frac{\overline{X}}{|X|} & 0 \end{pmatrix} , \qquad (4)$$

where $X = X_1 + \imath X_2$ (here the dimension is $D = 2$). It is clear that F is self-adjoint and satisfies $F^2 = \mathbf{1}$. An operator T on $\hat{\mathcal{H}}$ is said to be of degree 0 if it commutes with G and of degree 1 if it anticommutes with G. The graded commutator (or supercommutator) of two operators and the graded differential dT are defined by

$$[T, T']_S = TT' - (-)^{\deg(T)\deg(T')} T'T , \qquad\qquad dT = [F, T]_S .$$

Then, $d^2 T = 0$. The graded trace Tr_S (or supertrace) is defined by

$$\mathrm{Tr}_S(T) = \frac{1}{2}\mathrm{Tr}_{\hat{\mathcal{H}}}(GF[F, T]_S) = \mathrm{Tr}_{\mathcal{H}}(T_{++} - uT_{--}u^*) , \qquad (5)$$

where $u = X/|X|$ and T_{++} and T_{--} are the diagonal components of T with respect to the decomposition of $\hat{\mathcal{H}}$. It is a linear map on the algebra of operators such that $\mathrm{Tr}_S(TT') = \mathrm{Tr}_S(T'T)$. However, this trace is not positive. Observables in $\mathcal{A}$ will become operators of degree 0, namely $A \in \mathcal{A}$ will be represented by $\hat{A}_\omega = A_\omega \oplus A_\omega$.

Given a Hilbert space $\mathcal{H}$, the nth characteristic value μ_n of a compact operator T is the distance in norm of T to the set of operators of rank at most n. Equivalently, it is the nth eigenvalue of $|T| = (TT^*)^{1/2}$ labeled in decreasing order. The Mačaev ideals $\mathcal{L}^{p+}(\mathcal{H})$ are the set of compact operators on $\mathcal{H}$ with characteristic values satisfying

$$\|T\|_{p+} = \sup_{N \to \infty} \frac{1}{\ln N} \sum_{n=1}^{N} \mu_n^p < \infty .$$

Let Lim be a positive linear functional on the space of bounded sequences $l_+^\infty(\mathbf{N})$ of positive real numbers that is translation and scale invariant. For $T \in \mathcal{L}^{1+}(\mathcal{H})$ its Dixmier trace is defined by

$$\mathrm{Tr}_{\mathrm{Dix}}(T) = \mathrm{Lim}(\frac{1}{\ln N} \sum_{n=1}^{N} \mu_n) .$$

Note that $T \in \mathcal{L}^{1+}$ if and only if $\mathrm{Tr}_{\mathrm{Dix}}(|T|) < \infty$. Moreover, if the sequence $(\frac{1}{\ln N} \sum_{n=1}^{N} \mu_n)$ converges, then all functionals Lim of the sequence are equal to the limit and the Dixmier trace is given by this limit. From this definition, one can show that $\mathrm{Tr}_{\mathrm{Dix}}$ is a trace [12], [11].

The first important result is provided by a formula that was suggested by a result of Connes [9]. Namely, if $A \in \mathcal{C}^1(\mathcal{A})$ and if $\vec{\nabla} = (\partial_1, \partial_2)$ we have [7]:

$$\mathcal{T}_P(|\vec{\nabla} A|^2) = \frac{1}{\pi} \mathrm{Tr}_{\mathrm{Dix}}(|dA_\omega|^2) , \qquad \text{for } \mathbf{P}\text{-almost all } \omega . \tag{6}$$

Let now $\mathcal{S}$ denote the closure of $\mathcal{C}^1(\mathcal{A})$ under the noncommutative *Sobolev norm* $\|A\|_{\mathcal{S}}^2 = \mathcal{T}_P(A^*A) + \mathcal{T}_P(\vec{\nabla} A^* \vec{\nabla} A)$. Eq. (6) shows that for any element $A \in \mathcal{S}$, dA_ω belongs to $\mathcal{L}^{2+}(\hat{\mathcal{H}})$ $\mathbf{P}$-almost surely.

The following formula, valid for $A_0, A_1, A_2 \in \mathcal{C}^1(\mathcal{A})$, is the next important result proved in [8], [4], [2], [7]:

$$\int_\Omega d\mathbf{P}(\omega) \mathrm{Tr}_S(\hat{A}_{0,\omega} d\hat{A}_{1,\omega} d\hat{A}_{2,\omega}) = 2\imath\pi \mathcal{T}_P(A_0 \partial_1 A_1 \partial_2 A_2 - A_0 \partial_2 A_1 \partial_1 A_2) . \tag{7}$$

Thanks to eq. (6) this formula extends to $A_i \in \mathcal{S}$.

Applying these formulæ to the Fermi projection, the Chern character $\mathbf{Ch}(P_F)$ is well defined provided $P_F \in \mathcal{S}$ and

$$\mathbf{Ch}(P_F) = \int_\Omega d\mathbf{P}(\omega) \mathrm{Tr}_S(\hat{P}_{F,\omega} d\hat{P}_{F,\omega} d\hat{P}_{F,\omega}) . \tag{8}$$

The last step is a consequence of the Fedosov formula [13]; namely, the operator $P_\omega F^{+-}|_{P_\omega \mathcal{H}_-}$ is Fredholm and its index is an integer given by:

$$n(\omega) = \mathrm{Ind}(P_\omega F^{+-}|_{P_\omega \mathcal{H}_-}) = \mathrm{Tr}_S(\hat{P}_{F,\omega} d\hat{P}_{F,\omega} d\hat{P}_{F,\omega}) . \tag{9}$$

It remains to show that this index is $\mathbf{P}$-almost surely constant. By the covariance condition $P_{T^a\omega} F^{+-}|_{P_{T^a\omega}\mathcal{H}_-}$ and $P_\omega T(a)^{-1} F^{+-} T(a)|_{P_\omega \mathcal{H}_-}$ are unitarily equivalent, so that they have the same Fredholm index. Moreover, $P_\omega T(a)^{-1} F^{+-} T(a)$ $|_{P_\omega \mathcal{H}_-} - P_\omega F^{+-}|_{P_\omega \mathcal{H}_-}$ is easily seen to be compact so that $P_{T^a\omega} F^{+-}|_{P_{T^a\omega}\mathcal{H}_-}$ have the same index as $P_\omega F^{+-}|_{P_\omega \mathcal{H}_-}$. In other words, $n(\omega)$ is a $\mathcal{G}$-invariant function of ω. The probability $\mathbf{P}$ being $\mathcal{G}$-invariant and ergodic, $n(\omega)$ is $\mathbf{P}$-almost surely constant. Consequently, as $F^{+-} = u$, if $P_F \in \mathcal{S}$:

$$\mathbf{Ch}(P_F) = \mathrm{Ind}(P_{F,\omega} u|_{P_{F,\omega}\mathcal{H}}) \in \mathbf{Z} , \qquad \mathbf{P}\text{-almost surely} .$$

5. Localization

It remains to show how the condition $P_F \in \mathcal{S}$ is related to the Anderson localization. The easiest way to define the localization length consists of measuring the averaged square displacement of a wave packet on the long run. Let Δ

be an interval. We denote by P_Δ the eigenprojection of the Hamiltonian corresponding to energies in Δ. Then, if X is the position operator in $\mathcal{G}$, we set $X_{\Delta,\omega}(t) = e^{\frac{i}{\hbar}H_\omega t} P_{\Delta,\omega} X P_{\Delta,\omega} e^{-\frac{i}{\hbar}H_\omega t}$. Then we define the Δ-localization length as:

$$l^2(\Delta) = \limsup_{T\to\infty} \int_0^T \frac{dt}{T} \int_\Omega d\mathbf{P}(\omega) < 0|(X_{\Delta,\omega}(t) - X_{\Delta,\omega}(0))^2|0 > .$$

In [7] we have shown that equivalently

$$l^2(\Delta) = \limsup_{T\to\infty} \int_0^T \frac{dt}{T} \mathcal{T}_P(|\vec{\nabla}(e^{-\frac{i}{\hbar}Ht} P_\Delta)|^2) = \sup_{\mathcal{P}} \sum_{\Delta'\in\mathcal{P}} \mathcal{T}_P(|\vec{\nabla}P_{\Delta'}|^2) , \qquad (10)$$

where $\mathcal{P}$ runs in the set of finite partitions of Δ by Borel subsets. Moreover, we have also shown [7] that $l^2(\Delta) < \infty$ *implies that the spectrum of H_ω is pure point in Δ, $\mathbf{P}$-almost surely.*

The density of states is the positive measure $\mathcal{N}$ on $\mathbf{R}$ defined, for f a continuous function with compact support, by $\int_{-\infty}^{+\infty} d\mathcal{N}(E)f(E) = \mathcal{T}_P(f(H))$. It turns out [7] that if $l^2(\Delta) < \infty$ one can find a positive $\mathcal{N}$-square integrable function l on Δ such that

$$l^2(\Delta') = \int_{\Delta'} d\mathcal{N}(E)\, l(E)^2 , \qquad (11)$$

for any subinterval Δ' of Δ. We propose to call $l(E)$ the *localization length* at energy E.

We can now conclude. Thanks to eq. (10) the finiteness of the localization length in the interval Δ implies that [7]

(i) $P_F \in \mathcal{S}$ whenever the Fermi level E_F lies in Δ,

(ii) $E_F \in \Delta \mapsto P_F \in \mathcal{S}$ is continuous (for the Sobolev norm) at every regularity point of $\mathcal{N}$,

(iii) $\mathbf{Ch}(P_F)$ is constant on Δ, leading to the existence of plateaus for the transverse conductivity,

(iv) If the Hamiltonian is changed continuously (in the norm resolvent topology), $\mathbf{Ch}(P_F)$ stays constant as long as the localization length remains finite at the Fermi level.

As a Corollary, we notice that between two Hall plateaus with different indices, the localization length must diverge [15], [18]. The reader will find in [7] how to compute practically the Hall index using homotopy (property (iv)) and explicit calculation for simple models.

Acknowledgments. This work has benefited from many contacts during the last ten years. It is almost impossible to give the list of all colleagues who contributed to these discussions. I want to address my special thanks to Alain Connes for his outstanding contribution, continuous support and warm encouragements. Also I want to acknowledge Y. Avron, R. Seiler, and B. Simon for their constant interest in shading light on this difficult subject. Let me thank my recent and young collaborators Hermann Schulz-Baldes and Andreas van Elst, whose enthusiasm and competence permitted me to write the main report, which is the basis of this work.

References

[1] J. E. Avron, R. Seiler, and B. Simon, *Charge deficiency, charge transport and comparison of dimensions*, Comm. Math. Phys. **399** (1994).

[2] J. E. Avron, R. Seiler, and B. Simon, Phys. Rev. Lett **51**, 51 (1983).

[3] J. Bellissard, in Statistical Mechanics and Field Theory: Mathematical Aspects, Lecture Notes in Phys. **257**, edited by T. Dorlas, M. Hugenholtz, and M. Winnink (Springer-Verlag, Berlin and New York, 1986).

[4] J. Bellissard, in Proc. of the Bad Schandau Conference on Localization, edited by Ziesche and Weller (Teubner-Verlag, Leipzig, 1987).

[5] J. Bellissard, in Operator Algebras and Applications, vol. 2, edited by E. Evans and M. Takesaki (Cambridge University Press, Cambridge, 1988).

[6] J. Bellissard, in From Number Theory to Physics, edited by M. Waldschmidt, P. Moussa, J. Luck, and C. Itzykson (Springer-Verlag, Berlin and New York, 1991).

[7] J. Bellissard, A. van Elst, and H. Schulz-Baldes, *The non-commutative geometry of the quantum Hall effect*, J. Math. Phys. **35** (1994), 5373–5451.

[8] A. Connes, *Non-commutative differential geometry*, Publ. IHES **62**, 257 (1986).

[9] A. Connes, *The action functional in non-commutative geometry*, Commun. Math. Phys. **117**, 673 (1988).

[10] A. Connes, Géométrie non commutative (InterEditons, Paris, 1990)

[11] A. Connes, Non-commutative Geometry, Acad. Press, San Diego (1994).

[12] J. Dixmier, *Existence de traces non normales*, C. R. Acad. Sci., 1107 (1966).

[13] B. Fedosov, Functional Anal. Appl. **4**, 339 (1967).

[14] E. Hall, *On a new action of the magnet on electric currents*, Amer. J. Math. **2**, 287 (1879) and in Quantum Hall Effect: A Perspective, edited by A. Mac Donald (Kluwer Academic Publishers, Dordrecht, 1989).

[15] B. I. Halperin, *Quantized Hall conductance, current-carrying edge states and the existence of extended states in a two-dimensional disordered potential*, Phys. Rev. B **25**, 2185 (1982).

[16] R. Joynt and R. Prange, *Conditions for the quantum Hall effect*, Phys. Rev. B **29**, 3303 (1984).

[17] K. v. Klitzing, G. Dorda, and M. Pepper, *New method for high accuracy determination of the fine structure constant based on quantized Hall resistance*, Phys. Rev. Lett. **45**, 494 (1980).

[18] H. Kunz, *The quantized Hall effect for electrons in a random potential*, Comm. Math. Phys. **112**, 121 (1987).

[19] R. B. Laughlin, *Quantized Hall conductivity in two-dimension*, Phys. Rev. B **23**, 5632 (1981).

[20] R. E. Prange, *Quantized Hall resistance and the measurement of the fine structure constant*, Phys. Rev. B **23**, 4802 (1981).

[21] M. Stone, ed., The Quantum Hall Effect (World Scientific, Singapore, 1992).

[22] D. J. Thouless, J. Phys. C **14**, 3475 (1981).

[23] D. Thouless, M. Kohmoto, M. Nightingale, and M. den Nijs, *Quantized Hall conductance in two-dimensional periodic potential*, Phys. Rev. Lett. **49**, 405 (1982).

Conformal Field Theory and Integrable Systems Associated to Elliptic Curves

GIOVANNI FELDER[*]

Forschungsinstitut für Mathematik, ETH-Zentrum, CH-8092 Zürich, Switzerland	and	Department of Mathematics, University of North Carolina at Chapel Hill, Chapel Hill, NC 27599-3250, USA[†]

1. Introduction

It has become clear over the years that quantum groups (i.e., quasitriangular Hopf algebras, see [D]) and their semiclassical counterpart, Poisson Lie groups, are an essential algebraic structure underlying three related subjects: integrable models of statistical mechanics, conformal field theory, and integrable models of quantum field theory in 1+1 dimensions. Still, some points remain obscure from the point of view of Hopf algebras. In particular, integrable models associated with elliptic curves are still poorly understood. We propose here an elliptic version of quantum groups, based on the relation to conformal field theory, which hopefully will be helpful to complete the picture.

But before going to the elliptic case, let us review the relations between the three subjects in the simpler rational and trigonometric cases.

In integrable models of statistical mechanics (see [B],[F]), the basic object is an R-matrix, i.e., a meromorphic function of a spectral parameter $z \in \mathbb{C}$ with values in $\mathrm{End}(V \otimes V)$ for some vector space V, obeying the Yang-Baxter equation

$$R^{(12)}(z)R^{(13)}(z+w)R^{(23)}(w) = R^{(23)}(w)R^{(13)}(z+w)R^{(12)}(z),$$

in $\mathrm{End}(V \otimes V \otimes V)$. The notation is customary in this subject: $X^{(j)} \in \mathrm{End}(V \otimes \cdots \otimes V)$, for $X \in \mathrm{End}(V)$, means $\mathrm{Id} \otimes \cdots \otimes \mathrm{Id} \otimes X \otimes \mathrm{Id} \otimes \cdots \otimes \mathrm{Id}$, with X at the jth place, and if $R = \Sigma X_\nu \otimes Y_\nu$, $R^{(ij)} = \Sigma X_\nu^{(i)} Y_\nu^{(j)}$.

The Yang-Baxter equation implies the commutativity of infinitely many transfer matrices constructed out of R. Rational and trigonometric solutions of the Yang-Baxter equation appear naturally in the theory of quasitriangular Hopf algebras.

If R depends on a parameter $\hbar$ so that $R = \mathrm{Id} + \hbar r + O(\hbar^2)$, as $\hbar \to 0$, then the "classical r-matrix" r obeys the classical Yang-Baxter equation

$$[r^{(12)}(z), r^{(13)}(z+w) + r^{(23)}(w)] + [r^{(13)}(z+w), r^{(23)}(w)] = 0. \qquad (1)$$

This equation appears in the theory of Poisson-Lie groups, but has the following relation with conformal field theory. In the skew-symmetric case $r(z) = -r^{(21)}(-z)$,

---

[*]Supported in part by NSF grant DMS-9400841.   [†] Permanent address.

Proceedings of the International Congress
of Mathematicians, Zürich, Switzerland 1994
© Birkhäuser Verlag, Basel, Switzerland 1995

it is the compatibility condition for the system of equations

$$\partial_{z_i} u = \sum_{j:j \neq i} r^{(ij)}(z_i - z_j) u \qquad (2)$$

for a function $u(z_1, \ldots, z_n)$ on $\mathbb{C}^n - \cup_{i<j}\{z|z_i = z_j\}$, with values in $V \otimes \cdots \otimes V$. In the rational case, very simple skew-symmetric solutions are known: $r(z) = C/z$, where $C \in \mathfrak{g} \otimes \mathfrak{g}$ is a symmetric invariant tensor of a finite-dimensional Lie algebra $\mathfrak{g}$ acting on a representation space V. The corresponding system of differential equations is the Knizhnik-Zamolodchikov (KZ) equation for conformal blocks of the Wess-Zumino-Witten model of conformal field theory on the sphere. Solutions of (1) in $\mathfrak{g} \otimes \mathfrak{g}$, for simple Lie algebras $\mathfrak{g}$, were partially classified by Belavin and Drinfeld [BD], who in particular proved that, under a nondegeneracy assumption, solutions can be divided into three classes, according, say, to the lattice of poles: rational, trigonometric, and elliptic. Elliptic solutions are completely classified and exist only for sl_N.

Recently, Frenkel and Reshetikhin [FR] considered the "quantization" of the Knizhnik-Zamolodchikov equations based on the representation theory of Yangians (rational case) and affine quantum enveloping algebras (trigonometric case). These equations form a compatible system of *difference* equations that as $\hbar \to 0$, reduce to the differential equations (2). In an important special case, these difference equations had been introduced earlier by Smirnov [S] who derived them as equations for "form factors" in integrable quantum field theory, and gave relevant solutions. In the quantum field theory setting R has the interpretation of a two-particle scattering matrix, and is required to obey the "unitarity" relation $R(z)R^{(21)}(-z) = \mathrm{Id}$, as well as a "crossing symmetry" condition.

In the elliptic case, one knows solutions of the Yang-Baxter equation whose semiclassical limits are the sl_N solutions discussed above [Bel]. The relevant algebraic structure is here the Sklyanin algebra [Sk], [Ch], which however does not fall into the general quantum group theory. Although elliptic solutions related to other Lie algebras have not been found (and they could not have a semiclassical limit by the Belavin-Drinfeld theorem), many elliptic solutions of the Star-Triangle relation, a close cousin of the Yang-Baxter equation, are known (see [JMO], [JKMO], [DJKMO]). This is somewhat mysterious, as, in the trigonometric case, solutions of both equations are in one-to-one correspondence. Another apparent puzzle we want to point out is that conformal field theory can be defined on arbitrary Riemann surfaces [TUY] whereas r-matrices exist only up to genus one, and in genus one only for sl_N.

We will start from the solution to this last puzzle to arrive, via quantization and difference equations, to a notion of elliptic quantum group, which is the algebraic structure underlying the elliptic solutions of the Star-Triangle relation.

Let us mention some other recent progress in similar directions. In [FR] solutions of the Star-Triangle relation are obtained as connection matrices for the trigonometric quantum KZ equations. Very recently, Foda et al. [FIJKMY] have proposed an elliptic quantum algebra of nonzero level, by another modification of the Yang-Baxter equation.

2. Conformal field theory, KZB equations

Our starting point is the set of genus one Knizhnik-Zamolodchikov-Bernard (KZB) equations, obtained by Bernard [B1], [B2] as a generalization of the KZ equations. These equations have been studied recently in [FG], [EK], and [FW].

Let $\mathfrak{g}$ be a simple complex Lie algebra with invariant bilinear form normalized in such a way that long roots have square length 2. Fix a Cartan subalgebra $\mathfrak{h}$. The KZB equations are equations for a function $u(z_1, \ldots, z_n, \tau, \lambda)$ with values in the weight zero subspace (the subspace killed by $\mathfrak{h}$) of a tensor product of irreducible finite-dimensional representations of $\mathfrak{g}$. The arguments $z_1, \ldots z_n, \tau$ are complex numbers with τ in the upper half plane, and the z_i are distinct modulo the lattice $\mathbb{Z} + \tau\mathbb{Z}$, and $\lambda \in \mathfrak{h}$. Let us introduce coordinates $\lambda = \Sigma\lambda_\nu h_\nu$ in terms of an orthonormal basis (h_ν) of $\mathfrak{h}$. In the formulation of [FW], the KZB equations take the form

$$\kappa\partial_{z_j}u \;=\; -\sum_\nu h_\nu^{(j)}\partial_{\lambda_\nu}u + \sum_{l:l\neq j}\Omega^{(j,l)}(z_j - z_l, \tau, \lambda)u, \tag{3}$$

$$4\pi i\kappa\partial_\tau u \;=\; \sum_\nu \partial_{\lambda_\nu}^2 u + \sum_{j,l}\mathrm{H}^{(j,l)}(z_j - z_l, \tau, \lambda)u. \tag{4}$$

Here κ is an integer parameter that is large enough depending on the representations in the tensor product and Ω, $\mathrm{H} \in \mathfrak{g}\otimes\mathfrak{g}$ are tensors preserving the weight zero subspace that we now describe. Let $\mathfrak{g} = \mathfrak{h} + \sum_{\alpha\in\Delta}\mathfrak{g}_\alpha$ be the root decomposition of $\mathfrak{g}$, and $C \in S^2\mathfrak{g}$ be the symmetric invariant tensor dual to the invariant bilinear form on $\mathfrak{g}$. Write $C = \sum_{\alpha\in\Delta\cup\{0\}}C_\alpha$, where $C_0 = \sum_\nu h_\nu\otimes h_\nu$ and $C_\alpha \in \mathfrak{g}_\alpha\otimes\mathfrak{g}_{-\alpha}$. Let $\theta_1(t, \tau)$ be Jacobi's theta function

$$\theta_1(t|\tau) = -\sum_{j=-\infty}^{\infty}e^{\pi i(j+\frac{1}{2})^2\tau + 2\pi i(j+\frac{1}{2})(t+\frac{1}{2})}$$

and introduce functions ρ, σ:

$$\rho(t) \;=\; \partial_t \log\theta_1(t|\tau),$$

$$\sigma(w,t) \;=\; \frac{\theta_1(w - t|\tau)\partial_t\theta_1(0|\tau)}{\theta_1(w|\tau)\theta_1(t|\tau)}.$$

The tensor Ω is given by

$$\Omega(z, \tau, \lambda) = \rho(z)C_0 + \sum_{\alpha\in\Delta}\sigma(\alpha(\lambda), z)C_\alpha.$$

The tensor H has a similar form. We need the following special functions of $t \in \mathbb{C}$, expressed in terms of $\sigma(w, t)$, $\rho(t)$ and Weierstrass' elliptic function $\wp$ with periods $1, \tau$.

$$I(t) \;=\; \frac{1}{2}(\rho(t)^2 - \wp(t)),$$

$$J_w(t) \;=\; \partial_t\sigma(w,t) + (\rho(t) + \rho(w))\sigma(w,t).$$

These functions are regular at $t = 0$. The tensor H is then given by the formula

$$\mathrm{H}(t, \tau, \lambda) = I(t)C_0 + \sum_{\alpha \in \Delta} J_{\alpha(\lambda)}(t)C_\alpha.$$

As shown in [FW], the functions u from conformal field theory have a special dependence on the parameter λ. For fixed z, τ, the function u, as a function of λ, belongs to a finite-dimensional space of antiinvariant theta function of level κ. Therefore, the right way of looking at these equations is to consider u as a function of $z_1, \ldots, z_n, \tau$ taking values in a finite-dimensional space of functions of λ.

The tensors have the skew-symmetry property $\Omega(z) + \Omega^{(21)}(-z) = 0$, and $\mathrm{H}(z) - \mathrm{H}^{(21)}(-z) = 0$ and commute with $X^{(1)} + X^{(2)}$ for all $X \in \mathfrak{h}$. The compatibility condition of (3) is then the *modified classical Yang-Baxter equation* [FW]

$$\sum_\nu \partial_{\lambda_\nu} \Omega^{(1,2)} h_\nu^{(3)} + \sum_\nu \partial_{\lambda_\nu} \Omega^{(2,3)} h_\nu^{(1)} + \sum_\nu \partial_{\lambda_\nu} \Omega^{(3,1)} h_\nu^{(2)}$$

$$-[\Omega^{(1,2)}, \Omega^{(1,3)}] - [\Omega^{(1,2)}, \Omega^{(2,3)}] - [\Omega^{(1,3)}, \Omega^{(2,3)}] \quad = \quad 0 \qquad (5)$$

in $\mathfrak{g} \otimes \mathfrak{g} \otimes \mathfrak{g}$. In this equation, $\Omega^{(ij)}$ is taken at $(z_i - z_j, \tau, \lambda)$. Moreover, there are relations involving H, which we do not consider here, as we will consider only the first equation (3). The quantization of (4) is an important open problem related, for $n = 1$, to the theory of elliptic Macdonald polynomials [EK].

3. The quantization

Let $\mathfrak{h}$ be the complexification of a Euclidean space $\mathfrak{h}_r$ and extend the scalar product to a bilinear form on $\mathfrak{h}$. View $\mathfrak{h}$ as an abelian Lie algebra. We consider finite-dimensional diagonalizable $\mathfrak{h}$-modules V. This means that we have a weight decomposition $V = \oplus_{\mu \in \mathfrak{h}} V[\mu]$ such that $\lambda \in \mathfrak{h}$ acts as (μ, λ) on $V[\mu]$. Let $P_\mu \in \mathrm{End}(V)$ be the projection onto $V[\mu]$.

It is convenient to introduce the following notation. Suppose $V_1, \ldots, V_n$ are finite-dimensional diagonalizable $\mathfrak{h}$-modules. If $f(\lambda)$ is a meromorphic function on $\mathfrak{h}$ with values in $\otimes_i V_i = V_1 \otimes \cdots \otimes V_n$ or $\mathrm{End}(\otimes_i V_i)$, and η_i are complex numbers, we define a function on $\mathfrak{h}$

$$f(\lambda + \sum \eta_i h^{(i)}) = \sum_{\mu_1, \ldots, \mu_n} \prod_{i=1}^n P_{\mu_i}^{(i)} f(\lambda + \Sigma \eta_i \mu_i),$$

taking values in the same space as f.

Given $\mathfrak{h}$ and V as above, the quantization of (5) is an equation for a meromorphic function R of the spectral parameter $z \in \mathbb{C}$ and an additional variable $\lambda \in \mathfrak{h}$, taking values in $\mathrm{End}(V \otimes V)$:

$$R^{(12)}(z_{12}, \lambda + \eta h^{(3)}) R^{(13)}(z_{13}, \lambda - \eta h^{(2)}) R^{(23)}(z_{23}, \lambda + \eta h^{(1)})$$

$$= R^{(23)}(z_{23}, \lambda - \eta h^{(1)}) R^{(13)}(z_{13}, \lambda + \eta h^{(2)}) R^{(12)}(z_{12}, \lambda - \eta h^{(3)}). \qquad (6)$$

The parameter η is proportional to $\hbar$, and z_{ij} stands for $z_i - z_j$. This equation forms the basis for the subsequent analysis. Let us call it the modified Yang-Baxter equation (MYBE). Note that a similar equation, without spectral parameter, has appeared for the monodromy matrices in Liouville theory, see [GN], [B], [AF]. We supplement it by the "unitarity" condition

$$R^{(12)}(z_{12}, \lambda) R^{(21)}(z_{21}, \lambda) = \mathrm{Id}_{V \otimes V}, \qquad (7)$$

and the "weight zero" condition

$$[X^{(1)} + X^{(2)}, R(z, \lambda)] = 0, \qquad \forall X \in \mathfrak{h}. \qquad (8)$$

We say that $R \in \mathrm{End}(V \otimes V)$ is a *generalized quantum R-matrix* if it obeys (6), (7), and (8).

If we have a family of solutions parametrized by η in some neighborhood of the origin, and $R(z, \lambda) = \mathrm{Id}_{V \otimes V} - 2\eta\Omega(z, \lambda) + O(\eta^2)$ has a "semiclassical asymptotic expansion", then (6) reduces to the modified classical Yang-Baxter equation (5).

Here are examples of solutions. Take $\mathfrak{h}$ to be the abelian Lie algebra of diagonal N by N complex matrices, with bilinear form Trace(AB), acting on $V = \mathbb{C}^N$. Denote by E_{ij} the N by N matrix with a one in the ith row and jth column and zeros everywhere else. Then we have

PROPOSITION 3.1 *The function*

$$R(z, \lambda) = \sum_i E_{ii} \otimes E_{ii} + \sum_{i \neq j} \frac{\sigma(\gamma, \lambda_{ij})}{\sigma(\gamma, z)} E_{ii} \otimes E_{jj} + \sum_{i \neq j} \frac{\sigma(\lambda_{ij}, z)}{\sigma(\gamma, z)} E_{ij} \otimes E_{ji}$$

is a "unitary" weight zero solution of the modified Yang-Baxter equation, i.e., it is a generalized quantum R-matrix, with $\eta = \gamma/2$.

The proof is based on comparing poles and behavior under translation of spectral parameters by $\mathbb{Z} + \tau\mathbb{Z}$ on both sides of the equation. It uses unitarity, the $\mathbb{Z}$ periodicity of R, and the transformation property

$$R(z + \tau, \lambda) = e^{-4\pi i \eta} \exp(2\pi i(\eta C_0 + \lambda^{(1)})) R(z, \lambda) \exp(2\pi i(\eta C_0 - \lambda^{(1)})), \quad (9)$$

$$C_0 = \sum E_{ii} \otimes E_{ii}. \qquad (10)$$

Two limiting cases of this solution are of interest. First, if $\tau \to i\infty$ and $\lambda_j - \lambda_l \to i\infty$, if $j < l$, we recover the well-known trigonometric R matrix connected with the quantum enveloping algebra of $A_{N-1}^{(1)}$ (see [J]).

The semiclassical limit is more subtle. To obtain precisely Ω of the KZB equations, replace $\sigma(\gamma, \lambda_{ij})$ by $\sigma(\gamma, \lambda_{ij}) \exp(\rho(\lambda_{ij})\gamma)$, and multiply the resulting R-matrix by $\exp(\gamma\rho(z)(1 - N)/N)$. It turns out that these changes are compatible with the MYBE (but violate the assumption of meromorphy). Then R has a semiclassical asymptotic expansion with Ω (for sl_N) as coefficient of $-\gamma$.

Following the Leningrad school (see [F]), one associates a bialgebra with quadratic relations to each solution of the Yang-Baxter equation. In our case we have to slightly modify the construction. Let us consider an "algebra" $A(R)$ associated to a generalized quantum R-matrix R, generated by meromorphic functions on $\mathfrak{h}$

and the matrix elements (in some basis of V) of a matrix $L(z, \lambda) \in \text{End}(V)$ with noncommutative entries, subject to the relations

$$R^{(12)}(z_{12}, \lambda + \eta h)L^{(1)}(z_1, \lambda - \eta h^{(2)})L^{(2)}(z_2, \lambda + \eta h^{(1)})$$
$$= L^{(2)}(z_2, \lambda - \eta h^{(1)})L^{(1)}(z_1, \lambda + \eta h^{(2)})R^{(12)}(z_{12}, \lambda - \eta h).$$

Instead of giving a more precise definition of this algebra, let us define the more important notion (for our purposes) of representation of $A(R)$.

Definition: Let $R \in \text{End}(V \otimes V)$ be a meromorphic unitary weight zero solution of the MYBE (a generalized quantum R-matrix). A representation of $A(R)$ is a diagonalizable $\mathfrak{h}$-module W together with a meromorphic function $L(z, \lambda)$ (called the L-operator) on $\mathbb{C} \times \mathfrak{h}$ with values in $\text{End}(V \otimes W)$ such that the identity

$$R^{(12)}(z_{12}, \lambda + \eta h^{(3)})L^{(13)}(z_1, \lambda - \eta h^{(2)})L^{(23)}(z_2, \lambda + \eta h^{(1)})$$
$$= L^{(23)}(z_2, \lambda - \eta h^{(1)})L^{(13)}(z_1, \lambda + \eta h^{(2)})R^{(12)}(z_{12}, \lambda - \eta h^{(3)})$$

holds in $\text{End}(V \otimes V \otimes W)$, and so that L is of weight zero:

$$[X^{(1)} + X^{(2)}, L(z, \lambda)] = 0, \qquad \forall X \in \mathfrak{h}.$$

We have natural notions of homomorphisms of representations.

THEOREM 3.2 *(Existence and coassociativity of the coproduct) Let (W, L) and (W', L') be representations of $A(R)$. Then $W \otimes W'$ with $\mathfrak{h}$-module structure $X(w \otimes w') = Xw \otimes w' + w \otimes Xw'$ and L-operator*

$$L^{(12)}(z, \lambda + \eta h^{(3)})L^{(13)}(z, \lambda - \eta h^{(2)})$$

is a representation of $A(R)$. Moreover, if we have three representations W, W', W'', then the representations $(W \otimes W') \otimes W''$ and $W \otimes (W' \otimes W'')$ are isomorphic (with the obvious isomorphism).

Note also that if $L(z, \lambda)$ is an L-operator then also $L(z - w, \lambda)$ for any complex number w. Because the MYBE and the weight zero condition mean that (V, R) is a representation, we may construct representations on $V^{\otimes n} = V \otimes \cdots \otimes V$ by iterating the construction of Theorem 3.2. The corresponding L operator is the "monodromy matrix" with parameters $z_1, \ldots, z_n$:

$$\prod_{j=2}^{n+1} R^{(1j)}(z - z_j, \lambda - \eta \Sigma_{1<i<j}h^{(i)} + \eta \Sigma_{j<i\leq n+1}h^{(i)})$$

(the factors are ordered from left to right). Although the construction is very reminiscent of the quantum inverse scattering method [F], we cannot at this point construct commuting transfer matrices by taking the trace of the monodromy matrices. As will be explained below, one has to pass to IRF models.

4. Difference equations

We now give the quantum version of the KZB equation (3). As in the trigonometric case [S], [FR] it is a system of difference equations. The system is symmetric, i.e., there is an action of the symmetric group mapping solutions to solutions. In the trigonometric and rational case, symmetric meromorphic solutions with proper pole structure are "form factors" of integrable models of quantum field theory in two dimensions [S].

It is convenient to formulate the construction in terms of representation theory of the affine symmetric group. Let S_n be the symmetric group acting on $\mathbb{C}^n$ by permutations of coordinates, and s_j, $j = 1, \ldots, n-1$, be the transpositions $(j, j+1)$. These transpositions generate S_n with relations $s_j s_l = s_l s_j$, if $|j - l| \geq 2$, $s_j s_{j+1} s_j = s_{j+1} s_j s_{j+1}$, and $s_j^2 = 1$. Let also $P \in \operatorname{End}(V \otimes V)$ be the "flip" operator $P u \otimes v = v \otimes u$ and if R is a generalized quantum R-matrix, set $\hat{R} = RP$. The defining properties of a generalized quantum R-matrix imply:

PROPOSITION 4.1 *Suppose that R is a generalized quantum R-matrix. The formula*

$$s_j f(z, \lambda) = \hat{R}^{(j, j+1)}(z_{j,j+1}, \lambda - \eta \Sigma_{i<j} h^{(i)} + \eta \Sigma_{i>j+1} h^{(i)}) f(s_j z, \lambda) \qquad (11)$$

defines a representation of S_n on meromorphic functions on $\mathbb{C} \times \mathfrak{h}$ with values in $V^{\otimes n}$.

The (extended) affine symmetric group S_n^a is the semidirect product of S_n by $\mathbb{Z}^n$. It is generated by s_j and commuting generators e_j, $j = 1, \ldots, n$, with relations $s_j e_l = e_l s_j$, if $l \neq j, j+1$ and $s_j e_j = e_{j+1} s_j$. Let us introduce a parameter $a \in \mathbb{C}$ and let e_j act on $z \in \mathbb{C}^n$ as $e_j(z_1, \ldots, z_n) = (z_1, \ldots, z_j - a, \ldots, z_n)$. Note that S_n^a is actually generated by $s_1, \ldots, s_{n-1}$ and e_n, as the other e_j are constructed recursively as $e_j = s_j e_{j+1} s_j$.

THEOREM 4.2 *Suppose that R is a generalized quantum R-matrix. Let $T_j f(z, \lambda) = f(z, \lambda - 2\eta h^{(j)})$, $\Gamma_j f(z, \lambda) = f(e_j^{-1} z, \lambda)$, and $R^{(j,n)}$ denote the operator of multiplication by $R^{(j,n)}(z_{j,n}, \lambda - \eta \Sigma_{i<j} h^{(i)} + \eta \Sigma_{j<i<n} h^{(i)})$. Then (11) and*

$$e_n f = R^{(n-1,n)} \cdots R^{(2,n)} R^{(1,n)} \Gamma_n T_n f$$

define a representation of S_n^a on meromorphic functions on $\mathbb{C} \times \mathfrak{h}$ with values in $V^{\otimes n}$.

It is easy to calculate the action of the other generators e_j. One gets expressions similar to the ones in [FR], [S].

The compatible system of difference equations (quantum KZB equations) is then

$$e_j f = f, \qquad j = 1, \ldots n.$$

The symmetric group maps solutions to solutions.

Moreover, it turns out that, for special values of a, the representation of S_n^a for the solution of Proposition 3.1 preserves a space of theta functions, as in the classical case.

5. IRF models

In our setting, the relation between the generalized quantum R-matrix and the Boltzmann weights W of the corresponding interaction-round-a-face (IRF) model [B] is very simple. Let $R \in \mathrm{End}(V \otimes V)$ be a generalized quantum R-matrix, and let $V[\mu]$ be the component of weight $\mu \in \mathfrak{h}^*$ of V, with projection $E[\mu] : V \to V[\mu]$. Then for $a, b, c, d \in \mathfrak{h}^*$, such that $b - a$, $c - b$, $d - a$, and $c - d$ occur in the weight decomposition of V, define a linear map

$$W(a, b, c, d, z, \lambda) : V[d - a] \otimes V[c - d] \to V[c - b] \otimes V[b - a],$$

by the formula

$$W(a, b, c, d, z, \lambda) = E[c - b] \otimes E[b - a] R(z, \lambda - \eta a - \eta c)|_{V[d-a] \otimes V[c-d]}. \qquad (12)$$

Note that $W(a + x, b + x, c + x, d + x, z, \lambda + 2\eta x)$ is independent of $x \in \mathfrak{h} \simeq \mathfrak{h}^*$. Set $W(a, b, c, d, z) = W(a, b, c, d, z, 0)$.

THEOREM 5.1 *If R is a solution of the MYBE, then $W(a, b, c, d, z)$ obeys the Star-Triangle relation*

$$\sum_g W(b, c, d, g, z_{12})^{(12)} W(a, b, g, f, z_{13})^{(13)} W(f, g, d, e, z_{23})^{(23)}$$
$$= \sum_g W(a, b, c, g, z_{23})^{(23)} W(g, c, d, e, z_{13})^{(13)} W(a, g, e, f, z_{12})^{(12)},$$

on $V[f - a] \otimes V[e - f] \otimes V[d - e]$.

The familiar form of the Star-Triangle relation [B],[JMO] is recovered when the spaces $V[\mu]$ are 1-dimensional. Upon choice of a basis, the Boltzmann weights $W(a, b, c, d, z)$ are then numbers.

For example, if R is the solution of Proposition 3.1, we obtain the well-known $A_{n-1}^{(1)}$ solution (see [JMO], [JKMO], and references therein).

It is known that solutions of the Star-Triangle relations can be used to construct solvable models of statistical mechanics. Several elliptic solutions are known. It is to be expected that the representation theory of the algebra $A(R)$ above will give a more systematic theory of solutions. Also, the fact that these Boltzmann weights arise as connection matrices of the quantum KZ equation [FR] and the similarity of our Yang-Baxter equation with the triangle equation of [GN] suggest that our algebra is the quantum conformal field theory analogue of $U_q(\mathfrak{g})$, the algebra governing the monodromy of conformal field theory.

Acknowledgment. I wish to thank V. Kac, who asked a question that triggered the quantum part of this research.

References

[AF]   A. Alekseev and L. Faddeev, $(T^*G)_t$: *A toy model for conformal field theory*, Comm. Math. Phys. **159** (1994), 549–579.

[Ba]   O. Babelon, *Universal exchange algebra for Bloch waves and Liouville theory*, Comm. Math. Phys. **139** (1991), 619–643.

[B]   R. J. Baxter, *Exactly Solved Models of Statistical Mechanics*, Academic Press, London 1982.

[B1]   D. Bernard, *On the Wess-Zumino-Witten model on the torus*, Nuclear Phys. B, **303** (1988), 77–93.

[B2]   D. Bernard, *On the Wess-Zumino-Witten model on Riemann surfaces*, Nuclear Phys. B, **309** (1988),145–174.

[Bel]   A. Belavin, *Dynamical symmetry of integrable systems*, Nuclear Phys. B, **108** [**FS2**] (1981), 189–200.

[BD]   A. Belavin and V. Drinfeld, *Solutions of the classical Yang-Baxter equation for simple Lie algebras*, Functional Anal. Appl. **16** (1982), 159.

[Ch]   I. Cherednik, *Some finite dimensional representations of generalized Sklyanin algebras*, Functional Anal. Appl. **19** (1985), 77–79.

[D]   V. G. Drinfeld, *Quantum groups*, Proc. Internat. Congress Math. Berkeley 1986, Academic Press, San Diego, CA, and New York (1986), 798–820.

[DJKMO]   E. Date, M. Jimbo, T. Miwa, and M. Okado, *Exactly solvable SOS models II: Proof of the star-triangle relation and combinatorial identities*, Adv. Stud. Pure Math. **16** (1988), 17–22.

[EK]   P. Etingof and A. Kirillov, Jr., *On the affine analogue of Jack's and Macdonald's polynomials*, Yale preprint (1994).

[F]   L. Faddeev, *Integrable models in $(1+1)$-dimensional quantum field theory*, Proceedings of the Les Houches Summer School 1982, Elsevier 1984.

[FG]   F. Falceto and K. Gawędzki, *Chern-Simons states at genus one*, Comm. Math. Phys. **159** (1994), 549–579.

[FW]   G. Felder and C. Wieczerkowski, *Knizhnik-Zamolodchikov-Bernard equations and invariant theta functions*, preprint (1994).

[FIJKMY]   O. Foda, K. Iohara, M. Jimbo, R. Kedem, T. Miwa, and H. Yan, *An elliptic quantum algebra for $\widehat{sl}_2$*, preprint hep-th/9403094 (1994).

[FR]   I. Frenkel and N. Reshetikhin, *Quantum affine algebras and holonomic difference equations*, Comm. Math. Phys. **146** (1992), 1–60.

[GN]   J. L. Gervais and A. Neveu, *Novel triangle relation and absence of tachyons in Liouville theory*, Nuclear Phys. B, **238** (1984), 125.

[J]   M. Jimbo, *Quantum R matrix for generalized Toda systems*, Comm. Math. Phys. **102** (1986), 537–547.

[JKMO]   M. Jimbo, A. Kuniba, T. Miwa, and M. Okado, *The $A_n^{(1)}$ face models*, Comm. Math. Phys. **119** (1988), 543–565.

[JMO]   M. Jimbo, T. Miwa, and M. Okado, *Solvable lattice models related to the vector representation of classical simple Lie algebras*, Comm. Math. Phys. **116** (1988), 507–525.

[KZ]   V. Knizhnik and A. Zamolodchikov, *Current algebra and the Wess-Zumino model in two dimensions*, Nuclear Phys. B, **247** (1984), 83–103.

[S]   F. Smirnov, *Form Factors in Completely Integrable Systems of Quantum Field Theory*, World Scientific, Singapore, 1992.

[Sk]   E. Sklyanin, *Some algebraic structures connected with the Yang-Baxter equation*, Functional Anal. Appl. **16** (1982), 27–34; Functional Anal. Appl. **17** (1983), 273–234.

[TUY]   A. Tsuchiya, K. Ueno, and Y. Yamada, *Conformal field theory on a universal family of stable curves with gauge symmetries*, Adv. Stud. Pure Math. **19** (1989), 459–566.

Free Field Realizations in Representation Theory and Conformal Field Theory

EDWARD FRENKEL

Department of Mathematics, Harvard University
Cambridge, MA 02138, USA

Free field realization is a relatively new formalism that intertwines representation theory, conformal field theory, and the theory of nonlinear integrable equations. From the physics point of view, it provides realizations of two-dimensional conformal field theories via free bosonic theories. This allows one to find integrals of motion and compute explicitly correlation functions. Mathematically, free field realizations bring new insights into representation theory of conformal algebras, e.g., a new series of representations of affine Kac-Moody algebras, which can be considered as an analogue of the principal series of semi-simple groups over local fields. This connects representation theory of affine algebras with Langlands philosophy.

In this report, which is based mainly on my joint works with Feigin, we will focus on free field realizations of affine Kac-Moody algebras and W-algebras. We will give two constructions of free field realizations: geometric and Hamiltonian, and discuss their applications.

1. Geometric approach to free field realizations

In this section we will give a construction of a family of free field representations of affine algebras, which we call *Wakimoto modules*. These modules were defined by Wakimoto [W] for the simplest affine algebra $\widehat{\mathfrak{sl}}_2$ and by Feigin and the author [FF1] for an arbitrary affine algebra.

1.1. Finite-dimensional case.
Let us first consider the finite-dimensional analogue — the Borel-Weil-Bott construction of representations of semi-simple Lie algebras.

Let $\mathfrak{g}$ be a simple Lie algebra with the Cartan decomposition $\mathfrak{g} = \mathfrak{n}_+ \oplus \mathfrak{h} \oplus \mathfrak{n}_-$, and X be its flag manifold G/B_-, where G is the Lie group of $\mathfrak{g}$, and B_- is its Borel subgroup — the Lie algebra of $\mathfrak{n}_- \oplus \mathfrak{h}$. The *big cell* $\mathcal{U} = N_+ \cdot 1$ is an open subset of X that is isomorphic to the Lie group N_+ and hence to the Lie algebra $\mathfrak{n}_+$ via the exponential map.

The infinitesimal action of $\mathfrak{g}$ on X gives us an embedding of $\mathfrak{g}$ into the Lie algebra of vector fields on $\mathcal{U}$ and hence to the algebra $\mathcal{D}$ of differential operators on $\mathcal{U}$; here $\mathcal{D}$ is a *Heisenberg*, or Weyl, *algebra*. In fact, such an embedding is not unique: one can associate an embedding $\mathfrak{g} \to \mathcal{D}$ to an arbitrary $\lambda \in \mathfrak{h}^*$. By restricting the $\mathcal{D}$-module of regular functions on $\mathcal{U}$ to the image of this embedding we obtain a $\mathfrak{g}$-module. This module is contragradient to the Verma module M_λ over $\mathfrak{g}$.

Proceedings of the International Congress
of Mathematicians, Zürich, Switzerland 1994
© Birkhäuser Verlag, Basel, Switzerland 1995

1.2. Affine algebras. Our geometric construction of Wakimoto modules [FF1], [FF4] essentially exploits the same idea: we should find an appropriate homogeneous space of the Lie group of an affine algebra $\widehat{\mathfrak{g}}$ and try to embed $\widehat{\mathfrak{g}}$ into the algebra of differential operators on its big cell. We should then choose a module over this algebra in such a way that its restriction to $\widehat{\mathfrak{g}}$ lies in an appropriate category of $\widehat{\mathfrak{g}}$-modules.

Recall that the *affine Lie algebra* associated to $\mathfrak{g}$ is the extension $\widehat{\mathfrak{g}}$ of the loop algebra $L\mathfrak{g} = \mathfrak{g} \otimes \mathbb{C}[t, t^{-1}]$ by a one-dimensional center $\mathbb{C}K$ [K].

The Lie algebra $\widehat{\mathfrak{g}}$ has a *Cartan decomposition*: $\widehat{\mathfrak{g}} = \widetilde{\mathfrak{n}}_+ \oplus \widetilde{\mathfrak{h}} \oplus \widetilde{\mathfrak{n}}_-$, where $\widetilde{\mathfrak{n}}_{\pm} = (\mathfrak{n}_{\pm} \otimes \mathbb{C}1) \oplus (\mathfrak{g} \otimes t^{\pm 1}\mathbb{C}[t^{\pm 1}])$, and $\widetilde{\mathfrak{h}} = (\mathfrak{h} \otimes \mathbb{C}1) \oplus \mathbb{C}K$.

Using this decomposition we can define the *category* $\mathcal{O}$, which consists of $\widehat{\mathfrak{g}}$-modules, on which (1) the upper nilpotent subalgebra $\widetilde{\mathfrak{n}}_+$ acts locally nilpotently, and (2) the Cartan subalgebra $\widetilde{\mathfrak{h}}$ acts semi-simply [BGG], [RW], [DGK]. The Lie group of $\widetilde{\mathfrak{h}} \oplus \widetilde{\mathfrak{n}}_+$ is an analogue of the Iwahori subgroup of the group G over a local nonarchimedian field. Note also that elements of $\widetilde{\mathfrak{n}}_+$ annihilate the vacuum state of the corresponding quantum field theory. This motivates the definition of category $\mathcal{O}$.

The fundamental objects of the category $\mathcal{O}$ are Verma modules. Such a module is the induced representation $M_\lambda = U(\widehat{\mathfrak{g}}) \otimes_{U(\widetilde{\mathfrak{n}}_+ \oplus \widetilde{\mathfrak{h}})} \mathbb{C}_\lambda$, where $\mathbb{C}_\lambda$ is the one-dimensional $\widetilde{\mathfrak{n}} \oplus \widetilde{\mathfrak{h}}$-module, on which the first summand acts by 0, and the second summand acts according to its character $\lambda \in \widetilde{\mathfrak{h}}^*$; λ is called *highest weight*. We will write $\lambda = (\bar{\lambda}, k)$, where $\bar{\lambda} \in \mathfrak{h}^*$ is the restriction of λ to $\mathfrak{h} \subset \widetilde{\mathfrak{h}}$, and $k = \lambda(K)$; k is called *level*. All irreducible objects in $\mathcal{O}$ can be obtained as quotients of Verma modules.

The construction of the previous section carries over to the case of the standard flag manifold of $\widehat{\mathfrak{g}}$. This manifold is the quotient of the Lie group of $\widehat{\mathfrak{g}}$ by its standard Borel subgroup — the Lie group of $\widetilde{\mathfrak{n}}_- \oplus \widetilde{\mathfrak{h}}$. This gives a realization of the modules contragradient to the Verma modules over $\widehat{\mathfrak{g}}$ in the space of functions on the big cell of this flag manifold.

1.3. Semi-infinite flag manifold. In the affine case there are also other possibilities that have no analogues in the finite-dimensional picture. The reason is that in the affine algebra there are many different "Borel subalgebras" that are not conjugated to each other. One of them is $(\mathfrak{n}_+ \otimes \mathbb{C}[t, t^{-1}]) \oplus (\mathfrak{h} \otimes t\mathbb{C}[t])$, a Lie subalgebra of loops to the Borel subalgebra of $\mathfrak{g}$. To this subalgebra there corresponds the *semi-infinite flag manifold* $\widetilde{X}$, which is the quotient of the loop group LG by the connected component of the loop group of the Borel subgroup B_- of G. One can also describe $\widetilde{X}$ as the universal covering space of the loop space of the flag manifold X of G, cf. [FF4, Section 4].

Consider the big cell $\widetilde{\mathcal{U}} = LN_+ \cdot 1$ on $\widetilde{X}$, where LN_+ is the loop group of N_+. This orbit is isomorphic to LN_+, and hence to its Lie algebra $\widehat{\mathfrak{n}}_+ := \mathfrak{n}_+ \otimes \mathbb{C}[t, t^{-1}]$, because $N_+ \simeq \mathfrak{n}_+$ via the exponential map. Hence we obtain coordinates $x_\alpha(n) = x_\alpha \otimes t^n, \alpha \in \Delta_+, n \in \mathbb{Z}$, where Δ_+ is the set of positive roots of $\mathfrak{g}$, on $\widetilde{\mathcal{U}}$.

We can now identify the algebra of differential operators on $\widetilde{\mathcal{U}}$ with the Heisenberg algebra $\mathcal{H}$, which has generators $x_\alpha(n), \partial/\partial x_\alpha(n), \alpha \in \Delta_+, n \in \mathbb{Z}$, with the

standard commutation relations $[\partial/\partial x_\alpha(n), x_\beta(m)] = \delta_{\alpha,\beta}\delta_{n,m}$. In physics literature $\mathcal{H}$ is called a $\beta\gamma$-system.

The loop algebra $L\mathfrak{g}$ infinitesimally acts on $\widetilde{\mathcal{U}}$ by vector fields. These vector
fields are actually infinite, and therefore lie in a completion of the Lie algebra of
vector fields on $\widetilde{\mathcal{U}}$. If we could lift those vector fields to a completion of $\mathcal{H}$, we
would obtain a $L\mathfrak{g}$-module structure on a module over $\mathcal{H}$, on which the action of
the completion is well defined.

We could take as such a module, the space of functions on $\widetilde{\mathcal{U}}$, i.e. the module
generated by a vector v, such that $\partial/\partial x_\alpha(n) \cdot v = 0, n \in \mathbb{Z}$. But then the resulting
$L\mathfrak{g}$-module would not lie in the category $\mathcal{O}$, cf. [JK]. In order to obtain a module
from the category $\mathcal{O}$, we should instead take the space M of δ-functions on $\widetilde{\mathcal{U}}$ with
support on its subspace $\mathfrak{n}_+ \otimes \mathbb{C}[t] \subset \mathfrak{n}_+ \otimes \mathbb{C}[t, t^{-1}] = \widetilde{\mathcal{U}}$ of "semi-infinite dimension".
This module is therefore generated by a vector v, such that $\partial/\partial x_\alpha(n) \cdot v = 0, n \geq 0$,
and $x_\alpha(n) \cdot v = 0, n < 0$. This is the *Fock representation* of $\mathcal{H}$.

The module M carries a *vertex operator algebra* (VOA) structure. Recall
that a VOA structure is essentially a linear operation on a $\mathbb{Z}$-graded linear space
V that associates to any homogeneous vector $A \in V$ a formal power series, called
a *current*, $Y(A, z) = \sum_{m \in \mathbb{Z}} A_m z^m$, where $A_m : V \to V$ is a linear operator of
degree $\deg A + m$. These series satisfy certain axioms, cf. [B], [FLM].

Using the VOA structure on M we can define a *local completion* $\mathcal{H}_{\mathrm{loc}}$ of $\mathcal{H}$
[FF5], which consists of all Fourier coefficients of currents defined by M. These
currents have the form

$$: \partial_z^{m_1} a_{\alpha_1}[z] \cdot \ldots \cdot \partial_z^{m_k} a_{\alpha_k}[z] \cdot \partial_z^{n_1} a_{\beta_1}^*[z] \cdot \ldots \cdot \partial_z^{n_l} a_{\beta_l}^*[z] :, \qquad m_i, n_j \geq 0,$$

where columns stand for *normal ordering* and

$$a_\alpha[z] = \sum_{n \in \mathbb{Z}} \frac{\partial}{\partial x_\alpha(n)} z^{-n-1}, \qquad a_\alpha^*[z] = \sum_{n \in \mathbb{Z}} x_\alpha(n) z^n.$$

Clearly, the action of $\mathcal{H}_{\mathrm{loc}}$ on M is well defined and hence this is a suitable
completion of $\mathcal{H}$ into which to embed $L\mathfrak{g}$.

1.4. Wakimoto modules.

There is a filtration of $\mathcal{H}_{\mathrm{loc}}$ by powers of the generators
$\partial/\partial x_\alpha(n)$: $0 \subset \mathcal{H}_{\mathrm{loc}}^0 \subset \mathcal{H}_{\mathrm{loc}}^1 \subset \ldots$. We have the exact sequence:

$$0 \to \mathcal{H}_{\mathrm{loc}}^0 \to \mathcal{H}_{\mathrm{loc}}^1 \to \mathrm{Vect}\,\widetilde{\mathcal{U}}_{\mathrm{loc}} \to 0, \tag{1}$$

and an embedding $\epsilon : L\mathfrak{g} \to \mathrm{Vect}\,\widetilde{\mathcal{U}}_{\mathrm{loc}}$, where $\mathrm{Vect}\,\widetilde{\mathcal{U}}_{\mathrm{loc}}$ is a completion of $\mathrm{Vect}\,\widetilde{U}$.

In order to make M into a module over $L\mathfrak{g}$, we have to lift the map ϵ to a
map $\epsilon' : L\mathfrak{g} \to \mathcal{H}_{\mathrm{loc}}^1$. However, this cannot be done, because in contrast to the
finite-dimensional case, the exact sequence (1) *does not split*. It defines a class in
the cohomology group $H^2(\mathrm{Vect}\,\widetilde{\mathcal{U}}_{\mathrm{loc}}, \mathcal{H}_{\mathrm{loc}}^0)$ that is one dimensional [FF4, Section
5.1].

A miraculous fact is however that the extension of $L\mathfrak{g}$ by $\mathcal{H}_{\mathrm{loc}}^0$ defined by (1)
is cohomologically equivalent to its extension by $\mathbb{C} \subset \mathcal{H}_{\mathrm{loc}}^0$. It is possible therefore
to lift ϵ to a map from $\widehat{\mathfrak{g}}$ to $\mathcal{H}_{\mathrm{loc}}^1$.

THEOREM 1 [W], [FF1]. *(a) There exists a Lie algebra homomorphism* $\widehat{\mathfrak{g}} \to \mathcal{H}^1_{\mathrm{loc}}$, *which maps K to $-h^\vee$, where $h^\vee$ is the dual Coxeter number of $\mathfrak{g}$.*

(b) The space of homomorphisms $\widehat{\mathfrak{g}} \to \mathcal{H}^1_{\mathrm{loc}}$ *is a principal homogeneous space over* $\mathfrak{h}^* \otimes \mathbb{C}((z))dz$.

The Fock representation M now provides a family of *Wakimoto modules* over $\widehat{\mathfrak{g}}$ of level $-h^\vee$, which is called the *critical level*. Such a module $W_{\chi(z)}$ is attached to an arbitrary operator of the form $\partial_z + \chi(z)$, where $\chi(z) \in \mathfrak{h}^* \otimes \mathbb{C}((z))dz$. It can be considered as a connection on a principal H^L-bundle on the punctured formal disc, where H^L is the dual group of the Cartan subgroup H of G. One can write explicit formulas for the action of $\widehat{\mathfrak{g}}$ on the Wakimoto modules, cf. [W] in the case of $\widehat{\mathfrak{sl}}_2$ and [FF1] in the case of $\widehat{\mathfrak{sl}}_n$.

We see that the category $\mathcal{O}$ at the critical level is much larger than at other levels. In fact, one can show that all irreducible objects of this category can be constructed as quotients of Verma modules of critical level by characters of the center of a completion of $U(\widehat{\mathfrak{g}})$. The space of central characters is isomorphic to a space of G^L-connections on the formal disc satisfying a special transversality condition, cf. [FF5] and Section 3.5 below. Here G^L is the Langlands dual group of G, and these connections can be considered as "local Langlands parameters" of $\widehat{\mathfrak{g}}$-modules of critical level. This fact can also be used for constructing a global geometric *Langlands correspondence* for complex algebraic curves in the context of affine algebras (Drinfeld).

It is not difficult to generalize the construction of Wakimoto modules above to an arbitrary level.

The Lie subalgebra $\widehat{\mathfrak{h}} = \mathfrak{h} \otimes \mathbb{C}[t, t^{-1}] \oplus \mathbb{C}K$ of $\widehat{\mathfrak{g}}$ is a Heisenberg Lie algebra. It has generators $h_i(n) = h_i \otimes t^n, i = 1, \ldots, l; n \in \mathbb{Z}$, and K, and the commutation relations

$$[h_i(n), h_j(m)] = n(h_i, h_j)K, \qquad [K, h_i(n)] = 0,$$

where $(\cdot, \cdot)$ is the restriction of the Killing form of $\mathfrak{g}$.

Let λ be an element of $\mathfrak{h}^*$ and ν be a nonzero complex number. We define the Fock space representation π^ν_λ of $\widehat{\mathfrak{h}}$ as a module freely generated by $h_i(n), i = 1, \ldots, l; n < 0$, from a vector v_λ, such that $h_i(n)v_\lambda = 0, n > 0; h_i(0)v_\lambda = \lambda(h_i)v_\lambda$; and $Kv_\lambda = \nu v_\lambda$.

THEOREM 2 [W], [FF1]. *There is a structure of $\widehat{\mathfrak{g}}$-module of level k from the category $\mathcal{O}$ on* $W_{\chi,k} = M \otimes \pi^{k+h^\vee}_\chi$.

For generic values of λ the modules M_λ, M^*_λ and W_λ are irreducible and isomorphic to each other. When they are not irreducible, they may have different composition series, cf., e.g., [FF2] and [BeF] in the case of $\widehat{\mathfrak{sl}}_2$. A surprising fact [Fr2] is that if k is real and *less* than $-h^\vee$, then $W_{\chi,k} \simeq M^*_{\chi,k}$ for positive χ.

The integrable representation L_λ of $\widehat{\mathfrak{g}}$ with dominant integral weight λ is a subquotient of the Wakimoto module W_λ. One can construct an analogue R^*_λ of the Bernstein-Gelfand-Gelfand (BGG) resolution [BGG], [RW]. It consists of Wakimoto modules and its cohomology is concentrated in one dimension, where it is isomorphic to L_λ. In contrast to the usual resolution, R^*_λ is two-sided. In the case of $\widehat{\mathfrak{sl}}_2$ these resolutions were constructed explicitly in [FF4, Section 7.3] and in [BeF] (they are closely connected with similar resolutions over the Virasoro algebra constructed in [Fel]). In [BMP1], [BMP2] a remarkable connection between R_λ^* and resolutions over the quantum group $U_q(\mathfrak{g})$ with $q = \exp \pi i/(k + h^\vee)$ was found.

By construction, the modules $W_{\chi(z)}$ and $W_{\chi,k}$ are free over the Lie algebra $\widehat{\mathfrak{n}}_+ \cap \widetilde{\mathfrak{n}}_-$ and cofree over the Lie algebra $\widehat{\mathfrak{n}}_+ \cap \widetilde{\mathfrak{n}}_+$. Therefore they are *flat* over $\widehat{\mathfrak{n}}_+$ in the sense of semi-infinite cohomology [F]: $H^{\infty/2+i}(\widehat{\mathfrak{n}}_+, W_{\chi,k}) = \pi_\chi^{k+h^\vee}$ if $i = 0$, and 0 if $i \neq 0$ (note that $H^i(\widetilde{\mathfrak{n}}_+, M_\lambda^*) = \mathbb{C}_\lambda$ if $i = 0$, and 0 if $i \neq 0$, where H^i stands for the usual Lie algebra cohomology functor). Using this result and the two-sided resolution R_λ^*, we computed $H^{\infty/2+*}(\widehat{\mathfrak{n}}_+, L_\lambda)$ [FF4, Theorem 4].

1.5. Remarks. (1) The construction of Wakimoto modules is a semi-infinite version of the construction of induced and coinduced modules. We can define a $\widehat{\mathfrak{g}}$-bimodule $\widetilde{U}_k(\widehat{\mathfrak{g}})$ on which $\widehat{\mathfrak{g}}$ acts on the left with level k and on the right with level $-2h^\vee - k$, so that $W_{\chi,k} \simeq \mathrm{Tor}_{\infty/2}^{\widehat{\mathfrak{h}} \oplus \widehat{\mathfrak{n}}_-}(\widetilde{U}_k(\widehat{\mathfrak{g}}), \pi_\chi^{k+h^\vee})$, where $\mathrm{Tor}_{\infty/2+*}$ is the semi-infinite Tor functor (note that $M_\lambda = \mathrm{Tor}^{\widetilde{\mathfrak{h}} \oplus \widetilde{\mathfrak{n}}_+}(U(\widehat{\mathfrak{g}}), \mathbb{C}_\lambda)$).

(2) The semi-infinite flag manifold is stratified by orbits of the Lie group of $\widetilde{\mathfrak{n}}_+$, which are called Schubert cells. Wakimoto modules are related to these Schubert cells in the same way as Verma modules are related to the Schubert cells of the standard flag manifold, cf. [FF4]. In particular, the Floer cohomology of the semi-infinite flag manifold is the double of the semi-infinite cohomology of the Lie algebra $\widehat{\mathfrak{n}}_+$ (compare with the finite-dimensional case [Kos]).

(3) In [FF4] a more general construction is given that associates to an arbitrary parabolic subalgebra $\mathfrak{p}$ of $\mathfrak{g}$, a "Borel subalgebra" of $\widehat{\mathfrak{g}}$. These Borel subalgebras are not conjugated to each other and therefore lead to different flag manifolds. Generalized Wakimoto modules, which are flat with respect to the corresponding Borel subalgebras, can be defined as delta-functions supported on Schubert cells of these manifolds. In particular, $M_{\chi,k}^*$ corresponds to $\mathfrak{p} = \mathfrak{g}$ and $W_{\chi,k}$ corresponds to $\mathfrak{p} = \mathfrak{h} \oplus \mathfrak{n}_+$.

(4) One can show that $W_{\chi(z)}$ is irreducible for generic $\chi(z)$ [FF2]. This implies the Kac-Kazhdan conjecture [KK] on characters of irreducible modules at the critical level.

(5) It is possible to construct explicitly intertwining (or screening) operators acting between Wakimoto modules [FF2], [FF3], [BeF], [BMP1], [BMP2].

2. Solutions of the Knizhnik-Zamolodchikov equation

In this section we will outline the application of the Wakimoto realization to the computation of correlation functions (or conformal blocks) in the Wess-Zumino-Novikov-Witten (WZNW) model. It is known that in genus zero they satisfy a system of partial differential equations (PDEs) with regular singularities, called Knizhnik-Zamolodchikov (KZ) equations. Wakimoto realization allows one to express these correlation functions as integrals of much simpler correlation functions of free bosonic fields. This gives the Schechtman-Varchenko solutions of the KZ equations.

2.1. Genus zero conformal blocks.

Let us recall the definition of the space of conformal blocks in the WZNW model. Consider the projective line $\mathbb{CP}^1$ with a global coordinate t and N distinct finite points $z_1, \ldots, z_N \in \mathbb{CP}^1$. In the neighborhood of each point z_i we have the local coordinate $t - z_i$; denote $\widetilde{\mathfrak{g}}(z_i) = \mathfrak{g} \otimes \mathbb{C}((t - z_i))$. Let $\widehat{\mathfrak{g}}_N$ be the diagonal extension of the Lie algebra $\oplus_{i=1}^N \widetilde{\mathfrak{g}}(z_i)$ by one-dimensional center $\mathbb{C}K$. The Lie algebra $\widehat{\mathfrak{g}}_N$ naturally acts on N-fold tensor products of $\widehat{\mathfrak{g}}$-modules $\otimes_{i=1}^N M_i$ of a given level $k \neq -h^\vee$.

Let $\mathfrak{g}_{\mathbf{z}}$ be the Lie algebra of $\mathfrak{g}$-valued regular functions on $\mathbb{CP}^1 \backslash \{z_1, \ldots, z_N\}$, which vanish at ∞. By expanding elements of $\mathfrak{g}_{\mathbf{z}}$ in Laurent power series in the local coordinates $t - z_i$ at each point z_i, we obtain an embedding $\mathfrak{g}_{\mathbf{z}} \to \widehat{\mathfrak{g}}_N$. Denote by $H(M_1, \ldots, M_N)$ the space of $\mathfrak{g}_{\mathbf{z}}$-invariant linear functionals on $\otimes_{i=1}^N M_i$. This space is called the *space of conformal blocks*.

There is a canonical *flat connection* on the trivial bundle over the space $\mathbb{C}^N \backslash \{\text{diagonals}\}$ with the fiber $H(M_1, \ldots, M_N)$.

Now let us choose as the modules M_i, the modules $M^*_{\lambda_i, k}$ (recall that $M^*_{\lambda, k} \simeq W_{\lambda, k}$ for generic λ and k). Then $H(M_1, \ldots, M_N) = \otimes_{i=1}^N M_{\lambda_i}$ is the tensor product of Verma modules over $\mathfrak{g}$. In that case the flat connection is defined by the system of *KZ equations* [KZ]:

$$(k + h^\vee) \frac{\partial \psi(\mathbf{z})}{\partial z_i} = H_i \cdot \psi(\mathbf{z}), \qquad i = 1, \ldots, N, \tag{2}$$

where $H_i = \sum_{j \neq i} I_a^{(i)} I_a^{(j)} / (z_i - z_j)$, and $I_a^{(i)}$ denotes an element of an orthonormal basis $\{I_a\}$ of $\mathfrak{g}$ acting on the ith factor of $\otimes_{i=1}^N M_{\lambda_i}$, cf., e.g., [FFR].

2.2. Solutions.

In the same way as in Section 2.1 we can define spaces of conformal blocks with respect to the Heisenberg algebra $\widehat{\mathfrak{h}} \oplus \mathcal{H}$ [FFR]. Denote by $J_p(\mathbf{x})$ the space associated to the tensor product of Wakimoto modules $\otimes_{i=1}^p W_{\chi_i, k}$, where $\mathbf{x} = (x_1, \ldots, x_p)$ [FFR, Section 6]. This space is one dimensional, and the equation defining the flat connection takes the form:

$$(k + h^\vee) \frac{\partial \varphi}{\partial x_i} = \sum_{j \neq i} \frac{(\chi_i, \chi_j)}{x_i - x_j} \varphi, \qquad i = 1, \ldots, p. \tag{3}$$

These are "baby KZ" equations for the Heisenberg algebra. They are much simpler: the unique up to a constant factor solution is $\varphi_p = \prod_{i < j} (x_i - x_j)^{(\chi_i, \chi_j)/(k+h^\vee)}$. This is the correlation function of the scalar bosonic field.

Now set $p = N + m$, $x_i = z_i, \chi_i = \lambda_i, i = 1, \ldots, N$, and $x_{N+j} = w_j, \chi_{N+j} = -\alpha_{i_j}, j = 1, \ldots, m$. Then $\varphi_{N, m} =$

$$\prod_{i < j} (z_i - z_j)^{(\lambda_i, \lambda_j)/(k+h^\vee)} \prod_{i,j} (z_i - w_j)^{-(\lambda_i, \alpha_{i_j})/(k+h^\vee)} \prod_{s < j} (w_s - w_j)^{(\alpha_{i_s}, \alpha_{i_j})/(k+h^\vee)}.$$

Using Wakimoto modules, we can obtain solutions of the KZ equations (2) by integrating solutions of the "baby KZ" equations (3). We refer the reader to [ATY] and [FFR] for this computation and only give the final result.

Introduce the vector $|w_1^{i_1}, \ldots, w_m^{i_m}\rangle \in \otimes_{i=1}^N M_{\lambda_i}$ by the formula

$$|w_1^{i_1}, \ldots, w_m^{i_m}\rangle = \sum_{p=(I^1,\ldots,I^N)} \prod_{j=1}^N \frac{f_{i_1^j}^{(j)} f_{i_2^j}^{(j)} \cdots f_{i_{a_j}^j}^{(j)}}{(w_{i_1^j} - w_{i_2^j})(w_{i_2^j} - w_{i_3^j}) \ldots (w_{i_{a_j}^j} - z_j)} |0\rangle, \tag{4}$$

where the summation is taken over all *ordered* partitions $I^1 \cup I^2 \cup \ldots \cup I^N$ of the set $\{i_1, \ldots, i_m\}$, $I^j = \{i_1^j, i_2^j, \ldots, i_{a_j}^j\}$, $f_i^{(j)}$ denotes the generator $f_i \in \mathfrak{g}$ acting on the jth component of $\otimes_{i=1}^N M_{\lambda_i}$, and $|0\rangle$ is the highest weight vector.

Denote by $C_{m,\mathbf{z}}$ the space $\mathbb{C}^m$ with coordinates $w_1, \ldots, w_m$ without all diagonals $w_j = w_i$ and all hyperplanes of the form $w_j = z_i$. The multi-valued function $\varphi_{N,m}$ defines a one-dimensional local system $\mathcal{L}$ on the space $C_{m,\mathbf{z}}$.

THEOREM 3. *Let Δ be an m-dimensional cycle on $C_{m,\mathbf{z}}$ with coefficients in $\mathcal{L}^*$. The $\otimes_{i=1}^N M_{\lambda_i}$-valued function*

$$\int_\Delta \varphi_{N,m} |w_1^{i_1}, \ldots, w_m^{i_m}\rangle \, dw_1 \ldots dw_m$$

is a solution of the KZ equation.

Thus, we obtained solutions of the KZ equations in terms of generalized hypergeometric functions using Wakimoto modules. These solutions were first derived by Schechtman and Varchenko by other methods [SV1] (cf. also [L], [DJMM]).

2.3. Remarks. (1) The results of this section mean that a complicated $\mathcal{D}$-module on the space $\mathbb{C}^N \backslash \{\text{diagonals}\}$ defined by the KZ equations (2) can be embedded into the direct image of a much simpler $\mathcal{D}$-module on a larger space $\mathbb{C}^{N+m} \backslash \{\text{diagonals}\}$ defined by the equations (3). Wakimoto realization provides a natural explanation of this remarkable fact.

(2) Using Wakimoto modules at the *critical level* in a similar fashion, it was shown in [FFR] that the vector (4) is an eigenvector of the *Gaudin Hamiltonians* H_i, if w_j's satisfy a system of *Bethe ansatz* equations.

3. Free field realizations from the theory of nonlinear equations

Local integrals of motion of nonlinear integrable equations form Poisson algebras, which in many cases can be naturally embedded into larger Heisenberg-Poisson algebras. By quantizing this embedding we can obtain an embedding of the algebra of quantum integrals of motion into a Heisenberg algebra. This provides another source for free field realizations. We will describe this construction in the case of Toda equations following [FF6], [FF7].

3.1. Classical Toda field theory. Let $\mathfrak{g}$ be a simple Lie algebra and $\alpha_1, \ldots, \alpha_l \in \mathfrak{h}^*$ be the set of simple roots of $\mathfrak{g}$. The Toda equation associated to $\mathfrak{g}$ reads

$$\partial_\tau \partial_t \phi_i(t, \tau) = \sum_{j=1}^l (\alpha_i, \alpha_j) \exp[\phi_j(t, \tau)], \qquad i = 1, \ldots, l, \tag{5}$$

where each $\phi_i(t, \tau)$ is a family of functions on the circle with a coordinate t, depending on the time variable τ.

Let $\pi_0 = \mathbb{C}[u_i^{(n)}]_{1 \leq i \leq l, m \geq 0}$ be the algebra of *differential polynomials* in $u_1, \ldots ,$ u_l, where $u_i \equiv u_i^{(0)}$. It is $\mathbb{Z}$ graded according to $\deg u_i^{(n)} = n + 1$, and there is a derivation ∂ on π_0, such that $\partial u_i^{(n)} = u_i^{(n+1)}$. Let us formally introduce variables $\phi_i, i = 1, \ldots , l$, such that $\partial \phi_i = u_i$. Then we have an action of ∂ on the space $\pi_{\alpha_i} = \pi_0 \otimes e^{\phi_i}$.

There is a derivative on the space $\pi_\lambda \otimes \mathbb{C}[t, t^{-1}]$, where $\lambda = 0$ or α_i, given by $\partial \otimes 1 + 1 \otimes \partial_t$. Denote by $\mathcal{F}_\lambda$ the quotient of $\pi_\lambda \otimes \mathbb{C}[t, t^{-1}]$ by the subspace of total derivatives (and constants, if $\lambda = 0$) and let $\int$ be the projection $\pi_\lambda \otimes \mathbb{C}[t, t^{-1}] \to \mathcal{F}_\lambda$. The space $\mathcal{F}_0$ can be viewed as the space of *local functionals* in $u_1(t), \ldots , u_l(t) \in \mathfrak{h} \otimes \mathbb{C}[t, t^{-1}]$ of the form $\int P(u_i(t), \partial_t u_i(t), \ldots ; t) \, dt$, where P is a polynomial.

There is a unique partial Poisson bracket $\{\cdot, \cdot\} : \mathcal{F}_0 \times \mathcal{F}_\lambda \to \mathcal{F}_\lambda$, such that:

$$\{\int u_i t^n, \int u_j t^m\} = n(\alpha_i, \alpha_j)\delta_{n,-m}, \quad \{\int u_i t^n, \int e^{\phi_j} t^m\} = (\alpha_i, \alpha_j) \int e^{\phi_j} t^{n+m},$$

cf. [GD], [KW]. The restriction of this bracket to $\mathcal{F}_0$ makes it into a Lie algebra.

The equation (5) can be presented in the Hamiltonian form as $\partial_\tau u_i(t, \tau) = \{u_i(t, \tau), H\}, i = 1, \ldots , l$, where $H = \sum_{i=1}^{l} \int e^{\phi_i}$. This motivates the definition of the space $I(\mathfrak{g})$ of *local integrals of motion* of the Toda theory associated to $\mathfrak{g}$ as the intersection of kernels of the operators $\bar{Q}_i = \{\cdot, \int e^{\phi_i}\} : \mathcal{F}_0 \to \mathcal{F}_{\alpha_i}, i = 1, \ldots , l$. The bracket $\{\cdot, \cdot\}$ satisfies the Jacobi identity; therefore, $I(\mathfrak{g})$ is a Lie algebra.

3.2. Hidden nilpotent action. Introduce linear operators

$$\tilde{Q}_i = \sum_{1 \leq j \leq l; n \geq 0} (\alpha_i, \alpha_j)[\partial^n e^{\phi_i}]\frac{\partial}{\partial u_j^{(n)}}, \qquad i = 1, \ldots , l,$$

acting from π_0 to π_{α_i}. They commute with ∂, and the corresponding operators $\mathcal{F}_0 \to \mathcal{F}_{\alpha_i}$ coincide with $\bar{Q}_i$. We proved in [FF6, (2.2.8)], that *the operators $\tilde{Q}_i$ generate the nilpotent Lie subalgebra $\mathfrak{n}_+$ of $\mathfrak{g}$.*

By definition, $I(\mathfrak{g})$ is the 0th cohomology of the complex $\mathcal{F}_0 \xrightarrow{\oplus \bar{Q}_i} \oplus_{i=1}^{l} \mathcal{F}_{\alpha_i}$. Using the fact that $\tilde{Q}_i$'s generate $\mathfrak{n}_+$ and the BGG resolution of $\mathfrak{g}$ [BGG] we can extend this complex further to the right and relate $I(\mathfrak{g})$ to the cohomology of $\mathfrak{n}_+$ with coefficients in π_0 (the action of $\mathfrak{n}_+$ on π_0 is generated by the operators $e^{-\phi_i}\tilde{Q}_i$), cf. [FF6, Sections 2.3–2.4].

In [FF6] we proved that $H^i(\mathfrak{n}_+, \pi_0) = 0, i \neq 0$, and that there exist elements $W_i \in \pi_0$ of degrees $d_i + 1, i = 1, \ldots , l$, where d_i is the ith exponent of $\mathfrak{g}$, such that $H^0(\mathfrak{n}_+, \pi_0)$ is isomorphic to the algebra $\mathcal{W}(\mathfrak{g}) = \mathbb{C}[W_i^{(n)}]_{i=1,\ldots,l; n \geq 0}$ of differential polynomials in $W_1, \ldots , W_l$. This implies that the space $I(\mathfrak{g})$ is isomorphic to the quotient of $\mathcal{W}(\mathfrak{g}) \otimes \mathbb{C}[t, t^{-1}]$ by total derivatives and constants, i.e. the space of local functionals in $W_1(t), \ldots , W_l(t)$.

For example, $\mathcal{W}(\mathfrak{sl}_2) = \mathbb{C}[W^{(m)}]_{m \geq 0}$, where $W = \frac{1}{2}u^2 - \partial u$. Thus, for $\mathfrak{g} = \mathfrak{sl}_2$ integrals of motion are local functionals in $W(t) = \frac{1}{2}u(t)^2 - \partial_t u(t)$. They generate a classical limit of the Virasoro algebra [FF6, Section 2.1].

REMARK. The Lie algebra $I(\mathfrak{g})$ is called the *classical W-algebra* associated to $\mathfrak{g}$. It can be identified with the Poisson algebra of local functionals on an infinite-dimensional Hamiltonian space obtained from the dual space to $\widehat{\mathfrak{g}}$ by the *Drinfeld-Sokolov reduction* [DS]. This Hamiltonian space consists of connections on a G-bundle over the circle, which satisfy a certain transversality condition. For $\mathfrak{g} = \mathfrak{sl}_n$, $I(\mathfrak{g})$ was first defined by Adler and by Gelfand and Dickey.

3.3. Quantum integrals of motion. In the previous section we realized the space of integrals of motion of a Toda theory as a Lie subalgebra of $\mathcal{F}_0$, which lies in the kernel of the operators $\bar{Q}_i$. Now we want to quantize this embedding. In order to do that, we have to quantize the Lie algebra $\mathcal{F}_0$ and the operators $\bar{Q}_i$.

Let π_λ^ν be the Fock representation of $\widehat{\mathfrak{h}}$ defined in Section 3.1, and $\mathcal{F}_\lambda^\nu$ be the quotient of $\pi_\lambda^\nu \otimes \mathbb{C}[z, z^{-1}]$ by total derivatives (and constants, if $\lambda = 0$). The VOA structure on π_0^ν provides a Lie algebra structure on $\mathcal{F}_0^\nu$ [FF5]. This Lie algebra is a quantum deformation of the Lie algebra $\mathcal{F}_0$, cf. [FF6, Section 4.2].

Introduce *bosonic vertex operators*

$$V_\gamma^\nu[z] = \sum_{n \in \mathbb{Z}} V_\gamma^\nu(n) z^{-n} = T_\gamma \exp\left(-\sum_{n<0} \frac{\gamma(n) z^{-n}}{n}\right) \exp\left(-\sum_{n>0} \frac{\gamma(n) z^{-n}}{n}\right),$$

where $\gamma \in \mathfrak{h}^* \simeq \mathfrak{h}$ and $T_\gamma : \pi_0^\nu \to \pi_\gamma^\nu$ is such that $T_\gamma \cdot v_0 = v_\gamma$ and $[T_\gamma^\nu, b_i(n)] = 0, n < 0$. Thus, $V_\gamma^\nu(n), n \in \mathbb{Z}$, are well-defined linear operators $\pi_0^\nu \to \pi_\gamma^\nu$.

The operator $\widetilde{Q}_i^\nu = V_{\alpha_i}^\nu(1) : \pi_0^\nu \to \pi_{\alpha_i}^\nu$ is a quantum deformation of the operator $\widetilde{Q}_i : \pi_0 \to \pi_{\alpha_i}$ in the sense that $\widetilde{Q}_i^\nu = \nu \cdot \widetilde{Q}_i + \nu^2(\dots)$, [FF6, (4.2.4)]. Further, $\widetilde{Q}_i^\nu$ commutes with ∂, and hence provides an operator $\bar{Q}_i^\nu : \mathcal{F}_0^\nu \to \mathcal{F}_{\alpha_i}^\nu$, which is a quantum deformation of $\bar{Q}_i$.

We can now define the space $I_\nu(\mathfrak{g})$ of *quantum integrals of motion* of the Toda theory associated to $\mathfrak{g}$ as

$$I_\nu(\mathfrak{g}) = \bigcap_{i=1}^{l} \operatorname*{Ker}_{\mathcal{F}_0^\nu} \bar{Q}_i^\nu.$$

One can check that $I_\nu(\mathfrak{g})$ is a Lie subalgebra of $\mathcal{F}_0^\nu$, [FF6, (4.2.8)]. Thus, we *define* it through its embedding into $\mathcal{F}_0^\nu$, i.e. through its free field realization.

We also define the space $\mathcal{W}_\nu(\mathfrak{g})$ as $\quad \mathcal{W}_\nu(\mathfrak{g}) = \bigcap_{i=1}^{l} \operatorname*{Ker}_{\pi_0^\nu} \widetilde{Q}_i^\nu.$

One can check that $\mathcal{W}_\nu(\mathfrak{g})$ is a VOA [FF6, (4.2.8)].

3.4. W-algebras. Our computation of $I_\nu(\mathfrak{g})$ and $\mathcal{W}_\nu(\mathfrak{g})$ is based on the fact that the operators $\widetilde{Q}_i^\nu$ generate the quantized enveloping algebra $U_q(\mathfrak{n}_+)$ with $q = \exp(\pi i \nu)$ [BMP1] (cf. also [FF6, Section 4.5]. This fact follows from a remarkable connection between local systems on configuration spaces and quantum groups [SV2].

Using this and a quantum deformation of the BGG resolution [FF6, Section 4.4], we can construct a deformation $F_\nu^*(\mathfrak{g})$ of the extended complex used in the classical case ($\nu = 0$). We have: $F_\nu^j(\mathfrak{g}) = \oplus_{l(s)=j} \pi_{s(\rho)-\rho}^\nu$, where s runs over the Weyl group of $\mathfrak{g}$. From vanishing of higher cohomologies in the classical case we obtain the following result.

THEOREM 4 [FF6]. *(a) For generic ν higher cohomologies of the complex $F_\nu^*(\mathfrak{g})$ vanish. The 0th cohomology $\mathcal{W}_\nu(\mathfrak{g})$ is a VOA, in which there exist elements W_i^ν of degrees $d_i + 1$, $i = 1, \ldots, l$, where d_i is the ith exponent of $\mathfrak{g}$, such that $\mathcal{W}_\nu(\mathfrak{g})$ has a linear basis of lexicographically ordered monomials in the Fourier components $W_i^\nu(n_i)$, $1 \leq i \leq l$, $n_i < -d_i$, of the currents $Y(W_i^\nu, z) = \sum_{n \in \mathbb{Z}} W_i^\nu(n) z^{-n-d_i-1}$.*

(b) The Lie algebra $I_\nu(\mathfrak{g})$ of quantum integrals of motion of the Toda theory associated to $\mathfrak{g}$ consists of all Fourier components of currents of the VOA $\mathcal{W}_\nu(\mathfrak{g})$.

The Lie algebra $I_\nu(\mathfrak{g})$ is called the $\mathcal{W}$-*algebra* associated to $\mathfrak{g}$. Such $\mathcal{W}$-algebras are, along with the affine algebras, the main examples of algebras of symmetries of conformal field theories [BS].

Note that $\mathcal{W}_\nu(\mathfrak{sl}_2)$ is the VOA of the Virasoro algebra. Its embedding into π_0^ν has been known for a long time [FCT]. In was used by Feigin and Fuchs [FeFu] to study representations of the Virasoro algebra, and by Dotsenko and Fateev [DF] to obtain correlation functions of the minimal models, in the same way as in Section 2.

The VOA $\mathcal{W}_\nu(\mathfrak{sl}_3)$ was first constructed by Zamolodchikov [Z1], and the VOA $\mathcal{W}_\nu(\mathfrak{sl}_n), n > 3$, was first constructed by Fateev and Lukyanov [FL] (cf. also [BG]). The existence of $\mathcal{W}_\nu(\mathfrak{g})$ as a VOA "freely" generated by currents of degrees $d_i + 1, i = 1, \ldots, l$ (cf. part (a) of Theorem 4) for an arbitrary $\mathfrak{g}$ was an open question until [Fr1], [FF6].

To summarize, we defined the VOA of a $\mathcal{W}$-algebra as a vertex operator subalgebra of a VOA of free fields, subject to a set of constraints. These constraints satisfy certain algebraic relations, which make it possible to describe the structure of the $\mathcal{W}$-algebra: the operators $\widetilde{Q}_i^\nu$ generate the nilpotent part of $U_q(\mathfrak{g})$, so that $U_q(\mathfrak{g})$ and $\mathcal{W}_\nu(\mathfrak{g})$ form a "dual pair". The classical origin of these constraints is a nonlinear integrable equation — the Toda equation — and therefore the classical limit of a $\mathcal{W}$-algebra consists of local integrals of motion of that equation. In [FF8] the Wakimoto realization is derived in a similar fashion in connection with the non-linear Schrödinger equation.

3.5. Quantum Drinfeld-Sokolov reduction.

$\mathcal{W}$-algebras can also be defined through the quantum Drinfeld-Sokolov reduction [FF5], [Fr1]. Let $\mathcal{C}$ be the Clifford algebra with generators $\psi_\alpha(n), \psi_\alpha^*(n), \alpha \in \Delta_+, n \in \mathbb{Z}$, and anti-commutation relations

$$[\psi_\alpha(n), \psi_\beta(m)]_+ = [\psi_\alpha^*(n), \psi_\beta^*(m)]_+ = 0, \quad [\psi_\alpha(n), \psi_\beta^*(m)]_+ = \delta_{\alpha,\beta}\delta_{n,-m}.$$

Denote by $\bigwedge$ its Fock representation, generated by vector v, such that $\psi_\alpha(n)v = 0, n \geq 0, \psi_\alpha^*(n)v = 0, n > 0$. This is the super-VOA of $\mathcal{C}$. Introduce a $\mathbb{Z}$-grading on $\mathcal{C}$ and $\bigwedge$ by putting $\deg \psi_\alpha^*(n) = -\deg \psi_\alpha(n) = 1, \deg v = 0$.

Now consider the complex $(V_k \otimes \bigwedge, d)$, where V_k is the VOA of $\mathfrak{g}$ of level k, and $d = d_{\text{st}} + \chi$. Here d_{st} is the standard differential of semi-infinite cohomology of $\widehat{\mathfrak{n}}_+$ with coefficients in V_k [F], and $\chi = \sum_{i=1}^l \psi_{\alpha_i}^*(1)$ corresponds to the Drinfeld-Sokolov character of $\widehat{\mathfrak{n}}_+$ [DS]. The cohomology $H_k^*(\mathfrak{g}) = \oplus_{n \in \mathbb{Z}} H_k^n(\mathfrak{g})$ of this complex is a VOA [FF5]. This cohomology can be computed using the spectral sequence, in which the 0th differential is d_{st} and the first differential is χ.

THEOREM 5 [FF5], [Fr1]. *For generic $k \neq -h^\vee$ the spectral sequence degenerates into the complex $F^*_{1/(k+h^\vee)}(\mathfrak{g})$. Thus, $H^0_k(\mathfrak{g}) \simeq \mathcal{W}_{1/(k+h^\vee)}(\mathfrak{g})$ and $H^i_k(\mathfrak{g}) = 0, i \neq 0$.*

The second part of Theorem 5 was proved for an arbitrary k in [dBT] using the opposite spectral sequence.

For any module M from the category $\mathcal{O}$ of $\widehat{\mathfrak{g}}$, the cohomology of the complex $(M \otimes \bigwedge, d)$ is a module over the $\mathcal{W}$-algebra $I_{1/(k+h^\vee)}(\mathfrak{g})$. This defines a functor, which was studied in detail in [FKW].

The limit of the $\mathcal{W}$-algebra $I_{1/(k+h^\vee)}(\mathfrak{g})$ when $k = -h^\vee$ is isomorphic to the center $Z_{-h^\vee}(\widehat{\mathfrak{g}})$ of the local completion of $U_{-h^\vee}(\widehat{\mathfrak{g}})$. On the other hand it can be identified with $I(\mathfrak{g}^L)$, where $\mathfrak{g}^L$ is the Langlands dual Lie algebra to $\mathfrak{g}$. This proves Drinfeld's conjecture that $Z_{-h^\vee}(\widehat{\mathfrak{g}}) \simeq I(\mathfrak{g}^L)$ [FF5], [Fr1], which can be used in the study of geometric Langlands correspondence.

3.6. Affine Toda field theories. Our approach can be extended to *affine* Toda field theories. First consider the classical case [FF6], [FF7]. The Toda equation associated to an affine algebra $\widehat{\mathfrak{g}}$ is given by formula (5), in which the summation is over $i = 0, \dots, l$, and $\phi_0(t) = -(1/a_0) \sum_{i=1}^l a_i \phi_i(t)$, where a_i's are the labels of the Dynkin diagram of $\widehat{\mathfrak{g}}$ [K]; $\phi_0(t)$ corresponds to the extra root α_0 of $\widehat{\mathfrak{g}}$. Following the scheme of Section 3.1, we define the space $I(\widehat{\mathfrak{g}})$ of local integrals of motion as the intersection of kernels of the operators $\bar{Q}_i : \mathcal{F}_0 \to \mathcal{F}_{\alpha_i}$.

The operators $\widetilde{Q}_i : \pi_0 \to \pi_{\alpha_i}, i = 0, \dots, l$, generate the nilpotent subalgebra $\widetilde{\mathfrak{n}}_+$ of $\widehat{\mathfrak{g}}$. Using the BGG resolution of $\widehat{\mathfrak{g}}$, we can identify $I(\widehat{\mathfrak{g}})$ with $H^1(\widetilde{\mathfrak{n}}_+, \pi_0)$ [FF6], [FF7].

THEOREM 6 [FF6], [FF7]. *$\pi_0 \simeq \mathbb{C}[\widetilde{N}_+/A_+]$, where A_+ is the principal abelian subgroup of the Lie group $\widetilde{N}_+$ of $\widetilde{\mathfrak{n}}_+$, and $H^*(\widetilde{\mathfrak{n}}_+, \pi_0) \simeq \bigwedge^*(\mathfrak{a}^*_+)$, where $\mathfrak{a}_+$ is the Lie algebra of A_+. Thus $I(\widehat{\mathfrak{g}})$ is naturally isomorphic to $\mathfrak{a}^*_+$.*

Theorem 6 implies that local integrals of motion of the Toda theory associated to $\widehat{\mathfrak{g}}$ have degrees equal to the exponents of $\widehat{\mathfrak{g}}$ modulo the Coxeter number. The corresponding Hamiltonian equations form the modified KdV hierarchy [DS], [KW]. In [FF7] we showed that the corresponding flows on $\widetilde{N}_+/A_+$ coincide with those given by the right action of the opposite abelian subalgebra $\mathfrak{a}_-$ of $\widehat{\mathfrak{g}}$.

We can also define the space $I_\nu(\widehat{\mathfrak{g}})$ of quantum integrals of motion as

$$I_\nu(\widehat{\mathfrak{g}}) = \bigcap_{i=0}^l \mathop{\mathrm{Ker}}_{\mathcal{F}^\nu_0} \bar{Q}^\nu_i.$$

Using the BGG resolution over $U_q(\widehat{\mathfrak{g}})$ we proved in [FF6] that *all classical integrals of motion can be quantized*, so that $I_\nu(\widehat{\mathfrak{g}}) \simeq I(\widehat{\mathfrak{g}})$. For $\widehat{\mathfrak{g}} = \widehat{\mathfrak{sl}}_2$ this was conjectured in [G] and [Z2].

The quantum integrals of motion form an abelian subalgebra of the $\mathcal{W}$-algebra $I_\nu(\mathfrak{g})$. They can be viewed as conservation laws of massive perturbations of conformal field theories associated to $\mathcal{W}_\nu(\mathfrak{g})$ [Z2], [EY], [HM]. This is just one of many indications that the ideas of free field realizations are applicable beyond conformal field theory.

References

[ATY] H. Awata, A. Tsuchiya, and Y. Yamada, *Integral formulas for the WZNW correlation functions*, Nuclear Phys. B, **365** (1991), 680–698.

[BeF] D. Bernard and G. Felder, *Fock representations and BRST cohomology in SL(2) current algebra*, Comm. Math. Phys., **127** (1990), 145–168.

[BGG] I. N. Bernstein, I. M. Gelfand, and S. I. Gelfand, *Differential operators on the basic affine space and a study of $\mathfrak{g}$-modules*, in Lie Groups and Their Representations, ed. I. M. Gelfand, 21–64, Halsted Press, New York, 1975.

[BG] A. Bilal and L.-L. Gervais, *Extended $c = \infty$ conformal systems from classical Toda field theories*, Nuclear Phys. B, **314** (1989), 546–686; **318** (1989), 579–630.

[B] R. Borcherds, *Vertex algebras, Kac-Moody algebras, and the Monster*, Proc. Nat. Acad. Sci. U.S.A., **83** (1986), 3068–3071.

[BMP1] P. Bouwknegt, J. McCarthy, and K. Pilch, *Quantum group structure in the Fock space resolutions of $SL(n)$ representations*, Comm. Math. Phys., **131** (1990), 125–156.

[BMP2] P. Bouwknegt, J. McCarthy, and K. Pilch, *Free field approach to two-dimensional conformal field theory*, Progr. Theoret. Phys. Suppl., **102** (1990), 67–135.

[BS] P. Bouwknegt and K. Schoutens, *W-symmetry in conformal field theory*, Phys. Rep., **223** (1993), 183–276.

[dBT] J. de Boer and T. Tjin, *The relation between quantum W-algebras and Lie algebras*, Comm. Math. Phys., **160** (1994), 317–332.

[DJMM] E. Date, M. Jimbo, A. Matsuo, and T. Miwa, *Hypergeometric type integrals and the $SL(2, \mathbb{C})$ Knizhnik-Zamolodchikov equations*, Internat. J. Modern Phys. B, **4** (1990), 1049–1057.

[DGK] V. V. Deodhar, O. Gabber, and V. G. Kac, *Structure of some categories of representations of infinite-dimensional Lie algebras*, Adv. in Math., **45** (1982), 92-116.

[DF] V. Dotsenko and V. Fateev, *Conformal algebra and multipoint correlation functions in 2D statistical models*, Nuclear Phys. B, **240** (1984), 312–348.

[DS] V. Drinfeld and V. Sokolov, *Lie algebras and KdV type equations*, J. Soviet Math., **30** (1985), 1975–2036.

[EY] T. Eguchi and S.-K. Yang, *Deformations of conformal field theories and soliton equations*, Phys. Lett. B, **224** (1989), 373–378.

[FCT] D. Fairlie, unpublished; A. Chodos and C. Thorn, *Making the massless string massive*, Nuclear Phys. B, **72** (1974), 509–522.

[FL] V. Fateev and S. Lukyanov, *The models of two-dimensional conformal quantum field theory with $\mathbb{Z}_n$ symmetry*, Internat. J. Modern Phys. A, **3** (1988), 507–520.

[F] B. Feigin, *The semi-infinite cohomology of the Virasoro and Kac-Moody Lie algebras*, Uspekhi Mat. Nauk **39**:2 (1984), 195–196 [English translation: Russian Math. Surveys, **39**:2 (1984), 155–156].

[FF1] B. Feigin and E. Frenkel, *A family of representations of affine Lie algebras*, Uspekhi Mat. Nauk, **43**:5 (1988), 227–228 [English translation: Russian Math. Surveys, **43**:5 (1988), 221–222].

[FF2] B. Feigin and E. Frenkel, *Representations of affine Kac-Moody algebras and bosonization*, in Physics and Mathematics of Strings, V. G. Knizhnik Memorial Volume, eds. L. Brink, et al., 271–316. World Scientific, Singapore 1990.

[FF3] B. Feigin and E. Frenkel, *Representations of affine Kac-Moody algebras, bosonization, and resolutions*, Lett. Math. Phys., **19** (1990), 307–317.

[FF4] B. Feigin and E. Frenkel, *Affine Kac-Moody algebras and semi-infinite flag manifolds*, Comm. Math. Phys. **128** (1990), 161-189.

[FF5] B. Feigin and E. Frenkel, *Affine Kac-Moody algebras at the critical level and Gelfand-Dikii algebras*, Internat. J. Math. Phys. A, **7**, Supplement **1A** (1992), 197–215.

[FF6] B. Feigin and E. Frenkel, *Integrals of motion and quantum groups*, preprint YITP/K-1036, September 1993, to appear in Lecture Notes in Math.

[FF7] B. Feigin and E. Frenkel, *Kac-Moody groups and integrability of soliton equations*, Invent. Math. **120** (1995), 379–408.

[FF8] B. Feigin and E. Frenkel, *Wakimoto modules and non-linear Schrödinger equations*, preprint, October 1994.

[FFR] B. Feigin, E. Frenkel, and N. Reshetikhin, *Gaudin model, Bethe ansatz and critical level*, Comm. Math. Phys. **166** (1994), 27–62.

[FeFu] B. Feigin and D. Fuchs, *Representations of the Virasoro algebra*, in Representations of Lie Groups and Related Topics, eds. A. M. Vershik, and D. P. Zhelobenko, 465–554, Gordon & Breach, New York, 1990.

[Fel] G. Felder, *BRST approach to minimal models*, Nuclear Phys. B, **317** (1989), 215–236.

[Fr1] E. Frenkel, *W-algebras and Langlands-Drinfeld correspondence*, in New Symmetries in Quantum Field Theory, eds. J. Fröhlich, et al., 433–447, Plenum Press, New York, 1992.

[Fr2] E. Frenkel, *Determinant formulas for free field representations of the Virasoro and Kac-Moody algebras*, Phys. Lett. B, **286** (1992), 71–77.

[FKW] E. Frenkel, V. Kac, and M. Wakimoto, *Characters and fusion rules for W-algebras via quantized Drinfeld-Sokolov reduction*, Comm. Math. Phys., **147** (1992), 295–328.

[FLM] I. Frenkel, J. Lepowsky, and A. Meurman, Vertex Operator Algebras and the Monster, Academic Press, New York, 1988.

[GD] I. Gelfand and L. Dickey, *A Lie algebra structure in a formal variations calculus*, Functional Anal. Appl., **10** (1976), 16–22.

[G] J.-L. Gervais, *Infinite family of polynomial functions of the Virasoro generators with vanishing Poisson brackets*, Phys. Lett. B, **160** (1985), 277–278.

[HM] T. Hollowood and P. Mansfield, *Rational conformal field theories at, and away from, criticality as Toda field theories*, Phys. Lett. B, **226** (1989), 73–79.

[JK] H. Jacobsen and V. Kac, *A new class of unitarizable highest weight representations of infinite-dimensional Lie algebras*, Lecture Notes in Phys., **226** (1985), 1–20; II. J. Funct. Anal., **82** (1989), 69–90.

[K] V. Kac, Infinite-dimensional Lie algebras, Third edition, Cambridge University Press, London and New York, 1990.

[KK] V. Kac and D. Kazhdan, *Structure of representations with highest weight of infinite-dimensional Lie algebras*, Adv. in Math., **34** (1979), 97–108.

[KZ] V. Knizhnik and A. Zamolodchikov, *Current algebra and Wess-Zumino model in two dimensions*, Nuclear Phys. B, **247** (1984), 83–103.

[Kos] B. Kostant, *Lie algebra cohomology and generalized Shubert cells*, Ann. of Math. (2), **77** (1963), 72–144.

[KW] B. A. Kupershmidt and G. Wilson, *Conservation laws and symmetries of generalized sine-Gordon equations*, Comm. Math. Phys., **81** (1981), 189–202.

[L] R. Lawrence, *Homology representations of braid groups*, Ph.D. thesis, Oxford University, 1989.

[RW] A. Rocha-Caridi and N. Wallach, *Projective modules over graded Lie algebras*, Math. Z., **180** (1982), 151–177.

[SV1] V. Schechtman and A. Varchenko, *Arrangements of hyperplanes and Lie algebra cohomology*, Invent. Math., **106** (1991), 139–194.

[SV2] V. Schechtman and A. Varchenko, *Quantum groups and homology of local systems*, in Algebraic Geometry and Analytic Geometry, 182–191, Springer-Verlag, Berlin and New York, 1991.

[W] M. Wakimoto, *Fock representations of affine Lie algebra $A_1^{(1)}$*, Comm. Math. Phys., **104** (1986), 605–609.

[Z1] A. Zamolodchikov, *Infinite additional symmetries in two-dimensional conformal field theory*, Theor. Math. Phys., **65** (1985), 1205–1213.

[Z2] A. Zamolodchikov, *Integrable field theory from conformal field theory*, Adv. Stud. Pure Math., **19** (1989), 641–674.

Homogenized Models of Strongly Inhomogeneous Media

Eugene Ya. Khruslov

Institute for Low Temperature Physics and Engineering
Prospekt Lenina 47,
310164 Kharkov, Ukraine

Introduction

The paper is devoted to the homogenization theory for partial differential equations, which describe various physical processes in strongly inhomogeneous media, i.e. media whose local physical characteristics rapidly oscillate in space. Therefore such differential equations either have rapidly oscillating coefficients or are considered in highly perforated domains with complicated microstructures.

Direct solution of corresponding boundary value (or initial boundary value) problems is practically impossible either by analytical or by numerical methods. However, if the medium microstructure characteristic scale is much smaller than the size of the domain in which the process is studied, then there exists a possibility of homogenized (macroscopic) description of the behavior of the system. In these cases the medium often (but not always) has steady characteristics (such as heat conductivity, magnetic permeability, etc.) that, in general, essentially differ from the local characteristics. To determine them the homogenization of differential equations must be used. That is why they are called homogenized (or effective) characteristics.

Homogenization problems for partial differential equations have been studied by physicists since Maxwell's and Rayleigh's works, but for a long time they were out of the interest of mathematicians. Starting from the early 1960s the homogenization theory has been intensively developed by mathematicians. Such interest in the problem may be explained on one hand by numerous applications (first of all, in the theory of composite materials) and on other hand by the appearance of new deep ideas, significant for mathematics itself. At the present time there exist a great number of papers and books devoted to the problem. We shall note here only monographs ([1], [3], [9], [17], [18], [19]) containing extensive bibliographies.

Mathematical description of physical processes in strongly inhomogeneous media implies that their local characteristics depend on a small parameter ε, which is the characteristic scale of the microstructure. Therefore, for the construction of homogenized models of these processes an asymptotic analysis is carried out for $\varepsilon \to 0$. Namely, the asymptotic behavior of the solutions of the corresponding problems is studied, and the possible limits of the solutions are described by some modified equations having relatively smoothly varying coefficients and considered

Proceedings of the International Congress
of Mathematicians, Zürich, Switzerland 1994
© Birkhäuser Verlag, Basel, Switzerland 1995

in simple domains. These equations are called homogenized equations or homogenized (macroscopic) models.

The early papers considered strongly inhomogeneous media with local characteristics described by the functions of the form $a\left(\frac{x}{\varepsilon}\right)$, where $x \in \mathbf{R}^n$ and $a(x)$ is a periodic or almost periodic function or realization of a homogeneous random field (see [6], [9], [14], [15], [16]). In these cases the medium has the effective characteristics, which do not depend on x, and the homogenized model has the same form as the microscopic one.

However, there exist media with more complicated microstructure for which it is impossible to prescribe effective characteristics, completely determining the macroscopic model. In these cases the homogenized models essentially differ from microscopic ones: they may be nonlocal, multicomponent models or models with memory (see [7], [8], [10], [11], [12], [13]).

It is the purpose of this paper to characterize in general the strongly inhomogeneous media whose macroscopic description is provided by such unusual homogenized models.

1. Formulation of the problem. Main homogenized models

To be specific, we shall consider the equation describing the diffusion process in strongly inhomogeneous medium as the input model.

Let Ω be a bounded domain in the Euclidean space $\mathbf{R}^n$ $(n \geq 2)$ and $Q_T = \Omega \times (0, T)$, $S_T = \partial\Omega \times (0, T)$. Consider in Q_T the initial boundary value problem

$$\frac{\partial u^{(s)}}{\partial t} - \sum_{i,j=1}^{n} \frac{\partial}{\partial x_i}\left(a_{ij}^{(s)}(x)\frac{\partial u^{(s)}}{\partial x_j}\right) = 0, \quad (x,t) \in Q_T; \tag{1.1}$$

$$\left.\frac{\partial u^{(s)}}{\partial \nu_s}\right|_{S_T} = 0; \qquad u^{(s)}(x,0) = U(x), x \in \Omega; \tag{1.2}$$

where $U(x) \in H^1(\Omega)$ and $\partial/\partial\nu_s$ is the conormal derivative to $\partial\Omega$, corresponding to the coefficients $a_{ij}^{(s)}(x)$. We shall assume that the functions $a_{ij}^{(s)}(x)$ depend on the parameter $s \in \mathbf{N}$, which characterizes their space oscillations, and for any s and x the conditions of ellipticity and boundedness are valid. Namely for any vector $\xi \in \mathbf{R}^n$

$$a_s(x)|\xi|^2 \leq \sum_{i,j=1}^{n} a_{ij}^{(s)}(x)\xi_i\xi_j \leq A_s(x)|\xi|^2, (0 < a_s(x) \leq A_s(x) < \infty).$$

Then, as is known, there exists a unique generalized solution $u^{(s)}(x,t)$ of the problem (1.1), (1.2), depending on the parameter s. We study its asymptotic behavior as $s \to \infty$.

Consider at first the case when the conditions of ellipticity and boundedness are fulfilled uniformly with respect to s, i.e. $0 < a \leq a_s(x) \leq A_s(x) \leq A < \infty$. As is known, in this case the sequence of the solutions $\{u^{(s)}(x,t)\}$ is compact (for example, in the space $L^2(Q_T)$), and all its possible limits with respect to different subsequences are described by the initial boundary value problems of the form:

$$\frac{\partial u}{\partial t} - \sum_{i,j=1}^{n} \frac{\partial}{\partial x_i} \left(a_{ij}(x) \frac{\partial u}{\partial x_j} \right) = 0, \quad (x,t) \in Q_T; \tag{1.3}$$

$$\left. \frac{\partial u}{\partial \nu} \right|_{S_T} = 0; \quad u(x,0) = U(x), \quad x \in \Omega; \tag{1.4}$$

where $a_{ij}(x)$ are measurable bounded functions in Ω that satisfy the ellipticity condition and $\partial/\partial\nu$ is the corresponding conormal derivative to $\partial\Omega$. Equation (1.3) is the simplest homogenized model of the diffusion process. It is a local, one-component model without memory. The tensor $\{a_{ij}(x); i, j = 1, 2, \ldots, n\}$ is the effective conductivity tensor completely determining such a model.

However, if the condition of uniform ellipticity or boundedness is not fulfilled then the homogenized model is not determined solely by the effective conductivity tensor. Its form in the general case differs from the form of the input model and depends on the structure of the sets $\Omega_{0s} \subset \Omega$, where this condition is violated.

Let us assume at first that the uniform ellipticity condition is fulfilled, but the uniform boundedness condition is not, i.e. there exist such domains $\Omega_{0s} \subset \Omega$, that $\inf\{A_s(x), x \in \Omega_{0s}\} \to \infty$, as $s \to \infty$. In this case, the sequence of the solutions of the problem (1.1), (1.2) is also compact and the conditions can be indicated at which this sequence converges to the solution $u(x,t)$ of the following initial boundary value problem:

$$\frac{\partial u}{\partial t} - \sum_{i,j=1}^{n} \frac{\partial}{\partial x_i} \left(a_{ij}(x) \frac{\partial u}{\partial x_j} \right) + c(x)u - \int_{\Omega} R(x,y)u(y,t)\,dy = 0; \tag{1.5}$$

$$\left. \frac{\partial u}{\partial \nu} \right|_{S_T} = 0, \quad u(x,0) = U(x), x \in \Omega; \tag{1.6}$$

where $c(x)$ and $R(x,y)$ are nonnegative functions.

Note that if the set Ω_{0s} is fine-grained, i.e. consists of small grains located not very close to each other, then the functions $c(x), R(x,y)$ are equal to zero. In the general case, the presence of the integral term in equation (1.5) means that the homogenized equation is a nonlocal model of the diffusion process, though the input model was local.

Now we shall assume that the uniform boundedness condition is fulfilled, but the uniform ellipticity condition is not, i.e. there exist the domains Ω_{0s} such that $\sup\{a_s(x), x \in \Omega_{0s}\} \to 0$, as $s \to \infty$. In this case the sequence of the solutions $\{u^{(s)}(x,t)\}$ is not, generally speaking, compact in $L^2(Q_T)$. However, one can indicate the conditions, when the solutions $u^{(s)}(x,t)$ converge in some sense to the vector-function $u(x,t) = \{u_1(x,t), \ldots, u_m(x,t)\}$. The system of functions

$\{u_1(x,t),\ldots,u_m(x,t)\}$ is the solution of the following initial boundary value problem:

$$\frac{\partial u_r}{\partial t} - \frac{1}{m_r(x)} \sum_{i,j=1}^{n} \frac{\partial}{\partial x_i}\left(a_{ij}^r(x)\frac{\partial u_r}{\partial x_j}\right) + \frac{\partial}{\partial t}\int_0^t \sum_{q=1}^{m} B_{rq}(x,t-\tau)u_q(x,\tau)\,d\tau$$

$$= U(x)\sum_{q=1}^{m} B_{rq}(x,t); \quad (r=1,\ldots,m) \tag{1.7}$$

$$\left.\frac{\partial u_r}{\partial \nu_r}\right|_{S_T} = 0; \quad u_r(x,0) = U(x), x \in \Omega; \tag{1.8}$$

where $a_{ij}^r(x)$ $(r=1,\ldots,m)$ and $m_r(x)$ are measurable bounded functions $m_r(x) > 0$, $\{a_{ij}^r(x); i,j=1,\ldots,n\}$ are positive definite tensor and $\partial/\partial\nu_r$ is the conormal derivative, corresponding to $a_{ij}^r(x)$.

The system (1.7) is an m-component model of a diffusion process in strongly inhomogeneous medium, moreover the presence of the time-integral term means that this is a model with memory.

To characterize the structure of the sets Ω_{0s} we shall introduce the notions of strong and weak connectedness of domains.

2. Strongly and weakly connected sequences of domains

Consider in the domain Ω the sequence of subdomains $\{\Omega_s, s \in \mathbf{N}\}$. Let $m_s = \mathrm{mes}\,\Omega_s$ be the Lebesgue measure of Ω_s. We assume that for any sphere $B_\varepsilon \subset \Omega$ with radius $\varepsilon > 0$ for sufficiently large $s(s \geq \hat{s}(\varepsilon))$ the inequality holds: $C_1\varepsilon^n m_s \leq \mathrm{mes}[\Omega_s \bigcap B_\varepsilon] \leq C_2\varepsilon^n m_s$. In this case we say that the sequence of the domains $\{\Omega_s\}$ satisfies the density condition.

Let a function $v_s(x) \in L^2(\Omega_s)$ be determined in each domain Ω_s.

DEFINITION 1 We say that the sequence of functions $\{v_s(x) \in L^2(\Omega_s)\}$ $L^2(\Omega_s)$-converges to the function $v(x) \in L^2(\Omega)$ if there exists the sequence of functions $\{\tilde{v}_M(x), M \in \mathbf{N}\}$ belonging to the Lipschitz class $\mathrm{Lip}(M,\Omega)$ such that

$$\lim_{M\to\infty}\overline{\lim_{s\to\infty}}\left\{\|v - \tilde{v}_M\|_{L^2(\Omega)} + m_s^{-1}\|v_s - \tilde{v}_M\|_{L^2(\Omega_s)}\right\} = 0$$

The sequence of functions $\{v_s(x) \in L^2(\Omega_s)\}$ is called $L^2(\Omega_s)$-compact if from each of its subsequences one can separate a subsequence that $L^2(\Omega_s)$-converges to a function $v(x) \in L^2(\Omega)$.

When dealing with homogenization problems the first question arising is: Under what conditions on the subdomains $\Omega_s \subset \Omega$ is the sequence of functions $\{v_s \in H^1(\Omega_s)\}$, which satisfies the inequality

$$\|v_s(x)\|_{H^1(\Omega_s)}^2 \leq Cm_s, \tag{2.1}$$

$L^2(\Omega_s)$-compact?

DEFINITION 2 We say that the sequence of subdomains $\{\Omega_s \subset \Omega\}$ satisfies the condition of strong connectedness and denote $\{\Omega_s\} \in SC$ if for any sequence of functions $\{v_s \in C^1(\Omega_s)\}$ that satisfies inequality (2.1) and for arbitrary $M \in \mathbf{N}$ there exist subsets $\Omega_{sM} \subset \Omega_s$ and $\Omega_M^s = \Omega_s \setminus \Omega_{sM}$ such that $v_s(x) \in \text{Lip}(M, \Omega_{sM})$, and for $s > \hat{s}(M)$

$$\text{mes}(\Omega_M^s) + M^{-2}\|v_s\|_{L^2(\Omega_M^s)}^2 \leq \varphi(M)m_s,$$

where $\varphi(M)$ is a monotone function satisfying the conditions: $\lim_{M \to \infty} M^2\varphi(M) = 0$ and $\underline{\lim}_{M \to \infty} M^{2+\gamma}\varphi(M) > 0$ for some $\gamma > 0$.

The answer to the above question is given by the following theorem.

THEOREM 2.1 *If $\{\Omega_s\} \in SC$, then any sequence of the functions $\{v_s \in H^1(\Omega_s)\}$ that satisfies the inequality (2.1) is $L^2(\Omega_s)$-compact.*

Introduce the main quantative characteristic of strongly connected domains. Let $K_h^x \subset \Omega$ be a cube with the center at the point x and with the sides of length h, directed along the coordinate axes. Set $m(x, s, h) = h^{-n} \text{mes}(K_h^x \bigcap \Omega_s)$, $g(h) = \varphi^{-1/2}(h^{-1})$, and

$$a_s(x, p) = \sum_{i,j=1}^{n} a_{ij}^{(s)}(x)p_i p_j \quad (x \in \Omega, p \in \mathbf{R}^n).$$

Consider the functional

$$A_y^{sh}[l] = \inf_{v_s} \frac{1}{h^n} \int_{K_h^y \bigcap \Omega_s} \left\{ a_s(x, \nabla v_s) + \frac{g(h)}{m(y, s, h)}|v_s - (x - y, l)|^2 \right\} dx, \quad (2.2)$$

where $l = (l_1, \dots, l_n)$ is an arbitrary vector in $\mathbf{R}^n$, $(\cdot, \cdot)$ is the scalar product in $\mathbf{R}^n$, and the infimum is taken over the class of functions $v^{(s)} \in H^1(K_h^y \bigcap \Omega_s)$.

This functional is quadratic and can be represented as

$$A_y^{sh}[l] = \sum_{i,j=1}^{n} a_{ij}(y, s, h)l_i l_j,$$

where $a_{ij}(y, s, h)$ are measurable and bounded functions with respect to y, forming the positive definite tensor in $\mathbf{R}^n$. We shall call this tensor the homogenized conductivity tensor of the domains Ω_s.

Introduce now the notion of weakly connected domains. Let Ω_{rs} $(r = 1, \dots, m)$ be subdomains in Ω_s, satisfying the density condition in Ω. Denote by $m_r(x, s, h) = h^{-n} \text{mes}(K_h^x \bigcap \Omega_{rs})$, and by χ_{rs} and χ_{0s} the characteristic functions of domains Ω_{rs} $(r = 1, \dots, m)$ and $\Omega \setminus \bigcup_{r=1}^{m} \Omega_{rs}$, respectively.

Consider the functional

$$B_{g\lambda}^{sh}[e] = \inf_{v_s} \frac{1}{h^n} \int_{K_h^y \bigcap \Omega} \left\{ a_s(x, \nabla v_s) + \lambda|v_s|^2 \chi_0^{(s)} \right.$$

$$\left. + g(h) \sum_{r=1}^{m} |v_s - e_r|^2 m_r^{-1}(y, s, h)\chi_{rs} \right\} dx,$$

$$(2.3)$$

where $e = (e_1, \ldots, e_m)$ is an arbitrary vector in $\mathbf{R}^m$; λ is an arbitrary positive number; and the infimum is taken over the class of functions $v_s \in H^1(K_h^y \bigcap \Omega)$.

This functional is quadratic and can be represented as

$$B_{g\lambda}^{sh}[e] = \sum_{r,q=1}^{m} b_{rq}(y, \lambda, s, h) e_r e_q,$$

where $b_{rq}(y, \lambda, s, h)$ are measurable and bounded functions with respect to y. We shall call the matrix-function $\{b_{rq}(y, \lambda, s, h); \ r, q = 1, \ldots, m\}$ the connection matrix.

DEFINITION 3 The sequences of the subdomains $\{\Omega_{rs} \subset \Omega_s\}$ $(r = 1, \ldots, m)$ are called mutually a_s-weakly connected if

$$\lim_{h \to 0} \overline{\lim_{s \to \infty}} \sup \left\{ \sum_{q=1}^{m} b_{qq}(y, 0, s, h), \quad y \in \Omega \right\} < \infty.$$

THEOREM 2.2 Let in (2.3) $\Omega = \Omega_s$ and $a_s(x, p) = m_s^{-1} \chi_s(x) |p|^2$, where $\chi_s(x)$ is the characteristic function of the domain Ω_s. If $\{\Omega_s\} \in SC$ then in the domain Ω_s there do not exist the subdomains $\Omega_{rs} \subset \Omega_s$ $(r = 1, \ldots, m)$ such that $m_r(x, s, h) \geq Cm(x, s, h)$ and the sequences $\{\Omega_{rs}\}$ $(r = 1, \ldots, m)$ are mutually a_s-weakly connected.

3. Statements of main theorems

Let the coefficients $a_{ij}(x)$ satisfy the condition of uniform ellipticity (i.e. $a_s(x) \geq a > 0$ everywhere), and let the condition of uniform boundedness be violated only on the sets Ω_{0s}, i.e. $\sup\{A_s(x), \ x \in \Omega \setminus \Omega_{0s}\} \leq A$.

First we assume that for each s, Ω_{0s} consists of nonintersected components G_{qs} such that the distance r_{qs} from G_{qs} to $\bigcup_{r \neq q} G_{rs} \bigcup \partial\Omega$ satisfies the inequality $r_{qs} \geq c \, diam(G_{qs})$, for a constant $c > 0$ independent of s. Then the homogenized conductivity tensor $\{a_{ij}(x, s, h), \ i, j = 1, \ldots, n\}$ of the domain Ω (see (2.2) for $\Omega_s = \Omega$) is positive definite and uniformly bounded with respect to s and h.

THEOREM 3.1 The solution of the problem (1.1), (1.2) for any $U(x)$ converges in $L^2(Q_T)$ to the solution of the problem (1.3), (1.4) if and only if the functions $a_{ij}(x, s, h)$ converge to $a_{ij}(x)$ in $L^1(\Omega)$ as $s \to \infty$ and $h \to 0$, i.e.

$$\lim_{h \to 0} \overline{\lim_{s \to \infty}} \int_{\Omega} |a_{ij}(x, s, h) - a_{ij}(x)| \, dx = 0, \ (i, j = 1, \ldots, n). \tag{3.1}$$

Consider now the sets Ω_{0s} of another topological structure. Namely, assume that for any s, $\Omega = \Omega_{0s} \bigcup \Omega_{1s} \bigcup F_s$ and for $s \to \infty$ the conditions are fulfilled:
(i) $\inf\{a_s(x), \ x \in \Omega_{0s}\} \geq a_0 m_{0s}^{-1} \to \infty$, $\sup\{A_s(x), \ x \in \Omega_{0s} \bigcup F_s\} \leq A_0 m_{0s}^{-1}$, and $\sup\{A_s(x), \ x \in \Omega_{1s}\} \leq A$;
(ii) $m_{0s} = \text{mes} \, \Omega_{0s} \to 0$ and $\text{mes} \, F_s = o(m_{0s})$;

(iii) the sequences $\{\Omega_{0s}\}$ and $\{\Omega_{1s}\}$ satisfy the conditions of density and strong connectedness;

and (iv) the subdomains $\{\Omega_{0s}\}$ and $\{\Omega_{1s}\}$ are a_{0s}-weakly connected in Ω.

Denote by $\{a_{ij}^r(x,s,h),\ i,j=1,\ldots,n\}\ (r=0,1)$ the homogenized conductivity tensors of the subdomains Ω_{rs} (see (2.2) for $\Omega_s=\Omega_{rs}$) and by $b_{rq}(x,0,s,h)$ the elements of the connection matrix for $\lambda=0$ (see (2.3)).

THEOREM 3.2 *Let the functions* $\{a_{ij}^r(x,s,h),\ i,j=1,\ldots,n\}\ (r=0,1)$ *and* $b_{11}(x,0,s,h)$ *converge in* $L^1(\Omega)$ *to the functions* $a_{ij}^r(x)$ *and* $C(x)$ *as* $s\to\infty$ *and* $h\to 0$, *respectively (see (3.1)).*

Then the solution $u^{(s)}(x,t)$ *of the problem* (1.1), (1.2) *converges in* $L^2(\Omega)$ *uniformly with respect to* $t\in[0,T]$ *to the solution of the problem* (1.5), (1.6), *where* $R(x,y)=C(x)G(x,y)C(y)$ *and* $G(x,y)$ *is the Green function of the problem*

$$-\sum_{i,j=1}^n \frac{\partial}{\partial x_i} a_{ij}^0(x)\frac{\partial G}{\partial x_j} + C(x)G = \delta(x-y),\ x,y\in\Omega;$$

$$\frac{\partial G}{\partial \nu_0} = 0,\ x\in\partial\Omega.$$

Now we shall consider the case where the coefficients $a_{ij}^{(s)}(x)$ satisfy the condition of uniform boundedness (i.e. $A_s(x)\le A<\infty$ everywhere) but the condition of uniform ellipticity is violated. Namely, we shall assume that for any s,

$$\Omega = \Omega_{0s}\bigcup\Omega_s\bigcup F_s,\quad \Omega_s = \left(\bigcup_{r=1}^m \Omega_{rs}\right)\bigcup\left(\bigcup_{q=1}^{N_s} G_{qs}\right),$$

and for $s\to\infty$ the conditions are fulfilled: (i) $\sup\{A_s(x),x\in\Omega_{0s}\}\to 0$ and $\inf\{a_s(x),x\in\Omega_s\}\ge a>0$; (ii) mes $F_s\to 0$, $\max_q\{\text{diam}\,G_{qs}\}\to 0$; (iii) for each $r=1,\ldots,m$ the sequence $\{\Omega_{rs}\}$ satisfies the conditions of density and strong connectedness; (iv) the subdomains $\{\Omega_{rs}\}\ (r=1,\ldots,m)$ are mutually a_s-weakly connected; and (v) there exist the nonintersecting subdomains $\tilde{G}_{qs}\ (q=1,\ldots,N_s)$ such that $G_{qs}\subset\tilde{G}_{qs}\subset\Omega\setminus\bigcup_{r=1}^m\Omega_{rs}$, $\max_q\{\text{diam}\,\tilde{G}_{qs}\}\to 0$ as $s\to\infty$, and

$$\sum_{q=1}^{N_s}\inf_{\psi_{qs}}\int_\Omega\sum_{i,j=1}^n a_s(x,\nabla\psi_{qs})\,dx \le C<\infty,$$

where the infimum is taken over the class of the functions $\psi_{qs}\in H^1(\Omega)$, which are equal to 1 on G_{qs} and 0 outside $\tilde{G}_{qs}$, and the constant C does not depend on s.

THEOREM 3.3 *Let the functions* $a_{ij}^r(x,s,h)$, $m_r(x,s,h)$, *and* $b_{rq}(x,\lambda,s,h)$ *converge in* $L^1(\Omega)$ *to the functions* $a_{ij}^r(x)$, $m_r(x),b_{rq}(x,\lambda)$, *as* $s\to\infty$ *and* $h\to 0$. *Let* $m_r(x)\ge C>0$, *and* $b_{rq}(x,\lambda)$ *allows analytical continuation with respect to* λ *into the complex plane with the cut along the ray* $\arg\lambda=\pi$, *and let for* $|\lambda|\to\infty$ *and* $|\arg\lambda-\pi|\ge\alpha_0>0$ *the estimates* $|b_{rq}(x,\lambda)|=O(|\lambda|^\delta)$, *be valid where* $\delta<1$.

Then the solution $u^{(s)}(x,t)$ of the problem (1.1), (1.2) $L^2(\Omega_{rs})$-converges uniformly with respect to $t \in [0, T]$ to the functions $u_r(x,t)$ $(r = 1, \ldots, m)$ and the system of functions $\{u_1(x,t), \ldots, u_m(x,t)\}$ is the solution of initial boundary value problem (1.5), (1.6), where the matrix-function $\{B_{rq}(x,t); r, q = 1, \ldots, m\}$ is the inverse Laplace transform of the matrix-function $\{b_{rq}(x, \lambda)\lambda^{-1}m_r^{-1}(x); r, q = 1, \ldots, m\}$ with respect to t.

Simple examples of fulfillment of the conditions of Theorems 3.2 and 3.3 are considered in [13].

4. Homogenization in highly perforated domains and on Riemann manifolds

The analogues of Theorem 3.3 can also be proved for perforated domains Ω_s, which satisfy the condition of density in Ω. Moreover the domains with vanishing measure are permissible. Consider, for example, initial boundary value problem (1.1), (1.2) in the domains Ω_s with $a_{ij}^{(s)}(x) = \delta_{ij}$ $(x \in \Omega_s)$ and Neumann boundary condition on $\partial\Omega_s$. Set in (2.2) and (2.3), $a_s(x, p) = m_s^{-1}\chi_s(x)|p|^2$ and replace everywhere Ω by Ω_s and $\chi_{0s}(x)$ by $m_s^{-1}\chi_{0s}(x)$. In this case the statement of Theorem 3.3 holds as well.

As is seen from the examples considered in [13], in this case the weak connection between the domains Ω_{rs} $(r = 1, \ldots, m)$ is effected by narrow bridges or microcontacts, unlike the case of Section 3, where this connection was realized by the intermediate set Ω_{0s} with very weak conductivity. Note that both situations occur in the physics of high temperature superconductors, where a medium with volume distribution of weak connections is called the Josephson medium [2].

The notion of weak connection also arises in the problems of homogenization on Riemann manifolds M_s that depend on the parameter s. Consider, for example, the diffusion equation on the Riemann manifolds M_s of special form that consist of m copies Ω_{rs} $(r = 1, \ldots, m)$ of the Euclidean space $\mathbf{R}^n$ $(n \geq 2)$ with a large number of small "holes" F_{is} $(i = 1, \ldots, N_s; N_s \to \infty$ as $s \to \infty)$. The edges of the holes are attached either to each other or to the edges of the n-dimensional manifolds G_{rqs}^{ij} $(i, j = 1, \ldots, N_s; r, q = 1, \ldots, m)$, which are homotopic to finite length pipes or spherical caps. Thus, we have the manifold $M_s = (\bigcup \Omega_{rs})\bigcup(\bigcup G_{rqs}^{ij})$ whose topological type, generally speaking, increases as $s \to \infty$ (for $n = 2$ this is an m-sheeted surface with a large number of "handle", "bubbles" and "worm holes"). For the corresponding choice of the diameters of holes F_{is} and the distances between them, the subdomains Ω_{rs} $(r = 1, \ldots, m)$ and $\bigcup G_{rqs}^{ij}$ are "weakly connected", and as is shown in [4] the asymptotic behavior of the solutions of the diffusion equation on M_s is described by a system of equations that is an m-component nonlocal model with memory.

References

[1] Bakhvalov, N. S., and Panasenko, G. N.,*Averaging of processes in periodic media.* Mathematical Problems of Composite Material Mechanics, "Nauka", Moscow, 1984, p. 352 (Russian).

[2] Belous, N. A. et al., *Coherent properties and current-transfer in a Josephson medium the $BaPl_{1-x}Bi_xO_3$ ceramics*, JETP, **91**, 1986, pp. 274–286. (Russian).

[3] Bensoussan, A.; Lions J.-L.; and Papanicolau, G., Asymptotic analysis for periodic structures. Stud. Math. Appl., **5**, North-Holland, Amsterdam, 1978.

[4] Boutet de Monvel, L., and Khruslov, E. Ya., *Homogenization on Riemmanian manifolds*, preprint BiBoS N 560/93, Bielefeld, 1993, p. 37.

[5] Dal Maso, G., An introduction to Γ-convergence, Birkhäuser, Basel and Boston, MA, 1993, p. 400.

[6] De Giorgi, S. Spagnolo, *Sulla convergenza delli integrali dell energia per operatori ellittici del secondo ordine*, Boll. Un. Mat. Ital., **8**, 1973, pp. 391–411.

[7] Fenchenko, V. M., and Khruslov, E. Ya., *Asymptotics of solutions of differential equations with rapidly oscillating and degenerating matrix of coefficients*, Dokl. Akad. Nauk Ukrain. SSR Ser. A, N 4, 1980, pp. 25–29. (Russian).

[8] Fenchenko, V. M., and Khruslov, E. Ya., *Asymptotics of solutions of differential equations with strongly oscillating matrix of coefficients, which doesn't satisfy the condition of uniform boundedness*, Dokl. Akad. Nauk Ukrain. SSR Ser. A, N 4, 1981, pp. 23–27. (Russian).

[9] Freidlin, M. I., *Dirichlet problem for the equation with periodical coefficients, which depend on a small parameter*, Teor. Veroyatnost, **9**, N 1, 1964, pp. 133–139. (Russian).

[10] Khruslov, E. Ya., *On convergence of second boundary value problem solutions in weakly connected domains*, In: Theory of Operators in Functional Spaces and its Application, Naukova Dumka, Kiev (1981), pp. 129–173 (Russian).

[11] Khruslov, E. Ya., *Homogenized diffusion model in cracked-porous media*, Dokl. Akad. Nauk SSSR, **309**, N 2, 1989, pp. 332–335 (Russian).

[12] Khruslov, E. Ya., *Homogenized model of strongly inhomogeneous medium with memory*, Uspekhi Mat. Nauk, **45**, N 1, 1990, pp. 197–199. (Russian).

[13] Khruslov, E. Ya., *Homogenized models of composite media*, In: Composite Media and Homogenization Theory (edited by G. Dal Maso, and G. F. Dell'Antonio) Birkhäuser, Basel and Boston, MA, 1991, pp. 159–182.

[14] Kozlov, S. M., *Homogenization of differential operators with almost periodical coefficients*, Mat. Sb. (New Ser.), **107**, N 2, 1978, pp. 199–217. (Russian).

[15] Kozlov, S. M., *Homogenization of random operators*, Mat. Sb. (New Ser.), N 2, 1979, pp. 188–202. (Russian).

[16] Lions, J.-L., *Asymptotic expansions in perforated media with periodic structure*, Rocky Mountain J. Math., **10**, N 1, 1980, pp. 125–144.

[17] Marchenko, V. A., and Khruslov, E. Ya., *Boundary value problems in domains with fine-grained boundary*, Naukova Dumka, Kiev, 1974, p. 279 (Russian).

[18] Skrypnik, I. V., *Methods of investigations of nonlinear elliptic boundary value problems*, "Nauka", Moscow 1990, p. 448 (Russian).

[19] Zhikov, V. V.; Kozlov S. M.; and Oleinik, O. A., *Homogenization of differential operators*, Fiz.-Mat. Lit., Moscow, 1993, p. 461 (Russian).

Fluxes and Dimers in the Hubbard Model

ELLIOTT H. LIEB

Departments of Mathematics and Physics
Princeton University, Princeton, NJ, USA

The Hubbard model is the simplest conceivable example of interacting quantum-mechanical particles — yet it seems to have real world applications. Like the Ising model, it is "merely" a discrete lattice model, but despite four decades of research it is far from being understood rigorously, or even heuristically in many respects. The most amenable and symmetric situation is the "half-filled band", in which the number of particles equals the number of vertices in the lattice. Two conjectures about the model at half-filling have very recently been proved and are discussed here. The first can be found in [LN] and the second in [LE].

(A) The first is the old conjecture of Peierls, Fröhlich, and others that the *one-dimensional* version on a ring with an even number of vertices can exhibit symmetry breaking (called dimerization) from period one to period two, but no further, when the number of vertices is 2(mod4). More specifically, when the hopping amplitudes between adjacent vertices are allowed to vary freely, but where an energy penalty has to be paid to make these amplitudes large, then the optimal choice of the amplitudes will be either that they are all identical (period one) or else that they will be alternately large-small-large-small (period two), but nothing more chaotic than this.

To illustrate this first problem in more detail, consider a simple $(N \times N)$ Jacobi matrix T whose only nonzero entries are $t_{j+1,j} = t_{j,j+1} > 0$ (with $N+1 \equiv 1$ and N even). The eigenvalues of T come in opposite pairs $\mu, -\mu$ and we define $\lambda(T)$ to be the sum of the negative ones, i.e., $\lambda(T) = -\frac{1}{2}\mathrm{Tr}|T|$. Next, we take V to be a nice convex function on $\mathbf{R}^+$ with sufficiently fast increase at infinity, and define the energy

$$E(T) = \lambda(T) + \sum_{j=1}^{N} V(t_{j,j+1}). \tag{1}$$

The potential V thus acts as a penalty function for large $t_{j,j+1}$'s. The problem is to minimize $E(T)$ subject only to $t_{j+1,j} = t_{j,j+1} \geq 0$ for all j. The problem has the symmetry of the ring. The conjecture was that for a minimizer

$$t_{j,j+1} = a + (-1)^j b, \tag{2}$$

i.e., if the symmetry is broken it is broken only from period one to period two. This was shown to be the case by Kennedy and the author [KL] for quadratic V, i.e., when $V(t) = d(t - c)^2$.

Proceedings of the International Congress
of Mathematicians, Zürich, Switzerland 1994
© Birkhäuser Verlag, Basel, Switzerland 1995

The new result [LN] (obtained jointly with Nachtergaele) is that conclusion (2) is always correct for *general* V when $N = 2(\text{mod}4)$. When $N = 0(\text{mod}4)$ counterexamples can be found!

The model above describes free electrons on a ring, where the electron number equals the number of vertices in the ring, namely N. More important than the generalization to arbitrary V is the fact that the results in [LN] also include the case in which the electrons interact with each other by an attractive or repulsive Hubbard type "on-site" interaction. The conclusion does not change. Indeed the presence of the Hubbard interaction only seems to make it more difficult to find counterexamples.

(B) The second is the more recent "flux-phase" conjecture concerning a *planar, square lattice*. Here, the hopping elements are fixed in modulus (usually taken to be the elements of the adjacency matrix), but their arguments are *not* fixed. That is, if $t_{xy} = \bar{t}_{xy} = |t_{xy}| \exp[i\phi(x,y)]$ denotes the hopping matrix element from vertex x to vertex y in the lattice then $|t_{xy}|$ is fixed, but $\phi(x,y)$ is the variable quantity with respect to which an energy is to be minimized. The sum of the arguments of the t_{xy}'s around a circuit is called the flux, and the conjecture is that the value of the fluxes that minimizes the total energy is π in each square face of the lattice when the electron number equals the number of vertices. Even in the absence of a Hubbard interaction the conjecture is not obvious and previously unproved, and possibly has a differential-geometric significance. Partial results were obtained earlier by Loss and the author [LL] for special quasi-one-dimensional geometries without a Hubbard interaction. The new result [LE] goes beyond the original conjecture in several ways: (1) It does not assume, *a priori*, that all squares in the lattice must have the same flux; (2) a Hubbard type interaction can be included; (3) the conclusion holds for positive temperature as well as the ground state; (4) the results hold in $D \geq 2$ dimensions (e.g., the cubic lattice has the lowest energy if there is flux π in each square face).

References

[KL] T. Kennedy and E. H. Lieb, *Proof of the Peierls instability in one dimension*, Phys. Rev. Lett. **59** (1987), 1309–1312.

[LE] E. H. Lieb, *The flux phase of the half-filled band*, Phys. Rev. Lett. **73** (1994), 2158–2161.

[LL] E. H. Lieb and M. Loss, *Fluxes, Laplacians and Kasteleyn's theorem*, Duke Math. J. **71** (1993), 337–363.

[LN] E. H. Lieb and B. Nachtergaele, *The Stability of the Peierls Instability for Ring Shaped Molecules*, Phys. Rev. B. (3) **51** (1995), 4777–4791.

Von Neumann Algebras and Quantum Field Theory

ROBERTO LONGO

Dipartimento di Matematica
Università di Roma "Tor Vergata"
Via della Ricerca Scientifica
I-00133 Roma, Italy

and

Centro Linceo Interdisciplinare
Accademia Nazionale dei Lincei
Via della Lungara 10
I-00165 Roma, Italy

Von Neumann algebras were originally designed as a framework for various subjects, a major motivation coming from quantum theory. Although von Neumann algebras have since turned out to have unexpected connections with many different fields, the contact with quantum physics has remained constant and fruitful. I will try to give an account of a recent interplay between quantum field theory and inclusions of von Neumann algebras. The latter may arise for example as inclusions of local observable algebras, a structure analyzed earlier [10], [13], [39]–[42] and still productive. Here however I will deal with inclusions associated with superselection sectors [11], a setting related to index theory [36] along a line of research I have been following during about the last six years.

I will recall some of the general context. Let $\mathcal{H}$ be a Hilbert space that we always assume to be separable to simplify the exposition. With $B(\mathcal{H})$ the algebra of all bounded linear operators on $\mathcal{H}$ a von Neumann algebra $\mathcal{M}$ is a *-subalgebra of $B(\mathcal{H})$ containing the identity operator such that $\mathcal{M} = \mathcal{M}^-$ (weak closure) or equivalently $\mathcal{M} = \mathcal{M}''$ (double commutant): already at the foundation von Neumann density theorem manifests a typical double aspect, analytical and algebraic.

1 Tomita-Takesaki theory [54]

Let $\mathcal{M}$ be a von Neumann algebra and ω a normal faithful state on $\mathcal{M}$. The modular theory assigns to ω a canonical "evolution", i.e. a one-parameter group of *modular automorphisms* σ_t^ω of $\mathcal{M}$. σ^ω is a manifestation of the algebraic *-operation, but is characterized by the analytical Kubo-Martin-Schwinger (KMS) condition:

$$\omega(yx) = \underset{t \to i}{\text{anal.cont.}}\,\omega(\sigma_t^\omega(x)y), \qquad x, y \in \mathcal{M},$$

where the analytic continuation is realized by a bounded continuous function in the strip $0 \leq \text{Im}(z) \leq 1$ and analytic in its interior. Parallel to the modular theory the KMS condition was proposed by Haag, Hugenholtz, and Winnink [26] to characterize thermal equilibrium states in quantum statistical mechanics. The physical meaning of the modular theory remained however unclear in quantum field theory; see Section 7 (compare also with [6]).

Proceedings of the International Congress
of Mathematicians, Zürich, Switzerland 1994
© Birkhäuser Verlag, Basel, Switzerland 1995

By the GNS construction, we may assume that $\omega = (\cdot\,\Omega, \Omega)$ with Ω a cyclic and separating vector (namely $\mathcal{M}$ acts standardly). σ_t^ω is implemented by Δ^{it} with Δ the (positive, nonsingular) *modular operator*, the anti-unitary involution *modular conjugation* J implements an anti-isomorphism of $\mathcal{M}$ with $\mathcal{M}'$

$$JMJ = \mathcal{M}'.$$

2 Jones theory [36]

Let $\mathcal{N} \subset \mathcal{M}$ be an inclusion of factors (von Neumann algebras with trivial center). Assume $\mathcal{M}$ to be finite, namely there exists a tracial state ω on $\mathcal{M}$, i.e. σ^ω is trivial. As above we may assume that $\mathcal{M}$ acts standardly on $\mathcal{H}$. With e the projection onto the closure of $\mathcal{N}\Omega$, the von Neumann algebra generated by $\mathcal{M}$ and e

$$\mathcal{M}_1 = \langle \mathcal{M}, e \rangle = J_\mathcal{M} \mathcal{N}' J_\mathcal{M}$$

is a semifinite factor; if it is finite $\mathcal{N} \subset \mathcal{M}$ has finite *index* $\lambda = \omega(e)^{-1}$ with ω also denoting the trace of $\mathcal{M}_1$. Jones' theorem shows the possible values for the index:

$$\lambda \in \left\{ 4\cos^2\frac{\pi}{n}, n \geq 3 \right\} \cup [4, \infty]. \tag{2.1}$$

A probabilistic definition of the index was given by Pimsner and Popa through the inequality

$$\varepsilon(x) \geq \frac{1}{\lambda}x, \quad x \in \mathcal{M}^+,$$

where $\varepsilon : \mathcal{M} \to \mathcal{N}$ is the trace preserving conditional expectation [48].

A definition for the index $\mathrm{Ind}_\varepsilon(\mathcal{N}, \mathcal{M})$ of an arbitrary inclusion of factors $\mathcal{N} \subset \mathcal{M}$ with a normal conditional expectation $\varepsilon : \mathcal{M} \to \mathcal{N}$ was given by Kosaki [31] using Haagerup's dual weights. It depends on the choice of ε. The good properties are shared by the *minimal index* [29], [43]

$$\mathrm{Ind}(\mathcal{N}, \mathcal{M}) = \inf_\varepsilon \{ \mathrm{Ind}_\varepsilon(\mathcal{N}, \mathcal{M}) \} = \mathrm{Ind}_{\varepsilon_0}(\mathcal{N}, \mathcal{M})$$

where ε_0 is the unique *minimal conditional expectation*. I shall return to this from a different point of view in the next section.

3 Joint modular structure. Sectors

Let $\mathcal{N} \subset \mathcal{M}$ be an inclusion of infinite factors (no nonzero trace). We may assume that $\mathcal{N}'$ and $\mathcal{M}'$ are infinite so $\mathcal{M}$ and $\mathcal{N}$ act standardly. With $J_\mathcal{N}$ and $J_\mathcal{M}$ modular conjugations of $\mathcal{N}$ and $\mathcal{M}$, the unitary $\Gamma = J_\mathcal{N} J_\mathcal{M}$ implements a *canonical endomorphism* of $\mathcal{M}$ into $\mathcal{N}$

$$\gamma(x) = \Gamma x \Gamma^*, \qquad x \in \mathcal{M}.$$

γ depends on the choice of $J_\mathcal{N}$ and $J_\mathcal{M}$ only up to perturbations by an inner automorphism of $\mathcal{M}$ associated with a unitary in $\mathcal{N}$; γ is canonical as a sector of $\mathcal{M}$ as we define now.

Given the infinite factor $\mathcal{M}$, the *sectors of* $\mathcal{M}$ are given by

$$\mathrm{Sect}(\mathcal{M}) = \mathrm{End}(\mathcal{M})/\mathrm{Inn}(\mathcal{M}),$$

namely $\mathrm{Sect}(\mathcal{M})$ is the quotient of the semigroup of the endomorphisms of $\mathcal{M}$ modulo the equivalence relation: $\rho, \rho' \in \mathrm{End}(\mathcal{M})$, $\rho \sim \rho'$ iff there is a unitary $u \in \mathcal{M}$ such that $\rho'(x) = u\rho(x)u^*$ for all $x \in \mathcal{M}$.

$\mathrm{Sect}(\mathcal{M})$ is a *-semiring (there is an addition, a product and an involution) equivalent to the Connes correspondences (bimodules) on $\mathcal{M}$ up to unitary equivalence. We shall use the same symbol for an element of $\mathrm{End}(\mathcal{M})$ and for its class in $\mathrm{Sect}(\mathcal{M})$. The operations are:

Addition (direct sum): Let $\rho_1, \rho_2 \in \mathrm{End}(\mathcal{M})$; then

$$x \in \mathcal{M} \to \begin{bmatrix} \rho_1(x) & 0 \\ 0 & \rho_2(x) \end{bmatrix}$$

is an isomorphism of $\mathcal{M}$ into $\mathrm{Mat}_2(\mathcal{M})$. Because $\mathcal{M}$ is infinite, $\mathrm{Mat}_2(\mathcal{M})$ is (naturally up to inners) identified with $\mathcal{M}$ and we obtain a well-defined sector $\rho_1 \oplus \rho_2$ of $\mathcal{M}$.

Composition (monoidal product). The usual composition of maps

$$\rho_1 \circ \rho_2(x) = \rho_1(\rho_2(x)), \qquad x \in \mathcal{M},$$

defined on $\mathrm{End}(\mathcal{M})$ passes to the quotient $\mathrm{Sect}(\mathcal{M})$.

Conjugation. With $\rho \in \mathrm{End}(\mathcal{M})$, choose a canonical endomorphism $\gamma_\rho : \mathcal{M} \to \rho(\mathcal{M})$. Then

$$\bar{\rho} = \rho^{-1} \circ \gamma_\rho \tag{3.1}$$

well-defines a conjugation in $\mathrm{Sect}(\mathcal{M})$. By definition we thus have

$$\gamma_\rho = \rho \circ \bar{\rho}. \tag{3.2}$$

Moreover $\mathrm{End}(\mathcal{M})$ is a strict *tensor C^*-category with conjugates* [47], in particular the intertwiner linear space (ρ, ρ') between two objects $\rho, \rho' \in \mathrm{End}(\mathcal{M})$ is defined: $T \in (\rho, \rho')$ means $T \in \mathcal{M}$ and $\rho'(x)T = T\rho(x)$ for all $x \in \mathcal{M}$ (compare with [14]). The concepts of representation theory apply.

If ρ is irreducible (i.e. $\rho(\mathcal{M})' \cap \mathcal{M} = \mathbb{C}$) and has finite index, then $\bar{\rho}$ is the unique irreducible sector such that $\rho \circ \bar{\rho}$ contains the identity sector. More generally the objects $\rho, \bar{\rho} \in \mathrm{End}(\mathcal{M})$ are conjugate according to the analytic definition (3.1) and have finite index if and only if there exist isometries $v \in (\iota, \rho \circ \bar{\rho})$ and $\bar{v} \in (\iota, \bar{\rho} \circ \rho)$ such that

$$\bar{v}^* \bar{\rho}(v) = \frac{1}{d}, \qquad v^* \rho(\bar{v}) = \frac{1}{d}, \tag{3.3}$$

for some $d > 0$ (algebraic definition).

Formula (3.2) shows that given $\gamma \in \text{End}(\mathcal{M})$ the problem of deciding whether it is a canonical endomorphism with respect to some subfactor is essentially the problem of finding a "square root" ρ. If γ has finite index one finds that γ is canonical iff there exist isometries $T \in (\iota, \gamma)$, $S \in (\gamma, \gamma^2)$ satisfying the algebraic relations

$$\gamma(S)S = S^2 , \tag{3.4}$$

$$S^*\gamma(T) \in \mathbb{C}\backslash\{0\} , \quad T^*S \in \mathbb{C}\backslash\{0\}. \tag{3.5}$$

The minimal value of d in the formulas (3.3) is the *dimension* $d(\rho)$ of ρ; it is related to the minimal index by

$$\text{Ind}(\rho) = d(\rho)^2 \tag{3.6}$$

(with $\text{Ind}(\rho) = \text{Ind}(\rho(\mathcal{M}), \mathcal{M})$) and satisfies the character properties [33], [45]

$$d(\rho_1 \oplus \rho_2) = d(\rho_1) + d(\rho_2)$$
$$d(\rho_1 \circ \rho_2) = d(\rho_1)d(\rho_2)$$
$$d(\bar{\rho}) = d(\rho).$$

This structure has found applications to subfactor theory by Kosaki, Izumi, Popa, and others, see e.g. [9], [32], [34], [35], [49]. A related example is contained in [38].

The construction extends to the case of an abstract strict tensor C^*-category with conjugates $\mathcal{T}$ [47]. For each finite-dimensional object ρ there is an associated von Neumann algebra $\mathcal{M}_\rho$ and a functor

$$F : \mathcal{T}_\rho \to \text{End}(\mathcal{M}_\rho) \tag{3.7}$$

of the tensor C^*-category $\mathcal{T}_\rho$ of tensor powers of ρ into $\text{End}(\mathcal{M}_\rho)$. The functor F is full if ρ is amenable following Popa's work [49].

Examples of tensor C^*-categories with conjugates are provided by quantum groups or by (conformal) quantum field theory.

As an application one finds a duality for subfactors and for finite-dimensional complex semisimple Hopf algebras [46], [47]. An irreducible finite index inclusion of factors gives rise by (3.4), (3.5) to an irreducible *Q-system* i.e. a triple $(\mathcal{T}, \lambda, S)$, where $\mathcal{T}$ is a strict tensor C^*-category with conjugates, λ an object that generates $\mathcal{T}$ with the relations:

(a) (ι, λ) is one dimensional; namely there exists a unique element $T \in (\iota, \lambda)$, up to a phase; T is proportional to an isometry.

(b) there exists an arrow $S \in (\lambda, \lambda \otimes \lambda)$ proportional to an isometry such that

$$(b_1)\, 1_\lambda \otimes S \circ S = S \otimes 1_\lambda \circ S, \quad (b_2) \begin{cases} S^* \circ 1_\lambda \otimes T = 1_\lambda \\ T^* \otimes 1_\lambda \circ S = 1_\lambda . \end{cases}$$

Conversely a Q-system gives rise to a finite-index inclusion of von Neumann algebras. One sets up a bijection in the amenable case.

A finite-dimensional Hopf algebra is characterized by a Q-system with the additional distinguished property of the regular representation: $\lambda \otimes \lambda$ is equivalent to a finite multiple of λ.

4 Doplicher-Haag-Roberts theory [11], [12]

Haag and Kastler [27] have proposed long ago a framework to analyze a quantum field theory starting from first principles. With a region $\mathcal{O}$ of the space-time $\mathbb{R}^4$ (say $\mathcal{O} \in \mathcal{K}$, the family of double cones) one considers the von Neumann algebra $\mathcal{A}(\mathcal{O})$ generated by the observables localized in $\mathcal{O}$. The net $\mathcal{O} \to \mathcal{A}(\mathcal{O})$ satisfies the usual properties: *isotony*: $\mathcal{O}_1 \subset \mathcal{O}_2 \Rightarrow \mathcal{A}(\mathcal{O}_1) \subset \mathcal{A}(\mathcal{O}_2)$; *additivity*: $\mathcal{O} \subset \mathcal{O}_1 \cup \mathcal{O}_2 \cdots \cup \mathcal{O}_n \Rightarrow \mathcal{A}(\mathcal{O}) \subset (\cup_i \mathcal{A}(\mathcal{O}_i))''$; *locality*: $\mathcal{O}_1 \subset \mathcal{O}_2' \Rightarrow \mathcal{A}(\mathcal{O}_1) \subset \mathcal{A}(\mathcal{O}_2)'$, with $\mathcal{O}'$ denoting the space-like complement; this is strengthened to *Haag duality*: $\mathcal{A}(\mathcal{O}')' = \mathcal{A}(\mathcal{O})$, $\mathcal{O} \in \mathcal{K}$; *Poincaré covariance*: there exists a unitary representation U of the (connected) Poincaré group $\mathcal{P}_+^\uparrow$ on the underlying Hilbert space $\mathcal{H}$ with positive energy-momentum and a unique U-invariant vector Ω (*vacuum*) cyclic for the *quasi-local $C^* - algebra$* $\mathcal{A} = \cup_{\mathcal{O} \in \mathcal{K}} \mathcal{A}(\mathcal{O})^-$ (norm closure), with a covariant action $U(g)\mathcal{A}(\mathcal{O})U(g)^{-1} = \mathcal{A}(g\mathcal{O}), g \in \mathcal{P}_+^\uparrow, \mathcal{O} \in \mathcal{K}$.

A *superselection sector* in the sense of Wick, Wightman, and Wigner [58], i.e. a label for quantum "charges", is viewed in [25] as an equivalence class of representations of $\mathcal{A}$. To select physical representations Borchers has proposed to consider positive-energy covariant unitary representations ρ, namely there is a positive-energy representation U_ρ of the universal covering group $\tilde{\mathcal{P}}_+^\uparrow$ of the Poincaré group with

$$U_\rho(g)\rho(X)U_\rho(g)^{-1} = \rho(U(g)XU(g)^{-1}), \quad X \in \mathcal{A}, \quad g \in \tilde{\mathcal{P}}_+^\uparrow \qquad (4.1)$$

(where U is viewed as a representation of $\tilde{\mathcal{P}}_+^\uparrow$) [2].

An effective selection criterion was considered in [11]: for every double cone $\mathcal{O}$ the equivalence class of the representation contains an endomorphism ρ of $\mathcal{A}$ that acts identically on $\mathcal{A}(\mathcal{O}_1)$ for all $\mathcal{O}_1 \in \mathcal{K}$, $\mathcal{O}_1 \subset \mathcal{O}'$ (*localized endomorphism*). This criterion, suitable for short range interactions, was extended in [8] by relaxing $\mathcal{O}$ to be a space-like cone showing this class to exhaust all massive representations.

Let ρ be an endomorphism of $\mathcal{A}$ localized in $\mathcal{O} \in \mathcal{K}$ and $\rho_1 = u\rho(\cdot)u^*$ in the same class of ρ and localized in $\mathcal{O}_1 \subset \mathcal{O}'$. The statistics operator $\epsilon = u^*\rho(u)$ interchanges $\rho \circ \rho_1$ and $\rho_1 \circ \rho$. The $\epsilon_i = \rho^{i-1}(\epsilon)$, $i \in \mathbb{N}$, form a presentation of the permutation group $\mathbb{P}_\infty$ by transpositions

$$\epsilon_i^2 = 1, \quad \epsilon_i\epsilon_j = \epsilon_j\epsilon_i \text{ if } |i - j| \geq 2, \quad \epsilon_i\epsilon_{i+1}\epsilon_i = \epsilon_{i+1}\epsilon_i\epsilon_{i+1} \qquad (4.2)$$

and one obtains a unitary representation of $\mathbb{P}_\infty$, the *statistics* of ρ. There is a left inverse of ρ, a completely positive map $\Phi : \mathcal{A} \to \mathcal{A}$ with $\Phi \circ \rho = \text{id}$. If ρ is irreducible the *statistics parameter* $\lambda_\rho = \Phi(\epsilon)$ takes one of the values

$$\lambda_\rho = 0, \pm 1, \pm\frac{1}{2}, \pm\frac{1}{3}, \cdots$$

and classifies the statistics.

5 Index-statistics theorem

Let ρ be an endomorphism localized in the double cone $\mathcal{O}$. By duality ρ restricts to an endomorphism of $\mathcal{A}(\mathcal{O})$. A natural connection between the Jones and DHR theories is realized by the formula [43], [44]

$$\mathrm{Ind}(\rho) = d_{\mathrm{DHR}}(\rho)^2. \tag{5.1}$$

Here $\mathrm{Ind}(\rho)$ is $\mathrm{Ind}(\rho|_{\mathcal{A}(\mathcal{O})})$, the minimal index independently of the double cone $\mathcal{O}$ provided ρ is localized therein (however, in order that $\mathcal{A}(\mathcal{O})$ be a factor, one better replaces $\mathcal{O}$ by a wedge region $\mathcal{W}$, i.e. a Poincaré transformed of the region $\mathcal{W}_0 = \{\mathbf{x} : x_1 > x_0\}$; a straightforward comment on Driessler's theorem shows indeed that $\mathcal{A}(\mathcal{W})$ is a III_1-factor in Connes classification, see [39]) and $d_{\mathrm{DHR}}(\rho) = |\lambda_\rho|^{-1}$ is the DHR *statistical dimension* of ρ, in other words $d(\rho|_{\mathcal{A}(\mathcal{O})}) = d_{\mathrm{DHR}}(\rho)$. By formula (3.6) we will omit the suffix DHR. Because by duality $\rho(\mathcal{A}(\mathcal{O})) \subset \mathcal{A}(\mathcal{O})$ coincides with $\rho(\mathcal{A}(\mathcal{O})) \subset \rho(\mathcal{A}(\mathcal{O}'))'$ one may rewrite the index formula (5.1) directly in terms of the representation ρ.

The map $\rho \to \rho|_{\mathcal{A}(\mathcal{O})}$ is a faithful functor of tensor C^*-categories with conjugates from the endomorphisms localized in $\mathcal{O}$ into $\mathrm{End}(\mathcal{M})$ with $\mathcal{M} = \mathcal{A}(\mathcal{O})$ for any choice of $\mathcal{O} \in \mathcal{K}$. Passing to quotient one obtains a natural embedding

$$\text{Superselection sectors} \longrightarrow \mathrm{Sect}(\mathcal{M}). \tag{5.2}$$

The functor is full if $\mathcal{O}$ is a wedge.

6 Low-dimensional quantum field theory

The DHR theory is valid if the space-time dimension is 4, but not in low dimension. Low-dimensional statistics was considered in [20]. A first analysis was independently made in [17] and in [43]. On a two-dimensional space-time the formula $\epsilon_i^2 = 1$ is no longer true in general, but the other equations in (4.2) remain and provide a presentation of the Artin braid group $\mathbb{B}_\infty$.

The index-statistics theorem still applies, thus Jones' theorem (3.1) gives restrictions for the possible values of the statistical dimension

$$d(\rho) \in \{2\cos\frac{\pi}{n}, n \geq 3\} \cup [2, \infty].$$

If $d(\rho)$ is small and not 1 then either ρ^2 has two irreducible components (2-channel) or three irreducible components, one of which is an automorphism (3-channel essentially self-conjugate). The statistics is then classified in these cases. It is a braid group representation of Jones [37], [19] or Birman-Wenzl-Murakami [57] and is therefore described by the knot and link polynomial invariants of Jones and Kaufman.

Wenzl analysis [56], [57] is applicable. This shows in particular that, although there is a continuum of subfactor index values after 4, there are many gaps for the values of the square of the statistical dimension; in particular 4 is isolated on the right showing that subfactors arising from quantum field theory are of a particular kind, they are in fact "braided subfactors" [45]. Rehren [51] has refined this calculation to an optimal form in the interval $(4, 6)$ showing that only four index values are admissible in this range

$$4 < d(\rho)^2 < 6 \Rightarrow d(\rho)^2 = 5, 5.049\ldots, 5.236\ldots, 5.828\ldots$$

7 Relativistic invariance and the particle-antiparticle symmetry

By the Reeh-Schlieder theorem the vacuum vector Ω is cyclic and separating for any $\mathcal{A}(\mathcal{O})$, $\mathcal{O} \in \mathcal{K}$, as a consequence of the positivity of the energy. The Tomita-Takesaki theory faces naturally in this setting. Our analysis is based on the following

Modular covariance principle. With $\mathcal{W}$ any wedge region, the modular group of $\mathcal{A}(\mathcal{W})$ with respect to Ω acts covariantly as the rescaled pure Lorentz transformations $\Lambda_{\mathcal{W}}$ preserving $\mathcal{W}$:

$$\Delta_{\mathcal{W}}^{it} \mathcal{A}(\mathcal{O}) \Delta_{\mathcal{W}}^{-it} = \mathcal{A}(\Lambda_{\mathcal{W}}(2\pi t)\mathcal{O}), \quad \mathcal{O} \in \mathcal{K}, \quad t \in \mathbb{R}.$$

The main justification for this property comes from the Bisognano-Wichmann theorem [1] showing that indeed in a Wightman frame $\Delta_{\mathcal{W}}^{it} = U(\Lambda_{\mathcal{W}}(2\pi t))$.

On the physical side modular covariance enters in connection with an explanation of the Unruh effect [55] and the Hawking black hole thermal radiation [28], as noticed by Sewell [52] as follows. Modular covariance is equivalent to the KMS condition for the boosts as automorphisms of $\mathcal{A}(\mathcal{W})$ showing the vacuum to be a thermal equilibrium state for the system; on the other hand the boosts are the trajectory of a uniformly accelerated motion for which the "Rindler universe" $\mathcal{W}$ is a natural horizon; the equivalence principle in relativity theory then allows an interpretation of the thermal outcome manifested by the KMS property as a gravitational effect. On this basis Haag has proposed long ago to derive the Bisognano-Wichmann theorem.

Note that the formulation of the modular covariance property does not require a priori the Poincaré covariance of the vacuum sector, nor additivity, but the Reeh-Schlieder property for the vacuum vector, which we assume to hold for space-like cones; then the modular covariance implies the Poincaré covariance with positive energy of the vacuum sector and provides an intrinsic characterization of the vacuum state capable of extensions beyond the Minkoski space [4], [5].

In conformal quantum field theory the modular covariance principle is automatic [30], [5] and [21] (S^1 case), using a theorem of Borchers [3].

Let ρ be a localized endomorphism. Assuming the modular covariance property we have [22]:

$$\rho \text{ Poincaré covariant} \iff \exists \text{ conjugate sector } \bar{\rho}$$

(the implication $\Leftarrow$ needs a regularity condition for the net). Here the sector ρ may have infinite statistics and the definition of the conjugate sector $\bar{\rho}$ is provided by a consistent choice of a conjugate endomorphism (3.1) for $\rho|_{\mathcal{A}(\mathcal{W})}$ as $\mathcal{W}$ varies in the wedge regions. In [22] only wedges that are Euclidean transformed of $\mathcal{W}_0$ were considered and one obtained Euclidean covariance.

If ρ has finite statistical dimension a conjugate sector exists [11]. On the other hand in a massive theory a sector is localizable in a space-like cone and has finite statistical dimension [8], [16]; we thus have by the Buchholz-Fredenhagen theorem that, assuming the spectrum of $U|_{\mathbb{R}^4}$ has an isolated mass shell

$$\rho \text{ translation covariant} \implies \rho \text{ Poincaré covariant.}$$

The point is that the algebraic definition of the conjugate sector (3.1) makes the embedding (5.2) independent of $\mathcal{O}$. An interpretation of this fact according to the analytical definition of the conjugate sector and the geometric interpretation provided by the modular covariance property gives the results.

8 Algebraic spin-statistics theorem

The DHR theory shows the statistics, in particular the statistics parameter, to be intrinsically associated with a superselection sector ρ. The index-statistics theorem provides a new understanding of the absolute value of λ_ρ, but also

$$\kappa_\rho = \text{phase}(\lambda_\rho)$$

is intrinsic. On the other hand we have seen in Section 7 that, if $\text{Ind}(\rho) < \infty$, the Poincaré covariance property is also intrinsically associated with ρ and it is not an extra-structure. In particular the univalence $U_\rho(2\pi)$ of the representation U_ρ is also an intrinsic quantity that we are naturally led to relate to it. Now the spin-statistics theorem, familiar in the Wightman framework (see [51]), has an extension to the algebraic setting based on assumptions of finite multiplicity and strictly positive mass [12], [7]. The modular covariance principle provides the ground to obtain a general spin-statistics relation [23], [24]

$$\kappa_\rho = U_\rho(2\pi).$$

Moreover we have

$$\text{modular covariance} \Longrightarrow \text{PCT invariance}.$$

Instead of discussing the above form of the spin-statistics theorem, I will exemplify the result in the case of a conformal theory on S^1, where the modular covariance property is automatic (Section 7). We start then with a pre-cosheaf $\mathcal{A}$ of von Neumann algebras $\mathcal{A}(I)$ associated with (proper) intervals of S^1 ($\mathcal{A}$ is an inclusion preserving family, not a net), with locality $\mathcal{A}(S^1 \backslash I) \subset \mathcal{A}(I)'$, covariance with respect to the Möbius group $SL(2, \mathbb{R})/\{1, -1\}$ with positive (conformal) energy and uniqueness of the vacuum vector. Denote by I_1 the upper half-circle and I_2 the right half-circle and let the covariant (irreducible) superselection sector ρ be localized in $I_1 \cap I_2$. For simplicity we also assume that ρ is abelian ($\text{Ind}(\rho) = 1$). Then $\rho|_{\mathcal{A}(I_i)}$ is an automorphism and we consider the Araki-Connes-Haagerup unitary standard implementation V_i of $\rho|_{\mathcal{A}(I_i)}$ with respect to the vacuum vector.

It turns out that V_1 and V_2 commute up to a phase

$$V_1 V_2 = \mu V_2 V_1.$$

By construction the invariant μ has not only an algebraic and an analytic character, but also reflects a geometric aspect. Looking at it from these different points of view, one can show separately that

$$\mu = \kappa_\rho \qquad \text{and} \qquad \mu = U_\rho(2\pi),$$

where U_ρ is the unitary representation of the universal covering group of $SL(2, \mathbb{R})$ in the sector ρ. This is a starting point for more general analysis on curved space-time.

References

[1] Bisognano, J. and Wichmann, E., *On the duality condition for a Hermitian scalar field*, J. Math. Phys. **16** (1975), 985–1007.

[2] Borchers, H. J., *Local rings and the connection between spin and statistics*, Comm. Math. Phys. **1** (1965), 281–307.

[3] Borchers, H. J., *The CPT theorem in two-dimensional theories of local observables*, Comm. Math. Phys. **143** (1992), 315.

[4] Brunetti, R.; Guido, D.; and Longo, R., *Modular structure and duality in conformal quantum field theory*, Comm. Math. Phys. **156** (1993), 201–219.

[5] Brunetti, R.; Guido, D.; and Longo, R., *Group cohomology, modular theory and space-time symmetries*, Rev. Math. Phys., **7** (1995), 57–71.

[6] Buchholz, D.; D'Antoni, C.; and Longo, R., *Nuclear maps and modular structures. I*, J. Funct. Anal. **88** (1990), 223–250; *II*, Comm. Math. Phys. **129** (1990), 115–138.

[7] Buchholz, D., and Epstein, H., *Spin and statistics of quantum topological charges*, Fizika **3** (1985), 329–343.

[8] Buchholz, D., and Fredenhagen, K., *Locality and structure of particle states*, Comm. Math. Phys. **84** (1982), 1–54.

[9] Choda, M., *Extension algebras of II_1-factors via endomorphisms*, preprint.

[10] D'Antoni, C., and Longo, R., *Interpolation by type I factors and the flip automorphism*, J. Funct. Anal. **51** (1983), 361–371.

[11] Doplicher, S., Haag, R., and Roberts, J. E., *Local observables and particle statistics I*, Comm. Math. Phys. **23** (1971), 199–230.

[12] Doplicher, S., Haag, R., and Roberts, J. E., *Local observables and particle statistics II*, Commun. Math. Phys. **35** (1974), 49–85.

[13] Doplicher, S., and Longo, R., *Standard and split inclusions of von Neumann algebras*, Invent. Math. **73** (1984), 493–536.

[14] Doplicher, S., and Roberts, J. E., *Monoidal C^*-categories and a new duality for compact groups*, Invent. Math. **98** (1989), 157.

[15] Fidaleo, F., and Isola, T., *On the conjugate endomorphism for discrete and compact inclusions*, preprint.

[16] Fredenhagen, K., *On the existence of antiparticles*, Comm. Math. Phys. **79** (1981), 141.

[17] Fredenhagen, K., Rehren, K. H., and Schroer, B., *Superselection sectors with braid group statistics and exchange algebras. I*, Comm. Math. Phys. **125** (1989), 201–226.

[18] Fredenhagen, K., Rehren, K. H., and Schroer, B., *Superselection sectors with braid group statistics and exchange algebras. II*, Rev. Math. Phys. (1992), no. special issue, 111.

[19] Freyd, P., Yetter, D., Hoste, J., Lickorish, W., Millet, K., and Ocneanu, A., *A new polynomial invariant for knots and links*, Bull. Amer. Math. Soc. **12** (1985), 103–111.

[20] Fröhlich, J., *Statistics of fields, the Yang-Baxter equation, and the theory of knots and links*, G. t'Hooft et al. (eds)., Non-perturbative Quantum Field Theory, Plenum, New York, 1988.

[21] Fröhlich, J., and Gabbiani, F., *Operator algebras and Conformal Field Theory*, Comm. Math. Phys. **155** (1993), 569–640.

[22] Guido, D., and Longo, R., *Relativistic invariance and charge conjugation in quantum field theory*, Comm. Math. Phys. **148** (1992), 521–551.

[23] Guido, D., and Longo, R., *An algebraic spin and statistics theorem*, Comm. Math. Phys., to appear.

[24] Guido, D., and Longo, R., *The conformal spin and statistics theorem*, preprint.

[25] Haag, R., Local Quantum Physics, Springer-Verlag, Berlin and Heidelberg, 1992.

[26] Haag, R., Hugenoltz, N. M., and Winnink M., *On the equilibrium states in quantum statistical mechanics*, Comm. Math. Phys. **5** (1967), 215.

[27] Haag, R., and Kastler, D., *An algebraic approach to Quantum Field Theory*, J. Math. Phys. **5** (1964), 848–861.

[28] Hawking, S. W., *Particle creation by black holes*, Comm. Math. Phys. **43** (1975), 199.

[29] Hiai, F., *Minimizing index of conditional expectations onto subfactors*, Publ. Res. Inst. Math. Sci. **24** (1988), 673–678.

[30] Hislop, P., and Longo R., *Modular structure of the von Neumann algebras associated with the free scalar massless field theory*, Comm. Math. Phys. **84** (1982), 71–85.

[31] Kosaki, H., *Extension of Jones' theory on index to arbitrary subfactors*, J. Funct. Anal. **66** (1986), 123–140.

[32] Kosaki, H., *Sector theory and automorphisms for factor-subfactor pairs*, preprint.

[33] Kosaki, H., and Longo, R., *A remark on the minimal index of subfactors*, J. Funct. Anal. **107** (1992), 458–470.

[34] Izumi, M., *Application of fusion rules to classification of subfactors*, Publ. Res. Inst. Math. Sci. Kyoto Univ. **27** (1991), 953–994.

[35] Izumi, M., *Subalgebras of C^*-algebras with finite Watatani indices I. Cuntz algebras*, Comm. Math. Phys. **155** (1993), 157–182.

[36] Jones, V. R. F., *Index for subfactors*, Invent. Math. **72** (1983), 1–25.

[37] Jones, V. R. F., *Hecke algebras representations of braid groups and link polynomials*, Ann. Mat. **126** (1987), 335–338.

[38] Jones, V. R. F., *On a family of almost commuting endomorphisms*, J. Funct. Anal., to appear.

[39] Longo, R., *Algebraic and modular structure of von Neumann algebras of Physics*, Proc. Sympos. Pure Math. **38** (1982), Part 2, 551.

[40] Longo, R., *Solution of the factorial Stone-Weierstrass conjecture*, Invent. Math. **76** (1984), 145–155.

[41] Longo, R., *Simple injective subfactors*, Adv. in Math. **63** (1987), 152–171.

[42] Longo R., *Restricting a compact action to an injective subfactor*, Ergodic Theory Dynamical Systems **9** (1989), 127–135.

[43] Longo, R., *Index of subfactors and statistics of quantum fields. I*, Comm. Math. Phys. **126** (1989), 145–155.

[44] Longo, R., *Index of subfactors and statistics of quantum fields. II Correspondences, braid group statistics and Jones polynomial*, Comm. Math. Phys. **130** (1990), 285–309.

[45] Longo, R., *Minimal index and braided subfactors*, J. Funct. Anal. **109** (1992), 98–112.

[46] Longo, R., *A duality for Hopf algebras and for subfactors. I*, Comm. Math. Phys. **159** (1994), 133–150.

[47] Longo, R., and Roberts, J. E., *A theory of dimension*, K-theory (to appear).

[48] Pimsner, M., and Popa, M., *Entropy and index for subfactors*, Ann. Sci. École Norm. Sup. (4) **19** (1986), 57–106.

[49] Popa, S., *Classification of hyperfinite subfactors of type III_1*, CBMS lectures, preprint.

[50] Rehren, K. H., *Braid group statistics and their superselection rules*, D. Kastler, ed., The algebraic theory of superselection sectors. Introduction and results, World Scientific, Singapore and Teaneck, NJ, 1989.

[51] Rehren, K. H., *On the range of the index of subfactors*, J. Funct. Anal., to appear.

[52] Sewell, G. L., *Relativity of temperature and Hawking effect*, Phys. Lett. A **79** (1980), 23.

[53] Streater, R. F., and Wightman, A. S., PCT, Spin and Statistics, and all that, Benjamin, Reading, MA, 1964.

[54] Takesaki, M., Tomita's theory of modular Hilbert algebras and its applications, vol. 128, 1970.

[55] Unruh, W. G., *Notes on black hole evaporation*, Phys. Rev. D **14** (1976), 870.

[56] Wenzl, H., *Hecke algebras of type A_n and subfactors*, Invent. Math. **92** (1988), 349–383.

[57] Wenzl, H., *Quantum groups and subfactors of type B, C and D*, Comm. Math. Phys. **133** (1990), 1–49.

[58] Wick, G. C., Wightman, A. S., and Wigner, E. P., *The intrinsic parity of elementary particles*, Phys. Rev. **88** (1952), 101–105.

Two-dimensional Yang-Mills Theory and Topological Field Theory

GREGORY MOORE

Department of Physics, Yale University
New Haven, CT 06511, USA

1 Introduction

Two-dimensional Yang-Mills theory (YM_2) is often dismissed as a trivial system. In fact it is very rich mathematically and might be the source of some important lessons physically.

Mathematically, YM_2 has served as a tool for the study of the topology of the moduli spaces of flat connections on surfaces [2], [25], [38], [39]. Moreover, recent work has shown that it contains much information about the topology of Hurwitz spaces — moduli spaces of coverings of surfaces by surfaces.

Physically, YM_2 is important because it is the first example of a nonabelian gauge theory that can be reformulated as a string theory. Such a reformulation offers one of the few ways in which analytic results could be obtained for strongly coupled gauge theories. Motivations for a string reformulation include experimental "approximate duality" of strong interaction amplitudes, weak coupling expansions [34], strong coupling expansions [37], and loop equations [29]. The evidence is suggestive but far from conclusive. In [18] Gross proposed the search for a string formulation of Yang-Mills theory using the exact results of YM_2. This program has enjoyed some success. A successful outcome for YM_4 would have profound consequences, both mathematical and physical.

In order to describe the string interpretation of YM_2 properly, we will be led to a subject of broader significance: the construction of cohomological field theory (CohFT). This is reviewed in Section 6.

2 Exact Solution of YM_2

Let Σ_T be a closed 2-surface equipped with a Euclidean metric. Let G be a compact Lie group with Lie algebra $\mathbf{g}$, $P \rightarrow \Sigma_T$ a principal G-bundle, $\mathcal{G}(P) = \text{Aut}(P)$, $\mathcal{A}(P) =$ the space of connections on P. The action for YM_2 is the $\mathcal{G}(P)$-invariant function on $\mathcal{A}(P)$ defined by: $I_{\text{YM}} = \frac{1}{4e^2} \int_{\Sigma_T} \text{Tr}(F \wedge *F)$; $F = dA + A^2$, $* =$ Hodge dual, $e^2 =$ gauge coupling. I_{YM} is equivalent to a theory with action: $I(\phi, A) = -\frac{1}{2} \int_{\Sigma_T} i\text{Tr}(\phi F) + \frac{1}{2} e^2 \mu \text{Tr}\phi^2$; $\phi \in \Omega^0(M; \mathbf{g})$, $\mu = *1$, and Tr is normalized as in [38]: $\frac{1}{8\pi^2} \text{Tr} F^2$ represents the fundamental class of $H^4(B\tilde{G}; \mathbb{Z})$, where $\tilde{G}$ is the universal cover of G. Various definitions of the quantum theory will differ by a renormalization ambiguity $\Delta I = \alpha_1 \int \frac{R}{4\pi} + \alpha_2 e^2 \int \mu$. Equivalence to the theory

Proceedings of the International Congress
of Mathematicians, Zürich, Switzerland 1994
© Birkhäuser Verlag, Basel, Switzerland 1995

$I(\phi, A)$ shows that YM_2 is SDiff(Σ_T) invariant (no gluons!) and that amplitudes are functions only of the topology of Σ_T and $e^2 a$, where $a = \int \mu$.

The Hilbert space $\mathcal{H}_G$ is the space of class functions $L^2(G)^{Ad(G)}$ and has a natural basis given by unitary irreps: $\mathcal{H}_G = \oplus_R \mathbb{C} \cdot |R\rangle$. The Hamiltonian is essentially the quadratic Casimir: $C_2 + \alpha_2$. The amplitudes are nicely summarized using standard ideas from topological field theory. Let $\underline{\mathbf{S}}$ be the tensor category of oriented surfaces with area: Obj$(\underline{\mathbf{S}})=$ disjoint oriented circles, Mor$(\underline{\mathbf{S}})=$ oriented cobordisms, then:

THEOREM 2.1. *YM_2 amplitudes provide a representation of the geometric category $\underline{\mathbf{S}}$. The state associated to the cap of area a is:*

$$e^{\alpha_1} \sum_R \dim R e^{-e^2 a(C_2(R)+\alpha_2)} \, | R\rangle \qquad .$$

The morphism associated to the tube is

$$\sum_R e^{-e^2 a(C_2(R)+\alpha_2)} \, | R\rangle\langle R | \quad ,$$

and the trinion with two ingoing circles and one outgoing circle is:

$$e^{-\alpha_1} \sum_R (\dim R)^{-1} e^{-e^2 a(C_2(R)+\alpha_2)} \, | R\rangle\langle R | \otimes\langle R | \qquad .$$

Proof: The heat kernel defines a renormalization-group invariant plaquette action. $\qquad\square$

COROLLARY: *On a closed oriented surface Σ_T of area a and genus p the partition function is*

$$Z(e^2 a, p, G) = e^{\alpha_1(2-2p)} \sum_R (\dim R)^{2-2p} e^{-e^2 a(C_2(R)+\alpha_2)}. \tag{2.1}$$

These considerations go back to [28]. A clear exposition is given in [38].

3 YM_2 and the Moduli Space of Flat Bundles

At $e^2 a = 0$ the action $I(\phi, A)$ defines a topological field theory "of Schwarz type" [8]. In [38], [39] Witten applied YM_2 to the study of the topology of the space of flat G-connections on Σ_T: $\mathcal{M} \equiv \mathcal{M}(F = 0; \Sigma_T, P) = \{A \in \mathcal{A}(P) : F(A) = 0\}/\mathcal{G}(P).$[1]

Witten's first result is that, for appropriate choice of α_1, Z computes the symplectic volume of $\mathcal{M}$ [38]:

$$Z(0, p, G) = \frac{1}{\#Z(G)} \int_{\mathcal{M}} \exp \omega = \frac{1}{\#Z(G)} \text{vol}(\mathcal{M}) \tag{3.1}$$

---

[1]   We take a topologically trivial P for simplicity. $\mathcal{M}$ then has singularities, but the results extend to the case of twisted P, where $\mathcal{M}$ can be smooth [2].

where $Z(G)$ is the center, and ω is the symplectic form on $\mathcal{M}$ inherited from the 2-form on $\mathcal{A}$: $\omega(\delta A_1, \delta A_2) = \frac{1}{4\pi^2} \int_\Sigma \mathrm{Tr}(\delta A_1 \wedge \delta A_2)$. The argument uses a careful application of Faddeev-Popov gauge fixing and the triviality of analytic torsion on oriented 2-surfaces. The result extends to the unorientable case, and the constant α_1 can be evaluated by a direct computation of the Reidemeister torsion.

According to [38], (3.1) is the large k limit of the Verlinde formula [35]. Let $S_{RR'}(k)$ be the modular transformation matrix for the characters of integrable highest weight modules $R \in P_+^k$ of the affine Lie algebra $\mathbf{g}_k^{(1)}$ under $\tau \to -1/\tau$ [23]. At $e^2 a = 0$ we have:

$$Z = \lim_{k \to \infty} e^{\alpha_1 \chi(\Sigma_T)} \sum_{R \in P_+^k} \left(\frac{S_{00}(k)}{S_{0R}(k)} \right)^{2p-2}$$

where 0 denotes the basic representation. On the other hand, we may choose a complex structure J on Σ_T inducing a holomorphic line bundle $\mathcal{L} \to \mathcal{M}$ with $c_1(\mathcal{L}) = \omega$, and apply the Verlinde formula to get: $\lim_{k \to \infty} k^{-n} \sum_{P_+^k} (\frac{1}{S_{0R}})^{2p-2} = \lim_{k \to \infty} k^{-n} \dim H^0(\Sigma_T; \mathcal{L}^{\otimes k}) = \lim_{k \to \infty} k^{-n} \langle e^{kc_1(\mathcal{L})} Td\mathcal{M}, \mathcal{M} \rangle = \mathrm{vol}\,\mathcal{M}$, where $n = \frac{1}{2} \dim \mathcal{M}$. Using [23] one recovers (3.1) with

$$e^{\alpha_1} = (2\pi)^{\dim G} / (\sqrt{|P/L|}\,\mathrm{vol}\,G) = (\prod_{\alpha > 0} 2\pi(\alpha, \rho)) / \sqrt{|P/L|} \quad,$$

P is the weight lattice, L the long root lattice, and ρ the Weyl vector. The fact that the trinion is diagonal in the sum over representations is the large k limit of Verlinde's diagonalization of fusion rules.[2]

Witten's second result [39] gives the asymptotics of (2.1) for $e^2 a \to 0$ (set $a = 1$):

$$Z(e^2, p, G) \overset{e^2 \to 0}{\sim} \frac{1}{\#Z(G)} \int_\mathcal{M} e^{\omega + \epsilon \Theta} + \mathcal{O}(e^{-c/e^2}). \tag{3.2}$$

$e^2 = 2\pi^2 \epsilon$, $\alpha_2 = (\rho, \rho)$, and c is a constant. $\Theta \in H^4(\mathcal{M}; \mathbb{Q})$ is — roughly — the characteristic class obtained from $c_2(\mathcal{Q})$ where $\mathcal{Q} \to \Sigma_T \times \mathcal{M}^{\mathrm{irr}}$ is the universal flat G-bundle.[3] Θ is best thought of in terms of the $\mathcal{G}(P)$-equivariant cohomology of $\{A \in \mathcal{A}(P) : F(A) = 0\}$. In the Cartan model it is represented by $\frac{1}{8\pi^2} \mathrm{Tr} \phi^2$. The "physical argument" for (3.2) proceeds by writing the path integral as:

$$Z(e^2, p, G) = \frac{1}{\mathrm{vol}\,\mathcal{G}} \int d\phi\, dA\, d\psi \exp \left\{ \left[\frac{i}{4\pi^2} \int_\Sigma \mathrm{Tr}(\phi F - \frac{1}{2} \psi \wedge \psi) \right] \right.$$
$$\left. + \left[\epsilon \int_\Sigma \mu \frac{1}{8\pi^2} \mathrm{Tr} \phi^2 \right] \right\} \tag{3.3}$$

where ψ are the odd generators of the functions on the superspace $\Pi T\mathcal{A}$ and $dA\, d\psi$ is the Berezin measure. This path integral is the $t \to 0$ limit of the partition function of a cohomological field theory whose Q-exact action is $\Delta I = tQ \int \mu \mathrm{Tr} \psi^\alpha D_\alpha f$;

---

2)   This was first proved, using conformal field theoretic techniques, in [31].

3)   Precise definitions are given in [2].

$f = *F$, Q is the Cartan model differential for $\mathcal{G}$-equivariant cohomology of $\mathcal{A}$. The partition function is t-independent and localizes on the classical solutions of Yang-Mills. A clever argument maps the theory at $t \to \infty$ to "$D = 2$ Donaldson theory" and establishes the result. From a mathematical perspective the first term in the action of (3.3) is the $\mathcal{G}$-equivariant extension of the moment map on $\mathcal{A}$, the integral over A, ψ defines an equivariant differential form in $\Omega_{\mathcal{G}}(\mathcal{A})$, and the integral over ϕ defines equivariant integration of such forms. When the argument is applied to finite-dimensional integrals it leads to a rigorous result, namely the non-abelian localization theorem for equivariant integration of equivariant differential forms [39], [22].

4 Large N Limit: The Hilbert Space

The large N limit of YM_2 amplitudes is defined by taking $N \to \infty$ asymptotics for gauge group $G = SU(N)$, holding $e^2 a N \equiv \frac{1}{2}\lambda$ fixed. It is instructive to consider first the Hilbert space of the theory. In the large N limit the state space can be described by the conformal field theory (CFT) of free fermions [30], [13], [14]. Bosonization then provides the key to a geometrical reformulation in terms of coverings [19].

Nonrelativistic free fermions on S^1 enter the theory because class functions on $SU(N)$ can be mapped to totally antisymmetric functions on the maximal torus. The Slater determinants of N-body wavefunctions give the numerators of the Weyl character formula. The Fermi sea corresponds to the trivial representation with one-body states $\psi(\theta) = e^{in\theta}$ occupied for $\mid n \mid \leq \frac{1}{2}(N-1)$. In the representation basis the Hilbert space is: $\mathcal{H}_{SU(N)} = \oplus_{n \geq 0} \oplus_{Y \in \mathcal{Y}_n^{(N)}} \mathbf{C} \cdot \mid R(Y) \rangle$; $\mathcal{Y}_n$ = the set of Young diagrams with n boxes, $\mathcal{Y}_n^{(N)}$ is the subset of diagrams with $\leq N$ rows, $R(Y)$ is the $SU(N)$ representation corresponding to $Y \in \mathcal{Y}_n^{(N)}$. The naive $N \to \infty$ limit of $\mathcal{H}_{SU(N)}$ is $\mathcal{H}^+ = \oplus_{n \geq 0} \oplus_{Y \in \mathcal{Y}_n} \mathbf{C} \cdot \mid Y \rangle$. The space $\mathcal{H}^+$ is related to the state space of a $c = 1$ CFT. Excitations of energy $\ll N$ around the Fermi level $n_F = \frac{1}{2}(N-1)$ are described using the zero-charge sector $\mathcal{H}_{bc}^{Q=0}$ of a "$\lambda = 1/2$ bc CFT" [16], where $Q = \oint_{S^1} bc$. The point of this reformulation is that one can apply the well-known bosonization theorem, which relates the "representation basis" to the "conjugacy class basis." Focusing on one Fermi level we define fermionic oscillators $\{b_n, c_m\} = \delta_{n+m,0}$, a Heisenberg algebra $[\alpha_n, \alpha_m] = n\delta_{n+m,0}$ related by $\alpha_n = \sum b_{n-m} c_m$, and compare, at level $L_0 = n$, the fermionic basis: $\mid Y(h_1, \ldots, h_s) \rangle = c_{-h_1+1-\frac{1}{2}} \cdots c_{-h_s+s-\frac{1}{2}} b_{-v_1+1-\frac{1}{2}} \cdots b_{-v_s+s-\frac{1}{2}} \mid 0 \rangle$ where $Y(h_1, \ldots, h_s) \in \mathcal{Y}_n$ is a Young diagram with row lengths h_i, with the bosonic basis $\mid \vec{k} \rangle \equiv \prod_{j=1}^{\infty} (\alpha_{-j})^{k_j} \mid 0 \rangle$ where $\vec{k} = (k_1, k_2, \ldots)$ is a tuple of nonnegative integers, almost all 0. $\vec{k}$ specifies a partition of $n = \sum j k_j$ and a conjugacy class $C(\vec{k}) \subset S_n$. The fermi/bose overlap is given by the characters of the symmetric group representation $r(Y)$: $\langle \vec{k} \mid Y \rangle = \frac{1}{n!} \chi_{r(Y)}(C(\vec{k}))$.

When applying the above well-known technology to YM_2, one finds a crucial subtlety [19]: $\mathcal{H}^+$ is *not* the appropriate limit for YM_2. At $N \leq \infty$ there are two Fermi levels $n_F = \pm\frac{1}{2}(N-1)$; excitations around these different levels are related to tensor products of $N, \bar{N}$ representations, respectively. In the large N

limit one must consider representations occurring in the decomposition of tensor products $R \otimes \bar{S}$ where R, S are associated with Young diagrams with $n \ll N$ boxes, and $\bar{S}$ is the conjugate representation. That is, the correct limit for YM_2 is $\mathcal{H}_{SU(N)} \to \mathcal{H}^+ \otimes \mathcal{H}^-$. The two bc systems are naturally interpreted as left- and right-moving sectors of a $c = 1$ CFT.

Gross and Taylor provided an elegant interpretation of the $N \to \infty$ YM_2 Hilbert space in terms of covering maps [19]. The one-body string Hilbert space is identified with the group algebra $\mathbb{C}[\pi_1(S^1)]$. The state $| \vec{k} \rangle \in \mathcal{H}^+$ is identified with a state in the Fock space of strings defined by k_j j-fold oriented coverings $S^1 \to S^1$. The structure of the state space $\mathcal{H}^+ \otimes \mathcal{H}^-$ has a natural geometrical interpretation in terms of string states $| \vec{k} \rangle \otimes | \vec{l} \rangle$: $\vec{k}, \vec{l}$ describe orientation preserving/reversing coverings.

5 $1/N$ Expansion of Amplitudes

The $1/N$ expansion of YM_2 has a very interesting interpretation in terms of the mathematics of covering spaces of Σ_T. Heuristically, the worldsheet swept out by a j-fold cover $S^1 \to S^1$ defines a j-fold cover of a cylinder by a cylinder. Moreover, the Hamiltonian $H = C_2$ is not diagonal in the string basis. One finds a cubic interaction term describing the branched cover of a cylinder by a trinion [30], [13], [14].

To state a more precise relation we define the chiral partition function to be: $Z^+(\lambda, p) \equiv \sum_{n \geq 0} \sum_{Y \in \mathcal{Y}_n} (\dim R(Y))^{2-2p} e^{-\lambda C_2(R(Y))/(2N)}$. Z^+ exists as an asymptotic expansion in $1/N$. The $1/N$ expansion is related to topological invariants of Hurwitz spaces. To define these let $H(n, B, p, L)$ stand for the equivalence classes of connected branched coverings of Σ_T of degree n, branching number B, and L branch points. If $\mathcal{C}_L(\Sigma_T) \equiv \{(z_1, \ldots, z_L) \in \Sigma_T^L | z_i \in \Sigma_T, z_i \neq z_j\}/S_L$, then $H(n, B, p, L) \to \mathcal{C}_L(\Sigma_T)$ is an unbranched cover with discrete fiber above $S \in \mathcal{C}_L$ given by the equivalence classes of homomorphisms $\pi_1(\Sigma_T - S, y_0) \to S_n$, $y_0 \notin S$ [17]. Let $H(h, p) \equiv \amalg'_{n, B \geq 0} \amalg_{L=0}^B H(n, B, p, L)$ where the union on n, B is taken consistent with the Riemann-Hurwitz relation: $2h - 2 = n(2p - 2) + B$. We define the orbifold Euler characters of Hurwitz spaces by the formula

$$\chi_{\text{orb}}\{H(h, p)\} \equiv \sum_{n, B \geq 0}' \sum_{L=0}^{B} \chi(\mathcal{C}_L(\Sigma_T)) \sum_{\pi_0(H(n, B, p, L))} |\text{Aut } f|^{-1}.$$

THEOREM 5.1. ([19] and [10]). *For $p > 1$:*

$$Z^+(0, p) \overset{N \to \infty}{\sim} \exp\left[\sum_{h=0}^{\infty} \left(\frac{1}{N}\right)^{2h-2} \chi_{\text{orb}}\{H(h, p)\} \right].$$

Proof: In [19] Gross and Taylor used Schur-Weyl reciprocity to write $SU(N)$ representation-theoretic objects in terms of symmetric groups. A key step was the introduction of an element of the group algebra $\Omega_n = \sum_{v \in S_n} \left(\frac{1}{N}\right)^{n-K_v} v \in \mathbb{C}[S_n]$ where K_v is the number of cycles in v. Ω_n is invertible for $N > n$ and satisfies:

$$(\dim R(Y))^m = \left(\frac{N^n \dim r(Y)}{n!} \right)^m \frac{\chi_{r(Y)}(\Omega_n^m)}{\dim r(Y)}$$

for *all* integers m. Gross and Taylor showed that:

$$Z^+(\lambda, p) \sim \sum_{n,i,t,h=0}^{\infty} e^{-n\lambda/2}(-1)^i \frac{(\lambda)^{i+t+h}}{i!t!h!}\left(\frac{1}{N}\right)^{n(2p-2)+2h+i+2t}\frac{n^h(n^2-n)^t}{2^{t+h}}$$

$$\sum_{p_1,\ldots,p_i \in T_{2,n}} \sum_{s_1,t_1,\ldots,s_p,t_p \in S_n}\left[\frac{1}{n!}\delta(p_1\cdots p_i\Omega_n^{2-2p}\prod_{j=1}^{p}s_jt_js_j^{-1}t_j^{-1})\right].$$

$$(5.1)$$

$T_{2,n} \subset S_n$ is the conjugacy class of transpositions, and δ acts on an element of the group algebra by evaluation at 1. δ is nonvanishing when its argument defines a homomorphism $\psi : \pi_1(\Sigma_T - S, y_0) \to S_n$ for some subset $S \subset \Sigma_T$. One now uses Riemann's theorem identifying equivalence classes of degree n branched covers branched at $S \in \mathcal{C}_L(\Sigma_T)$ with equivalence classes of homomorphisms $\psi : \pi_1(\Sigma_T - S, y_0) \to S_n$ to interpret (5.1) as a sum over branched covers. Expanding the Ω^{-1} points to obtain the coefficients of the $1/N$ expansion gives the orbifold Euler characters of Hurwitz spaces. $\qquad\square$

The significance of this theorem is that it relates YM_2 to CohFT. To see this note that branched covers are related to holomorphic maps. Indeed, let $\tilde{\mathcal{M}}(\Sigma_w, \Sigma_T)$ $= \mathcal{C}^{\infty}(\Sigma_w, \Sigma_T) \times \text{Met}(\Sigma_w)$; $\text{Met}(\Sigma_w)$ is the space of smooth Riemannian metrics on a 2-surface Σ_w of genus h. The moduli space of holomorphic maps is $\text{Hol}(\Sigma_w, \Sigma_T) \equiv \{(f,g) \in \tilde{\mathcal{M}}(\Sigma_w, \Sigma_T) : df\epsilon(g) = Jdf\}/(\text{Diff}^+ \times \text{Weyl}(\Sigma_w))$; $\epsilon(g)$ is the complex structure on Σ_w inherited from g, $\text{Weyl}(\Sigma_w)$ is the group of local conformal rescalings acting on $\text{Met}(\Sigma_w)$. The definition of orbifold Euler character above is thus natural because the action by $\text{Diff}^+(\Sigma_w)$ on $\tilde{\mathcal{M}}$ has fixed points at maps with automorphism: $\chi_{\text{orb}}(H(h,p)) = \chi_{\text{orb}}(\text{Hol}(\Sigma_w, \Sigma_T))$. As we explain in the next section, CohFT partition functions are Euler characters of vector bundles over moduli spaces.

Theorem 5.1 has been extended in many directions to cover other correlation functions of YM_2[19], [10]. The results are not yet complete but are all in harmony with the identification of YM_2 as a CohFT. Wilson loop amplitudes are accounted for by Hurwitz spaces for coverings $\Sigma_w \to \Sigma_T$ by manifolds with boundary.[4] A formula analogous to (5.1) for the full, nonchiral theory has been given in [19]. The proof is not as rigorous as one might wish, but we do not doubt the result. The analog of Theorem 5.1 involves "coupled covers" [10]. A *coupled cover* $f : \Sigma_w \to \Sigma_T$ of Riemann surfaces is a map such that on the normalization of $N(\Sigma_w) = N^+(\Sigma_w) \amalg N^-(\Sigma_w)$ along the double points $\{Q_1, \ldots, Q_d\}$ of Σ_w, $N(f) = f^+ \amalg f^-$ where $f^+ : N^+(\Sigma_w) \to \Sigma_T$ is holomorphic and $f^- : N^-(\Sigma_w) \to \Sigma_T$ is antiholomorphic and $\forall i$, ramification indices match: $\text{Ram}(f^+, Q_i^+) = \text{Ram}(f^-, Q_i^-)$. One may define a "coupled Hurwitz space" $\mathcal{CH}(\Sigma_w, \Sigma_T)$ along the lines of the purely holomorphic theory. The $1/N$ expansion of the partition function again generates the Euler characters of $\mathcal{CH}(\Sigma_w, \Sigma_T)$, *if* coupled covers with ramified double points receive a weighting

---

4)   The $\text{SDiff}(\Sigma_T)$ invariance of YM_2 implies that Wilson loop averages define infinitely many invariants of immersions $S^1 \to \Sigma_T$.

factor $\prod_Q \mathrm{Ram}(f^+, Q^+)$ in the calculation of the Euler characteristic [10]. Put differently, the proper definition of "coupled Hurwitz space" involves a covering of the naive moduli space of coupled covers. This point has *not* been properly understood from the string viewpoint.

Finally, the results need to be extended to the case of nonzero area. When $\lambda \neq 0$ the expansion (5.1) and its nonchiral analog have the form:

$$Z^+(\lambda, p) = \sum_{h \geq 0} (1/N)^{2h-2} Z_{h,p}^+(\lambda) = \sum_{h \geq 0} (1/N)^{2h-2} \sum_n e^{-n\lambda/2} Z_{n,h,p}^+(\lambda).$$

For $p > 1$, $Z_{n,h,p}^+(\lambda)$ is polynomial in λ, of degree at most $(2h-2) - n(2p-2) = B$. The string interpretation described below shows that these polynomials are related to intersection numbers in $H(h, p)$. For $p = 1$, $Z_{h,1}^+(\lambda)$ are infinite sums that can be calculated using the relation to CFT described above [14]. These functions may be expressed in terms of Eisenstein series and hence satisfy modular properties in $\tau = i\lambda/(4\pi)$. For example: $Z_{1,1}^+ = e^{\lambda/48}\eta(i\lambda/(4\pi))$ (η is the Dedekind function) [18]. The modularity in the coupling constant might be an example of the phenomenon of "S-duality" which is currently under intensive investigation in other theories. For the case of a sphere: $Z_{0,0}(\lambda)$ has finite radius of convergence. At $\lambda = \pi^2$ there is a third order phase transition (=discontinuity in the third derivative of the free energy) [15]. The existence of such large N phase transitions might present a serious obstacle to a higher-dimensional Yang-Mills string formulation.

The λ-dependence of (5.1) has been interpreted geometrically in [19]. Contributions with $h, t > 0$ are related to degenerate Σ_w. In the framework of topological string theory the $h > 0$ contributions are probably related to the phenomenon of bubbling [5].

6 Cohomological Field Theory

CohFT is the study of intersection theory on moduli spaces using quantum field theory. Reviews include [40], [8], [9], [11]. The following discussion is a summary of the point of view explained at length in [11]. In physics the moduli spaces are presented as $\mathcal{M} = \{f \in \mathcal{C} : Df = 0\}/\mathcal{G}$ where $\mathcal{C}$ is a space of fields, D is a differential operator, and $\mathcal{G}$ is a group of local transformations. The action is an exact form in a model for the $\mathcal{G}$-equivariant cohomology of a vector bundle over $\mathcal{C}$. The path integral localizes to the fixed points of the differential Q of equivariant cohomology.

More precisely, the following construction of CohFT actions can be extracted from the literature [40], [44], [45], [7], [6], [3], [32], [24]. We begin with the basic data:

(1) $\mathcal{E} \to \mathcal{C}$, a vector bundle over a field space that is a sum of three factors: $\mathcal{E} = \Pi\mathcal{E}_{\mathrm{loc}} \oplus \mathcal{E}_{\mathrm{proj}} \oplus \Pi\mathcal{E}_{\mathrm{g.f.}}$ (the Π means the fiber is considered odd).

(2) $\mathcal{G}$-invariant metrics on $\mathcal{C}$ and $\mathcal{E}$.

(3) a $\mathcal{G}$-equivariant section $s : \mathcal{C} \to \mathcal{E}_{\mathrm{loc}}$, a $\mathcal{G}$-equivariant connection $\nabla s = ds + \theta s \in \Omega^1(\mathcal{C}; \mathcal{E}_{\mathrm{loc}})$, and a $\mathcal{G}$-*nonequivariant* section $\mathcal{F} : \mathcal{C} \to \mathcal{E}_{\mathrm{g.f.}}$ whose zeros determine local cross-sections for $\mathcal{C} \to \mathcal{C}/\mathcal{G}$.

The observables and action are best formulated using the "BRST model" of $\mathcal{G}$-equivariant cohomology [32], [24], [33]. To any Lie algebra $\mathbf{g}$ there is an associated differential graded Lie algebra (DGLA) $\mathbf{g}[\theta] \equiv \mathbf{g} \otimes \Lambda^* \theta;\ \theta^2 = 0,\ \deg \theta = -1,$ $\deg \mathbf{g} = 0,\ \partial\theta = 1$. Moreover, if M is a superspace with a $\mathbf{g}$-action then $\Omega^*(M)$ is a differential graded $\mathbf{g}[\theta]$ module, with $X \in \mathbf{g} \to \mathcal{L}_X,\ \bar{X} \otimes \theta \to \iota_X$. In our case $\mathbf{g} \to \mathrm{Lie}(\mathcal{G})$ and M is the total space of $\mathcal{E}$. The BRST complex is $\hat{\mathcal{E}} \equiv \Lambda^*\Sigma(\overline{\mathrm{Lie}(\mathcal{G})[\theta]})^* \otimes \Omega^*(\mathcal{E})$ where Σ is the suspension, increasing grading by 1. The differential on the complex is $Q = (d_{\mathcal{E}} + \partial') + d_{\mathrm{C.E.}}$ where ∂' is dual to ∂ and $d_{\mathrm{C.E.}}$ is the Chevalley-Eilenberg differential for the DGLA $\mathrm{Lie}(\mathcal{G})[\theta]$ acting on $\Omega^*(\mathcal{E})$. Physical observables $\hat{\mathcal{O}}_i$ are Q-cohomology classes of the "basic" ($\mathrm{Lie}(\mathcal{G})$-relative) subcomplex and correspond to basic forms $\mathcal{O}_i \in \Omega^*(\mathcal{C})$ that descend and restrict to cohomology classes $\omega_i \in H^*(\mathcal{M})$.

The Lagrangian is $I = Q\Psi$, the gauge fermion is a sum of three terms: $\Psi = \Psi_{\mathrm{loc}} + \Psi_{\mathrm{proj}} + \Psi_{\mathrm{g.f.}}$ for localization, projection, and gauge-fixing, respectively. Denoting antighosts (= generators of the functions on the fibers of $\mathcal{E}$) by $\rho + \theta\pi, \lambda + \theta\eta, \bar{c} + \theta\bar{\pi}$, of degrees $-1, -2, -1$, respectively, and taking, for definiteness, $\mathcal{E}_{\mathrm{g.f.}}|_f = \mathcal{E}_{\mathrm{proj}}|_f = \mathrm{Lie}(\mathcal{G})$ we have:

$$\begin{aligned}
\Psi_{\mathrm{loc}} &= -i\langle \rho, s \rangle - (\rho, \theta \cdot \rho)_{\mathcal{E}^*_{\mathrm{loc}}} + (\rho, \pi)_{\mathcal{E}^*_{\mathrm{loc}}} \\
\Psi_{\mathrm{proj}} &= i(\lambda, C^\dagger)_{\mathrm{Lie}(\mathcal{G})} \\
\Psi_{\mathrm{g.f.}} &= \langle \bar{c}, \mathcal{F}[A] \rangle - (\bar{c}, \bar{\pi})_{\mathrm{Lie}(\mathcal{G})}
\end{aligned} \tag{6.1}$$

where $C^\dagger = (dR_f)^\dagger \in \Omega^1(\mathcal{C}; \mathrm{Lie}(\mathcal{G}))$, is obtained, using the metrics, from the right $\mathcal{G}$ action through f, $R_f : \mathcal{G} \to \mathcal{C}$.

The main result of the theory is a path-integral representation for intersection numbers on $\mathcal{M}$ as correlation functions in the cohomological field theory:

$$\int_{\hat{\mathcal{E}}} \hat{\mu} e^{-I} \hat{\mathcal{O}}_1 \cdots \hat{\mathcal{O}}_k = \int_{\mathcal{M}=\mathcal{Z}(s)/\mathcal{G}} \chi[\mathrm{cok}(\mathbb{O})/\mathcal{G}] \wedge \omega_1 \wedge - \wedge \omega_k \tag{6.2}$$

where $\hat{\mu}$ is the Berezin measure on $\hat{\mathcal{E}}$ and $\mathbb{O} = \nabla s \oplus C^\dagger \in \Omega^1(\mathcal{C}; \mathcal{V} \oplus \mathrm{Lie}(\mathcal{G}))$ is Fredholm with $T\mathcal{M} \cong \ker\mathbb{O}/\mathcal{G}$. The argument for (6.2) may be sketched as follows. The equations $Df = 0$ define the vanishing locus of a cross-section $s(f) = Df \in \Gamma[\mathcal{E}_{\mathrm{loc}} \to \mathbb{CC}]$. Using the data of a metric and connection ∇ on a vector bundle E, one constructs the Mathai-Quillen representative $\Phi(E, \nabla)$ of the Thom class of E [27]. This construction can be applied — formally — in infinite dimensions to write the Thom class for $\mathcal{E}_{\mathrm{loc}}/\mathcal{G}$. When pulled back by a section $\bar{s} : \mathcal{C}/\mathcal{G} \to \mathcal{E}_{\mathrm{loc}}/\mathcal{G}$, $\bar{s}^*(\Phi(\mathcal{E}_{\mathrm{loc}}/\mathcal{G}, \nabla))$ is Poincaré dual to the zero locus $\mathcal{Z}(\bar{s}) = \mathcal{Z}(s)/\mathcal{G}$. The natural connection on $\mathcal{E}_{\mathrm{loc}}/\mathcal{G}$ is nonlocal in spacetime. In order to find a useful field-theoretic representation of the integral over $\mathcal{C}/\mathcal{G}$ one uses the "projection gauge fermion" Ψ_{proj} to rewrite the expression as an integral over $\mathcal{C}$. Finally, one must divide by the volume of the gauge group $\mathrm{vol}\,\mathcal{G}$, necessitating the introduction of $\Psi_{\mathrm{g.f.}}$. The "extra" factor of $\chi(\mathrm{cok}(\mathbb{O})/\mathcal{G})$ follows from a general topological argument [41] or from a careful evaluation of the measure near the Q-fixed points. $\square$

Two remarks are in order. First, the factor $\chi(\mathrm{cok}(\mathbb{O})/\mathcal{G})$ is crucial in studies of mirror symmetry [1], [42] and is also crucial to the formulation of the YM_2 string. Second, the formula (6.2) ignores the (important) singularities in $\mathcal{M}$. We conclude with four examples.

1. Donaldson Theory [44], [7], [3]: $P \to M$ is a principal G-bundle over a 4-fold M. $\mathcal{C} = \mathcal{A}(P)$, $\mathcal{G} = \mathrm{Aut}(P)$, $s(A) = F_+ \in \mathcal{E}_{\mathrm{loc}} = \mathcal{A} \times \Omega^{2,+}(M;\mathbf{g})$. $\mathrm{cok}\mathbb{O} = \{0\}$ (at irreducible connections). Observables are $\int_\gamma \Phi^*(\xi)$; $\gamma \in H_*(M), \xi \in H^*(BG)$, $\Phi : (P \times \mathcal{A}(P))/(G \times \mathcal{G}(P)) \to BG$ is the classifying map of the G-bundle $\mathcal{Q} \to \mathcal{A}/\mathcal{G} \times M$ of Atiyah-Singer [4]. (6.2) becomes Witten's path integral representation of the Donaldson polynomials.

2. Topological σ Model, $T\sigma(X)$ [45], [6]: $X = $ a compact, almost Kähler manifold with almost complex structure J. $\mathcal{C} = \mathrm{Map}(\Sigma_w, X)$. Σ_w has complex structure ϵ and $s(f) = df + Jdf\epsilon \in \mathcal{E}_{\mathrm{loc,f}} = \Gamma(T^*\Sigma \otimes f^*TX)$. Choosing a natural connection on $\mathcal{E}_{\mathrm{loc}}$ one finds $\mathrm{cok}(\mathbb{O}) \cong H^1(\Sigma, f^*(TX))$. Observables are the Gromov-Witten classes: $\int_\gamma \Phi^*(\xi)$; $\gamma \in H_*(\Sigma_w), \xi \in H^*(X)$, $\Phi : \Sigma_w \times \mathcal{C} \to X$ is the universal map.

3. Topological String Theory, $TS(X)$ [43], [36], [12]: $X = $ compact, Kähler, $\mathbb{F} = (f,h) \in \mathcal{C} = \hat{\mathcal{M}}(\Sigma_w, X) = \mathrm{Map}(\Sigma_w, X) \times \mathrm{Met}(\Sigma_w)$, $\mathcal{G} = \mathrm{Diff}^+(\Sigma_w)$, $s(\mathbb{F}) = (R(h)+1, df + Jdf\epsilon(h))$, the first equation eliminating the Weyl mode of the metric. Observables are products of Gromov-Witten classes and Mumford-Morita-Miller classes on the moduli space of curves.

4. Euler σ Model, $\mathcal{E}\sigma(X)$ [10], [11]: $X = $ compact, Kähler. If, in $TS(X)$, $\mathrm{cok}\mathbb{O} = \{0\}$, $\mathcal{E}\sigma(X)$ computes the Euler character of $\mathrm{Hol}(\Sigma_w, X)$. The fieldspace $\mathcal{C} \to \hat{\mathcal{M}}(\Sigma_w, X)$ is a vector bundle with fiber $\mathcal{E}_{\mathrm{loc}}^* \oplus \mathrm{Lie}(\mathcal{G})$. The section is: $s(\mathbb{F}, \hat{\mathbb{F}}) = (s(\mathbb{F}), \mathbb{O}^\dagger\hat{\mathbb{F}})$, and, by construction, the partition function is: $Z = \chi_{\mathrm{orb}}(\mathrm{Hol}(\Sigma_w, X))$, so $\mathcal{E}\sigma(\Sigma_T)$ is the string theory of (chiral) YM_2. The area dependence is obtained by perturbing the action by $\Delta I = \frac{1}{2}\int f^*\mathbf{k}$ where $\mathbf{k}$ is the Kähler class of Σ_T (this is only partially proven). A similar construction reproduces the nonchiral amplitudes but introduces an action that is fourth order in derivatives and requires further investigation. An alternative proposal for a string interpretation of YM_2 was made in [21]. This approach certainly deserves further study.

7 Application and a Guess

The original motivation for the program of Gross was to find a string interpretation of YM_4. Have we made any progress towards this end? The answer is not clear at present. We offer one suggestion here in the form of a guess.

Combining (3.2) with the $1/N$ asymptotics of the YM_2 partition function we expect[5] an intriguing relation between intersection theory on $\mathcal{M}(F = 0, \Sigma_T)$ for

---

5) To make this statement rigorous one must (a.) take care of the singularities in $\mathcal{M}$ and (b.) ensure that the corrections $\sim \mathcal{O}(e^{-2Nc/\lambda})$ from (3.2) are not overwhelmed by the "entropy of unstable solutions" [00]. The absence of phase transitions as a function of λ for $G > 1$ suggests that, for $G > 1$, these terms are indeed $\sim \mathcal{O}(e^{-Nc'})$ for some constant c'.

$G = SU(N)$ and the moduli spaces of holomorphic maps $\Sigma_w \to \Sigma_T$:

$$\left\langle exp\left[\omega + \frac{\lambda}{4\pi^2 N}\Theta\right]\right\rangle \overset{N\to\infty}{\sim}$$

$$C_N \sum_{h\geq 0}\left(\frac{1}{N}\right)^{2h-2} \sum_{d\geq 0} e^{-\frac{1}{2}d\lambda} P_d(\lambda)\chi_{\mathrm{orb}}(\mathcal{CH}(\Sigma_w, \Sigma_T, d)).$$

$$(7.1)$$

$C_N = N e^{\alpha_1(2-2p)-\lambda\alpha_2/(2N)}$, $\mathcal{CH}(\Sigma_w, \Sigma_T, d)$ is the coupled Hurwitz space for maps of total degree d, and P_d is a polynomial with $P_d(0) = 1$.

Now, the string theory of YM_2 *does* have a natural extension to four-dimensional target spaces: $I = I(\mathcal{E}\sigma(X))$ for X a compact Kähler 4-fold. Let e^α be a basis of $H_2(X, \mathbb{Z})$ with Poincaré dual basis $\mathbf{k}_\alpha$. The action may be perturbed by $\Delta I = t^\alpha \int f^* \mathbf{k}_\alpha$. Defining degrees d_α by: $f(\Sigma) = \sum d_\alpha e^\alpha \in H_2(X, \mathbb{Z})$, the partition function of the theory should have the form

$$Z(\mathcal{E}\sigma(X)) \sim \sum_{h\geq 0}\kappa^{2h-2} \sum_{d_\alpha \geq 0} e^{-t^\alpha d_\alpha} P_{d_\alpha}(t^\alpha)\chi(\mathcal{CH}(\Sigma_w, X; d_\alpha)),$$

more or less by construction, where κ is a string coupling constant and $P_{d_\alpha}(t^\alpha)$ is a polynomial whose value at zero is one. Our guess is that a formula analogous to (7.1) holds in four dimensions, and that the asymptotic expansion of $Z(\mathcal{E}\sigma(X))$ in κ is closely related to the large N asymptotics of intersection numbers of the classes $\mathcal{O}_2^{(2)}(e_\alpha) = \int_{e_\alpha} c_2(\mathcal{Q})$ on the moduli space of anti-self-dual instantons on X:

$$\left\langle e^{r^\alpha \mathcal{O}_2^{(2)}(e_\alpha)}\right\rangle_{\mathcal{M}_+(X;SU(N))}$$

where $\kappa \sim 1/N$ and r^α are analytic functions of the t^α.

Acknowledgments

The author is grateful to many colleagues for essential discussions and collaboration on the above issues; he thanks especially S. Cordes, R. Dijkgraaf, M. Douglas, E. Getzler, S. Ramgoolam, W. Taylor, and G. Zuckerman. He thanks the theory group at CERN for hospitality while the manuscript was being completed. He also thanks I. Frenkel for useful comments on the manuscript. This work is supported by DOE grants DE-AC02-76ER03075 and DE-FG02-92ER25121, and by a Presidential Young Investigator Award.

References

[1]   P. Aspinwall and D. Morrison, *Topological field theory and rational curves*, Comm. Math. Phys. **151** (1993), 245, hep-th/9110048.

[2]   M. F. Atiyah and R. Bott, *The Yang-Mills equations over Riemann surfaces*, Philos. Trans. Roy. Soc. London **A308** (1982), 523.

[3]   M.F. Atiyah and L. Jeffrey, *Topological Lagrangians and cohomology*, Jour. Geom. Phys. **7** (1990), 119.

[4]   M.F. Atiyah and I.M. Singer, *Dirac operators coupled to vector potentials*, Proc.
        Natl. Acad. Sci. **81** (1984), 2597.

[5]   M. Audin and J. Lafontaine, Holomorphic Curves in Symplectic Geometry, Birk-
        häuser, Basel and Boston, 1994.

[6]   L. Baulieu and I. Singer, *The Topological sigma model*, Comm. Math. Phys. **125**
        (1989), 227–237.

[7]   L. Baulieu and I. Singer, *Topological Yang-Mills Symmetry*, Nuclear Physics B.
        Proc. Suppl. **5B** (1988), 12.

[8]   Birmingham, Blau, Rakowski, and Thompson, *Topological field theories*, Phys. Rep.
        209 (1991), 129.

[9]   M. Blau, *The Mathai-Quillen formalism and topological field theory*, Notes of lec-
        tures given at the Karpacz Winter School on Infinite Dimensional Geometry in
        Physics.

[10]   S. Cordes, G. Moore, and S. Ramgoolam, *Large N 2D Yang-Mills theory and topo-
        logical string theory*, hep-th/9402107.

[11]   S. Cordes, S. Ramgoolam, and G. Moore, *Lectures on 2D Yang-Mills theory, equiv-
        ariant cohomology and topological field theory*, hep-th/9411210.

[12]   R. Dijkgraaf, E. Verlinde, and H. Verlinde, *Loop equations and Virasoro constraints
        in nonperturbative 2d quantum gravity; Topological strings in d < 1*, Nuclear Physics
        B **348** (1991) 435; B **352** (1991) 59; in: String Theory and Quantum Gravity, Proc.
        Trieste Spring School April 1990, World Scientific, Singapore, 1991.

[13]   M. R. Douglas, *Some comments on QCD string*, to appear in the proceedings of
        the Strings '93 Berkeley conference.

[14]   M.R. Douglas, *Conformal field theory techniques in large N Yang-Mills theory*, hep-
        th/9311130, to be published in the proceedings of the May 1993 Cargèse workshop
        on Strings, Conformal Models and Topological Field Theories.

[15]   M. Douglas and Kazakov, *Large N phase transition in continuum QCD₂*, hep-
        th/9305047.

[16]   D. Friedan, E. Martinec, and S. Shenker, *Conformal invariance, supersymmetry,
        and string theory*, Nuclear Physics B **271** (1986), 93.

[17]   W. Fulton, *Hurwitz schemes and irreducibility of moduli of algebraic curves*, Ann.
        of Math. (2) **90** (1969), 542.

[18]   D. Gross, *Some new/old approaches to QCD*, hep-th/9212148; *Two-dimensional
        QCD as a string theory*, hep-th/9212149, Nuclear Physics B **400** (1993), 161–180.

[19]   D. Gross and W. Taylor, *Two-dimensional QCD is a string theory*, hep-th/9301068;
        Twists and loops in the string theory of two-dimensional QCD, hep-th/9303046.

[20]   D.J. Gross and W. Taylor, *Two-dimensional QCD and strings*, talk presented at
        Strings '93, Berkeley, 1993; hep-th/9311072.

[21]   P. Horava, *Topological strings and QCD in two dimensions*, EFI-93-66,
        hep-th/9311156; to appear in the proc. of the Cargese Workshop, 1993.

[22]   L. Jeffrey and F. Kirwan, *Localization for nonabelian group actions*,
        alg-geom/9307001.

[23]   V.G. Kac and D.H. Peterson, Adv. in Math. **53** (1984), 125.

[24]   J. Kalkman, *BRST Model for equivariant cohomology and representatives for the
        equivariant Thom class*, Comm. Math. Phys. **153** (1993), 447; BRST model applied
        to symplectic geometry, hep-th/9308132.

[25]   F. Kirwan, Cohomology of Quotients in Symplectic and Algebraic Geometry,
        Princeton University Press, Princeton, 1984.

[26]   J.M.F. Labastida, M. Pernici, and E. Witten, *Topological gravity in two dimensions*,
        Nuclear Physics B **310** (1988), 611.

[27] V. Mathai and D. Quillen, *Superconnections, Thom classes, and equivariant differential forms*, Topology **25** (1986), 85.

[28] A. Migdal, *Recursion relations in Gauge theories*, Zh. Eksper. Teoret. Fiz. **69** (1975), 810 (Soviet Phys. JETP. **42**, 413).

[29] A. A. Migdal, *Loop equations and $1/N$ expansion*, Phys. Reports **102**, (1983), 199–290.

[30] J. Minahan and A. Polychronakos, *Equivalence of two-dimensional QCD and the $c = 1$ matrix model*, hep-th/9303153.

[31] G. Moore and N. Seiberg, *Polynomial equations for rational conformal field theories*, Phys. Lett. **212B** (1988), 451; *Classical and quantum conformal field theory*, Comm. Math. Phys. **123** (1989), 177; *Lectures on rational conformal field theory*, in: Strings 90, Proceedings of the 1990 Trieste Spring School on Superstrings.

[32] S. Ouvry, R. Stora, and P. Van Baal, Phys. Lett. **220B** (1989), 159.

[33] The following construction goes back to D. Quillen, *Rational homotopy theory*, Ann. of Math. (2) **90** (1969), 205. The application to the present case was developed with G. Zuckerman.

[34] G. 't Hooft, *A planar diagram theory for strong interactions*, Nuclear Physics B **72** (1974), 461.

[35] E. Verlinde, *Fusion rules and modular transformations in 2D conformal field theory*, Nuclear Physics B **300** (1988), 360.

[36] E. Verlinde and H. Verlinde, *A solution of two-dimensional topological quantum gravity*, Nuclear Physics B **348** (1991), 457.

[37] K. G. Wilson, *Confinement of quarks*, Phys. Rev. **D10** (1974), 2445.

[38] E. Witten, *On Quantum gauge theories in two dimensions*, Comm. Math. Phys. **141**, **153** (1991).

[39] E. Witten, *Two-dimensional gauge theories revisited*, hep-th/9204083, J. Phys. **G9** (1992), 303.

[40] E. Witten, *Introduction to cohomological field theory*, in: Trieste Quantum Field Theory 1990, 15–32 (QC174.45:C63:1990).

[41] E. Witten, *The N-matrix model and gauged WZW models*, Nuclear Physics B **371** (1992), 191.

[42] E. Witten, *Mirror manifolds and topological field theory*, hep-th/9112056, in: Essays on Mirror Manifolds, International Press, 1992.

[43] E. Witten, Nuclear Physics B **340** (1990), 281.

[44] E. Witten, *Topological quantum field theory*, Comm. Math. Phys. **117** (1988), 353.

[45] E. Witten, *Topological sigma models*, Comm. Math. Phys. **118** (1988), 411–419.

Mirror Symmetry and Moduli Spaces
of Superconformal Field Theories

DAVID R. MORRISON[*]

Department of Mathematics
Duke University
Durham, NC 27708-0320, USA

Mirror symmetry is the remarkable discovery in string theory that certain "mirror pairs" of Calabi-Yau manifolds apparently produce isomorphic physical theories — related by an isomorphism that reverses the sign of a certain quantum number — when used as backgrounds for string propagation [13], [19], [10], [16]. The sign reversal in the isomorphism has profound effects on the geometric interpretation of the pair of physical theories. This leads to startling predictions that certain geometric invariants of one Calabi-Yau manifold (essentially the numbers of holomorphic 2-spheres of various degrees) should be related to a completely different set of geometric invariants of the mirror partner ("period" integrals of holomorphic forms).

We will discuss the applications of this mirror symmetry principle to the study of the moduli spaces of two-dimensional conformal field theories with $N=(2,2)$ supersymmetry. Such theories depend on finitely many parameters, and for a large class of these theories the parameters admit a clear geometric interpretation. To circumvent the difficulties of trying to treat path integrals in a mathematically rigorous manner, we shall simply *define* the moduli spaces in terms of these geometric parameters. Other interesting physical quantities — the "topological" correlation functions — can then also be defined as asymptotic series whose coefficients have geometric meaning. The precise forms of the definitions are motivated by path integral arguments.

Mirror symmetry predicts some unexpected identifications between these moduli spaces, and serves as a powerful tool for understanding their structure. Perhaps the most striking consequence is the prediction that the moduli spaces can be analytically continued beyond the original domain of definition, into new regions, some of which parameterize conformal field theories that are related not to the original Calabi-Yau manifold, but rather to close cousins of it, which differ by simple topological transformations.

In preparing this report, I have drawn on a considerable body of earlier work [1]–[5], [21]–[24], much of which was collaborative. I would like to thank

---

[*]   Research partially supported by National Science Foundation Grants DMS-9103827, DMS-9304580, and DMS-9401447, and by an American Mathematical Society Centennial Fellowship.

Proceedings of the International Congress
of Mathematicians, Zürich, Switzerland 1994
© Birkhäuser Verlag, Basel, Switzerland 1995

my colleagues and collaborators Paul Aspinwall, Robert Bryant, Brian Greene, Sheldon Katz, Ronen Plesser, and Edward Witten for their contributions.

1 The physics of nonlinear σ-models

We begin by describing nonlinear σ-models from the point of view of physics (see [15] and the references therein), and giving a geometric interpretation to the parameters that appear in the theory. The starting data for constructing a nonlinear σ-model consists of a compact manifold X, a Riemannian metric g_{ij} on X, and a class $B \in H^2(X, \mathbb{R}/\mathbb{Z})$ (which we represent as a closed, $\mathbb{R}/\mathbb{Z}$-valued 2-form; i.e., a collection of closed, $\mathbb{R}$-valued 2-forms on the sets of an open cover of X that differ by $\mathbb{Z}$-valued forms on overlaps). The bosonic version of the nonlinear σ-model is then specified, in the Lagrangian formulation, by the $\mathbb{C}/\mathbb{Z}$-valued (Euclidean) action which assigns to each sufficiently smooth map ϕ from an oriented Riemannian 2-manifold Σ to X the quantity[1]

$$\mathcal{S}[\phi] := i \int_\Sigma \|d\phi\|^2 \, d\mu + \int_\Sigma \phi^*(B), \tag{1}$$

where the norm $\|d\phi\|$ of $d\phi \in \mathrm{Hom}(T_\Sigma, \phi^*(T_X))$ is determined from the Riemannian metrics on X and on Σ.

There is a variant of this theory in which additional fermionic terms are added to (1) to produce an action that is invariant under at least one supersymmetry transformation. (We will not write the fermionic terms in the action explicitly, as they do not enter into our analysis of the parameters.) The supersymmetric form of the action is also invariant under *additional* supersymmetry transformations when the geometry is restricted in certain ways — if the metric is Kähler then the theory has what is called $N{=}(2,2)$ supersymmetry, whereas if the metric is hyper-Kähler then the supersymmetry algebra is extended to $N{=}(4,4)$.

A nonlinear σ-model describes a consistent background for string propagation only if it is conformally invariant. The possible failure of conformal invariance is measured by the so-called "β-function" of the theory, and a perturbative calculation yields the result that the one-loop contribution to this β-function is proportional to the Ricci tensor of the metric. This makes Ricci-flat metrics — those with vanishing Ricci tensor — good candidates for producing conformally invariant σ-models. In fact, supersymmetric σ-models whose Ricci-flat metric is in addition hyper-Kähler are believed to be conformally invariant, as are bosonic σ-models whose metric is flat.

When the supersymmetry algebra of the theory cannot be extended as far as $N{=}(4,4)$, the Ricci-flat theories fail to be conformally invariant. However, when the Ricci-flat metrics are Kähler (i.e., when the theory has $N{=}(2,2)$ supersymmetry), we can deduce some of the properties of the conformally invariant theory by a careful study of the Ricci-flat theories. This works as follows: renormalization produces a flow on the space of metrics, and along a trajectory that begins at a

---

[1]We suppress the string coupling constant, and use a normalization in which the action appears as $\exp(2\pi i \mathcal{S})$ in the path integrals for correlation functions.

Ricci-flat Kähler metric, the metric is expected to remain Kähler with respect to a fixed complex structure on X, and the Kähler class of the metric is not expected to change. Thus, if there is a conformally invariant theory in the same universality class as this trajectory, i.e., if there is a fixed point of the flow that lies in the trajectory's closure, then any property of the conformal theory that depends only on the complex structure, the Kähler class, and the 2-form B can be calculated anywhere along the trajectory, including the initial, Ricci-flat theory. Furthermore, every Ricci-flat Kähler metric whose Kähler class is sufficiently deep within the Kähler cone is expected to determine a unique conformally invariant theory (which lies in the same universality class).

We can thus define a first approximation to the parameter space for $N{=}(2,2)$ superconformal field theories as follows (cf. [22]). Fix a compact manifold X, and define the *one-loop semiclassical nonlinear σ-model moduli space* of X to be

$$\mathcal{M}_\sigma := \{(g_{ij}, B)\} / \operatorname{Diff}(X), \tag{2}$$

where g_{ij} runs over the set of Ricci-flat metrics which are Kähler for some complex structure on X, B is an element of $H^2(X, \mathbb{R}/\mathbb{Z})$, and $\operatorname{Diff}(X)$ denotes the diffeomorphism group of X. Manifolds for which $\mathcal{M}_\sigma$ is nonempty (that is, those that admit a Ricci-flat Kähler metric) are called *Calabi-Yau manifolds*. The 2-form B should be regarded as some sort of "extra structure" (cf. [21]) that supplements the choice of metric.

It is important to keep in mind that the space $\mathcal{M}_\sigma$ is only an approximation to the moduli space of conformal field theories, for several reasons:

- As already mentioned, not every pair (g_{ij}, B) is expected to determine a conformal field theory, only those whose Kähler class is sufficiently deep within the Kähler cone.
- There may be analytic continuations of the space of conformal field theories beyond the domain where the theories have a σ-model interpretation. (We will see this in more detail in Section 6.)
- There may be points of $\mathcal{M}_\sigma$ that define isomorphic conformal field theories, even though they do not define isomorphic σ-models. This phenomenon was first observed in the case in which X is a torus of real dimension $2d$, and g_{ij} is a flat metric [25], [26]: in this case, $\mathcal{M}_\sigma = \Gamma_0 \backslash \mathcal{D}$, where $\mathcal{D}$ is a certain symmetric space and $\Gamma_0 = \Lambda^2 \mathbb{Z}^{2d} \rtimes \operatorname{GL}(2d, \mathbb{Z})$, while the actual moduli space of conformal field theories takes the form $\Gamma \backslash \mathcal{D}$ for some Γ containing the integral orthogonal group $\operatorname{O}(\mathbb{Z}^{2d,2d})$ (in which Γ_0 is a parabolic subgroup).

In spite of these limitations, $\mathcal{M}_\sigma$ provides a good arena for formulating a mathematical version of the theory, based on definitions using asymptotic expansions.

2 The correlation functions

The correlation functions of these quantum field theories will depend on the parameters in the action functional. If we construct a vector bundle over the moduli space whose fiber over a particular point is the Hilbert space of operators in the theory labeled by that point, then the correlation functions can be regarded as multilinear maps from this bundle to the complex numbers. These maps and their dependence on parameters can be studied by means of a semiclassical analysis, at least in a certain "topological" sector of the theory. (In this sector, the dependence of the correlation functions on the metric will always be a dependence on the Kähler class alone.)

The semiclassical properties of the $N=(2,2)$ theory are calculated in terms of the set of stationary values for the action (1). To find these, we pick a complex structure on Σ that makes its Riemannian metric Kähler, and that is compatible with its orientation. Then the first term in the action (1) can be rewritten using the formula:

$$\int_\Sigma \|d\phi\|^2 \, d\mu = \int_\Sigma \|\bar\partial\phi\|^2 \, d\mu + \int_\Sigma \phi^*(\omega), \qquad (3)$$

where $\bar\partial\phi \in \mathrm{Hom}(T_\Sigma^{(1,0)}, \phi^*(T_X^{(0,1)}))$ is determined by the complex structures, and where ω is the Kähler form of the metric g_{ij} on X. From this formula it is clear that the stationary values are the holomorphic maps; i.e., those with $\bar\partial\phi \equiv 0$. Furthermore, the action (1) evaluated on such a stationary value is the quantity

$$i\int_\Sigma \phi^*(\omega) + \int_\Sigma \phi^*(B) \in \mathbb{C}/\mathbb{Z}, \qquad (4)$$

which depends only on the homology class η of the map ϕ.

The path integral describing this quantum field theory has bosonic part

$$\int \mathcal{D}\phi \, e^{2\pi i \, \mathcal{S}[\phi]}, \qquad (5)$$

and the correlation functions are calculated by inserting operators into this expression (see for example Witten's address at the Berkeley ICM [31]). Such path integrals are of course problematic for mathematicians, but it is possible to use the outcome of the path integral manipulations as a basis for mathematical definitions.

To analyze these correlation functions, we break the path integral into a sum over homology classes. This produces an asymptotic expansion that is expected to converge for metrics whose Kähler class is sufficiently deep within the Kähler cone. The terms in the asymptotic expansion are themselves path integrals whose bosonic parts are the integrals of $\exp(2\pi i \int_\Sigma \|\bar\partial\phi\|^2 \, d\mu)$ over all maps of class η (with operators inserted), weighted by the exponential of $2\pi i$ times the classical action (4). For certain of the correlation functions, these "coefficient" path integrals can in turn be evaluated by the methods of topological field theory (cf. [32], [34]): upon modifying the fermionic terms in the action and introducing a parameter t, the path integral with bosonic part

$$\int_{[\phi]=\eta} \mathcal{D}\phi \, e^{2\pi i \, t \int_\Sigma \|\bar\partial\phi\|^2 \, d\mu} \qquad (6)$$

and "topological" operator insertions becomes independent of t. This integral can then be evaluated by the method of stationary phase, which reduces it to a finite-dimensional integral over the set of stationary maps in class η. Rigorous mathematical definitions for such "topological" correlation functions can be based on these finite-dimensional integrals, following ideas of Gromov [17] and Witten [32], [33]. See [20], [27], or Kontsevich's address at this congress for an account of these definitions and their properties.

In short, the physical quantities that can be calculated (by physicists) or defined (by mathematicians) using topological field theory will take the general form

$$\sum_{\eta \in H_2(X,\mathbb{Z})} c_\eta \, e^{2\pi i \, \langle B+i\omega, \eta \rangle}. \tag{7}$$

Notice that the only dependence on the metric is through the complex structure and the Kähler class ω. The coefficient c_η will depend on the set of all holomorphic maps in class η, and may well depend on the complex structure of X. (It also depends on the behavior of the fermionic terms in the action that we have suppressed.) The key property of interest here is the holomorphic dependence of these functions on parameters: the coefficients c_η depend holomorphically on the complex structure, and the dependence of (7) on $B + i\omega$ is also holomorphic (provided that the series converges and that $H^{2,0}(X_{\mathcal{J}}) = \{0\}$).

3 Mirror symmetry

The analysis of the previous sections ultimately derives from the specific form of our physical theory, which is based on the geometry of Ricci-flat metrics on X. We now adopt a somewhat more abstract point of view, and consider the structure of $N=(2,2)$ superconformal field theories *per se*.

The algebraic approach to conformal field theories — which treats them as unitary representations of the Virasoro algebra — has been extensively studied in the mathematics literature (cf. [14], for example). When the theories are supersymmetric, the algebra that acts on the representation can be enlarged. The enlargement relevant here is the $N=2$ superconformal algebra (for which a convenient reference is [19]). This is a super extension of the Virasoro algebra whose even part contains a $\mathfrak{u}(1)$-subalgebra in addition to the Virasoro algebra itself. From this algebraic point of view, an $N=(2,2)$ superconformal field theory is simply a unitary representation of two commuting copies of this algebra; there is thus an induced representation of the subalgebra $\mathfrak{u}(1) \times \mathfrak{u}(1)$.

The deformations of these representations have been analyzed in the physics literature [12], [13]. The infinitesimal deformations can be identified with the finite-dimensional kernel V of a certain operator, and it is argued in [12], [13] that there should be no obstructions to deforming in the directions corresponding to V.[2]

The $\mathfrak{u}(1) \times \mathfrak{u}(1)$ manifests itself on V in the following way: there are two commuting complex structures $\mathcal{J}$ and $\mathcal{J}'$ on V, each of which determines a natural representation of $\mathfrak{u}(1)$ on $V \otimes \mathbb{C}$ (with respect to which half of the charges[3] are $+1$ and half are -1). The two complex structures together determine a representation

---

[2]The arguments in [12] and [13] involve more of the physical structure than is present in the purely algebraic formulation we are discussing here. It would be desirable to have a purely algebraic proof of this statement.

[3]For a representation ρ of $\mathfrak{u}(1) \cong i\mathbb{R}$, the eigenvalues of $\rho(i)$ are called the *charges* of the representation.

of $\mathfrak{u}(1) \times \mathfrak{u}(1)$ on $V \otimes \mathbb{C}$, and we can decompose $V \otimes \mathbb{C}$ into four complex subspaces $V^{\pm 1, \pm 1}$ according to the $\mathfrak{u}(1) \times \mathfrak{u}(1)$ charges.

If we use $\mathcal{J}$ to put a complex structure on V and call the resulting space $V_{\mathcal{J}}$, then we can write $V_{\mathcal{J}} = V^{1,1} \oplus V^{1,-1}$. From this point of view, because $\mathcal{J}'$ has eigenvalues $\pm i$, respectively, on the two summands whereas $\mathcal{J}$ is simply multiplication by i, we can identify the summands as $V^{1,1} = \ker(\mathcal{J}\mathcal{J}' - \mathrm{Id}) \subset V_{\mathcal{J}}$ and $V^{1,-1} = \ker(\mathcal{J}\mathcal{J}' + \mathrm{Id}) \subset V_{\mathcal{J}}$.

A *mirror isomorphism* between two $N=(2,2)$ superconformal field theories is an isomorphism that reverses the sign of *one* of the $\mathfrak{u}(1)$ charges. If it is the second $\mathfrak{u}(1)$ charge that is reversed, then the isomorphism will map $V^{\pm 1, \pm 1}$ to $V^{\pm 1, \mp 1}$, and will interchange the factors in the decomposition $V_{\mathcal{J}} = V^{1,1} \oplus V^{1,-1}$.

A mirror isomorphism must preserve *all* correlation functions, not just the topological ones. It particular, it preserves the bilinear form on V that corresponds to the so-called Zamolodchikov metric on $\mathcal{M}_\sigma$. Thanks to the preservation of this metric, a mirror isomorphism at a single point can always be extended to a local isometry between the moduli spaces. There will also be a compatible isomorphism of the bundles of Hilbert spaces that maps the topological correlation functions from one theory to those of the other, but because of the sign change in the $\mathfrak{u}(1)$ charge, the geometric interpretations of these correlation functions may be rather different. For example, Candelas, de la Ossa, Green, and Parkes [11] used a mirror isomorphism to assert that a correlation function that they could compute exactly (using period integrals) as

$$5 + 2875 \, \frac{q}{1-q} + 609250 \, \frac{2^3 q^2}{1-q^2} + 317206375 \, \frac{3^3 q^3}{1-q^3} + 242467530000 \, \frac{4^3 q^4}{1-q^4} + \cdots$$

should coincide with a generating function of the form (7) in which the coefficients represent the numbers of holomorphic 2-spheres of various degrees on a quintic hypersurface in $\mathbb{CP}^4$. (See [21] or the Givental's address at this congress for some of the mathematical aspects of this generating function.)

4 Local analysis of the σ-model moduli space

The abstract description of the deformations of $N=(2,2)$ theories can be made very concrete for σ-models, where it reveals the local structure of the space $\mathcal{M}_\sigma$. The set of first-order variations δg of a fixed Riemannian metric g_{ij} on X can be identified with the space of symmetric contravariant 2-tensors $\Gamma(\mathrm{Sym}^2 T_X^*)$. If X is compact and g_{ij} is Ricci-flat, then according to a theorem of Berger and Ebin [6] the space of first-order variations[4] δg (modulo $\mathrm{Diff}(X)$) that preserve the Ricci-flat condition can be identified with the kernel of the Lichnérowicz Laplacian Δ_L acting on $\Gamma(\mathrm{Sym}^2 T_X^*)$. On the other hand, the set of first-order variations δB of the 2-form B can be identified with the space of harmonic 2-forms $\ker \Delta \subset \Gamma(\Lambda^2 T_X^*)$. Because the Lichnérowicz Laplacian on 2-forms coincides with the ordinary Laplacian, the combined contravariant 2-tensor $\delta g + \delta B \in \Gamma(\bigotimes^2 T_X^*)$ satisfies $\Delta_L(\delta g + \delta B) = 0$. We can thus identify the tangent space to $\mathcal{M}_\sigma$ at (g_{ij}, B) with $\ker \Delta_L \subset \Gamma(\bigotimes^2 T_X^*)$.

---

[4]It follows from the theorem of Bogomolov [8], Tian [28], and Todorov [29] that first-order variations can always be extended to deformations of the metric.

Let us assume that the holonomy of g_{ij} takes its "generic" value for Ricci-flat Kähler metrics, namely $\mathrm{SU}(n)$, $n \geq 3$ (where $n := \dim_{\mathbb{C}} X$). In this case, the two complex structures that we are expecting from our abstract analysis can be described as follows. First, if we fix a complex structure[5] $\mathcal{J}$ on X with respect to which g_{ij} is Kähler, there is an induced operator $\mathcal{J}$ on $\Gamma(\bigotimes^2 T_X^*)$ defined by $\mathcal{J}h(x,y) := h(x, \mathcal{J}y)$. This new operator $\mathcal{J}$ commutes with Δ_L, and so induces an operator on the tangent space $\ker \Delta_L$ of $\mathcal{M}_\sigma$ whose square is $-\mathrm{Id}$, that is, a complex structure on $\ker \Delta_L$.

The second complex structure $\mathcal{J}'$ on $\ker \Delta_L$ is much less obvious. It can be characterized by the property that the product $\mathcal{J}\mathcal{J}'$ acts as $-\mathrm{Id}$ on the space of symmetric, skew-Hermitian tensors, and as $+\mathrm{Id}$ on the space of tensors that are either Hermitian or skew-symmetric. Explicitly, $\mathcal{J}'$ can be defined by the formula

$$\mathcal{J}'h(x,y) := \frac{1}{2}\left(-h(x, \mathcal{J}y) + h(y, \mathcal{J}x) + h(\mathcal{J}x, y) + h(\mathcal{J}y, x) \right). \tag{8}$$

Using $\mathcal{J}$ to put a complex structure on $\ker \Delta_L$, we can identify $V^{1,-1}$ with the space of symmetric, skew-Hermitian tensors in $\ker \Delta_L$. This space corresponds to that part of the moduli space of metrics that is obtained by varying the complex structure (cf. [7, Chapter 12]). The operator $\mathcal{J}$ preserves that space, and induces the usual complex structure on it. In fact, under our assumptions about the holonomy, the complex structure can be varied freely and we have $V^{1,-1} \cong H^1(T_X^{(1,0)})$, the latter being the space of first-order variations of complex structure.

We can similarly identify $V^{1,1}$ as the space consisting of tensors that are either Hermitian or skew-symmetric; on this space, the operator $\mathcal{J}$ mixes symmetric and skew-symmetric forms, so it does not have a classical interpretation in terms of metrics alone. The parameters associated to this part of the deformation space are of the form $B + i\omega$, and $V^{1,1} \cong H^{1,1}(X_{\mathcal{J}}) \cong H^2(X, \mathbb{C})$ (under our assumption that the holonomy is $\mathrm{SU}(n)$, $n \geq 3$).

A mirror isomorphism between Calabi-Yau manifolds X and Y thus identifies the space of complex deformations of X with the space of complexified Kähler deformations of Y, and vice versa (at least when the holonomy is "generic").

5 Global analysis of the σ-model moduli space

The moduli space of Ricci-flat metrics (and hence the nonlinear σ-model moduli space) can be analyzed globally as well as locally. To carry this out, we introduce a related space, which includes a choice of complex structure. Define

$$\mathcal{M}_{N=2} := \{(g_{ij}, B, \mathcal{J})\} / \mathrm{Diff}(X), \tag{9}$$

where $\mathcal{J}$ ranges over the complex structures on X with respect to which g_{ij} is Kähler. The holonomy group of the metric g_{ij} is necessarily contained in the $\mathrm{SU}(n)$ specified by $\mathcal{J}$. The fibers of the natural map $\mathcal{M}_{N=2} \to \mathcal{M}_\sigma$ depend on

---

[5]When the holonomy is $\mathrm{SU}(n)$, $n \geq 3$, there are precisely two such complex structures: $\mathcal{J}$ and $-\mathcal{J}$.

this holonomy group, and can be described as the set of U(n)'s that lie between the holonomy group and O($2n$). Some examples:

1. If the holonomy is SU(n), $n \geq 3$, then the fiber consists of two points. (This is the "generic" case.)
2. If the holonomy is Sp($n/2$), then the fiber is $\mathbb{CP}^1$. (This is the case of an indecomposable hyper-Kähler manifold, such as a K3 surface.)[6]

The real dimension of the fiber is always $\dim_{\mathbb{R}} H^{2,0}(X_{\mathcal{J}})$.

The structure of the space $\mathcal{M}_{N=2}$ can be determined from the natural map $\mathcal{M}_{N=2} \to \mathcal{M}_{\text{complex}} := \{\mathcal{J}\}/\operatorname{Diff}(X)$. By the theorems of Calabi [9] and Yau [36], the fibers of this map take the form $\mathcal{K}_{\mathbb{C}}(X_{\mathcal{J}})/\operatorname{Aut}(X_{\mathcal{J}})$, where $\mathcal{K}_{\mathbb{C}}(X_{\mathcal{J}})$ is the *complexified Kähler cone*[7]

$$\mathcal{K}_{\mathbb{C}}(X_{\mathcal{J}}) := \{B + i\omega \in H^2(X, \mathbb{C}/\mathbb{Z}) \mid \omega \in \mathcal{K}_{\mathcal{J}}\}, \tag{10}$$

$\mathcal{K}_{\mathcal{J}}$ being the set of Kähler classes on $X_{\mathcal{J}}$, and $\operatorname{Aut}(X_{\mathcal{J}})$ being the group of holomorphic automorphisms. It is this fact that gives us access to global information about the conformal field theory moduli space, as the moduli space of complex structures can be studied by the methods of algebraic geometry. For example, by a theorem of Viehweg [30] the subspace $\mathcal{M}_{\text{complex}}^{\mathcal{L}} \subset \mathcal{M}_{\text{complex}}$ consisting of all complex structures polarized with respect to a fixed class $\mathcal{L}$ is a quasi-projective variety, i.e., the complement of a finite number of compact subvarieties in a compact complex manifold. (And the spaces $\mathcal{M}_{\text{complex}}^{\mathcal{L}}$ are open subsets of $\mathcal{M}_{\text{complex}}$ when $H^{2,0}(X_{\mathcal{J}}) = \{0\}$.) In contrast, although $\mathcal{K}_{\mathbb{C}}$ has a canonical complex structure when $H^{2,0}(X_{\mathcal{J}}) = \{0\}$, it is typically a rather small domain.

Note that the expected condition for a given pair (g_{ij}, B) to determine a conformal field theory was stated in terms of the Kähler class only and was valid for every choice of complex structure. Thus, the global description of the complex structures should be valid for the conformal field theory moduli space itself. On the other hand, the complexified Kähler directions are subject to modification.

6 Beyond the Kähler cone

We now apply the mirror symmetry principle to study the moduli space in the case in which the holonomy of the Ricci-flat metrics on X is SU(n), $n \geq 3$.

Suppose that a mirror partner Y is known for X. The mirror map between the moduli spaces $\mathcal{M}_{\sigma}(X)$ and $\mathcal{M}_{\sigma}(Y)$ will certainly be well defined at points corresponding to metrics whose Kähler class is sufficiently deep within the Kähler cone, but in general we can only expect a partially defined, local isomorphism between these spaces. However, because of the global nature of the complex structure space $\mathcal{M}_{\text{complex}}(Y)$, we can deduce the structure of the Kähler moduli space $\mathcal{K}_{\mathbb{C}}(X)$ from even a *local* knowledge of the mirror map. In principle, the mirror map should be determined essentially uniquely from the structure of the Zamolodchikov

---

[6]K3 surfaces are "self-mirror," and the mirror map induces an automorphism of $\mathcal{M}_{\sigma}$. Thus, as in the case of a torus, the moduli space of conformal field theories of this type is a nontrivial quotient of $\mathcal{M}_{\sigma}$ (cf. [5], where this quotient is determined precisely).

[7]This definition differs slightly from ones we have given elsewhere [22], [23].

metric, once the derivative of the map is known at a single point. In practice, it is easier to approach the construction of the mirror map in other ways (based on the topological correlation functions), which determine it up to a finite number of unknown parameters. Even those parameters can often be determined. (See [24] for a recent review of this problem.)

This comparison of structure between Kähler and complex moduli spaces has been carried out in [1], [2] for cases in which a mirror partner is known (to physicists), thanks to some explicit constructions using the discrete series representation of the $N=(2,2)$ superconformal algebra [16]. The results are quite illuminating: on the one hand, the locally defined map

$$\mathcal{K}_{\mathbb{C}}(X) \dashrightarrow \mathcal{M}_{\text{complex}}(Y) \tag{11}$$

does *not* in general extend throughout $\mathcal{K}_{\mathbb{C}}(X)$, but instead there are points where the theories become singular, and the map encounters difficulties beyond those points.[8] On the other hand, the image of (11) is *not* all of $\mathcal{M}_{\text{complex}}(Y)$ — as we have already suggested, $\mathcal{K}_{\mathbb{C}}(X)$ is much smaller than $\mathcal{M}_{\text{complex}}(Y)$. This means that there must be a way to analytically continue the conformal field theories on X beyond the theories specified by $\mathcal{K}_{\mathbb{C}}(X)$ (because such theories occur in $\mathcal{M}_{\text{complex}}(Y)$). This second conclusion was independently reached by Witten [35] on somewhat different grounds.

What, then, lies beyond the Kähler cone for such theories? In some cases, the conformal field theories are σ-models on other Calabi-Yau manifolds that are obtained by a simple topological surgery from X (see [1], [2], and [35], or for a more mathematical account, [23]). In these cases, as the Kähler class is varied and allowed to approach a wall of the Kähler cone, a finite number of holomorphic 2-spheres have their areas approach 0. When the Kähler class is pushed beyond that wall, the areas of those 2-spheres would apparently become negative. However, the analytically continued σ-model should instead be formulated as a σ-model on a modified manifold X', which is obtained from X by a surgery along the 2-spheres in such a way that the sign of their (common) homology class has been reversed (cf. [18]).

The collection of complexified Kähler cones of the various topological models produces a rich combinatorial structure of regions in the moduli space corresponding to the different models. But even these do not fill up the entire conformal field theory moduli space — there are additional regions whose associated conformal field theories must be described by constructions other than σ-models [35], [2]. These theories are currently under active study.

---

[8]This phenomenon is already visible in the example considered in [11].

References

[1] P. S. Aspinwall, B. R. Greene, and D. R. Morrison, *Calabi-Yau moduli space, mirror manifolds and spacetime topology change in string theory*, Nuclear Phys. B **416** (1994), 414–480.

[2] _____, *Measuring small distances in N=2 sigma models*, Nuclear Phys. B **420** (1994), 184–242.

[3] P. S. Aspinwall and D. R. Morrison, *Topological field theory and rational curves*, Comm. Math. Phys. **151** (1993), 245–262.

[4] _____, *Chiral rings do not suffice: N=(2,2) theories with nonzero fundamental group*, Phys. Lett. B **334** (1994), 79–86.

[5] _____, *String theory on K3 surfaces*, Essays on Mirror Manifolds II (B. R. Greene and S.-T. Yau, eds.), International Press, Hong Kong, to appear.

[6] M. Berger and D. Ebin, *Some decompositions of the space of symmetric tensors on a Riemannian manifold*, J. Differential Geom. **3** (1969), 379–392.

[7] A. L. Besse, Einstein Manifolds, Springer-Verlag, Berlin, Heidelberg, and New York, 1987.

[8] F. A. Bogomolov, *Hamiltonian Kähler manifolds*, Dokl. Akad. Nauk SSSR **243**, no. 5 (1978), 1101–1104.

[9] E. Calabi, *On Kähler manifolds with vanishing canonical class*, Algebraic Geometry and Topology, A Symposium in Honor of S. Lefschetz (R. H. Fox et al., eds.), Princeton University Press, Princeton, NJ, 1957, pp. 78–89.

[10] P. Candelas, M. Lynker, and R. Schimmrigk, *Calabi-Yau manifolds in weighted* $\mathbb{P}_4$, Nuclear Phys. B **341** (1990), 383–402.

[11] P. Candelas, X. C. de la Ossa, P. S. Green, and L. Parkes, *A pair of Calabi-Yau manifolds as an exactly soluble superconformal theory*, Nuclear Phys. B **359** (1991), 21–74.

[12] M. Dine and N. Seiberg, *Microscopic knowledge from macroscopic physics in string theory*, Nuclear Phys. B **301** (1988), 357–380.

[13] L. J. Dixon, *Some world-sheet properties of superstring compactifications, on orbifolds and otherwise*, Superstrings, Unified Theories, and Cosmology 1987 (G. Furlan et al., eds.), World Scientific, Singapore and Teaneck, NJ, 1988, pp. 67–126.

[14] I. Frenkel, J. Lepowsky, and A. Meurman, Vertex Operator Algebras and the Monster, Academic Press, New York and San Diego, 1988.

[15] M. B. Green, J. H. Schwarz, and E. Witten, Superstring Theory, 2 vols., Cambridge University Press, London and New York, 1987.

[16] B. R. Greene and M. R. Plesser, *Duality in Calabi-Yau moduli space*, Nuclear Phys. B **338** (1990), 15–37.

[17] M. Gromov, *Soft and hard symplectic geometry*, Proc. Internat. Congress Math. Berkeley 1986, vol. 1, Amer. Math. Soc., Providence, RI, 1987, pp. 81–98.

[18] V. Guillemin and S. Sternberg, *Birational equivalence in the symplectic category*, Invent. Math. **97** (1989), 485–522.

[19] W. Lerche, C. Vafa, and N. P. Warner, *Chiral rings in* $N = 2$ *superconformal theories*, Nuclear Phys. B **324** (1989), 427–474.

[20] D. McDuff and D. Salamon, J-holomorphic Curves and Quantum Cohomology, University Lecture Series, vol. 6, Amer. Math. Soc., Providence, RI, 1994.

[21] D. R. Morrison, *Mirror symmetry and rational curves on quintic threefolds: A guide for mathematicians*, J. Amer. Math. Soc. **6** (1993), 223–247.

[22] _____, *Compactifications of moduli spaces inspired by mirror symmetry*, Journées de Géométrie Algébrique d'Orsay (Juillet 1992), Astérisque, vol. 218, Société Mathématique de France, 1993, pp. 243–271.

[23] _____, *Beyond the Kähler cone*, Proc. Hirzebruch's 65th Birthday Workshop in Algebraic Geometry, to appear.

[24] _____, *Making enumerative predictions by means of mirror symmetry*, Essays on Mirror Manifolds II (B. R. Greene and S.-T. Yau, eds.), International Press, Hong Kong, to appear.

[25] K. S. Narain, *New heterotic string theories in uncompactified dimensions* < 10, Phys. Lett. B **169** (1986), 41–46.

[26] K. S. Narain, M. H. Sarmadi, and E. Witten, *A note on toroidal compactification of heterotic string theory*, Nuclear Phys. B **279** (1987), 369–379.

[27] Y. Ruan and G. Tian, *A mathematical theory of quantum cohomology*, Math. Res. Lett. **1** (1994), 269–278.

[28] G. Tian, *Smoothness of the universal deformation space of compact Calabi-Yau manifolds and its Petersson-Weil metric*, Mathematical Aspects of String Theory (S.-T. Yau, ed.), World Scientific, Singapore, 1987, pp. 629–646.

[29] A. N. Todorov, *The Weil-Petersson geometry of the moduli space of $SU(n \geq 3)$ (Calabi-Yau) manifolds, I*, Comm. Math. Phys. **126** (1989), 325–346.

[30] E. Viehweg, *Weak positivity and the stability of certain Hilbert points, III*, Invent. Math. **101** (1990), 521–543.

[31] E. Witten, *Geometry and physics*, Proc. Internat. Congress Math. Berkeley 1986, vol. 1, Amer. Math. Soc., Providence, RI, 1987, pp. 267–303.

[32] _____, *Topological sigma models*, Comm. Math. Phys. **118** (1988), 411–449.

[33] _____, *On the structure of the topological phase of two-dimensional gravity*, Nuclear Phys. B **340** (1990), 281–332.

[34] _____, *Mirror manifolds and topological field theory*, Essays on Mirror Manifolds (S.-T. Yau, ed.), International Press, Hong Kong, 1992, pp. 120–159.

[35] _____, *Phases of N=2 theories in two dimensions*, Nuclear Phys. B **403** (1993), 159–222.

[36] S.-T. Yau, *The role of partial differential equations in differential geometry*, Proc. Internat. Congress Math. Helsinki 1978, vol. 1, Academia Scientiarum Fennica, 1980, pp. 237–250.

The Critical Behavior of Random Systems

GORDON SLADE[*]

Department of Mathematics and Statistics, McMaster University,
Hamilton, ON, Canada L8S 4K1

1. Introduction

Self-avoiding walks, lattice trees and lattice animals, and percolation are among
the simplest models exhibiting the general features of critical phenomena. A basic
problem is to prove the existence of critical exponents governing their behavior
near the critical point. This problem gains importance from interrelations between
these models and models of ferromagnetism such as the Ising model, and from their
role in the theory of polymer molecules.

In low spatial dimensions, the principal mathematical questions about critical
exponents for these models remain unsolved. This is true in spite of significant
progress on such questions in the physics, chemistry, and numerical literature.
Proofs of existence of critical exponents are currently restricted to high spatial
dimensions, above the so-called upper critical dimensions. These proofs are based
on joint work with Hara, and for percolation, also on results of [4], [6]. The proofs
make use of an expansion method first introduced by Brydges and Spencer [9] and
known as the lace expansion. The results are summarized below. More extensive
summaries, with further references, can be found in [22], [26].

The results described below form a natural sequel to those for ferromagnetic
spin systems [1], [13] presented in Aizenman's ICM 1983 lecture [2], and provide
a resolution of an issue raised there for percolation.

2. Self-avoiding walks

An n-step self-avoiding walk on the hypercubic (integer) lattice is a mapping
$\omega : \{0, 1, \ldots, n\} \to \mathbb{Z}^d$, with $|\omega(i+1) - \omega(i)| = 1$ for all i (Euclidean distance) and
$\omega(i) \neq \omega(j)$ when $i \neq j$. Let $c_n(x, y)$ denote the number of n-step self-avoiding
walks with $\omega(0) = x$ and $\omega(n) = y$, and set $c_0(x, y) = \delta_{x,y}$. Let $c_n = \sum_y c_n(0, y)$
be the number of n-step self-avoiding walks that begin at the origin and end any-
where. Forty years ago, it was observed [15] that the elementary submultiplicativity inequality $c_{n+m} \leq c_n c_m$ implies the existence of the *connective constant*
$\mu = \lim_{n \to \infty} c_n^{1/n}$, with $c_n \geq \mu^n$ for all n. Accurate bounds on the value of μ have
been obtained [5], [11], [23].

[*]Research supported in part by NSERC grant A9351.

Proceedings of the International Congress
of Mathematicians, Zürich, Switzerland 1994
© Birkhäuser Verlag, Basel, Switzerland 1995

The sequence c_n is a primary object of study. It is believed that asymptotically

$$c_n \sim A\mu^n n^{\gamma-1} \quad \text{as } n \to \infty, \tag{2.1}$$

where the amplitude A and the critical exponent γ are dimension-dependent positive numbers. The numerical values of γ are believed to be $\frac{43}{32}$ for $d = 2$, about 1.162 for $d = 3$, and 1 for $d \geq 4$, with (2.1) replaced by $c_n \sim A\mu^n[\log n]^{1/4}$ in four dimensions. Attention is focused on the critical exponent γ because it is believed to be *universal*. Universality means that γ should depend only on the spatial dimension and be independent of such details as whether self-avoiding walks are defined on the square lattice or the hexagonal lattice in two dimensions, or whether we consider self-avoiding walks taking only nearest-neighbor steps or more general symmetric steps. The same will not be true of A or μ; for example, μ can be thought of roughly as the average number of possible next steps available to a long self-avoiding walk and hence depends on the specific lattice and the nature of the allowed steps.

A proof of universality for γ has not been found, and the very existence of γ has not yet been established in low dimensions. The current best bounds are

$$\mu^n \leq c_n \leq \begin{cases} \mu^n \exp[Cn^{1/2}] & d = 2 \\ \mu^n \exp[Cn^{2/5}\log n] & d = 3 \\ \mu^n \exp[Cn^{1/3}\log n] & d = 4, \end{cases} \tag{2.2}$$

with the lower bound due to submultiplicativity, and the upper bounds, which are based on submultiplicativity, due to [16], [25]. The bounds (2.2) are a long way from (2.1), but the situation is better in high dimensions.

THEOREM 2.1 *For any $d \geq 5$, there is a positive constant A such that as $n \to \infty$*

$$c_n = A\mu^n[1 + O(n^{-\epsilon})] \quad \text{for any } \epsilon < \tfrac{1}{2}.$$

For $d = 5$, $1 \leq A \leq 1.493$.

This theorem, due to [21], [19], implies that $\gamma = 1$ for $d \geq 5$. This is the same value as for simple random walks having no self-avoidance constraint, as n-step simple random walks are $(2d)^n$ in number.

A second quantity of interest is the mean-square displacement

$$\langle |\omega(n)|^2 \rangle = \frac{1}{c_n} \sum_\omega |\omega(n)|^2, \tag{2.3}$$

where the sum is over all n-step self-avoiding walks beginning at the origin. For simple random walks, the analogue of the mean-square displacement is equal to n. It is believed that asymptotically

$$\langle |\omega(n)|^2 \rangle \sim Dn^{2\nu}, \tag{2.4}$$

with positive D and ν. The critical exponent ν is believed to be equal to $\frac{3}{4}$ for $d = 2$, about 0.588 for $d = 3$, and $\frac{1}{2}$ for $d \geq 4$, with a logarithmic correction $\langle |\omega(n)|^2 \rangle \sim Dn[\log n]^{1/4}$ when $d = 4$. This has been proved for $d \geq 5$ [21], [19].

THEOREM 2.2 *For any $d \geq 5$, there is a positive constant D such that as $n \to \infty$*

$$\langle |\omega(n)|^2 \rangle = Dn[1 + O(n^{-\epsilon})] \quad \text{for any } \epsilon < \tfrac{1}{4}. \tag{2.5}$$

For $d = 5$, $1.098 \leq D \leq 1.803$.

For dimensions $d = 2, 3, 4$, it has not yet been proved either that $\langle |\omega(n)|^2 \rangle \geq O(n)$ nor $\langle |\omega(n)|^2 \rangle \leq O(n^{2-a})$ with $a > 0$, even though it appears implausible that self-avoiding walks typically remain closer to the origin than simple random walks, or move ballistically ($\nu = 1$) when $d > 1$.

The above two theorems show that in some aspects, self-avoiding walks behave like simple random walks above four dimensions. This is true in the following general sense. Given an n-step self-avoiding walk ω, define a continuous path X_n in $\mathbb{R}^d$ by setting $X_n(j/n) = (Dn)^{-1/2} \omega(j)$ for $j = 0, 1, \ldots, n$ and taking $X_n(t)$ to be the linear interpolation of this. This is a random path, via the uniform measure on n-step self-avoiding walks. For $d \geq 5$, a theorem of [21], [19] states that X_n converges in distribution to Brownian motion.

The fact that self-avoiding walks behave like simple random walks above four dimensions, but do not below four dimensions, is summarized by the statement that the upper critical dimension is equal to four. This can be partially understood from the fact that intersection properties of simple random walks change dramatically at $d = 4$. For example, the probability that two independent n-step simple random walks do not intersect remains bounded away from zero for $d > 4$, but not for $d \leq 4$. Also, two independent Brownian motion paths in $\mathbb{R}^d$ intersect each other with probability 1 if $d < 4$, but have empty intersection with probability 1 if $d \geq 4$. This is consistent with the fact that Brownian motion paths have Hausdorff dimension 2, as two 2-dimensional sets generically do not intersect above $d = 4$.

The proofs of the above theorems use generating functions. To define these generating functions, we let z denote a complex parameter and first define the *two-point function* by

$$G_z(x, y) = \sum_{n=0}^{\infty} c_n(x, y) z^n. \tag{2.6}$$

We define the *susceptibility* $\chi(z)$ by

$$\chi(z) = \sum_{x \in \mathbb{Z}^d} G_z(0, x) = \sum_{n=0}^{\infty} c_n z^n \tag{2.7}$$

and the *correlation length of order two* $\xi_2(z)$ by

$$\xi_2(z) = \left[\frac{\sum_{x \in \mathbb{Z}^d} |x|^2 G_z(0, x)}{\sum_{x \in \mathbb{Z}^d} G_z(0, x)} \right]^{1/2}. \tag{2.8}$$

These all have a radius of convergence equal to the *critical point* $z_c = \mu^{-1}$, and the manner of divergence of the susceptibility and correlation length of order two at z_c reflects the large-n asymptotics of c_n and the mean-square displacement. Theorems 2.1 and 2.2 are obtained using the following theorem together with contour integration methods.

THEOREM 2.3 *For any $d \geq 5$, and uniformly in complex z satisfying $|z| < z_c$,*

$$\chi(z) = \frac{Az_c}{z_c - z} + O(|z_c - z|^{-1+\epsilon}) \quad \text{for any } \epsilon < \tfrac{1}{2} \tag{2.9}$$

$$\xi_2(z) = \left(\frac{Dz_c}{z_c - z}\right)^{1/2} + O(|z_c - z|^{\epsilon-1/2}) \quad \text{for any } \epsilon < \tfrac{1}{4} \tag{2.10}$$

with the same constants A and D as in Theorems 2.1 and 2.2.

An important ingredient in the proof of the above theorems, as it was for the ferromagnetic models discussed in [1], [2], [13], is an infrared bound. This bound reflects the long-distance behavior of the critical two-point function indirectly through the behavior of its Fourier transform near the origin. In general, the Fourier transform of a summable function f on $\mathbb{Z}^d$ is defined by

$$\hat{f}(k) = \sum_{x \in \mathbb{Z}^d} f(x)e^{ik \cdot x}, \qquad k = (k_1, \ldots, k_d) \in [-\pi, \pi]^d, \tag{2.11}$$

where $k \cdot x = \sum_{j=1}^d k_j x_j$. The conjectured behavior of the critical two-point function is

$$\hat{G}_{z_c}(k) \sim \text{const.}\frac{1}{k^{2-\eta}}, \quad \text{as } k \to 0. \tag{2.12}$$

Scaling theory predicts that the critical exponent η is given in terms of γ and ν by Fisher's scaling relation $\gamma = (2 - \eta)\nu$ [12]. According to the conjectured values of γ and ν, η is nonnegative in all dimensions. This is a statement of the infrared bound, which can also be stated in the form $\hat{G}_{z_c}(k) \leq O(k^{-2})$. The infrared bound is believed to be true in all dimensions, but remains unproven for dimensions 2, 3, and 4. This k^{-2} behavior is the same as that for simple random walks, for which the analogue of $\hat{G}_{z_c}(k)$ is the massless lattice Green function $[1 - d^{-1}\sum_{j=1}^d \cos k_j]^{-1} \sim (2d)k^{-2}$. The following theorem gives an infrared bound for self-avoiding walks when $d \geq 5$.

THEOREM 2.4 *For $d \geq 5$, $\hat{G}_{z_c}(k)^{-1} = (2d)^{-1}k^2[DA^{-1} + O(k^\epsilon)]$ for any $\epsilon < \tfrac{1}{2}$.*

Proofs of these theorems, with further results along these lines, can be found in [21], [19]. The proofs all rely on the lace expansion, which is an expansion for the two-point function which treats self-avoiding walks as a perturbation of simple random walks. This perturbation is expected only to be small for $d > 4$, as self-avoiding walks and simple random walks do not have similar behavior in lower dimensions. Thus, $d > 4$ appears as a necessary condition for convergence of the lace expansion. But it is not the only condition, and a small parameter is needed to ensure convergence.

The small parameter turns out to be $B(z_c) - 1$, where the "critical bubble diagram" $B(z_c)$ is defined by

$$B(z) = \sum_{x \in \mathbb{Z}^d} G_z(0, x)^2. \tag{2.13}$$

For simple random walks, the analogue of the critical bubble diagram is the expected number of intersections of two independent simple random walks beginning at the origin. This is infinite for $d \leq 4$, but is finite for $d > 4$. For self-avoiding walks, by the Parseval relation,

$$B(z_c) = \int_{[-\pi,\pi]^d} \hat{G}_{z_c}(k)^2 \frac{d^d k}{(2\pi)^d}. \tag{2.14}$$

The infrared bound implies that this is finite for $d > 4$, whereas the conjectured values of the critical exponent η give an infinite value in lower dimensions.

The first attempts at theorems like the above worked in asymptotic regimes where the small parameter could be made as small as desired. Brydges and Spencer [9] analyzed the weakly self-avoiding walk, which is a measure on simple random walk paths where self-intersections are not prohibited, but rather give rise to a small penalty λ in the measure of a path. The weakly self-avoiding walk is believed on the basis of renormalization group considerations to be in the same universality class as the strictly self-avoiding walk, and hence to have the same critical exponents. For the weakly self-avoiding walk, $B(z_c) - 1$ enjoys a multiplicative factor λ^2, and for $d > 4$, can be made as small as desired by taking λ small. This simplifies convergence issues for the lace expansion, and has led to results for the weakly self-avoiding walk for dimensions $d > 4$.

A second attempt arose in a series of papers initiated in [29], where the strictly self-avoiding walk was studied in very high dimensions. As $d \to \infty$, the bubble diagram satisfies $B(z_c) - 1 = O(d^{-1})$, so by taking d sufficiently large, convergence of the lace expansion can be ensured. This allows for the treatment of the strictly self-avoiding walk, but the critical nature of $d = 4$ becomes obscured. Alternately, one can consider spread-out models. An example of a spread-out model is to consider self-avoiding walks in which the condition that a walk take nearest-neighbor steps is relaxed to allow any steps that change all coordinates by at most an amount L. For such walks, when $d > 4$, $B(z_c) - 1 \to 0$ as $L \to \infty$ and yields a small parameter. For L sufficiently large, the analogue of the theorems stated above can then be proved for all $d > 4$ [26]. This provides an example of universality, with the value of critical exponents independent of the value of (large) L, or indeed of the precise nature of the definition of the spread-out model.

Remarkably, even for the usual strictly self-avoiding walk in $d \geq 5$, $B(z_c) - 1$ turns out to be small enough to allow for a proof of convergence of the lace expansion. Our bound for $d = 5$ is $B(z_c) - 1 \leq 0.493$. This value is large enough that very detailed estimates were required to prove convergence, and computer assistance was necessary [19]. This is unsatisfactory, because the precise numerical value of the bubble should not be crucial. What should matter is that it is finite. The fact that the proof works for $d = 5$ is fortuitous; presumably the bubble diagram diverges when $d = 4$, so its value at $d = 5$ might well have been larger than 0.493. However, a proof whose driving force is the finiteness of the bubble, rather than its smallness, has not been found.

Finally, it should be mentioned that for $d = 4$ there has been recent progress towards computing the logarithmic corrections mentioned above, in work on related models involving a small parameter [8], [24].

3. Lattice trees and lattice animals

We will define both nearest-neighbor and spread-out versions of lattice trees and lattice animals. In the nearest-neighbor model, a *bond* is defined to be a pair $\{x, y\}$ of sites in $\mathbb{Z}^d$ separated by Euclidean distance 1. In the spread-out model, a bond is a pair $\{x, y\}$ with $|x_i - y_i| \le L$ $(i = 1, \dots, d)$. A *lattice tree* is defined to be a finite connected set of bonds without cycles (closed loops). Although a tree T is defined as a set of bonds, we will write $x \in T$ if x is an element of a bond in T. The number of bonds in T will be denoted $|T|$. A *lattice animal* is a finite connected set of bonds that may contain closed loops. Trees and animals are believed to be in the same universality class.

Let t_n denote the number of n-bond trees modulo translation, and let a_n denote the number of n-bond animals modulo translation. By a supermultiplicativity argument, $t_n^{1/n}$ and $a_n^{1/n}$ both converge to finite positive limits λ and λ_a as $n \to \infty$, with $t_n \le \lambda^n$ and $a_n \le \lambda_a^n$. The asymptotic behavior of both t_n and a_n as $n \to \infty$ is believed to be governed by the same universal critical exponent γ:

$$t_n \sim \text{const.}\, \lambda^n n^{\gamma-3}, \quad a_n \sim \text{const.}\, \lambda_a^n n^{\gamma-3}. \tag{3.1}$$

The typical size of a lattice tree or animal is characterized by the average radius of gyration. Let $\bar{x}_T = (|T| + 1)^{-1} \sum_{x \in T} x$ denote the center of mass of T, and let $R_T^2 = (|T| + 1)^{-1} \sum_{x \in T} |x - \bar{x}_T|^2$ be the squared radius of gyration of T. The average radius of gyration is then given by

$$R(n) = \left[\frac{1}{t_n} \sum_{T : |T| = n} R_T^2 \right]^{1/2}, \tag{3.2}$$

where the summation is over one tree from each equivalence class modulo translation. It is believed that asymptotically

$$R(n) \sim \text{const.}\, n^\nu \tag{3.3}$$

for a universal critical exponent ν, which is the same for both trees and animals. The following theorem [20] gives results for these critical exponents for trees in high dimensions. Related results have been obtained for lattice animals, at the level of generating functions [18].

THEOREM 3.1 *For nearest-neighbor trees with d sufficiently large, or for spread-out trees with $d > 8$ and L sufficiently large, there are positive constants such that for every $\epsilon < \min\{\frac{1}{2}, \frac{d-8}{4}\}$,*

$$\begin{aligned} t_n &= \text{const.}\, \lambda^n n^{-5/2}[1 + O(n^{-\epsilon})] \\ R(n) &= \text{const.}\, n^{1/4}[1 + O(n^{-\epsilon})]. \end{aligned}$$

Some hint can be gleaned from this theorem as to why the upper critical dimension should be 8. The fact that $n \approx R(n)^4$ is a sign that in some sense trees in high dimensions are 4-dimensional objects, and hence two trees will typically not intersect above 8 dimensions. This suggests that for $d > 8$ lattice trees will have similar behavior to the "mean-field" model of abstract trees embedded in the lattice with no constraint that the embedding be a tree, whereas for $d \leq 8$ their behavior will be different. The upper critical dimension for both trees and animals is believed to be 8.

As was the case for self-avoiding walks, the proof proceeds first by studying generating functions near their closest singularity to the origin, and then uses contour integration to extract the large-n asymptotics of t_n and the radius of gyration. Let $z_c = \lambda^{-1}$, and for $|z| \leq z_c$ define the two-point function

$$G_z(x, y) = \sum_{T:T \ni x, y} z^{|T|}. \tag{3.4}$$

It is believed that $\hat{G}_{z_c}(k)$ is asymptotic to a multiple of $k^{\eta-2}$ as $k \to 0$, with η determined by Fisher's relation $\gamma = (2-\eta)\nu$. The lace expansion is used to compare the two-point function with that for a simple random walk. An infrared bound for $\hat{G}_z(k)$ is obtained under the hypotheses of Theorem 3.1. Here, the square diagram $\mathsf{S}(z_c) = (2\pi)^{-d} \int_{[-\pi,\pi]^d} \hat{G}_{z_c}(k)^4 d^d k$ (minus 1) plays a role as a small parameter analogous to the bubble diagram for self-avoiding walks. In contrast to self-avoiding walks, it has been conjectured [7] that the infrared bound fails for lattice trees and animals below 8 dimensions.

4. Percolation

Percolation is a simple probabilistic model that exhibits a phase transition. We consider bond percolation on $\mathbb{Z}^d$, either nearest-neighbor or spread-out, assigning to each bond $\{x, y\}$ an independent Bernoulli random variable $n_{\{x,y\}}$ that takes the value 1 with probability p and the value 0 with probability $1 - p$, where p is a parameter in $[0, 1]$. If $n_{\{x,y\}} = 1$ then we say that the bond $\{x, y\}$ is *occupied*, and otherwise we say that it is *vacant*. Given a realization of the bond variables, and any two sites x and y, we say that x and y are *connected* if there is a self-avoiding walk from x to y whose steps are occupied bonds, or if $x = y$. We denote by $C(x)$ the random set of sites connected to x, and denote its cardinality by $|C(x)|$. For $d \geq 2$, it is known that there is a phase transition, in the sense that there is a critical value $p_c \in (0, 1)$ such that the probability $\theta(p)$ that $|C(0)| = \infty$ is zero for $p < p_c$ and strictly positive for $p > p_c$. A general reference is [14].

We denote the joint distribution of the Bernoulli random variables $n_{\{x,y\}}$ by P_p and expectation with respect to this distribution by $\langle \cdot \rangle_p$. The two-point function $\tau_p(x, y)$ is defined to be the probability that x and y are connected, and is analogous to the functions $G_z(x, y)$ defined previously for self-avoiding walks and for lattice trees and animals. The *susceptibility*, or expected cluster size, is defined by

$$\chi(p) = \sum_{x \in \mathbb{Z}^d} \tau_p(0, x) = \langle |C(0)| \rangle_p. \tag{4.1}$$

The susceptibility is known to be finite for $p < p_c$ and to diverge as $p \uparrow p_c$. The *magnetization*, defined by

$$M(p,h) = 1 - \sum_{n=1}^{\infty} e^{-hn} P_p[|C(0)| = n], \qquad (4.2)$$

generalizes the percolation probability $\theta(p)$, as $M(p,0) = \theta(p)$.

The following power laws are believed to hold:

$$\begin{aligned}
\chi(p) &\sim A_1(p_c - p)^{-\gamma} & \text{as } p \uparrow p_c, \\
\theta(p) &\sim A_2(p - p_c)^{\beta} & \text{as } p \downarrow p_c, \\
M(p_c, h) &\sim A_3 h^{1/\delta} & \text{as } h \downarrow 0,
\end{aligned}$$

for some amplitudes A_i and universal critical exponents γ, β, δ. Such behavior is not difficult to establish for percolation on a tree, with the "mean-field values" $\gamma = 1$, $\beta = 1$, $\delta = 2$. With more work, it can be shown that these values provide the following lower bounds for percolation on $\mathbb{Z}^d$ for all $d \geq 2$: $\chi(p) \geq c_1(p_c - p)^{-1}$, $\theta(p) \geq c_2(p - p_c)$, $M(p_c, h) \geq c_3 h^{1/2}$ [3], [4], [10]. The next theorem gives complementary upper bounds, above the upper critical dimension 6.

THEOREM 4.1 *For the nearest-neighbor model with d sufficiently large ($d \geq 19$ is large enough), or for the spread-out model with $d > 6$ and L sufficiently large, there are constants a_i such that*

$$\begin{aligned}
a_1(p_c - p)^{-1} &\leq & \chi(p) &\leq a_2(p_c - p)^{-1} & \text{as } p \uparrow p_c, \\
a_3(p - p_c)^1 &\leq & \theta(p) &\leq a_4(p - p_c)^1 & \text{as } p \downarrow p_c, \\
a_5 h^{1/2} &\leq & M(p_c, h) &\leq a_6 h^{1/2} & \text{as } h \downarrow 0.
\end{aligned}$$

The proof of the upper bounds of Theorem 4.1 is a combination of several results that center on the triangle condition. The triangle condition is the statement that the triangle diagram is finite at $p = p_c$, with the triangle diagram given by

$$\mathsf{T}(p) = \sum_{x,y \in \mathbb{Z}^d} \tau_p(0,x)\tau_p(x,y)\tau_p(y,0) = \int_{[-\pi,\pi]^d} \hat{\tau}_p(k)^3 \frac{d^d k}{(2\pi)^d}. \qquad (4.3)$$

Aizenman and Newman [4] introduced the triangle condition and showed that it implies $\gamma = 1$, arguing that the upper critical dimension is 6. Barsky and Aizenman [6] used differential inequalities to prove that the triangle condition implies $\beta = 1$, $\delta = 2$. Then in [17], the lace expansion was used to show that the triangle condition holds under the hypotheses of the theorem. In fact, $\mathsf{T}(p_c) - 1$ serves as a small parameter for convergence of the lace expansion, and the limitation to $d \geq 19$ for the nearest-neighbor model arises to ensure it is small enough.

The proof that the triangle condition holds above 6 dimensions involves proving the infrared bound $\hat{\tau}_p(k) \leq \text{const.} k^{-2}$, with a constant that is uniform in $p < p_c$. The conjectured behavior in general dimensions is again $\hat{\tau}_{p_c}(k) \sim \text{const.} k^{\eta-2}$. However, for percolation it has been conjectured that the infrared bound is violated ($\eta < 0$) for some dimensions below 6. The triangle condition is expected not to hold for any $d \leq 6$.

It follows from the behavior of $\theta(p)$ given in Theorem 4.1 that the percolation probability is zero at the critical point: $\theta(p_c) = 0$. Although this is strongly believed to be true in all dimensions, it has otherwise been proven only for the nearest-neighbor model in two dimensions and remains a major open problem in general dimensions.

Related results for oriented bond percolation models can be found in the work of Nguyen and Yang [27], [28].

References

[1] M. Aizenman, *Geometric analysis of φ^4 fields and Ising models, Parts I and II*, Comm. Math. Phys., **86**:1–48, (1982).

[2] M. Aizenman, *Stochastic geometry in statistical mechanics and quantum field theory*, in Z. Ciesielski and C. Olech, editors, Proc. Internat. Congress Math., August 16-24, 1983, Warszawa, Warsaw, (1984). Polish Scientific Publishers.

[3] M. Aizenman and D. J. Barsky, *Sharpness of the phase transition in percolation models*, Comm. Math. Phys., **108**:489–526, (1987).

[4] M. Aizenman and C. M. Newman, *Tree graph inequalities and critical behaviour in percolation models*, J. Statist. Phys., **36**:107–143, (1984).

[5] S. E. Alm, *Upper bounds for the connective constant of self-avoiding walks*, Combinatorics Probab. Comput., **2**:115–136, (1993).

[6] D. J. Barsky and M. Aizenman, *Percolation critical exponents under the triangle condition*, Ann. Probab., **19**:1520–1536, (1991).

[7] A. Bovier, J. Fröhlich, and U. Glaus, *Branched polymers and dimensional reduction*, in K. Osterwalder and R. Stora, editors, Critical Phenomena, Random Systems, Gauge Theories, Amsterdam, (1986). North-Holland. Les Houches 1984.

[8] D. Brydges, S. N. Evans, and J. Z. Imbrie, *Self-avoiding walk on a hierarchical lattice in four dimensions*, Ann. Probab., **20**:82–124, (1992).

[9] D. C. Brydges and T. Spencer, *Self-avoiding walk in 5 or more dimensions*, Comm. Math. Phys., **97**:125–148, (1985).

[10] J. T. Chayes and L. Chayes, *The mean field bound for the order parameter of Bernoulli percolation*, in H. Kesten, editor, Percolation Theory and Ergodic Theory of Infinite Particle Systems. Springer, Berlin and New York, (1987).

[11] A. R. Conway and A. J. Guttmann, *Lower bound on the connective constant for square lattice self-avoiding walks*, J. Phys. A: Math. Gen., **26**:3719–3724, (1993).

[12] M. E. Fisher, *Correlation functions and the critical region of simple fluids*, J. Math. Phys., **5**:944–962, (1964).

[13] J. Fröhlich, *On the triviality of φ_d^4 theories and the approach to the critical point in $d \geq 4$ dimensions*, Nuclear Phys. B, **200** [FS4]:281–296, (1982).

[14] G. Grimmett, Percolation, Springer, Berlin and New York, (1989).

[15] J. M. Hammersley and K. W. Morton, *Poor man's Monte Carlo*, J. Roy. Statist. Soc. Ser. B, **16**:23–38, (1954).

[16] J. M. Hammersley and D. J. A. Welsh, *Further results on the rate of convergence to the connective constant of the hypercubical lattice*, Quart. J. Math. Oxford Ser. (2), **13**:108–110, (1962).

[17] T. Hara and G. Slade, *Mean-field critical behaviour for percolation in high dimensions*, Comm. Math. Phys., **128**:333–391, (1990).

[18] T. Hara and G. Slade, *On the upper critical dimension of lattice trees and lattice animals*, J. Statist. Phys., **59**:1469–1510, (1990).

[19] T. Hara and G. Slade, *The lace expansion for self-avoiding walk in five or more dimensions*, Rev. Math. Phys., **4**:235–327, (1992).

[20] T. Hara and G. Slade, *The number and size of branched polymers in high dimensions*, J. Statist. Phys., **67**:1009–1038, (1992).

[21] T. Hara and G. Slade, *Self-avoiding walk in five or more dimensions. I. The critical behaviour*, Comm. Math. Phys., **147**:101–136, (1992).

[22] T. Hara and G. Slade, *Mean-field behaviour and the lace expansion*, in G. R. Grimmett, editor, Probability and Phase Transition, Dordrecht, (1994). Kluwer.

[23] T. Hara, G. Slade, and A. D. Sokal, *New lower bounds on the self-avoiding-walk connective constant*, J. Statist. Phys., **72**:479–517, (1993). Erratum, J. Statist. Phys., **78**:1187–1188, (1995).

[24] D. Iagolnitzer and J. Magnen, *Polymers in a weak random potential in dimension four: rigorous renormalization group analysis*, Comm. Math. Phys., **162**:85–121, (1994).

[25] H. Kesten, *On the number of self-avoiding walks. II*, J. Math. Phys., **5**:1128–1137, (1964).

[26] N. Madras and G. Slade, The Self-Avoiding Walk, Birkhäuser, Basel and Boston, (1993).

[27] B. G. Nguyen and W-S. Yang, *Gaussian limit for critical oriented percolation in high dimensions*, J. Statist. Phys., **78**:841–876, (1995).

[28] B. G. Nguyen and W-S. Yang, *Triangle condition for oriented percolation in high dimensions*, Ann. Probab., **21**:1809–1844, (1993).

[29] G. Slade, *The diffusion of self-avoiding random walk in high dimensions*, Comm. Math. Phys., **110**:661–683, (1987).

A Rigorous (Renormalization Group) Analysis of Superconducting Systems

EUGENE TRUBOWITZ

ETH Zentrum
CH-8092 Zürich, Switzerland

The work I'll discuss today is part of a larger program carried out in collaboration with Feldman, Knörrer, Lehmann, Magnen, Rivasseau, Salmhofer and Sinclair. The parenthesized term, "Renormalization Group" in the title refers to a technique for analyzing phenomena that are characterized by an intrinsic, infinite hierarchy of different energy or length scales. Superconductivity is a phenomenon of this kind.

The electrical resistivity of aluminum is strictly positive at temperatures above 1.2 Kelvins. It drops to *zero* at all smaller temperatures. One says that there is a transition between the "normal" and "superconducting" states of aluminum at the "critical temperature" $T_c = 1.2\,\mathrm{K}$. Such a transition has never been observed in silver. Here are

Four Metallic Elements and their Critical Temperatures

Aluminum	(Al)	$T_c = 1.19\,\mathrm{K}$
Silver	(Ag)	$T_c = ?$
Mercury	(Hg)	$T_c = 4.15\,\mathrm{K}$
Lead	(Pb)	$T_c = 7.19\,\mathrm{K}$

For comparison, helium liquifies at $4.2\,\mathrm{K}$. Al, Hg, and Pb are called "conventional" superconductors.

The superconducting phase transition was discovered by H. Kamerlingh Onnes in 1911 when he cooled mercury below its critical temperature of 4.15 K. Between 1911 and 1972 materials were found with higher transition temperatures and the highest attainable transition temperature was gradually increased by about 19 Kelvins from $4.15\,\mathrm{K}$ for Hg to $23\,\mathrm{K}$ for the compound Nb_3Ge. The highest transition temperature remained at $23\,\mathrm{K}$ for another 14 years until 1986 when Bednorz and Müller at IBM in Rüschlikon reported superconductivity in mixtures of La and Ba copper oxides at $30\,\mathrm{K}$. There has been astonishing progress since then. Here are two compounds with spectacularly high transition temperatures.

Two Compounds and their Critical Temperatures

$Y\,Ba_2\,Cu_3\,O_{6.9}$	$T_c = 92.5\,\mathrm{K}$
$Hg\,Ba_2\,Ca_3\,O_8$	$T_c = 134\,\mathrm{K}$

Proceedings of the International Congress
of Mathematicians, Zürich, Switzerland 1994
© Birkhäuser Verlag, Basel, Switzerland 1995

They are "unconventional" superconductors.

134 K is presently (August, 1994) the world's record transition temperature obtained by Ott at ETH-Zürich. For comparison, nitrogen liquifies at 77 K.

There is a very good physical theory (see, for example [deG]) of "conventional" superconductivity in, for example, metallic elements. Here, as we shall explain in more detail, the underlying mechanism is an indirect electron-electron interaction that is generated by the vibrations of the ions. It works especially well for aluminum. However, at this time, there are many conflicting explanations (see, for example [H]), by many physicists, of "unconventional high temperature superconductivity". We shall state a rigorous mathematical theorem about "conventional" superconductivity in the standard model of a weakly coupled electron-ion system.

Superconductivity is a macroscopic quantum phenomenon that reflects the collective behavior of many, many microscopic electrons and ions. It is a problem of quantum statistical mechanics and therefore a problem of nonrelativistic quantum field theory. Unfortunately, it is not possible to explain in a few words how to express an electron-ion system in terms of quantum fields unless one takes for granted, among other things, the formalism of Green's functions and fermionic functional integration. Similarly, the methods that we have developed to rigorously construct and control the relevant quantum fields are complicated and as a result hard to explain in a short time. Nevertheless, I will try to give an *heuristic* introduction to the physics and mathematics of superconductivity that leads to the statement of a rigorous theorem. Let's start with one free electron.

Let $d = 2, 3$ and fix $L > 0$. The state space for a free electron with position $\mathbf{x} = (x_1, \ldots, x_d)$ and spin $\sigma \in \{\uparrow, \downarrow\}$ moving in the periodic box $\mathbb{R}^d/L\mathbb{Z}^d = [-L/2, L/2)^d$ is the Hilbert space $L^2\left(\mathbb{R}^d/L\mathbb{Z}^d \times \{\uparrow, \downarrow\}\right)$. The Hamiltonian for a free electron is the kinetic energy operator

$$-\tfrac{1}{2m}\Delta_{\mathbf{x}} \;=\; -\tfrac{1}{2m} \sum_{i=1}^{d} \frac{\partial^2}{\partial x_i^2}$$

You should think of the two-dimensional torus as a thin metallic film of area L^2. The three-dimensional torus is a model for the usual bulk sample of volume L^3. The torus imposes periodic boundary conditions on the single particle states. Any other physical boundary conditions will do.

There is a special basis of single electron states — the plane waves. Let $\mathbf{k} \in \frac{2\pi}{L}\mathbb{Z}^d$ and $\tau \in \{\uparrow, \downarrow\}$. The plane wave

$$\phi_{\mathbf{k},\tau}(\mathbf{x}, \sigma) = \frac{e^{i\langle \mathbf{k}, \mathbf{x}\rangle}}{L^{d/2}}\, \delta_{\tau,\sigma}$$

is a single electron state with energy $|\mathbf{k}|^2$, momentum $\mathbf{k}$, and spin τ because it is an eigenvector of the energy and momentum operators. That is,

$$-\Delta_{\mathbf{x}}\phi_{\mathbf{k},\tau} \;=\; |\mathbf{k}|^2\, \phi_{\mathbf{k},\tau}$$
$$-i\nabla_{\mathbf{x}}\phi_{\mathbf{k},\tau} \;=\; \mathbf{k}\, \phi_{\mathbf{k},\tau}$$

The momentum of an electron in a plane wave state is $\mathbf{k}$ with probability one. Its position probability density, given by the modulus squared of the plane wave, is consistent with the Heisenberg uncertainty principle, completely smeared out.

To pass from one to many free electrons we must implement the Pauli exclusion principle. The state space $\mathcal{F}_{n(L)}$ for n free electrons moving on the torus $\mathbb{R}^d/L\mathbb{Z}^d$ is the Hilbert space of all *antisymmetric* functions $\psi(\mathbf{x}_1, \sigma_1, \ldots, \mathbf{x}_n, \sigma_n)$ in $L^2\left(\left(\mathbb{R}^d/L\mathbb{Z}^d \times \{\uparrow, \downarrow\}\right)^n\right)$. If the position, spin pair $\mathbf{x}_i, \sigma_i$ is interchanged with the pair $\mathbf{x}_j, \sigma_j$, then, by antisymmetry, the wave function ψ changes sign. In particular, the configuration of two electrons at the same place with the same spin is excluded because the wave function has to vanish. The Hamiltonian is

$$H_0(n,L) = \sum_{i=1}^{n} -\Delta_{\mathbf{x}_i}$$

There is also a special basis of n-electron states; the wedge products. The wedge product

$$\phi_{\mathbf{k}_1, \tau_1} \wedge \cdots \wedge \phi_{\mathbf{k}_n, \tau_n} = \frac{1}{n!} \det\left(\phi_{\mathbf{k}_i, \tau_i}(\mathbf{x}_j, \sigma_j)\right)$$

of plane waves is an n-electron state with energy $E = |\mathbf{k}_1|^2 + \cdots + |\mathbf{k}_n|^2$, meaning

$$H_0(n,L)\, \phi_{\mathbf{k}_1, \tau_1} \wedge \cdots \wedge \phi_{\mathbf{k}_n, \tau_n} = E\, \phi_{\mathbf{k}_1, \tau_1} \wedge \cdots \wedge \phi_{\mathbf{k}_n, \tau_n}.$$

By construction, a wedge product vanishes when any two of the plane waves have the same momentum and spin. The matrix of the Hamiltonian $H_0(n, L)$ is diagonal in this basis. For this reason, it is easy to construct the ground state for n free electrons, that is, the normalized wave function with the smallest energy.

Suppose, for simplicity, that n is even. To minimize the energy, let $\mathbf{s}_1, \ldots, \mathbf{s}_{\frac{n}{2}}$ be the $\frac{n}{2}$ shortest vectors in the dual lattice $\frac{2\pi}{L}\mathbb{Z}^d$; that is, the lattice dual to the torus of side L. Then the wedge product

$$\textbf{Fermi sea} \; = \; \sqrt{n!}\, \phi_{\mathbf{s}_1, \uparrow} \wedge \phi_{\mathbf{s}_1, \downarrow} \wedge \cdots \wedge \phi_{\mathbf{s}_{\frac{n}{2}}, \uparrow} \wedge \phi_{\mathbf{s}_{\frac{n}{2}}, \downarrow}$$

is the ground state wave function of $H_0(n, L)$. It is somewhat poetically called the Fermi sea because $\mathbf{s}_1, \ldots, \mathbf{s}_{\frac{n}{2}}$ are the points of the dual lattice that lie inside the Fermi sphere of radius

$$\rho^{\frac{1}{d}} \; = \; \left(\frac{n}{L^d}\right)^{\frac{1}{d}}$$

where ρ is the number of electrons per unit volume, that is, the density. I have ignored in the expression for the radius of the Fermi sphere overall constants and error terms that tend to zero as L tends to infinity.

It is useful to introduce operators that "create" and "annihilate" electrons with momentum $\mathbf{k}$ and spin τ. They are a first hint of the underlying quantum fields. The "creation operator" $a^\dagger_{\mathbf{k}, \tau}$ is defined for all $\psi \in \mathcal{F}_{\ell(L)}$ by

$$a^\dagger_{\mathbf{k}, \tau}\psi \; = \; L^{\frac{d}{2}} \sqrt{\ell+1}\; \phi_{\mathbf{k}, \tau} \wedge \psi \; \in \; \mathcal{F}_{\ell+1}(L).$$

We have

$$\textbf{Fermi sea} \;=\; L^{-\frac{d}{2}n} \prod_{i=1}^{\frac{n}{2}} a_{s_i,\uparrow}^{\dagger} a_{s_i,\downarrow}^{\dagger} \, \textbf{1}$$

where $\textbf{1}$ is the constant function "one" on the torus. The adjoint $a_{\mathbf{k},\tau}$ of $a_{\mathbf{k},\tau}^{\dagger}$ "annihilates" an electron with momentum $\mathbf{k}$ and spin τ. For example,

$$a_{\mathbf{k},\tau} \, \phi_{\mathbf{k},\tau} \wedge \phi_{\mathbf{k}_1,\tau_1} \wedge \cdots \wedge \phi_{\mathbf{k}_\ell,\tau_\ell} \;=\; \frac{L^{\frac{d}{2}}}{\sqrt{\ell+1}} \, \phi_{\mathbf{k}_1,\tau_1} \wedge \cdots \wedge \phi_{\mathbf{k}_\ell,\tau_\ell}.$$

The next step is to introduce interactions. The state space for n electrons, moving on the torus $\mathbb{R}^d/L\mathbb{Z}^d$, that interact with each other and with ions vibrating around their equilibrium positions in the "crystal lattice $\mathbb{Z}^d$" is the Hilbert space of all functions $\psi(\mathbf{x}_1, \sigma_1, \ldots, \mathbf{x}_n, \sigma_n; \text{ion positions})$ that are antisymmetric in the electron variables. The electron-ion Hamiltonian is

$$H \;=\; \sum_{i=1}^{n} -\Delta_{\mathbf{x}_i} + \tfrac{1}{2} \sum_{i \neq j} V(\mathbf{x}_i - \mathbf{x}_j) + \; \textbf{ionic kinetic energy} + \; \textbf{ion/ion interaction}$$

$$+ \; \textbf{electron/ion interaction}.$$

The direct interaction through the potential V between an electron at $\mathbf{x}_i$ and an electron at $\mathbf{x}_j$ is repulsive. There is also an indirect electron/ion/ion/electron interaction generated by the last three terms of the Hamiltonian. To visualize the indirect interaction, imagine that an electron at $\mathbf{x}_i$ interacts through the last term of the Hamiltonian with an ion that interacts through the second and third terms with another ion that finally interacts with an electron at $\mathbf{x}_j$. It is a basic observation in the theory of conventional superconductivity that the indirect electron/ion/ion/electron interaction is attractive.

To make a rough but ready analogy, imagine that the first of a couple gets into bed. He or she makes a dent. The dent represents the screening of a single negatively charged electron by the lattice of positively charged ions. When the second person gets in bed he or she also makes a dent. As some of you may have noticed, two dents in a soft mattress tend to attract each other into the center. For this reason the indirect attraction between electrons is sometimes called the mattress effect.

We show rigorously, under physically reasonable hypotheses, that the indirect interaction is indeed attractive and moreover dominates the repulsive direct interaction so that the total force between two electrons is attractive for a small but crucial range of momenta. We also show rigorously, using this attraction, that the electron-ion Hamiltonian exhibits superconductivity.

The total attractive force loosely binds special pairs of electrons. They are the Cooper pairs. The notion of Cooper pair is basic for superconductivity. I'll now describe a remarkable experiment that "displays" Cooper pairs.

Construct a silver film-lead film interface. If the temperature of the interface is above $7.2\,\mathrm{K}$, the critical temperature for lead, we have a normal metal-normal metal junction. Inject an electron with momentum $\mathbf{k}$ and spin $\uparrow$ into the silver film. It may be elastically scattered off the silver film or it may tunnel into the lead. Not very exciting! Just what you would expect.

To see something more interesting, cool the interface to any temperature below $7.2\,$K. We have a normal-superconducting junction, because the silver is still in a normal state. Now, at temperature $T \approx 0$ inject an electron with momentum $\mathbf{k}$ and spin $\uparrow$ and energy

$$0 \; < \; e(\mathbf{k}) = |\mathbf{k}|^2 - \rho_{\mathbf{Ag}}^{\frac{2}{d}} \; < \; \Delta_{\mathbf{Pb}}$$

into the silver film. Here, $\rho_{\mathbf{Ag}}$ is the electron density of silver and

$$\Delta_{\mathbf{Pb}} \approx 27.3 \times 10^{-4} eV.$$

In other words, the momentum $\mathbf{k}$ lies in a thin shell outside the Fermi sphere of silver. Then, with probability one a *positively* charged "particle" with momentum

$$\mathbf{k}_{\mathrm{R}} = -\sqrt{2\rho^{\frac{2}{d}} - |\mathbf{k}|^2} \; \frac{\mathbf{k}}{|\mathbf{k}|} \; \approx \; -\mathbf{k}$$

and spin $\downarrow$ is reflected. The momentum of the reflected particle lies inside the Fermi sphere of silver. Note that $\mathbf{k}_{\mathrm{R}} = -\mathbf{k}$ when $\mathbf{k}$ lies on the Fermi sphere.

If, on the other hand, the momentum of the injected electron lies outside the thin shell, that is, $\Delta_{\mathbf{Pb}} < e(\mathbf{k}) = |\mathbf{k}|^2 - \rho_{\mathbf{Ag}}^{\frac{2}{d}}$, then the probability that a positively charged particle is reflected goes to zero as $\frac{1}{4}\left(\frac{\Delta_{\mathbf{Pb}}}{e(\mathbf{k})}\right)^2$.

We first conclude from the experiment described above that there are no single particle states with energy $|e(\mathbf{k})| < \Delta_{\mathbf{Pb}}$ in superconducting lead, because, with probability one, any incident electron with energy $e(\mathbf{k})$ is "Andreev reflected". In other words, there is a gap of size $\Delta_{\mathbf{Pb}}$ in the single particle spectrum of superconducting lead.

What really happened? Let's make the naive, but quite reasonable, approximation that the ground state of silver is a Fermi sea with radius $\rho_{\mathbf{Ag}}^{\frac{1}{d}}$. Then, the ground state for the normal-superconducting interface is the tensor product

$$\mathbf{Fermi\ sea\,(Ag)} \otimes \mathbf{SC(Pb)}$$

where $\mathbf{SC(Pb)}$ is the superconducting ground state of lead, whatever that is. In Andreev scattering, we see the transition from the tensor product state

$$(a_{\mathbf{k},\uparrow}^{\dagger}\mathbf{Fermi\ sea\,(Ag)}) \otimes \mathbf{SC(Pb)}$$

to the tensor product state

$$(a_{-\mathbf{k},\downarrow}\mathbf{Fermi\ sea\,(Ag)}) \otimes \; a_{\mathbf{k},\uparrow}^{\dagger}a_{-\mathbf{k},\downarrow}^{\dagger}\mathbf{SC(Pb)}$$

in which the incident electron in the plane wave state $\phi_{\mathbf{k}\uparrow}$ pairs with the plane wave $\phi_{\mathbf{k}_{\mathrm{R}}\downarrow}$ in the Fermi sea and despite having arbitrarily small energy enters the lead. A "hole" $a_{-\mathbf{k},\downarrow}\mathbf{Fermi\ sea\,(Ag)}$ in the Fermi sea of the silver film is created and the

"Cooper pair operator" $a^{\dagger}_{\mathbf{k},\uparrow} a^{\dagger}_{-\mathbf{k},\downarrow}$ adds a Cooper pair to $\mathbf{SC(Pb)}$. In particular, there is no gap in the two particle excitation spectrum of superconducting lead.

It is now easy to explain the experiment. The hole is the "reflected" positively charged "particle". It is positively charged because one negatively charged electron has been removed from the electrically neutral silver. We also conclude that the superconducting ground state is a "soup" of Cooper pairs. Observe that the presence of Cooper pairs can be detected by the expected value of the Cooper pair operator. For example,

$$\left\langle \mathbf{Fermi\,sea}, a^{\dagger}_{\mathbf{k},\uparrow} a^{\dagger}_{-\mathbf{k},\downarrow} \mathbf{Fermi\,sea} \right\rangle \;=\; 0$$

as it should be, as there are no Cooper pairs in the Fermi sea.

What does the Cooper pair soup look like? Bardeen, Cooper, and Schrieffer had a seminal idea. They minimized the energy

$$\min_{\{\theta_{\mathbf{k}}\}} \left\langle \Phi_{\{\theta_{\mathbf{k}}\}}, H_{\mathrm{BCS}} \Phi_{\{\theta_{\mathbf{k}}\}} \right\rangle$$

of a "toy" Hamiltonian H_{BCS} over the class of "pairing states"

$$\Phi_{\{\theta_{\mathbf{k}}\}} \;=\; \prod_{|e(\mathbf{k})|<\varepsilon} \left(\cos\theta_{\mathbf{k}} + \sin\theta_{\mathbf{k}}\, a^{\dagger}_{\mathbf{k},\uparrow} a^{\dagger}_{-\mathbf{k},\downarrow} \right) \mathbf{1}.$$

Miraculously, the minimum exhibits, to a good approximation, the correct phenomenology. It is a crucial observation that the Fermi sea is homogeneous — a single wedge product — whereas a pairing state is inhomogeneous. It is a sum of wedge products of different degrees. Notice that the expected value

$$\left\langle \Phi_{\{\theta_{\mathbf{k}}\}}, a^{\dagger}_{\mathbf{q},\uparrow} a^{\dagger}_{-\mathbf{q},\downarrow} \Phi_{\{\theta_{\mathbf{k}}\}} \right\rangle \;=\; \frac{1}{2}\sin 2\theta_{\mathbf{q}}$$

of finding a Cooper pair whose component electrons have momenta $\pm\mathbf{q}$ in the BCS pairing state $\Phi_{\{\theta_{\mathbf{k}}\}}$ is not zero unless $2\theta_{\mathbf{q}}$ is a multiple of π.

Bardeen, Cooper, and Schrieffer used their physical intuition to whittle the exact electron-ion Hamiltonian into a toy Hamiltonian and the exact Hilbert space down to a small select set of pairing states. We now state a theorem, a fragment of a more complete theorem, that demonstrates superconductivity for the exact electron-ion Hamiltonian H.

THEOREM (FELDMAN, MAGNEN, RIVASSEAU, TRUBOWITZ) *Let Ω be the ground state of the electron-ion Hamiltonian H. Suppose the total coupling λ between the electrons is sufficiently small and that the direct electron-electron interaction is short range. Fix $\rho = \frac{n}{L^d}$. Then, under physically reasonable assumptions on the ion-ion and electron-ion interactions, there is a*

$$\Delta \;\approx\; \mathrm{const}\, e^{-\mathrm{const}' \frac{1}{|\lambda|}}$$

and a constant $\mathbf{c} > 1$ and such that

(1) *If* $L \leq \frac{1}{\mathbf{c}} \frac{1}{\Delta}$,

$$\left\langle \Omega, a^{\dagger}_{\mathbf{k},\uparrow} a^{\dagger}_{-\mathbf{k},\downarrow} \Omega \right\rangle = 0$$

where L^d *is the volume of the torus. In other words, the expected value of the Cooper pair operator* $a^{\dagger}_{\mathbf{k},\uparrow} a^{\dagger}_{-\mathbf{k},\downarrow}$ *in the ground state* Ω *for a box of volume less than* $\left(\frac{1}{\mathbf{c}} \frac{1}{\Delta} \right)^d$ *is zero. There are "no Cooper pairs".*

(2) *If* $L \geq \mathbf{c} \frac{1}{\Delta}$,

$$\left\langle \Omega, a^{\dagger}_{\mathbf{k},\uparrow} a^{\dagger}_{-\mathbf{k},\downarrow} \Omega \right\rangle = \frac{\Delta}{2\sqrt{e(\mathbf{k})^2 + \Delta^2}}$$

where

$$e(\mathbf{k}) = |\mathbf{k}|^2 - \rho^{\frac{2}{d}}.$$

That is, Cooper pairs appear when the volume of the box is greater than $\left(\mathbf{c} \frac{1}{\Delta} \right)^d$.

(3) *If* $L \gg \frac{1}{\Delta}$, *Cooper pairs interact through a long range interaction that is mediated by a "massless particle", the Goldstone boson.*

In the statement of the theorem, for simplicity, we have ignored, among other things, boundary conditions that force Δ to be real.

First, observe that "the gap" Δ is nonperturbatively small. One cannot see it by expanding to any finite order in the coupling constant λ. The second point is that there are three distinct physical regimes determined by the size L of the torus. A different mathematical technique is required for each of the three regimes. There is also the problem of continuing the analysis through the transition region from the first to the second regime in which a fraction of the electrons condense into a soup of Cooper pairs.

Unfortunately, the proof is too complicated to describe here. However, for a pedagogical introduction see [FT1] and [FMRT1], [FMRT2].

References

[deG]   P.G. deGennes, Superconductivity of Metals and Alloys, Benjamin, New York, 1966.

[F]   J. Feldman, *Introduction to constructive quantum field theory*, Proc. Internat. Congress Math. Kyoto (1990), 1335–1341.

[FT1]   J. Feldman and E. Trubowitz, *Renormalization in classical mechanics and many body quantum field theory*, J. Analyse Math. **58** (1992), 213–247.

[FT2]   J. Feldman and E. Trubowitz, *Perturbation theory for many fermion systems*, Helv. Phys. Acta **63** (1990), 156–260.

[FT3]   J. Feldman and E. Trubowitz, *The flow of an electron-phonon system to the superconducting state*, Helv. Phys. Acta **64** (1991), 214–357.

[FKLT]   J. Feldman, H. Knörrer, D. Lehmann, and E. Trubowitz, *Two dimensional Fermi liquids*, in preparation.

[FKST]   J. Feldman, H. Knörrer, M. Salmhofer and E. Trubowitz, *Renormalization of Fermi surfaces*, in preparation.

[FMRT1]   J. Feldman, J. Magnen, V. Rivasseau, and E. Trubowitz, *Constructive many-body theory*, in: The State of Matter (M. Aizenmann and H. Araki, eds.), Adv. Ser. Math. Phys. **20**, World Scientific, Singapure (1994).

[FMRT2]   J. Feldman, J. Magnen, V. Rivasseau, and E. Trubowitz, *Fermionic many-body models*, in: Mathematical Quantum Theory I: Field Theory and Many-Body Theory (J. Feldman, R. Froese, and L. Rosen, eds.), CRM Proc. and Lecture Notes.

[FMRT3]   J. Feldman, J. Magnen, V. Rivasseau, and E. Trubowitz, *An infinite volume expansion for many Fermion Green's functions*, Helv. Phys. Acta **65** (1992), 679–721.

[FMRT4]   J. Feldman, J. Magnen, V. Rivasseau, and E. Trubowitz, *Ward identities and a perturbative analysis of a U(1) Goldstone boson in a many Fermion system*, Helv. Phys. Acta **66** (1993), 498–550.

[FMRT5]   J. Feldman, J. Magnen, V. Rivasseau, and E. Trubowitz, *An intrinsic 1/N expansion for many-Fermion systems*, Europhys. Lett. **24** (6) (1993), 437–442.

[FMRT6]   J. Feldman, J. Magnen, V. Rivasseau, and E. Trubowitz, *Two-dimensional many-Fermion systems as vector models*, Europhys. Lett. **24** (7) (1993), 521–526.

[H]   J. Woods Halley, ed., Theories of High Temperature Superconductivity, Addison-Wesley, Reading, MA, 1988.

Eigenvalues of Graphs

FAN R. K. CHUNG

Department of Mathematics, University of Pennsylvania
Philadelphia, PA 19104, USA

1. Introduction

The study of eigenvalues of graphs has a long history. Since the early days, representation theory and number theory have been very useful for examining the spectra of strongly regular graphs with symmetries. In contrast, recent developments in spectral graph theory concern the effectiveness of eigenvalues in studying general (unstructured) graphs. The concepts and techniques, in large part, use essentially geometric methods. (Still, extremal and explicit constructions are mostly algebraic [20].) There has been a significant increase in the interaction between spectral graph theory and many areas of mathematics as well as other disciplines, such as physics, chemistry, communication theory, and computer science.

In this paper, we will briefly describe some recent advances in the following three directions.

1. The connections of eigenvalues to graph invariants such as diameter, distances, flows, routing, expansion, isoperimetric properties, discrepancy, containment, and, in particular, the role eigenvalues play in the equivalence classes of so-called quasi-random properties;
2. The techniques of bounding eigenvalues and eigenfunctions, with special emphasis on the Sobolev and Harnack inequalities for graphs;
3. Eigenvalue bounds for special families of graphs, such as the convex subgraphs of homogeneous graphs, with applications to random walks and efficient approximation algorithms.

This paper is organized as follows. Section 2 includes some basic definitions. In Section 3 we discuss the relationship of eigenvalues to graph invariants. In Section 4 we describe the consequences and limitations of the Sobolev and Harnack inequalities. In Section 5 we use the heat kernel to derive eigenvalue lower bounds that are especially useful for the case of convex subgraphs. In Section 6 some examples and applications are illustrated. All proofs will not be included here and the statements can sometimes be very brief; thus, the reader is referred to [7] for more discussion and details.

Proceedings of the International Congress
of Mathematicians, Zürich, Switzerland 1994
© Birkhäuser Verlag, Basel, Switzerland 1995

2. Preliminaries

In a graph G with vertex set $V = V(G)$ and edge set $E = E(G)$, we define the Laplacian $\mathcal{L}$ as a matrix with rows and columns indexed by V as follows:

$$\mathcal{L}(u, v) = \begin{cases} 1 & \text{if } u = v \\ -\dfrac{1}{\sqrt{d_u d_v}} & \text{if } u \text{ and } v \text{ are adjacent } (u_u \sim v_v) \\ 0 & \text{otherwise} \end{cases}$$

where d_v denotes the degree of v. Here we consider simple, loopless graphs (because all results can be easily extended to general weighted graphs with loops [7]). For k-regular graphs, it is easy to see that

$$\mathcal{L} = I - \frac{1}{k} A$$

where A is the adjacency matrix. For a general graph, we have

$$\mathcal{L} = I - T^{-\frac{1}{2}} A T^{-\frac{1}{2}}$$

where T is the diagonal matrix with value d_v at the (v, v)-entry. The eigenvalues of $\mathcal{L}$ are denoted by

$$0 = \lambda_0 \leq \lambda_1 \leq \cdots \leq \lambda_{n-1}$$

and

$$\lambda_G := \lambda_1 \quad = \quad \inf_{\substack{f \\ \sum f(v) d_v = 0}} \frac{\displaystyle\sum_{u \sim v} (f(u) - f(v))^2}{\displaystyle\sum_v f(v)^2 d_v}$$

$$= \quad \inf_{\substack{h \\ \sum h(v) \sqrt{d_v} = 0}} \frac{\langle h, \mathcal{L} h \rangle}{\langle h, h \rangle} .$$

In a way, the eigenvalues λ_i can be viewed as the discrete analogues of the Laplace-Beltrami operator for Riemannian manifolds

$$\lambda_M = \inf_f \frac{\displaystyle\int_M ||\nabla f||^2}{\displaystyle\int_M ||f||^2}$$

where f ranges over functions satisfying $\int_M f = 0$. For a connected graph G, we have $\lambda_G > 0$ and in general $0 \leq \lambda_G \leq 1$, with the exception of $G = K_n$, the complete graph (in which case $\lambda_G = n/(n-1)$). Also $1 < \lambda_{n-1} \leq 2$, with equality holding for bipartite graphs.

3. Eigenvalues and graph properties

In a graph G on n vertices, the distance between two vertices u and v, denoted by $d(u, v)$, is the length of a shortest path joining u and v. The diameter of G, denoted by $D(G)$, is the maximum distance over all pairs of vertices: a lower bound for λ_1

implies an upper bound for $D(G)$. Namely, in [6], it was shown that for regular graphs, we have

$$D(G) \leq \left\lceil \frac{\log n - 1}{\log \frac{1}{1-\lambda_1}} \right\rceil \tag{1}$$

with the exception of the complete graph K_n. (We assume in this section that $G \neq K_n$ and G is connected.) The proof is based on the simple observation that $D(G) \leq t$ if, for some polynomial P_t of degree t and some $(n \times n)$-matrix M with $M(u,v) = 0$ for $u \not\sim v$, we have all entries of $P_t(M)$ nonzero. The above inequality can be further extended for distances between any two subsets X, Y of vertices in G. Here we denote the distance $d(X,Y)$ to be the minimum distance between a vertex in X and a vertex in Y:

$$d(X,Y) \leq \left\lceil \frac{\log \frac{vol\ V}{vol\ X\ vol\ Y}}{\log \frac{1}{1-\lambda'}} \right\rceil \tag{2}$$

where the volume of a subset X is defined to be $vol\ X = \sum_{v \in X} d_v$, and λ' is equal to λ_1 if $1 - \lambda_1 \geq \lambda_{n-1} - 1$, or else $\lambda' = 2\lambda_1/(\lambda_1 + \lambda_{n-1})$. The above inequalities have several generalizations. For example, the distances among $k+1$ subsets $X_1, \ldots, X_{k+1}$ of V are related to the kth eigenvalue λ_k for $k \geq 2$:

$$\min_{i \neq j} d(X_i, X_j) \leq \max_{i \neq j} \left\lceil \frac{\log \frac{vol\ V}{vol\ X_i\ vol\ X_j}}{\log \frac{1}{1-\lambda_k}} \right\rceil \tag{3}$$

if $1 - \lambda_k \geq \lambda_{n-1} - 1$; otherwise replace λ_k by $\frac{2\lambda_k}{\lambda_k + \lambda_{n-1}}$ in (3).

This can be further generalized to eigenvalue bounds for a Laplace operator on a smooth, connected, compact Riemannian manifold M [11]:

$$\lambda_k \leq \frac{4}{t^2} \max_{i \neq j} \left(\log \frac{2\ vol\ M}{\sqrt{vol\ X_i\ vol\ X_j}} \right)^2 \tag{4}$$

if there are $k+1$ disjoint subsets $X_1, \ldots, X_{k+1}$ such that the geodesic distance between any pair of them is at least t.

The above inequalities can be used to derive isoperimetric inequalities in the following way. For a subset X of vertices, we define the t-boundary $\delta_t(X) = \{u \notin X : d(u,v) \leq t$ for some $v \in X\}$. By substituting $Y = V - \delta_t(X) - X$ in (2) we deduce

$$\frac{vol(\delta_t(X))}{vol\ (X)} \geq (1 - (1-\lambda')^{2t})(1 - \frac{vol\ X}{vol\ V}). \tag{5}$$

We remark that the special case of (5) for regular graphs was proved by Alon [1] and Tanner [24].

Another type of boundary for a subset X is

$$\partial(X) = \{\{x, x'\} \in E : x \in X, x' \notin X\}$$

The Cheeger constant h_G is defined to be

$$h_G = \min_{\substack{x \\ |\text{vol } X| \le \frac{1}{2} \text{vol } V}} \frac{|\partial(X)|}{\text{vol } X}$$

and Cheeger's inequality states

$$2h_G \ge \lambda_1 \ge \frac{h_G^2}{2}.$$

The discrete version of Cheeger's inequality was considered in [18], [3] with proof techniques quite similar to those used for the continuous case by Cheeger [5], and can be traced back to the early work of Polya and Szego [22].

The implications of the above isoperimetric inequalities can be summarized as follows: when λ_1 is bounded away from 0, i.e., $\lambda_1 \ge c > 0$ for some absolute constant c, the diameter is "small" and the boundary of a subset X is "large" (proportional to the volume of the subset). As an immediate consequence of the isoperimetric inequalities, there are many paths with "small" overlap simultaneously joining all pairs of vertices. In fact, the following dynamic version of routing can be achieved efficiently (in logarithmic time in n). Namely, in a regular graph G suppose pebbles p_i are placed on vertices v_i with destination $v_{\pi(i)}$ for some permutation $\pi \in S_n$. At each step, every pebble is allowed to move along some edge to a neighboring vertex provided that no two pebbles can be placed at the same vertex simultaneously. Then there is a routing scheme to move all pebbles to their destinations in $O(\frac{1}{\lambda^2} \log^2 n)$ time (see [2] and [7]).

When both λ_1 and λ_{n-1} are close to 1, the graph G satisfies additional properties. For example, for two subsets of vertices, say X and Y, the number $e(X, Y)$ of pairs $(x, y), x \in X, y \in Y$ and $\{x, y\} \in E$ is close to the expected value. Here by "expected" value, we mean the expected value for a random graph with the same edge density. To be precise, we have the following inequality:

$$\left| e(X, Y) - \frac{\text{vol } X \, \text{vol } Y}{\text{vol } V} \right| \le \max_{i \ne 0} |1 - \lambda_i| \sqrt{\text{vol } X \, \text{vol } Y}.$$

When $X = Y$, the left-hand side of the above inequality is called the discrepancy of X.

For sparse graphs, say k-regular graphs for some fixed k, $1 - \lambda_1$ cannot be too small. In fact, $1 - \lambda_1 \ge \frac{1}{\sqrt{k}}$. However for dense graphs, $1 - \lambda_1$ can be very close to zero. For example, almost all graphs have $1 - \lambda_1$ at most $\frac{c}{\sqrt{n}}$. For graphs with constant edge density, say $\rho = \frac{1}{2}$, the condition of $1 - \lambda_1 = o(1)$ implies many strong graph properties. Here we will use descriptions of graph properties containing the $o(1)$ notation so that $P(o(1)) \to P'(o(1))$ means that for any $\epsilon > 0$, there exists δ such that $P(\delta) \to P'(\epsilon)$. Two properties P and P' are equivalent if $P \to P'$ and $P' \to P$. The following class of properties for an almost regular graph G, with edge density $\frac{1}{2}$, have all been shown to be equivalent [10] (also see [7]) and this class of graph properties is termed "quasi-random" because a random graph shares these properties.

P_1:: $\max_{i>0} |1 - \lambda_i| = o(1)$

P_2:: For any subset X of vertices, the discrepancy of $X = o(1) \cdot \text{vol } X$.

For a fixed $s \geq 4$,

$P_3(s)$:: For any graph H on s vertices, the number of occurrences of H as an induced subgraph of G is $(1 + o(1))$ times the expected number.

P_4 :: For almost all pairs x, y of vertices, the number of vertices w satisfies $(w \sim x$ and $w \sim y)$ or $(w \not\sim x$ and $w \not\sim y)$ is $(1 + o(1))$ times the expected number.

We remark that the $o(1)$ terms in the above properties represent the estimates of deviations from the expectation. The problems of determining the order and the behavior of these deviations and the relations between various estimates touch many aspects of extremal graph theory and random graph theory. Needless to say, many intriguing questions remain open. We remark that quasi-random classes for hypergraphs have also been established and examined in [9].

4. Sobolev and Harnack inequalities

In this section, we will describe the Sobolev inequalities and Harnack inequalities for eigenfunctions of graphs that then lead to eigenvalue bounds. The ideas and proof techniques are quite similar to various classical methods in treating the eigenvalues of connected smooth compact Riemannian manifolds. In general, there are often various obstructions to applying continuous methods in the discrete domain. For example, many differential techniques can be quite hard to utilize because the eigenfunctions for graphs are defined on a finite number of vertices and the task of taking derivatives can therefore be difficult (if not impossible). Furthermore, general graphs usually represent all possible configurations of edges, and, as a consequence, many theorems concerning smooth surfaces are simply not true for graphs. Nevertheless, there are many common concepts that provide connections and interactions between spectral graph theory and Riemannian geometry. As a successful example, the Sobolev inequalities for graphs can be proved almost entirely by classical techniques that can be traced back to Nash [26]. The situation for the Harnack inequalities for graphs is somewhat different because discrete versions of the statement for the continuous cases do not hold in general. However, we will describe a Harnack inequality that works for eigenfunctions of homogeneous graphs and some special subgraphs that we call "strongly convex."

We first consider Sobolev inequalities that hold for all general graphs. To start with, we define a graph invariant, the so-called isoperimetric dimension, which is involved in the Sobolev inequality.

We say that a graph G has isoperimetric dimension δ with an isoperimetric constant c_δ if for all subsets X of $V(G)$, the number of edges between X and the complement $\bar{X}$ of X, denoted by $e(X, \bar{X})$, satisfies

$$e(X, \bar{X}) \geq c_\delta (\text{vol } X)^{\frac{\delta-1}{\delta}}$$

where $\text{vol } X \leq \text{vol } \bar{X}$ and c_δ is a constant depending only on δ. Let f denote an arbitrary function $f : V(G) \to \mathbf{R}$. The following Sobolev inequalities hold:

(i) For $\delta > 1$,

$$\sum_{u \sim v} |f(u) - f(v)| \geq c_\delta \frac{\delta - 1}{\delta} \min_m \left(\sum_v |f(v) - m|^{\frac{\delta}{\delta-1}} d_v \right)^{\frac{\delta-1}{\delta}}$$

(ii) For $\delta > 2$,

$$(\sum_{u \sim v} |f(u) - f(v)|^2)^{\frac{1}{2}} \geq \sqrt{c_\delta} \frac{\delta - 1}{\sqrt{2\delta}} \min_m (\sum_v |(f(v) - m)^\alpha d_v)^{\frac{1}{\alpha}}$$

where $\alpha = \frac{2\delta}{\delta - 2}$.

The above two inequalities can be used to derive the following eigenvalue inequalities for a graph G (see [13]):

$$\sum_{i > 0} e^{-\lambda_i t} \leq c \frac{\text{vol } V}{t^{\frac{\delta}{2}}} \tag{6}$$

$$\lambda_k \geq c'(\frac{k}{\text{vol } V})^{\frac{2}{\delta}} \tag{7}$$

for suitable contants c and c' that depend only on δ.

In a way, a graph can be viewed as a discretization of a Riemannian manifold in $\mathbf{R}^n$ where n is roughly equal to δ. The eigenvalue bounds in (7) are analogues of the Polya conjecture for Dirichlet eigenvalues of a regular domain M,

$$\lambda_k \geq \frac{2\pi}{w_n}(\frac{k}{\text{vol } M})^{\frac{2}{n}}$$

where w_n is the volume of the unit disc in $\mathbf{R}^n$.

From now on, we assume that f is an eigenfunction of the Laplacian of G. The usual Harnack inequality concerns establishing an upper bound for the quantity $\max_{x \sim y}(f(x) - f(y))^2$ by a multiple of λ and $\max_x f^2(x)$. Such an inequality does not hold for general graphs (for example, for the graph formed by joining two complete graphs K_n by an edge). We will show that we can have a Harnack inequality for certain homogeneous graphs and some of their subgraphs.

A homogeneous graph is a graph Γ together with a group H acting on the vertices satisfying:

1. For any $g \in H, u \sim v$ if and only if $gu \sim gv$.
2. For any $u, v \in V(\Gamma)$ there exists $g \in H$ such that $gu = v$.

In other words, Γ is vertex transitive under the action of H and the vertices of Γ can be labelled by cosets H/I where $I = \{g | gv = v\}$ for a fixed v. Also, there is an edge generating set $K \subset H$ such that for all vertices $v \in V(\Gamma)$ and $g \in K$, we have $\{v, gv\} \in E(\Gamma)$.

A homogeneous graph is said to be invariant if K is invariant as a set under conjugation by elements of K, i.e., for all $a \in K$, $aKa^{-1} = K$.

Let f denote an eigenfunction in an invariant homogeneous graph with edge generating set K consisting of k generators. Then it can be shown [14]

$$\frac{1}{k} \sum_{a \in K} (f(x) - f(ax))^2 \leq 8\lambda \sup_y f^2(y).$$

An induced subgraph S of a graph Γ is said to be *strongly convex* if for all pairs of vertices u and v in S, all shortest paths joining u and v in Γ are contained in S. The main theorem in [14] asserts that the following Harnack inequality holds.

Suppose S is a strongly convex subgraph in an abelian homogeneous graph with edge generating set K consisting of k generators. Let f denote an eigenfunction of S associated with the Neumann eigenvalue λ. Then for all $x \in S$, $x \sim y$,

$$|f(x) - f(y)|^2 \leq 8k\lambda \sup_{z \in S} f^2(z).$$

The Neumann eigenvalues for subgraphs will be defined in the next section. A direct consequence of the Harnack inequalities is the following lower bound for the Neumann eigenvalue λ of S:

$$\lambda \geq \frac{1}{8kD^2}$$

where k is the maximum degree and D is the diameter of S. Such eigenvalue bounds are particularly useful for deriving polynomial approximation algorithms when enumeration problems of combinatorial structures can often be represented as random walk problems on "convex" subgraphs of appropriate homogeneous graphs. However, the condition of a strongly convex subgraph poses quite severe constraints, which will be relaxed in the next section.

5. Eigenvalue inequalities for subgraphs and convex subgraphs

Let S denote a subset of vertices in G. An induced subgraph on S consists of all edges with both end points in S. Whereas a graph corresponds to a manifold with no boundary, an induced subgraph on S can be associated with a submanifold with a boundary. Next, we define the Neumann eigenvalue for an induced subgraph on S. Let $\hat{S}$ denote the extension of S formed by all edges with at least one end point in S. The Neumann eigenvalue λ_S for S is defined to be

$$\lambda_S = \inf_f \frac{\displaystyle\sum_{\{x,y\} \in \hat{S}} (f(x) - f(y))^2}{\displaystyle\sum_{x \in S} f^2(x) d_x} = \inf_g \frac{\langle g, \mathcal{L}g \rangle}{\langle g, g \rangle}$$

where f ranges over all functions $f : S \cup \delta S \to \mathbf{R}$ satisfying $\displaystyle\sum_{x \in S} f(x) d_x = 0$, $g(x) = f(x)\sqrt{d_x}$, and $\mathcal{L}$ denotes the Laplacian of S.

Let ϕ_i denote the eigenfunction for the Laplacian corresponding to eigenvalue λ_i. Then ϕ_i satisfies

$$\mathcal{L}\phi_i(x) = \begin{cases} \lambda_i \phi_i(x) & \text{if } x \in S \\ 0 & \text{if } x \in \delta S. \end{cases}$$

We now define the heat kernel of S as an $(n \times n)$-matrix

$$H_t = \sum e^{\lambda_i t} P_i = e^{-t\mathcal{L}} = I - t\mathcal{L} + \frac{\Sigma^2}{2}\mathcal{L} + \cdots$$

where $\mathcal{L} = \sum \lambda_i P_i$ is the decomposition of the Laplacian $\mathcal{L}$ into projections on its eigenspaces. In particular, we have

- $H_0 = I$
- $F(x, t) = \displaystyle\sum_{y \in S \cup \delta S} H_t(x, y) f(y) = (H_t f)(x)$

- $F(x, 0) = f(x)$
- F satisfies the heat equation $\frac{\partial F}{\partial t} = -\mathcal{L}F$
- $H_t(x, y) \geq 0$.

By using the heat kernel, the following eigenvalue inequality can be derived, for all $t > 0$:

$$\lambda_S \geq \frac{\displaystyle\sum_{x \in S} \inf_{y \in S} H_t(x, y) \frac{\sqrt{d_x}}{\sqrt{d_y}}}{2t}.$$

One way to use the above theorem is to bound the heat kernel of a graph by the (continuous) heat kernel of the Riemannian manifolds, for certain graphs that we call convex subgraphs. We say Γ is a lattice graph if Γ is embedded into a d-dimensional Riemannian manifold $\mathcal{M}$ with a metric μ such that $\epsilon = \mu(x, gx) = \mu(y, g'y)$ for all $g, g' \in K$. An induced subgraph of a homogeneous graph Γ is said to be convex if the following conditions are satisfied:

1. There is a submanifold $M \subset \mathcal{M}$ with a convex boundary such that

$$V(P)(\Gamma) \cap M - \partial M = S.$$

2. For any $x \in S$, the ball centered at x of S of radius $\varepsilon/2$ is contained in M.

$$\mu(x, \widetilde{S}) \geq cM(x, gx) \text{ for some } g \in K$$

where $\widetilde{S}$ denotes some convex submanifold of M containing all vertices in S.

We need one more condition to apply our theorem on convex subgraphs. Basically, ϵ has to be "small" enough so that the count of vertices in S can be used to approximate the volume of the manifold M. Namely, let us define

$$r = \frac{U |S|}{\text{vol } M}. \tag{8}$$

where U denotes the volume of *Vononoi* region which consists of all points in $\mathcal{M}$ closest to a lattice point. Then the main result in [16] states that the Neumann eigenvalue of S satisfies the following inequality:

$$\lambda_1 \geq \frac{c \, r\varepsilon^2}{dD^2(M)}$$

for some absolute constant c which depends only on Γ; and $D(M)$ denotes the diameter of the manifold M. We note that r in (8) can be lower bounded by a constant if the diameter of M measured in L_1 norm is at least as large as ϵd.

6. Applications to random walks and rapidly mixing Markov chains

As an application of the eigenvalue inequalities in the previous sections, we consider the classical problem of sampling and enumerating the family S of $(n \times n)$-matrices with nonnegative integral entries with given row and column sums. Although the problem is presumed to be computationally intractable (in the so-called #P-complete class), the eigenvalue bounds in the previous section can be used to obtain a polynomial approximation algorithm. To see this, we consider the homogeneous graph Γ with the vertex set consisting of all $(n \times n)$-matrices with integral entries (possibly negative) with given row and column sums. Two vertices u and

v are adjacent if u and v differ at just the four entries of a (2×2)-submatrix with entries $u_{ik} = v_{ik} + 1, u_{jk} = v_{jk} - 1, u_{im} = v_{im} - 1, u_{jm} = v_{jm} + 1$. The family S of matrices with all nonnegative entries is then a convex subgraph of Γ.

On the vertices of S, we consider the following random walk. The probability $\pi(u,v)$ of moving from a vertex u in S to a neighboring vertex v is $\frac{1}{k}$ if v is in S where k is the degree of Γ. If a neighbor v of u (in Γ) is not in S, then we move from u to each neighbor z of v, z in S, with the (additional) probability $\frac{1}{d'_v}$ where $d'_v = |\{z \in S : z \sim v \text{ in } \Gamma\}|$ for $v \notin S$. In other words, for $u, v \in S$,

$$\pi(u,v) = \frac{w_{uv}}{d_u} + \sum_{\substack{z \notin S \\ u \sim z, v \sim z}} \frac{w_{uz}}{d_v d'_z} w_{zv}$$

where w_{uv} denotes the weight of the edge $\{u, v\}$ ($w_{uv} = 1$ or 0 for simple graphs) and $d_u = \sum_{u \sim v} d_{uv}$.

The stationary distribution for this walk is uniform. Let λ_π denote the second largest eigenvalue of π. It can be shown [15] that

$$1 - \lambda_\pi \geq \lambda_S .$$

In particular, if the total row sum (minus the maximum row sum) is $\geq c' \, n^2$, we have $1 - \lambda_\pi \geq \frac{c}{kD^2}$. This implies that a random walk converges to the uniform distribution in $O(\frac{1}{1-\lambda_\pi}) = O(kD^2)$ steps (measured in L_2 norm) and in $O(kD^2(\log n))$ steps for relative pointwise convergence.

It is reasonable to expect that the above techniques can be useful for developing approximation algorithms for many other difficult enumeration problems by considering random walk problems in appropriate convex subgraphs. Further applications using the eigenvalue bounds in previous sections can be found in [11].

References

[1] N. Alon, *Eigenvalues and expanders*, Combinatorica **6** (1986) 86–96.

[2] N. Alon, F. R. K. Chung and R. L. Graham, *Routing permutations on graphs via matchings*, SIAM J. Discrete Math. **7** (1994) 513–530.

[3] N. Alon and V. D. Milman, λ_1 *isoperimetric inequalities for graphs and superconcentrators*, J. Combin. Theory B **38** (1985), 73–88.

[4] L. Babai and M. Szegedy, *Local expansion of symmetrical graphs*, Combinatorics, Probab. Comput. **1** (1991), 1–12.

[5] J. Cheeger, *A lower bound for the smallest eigenvalue of the Laplacian*, Problems in Analysis, (R. C. Gunning, ed.), Princeton Univ. Press, Princeton, NJ (1970) 195–199.

[6] F. R. K. Chung, *Diameters and eigenvalues*, J. Amer. Math. Soc. **2** (1989) 187–196.

[7] F. R. K. Chung, Lectures on Spectral graph theory, CBMS Lecture Notes, 1995, AMS Publications, Providence, RI.

[8] F. R. K. Chung, V. Faber, and T. A. Manteuffel, *An upper bound on the diameter of a graph from eigenvalues associated with its Laplacian*, SIAM J. Discrete Math. **7** (1994) 443–457.

[9] F. R. K. Chung and R. L. Graham, *Quasi-random set systems*, J. Amer. Math. Soc. **4** (1991), 151–196.

[10] F. R. K. Chung, R. L. Graham, and R. M. Wilson, *Quasi-random graphs*, Combinatorica **9** (1989) 345–362.

[11] F. R. K. Chung, R. L. Graham, and S.-T. Yau, *On sampling with Markov chains, random structures and algorithms*, to appear.

[12] F. R. K. Chung, A. Grigor'yan, and S.-T. Yau, *Eigenvalues and diameters for manifolds and graphs*, Adv. in Math., to appear.

[13] F. R. K. Chung and S.-T. Yau, *Eigenvalues of graphs and Sobolev inequalities*, to appear in Combinatorics, Probab. Comput.

[14] F. R. K. Chung and S.-T. Yau, *A Harnack inequality for homogeneous graphs and subgraphs*, Communications in Analysis and Geometry **2** (1994) 628–639.

[15] F. R. K. Chung and S.-T. Yau, *The heat kernels for graphs and induced subgraphs*, preprint.

[16] F. R. K. Chung and S.-T. Yau, *Heat kernel estimates and eigenvalue inequalities for convex subgraphs*, preprint.

[17] P. Diaconis and D. Stroock, *Geometric bounds for eigenvalues of Markov chains*, Ann. Appl. Prob. **1** 36–61.

[18] M. Jerrum and A. Sinclair, *Approximating the permanent*, SIAM J. Comput. **18** (1989) 1149–1178.

[19] Peter Li and S. T. Yau, *On the parabolic kernel of the Schrödinger operator*, Acta Mathematica **156**, (1986) 153–201.

[20] A. Lubotsky, R. Phillips, and P. Sarnak, *Ramanujan graphs*, Combinatorica **8** (1988) 261–278.

[21] G. A. Margulis, *Explicit constructions of concentrators*, Problemy Peredachi Informasii **9** (1973) 71–80 (English transl. in Problems Inform. Transmission **9** (1975) 325–332.

[22] G. Polya and S. Szego, Isoperimetric inequalities in mathematical physics, Ann. of Math. Stud., no. **27**, Princeton University Press, Princeton, NJ (1951).

[23] P. Sarnak, Some Applications of Modular Forms, Cambridge University Press, London and New York (1990).

[24] R. M. Tanner, *Explicit construction of concentrators from generalized N-gons*, SIAM J. Algebraic Discrete Methods **5** (1984) 287–294.

[25] L. G. Valiant, *The complexity of computing the permanent*, Theoret. Comput. Sci. **8** (1979) 189–201.

[26] S.-T. Yau and R. M. Schoen, Differential Geometry, International Press, Cambridge Massachusetts (1994).

Extremal Hypergraphs and Combinatorial Geometry

ZOLTÁN FÜREDI[*]

University of Illinois
at Urbana-Champaign   and
Urbana, IL 61801, USA

Math. Inst.
Hungarian Academy of Sciences
1364 Budapest, POB 127, Hungary

ABSTRACT. Here we overview some of the methods and results of extremal graph and hypergraph theory. A few geometric applications are also given.

1. Introduction and notation

Most combinatorial problems can be formulated as (extremal) hypergraph problems. Extremal hypergraph theory applies a broad array of tools and results from other fields like number theory, linear and commutative algebra, probability theory, geometry, and information theory. On the other hand, it has a number of interesting applications in all parts of combinatorics, and in geometry, integer programming, and computer science. Some recent successes include: the best upper bound for the number of unit distances in a convex polygon [40]; the first nontrivial upper bound for the number of halving hyperplanes [3]; and the counterexample to the longstanding Borsuk's conjecture by Kahn and Kalai [48].

We overview some of the methods used in extremal graph and hypergraph theory and illustrate them by Turán-type problems. Some geometric applications are also given; more can be found in the recent monograph [60].

A *hypergraph* H is a pair $H = (V, \mathcal{E})$, where V is a finite set, the set of *vertices*, and $\mathcal{E}$ is a family of subsets of V, the set of *edges*. If all the edges have r elements, then H is called an *r-graph*, or r-uniform hypergraph. The complete r-partite hypergraph $\mathcal{K}_{t_1, t_2, \dots, t_r}$ has a partition of its vertex set $V = V_1 \cup \dots \cup V_r$, such that $|V_i| = t_i$, and $\mathcal{E} = \{E : |E \cap V_i| = 1\}$ for all $1 \leq i \leq r$. The set $\{1, 2, \dots, n\}$ is abbreviated as $[n]$.

2. The Turán problem

Given a graph F, what is $\mathrm{ex}(n, F)$, the maximum number of edges of a graph with n vertices not containing F as a subgraph? This problem was proposed for $F = C_4$ by Erdős [19] in 1938 and in general by Turán [72]. For example, $\mathrm{ex}(n, K_3) = \lfloor n^2/4 \rfloor$ (Mantel [57], Turán [72]). The Erdős-Stone-Simonovits [29], [26] theorem says that the order of magnitude of $\mathrm{ex}(n, F)$ depends only on the chromatic number, $\lim_{n \to \infty} \mathrm{ex}(n, F)/\binom{n}{2} = 1 - (\chi(F) - 1)^{-1}$. This gives a sharp estimate, except for bipartite graphs.

[*] *1991 Mathematics Subject Classification.* Primary 05D05; secondary 05B25, 05C65, 52C10.
Key words and phrases. Turán problems, intersecting hypergraphs, embeddings.

Proceedings of the International Congress
of Mathematicians, Zürich, Switzerland 1994
© Birkhäuser Verlag, Basel, Switzerland 1995

Very little is known even about simple cases when F is a fixed even cycle C_{2k} or a fixed complete bipartite graph $\mathcal{K}_{k,k}$. For a survey of extremal graph problems, see Bollobás' book [5], or Simonovits [67], [66]. For Turán problems for hypergraphs see [41].

3. Minimum graphs of given girth

Erdős proved in 1959 that for any $\chi \geq 2$ and $g \geq 3$ there exists a graph of chromatic number χ and girth g. (The *girth* is the length of the shortest cycle.) Known elementary constructions yield graphs with an enormous number of vertices. Recently, deep results in number theory combined with the eigenvalue methods in graph theory have been invoked with success to explicitly construct relatively small graphs, called *Ramanujan graphs*, with large chromatic number and girth (Margulis [58], Imrich [47], and Lubotzky, Phillips, and Sarnak [56]). These graphs give the lower bound in the following inequality:

$$\Omega(n^{1+(3/(4k+21))}) \leq \text{ex}(n, C_{2k}) \leq 90kn^{(k+1)/k}. \tag{1}$$

The first nontrivial lower bound, $\Omega(n^{1+(1/2k)})$, was given by Erdős (see in [28]) using probabilistic methods. The upper bound is due to Bondy and Simonovits [6] and is believed to give the correct order of magnitude.

Constructions giving $\Omega(n^{1+(1/k)})$ are known only for $k = 2, 3$, and 5 (Benson [4]). Wenger [74] simplified these cases. Recently Lazebnik, Ustimenko, and Woldar gave new algebraic constructions [53] for all k.

4. Bipartite graphs

For every bipartite graph F that is not a forest there is a positive constant c (not depending on n) such that $\Omega(n^{1+c}) \leq \text{ex}(n, F) \leq O(n^{2-c})$. The lower bound follows from (1). The upper bound is provided by the following result of Kővári, Sós, and Turán [50] concerning the complete bipartite graph.

$$\text{ex}(n, \mathcal{K}_{t,t}) < \frac{1}{2}(t-1)^{1/t}n^{2-(1/t)} + (t-1)n/2 = O(n^{2-(1/t)}). \tag{2}$$

This bound gives the right order of magnitude of $\text{ex}(n, \mathcal{K}_{t,t})$ for $t = 2$ and $t = 3$ and probably for all t. For $t > 3$ the best lower bound, $\text{ex}(n, \mathcal{K}_{t,t}) \geq \Omega(n^{2-2/(t+1)})$, is due to Erdős and Spencer [28]. Until now the only asymptotic for a bipartite graph that is not a forest, $\text{ex}(n, C_4) = \frac{1}{2}(1 + o(1))n^{3/2}$, was due to Erdős, Rényi, and T. Sós [25] and to Brown [8]. This has recently been generalized [43]:

THEOREM 1 *For any fixed $t \geq 1$* $\quad \text{ex}(n, \mathcal{K}_{2,t+1}) = \frac{1}{2}\sqrt{t}n^{3/2} + O(n^{4/3}).$

A large graph with no $\mathcal{K}_{2,t+1}$. The following algebraic construction is closely related to the examples for C_4-free graphs and is inspired by an example of Hyltén-Cavallius [46] and Mörs [59] given for Zarankiewicz's problem [76]. Let q be a prime power such that $(q-1)/t$ is an integer. We construct a $\mathcal{K}_{2,t+1}$-free graph G on $(q^2 - 1)/t$ vertices such that every vertex has degree q or $q - 1$. Let $\mathbf{F}$ be the

q-element field, $h \in \mathbf{F}$ an element of order t, $H = \{1, h, h^2, \ldots, h^{t-1}\}$. The vertices of G are the t-element orbits of $(\mathbf{F} \times \mathbf{F}) \setminus (0, 0)$ under the action of multiplication by powers of h. Two classes $\langle a, b \rangle$ and $\langle x, y \rangle$ are joined by an edge in G if $ax + by \in H$.

Note that the sets $N\langle a, b \rangle = \{\langle x, y \rangle : ax + by \in H\}$ form a q-uniform, symmetric, solvable, group divisible t-design.

Brown [8] gave an algebraic construction to show $\mathrm{ex}(n, \mathcal{K}_{3,3}) \geq (1/2 - o(1))n^{5/3}$. Very recently, it was shown to be asymptotically optimal [44].

THEOREM 2 $\mathrm{ex}(n, \mathcal{K}_{t,t}) \leq \dfrac{1}{2}(1 + o(1))n^{2-(1/t)}$.

5. The number of unit distances

What is the maximum number of times, $f^{(d)}(n)$, that the same distance can occur among pairs of n points in the d-dimensional space $\mathbf{R}^d$? The complete bipartite graph $\mathcal{K}_{2,3}$ cannot be realized on the plane, so $f^{(2)}(n) \leq \mathrm{ex}(n, \mathcal{K}_{2,3}) = O(n^{3/2})$. Erdős [20] conjectured in 1945 that the grid gave the best value, $f^{(2)}(n) = O(n^{1+C/\log\log n})$. Spencer, Szemerédi, and Trotter [69] proved $f^{(2)}(n) \leq O(n^{4/3})$. A new proof appeared in Clarkson et al. [13]. Erdős observed that for the 3-space $n^{4/3} \log \log n \leq f^{(3)}(n) \leq \mathrm{ex}(n, \mathcal{K}_{3,3}) = O(n^{5/3})$. The best upper bound is due to Clarkson et al. [13], $f^{(3)}(n) \leq O(n^{3/2}\beta(n))$, where $\beta(n)$ is an extremely slowly growing function related to the inverse of Ackermann's function.

It is proved in [40], using the Turán theory of matrices resembling the Davenport-Schinzel problem solved by Sharir [64], that the maximum number of unit distances in a *convex* n-gon, $g^{(2)}(n)$, is at most $7n \log n$. Erdős and Moser [24] conjecture that $g^{(2)}(n)$ is linear. Edelsbrunner and Hajnal [17] showed that $g^{(2)}(n) \geq 2n - 4$.

6. The number of halving planes

In most problems an estimate on the number of sub(hyper)graphs isomorphic to a given structure F is more applicable than the information about the Turán number $\mathrm{ex}(n, F)$. Rademacher proved in 1941 that a graph with $\lfloor n^2/4 \rfloor + 1$ edges has at least $\lfloor n/2 \rfloor$ triangles (see Lovász and Simonovits [55]). The best, in most cases almost optimal, lower bound for the number of triangles in a graph of n vertices and e edges was given by Fisher [31]. The following theorem was proved in [21] in an implicit form. For more explicit formulations see Erdős and Simonovits [27] or Frankl and Rödl [37].

THEOREM 3 (Erdős [21]) *For any positive integers r and $t_1 \leq \cdots \leq t_r$ there exist positive constants c' and c'' such that the following holds. If an r-graph has n vertices and $e \geq c'n^{r-\alpha}$ edges, where $\alpha = 1/(t_1 t_2 \cdots t_{r-1})$, then it contains at least*

$$c''(e/n^r)^{t_1 t_2 \cdots t_r} n^{t_1 + \cdots + t_r}$$

copies of the complete r-hypergraph $\mathcal{K}_{t_1, \ldots, t_r}$.

Let $S \subset \mathbf{R}^3$ be an n-set in general position. A plane containing three of the points is called a *halving plane* if it dissects S into two parts of (almost) equal cardinality. In [3] it was proved that the number of halving planes is at most $O(n^{2.998})$. As a main tool, for every set Y of n points in the plane, a set N of size $O(n^4)$ is constructed such that the points of N are distributed almost evenly in the triangles determined by Y. The proof is a combined application of Turán theory (Theorem 3), the random method, and fractional hypergraph coverings.

A generalization of Tverberg's theorem [73], conjectured in [3], was proved by Živaljević and Vrećica [77] by Lovász' topological method [54]. The exponent 2.998 was improved most recently by Dey and Edelsbrunner [14] to 8/3. The best lower bound is $\Omega(n^2 \log n)$. The best 2-dimensional upper bound is due to Pach, Szemerédi, and Steiger [61].

7. Intersecting hypergraphs

Here we consider the more general hypergraph problems, where the forbidden configurations are k-uniform hypergraphs. For example, if the excluded hypergraph consists of two disjoint edges; i.e., the family $\mathcal{H}$ of k-sets is intersecting, then $|\mathcal{H}| \leq \binom{n-1}{k-1}$ for $n \geq 2k$, where n stands for the number of vertices. If $\mathcal{G}$ is a family of k-sets of $[n]$ such that any two members intersect in at least t elements, then $|\mathcal{G}| \leq \binom{n-t}{k-t}$, provided n is sufficiently large, $n > n_0(k,t)$. Equality holds if and only if $\mathcal{G}$ consists of all k-element subsets of $[n]$ containing a fixed t-element subset (Erdős, Ko, and Rado [23]). The exact value of $n_0(k,t) = (k-t+1)(t+1)$ was determined by Frankl [32] (for $t \geq 15$), and by Wilson [75] (for all t, using association schemes). Define

$$\mathcal{A}^r = \left\{ G \in \binom{[n]}{k} : |G \cap [t+2r]| \geq t+r \right\}.$$

$|\mathcal{A}^r|$ is the largest among the $\mathcal{A}^i$'s if $(k-t+1)(2+\frac{t-1}{r+1}) \leq n < (k-t+1)(2+\frac{t-1}{r})$.

CONJECTURE 1 (Erdős, Ko, and Rado [23]; Frankl [32])   *If $\mathcal{G}$ is a t-intersecting family of maximum cardinality, then $\mathcal{G}$ is isomorphic to $\mathcal{A}^r$ for some r.*

This conjecture was proved [35] for $r < c\sqrt{t \log t}$, where $c > 0.02$ is an absolute constant. The proof is a triumph of the transformation method (left shifting).

THEOREM 4 [34]   *Suppose that a k-uniform hypergraph on n vertices has more than $\binom{n-t-1}{k-t-1}$ edges, $k \geq 2t+2$, $n > n_1(k)$. Then it contains two edges F, F' such that $|F \cap F'| = t$.*

8. Prescribed intersections

Let $0 \leq \ell_1 < \ell_2 < \cdots < \ell_s < k \leq n$ be integers. The family $\mathcal{G} \subseteq \binom{V}{k}$ is an $(n, k, \{\ell_1, \ldots, \ell_s\})$-system if $|G \cap G'| \in \{\ell_1, \ldots, \ell_s\}$ holds for every $G, G' \in \mathcal{G}$, $G \neq G'$. Denote $\{\ell_1, \ldots, \ell_s\}$ by L and let us denote by $m(n, k, L)$ the maximum cardinality of an (n, k, L)-system. The determination of $m(n, k, L)$ is the simplest looking Turán-type problem; the family of forbidden configurations consists only of hypergraphs of size two. The most well-known result of this type is the Erdős-Ko-Rado theorem dealing with the case $L = \{t, t+1, \ldots, k-1\}$ (see above).

The problem of determining $m(n, k, L)$ for general L was proposed by Larman [51], and first studied by Deza, Erdős, and Frankl [15]. A few years earlier yet, Ray-Chaudhuri and Wilson [10] proved a very general upper bound, namely that $m(n, k, L) \leq \binom{n}{s}$ holds for all $n \geq k$ and $|L| = s$. The proof uses linear algebraic independence of some higher order incidence matrices over the reals. This was generalized for finite fields by Frankl and Wilson [39]. Very recently Frankl, Ota, and Tokushige have determined almost all the 8192 exponents of the $m(n, k, L)$'s up to $k \leq 12$. The complexity of these questions can be seen in the following result of Frankl [33]. For every rational $r \geq 1$ there exist k and L such that $m(n, k, L) = \Theta(n^r)$. The proof of this combines the Δ-system method, and algebraic and geometric constructions. A similar conjecture of Erdős and Simonovits [27] for graphs is still open: for every rational $1 < p/q < 2$ there exists a bipartite graph G with $ex(n, G) = \Theta(n^{p/q})$, and every bipartite graph has a rational exponent r with $ex(n, G) = \Theta(n^r)$.

Improving a result of Babai and Frankl [1] a necessary and sufficient condition for $m(n, k, L) = \Theta(n)$ has been found. We say that the numbers $\ell_1, \ldots, \ell_s$ and k satisfy property $(*)$ if there exists a family $\mathcal{I} \subset 2^{[k]}$, closed under intersection, such that $\cup \mathcal{I} = [k]$ and $|I| \in L$ for all $I \in \mathcal{I}$.

THEOREM 5 [41] *If $(*)$ is satisfied, then $m(n, k, L) > (1/8k)n^{k/(k-1)}$. On the other hand, if $(*)$ does not hold, then $m(n, k, L) \leq (2^{k^k})n$.*

9. The chromatic number of the space

The following problem was proposed by Hadwiger [45]. What is the minimum number $c(n)$ such that $\mathbf{R}^n$ can be divided into $c(n)$ subsets $\mathbf{R}^n = C_1 \cup \ldots \cup C_{c(n)}$ such that no pair of points within the same C_i is at unit distance? In other words, what is the chromatic number of the unit distance graph? This problem is wide open even in the plane, we have only $4 \leq c(2) \leq 7$. The regular simplex shows $c(n) \geq n + 1$; the first nonlinear lower bound $\Omega(n^2)$ was given by Larman and Rogers [52]. They also gave an exponential upper bound of 3^n. The above-mentioned forbidden intersection theorems of Frankl and Wilson [39] easily lead to a lower bound of 1.2^n.

Sixty years ago Borsuk [7] raised the following question. Is it true that every set of diameter one in $\mathbf{R}^d$ can be partitioned into $d + 1$ sets of diameter smaller than one? The following theorem of Frankl and Rödl [38] led to the counterexample given by Kahn and Kalai [48]. Let n be an integer divisible by four, and let $\mathcal{F}$ be a family of subsets of an n-element underlying set such that no two sets in the family have intersection of size $n/4$. Then $|\mathcal{F}| < 1.99^n$.

10. Szemerédi's regularity lemma

This is a powerful graph-approximation method. We need some notation. Let G be an arbitrary, fixed graph. For two disjoint subsets $V_1, V_2 \subset V(G)$, let $E(V_1, V_2)$ denote the set of edges of G with one endpoint in V_1 and the other in V_2. The *edge-density* between these sets is

$$\delta(V_1, V_2) = \frac{|E(V_1, V_2)|}{|V_1| \cdot |V_2|}.$$

The pair (V_1, V_2) is called ε-*regular*, if $|\delta(V_1', V_2') - \delta(V_1, V_2)| < \varepsilon$ holds for all $V_1' \subset V_1$ and $V_2' \subset V_2$ whenever $|V_1'| \geq \varepsilon|V_1|$ and $|V_2'| > \varepsilon|V_2|$.

THEOREM 6 (Szemerédi's regularity lemma [70]) *For every $0 < \varepsilon < 1$ and for every integer r there exists an $M(\varepsilon, r)$ such that the following is true for every graph G. The vertex set of G can be partitioned into ℓ classes $V_1, \ldots, V_\ell$ for some $r \leq \ell \leq M(\varepsilon, r)$ so that these classes are almost equal (i.e., $||V_i| - |V(G)|/\ell| < 1$), and all but at most $\varepsilon \ell^2$ pairs (V_i, V_j) are ε-regular.*

The main feature of Theorem 6 is that it allows us to handle any given graph as if it were a random one. Even the most chaotic graph can be decomposed into a relatively small number of almost regular systems. Rödl [62] and Elekes [18] showed that one cannot require all pairs (V_i, V_j) to be ε-regular. Sós and Simonovits [68] (joining to works of Thomason [71] and Chung, Graham, and Wilson [12]) used Theorem 6 to describe the so-called *quasi-random* sequences of graphs. This connection is illuminated in the next section.

11. Graphs with a small number of triangles

This is an application of Szemerédi's regularity lemma. Let F be a fixed graph on the k-element vertex set $\{u_1, \ldots, u_k\}$, and suppose that the graph G on n vertices contains only $o(n^k)$ copies of F. We will prove that one can delete $o(n^2)$ edges from G to eliminate all copies of F. Reformulating this statement without o's for the special case $F = K_3$ we get

THEOREM 7 *For every $\varepsilon > 0$, there exists a $\delta = \delta(\varepsilon) > 0$ such that the following holds: For every graph G on n vertices with at most δn^3 triangles, one can find a set E' with at most εn^2 edges, such that $G \setminus E'$ is triangle-free.*

The theorem says that it is impossible to distribute *evenly* a small number of triangles in a graph with a large number of edges.

Let $V_1, \ldots, V_k$ be disjoint m-element sets and let $0 \leq \delta \leq 1$. A random graph on the vertex set $V_1 \cup \ldots \cup V_k$ is defined by choosing every pair of vertices $u \in V_i$, $v \in V_j$ with probability δ. The expected edge-density between V_i and V_j is δ. Moreover, the expected number of copies of F such that $v_i \in V_i$, and $v_i v_j$ is connected for $u_i u_j \in E(F)$ is $\delta^{|E(F)|} m^k$. The next lemma is used (sometimes in implicit form) in most applications of the Regularity Lemma.

LEMMA 1 *Let $a_1, \ldots, a_k$ be natural numbers, $\sum a_i = p$, let F be a graph on the p-element vertex set $\{u_{ij} : 1 \leq i \leq k, 1 \leq j \leq a_i\}$, and let $0 < \varepsilon < p^{-p}$, $\varepsilon^{1/p} \leq \delta < 1/2$. Suppose that the graph G has k pairwise disjoint subsets $V_1, \ldots, V_k \subset V(G)$, $|V_i| \geq m_i$ for all $1 \leq i \leq k$, and $\delta(V_i', V_j') \geq \delta$ hold for all $V_i' \subset V_i$, $V_j' \subset V_j$ if for some $1 \leq i < j \leq k$ there is an edge $u_{ia} u_{jb} \in E(F)$ and $|V_i'| \geq \varepsilon|V_i|$, $|V_j'| \geq \varepsilon|V_j|$. Then the subgraph of G induced by $V_1 \cup \ldots \cup V_k$ contains at least $\delta^{|E(F)|} 2^{-p} \prod(m_i)^{a_i}$*

copies (embeddings) of F with vertex sets $\{v_{ia}\}$ such that $v_{ia} \in V_i$, $v_{jb} \in V_j$, and $v_{ia}v_{jb} \in E(G)$ for $u_{ia}u_{jb} \in E(F)$.

Proof. The case $a_1 = \cdots = a_k = 1$ implies the general case. Indeed, choose a_i disjoint (m_i/a_i)-element sets from V_i, and apply the lemma for this new partition $(\varepsilon^* = \varepsilon/(\max a_i),\ \delta^* = \delta)$. Then do this for all possible partitions. Finally, the case $p = k$ follows by induction on k, as was done for $F = K_k$ in [60]. $\qquad\Box$

Proof of Theorem 7. Let $\varepsilon_0 = (\varepsilon/3)^k$. Suppose that $\varepsilon_0 < k^{-k}/3$, and define $r = \lceil 3/\varepsilon_0 \rceil$. We claim that $\delta = (2M(r, \varepsilon_0))^{-k}(\varepsilon_0)^{|E(F)|/k}$ will suffice. Let G be an arbitrary graph with at most δn^k copies of F. Apply Szemerédi's lemma with the above r and ε_0. We get a partition $V_1, \ldots, V_\ell$. Delete all edges covered by any V_i, then delete all edges connecting V_i and V_j if the pair (V_i, V_j) is not ε_0-regular, or if its density is less than $\varepsilon_0^{1/k}$. We have deleted at most $n^2/\ell + \varepsilon_0 n^2 + \varepsilon_0^{1/k} n^2$ edges. Then the rest of the graph is F-free; otherwise, the lemma would provide us at least $\varepsilon_0^{|E(F)|/k}(n/2\ell)^k$ copies. $\qquad\Box$

12. The maximum number of edges in a minimal graph of diameter 2

A graph G of diameter 2 is minimal if the deletion of any edge increases its diameter. Murty and Simon (see in [9]) conjecture that such a G cannot have more than $n^2/4$ edges. This was proved for $n > n_0$ in [42] in the following slightly stronger form: the only extremum is the complete bipartite graph. The value of n_0 is explicitly computable, but the proof gives a vastly huge number, a tower of 2's of height about 10^{14}.

This theorem is the first application of Szemerédi's regularity lemma yielding an exact answer (at least for $n > n_0$). Bounds were given by Caccetta and Häggkvist [9] and Fan [30]. It is easy to see that the theorem gives a direct generalization of Turán's triangle theorem. (If every edge in G that is contained in a triangle is also contained in some minimal path of length 2, then $|E(G)| \leq n^2/4$.)

The proof utilizes the following Turán type result of Ruzsa and Szemerédi [63]: if $\mathcal{F}$ is a triangle-free, 3-uniform hypergraph on n vertices (that means that no 6 vertices carry more than 2 triples), then $|\mathcal{F}| = o(n^2)$. In almost all other applications of Theorem 6 one only needs the Ruzsa-Szemerédi theorem. Note that it is an easy corollary of Theorem 7. (Replacing each triple by the 3 pairs contained in it one gets a graph with $3|\mathcal{F}|$ edges and only $|\mathcal{F}| \leq \binom{n}{2} = o(n^3)$ triangles.) Other short proofs and generalizations for r-uniform hypergraphs (also based on Theorem 6) were given by Erdős, Frankl, and Rödl [22] and by Duke and Rödl [16].

References

[1] L. Babai and P. Frankl, *Note on set intersections*, J. Combin. Theory Ser. A **28** (1980), 103–105.

[2] L. Babai and P. Frankl, *Linear algebra methods in combinatorics*, (Preliminary version 2.), Dept. Comp. Sci., The University of Chicago, 1992.

[3] I. Bárány, Z. Füredi, and L. Lovász, *On the number of halving planes*, Combinatorica **10** (1990), 175–183.

[4] C. T. Benson, *Minimal regular graphs of girth eight and twelve*, Canad. J. Math. **26** (1966), 1091–1094.

[5] B. Bollobás, Extremal Graph Theory, Academic Press, London, 1978.

[6] A. Bondy and M. Simonovits, *Cycles of even length in graphs*, J. Combin. Theory Ser. B **16** (1974), 97–105.

[7] K. Borsuk, *Drei Sätze über die n-dimensionale euklidische Sphäre*, Fund. Math. **20** (1933), 177–190.

[8] W. G. Brown, *On graphs that do not contain a Thomsen graph*, Canad. Math. Bull. **9** (1966), 281–289.

[9] L. Caccetta and R. Häggkvist, *On diameter critical graphs*, Discrete Math. **28** (1979), 223–229.

[10] D. K. Ray-Chaudhuri and R. M. Wilson, *On t-designs*, Osaka J. Math. **12** (1975), 737–744.

[11] F. R. K. Chung and P. Erdős, *On unavoidable graphs*, Combinatorica **3** (1983), 167–176; *On unavoidable hypergraphs*, J. Graph. Theory **11** (1987), 251–263.

[12] F. R. K. Chung, R. L. Graham, and R. M. Wilson, *Quasi-random graphs*, Combinatorica **9** (1989), 345–362.

[13] L. K. Clarkson, H. Edelsbrunner, L. Guibas, M. Sharir, and E. Wetzl, *Combinatorial complexity bounds for arrangements of curves and spheres*, Discrete Comput. Geom., **5** (1990), 99–160.

[14] Tamal K. Dey and H. Edelsbrunner, *Counting triangle crossings and halving planes*, manuscript Nov. 1992.

[15] M. Deza, P. Erdős, and P. Frankl, *Intersection properties of systems of finite sets*, Proc. London Math. Soc. (3) **36** (1978), 368–384.

[16] R. Duke and V. Rödl, *The Erdős-Ko-Rado theorem for small families*, J. Combin. Theory Ser. A **65** (1994), 246–251.

[17] H. Edelsbrunner and P. Hajnal, *A lower bound on the number of unit distances between the vertices of a convex polygon*, J. Combin. Theory Ser. A **55** (1990), 312–314.

[18] G. Elekes, *Irregular pairs are necessary in Szemerédi's regularity lemma*, manuscript (1992).

[19] P. Erdős, *On sequences of integers no one of which divides the product of two others and some related problems*, Izv. Naustno-Issl. Inst. Mat. i Meh. Tomsk **2** (1938), 74–82. (Zbl. **20**, p. 5)

[20] P. Erdős, *On sets of distances of n points*, Amer. Math. Monthly **53** (1946), 248–250.

[21] P. Erdős, *On extremal problems of graphs and generalized graphs*, Israel J. Math. **2** (1964), 183–190.

[22] P. Erdős, P. Frankl, and V. Rödl, *The asymptotic number of graphs not containing a fixed subgraph and a problem for hypergraphs having no exponent*, Graphs Combin. **2** (1986), 113–121.

[23] P. Erdős, Chao Ko, and R. Rado, *An intersecting theorem for finite sets*, Quart. J. Math. Oxford **12** (1961), 313–320.

[24] P. Erdős and L. Moser, *Problem 11*, Canad. Math. Bull. **2** (1959), 43.

[25] P. Erdős, A. Rényi, and V. T. Sós, *On a problem of graph theory*, Studia Sci. Math. Hungar. **1** (1966), 215–235.

[26] P. Erdős and M. Simonovits, *A limit theorem in graph theory*, Studia Sci. Math. Hungar. **1** (1966), 51–57.

[27] P. Erdős and M. Simonovits, *Supersaturated graphs and hypergraphs*, Combinatorica **3** (1983), 181–192.

[28] P. Erdős and J. Spencer, *Probabilistic Methods in Combinatorics*, Akadémiai Kiadó, Budapest, 1974.

[29] P. Erdős and A. H. Stone, *On the structure of linear graphs*, Bull. Amer. Math. Soc. **52** (1946), 1087–1091.

[30] G. Fan, *On diameter 2-critical graphs*, Discrete Math. **67** (1987), 235–240.

[31] D. C. Fisher, *Lower bounds on the number of triangles in a graph*, J. Graph Theory **13** (1989), 505–512.

[32] P. Frankl, *The Erdős-Ko-Rado theorem is true for n = ckt*, Combinatorics, Proc Fifth Hungarian Colloq. Combin., Keszthely, Hungary, 1976, (A. Hajnal et al., eds.), Proc. Colloq. Math. Soc. J. Bolyai **18** (1978), 365–375, North-Holland, Amsterdam.

[33] P. Frankl, *All rationals occur as exponents*, J. Combin. Theory Ser. A **42** (1986), 200–206.

[34] P. Frankl and Z. Füredi, *Exact solution of some Turán-type problems*, J. Combin. Theory Ser. A **45** (1987), 226–262.

[35] P. Frankl and Z. Füredi, *Beyond the Erdős-Ko-Rado theorem*, J. Combin. Theory Ser. A **56** (1991), 182–194.

[36] P. Frankl, K. Ota, and N. Tokushige, *Exponents of uniform L-systems*, manuscript, 1994.

[37] P. Frankl and V. Rödl, *Hypergraphs do not jump*, Combinatorica **4** (1984), 149–159.

[38] P. Frankl and V. Rödl, *Forbidden intersections*, Trans. Amer. Math. Soc. **300** (1987), 259–286.

[39] P. Frankl and R. M. Wilson, *Intersection theorems with geometric consequences*, Combinatorica **1** (1981), 357–368.

[40] Z. Füredi, *The maximum number of unit distances in a convex n-gon*, J. Combin. Theory Ser. A **55** (1990), 316–320.

[41] Z. Füredi, *Turán type problems*, in Surveys in Combinatorics, Proc. of the 13th British Combinatorial Conference, (ed. A. D. Keedwell), Cambridge Univ. Press, London Math. Soc. Lecture Note Series **166** (1991), 253–300.

[42] Z. Füredi, *The maximum number of edges in a minimal graph of diameter 2*, J. Graph Theory **16** (1992), 81–98.

[43] Z. Füredi, *New asymptotics for bipartite Turán numbers*, J. Combin. Theory Ser. A, to appear.

[44] Z. Füredi, *An upper bound on Zarankiewicz' problem*, Comb. Prob. and Comput., to appear.

[45] H. Hadwiger, *Überdeckungssätze für den Euklidischen Raum*, Portugal. Math. **4** (1944), 140–144.

[46] C. Hyltén-Cavallius, *On a combinatorial problem*, Colloq. Math. **6** (1958), 59–65.

[47] W. Imrich, *Explicit construction of graphs without small cycles*, Combinatorica **4** (1984), 53–59.

[48] J. Kahn and G. Kalai, *A counterexample to Borsuk's conjecture*, Bull. Amer. Math. Soc. **29** (1993), 60–63.

[49] Gy. Katona, *Intersection theorems for systems of finite sets*, Acta Math. Hungar. **15** (1964), 329–337.

[50] T. Kővári, V. T. Sós, and P. Turán, *On a problem of K. Zarankiewicz*, Colloq. Math. **3** (1954), 50-57.

[51] D. C. Larman, *A note on the realization of distances within sets in Euclidean space*, Comment. Math. Helv. **53** (1978), 529–535.

[52] D. Larman and C. Rogers, *The realization of distances within sets in Euclidean space*, Mathematika **19** (1972), 1–24.

[53] F. Lazebnik, V. A. Ustimenko, and A. J. Woldar, *A new series of dense graphs of high girth*, Bull. Amer. Math. Soc., **32** (1995), 73–79.

[54] L. Lovász, *Kneser's conjecture, chromatic number and homotopy*, J. Combin. Theory Ser. A **25** (1978), 319–324.

[55] L. Lovász and M. Simonovits, *On the number of complete subgraphs of a graph II*, in *Studies in Pure Math.*, Birkhäuser Verlag, Basel, (1983), 459–495.

[56] A. Lubotzky, R. Phillips, and P. Sarnak, *Ramanujan graphs*, Combinatorica **8** (1988), 261–277.

[57] W. Mantel, *Problem 28*, Wiskundige Opgaven **10** (1907), 60–61.

[58] G. A. Margulis, *Explicit construction of graphs without short cycles and low density codes*, Combinatorica **2** (1982), 71–78.

[59] M. Mörs, *A new result on the problem of Zarankiewicz*, J. Combin. Theory Ser. A **31** (1981), 126–130.

[60] J. Pach and P. K. Agarwal, Combinatorial Geometry, Wiley, New York, 1995.

[61] J. Pach, W. Steiger, and E. Szemerédi, *An upper bound on the number of planar k-sets*, Discrete Comput. Geom. **7** (1992), 109–123.

[62] V. Rödl, personal communication

[63] I. Z. Ruzsa and E. Szemerédi, *Triple systems with no six points carrying three triangles*, Combinatorics (Keszthely, 1976), Proc. Colloq. Math. Soc. J. Bolyai **18**, vol. II., 939–945, North-Holland, Amsterdam and New York, 1978.

[64] M. Sharir, *Almost linear upper bounds on the length of generalized Davenport-Schinzel sequences*, Combinatorica **7** (1987), 131–143.

[65] A. F. Sidorenko, *On the maximal number of edges in a uniform hypergraph that does not contain prohibited subgraphs*, Math. Notes **41** (1987), 247–259.

[66] M. Simonovits, *Extremal graph theory*, in Selected Topics in Graph Theory 2, (eds. L. W. Beineke and R. J. Wilson), Academic Press, New York, 1983, 161–200.

[67] M. Simonovits, *Extremal graph problems, degenerate extremal problems, and supersaturated graphs*, Progress in Graph Theory (Waterloo, Ont. 1982), 419–437, Academic Press, Toronto, Ont., 1984.

[68] V. T. Sós and M. Simonovits, *Szemerédi's partition and quasi-randomness*, Random Structures and Algorithms **2** (1991), 1–10.

[69] J. Spencer, E. Szemerédi, and W. T. Trotter, *Unit distances in the Euclidean plane*, Graph Theory and Combinatorics, (ed. B. Bollobás), Academic Press, London, 1984, pp. 293–303.

[70] E. Szemerédi, *Regular partitions of graphs*, Problèmes Combinatoires et Théorie des Graphes, Proc. Colloq. Int. CNRS **260**, (1978), 399–401, Paris.

[71] A. Thomason, *Pseudo random graphs*, Proceedings on Random Graphs, Poznan, 1985, (M. Karoński, ed.), Ann. Discrete Math. **33** (1987), 307–331.

[72] P. Turán, *On an extremal problem in graph theory*, Mat. Fiz. Lapok **48** (1941), 436–452 (in Hungarian). (Also see Colloq. Math. **3** (1954), 19–30.)

[73] H. Tverberg, *A generalization of Radon's theorem*, J. London Math. Soc. **41** (1966), 123–128.

[74] R. Wenger, *Extremal graphs with no C^4's, C^6's or C^{10}'s*, J. Combin. Theory Ser. B **52** (1991), 113–116.

[75] R. M. Wilson, *The exact bound in the Erdős-Ko-Rado theorem*, Combinatorica **4** (1984), 247–257.

[76] K. Zarankiewicz, *Problem of P101*, Colloq. Math. **2** (1951), 301.

[77] Živaljević and Vrećica, *The colored Tverberg's problem and complexes of injective functions*, J. Combin. Theory Ser. A **61** (1992), 309–318.

Asymptotics of Hypergraph Matching, Covering and Coloring Problems

JEFF KAHN*

Department of Mathematics and RUTCOR,
Rutgers University, New Brunswick, NJ 08903, USA

1. Introduction

A *hypergraph* $\mathcal{H}$ is simply a collection of subsets of a finite set, which we will always denote by V. Elements of V are called *vertices* and elements of $\mathcal{H}$ *edges*. A hypergraph is *k-uniform (k-bounded)* if each of its edges has size k (at most k).

The *degree* in $\mathcal{H}$ of a vertex x is the number of edges containing x, and is denoted by $d(x)$. Similarly, $d(x,y)$ denotes the number of edges containing both of the vertices x, y. A hypergraph is *d-regular* if each of its vertices has degree d and *simple* if $d(x,y) \leq 1$ for all x,y. We use $D(\mathcal{H})$ and $\delta(\mathcal{H})$ for the maximum and minimum degrees of $\mathcal{H}$. A simple 2-uniform hypergraph is a *graph*, usually denoted G, whereas a general 2-uniform hypergraph is a *multigraph*.

Principal objects associated with a hypergraph are matchings, covers, and colorings. A *matching* is a collection of pairwise disjoint edges. We write $\mathcal{M} = \mathcal{M}(\mathcal{H})$ for the set of matchings of $\mathcal{H}$ and $\nu(\mathcal{H})$ for the *matching number*, the maximum size of a matching in $\mathcal{H}$. A *cover* is a collection of edges whose union is V, and an *(edge-)coloring* is $\sigma : \mathcal{H} \to S$ (S a set) with $A \cap B \neq \emptyset \Rightarrow \sigma(A) \neq \sigma(B)$ (so a partition of $\mathcal{H}$ into matchings). For these the parameters analogous to ν are $\rho(\mathcal{H})$, the minimum size of a cover, and $\chi'(\mathcal{H})$, the minimum number of matchings in a coloring. For further background see e.g. [16].

Much of this talk deals with hypergraphs in which edge sizes are fixed or bounded and degrees are large. In contrast to the familiar intractability of hypergraph problems, a central message here is that under the restrictions just stated one does often have, or at least seems to have, good *asymptotic* behavior. This is tied to notions of approximate independence (Sections 3., 4.) and relations between integer and linear programs (Section 5.). For a somewhat less compressed account of most of what is covered here, see [29]; as discussed there (and apparent here), much of this material had its beginnings in problems of Paul Erdős.

2. Some coloring problems

We begin with a few problems intended both to give some flavor of the subject and to form a basis for later discussion. For many more coloring problems see [43].

*Supported in part by NSF.

Proceedings of the International Congress
of Mathematicians, Zürich, Switzerland 1994
© Birkhäuser Verlag, Basel, Switzerland 1995

Vizing, Pippenger, Pippenger-Spencer

The classic result on edge-colorings of graphs is Vizing's theorem [44]: for any multigraph G, $\chi'(G) \leq D(G) + \max\{d(x,y) : x,y \in V, x \neq y\}$. In particular, for simple, d-regular G we have

$$\chi'(G) \leq d+1 \quad \text{and consequently} \quad \nu(G) \geq (1 - 1/(d+1))v(G)/2. \qquad (1)$$

Here the bound on ν is not hard, but that on χ' is far from obvious. The next two results, due respectively to Pippenger (unpublished; see [16]) and Pippenger and Spencer [38], give analogues of the bounds in (1) for hypergraphs of fixed edge size, and elegantly illustrate the "central message" mentioned in Section 1.. (We omit corresponding statements for covers.)

THEOREM 2.1 *Let k be fixed and $\mathcal{H}$ a k-uniform, d-regular hypergraph on n vertices satisfying*

$$d(x,y) < o(d) \text{ for all distinct vertices } x,y. \qquad (2)$$

Then $\nu(\mathcal{H}) \sim n/k \quad (d \to \infty)$.

THEOREM 2.2 *Under the hypotheses of Theorem 2.1,* $\chi'(\mathcal{H}) \sim d \quad (d \to \infty)$.

Note that convergence in these statements is uniform in $\mathcal{H}$. A similar "uniformity convention" is assumed to be in force whenever appropriate in what follows.

Of course Theorem 2.2 contains Theorem 2.1. (Its proof does require Theorem 2.1, but this is not true of the stronger Theorem 3.1 below.)

Part of the appeal of hypergraph problems is that they usually are not vulnerable to traditional graph-theoretic methods (e.g. recoloring as for Vizing's theorem); thus, they require the development of new (and perhaps more interesting) approaches. Thus, for $k \geq 3$, even Theorem 2.1 is quite a deep result (the theorems do not become easier if one substitutes simplicity for (2)). Its proof is achieved by a "semirandom" method that constructs the desired matching in small random increments. The proof of Theorem 2.2 is in the same vein, but requires a second level of ideas: the "increments" are now small sets of matchings chosen at random, and the choice of an appropriate probability distribution on $\mathcal{M}$ is a subtle matter.

A related semirandom approach was pioneered by Ajtai, Komlós, and Szemerédi [1]. Something closer to what is needed for Theorem 2.1 appears in a breakthrough paper of Rödl [40], proving the "Erdős-Hanani conjecture" [11]. Theorem 2.1 strengthens a result of Frankl and Rödl [14], which was itself a broad generalization of Rödl's original work. We will later give a result (Theorem 3.1) even more general than Theorem 2.2, together with just a hint at the workings of the semirandom method. For more serious discussions see [16], [25].

The Erdős-Faber-Lovász conjecture

This celebrated conjecture from 1972 may be stated as follows:

CONJECTURE 2.3 *Any simple n-vertex hypergraph has chromatic index at most n.*

(To avoid trivialities, hypergraphs here are assumed to have no singleton edges.)

Erdős (e.g. [9]) has for many years listed Conjecture 2.3 as one of his "three favorite problems." It is sharp when $\mathcal{H}$ is a projective plane or complete graph K_n with n odd, and also in a few related cases, but there ought to be some slack in the bound away from these extremes. A natural strengthening, suggested by Meyniel (unpublished), Berge [3], and Füredi [15] would include Vizing's theorem:

CONJECTURE 2.4 *For $\mathcal{H}$ simple on vertex set V, $\chi'(\mathcal{H}) \le \max_{x \in V} |\cup_{x \in A \in \mathcal{H}} A|$.*

Notice that for $\mathcal{H}$ k-uniform ($k \ge 3$ fixed) and n large, Theorem 2.2 implies a much better bound than that of Conjecture 2.3, namely $\chi'(\mathcal{H}) \sim n/(k-1)$. From a relatively modest extension of Theorem 2.2 — or, more naturally, from Theorem 3.1 below — one may derive an asymptotic version of Conjecture 2.4 ([24]; the proof given there for $\chi' < n + o(n)$ applies here as well):

THEOREM 2.5 *For $\mathcal{H}$ simple on V, $\chi'(\mathcal{H}) < (1 + o(1)) \max_{x \in V} |\cup_{x \in A \in \mathcal{H}} A|$.*

It seems clear that the random methods used here will not by themselves suffice to settle the above conjectures, though it does not seem impossible that some combination of random and constructive techniques might prove effective. (The reader could try proving Conjecture 2.3 when edge sizes are at most 3, a case that already seems to capture much of the difficulty of the problem.)

List-colorings

The *list-chromatic index*, $\chi'_l(\mathcal{H})$, of $\mathcal{H}$ is the least t such that if $S(A)$ is a set ("list") of size t for each $A \in \mathcal{H}$, then there exists a coloring σ of $\mathcal{H}$ with $\sigma(A) \in S(A)$ for each $A \in \mathcal{H}$. One natural reason for considering such a notion is that an ordinary coloring problem in which some colors have already been assigned is a list-coloring problem. See [2] for an (already somewhat out of date) survey of recent developments in this area.

Of course one always has $\chi'_l \ge \chi'$. The intuition that coloring should be most difficult when all lists are equal is specious in general (see [13], [45]), but seems correct for edge-colorings of graphs. The following "list-chromatic" or "list-coloring" conjecture was proposed several times, probably first by Vizing in 1975 (see e.g. [20] for more on this story):

CONJECTURE 2.6 *For every multigraph G, $\chi'_l(G) = \chi'(G)$.*

The case $G = K_{n,n}$ — the *Dinitz conjecture* — was proposed by Dinitz in about 1978 (see [8]) in the context of Latin squares. This version is particularly appealing, and seems to have provided much of the initial stimulus for western interest in such questions.

Conjecture 2.6, and the Dinitz conjecture in particular, received considerable attention, especially in the last five years (see e.g. [2], [20], [17] for discussion and references). The Dinitz conjecture was finally given a beautiful and wholly elementary proof by Galvin [17], who in fact proved Conjecture 2.6 for all *bipartite* multigraphs. The nonbipartite case is still open. For G with $d(x,y) < o(D(G)) \; \forall x, y$ the bound $\chi'_l < (1 + o(1))D(G)$ is contained in Theorem 3.1.

Borsuk's conjecture

A classic problem of elementary geometry that is also a coloring problem (though not an edge-coloring problem) is "Borsuk's conjecture" stating that every bounded set in $\mathbf{R}^d$ is the union of $d+1$ sets of smaller diameter ([5]; see [19], [4], [6] for further discussion). A conjecture of Larman [36] extracts something of the combinatorial essence of Borsuk's conjecture and puts it a little closer to some of our other problems (note the formal similarity to Conjecture 2.3):

CONJECTURE 2.7 *If $\mathcal{H}$ is a t-intersecting hypergraph on n vertices, then there are $(t+1)$-intersecting $\mathcal{H} = \mathcal{H}_1, \ldots, \mathcal{H}_n$ with $\mathcal{H} = \cup \mathcal{H}_i$.*

(A hypergraph is *t-intersecting* if any two of its edges share at least t vertices. Note that for uniform $\mathcal{H}$ Larman's conjecture is included in Borsuk's.)

Conjecture 2.7 and Borsuk's conjecture were recently disproved in [30]. This does not have much to do with the rest of our discussion, though at least "thinking big" was again a key: for moderately large d, even $(1.2)^{\sqrt{d}}$ (vs. $d+1$) sets are not enough, and the smallest counterexamples known are in the vicinity of $d = 1000$.

The case $t = 1$ of Conjecture 2.7 remains open (and interesting). Here Füredi and Seymour (see [10], [34]) proposed the stronger conjecture that one may use $\mathcal{H}_i$'s of the form $\{A \in \mathcal{H} : A \supseteq \{x, y\}\}$. This too turns out to be false [31], though a simple disproof would still be welcome. See Section 5. for a little more on this.

3. List-colorings and the semirandom method again

A rough rationale for the good asymptotics of the types of hypergraphs appearing in Theorems 2.1 and 2.2 is that we are in a probability space — namely $\mathcal{M}(\mathcal{H})$ with an appropriate measure — endowed with considerable approximate (though usually no *exact*) independence.

Different manifestations of this idea appear in the workings of the semirandom method and in the material of the next section. Here we give just a hint at the former in the context of the list-coloring version of Theorem 2.2 [27]:

THEOREM 3.1 *Under the hypotheses of Theorem 2.1, $\chi'_l(\mathcal{H}) \sim d$ $(d \to \infty)$.*

The basic idea of the proof is quite natural, though a little strange in that it initially seems doomed to failure. We present a thumbnail sketch in the "standard" case that all the $S(A)$'s are the same. (The general case is not essentially different. Lest we create false impressions, it should be stressed that implementation of the following is reasonably delicate.)

We color the hypergraph in stages. At each stage we tentatively assign each as yet uncolored edge A a random color from its current list of legal colors. In some (most) cases, the color tentatively assigned to A will also be assigned to one or more edges meeting A. Such edges A are simply returned to the pool of uncolored edges. The remaining edges are permanently colored with their tentative colors and removed from the hypergraph. We then modify the lists of legal colors (mainly meaning that we delete from $S(A)$ all colors already assigned to edges that meet A) and repeat the process.

Martingale concentration results together with the Lovász local lemma [12] are used to show that this procedure can be repeated many times, leaving after each stage a hypergraph and lists of legal colors that are reasonably well behaved. (Finding the correct definition of "well behaved" is crucial.) Eventually our control here does deteriorate, but by the time this happens the degrees in the remaining hypergraph are small relative to the number of colors still admissible at an edge, and the remaining edges can be colored greedily. The strange feature alluded to above (this is not shared by proofs of earlier results) is that the lists of legal colors initially shrink much faster than the degrees. (Roughly, when the degrees have shrunk to βD, with β not too small, the lists will have size about $\beta^k D$.) This at first seems unpromising, because we are accustomed to thinking of the degree as a trivial lower bound on chromatic index. The situation is saved by approximate independence (though we can only hint at how this goes): the lists $S^i(A)$ (of legal colors for A at the end of stage i) tend to evolve fairly independently, except where obviously dependent. So, for example, for a color γ that through stage i has not been permanently assigned to any edge meeting $A \cap B$ (that is, we condition on this being so), the probability that γ belongs to $S^i(B)$ is not much affected by its membership or nonmembership in $S^i(A)$.

4.   Random matchings

We now consider a matching M drawn uniformly at random from $\mathcal{M}(G)$ or $\mathcal{M}(\mathcal{H})$. (It is also sometimes worthwhile to consider more general "normal" distributions on $\mathcal{M}$; see e.g. [22], [35], [39], [32]. The last of these was the link between Pippenger's theorem and the material of this section; see [29].)

Suppose first that M is uniform from $\mathcal{M} = \mathcal{M}(G)$. Set $\xi = \xi(G) = |M|$, $p_k(G) = Pr(\xi = k)$, and let $\mu = \mu(G)$ and $\sigma = \sigma(G)$ denote the mean and standard deviation of ξ. For a vertex x write $p(\overline{x})$ for the probability that x is not contained in any edge of M.

For a sequence $\{G_n\}$ of graphs, we abbreviate $\xi(G_n), \mu(G_n), \ldots$ to $\xi_n, \mu_n, \ldots$. To avoid trivialities we assume $|V(G_n)| \to \infty$ ($n \to \infty$). The sequence $\{p_k(G_n)\}_{k \geq 0}$ is *asymptotically normal* if for each $x \in \mathbf{R}$

$$\Pr\left(\frac{\xi_n - \mu_n}{\sigma_n} < x\right) \to \frac{1}{\sqrt{2\pi}} \int_{-\infty}^{x} e^{-t^2/2} dt \quad (n \to \infty).$$

The following results and conjectures are again motivated by — and their proofs (if any) depend on — the idea of approximate independence mentioned at the beginning of the preceding section. (But the proofs are not related to the "semirandom" method.)

THEOREM 4.1 *[28] The distribution $\{p_k(G_n)\}_{k \geq 0}$ is asymptotically normal if and only if*

$$\nu_n - \mu_n \to \infty \quad (n \to \infty). \tag{3}$$

THEOREM 4.2 *[33] For d-regular graphs G,*
(a) $p(\overline{x}) \sim d^{-1/2}$ $\forall x \in V$, *and in particular* $\mu(G) \sim |V(G)|/2$,
(b) $\sigma^2(G) \sim |V(G)|/(4\sqrt{d})$.

Again, limits in Theorem 4.2 are taken as $d \to \infty$; of course, what is surprising is that the values of the parameters are hardly affected by G, x.

It follows from results of [22], [35], and an observation of Harper [21] that Theorem 4.1 is true if we replace (3) by "$\sigma_n \to \infty$." The advantages of Theorem 4.1 over this may not be apparent, but for example it fairly easily gives

COROLLARY 4.3 *Each of the following implies asymptotic normality:*

(a) $\nu_n/D_n \to \infty$,

(b) $\delta_n < (1 - o(1))D_n$,

(c) $\nu_n > (1 - o(1))|V(G_n)|/2$.

In particular, Theorem 4.1 gives the first proof of asymptotic normality for sequences of regular graphs. (Theorem 4.2 (b) gives a second.) That (a) implies asymptotic normality was shown in [41], improving a result of [18].

For hypergraphs we have only conjectures. (We extend the above notation in the obvious ways.)

CONJECTURE 4.4 *Fix k. If $\mathcal{H}_n$ is a sequence of simple, k-bounded hypergraphs with $\delta_n \to \infty$, then the following are equivalent:*

(a) $\{p_k(\mathcal{H}_n)\}$ *is asymptotically normal,*

(b) $\sigma_n \to \infty$,

(c) $\nu_n - \mu_n \to \infty$.

In contrast to the situation for graphs, this is not true if $\delta_n \nrightarrow \infty$.

CONJECTURE 4.5 *For fixed k and simple, k-uniform, d-regular $\mathcal{H}$,*

$$p(\overline{x}) \sim d^{-1/k} \qquad \forall x \in V(\mathcal{H}).$$

These conjectures may be nonsense, but they are certainly very interesting if true. Note that Conjecture 4.5 is far stronger than Theorem 2.1 for simple $\mathcal{H}$ (we omit speculation on what happens under (2)). For a little more in this direction and possible extensions of Theorem 4.1 to multigraphs, see [28].

It would be very interesting to have analogues of Theorem 4.1 in other combinatorial situations; for example, is Conjecture 4.4 true if we replace ξ_n by the size of a uniformly chosen forest from a graph G_n? (And ν_n by the size of a largest forest in G_n. Compare here the well-known unimodal and log concavity conjectures for independent sets in matroids [46], [37].)

5. Fractional vs. integer

Finally, we want to say a little about connections with linear programming. Matching, covering, and coloring problems are integer programming problems, so it is often edifying to compare them with their "fractional" versions (linear relaxations). For example, the fractional version of ν is the *fractional matching number*

$$\nu^* = \max\{\sum_{A\in\mathcal{H}} f(A) : f \text{ a fractional matching}\},$$

where a *fractional matching* is $f : \mathcal{H} \to \mathbf{R}^+$ satisfying $\sum_{A\ni v} f(A) \le 1$ for every $v \in V$; and the fractional version of chromatic index is

$$\chi'^*(\mathcal{H}) = \min\{\sum_{M\in\mathcal{M}} f(M) : f : \mathcal{M} \to \mathbf{R}^+, \sum_{A\in M\in\mathcal{M}} f(M) \ge 1 \ \forall A \in \mathcal{H}\}.$$

We usually consider linear programs tractable, integer programs intractable, and expect the behavior of an IP to be quite different from that of its associated LP. Nonetheless, as will be evident in several of our examples, fractional behavior is often a good guide in combinatorial situations. In particular, one concretization of "nice behavior" is asymptotic agreement of fractional and integer versions of our various parameters. (The problems discussed here are probably special cases of such agreement in more general IP/LP situations, but this has yet to be explored.)

On the other hand, again as illustrated below, *disagreement* of fractional and integer is often a hallmark of difficult combinatorial problems.

There are interesting connections between "normal" distributions on $\mathcal{M}(G)$ and Edmonds' matching polytope theorem [7], but space does not permit; see [39], [32].

Borsuk again
Borsuk's conjecture is false fractionally (though we omit definitions): [30] gives a finite $X \subseteq \mathbf{R}^d$ such that any subset of smaller diameter has size less than $(1.2)^{-\sqrt{d}}|X|$. On the other hand, as shown by Füredi and Seymour (see [34]), the conjecture disproved in [31] (see discussion following Conjecture 2.7) is *true* fractionally, which may make the difficulty of a counterexample less surprising.

Incidentally, Borsuk's *theorem* (a ball in $\mathbf{R}^d$ cannot be covered by d sets of smaller diameter [5]) is, though not in this language, a classic example where fractional and integer disagree.

Pippenger again
It is easy to see that any k-uniform, regular $\mathcal{H}$ satisfies $\nu^*(\mathcal{H}) = |V(\mathcal{H})|/k$. Thus, the conclusion of Theorem 2.1 is $\nu(\mathcal{H}) \sim \nu^*(\mathcal{H})$. That this is the proper interpretation of Pippenger's theorem is the contention of [23], where a similar conclusion is proved assuming only that $\mathcal{H}$ is k-bounded and satisfies something like (2). (The statement of this requires a little care and necessarily refers to a given fractional matching of $\mathcal{H}$.) A more subtle extension of Theorem 2.1, again involving the interplay of IPs and LPs, is given in [32].

A problem of Erdős and Lovász
Another of Erdős' favorite problems (again see [9]) may be stated as follows (see [26] for some motivation).

QUESTION 5.1 *[12] Fix c. Suppose $\mathcal{H}$ is an r-regular hypergraph with $n = |V| \leq cr$ and $d(x, y) \geq 1$ for all $x, y \in V$. Is it true that for large enough r, $\rho(\mathcal{H}) < r$?*

Of course $\rho(\mathcal{H}) \leq r$ because the edges containing any given vertex form a cover. On the other hand, the *fractional* cover number is considerably smaller. The function $t : \mathcal{H} \to \mathbf{R}^+$ given by $t(A) = |A|/(n + r - 1)$ is a fractional cover (we omit the obvious definition) with

$$\sum_{A \in \mathcal{H}} t(A) = nr/(n + r - 1) \approx nr/(n + r) \leq cr/(c + 1).$$

But surprisingly (that is, despite what seems to have been a general expectation to the contrary),the answer to Question 5.1 is *no* [26].

So here we have disagreement of fractional and integer (but support for the idea that such disagreement signals combinatorial difficulties). It was the understanding provided by [23] of situations where one *cannot* have such disagreement that gave the first clue to [26]; see [26], [29].

Erdős-Faber-Lovász again
The fractional version of Conjecture 2.3 was conjectured in [42] and proved in [34]:

THEOREM 5.2 *If $\mathcal{H}$ is simple on n vertices, then $\chi'^* \leq n$.*

Due to its greater generality,this (eventually) turned out to have a proof considerably easier than that of the implied $\nu(\mathcal{H}) \geq |\mathcal{H}|/n$, which is the main result of [42]. The fractional version of Conjecture 2.4 is still unresolved.

"Nice" again
We close with our favorite realization of the idea that hypergraphs of bounded edge size and large degree are asymptotically well ehaved [25], [29]:

CONJECTURE 5.3 *For fixed k and k-bounded $\mathcal{H}$, $\chi'_l(\mathcal{H}) \sim \chi'(\mathcal{H}) \sim \chi'^*(\mathcal{H})$.*

Here even $\chi' \sim \chi'^*$ for multigraphs is open, though more precise results have been conjectured for more than twenty years (see [43, Problem 63]).

Conjecture 5.3 goes far beyond Theorem 3.1, and may be regarded as providing a complete understanding of the asymptotics of chromatic and list-chromatic indices of k-bounded hypergraphs, even in the absence of anything like (2). (That it implies Theorems 2.2, and 3.1 is not obvious, but follows from the version of Theorem 2.2 given in [38].)

References

[1] M. Ajtai, J. Komlós, and E. Szemerédi, *A dense infinite Sidon sequence*, European J. Combin. **2** (1981), 1–11.

[2] N. Alon, *Restricted colorings of graphs*, pp. 1–33 in Surveys in Combinatorics, 1993 (Proc. 14th British Combinatorial Conf.), Cambridge Univ. Press, Cambridge, 1993.

[3] C. Berge, *On the chromatic index of a linear hypergraph and the Chvátal conjecture*, pp. 40–44 in Combinatorial Mathematics: Proc. 3rd Int'l. Conf (New York, 1985), Ann. N.Y. Acad. Sci. **555**, New York Acad. Sci., New York, 1989.

[4] V. Boltjansky and I. Gohberg, Results and Problems in Combinatorial Geometry, Cambridge Univ. Press, Cambridge, 1985.

[5] K. Borsuk, *Drei Sätze über die n-dimensionale euklidische Sphäre*, Fund. Math. **20** (1933), 177–190.

[6] H. Croft, K. Falconer, and R. Guy, Unsolved Problems in Geometry, Springer-Verlag, Berlin and New York, 1991.

[7] J. Edmonds, *Maximum matching and a polyhedron with 0, 1-vertices*, J. Res. Nat. Bur. Standards (B) **69** (1965), 125–130.

[8] P. Erdős, *Some old and new problems in various branches of combinatorics*, Congr. Numer. **23** (1979), 19–37.

[9] P. Erdős, *On the combinatorial problems which I would most like to see solved*, Combinatorica **1** (1981), 25–42.

[10] P. Erdős, *Some of my old and new combinatorial problems*, pp. 35–46 in Paths, Flows and VLSI-Layout (B. Korte, L. Lovász, H.-J. Promel and A. Schrijver, eds.), Springer-Verlag, Berlin and New York, 1990.

[11] P. Erdős and H. Hanani, *On a limit theorem in combinatorial analysis*, Publ. Math. Debrecen **10** (1963), 10–13.

[12] P. Erdős and L. Lovász, *Problems and results on 3-chromatic hypergraphs and some related questions*, Colloq. Math. Soc. János Bolyai **10** (1974), 609–627.

[13] P. Erdős, A. Rubin, and H. Taylor, *Choosability in graphs*, Congr. Numer. **26** (1979), 125–157.

[14] P. Frankl and V. Rödl, *Near-perfect coverings in graphs and hypergraphs*, Europ. J. Combin. **6** (1985), 317–326.

[15] Z. Füredi, *The chromatic index of simple hypergraphs*, Graphs Combin. **2** (1986), 89–92.

[16] Z. Füredi, *Matchings and covers in hypergraphs*, Graphs Combin. **4** (1988), 115–206.

[17] F. Galvin, *The list chromatic index of a bipartite multigraph*, manuscript, 1994.

[18] C. D. Godsil, *Matching behavior is asymptotically normal*, Combinatorica **1** (1981), 369–376.

[19] B. Grünbaum, Borsuk's problem and related questions, Proc. Sympos. Pure Math. **7** (Convexity), Amer. Math. Soc., 1963.

[20] R. Häggkvist and J. C. M. Janssen, *On the list-chromatic index of bipartite graphs*, manuscript, 1993.

[21] L. H. Harper, *Stirling behavior is asymptotically normal*, Ann. Math. Stat. **38** (1967), 410–414.

[22] O. J. Heilmann and E. H. Lieb, *Theory of monomer-dimer systems*, Comm. Math. Phys. **25** (1972), 190–232.

[23] J. Kahn, *On a theorem of Frankl and Rödl*, in preparation.

[24] J. Kahn, *Coloring nearly-disjoint hypergraphs with $n + o(n)$ colors*, J. Combin. Theory Ser. A **59** (1992), 31–39.

1362 Jeff Kahn

[25] J. Kahn, *Recent results on some not-so-recent hypergraph matching and covering problems*, Proc. 1st Int'l Conference on Extremal Problems for Finite Sets, Visegrád, June 1991, to appear.

[26] J. Kahn, *On a problem of Erdős and Lovász II: $n(r) = O(r)$*, J. Amer. Math. Soc. **7** (125–143), 1994.

[27] J. Kahn, *Asymptotically good list-colorings*, J. Combin. Theory Ser. A, to appear.

[28] J. Kahn, *A normal law for matchings*, in preparation.

[29] J. Kahn, *On some hypergraph problems of Paul Erdős and the asymptotics of matchings, covers and colorings*, to appear.

[30] J. Kahn and G. Kalai, *A counterexample to Borsuk's Conjecture*, Bull. Amer. Math. Soc. **29** (1993), 60–62.

[31] J. Kahn and G. Kalai, *A problem of Füredi and Seymour on covering intersecting families by pairs*, J. Combin. Theory, to appear.

[32] J. Kahn and M. Kayll, *Fractional vs. integer covers in hypergraphs of bounded edge size*, in preparation.

[33] J. Kahn and J. H. Kim, *Random matchings in regular graphs*, in preparation.

[34] J. Kahn and P. D. Seymour, *A fractional version of the Erdős-Faber-Lovász Conjecture*, Combinatorica **12** (1992), 155–160.

[35] H. Kunz, *Location of the zeros of the partition function for some classical lattice systems*, Phys. Lett. (A) (1970), 311–312.

[36] D. G. Larman, in Convexity and Graph Theory (Rosenfeld and Zaks, eds.), Ann. Discrete Math. **20** (1984), 336.

[37] J. H. Mason, *Matroids: Unimodal conjectures and Motzkin's theorem*, pp. 207–221 in Combinatorics (D. J. A. Welsh and D. R. Woodall, eds.), Inst. Math. Appl., Oxford Univ. Press, New York, 1972.

[38] N. Pippenger and J. Spencer, *Asymptotic behavior of the chromatic index for hypergraphs*, J. Combin. Theory Ser. A **51** (1989), 24–42.

[39] Y. Rabinovich, A. Sinclair, and A. Wigderson, *Quadratic dynamical systems*, Proc. 33rd IEEE Symposium on Foundations of Computer Science (1992), 304–313.

[40] V. Rödl, *On a packing and covering problem*, European J. Combin. **5** (1985), 69–78.

[41] A. Ruciński, *The behaviour of $\binom{n}{k,,\ldots,,k,n-ik}c^i/i!$ is asymptotically normal*, Discrete Math. **49** (1984), 287–290.

[42] P. D. Seymour, *Packing nearly-disjoint sets*, Combinatorica **2** (1982), 91–97.

[43] B. Toft, 75 *graph-colouring problems*, pp. 9–35 in Graph Colourings (R. Nelson and R.J. Wilson, eds.), Wiley, New York, 1990.

[44] V. G. Vizing, *On an estimate of the chromatic class of a p-graph* (in Russian), *Diskret. Analiz* **3** (1964), 25–30.

[45] V. G. Vizing, *Coloring the vertices of a graph in prescribed colors* (in Russian), Diskret. Analiz No 29 Metody Diskret. Anal. v Teorii Kodov i Shem (1976), 3–10, 101 (MR58 #16371).

[46] D. J. A. Welsh, *Combinatorial problems in matroid theory*, pp. 291–307 in Combinatorial Mathematics and its Applications, Academic Press, New York, 1971.

Combinatorics and Convexity

GIL KALAI[*]

Hebrew University
Givat Ram, Jerusalem 91904, Israel

Connections between Euclidean convex geometry and combinatorics go back to Euler, Cauchy, Minkowski, and Steinitz. The theory was advanced greatly since the 1950's and was influenced by the discovery of the simplex algorithm, the connections with extremal combinatorics, the introduction of methods from commutative algebra, and the relations with complexity theory.

The first part of this paper deals with convexity in general, and the second part deals with the combinatorics of convex polytopes. There are many excellent surveys [21], [10] and collections of open problems [14], [30]. I try to discuss several specific topics and to zoom in on issues with which I am more familiar.

1. Convex sets in general

1.1 Covering, packing, and tiling

Borsuk conjectured (1933) that every bounded set in $\mathbb{R}^d$ can be covered by $d+1$ sets of smaller diameter. Kahn and Kalai [23] showed that Borsuk's conjecture is very false in high dimensions.

Here is the disproof of Borsuk's conjecture. Let $f(d)$ be the smallest integer such that every bounded set in $\mathbb{R}^d$ can be covered by $f(d)$ sets of smaller diameter. For a bounded metric space X, let $b(X)$ be the minimum number of sets of smaller diameter needed to cover X. Consider $\mathbb{P}^{d-1}$ the space of lines through the origin in $\mathbb{R}^d$ where the metric is given by the angle between two lines. The diameter of $\mathbb{P}^{d-1}$ is $\pi/2$ and the distance between two lines is $\pi/2$ iff they are orthogonal. Let $d = 4p$, p a prime. Frankl and Wilson [18], see also [40], [17], proved that there are at most 1.8^d vectors in $\{-1,+1\}^d$ such that no two are orthogonal. This yields $b(\mathbb{P}^{d-1}) > 1.1^d$, because if $\mathbb{P}^{d-1}$ is covered by t sets of smaller diameter, each such set contains at most 1.8^d of the lines spanned by the vectors in $\{-1,+1\}^d$. But there are 2^{d-1} such lines and therefore $t \geq (2/1.8)^d$. Now, embed $\mathbb{P}^{d-1}$ into $\mathbb{R}^{d^2}$ by the map $x \rightarrow x \otimes x$, where x is a vector of norm 1 in $\mathbb{R}^d$. Note[1] that $< x \otimes x, y \otimes y > = < x, y >^2$. Therefore, the order relation between distances is preserved, and the image of $\mathbb{P}^{d-1}$ is the required counterexample. This example gives $f(d) > 1.2^{\sqrt{d}}$, for sufficiently large d.

---

[*]e-mail: kalai@math.huji.ac.il

[1]If $x = (x_1, x_2, \ldots, x_d)$ and $y = (y_1, y_2, \ldots, y_k)$, you can regard $x \otimes y$ as the $(d \times k)$-matrix whose (i,j)-entry is $x_i \cdot y_j$.

Proceedings of the International Congress
of Mathematicians, Zürich, Switzerland 1994
© Birkhäuser Verlag, Basel, Switzerland 1995

Betke, Henk, and Wills [7] proved for sufficiently high dimensions Fejes Toth's sausage conjecture. They showed that the minimum volume of the convex hull of n nonoverlapping congruent balls in $\mathbb{R}^d$ is attained when the centers are on a line.

Keller conjectured (1930) that in every tiling of $\mathbb{R}^d$ by cubes there are two cubes that share a complete facet. Lagarias and Shor [31] showed this to be false for $d \geq 10$. They used a reduction to a purely combinatorial problem, which was found by Corŕadi and Szabó.

Some problems
There are many problems on packing, covering, and tiling, and the most famous are perhaps the sphere packing problem in $\mathbb{R}^3$ and the (asymptotic) sphere packing problem in $\mathbb{R}^d$. There are several open problems around Borsuk's problem. What is the asymptotic behavior of $f(d)$? What is the situation in low dimensions? What is the behavior of $b(\mathbb{P}^n)$? Witsenhausen conjectured (see [17]) that if A is a subset of the unit sphere without two orthogonal vectors, then $\mathrm{vol}(A) \leq 2v_{\pi/4}$, where $v_{\pi/4}$ is the volume of a spherical cap of radius $\pi/4$. This would imply that $b(\mathbb{P}^n) \leq (\sqrt{2} + o(1))^n$. Perhaps the algebraic methods used for the Frankl-Wilson theorem can be of help.

Schramm [42] proved an upper bound $f(d) \leq s(d) = (\sqrt{3/2} + o(1))^d$. He showed that every set of constant width can be covered by $s(d)$ smaller homothets. Bourgain and Lindenstrauss [13] proved the same bound by covering every bounded set by $s(d)$ balls of the same diameter. (Danzer already showed that an exponential number of balls is sometimes necessary.) In his proof Schramm related the value of $f(d)$ to another classical problem in convexity, that of finding or estimating the minimal volume of Euclidean (and more generally spherical) sets of constant width.

It is not known if there are sets of constant width 1 in $\mathbb{R}^d$ whose volume is *exponentially* smaller than the volume of a ball of radius $1/2$. Perhaps the following series of examples (suggested by Schramm) $K_d \subset \mathbb{R}^d$ will do, but we do not know to compute or estimate their volumes. $K_0 = 0$ and K_{d+1} is obtained as follows. Consider K_d as sitting in the hyperplane given by $x_{d+1} = 0$ in $\mathbb{R}^{d+1}$. Now, take $K_{d+1} = A_{d+1} \cup B_{d+1}$ where A_{d+1} is the set of all points z with $x_{d+1} \geq 0$ such that the ball of radius 1 around z contains K_d and B_{d+1} is the set of all points z with $x_{d+1} \leq 0$ that belong to every ball of radius 1 that contains K_d. Schramm also conjectured that the minimal volume of a spherical set of constant width $\pi/4$ is obtained for an orthant.

Finally, what is the minimal diameter d_n such that the unit n-ball can be covered by $n + 1$ sets of diameter d_n? It is known that $2 - O(\log n/n) \leq d_n \leq 2 - O(1/n)$, see [32]. Hadwiger conjectured that the upper bound (which corresponds to the standard symmetric decomposition of the ball to $n + 1$ regions) is the truth. Perhaps also here the natural conjecture is false?

1.2 Helly-type theorems
Tverberg's theorem
Sarkaria [41] found a striking simple proof of the following theorem of Tverberg [50]:

Every $(d + 1)(r - 1) + 1$ points in $\mathbb{R}^d$ can be partitioned into r parts such that the convex hulls of these parts have nonempty intersection.

He used the following result of Bárány [2]. *Let $A_1, A_2, \ldots, A_{d+1}$ be sets in $\mathbb{R}^d$ such that $x \in \text{conv}(A_i)$ for every i. Then it is possible to choose $a_i \in A_i$ such that $x \in \text{conv}(a_1, a_2, \ldots, a_{d+1})$.* (To prove this consider the minimal distance t between x and such $\text{conv}(a_1, a_2, \ldots, a_{d+1})$ and show that if $t > 0$ one of the a_i's can be replaced to decrease t.)

Now consider $m = (d+1)(r-1) + 1$ points $a_1, a_2, \ldots, a_m$ in $\mathbb{R}^d$ and regard them as points in $V = \mathbb{R}^{d+1}$ whose sum of coordinates is 1. Sarkaria's idea was to consider the tensor product $V \otimes W$ where W is a $(r-1)$-dimensional space spanned by r vectors $w_1, w_2, \ldots, w_r$ whose sum is zero. Next define m $(= \dim V \otimes W - 1)$ sets in $V \otimes W$ as follows:

$$A_i = \{a_i \otimes w_1, a_i \otimes w_2, \ldots, a_i \otimes w_r\}.$$

Note that 0 is in the convex hull of each A_i and by Bárány's theorem $0 \in \text{conv}\{a_1 \otimes w_{i_1}, a_2 \otimes w_{i_2}, \ldots, a_m \otimes w_{i_m}\}$, for some choices of $i_1, i_2, \ldots, i_m$. The required partition of the points is given by $\Omega_j = \{a_k : i_k = j\}$, $j = 1, 2, \ldots, r$. To see this write $0 = \sum \lambda_k a_k \otimes w_{i_k}$, where the coefficients λ_k are nonnegative and sum to 1. Deduce that the vectors $v_j = \sum_{k \in \Omega_j} \lambda_k a_k$, $1 \leq j \leq r$, are all equal, as are the scalars $\alpha_j = \sum_{k \in \Omega_j} \lambda_k$.

There are many beautiful problems and results concerning Tverberg's theorem, see [16]. Topological versions were found for the case where r is a prime [3] and were extended to derive colored versions of Tverberg's theorem [53]. Sierksma conjectured, see [51], that the number of Tverberg partitions is at least $(r-1)!^d$. For a finite set A in $\mathbb{R}^d$ let $f(A, r) = \max\{\dim \cap_{i=1}^r \text{conv}(\Omega_i)\}$, where the maximum is taken over all partitions $(\Omega_1, \Omega_2, \ldots, \Omega_r)$ of A.

CONJECTURE: $\sum_{r=1}^{|A|} f(A, r) \geq 0$. (Note: $\dim \emptyset = -1$.)

This extension of Tverberg's theorem was proved by Kadari for planar sets.

Hadwiger-Debrunner's piercing conjecture

Alon and Kleitman [1] proved the Hadwiger-Debrunner piercing conjecture.

For every d and every $p \geq d + 1$ there is a $c = c(p, d) < \infty$ such that the following holds. For every family H of compact, convex sets in $\mathbb{R}^d$ in which any set of p members of the family contains a subset of cardinality $d + 1$ with a nonempty intersection there is a set of c points in $\mathbb{R}^d$ that intersects each member of H.

Helly's theorem asserts that $c(d+1, d+1) = 1$ and it is not difficult to see that $c(p, 1) = p - 1$. We describe the proof for the first (typical) case $d = 2, p = 4$. We are given a family of n planar convex sets and out of every four sets in the family we can nail three with a point. We want to nail the entire family with a fixed number of points. The first step is to show that there is a way to nail a constant fraction (independent from n) of the sets with one point. This follows from a "fractional Helly theorem" of Katcalski and Liu. A more sophisticated use of the Katcalski-Liu theorem shows that for every assignment of nonnegative weights to the sets in the family we can nail with one point sets representing a constant proportion of the entire weight. Using linear programming duality Alon and Kleitman proceeded to show that there is a collection Y of points (their number may depend on n) such that every set in the family is nailed by a constant fraction of the points in

Y. The final step, replacing Y with a set of bounded cardinality that meets all the sets in the family, is done using the theorems of Bárány and Tverberg mentioned above.

2. Convex polytopes

2.1 Polytopes, spheres, and the Steinitz Theorem

Convex polytopes are among the most ancient mathematical objects of study. The combinatorial theory of polytopes is the study of their face structure and in particular their face numbers. There is also a developed metric theory of polytopes (problems concerning volume, width, sections, projections, etc.) and arithmetic theory (lattice points in polytopes). These three aspects of convex polytopes are related and some of the algebraic tools mentioned below are relevant to all of them.

A convex d-dimensional polytope (briefly, a d-polytope) is the convex hull of a finite set of points that affinely span $\mathbb{R}^d$. A (nontrivial) face of a d-polytope P is the intersection of P with a supporting hyperplane. The empty set and P itself are regarded as trivial faces. 0-faces are called *vertices*, 1-faces are called *edges*, and $(d-1)$-faces are called *facets*. The set of faces of a polytope is a graded lattice. Two polytopes P and Q are *combinatorially isomorphic* if there is an order preserving bijection between their face lattices. P and Q are *dual* if there is an order reversing bijection between their face lattices.

Simplicial polytopes are polytopes all of whose proper faces are simplices. Duals of simplicial polytopes are called *simple* polytopes. A d-polytope P is simple iff every vertex of P belongs to d edges. Denote by $f_i(P)$ the number of i-faces of P. The vector $(f_0(P), f_1(P), \ldots, f_d(P))$ is called the f-vector of P. Euler's famous formula $V - E + F = 2$ is the beginning of a rich theory on face numbers of convex polytopes and related combinatorial structures.

The wild behavior for $d \geq 4$

The boundary of every simplicial d-polytope is a triangulation of a $(d-1)$-sphere, but there are triangulations of $(d-1)$-spheres that cannot be realized as boundary complexes of simplicial polytopes (for $d \geq 4$). Goodman and Pollack [20] proved that the number of combinatorial types of polytopes is surprisingly small. The number of d-polytopes with 1,000,000 vertices (in any dimension) is bounded above by $2^{2^{70}}$ whereas the number of triangulations of spheres with 1,000,000 vertices is between $2^{2^{692,225\pm25}}$. (This is achieved for $d \sim 552,786$.) There are combinatorial types of convex polytopes that cannot be realized by points with rational coordinates [22], [52] and there are polytopes that have a combinatorial automorphism that cannot be realized geometrically and whose realization space is not connected. Mnev [39] showed that for every simplicial complex C, there is a polytope whose realization space is homotopy equivalent to C. Recently, Richter announced that all these phenomena occur already in dimension 4, that all algebraic numbers are needed to coordinatize all 4-polytopes, and that there is a nonrational 4-polytope with 34 vertices.

The tame behavior for $d = 3$

All these "pathologies" do not occur for 3-polytopes by a deep theorem of Steinitz asserting that every 3-connected planar graph is the graph of a polytope and related theorems. Relatives of the Koebe-Andreev-Thurston circle packing theorem provide a new approach to the Steinitz Theorem, see [43]. Andreev and Thurston proved that there is a realization of every 3-polytope P such that all its edges are tangent to the unit ball, and this realization is unique up to projective transformations preserving the unit sphere. Schramm observed that by choosing the realization so that the hyperbolic center of the tangency points of edges with the unit sphere is at the origin, you get the following result (this answers a question of Grünbaum, and extends a result of Mani):

Let P be a 3-polytope, and let Γ be the group of combinatorial isomorphisms of the pair (P, P^*), where P^* is the dual of P. (In other words, each element of Γ is either a combinatorial automorphism of P or an isomorphism from P to P^*.) Then there is a realization of the polyhedron so that every element of Γ is induced by a congruence.

An open problem of Perles is whether every combinatorial automorphism ϕ of a centrally symmetric d-polytope (P is centrally symmetric if $x \in P$ implies $-x \in P$) satisfies $\phi(-v) = -\phi(v)$.

2.1 Face numbers and h-numbers of simplicial polytopes

The upper bound theorem and the lower bound theorem

Motzkin conjectured in 1957 and McMullen proved in 1970 [38] the *upper bound theorem*: Among all d-polytopes with n vertices the *cyclic* polytope has the maximal number of k-faces for every k. The cyclic d-polytope with n vertices is the convex hull of n points on the *moment curve* $x(t) = (t, t^2, \ldots, t^d)$. Cyclic d-polytopes have the remarkable property that every set of k vertices determines a $(k-1)$-face for $1 \le k \le [d/2]$.

Klee proved in 1964 the assertion of the upper bound theorem when n is large w.r.t. d for arbitrary *Eulerian complexes*, namely $(d-1)$-dimensional simplicial complexes such that the link of every r-face has the same Euler characteristics as a $(d - r - 1)$-sphere. The assertion of the upper bound theorem for arbitrary Eulerian complexes (even manifolds) is still open.

Brückner conjectured in 1909 and Barnette [4] proved in 1970 the *lower bound theorem*: The minimal number of k-faces for simplicial d-polytopes with n vertices is attained for *stacked* polytopes. Stacked polytopes are those polytopes built by gluing simplices along facets.

The g-theorem

Let $d > 0$ be a fixed integer. Given a sequence $f = (f_0, f_1, \ldots, f_{d-1})$ of nonnegative integers, set $f_{-1} = 1$ and define $h[f] = (h_0, h_1, \ldots, h_d)$ by the relation

$$\sum_{k=0}^{d} h_k x^{d-k} = \sum_{k=0}^{d} f_{k-1}(x-1)^{d-k}.$$

If $f = f(K)$ is the f-vector of a $(d-1)$-dimensional simplicial complex K then $h[f] = h(K)$ is called the h-*vector* of K. The h-vectors are of great importance in the combinatorial theory of simplicial polytopes. The upper bound theorem and the lower bound theorem have simple forms in terms of the h-numbers. The upper bound theorem follows from the inequality $h_k \leq \binom{n-d+k-1}{k}$. The lower bound theorem amounts to the relation $h_1 \leq h_2$. The Dehn-Sommerville relations for the face numbers of simplicial polytopes assert that $h_k = h_{d-k}$.

In 1970 McMullen proposed a complete characterization of f-vectors of boundary complexes of simplicial d-polytopes. McMullen's conjecture was settled in 1980. Billera and Lee [8] proved the sufficiency part of the conjecture and Stanley [45] proved the necessity part. Recently, McMullen [36], [37] found an elementary proof of the necessity part of the g-theorem.

The McMullen conjecture, now called the *g-theorem*, asserts that $(h_0, h_1, \ldots, h_d)$ *is the h-vector of a simplicial d-polytope if and only if the following conditions hold: (a)* $h_i = h_{d-i}$, *(b) there is a graded standard algebra* $M = \oplus_{i=0}^{d/2} M_i$ *such that* $\dim M_i = h_i - h_{i-1}$, *for* $0 \leq i \leq [d/2]$. (A graded algebra is standard if it is generated as an algebra by elements of degree 1.)

The second condition was originally given in purely combinatorial terms, which is equivalent to the formulation given here by an old theorem of Macaulay. In the rest of this section we will describe methods used to attack the upper and lower bound theorems and the g-conjecture.

It is conjectured that the assertion of the g-theorem applies to *arbitrary simplicial spheres*.

Shellability and the h-vector

A *shelling* of a simplicial sphere is a way to introduce the facets (maximal faces) one by one so that at each stage you have a topological ball until the last facet is introduced and you get the entire sphere. Let P be a simplicial polytope and let Q be its polar (which is a simple polytope). A shelling order for the facets of P is obtained simply by ordering the vertices of Q according to some linear objective function ϕ. The number $h_k(P)$ has a simple interpretation as the number of vertices v of Q of degree k where the degree of a vertex is the number of its neighboring vertices with lower value of the objective function. Switching from ϕ to $-\phi$ we get the Dehn-Sommerville relations $h_k = h_{d-k}$ (including the Euler relation for $k = 0$). Put $h_k^*(Q) = h_k(P)$.

We are ready to describe McMullen's proof of the upper bound theorem (in a dual form). Let Q be a simple d-polytope with n facets. Consider a linear objective function ϕ that gives higher values to vertices in a facet F than to all other vertices to obtain that (*) $h_{k-1}^*(F) \leq h_k^*(Q)$. Next,

$$(**) \qquad \sum h_k^*(F) = (k+1)h_{k+1}^*(Q) + (d-k)h_k^*(Q),$$

where the sum is over all facets F of Q. To see this note that every vertex of degree k in Q has degree $k - 1$ in k facets containing v and degree k in the remaining $d - k$ facets. (*) and (**) give the upper bound relations $h_{d-k}^*(Q) \leq \binom{n-d+k-1}{k}$ by induction on k.

Cohen-Macaulay rings

Stanley, see [46], proved the upper bound theorem for arbitrary simplicial spheres using the theory of Cohen-Macaulay rings. Let K be a $(d-1)$-dimensional simplicial complex on n vertices $x_1, \ldots, x_n$. The *face ring* $R(K)$ of K is the quotient

$\mathbb{R}[x_1, x_2 \ldots x_n]/I$ where I is the ideal generated by non faces of K (i.e., I is generated by monomials of the form $x_{i_1} \cdot x_{i_2} \cdots x_{i_m}$ where $[x_{i_1}, x_{i_2}, \ldots, x_{i_m}]$ is *not* a face of K). $R(K)$ is a *Cohen-Macaulay* ring if it decomposes into a direct sum of (translation of) polynomial rings as follows: there are elements of $R(K)$, $\theta_1, \theta_2, \ldots, \theta_d$ and $\eta_1, \eta_2, \ldots, \eta_t$ such that

$$R(K) = \oplus_{i=1}^{t} \eta_i \mathbb{R}[\theta_1, \theta_2, \ldots, \theta_d].$$

It turns out that the θ's can be chosen as linear combinations of the variables and then the number of η's of degree i is precisely h_i. Reisner found topological conditions for the Cohen-Macaulayness of $R(K)$ that imply that $R(K)$ is Cohen-Macaulay when K is a simplicial sphere. All this implies the upper bound inequalities for the h numbers because after moding out by d linear forms the dimension of the space of homogeneous polynomials of degree k (from which the $\eta's$ are taken) is $\binom{n-d+k-1}{k}$.

Toric varieties

For every rational d-polytope P one associates an algebraic variety $T(P)$ of dimension $2d$. If P has n vertices $v_1, v_2, \ldots, v_n$ then consider n complex variables $z_1, \ldots, z_n$ and replace each affine relation with integer coefficients $\sum n_i v_i = 0$, where $\sum n_i = 0$, by the polynomial relation $\prod z_i^{n_i} = 1$. When P is simplicial Danilov proved that the $2i$th Betti number of $T(P)$ is h_i. This enabled Stanley [45] to prove the necessity part of the g-conjecture via the Hard-Lefschetz theorem for $T(P)$.

Rigidity

Let P be a simplicial d-polytope, $d \geq 3$. then P is *rigid*. Namely, every small perturbation of the vertices of P that does not change the length of the edges of P is induced by an affine rigid motion of $\mathbb{R}^d$. The rigidity of simplicial 3-polytopes follows from Cauchy's rigidity theorem, which asserts that if two combinatorially isomorphic convex polytopes have pairwise congruent 2-faces then they are congruent. (It follows also from Dehn's infinitesimal rigidity theorem for simplicial 3-polytopes.) There is a simple inductive argument on the dimension to prove rigidity of simplicial d-polytopes starting with the case $d = 3$. If P is a simplicial d-polytope with n vertices, there are dn degrees of freedom to move the vertices and the dimension of the group of rigid motions of $\mathbb{R}^d$ is $\binom{d+1}{2}$. Therefore the rigidity of P implies the lower bound inequality $f_1(P) \geq dn - \binom{d+1}{2}$. This observation also gives various extensions of the lower bound theorem, see Kalai [25]. Lee [34] extended this idea to higher h-numbers and found relations to the face ring.

The algebra of weights

A remarkable recent development is McMullen's elementary proof of the necessity part of the g-theorem [36], [37]. McMullen proved, in fact, the assertion of the Hard-Lefschetz theorem and his proof applies to non-rational simplicial polytopes. (There, the toric varieties do not exist but the assertion of the Hard-Lefschetz theorem in terms of the face ring still makes sense.) McMullen defines r-weights of simple d-polytopes to be an assignment of weights $w(F)$ to each r-face F such that in each $(r+1)$-face G, $\sum w(F)u_{F,G} = 0$, where the sum is taken over all

r-faces F of G and $u_{F,G}$ is the outer normal of F in G. Let $\Omega_r(P)$ denote the space of r-weights of the polytope P. A well-known theorem of Minkowski asserts that assigning to an r-face its r-dimensional volume is an r-weight. These special weights have a central role in the proof.

McMullen's proof proceeds in the following steps: (1) He defines an algebra structure on weights and shows that this algebra is generated by 1-weights. (2) He proves that $\dim \Omega_r(P) = h_r(P)$. (3) He considers the special 1-weight ω, which assigns to each edge its length, and proves that $\omega^{d-2r} : \Omega_r \to \Omega_{d-r}$ is an isomorphism. To show this McMullen computes the signature of the quadratic form $\omega^{d-2r}x^2$ on $\Omega_r(P)$. This is achieved via new geometric inequalities of Brunn-Minkowski type.

Algebraic shifting

Algebraic shifting, introduced by Kalai in [24], is a way to assign to every simplicial complex K an auxiliary simplicial complex $\Delta(K)$ of a special type. The vertices of $\Delta(K)$ are $v_1, v_2, v_3, \ldots$ and the r-faces of $\Delta(K)$ respect a certain partial order. Namely, if $S = (v_{i_0}, v_{i_1}, \ldots, v_{i_r})$ form an r-face of $\Delta(K)$ then if one of the vertices v_j of S is replaced with a vertex v_i with $i < j$ this results also with a face of $\Delta(K)$. (For example, if (v_3, v_7) is a 1-face of $\Delta(K)$ then so is (v_3, v_5).) The definition of $\Delta(K)$ is given by a certain generic change of basis for the cochain groups of K, see [11].

Algebraic shifting complements the classical notion of shifting in extremal combinatorics which was introduced by Erdős, Ko and Rado. It is also closely related to the notion of "generic initial ideals" in commutative algebra.

Various combinatorial and topological properties of simplicial complexes are preserved by the operation $K \to \Delta(K)$. $\Delta(K)$ has the same f-vector as K. $\Delta(K)$ also have the same Betti numbers as K but other homotopical information is eliminated as $\Delta(K)$ has the homotopy type of a wedge of spheres. K has the Cohen-Macaulay property (its face ring is Cohen-Macaulay) iff $\Delta(K)$ has.

What is still missing is the relation of algebraic shifting with embeddability in $\mathbb{R}^n$. It is a well-known fact that K_5, the complete graph with five vertices, cannot be embedded in the plane. More generally, van Kampen and Flores proved that σ_r^{2r+2}, the r-skeleton of the $(2r+2)$-simplex, cannot be embedded in $\mathbb{R}^{2r}$. Kalai and Sarkaria propose

CONJECTURE: σ_r^{2r+2} is not contained in $\Delta(K)$ whenever K is embeddable in $\mathbb{R}^{2r}$.

This conjecture would imply the assertion of the g-theorem for arbitrary simplicial spheres.

2.3 Other topics
Flag numbers and and other invariants of general polytopes

Flag numbers of polytopes count chains of faces of prescribed dimensions. There are 2^d flag numbers but Bayer and Billera [5] showed that the affine space of flag numbers of d-polytopes has dimension $c_d - 1$ where c_d is the dth Fibonacci number. A significant basis of the space of flag numbers is Fine's CD-index, see [6], [49]. Toric varieties supply interesting invariants for general polytopes P. The dimensions of the (middle perversity) intersection homology groups of $T(P)$ are linear combinations of flag numbers, see [47], [26]. There are mysterious connections between these invariants of a polytope P and its dual P^* (see [25, Section 12], [6], [48]).

The following very simple problem is open: *Show that a centrally symmetric polytope P in $\mathbb{R}^d$ must have at least 3^d nonempty faces.*

Reconstruction theorems

Whitney proved that the graph of a 3-polytope determines its face structure. The 2-faces of the polytope are given by the induced cycles, which do not separate the graph. This can be extended to show that the $(d-2)$-skeleton of a d-polytope determines the face structure and for general polytopes this cannot be improved. (See [22, Chapter 12].) Perles proved that the $[d/2]$-skeleton of a *simplicial d-polytope* determines the face structure, and Dancis [15] extended this result to arbitrary simplicial spheres. Perles conjectured and Blind and Mani [12] proved that the face structure of every simple d-polytope is determined by the graph (1-skeleton) of the polytope. For a simple proof see Kalai [27]. Consider a simplicial $(d-1)$-dimensional sphere and a *puzzle* in which the pieces are the facets and for each piece there is a list of the d neighboring pieces. The Blind-Mani theorem asserts that for boundary complexes of simplicial polytopes (and for a certain class of shellable spheres) the puzzle has only one solution. Conjecture: *For an arbitrary simplicial sphere the puzzle has a unique solution.* Perhaps the machinery of Cohen-Macaulay rings can be of help.

Polytopes of triangulations

Lee [33] and Haiman proved that the set of triangulations of the regular n-gon with noncrossing diagonals corresponds to the vertices of an $(n-3)$-dimensional polytope. The r-faces of this polytope correspond to all triangulations containing a given set of $n-3-r$ diagonals. Independently (as part of a theory of generalized hypergeometric functions), Gelfand, Kapranov, and Zelevinskii [19] defined much more general objects called "secondary polytopes", which correspond to certain triangulations of arbitrary polytopes. Further extensions were given by several authors, including the Billera and Sturmfels "fiber polytopes" [9]. It appears now that these constructions are quite fundamental in convex polytope theory and the reader is referred to Zeigler's book [52]. In another independent development, Slater, Tarjan, and Thurston [44] proved a sharp lower bound on the (combinatorial) diameter of the associahedron using volume estimates of hyperbolic polytopes.

2.4 The simplex algorithm and the diameter of graphs of polytopes

The simplex algorithm solves linear programming problems by moving from vertex to vertex of a polytope (the set of feasible solutions) along its edges. Let $\Delta(d, n)$ be the maximum diameter of the graphs of d-polytopes P with n facets. It is not known if $\Delta(d, n)$ is bounded above by a linear function of d and n, or even by a polynomial function of d and n. In 1970 Larman proved that $\Delta(d, n) \leq 2^{d-3}n$. Recently, quasi-polynomial bounds were found, see Kalai and Kleitman [29] for a simple proof for $\Delta(d, n) \leq n^{\log d + 1}$. All the known upper bounds use only the facts that the intersection of faces of a polytope is a face and that the graph of every face is connected.

Consider a linear programming problem with d variables and n constraints. Given the fact that the diameter of the feasible polytope is relatively small, the next step would be to find a pivot rule for linear programming that requires for *every* linear programming problem a subexponential number of pivot steps. Here, we assume that each individual pivot step should be performed by a polynomial number of arithmetic operations in d and n. However, no such pivot rule is known. Recently, Kalai [28] and independently Matouŝek, Sharir, and Welzl [35] found a *randomized* pivot rule such that the *expected* number of pivot steps needed is at most $\exp(c\sqrt{d\log n})$.

Added in proof: Amenta [Discr. Comp. Geometry, to appear] proved an old conjecture of Grünbaum and Motzkin that the Helly *order* of families whose members are disjoint unions of t convex sets is $t(d+1)$; that is, she proved that given a finite family $\mathcal{K}$ so that every intersection of members of $\mathcal{K}$ is the union of t pairwise-disjoint convex sets, and every intersection of at most $t(d+1)$ members is non-empty then the intersection of all members is non-empty as well. Sarkaria proved Sierksma's conjecture on the number of Tverberg's partitions using certain computations of Chern classes. Sheftel verified the assertion of the upper bound theorem (for f-vectors) for arbitrary odd dimensional manifolds. She relies on a result of Schenzel on h-vectors of Buchsbaum complexes.

References

[1] N. Alon and D. Kleitman, *Piercing convex sets and the Hadwiger Debrunner (p,q)-problem*, Adv. in Math. 96 (1992), 103–112.

[2] I. Bárány, *A generalization of Caratheodory's theorem*, Discrete Math. 40 (1982), 141–152.

[3] I. Bárány, S. Shlosman, and A. Szücs, *On a topological generalization of a theorem of Tverberg*, J. London Math. Soc. 23 (1981), 158–164.

[4] D. Barnette, *A proof of the lower bound conjecture for convex polytopes*, Pacific J. Math. 46 (1971), 349–354.

[5] M. Bayer and L. Billera, *Generalized Dehn-Sommerville relation for polytopes, spheres and Eulerian partially ordered sets*, Invent. Math. 79 (1985), 143–157.

[6] M. Bayer and A. Klapper, *A new index for polytopes*, Discrete Comput. Geom. 6(1991), 33–47.

[7] U. Betke, M. Henk, and J. Wills, *Finite and infinite packings*, J. Reine Angew. Math. 453(1994), 165–191.

[8] L. Billera and C. Lee, *A proof of the sufficiency of McMullen's conditions for f-vectors of simplicial convex polytopes*, J. Combin. Theory Ser. A 31 (1981), 237–255.

[9] L. Billera and B. Sturmfels, *Fiber polytopes*, Ann. of Math. (2) 135(1992), 527–549.

[10] T. Bistritsky, P. McMullen, R. Schneider, and A. Weiss (eds.), Polytopes — Abstract, Convex and Computational, Kluwer, Dordrecht, 1994.

[11] A. Björner and G. Kalai, *Extended Euler Poincaré relations*, Acta Math. 161 (1988), 279–303.

[12] R. Blind and P. Mani, *On puzzles and polytope isomorphism*, Aequationes Math., 34 (1987), 287–297.

[13] J. Bourgain and J. Lindenstrauss, *On covering a set in* $\mathbb{R}^N$ *by balls of the same diameter*, in Geometric Aspects of Functional Analysis (J. Lindenstrauss and V. Milman, eds.), Lecture Notes in Math. 1469, Springer-Verlag, Berlin and New York, 1991, 138–144.

[14] H. Croft, K. Falconer, and R. Guy, Unsolved Problems in Geometry, Springer-Verlag, Berlin and New York, 1991, 123–125.

[15] J. Dancis, *Triangulated n-manifolds are determined by their* $[n/2] + 1$*-skeletons*, Topology Appl. 18(1984), 17–26.

[16] J. Eckhoff, *Helly, Radon, and Carathèodory type theorems*, in [21], 389–448.

[17] P. Frankl and V. Rödl, *Forbidden intersections*, Trans. Amer. Math. Soc. 300(1987), 259–286.

[18] P. Frankl and R. Wilson, *Intersection theorems with geometric consequences*, Combinatorica 1 (1981), 259–286.

[19] I. Gelfand, A. Zelevinskii, and M. Kapranov, *Newton polytopes of principal A-determinants*, Soviet Math. Dokl., 40(1990), 278–281; Consequences, Combinatorica 1 (1981), 357–368.

[20] J. Goodman and R. Pollack, *There are asymptotically far fewer polytopes than we thought*, Bull. Amer. Math. Soc. 14(1986), 127–129.

[21] P. Grüber and J. Wills (eds.), Handbook of Convex Geometry, North-Holland, Amsterdam, 1993.

[22] B. Grünbaum, Convex Polytopes, Wiley Interscience, London, 1967.

[23] J. Kahn and G. Kalai, *A counterexample to Borsuk's conjecture*, Bull. Amer. Math. Soc. 29(1993), 60–62.

[24] G. Kalai, *A characterization of f-vectors of families of convex sets in* $\mathbb{R}^d$*, Part I: Necessity of Eckhoff's conditions*, Israel J. Math. 48 (1984), 175–195.

[25] G. Kalai, *Rigidity and the lower bound theorem I*, Invent. Math. 88(1987), 125–151.

[26] G. Kalai, *A new basis of polytopes*, J. Comb. Theory Ser. A 49(1988), 191–209.

[27] G. Kalai, *A simple way to tell a simple polytope from its graph*, J. Comb. Theory Ser. A 49(1988), 381–383.

[28] G. Kalai, *A subexponential randomized simplex algorithm*, Proceedings of the 24th Ann. ACM Symp. on the Theory of Computing, 475–482, ACM Press, Victoria, 1992.

[29] G. Kalai and D. J. Kleitman, *A quasi-polynomial bound for diameter of graphs of polyhedra*, Bull. Amer Math. Soc. 26(1992), 315–316.

[30] V. Klee and S. Wagon, Old and New Unsolved Problems in Plane Geometry and Number Theory, Math. Assoc. of Amer., Washington, D.C., 1991.

[31] J. Lagarias and P. Shor, *Keller's cube tiling conjecture is false in high dimensions*, Bull. Amer. Math. Soc. 27(1992), 279–283.

[32] D. Larman and N. Tamvakis, *The decomposition of the n-sphere and the boundaries of plane convex domains*, Ann. Discrete Math. 20 (1984), 209–214.

[33] C. Lee, *The associahedron and triangulations of the n-gon*, European J. Combin., 10 (1989), 551–560.

[34] C. Lee, *Generalized stress and motions*, in [10], 249–271.

[35] J. Matoušek, M. Sharir, and E. Welzl, *A subexponential bound for linear programming*, Proc. 8th Annual Symp. on Computational Geometry, 1992, 1–8.

[36] P. McMullen, *On simple polytopes*, Invent. Math. 113(1993), 419–444.

[37] P. McMullen, *Weights on polytopes*, Discrete Comput. Geom., to appear.

[38] P. McMullen and G. C. Shephard, Convex Polytopes and the Upper Bound Conjecture, Cambridge Univ. Press, London and New York, 1971.

[39] N. Mnëv, *The universality theorems on the classification problem of configuration varieties and convex polytopes varieties*, in Topology and Geometry — Rohlin Seminar, (O. Ya. Viro, ed.), Lecture Notes in Math., 1346, Springer-Verlag, Berlin, Heidelberg, and New York, 1988, 527–544.

[40] A. Nilli, *On Borsuk problem*, in Jerusalem Combinatorics 1993 (H. Barcelo and G. Kalai, eds.) 209–210, Contemp. Math. 178, Amer. Math. Soc., Providence, RI, 1994.

[41] K. Sarkaria, *Tverberg's theorem via number fields*, Israel J. Math. 79 (1992), 317–320.

[42] O. Schramm, *Illuminating sets of constant width*, Mathematica 35(1988), 180–199.

[43] O. Schramm, *How to cage an egg*, Invent. Math. 107(1992), 543–560.

[44] D. Slater, R. Tarjan, and W. Thurston, *Rotation distance, triangulations and hyperbolic geometry*, J. Amer. Math. Soc. 1(1988), 647–681.

[45] R. Stanley, *The number of faces of simplicial convex polytopes*, Adv. in Math. 35(1980), 236–238.

[46] R. Stanley, Combinatorics and Commutative Algebra, Birkhäuser, Basel and Boston, 1983.

[47] R. Stanley, *Generalized h-vectors, intersection cohomology of toric varieties, and related results*, in Commutative Algebra and Combinatorics (M. Nagata and H. Matsumura, eds.), Adv. Stud. Pure Math. 11, Kinokuniya, Tokyo, and North-Holland, Amsterdam/New York, 1987, 187–213.

[48] R. Stanley, *Subdivisions and local h-vectors*, J. Amer. Math. Soc. 5(1992), 805–851.

[49] R. Stanley, *Flag vectors and the CD-index*, Math. Z. 216(1994), 483–499.

[50] H. Tverberg, *A generalization of Radon's Theorem*, J. London Math. Soc. 41 (1966), 123–128.

[51] A. Vučić and R. Živaljevic, *Note on a conjecture by Seirksma*, Discrete Comput. Geom., 9(1993), 339–349.

[52] G. Ziegler, Lectures on Polytopes, Springer-Verlag, Berlin and New York, 1994.

[53] R. Živaljević and S. Vrećica, *The colored Tverberg's problem and complexes of injective functions*, J. Combin. Theory Ser. A 61(1992), 309–318.

Probabilistic Methods in Combinatorics

JOEL SPENCER

Courant Institute
251 Mercer Street
New York, NY 10012, USA

In 1947 Paul Erdős [8] began what is now called the probabilistic method. He showed that if $\binom{n}{k}2^{1-\binom{k}{2}} < 1$ then there *exists* a graph G on n vertices with clique number $\omega(G) < k$ and independence number $\alpha(G) < k$. (In terms of the Ramsey function, $R(k,k) > n$.) In modern language he considered the random graph $G(n, .5)$ as described below. For each k-set S let B_S denote the "bad" event that S is either a clique or an independent set. Then $\Pr[B_S] = 2^{1-\binom{k}{2}}$ so that $\sum \Pr[B_S] < 1$, hence $\wedge \overline{B}_S \neq \emptyset$ and a graph satisfying $\wedge \overline{B}_S$ must exist.

In 1961 Erdős with Alfred Rényi [11] began the systematic study of random graphs. Formally $G(n,p)$ is a probability space whose points are graphs on a fixed labelled set of n vertices and where every pair of vertices is adjacent with independent probability p. A graph theoretic property A becomes an event. Whereas in the probabilistic method one generally requires only $\Pr[A] > 0$ from which one deduces the existence of the desired object, in random graphs the estimate of $\Pr[A]$ is the object itself. Let A denote connectedness. In their most celebrated result Erdős and Rényi showed that if $p = p(n) = \frac{\ln n}{n} + \frac{c}{n}$ then $\Pr[A] \to \exp(-e^{-c})$. We give [2], [6] as general references for these topics.

Although pure probability underlies these fields, most of the basic results use fairly straightforward methods. The past ten years (our emphasis here) have seen the use of a number of more sophisticated probability results. The Chernoff bounds have been enhanced by inequalities of Janson and Talagrand and new appreciation of an inequality of Azuma. Entropy is used in new ways. In its early days the probabilistic method had a magical quality — where is the graph that Erdős in 1947 proved existed? With the rise of theoretical computer science these questions take on an algorithmic tone — having proven the existence of a graph or other structure, can it be constructed in polynomial time? A recent success of Beck allows the Lovász local lemma to be derandomized. Sometimes. We close with two forays into a land dubbed Asymptopia by David Aldous. There the asymptotic behavior of random objects is given by an infinite object, allowing powerful noncombinatorial tools to be used.

1. Chernoff, Azuma, Janson, Talagrand

Let $X = X_1 + \cdots + X_m$ with the X_i mutually independent and normalized so that $E[X] = E[X_i] = 0$. The so-called Chernoff bounds (Bernstein or antiquity might

Proceedings of the International Congress
of Mathematicians, Zürich, Switzerland 1994
© Birkhäuser Verlag, Basel, Switzerland 1995

be more accurate attributions) bound the "large deviation"

$$\Pr[X > a] < e^{-\lambda a} E[e^{\lambda X}] = e^{-\lambda a} \prod_i E[e^{\lambda X_i}]$$

(See, e.g., the appendix of [2].) The power in the inequality is that it holds for all $\lambda > 0$ and one chooses $\lambda = \lambda(a)$ for optimal results. Suppose, for example, that $|X_i| \leq 1$. One can show $E[e^{\lambda X_i}] \leq \cosh(\lambda) \leq \exp(\lambda^2/2)$, the extreme case when $X_i = \pm 1$ uniformly. Then $\Pr[X > a] < \exp(-\lambda a + \lambda^2 m/2) = \exp(-a^2/2m)$ by the optimal choice $\lambda = a/m$. Intuition is guided by comparison to the Gaussian, in the above example $\mathrm{Var}(X_i) \leq 1$ so $\mathrm{Var}(X) \leq m$ and the probability of being more than $a = \sigma\sqrt{m}$ of the mean should, and here does, drop like the chance of being σ standard deviations off the mean, like $\exp(-\sigma^2/2)$.

The new inequalities are used when the X_i exhibit slight dependencies. To illustrate them, let $G \sim G(n, .5)$. Let $f(x) = \binom{n}{x} 2^{-\binom{x}{2}}$ be the expected number of x-cliques and let $k_0 = k_0(n)$ satisfy $f(k_0) > 1 > f(k_0 + 1)$. Calculation gives $k_0 \sim 2\log_2 n$ and it has long been known that $\omega(G)$ is almost surely very close to k_0. Now set $k = k_0 - 4$ so that $f(k) > n^{3+o(1)}$ is large. We show thrice that

$$\Pr[\omega(G) < k] < 2^{-n^2 \ln^{-c} n}$$

(As G may be empty the probability is at least 2^{-cn^2}.) The proof via Azuma's Inequality, given below, was given by Béla Bollobás [7] and was essential to his discovery that the chromatic number $\chi(G) \sim n/(2\log_2 n)$ almost surely.

Azuma's Inequality: Let $\mu = X_0, X_1, \ldots, X_m = X$ be a martingale in which $|X_{i+1} - X_i| \leq 1$. Then $\Pr[X > \mu + a] < \exp(-a^2/2m)$.

In application we use an isoperimetric version. Let $\Omega = \prod_{i=1}^m \Omega_i$ be a product probability space and X a random variable on it. Call X *Lipschitz* if whenever $\omega, \omega' \in \Omega$ differ on only one coordinate $|X(\omega) - X(\omega')| \leq 1$. Set $\mu = E[X]$.

Azuma's Perimetric Inequality: $\Pr[X \geq \mu + a] < e^{-a^2/2m}$.

The connection is via the Doob martingale, $X_i(\omega)$ being the conditional expectation of X given the first i coordinates of ω. The same inequality holds for $\Pr[X \leq \mu - a]$. The random graph $G(n, .5)$ can be viewed as the product of its $m = \binom{n}{2}$ coin flips. Bollobás set X equal to the maximal number of edge disjoint k-cliques. From probabilistic methods he showed $E[X] > cn^2 k^{-4}$. (One may conjecture that the true value is $\Theta(n^2 k^{-2})$.) Then $\omega(G) < k$ if and only if $X = 0$ and

$$\Pr[X = 0] = \Pr[X \leq \mu - \mu] < e^{-\mu^2/2m} = e^{-\Theta(n^2 \ln^{-8} n)}$$

For Janson's inequalities let Ω be a fixed set and $Y \subseteq \Omega$ a random subset (so, formally, 2^Ω is the probability space) where the events $y \in Y$ are mutually independent over $y \in \Omega$. $G(n, p)$ fits this perfectly with $\Omega = [n]^2$ the set of potential edges and $\Pr[y \in G(n, p)] = p$ for every $y \in \Omega$. Let $A_1, \ldots, A_m \subseteq \Omega$. Let B_i be the event $Y \supseteq A_i$, I_i its characteristic function, $X = \sum I_i$, and $\mu = E[X]$. Write $i \sim j$ if $i \neq j$ and $A_i \cap A_j \neq \emptyset$. Roughly $\sim$ represents dependence of the corresponding B_i. Let ϵ be an upper bound for all $\Pr[B_i]$. Set

$$M = \prod \Pr[\overline{B_i}] \quad \text{and} \quad \Delta = \sum_{i \sim j} \Pr[B_i \wedge B_j]$$

Janson's Inequality:

$$M \leq \Pr[\wedge \overline{B}_i] \leq M e^{\frac{1}{1-\epsilon} \frac{\Delta}{2}}$$

Generalized Janson Inequality: If $\Delta \geq \mu(1 - \epsilon)$ then

$$\Pr[\wedge \overline{B}_i] \leq e^{-\mu^2(1-\epsilon)/\Delta}$$

In many cases $\epsilon \to 0$, $\Delta \to 0$, and $M \sim e^{-\mu}$ so that Janson's inequality gives $\Pr[X = 0] \sim e^{-\mu}$. In this sense Janson's Inequality acts as a Poisson approximation for X, though with particular emphasis at $X = 0$. For example, when $p = c/n$ and $A_{ijk} = \{\{i, j\}, \{i, k\}, \{j, k\}\}$ range over all triangles these conditions hold and $G(n, p)$ is triangle free with probability $\sim \exp(-c^3/6)$, as known to Erdős and Rényi. Sweeping generalizations of this are given in [13] where the first proof of Janson's inequality may be found. Other proofs and generalizations are given in [12], [2].

Applying Janson to $\Pr[\omega(G(n, .5)) < k]$ we let $A_S = [S]^2$, S ranging over the k-sets of vertices. Then $\epsilon \to 0$, $\mu = f(k)$. Δ is the expected number of edge overlapping k-cliques, calculation gives domination by cliques overlapping in a single edge, and $\Delta \sim \mu^2(2k^4 n^{-2})$. The Poisson approximation does *not* apply but the extended Janson inequality gives

$$\Pr[\omega(G(n, .5)) < k] < e^{-c\mu^2/\Delta} = e^{-c'n^2 \ln^{-4} n}$$

The newest result, Talagrand's inequality, has a similar framework to that of Azuma. Let $\Omega = \prod_1^m \Omega_i$ be a product probability space. For $A \subseteq \Omega$, $x = (x_1, \ldots, x_t) \in \Omega$ define a "distance" $\rho(A, x)$ as the least t so that for any real $\alpha_1, \ldots, \alpha_m$ with $\sum \alpha_i^2 = 1$ there exists $y = (y_1, \ldots, y_t) \in A$ with $\sum_{x_i \neq y_i} \alpha_i \leq t$. Note critically that y may depend on $\alpha_1, \ldots, \alpha_m$. Set A_t equal to the set of all $x \in \Omega$ with $\rho(A, x) \leq t$.

Talagrand's Inequality [20]:

$$\Pr[A] \Pr[\overline{A}_t] \leq e^{-t^2/4}$$

Call $X : \Omega \to R$ f-certifiable ($f : N \to N$) if whenever $X(x) \geq s$, $x = (x_1, \ldots, x_m)$, there is a set of at most $f(s)$ indices I that certify $X \geq s$ in that if $y = (y_1, \ldots, y_m)$ has $y_i = x_i$ for $i \in I$ then $X(y) \geq s$.

COROLLARY 1 *If X is Lipschitz and f-certifiable then for all $t \geq 0$, b*

$$\Pr[X \leq b - t\sqrt{f(b)}] \Pr[X \geq b] \leq e^{-t^2/4}$$

Proof. Set $A = \{x : h(x) < b - t\sqrt{f(b)}\}$. Now suppose $h(y) \geq b$. We claim $y \notin A_t$. Let I be a set of indices of size at most $f(b)$ that certifies $h(y) \geq b$ as given above. Define $\alpha_i = 0$ when $i \notin I$, $\alpha_i = |I|^{-1/2}$ when $i \in I$. If $y \in A_t$ there exists a $z \in A$ that differs from y in at most $t\sqrt{f(b)}$ coordinates of I though at arbitrary coordinates outside of I. Let y' agree with y on I and agree with z outside of I. By the certification $h(y') \geq b$. Now y', z differ in at most $t\sqrt{f(b)}$ coordinates and so, by Lipschitz,

$$h(z) > h(y') - t\sqrt{f(b)} \geq b - t\sqrt{f(b)}$$

but then $z \notin A$, a contradiction. So $\Pr[X > b] \leq 1 - \Pr[A_t]$, thus

$$\Pr[X < b - t\sqrt{f(b)}]\Pr[X \geq b] \leq e^{-t^2/4}$$

As the right-hand side is continuous in t we may replace $<$ by $\leq$ giving the Corollary. $\square$

Letting b (or $b - t\sqrt{f(b)}$) be the median of X the Corollary gives a sharp concentration result. For example, let $\Omega = [0,1]^n$ with uniform distribution and let $X(x_1, \ldots, x_n)$ be the length of the longest monotone subsequence of $x_1, \ldots, x_n$. X is Lipschitz and f-certifiable with $f(s) = s$ as a monotone subsequence certifies itself. It is known that $X \sim 2\sqrt{n}$ almost surely. Therefore X almost surely lies within $n^{1/4}\omega(n)$ ($\omega(n) \to \infty$) of its median.

In $G(n, .5)$ let X be, as before, the maximal number of edge disjoint k-cliques. X is Lipschitz and f-certifiable with $f(s) = \binom{k}{2}s$ as the s k-cliques certify themselves. Although medians are notoriously difficult to calculate, tight concentration yields that the median $b \sim \mu > cn^2 k^{-4}$ as previously discussed. Setting $t = bf(b)^{-1/2}$

$$\Pr[\omega(G) < k] = \Pr[X = 0] = \Pr[X \leq b - t\sqrt{f(b)}] < 2e^{-t^2/4} < ce^{-c'n^2\ln^{-6}n}$$

2. Entropy

Let $\mathcal{F}$ be a family of subsets of Ω. A two-coloring is a map $\chi : \Omega \to \{-1, +1\}$. For $A \subseteq \Omega$ define $\chi(A) = \sum_{a \in A} \chi(a)$ so that $|\chi(A)|$ is small if the coloring is "nearly balanced" on A. An object of discrepancy theory is to find χ so all $|\chi(A)|$, $A \in \mathcal{F}$, are small. It is convenient to also define partial colorations as maps $\chi : \Omega \to \{-1, 0, 1\}$, a is called colored when $\chi(a) \neq 0$, $\chi(A)$ is as before.

Under random coloring of an n-set A, $\chi(A)$ has distribution S_n, roughly Gaussian with zero mean and standard deviation $n^{1/2}$. Chernoff bounds give $\Pr[|\chi(A)| > \lambda n^{-1/2}] < 2e^{-\lambda^2/2}$. When $\mathcal{F}$ consists of m sets, each of size n, one sets $\lambda = (2\ln(2m))^{1/2}$ so these "failure events" each have probability less than $\frac{1}{m}$ and thus there exists χ with all $|\chi(A)| \leq \lambda\sqrt{n}$. With entropy we can sometimes do better.

Define the roundoff function $R_b(x)$ as that integer i with $2bi$ closest to x. Note that $R_b(S_n) = 0$ when $|S_n| < b$. Define $\text{ENT}(n, b)$ to be the entropy of the random variable $R_b(S_n)$.

THEOREM 1 Let $\mathcal{F} = \{S_1, \ldots, S_v\}$ with $|\Omega| = n$ and $|S_i| = n_i$. Suppose b_i, ϵ, and $\gamma < \frac{1}{2}$ are such that

$$\sum_{i=1}^{v} \text{ENT}(n_i, b_i) \leq \epsilon n \qquad and \qquad \sum_{j=0}^{\gamma n} \binom{n}{j} < 2^{n(1-\epsilon)}$$

Then there is a partial coloring χ of Ω with

$$|\chi(S_i)| \leq b_i \text{ for all } i$$

and more than $2\gamma n$ points $x \in \Omega$ colored.

Proof. Let $\chi : \Omega \to \{-1, +1\}$ and define

$$L(\chi) = (R_{b_1}(\chi(S_1)), \ldots, R_{b_v}(\chi(S_v)))$$

Entropy, critically, is subadditive so L has entropy at most ϵn. Therefore some value of L obtained with probability at least $2^{-\epsilon n}$, and some $2^{(1-\epsilon)n}$ colorings χ have the same L-value. Colorings χ can be considered points on the Hamming cube $\{-1, +1\}^n$. A classic result of Kleitman [14] gives that some two χ_1, χ_2 of these must differ in at least $2\gamma n$ coordinates. Then $\chi = (\chi_1 - \chi_2)/2$ gives the desired partial coloring. $\square$

It is best to consider $\mathrm{ENT}(n, b)$ under the parametrization $b = \lambda n^{1/2}$. Then $R_b(S_n)$ is roughly $R_\lambda(N)$, with N standard Gaussian. For λ large $\mathrm{ENT}(n, b) < e^{-c\lambda^2}$, the terms $R = 0, \pm 1$ dominating. In particular, for λ a large constant $\mathrm{ENT} < \epsilon$. For λ small $\mathrm{ENT}(n, b) < c \ln(\lambda^{-1})$, the dominating factor being that R is roughly uniform for $|i| = O(\lambda^{-1})$.

Suppose $\mathcal{F}$ consists of n sets on an n-set Ω, so all sets have size at most n. For λ a large constant (six will suffice) the Theorem gives a coloring with only a small (but fixed) fraction of the points uncolored and all $|\chi(A)| \leq \lambda n^{1/2}$. Appropriately iterating, this author [19] showed that for suitable constant λ one can find χ as above with *no* points uncolored.

Let $\Omega = [n]$ and $\mathcal{F}$ be the arithmetic progressions on $[n]$. The discrepancy $disc(\mathcal{F})$ is the least $g(n)$ for which there is a $\chi : \Omega \to \{-1, +1\}$ with $|\chi(A)| \leq g(n)$ for all $A \in \mathcal{F}$. In 1964 Roth [18] used analytic methods to show $disc(\mathcal{F}) > cn^{1/4}$. The upper bound has been lowered from $n^{.5+o(1)}$ to $n^{1/3+o(1)}$ to $n^{1/4} \ln^c n$ [3] over the decades and just recently to $c'n^{1/4}$ by Matoušek and this author [16]. Beck [3] provided a key decomposition. For each $d \leq n$, $0 \leq i < d$, and $j \geq 0$ with $2^j \leq n$, split $\{x \in [n] : x \equiv i \bmod d\}$ into consecutive intervals of length 2^j, leaving out the excess. Let $\mathcal{G}$ be the family of sets obtained. Any $A \in \mathcal{F}$ can be written $A = B - C$ with $C \subset B$ and both B, C the disjoint union of $S \in \mathcal{G}$ of distinct cardinalities. Thus, a coloring χ for which all $S \in \mathcal{G}$ with $|S| = 2^j$ have $|\chi(S)| \leq f(2^j)$ would have the property that $|\chi(A)| \leq 2 \sum_j f(2^j)$ for all $A \in \mathcal{F}$. Calculation gives that $\mathcal{G}$ has roughly $n^2 s^{-2}$ sets of size $s = 2^j$. To get a substantial partial coloring with $|\chi(A)| \leq f(|A|)$ for $A \in \mathcal{G}$ the entropy requirement becomes

$$\sum n^2 s^{-2} \mathrm{ENT}(s, f(s)) \leq \epsilon n$$

When $s \sim n^{1/2}$ we may take $f(s) = kn^{1/4}$. For larger s the savings in s^{-2} allow for a smaller $f(s)$ and for smaller s the savings in $s^{1/2}$ also allow for a smaller $f(s)$. With care we can ensure the entropy requirement and that $2 \sum f(s) = O(n^{1/4})$. This gives a substantial partial coloring of $[n]$ with $|\chi(A)| = O(n^{1/4})$ for all $A \in \mathcal{F}$. The iteration of this method to get a full coloring χ (without losing a logarithmic factor!) uses interesting but noncombinatorial ideas.

Matoušek [15] applied entropy to discrepancy of halfplanes. Let P be a set of n points in the plane and $\mathcal{F}$ the family of $H \cap P$, H a halfplane. Here the decomposition is more difficult, the end result again being a family $\mathcal{G}$ so that all $A \in \mathcal{F}$ are expressible in terms of $B \in \mathcal{G}$ of distinct cardinalities 2^j. Again $\mathcal{G}$ has $\sim n^2 s^{-2}$ sets of size s and the entropy argument gives a partial coloring χ —

which again can be extended to a full coloring χ — with $|\chi(A)| = O(n^{1/4})$ for all $A \in \mathcal{F}$. This result is best possible up to constants and the method works for halfspaces in R^d for any constant d. Indeed the discrepancy of halfplanes came first and motivated the reinvestigation of Roth's result.

Let $\vec{v}_i = (a_{i1}, \ldots, a_{in}) \in R^n$, $1 \leq i \leq n$. For $\chi : [n] \to \{-1, +1\}$ set

$$\vec{S} = \sum_{i=1}^{n} \chi(i)\vec{v}_i = (L_1, \ldots, L_n)$$

with $L_j = \sum_i \chi(i)a_{ij}$. Entropy methods give that if $|\vec{v}_i|_\infty \leq 1$ there exists χ with $|\vec{S}|_\infty \leq cn^{1/2}$. (When $a_{ij} \in \{0,1\}$ this reduces to n sets on n points and the same proof applies.) Linear algebra methods [5] give that if $|\vec{v}_i|_1 \leq 1$ there exists χ with $|\vec{S}|_\infty \leq 2$. Assume now $|\vec{v}_i|_2 \leq 1$. Set $\sigma_j^2 = \sum_i a_{ij}^2$ so $\sum \sigma_j^2 = \sum\sum a_{ij}^2 \leq n$. Let χ be random, L_i acts like $\sigma_i N$. For k large $\text{ENT}(\sigma N, k) < \epsilon$ when $\sigma \sim 1$. Further $\text{ENT}(\sigma N, k) < \epsilon\sigma^2$ for *all* σ. One calculates $\sum \text{ENT}(L_i, k) < \epsilon n$ so there exists $\chi : [n] \to \{-1, 0, +1\}$ with many $\chi(i) \neq 0$ and $|\vec{S}|_\infty \leq K$. Here iteration fails! More precisely, one may [19] iterate the process $O(\ln n)$ times to give χ with all $\chi(i) = \pm 1$ and $|\vec{S}|_\infty = O(\ln n)$. Still open is a challenging conjecture of Komlós that such χ exists with $|\vec{S}|_\infty \leq K$.

3. Algorithmic Sieve

Let $B_i, i \in I$ be events, I finite. Let $\sim$ be a symmetric relation on I so that B_i is mutually independent of all B_j with $i \nsim j$. This includes the Janson scenario of Section 1 but is far more general.

Lovász Local Lemma [10] (symmetric case): If all $\Pr[B_i] \leq p$ and, for each $i \in I$, $i \sim j$ for at most d $j \in I$ and if $p < d^d(d+1)^{-(d+1)}$ then $\wedge_{i \in I} \overline{B}_i \neq \emptyset$.

The strength of the LLL is that I may be of arbitrary size. With B_i as bad events it sieves out a good outcome. We will concentrate on one example. Let $S_1, \ldots, S_n \subseteq [n]$ with all $|S_i| = k$ and all j in precisely $k + 1$ sets S_i. We want a coloring $\chi : [n] \to \{Red, Blue\}$ so that no S_i is monochromatic. Let χ be random and let B_i be the event that S_i is monochromatic. We naturally define $i \sim i'$ when $S_i \cap S_{i'} \neq \emptyset$. Then $p = 2^{1-k}$ and $d = k^2$. For k large ($k = 10$ suffices) the LLL conditions hold and χ exists.

The probabilistic method has always had a magical quality — just where is the coloring, graph, tournament or whatever that we have proved exists? Here we can ask a precise question. Fix $k = 10$. Given $S_1, \ldots, S_n \subseteq [n]$ as above, can the desired χ be found with a polynomial (in n) time algorithm? Even allowing randomized algorithms the answer is not clear. Though LLL guarantees $\Pr[\wedge_{i \in I} \overline{B}_i] \neq 0$ it will be exponentially small in n so checking random χ would take expected exponential time. As stated, the problem remains open. But a recent breakthrough by Beck [4] gives an algorithm when k is somewhat larger.

We outline Beck's idea as a randomized algorithm though it can be, and originally was, expressed in deterministic fashion. Fix $k = 100$ for definiteness. First $[n]$ is colored randomly. Any S_i with more than 80 (say) points in one color is considered dangerous. All points in dangerous sets are uncolored. If S_i still has red and blue colors, fine. Otherwise, we say S_i survives and let S_i^* be the set of uncolored points. Then $|S_i^*| \geq 20$ for otherwise it had had more than 80 points all one color, so it was dangerous and all points were uncolored. Let $\mathcal{F}^*$ be the family of S_i^*. We want a 2-coloring χ of $\mathcal{F}^*$ with no S_i^* monochromatic. Having picked $100, 80, 20$ appropriately LLL applies and χ exists. But isn't this begging the question? Surprisingly, no. The family $\mathcal{F}^*$ has, almost surely, a quite simple structure. Make a graph G with vertices the indices $1 \leq i \leq n$ and adjacency $i \sim j$ if $S_i \cap S_j \neq \emptyset$. Each i has at most 10^4 neighbors. For S_i to survive one of its neighbors must be dangerous, and this occurs with probability at most a very small constant ϵ. Let G^* be the restriction of G to the surviving i. Imagine that each i survived with independent probability ϵ. When i survived it would have in G^* on average $\gamma = 10^4 \epsilon$ surviving neighbors who would have on average γ^2 further neighbors, etc. With $\gamma < 1$ the neighborhood of i looks locally like a birth process that will almost surely die. An even better analogy is to components of the random graph $G(n, \frac{\gamma}{n})$ with $\gamma < 1$. There, as discussed in Section 4.1, all components are of size $O(\ln n)$. Of course, the i do not survive independently, when $i \sim j$ the dependence can be quite strong. Nonetheless Beck showed that G^* almost surely has all components of size $O(\ln n)$. The coloring of $\mathcal{F}^*$ then breaks into coloring the at most n components separately. Each component has $O(\ln n)$ sets hence $O(\ln n)$ vertices. On each component a coloring χ exists. Beck finds it by using exhaustive search! This takes exponential time but the problem has only logarithmic size so the time is polynomial in n. Alon [1] has given an alternate, parallelizable version of this algorithm and many applications. Still, the general, if ill-formed, question of whether LLL always admits an algorithmic implementation remains open. More likely the opposite is true. A class of problems may well be found where the existence of solutions is guaranteed by LLL but a polynomial time algorithm to find them would violate usual assumptions in complexity theory.

4.   Adventures in Asymptopia

4.1.   Inside the Double Jump.   In their original [11] Erdős and Rényi discovered what they called the "double jump" in the evolution of the random graph $G(n, p)$ around $p = n^{-1}$. When $p = \gamma n^{-1}$, $\gamma < 1$, all components of G are small, the largest of size $\Theta(\ln n)$, but when $\gamma > 1$ a giant component of size $\Theta(n)$ has been created. We now know that the proper magnification with which to slow down the double jump is

$$p = \frac{1}{n} + \frac{\lambda}{n^{4/3}}$$

This narrower range of p is called the phase transition. When $\lambda = \lambda(n) \to -\infty$ the largest components are all of size $o(n^{2/3})$; they are all almost the same size and they are all trees. The phase transition has not started. By the time $\lambda = \lambda(n) \to +\infty$ there is a dominant component whose size is $\gg n^{2/3}$ while all other components have size $o(n^{2/3})$. Moreover the complexity (defined as edges minus vertices) of the dominant component goes to infinity. In Asymptopia the situation at λ constant is given by an infinite sequence $c_1 > c_2 > \ldots$, representing components of sizes $c_1 n^{2/3}, c_2 n^{2/3}, \ldots$ in $G(n, p)$. We think of this as an infinite asteroid belt with asteroids of these sizes. The distribution of these sequences is complex. But the dynamic situation, moving from time λ to time $\lambda + d\lambda$ is easy to describe. Given components of sizes $c_i n^{2/3}, c_j n^{2/3}$ there are $c_i c_j n^{4/3}$ potential edges between them and $n^{2/3} d\lambda/2$ random edges are being selected so they are joined with probability $\sim c_1 c_2 d\lambda$. In Asymptopia we have a peculiar physics in which with probability $c_1 c_2 d\lambda$ asteroids of sizes c_1, c_2 merge to form a new asteroid of size $c_1 + c_2$. Each asteroid further has a complexity x_i, the complexity of the component. For λ large negative most of the components will be trees, so $x_i = -1$. When asteroids of complexities x_i, x_j merge, the merged asteroid has complexity $x_i + x_j - 1$. With λ large negative the asteroids are all tiny, but as λ increases moderate size asteroids are created. This physics favors the rich; a larger asteroid is more likely to merge with others and so become still larger. Computer experiments reveal the process quite strikingly, when $\lambda = -4$ the sizes are small, whereas by $\lambda = +4$ in over 90% of the cases a clear dominant component has emerged.

4.2. Asymptotic Packing. For $2 \le l < k < n$ let $m(n, k, l)$ denote the maximal size of a family F of k-element subsets of $\{1, \ldots, n\}$ so that no l-set E is contained in more than one $A \in F$. We set $Q = \binom{k}{l}$ for notational convenience. Elementary counting gives $m(n, k, l) \le \binom{n}{l}/Q$, with equality holding if and only if there is an appropriate tactical configuration. (For $l = 2, k = 3$, these are the Steiner Triple Systems.) In 1963 Erdős and Hanani [9] conjectured that for all $2 \le l < k$

$$\lim_{n \to \infty} m(n, k, l) Q / \binom{n}{l} = 1$$

This was first proven by Rödl [17]. Here we outline a new proof. Indeed, we show that a random greedy algorithm gives F of desired size.

We describe a greedy algorithm with a handy parametrization. Assign to each k-set A a random real $x_A \in [0, \binom{n-l}{k-l})]$. This orders the k-sets. Consider them in order accepting A if no B with $|A \cap B| \ge l$ has already been accepted. Let F_c be the family of A accepted with $x_A < c$. An l-set E is said to survive at "time" c if no $A \in F_c$ contains E.

To determine if E survives at time c we create a tree with root E. If $A \supset E$ and $x_A < c$ we consider the $Q - 1$ l-sets $E' \subset A$, $E' \ne E$, as a brood of children of E, born at time x. If all these E' survive at time x then either A is placed in F at time x or some $A' \supset E$ had already been placed in F. Either way E does not survive at time c.

In Asymptopia this becomes a continuous time birth process. E, now Eve, has birthdate c. Time goes backwards. Eve gives birth to broods of size $Q - 1$ by a Poisson process with unit density. Children with birthdate x in turn have broods in $[0, x)$ by the same process. With probability one a finite tree T is produced. Survival is defined inductively. Childless E' survive and E' does not survive if and only if she has a brood all of whom survive. Let $f(c)$ be the probability Eve survives. Some technical work gives $\lim_{n \to \infty} f_n(c) = f(c)$.

In Asymptopia we estimate $f(c) - f(c + \Delta c)$. The difference for Eve is if she has no surviving broods born in $[0, c)$, a brood born in $[c, c + \Delta c)$, and that brood all survive. For Δc small $f(c) - f(c + \Delta c) \sim f(c)(\Delta c)f(c)^{Q-1}$.

Here we bring out the most powerful tool of all, calculus! In the limit the derivative $f'(c) = -f(c)^Q$. Eve born at $c = 0$ is always childless so $f(0) = 1$. We solve the differential equation

$$f(c) = [1 + (Q - 1)c]^{-1/(Q-1)}$$

For any $\epsilon > 0$ we find c and then n so that on average fewer than $\epsilon\binom{n}{l}$ E survive at time c. Thus, there *exists* an outcome for which fewer than $\epsilon\binom{n}{l}$ E survive. Then the $A \in F_c$ must cover $(1 - \epsilon)\binom{n}{l}$ sets E and

$$|F| \geq |F_c| \geq (1 - \epsilon)\binom{n}{l}\bigg/\binom{k}{l} \qquad \text{as desired.}$$

References

[1] N. Alon, *A parallel algorithmic version of the local lemma*, Random Stuctures & Algorithms 2:367–378, 1991.

[2] N. Alon and J. Spencer, The Probabilistic Method, J. Wiley & Sons, New York, 1992.

[3] J. Beck, *Roth's estimate on the discrepancy of integer sequences is nearly sharp*, Combinatorica 1(4):319–325, 1981.

[4] J. Beck, *An algorithmic approach to the Lovász Local Lemma I*, Random Stuctures & Algorithms 2:343–365, 1991.

[5] J. Beck and T. Fiala, *Integer-making theorems*, Discrete Appl. Math. 3:1–8, 1981.

[6] B. Bollobás. Random Graphs, Academic Press, New York and San Diego, 1985.

[7] B. Bollobás, *The chromatic number of random graphs*, Combinatorica 8:49–55, 1988.

[8] P. Erdős, *Some remarks on the theory of graphs*, Bull. Amer. Math. Soc. 53:292–294, 1947.

[9] P. Erdős and H. Hanani, *On a limit theorem in combinatorial analysis*, Publ. Math. Debrecen 10:10–13, 1963.

[10] P. Erdős and L. Lovász, *Problems and results on 3-chromatic hypergraphs and some related questions*, in Infinite and Finite Sets, A. Hajnal et al., eds., North-Holland, Amsterdam, 1975, 609–628.

[11] P. Erdős and A. Rényi, *On the evolution of random graphs*, Magyar Tud. Akad. Mat. Kut. Int. Közl 5:17–61, 1960

[12] S. Janson, *Poisson approximation for large deviations*, Random Structures & Algorithms 1:221–230, 1990.

[13] S. Janson, T. Łuczak, and A. Rucinski, *An exponential bound for the probability of nonexistence of specified subgraphs of a random graph*, in Proceedings of Random Graphs '87, M. Karonski et al., eds, J. Wiley, New York, 1990, 73–87.

[14] D. J. Kleitman, *On a combinatorial problem of Erdős*, J. Combin. Theory 1:209–214, 1966.

[15] J. Matoušek, *Tight upper bounds for the discrepancy of halfspaces*, KAM Series (Tech. Report), Charles University, Prague 1994.

[16] J. Matoušek and J. Spencer, *Discrepancy in arithmetic Progressions*, to appear.

[17] V. Rödl, *On a packing and covering problem*, European J. Combin., 5 (1985), 69–78.

[18] K. F. Roth, *Remark concerning integer sequences*, Acta Arith. 9:257–260, 1964.

[19] J. Spencer, *Six standard deviations suffice*, Trans. Amer. Math. Soc. 289:679–706, 1985.

[20] M. Talagrand, *A new isoperimetric inequality for product measure and the tails of sums of independent random variables*, Geom. Functional Anal. 1:211–223, 1991.

Asymptotic Combinatorics and Algebraic Analysis

ANATOLY M. VERSHIK[*]

Steklov Mathematical Institute
St. Petersburg Branch
Fontanka 27, St. Petersburg 191011, Russia

1 Asymptotic problems in combinatorics and their algebraic equivalents

A large number of asymptotic questions in mathematics can be stated as combinatorial problems. I can give examples from algebra, analysis, ergodic theory, and so on. Therefore the study of asymptotic problems in combinatorics is stimulated enormously by taking into account the various approaches from different branches of mathematics. Recently we found many new aspects of this development of combinatorics. The main question in this context is: *What kind of limit behavior can have a combinatorial object when it "grows" ?*

One of the recent examples we can find in a very old area, namely the theory of symmetric and other classical groups and their representations. Let me quote the remarkable words of Weyl from his book *Philosophy of mathematics and natural science* (1949): *Perhaps the simplest combinatorial entity is the group of permutations of n objects. This group has a different constitution for each individual number n. The question is whether there are nevertheless some asymptotic uniformities prevailing for large n or for some distinctive class of large n.* He continued: *Mathematics has still little to tell about such a problem.*

In the meantime, a lot of progress has been made in this direction. We should mention the names of some persons who have made important contributions to this area, namely P. Erdős, V. Goncharov, P. Turan, A. Khinchin, W. Feller, and others. In the more general context of what is called nowadays the asymptotic theory of representations, I want to mention the names of H. Weyl and J. von Neumann.

2 Typical objects in asymptotic combinatorial theory

Besides the symmetric groups there are other classical objects in mathematics and in combinatorics, namely partitions of natural numbers. They provide another source of extremely important asymptotic problems that are also closely related to analysis, algebra, number theory, measure theory, and statistical physics.

The third class of objects, which plays the role of a link between combinatorics on one side and algebra and analysis on the other side, is a special kind of graphs,

*)   email: vershik@pdmi.ras.ru

Proceedings of the International Congress
of Mathematicians, Zürich, Switzerland 1994
© Birkhäuser Verlag, Basel, Switzerland 1995

the so-called Bratteli diagrams, i.e. $\mathbb{Z}_+$-graded locally finite graphs. These are the combinatorial analogues of locally semisimple algebras. This important class of algebras arises in asymptotic theory of finite and locally finite groups, and can be considered as an algebraic equivalent of asymptotic theory in analysis.

We now have described some of the objects that used to be basic in the theory of asymptotic combinatorial problems.

3 Problems

Next we will formulate the typical problems for these objects. We will start with the problems related to symmetric groups. It is important to emphasize that the same problems can also be stated for any other series of classical groups like Coxeter groups, $\mathrm{GL}(n, \mathbb{F}_p)$, and so on.

In all the considerations we use some probability measure. For example it is natural to provide the symmetric group S_n ($n \in \mathbb{N}$) with the uniform distribution (Haar measure).

PROBLEM. *Describe the asymptotic behavior (on n) of conjugacy classes; more precisely, find the common limit distribution of the numerical invariants of the classes.*

Now let us consider linear representations of those groups or their dual objects $\hat{S}_n$ provided with the Plancherel measure. (Then the measure of a representation is the normalized square of its dimension; this is the right analogue of the Haar measure for the dual space. The deep connection between these two measures is given by the RSK-Robinson-Shensted-Knuth-correspondence.)

PROBLEM. *Describe the asymptotic behavior, i.e. find the common limit distribution of a complete system of invariants of the representations.*

For the symmetric groups there are natural parameters both for conjugacy classes, namely the lengths of cycles, and for representations, namely Young diagrams. So we have to study asymptotic combinatorial problems about random partitions or random Young diagrams. Both of these problems were posed by the author in the early 1970s and were solved in the 1970s in joint papers of the author with Kerov (see [KV1]) and Schmidt (see [SV]), and also partially in papers of Shepp and Logan (see [LS]).

Now let us consider partitions of natural numbers $\mathcal{P}(n)$. As we mentioned before, the previous problems can be reduced to problems about partitions. Roughly speaking, all the questions concern the following problem: suppose we have some statistics on the space of partitions $\mathcal{P}(n)$ for all n, say μ_n; how do we scale the space $\mathcal{P}(n)$ in order to obtain the true nontrivial limit distribution of the measures μ_n? The same question can be asked for Young diagrams, graphs, configurations, and higher-dimensional objects of such a type.

A possible kind of answer can be a *limit-shape theorem*, which asserts that the limit distribution is a δ-measure concentrated at one configuration, called the *limit shape* of the random partition diagram, configuration, etc. In the problems that we discuss below, examples are Plancherel statistics, uniform statistics, and convex problems. Other examples for the same situation are the so-called Richardson-Eden model in the probability theory of a many-particles system, one-dimensional hydrodynamics, Maxwell (Bell) statistics on partitions, etc. We obtain rich information about asymptotics from the properties of the limit shape.

In the opposite (nonergodic) case the limit distribution is a nondegenerate distribution. Examples for this case are conjugacy classes in symmetric groups (see above) and other series of classical groups, and harmonic measures on Young diagrams.

In all these examples we have a completely different scaling in comparison with the first case. The dichotomy of the two cases can be compared with the dichotomy of trivial and nontrivial Poisson boundary in probability theory — there is a deep analogy. A systematic theory as well as general criteria for the two cases are still unknown.

4 Results

Here we will list the main results that have been obtained in this direction. Let us first define some of the important statistics on partitions, some of which we have mentioned shortly above. We use the bijection between Young diagrams with n cells and the set of partitions $\mathcal{P}(n)$ of n (see Figure 1).

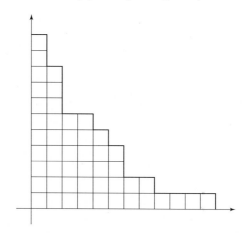

$$\lambda(t) = \sum_{i \geq t} \kappa_i; \qquad \kappa_i = \#\{j : \lambda_j = i\};$$

$$\tilde{\lambda}(t) = \frac{1}{\varphi_n} \sum_{i \geq t\psi_n} \kappa_i; \qquad \varphi_n \psi_n = n.$$

Figure 1

(a) Haar statistics (for conjugacy classes in symmetric groups): Let $\lambda \in \mathcal{P}(n)$ and let $k_1, \ldots, k_n$ be the multiplicities of the summands $1, 2, \ldots, n$ respectively. Then

$$\mu_h^n(\lambda) = \prod_{m=1}^{n} \frac{1}{k_m! \, m^{k_m}} \ .$$

(b) Uniform statistics:

$$\mu_u^n(\lambda) = \frac{1}{p(n)} \ ,$$

where $p(n)$ is the Euler-Hardy-Ramanujan function.

(c) Maxwell or Bell statistics:

$$\mu_b^n(\lambda) = \prod_{m=1}^{n} \frac{1}{k_m!\,(m!)^{k_m}} \ .$$

This is the image on $\mathcal{P}(n)$ of the uniform distribution on *partitions of n distinct objects*.

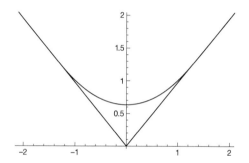

$$\Omega(s) = \begin{cases} \frac{2}{\pi}\left(s\arcsin s + \sqrt{1-s^2}\right) & |s| \le 1, \\ |s| & |s| > 1. \end{cases}$$

Figure 2

(d) Plancherel statistics on Young diagrams:

$$\mu_p^n(\lambda) = n!/(\prod_{\alpha} h_\alpha)^2 \ ,$$

where α is a cell of the Young diagram, h_α is the *hooklength* of the cell α, and the product here is taken over all cells of the diagram. This is the probability of the diagram (or partition) as a representation of the symmetric group: the probability is proportional to the square of the dimension of the representation.

(e) Uniform statistics on Young diagrams which sit inside a given rectangle.

(f) Fermi statistics on Young diagrams (no rows with equal length).

(g) Uniform statistics on convex diagrams: This is one of the first two-dimensional problems. We consider the set of all diagrams inside a given square of a lattice whose border is convex (or concave). This set can be considered as the set of *vector partitions* of the vector (n, n). The correspondence with the previous description is established by considering the slopes of the edges (see Figure 3), etc.

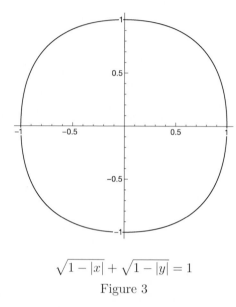

$$\sqrt{1 - |x|} + \sqrt{1 - |y|} = 1$$

Figure 3

All the measures are defined for all natural numbers n. In order to be able to speak about convergence we have to normalize or rescale the axes of the diagrams (partitions), dividing by appropriate sequences of numbers ϕ_n and ψ_n which depend on the cases; the choice of those numbers is unique (and they exist). Suppose the rescaling is done and we can consider the measures μ^n for all n in the same limit topological space of normalized diagrams. Let us say that we have the *ergodic case* if the weak limit of the measures μ^n is a δ-measure at some point of the space — this is usually some curve — "continuous" diagram. If the limit of the measures μ^n does exist but is a nondegenerate measure we will say that the case is *nonergodic*.

THEOREM 1. *The case (a) is nonergodic, the cases (b)–(e) are ergodic. The normalizations of the axes are the following:*

(a) *For $\lambda \in \mathcal{P}(n)$, $\lambda = (\lambda_1, \ldots, \lambda_n)$: $\lambda_i \to \lambda_i/n$ for $i = 1, \ldots, n$, there is no normalization along the second axis.*

(b), (d), (e), (f), (g) *The normalization along both axes is $1/\sqrt{n}$.*

(c) *For $\lambda \in \mathcal{P}(n)$, the normalization of the values of $\lambda - s$ is $1/\ln n$ and the normalization of the indices is $n/\ln n$.*

Now we will give the precise answer to the questions about limit measures or limit shapes.

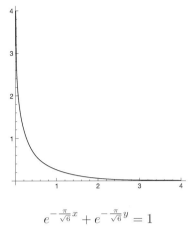

$$e^{-\frac{\pi}{\sqrt{6}}x} + e^{-\frac{\pi}{\sqrt{6}}y} = 1$$

Figure 4

THEOREM 2. *The following curves are the limit shapes:*
Case (b) $\exp[-(\pi/\sqrt{6})x] + \exp[-(\pi/\sqrt{6})y] = 1$ *(see Figure 4),*
Case (c) $y(x) \equiv 1$.
Case (d)   *Let* $w = (x+y)/2$, $s = (x-y)/2$, *then*

$$\omega \equiv \Omega(s) = \begin{cases} (2/\pi)(s\arcsin s + \sqrt{(1-s^2)}) & \text{for } |x| \le 1 \\ |s| & \text{for } |x| \ge 1 \end{cases}$$

(Kerov and Vershik [KV1], Logan and Shepp [LS]), see Figure 2.
Case (e) $(1 - \exp[-c\lambda])\exp[-cy] + (1 - \exp[-c\mu])\exp[-cx] = 1 - \exp[-c(\lambda+\mu)]$.
        λ, μ *are the size of the rectangle,* $c = c(\lambda, \mu)$. *(See Figure 5 for the case*
        $\lambda = \mu = 2$).
Case (f) $\exp[-\pi/\sqrt{12}y] - \exp[-\pi/\sqrt{12}x] = 1$.
Case (g) $\sqrt{1 - |x|} + \sqrt{1 - |y|} = 1$
        (Barany [B], Sinai [S], Vershik [V4]), see Figure 3.

This means that in each of those cases the following is true: for any $\epsilon > 0$
there exists an N such that if $n > N$ then the measure μ^n on $\mathcal{P}(n)$ has the property
$\mu^n\{\lambda:$ the normalized $\lambda \in V_\epsilon(\Gamma)\} > 1 - \epsilon$, where Γ is the limit shape curve, V_ϵ is
the ϵ-neighborhood of the curve Γ in the uniform topology, and the normalization
of the diagram for the cases is as above.

A completely different situation occurs in case (a), i.e. the limit distributions
of normalized lengths of cycles. The complete answer for this case was given in
a joint paper with Schmidt [SV]. In this case the limit measure is concentrated
in the space of positive series with sum 1. This remarkable measure also appears
in the context of number theory (the distribution of logarithms of prime divisors
of natural numbers) as was recently described by the author [V1] (see also P.
Billingsley, D. Knuth, and Trab-Pardo).

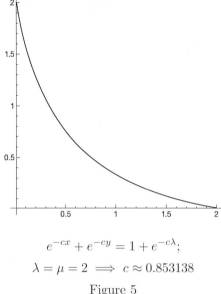

$$e^{-cx} + e^{-cy} = 1 + e^{-c\lambda};$$

$$\lambda = \mu = 2 \implies c \approx 0.853138$$

Figure 5

Let me give some examples of the applications of Theorem 2.

For $\lambda \in \mathcal{P}(n)$ let $\dim \lambda = n! / \prod h_\alpha$ (hook formula).

PROBLEM. *Find* $\max\{\dim \lambda : \lambda \in \mathcal{P}\}$.

This functional is rather complicated and, as H. Weyl suspected, the optimal diagram has a different feature for each n, hardly depending on the arithmetic of n. But it happens that *asymptotically* there is a prevailing form of diagram: this is again the limit shape Ω that I mentioned above. We obtain:

THEOREM 3 (Kerov and Vershik [KV3]). *There are constants c_1 and c_2 such that*

$$0 < c_1 < -\frac{1}{\sqrt{n}} \ln(\max \frac{\dim \lambda}{\sqrt{n!}}) < c_2 < \infty.$$

It is an important observation that the average diagram with respect to the Plancherel measure asymptotically coincides with the diagram of maximal dimension. Another important fact is that the limit shape Ω arises in many different contexts such as the asymptotic of the spectrum of random matrices, zeros of the orthogonal polynomials, etc.

For the case of Maxwell-Bell statistics the limit shape is not interesting — then the generic block of the partition has size r where r is a solution of the equation $x \exp x = n$. But it is possible to refine this answer in the spirit of CLT: Let $\varphi(i) = (b(i) - r)/\sqrt{r}$, where $b(i)$ is the length of the block containing i.

THEOREM 4.

$$\lim \mu_b^n \left\{ \lambda \in \mathcal{P}(n) : \left| \frac{1}{n} \#\{i : \varphi(i) < a\} - \mathrm{Erf}(a) \right| < \epsilon \right\} = 1$$

for all $a \in \mathbb{R}$ and all $\epsilon > 0$.

This theorem together with case (c) of Theorem 2, proven by my students Yu. Yakubovich and D. Alexandrovsky, describes the limit structure of generic finite partitions (see [Ya]).

5 Techniques

Now we will discuss some technical aspects that are important by themselves. We emphasize four tools:

(i) generating functions and the Hardy method, the saddle-point method;
(ii) the variational principle in combinatorics;
(iii) functional equations and the ergodic approach;
(iv) methods from statistical physics: big canonical set and the local limit theorem of probability theory.

The classical method for studying enumeration problems uses generating functions. For our goals we also can use them, but with some modification: we need to consider the generating function for the number of combinatorial objects with some special properties. For example, instead of the Euler function for partitions $F(z) = \{\prod_{k=1}^{\infty}(1 - z^k)\}^{-1}$ we have to use the generating function for the number of partitions with a given number of blocks whose lengths are less than a constant. In a different context such a method has been applied by Turan and Szalay.

For the higher-dimensional case (g) of convex diagrams or convex lattice polygons we introduce a new kind of generating function of two variables:

$$F(t, s) = \prod_{(k,r) \text{ coprime}} (1 - t^k s^r)^{-1}.$$

LEMMA. *Let* $p(n, m) = \mathrm{Coeff}(t^n s^m; F)$; *then this is the number of convex lattice polygons in the rectangle* $(0, n) \times (0, m)$ *which meet the points* $(0, 0)$ *and* (n, m). *The formula*

$$\ln(p(n, n))/n^{2/3} = 3\sqrt[3]{\zeta(3)/\zeta(2)}(1 + o(1))$$

holds. This formula also gives us the number of vector partitions without collinear summands.

The two-dimensional saddle point method applied to this function is the main ingredient to obtain the limit-shape theorem for this case. But in addition we need considerations on generating functions (see Barany [B] and Vershik [V4]). We can combine this approach with an approach from statistical physics. For case (g) this was done by Sinai [S]. In the early 1950s A. Khinchin [Kh] used this method in statistical physics.[*] Here we present a general context that covers all these papers.

The main idea is the following. Instead of studying the asymptotics of the coefficients of the generating function with the help of methods from the theory of

---

[*] We want to emphasize that the difference between the saddle point method (Darwin and Fauler's method) and the local limit theorem (Khinchin's approach) is not so big: technically to find a saddle point is the same as to find the value of a parameter which realizes a needed mathematical expectation. (See [V5]).

complex variables we can introduce one-parametric families of measures for which the natural coordinates, say the number of rows of given length, are independent. After this we can easily find the distribution of the functionals and then (the hardest part) prove that the distributions are the same as in the initial problem. This is completely analogous to the method of equivalence of great and small canonical sets.

The general definition: let $F(t) = \prod f_i(t) = 1 + b_1 z + b_2 z^2 + \cdots$ and let $f_i(t) = 1 + a_{i1}t + a_{i2}t^2 + \cdots$ be series converging in a circle and with nonnegative coefficients a_{ij}, $i = 1, 2, \ldots$. Now we introduce two sets of measures on the set of partitions: the first set consists of the measures μ^n on the sets $\mathcal{P}(n)$, defined by $\mu(\lambda) = c_n \prod a_{ij(i)}$, where a partition $\lambda = (j(1), \ldots, j(n)) \in \mathcal{P}(n)$ has $j(s)$ summands equal to s, and c_n is a constant; the second set is a one-parametric set of measures ν_t ($t \in (0,1)$) on the big canonical set $\bigcup \mathcal{P}(n)$, $n \in \mathbb{N}$, and defined as follows:

$$\nu_t(\lambda) = (F(t))^{-1} b_n t^n, \text{ where } \lambda \in \mathcal{P}(n).$$

The simple observation is that the number of summands of a given size is independent of ν_t and consequently we can use powerful methods of probability theory such as large deviations and so on. The main fact is contained in the following theorem.

THEOREM 5. *There exists a sequence t_n such that the main terms of the asymptotics of the expected smooth functionals on $\mathcal{P}(n)$ with respect to the measures μ^n and ν_{t_n} coincide.*

But we cannot claim that the expected distributions of the functionals also coincide. For this we need some additional assumptions. In particular, it is true in the cases (b)–(e) above. This assertion is essentially the above-mentioned equivalence between two sets. The technique is slightly simpler but parallel to the saddle point method.

Variational principles in these problems are very useful — we will give some examples. From the combinatorial point of view the variational principle defines a functional on the configuration (or diagram, partitions) which gives the main term in the asymptotics (like energy or entropy).

Suppose the continuous diagram Γ is fixed (see Figure 3) and Γ is the graph of a differentiable function $\gamma(\cdot)$. We want to find the asymptotic of the number of convex diagrams T_n that are close to Γ in the uniform metric.

THEOREM 6. [V4]

$$n^{-\frac{2}{3}} \ln T_n = 3\sqrt[3]{\frac{\zeta(3)}{4\zeta(2)}} \left(\int_\Gamma \kappa^{1/3} ds + o(1) \right)$$

where κ is a curvature of Γ.

This means that in case (g) the integral takes its maximal value on the class of all monotone differentiable curves in the limit shape curve.

6 Combinatorics of infinite objects

Our approach can be extended by considering certain limit objects preserving the combinatorial structure. That gives us an interpretation of some limit distributions that previously appeared as pure limit objects.

The best example of such an extension is the notion of *virtual permutations*. This theory is developed in a recent paper by Kerov, Ol'shansky, and myself [KOV]. The main definition is the following.

LEMMA.   *There is a unique projection* $p_n \colon S_n \to S_{n-1}$ *that commutes with the two-sided action of* S_{n-1}.

The inverse limit $X = \varprojlim(S_n, p_n)$ is called the space of *virtual permutations*. On this space X a two-sided action of the infinite symmetric group is well defined. The space of virtual permutations does not form a group, but it is possible to define the notion of cycles, and their normalized length, Haar measure, and other group-like notions. We will not give the precise definitions for those notions here (see [KOV]).

THEOREM 7.   *The common distribution of normalized lengths of virtual permutations coincides with the above-defined limit distribution for normalized lengths of cycles of ordinary permutations.*

The main application of the space of virtual permutations consists in the construction of some new types of representations of the infinite symmetric group like the regular representation based on the analogue of the Haar measure and its one-parametric deformation.

One of the main problems in this area is the problem of limit shapes for multidimensional configurations, Young diagrams etc. Perhaps variational principles with ideas coming from statistical physics will help to solve them (see [V5]).

References

[B]   I. Barany, *The limit shape theorem of convex lattice polygons*, Discrete Comput. Geom. **13** (1995), 279–295.

[K1]   S. Kerov, *Gaussian limit for the Plancherel measure of the symmetric groups*, C. R. Acad. Sci. **316** (1993), 303–308.

[K2]   S. Kerov, *Transition probabilities of continuous Young diagrams and the Markov moment problem*, Functional Anal. Appl. **27** (3) (1993), 32–49.

[KOV]   S. Kerov, G. Ol'shansky, and A. Vershik, *Harmonic analysis on the infinite symmetric group*, C. R. Acad. Sci. **316** (1993), 773–778.

[KV1]   S. Kerov and A. Vershik, *Asymptotics of the Plancherel measure of the symmetric group and limit shapes of Young diagrams*, Soviet Math. Dokl. **18** (1977), 527–531.

[KV2]   S. Kerov and A. Vershik, *Characters and factor representations of the infinite symmetric group*, Soviet Math. Dokl. **23** (5) (1981), 1037–1040.

[KV3]   S. Kerov and A. Vershik, *The asymptotic of maximal and typical dimension of irreducible representations of the symmetric group*, Functional Anal. Appl. **19** (1) (1985), 21–31.

[Kh]   A. Khinchin, Mathematical Foundations of Quantum Statistics, Moscow, 1951, (in Russian).

[LS]   B. Logan and L. Shepp, *A variational problem for random Young tableaux*, Adv. in Math. **26** (1977), 206–292.

[SV]   A. Schmidt and A. Vershik, *Limit measures that arise in the asymptotic theory of symmetric groups* I, Teor. Veroyatnast. i. Primenen **22** (1) (1977), 72–88; II, Teor. Veroyatnast. i. Primenen. **23** (1) (1978), 42–54.

[S]   Ya. Sinai, *Probabilistic approach to analysis of statistics of convex polygons*, Functional Anal. Appl. **28** (2) (1994), 41–48.

[V1]   A. Vershik, *Asymptotic distribution of decompositions of natural numbers into prime divisors*, Soviet Math. Dokl. **289** (2) (1986), 269–272.

[V2]   A. Vershik, *Statistical sum associated with Young diagrams*, J. Soviet Math. **47** (1987), 2379–2386.

[V3]   A. Vershik, *Asymptotic theory of the representation of the symmetric group*, Selecta Math. Soviet. **19** (2) (1992), 281–301.

[V4]   A. Vershik, *Limit shape of convex lattice polygons and related questions*, Functional Anal. Appl. **28** (1) (1994), 16–25.

[V5]   A. Vershik, *Statistical physics of combinatorial partitions and its limit configuration*, Func. Anal. **30** (1) (1996).

[Ya]   Yu. Yakubovich, *Asymptotics of a random partitions of a set*, Zap. nauch. sem. POMI, Representation theory, dynamical systems, combinatorial and algorithmic methods I **223** (1995), 227–249 (in Russian).

Probabilistic Proof Systems

ODED GOLDREICH*

Department of Applied Mathematics and Computer Science
Weizmann Institute of Science, Rehovot, Israel

ABSTRACT. Various types of *probabilistic* proof systems have played a central role in the development of computer science in the last decade. In this exposition, we concentrate on three such proof systems — *interactive proofs*, *zero-knowledge proofs*, and *probabilistic checkable proofs* — stressing the essential role of randomness in each of them.

1. Introduction

The glory given to the creativity required to find proofs makes us forget that it is the less glorified procedure of verification that gives proofs their value. Philosophically speaking, proofs are secondary to the verification procedure; whereas technically speaking, proof systems are defined in terms of their verification procedures.

The notion of a verification procedure assumes the notion of computation and furthermore the notion of efficient computation. This implicit assumption is made explicit in the definition of $\mathcal{NP}$, in which efficient computation is associated with (deterministic) polynomial-time algorithms.

DEFINITION 1 (NP-proof systems): *Let $S \subseteq \{0,1\}^*$ and $\nu : \{0,1\}^* \times \{0,1\}^* \mapsto \{0,1\}$ be a function so that $x \in S$ if and only if there exists a $w \in \{0,1\}^*$ such that $\nu(x,w) = 1$. If ν is computable in time bounded by a polynomial in the length of its first argument then we say that S is an* NP-set *and ν defines an* NP-proof system.

Traditionally, $\mathcal{NP}$ is defined as the class of NP-sets (cf. [14]). Yet, each such NP-set can be viewed as a proof system. For example, consider the set of satisfiable Boolean formulae. Clearly, a satisfying assignment π for a formula ϕ constitutes an NP-proof for the assertion "ϕ is satisfiable" (the verification procedure consists of substituting the variables of ϕ by the values assigned by π and computing the value of the resulting Boolean expression).

The formulation of NP-proofs restricts the "effective" length of proofs to be polynomial in length of the corresponding assertions (as the running time of the verification procedure is restricted to be polynomial in the length of the assertion). However, longer proofs may be allowed by padding the assertion with sufficiently

*Supported by grant No. 92-00226 from the United States - Israel Binational Science Foundation (BSF), Jerusalem, Israel.

many blank symbols. So it seems that NP gives a satisfactory formulation of proof systems (with efficient verification procedures). This is indeed the case if one associates efficient procedures with *deterministic* polynomial-time algorithms. However, we can gain a lot if we are willing to take a somewhat nontraditional step and allow *probabilistic* verification procedures. In particular,

- Randomized and interactive verification procedures, giving rise to *interactive proof systems*, seem much more powerful than their deterministic counterparts.

- Such randomized procedures allow the introduction of *zero-knowledge proofs*, which are of great theoretical and practical interest.

- NP-proofs can be efficiently transformed into a (redundant) form that offers a trade-off between the number of locations examined in the NP-proof and the confidence in its validity (see *probabilistically checkable proofs*).

In all above-mentioned types of probabilistic proof systems, explicit bounds are imposed on the computational complexity of the verification procedure, which in turn is personified by the notion of a verifier. Furthermore, in all these proof systems, the verifier is allowed to toss coins and rule by statistical evidence. Thus, all these proof systems carry a probability of error; yet, this probability is explicitly bounded and, furthermore, can be reduced by successive application of the proof system.

1.1. Basic background from computational complexity.
The following are standard complexity classes.

- $\mathcal{P}$ denotes the class of sets in which membership can be decided in (deterministic) polynomial time. Namely, for every $S \in \mathcal{P}$ there exists a (deterministic) polynomial-time algorithm A so that $x \in S$ iff $A(x) = 1$, for all $x \in \{0, 1\}^*$.

- $\mathcal{RP}$ (resp., $\mathcal{BPP}$) denotes the class of sets in which membership can be decided in probabilistic polynomial time with one-sided (resp., two-sided) error probability. Specifically, for every $S \in \mathcal{RP}$ (resp., $S \in \mathcal{BPP}$) there exists a probabilistic polynomial-time algorithm A so that $x \in S$ implies $\text{Prob}(A(x) = 1) \geq \frac{1}{2}$ (resp., $\text{Prob}(A(x) = 1) \geq \frac{2}{3}$) whereas $x \notin S$ implies $\text{Prob}(A(x) = 1) = 0$ (resp., $\text{Prob}(A(x) = 1) \leq \frac{1}{3}$). The class $co\mathcal{RP} \overset{\circ}{=} \{\overline{S} : S \in \mathcal{RP}\}$ has one-sided error in the "other direction".

- $\mathcal{NP}$ denotes the class of NP-sets and $co\mathcal{NP}$ denotes the class of their complements (i.e., $S \in co\mathcal{NP}$ iff $\overline{S} \in \mathcal{NP}$).

- A set S is *polynomial-time reducible* to a set T if there exists a polynomial-time computable function f so that $x \in S$ iff $f(x) \in S$ (for every x). A set is *NP-hard* if every NP-set is polynomial-time reducible to it. A set is *NP-complete* if it is both NP-hard and in $\mathcal{NP}$.

- $\mathcal{PSPACE}$ denotes the class of sets in which membership can be decided in polynomial space (i.e., the work space taken by the decider is polynomial in length of the input).

Obviously, $P \subseteq \mathcal{RP} \subseteq \mathcal{BPP} \subseteq \mathcal{PSPACE}$. It is not hard to see that $\mathcal{RP} \subseteq \mathcal{NP}$ and that $\mathcal{NP} \subseteq \mathcal{PSPACE}$. It is widely believed that $P \neq \mathcal{NP}$ and $\mathcal{NP} \neq \mathcal{PSPACE}$. Furthermore, it is also believed that $\mathcal{NP} \neq \mathrm{co}\mathcal{NP}$. NP-hard sets (or tasks) are assumed to be infeasible, because if an NP-hard set is in P then $\mathcal{NP} = P$.

1.2. Conventions. When presenting a proof system, we state all complexity bounds in terms of the length of the assertion to be proven (which is viewed as an input to the verifier). Namely, polynomial time means time polynomial in the length of this assertion. Note that this convention is consistent with our definition of NP-proofs.

Denote by `poly` the set of all polynomials and by `log` the set of all integer functions bounded by $O(\log n)$.

1.3. Basic background from combinatorics. Two graphs, $G_1 = (V_1, E_1)$ and $G_2 = (V_2, E_2)$, are called *isomorphic* if there exists a 1-1 and onto mapping ϕ from the vertex set V_1 to the vertex set V_2 so that $(u, v) \in E_1$ if and only if $(\phi(v), \phi(u)) \in E_2$. The mapping ϕ, if existing, is called an *isomorphism* between the graphs.

A graph $G = (V, E)$ is said to be *3-colorable* if there exists a function $\pi : V \mapsto \{1, 2, 3\}$ so that $\pi(v) \neq \pi(u)$ for every $(u, v) \in E$. Such a function π is called a *3-coloring* of the graph.

2. Interactive Proof Systems

In light of the growing acceptability of randomized and distributed computations, it is only natural to associate the notion of efficient computation with probabilistic and interactive polynomial-time computations. This leads naturally to the notion of interactive proof systems in which the verification procedure is interactive and randomized, rather than being noninteractive and deterministic. Thus, a "proof" in this context is not a fixed and static object but rather a randomized (dynamic) process in which the verifier interacts with the prover. Intuitively, one may think of this interaction as consisting of "tricky" questions asked by the verifier to which the prover has to reply "convincingly". The above discussion, as well as the following definition, makes explicit reference to a prover, whereas a prover is only implicit in the traditional definitions of proof systems (e.g., NP-proofs).

2.1. Definition. Loosely speaking, an interactive proof is a game between a computationally bounded verifier and a computationally unbounded prover whose goal is to convince the verifier of the validity of some assertion. Specifically, the verifier is probabilistic polynomial time. It is required that if the assertion holds then the verifier always accepts (i.e., when interacting with an appropriate prover strategy). On the other hand, if the assertion is false then the verifier must reject with probability at least $\frac{1}{2}$, no matter what strategy is being employed by the prover. A sketch of the formal definition is given in Item (1) below. Items (2) and

(3) introduce additional complexity measures, which can be ignored in the first reading.

DEFINITION 2 (Interactive Proofs – IP) [19]:

(1) An interactive proof system *for a set S is a two-party game between a* verifier, *executing a probabilistic polynomial-time strategy (denoted V), and a* prover, *which executes a computationally unbounded strategy (denoted P), satisfying*

- Completeness*: For every $x \in S$ the verifier V always accepts after interacting with the prover P on common input x.*
- Soundness*: For every $x \notin S$ and every potential strategy P^*, the verifier V rejects with probability at least $\frac{1}{2}$, after interacting with P^* on common input x.*

(2) Let m be an integer function. The complexity class $\mathcal{IP}(m(\cdot))$ consists of sets having an interactive proof system in which, on common input x, the total number of messages exchanged between the parties is bounded by $m(|x|)$.

(3) Let M be a set of integer functions. Then, $\mathcal{IP}(M)$ denotes $\cup_{m \in M}\mathcal{IP}(m(\cdot))$. Finally, $\mathcal{IP} \stackrel{\text{def}}{=} \mathcal{IP}(\text{poly})$.

In Item (1), we have followed the standard definition, which specifies strategies for both the verifier and the prover. An alternative presentation only specifies the verifier's strategy while rephrasing the completeness condition as follows:

> There exists a prover strategy P so that, for every $x \in S$, the verifier V always accepts after interacting with P on common input x.

Arthur-Merlin games[1] introduced in [4] are a special case of interactive proofs; yet, as shown in [20], this restricted case has essentially[2] the same power as the general case previously introduced in [19]. Also, in some sources interactive proofs are defined so that two-sided error probability is allowed; yet, essentially this does not increase their power [13].

2.2. The role of randomness. Randomness is essential to the formulation of interactive proofs; if randomness is not allowed (or if it is allowed but zero error is required in the soundness condition) then interactive proof systems collapse to NP-proof systems. The reason for this is that the prover can predict the verifier's part of the interaction and thus it suffices to let the prover send the full transcript of the interaction and let the verifier check that the interaction is indeed valid. (In case the verifier is not deterministic, the transcript sent by the prover may not match the outcome of the verifier coin tosses.) The moral is that there is no point to interact with predictable parties which are also computationally weaker.[3]

---

[1] In Arthur-Merlin games, the verifier must send the outcome of any coin it tosses (and thus need not send any other information).

[2] Here and in the next sentence, not only $\mathcal{IP}$ remains invariant under the various definitions, but also $\mathcal{IP}(m(\cdot))$, for every integer function satisfying $m(n) \geq 2$ for every n.

[3] This moral represents the prover's point of view. Certainly, from the verifier's point of view it is beneficial to interact with the prover, as it is computationally stronger.

2.3. The power of interactive proofs. A simple example demonstrating the power of interactive proofs follows. Specifically, we present an interactive proof for proving that two graphs are not isomorphic. It is not known whether such a statement can be proven via an NP-proof system.

CONSTRUCTION 1 (Interactive proof system for graph nonisomorphism) [15]:

- Common input: *A pair of two graphs, $G_1 = (V_1, E_1)$ and $G_2 = (V_2, E_2)$. Suppose, without loss of generality, that $V_1 = \{1, 2, ..., |V_1|\}$, and similarly for V_2.*

- Verifier's first step (V1): *The verifier selects at random one of the two input graphs, and sends to the prover a random isomorphic copy of this graph. Namely, the verifier selects uniformly $\sigma \in \{1, 2\}$, and a random permutation π from the set of permutations over the vertex set V_σ. The verifier constructs a graph with vertex set V_σ and edge set $F \stackrel{\text{def}}{=} \{(\pi(u), \pi(v)) : (u, v) \in E_\sigma\}$ and sends (V_σ, F) to the prover.*

- Motivating remark: *If the input graphs are nonisomorphic, as the prover claims, then the prover should be able to distinguish (not necessarily by an efficient algorithm) isomorphic copies of one graph from isomorphic copies of the other graph. However, if the input graphs are isomorphic then a random isomorphic copy of one graph is distributed identically to a random isomorphic copy of the other graph.*

- Prover's step: *Upon receiving a graph, $G' = (V', E')$, from the verifier, the prover finds a $\tau \in \{1, 2\}$ so that the graph G' is isomorphic to the input graph G_τ. (If both $\tau = 1, 2$ satisfy the condition then τ is selected arbitrarily. In case no $\tau \in \{1, 2\}$ satisfies the condition, τ is set to 0). The prover sends τ to the verifier.*

- Verifier's second step (V2): *If the message τ received from the prover equals σ (chosen in Step V1) then the verifier outputs 1 (i.e., accepts the common input). Otherwise the verifier outputs 0 (i.e., rejects the common input).*

The verifier strategy presented above is easily implemented in probabilistic polynomial time. We do not know of a probabilistic polynomial-time implementation of the prover's strategy, but this is not required. The motivating remark justifies the claim that Construction 1 constitutes an interactive proof system for the set of pairs of nonisomorphic graphs, which is a coNP-set (not known to be in $\mathcal{NP}$).

Interactive proofs are powerful enough to prove any coNP assertion (e.g., that a graph is not 3-colorable) [23]. Furthermore, the class of sets having interactive proof systems coincides with the class of sets that can be decided using a polynomial amount of work space [28].

THEOREM 1 [28]: $\mathcal{IP} = \mathcal{PSPACE}$.

Recall that it is widely believed that $\mathcal{NP} \subset \mathcal{PSPACE}$. Thus, under this conjecture, interactive proofs are more powerful than NP-proofs.

Concerning the finer structure of the IP hierarchy it is known that this hierarchy has a "linear speed-up" property [7]. Namely, for every integer function f so that $f(n) \geq 2$ for all n, the class $\mathcal{IP}(O(f(\cdot)))$ collapses to the class $\mathcal{IP}(f(\cdot))$. In particular, $\mathcal{IP}(O(1))$ collapses to $\mathcal{IP}(2)$. It is conjectured that co$\mathcal{NP}$ is *not* contained in $\mathcal{IP}(2)$, and consequently that interactive proofs with an unbounded number of message exchanges are more powerful than interactive proofs in which only a bounded (i.e., constant) number of messages is exchanged. Yet, the class $\mathcal{IP}(2)$ contains sets not known to be in $\mathcal{NP}$; e.g., graph nonisomorphism (as shown above).

3. Zero-Knowledge Proof Systems

Zero-knowledge proofs, introduced in [19], are central to cryptography. Furthermore, zero-knowledge proofs are very intriguing from a conceptual point of view, because they exhibit an extreme contrast between being convinced of the validity of a statement and learning anything in addition while receiving such a convincing proof. Namely, zero-knowledge proofs have the remarkable property of being convincing while also yielding nothing to the verifier, beyond the fact that the statement is valid. Formally, the fact that "nothing is gained by the interaction" is captured by stating that whatever the verifier can efficiently compute after interacting with a zero-knowledge prover can be efficiently computed from the assertion itself without interacting with anyone.

3.1. Sketch of definition. Zero-knowledge is a property of some interactive proof systems, or, more accurately, of some specified prover strategies. The formulation of the zero-knowledge condition considers two ensembles of probability distributions; each ensemble associates to each valid assertion a probability distribution. The first ensemble represents the output distribution of the verifier after interacting with the prover strategy P, where the verifier is not necessarily employing the specified strategy (i.e., V) — but rather any efficient strategy. The second ensemble represents the output distribution of some probabilistic polynomial-time algorithm (which does not interact with anyone). The basic paradigm of zero-knowledge asserts that for every ensemble of the first type there exists a "similar" ensemble of the second type.

The specific variants differ by the interpretation given to "similarity". The most strict interpretation, leading to *perfect zero-knowledge*, is that similarity means equality. A somewhat relaxed interpretation, leading to *almost-perfect zero-knowledge*, is that similarity means statistical closeness (i.e., negligible difference between the ensembles). The most liberal interpretation, leading to the standard usage of the term zero-knowledge (and sometimes referred to as *computational zero-knowledge*), is that similarity means computational indistinguishability (i.e., failure of any efficient procedure to tell the two ensembles apart — cf. [18] and [29]).

3.2. The power of zero-knowledge. A simple example, demonstrating the power of zero-knowledge proofs, follows. Specifically, we will present a simple zero-knowledge proof for proving that a graph is 3-colorable. The interactive proof will be described using "boxes" in which information can be hidden and later revealed. Such "boxes" can be implemented using one-way functions (see below).

CONSTRUCTION 2 (Zero-knowledge proof of 3-colorability) [15]:

- Common input: *A simple graph $G = (V, E)$.*

- Prover's first step: *Let ψ be a 3-coloring of G. The prover selects a random permutation π over $\{1, 2, 3\}$, and sets $\phi(v) \stackrel{\text{def}}{=} \pi(\psi(v))$, for each $v \in V$. Hence, the prover forms a random relabelling of the 3-coloring ψ. The prover sends the verifier a sequence of $|V|$ locked and nontransparent boxes so that the vth box contains the value $\phi(v)$.*

- Verifier's first step: *The verifier uniformly selects an edge $(u, v) \in E$, and sends it to the prover.*

- Motivating remark: *The verifier asks to inspect the colors of vertices u and v.*

- Prover's second step: *The prover sends to the verifier the keys to boxes u and v;*

- Verifier's second step: *The verifier opens boxes u and v, and accepts if and only if they contain two different elements in $\{1, 2, 3\}$.*

The verifier strategy presented above is easily implemented in probabilistic polynomial time. The same holds with respect to the prover's strategy, provided it is given a 3-coloring of G as auxiliary input. Clearly, if the input graph is 3-colorable then the prover can cause the verifier to accept always. On the other hand, if the input graph is not 3-colorable then any contents put in the boxes must be invalid on at least one edge, and consequently the verifier will reject with probability at least $\frac{1}{|E|}$. Hence, the above game exhibits a nonnegligible gap in the accepting probabilities between the case of 3-colorable graphs and the case of non-3-colorable graphs. To increase the gap, the game may be repeated sufficiently many times (of course, using independent coin tosses in each repetition). The zero-knowledge property follows easily, in this abstract setting, as one can simulate the real interaction by placing a random pair of different colors in the boxes indicated by the verifier. This indeed demonstrates that the verifier learns nothing from the interaction (because it expects to see a random pair of different colors and indeed this is what it sees). We stress that this simple argument is not possible in the digital implementation because the boxes are not totally unaffected by their contents (they are affected, yet in an indistinguishable manner).

As stated above, the "boxes" need to be implemented digitally, and this is done using an adequately defined "commitment scheme". Loosely speaking, such a scheme is a two-phase game between a sender and a receiver so that after the first phase the sender is "committed" to a value and yet, at this stage, it is infeasible for

the receiver to find out the committed value. The committed value will be revealed
to the receiver in the second phase and it is guaranteed that the sender cannot
reveal a value other than the one committed. Such commitment schemes can be
implemented assuming the existence of one-way functions (i.e., loosely speaking,
functions that are easy to compute but hard to invert, such as multiplication of
two large primes) — cf. [25] and [21].

Using the fact that 3-colorability is NP-complete, one gets zero-knowledge proofs
for any NP-set.

THEOREM 2 [15]: *Assuming the existence of one-way functions, any NP-proof can
be efficiently transformed into a* (computational) *zero-knowledge interactive proof.*

Theorem 2 has a dramatic effect on the development of cryptographic protocols (cf.
[15] and [16]). For the sake of elegance, we mention that, using further ideas and
under the same assumption, any interactive proof can be efficiently transformed
into a zero-knowledge one (cf. [22] and [8]).

The above results may be contrasted with the results regarding the complexity of *almost-perfect* zero-knowledge proof systems; namely, that almost-perfect
zero-knowledge proof systems exist only for sets in $\mathcal{IP}(2) \cap \mathrm{co}\mathcal{IP}(2)$ [11], [1],
and thus are unlikely to exist for all NP-sets. Also, a very recent result indicates
that one-way functions are essential for the existence of zero-knowledge proofs for
"hard" sets (i.e., sets that cannot be decided in average polynomial time)[26].

3.3. The role of randomness.   Again, randomness is essential to all of the above-
mentioned (positive) results. Namely, if either verifier or prover is required to be
deterministic then only BPP-sets can be proven in a zero-knowledge manner [17].
However, BPP-sets have trivial zero-knowledge proofs in which the prover sends
nothing and the verifier just tests the validity of the assertion by itself.[4] Thus,
randomness is essential to the usefulness of zero-knowledge proofs.

4. Probabilistically Checkable Proof Systems

When viewed in terms of an interactive proof system, the probabilistically checkable proof setting consists of a prover which is memoryless. Namely, one can think
of the prover as being an oracle and of the messages sent to it as being queries.
A more appealing interpretation is to view the probabilistically checkable proof
setting as an alternative way of generalizing $\mathcal{NP}$. Instead of receiving the entire proof and conducting a deterministic polynomial-time computation (as in the
case of $\mathcal{NP}$), the verifier may toss coins and query the proof only at locations of
its choice. Potentially, this allows the verifier to utilize very long proofs (i.e., of
super-polynomial length) or alternatively examine very few bits of an NP-proof.

---

[4]Actually, this is slightly inaccurate because the resulting "interactive proof" may have a
two-sided error, whereas we have required interactive proofs to have only a one-sided error. Yet,
because the error can be made negligible by successive repetitions, this issue is insignificant.
Alternatively, one can use ideas in [13] to eliminate the error by letting the prover send some
random-looking help.

4.1. Definition. Loosely speaking, a probabilistically checkable proof system consists of a probabilistic polynomial-time verifier having access to an oracle that represents a proof in redundant form. Typically, the verifier accesses only a few of the oracle bits, and these bit positions are determined by the outcome of the verifier's coin tosses. Again, it is required that if the assertion holds then the verifier always accepts (i.e., when given access to an adequate oracle); whereas, if the assertion is false then the verifier must reject with probability at least $\frac{1}{2}$, no matter which oracle is used. The basic definition of the PCP setting is given in Item (1) below. Yet, the complexity measures introduced in Items (2) and (3) are of key importance for the subsequent discussions, and should not be ignored.

DEFINITION 3 (Probabilistic Checkable Proofs – PCP):

(1) A probabilistic checkable proof (PCP) system *for a set* S *is a probabilistic polynomial-time oracle machine* (called verifier), *denoted* V, *satisfying*

- Completeness: *For every* $x \in S$ *there exists an oracle set* π_x *so that* V, *on input* x *and access to oracle* π_x, *always accepts* x.

- Soundness: *For every* $x \notin S$ *and every oracle set* π, *machine* V, *on input* x *and access to oracle* π, *rejects* x *with probability at least* $\frac{1}{2}$.

(2) Let r *and* q *be integer functions. The complexity class* $\mathcal{PCP}(r(\cdot), q(\cdot))$ *consists of sets having a probabilistic checkable proof system in which the verifier, on any input of length* n, *makes at most* $r(n)$ *coin tosses and at most* $q(n)$ *oracle queries.*

(3) Let R *and* Q *be sets of functions. Then* $\mathcal{PCP}(R, Q)$ *denotes* $\cup_{r \in R, q \in Q} \mathcal{PCP}(r(\cdot), q(\cdot))$.

The above model was suggested in [12] and shown to be related to a multi-prover model introduced previously in [9]. The fine complexity measures were introduced and motivated in [10], and the notation is due to [3]. A related model was presented in [5], stressing the applicability to program checking.

We stress that the oracle π_x in a PCP system constitutes a proof in the standard mathematical sense.[5] Yet, this oracle has the extra property of enabling a lazy verifier, to toss coins, take its chances, and "assess" the validity of the proof without reading all of it (but rather by reading a tiny portion of it).

4.2. The power of probabilistically checkable proofs. Clearly, $\mathcal{PCP}(\text{poly}, 0)$ equals $\mathcal{BPP}$, whereas $\mathcal{PCP}(0, \text{poly})$ equals $\mathcal{NP}$. It is easy to prove an upper bound on the nondeterministic time complexity of sets in the PCP hierarchy. In particular,

PROPOSITION 3 $\mathcal{PCP}(\text{log}, \text{poly})$ *is contained in* $\mathcal{NP}$.

---

[5]Jumping ahead, the oracles in PCP systems characterizing $\mathcal{NP}$ have the property of being NP-proofs themselves.

These upper bounds turn out to be tight, but proving this is much more difficult (to say the least). The following result is a culmination of a sequence of works [6], [5], [10], [3], [2].[6]

THEOREM 4 [2]: $\mathcal{NP}$ is contained in $\mathcal{PCP}(\log, O(1))$.

Thus, probabilistically checkable proofs in which the verifier tosses only logarithmically many coins and makes only a constant number of queries exist for every set in the complexity class $\mathcal{NP}$. It follows that NP-proofs can be transformed into NP-proofs that offer a trade-off between the portion of the proof being read and the confidence it offers. Specifically, if the verifier is willing to tolerate an error probability of ϵ then it suffices to let it examine $O(\log(1/\epsilon))$ bits of the (transformed) NP-proof. These bit locations need to be selected at random.

The characterization of $\mathcal{NP}$ in terms of probabilistically checkable proofs plays a central role in recent developments concerning the difficulty of approximation problems (cf., [10], [2], and [24]). To demonstrate this relationship, we first note that Theorem 4 can be rephrased without mentioning the class $\mathcal{PCP}$ altogether. Instead, a new type of polynomial-time reductions, which we call *amplifying*, emerges.

THEOREM 5 (Theorem 4 — Rephrased): *There exists a constant $\epsilon > 0$, and a polynomial-time computable function f, mapping the set of 3CNF formulae[7] to itself so that*

- *As usual, f maps satisfiable 3CNF formulae to satisfiable 3CNF formulae; and*

- *f maps nonsatisfiable 3CNF formulae to (nonsatisfiable) 3CNF formulae for which every truth assignment satisfies at most a $1 - \epsilon$ fraction of the clauses.*

The function f is called an amplifying reduction.

Proof sketch (Theorem 4 $\Rightarrow$ Theorem 5): Start by considering the PCP for a satisfiable 3CNF formula (guaranteed by Theorem 4). Use the fact that the PCP system used in the proof of Theorem 4 is nonadaptive[8] (i.e., the queries are determined as a function of the input and the random-tape — and do not depend on answers to previous queries). Next, associate the bits of the oracle with Boolean variables and introduce a (constant size) Boolean formula for each possible outcome of the sequence of $O(\log \cdot)$ coin tosses, describing whether the verifier would have accepted given this outcome. Finally, using auxiliary variables, convert each

---

[6]The sequence has started with the characterization of $\mathcal{PCP}(\texttt{poly}, \texttt{poly})$ as equal nondeterministic exponential-time [6], and continued with its scaled-down version in [5] and [10], which led to the $\mathcal{NP} \subseteq \mathcal{PCP}(\texttt{polylog}, \texttt{polylog})$ result of [10]. The first PCP-characterization of $\mathcal{NP}$, by which $\mathcal{NP} = \mathcal{PCP}(\log, \log)$, has appeared in [3] and the cited result was obtained in [2].

[7]A 3CNF formula is a Boolean formula consisting of a conjunction of clauses, where each clause is a disjunction of up to 3 literals. (A literal is a variable or its negation.)

[8]Actually, this is not essential as one can convert an adaptive system into a nonadaptive one, while incurring an exponential blowup in the query complexity (which in our case is a constant).

of these formulae into a 3CNF formula and obtain (as the output of the reduction) the conjunction of all these polynomially many clauses. $\square$

As an immediate corollary one gets results concerning the intractability of approximation. For example,

COROLLARY 6 *There exists a constant $\epsilon > 0$, so that the following approximation problem* (known as Max3Sat) *is "NP-hard"* (i.e., cannot be solved in polynomial-time unless $\mathcal{P} = \mathcal{NP}$):

> *Given a satisfiable 3CNF formula, find a truth assignment that satisfies at least a $1 - \epsilon$ fraction of its clauses.*

4.3. The role of randomness. No trade-off between the number of bits examined and the confidence is possible if one requires the verifier to be deterministic. In particular, $\mathcal{PCP}(0, q(\cdot))$ contains only sets that are decidable by a deterministic algorithm of running time $2^{q(n)} \cdot \text{poly}(n)$. It follows that $\mathcal{PCP}(0, \log) = \mathcal{P}$. Furthermore, as it is unlikely that all NP-sets can be decided by (deterministic) algorithms of running time, say, $2^{\sqrt{n}} \cdot \text{poly}(n)$, it follows that $\mathcal{PCP}(0, \sqrt{n})$ cannot contain $\mathcal{NP}$.

Acknowledgment: I am grateful to Shafi Goldwasser for suggesting the essential role of randomness as the unifying theme for this exposition.

References

[1] W. Aiello and J. Håstad, *Perfect zero-knowledge languages can be recognized in two rounds*, 28th FOCS, pages 439–448, 1987.

[2] S. Arora, C. Lund, R. Motwani, M. Sudan, and M. Szegedy, *Proof verification and intractability of approximation problems*, 33rd FOCS, pages 14–23, 1992.

[3] S. Arora and S. Safra, *Probabilistic checkable proofs: A new characterization of NP*, In 33rd FOCS, pages 1–13, 1992.

[4] L. Babai, *Trading group theory for randomness*, 17th STOC, pages 421–420, 1985.

[5] L. Babai, L. Fortnow, L. Levin, and M. Szegedy, *Checking computations in poly-logarithmic time*, 23rd STOC, pages 21–31, 1991.

[6] L. Babai, L. Fortnow, and C. Lund, *Non-deterministic exponential time has two-prover interactive protocols*, 31st FOCS, pages 16–25, 1990.

[7] L. Babai and S. Moran, *Arthur-Merlin games: A randomized proof system and a hierarchy of complexity classes*, J. Comput. System Sci., vol. 36, pp. 254–276, 1988.

[8] M. Ben-Or, O. Goldreich, S. Goldwasser, J. Håstad, J. Kilian, S. Micali, and P. Rogaway, *Everything provable is probable in zero-knowledge*, Crypto88, Springer-Verlag, Berlin and New York, Lecture Notes in Comput. Sci. vol. 403, pages 37–56, 1990.

[9] M. Ben-Or, S. Goldwasser, J. Kilian, and A. Wigderson, *Multi-prover interactive proofs: How to remove intractability*, 20th STOC, pages 113–131, 1988.

[10] U. Feige, S. Goldwasser, L. Lovász, S. Safra, and M. Szegedy, *Approximating clique is almost NP-complete*, 32nd FOCS, pages 2–12, 1991.

[11] L. Fortnow, *The complexity of perfect zero-knowledge*, 19th STOC, pages 204–209, 1987.

[12] L. Fortnow, J. Rompel, and M. Sipser, *On the power of multi-prover interactive protocols*, Proc. 3rd IEEE Symp. on Structure in Complexity Theory, pages 156–161, 1988.

[13] M. Furer, O. Goldreich, Y. Mansour, M. Sipser, and S. Zachos, *On Completeness and Soundness in Interactive Proof Systems*, Advances in Computing Research: A Research Annual, vol. 5 (Randomness and Computation, S. Micali, ed.), pp. 429–442, 1989.

[14] M. R. Garey and D. S. Johnson, Computers and Intractability: A Guide to the Theory of NP-Completeness. W. H. Freeman, San Francisco, CA, 1979.

[15] O. Goldreich, S. Micali, and A. Wigderson, *Proofs that yield nothing but their validity or all languages in NP have zero-knowledge proof systems*, J. Assoc. Comput. Math., vol. 38, No. 1, pages 691–729, 1991. Extended abstract in 27th FOCS, 1986.

[16] O. Goldreich, S. Micali, and A. Wigderson, *How to play any mental game or a completeness theorem for protocols with honest majority*, 19th STOC, pages 218–229, 1987.

[17] O. Goldreich and Y. Oren, *Definitions and properties of zero-knowledge proof systems*, J. Cryptology, vol. 7, No. 1, pages 1–32, 1994.

[18] S. Goldwasser and S. Micali, *Probabilistic encryption*, J. Comput. System Sci., vol. 28, No. 2, pages 270–299, 1984. Extended abstract in 14th STOC, 1982.

[19] S. Goldwasser, S. Micali, and C. Rackoff, *The knowledge complexity of interactive proof systems*, SIAM J. Comput., vol. 18, pages 186–208, 1989. Extended abstract in 17th STOC, 1985.

[20] S. Goldwasser and M. Sipser, *Private coins versus public coins in interactive proof systems*, 18th STOC, pages 59–68, 1986.

[21] J. Håstad, R. Impagliazzo, L.A. Levin, and M. Luby, *Construction of pseudorandom generator from any one-way function*, manuscript, 1993. See preliminary versions by Impagliazzo et al. in 21st STOC and Håstad in 22nd STOC.

[22] R. Impagliazzo and M. Yung, *Direct zero-knowledge computations*, Crypto87, Springer-Verlag, Berlin and New York, Lecture Notes in Comput. Sci. vol. 293, pages 40–51, 1987.

[23] C. Lund, L. Fortnow, H. Karloff, and N. Nisan, *Algebraic methods for interactive proof systems*, 31st FOCS, pages 2–10, 1990.

[24] C. Lund and M. Yannakakis, *On the hardness of approximating minimization problems*, 25th STOC, pages 286–293, 1993.

[25] M. Naor, *Bit commitment using pseudorandom generators*, Crypto89, pages 123–132, 1990.

[26] R. Ostrovsky and A. Wigderson, *One-way functions are essential for non-trivial zero-knowledge*, Proc. 2nd Israel Symp. on Theory of Computing and Systems (ISTCS93), IEEE Computer Society Press, pages 3–17, 1993.

[27] C. H. Papadimitriou and M. Yannakakis, *Optimization, approximation, and complexity classes*, 20th STOC, pages 229–234, 1988.

[28] A. Shamir, *IP=PSPACE*, 31st FOCS, pages 11–15, 1990.

[29] A.C. Yao, *Theory and application of trapdoor functions*, 23st FOCS, pages 80–91, 1982.

The Computational Complexity of Counting

MARK JERRUM[*]

Department of Computer Science
University of Edinburgh,
The King's Buildings
Edinburgh EH9 3JZ, United Kingdom

ABSTRACT. The complexity theory of counting contrasts intriguingly with
that of existence or optimization.

1. Counting versus existence

The branch of theoretical computer science known as *computational complexity* is
concerned with quantifying the computational resources required to achieve specified computational goals. Classically, the goal is often to decide the existence of
a certain combinatorial structure, for example, whether a given graph G contains
a Hamilton cycle. Alternatively, the goal might be to find an occurrence of the
structure that is optimal with respect to a certain measure; in the context of the
structure "Hamilton cycle," the notorious *Travelling Salesman Problem* may be
cited as an example. Less well studied, and somewhat less well understood, are
counting problems, such as determining how many Hamilton cycles a graph G contains. In some areas, such as statistical physics, counting problems arise directly;
in many others they appear in the guise of discrete approximations to continuous
problems involving multivariate integration. This article aims to sketch the complexity theory of counting, highlighting the ways in which it diverges from that of
existence or optimization.

Let Σ be an alphabet, possibly the binary alphabet, in which the objects
of interest (e.g., graphs and Hamilton cycles) may be encoded. A *witness-testing
predicate* for some combinatorial structure $\mathcal{S}$ is a predicate $\psi : \Sigma^* \times \Sigma^* \to \{0, 1\}$,
where the truth of $\psi(x, y)$ is to be interpreted as "y is an occurrence of structure $\mathcal{S}$
within instance x." Specializing to the structure "Hamilton cycle," $\psi(x, y)$ would
be true precisely if the words x and y encode (respectively) a graph G and a
subgraph H of G, and H is a Hamilton cycle in G. The existence predicate $\phi(x)$
for the structure $\mathcal{S}$ may be expressed as

$$\phi(x) \Leftrightarrow \exists y \in \Sigma^* \left[|y| = p(|x|) \wedge \psi(x, y) \right], \tag{1}$$

---
[*]The author is a Nuffield Foundation Science Research Fellow, and is supported in part by
grant GR/F 90363 of the UK Science and Engineering Research Council, and by Esprit Working
Group No. 7097, "RAND."

Proceedings of the International Congress
of Mathematicians, Zürich, Switzerland 1994
© Birkhäuser Verlag, Basel, Switzerland 1995

where p is a polynomial that depends on the exact encodings used. In our running example, ϕ is the "Hamiltonicity predicate," so that $\phi(x)$ expresses the situation that x encodes a Hamiltonian graph. For convenience, we abbreviate the relationship (1) between ϕ and ψ to $\phi(x) \Leftrightarrow \exists^p y \, \psi(x, y)$.

In general, the witness-testing predicate $\psi(x, y)$ belongs to the class P of polynomial-time predicates, i.e., those that are computable in time polynomial in the lengths of their arguments. (Observe, for example, that it is easy to determine whether a given subgraph H of a graph G is a Hamilton cycle of G, and that the same is true of most other combinatorial structures we might consider.) The class NP (which might more suggestively be denoted $\exists$P) contains precisely those predicates ϕ that may be derived from polynomial-time two-place predicates ψ via (1). It is clear that if $\phi \in$ NP then $\phi(x)$ may be decided in time exponential in $|x|$ by an exhaustive search over all words $y \in \Sigma^*$ of length polynomial in $|x|$. The key issue from a complexity-theoretic viewpoint is whether $\phi(x)$ may be decided in much less than exponential time, in particular, whether $\phi \in$ P.

It is important to note that the *complexity classes* P and NP (and indeed all the others we shall encounter in this article) are invariant under "reasonable" variation in the computational model and in the choice of encodings. Thus, it is of no great significance whether computation is modelled by a Turing machine or something more akin to existing computers, nor whether graphs are encoded as adjacency matrices or as incidence lists. A polynomial-time computation will remain polynomial-time, even though the degree of the polynomial may change.

It is clear that P $\subseteq$ NP, but strictness of the inclusion remains a major open question. Under the assumption P $\subset$ NP — which is widely conjectured to be the case — it is possible to exhibit "natural" predicates in NP $-$ P using the notion of polynomial-time reducibility. A predicate ϕ' is said to be *polynomial-time (many-one or Cook-Karp) reducible to* ϕ if there exists a polynomial-time-computable function $g : \Sigma^* \to \Sigma^*$ satisfying $\phi'(x) \Leftrightarrow \phi(g(x))$. Many natural predicates $\phi \in$ NP, including Hamiltonicity, have the property that every predicate in NP is polynomial-time reducible to ϕ, and such predicates are said to be *NP-complete*. It is an easy consequence of the definitions that no NP-complete predicate can be in P, unless P $=$ NP. Thus, NP-completeness of a predicate such as Hamiltonicity can be regarded as evidence of computational intractability.

The framework just described extends smoothly to counting problems. The counting function $f : \Sigma^* \to \mathbb{N}$ associated with witness-testing predicate ψ is

$$f(x) = \left| \left\{ y \in \Sigma^* : |y| = p(|x|) \wedge \psi(x, y) \right\} \right|; \tag{2}$$

for convenience, we abbreviate this relationship to $f(x) = \#^p y \, \psi(x, y)$. The class #P, the counting analogue of NP, contains precisely those functions f that can be derived from polynomial-time two-place predicates ψ via (2). As with NP, the class #P contains complete problems, which are computationally the hardest problems in #P relative to polynomial-time reducibility. However, to obtain a rich class of #P-complete problems it is necessary to employ a somewhat more liberal notion of reducibility known as Turing reducibility. A function f' is *Turing reducible to* f if there is a polynomial-time algorithm that computes f' given an oracle for f.[1] A function $f \in \#P$ is #P-*complete* if every function in #P is Turing reducible to f.

It is clear that a procedure for counting witnesses must in particular decide between no witnesses and some, and hence counting is computationally at least as hard as deciding existence. The first evidence that counting can be harder than deciding existence for significant natural problems was provided by Valiant [18], who exhibited a natural witness-checking predicate for which the associated counting function is #P-complete, even though the associated existence predicate is in P. A *perfect matching* in a graph G is a subset M of the edge set of G such that every vertex of G is the endpoint of precisely one edge in M. Valiant showed that counting the number of perfect matchings in a bipartite graph (equivalently, evaluating the permanent of a 0,1-matrix) is #P-complete and hence likely to be computationally intractable, whereas deciding whether a bipartite graph contains a perfect matching (equivalently, deciding whether the permanent of a 0,1-matrix is nonzero) is in P, by virtue of the classical "augmenting path" algorithm.

From Valiant's initial collection [19], the catalogue of #P-complete counting problems has grown steadily. A significant recent contribution is Brightwell and Winkler's proof [2] that counting linear extensions of a partial order is #P-complete, settling a decade-old question. Perhaps the most extensive investigation in this direction was initiated by Jaeger, Vertigan, and Welsh [6], who considered the computational complexity of evaluating the two-variable Tutte polynomial $T(M; x, y)$ of a matroid M, at all points and along significant curves in the x, y-plane. An astonishing variety of counting problems connected with matroids in general, and graphs in particular, may by viewed as evaluations of the Tutte polynomial at certain points and along certain curves. Many complexity-theoretic results may be read off from the classification developed in [6] and succeeding articles.

Empirically it appears to be the case that counting occurrences of a combinatorial structure is #P-complete whenever deciding existence of the structure is NP-complete. This apparent entailment may be formalized, leading to the following precise open question. Let ψ be any polynomial-time witness-testing predicate; is it the case that the function $f(x) = \#^p y\, \psi(x, y)$ is #P-complete whenever the predicate $\phi(x) = \exists^p y\, \psi(x, y)$ is NP-complete? There seems to be no good reason why the answer should be yes in general, even though it is for all "natural" witness-checking predicates ψ that have been considered. Note that Valiant's result [18] assures us that the converse cannot hold unless P = NP.

2. Structural considerations

From the point of view of "structural complexity" it is natural to consider where the counting class #P lies in relation to other complexity classes. Certain relationships, for example that #P is contained in the class FPSPACE of functions computable in polynomial space, follow immediately from the definitions. A far from obvious and more enlightening containment was discovered about five years

---

[1]An *oracle* for a function f is a black box that takes as input a word $x \in \Sigma^*$ and in one time-step produces as output $f(x)$.

ago. Let us first extend the notation of Section 1. in a natural way. For $\mathcal{C}$ a class of predicates, let $\exists\mathcal{C}$ denote the derived class obtained by the following construction. A k-place predicate ϕ is contained in $\exists\mathcal{C}$ if there is a $(k+1)$-place predicate $\psi \in \mathcal{C}$ and a polynomial p such that

$$\phi(x_1,\ldots,x_k) \Leftrightarrow \exists y \in \Sigma^* \left[|y| = p(|x_1|,\ldots,|x_k|) \wedge \psi(x_1,\ldots,x_k,y) \right].$$

The derived class $\forall\mathcal{C}$ is defined analogously with $\forall$ replacing $\exists$. The *polynomial hierarchy* PH of Meyer and Stockmeyer is the class

$$\text{PH} = \text{P} \cup \exists\text{P} \cup \forall\exists\text{P} \cup \exists\forall\exists\text{P} \cup \cdots \tag{3}$$

of predicates obtained from the base class P by finite quantification. (Note the analogy with Kleene's arithmetic hierarchy from classical recursion theory.) At the base of the hierarchy we find the classes P and NP $= \exists$P that were encountered in the previous section. Toda [17] has shown the following.

THEOREM 1 *Every predicate in* PH *is polynomial-time Turing reducible to a function in* #P.

In other words, the class #P essentially contains the entire polynomial hierarchy. The power of the counting operator # to simulate arbitrary finite alternations of the operators $\exists$ and $\forall$ is, I think, surprising.

To give the flavor of the proof of Theorem 1, which relies on a subtle "quantifier swapping" argument, it is necessary to define two further operators of independent interest, $\oplus$ and B. A k-place predicate ϕ is contained in $\oplus\mathcal{C}$ if there is a $(k+1)$-place predicate $\psi \in \mathcal{C}$ and a polynomial p such that the number of words $y \in \Sigma^*$ of length $p(|x_1|,\ldots,|x_k|)$ satisfying $\psi(x_1,\ldots,x_k,y)$ is odd. A k-place predicate ϕ is contained in B$\mathcal{C}$ if there is a $(k+1)$-place predicate $\psi \in \mathcal{C}$, a polynomial p, and a number $\alpha > 0$, such that $\Pr\left(\phi(x_1,\ldots,x_k) \Leftrightarrow \psi(x_1,\ldots,x_k,y)\right) \geq \frac{1}{2} + \alpha$, where $y \in \Sigma^*$ is selected uniformly at random (u.a.r.) from words of length $p(|x_1|,\ldots,|x_k|)$.

Toda showed that for predicate classes $\mathcal{C}$ satisfying a certain technical condition, which is satisfied by the classes forming the polynomial hierarchy (3),

(i) $\exists\mathcal{C} \subseteq \text{B}\oplus\mathcal{C}$,   (ii) $\oplus\text{B}\oplus\text{P} \subseteq \text{B}\oplus\text{P}$,   and   (iii) $\text{BB}\oplus\text{P} \subseteq \text{B}\oplus\text{P}$.

Inclusions (i) and (ii) are nontrivial, decidedly so in the case of (i), which rests on a universal hashing technique of Valiant and Vazirani [20]. If $\mathcal{C} \subseteq \text{B}\oplus\text{P}$ satisfies the technical condition, then

$$\exists\mathcal{C} \subseteq \text{B}\oplus\mathcal{C} \subseteq \text{B}\oplus\text{B}\oplus\text{P} \subseteq \text{BB}\oplus\text{P} \subseteq \text{B}\oplus\text{P},$$

using containments (i)–(iii). Thus, by induction on quantifier depth, PH $\subseteq$ B$\oplus$P. The proof is completed by demonstrating that every predicate in the class B$\oplus$P is polynomial-time Turing reducible to a function in #P.

3.   Random instances

Under the assumption P $\neq$ NP, there can be no polynomial-time algorithm for deciding Hamiltonicity or any other NP-complete predicate. This is a "worst-case" result: for any NP-complete predicate ϕ, and algorithm $\mathcal{A}$ for deciding ϕ, there exist instances $x \in \Sigma^*$ that cause $\mathcal{A}$ to run for superpolynomially many steps. However, these difficult instances may be sparse, so there is no reason to suppose that $\mathcal{A}$ cannot decide $\phi(x)$ quickly on most inputs. The same remarks hold good for #P-complete functions.

Empirically, most NP-complete predicates $\phi(x)$ seem easy to decide when the instance x is chosen at random, provided the probability distribution on instances is fairly natural. For example, deciding Hamiltonicity of a random graph seems in practice to be easy, whichever of the established random graph models is used; indeed, good results may be obtained using fairly simple heuristics.[2] In contrast, again from an empirical standpoint, counting seems to be hard for random instances. Suppose, for example, we are presented with an n-vertex graph G and asked to compute the number of Hamilton cycles in G. Whether the graph G is provided by an adversary or is selected u.a.r. seems to make no difference to the computational difficulty of the task. We can gain an understanding of why this is so by considering a phenomenon that has become known as "random self-reducibility." Following Lipton [13], we find it convenient to investigate this phenomenon in the context of the permanent function.

Let $A = (a_{ij} : 0 \leq i, j \leq n - 1)$ be an $(n \times n)$-matrix with entries in some ring. The *permanent* of A is defined as

$$\operatorname{per} A = \sum_{\pi} \prod_{i=0}^{n-1} a_{i,\pi(i)},$$

where the sum is over all permutations π of $\{0, \ldots, n - 1\}$. As noted in Section 1, the permanent is a #P-complete function, even when restricted to 0,1-matrices. Our aim is to show that evaluating the permanent of a random matrix R is as hard as evaluating the permanent of an arbitrary matrix. The first notion of "random matrix" that comes to mind is that of a random 0,1-matrix, because per R then has a natural combinatorial interpretation, namely, the number of perfect matchings in the bipartite graph with adjacency matrix R. However, the argument we present requires that the matrix elements take on a wider range of values, and we accordingly work with matrices over the finite field GF(p), where p is a prime not less than $n + 2$.

Let A be an arbitrary $(n \times n)$-matrix over GF(p), which we may imagine to be chosen by some adversary. Our aim is to reduce the computation of per A to the computation of the permanents of a number of random matrices. Choose a random $(n \times n)$-matrix R, i.e, one with each entry selected independently, and u.a.r. from GF(p). The function $f(z) = \operatorname{per}(A + Rz)$ is a polynomial of degree n

---

[2]This rather sweeping statement deserves more substantial qualification than space here allows. Suffice it to say that the task of exhibiting algorithms that are provably efficient for almost every instance presents a considerable challenge.

in the indeterminate z. If we knew the $n + 1$ values $f(1), f(2), \ldots, f(n+1)$ then we could interpolate to find $f(0) = \operatorname{per} A$, which is precisely the quantity we wish to compute. But $f(i) = \operatorname{per}(A + iR)$, and $A + iR$ is a random matrix, provided $i \not\equiv 0 \pmod{p}$. It follows that the existence of a polynomial-time algorithm that correctly computes $\operatorname{per} A$ for all but a fraction $1/3(n+1)$ of $(n \times n)$-matrices A over $\mathrm{GF}(p)$ would imply the existence of a randomized polynomial-time algorithm that computes the permanent correctly, with high probability, on all inputs.

Using the theory of error correcting codes, it is possible to interpolate polynomials reliably from "noisy" data. This fact — which is implicit in a decoding procedure due to Berlekamp and Welch — allows the fraction $1/3(n+1)$ mentioned above to be increased to $\frac{1}{2} - \delta$ for any $\delta > 0$, which is clearly as much as can be achieved using a simple-minded application of random self-reducibility. Using more sophisticated techniques, Feige and Lund [5] show that the existence of a polynomial-time algorithm that correctly computes the permanent of even a tiny proportion of all matrices would have a surprising consequence: that the polynomial hierarchy collapses to the second level. In the notation introduced in Section 2.:

THEOREM 2 *If there exists a polynomial-time algorithm that correctly computes the permanent on a fraction $100n^3/\sqrt{p}$ of all $(n \times n)$ matrices over $\mathrm{GF}(p)$, then $\exists \forall \mathrm{P} = \mathrm{PH}$.*

Note, however, that to obtain a dramatic effect it is necessary to work over rather large finite fields. Their proof exploits ideas from the theory of "interactive proofs."

Existing results concerning hardness of counting problems on random instances rely on polynomial interpolation, which can only work over a sufficiently large field. Intuitively, evaluating the permanent of a random 0,1-matrix ought to be as hard as evaluating the permanent of an arbitrary 0,1-matrix, and the same ought to be true of counting Hamilton cycles in a random graph, or of counting many other combinatorial structures in random graphs. (Note that these problems are "essentially combinatorial," in contrast to that of permanent evaluation over a finite field, which has an algebraic flavor.) There is no complexity-theoretic evidence that the cited problems are hard with respect to the uniform probability distribution on instances, and remedying this deficiency would be a significant advance.

4. Approximation algorithms

The familiar dichotomy within the class NP, namely P versus NP-complete, has a counterpart in the complexity class #P: almost all the "natural" functions in #P that have been considered have been shown either to be polynomial-time computable or to be #P-complete. Unfortunately, very few counting problems fall into the former category, notable examples being (a) spanning trees in a graph, which can be counted via Kirchhoff's matrix-tree theorem, (b) perfect matchings in a planar graph (including dimer coverings of a two-dimensional lattice as a special case), the solution of which is a classical result of Kasteleyn [12], and (c) a few other structures — such as Eulerian circuits in a directed graph or Ising configurations in a planar lattice — that can be handled by fairly direct reduction to either (a) or (b). Valiant has made the intriguing observation that all the cited algorithms ultimately rely on linear algebra, in particular, the well-known fact that the determinant of a matrix may be computed in polynomial time by Gaussian elimination.

Given the paucity of positive results, it is tempting to weaken the requirements somewhat, and consider whether *approximate* counting is feasible; for example, whether it is possible, in polynomial time, to evaluate the permanent of a 0,1-matrix with relative error at most 1%. The reader who has followed the recent dramatic advances in the field of approximation algorithms for combinatorial optimization will be aware of the emerging classification of optimization problems according to degrees of approximability: problems, such as chromatic number of an n-vertex graph, that cannot be approximated within ratio n^α for some $\alpha > 0$ unless P = NP; problems, such as "MAXSAT," that can be approximated within ratio α but not within ratio β, for some $0 < \alpha < \beta < 1$, and so on. In contrast, Sinclair and the author [16] have demonstrated that no correspondingly rich classification can be expected in the domain of approximate counting. A witness-checking predicate ψ is (downward) self-reducible if, roughly, the set $\{y : \psi(x,y)\}$ has a simple recursive expression in terms of some collection of sets $\{y : \psi(x_1,y)\}, \ldots, \{y : \psi(x_k,y)\}$ with $|x_i| < |x|$ for all $1 \leq i \leq k$. Almost all naturally occurring predicates are self-reducible in this sense.

THEOREM 3 *Suppose ψ is a self-reducible predicate, and let A be the set of $\alpha \in \mathbb{R}$ for which there exist $\delta > 0$ and a polynomial-time randomized algorithm that approximates the function $f(x) = \#^p y\, \psi(x,y)$ within ratio $1 + |x|^\alpha$, with probability $\frac{1}{2} + \delta$. Then A is either empty or the whole of $\mathbb{R}$.*

In the absence of differing degrees of approximability, a single notion of efficient approximation algorithm has assumed considerable importance. Suppose $f : \Sigma^* \to \mathbb{N}$ is a function mapping problem instances to natural numbers. A *randomized approximation scheme* for f is a randomized algorithm that takes as input a word (instance) $x \in \Sigma^*$ and $\varepsilon > 0$, and produces as output a number Y (a random variable) such that $\Pr\big((1 - \varepsilon)f(x) \leq Y \leq (1 + \varepsilon)f(x)\big) \geq \frac{1}{2} + \delta$, for some $\delta > 0$.[3] A randomized approximation scheme is said to be *fully polynomial* if it runs in time polynomial in $|x|$ and ε^{-1}. The rather unwieldy phrase "fully polynomial randomized approximation scheme" is often abbreviated to *fpras*.

The above definition is due to Karp and Luby, who presented a simple but elegant fpras for counting satisfying assignments to a Boolean formula in DNF, a #P-complete problem. Unfortunately, the classical Monte Carlo approach adopted by Karp and Luby appears to have only limited application in this area. An alternative approach that has proved very fruitful exploits the close relationship that exists between approximate counting and (almost) uniform sampling. Subject to self-reducibility, the existence of an efficient (i.e., polynomial-time) almost uniform

---

[3] Any success probability strictly greater than $\frac{1}{2}$ may be boosted to a value arbitrarily close to 1 by making a small number of trials and taking the median of the results [10].

sampling procedure for a certain combinatorial structure entails the existence of an efficient approximate counting procedure (i.e., fpras) for that structure, and vice versa. This connection was investigated by Valiant, Vazirani, and the author [10], and further elucidated in [16].

For the sake of definiteness, let us suppose that the goal is to estimate the permanent of a 0,1-matrix. We have noted that this task is equivalent to estimating the number of perfect matchings in a bipartite graph G, and hence to sampling perfect matchings in G almost uniformly. Broder [3] suggested the following approach to sampling matchings. Let M denote the set of all perfect matchings in G and M^- the set of "near-perfect" matchings that leave exactly two vertices uncovered. Construct a Markov chain (X_t) whose state space $\Omega = M \cup M^-$ consists of all perfect and near-perfect matchings. A transition from state $X_t \in \Omega$ to $X_{t+1} \in \Omega$ is possible if $X_t \oplus X_{t+1}$ — the symmetric difference of X_t and X_{t+1} — contains at most two edges; all allowed transitions occur with equal probability.

It can be shown that the Markov chain (X_t) is ergodic and has uniform stationary distribution. This observation suggests the following sampling procedure for perfect matchings: simulate the Markov chain from a fixed initial state X_0 for T steps; return as result the final state X_T if $X_T \in M$, otherwise repeat the simulation. If T is chosen large enough, the result of this procedure is a perfect matching chosen almost u.a.r. For this "Markov chain Monte Carlo" sampling procedure to be efficient, two conditions must be satisfied: (a) the ratio $|M|/|\Omega|$ must be nonnegligible, and (b) the Markov chain (X_t) must be "rapidly mixing," that is to say, must be close to stationarity after a number of steps polynomial in n, the size of G. It is in verifying condition (b) that the work lies.

The classical approach to establishing that a Markov chain is rapidly mixing is via a "coupling argument." Broder attempted to use a coupling argument to prove that his perfect matching Markov chain is rapidly mixing; however, the technique appears ill suited to the irregular Markov chains that typically arise in this context. A modest exception to this general rule is an algorithm of the author's for estimating the number of colorings of a low-degree graph [7], the correctness of which rests on a surprisingly simple coupling argument.

The failure of coupling to produce results has spurred the development of two new approaches to analyzing rates of convergence of ergodic Markov chains. The first of these is the "canonical paths" argument, which was introduced by Sinclair and the author, and used by them to analyze Broder's perfect matching chain [8]. The canonical paths argument provides a lower bound on a quantity known as the *conductance* of the Markov chain, which is a weighted version of the familiar notion of expansion of a graph. Sinclair and the author [16] showed that conductance and mixing time (roughly, number of steps until the distribution on states is close to stationarity) are related, so that large conductance entails a short mixing time; similar results were obtained independently by Alon and Aldous, and can be regarded as discrete analogues of Cheeger's inequality.

The idea is to specify, for every pair of states $I, F \in \Omega$, a *canonical path*, starting at I and ending at F, and composed of valid transitions in the Markov chain. If the canonical paths can be chosen so that the total number of paths using any single transition is relatively small, then the conductance of the Markov chain must be relatively large, and the mixing time correspondingly short. This line of argument has been used to demonstrate the existence of an fpras for the permanent of a wide class of 0,1-matrices, including all matrices that are sufficiently dense [8], and for the partition function of a ferromagnetic Ising system [9].

The second approach is geometric in nature. The idea is to assign a geometric interpretation to the state space Ω, from which a lower bound on conductance can be read off using (enhancements of) classical isoperimetric inequalities. This was the approach used by Dyer, Frieze, and Kannan [4] in their seminal work on estimating the volume of a convex body in high-dimensional Euclidean space, which has been variously refined by the original authors and by Lovász and Simonovits [14], and extended to the integration of log-concave functions over convex regions by Applegate and Kannan [1]. The approach was also used by Karzanov and Khachiyan [11] to validate an fpras for the number of linear extensions of a partial order.

It would be interesting to know which #P-complete functions can be approximated in the fpras sense, and which cannot. As we have seen, definite progress has been made in the last few years, at least on the positive side of the classification. However, many natural #P-complete functions have so far resisted analysis. There seem to be difficulties on both sides. On the one hand, several problems seem amenable to the Markov chain Monte Carlo method, but the Markov chains that have been proposed for their solution are not provably rapidly mixing. Further progress may be predicated on the development of new techniques for bounding the mixing time of irregular Markov chains.

On the other hand, negative results are hard to come by, even though there are no doubt many natural #P-complete problems for which no fpras exists. Current techniques for establishing negative results are primitive. It is easy to show that if deciding existence of some combinatorial structure is NP-complete, then there can be no fpras for the number of occurrences of that structure unless RP = NP.[4] Furthermore, a simple argument [15, Theorem 1.17] suffices to rule out the existence of an fpras for certain problems — including counting independent sets of all sizes in graph — again unless RP = NP. However, the current techniques seem to fall well short of resolving whether (for example) there is an fpras for the permanent of an *arbitrary* 0,1-matrix.

Acknowledgments. I thank Sigal Ar, Alistair Sinclair, and Leslie Valiant for comments on a preliminary draft of this article.

References

[1] David Applegate and Ravi Kannan, *Sampling and integration of near log-concave functions*, Proceedings of the 23rd ACM Symposium on Theory of Computing, ACM Press, New York 1991, 156–163.

[2] Graham Brightwell and Peter Winkler, *Counting linear extensions*, Order **8** (1991), 225–242.

---

[4]RP is is the class of predicates that can be decided by a polynomial-time *randomized* algorithm.

[3] Andrei Z. Broder, *How hard is it to marry at random? (On the approximation of the permanent)*, Proceedings of the 18th ACM Symposium on Theory of Computing, ACM Press, New York 1986, 50–58. Erratum in Proceedings of the 20th ACM Symposium on Theory of Computing, 1988.

[4] M. Dyer, A. Frieze, and R. Kannan, *A random polynomial time algorithm for approximating the volume of convex bodies*, Journal of the J. Assoc. Comput. Math. **38** (1991), 1–17.

[5] Uriel Feige and Carsten Lund, *On the hardness of computing the permanent of random matrices*, Proceedings of the 24th ACM Symposium on Theory of Computing, ACM Press, New York 1992, 643–654.

[6] F. Jaeger, D. L. Vertigan, and D. J. A. Welsh, *On the computational complexity of the Jones and Tutte polynomials*, Math. Proc. Cambridge Philos. Soc. **108** (1990), 35–53.

[7] Mark Jerrum, *A very simple algorithm for estimating the number of k-colourings of a low-degree graph*, Report ECS-LFCS-94-290, Department of Computer Science, University of Edinburgh, April 1994.

[8] Mark Jerrum and Alistair Sinclair, *Approximating the permanent*, SIAM J. Comput. **18** (1989), 1149–1178.

[9] Mark Jerrum and Alistair Sinclair, *Polynomial-time approximation algorithms for the Ising model*, SIAM J. Comput. **22** (1993), 1087–1116.

[10] Mark R. Jerrum, Leslie G. Valiant, and Vijay V. Vazirani, *Random generation of combinatorial structures from a uniform distribution*, Theoret. Comput. Sci. **43** (1986), 169–188.

[11] A. Karzanov and L. Khachiyan, *On the conductance of order Markov chains*, Technical Report DCS 268, Rutgers University, June 1990.

[12] P. W. Kasteleyn, *Dimer statistics and phase transitions*, J. Math. Phys. **4** (1963), 287–293.

[13] Richard J. Lipton, *New directions in testing*, Distributed Computing and Cryptography (J. Feigenbaum and M. Merritt, eds), DIMACS Series in Discrete Mathematics and Theoretical Computer Science **2**, Amer. Math. Soc., Providence RI, 1991, 191–202.

[14] L. Lovász and M. Simonovits, *Random walks in a convex body and an improved volume algorithm*, Random Structures and Algorithms, to appear.

[15] Alistair Sinclair, Algorithms for random generation and counting: A Markov chain approach, Progr. Theoret. Comput. Sci., Birkhäuser, Basel and Boston, MA, 1993.

[16] Alistair Sinclair and Mark Jerrum, *Approximate counting, uniform generation and rapidly mixing Markov chains*, Inform. and Comput. **82** (1989), 93–133.

[17] Seinosuke Toda, *On the computational power of PP and ⊕P*, Proceedings of the 30th Annual IEEE Symposium on Foundations of Computer Science, Computer Society Press (1989), 514–519.

[18] L. G. Valiant, *The complexity of computing the permanent*, Theoret. Comput. Sci. **8** (1979), 189–201.

[19] L. G. Valiant, *The complexity of enumeration and reliability problems*, SIAM J. Comput. **8** (1979), 410–421.

[20] L. G. Valiant and V. V. Vazirani, *NP is as easy as detecting unique solutions*, Theoret. Comput. Sci. **47** (1986), 85–93.

Methods for Message Routing in Parallel Machines

Tom Leighton

Mathematics Department and Laboratory for Computer Science
Massachusetts Institute of Technology
Cambridge, MA 02139, USA

Abstract

The problem of getting the right data to the right place within a reasonable amount of time is one of the most challenging and important tasks facing the designer (and, in some cases, the user) of a large-scale general-purpose parallel machine. This is because the processors comprising a parallel machine need to communicate with each other (or with a common shared memory) in a tightly constrained fashion in order to solve most problems of interest in a timely fashion. Supporting this communication is often an expensive task, in terms of both hardware and time. In fact, most parallel machines devote a significant portion of their resources to handling communication between the processors and the memory.

In the talk, we surveyed several of the ideas and approaches that have been proposed for solving communication problems in parallel machines. Particular emphasis was placed on recent work involving randomly wired networks (known as multibutterflies). Results were presented that indicate that randomly wired networks significantly outperform traditional networks such as the butterfly in terms of both speed and fault tolerance.

References

[1] F. T. Leighton, *Methods for message routing in parallel machines*, Theoret. Comput. Sci. **128** (1994), 31–62.

[2] T. Leighton and B. Maggs, *The role of randomness in the design of interconnection networks*, in: Algorithms, Software, Architecture (J. van Leeuwen, ed.), Information Processing 92, Vol. I, Elsevier Science Publishers, BV, 1992.

Proceedings of the International Congress
of Mathematicians, Zürich, Switzerland 1994
© Birkhäuser Verlag, Basel, Switzerland 1995

Randomness and Nondeterminism

LEONID A. LEVIN*

Computer Science Department, Boston University
111 Cummington Street, Boston, MA 02215, USA

Exponentiation makes the difference between the bit size of this line and the number ($\ll 2^{300}$) of particles in the known universe. The expulsion of exponential time algorithms from computer theory in the 1960s created a deep gap between deterministic computation and — formerly its unremarkable tools — randomness and nondeterminism. These two "freedoms" of computation preserved their reputation as some of the most mysterious phenomena in science and seem to play an ever more noticeable role in computer theory. We have learned little in the past decades about the power of either, but a vague pattern is emerging in their relationships.

A nondeterministic task is to invert an easy computable function. It is widely believed to require, in general, exponential computing time. Great efforts, however, have failed to prove the existence of such one-way (easy to compute, infeasible to invert) functions (owf). Randomness too preserves its mysterious reputation: it seems now, it can be generated deterministically. A simple program can transform a short random "seed" into an unlimited array of random bits. As long as the seed is not disclosed, no computational power that may fit in the whole universe can distinguish this array from truly random ones. This conjecture is proven equivalent to the existence of owf. I will survey a number of papers that have resulted in this theorem.

Although fundamental in many areas of science, randomness is really "native" to computer science. Its computational nature was clarified by Kolmogorov. He and his followers (see survey [Kolmogorov Uspenskii 1987]) built in the 1960s–1970s the first successful theory of random objects, defined roughly as those that cannot be computed from short descriptions. Kolmogorov also suggested in the 1960s that randomness may have an important relationship with nondeterminism; namely, that the task of finding a "nonrandomness" witness (i.e. short fast program generating a given string) may be a good candidate to prove that exhaustive search cannot be avoided (in today's terms, P<NP).

The next step came from cryptography of the 1980s. Many applications require only few properties of random strings. So, they do not care about the fundamental issues of randomness, as long as some laws (e.g., of big numbers) hold. Cryptographers, in contrast, need to deprive the adversary from taking advantage of *any* regularities in their random strings, no matter how peculiar, making "perfect" randomness an important practical concern. [Blum Micali], [Yao][1] used

---

*Supported by NSF grant CCR-9015276. The author's e-mail address: Lnd@bu.edu.
[1]See some other details also in [Levin-ow]

the idea of a *hard core* or *hidden* bit. Suppose from a length-preserving one-way function $f(x)$ it is hard to compute not only x but even its one bit $b(x)$ (easily computable from x). Assume that even guessing $b(x)$ with any noticeable correlation is infeasible. If f is bijective, $f(x)$ and $b(x)$ are both random and *appear to be* independent to any feasible test, thus increasing the initial amount $|x|$ of randomness by one bit. Then, a short random seed x can be transformed into an arbitrary long string $b(f^{(i)}(x))$, $i = 1, 2, \ldots$, which passes any feasible randomness test. [Blum Micali] established such a hard-core b assuming a particular function (discrete log) to be one way. But if its inversion turns out to be feasible, all is lost. [Yao] uses a more general one-way candidate. It is less likely to be technically in P, but is, nevertheless, easy to invert for reasonable $|x|$ (it breaks input into small pieces and applies an owf to each of them).

In [Goldreich Levin] and [Levin] it is proven that a boolean inner product provides every owf f with a hidden bit b of the same security. Actually, the (very efficient) construction used in [Levin] is quite simple and can be understood by non-experts. It yields "perfect" pseudo-random generators from any one-way bijection. The bijection requirement was lifted in [Hastad Impagliazzo Levin Luby]. It showed that the possibility of deterministic generation of pseudo-randomness is exactly equivalent to the existence of owf. Thus, Kolmogorov's intuition that nondeterministic phenomena are intimately related with randomness proved to be accurate.

References

[Alexi Chor Goldreich Schnorr] W. Alexi, B. Chor, O. Goldreich and C.P. Schnorr, *RSA and Rabin functions: Certain parts are as hard as the whole*, SIAM J. Comput. **17** (1988), 194–209.

[BFLS] L. Babai, L. Fortnow, L. Levin, and M. Szegedy, *Checking computations in polylogarithmic time*, ACM Sympos. on Theory of Computing (1991), 21–31.

[Blum Micali] M. Blum, S. Micali, *How to generate cryptographically strong sequences of pseudo-random bits*, SIAM J. Comput. **13** (1984), 850–864.

[GGM] O. Goldreich, S. Goldwasser, S. Micali, *How to construct random functions*, FOCS-84, J. ACM **33/4** (1986), 792–807.

[Goldreich Levin] O. Goldreich, L. Levin, *A hard-core predicate for all one-way functions*, ACM Sympos. on Theory of Computing (1989), 25–32.

[Hastad Impagliazzo Levin Luby] J. Hastad, R. Impagliazzo, L. Levin, and Michael Luby, *Construction of a pseudo-random generator from any one-way function*, Internat. Comp. Sci. Inst. (Berkeley). Tech. Rep. 91-068, (12/1991), 1–36; to appear in SICOMP.

[Kolmogorov Uspenskii] A.N. Kolmogorov, and V.A. Uspenskii, *Algorithms and Randomness*, Theor. Veroyatnost. i Primenen. **3** (32) (1987), 389–412.

[Levin-ow] L. Levin, *One-way functions and pseudorandom generators*, Combinatorica **7** (4) (1987), 357–363.

[Levin] L. Levin, *Randomness and nondeterminism*, J. of Symb. Logic **58** (3) (1993), 1102–1103.

[Vazirani] U. Vazirani, *Efficiency considerations in using semi-random sources*, ACM Sympos. on Theory of Computing (1987), 160–168.

[Yao] A.C. Yao, *Theory and applications of trapdoor functions*, in: Proc. of IEEE Symp. on Foundations of Computer Sci. (1982), 80–91.

Mathematical Modeling and Numerical Analysis of Linearly Elastic Shells

Philippe G. Ciarlet

Université Pierre et Marie Curie
Analyse Numérique
4 Place Jussieu, F-75005 Paris
France

1 Introduction

Shells and their assemblages are found in a wide variety of elastic structures: cooling towers, aircrafts, car bodies, sails, hulls of vessels, etc. These shells are, and therefore should be *in principle* studied as, *three-dimensional* bodies. The *"small" thickness* of a shell makes it however natural to "replace" the genuine three-dimensional model by a simpler *two-dimensional* model, i.e., one that is posed over the middle surface of the shell.

Not only is this replacement intuitively "natural", but it becomes a *necessity* when *numerical methods* must be devised for computing approximate displacements and stresses. Any reasonably accurate *three-dimensional* discretization necessarily involves an outstandingly large number of unknowns, which renders it prohibitively expensive, and at any rate its implementation requires extreme care. By contrast, the situation is on much safer ground as regards the application of approximate methods to *two-dimensional shell models* [2], although there remain challenging problems, such as that of efficiently coping with the onset of *locking* [5].

As two-dimensional shell models are for these reasons by and large preferred, two major, and closely related, questions naturally arise:

How does one derive two-dimensional shell models in a rational manner from three-dimensional elasticity? This first question is one in *asymptotic analysis*; it consists of analyzing the behavior of the three-dimensional solution as the thickness (the "small" parameter) approaches zero.

How does one choose among the various available two-dimensional shell models in a given physical situation, so that the chosen one is indeed a "good" approximation of the three-dimensional model it "replaces"? This second question is of paramount *practical* importance, for it makes no sense to devise methods for accurately approximating the solution of a "wrong" model!

Our purpose is to describe recent progress on these questions. Our discussion will focus in particular on three well-known two-dimensional shell models: the *bending* and *membrane* models, and *Koiter's model*.

Proceedings of the International Congress
of Mathematicians, Zürich, Switzerland 1994
© Birkhäuser Verlag, Basel, Switzerland 1995

2 The three-dimensional problem of a linearly elastic shell

Greek indices and exponents (except ε) belong to $\{1,2\}$; Latin indices and exponents belong to $\{1,2,3\}$; the summation convention is used. The Euclidean norm and inner product, and the vector product of $\boldsymbol{u}, \boldsymbol{v} \in \mathbf{R}^3$ are denoted by $|\boldsymbol{u}|, \boldsymbol{u} \cdot \boldsymbol{v}, \boldsymbol{u} \times \boldsymbol{v}$.

Let ω be a bounded, open, connected subset of $\mathbf{R}^2$, with a Lipschitz-continuous boundary γ, the set ω being locally on one side of γ. We let $y = (y_\alpha)$ denote a generic point in $\overline{\omega}$, $\partial_\alpha = \partial/\partial y_\alpha, \partial_{\alpha\beta} = \partial_\alpha\partial_\beta$, and ∂_ν denote the outer normal derivative along γ. Let $\boldsymbol{\varphi} : \overline{\omega} \to \mathbf{R}^3$ be an injective mapping of class $\mathcal{C}^3$ such that the two vectors $\boldsymbol{a}_\alpha = \partial_\alpha\boldsymbol{\varphi}$ span the tangent plane to the *surface* $S = \boldsymbol{\varphi}(\overline{\omega})$ at all points in $\overline{\omega}$. We also define the vectors $\boldsymbol{a}^\beta$ of the tangent plane by $\boldsymbol{a}^\beta \cdot \boldsymbol{a}_\alpha = \delta_\alpha^\beta$, the normal vector $\boldsymbol{a}^3 = (\boldsymbol{a}_1 \times \boldsymbol{a}_2)/|\boldsymbol{a}_1 \times \boldsymbol{a}_2|$, the *Christoffel symbols* $\Gamma_{\alpha\beta}^\sigma = \boldsymbol{a}^\sigma \cdot \partial_\alpha\boldsymbol{a}_\beta$, the *metric tensor* by its covariant and contravariant components $a_{\alpha\beta} = \boldsymbol{a}_\alpha \cdot \boldsymbol{a}_\beta$ and $a^{\alpha\beta} = \boldsymbol{a}^\alpha \cdot \boldsymbol{a}^\beta$, the *area element* $\sqrt{a}\, dy$ with $a = \det(a_{\alpha\beta})$, the *curvature tensor* by its covariant and mixed components $b_{\alpha\beta} = \boldsymbol{a}^3 \cdot \partial_\alpha\boldsymbol{a}_\beta$ and $b_\alpha^\beta = a^{\beta\sigma}b_{\sigma\alpha}$. Finally, we let $c_{\alpha\beta} = b_\alpha^\sigma b_{\sigma\beta}$.

For each $\varepsilon > 0$, define the sets $\Omega^\varepsilon = \omega \times\,]-\varepsilon, \varepsilon[$ and $\Gamma_0^\varepsilon = \gamma_0 \times\,]-\varepsilon, \varepsilon[$, where $\gamma_0 \subset \gamma$ and *length* $\gamma_0 > 0$, let $x^\varepsilon = (x_i^\varepsilon)$ denote a generic point in $\overline{\Omega}^\varepsilon$ (hence $x_\alpha^\varepsilon = y_\alpha$), and let $\partial_i^\varepsilon = \partial/\partial x_i^\varepsilon$. For each $\varepsilon > 0$, we consider an *elastic shell* with *middle surface* S and *thickness* 2ε, whose *reference configuration* is thus the set $\boldsymbol{\Phi}(\overline{\Omega}^\varepsilon) \subset \mathbf{R}^3$, where $\boldsymbol{\Phi}(x^\varepsilon) = \boldsymbol{\varphi}(y) + x_3^\varepsilon\,\boldsymbol{a}^3(y)$ for all $x^\varepsilon = (y, x_3^\varepsilon) \in \overline{\Omega}^\varepsilon$. For $\varepsilon > 0$ small enough, the vectors $\boldsymbol{g}_i^\varepsilon = \partial_i^\varepsilon\boldsymbol{\Phi}$ are linearly independent at all points in $\overline{\Omega}^\varepsilon$. We also define the vectors $\boldsymbol{g}^{j,\varepsilon}$ by $\boldsymbol{g}^{j,\varepsilon} \cdot \boldsymbol{g}_i^\varepsilon = \delta_i^j$, the *metric tensor* by its covariant and contravariant components $g_{ij}^\varepsilon = \boldsymbol{g}_i^\varepsilon \cdot \boldsymbol{g}_j^\varepsilon$ and $g^{ij,\varepsilon} = \boldsymbol{g}^{i,\varepsilon} \cdot \boldsymbol{g}^{j,\varepsilon}$, the *volume element* $\sqrt{g^\varepsilon}\, dx^\varepsilon$ with $g^\varepsilon = \det(g_{ij}^\varepsilon)$, and the *Christoffel symbols* $\Gamma_{ij}^{p,\varepsilon} = \boldsymbol{g}^{p,\varepsilon} \cdot \partial_i^\varepsilon\boldsymbol{g}_j^\varepsilon$.

The *unknown* is the vector field $\boldsymbol{u}^\varepsilon = (u_i^\varepsilon) : \overline{\Omega}^\varepsilon \to \mathbf{R}^3$, where the functions $u_i^\varepsilon = \overline{\Omega}^\varepsilon \to \mathbf{R}$ are the *covariant components* of the *displacement field of the shell*; this means that, for each $x^\varepsilon \in \overline{\Omega}^\varepsilon$, the vector $u_i^\varepsilon(x^\varepsilon)\boldsymbol{g}^{i,\varepsilon}(x^\varepsilon)$ is the displacement of the point $\boldsymbol{\Phi}(x^\varepsilon)$. In *linearized elasticity*, the unknown solves the variational problem:

$$\boldsymbol{u}^\varepsilon \in \boldsymbol{V}(\Omega^\varepsilon) := \{\boldsymbol{v}^\varepsilon = (v_i^\varepsilon) \in \boldsymbol{H}^1(\Omega^\varepsilon); \boldsymbol{v}^\varepsilon = \boldsymbol{0} \text{ on } \Gamma_0^\varepsilon\}, \tag{1}$$

$$\int_{\Omega^\varepsilon} A^{ijk\ell,\varepsilon} e_{k\|\ell}^\varepsilon(\boldsymbol{u}^\varepsilon) e_{i\|j}^\varepsilon(\boldsymbol{v}^\varepsilon)\sqrt{g^\varepsilon}\, dx^\varepsilon = \int_{\Omega^\varepsilon} f^{i,\varepsilon} v_i^\varepsilon \sqrt{g^\varepsilon}\, dx^\varepsilon \tag{2}$$

for all $\boldsymbol{v}^\varepsilon \in \boldsymbol{V}(\Omega^\varepsilon)$, where

$$A^{ijk\ell,\varepsilon} := \lambda g^{ij,\varepsilon} g^{k\ell,\varepsilon} + \mu(g^{ik,\varepsilon} g^{j\ell,\varepsilon} + g^{i\ell,\varepsilon} g^{jk,\varepsilon}), \tag{3}$$

$$e_{i\|j}^\varepsilon(\boldsymbol{v}^\varepsilon) := \frac{1}{2}(\partial_i^\varepsilon v_j^\varepsilon + \partial_j^\varepsilon v_i^\varepsilon) - \Gamma_{ij}^{p,\varepsilon} v_p^\varepsilon, \tag{4}$$

where $\lambda > 0$ and $\mu > 0$ denote the *Lamé constants*, assumed to be *independent of ε*, of the material constituting the shell, and $f^{i,\varepsilon} \in L^2(\Omega^\varepsilon)$ are the *covariant components* of the *applied body force* density. *Problem* (1)–(2) *has one and only one solution*; this either follows from the classical Korn inequality [15], or from Korn's inequality in curvilinear coordinates [7].

In the ensuing asymptotic analysis, the following "two-dimensional" analogues of the "three-dimensional" functions (3)–(4) will naturally arise:

$$a^{\alpha\beta\sigma\tau} = \frac{4\lambda\mu}{\lambda+2\mu}\, a^{\alpha\beta} a^{\sigma\tau} + 2\mu(a^{\alpha\sigma}a^{\beta\tau} + a^{\alpha\tau}a^{\beta\sigma}), \tag{5}$$

$$\gamma_{\alpha\beta}(\boldsymbol{\eta}) = \frac{1}{2}(\partial_\alpha\eta_\beta + \partial_\beta\eta_\alpha) - \Gamma^\sigma_{\alpha\beta}\eta_\sigma - b_{\alpha\beta}\eta_3, \tag{6}$$

$$\rho_{\alpha\beta}(\boldsymbol{\eta}) = \partial_{\alpha\beta}\eta_3 - \Gamma^\sigma_{\alpha\beta}\partial_\sigma\eta_3 + b^\sigma_\beta(\partial_\alpha\eta_\sigma - \Gamma^\tau_{\alpha\sigma}\eta_\tau) + b^\sigma_\alpha(\partial_\beta\eta_\sigma - \Gamma^\tau_{\beta\sigma}\eta_\tau) \\ + (\partial_\beta b^\sigma_\alpha + \Gamma^\sigma_{\beta\tau}b^\tau_\alpha - \Gamma^\tau_{\alpha\beta}b^\sigma_\tau)\eta_\sigma - c_{\alpha\beta}\eta_3. \tag{7}$$

They respectively denote the contravariant components of the *elasticity tensor of the surface S*, and the covariant components of the *linearized strain tensor* and *linearized change of curvature tensor* associated with an arbitrary displacement field $\eta_i \boldsymbol{a}^i$ of the surface S. In what follows, $\|\cdot\|_{0,A}$ and $\|\cdot\|_{1,A}$ denote the $L^2(A)$- and $H^1(A)$-norms of real or vector-valued functions.

3 Asymptotic analysis of bending-dominated shells

Let $\Omega = \omega\times]-1,1[$ and $\Gamma_0 = \gamma_0 \times [-1,1]$. With a point $x^\varepsilon = (x^\varepsilon_i) \in \overline{\Omega}^\varepsilon$, we associate the point $x = (x_i) \in \overline{\Omega}$, defined by $x^\varepsilon_\alpha = x_\alpha$, $x^\varepsilon_3 = \varepsilon x_3$ as in (8), (9), (12), (16), (24), (25), and we let $\partial_i = \partial/\partial x_i$. We assume that there exist functions $f^i \in L^2(\Omega)$ *independent of* ε such that the components of the applied body force density satisfy

$$f^{i,\varepsilon}(x^\varepsilon) = \varepsilon^2 f^i(x) \quad \text{for all } x^\varepsilon \in \Omega^\varepsilon \tag{8}$$

(a definition of the otherwise loose statement that "the body force is $O(\varepsilon^2)$"). Then the *scaled unknown* $\boldsymbol{u}_f(\varepsilon) : \overline{\Omega} \to \mathbf{R}^3$ defined by

$$\boldsymbol{u}^\varepsilon(x^\varepsilon) = \boldsymbol{u}_f(\varepsilon)(x) \quad \text{for all } x^\varepsilon \in \overline{\Omega}^\varepsilon, \tag{9}$$

satisfies (compare with (1)–(4)):

$$\boldsymbol{u}_f(\varepsilon) \in \boldsymbol{V}(\Omega) := \{\boldsymbol{v} = (v_i) \in \boldsymbol{H}^1(\Omega); \boldsymbol{v} = \boldsymbol{0} \text{ on } \Gamma_0\}, \tag{10}$$

$$\int_\Omega A^{ijk\ell}(\varepsilon)e_{k\|\ell}(\varepsilon; \boldsymbol{u}_f(\varepsilon))e_{i\|j}(\varepsilon; \boldsymbol{v})\sqrt{g(\varepsilon)}\, dx = \varepsilon^2 \int_\Omega f^i v_i \sqrt{g(\varepsilon)}\, dx \tag{11}$$

for all $\boldsymbol{v} \in \boldsymbol{V}(\Omega)$, where

$$A^{ijk\ell}(\varepsilon)(x) = A^{ijk\ell,\varepsilon}(x^\varepsilon), \quad g(\varepsilon)(x) = g^\varepsilon(x^\varepsilon) \quad \text{for all } x^\varepsilon \in \Omega^\varepsilon, \tag{12}$$

$$e_{\alpha\|\beta}(\varepsilon; \boldsymbol{v}) = \frac{1}{2}(\partial_\alpha v_\beta + \partial_\beta v_\alpha) - \Gamma^p_{\alpha\beta}(\varepsilon)v_p, \tag{13}$$

$$e_{\alpha\|3}(\varepsilon; \boldsymbol{v}) = \frac{1}{2}(\partial_\alpha v_3 + \frac{1}{\varepsilon}\partial_3 v_\alpha) - \Gamma^\sigma_{\alpha3}(\varepsilon)v_\sigma, \tag{14}$$

$$e_{3\|3}(\varepsilon; \boldsymbol{v}) = \frac{1}{\varepsilon}\partial_3 v_3, \tag{15}$$

$$\Gamma^p_{ij}(\varepsilon)(x) = \Gamma^{p,\varepsilon}_{ij}(x^\varepsilon) \text{ for all } x^\varepsilon \in \Omega^\varepsilon. \tag{16}$$

Our objective is to study the behavior of the unique solution $\boldsymbol{u}_f(\varepsilon)$ of problem (10)–(11) as $\varepsilon \to 0$. Note that eq. (11) is *not* defined for $\varepsilon = 0$ (cf. (14)–(15)); it constitutes an instance of a *singular perturbation problem* in the sense of [19].

THEOREM 1 [10]. *Assume that the "space of inextensional displacements" (cf. (6)):*

$$\begin{aligned} \boldsymbol{V}_f(\omega) = \{\boldsymbol{\eta} = (\eta_i) &\in H^1(\omega) \times H^1(\omega) \times H^2(\omega); \\ \eta_i &= \partial_\nu \eta_3 = 0 \quad on \quad \gamma_0, \ \gamma_{\alpha\beta}(\boldsymbol{\eta}) = 0 \quad in \quad \omega\} \end{aligned} \tag{17}$$

does not reduce to $\{\boldsymbol{0}\}$. Then the scaled unknown defined in (9) satisfies

$$\boldsymbol{u}_f(\varepsilon) \to \boldsymbol{u}_f \text{ in } \boldsymbol{H}^1(\Omega) \text{ as } \varepsilon \to 0, \tag{18}$$

where the limit $\boldsymbol{u}_f \in \boldsymbol{V}(\Omega)$ is independent of the "transverse" variable x_3. Furthermore, the function $\boldsymbol{\zeta}_f := \dfrac{1}{2} \displaystyle\int_{-1}^{1} \boldsymbol{u}_f \, dx_3$ belongs to the space $\boldsymbol{V}_f(\omega)$ and satisfies the two-dimensional equations of a "bending-dominated shell" (cf. (5) and (7))

$$\frac{\varepsilon^3}{3} \int_\omega a^{\alpha\beta\sigma\tau} \rho_{\sigma\tau}(\boldsymbol{\zeta}_f) \rho_{\alpha\beta}(\boldsymbol{\eta}) \sqrt{a} \, dy = \int_\omega \left\{ \int_{-\varepsilon}^{\varepsilon} f^{i,\varepsilon} dx_3^\varepsilon \right\} \eta_i \sqrt{a} \, dy \tag{19}$$

for all $\boldsymbol{\eta} \in \boldsymbol{V}_f(\omega)$.

The space $\boldsymbol{V}_f(\omega)$ was introduced by Sanchez-Palencia [23], who also noted [24] that $\boldsymbol{V}_f(\omega) \neq \{\boldsymbol{0}\}$ when S is a portion of a cylinder and $\boldsymbol{\varphi}(\gamma_0)$ is contained in a generatrix of S. The proof of the convergence (18) hinges on *a priori estimates* on the family $(\boldsymbol{u}(\varepsilon))_{\varepsilon>0}$, which themselves crucially rely on a *"first" generalized Korn inequality* (20) valid for an *arbitrary* surface $S = \boldsymbol{\varphi}(\omega)$ with $\boldsymbol{\varphi} \in \mathcal{C}^3(\overline{\omega})$, irrespective of whether the space $\boldsymbol{V}_f(\omega)$ reduces to $\{\boldsymbol{0}\}$ or not:

THEOREM 2 [10]. *There exist $\varepsilon_1 > 0$ and $C_1 > 0$ such that (cf. (13)–(15))*

$$\|\boldsymbol{v}\|_{1,\Omega} \leq \frac{C_1}{\varepsilon} \left\{ \sum_{i,j} \|e_{i\|j}(\varepsilon; \boldsymbol{v})\|^2_{0,\Omega} \right\}^{\frac{1}{2}} \tag{20}$$

for all $0 < \varepsilon \leq \varepsilon_1$ and all $\boldsymbol{v} \in \boldsymbol{V}(\Omega)$ (cf. (10)).

4 Asymptotic analysis of membrane-dominated shells

It is remarkable that, *in some cases* (examples are given in Theorem 4), the "constant" C_1/ε appearing in (20) can be replaced by a constant *independent of ε*, at the expense however of "replacing $\|v_3\|_{1,\Omega}$ by $\|v_3\|_{0,\Omega}$" on the left-hand side.

THEOREM 3 [9]. *Define the space*

$$\boldsymbol{V}_m(\omega) = \{\boldsymbol{\eta} = (\eta_i); \eta_\alpha \in H^1_0(\omega), \eta_3 \in L^2(\omega)\} = H^1_0(\omega) \times H^1_0(\omega) \times L^2(\omega), \tag{21}$$

and assume that there exists a constant $c > 0$ such that (cf. (6))

$$\left\{\sum_\alpha \|\eta_\alpha\|_{1,\omega}^2 + \|\eta_3\|_{0,\omega}^2\right\}^{1/2} \le c\left\{\sum_{\alpha,\beta} \|\gamma_{\alpha\beta}(\boldsymbol{\eta})\|_{0,\omega}^2\right\}^{1/2} \text{ for all } \boldsymbol{\eta} \in \boldsymbol{V}_m(\omega). \quad (22)$$

Then there exist $\varepsilon_2 > 0$ and $C_2 > 0$ such that

$$\left\{\sum_\alpha \|v_\alpha\|_{1,\Omega}^2 + \|v_3\|_{0,\Omega}^2\right\}^{1/2} \le C_2\left\{\sum_{i,j} \|e_{i\|j}(\varepsilon; \boldsymbol{v})\|_{0,\Omega}^2\right\}^{1/2} \quad (23)$$

for all $0 < \varepsilon \le \varepsilon_2$ and all $\boldsymbol{v} \in \boldsymbol{V}(\Omega)$ where $\boldsymbol{V}(\Omega)$ is the space of (10) with $\Gamma_0 = \gamma \times [-1, 1]$.

THEOREM 4 [8], [13]. *Assume either that γ is of class $\mathcal{C}^3$ and $\boldsymbol{\varphi}$ is analytic in an open set containing $\overline{\omega}$, or that γ is of class $\mathcal{C}^4$ and $\boldsymbol{\varphi} \in \mathcal{C}^5(\overline{\omega})$; in addition, assume that the surface $S = \boldsymbol{\varphi}(\overline{\omega})$ is "uniformly elliptic", in the sense that the two principal radii of curvature are either both > 0, or both < 0, at all points of S. Then relation (22) holds.*

We assume now that there exist functions $f^i \in L^2(\Omega)$ *independent of ε* such that (compare with (8)):

$$f^{i,\varepsilon}(x^\varepsilon) = f^i(x) \text{ for all } x^\varepsilon \in \Omega^\varepsilon. \quad (24)$$

Then the *scaled unknown* $\boldsymbol{u}_m(\varepsilon) : \overline{\Omega} \to \mathbf{R}^3$, defined by

$$\boldsymbol{u}^\varepsilon(x^\varepsilon) = \boldsymbol{u}_m(\varepsilon)(x) \text{ for all } x^\varepsilon \in \overline{\Omega}^\varepsilon, \quad (25)$$

satisfies another *singular perturbation problem* (compare with (10)–(11)):

$$\boldsymbol{u}_m(\varepsilon) \in \boldsymbol{V}(\Omega), \quad (26)$$

$$\int_\Omega A^{ijk\ell}(\varepsilon)e_{k\|\ell}(\varepsilon; \boldsymbol{u}_m(\varepsilon))e_{i\|j}(\varepsilon; \boldsymbol{v})\sqrt{g(\varepsilon)} \, dx = \int_\Omega f^i v_i \sqrt{g(\varepsilon)} \, dx \quad (27)$$

for all $\boldsymbol{v} \in \boldsymbol{V}(\Omega)$, where $\boldsymbol{V}(\Omega)$ is the space of (10) with $\Gamma_0 = \gamma \times [-1, 1]$.

THEOREM 5 [9]. *Assume that $\gamma_0 = \gamma$ and that relation (22) holds. Then the scaled unknown defined in (25) satisfies*

$$\boldsymbol{u}_m(\varepsilon) \to \boldsymbol{u}_m \text{ in } H^1(\Omega) \times H^1(\Omega) \times L^2(\Omega) \text{ as } \varepsilon \to 0, \quad (28)$$

where the limit $\boldsymbol{u}_m$ is independent of the "transverse" variable x_3. Furthermore, the function $\boldsymbol{\zeta}_m := \dfrac{1}{2}\displaystyle\int_{-1}^1 \boldsymbol{u}_m \, dx_3$ belongs to the space $\boldsymbol{V}_m(\omega)$ of (21) and satisfies the two-dimensional equations of a "membrane-dominated shell" (cf. (5)–(6)):

$$\varepsilon \int_\omega a^{\alpha\beta\sigma\tau}\gamma_{\sigma\tau}(\boldsymbol{\zeta}_m)\gamma_{\alpha\beta}(\boldsymbol{\eta})\sqrt{a} \, dy = \int_\omega \left\{\int_{-\varepsilon}^\varepsilon f^{i,\varepsilon} \, dx_3^\varepsilon\right\}\eta_i \sqrt{a} \, dy \quad (29)$$

for all $\boldsymbol{\eta} \in \boldsymbol{V}_m(\omega)$.

The proof of the convergence (28) hinges again on *a priori estimates*, which now crucially rely on the *"second" generalized Korn inequality* (23). That the constant in (23) is independent of ε now allows the body force to be $O(1)$, as compared to $O(\varepsilon^2)$ in Section 3.

5 Koiter's shell model

For each $\varepsilon > 0$, consider the same linearly elastic shell as discussed in Section 2. Following a fundamental work of Fritz John, Koiter [16] has proposed a *two-dimensional* shell model, whose *unknown* is the vector field $\boldsymbol{\zeta}^\varepsilon = (\zeta_i^\varepsilon) : \overline{\omega} \to \mathbf{R}^3$ where the functions $\zeta_i^\varepsilon : \overline{\omega} \to \mathbf{R}$ are the *covariant components* of the *displacement field of the middle surface of the shell*; this means that, for each $y \in \overline{\omega}$, the vector $\zeta_i^\varepsilon(y)\boldsymbol{a}^i(y)$ is the displacement of the point $\boldsymbol{\varphi}(y)$. The unknown solves the variational problem (cf. (5)–(7)):

$$\boldsymbol{\zeta}^\varepsilon \in \boldsymbol{V}_K(\omega) := \{\boldsymbol{\eta} = (\eta_i) \in H^1(\omega) \times H^1(\omega) \times H^2(\omega); \eta_i = \partial_\nu \eta_3 = 0 \text{ on } \gamma_0\}, \quad (30)$$

$$\varepsilon \int_\omega a^{\alpha\beta\sigma\tau} \gamma_{\sigma\tau}(\boldsymbol{\zeta}^\varepsilon)\gamma_{\alpha\beta}(\boldsymbol{\eta})\sqrt{a}\,dy + \frac{\varepsilon^3}{3}\int_\omega a^{\alpha\beta\sigma\tau}\rho_{\sigma\tau}(\boldsymbol{\zeta}^\varepsilon)\rho_{\alpha\beta}(\boldsymbol{\eta})\sqrt{a}\,dy$$
$$= \int_\omega \left\{\int_{-\varepsilon}^\varepsilon f^{i,\varepsilon}\,dx_3^\varepsilon\right\}\eta_i\sqrt{a}\,dy \text{ for all } \boldsymbol{\eta} \in \boldsymbol{V}_K(\omega). \quad (31)$$

For each $\varepsilon > 0$, problem (30)–(31) has one and only one solution $\boldsymbol{\zeta}^\varepsilon$ [3], [4]. Assume first that $\boldsymbol{V}_f(\omega) \neq \{\boldsymbol{0}\}$, as in Theorem 1. Then Sanchez-Palencia [23, Theorem 2.1] has shown that, as $\varepsilon \to 0$, $\boldsymbol{\zeta}^\varepsilon$ *weakly converges* (in fact, strongly; cf. [11]) *in the space* $\boldsymbol{V}_K(\omega)$ *to the solution* $\boldsymbol{\zeta}_f \in \boldsymbol{V}_f(\omega)$ of (19). Combining this result with the convergence (18) and the scaling (9), we obtain:

THEOREM 6 [11]. *Assume that* $\boldsymbol{V}_f(\omega) \neq \{\boldsymbol{0}\}$ *and that assumption* (8) *holds. Then*

$$\frac{1}{2\varepsilon}\int_{-\varepsilon}^\varepsilon \boldsymbol{u}^\varepsilon\,dx_3^\varepsilon = \boldsymbol{\zeta}_f + o(1) \text{ in } \boldsymbol{H}^1(\omega), \quad (32)$$

$$\boldsymbol{\zeta}^\varepsilon = \boldsymbol{\zeta}_f + o(1) \text{ in } \boldsymbol{H}^1(\omega), \quad (33)$$

as $\varepsilon \to 0$, *where* $\boldsymbol{u}^\varepsilon$ *denotes the solution of the three-dimensional problem* (1)–(2) *and* $\boldsymbol{\zeta}^\varepsilon$ *denotes the solution of Koiter's model.*

Assume next that $\gamma_0 = \gamma$ and that relation (22) holds, as in Theorem 5. Then Sanchez-Palencia has shown [24, Theorem 4.1] that, *as* $\varepsilon \to 0$, $\boldsymbol{\zeta}^\varepsilon$ *converges in the space* $\boldsymbol{V}_m(\omega)$ *of* (21) *to the solution* $\boldsymbol{\zeta}_m \in \boldsymbol{V}_m(\omega)$ *of* (29). Combining this result with the convergence (28) and the scaling (25), we obtain:

THEOREM 7 [11]. *Assume that* $\gamma_0 = \gamma$, *that relation* (22) *holds, and that assumption* (24) *holds. Then*

$$\frac{1}{2\varepsilon}\int_{-\varepsilon}^\varepsilon \boldsymbol{u}^\varepsilon\,dx_3^\varepsilon = \boldsymbol{\zeta}_m + o(1) \text{ in } H^1(\omega) \times H^1(\omega) \times L^2(\omega), \quad (34)$$

$$\boldsymbol{\zeta}^\varepsilon = \boldsymbol{\zeta}_m + o(1) \text{ in } H^1(\omega) \times H^1(\omega) \times L^2(\omega), \quad (35)$$

as $\varepsilon \to 0$, *where* $\boldsymbol{u}^\varepsilon$ *denotes the solution of the three-dimensional problem* (1)–(2) *and* $\boldsymbol{\zeta}^\varepsilon$ *denotes the solution of Koiter's model.*

6 Conclusions and comments

6.1. The convergences (18) and (28) constitute, under the respective assumptions of Theorems 1 and 5, a *mathematical justification of the two-dimensional equations of a "bending-dominated shell"* (19) *and of a "membrane-dominated shell"* (29). They justify the *formal* asymptotic analysis of Sanchez-Palencia [25] (see also [22]).

6.2. Under the same respective assumptions, relations (32) and (33) on the one hand, and relations (34) and (35) on the other, likewise constitute a *mathematical justification of Koiter's shell model*, because they show that, in each case, its solution $\boldsymbol{\zeta}^\varepsilon$ and the average $\dfrac{1}{2\varepsilon} \displaystyle\int_{-\varepsilon}^{\varepsilon} \boldsymbol{u}^\varepsilon \, dx_3^\varepsilon$ obtained from three-dimensional elasticity both have the *same principal part with respect to powers of ε.*

One can therefore use Koiter's model without knowing "in advance" whether the shell is "bending-dominated" (in the sense that $\boldsymbol{V}_f(\omega) \neq \{\boldsymbol{0}\}$) *or "membrane-dominated"* (in the sense that $\gamma_0 = \gamma$ and relation (22) holds). This remarkable feature certainly explains why Koiter's model is blithely, and successfully, used in computational mechanics.

6.3. For a linearly elastic *plate*, or for a linearly elastic *shallow shell*, the asymptotic analysis yields a two-dimensional limit model where *both* the "bending" and "membrane" terms *simultaneously* appear (cf. [6, Theorem 3.3-1] and [12]). For a linearly elastic *shell*, by contrast, the asymptotic analysis yields a two-dimensional limit model that is *either* of the bending-dominated type, *or* of the membrane-dominated type.

6.4. The present analysis covers two cases: *either* $\boldsymbol{V}_f(\omega) \neq \{\boldsymbol{0}\}$, *or* $\gamma_0 = \gamma$ and relation (22) holds. It therefore remains to study *intermediary cases* (which correspond to surfaces that are "not well inhibited for the admissible displacements", in the terminology of Sanchez-Palencia [24]). For instance, the space

$$\mathbf{R}(\omega) := \{\boldsymbol{\eta} = (\eta_i) \in H^1(\omega) \times H^1(\omega) \times L^2(\omega); \eta_\alpha = 0 \text{ on } \gamma_0, \gamma_{\alpha\beta}(\boldsymbol{\eta}) = 0 \text{ in } \omega\}$$

may reduce to $\{\boldsymbol{0}\}$, but the equivalence of norm (22) does not hold; or the space $\boldsymbol{V}_f(\omega)$ may reduce to $\{\boldsymbol{0}\}$, but $\mathbf{R}(\omega) \neq \{\boldsymbol{0}\}$, etc.

6.5. An earlier convergence result was obtained by Destuynder for *membrane-dominated shells*. He showed in particular [14, Theorem 7.9, p. 305] that, under the assumption of uniform ellipticity of the middle surface S, $(\boldsymbol{u}_m(\varepsilon))_\alpha \to (u_m)_\alpha$ in $L^2(\Omega)$, $\dfrac{1}{2} \displaystyle\int_{-1}^{1} (\boldsymbol{u}_m(\varepsilon))_\alpha \, dx_3 \to \dfrac{1}{2} \displaystyle\int_{-1}^{1} (u_m)_\alpha \, dx_3$ in $H^1(\omega)$, $\varepsilon(\boldsymbol{u}_m(\varepsilon))_3 \to 0$ in $L^2(\Omega)$ (the notation is as in Theorem 5). However, his analysis was still "partially formal" in that it assumed the existence of a formal series expansion of $(\boldsymbol{u}_m(\varepsilon))_3$ as powers of ε.

Using *Γ-convergence techniques*, Acerbi, Buttazzo, and Percivale [1] have also obtained earlier convergence theorems for shells viewed as "thin inclusions" in a larger, surrounding elastic body; as a consequence, they do not seem to relate the distinction between the bending-dominated and membrane-dominated cases to the "geometry" of the middle surface and the boundary conditions.

6.6. For *nonlinearly elastic shells*, two-dimensional *"membrane-dominated"* or *"bending-dominated"* models can be likewise identified through a *formal asymptotic expansion* of the scaled three-dimensional solution [20], [21]. As for nonlinearly elastic plates [17], Γ-*convergence techniques* can also be successfully applied that yield a *convergence theorem* to the solution of a *"large deformation"* membrane shell model [18].

References

[1] E. Acerbi, G. Buttazzo, and D. Percivale, *Thin inclusions in linear elasticity: a variational approach*, J. Reine Angew. Math., **386** (1988), 99–115.

[2] M. Bernadou, Méthodes d'Éléments Finis pour les Problèmes de Coques Minces, Masson, Paris, 1994.

[3] M. Bernadou and P. G. Ciarlet, *Sur l'ellipticité du modèle linéaire de coques de W.T. Koiter*, Computing Methods in Applied Sciences and Engineering (R. Glowinski and J.-L. Lions, eds.), pp. 89–136, Springer-Verlag, Berlin and New York, 1976.

[4] M. Bernadou, P. G. Ciarlet, and B. Miara, *Existence theorems for two-dimensional linear shell theories*, J. Elasticity, **34** (1994), 111–138.

[5] D. Chenais and J.-C. Paumier, *On the locking phenomenon for a class of elliptic problems*, Numer. Math., **67** (1994), 427–440.

[6] P. G. Ciarlet, Plates and Junctions in Elastic Multi-Structures: An Asymptotic Analysis, Masson, Paris, and Springer-Verlag, Heidelberg, Berlin, and New York, 1990.

[7] P. G. Ciarlet, Mathematical Elasticity, Vol. II: Plates and Shells, North-Holland, Amsterdam, 1996.

[8] P. G. Ciarlet and V. Lods, *Ellipticité des équations membranaires d'une coque uniformément elliptique*, C. R. Acad. Sci. Paris, Série I, **318** (1994), 195–200.

[9] P. G. Ciarlet and V. Lods, *Analyse asymptotique des coques linéairement élastiques. I. Coques "membranaires"*, C. R. Acad. Sci. Paris, Série I, **318** (1994), 863–868.

[10] P. G. Ciarlet, V. Lods, and B. Miara, *Analyse asymptotique des coques linéairement élastiques. II. Coques "en flexion"*, C. R. Acad. Sci. Paris, Série I, **319** (1994), 95–100.

[11] P. G. Ciarlet and V. Lods, *Analyse asymptotique des coques linéairement élastiques. III. Une justification du modèle de W.T. Koiter*, C. R. Acad. Sci. Paris, Série I, **319** (1994), 299–304.

[12] P. G. Ciarlet and B. Miara, *Justification of the two-dimensional equations of a linearly elastic shallow shell*, Comm. Pure Appl. Math., **XLV** (1992), 327–360.

[13] P. G. Ciarlet and E. Sanchez-Palencia, *Un théorème d'existence et d'unicité pour les équations des coques membranaires*, C. R. Acad. Sci. Paris, Série I, **317** (1993), 801–805.

[14] P. Destuynder, Sur une justification des modèles de plaques et de coques par les méthodes asymptotiques, Thèse d'État, Université Pierre et Marie Curie, Paris, 1980.

[15] G. Duvaut and J.-L. Lions, Les Inéquations en Mécanique et en Physique, Dunod, Paris, 1972.

[16] W. T. Koiter, *On the foundation of the linear theory of thin elastic shells*, in: Proc. Kon. Nederl. Akad. Wetensch., **B73** (1970), 169–195.

[17] H. Le Dret and A. Raoult, *Le modèle de membrane non linéaire comme limite variationnelle de l'élasticité non linéaire tridimensionnelle*, C. R. Acad. Sci. Paris, Série II, **317** (1993), 221–226.

[18] H. Le Dret and A. Raoult, *Dérivation variationnelle du modèle non linéaire de coque membranaire*, C. R. Acad. Sci. Paris, Série I, **320** (1995), 511–516.

[19] J.-L. Lions, Perturbations Singulières dans les Problèmes aux Limites et en Contrôle Optimal, Springer-Verlag, Heidelberg, Berlin, and New York, 1973.

[20] V. Lods and B. Miara, *Analyse asymptotique des coques "en flexion" non linéairement élastiques*, C. R. Acad. Sci. Paris, Série I (1995), to appear.

[21] B. Miara, *Analyse asymptotique des coques membranaires non linéairement élastiques*, C. R. Acad. Sci. Paris, Série I, **318** (1994), 689–694.

[22] B. Miara and E. Sanchez-Palencia, *Asymptotic analysis of linearly elastic shells*, Asymptotic Anal., to appear.

[23] E. Sanchez-Palencia, *Statique et dynamique des coques minces. I. Cas de flexion pure non inhibée*, C. R. Acad. Sci. Paris, Série I, **309** (1989), 411–417.

[24] E. Sanchez-Palencia, *Statique et dynamique des coques minces. II. Cas de flexion pure inhibée*, C. R. Acad. Sci. Paris, Série I, **309** (1989), 531–537.

[25] E. Sanchez-Palencia, *Passage à la limite de l'élasticité tri-dimensionnelle à la théorie asymptotique des coques minces*, C. R. Acad. Sci. Paris, Série II, **311** (1990), 909–916.

Multiscale Techniques — Some Concepts and Perspectives

WOLFGANG DAHMEN

Institut für Geometrie und Praktische Mathematik
RWTH Aachen, Templergraben 55, 52056 Aachen, Germany

1. Introduction

The success of numerical simulation hinges to a great extent on the ability to handle larger and larger (ultimately) linear systems of equations that typically arise from discretizing integral or differential equations.

There are different possible responses to the challange posed by such large scale problems. One is mainly *data oriented*, relying on increased computing power and employment of parallel techniques. However, it seems to be generally agreed upon that this by itself may ultimately not suffice. The goal must be to develop *asymptotically optimal* (a.o.) schemes, which yield a solution within a given tolerance of accuracy, at the expense of storage and floating point operations, which remain proportional to the number of unknowns. Instead of isolating the linear algebra problem from its origin, it seems to be necessary to exploit as much information about the underlying analytical problem as possible. Such information cannot be extracted from a single scale of discretization, no matter how fine this scale might be. One key principle is to extract asymptotic information from the *interaction* of several levels of resolution.

The perhaps most prominent representative of such *multiscale techniques* is the *multigrid method* for solving elliptic boundary value problems, which under certain circumstances is known to be a.o. [17]. In addition during the past few years so-called *multilevel preconditioning* techniques have gained considerable importance. Only recently have they also been shown to give rise to a.o. schemes even under minimal regularity assumptions [2], [10], [24], [25]. Both methods can be interpreted as *correction schemes* [32], where successive corrections are realized through adding finer details on higher levels of discretization.

Adding details from successively finer scales is also the essence of the concept of (discrete) *wavelet transforms*, which, during the past decade, has become a completely independent, rapidly expanding and extremely active area of research [15], [16], [22], [23]. Although primary applications were mainly concerned with signal analysis, image processing, and data compression, there have been recent attempts to also apply wavelet-type concepts to integral or differential equations (cf. e.g. [1], [10], [14], [19, 20]). On the other hand, there is an important difference in the point of views of both methodologies. In contrast to multigrid and multilevel methods developed in the finite element community, wavelet-type concepts are strictly *basis oriented*. Such schemes rely on the explicit availability of bases with certain nice properties. The construction of such bases, however, is expected to

Proceedings of the International Congress
of Mathematicians, Zürich, Switzerland 1994
© Birkhäuser Verlag, Basel, Switzerland 1995

depend, in particular, on the underlying domain. So the interest stirred up by striking examples of such bases (cf. [15]) may actually have nourished hopes that ultimately are hard to realize. Also one could argue that for every presently known wavelet-based method there already exists a method (tuned to that particular problem at hand) that performs at least as well. Nevertheless, I think it is fair to say that the wavelet point of view has already been of significant help in providing a more unified analytic platform upon which one can draw from other areas such as approximation theory and theory of function spaces, to develop a.o. numerical linear algebra schemes.

In this paper we will illustrate this by focusing mainly on two issues:

- *Preconditioning*
- *Matrix Compression*

Our objective is to bring out some key principles and their consequences that are relevant in this context. They center upon the interplay between certain notions of *stability* and *norm equivalences* for Sobolev spaces. This should help to isolate those conditions that have to be satisfied in any particular realization. It also sheds some light on corresponding difficulties and limitations.

2.   Multiscale Decompositions

In this section we will outline a few basic concepts and ideas that are relevant to subsequent applications. We will choose a setting that is sufficiently general to also cover *multiresolution* of functions defined on bounded domains in $\mathbb{R}^n$ or on more general objects such as closed manifolds. We will not confine the discussion to orthogonal decompositions because they may often be too hard to realize and may sometimes not even be the best choice.

2.1.   Multiscale Transformations, Stability.   To *approximate* a function f in some Banach function space $\mathcal{F}$ one usually considers a *dense* sequence $\mathcal{S}$ of closed subspaces $S_j \subset \mathcal{F}$; i.e., the closure of the union of the S_j is all of $\mathcal{F}$. Approximants that possibly convey much information about the limit can often be derived from *representations* of the elements of $\mathcal{F}$. Suppose that $\mathcal{S}$ actually consists of nested spaces $S_j \subset S_{j+1}$ and assume that $\mathcal{Q}$ is an associated uniformly bounded sequence of linear projectors Q_j from $\mathcal{F}$. Then the telescoping expansion

$$f = \sum_{j=0}^{\infty}(Q_j - Q_{j-1})f, \qquad (2.1)$$

where $Q_{-1} = 0$, converges strongly in $\mathcal{F}$. $(Q_j - Q_{j-1})f$ represents the *detail* added on each scale. The detail should have no overlap with preceding scales, which means that the spaces $W_{j-1} := (Q_j - Q_{j-1})S_j = (Q_j - Q_{j-1})\mathcal{F}$ should be direct summands; i.e., $S_j = S_{j-1} \oplus W_{j-1}$. Thus, the above telescoping expansion corresponds to such direct sum decompositions if the mappings $Q_j - Q_{j-1}$ are also projectors, which is equivalent to the *commutator condition*

$$Q_j Q_n = Q_j \quad \text{for} \ \ j \leq n. \qquad (2.2)$$

Orthogonal projectors or Lagrange interpolation projectors satisfy (2.2).

One way of making practical use of the expansion (2.1) is to determine explicit representations

$$(Q_{j+1} - Q_j)f = \sum_{k \in J_j} d_{j,k}(f)\psi_{j,k} \tag{2.3}$$

of the details relative to stable *complement bases* $\Psi^j = \{\psi_{j,k} : k \in J_j\}$ of $W_{j-1} = (Q_j - Q_{j-1})\mathcal{F}$. To explain this, let $\mathcal{F} = L_2(\Omega)$, where in the following Ω is some measure space and could stand for domains in $\mathbb{R}^n$ or even for more general (possibly closed) manifolds. For convenience we will use the notation $a \lesssim b$ to express that a can be bounded by some constant multiple of b, where the constant is independent of the parameters a and b may depend on. In particular, $a \sim b$ means that $a \lesssim b$ and $b \lesssim a$. Then Ψ^j is called *stable* if

$$\|\mathbf{d}\|_{l_2(J_j)} \sim \| \sum_{k \in J_j} d_k \psi_{j,k} \|_{L_2(\Omega)}.$$

However, in practical applications the elements of S_j are usually defined through a single scale basis $\Phi_j = \{\varphi_{j,k} : k \in I_j\}$ that is stable in the above sense and consists of compactly supported functions such that $\mathrm{diam}\,(\mathrm{supp}\,\varphi_{j,k}) \sim \rho^{-j}$ for some $\rho > 1$. Examples are B-spline or box-spline bases, finite element bases, or translates of scaling functions for $\rho = 2$.

The core of any basis oriented multiscale scheme is then the transformation $\mathbf{T}_n$ that takes the *multiscale coefficients* $\mathbf{d}$ of some $f_n = \sum_{j=0}^{n-1} \sum_{k \in J_j} d_{j,k}\psi_{j,k} \in S_n$ into the coefficients $\mathbf{c}$ of the *single scale representation* $f_n = \sum_{k \in I_n} c_k\varphi_{n,k}$, where $I_{j+1} \simeq I_j \cup J_j$, $I_j \cap J_j = \emptyset$.

The structure of $\mathbf{T}_n$ is analogous to the wavelet transform (see e.g. [16]) except that the filters may depend on the scale.

To describe this for the present generality, it is convenient to view Φ_j as a column vector. Note that the nestedness of S_j and stability of Φ_j imply the existence of a matrix $\mathbf{M}_{j,0}$, defining a bounded linear map from $l_2(I_j)$ into $l_2(I_{j+1})$, such that $\Phi_j' = \Phi_{j+1}'\mathbf{M}_{j,0}$. Here Φ_j' is the transpose of Φ_j. The complement bases must then have the form $\Psi_j' = \Phi_{j+1}'\mathbf{M}_{j,1}$ for some matrix $\mathbf{M}_{j,1}$ such that $\mathbf{M}_j := (\mathbf{M}_{j,0}, \mathbf{M}_{j,1})$ establishes a change of bases between Φ_{j+1} and $\Phi_j \cup \Psi^j$. Thus, there must exist matrices $\mathbf{G}_j := (\mathbf{G}_{j,0}, \mathbf{G}_{j,1})$ such that

$$\mathbf{M}_j'\mathbf{G}_j = \mathbf{G}_j\mathbf{M}_j' = I. \tag{2.4}$$

The following observation is easily verified [4].

REMARK 2.1 *The complement bases Ψ^j are uniformly stable for each scale if and only if the $\mathbf{M}_j$ and $\mathbf{G}_j$ are uniformly bounded mappings on the spaces $l_2(I_{j+1})$.*

One easily checks that $\mathbf{T}_n$ has the form

$$\mathbf{T}_n = \hat{\mathbf{M}}_{n-1} \cdots \hat{\mathbf{M}}_0,$$

where

$$\hat{\mathbf{M}}_j := \begin{pmatrix} \mathbf{M}_j & 0 \\ 0 & I \end{pmatrix}.$$

Thus, $\mathbf{T}_n$ can still be represented by a *pyramid*-type scheme. The inverse operation has a similar structure but involves only the matrices $\mathbf{G}_j$.

There are two essential requirements to be satisfied by $\mathbf{T}_n$.

(1) The number of operations needed to carry out $\mathbf{T}_n$ should be at most of the order of $\dim S_n$.

(2) The transformations $\mathbf{T}_n$ should be *uniformly stable*; i.e.,

$$\|\mathbf{T}_n\|,\ \|\mathbf{T}_n^{-1}\| = O(1), \quad n \to \infty. \tag{2.5}$$

Typically $\#I_j$ behaves like a fixed fraction of $\#I_{j+1}$. Then (1) holds for $\mathbf{T}_n$ and $\mathbf{T}_n^{-1}$ if the matrices $\mathbf{M}_j, \mathbf{G}_j$ are all uniformly *sparse*. This is obviously a very stringent constraint, which explains the difficulty one usually encounters when constructing suitable multiscale bases.

We will concentrate in the following on (2.5). It will be seen that the uniform boundedness of the condition numbers of the transformations $\mathbf{T}_n$ is not only important for avoiding loss of numerical accuracy when applying $\mathbf{T}_n$ but entails further important properties that turn out to be crucial for subsequent applications.

2.2. Riesz Bases. It is not hard to see that stability in the sense of (2.5) is equivalent to the fact that $\Psi := \bigcup_{j=-1}^{\infty} \Psi^j$, $\Psi^{-1} := \Phi_0$ forms a *Riesz basis*. This means that every $f \in L_2(\Omega)$ has a unique expansion

$$f = \sum_{j=-1}^{\infty} \sum_{k \in J_j} d_{j,k}(f)\psi_{j,k}$$

such that

$$\|f\|_{L_2(\Omega)} \sim \left(\sum_{j=-1}^{\infty} \sum_{k \in J_j} |d_{j,k}(f)|^2 \right)^{1/2} := \|\mathbf{d}(f)\|_{\ell_2(\mathcal{J})}. \tag{2.6}$$

By the Riesz representation theorem each $d_{j,k}(f)$ may be represented by $\langle f, \tilde{\psi}_{j,k} \rangle$ for some $\tilde{\psi}_{j,k} \in L_2(\Omega)$ and it is not hard to show that the $\tilde{\psi}_{j,k}$ form another Riesz basis $\tilde{\Psi}$ that is *biorthogonal* to Ψ. In terms of the projectors Q_j one has

$$Q_n f := \sum_{j=-1}^{n-1} \sum_{k \in J_j} \langle f, \tilde{\psi}_{j,k} \rangle \psi_{j,k}, \quad Q_n^* f := \sum_{j=-1}^{n-1} \sum_{k \in J_j} \langle f, \psi_{j,k} \rangle \tilde{\psi}_{j,k}, \tag{2.7}$$

where Q_j^* is the adjoint of Q_j. This means that $\mathcal{Q}$ and $\mathcal{Q}^*$ are both uniformly bounded and satisfy (2.2), or, equivalently, that their ranges $\mathcal{S}$ and $\tilde{\mathcal{S}}$ are both dense and nested.

Note that the task of establishing stability in the sense of (2.6) can be split into two steps. First one has to ensure (uniform) stability of each complement basis Ψ^j in the sense of Remark 2.1, which is usually the easier part. Then it remains to establish stability across scales, which, except when dealing with orthogonal complements, is usually the more delicate problem. It can be expressed solely in terms of the Q_j without specifying the bases Ψ^j explicitly by

$$\|f\|_{L_2(\Omega)} \sim \left(\sum_{j=0}^{\infty} \|(Q_j - Q_{j-1})f\|_{L_2(\Omega)}^2 \right)^{1/2}. \tag{2.8}$$

In the context of biorthogonal wavelets on all of $\mathbb{R}^n$ for the shift invariant setting a complete characterization of the Riesz basis property based on Fourier techniques is known [6], [5].

However, one can show that, even in a much more general Hilbert space context, the condition (2.2) together with the validity of certain direct and inverse estimates for the spaces S_j and for the range $\tilde{S}_j$ of the adjoint Q_j^* suffice to ensure the validity of a corresponding analog to (2.8) [8].

The construction of appropriate projectors Q_j can be based on finding sufficiently regular dual bases $\tilde{\Phi}_j$ to Φ_j. A general strategy for constructing the $\tilde{\Phi}_j$ is to search for such left inverses $\mathbf{G}_{j,0}$ for which the products $\mathbf{G}_{j+m,0} \cdots \mathbf{G}_{j,0}$ converge in an appropriate sense (cf. [7]). This is expected to be generally rather difficult. Current work indicates that for important special cases like piecewise linear functions on triangulations or polyhedral manifolds stable multiscale bases are available.

2.3. Sobolev Norms. Once an equivalence of the form (2.8) has been established, introducing suitable weights in the expression on the right-hand side of (2.8) generates Sobolev- and Besov-norms for certain ranges of positive and negative exponents depending on certain approximation and regularity properties, expressed in terms of Jackson and Bernstein inequalities, of both the ranges $\mathcal{S}$ and $\tilde{\mathcal{S}}$ of the Q_j and Q_j^*, respectively:

$$\inf_{f_n \in S_n} \|f - f_n\|_{L_2(\Omega)} \le c \, \rho^{-nt} \|f\|_{H^t(\Omega)}, \quad f \in H^t(\Omega), \tag{2.9}$$

and

$$\|f_n\|_{H^t(\Omega)} \le c \, \rho^{nt} \|f_n\|_{L_2(\Omega)}, \quad f_n \in S_n. \tag{2.10}$$

Suppose that analogous relations hold for $\tilde{\mathcal{S}}$ and $t^* > 0$. Then the equivalence

$$\|f\|_{H^s(\Omega)} \sim \left(\sum_{j \in \mathbb{N}_0} \rho^{2sj} \|(Q_j - Q_{j-1})f\|_{L_2(\Omega)}^2 \right)^{1/2} \tag{2.11}$$

holds for $-t^* < s < t$. In other words, the operator $\Lambda_s f := \sum_{j=0}^{\infty} \rho^{js}(Q_j - Q_{j-1})f$ shifts between Sobolev scales (see e.g. [8], [21]); i.e.,

$$\|\Lambda_s f\|_{H^r(\Omega)} \sim \|f\|_{H^{s+r}(\Omega)}, \quad -t^* < s + r < t. \tag{2.12}$$

Moreover, one easily deduces from (2.9), (2.10), and (2.11) that

$$\|f\|_{H^s(\Omega)} \sim \left(\sum_{j \in \mathbb{N}_0} \|(Q_j - Q_{j-1})f\|_{H^s(\Omega)}^2 \right)^{1/2}, \quad -t^* < s < t. \tag{2.13}$$

3. Numerical Solution of Operator Equations

In this section we indicate the relevance of the above results for the numerical treatment of various types of linear operator equations. In the following H^s will stand for a Sobolev space defined on some domain Ω as above. To explain the principles it will not be essential to specify here precisely the nature of Ω nor the type of boundary conditions incorporated in H^s. Moreover, suppose that for some $r \in \mathbb{R}$, $A : H^s \to H^{s-r}$ is a *boundedly invertible* linear operator; i.e.,

$$\|Au\|_{H^{s-r}} \sim \|u\|_{H^s}, \quad u \in H^s. \tag{3.1}$$

Given an ascending sequence of trial spaces $S_j \subset H^s \cap L_2(\Omega)$, a standard approach to solving

$$Au = f \tag{3.2}$$

in H^s approximately is to determine $u_j \in S_j$ such that

$$\langle Au_j, v \rangle = \langle f, v \rangle, \quad v \in S_j. \tag{3.3}$$

For a given sequence of L_2-uniformly bounded projectors $Q_j : L_2(\Omega) \to S_j$ this can be reformulated as a *projection method*

$$Q_j^* A u_j = Q_j^* f. \tag{3.4}$$

In the following, let d, d^* denote the largest integers for which (2.9) holds relative to the spaces S_j and $\tilde{S}_j$. Moreover, we will always assume that $t, t^* > |r/2|$, and, for simplicity, that $\rho = 2$, which corresponds to halving the meshsize when progressing to the next level of discretization.

3.1. Stability and Convergence. Under the above assumptions, one can show for a wide range of cases that the Galerkin scheme (3.3) is $(s, s - r)$-*stable* for $r - d \leq s \leq r/2$; i.e.,

$$\|Q_j^* A v_j\|_{H^{s-r}} \gtrsim \|v_j\|_{H^s}, \quad v_j \in S_j. \tag{3.5}$$

Moreover, one has for the solutions u, u_j of (3.2), (3.3), respectively,

$$\|u - u_j\|_{H^\tau} \lesssim 2^{j(\tau - \nu)} \|u\|_{H^\nu}, \tag{3.6}$$

for $r - d \leq \tau < t$, $\tau \leq \nu$, $r/2 \leq \nu \leq d$ [14].

3.2. Preconditioning. (3.3) amounts to finding a matrix representation $\mathbf{A}_j$ of the operator $Q_j^* A Q_j$. A standard approach is to take $\mathbf{A}_j = \mathbf{A}_j^\Phi$ as the stiffness matrix of $\mathbf{A}$ relative to the stable basis Φ_j of S_j and solve the linear system of equations

$$\mathbf{A}_j^\Phi \mathbf{c} = \mathbf{f}_j^\Phi, \tag{3.7}$$

where $(\mathbf{f}_j^\Phi)_k = \langle f, \varphi_{j,k} \rangle$, $k \in I_j$. Let us pause briefly to consider the important special case when $r/2 \in \mathbb{N}$ and A is a partial differential operator of order r

such that (via Green's theorem) $(Au, v) := \int_\Omega Au(x)v(x)\, dx$ leads to the *symmetric* H^s-*elliptic* bilinear form $a(\cdot, \cdot)$, i.e.,

$$\|u\|_{H^{r/2}}^2 \sim a(u, u), \quad u \in H^{r/2}. \tag{3.8}$$

It is clear that then (3.1) holds for $s = r/2$. In this case Φ_j is usually arranged to consist of compactly supported functions so that the symmetric positive definite matrix $\mathbf{A}_j^\Phi$ is possibly very large but *sparse*. Therefore the only candidates for a.o. schemes are iterative methods. However, because the spectral condition numbers $\kappa_2(\mathbf{A}_j^\Phi)$ usually grow like 2^{rj} as j tends to infinity, significant *preconditioning* is of vital importance. In fact, a scheme can only be a.o. if the preconditioned matrix has *uniformly bounded* condition numbers. In principle, this is accomplished by multigrid methods [17]. Moreover, certain multilevel preconditioned conjugate gradient schemes have recently turned out to also be a.o. (see e.g. [2], [10], [20], [24], [25]). The latter fact is closely related to the considerations of the previous section. To explain this, let Q_j be the L_2-*orthogonal projector* onto S_j (so that $Q_j^* = Q_j$, $S_j = \tilde{S}_j$) and suppose that (2.9) and (2.10) hold for some $t = t^* > r/2$. In view of the min-max characterization of the smallest and largest eigenvalue of a symmetric positive definite matrix and $H^{r/2}$-ellipticity the norm equivalence (2.11) means that the operator $\mathbf{C}_j$ defined by

$$\mathbf{C}_j^{-1} := \sum_{l=0}^j 2^{rl}(Q_l - Q_{l-1}) \tag{3.9}$$

is an optimal preconditioner; i.e., $\kappa_2(\mathbf{C}_j^{1/2}\mathbf{A}_j\mathbf{C}_j^{1/2}) = \mathcal{O}(1)$, $j \to \infty$. Of course, when dealing with classical finite elements, the evaluation of $(Q_l - Q_{l-1})v$ would be by far too expensive. Fortunately, this can be avoided [29]. In fact, because $\mathbf{C}_j = \sum_{l=0}^j 2^{-rl}(Q_l - Q_{l-1})$ and r is positive, $\mathbf{C}_j$ is readily seen to be spectrally equivalent to $\sum_{l=0}^j 2^{-rl}Q_l$, which in turn is spectrally equivalent to the operator

$$\hat{\mathbf{C}}_j v := \sum_{l=0}^j 2^{-lr}\sum_{k \in I_l}\langle v, \varphi_{l,k}\rangle\varphi_{l,k} \tag{3.10}$$

if and only if the Φ_l are stable. $\hat{\mathbf{C}}_j$ now does give rise to an a.o. scheme. Thus, in the above situation one gets away without explicit knowledge of the complement bases Ψ^j corresponding to $Q_{j+1} - Q_j$.

Of course, many cases of significant importance are not covered by the above reasoning. A could be unsymmetric or could be an operator of negative order $r < 0$. This occurs, for example, in connection with *boundary integral equations* when an elliptic *exterior domain* boundary value problem is transformed into an integral equation. The following summarizes what can still be said [14].

THEOREM 3.1 *Let Q_j be uniformly L_2-bounded linear projectors onto S_j satisfying (2.2) and let Ψ^j be stable complement bases corresponding to $Q_{j+1} - Q_j$ (see (2.3)). Assume that the spaces S_j and $\tilde{S}_j := \mathrm{range}\, Q_j^*$ satisfy (2.9), (2.10)* with $t, t^* > |r/2|$. Let $\mathbf{D}_j^s$ be the diagonal matrix with entries $(\mathbf{D}_j^s)_{(l,k),(l',k')} = 2^{sl}\delta_{(l,k),(l',k')}$. Finally, let $\mathbf{A}_j^\Psi$ denote the stiffness matrix relative to the multiscale basis $\cup_{-1\le l<j}\Psi^l$. Then

$$\kappa_2(\mathbf{D}_j^{-r/2}\mathbf{A}_j^\Psi\mathbf{D}_j^{-r/2}) = \mathcal{O}(1), \quad j \to \infty.$$

Proof: Let $w_j = \Lambda_{r/2}v_j$. By (2.12), (3.1), and (3.5), one obtains

$$\|w_j\|_{L_2(\Omega)} \sim \|v_j\|_{H^{r/2}} \sim \|Q_j^* A v_j\|_{H^{-r/2}} \sim \|\Lambda_{-r/2}^* Q_j^* A Q_j \Lambda_{-r/2} w_j\|_{L_2(\Omega)}.$$

The matrix representation of $\Lambda_{-r/2}^* Q_j^* A Q_j \Lambda_{-r/2}$ is $\mathbf{B}_j := \mathbf{D}_j^{-r/2}\mathbf{A}_j^\Psi\mathbf{D}_j^{-r/2}$. The stability of the multiscale basis yields then $\|\mathbf{B}_j\|\,\|\mathbf{B}_j^{-1}\| = \mathcal{O}(1)$, $j \to \infty$, which proves the assertion. □

Because $\mathbf{A}_j^\Psi = \mathbf{T}_j'\mathbf{A}_j^\Phi\mathbf{T}_j$, the preconditioning is realized up to a diagonal scaling by a change of bases (see also [30]). In particular, $\mathbf{C}_j := (\mathbf{D}_j^{-r/2}\mathbf{T}_j')(\mathbf{D}_j^{-r/2}\mathbf{T}_j')'$ is an optimal preconditioner. Note that only $\mathbf{T}_j$ and not $\mathbf{T}_j^{-1}$ is needed. When the $\varphi_{j,k}, \psi_{j,k}$ are all compactly supported and the matrices $\mathbf{M}_j$ are uniformly sparse, the application of $\mathbf{C}_j$ requires only $\mathcal{O}(\dim S_j)$ operations. When dealing with differential operators one therefore does not have to store $\mathbf{A}_j^\Psi$ but only the sparse matrix $\mathbf{A}_j^\Phi$.

3.3. Matrix Compression. When $r \le 0$ the matrix representations of $Q_j^* A Q_j$ are generally *not* sparse. Therefore various strategies have been developed to facilitate fast (approximate) matrix-vector multiplication as the main ingredient of iterative solvers. The best known are perhaps multipole expansion [27], panel clustering [18], or the scheme in [3], which in turn is closely related to the method proposed in [1]. There it is shown that the stiffness matrix of certain (zero order) periodic integral operators relative to wavelet bases can be well approximated by sparse matrices. It is this point of view that seems to exploit the previous considerations best even under more general circumstances.

We will assume throughout the following that the hypotheses of Theorem 3.1 are satisfied. The strategy pursued in [14] may be sketched as follows. The first step is to estimate the entries of the stiffness matrix $\mathbf{A}_j^\Psi$. We will comment on this point later in more detail. Based on these estimates a truncation rule is chosen by which certain entries are replaced by zero. A Schur lemma argument (see e.g. [23]) is then employed to estimate the spectral norm of the difference between $\mathbf{A}_j^\Psi$ and the compressed version. To be able to work with the spectral norm also involves preconditioning techniques as used in the proof of Theorem 3.1. With a judicious choice of truncation rules one can then prove optimal consistency arguments. A perturbation argument then ensures stability of the compressed operators. Finally, this leads (in certain cases) to asymptotically optimal convergence estimates for the solutions of the perturbed systems. One can show that the number of nonvanishing entries in the compressed matrices is of the order $\mathcal{O}((\log \#I_j)^a\#I_j)$, where a is some nonnegative real number (which in some cases can be shown to be zero). A detailed study of periodic problems can be found in [12], [13] including a stability and convergence analysis for generalized Petrov-Galerkin schemes. In particular, it is shown in [13], [14] that the solutions of the compressed systems approximate the solution of (3.2) at the same asymptotic rate as the solutions of the uncompressed systems (3.4) (see also [26]).

The above line of arguments uses the stability properties and norm equivalences described in Section 2 combined with the estimates for the entries of $\mathbf{A}_j^{\Psi}$. In [1], [13], [9], [14], [26] these estimates are based on *moment conditions* to be satisfied by the $\psi_{j,k}$ combined with the following assumptions on the Schwartz kernel $K(\cdot,\cdot)$ of A that

$$|\partial_x^\alpha \partial_y^\beta K(x,y)| \lesssim \operatorname{dist}(x,y)^{-(n+r+|\alpha|+|\beta|)}, \quad n+r+|\alpha|+|\beta| > 0, \ x \neq y, \quad (3.11)$$

where n is the spatial dimension of Ω, and that K is smooth off the diagonal.

The decay of the entries of $\mathbf{A}_j^{\Psi}$ is a consequence of the approximation properties of the adjoint operators $\mathcal{Q}^*$. To see this, let $R_j := Q_{j+1} - Q_j$ and note that $R_j \psi_{j,k} = \psi_{j,k}$. Therefore one has

$$\begin{aligned}
|\langle A\psi_{l,k}, \psi_{l',k'}\rangle| &= \left| \int_{\Omega \times \Omega} (R_{l'}^* \otimes R_l^* K)(x,y)\psi_{l',k'}(x)\psi_{l,k}(y)\,dx\,dy \right| \\
&\leq \|R_{l'}^* \otimes R_l^* K\|_{L_2(\Omega_{l',k'} \times \Omega_{l,k})}, \quad (3.12)
\end{aligned}$$

where the first (second) factor in $R_{l'}^* \otimes R_l^*$ is meant to act on the first (second) group of variables in $K(\cdot,\cdot)$ and $\Omega_{l,k} := \operatorname{supp}\psi_{l,k}$. Next note that $\|R_j^* u\|_{L_2(\Omega)} \lesssim \inf_{v \in \tilde{S}_{j-1}} \|u - v\|_{L_2(\Omega)}$. If the diameters of the supports of $\varphi_{j,k}, \psi_{j,k}, \tilde\varphi_{j,k}, \tilde\psi_{j,k}$ behave like 2^{-j}, Q_j and Q_j^* are local projectors so that one expects to have the following local version of (2.9)

$$\|Q_j^* u - u\|_{L_2(D)} \lesssim 2^{-d^* j}|u|_{H^{d^*}(\hat{D})}, \quad u \in H^{d^*}, \quad (3.13)$$

where $\hat{D}$ is a domain containing D whose diameter exceeds that of D at most by a fixed multiple of 2^{-j}. If $\Omega_{l,k} \cap \Omega_{l',k'} = \emptyset$ the kernel K is by assumption smooth on $\Omega_{l,k} \times \Omega_{l',k'}$. Thus, applying (3.13) to (3.12) yields, on account of (3.11) [26],

$$|\langle A\psi_{l,k}, \psi_{l',k'}\rangle| \lesssim 2^{-(l+l')\frac{n}{2}-2d^*} \operatorname{dist}(\Omega_{l,k}, \Omega_{l',k'})^{-(n+r+2d^*)}.$$

Note that the larger d^* the better the effect of compression. Choosing $d^* > d$ appears to be essential for obtaining optimal convergence rates in high norms for the solutions of the compressed schemes [14]. This flexibility seems to be an advantage of the biorthogonal setting over orthogonal decompositions.

Aside from the above-mentioned important advantages of stable multiscale bases, one expects that corresponding multiscale expansions will provide information about local regularity properties and lead, in view of (3.8), naturally to local error estimators. Even when dealing with bounded domains one can often still exploit constructions for the shift-invariant case [21]. This can be used to adapt the multiresolution setting for instance to saddle-point problems [28]. It also admits the efficient computation of inner products of wavelets and scaling functions [11].

References

[1] G. Beylkin, R. Coifman, and V. Rokhlin, *The fast wavelet transform and numerical algorithms*, Comm. Pure Appl. Math., **44**(1991), 141–183.

[2] J. H. Bramble, J. E. Pasciak, and J. Xu, *Parallel multilevel preconditioners*, Math. Comp., **55**(1990), 1–22.

[3] A. Brandt and A. A. Lubrecht, *Multilevel matrix multiplication and fast solution of integral equations*, J. Comput. Phys., **90**(1991), 348–370.

[4] J. M. Carnicer, W. Dahmen, and J. M. Peña, *Local decompositions of nested spaces*, preprint, 1994.

[5] A. Cohen, I. Daubechies, and J.-C. Feauveau, *Biorthogonal bases of compactly supported wavelets*, Comm. Pure Appl. Math., **45**(1992), 485–560.

[6] A. Cohen and I. Daubechies, *Non-Separable Bidimensional Wavelet Bases*, Rev. Mat. Iberoamericana, **9**(1993), 51–137.

[7] W. Dahmen, *Some remarks on multiscale transformations, stability and biorthogonality*, in Curves and Surfaces II, P. J. Laurent, A. Le Méhauté, and L. L. Schumaker (eds.), AKPeters, Boston, (1994), 157–188.

[8] W. Dahmen, *Stability of multiscale transformations*, Preprint, 1994.

[9] W. Dahmen, B. Kleemann, S. Prössdorf, and R. Schneider, *A multiscale method for the double layer potential equation on a polyhedron*, in Advances in Computational Mathematics, H. P. Dikshit and C. A. Micchelli (eds.), World Scientific, Singapore and Raneck, NJ, (1994),15–57.

[10] W. Dahmen and A. Kunoth, *Multilevel preconditioning*, Numer. Math., **63**(1992), 315–344.

[11] W. Dahmen and C.A. Micchelli, *Using the refinement equation for evaluating integrals of wavelets*, SIAM J. Numer. Anal., **30**(1993), 507–537.

[12] W. Dahmen, S. Prössdorf, and R. Schneider, *Wavelet approximation methods for pseudodifferential equations I: Stability and convergence*, Math. Z., **215**(1994), 583–620.

[13] W. Dahmen, S. Prössdorf, and R. Schneider, *Wavelet approximation methods for pseudodifferential equations II: Matrix compression and fast solution*, Adv. Comput. Math., **1**(1993), 259–335.

[14] W. Dahmen, S. Prössdorf, and R. Schneider, *Multiscale methods for pseudodifferential equations on smooth manifolds*, in Wavelets: Theory, Algorithms, and Applications, C. K. Chui, L. Montefusco, and L. Puccio (eds.), Academic Press, 1994, 385–424.

[15] I. Daubechies, *Orthonormal bases of wavelets with compact support*, Comm. Pure Appl. Math., **41**(1987), 909–996.

[16] I. Daubechies, *Ten Lectures on Wavelets*, CBMS-NSF Regional Conf. Ser. in Appl. Math., **61**, SIAM, Philadelphia, PA, 1992.

[17] W. Hackbusch, Multi-Grid Methods and Applications, Springer; Berlin and New York, 1985.

[18] W. Hackbusch and Z. P. Nowak, *On the fast matrix multiplication in the boundary element method by panel clustering*, Numer. Math., **54**(1989), 463–491.

[19] A. Harten, *Discrete multi-resolution analysis and generalized wavelets*, Appl. Numer. Math., **12** (1993), 153–192.

[20] S. Jaffard, *Wavelet methods for fast resolution of elliptic problems*, SIAM J. Numer. Anal., **29**(1992), 965–986.

[21] A. Kunoth, *Multilevel preconditioning*, Dissertation, Fachbereich Mathematik und Informatik, Freie Universität Berlin, January 1994.

[22] Y. Meyer, Ondelettes et opérateurs 1: Ondelettes, Hermann, Paris, 1990.

[23] Y. Meyer, Ondelettes et opérateurs 2: Opérateur de Caldéron-Zygmund, Hermann, Paris, 1990.

[24] P. Oswald, *On function spaces related to finite element approximation theory*, Z. Anal. Anwendungen, **9**(1990), 43–64.

[25] P. Oswald, *On discrete norm estimates related to multilevel preconditioners in the finite element method*, in Constructive Theory of Functions, Proc. Int. Conf. Varna 1991, K. G. Ivanov, P. Petrushev, and B. Sendov (eds.), Bulg. Acad. Sci., Sofia (1992), 203–214.

[26] T. von Petersdorf and C. Schwab, *Wavelet approximation for first kind boundary integral equations on polygons*, Techn. Note BN-1157, Institute for Physical Science and Technology, University of Maryland at College Park, February 1994.

[27] V. Rokhlin, *Rapid solution of integral equations of classical potential theory*, J. Comput. Phys., **60**(1985), 187–207.

[28] K. Urban, *On divergence-free wavelets*, in Adv. Comput. Math., **4** (1995), 51–81.

[29] J. Xu, *Theory of multilevel methods*, Report AM 48, Department of Mathematics, Pennsylvania State University, 1989.

[30] H. Yserentant, *On the multilevel splitting of finite element spaces*, Numer. Math., **49**(1986), 379–412.

[31] H. Yserentant, *Two preconditioners based on the multilevel splitting of finite element spaces*, Numer. Math., **58** (1990), 163–184.

[32] H. Yserentant, *Old and new convergence proofs for multigrid methods*, Acta Numerica, **2**, A. Iserles (ed.), Cambridge, 1993.

Matrix Computation and the Theory of Moments

GENE H. GOLUB

Computer Science Department, Stanford University,
Stanford, CA 94305 USA

ABSTRACT. We study methods to obtain bounds or approximations to $u^T f(A)v$ where A is a symmetric, positive definite matrix and f is a smooth function. These methods are based on the use of quadrature rules and the Lanczos algorithm. We give some theoretical results on the behavior of these methods based on results for orthogonal polynomials as well as analytical bounds and numerical experiments on a set of matrices for several functions f. We discuss the effect of rounding error in the quadrature calculation.

1. Introduction

The classical theory of moments plays a vital role in numerical linear algebra. It has long been recognized that there is a strong connection between the theory of moments, Gauss quadrature, orthogonal polynomials, and the conjugate gradient method and the Lanczos process. In this paper, we will be exploring these connections in order to obtain bounds for various matrix functions that arise in applications.

Let A be a real symmetric positive definite matrix of order n. We want to find upper and lower bounds (or approximations, if bounds are not available) for the entries of a function of a matrix. We shall examine analytical expressions as well as numerical iterative methods that produce good approximations in a few steps. This problem leads us to consider

$$u^T f(A)v, \qquad (1.1)$$

where u and v are given vectors and f is some smooth (possibly C^∞) function on a given interval of the real line. As an example, if $f(x) = \frac{1}{x}$ and $u^T = e_i^T = (0, \dots, 0, 1, 0, \dots, 0)$, the nonzero element being in the ith position and $v = e_j$, we will obtain bounds on the elements of the inverse A^{-1}.

Some of the techniques presented in this paper have been used (without any mathematical justification) to solve problems in solid state physics, particularly to compute elements of the resolvent of a Hamiltonian modeling the interaction of atoms in a solid, see [9], [11], [12]. In these studies the function f is the inverse of its argument.

The outline of the paper is as follows: Section 2 considers the problem of characterizing the elements of a function of a matrix. The theory is developed in Section 3, and Section 4 deals with the construction of the orthogonal polynomials that are needed to obtain a numerical method for computing bounds. The Lanczos

Proceedings of the International Congress
of Mathematicians, Zürich, Switzerland 1994
© Birkhäuser Verlag, Basel, Switzerland 1995

method used for the computation of the polynomials is presented there. Applications are described in Section 5 where very simple iterative algorithms are given to compute bounds. In Section 6 we discuss some extensions and recent work.

2.   Elements of a function of a matrix

Because $A = A^T$, we write A as

$$A = Q \Lambda Q^T,$$

where Q is the orthonormal matrix whose columns are the normalized eigenvectors of A and Λ is a diagonal matrix whose diagonal elements are the eigenvalues λ_i which we order as

$$\lambda_1 \leq \lambda_2 \leq \ldots \leq \lambda_n.$$

By definition, we have

$$f(A) = Q f(\Lambda) Q^T.$$

Therefore,

$$\begin{aligned} u^T f(A) v &= u^T Q f(\Lambda) Q^T v \\ &= \alpha^T f(\Lambda) \beta, \\ &= \sum_{i=1}^{n} f(\lambda_i) \alpha_i \beta_i. \end{aligned}$$

This last sum can be considered as a Riemann-Stieltjes integral

$$I[f] = u^T f(A) v = \int_a^b f(\lambda) \, d\alpha(\lambda), \qquad (2.1)$$

where the measure α is piecewise constant and defined by

$$\alpha(\lambda) = \begin{cases} 0 & \text{if} \quad \lambda < a = \lambda_1 \\ \sum_{j=1}^{i} \alpha_j \beta_j & \text{if} \quad \lambda_i \leq \lambda < \lambda_{i+1} \\ \sum_{j=1}^{n} \alpha_j \beta_j & \text{if} \quad b = \lambda_n \leq \lambda. \end{cases}$$

In this paper, we are looking for methods to obtain upper and lower bounds L and U for $I[f]$,

$$L \leq I[f] \leq U.$$

In the next section, we review and describe some basic results from Gauss quadrature theory as this plays a fundamental role in estimating the integrals and computing bounds.

3.   Bounds on matrix functions as integrals

One way to obtain the bounds on the integral $I[f]$ is to match the moments associated with the distribution $\alpha(\lambda)$. Thus, we seek to compute quadrature rules so that

$$I[\lambda^r] = \int_a^b \lambda^r d\alpha(\lambda) = \sum_{j=1}^{N} w_j t_j^r + \sum_{k=1}^{M} v_k z_k^r$$

for $r = 0, 1, \ldots, 2N + M - 1$.

The quantity $I[\lambda^r]$ is the rth moment associated with the distribution $\alpha(\lambda)$. Note that this can be easily calculated because

$$\mu_r \equiv I[\lambda^r] = u^T A^r v \quad (r = 0, 1, \ldots, 2N + M - 1).$$

The general form of the Gauss, Gauss-Radau, and Gauss-Lobatto quadrature formulas is given by

$$\int_a^b f(\lambda)\, d\alpha(\lambda) = \sum_{j=1}^N w_j f(t_j) + \sum_{k=1}^M v_k f(z_k) + R[f], \qquad (3.1)$$

where the weights $[w_j]_{j=1}^N, [v_k]_{k=1}^N$ and the nodes $[t_j]_{j=1}^N$ are unknowns and the nodes $[z_k]_{k=1}^M$ are prescribed, see [1], [2], [3], [8].

When $u = v$, the measure is a positive increasing function and it is known (see for instance [13]) that

$$R[f] = \frac{f^{(2N+M)}(\eta)}{(2N+M)!} \int_a^b \prod_{k=1}^M (\lambda - z_k) \left[\prod_{j=1}^N (\lambda - t_j) \right]^2 d\alpha(\lambda), \qquad (3.2)$$

$$a < \eta < b.$$

If $M = 0$, this leads to the Gauss rule with no prescribed nodes. If $M = 1$ and $z_1 = a$, or $z_1 = b$ we have the Gauss-Radau formula. If $M = 2$ and $z_1 = a, z_2 = b$, this is the Gauss-Lobatto formula.

Let us recall briefly how the nodes and weights are obtained in the Gauss, Gauss-Radau, and Gauss-Lobatto rules. For the measure α, it is possible to define a sequence of polynomials $p_0(\lambda), p_1(\lambda), \ldots$ that are orthonormal with respect to α:

$$\int_a^b p_i(\lambda) p_j(\lambda)\, d\alpha(\lambda) = \begin{cases} 1 & \text{if } i = j \\ 0 & \text{otherwise} \end{cases}$$

and p_k is of exact degree k. Moreover, the roots of p_k are distinct, real, and lie in the interval $[a, b]$. We will see how to compute these polynomials in the next section. This set of orthonormal polynomials satisfies a three-term recurrence relationship (see [15]):

$$\gamma_j p_j(\lambda) = (\lambda - \omega_j) p_{j-1}(\lambda) - \gamma_{j-1} p_{j-2}(\lambda), \quad j = 1, 2, \ldots, N \qquad (3.3)$$

$$p_{-1}(\lambda) \equiv 0, \quad p_0(\lambda) \equiv 1, \text{if} \int d\alpha = 1.$$

In matrix form, this can be written as

$$\lambda p(\lambda) = J_N p(\lambda) + \gamma_N p_N(\lambda) e_N,$$

where

$$p(\lambda)^T = [p_0(\lambda) p_1(\lambda) \ldots p_{N-1}(\lambda)], \quad e_N^T = (0, 0, \ldots 0, 1),$$

$$J_N = \begin{pmatrix} \omega_1 & \gamma_1 & & & & \\ \gamma_1 & \omega_2 & \gamma_2 & & & \\ & \ddots & \ddots & \ddots & & \\ & & \gamma_{N-2} & \omega_{N-1} & \gamma_{N-1} & \\ & & & \gamma_{N-1} & \omega_N \end{pmatrix}. \tag{3.4}$$

The eigenvalues of J_N (which are the zeros of p_N) are the nodes of the Gauss quadrature rule (i.e., $M = 0$). The weights are the squares of the first elements of the normalized eigenvectors of J_N, cf. [8]. We note that all the eigenvalues of J_N are real and simple.

For the Gauss quadrature rule (renaming the weights and nodes w_j^G and t_j^G), we have

$$\int_a^b f(\lambda)\, d\alpha(\lambda) = \sum_{j=1}^{N} w_j^G f\left(t_j^G\right) + R_G[f],$$

with

$$R_G[f] = \frac{f^{(2N)}(\eta)}{(2N)!} \int_a^b \left[\prod_{j=1}^{N} (\lambda - t_j^G)\right]^2 d\alpha(\lambda),$$

and the next theorem follows.

THEOREM 1 *Suppose $u = v$ in (2.1) and f is such that $f^{(2n)}(\xi) > 0$, $\forall n$, $\forall \xi$, $a < \xi < b$, and let*

$$L_G[f] = \sum_{j=1}^{N} w_j^G f\left(t_j^G\right).$$

Then, $\forall N, \exists \eta \in [a, b]$ such that

$$L_G[f] \leq I[f], \qquad I[f] - L_G[f] = \frac{f^{(2N)}(\eta)}{(2N)!}.$$

A proof of this is given in [13]. To obtain the Gauss-Radau rule ($M = 1$ in (3.1)–(3.2)), we extend the matrix J_N in (3.4) in such a way that it has one prescribed eigenvalue, see [4].

For Gauss-Radau, the remainder R_{GR} is

$$R_{GR}[f] = \frac{f^{(2N+1)}(\eta)}{(2N+1)!} \int_a^b (\lambda - z_1) \left[\prod_{j=1}^{N} (\lambda - t_j)\right]^2 d\alpha(\lambda).$$

Therefore, if we know the sign of the derivatives of f, we can bound the remainder. This is stated in the following theorem.

THEOREM 2 *Suppose $u = v$ and f is such that $f^{(2n+1)}(\xi) < 0, \forall n, \forall \xi, a < \xi < b$. Let U_{GR} be defined as*

$$U_{GR}[f] = \sum_{j=1}^{N} w_j^a f\left(t_j^a\right) + v_1^a f(a),$$

w_j^a, v_1^a, t_j^a being the weights and nodes computed with $z_1 = a$ and let L_{GR} be defined as

$$L_{GR}[f] = \sum_{j=1}^{N} w_j^b f\left(t_j^b\right) + v_1^b f(b),$$

w_j^b, v_1^b, t_j^b being the weights and nodes computed with $z_1 = b$. Then $\forall N$ we have

$$L_{GR}[f] \leq I[f] \leq U_{GR}[f],$$

and

$$I[f] - U_{GR}[f] = \frac{f^{(2N+1)}(\eta)}{(2N+1)!} \int_a^b (\lambda - a) \left[\prod_{j=1}^{N} \left(\lambda - t_j^a\right) \right]^2 d\alpha(\lambda),$$

$$I[f] - L_{GR}[f] = \frac{f^{(2N+1)}(\eta)}{(2N+1)!} \int_a^b (\lambda - b) \left[\prod_{j=1}^{N} \left(\lambda - t_j^b\right) \right]^2 d\alpha(\lambda).$$

We remark that we need not always compute the eigenvalues and eigenvectors of the tridiagonal matrix. Let Y_N be the matrix of the eigenvectors of J_N (or $\hat{J}_N$) whose columns we denote by y_i and T_N be the diagonal matrix of the eigenvalues t_i that give the nodes of the Gauss quadrature rule. It is well known that the weights w_i are given by (cf. [16])

$$\frac{1}{w_i} = \sum_{l=0}^{N-1} p_l^2(t_i).$$

It can be easily shown that

$$w_i = \left(\frac{y_i^1}{p_0(t_i)} \right)^2,$$

where y_i^1 is the first component of y_i. But, as $p_0(\lambda) \equiv 1$, we have

$$w_i = \left(y_i^1\right)^2 = \left(e_1^T y_i\right)^2.$$

THEOREM 3

$$\sum_{l=1}^{N} w_l f(t_l) = e_1^T f(J_N) e_1.$$

Proof:

$$
\begin{aligned}
\sum_{l=1}^{N} w_l f(t_l) &= \sum_{l=1}^{N} e_1^T y_l f(t_l) y_l^T e_1 \\
&= e_1^T \left(\sum_{l=1}^{N} y_l f(t_l) y_l^T \right) e_1 \\
&= e_1^T Y_N f(T_N) Y_N^T e_1 \\
&= e_1^T f(J_N) e_1.
\end{aligned}
$$

The same statement is true for the Gauss-Radau and Gauss-Lobatto rules. Therefore, in some cases where $f(J_N)$ (or the equivalent) is easily computable (for instance, if $f(\lambda) = 1/\lambda$, see Section 5.), we do not need to compute the eigenvalues and eigenvectors of J_N.

4. Construction of the orthogonal polynomials

In this section we consider the problem of computing the orthonormal polynomials or equivalently the tridiagonal matrices that we need. A very natural and elegant way to do this is to use Lanczos algorithms. When $u = v$, we use the classical Lanczos algorithm.

Let $x_{-1} = 0$ and x_0 be given such that $||x_0|| = 1$. The Lanczos algorithm is defined by the following relations:

$$\gamma_j x_j = r_j = (A - \omega_j I) x_{j-1} - \gamma_{j-1} x_{j-2}, \quad j = 1, \ldots$$

$$\omega_j = x_{j-1}^T A x_{j-1},$$

$$\gamma_j = ||r_j||.$$

The sequence $\{x_j\}_{j=0}^l$ is an orthonormal basis of the Krylov space

$$\mathrm{span}\{x_0, Ax_0, \ldots, A^l x_0\}.$$

PROPOSITION 1 *The vector x_j is given by*

$$x_j = p_j(A)x_0,$$

where p_j is a polynomial of degree j defined by the three-term recurrence (identical to (3.3))

$$\gamma_j p_j(\lambda) = (\lambda - \omega_j) p_{j-1} j(\lambda) - \gamma_{j-1} p_{j-2}(\lambda), \quad p_{-1}(\lambda) \equiv 0, \quad p_0(\lambda) \equiv 1.$$

THEOREM 4 *If $x_0 = u$, we have*

$$x_k^T x_l = \int_a^b p_k(\lambda)p_l(\lambda) \, d\alpha(\lambda).$$

Proof: As the x_j's are orthonormal, we have

$$
\begin{aligned}
x_k^T x_l &= x_0^T P_k(A)^T P_l(A) x_0 \\
&= x_0^T Q P_k(\Lambda) Q^T Q P_l(\Lambda) Q^T x_0 \\
&= x_0^T Q P_k(\Lambda) P_l(\Lambda) Q^T x_0 \\
&= \sum_{j=1}^n p_k(\lambda_j) p_l(\lambda_j) \hat{x}_j^2,
\end{aligned}
$$

where $\hat{x} = Q^T x_0$. Therefore, the p_j's are the orthonormal polynomials related to α that we have referred to in (3.3).

5. Applications

The applications are explained at length in [5], [6], and [7].

5.1. Error bounds for linear systems.

Suppose we solve a system of equations $Ax = b$ and obtain an approximation ξ to the solution. We desire to estimate the vector e where $x = \xi + e$. Note that $r = b - A\xi = A(x - \xi) = Ae$. Hence, $||e||^2 = r^T A^{-2} r$. Thus, $u = r$, and $f(\lambda) = \lambda^{-2}$.

5.2. Minimizing a quadratic form with a quadratic constraint.

Consider the problem of determining x such that $x^T Ax - 2b^T x = \min$ and $||x||^2 = \alpha^2$. Consider the Lagrangian: $\varphi(x; \mu) = x^T Ax - 2b^T x + \mu \left(x^T x - \alpha^2 \right)$. Then grad $\varphi(x; \mu) = 0$ when $(A + \mu I)x = b$. This implies $b^T (A + \mu I)^{-2} b = \alpha^2$. We can approximate the quadratic form $b^T (A + \mu I)^{-2} b$ by using the Lanczos algorithm with the initial vector $b/||b||_2$. This procedure has been extensively studied in [7].

5.3. Inverse elements of a matrix.

The elements of the inverse of a matrix are given by $e_j^T A^{-1} e_j$ where e_j is the jth unit vector. Hence, $f(\lambda) = \lambda^{-1}$. Thus, using the Lanczos process with the initial vector e_j will produce upper and lower bounds on a_{jj}, providing a lower bound is known for the smallest eigenvalue and an upper bound for the largest eigenvalue of A. It is desirable to compute the diagonal of the inverse for the Vičsek Fractal Hamiltonian matrix. The matrices are defined as follows.

$$H_1 = \begin{bmatrix} -4 & 1 & 1 & 1 & 1 \\ 1 & -2 & 0 & 0 & 0 \\ 1 & 0 & -2 & 0 & 0 \\ 1 & 0 & 0 & -2 & 0 \\ 1 & 0 & 0 & 0 & -2 \end{bmatrix},$$

$$H_n = \begin{bmatrix} H_{n-1} & V_1^T & V_2^T & V_3^T & V_4^T \\ V_1 & H_{n-1} & 0 & 0 & 0 \\ V_2 & 0 & H_{n-1} & 0 & 0 \\ V_3 & 0 & 0 & H_{n-1} & 0 \\ V_4 & 0 & 0 & 0 & H_{n-1} \end{bmatrix},$$

where $H_n \in R^{N_n \times N_n}$ and $N_{n+1} = 5N_n$.

The following tables show the "exact" values of a_{ii} for some chosen i, and estimated bounds of a_{ii} by using Gauss quadrature rule and Gauss-Radau rule. The "exact" values are computed using the Cholesky decomposition and then triangular inversion. It is a dense matrix method, with storage $O(N^2)$ and flops $O(N^3)$. The Gauss and the Gauss-Radau rule are sparse matrix methods, and both storage and flop only $O(N)$ because of the structure of the matrix H_n. From these two tables, we see that the error between the "exact" and estimated value is at $O(10^{-5})$, which is generally satisfactory and also is the stopping criterion used in the inner loop of the Gauss rule and the Gauss-Radau rule.

Table 1 $N = 125$

i	"exact"	Gauss		Gauss-Radau		
		iter	lower bound	iter	lower bound	upper bound
1	$9.480088e - 01$	15	$9.480088e - 01$	12	$9.479939e - 01$	$9.480112e - 01$
10	$6.669905e - 01$	13	$6.669846e - 01$	13	$6.669864e - 01$	$6.669969e - 01$
20	$1.156877e + 00$	14	$1.156848e + 00$	14	$1.156868e + 00$	$1.156879e + 00$

Table 2 $N = 625$

i	"exact"	Gauss		Gauss-Radau		
		iter	lower bound	iter	lower bound	upper bound
1	$9.480142e - 0$	15	$9.480123e - 01$	13	$9.480026e - 01$	$9.480197e - 01$
100	$1.100525e + 0$	14	$1.100512e + 00$	15	$1.100520e + 00$	$1.100527e + 00$
301	$9.243102e - 0$	14	$9.243074e - 01$	12	$9.242992e - 01$	$9.243184e - 01$
625	$6.440025e - 0$	12	$6.439994e - 01$	13	$6.440017e - 01$	$6.440054e - 01$

6.   Extensions

These methods, though simple, can be used in many situations involving large scale computations. We have extended these results to bilinear forms and to the situation where one wishes to estimate $W^T f(A)W$ where W is an $(N \times p)$-matrix ([5]).

It is well known that the *numerical* Lanczos process will produce sequences different than that defined by the mathematical sequence. Nevertheless, it has been shown in [6] that robust estimates of the quadratic form are obtained even in the presence of roundoff.

Acknowledgments. I want to thank G. Meurant and Z. Strakoš with whom I have had an active and stimulating collaboration. I also want to thank Z. Bai who provided the example given in Section 5.

References

[1] P. Davis and P. Rabinowitz, Methods of numerical integration, Second edition (1984) Academic Press, New York and San Diego, CA.

[2] W. Gautschi, *Construction of Gauss-Christoffel quadrature formulas*, Math. Comp. 22 (1968) pp. 251–270.

[3] W. Gautschi, *Orthogonal polynomials — constructive theory and applications*, J. Comp. Appl. Math. 12 and 13 (1985) pp. 61–76.

[4] G. H. Golub, *Some modified matrix eigenvalue problems*, SIAM Rev. vol. 15 no. 2 (1973) pp. 318–334.

[5] G. H. Golub and G. Meurant, *Matrices, moments, and quadrature*, Proceedings of the 15-th Dundee Conference, June–July, 1993, D. F. Sciffeths and G.A. Watson, Ests., Longman Scientific & Technical, 1994.

[6] G. H. Golub and Z. Strakoš, *Estimates in quadratic formulas*, accepted for publication in Numer. Algorithms.

[7] G. H. Golub and U. von Matt, *Quadratically constrained least squares and quadratic problems*, Numer. Math. (59), 561–580, 1991.

[8] G. H. Golub, J. H. Welsch, *Calculation of Gauss quadrature rule*, Math. Comp. 23 (1969) pp. 221–230.

[9] R. Haydock, *Accuracy of the recursion method and basis non-orthogonality*, Comput. Phys. Comm. 53 (1989) pp. 133–139.

[10] G. Meurant, *A review of the inverse of tridiagonal and block tridiagonal matrices*, SIAM J. Matrix Anal. Appl. vol. 13, no. 3 (1992) pp. 707–728.

[11] C. M. Nex, *Estimation of integrals with respect to a density of states*, J. Phys. A, vol. 11, no. 4 (1978) pp. 653–663.

[12] C. M. Nex, *The block Lanczos algorithm and the calculation of matrix resolvents*, Comput. Phys. Comm. 53 (1989) pp. 141–146.

[13] J. Stoer and R. Bulirsch, Introduction to Numerical Analysis, Second edition (1983) Springer-Verlag, Berlin and New York.

[14] G. W. Struble, *Orthogonal polynomials: variable-signed weight functions*, Numer. Math. vol. 5 (1963) pp. 88–94.

[15] G. Szegö, Orthogonal Polynomials, Third edition (1974) Amer. Math. Soc., Providence, RI.

[16] H. S. Wilf, Mathematics for the Physical Sciences, (1962) Wiley, New York.

Subscale Capturing in Numerical Analysis

STANLEY OSHER

Mathematics Department,
University of California,   and
Los Angeles, CA, USA

Cognitech, Inc.
Santa Monica,
CA, USA

1. Introduction

In this paper we shall describe numerical methods that were devised for the purpose of computing small scale behavior without either fully resolving the whole solution or explicitly tracking certain singular parts of it. Techniques developed for this purpose include shock capturing, front capturing, and multiscale analysis. Areas in which these methods have recently proven useful include image processing, computer vision, and differential geometry, as well as more traditional fields of physics and engineering.

Shock capturing methods were devised for the numerical solution of nonlinear conservation laws. At the 1990 meeting of the ICM, Harten [16] gave an overview of recent developments in that area, culminating in the construction of essentially nonoscillatory (ENO) schemes [17], [18]. We shall describe some of the ideas and results relating to this subject in Section 3.

Rudin, in his Ph.D. thesis [36] noted that the ideas and techniques from the theory of hyperbolic conservation laws and their numerical solution are relevant to the field of image processing. Images have features such as edges, lines, and textures and shock capturing is therefore an appropriate tool. Later developments [38], [39], [26], [27], [35] indicate that subscale capturing contains a great number of relevant tools for both image and video processing, as well as computer vision. We shall discuss this in Section 4.

In 1987, together with J.A. Sethian [31] we devised a new numerical procedure for capturing fronts and applied it to curves and surfaces whose speeds depend on local curvature. The method uses a fixed (Eulerian) grid and finds the front as a particular level set (moving with time) of a scalar function. The method applies to a very general class of problems.

The technique handles topological merging and breaking, works in any number of space dimensions, does not require that the moving surface be written as a function, captures sharp gradients and cusps in the front, and is relatively easy to program. Theoretical justification, involving the concept of viscosity solutions, has been given in [13], [7].

Many applications and extensions have recently been found. We shall describe the method and some applications in Section 2. We also note that the motion of multiple junctions using related ideas has been studied in [28]. A particularly novel application and extension (done with Harabetian) is to the numerical study of

Proceedings of the International Congress
of Mathematicians, Zürich, Switzerland 1994
© Birkhäuser Verlag, Basel, Switzerland 1995

unstable fronts — e.g. vortex sheets, in [15]. This will also be described in Section 2. The level set formulation allows for the capturing of the front with minimal regularization because the zero level set of a continuous function can become quite complicated, even though the function itself is easy to compute.

Our last example of subscale capturing involves wavelet based algorithms for linear initial value problems. Using ideas of Beylkin, Coifman, and Rokhlin [2], we have with Engquist, Zhong, and Jiang [11], [19] devised very fast algorithms for evaluating the solution of linear initial value problems with time independent coefficients. This will be described in Section 5.

2.   The Level Set Method for Capturing Moving Fronts

In a variety of physical phenomena, one wishes to follow the motion of a front whose speed is a function of the local geometry and an underlying flow field. Generally the location of the interface or front affects the flow field. Typically there have been two types of numerical algorithms employed in the solution of such problems. The first parameterizes the moving front by some variable and discretizes this parameterization into a set of marker points. The positions of these marker points are updated according to approximations of the equations of motion. For large complex motion, several problems occur. First, marker particles come together in regions where the curvature builds, causing numerical instability unless regridding is used. The regridding mechanism often dominates the real effects. Moreover the numerical methods tend to become quite stiff in these regions — see e.g. [41]. Second, such methods suffer from topological problems; e.g., when two regions merge or a single region splits, ad hoc technologies are required.

Other algorithms commonly employed fall under the category of "volume of fluid" techniques, which track the motion of the interior region; e.g., [29], [3]. These are somewhat more adaptable to topological changes than the tracking methods but still lack the ability to easily compute geometrical quantities such as curvature of the front.

Both methods are difficult to implement in three space dimensional problems. Our idea, as first developed with Sethian in [31] is as follows. Given a region Ω in R^2 or R^3 (which could be multiply connected), and whose boundary is moving with time, we construct an auxiliary function $\varphi(\bar{x}, t)$ that is Lipschitz continuous and has the property

$$\varphi(\bar{x}, t) > 0 \quad \Leftrightarrow \quad \bar{x} \; \varepsilon \; \Omega \quad \text{at time } t \tag{2.1}$$

$$\varphi(\bar{x}, t) < 0 \quad \Leftrightarrow \quad \bar{x} \; \varepsilon \; \Omega^c \quad \text{at time } t \tag{2.2}$$

$$\varphi(\bar{x}, t) = 0 \quad \Leftrightarrow \quad \bar{x} \; \varepsilon \; \partial\Omega \quad \text{at time } t. \tag{2.3}$$

On any level set of φ we have

$$\varphi_t + \vec{u} \cdot \nabla\varphi = 0 \tag{2.4}$$

where $\vec{u} = (\dot{x}(t), \dot{y}(t))$, the motion of the front and the set $\varphi \equiv 0$ characterizes $\partial\Omega$ at time t.

Generally, if the normal velocity $\vec{u} \cdot \vec{n}$ is a given function f of the geometry, the level set motion is governed by

$$\varphi_t + |\nabla\varphi| f = 0. \tag{2.5}$$

Typically (in two dimensions) f is a function of the curvature of the front, $f = f(\kappa) = f\left(\nabla \cdot \left(\frac{\nabla\varphi}{|\nabla\varphi|}\right)\right)$. In this case we can replace (2.4) by an equation involving φ only

$$\varphi_t + |\nabla\varphi| f\left(\nabla \cdot \left(\frac{\nabla\varphi}{|\nabla\varphi|}\right)\right) = 0. \tag{2.6}$$

Our algorithm is merely to extend (2.6) to be valid throughout space and just pick out the zero level set as the front at all later times. Equations of this type for $f'(0) < 0$ have been analyzed in [13], [7] using the theory of viscosity solutions. In addition to well-posedness, it was shown that modulo a few exceptions, the level set method works. This means that the zero level set agrees with the classical motion for smooth, noninteracting curves. Moreover, the asymptotic behavior of certain fronts arising in reaction diffusion equations leads to this motion as the small parameter goes to zero [12].

In many applications involving multiphase flow in fluid dynamics the interface between any two regions can be represented by judiciously using delta functions as source terms in the equations of motion. This is true in particular for computing rising air bubbles in water, falling water drops in air, and in numerous other applications — see e.g. [46],[5]. In fact surface tension often plays a role and this quantity is just proportional to curvature, here easy to compute. Thus, an Eulerian framework is easily set up, using the level set approach, allowing phenomena such as merging of water drops, resulting in surface tension driven oscillations, and drops hitting the base and deforming [46].

A key requirement here and elsewhere is that the level set function φ stay well behaved; i.e., $0 < c \leq |\nabla\varphi| \leq C$ for fixed constants (except for isolated points). In fact it would be desirable to set

$$|\nabla\varphi| = 1 \tag{2.7}$$

with the additional criteria (2.1), (2.2), (2.3). In other words, we wish to replace (at least near $\partial\Omega$) φ by d, the signed distance to the boundary.

We can do this as described in [46], through reinitialization after every discrete update of the system, in a very fast way by obtaining the viscosity solution of

$$d_\tau + (|\nabla d| - 1)H(\varphi) = 0 \tag{2.8}$$

for $\tau > 0$, in fact as $\tau \uparrow \infty$, with $d(\bar{x},0) = \varphi(x,t)$. Here $H(\varphi)$ is any smooth monotone function of φ with $H(0) = 0$.

ENO schemes for Hamilton-Jacobi equations, as defined in [31],[32] may be used to solve this. By the method of characteristics it is clear that, near $\partial\Omega$, which is the zero level set of φ, the steady state is achieved very quickly. We thus have a fast method of computing signed distance to an arbitrary set of closed curves in R^2 or surfaces in R^3.

Another example of the use of this method in fluid dynamics involves area (or volume) preserving motion by mean curvature. This represents the simplified motion of foam and can be modelled simply by finding the zero level set of

$$\varphi_t = |\nabla\varphi| \left(\nabla \cdot \left(\frac{\nabla\varphi}{|\nabla\varphi|} \right) - \bar{\kappa} \right) \tag{2.9}$$

where $\bar{k}$ is the average curvature of the interface. This last can be easily computed

$$\bar{\kappa} = \frac{\int\int_\Omega \left(\nabla \cdot \left(\frac{\nabla\varphi}{|\nabla\varphi|} \right) \right) \delta(\varphi)|\nabla\varphi|}{\int\int_\Omega \delta(\varphi)|\nabla\varphi|}. \tag{2.10}$$

The distance reinitialization is used and the method easily yields merging and topological breaking, see [20]. More realistic models involving volume preserving acceleration by mean curvature are being developed and analyzed with the same group of people.

Another interesting example concerns Stefan problems. Earlier work was done using the level set formulation [42]. Our formulation seems to be quite simple and flexible. We solve for the temperature (in two or three dimensions)

$$
\begin{align}
T_t &= \nabla \cdot k(\bar{x})\nabla T \tag{2.11}\\
k(\bar{x}) &= k_1 \text{ if } \bar{x} \,\varepsilon\, \Omega \tag{2.12}\\
k(\bar{x}) &= k_2 \text{ if } \bar{x} \,\varepsilon\, \Omega^c \tag{2.13}\\
T &= 0 \text{ for } \bar{x} \,\varepsilon\, \partial\Omega \tag{2.14}
\end{align}
$$

and the boundary of Ω moves with normal velocity

$$\vec{v} \cdot \vec{n} = \left[\frac{\partial T}{\partial n} \right] c_1 + c_2\kappa \tag{2.15}$$

where κ = curvature of the front.

We solve this using φ, the level set function, with reinitialization, by using

$$\varphi_t + \vec{u} \cdot \nabla\varphi = 0 \tag{2.16}$$

for u defined semi-numerically as

$$
\begin{align}
\vec{u} &= c_1 \left[\Delta x \Delta_+^x \Delta_-^x T, \ \Delta y \Delta_+^y \Delta_-^y T \right] \tag{2.17}\\
&+ c_2 \left[\nabla \cdot \left(\frac{\nabla\varphi}{|\nabla\varphi|} \right) \right] \frac{\nabla\varphi}{|\nabla\varphi|}
\end{align}
$$

for Δ_+, Δ_- the usual undivided difference operators. The first term on the right is $O(\Delta x, \Delta y)$ except at the front.

We solve (2.11) by using the piecewise constant values k_1 or k_2 except when the discrete operators above cross the level set $\varphi = 0$. At such points we merely interpolate using the distance function to find the x and/or y value at which $T = 0$. We thus can get a one-sided arbitrary high order approximation to $\Delta x T_{xx}$ and/or

$\Delta y T_{yy}$ there. This is also used in (2.17). The results appear to be state of the art for this simple method. This is the result of joint work with Chen, Merriman, and Smereka [6].

Next, with Harabetian [15] we consider an extension of the level set method where the normal velocity need not be intrinsic (solely geometry or position based) and for which the problem written in Lagrangian (moving) coordinates is Hadamard ill posed. The main observation is that our approach provides an automatic regularization. There appear to be at least two reasons for this. The first is topological: a level set of a function cannot change its winding number — certain topological shapes based on the curve crossing itself are impossible. The second is analytical: the linearized problem is well posed in the direction of propagation normal to the level set in this formulation; however it is ill posed overall. The method was developed in [15] in R^2. The three-dimensional extension is relatively straightforward. Our two paradigms were: (1) the initial value problem for the Cauchy-Riemann equations and (2) the motion of a vortex sheet in two-dimensional, incompressible, inviscid fluid flow.

For the latter, we obtained roll-up past time of singularities even though we did not do any explicit filtering in the Fourier frequencies, nor did we use blobs to smooth out the flow as in [23].

Finally we mention that complicated motion of multiple junctions can be rather simply implemented by using as many level set functions as there are regions, see [28]. Also, in the special case of mean curvature motion, the simple heat equation together with a projection may be used [28].

3. Shock Capturing Methods

There is a vast amount of literature on this subject, also see [16] for a recent review article at the 1990 ICM. The fundamental problem is that the solution to the initial value problem for a system of hyperbolic conservation laws generally develops discontinuities (shocks) in finite time, no matter how smooth the initial data is. Weak solutions must be computed. The goal is to develop numerical methods that "capture" shocks automatically. Reasonable design principles are:

(1) Conservation form (defines shock capturing, see [14], [25]).

(2) No spurious overshoots, wiggles near discontinuities, yet sharp discrete shock profiles.

(3) High accuracy in smooth regions of the flow.

(4) Correct physical solution; i.e., satisfaction of the entropy conditions in the convergent limit [24].

Conventional methods had trouble with combining (1) and (3). It should be noted that wiggles can pollute the solution causing e.g. negative densities and pressures and other instabilities.

We have developed with Harten, Engquist, and Chakravarthy [18], [17] and later simplified with Shu [44], [45] a class of shock capturing algorithms designed to satisfy principles (1)–(4).

These methods are called essentially nonoscillatory (ENO) schemes. They resemble their predecessors — total variation diminishing (TVD) schemes — in that the stencil is adaptive; however, the total variation of the solution of the approximation to a one space dimensional scalar model might increase, but only at a rate $O((\text{grid size})^p)$, for p the order of the method, up to discontinuities, and the order of accuracy can be made arbitrary in regions of smoothness. TVD schemes traditionally degenerate to first order at isolated extrema (see [40] for extensions up to second order).

The basic idea is to extend Godunov's [14] ingenious idea past first order accuracy. This was first done up to second order accuracy by van Leer [47]. A key step, and the only one we have time to describe here, is the construction of a piecewise polynomial of degree m, which interpolates discrete data w given at grid x_j. In each cell $d_j = \{(x)x_j \leq x \leq x_{j+1}\}$ we construct a polynomial of degree m that interpolates $w(x)$ at $m+1$ successive points $\{x_i\}$ including x_j and x_{j+1}.

The idea is to avoid creating oscillations by choosing the points using the "smoothest" values of w. (This is a highly nonlinear choice, as it must be.) One way of doing this is to use the Newton interpolating polynomials and the associated coefficients. We start with a linear interpolant in each cell

$$q_{1,j+\frac{1}{2}} = w[x_j] + (x - x_j)w[x_j, x_{j+1}] \tag{3.1}$$

using the Newton coefficients

$$w[x_i] = w(x_i) \tag{3.2}$$
$$w[x_i, \ldots, x_{i+k}] = (x_{i+k} - x_i)^{-1}(w[x_{i+1}, \ldots, x_k] - w[x_i, \ldots, x_{k-1}]). \tag{3.3}$$

We get two candidates for $q_{2,j+\frac{1}{2}}$, which interpolate w at x_j, x_{j+1} and either x_{j-1}, or x_{j+2}

$$q_{2,j+\frac{1}{2}} = q_{1,j+\frac{1}{2}} + (x - x_j)(x - x_{j+1})[w[x_{j-1}, x_j, x_{j+1}] \text{ or } w[x_j, x_{j+1}, x_{j+2}]]. \tag{3.4}$$

Because we are trying to minimize oscillations by taking information from regions of smoothness, we pick the coefficient that is *smaller* in magnitude. We store this choice and proceed inductively up to degree m. The result is a method that is exact for piecewise polynomials of degree $\leq m$ and that is nonoscillatory (i.e. essentially monotone) across jumps. See [17] for further discussions.

Other choices are possible, in fact it seems advantageous to minimize truncation error by biasing the choice of stencil towards the center, see [43], [33].

4.   Image Processing

In his 1987 Ph.D. thesis, Rudin [36] made the connection between various tasks in image processing and the numerical solution of nonlinear partial differential equations whose solutions develop steep gradients. Images are characterized by edges and other singularities, thus the techniques used in shock capturing are relevant here. There are now many examples of this connection. We shall discuss only a few here.

We extend the notion of "shock filter" described first in [36] to enhance images that were first blurred by a mild smoothing process. Consider the (apparently ill-posed) initial value problem

$$u_t = -|\nabla u| F[(D^2 u \cdot \nabla u, \ \nabla u)] \tag{4.1}$$
$$u(x, y, 0) = u_0(x, y) \tag{4.2}$$

where $F(A)$ is an increasing function with $F(0) = 0$.

Here, $u_0(x, y)$ is the blurry image to be processed. Intuitively if, for example, $F(A)$ is the sign function, then the process involves propagating data towards blurred out edges, (zeros of the edge detector $((D^2 u)\nabla u, \nabla u)))$. The apparent ill-posedness is taken care of by the choice of finite difference approximation, which has the effect of turning off the motion at isolated extrema. See [30] for a further discussion of this. We note here that the resulting motion satisfies a local maximum principle and, in one space dimension, preserves the total variation of the original image.

An important extension of these ideas comes in the development of a total variation based restoration algorithm [38],[39]. We are given a blurry noisy image

$$u_0(x, y) = (Au)(x, y) + \bar{n}(x, y) \tag{4.3}$$

where A is a linear integral operator and $\bar{n}$ is additive noise. Also u_0 is the observed intensity function and u is the image to be restored. The method is quite general — A needs only to be a compact operator.

We minimize the total variation

$$\text{minimize} \int_\Omega \sqrt{u_x^2 + u_y^2} \, dx \, dy \tag{4.4}$$

subject to constraints on u involving the mean and variance of the noise

$$\int_\Omega u \, dx \, dy = \int_\Omega u_0 \, dx \, dy \tag{4.5}$$
$$\int (Au - u_0)^2 \, dx \, dy = \sigma^2. \tag{4.6}$$

We use the gradient projection method of Rosen [34], which in this case becomes the interesting "constrained" time dependent partial differential equation

$$u_t = \nabla \cdot \left(\frac{\nabla u}{|\nabla u|} \right) - \lambda A^*(Au - u_0) \tag{4.7}$$

for $t > 0$, $(x, y) \ \varepsilon \ \Omega$ with boundary conditions

$$\frac{\partial u}{\partial n} = 0 \ \text{ on } \partial\Omega \tag{4.8}$$

and $u(x, y, 0)$ given so that (4.5), (4.6) are satisfied.

The Lagrange multiplier is chosen so as to preserve (4.6), (the constraint (4.5) is automatic).

The method generalizes to multiplicative and other types of noise, and to localized constraints (suggested and implemented by Rudin). Theoretical justification and results on multiplicative noise are presented in [26]. The important observation is that noisy edges can be recovered to be crisp (reminiscent of shock capturing) without smearing or oscillations. See [38],[39],[26] for successful restoration of images using this approach and [9] for applications to different inverse problems. From a geometric point of view, (4.7) represents the motion of each level set of u normal to itself with normal velocity equal to its curvature divided by the magnitude of the gradient of u. The constraint term just acts to project the motion back so that (4.6) is satisfied. We note here that Alvarez-Guichard-Lions-Morel in a very important paper [1] demonstrate that the axioms of multiscale analysis lead inexorably to motion by mean curvature and variants, as in [31]. This sort of motion is also important in computer vision and shape recognition, see e.g. [4],[21].

The notions of subscale resolution also appear in segmentation [22], decluttering [37], reconstruction of shapes-from-shading [35], [27], etc.

5. Fast Wavelet Based Algorithms for Linear Initial Value Problems

This is joint work with Engquist and Zhong [11] based on results in [2]. We are interested in the fast numerical solution of a system of evolution equations

$$u_t + L(x, \partial_x)u = f(x), \quad x \in \Omega \subset R^d, \ t > 0 \tag{5.1}$$
$$u(x, 0) = u_0(x)$$

+ boundary conditions.

Here $L(x, \frac{\partial}{\partial x})$ is a linear differential operator.

We shall take an explicit discretization

$$u_j^n = u(x_j, t_n), \quad t_n = n\Delta t \tag{5.2}$$
$$x_j = (j_1 \Delta x_1, \ldots, j_d \Delta x_d)$$
$$u^{n+1} = Au^n + F \tag{5.3}$$
$$u^0 = u_0$$
$$u_0, F \in R^{N^d}, \quad \Delta t = \text{const } (\Delta x)^r.$$

The u^n vector contains all u_j^n at time level t_n.

The matrix A is $(N^d \times N^d)$ with the number of nonzero elements in each row or column bounded by a constant.

Each time step requires $O(N^d)$ arithmetic operations. The overall complexity for $t = 0(1)$ is $O(N^{d+r}) = $ (number of unknowns).

We proposed a general approach to speed up this calculation that works extraordinarily well for parabolic equations and is quite promising for hyperbolic equations.

We solve the discretization:

$$u^n = A^n u_0 + \sum_{\nu=0}^{n-1} A^\nu F. \tag{5.4}$$

We compute the solution for $F = 0$ in $\log_2 n$ steps ($n = 2^m$, $m =$ integer). Repeatedly square A

$$A^2, A^4, \ldots, A^{2^m}.$$

(This is why the equation needs to have time independent coefficients.)

Unfortunately, the later squarings involve almost dense matrices so the overall complexity is $O(N^{3d} \log N)$, which is worse than the straightforward approach.

Observation based on [2]: for the representation of A in a wavelet basis, all of the powers of A^r may be approximated by uniformly sparse matrices, and the algorithm using repeated squaring is advantageous.

Algorithm:

$$
\begin{aligned}
B &= SAS^{-1} \\
C &= I \\
C &= TRUNC(C + BC, \epsilon) \\
B &= TRUNC(BB, \epsilon) \\
u^n &= S^{-1}(BSu^0 + CSF) \\
S &= \text{fast wavelet transform}
\end{aligned}
$$

$$
TRUNC(A, \epsilon) \begin{cases} \tilde{a}_{ij} = a_{ij} \text{ if } |a_{ij}| > \epsilon \\ \tilde{a}_{ij} = 0, \text{ if } |a_{ij}| < \epsilon. \end{cases} \tag{5.5}
$$

If $\epsilon = 0$ we get the usual operator (up to similarity).

For a fixed accuracy predetermined, the computational complexity to compute a one-dimensional hyperbolic equation can be reduced from $O(N^2)$ to $O(N(\log N)^3)$ with small constant.

For parabolic d-dimensional an explicit calculation with standard complexity $O(N^{d+2})$ can be reduced to $O(N^d(\log N)^3)$.

Extensions to periodic in time sources $f(x, t)$ are easy.

Together with Jiang [19] we have shown the following: if we wish to evaluate the solution only in the neighborhood of one point $x = x^*$ at $t = t^n$, the complexity decreases tremendously; e.g., for a one-dimensional parabolic equation it becomes $O(\log^4 N)$ as opposed to $O(N(\log^3 N))$.

For a general multidimensional parabolic equation, the complexity is again only $O(\log^4 N)$.

For a d-dimensional hyperbolic system the complexity is $O(N^{(2d-2)} \log^3 N)$. This is advantageous for dimension $d = 1, 2$. We expect to do better using more localized basis functions of Coifman and Meyer see ([8]), and using a nonlinear partial differential equation based replacement for ray tracing see ([10]).

References

[1] L. Alvarez, F. Guichard, P. L. Lions, and J. M. Morel, *Axioms and fundamental equations of image processing*, Arch. Rational Mech. Anal., **123**, (1993), 199–258.

[2] G. Beylkin, R. Coifman, and V. Rokhlin, *Fast wavelet transforms and numerical algorithms, I*, Comm. Pure Appl. Math., **64**, (1991), 141–184.

[3] J. Brackbill, D. Kothe, and C. Zemach, *A continuum method for modeling surface tension*, J. Comput. Phys., **100**, (1992), 335–353.

[4] V. Caselles, F. Catte, T. Coll, and F. Dibos, *A geometric model for active contours in image processing*, Numer. Math., **66**, (1993), 1–31.

[5] Y. C. Chang, T. Y. Hou, B. Merriman, and S. Osher, *A level set formulation of Eulerian interface capturing methods for incompressible fluid flows*, to appear, J. Comput. Phys., (1994).

[6] S. Chen, B. Merriman, P. Smereka, and S. Osher, *A fast level set based algorithm for Stefan problems*, preprint, (1994).

[7] Y. G. Chen, Y. Giga, and S. Goto, *Uniqueness and existence of viscosity solutions of generalized mean curvature flow equations*, J. Differential Geom., **23**, (1986), 749–785.

[8] R. Coifman and Y. Meyer, *Remarques sur l'analyse de Fourier à fenêtre, série I*, C.R. Acad. Sci. Paris, **312**, (1991), 259–261.

[9] D. Dobson and F. Santosa, *An image enhancement for electrical impedance tomography, inverse problems*, to appear, (1994).

[10] B. Engquist, E. Fatemi, and S. Osher, *Numerical Solution of the High Frequency Asymptotic Expansion for Hyperbolic Equations*, in Proceedings of the 10th Annual Review of Processers in Applied Computational Electromagnetics, Monterey, CA, (1994), A. Terzuoli, ed., vol. 1, 32–44.

[11] B. Engquist, S. Osher, and S. Zhong, *Fast wavelet algorithms for linear evolution equations*, SIAM J. Sci. Statist. Comput., **15**, (1994), 755–775.

[12] L. C. Evans, M. Soner, and P. Souganidis, *Phase transitions and generalized motion by mean curvature*, Comm. Pure Appl. Math., **45**, (1992), 1097–1123.

[13] L. C. Evans and J. Spruck, *Motion of level sets by mean curvature, I*, J. Differential Geom., **23**, (1986), 69–96.

[14] S. Godunov, *A difference scheme for computation of discontinuous solutions of equations of fluid dynamics*, Math. Sbornik, **47**, (1959), 271–306.

[15] E. Harabetian and S. Osher, *Stabilizing ill-posed problems via the level set approach*, preprint, (1994).

[16] A. Harten, *Recent developments in shock capturing schemes*, Proc Internat. Congress Math., Kyoto 1990, (1990), 1549–1573.

[17] A. Harten, B. Engquist, S. Osher, and S. R. Chakravarthy, *Uniformly high order accurate essentially nonoscillatory schemes, III*, J. Comput. Phys., **71**, (1987), 231–303.

[18] A. Harten and S. Osher, *Uniformly high-order accurate nonoscillatory schemes, I*, SINUM, **24**, (1987), 279–304.

[19] A. Jiang, *Fast Wavelet algorithms for solving linear equations*, Ph.D. Prospectus, UCLA Math., (1993).

[20] M. Kang, P. Smereka, B. Merriman, and S. Osher, *On moving interfaces by volume preserving velocities or accelerations*, preprint, (1994).

[21] R. Kimmel, N. Kiryati, and A. Bruckstein, *Sub-pixel distance maps and weighted distance transforms*, JMIV, to appear, (1994).

[22] G. Koepfler, C. Lopez, and J. M. Morel, *A multiscale algorithm for image segmentation by variational method*, SINUM, **31**, (1994), 282–299.

[23] R. Krasny, *Computing vortex sheet motion*, Proc. Internat. Congress Math., Kyoto 1990, (1990), 1573–1583.

[24] P. D. Lax, *Weak solutions of nonlinear hyperbolic equations and their numerical computation*, Comm. Pure Appl. Math., **7**, (1954), 159–193.

[25] P. D. Lax and B. Wendroff, *Systems of conservation laws*, Comm. Pure Appl. Math., **13**, (1960), 217–237.

[26] P. L. Lions, S. Osher, and L. Rudin, *Denoising and deblurring images with constrained nonlinear partial differential equations*, submitted to SINUM.

[27] P. L. Lions, E. Rouy, and A. Tourin, *Shape from shading, viscosity solutions, and edges*, Numer. Math., **64**, (1993), 323–354.

[28] B. Merriman, J. Bence, and S. Osher, *Motion of multiple junctions: A level set approach*, J. Comput. Phys., **112**, (1994), 334–363.

[29] W. Noh and P. Woodward, *A simple line interface calculation*, Proceeding, Fifth Int'l. Conf. on Fluid Dynamics, A.I. van de Vooran and D. J. Zandberger, eds., Springer-Verlag, (1970).

[30] S. Osher and L. I. Rudin, *Feature-oriented image enhancement using shock filters*, SINUM, **27**, (1990), 919–940.

[31] S. Osher and J. A. Sethian, *Fronts propagating with curvature dependent speed, algorithms based on a Hamilton-Jacobi formulation*, J. Comput. Phys., Vol 79, (1988), 12–49.

[32] S. Osher and C.-W. Shu, *High-order essentially nonoscillatory schemes for Hamilton-Jacobi equations*, SINUM, **28**, (1991), 907–922.

[33] A. Rogerson and E. Meiburg, *A numerical study of convergence properties of ENO schemes*, J. Sci. Comput., **5**, (1990), 151–167.

[34] J. G. Rosen, *The gradient projection method for nonlinear programming, II, Nonlinear constraints*, J. SIAM, **9**, (1961), 514–532.

[35] E. Rouy and A. Tourin, *A viscosity solution approach to shape from shading*, SINUM, **27**, (1992), 867–884.

[36] L. Rudin, *Images, numerical analysis of singularities, and shock filters*, Caltech Comp. Sc. Dept. Report # TR 5250:87, (1987).

[37] L. Rudin, G. Koepfler, F. Nordby, and J. M. Morel, *Fast variational algorithm for clutter removal through pyramidal domain decomposition*, Proceedings SPIE Conference, San Diego, CA, July, 1993.

[38] L. Rudin, S. Osher, and E. Fatemi, *Nonlinear total variation based noise removal algorithms*, Phys. D, **60**, (1992), 259–268.

[39] L. Rudin, S. Osher, and C. Fu, *Total variation based restoration of noisy, blurred images*, SINUM, to appear.

[40] R. Sanders, *A third order accurate variation nonexpansive difference scheme for a single nonlinear conservation*, Math. Comp., 51, (1988), 535–558.

[41] J. A. Sethian, *Curvature and the evolution of fronts*, Comm. Math. Phys., **101**, (1985), 487–499.

[42] J. Sethian and J. Strain, *Crystal growth and dendrite solidification*, J. Comput. Phys., **98**, (1992), 231–253.

[43] C.-W. Shu, *Numerical experiments on the accuracy of ENO and modified ENO schemes*, J. Sci. Comput., **5**, (1990), 127–150.

[44] C.-W. Shu and S. Osher, *Efficient implementation of essentially nonoscillatory schemes I*, J. Comput. Phys., **77**, (1988), 439–471.

[45] C.-W. Shu and S. Osher, *Efficient implementation of essentially nonoscillatory schemes II*, J. Comput. Phys., **83**, (1989), 32–78.

[46] M. Sussman, P. Smereka, and S. Osher, *A level set approach for computing solutions to incompressible two phase flow*, to appear, J. Comput. Phys., (1994).

[47] B. Van Leer, *Towards the ultimate conservative difference scheme V. A second order sequel to Godunov's method*, J. Comput. Phys., **32**, (1979), 101–136.

Analysis-Based Fast Numerical Algorithms of Applied Mathematics

VLADIMIR ROKHLIN

Computer Science Department, Yale University
New Haven, CT 06520, USA

1. Introduction

One of the principal problems addressed by applied mathematics is the application of various linear operators (or rather, their discretizations) to more or less arbitrary vectors. As is well known, applying directly a dense $(N \times N)$-matrix to a vector requires roughly N^2 operations, and this simple fact is a cause of serious difficulties encountered in large-scale computations. For example, the main reason for the limited use of integral equations as a numerical tool in large-scale computations is that they normally lead to dense systems of linear algebraic equations, and the latter have to be solved, either directly or iteratively. Most iterative methods for the solution of systems of linear equations involve the application of the matrix of the system to a sequence of recursively generated vectors, which tends to be prohibitively expensive for large-scale problems. The situation is even worse if a direct solver for the linear system is used, as such solvers normally require $O(N^3)$ operations. As a result, in most areas of computational mathematics dense matrices are simply avoided whenever possible. For example, finite difference and finite element methods can be viewed as devices for reducing a partial differential equation to a sparse linear system. In this case, the cost of sparsity is the inherently high condition number of the resulting matrices.

For translation invariant operators, the problem of excessive cost of applying (or inverting) the dense matrices has been met by the Fast Fourier Transform (FFT) and related algorithms (fast convolution schemes, etc.). These methods use algebraic properties of a matrix to apply it to a vector in order $N \log(N)$ operations. Such schemes are exact in exact arithmetic, and are fragile in the sense that they depend on the exact algebraic properties of the operator for their applicability. During the last several years, a group of algorithms has been introduced for the rapid application to arbitrary vectors of matrices resulting from the discretization of integral equations from several areas of applied mathematics. The schemes include the Fast Multipole Method (FMM) for the Laplace equation in two and three dimensions (see, for example, [11]), the fast Gauss transform (see [12]), the fast Laplace transform (see [16], [20]), and several other schemes. In all cases, the resulting algorithms have asymptotic CPU time estimates of either $O(n)$ or $O(n \cdot \log(n))$, and are a dramatic improvement over the classical ones for large-scale problems.

Proceedings of the International Congress
of Mathematicians, Zürich, Switzerland 1994
© Birkhäuser Verlag, Basel, Switzerland 1995

Each of such schemes is based on one of two approaches.

(1) The first approach utilizes the fact that the kernel of the integral operator to be applied is smooth (away from the diagonal or some other small part of the matrix), and decomposes it into some appropriately chosen set of functions (Chebyshev polynomials in [16] and [3], wavelets in [4], wavelet-like objects in [2], etc.). This approach is extremely general and easy to use, as a single scheme is applicable to a wide class of operators.

(2) The second approach is restricted to the cases when the integral operator has some special analytical structure, and uses the corresponding special functions (multipole expansions for the Laplace equation in [11], Hermite polynomials in [12], Laguerre polynomials in [20], etc.

In this approach, a special-purpose algorithm has to be constructed for each narrow class of kernels, and in each case their appropriate special functions and translation operators (historically known as addition theorems) have to be available. However, once constructed, such algorithms tend to be extremely efficient. In addition, there are several important situations where the first approach fails, but the second can be used (a typical example is the n-body gravitational problem with a highly nonuniform distribution of particles, as in [7]).

All of the above algorithms are in fact based on the simple observation that many matrices of applied mathematics are smooth functions of their indices away from the main diagonal, or from some other small part of the matrix. Thus, their large submatrices can be approximated to any prescribed accuracy with matrices of low rank (sometimes the rank of such matrices depends only on the accuracy of the approximation; in other cases, it is proportional to the logarithm of the dimensionality of the submatrix). As a result, algorithms of this type are conceptually quite straightforward, and tend to require only a very limited mathematical apparatus. Both of the above approaches fail when the kernel is highly oscillatory, and simple counter-examples show that it is impossible to construct a scheme that would work in the general oscillatory case (the Nyquist theorem being the basic obstacle). However, several oscillatory problems are of sufficient importance that it is worthwhile to construct special-purpose methods for them. A typical example is kernels satisfying the Helmholtz equation in two and three dimensions, since this is the equation controlling the propagation of acoustic and electromagnetic waves, and many quantum-mechanical phenomena. Unlike the nonoscillatory case, the oscillatory one requires a fairly subtle mathematical apparatus, and for the Helmholtz equation such an apparatus is constructed in [17] in two dimensions and in [18] in the three-dimensional case.

The purpose of this article is to give a brief and complete, but not rigorous, exposition of the Fast Multipole Method (FMM) for the Helmholtz equation. For purposes of demonstration, we consider the scalar wave equation with Dirichlet boundary conditions on the surface of a scatterer. The FMM provides an efficient mechanism for the numerical convolution of the Green function for the Helmholtz equation with a source distribution and can be used to radically accelerate the iterative solution of boundary integral equations. In the simple single-stage form presented here, it reduces the computational complexity of the convolution from order N^2 to order $N^{3/2}$, where N is the number of nodes in the discretization of the problem. By implementing a multistage FMM (see [17], [18]), the complexity can be further reduced to order $N \log N$. However, even for problems that have an order of magnitude more variables than those currently tractable using dense matrix techniques ($N \approx 10^5$), we estimate that the performance of a carefully implemented single-stage algorithm should be near optimal.

The structure of this article is as follows. In Section 2, we define notation and introduce the analytical apparatus of the FMM. A detailed prescription for FMM implementation, except for the choice of some important parameters of the algorithm, is given in Section 3. After the structure of the method is exhibited, these parameters (the number of terms used in the multipole expansion, and the directions at which far-field quantities are tabulated) are analyzed in Section 4.

2. Analytical Apparatus

2.1. Notation. Vectors in three-dimensional space are represented by boldface ($\mathbf{x}$). The magnitude of a vector $\mathbf{x}$ is written as $x \equiv |\mathbf{x}|$, unit vectors as $\hat{x} \equiv \mathbf{x}/x$, and integrals over the unit sphere as $\int d^2\hat{x}$. The imaginary unit is denoted by i.

2.2. Time Independent Scattering and the Nyström Algorithm. We will be considering a scattering problem defined by the scalar wave equation

$$(\nabla^2 + k^2)\psi = 0\,, \tag{1}$$

a Dirichlet boundary condition

$$\psi(\mathbf{x}) = 0 \tag{2}$$

on the surface S of a bounded scatterer, and the radiation condition at infinity. The Nyström method provides a discretization of the first kind integral equation associated with this problem, giving a set of linear equations with a dense coefficient (impedance) matrix:

$$Z_{nn'} = w_i \cdot \frac{e^{ik|\mathbf{X}_n - \mathbf{X}_{n'}|}}{4\pi|\mathbf{x}_n - \mathbf{x}_{n'}|}, \tag{3}$$

with the nodes $\{\mathbf{x}_n\}$, $n = 1, 2, \ldots, N$, distributed more or less uniformly on the boundary of the scatterer, and the coefficients $\{w_n\}$, $n = 1, 2, \ldots, N$ chosen to be the weights of an appropriate quadrature formula (see, for example, [13]). The FMM provides a prescription for the rapid computation of the matrix-vector product

$$B_n = \sum_{n'=1}^{N} Z_{nn'} I_{n'} \tag{4}$$

for an arbitrary vector I. This rapid computation can then be used in an iterative (e.g. conjugate gradient) solution of the discretized integral equation $Z \cdot I = V$, where, for an incident wave with wave vector $\mathbf{k}$,

$$V_n(\mathbf{k}) = e^{i\mathbf{k}\cdot\mathbf{x}}. \tag{5}$$

REMARK 2.1. Although in this article we use a first kind integral equation to solve the Dirichlet problem (1), (2), we do so because of the simplicity of the formulation, not because we prefer it as a numerical tool. Similarly, we solve the

Dirichlet problem simply as an illustration of the technique, not because of any deep preference. In fact, many kinds of boundary value problems for the Helmholtz equation are encountered (see, for example, [10]), and our preferred technique is the **second** kind integral equations, whenever such formulations are available.

2.3. Identities. The FMM, as presented here, rests on two elementary identities. These are found in many texts and handbooks on special functions, such as [1]. The first identity is an expansion of the kernel in the formula (3) for the impedance matrix elements, and is a form of Gegenbauer's addition theorem,

$$\frac{e^{ik|\mathbf{X}+\mathbf{d}|}}{|\mathbf{X}+\mathbf{d}|} = ik \sum_{l=0}^{\infty} (-1)^l (2l+1) j_l(kd) h_l(kX) P_l(\hat{d} \cdot \hat{X}), \tag{6}$$

where j_l is a spherical Bessel function of the first kind, h_l is a spherical Hankel function of the first kind, P_l is a Legendre polynomial, and $d < X$. When using this expansion to compute the field at $\mathbf{x}$ from a source at $\mathbf{x}'$, $\mathbf{X}$ will be chosen to be close to $\mathbf{x} - \mathbf{x}'$ so that d will be small.

The second identity is an expansion of the product $j_l P_l$ in propagating plane waves:

$$4\pi i^l j_l(kd) P_l(\hat{d} \cdot \hat{X}) = \int d^2 \hat{k} \, e^{i\mathbf{k}\cdot\mathbf{d}} P_l(\hat{k} \cdot \hat{X}). \tag{7}$$

Substituting (7) into (6), we get

$$\frac{e^{ik|\mathbf{X}+\mathbf{d}|}}{|\mathbf{X}+\mathbf{d}|} = \frac{ik}{4\pi} \int d^2 \hat{k} \, e^{i\mathbf{k}\cdot\mathbf{d}} \sum_{l=0}^{\infty} i^l (2l+1) h_l(kX) P_l(\hat{k} \cdot \hat{X}), \tag{8}$$

where we have performed the illegitimate but expedient interchange of summation and integration. The key point is that we intend to precompute the function

$$\mathcal{T}_L(\kappa, \cos\theta) = \sum_{l=0}^{L} i^l (2l+1) h_l(\kappa) P_l(\cos\theta) \tag{9}$$

for various values of κ. This is not a function in the limit $L \to \infty$ but that need not concern us, as we obviously intend to truncate the sum in numerical practice. The number of kept terms $L+1$ will depend on the maximum allowed value of kd, as well as the desired accuracy. The choice of L is discussed in Section 4. It suffices for the present to note that, in order to obtain accuracy from (6), it must be slightly greater than kD, where D is the maximum value of d for which the expansion will be used.

Ignoring this question for now (except for noting that the required number of terms becomes small as $D \to 0$), we have

$$\frac{e^{ik|\mathbf{X}+\mathbf{d}|}}{|\mathbf{X}+\mathbf{d}|} \approx \frac{ik}{4\pi} \int d^2 \hat{k} \, e^{i\mathbf{k}\cdot\mathbf{d}} \mathcal{T}_L(kX, \hat{k} \cdot \hat{X}). \tag{10}$$

Using this, the impedance matrix element (3) is given by the formula

$$Z_{nn'} \approx \frac{k}{(4\pi)^2} \int d^2 \hat{k} \, e^{i\mathbf{k}\cdot(\mathbf{x}-\mathbf{x}'-\mathbf{X})} \mathcal{T}_L(kX, \hat{k} \cdot \hat{X}). \tag{11}$$

In infinite precision arithmetic and in the limit of large L, this result would be independent of the choice of $\mathbf{X}$ (for $X > |\mathbf{x} - \mathbf{x}' - \mathbf{X}|$). In practice, one chooses $\mathbf{x}$ to make $\mathbf{x} - \mathbf{x}' - \mathbf{X}$ relatively small, so that excellent accuracy can be obtained with a modest value of L. Notice that (11) gives the impedance matrix element (for well-separated interactions) in terms of the Fourier transforms *with wave number k* of the basis functions; i.e., the basis functions' far fields. The acceleration provided by the FMM comes from the fact that these far fields can be grouped together *before* the integral over $\hat{k}$ is performed.

REMARK 2.2: Physically, (10) can be interpreted as a decomposition of the Green's function for the Helmholtz equation into a collection of plane waves. From this point of view, it is immediately clear that $\mathcal{T}_L$ can not have a limit as $L \to \infty$, as then $e^{ik|\mathbf{X}+\mathbf{d}|}/|\mathbf{X} + \mathbf{d}|$ would be a singular function whose Fourier transform has compact support. However, with proper choice of L, the algorithm produces arbitrarily high precision.

3. Algorithmic Prescription

3.1. Setup.

1. Divide the N nodes into M localized groups, labeled by an index m, each supporting about N/M nodes. (For now, M is a free parameter. Later it will be seen that the best choice will be $M \sim \sqrt{N}$.) Thus, establish a correspondence between the node's index n and a pair of indices (m, α), where α labels the particular node within the mth group. Denote the center of the smallest sphere enclosing each group as $\mathbf{x}_m$.

   It is assumed for purposes of illustration only that each patch supports only one node.

2. For group pairs (m, m') that contain "nearby" nodes (defined for now as those separated by a distance comparable to or smaller than a wavelength $(2\pi/k)$, so that (11) is valid) construct the sparse matrix Z', with elements

$$Z'_{m\alpha m'\alpha'} = Z_{n(m,\alpha)n'(m',\alpha')} \tag{12}$$

   by direct numerical evaluation of the matrix elements (3). For all other pairs, $Z'_{m\alpha m'\alpha'} = 0$.

   This part of the matrix computation is identical to what is conventionally done. All matrix elements whose computations require subtraction of singularities belong to Z'. If the large N limit is taken with a fixed discretization interval and nearness criterion, this step would require $O(N)$ computations. In Section 4, we define nearby regions precisely, and it turns out that their volume increases as $\sqrt{N}$, so that this step requires $O(N^{3/2})$ computations.

3. For K directions $\hat{k}$, compute the "excitation vectors"

$$V_{m\alpha}(\hat{k}) = e^{i\mathbf{k} \cdot (\mathbf{x} - \mathbf{x}_m)}, \tag{13}$$

   where k is considered to be a parameter of the problem, not a variable. Because K needs to be chosen to give accurate numerical quadrature for all harmonics to some order $\propto L \sim kD$, $K \propto L^2 \sim (kD)^2$, and because (from geometrical considerations) $kD \propto \sqrt{N/M}$, this step requires order N^2/M computations.

4. For each pair (m, m') for which $Z'_{m\alpha m'\alpha'} = 0$ (regions that are not nearby), compute the matrix elements

$$T_{mm'}(\hat{k}) = \frac{k}{(4\pi)^2} \sum_{l=0}^{L} i^l (2l+1) h_l(kX_{mm'}) P_l(\hat{k} \cdot \hat{X}_{mm'}), \qquad (14)$$

for the same K directions $\hat{k}$ as in the previous step, where $L \propto \sqrt{K}$. If done in a naive manner, this computation requires order $KLM^2 \sim M^{1/2}N^{3/2}$ operations, but can be accomplished more rapidly in a number of ways, the most straightforward being the fast Legendre transform (see [3]).

3.2. Fast Matrix-Vector Multiplication.   Rapid computation of the vector elements

$$B_{m\alpha} = \sum_{m'\alpha'} Z_{m\alpha m'\alpha'} I_{m'\alpha'} \qquad (15)$$

is accomplished by the following steps:

1. Compute the KM quantities

$$s_m(\hat{k}) = \sum_{\alpha} V_{m\alpha} \star (\hat{k}) I_{m\alpha}, \qquad (16)$$

which represent the far fields of each group of N/n nodes. This step requires order $KN \sim N^2/M$ operations.

2. Compute the KM quantities

$$g_m(\hat{k}) = \sum_{m'} T_{mm'}(\hat{k}) s_{m'}(\hat{k}). \qquad (17)$$

These represent the Fourier components of the field in the neighborhood of group m generated by the sources in the groups that are not nearby. This step requires order $KM^2 \sim MN$ operations.

3. Finally, compute

$$B_{m\alpha} = \sum_{m'\alpha'} Z'_{m\alpha m'\alpha'} I_{m'\alpha'} + \int d^2\hat{k} \, V_{m\alpha}(\hat{k}) g_m(\hat{k}). \qquad (18)$$

The first term is the standard evaluation of near interactions, and the second term gives the far interactions in terms of the far fields generated by each group. Obviously, this step requires order $KN \sim N^2/M$ operations.

Straightforward substitution of Eqs. (13), (14), (16), and (17) into (18) and (9), (11) into (15) shows that the two expressions for the vector B, Eqs. (18) and (15), give equal results. Thus, computation of the vector B requires $aNM + bN^2/M$ operations, where a and b are machine and implementation dependent. The total operation count is minimized by choosing $M = \sqrt{bN/a}$; the result is an order $N^{3/2}$ algorithm.

4. Required Number of Multipoles and Directions

In this section, we show how to choose the summation limit in the transfer function $T_{mm'}(\hat{k})$ (see (14)) to achieve the desired accuracy (in the process giving a precise definition of nearby regions). We also discuss how to choose the K directions $\hat{k}$ for the tabulation of angular functions.

One must choose L sufficiently large so that the multipole expansion (6) of the Green function converges to the desired accuracy. As a function of l, the Bessel functions $j_l(z)$ and $h_l(z)$ are of roughly constant magnitude for $l < z$. For $l > z$, $j_l(z)$ decays rapidly and $h_l(z)$ grows rapidly. Although one must choose $L > kd = k|\mathbf{x} - \mathbf{x}' - \mathbf{x}_{mm'}|$ (so that the partial wave expansion has converged), L cannot be taken to be much larger than $kX_{mm'}$, because the transfer function (9) will grow rapidly while oscillating wildly. This will first cause inaccuracies in the numerical angular integrations in (10), (18), and then result in catastrophic round-off errors. This condition is a consequence of the interchange of summation and integration in (8). An excellent semi-empirical fit to the number of multipoles required for single precision (32-bit reals) is

$$L_s(kD) = kD + 5\ln(kD + \pi), \tag{19}$$

where $D \geq 1/k$ is the maximum d that will be required (the diameter of the node groups). For double precision (64-bit reals), a good estimate is

$$L_d(kD) = kD + 10\ln(kD + \pi). \tag{20}$$

If the L dictated by the appropriate formula exceeds $kX_{mm'}$, then the groups are too close to use the FMM, and their interaction must be included in the sparse matrix Z'.

The K directions $\hat{k}$ at which the angular functions are tabulated must be sufficient to give a quadrature rule that is exact for all spherical harmonics of order $l < 2L$. A simple method (see [18]) for accomplishing this is to pick polar angles θ such that they are zeros of $P_L(\cos\theta)$ and azimuthal angles ϕ to be $2L$ equally spaced points. Thus, for this choice of $\hat{k} = (\sin\theta\cos\phi, \sin\theta\sin\phi, \cos\theta)$, $K = 2L^2$. If more efficient quadrature rules for the sphere of the type described by McLaren (see [14]) are used, then $K \approx (4/3)L^2$.

5. Conclusions

During the last several years, analysis-based "fast" algorithms have been added to the classical algebraic "fast" techniques. Although the currently popular analytical techniques are inherently incapable of handling highly oscillatory matrices, for many important types of operators it is possible to construct extremely effective special-purpose algorithms. In this article, we outline one such scheme for the Helmholtz equation in three dimensions; the reader is referred to [17], [18] for a detailed exposition of this technique, and to [8], [5], [6], for related approaches to the rapid evaluation of integral operators with oscillatory kernels.

Acknowledgments. The author would like to thank Professor R. Coifman and Dr. S. Wandzura for their help in writing this article.

References

[1] M. Abramovitz and I. Stegun, Handbook of Mathematical Functions, Applied Math. Series (National Bureau of Standards), Washington, D.C., 1964.

[2] B. Alpert, G. Beylkin, R. Coifman, and V. Rokhlin, *Wavelet-like bases for the fast solution of second-kind integral equations*, SIAM J. Sci. Statist. Comput., 14(1):159–184 (January 1993).

[3] B. Alpert and V. Rokhlin, *A fast algorithm for the evaluation of Legendre expansions*, SIAM J. Sci. Statist. Comput., 12(1):158–179 (1991).

[4] G. Beylkin, R. Coifman, and V. Rokhlin, *Fast wavelet transforms and numerical logorithms I*, Comm. Pure and Appl. Math., 14:141–183 (1991).

[5] B. Bradie, R. Coifman, and A. Grossman, *Fast Numerical Computations of Oscillatory Integrals Related to Acoustic Scattering*, I, Applied and Computational Harmonic Analysis, 1:94–99 (1993).

[6] F. X. Canning, *Improved impedance matrix localization method*, IEEE Trans. Antennas and Propagation, 41(5):658–667 (1993).

[7] J. Carrier, L. Greengard, and V. Rokhlin, *A fast adaptive multipole algorithm for particle simulations*, SIAM J. Sci. Statist. Comput., 9(4), (1988).

[8] R. Coifman and Y. Meyer, *Remarques sur l'analyse de Fourier á fenêtre*, C. R. Acad. Sci. Paris, 312, Série 1:259–261 (1991).

[9] R. Coifman V. Rokhlin, and S. Wandzura, *The fast multipole method for the wave equation: A pedestrian prescription*, IEEE Trans. Antennas and Propagation, 35(3):7–12 (1993).

[10] D. Colton and R. Kress, Integral Equation Methods in Scattering Theory, Wiley, New York, 1983.

[11] L. Greengard and V. Rokhlin, *A fast algorithm for particle simulations*, J. Comput. Phys., 73:325–348 (1987).

[12] L. Greengard and J. Strain, *The fast Gauss transform*, SIAM J. Sci. Statist. Comput., 12:79–94 (1991).

[13] R. Kress, Linear Integral Equations, Springer, Berlin and New York, 1989.

[14] A. D. McLaren, *Optimal numerical integration on a sphere*, Math. Comp., 17:361–383 (1963).

[15] V. Rokhlin, *Rapid solution of integral equations of classical potential theory*, J. Comput. Phys., 60(2):187 (1985).

[16] V. Rokhlin, *A fast algorithm for the discrete Laplace transformation*, J. Complexity, 4:12–32 (1988).

[17] V. Rokhlin, *Rapid solution of integral equations of scattering theory in two dimensions*, J. Comput. Phys., 86(2):414 (1990).

[18] V. Rokhlin, *Diagonal forms of translation operators for the Helmholtz equation in three dimensions*, Appl. Comput. Harmonic Anal., 1:82–93 (1993).

[19] J. Strain, *The fast Gauss transform with variable scales*, SIAM J. Sci. Statist. Comput., 12:1131–1139 (1991).

[20] J. Strain, *The fast Laplace transform based on Laguerre functions*, to appear in Math. Comp.

Solving Numerically Hamiltonian Systems

J. M. Sanz-Serna

Departamento de Matemática Aplicada y Computación, Facultad de Ciencias, Universidad de Valladolid, Valladolid, Spain

1. Introduction

This is a short summary of my oral presentation at the Zurich International Congress of Mathematicians. The presentation was aimed at providing an easy introduction to the field of symplectic numerical integrators for Hamiltonian problems. Some sacrifices in rigor and precision were deliberately made.

We are concerned with initial value problems for systems of ordinary differential equations

$$\frac{dy}{dt} = f(y), \quad 0 \leq t \leq T, \qquad y(0) = \alpha \in \mathcal{R}^D, \tag{1}$$

where f is a smooth function. The basic theory of numerical methods for (1) has been known for more than thirty years, see e.g. [8]. This theory, in tandem with practical experimentation, has led to the development of general software packages for the efficient solution of (1). It is perhaps remarkable that both the theory and the packages do not take into account any structure the problem may have and work under virtually no assumption on the (smooth) vector field f. This contributes to the elegance of the theory and to the versatility of the software.

However, it is clear that a method that can solve "all" problems is bound to be inefficient in some problems. Stiff problems [9], frequent in many applications, provide an example of problems of the format (1) where general packages are very inefficient. Accordingly, a special theory and special software have been created to cope with stiff problems.

Are there other classes of problems of the form (1) that deserve a separate study? In recent years much work has been done on special methods for *Hamiltonian problems*. Of course, Hamiltonian problems [11] play a crucial role as mathematical models of situations where dissipative effects are absent or may be ignored. Most special methods for Hamiltonian problems are *symplectic* methods; other possibilities, not discussed here, include reversible and energy-conserving methods [16]. Early references on symplectic integration are Channell [4], Feng [5], and Ruth [12]. In the last ten years the growth of the "symplectic" literature has been impressive, both in mathematics and in the various application fields. The monograph [16] contains over a hundred references from the mathematical literature. The second edition of the excellent treatise by Hairer, Nørsett, and Wanner [8] includes a section on symplectic integration.

Proceedings of the International Congress
of Mathematicians, Zürich, Switzerland 1994
© Birkhäuser Verlag, Basel, Switzerland 1995

In the talk I presented several examples, taken from mathematics [16], astronomy [20], and molecular dynamics [6] that illustrated the practical advantages of symplectic integrators when compared with general software.

2. Symplecticness

For our purposes here, a Hamiltonian problem is a problem of the form (1) where the dimension D is even, $D = 2d$, and the components f_i of f are given by

$$f_i = -\frac{\partial H}{\partial y_{d+i}}, \quad f_{d+i} = +\frac{\partial H}{\partial y_i}, \quad i = 1, \ldots, d, \tag{2}$$

for a suitable real-valued function $H = H(y)$ (the Hamiltonian). It is standard notation to set $p_i = y_i$, $q_i = y_{d+i}$, $i = 1, \ldots, d$, and then the Hamiltonian system with Hamiltonian function H reads

$$\frac{dp_i}{dt} = -\frac{\partial H}{\partial q_i}, \quad \frac{dq_i}{dt} = +\frac{\partial H}{\partial p_i}, \quad i = 1, \ldots, d. \tag{3}$$

Whether a system of the form (1) with $D = 2d$ is Hamiltonian or otherwise can be decided [1] by the symplecticness of its flow. Recall that, for each real t, the t-flow ϕ_t of the differential system in (1) is the mapping in $\mathcal{R}^D$ that maps each $\alpha \in \mathcal{R}^D$ into the value $y(t)$ at time t of the solution y of the initial value problem (1). A symplectic transformation Φ in $\mathcal{R}^{2d}$ is a transformation that preserves the differential form

$$\omega = dp_1 \wedge dq_1 + \cdots + dp_d \wedge dq_d.$$

When $d = 1$ preservation of ω is simply preservation of oriented area: a smooth Φ is symplectic if and only if for each oriented domain D in $\mathcal{R}^2$, $\Phi(D)$ possesses the same area and orientation as D. For $d > 1$, preservation of ω means preservation of the sum of the two-dimensional oriented areas of the projections onto the planes (p_i, q_i), $i = 1, \ldots, d$, of oriented two-dimensional surfaces D in $\mathcal{R}^{2d}$.

For each t, the flow ϕ_t of (3) is a symplectic transformation. Conversely if (1) (with $D = 2d$) is such that, for each t, ϕ_t is symplectic then (1) is a Hamiltonian problem, in the sense that a scalar function H may be found such that (2) holds. The conclusion is that the symplecticness of the flow characterizes Hamiltonian problems. In fact, all qualitative properties of the solutions of Hamiltonian systems derive from the symplecticness of the flow.

When solving (1) with a one-step numerical method, the true flow $\phi_{\Delta t}$ is replaced by a computable approximation $\psi_{\Delta t}$. For instance, for the standard Euler rule $\psi_{\Delta t}(y) = y + \Delta t f(y)$. For an order r method $\psi_{\Delta t}$ is an $O(\Delta t^{r+1})$ perturbation of $\phi_{\Delta t}$ as $\Delta t \to 0$. The numerical approximation y^n at time $t_n = n\Delta t$, $n = 1, 2, \ldots$, is computed by iterating the map $\psi_{\Delta t}$, i.e. $y^{n+1} = \psi_{\Delta t}(y^n)$, $n = 0, 1, \ldots$. Then $y^n - y(n\Delta t)$ is $O(\Delta t^r)$ as $\Delta t \to 0$, uniformly in bounded intervals of the variable $t = n\Delta t$.

If (1) is a Hamiltonian problem, there is no guarantee that a given numerical method yields a mapping $\psi_{\Delta t}$ that is symplectic. Therefore, in general, numerical methods do not share the property of symplecticness that is the hallmark of

Hamiltonian problems. A numerical method is said to be symplectic, if, whenever it is applied to a Hamiltonian problem (3), it produces a mapping $\psi_{\Delta t}$ that is symplectic for each Δt.

3. Available symplectic methods

The available symplectic methods can be grouped broadly into three classes.

The earliest symplectic methods were based on the fact that symplectic transformations in $\mathcal{R}^{2d}$ can be expressed in terms of the partial derivatives of a real-valued *generating function*. For the true flow $\phi_{\Delta t}$, the generating function is a solution of the Hamilton-Jacobi equation, and by approximately solving the Hamilton-Jacobi equation one constructed the generating function of the numerical method $\psi_{\Delta t}$. The methods obtained in this way require the knowledge of higher derivatives of H and tend to be cumbersome.

Lasagni [10], Suris [19], and I [13] discovered independently that standard classes of methods, like *Runge-Kutta* methods, include schemes that just 'happen' to be symplectic.

The third class of symplectic methods is built around the idea of *splitting*. It is required that the Hamiltonian H of interest may be decomposed as a sum $H = H_1 + \cdots + H_s$ such that the Hamiltonian systems with Hamiltonians H_i may be integrated in closed form, so that the corresponding flows $\phi_{\Delta t, H_i}$, $i = 1, \ldots, s$, are explicitly available. These "fractional" flows are then combined to produce an approximation $\psi_{\Delta t, H}$ to $\phi_{\Delta t, H}$. When $s = 2$, the simplest possibility is to set

$$\psi_{\Delta t, H} = \phi_{\Delta t, H_1} \phi_{\Delta t, H_2}.$$

This provides a first-order method that is symplectic: $\psi_{\Delta t, H}$ is a composition of two Hamiltonian flows, and hence of two symplectic mappings. Higher-order splittings exist; for instance the second-order recipe

$$\psi_{\Delta t, H} = \phi_{\Delta t/2, H_1} \phi_{\Delta t, H_2} \phi_{\Delta t/2, H_1}$$

goes back to Strang [18], and Yoshida [21] has developed a way of constructing splittings of arbitrarily high orders.

4. Discussion

In which way are symplectic methods better than their conventional counterparts? The standard criterion for determining the merit of numerical methods for (1) is as follows. One measures the error $|y^n - y(n\Delta t)|$ (numerical minus exact) at some prescribed time $t = n\Delta t$; method A is then an improvement on method B if A attains a prescribed error size with less work than B. There is some evidence suggesting that symplectic methods may be advantageous when this standard comparison criterion is used. For instance, it is possible to show [3] that, in the integration of the classical two-body problem, symplectic integrators have errors whose leading terms in the asymptotic expansion grow linearly with t; the error in conventional methods grows like t^2.

However the standard criterion described above may not be a sensible choice in many instances. Sometimes numerical integrators are used to get an indication of the long-time behavior of a differential system [14]. When t is large all numerical methods are likely to produce approximations y^n that differ significantly from $y(n\Delta t)$; therefore all methods would be regarded as bad with the standard criterion. This is particularly clear in cases, including chaotic regimes, where neighboring solutions of the system diverge exponentially as t increases and hence numerical errors also increase exponentially. It is then useful to derive new alternative criteria to judge the goodness of numerical methods in long-time integrations, see e.g. [17].

An idea that has recently attracted much attention [16], [2], [15], [7] is that of *backward error analysis*. In numerical analysis, given a problem $\mathcal{P}$ with true solution $\mathcal{S}$ and given an approximate solution $\tilde{\mathcal{S}}$, forward error analysis consists of estimating the distance between $\mathcal{S}$ and $\tilde{\mathcal{S}}$. Traditionally, error analyses in numerical differential equations are forward error analyses. Backward error analysis consists of showing that $\mathcal{S}$ is the exact solution of a problem $\tilde{\mathcal{P}}$ that is close to $\mathcal{P}$. Let $\psi_{t,f}$ be a numerical method of order $r \geq 1$ for the integration of (1). Given any *large* integer N, there is an autonomous vector field $\tilde{f}$, that depends on N and Δt, such that $\psi_{\Delta t,f} - \phi_{\Delta t,\tilde{f}} = O(\Delta t^{N+1})$ as $\Delta t \to 0$. This means that the numerical solution, that is an approximation of order r to the solution of the problem (1) we are trying to solve, is an approximation of order $N >> r$ to the perturbed problem

$$\frac{dy}{dt} = \tilde{f}(y), \quad 0 \leq t \leq T, \qquad y(0) = \alpha \in \mathcal{R}^D. \tag{4}$$

Ignoring $O(\Delta t^N)$ terms, the numerical solution is really solving the modified problem (4). Here $\tilde{f} = f + O(\Delta t^r)$, so that the higher the order of the method, the closer the modified problem is to the true problem (1). In any case, if the discrepancy between f and $\tilde{f}$ is of the same size as the uncertainty in f that results from modelling errors, experimental errors in measuring the constants that may feature in f, etc., then we are sure that there is nothing seriously wrong in solving (1) numerically.

In this connection, it turns out that if the numerical method is symplectic and (1) is Hamiltonian, then (4) is also a Hamiltonian problem. In this sense, a symplectic integrator changes the problem being solved by slightly altering the Hamiltonian function H; a general integrator changes the problem being solved by introducing a non-Hamiltonian perturbation.

Acknowledgments. I have been supported by grant DGICYT PB92-254.

References

[1] V. I. Arnold, Mathematical Methods of Classical Mechanics, 2nd. ed., Springer, Berlin and New York, 1989.

[2] M. P. Calvo, A. Murua and J. M. Sanz-Serna, *Modified Equations for ODEs*, Chaotic Dynamics (K. Palmer and P. Kloeden, eds.), Contemp. Math., Amer. Math. Soc. (1994), in press.

[3] M. P. Calvo and J. M. Sanz-Serna, *The development of variable-step symplectic integrators, with application to the two-body problem*, SIAM J. Sci. Statist. Comput. **14** (1993), 936–952.

[4] P. J. Channell, *Symplectic Integration Algorithms*, Los Alamos National Laboratory Report AT-6:ATN 83-9 (1983).

[5] K. Feng, *Difference schemes for Hamiltonian formalism and symplectic geometry*, J. Comput. Math. **4** (1986), 279–289.

[6] S. K. Gray, D. W. Noid, and B. G. Sumpter, *Symplectic integrators for large scale molecular dynamics simulations: A comparison of several explicit methods*, preprint.

[7] E. Hairer, *Backward error analysis of numerical integrators and symplectic schemes*, preprint.

[8] E. Hairer, S. P. Nørsett, and G. Wanner, Solving Ordinary Differential Equations I, Nonstiff Problems, 2nd. ed., Springer, Berlin and New York, 1993.

[9] E. Hairer and G. Wanner, Solving Ordinary Differential Equations II, Stiff and Differential-Algebraic Problems, Springer, Berlin and New York, 1991.

[10] F. M. Lasagni, *Canonical Runge-Kutta methods*, Z. Angew. Math. Phys. **39** (1988), 952–953.

[11] R. S. MacKay and J. D. Meiss, Hamiltonian Dynamical Systems, Adam Hilger, Bristol, 1987.

[12] R. D. Ruth, *A Canonical Integration Technique*, IEEE Trans. Nucl. Sci. **30** (1983), 1669–2671.

[13] J. M. Sanz-Serna, *Runge-Kutta schemes for Hamiltonian problems*, BIT **28** (1988), 539–543.

[14] J. M. Sanz-Serna, *Numerical Ordinary Differential Equations vs. Dynamical Systems*, The Dynamics of Numerics and the Numerics of Dynamics (D.S. Broomhead and A. Iserles, eds.) Clarendon Press, Oxford, 1992, pp. 81–106.

[15] J. M. Sanz-Serna, *Backward Error Analysis of Symplectic Integrators*, Applied Mathematics and Computation Reports, Report 1994/1, Universidad de Valladolid.

[16] J. M. Sanz-Serna and M. P. Calvo, Numerical Hamiltonian Problems, Chapman & Hall, London, 1994.

[17] J. M. Sanz-Serna and S. Larsson, *Shadows, chaos and saddles*, Appl. Numer. Math. **13** (1993), 181–190.

[18] G. Strang, *On the construction and comparison of difference schemes*, SIAM J. Numer. Anal. **5** (1968), 506–517.

[19] Y. B. Suris, *The canonicity of mappings generated by Runge-Kutta type methods when integrating the systems $\ddot{x} = -\partial U/\partial x$*, U.S.S.R. Comput. Math. and Math. Phys. **29** (1989), 138–144.

[20] J. Wisdom and M. Holman, *Symplectic maps for the N-body problem*, Astronom. J. **102** (1991), 1528–1538.

[21] H. Yoshida, *Construction of high order symplectic integrators*, Phys. Lett. A **150** (1990), 262–268.

Methods of Control Theory in Nonholonomic Geometry

ANDREI A. AGRACHEV*

Steklov Mathematical Institute
ul. Vavilova 42
Moscow 117 966, Russia

1 Introduction

Let M be a C^∞-manifold and TM the total space of the tangent bundle. A control system is a subset $V \subset TM$. Fix an initial point $q_0 \in M$ and a segment $[0, t] \subset \mathbb{R}$. Admissible trajectories are Lipschitzian curves $q(\tau)$, $0 \le \tau \le t$, $q(0) = q_0$, satisfying a differential equation of the form

$$\dot{q} = v_\tau(q), \tag{1}$$

where $v_\tau(q) \in V \cap T_q M$, $\forall q \in M$, $v_\tau(q)$ is smooth in q, bounded and measurable in τ. The mapping $q(\cdot) \mapsto q(t)$, which maps admissible trajectories in their end points, is called an end-point mapping.

Control theory is in a sense a theory of end-point mappings. This point of view is rather restrictive but sufficient for our purposes. For instance, attainable sets are just images of end-point mappings. Geometric control theory tends to characterize properties of these mappings in terms of iterated Lie brackets of smooth vector fields on M with values in V. A number of researchers have shown a remarkable ingenuity in this regard leading to encouraging results. See, for instance, books [5], [7], [10] to get an idea of various periods in the development of this domain and for other references. A complete list of references would probably run to thousands of items.

A great part of the theory is devoted to the case of nonsmooth V such that $V \cap T_q M$ are polytopes or worse. There is a widespread view that such a nonsmoothness is the essence of control theory. This is not my opinion, and I am making the following radical assumption.

Let us assume that V forms a smooth locally trivial bundle over M with fibers V_q — smooth closed convex submanifolds in $T_q M$ of positive dimension, symmetric with respect to the origin. So we consider a very special class of control systems.

EXAMPLES. 1) V_q is an ellipsoid centered at the origin. This is the case of Riemannian geometry.

*)   Partially supported by Russian fund for fundamental research grant 93-011-1728 and ISF grant MSD000.

Proceedings of the International Congress
of Mathematicians, Zürich, Switzerland 1994
© Birkhäuser Verlag, Basel, Switzerland 1995

2) V_q is a proper linear subspace of $T_q M$. This case includes nonholonomic geometry.

3) V_q is the intersection of an ellipsoid and a subspace. This is sub-Riemannian geometry.

This paper essentially deals with cases 2) and 3).

2 Extremals

Denote by $\Omega_{q_0}(t)$ the space of all admissible trajectories on $[0,t]$ equipped with $W_{1,\infty}$-topology, i.e. the topology of uniform convergence for curves and their velocities. Under our assumptions for V, the space $\Omega_{q_0}(t)$ possesses the natural structure of a smooth Banach manifold, and the end-point mapping

$$f_t : \Omega_{q_0}(t) \to M, \quad f_t(q(\cdot)) = q(t)$$

is a smooth mapping. We will denote by $D_q f_t : T_q \Omega_{q_0}(t) \to T_{f_t(q)} M$ the differential of f_t at $q(\cdot)$.

A trajectory $q(\cdot)$ is a critical point for f_t iff $\exists \lambda \in T^*_{f_t(q)} M$, $\lambda \neq 0$, such that $\lambda D_q f_t = 0$, i.e. λ is orthogonal to the image of the linear mapping $D_q f_t$. It is a natural thing that critical points of f_t are the main object of our investigation. We study critical levels of f_t and restrictions of f_t to the sets of their critical points.

The cotangent bundle $T^* M$ possesses the canonical symplectic structure. We will denote by $\overrightarrow{\phi}$ the Hamiltonian vector field on $T^* M$ associated to the Hamiltonian $\phi \in C^\infty(T^* M)$. Let v be a smooth vector field on M, then $v^* : \lambda \mapsto \langle \lambda, v(q) \rangle$, $\lambda \in T^*_q M$, $q \in M$ is a Hamiltonian on $T^* M$, which is linear on fibers, and $\overrightarrow{v^*}$ is a lift of the vector field v on $T^* M$.

Set $\Omega_{q_0} = \Omega_{q_0}(1)$, $f = f_1$. Let $q(\cdot) \in \Omega_{q_0}$ be a critical point of f. Then $q|_{[0,t]}$ is obviously a critical point of f_t, $\forall t \in (0,1]$. Moreover, let $q(\cdot)$ satisfy the equation (1). If λ_t, $0 \leq t \leq 1$, is a solution of the nonstationary Hamiltonian system

$$\dot{\lambda} = \overrightarrow{v^*_t}(\lambda) \quad \text{and} \quad \lambda_1 D_q f = 0, \quad \lambda_1 \neq 0, \tag{2}$$

then $\lambda_t D_{q|_{[0,t]}} f_t = 0$, $\lambda_t \neq 0$, $\forall t \in [0,1]$.

The curves in $T^* M$ that satisfy (2) for some v_t are called extremals associated with $q(\cdot)$. Let $q \in M$, $\lambda \in T^*_q M$, $\lambda \neq 0$. Set $h(\lambda) = \max_{v \in V_q} \langle \lambda, v \rangle$ if the maximum exists. The function h is defined on a subset of $T^* M$. It is convex and positively homogeneous on fibers. We call h the Hamiltonian of the control system.

Let σ be the canonical symplectic structure on $T^* M$. The following proposition is a corollary of the Pontryagin Maximum Principle.

PROPOSITION 1. (a) $h(\lambda_t) = const$, $0 \leq t \leq 1$, for arbitrary extremal λ_t.

(b) Let a level set $h^{-1}(c)$ be a smooth submanifold of $T^* M$. Then any extremal $\lambda_t \in h^{-1}(c)$, $0 \leq t \leq 1$, is a characteristic of the differential form $\sigma|_{h^{-1}(c)}$ (i.e. $\dot{\lambda} \rfloor \sigma|_{h^{-1}(c)} = 0$), and any properly parametrized characteristic of this form started at $T^*_{q_0} M$ is the extremal.

Note that level sets of h are smooth in the above Examples 1)–3).

3 Distributions

Distributions are just smooth vector subbundles of the tangent bundle. Let Δ be the space of smooth sections of a distribution, and $\Delta_q \subset T_q M$ be the fiber at $q \in M$ of the corresponding subbundle.

Set $\Delta^1 = \Delta$, $\Delta^n = [\Delta, \Delta^{n-1}]$, $n = 2, 3, \ldots$, where the Lie bracket of spaces of vector fields is, by definition, the linear hull of the pairwise brackets of their elements. The distribution is called bracket generating if a number n_q exists for $\forall q \in M$ such that $\Delta_q^{n_q} = T_q M$. We will consider only bracket generating distributions in this paper. The distribution defines a control system $V = \bigcup_{q \in M} \Delta_q$.

The well-known Rashevskij-Chow theorem asserts that the end-point mapping $f : \Omega_{q_0} \to M$ is a surjective one. It is not a submersion, however, if $\Delta_q \neq T_q M$. Critical points of f are called singular or abnormal geodesics for Δ.

Let $\Delta_q^\perp \subset T_q^* M$ be the set of all nonzero covectors that are orthogonal to Δ_q, $\Delta^\perp = \bigcup_{q \in M} \Delta_q^\perp$. The manifold $\Delta^\perp$ is the domain of the Hamiltonian of the control system V. This Hamiltonian is identical to zero in its domain. Singular geodesics are exactly projections on M of the characteristics of the form $\sigma|_{\Delta^\perp}$, started at $\Delta_{q_0}^\perp$.

4 Rigidity

Let $\Omega_{q_0, q_1} = f^{-1}(q_1)$ be the set of admissible trajectories that connect q_0 with q_1. An admissible trajectory $q(\cdot)$ is called rigid if there exists a neighborhood of $q(\cdot)$ in $\Omega_{q_0, q(1)}$ that contains only reparametrizations of $q(\cdot)$. It is called locally rigid if its small enough pieces are rigid, cf. [6].

THEOREM 2. (see [3]) Let $q(\cdot) \in \Omega_{q_0}$, $\dot{q} = v(q)$, $v \in \Delta$.

(a) If $q(\cdot)$ is a locally rigid trajectory, then there exists an extremal λ_t associated with $q(\cdot)$ such that

$$\lambda_t \perp \Delta_{q(t)}^2, \quad \langle \lambda_t, [[v, w], w](q(t)) \rangle \geqslant 0, \quad \forall w \in \Delta, \quad 0 \leq t \leq 1. \tag{3}$$

(b) Let there exist an extremal λ_t that satisfies (3) and

$$\langle \lambda_t, [[v, w], w](q(t)) \rangle > 0, \quad \forall w \in \Delta, \quad w(q(t)) \nparallel \dot{q}(t), \quad 0 \leq t \leq 1. \tag{4}$$

Then $q(\cdot)$ is indeed a locally rigid trajectory.

We call $q(\cdot)$ the singular geodesic of the first order if there exists a unique up to a positive multiplier extremal λ_t associated with $q(\cdot)$ that satisfies (3), (4).

5 Jacobi curves

Let λ_t be an extremal that satisfies (3), and $Q_\tau : M \to M$ be the flow generated by v. Set $\Gamma_t = \{\overrightarrow{w^*}|_{\lambda_t} : w \in \Delta\}$ — an isotopic subspace of the symplectic space $T_{\lambda_t}(T^* M)$. Let $0 = t_0 < t_1 < \cdots < t_{k+1} = 1$ be a subdivision of the segment $[0, 1]$, $I = \{t_1, \ldots, t_k\}$. Let us identify $T_\lambda(T_q^* M) = T_q^* M$ for $\lambda \in T_q^* M$ and set

$$\Lambda_0(I) = T_{q_0}^* M, \quad \Lambda_t(I) = Q_{t_i - t}^* (\Lambda_{t_i}(I)^{\Gamma_{t_i}}) \text{ for } t_i < t \leq t_{i+1}, \quad i = 0, \ldots, k,$$

where Λ^Γ denotes the intersection of $\Lambda + \Gamma$ with the skew-orthogonal complement to Γ. Then $\Lambda_t(I) \subset T_{\lambda_t}(T^* M)$ is a piecewise smooth family of Lagrangian subspaces.

PROPOSITION 3. *Let $q(\cdot)$ be a locally rigid trajectory, $\dot q = v(q)$, $v \in \Delta$. Then there exists an extremal λ_t that satisfies (3) and such that*

$$\exists \, \mathcal{I}\text{-}\lim \Lambda_t(I) = \Lambda_t, \ \forall t \in [0,1], \ \text{where } \mathcal{I} = \{I \subset (0,1) : \#I < \infty\}.$$

We call Λ_t the Jacobi curve associated with λ_t. The Jacobi curve is smooth in $t \in (0,1]$ and satisfies a simple Hamiltonian equation if $q(\cdot)$ is a singular geodesic of the first order. See details in [1], [2], [3].

THEOREM 4. *Let $q(\cdot)$ be a singular geodesic of the first order, and Λ_t be the Jacobi curve associated with the corresponding extremal λ_t. Suppose that $\Lambda_1 \cap T^*_{q(1)} M = \mathbb{R}\lambda_1$. Then there exists an integer $d \geq 0$ and a neighborhood $\mathcal{O}_q$ of q in $\Omega_{q_0,q(1)}$ such that $\mathcal{O}_q \backslash \{q\}$ is homotopy equivalent to the sphere S^{d-1}. If $\Lambda_t \cap T^*_{q(t)} M = \mathbb{R}\lambda_t$, $\forall t \in (0,1]$, then $d = 0$.*

We write $d = \mathrm{ind}q(\cdot)$. This index has an explicit expression in terms of the Maslov cocycle on T^*M, cf. [2].

6 Low dimensions

Let Δ be a rank 2 distribution and $\dim M = 3$. Then $N = \{q \in M : \Delta_q = \Delta_q^2 \neq \Delta_q^3\}$ is a smooth 2-dimensional submanifold in M (maybe empty), and $\Delta_q \pitchfork N \ \forall q \in N$. Integral curves of the rank 1 distribution $\Delta_q \cap T_q N$ on N are singular geodesics of the first order and all of them are *rigid*.

One may say more about generic distributions using local normal forms, see [13], [14]. The closure $\bar N$ is a smooth submanifold in M for generic Δ, and $\bar N \backslash N$ consists of isolated points. These points are singularities of the foliation on N generated by rank 1 distribution $\Delta_q \cap T_q N$. They may be saddles or focuses. We obtain a nonsmooth rigid trajectory pasting together two neighboring separatrixes of the saddle, and a smooth but not a rigid singular geodesic if we paste together separatrixes lying opposite each other. One more interesting phenomenon: any neighborhood of the focus contains rigid trajectories of arbitrary length! This happens because the foliation is never generated by a linearizable vector field in a neighborhood of our focus.

Let $\mathrm{rank}\Delta = 2$, $\dim M = 4$, and $\Delta_q \neq \Delta_q^2 \neq \Delta_q^3$, $\forall q \in M$. Such a distribution is called the Engel distribution. A characteristic rank 1 subdistribution $K \subset \Delta$ is defined by the relation $[K, \Delta^2] \subset \Delta^2$. Singular geodesics for Δ are exactly parametrizations of integral curves of K. These integral curves are singular geodesics of the first order. Let $q(\cdot) \in \Omega_{q_0}$ be a piece of one of them without self-intersections, and $\mathcal{K}$ be the foliation generated by K. Replace M by a neighborhood M_0 of $\{q(t) : 0 \leq t \leq 1\}$ such that a factor-manifold $M_0/\mathcal{K}$ is well defined. Let $\kappa : M_0 \to M_0/\mathcal{K}$ be the canonical projection. Then $\kappa_* \Delta_{q(t)}^2$ is a 2-dimensional subspace in $T_q(M_0/\mathcal{K})$ that does not depend on t. Let $\bar\kappa_*$ denote the composition of the κ_* and the projectivization of $T_q(M_0/\mathcal{K})$. Hence $\bar\kappa_* \Delta_{q(t)}^2$ is a projective line.

PROPOSITION 5. *The singular geodesic $q(\cdot)$ satisfies conditions of Theorem 4 iff $\bar\kappa_* \Delta_{q(1)} \neq \bar\kappa_* \Delta_{q_0}$, and*

$$\mathrm{ind}q(\cdot) = \#\{t \in (0,1) : \bar\kappa_* \Delta_{q(t)} = \bar\kappa_* \Delta_{q_0}\}.$$

See also [3], [6].

Let $\operatorname{rank}\Delta = 2$, $\dim M$ is arbitrary. If $\Delta_{q_0}^3 \neq \Delta_{q_0}^2$, then Ω_{q_0} contains a smooth rigid trajectory. If $\dim(\Delta_{q_0}^3 / \Delta_{q_0}^2) = 2$, then there exists a smooth rigid trajectory $q(\cdot) \in \Omega_{q_0}$ such that $\dot{q}(0) = \xi$, for $\forall \xi \in \Delta_{q_0} \backslash \{0\}$, see [3], [6]. Note that 2 is the maximum possible dimension for $\Delta_{q_0}^3 / \Delta_{q_0}^2$.

REMARK. Recall that the spaces Ω_{q_0,q_1} have the $W_{1,\infty}$-topology. Homotopy types may change dramatically and become independent on the distribution if we replace this topology by a weaker $W_{1,s}$-topology (i.e. the L_s-topology for velocities), $1 \leq s < \infty$. The embedding of Ω_{q_0,q_1} in the space of *all* Lipschitzian curves in M connecting q_0 with q_1 is a homotopy equivalence in the $W_{1,s}$-topology, $1 \leq s < \infty$, see [9].

7 Sub-Riemannian geodesics

Let V_q^1 be the intersection of Δ_q with an ellipsoid in $T_q M$ centered at the origin and smoothly depending on $q \in M$. Set $V_q^l = l V_q^1$, $V^l = \bigcup_{q \in M} V_q^l$, $l > 0$. The family of control systems V^l is called the sub-Riemannian structure on M coordinated with Δ. We will denote by $\Omega_{q_0}^l$ the space of admissible trajectories for V^l on $[0, 1]$ equipped with the $W_{1,1}$-topology. Note that all $W_{1,s}$-topologies, $1 \leq s < \infty$, are equivalent in the sub-Riemannian case as V_q^l are compact. The number l is, by definition, the length of any curve in $\Omega_{q_0}^l$. Set $\Omega_{q_0,q_1}^l = \{q(\cdot) \in \Omega_{q_0}^l : q(1) = q_1\}$ — a subspace in $\Omega_{q_0}^l$.

We call $q(\cdot) \in \Omega_{q_0}^l$ the strong length minimizer if it is a $W_{1,1}$-isolated point in $\bigcup_{l' \leqslant l} \Omega_{q_0,q(1)}^{l'}$. We call $q(\cdot)$ the global length minimizer if $\Omega_{q_0,q(1)}^{l'} = \emptyset$, $\forall l' < l$.

PROPOSITION 6. *Let $q(\cdot)$ be an isolated point in $\Omega_{q_0,q(1)}^l$. Then $q(\cdot)$ is a strong length minimizer and its small enough pieces (reparametrized in the obvious way) are global length minimizers.*

Critical points of the end-point mappings for control systems V^l are called sub-Riemannian geodesics. Let h^l be the Hamiltonian of V^l. The function $h^l = l h_1$ is smooth outside its zero level set, which is equal to $\Delta^\perp$.

Let λ_t be an extremal associated with a sub-Riemannian geodesic. The extremal is called normal if $h^l(\lambda_t) \neq 0$, otherwise it is called abnormal. Normal extremals are exactly trajectories of the Hamiltonian system $\dot{\lambda} = \overrightarrow{h^l}(\lambda)$, $h(\lambda) \neq 0$, started at $T_{q_0}^* M$. Abnormal extremals are just extremals associated with properly parametrized singular geodesics for Δ.

A sub-Riemannian geodesic $q(\cdot)$ is called regular if there exists a unique up to a positive multiplier normal extremal associated with $q(\cdot)$, otherwise it is called singular or abnormal. It is easy to show that an abnormal extremal is associated with any singular geodesic $q(\cdot)$. If all extremals associated with $q(\cdot)$ are abnormal, then $q(\cdot)$ is called strictly abnormal.

Let λ_t be a normal sub-Riemannian extremal and $H_t^l : T^* M \to T^* M$ be the Hamiltonian flow generated by the vector field $\overrightarrow{h^l}$. Set

$$\Lambda_0^l = T_{\lambda_0}(T_{q_0}^* M), \ \Lambda_t^l = H_{t^*}^l(\Lambda_0^l), \ 0 \leq t \leq 1.$$

Then $\Lambda_t^l \subset T_{\lambda_t}(T^*M)$ is a smooth family of Lagrangian subspaces. We call Λ_t^l the Jacobi curve associated with λ_t.

THEOREM 4^l. *Let* $q(\cdot)$ *be a regular sub-Riemannian geodesic. The statement of Theorem 4 remains true if symbols* Λ *and* Ω *are replaced by* Λ^l *and* Ω^l *everywhere in its formulation.*

THEOREM 7. *Let* $\mathrm{rank}\Delta = 2$ *and* $q(\cdot) \in \Omega_{q_0}^l$ *be a singular geodesic meeting conditions of Theorem 4.*

(a) *If* $q(\cdot)$ *is rigid, then it is a strong length minimizer.*

(b) *If* $q(\cdot)$ *is strictly abnormal, then* $\mathcal{O}_q^l \backslash \{q\}$ *is homotopy equivalent to* $\mathcal{O}_q \backslash \{q\}$ *for some neighborhood* $\mathcal{O}_q \subset \Omega_{q_0,q(1)}$, $\mathcal{O}_q^l \subset \Omega_{q_0,q(1)}^l$.

In particular, smooth rigid trajectories described in the previous section are strong length minimizers for an arbitrary sub-Riemannian structure coordinated with Δ. It turns out however that nonsmooth rigid curves constructed there for typical rank 2 distributions on the 3-dimensional manifold are never strong length minimizers. Recall that a strong minimum is a local minimum in the $W_{1,1}$-topology (see the remark at the end of Section 6.) See also [4], [8], [11].

8 The Lie group case

In this section we consider examples of sub-Riemannian geodesics that are neither regular nor strictly abnormal. Although most likely nongeneric, these geodesics are common in symmetric situations.

Let $M = G$ be a compact semisimple Lie group with the Lie algebra g of left-invariant vector fields and a bi-invariant Riemannian structure $(v_1|v_2)$, $v_1, v_2 \in T_qG$, $q \in G$. Any left-invariant corank 1 distribution on G has a form $\Delta(a)$, where $a \in g$, $(a|a) = 1$, $\Delta_q(a) = \{v \in T_qG : (v|a(q)) = 0\}$. Consider a sub-Riemannian structure

$$V^l = \{v \in \Delta_q(a) : (v|v) = l^2, \ q \in G\}.$$

Sub-Riemannian geodesics for V^l that are not strictly abnormal are exactly the curves

$$q(t) = q_0 e^{tb} e^{-t(b|a)a}, \ b \in g, \ (b|b) - (b|a)^2 = l^2. \tag{5}$$

Let a be a regular element of g. The geodesic (5) is regular iff $[b, a] \neq 0$, otherwise it is neither regular nor strictly abnormal. Let $A = \{v \in g : [v, a] = 0\}$ be a Cartan subalgebra in g. Fix $c \in A$, $(c|a) = 0$, $(c|c) = 1$. Let $\pm\rho_i \in A^*$, $i = 1, \ldots, m$, be all roots of g (relative to A), $\langle \rho_i, c \rangle \geq 0$.

Let $q^l(t) = q_0 e^{tlc}$. Set $\mathbf{H}_n = H_n (\Omega_{q_0,q_1}^l, \Omega_{q_0,q_1}^l \backslash \{q^l\})$; these homology groups are determined by the disposition of the affine line $lc + \mathbb{R}a$ with respect to the Stifel diagram, i.e. the maximal triangulation of the complex $\{v \in A : \exists i \ \mathrm{s.t.} \langle \rho_i, v \rangle \in \mathbb{Z}\}$.

PROPOSITION 8. *Suppose that* $lc + \mathbb{R}a$ *is transversal to the Stifel diagram and* c *belongs to the interior of a Weyl chamber. Let* E *be the intersection of this Weyl chamber with* $lc + \mathbb{R}a$,

$$E_k = \{e \in E : 2\sum_{i=1}^{m}[\langle \rho_i, e \rangle] \leq k\}, \ k = 0, 1, 2, \ldots,$$

where $[\cdot]$ is the integral part of the number in brackets. Then

$$\mathbf{H}_n = H_0(E_n, E_{n-1}) \oplus H_1(E_{n+1}, E_n), \ n \geq 0.$$

EXAMPLE. Let $G = SU(3)$, then $\dim A = 2$, $m = 3$. Let $0 < \langle \rho_1, c \rangle < \langle \rho_2, c \rangle < \langle \rho_3, c \rangle$. A possible disposition of $lc + \mathbb{R}a$ is shown in Figure 1.

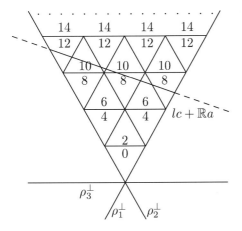

n-th Betti number
equals 1 for $n = 6, 10, 11$
and equals 0 for
the other n

Figure 1

    There is a rather involved explicit expression for the Betti numbers of the pair $(\Omega_{q_0}^l, \Omega_{q_0}^l \backslash \{q^l\})$ via $\langle \rho_i, lc \rangle$, but some asymptotic relations for $l \to \infty$ are transparent. Let $d(lc) = \min\{n : \mathbf{H}_n \neq 0\}$, $D(lc) = \max\{n : \mathbf{H}_n \neq 0\}$. Then

$$\lim_{l \to \infty} \frac{d(lc)}{D(lc)} = \frac{\langle \rho_1 + \rho_3, c \rangle}{\langle \rho_2 + \rho_3, c \rangle}.$$

This limit is a rough homological invariant of the sub-Riemannian structure and it is a rational function of a!

9 Contact structures

Our next topic is exponential mappings, i.e. the restrictions of the end-point mappings to the sets of sub-Riemannian geodesics. Following the philosophy of this paper, we deal with the most "smooth" case.

    Let Δ be a contact structure, i.e. a corank 1 distribution such that $[v, \Delta]_q = T_q M$, $\forall v \in \Delta$, $v(q) \neq 0$, $q \in M$. Hence the dimension of M is odd, $\dim M = 2m + 1$. We will consider a sub-Riemannian structure V^l, $l > 0$, coordinated with Δ. All geodesics for such a structure are regular except the constant trajectory $q(t) \equiv q_0$. While all nontrivial geodesics are regular, they form a smooth manifold $\mathcal{Q} = \bigcup_{l>0} \mathcal{Q}^l$ naturally diffeomorphic to an open subset of $T_{q_0}^* M \backslash \Delta_{q_0}^\perp$. We obtain a

desired diffeomorphism just by identifying a geodesic for V^l with the initial point
of the extremal λ_t associated with this geodesic and normalized by the relation
$h^l(\lambda_t) = 1$. Then $\mathcal{Q}^l$ is identified with $(h^l_{q_0})^{-1}(1)$ for all l small enough and with
an open subset in $(h^l_{q_0})^{-1}(1)$ for the arbitrary $l > 0$, where $h^l_{q_0} = h^l|_{T^*_{q_0}M}$.

A dilation $\delta_\tau : \mathcal{Q} \to \mathcal{Q}$, $0 < \tau \leq 1$, is defined by the relation $(\delta_\tau q)(t) = q(\tau t)$, $q(\cdot) \in \mathcal{Q}$, $t \in [0, 1]$. Then $\delta_\tau(\mathcal{Q}^l) = \mathcal{Q}^{\tau l}$ for $l > 0$ small enough. In other
words, $\mathcal{Q}^{l'}$ consists of reparametrized pieces of curves from $\mathcal{Q}^l$ if $l' < l$.

Consider "the exponential mapping" $ex :\ q(\cdot) \mapsto q(1)$, $q(\cdot) \in \mathcal{Q}$. Let us
denote by $\mathcal{C}$ the set of critical points of ex.

PROPOSITION 9. (a) $\#\{\tau \in (0, 1) : \delta_\tau q \in \mathcal{C}\} < \infty$, $\forall q \in \mathcal{Q}$.
(b) Let $q \in \mathcal{Q}\backslash\mathcal{C}$, then
q is a strong length minimizer $\Longleftrightarrow \delta_\tau q \notin \mathcal{C}$, $\forall \tau \in (0, 1)$.
(c) For any $K \Subset \mathcal{Q}$ there exists $\tau_K > 0$ such that $\delta_\tau(K) \cap \mathcal{C} = \emptyset$ $\forall \tau \leq \tau_K$.
(d) $\mathcal{Q}^l \cap \mathcal{C} \neq \emptyset$ for any small enough $l > 0$.

Properties (a)–(c) of the exponential mapping are similar to the case of Riemannian geometry but (d) is the exact opposite of the Riemannian case. It follows from (a), (b), (d) that there exist arbitrarily short geodesics started at q_0
that are not strong length minimizers. A formal reason is the noncompactness of
$\mathcal{Q}^l \approx (h^l_{q_0})^{-1}(1)$, as opposed to the Riemannian geometry. Actually, this phenomenon is easily predictable because arbitrarily short geodesics cover a neighborhood
of q_0, although all of them are tangent to the hyperplane Δ_{q_0}.

The set

$$\mathbf{C} = \{q(1) : q(\cdot) \in \mathcal{C},\ \delta_\alpha q \notin \mathcal{C},\ \forall \alpha \in (0, 1)\} \subset M$$

is called the sub-Riemannian caustic. It is an "envelope" of the family of geodesics.
Initial point q_0 belongs to the closure of $\mathbf{C}$. We need more notation to say more.

The sub-Riemannian structure V^l, $l > 0$, induces a Euclidean structure on
Δ_q, $q \in M$, such that the Euclidean length of $\forall v \in V^l_q$ is equal to l. Let ω be
a differential one-form that is orthogonal to Δ and normalized by the following
condition: $2m$-form $(d_q\omega)^m|_{\Delta_q}$ is the volume form for the Euclidean structure
induced by V^l. The form ω is defined up to a sign in a neighborhood of q_0, it
is defined globally iff contact structure Δ is coorientable. Our considerations are
local and we fix a sign of ω.

Set $h = \frac{1}{2}(h^1)^2 = \frac{1}{2l^2}(h^l)^2$ a Hamiltonian that is quadratic on the fibers
of T^*M. Relations $e \rfloor \omega = 1$, $e \rfloor d\omega = 0$ define a vector field e and a Hamilton-
ian $e^* : \lambda \mapsto \langle \lambda, e(q) \rangle$, $\lambda \in T_q M$, which is linear on fibers. Let us consider the
Poisson bracket $\{e^*, h\}$. It is one more Hamiltonian that is quadratic on fibers.
It is possible to show that $\Delta_q^\perp$ is contained in the kernel of the quadratic form
$\{e^*, h\}_q = \{e^*, h\}|_{T^*_q M}$. Hence we may consider $\{e^*, h\}_q$ as a quadratic form on
$\Delta_q^* = T^*_q M/\Delta_q^\perp$. Moreover, the Euclidean structure on Δ_q permits us to identify
Δ_q^* with Δ_q and to consider $\{e^*, h\}_q$ as a quadratic form on Δ_q or, in other words,
as a symmetric operator on the Euclidean space Δ_q. In particular, the trace and
the determinant of $\{e^*, h\}_q$ are well defined. It turns out that $\mathrm{tr}\{e^*, h\}_q = 0$ but
the determinant doesn't vanish, generally speaking.

If M is the total space of a principle bundle with one-dimensional fibers transversal to Δ, and V is invariant under the action of structure group (so that Δ is just a connection on the principle bundle), then e is a "vertical" vector field and $\{e^*, h\} = 0$. Conversely, if $\{e^*, h\} = 0$ for a contact sub-Riemannian structure, then the structure is invariant under the one-parametric group generated by e.

We have $T^*_{q_0} M = \mathbb{R}\omega_{q_0} + \Delta^*_{q_0}$. Let $\nu \in \mathbb{R}$ and $\eta \in \Delta^*_{q_0}$, $\eta \neq 0$. We will denote by $q(\cdot; \nu, \eta)$ the geodesic that is the projection on M of the extremal, starting at $(\nu\omega_{q_0} + \eta) \in T^*_{q_0} M$. It turns out that the mapping $\nu \mapsto q(\frac{1}{\nu}; \nu, \eta)$ possesses an asymptotic expansion for $\nu \to \infty$ in the power series in $\frac{1}{\nu}$ with coefficients that are elementary functions of η. It was the study of this expansion that made it possible to obtain fundamental invariants of the contact sub-Riemannian structures for $m = 1$ and to understand the form of the caustic near q_0 in the generic situation.

Dimension 3. Let $\dim M = 3$. Interesting calculations were made by various authors in this minimal possible dimension for a symmetric (Lie group) case where geodesics have a simple explicit expression (see especially [12]). We'll see, however, that principal invariants vanish in that symmetric case.

Figure 2 shows the form of the caustic $\mathbf{C}$ near q_0 if $\{e^*, h\}_{q_0} \neq 0$. "Horizontal" sections have 4 maps.

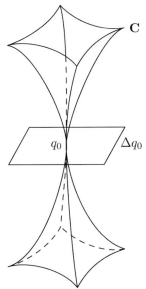

Figure 2

We don't use below a special notation for the standard identification of Δ_{q_0} and $\Delta^*_{q_0}$, and just put elements of Δ_{q_0} instead of $\Delta^*_{q_0}$ in formulas. Thus, $q(\cdot; \nu, v)$, $v \in \Delta_{q_0}$ is a geodesic whose velocity equals v at the starting point. The form $d_{q_0}\omega$ induces an orientation of Δ_{q_0} and of $V^1_{q_0}$ that is the unit circle in the Euclidean plane Δ_{q_0}. We will denote by $d\theta_\xi$, $\xi \in V^1_{q_0}$, the angle differential form on the oriented circle.

Let $\nu \in \mathbb{R}$, $v \in V_{q_0}^1$; set

$$l_c(\nu, v) = \min\{l > 0 : q(\cdot; \nu, lv) \in \mathcal{C}\}, \quad q_c(\nu, v) = q(l_c(\nu, v); \nu, v).$$

Then $l_c(\nu, v)$ is the supremum of the length of strong length minimizing pieces of the geodesic $q(\cdot; \nu, v)$, and $q_c(\nu, v)$ is the point of the caustic $\mathbf{C}$ where this geodesic ceases to be a strong length minimizer.

THEOREM 10. *The following asymptotic expansions hold for* $\nu \to \pm\infty$, $v \in V_{q_0}^1$:

$$q_c(\nu, v) = \pm\nu^{-2}\pi e(q_0) - \nu^{-3}\frac{3\pi}{2}\int_v^{-v}\{e^*, h\}_{q_0}(\xi)\,\xi\,d\theta_\xi + O(\nu^{-4})$$

$$l_c(\nu, v) = |\nu|^{-1}2\pi - |\nu|^{-3}\pi\rho(q_0) + O(\nu^{-4}),$$

where $\rho(q_0)$ *is a constant.*

The curve $v \mapsto \frac{3}{2}\int_v^{-v}\{e^*, h\}_{q_0}\,\xi\,d\theta_\xi$, $v \in V_{q_0}^1$, is a symmetric astroid in Δ_{q_0}; its radius equals $(-\det\{e^*, h\}_{q_0})^{\frac{1}{2}}$, and its cuspidal points belong to the isotropic lines of the form $\{e^*, h\}_{q_0}$.

The invariant $\rho(q)$, $q \in M$, is, in fact, a nonholonomic analog of the Gaussian curvature of a surface. Let $v_1(q)$, $v_2(q)$ be an orthonormal frame in Δ_q with a right orientation. Then $[v_1, v_2] = \alpha_1 v_1 + \alpha_2 v_2 - e$, $[e, v_j] \in \Delta$, $j = 1, 2$, where α_j are smooth functions, and

$$\rho = v_1\alpha_2 - v_2\alpha_1 - \alpha_1^2 - \alpha_2^2 + \frac{1}{2}(\langle[e, v_2], v_1\rangle - \langle[e, v_1], v_2\rangle).$$

A simple count of parameters shows that sub-Riemannian structures on a 3-dimensional manifold should have two "functional invariants". We already have two: $\det\{e^*, h\}$ and ρ.

THEOREM 11. *Let* $d\rho = 0$, $\det\{e^*, h\}_q = 0$ $\forall q \in M$, *where* M *is a parallelizable manifold, and* $H^1(M; \mathbb{R}) = 0$.

Then there exists an orthonormal frame v_1, v_2 *in* Δ *such that*

$$[v_2, v_1] = e, \quad [v_1, e] = \rho v_2, \quad [v_2, e] = -\rho v_1.$$

So a contact sub-Riemannian structure on a 3-dimensional manifold, with the identically vanishing $\{e^*, h\}$ and ρ, is locally equivalent to the Heisenberg group with the standard "symmetric" sub-Riemmanian structure — the most popular example in nonholonomic geometry. We obtain a model of the sub-Riemannian manifold with the identically vanishing $\{e^*, h\}$ and constant positive (negative) ρ if we consider the group $SU(2)$ ($\widetilde{SL}(2; \mathbb{R})$) with the sub-Riemannian structure that is defined by the restriction on a left-invariant distribution of the bi-invariant (pseudo-)Riemannian structure on $SU(2)$ ($\widetilde{SL}(2; \mathbb{R})$).

Acknowledgments

I wish to express my gratitude to Revaz V. Gamkrelidze, my teacher, and to Andrei Sarychev; some of the results presented in the paper were obtained jointly with them. I would also like to thank Richard Montgomery, Hector Sussmann, André Bellaiche, Fernand Pelletier, and Michail Zhitomirskii for stimulating discussions.

Note added in proof

I would like to thank Dr. Ge Zhong who kindly sent me his papers after the Congress. The papers contain, among other things, an essential information about the structure of sub-Riemannian caustics. He has also informed me that invariants of 3-dimensional contact sub-Riemannian structures we considered earlier by S. Webster.

References

[1] A. A. Agrachev and R. V. Gamkrelidze, *Symplectic geometry and necessary conditions for optimality*, Mat. Sb. **182** (1991), 36–54 (Russian); English transl. in Math. USSR-Sb. **72** (1992).

[2] A. A. Agrachev and R. V. Gamkrelidze, *Symplectic methods for optimization and control*, to appear, in Geometry of Feedback and Optimal Control, Marcel Dekker, New York and Basel.

[3] A. A. Agrachev and A. V. Sarychev, *Abnormal Sub-Riemannian geodesics: Morse index and rigidity*, to appear, Ann. Inst. H. Poincaré Anal. Non Linéaire.

[4] A. A. Agrachev and A. V. Sarychev, *Strong minimality of abnormal geodesics for 2-distributions*, J. Dynam. Contr. Syst. **1** (1995).

[5] R. W. Brockett, R. S. Millman, and H. J. Sussmann (eds.), Differential Geometric Control Theory, Birkhäuser, Boston, Basel, and Berlin, 1983.

[6] R. Bryant and L. Hsu, *Rigid trajectories of rank 2 distributions*, Invent. Math **114** (1993).

[7] B. Jakubczyk and W. Respondek (eds.), Geometry of Feedback and Optimal Control, Marcel Dekker, New York and Basel, to appear.

[8] R. Montgomery, *Abnormal Minimizers*, SIAM J. Control Optim. **32** (1994).

[9] A. V. Sarychev, *On homotopy properties of the space of trajectories of a completely nonholonomic differential system*, Dokl. Akad. Nauk SSSR **314** (1990) (Russian); English transl. in Soviet Math. Dokl. **42** (1991).

[10] H. J. Sussmann (ed.), Nonlinear Controllability and Optimal Control, Marcel Dekker, New York and Basel, 1990.

[11] H. J. Sussmann and W. Liu, *Shortest paths for sub-Riemannian metrics on rank-2 distributions*, preprint (1993).

[12] A. M. Vershik and V. Gershkovich, *Nonholonomic dynamical systems. Geometry of distributions and variational problems*, Itogi Nauki: Sovr. Probl. Mat.: Fund.Napr., vol. 16, VINITI, Moscow, 1987, pp. 5–85; English transl. in Encyclopedia of Math. Sci. vol. 16, (Dynamical Systems, VII), Springer-Verlag, Berlin and New York.

[13] I. Zelenko and M. Zhitomirskii, *Rigid curves of generic 2-distributions on 3-manifolds*, preprint (1994).

[14] M. Zhitomirskii, *Typical singularities of differential 1-forms and Pfaffian equations*, vol. 113, Transl. Math. Monographs, Amer. Math. Soc., Providence, RI, 1992.

Whiskered Tori and Chaotic Behavior in Nonlinear Waves

DAVID W. MCLAUGHLIN[*]

Courant Institute of Mathematical Sciences
New York University, New York, NY 10012 USA

1. Introduction and Overview

Chaotic behavior in deterministic dynamical systems is an important phenomenon with many physical ramifications. In finite-dimensional settings this behavior is well understood, because of several fundamental mathematical results and many numerical studies that unveil phenomena beyond the reach of current analytical methods.

However, most real physical applications are modeled by partial differential equations; therefore, to be really applicable, chaotic behavior in deterministic systems must be extended to the infinite-dimensional setting of pde's. In this lecture I will describe one initial step in this direction by summarizing some results for an admittedly idealized class of pde's — those nonlinear waves that are described by equations that are perturbations of completely integrable soliton equations.

The damped, driven pendulum,

$$u_{tt} + \sin u = \epsilon[-\alpha u_t + \Gamma \cos \omega t], \tag{1.1}$$

is a prototypic example of a three-dimensional deterministic system that admits chaotic behavior. When the perturbation is absent ($\epsilon = 0$), the pendulum has an unstable fixed point $(u, u_t) = (\pi, 0)$ representing its inverted position, and orbits homoclinic to this inverted position that separate the two-dimensional pendulum phase space into oscillating and rotating components. In the presence of a perturbation, one can use these integrable homoclinic orbits to construct a "horseshoe" with which one can establish the existence of an invariant set on which the perturbed dynamics is topologically equivalent to a Bernoulli shift on two symbols; moreover, these two symbols admit a natural physical interpretation as the pendulum falling from its inverted position by swinging to the right or to the left.

Knowing this behavior for the damped-driven pendulum, we began our study of chaotic behavior in pde's with a damped-driven perturbation of the "sine-Gordon" equation,

$$u_{tt} - u_{xx} + \sin u = \epsilon[-\alpha u_t + \Gamma \cos \omega t], \tag{1.2}$$

---

[*]Funded in part by AFOSR-90-0161 and by NSF DMS 8922717 A01.

which can be interpreted as the continuum limit of a chain of coupled pendula. Under periodic, even boundary conditions,

$$u(x + L, t) = u(x, t), \quad u(-x, t) = u(x, t),$$

the unperturbed ($\epsilon = 0$) problem is a completely integrable Hamiltonian system, as can be established with the "spectral transform" of soliton mathematics. In contrast to the trivial two-dimensional "phase plane" portrait of the unperturbed pendulum, the infinite-dimensional "phase portraits" for the sine-Gordon pde seem very complicated. However, through the spectral transform, one can obtain very detailed information about these portraits for this integrable wave equation.

First, however, we performed numerical experiments on the perturbed sine-Gordon equation. In these experiments we fixed the parameters ($\epsilon \alpha, \omega, L$), as well as the initial data, at judiciously chosen values, and we varied the amplitude ($\epsilon \Gamma$) of the sinusoidal driver. Although the detailed results varied depending upon the particular choices of parameter values, a typical sequence was as follows: as the "bifurcation parameter" ($\epsilon \Gamma$) was increased, the long time behavior of the observed solution changed as follows:

(1) **temporally periodic (with period of the driver), with no spatial structure (independent of x);**

(2) **temporally periodic (with period of the driver), with one spatially localized excitation — a solitary wave;**

(3) **temporally quasi-periodic (with the frequency of the driver, together with a second shorter frequency), with one spatially localized excitation, together with a significant long wavelength background;**

(4) **irregularly temporal, together with interesting interactions between the spatial structures.**

Detailed descriptions of these numerical experiments, together with descriptions of the numerical algorithms, validation procedures, and the specific "chaotic diagnostics" that were employed may be found in [3],[4], [20]. My point here is that the pde "route to chaos" is very different from the route for the single pendulum. Spatial structure develops and stabilizes at values of the driving parameter that are very small, certainly smaller than those required for the single pendulum to evolve chaotically. Chaotic behavior does indeed develop in the sine-Gordon experiments, but interactions between spatial structures play a major role in the chaos. Moreover, the amplitudes of the waves are small and far from the inverted unstable position of the pendula. Some other instability must be behind this chaotic behavior for the near integrable pde.

At this point in our study we have a choice — (i) to continue the analysis of the perturbed sine-Gordon equation [6], [20], or (ii) to turn to a similar study for a perturbed nonlinear Schrödinger (NLS) equation, which, because of the small amplitude nature of the chaotic state, should behave very similarly. With either option, the methods and results will be similar; however, as the NLS framework is simpler, I will restrict the rest of this lecture to the NLS case.

NLS describes the envelope of a small $O(\epsilon)$ amplitude, nearly monochromatic sine-Gordon wave, an approximation that is valid over long time scales of $O(\epsilon^{-2})$. Unfortunately, these time scales are not long enough to capture, reliably, the long time chaotic behavior of the sine-Gordon equation. Thus, we first repeated the numerical experiments for the NLS case [20] and showed that the "routes to chaos" for the NLS equation were very similar to those of the sine-Gordon equation.

Our numerical experiments immediately lead to the following analytical questions:

(1) Which instabilities in the integrable pde replace the inverted position of the pendulum?

(2) Which homoclinic structures in the integrable pde replace the pendulum's separatrix? In the integrable cases, how can such hyperbolic structures be constructed? In the perturbed cases, how can these hyperbolic structures be identified and monitored?

(3) In the pde cases, what replaces the "left-right" symbol dynamics that is so natural for the pendulum?

(4) Which parts of these integrable hyperbolic structures persist after perturbation? Can homoclinic orbits, horseshoes, and a symbol dynamics be constructed for the pde?

In the remainder of this lecture, I will describe the status of our efforts to answer these questions.

2. Hyperbolic Structure — Integrable Case

We consider a normalized form of the NLS equation:

$$- iq_t + q_{xx} + 2\,|q|^2 q = 0, \tag{2.1}$$

in H^1, under periodic, even boundary conditions. NLS is a completely integrable Hamiltonian system, as can be established through the "inverse spectral transform", that begins from the Zakharov-Shabat linear system [25]

$$\left[- i\sigma_3 \tfrac{d}{dx} - \begin{pmatrix} 0 & q \\ -\bar{q} & 0 \end{pmatrix} \right] \vec{\psi} = \lambda \vec{\psi}$$

$$\left[- i\sigma_3 \tfrac{d}{dt} + q\bar{q} + \begin{pmatrix} 0 & -2\lambda q + iq_x \\ 2\lambda\bar{q} + i\bar{q}_x & 0 \end{pmatrix} \right] \vec{\psi} = 2\lambda^2 \vec{\psi}, \tag{2.2}$$

where $\sigma_3 \equiv \mathrm{diag}[-1, 1]$. Compatibility of this overdetermined system ensures that the coefficient q satisfies the NLS equation.

2.1. Floquet Spectral Theory. The integration of the NLS equation is accomplished through the Floquet spectral theory of the differential operator $\hat{L} = \hat{L}(q)$,

$$\hat{L} = -i\sigma_3 \frac{d}{dx} - \begin{pmatrix} 0 & q \\ -\bar{q} & 0 \end{pmatrix}.$$

The difficulty with this spectral theory is that the operator $\hat{L}$ is not self-adjoint. Nevertheless, it can be controlled with the aid of certain "counting lemmas" [23], [15], and, as described in [20], the spectrum has a fascinating structure in the complex plane.

Certain properties of this spectrum follow as in the standard Floquet theory of Hill's operator [18]. The spectrum occurs on curves, not necessarily real, that terminate at periodic or antiperiodic eigenvalues. *Multiple points* are defined as points where these eigenvalues coalesce. All but a finite number of these multiple points must be real. The *complex multiple points* are especially important for temporal instabilities.

2.2. Temporal Instabilities. Fix a solution $q(x, t)$ of the NLS equation that is periodic in x and quasiperiodic in t; more precisely, fix a q on one of the finite dimensional invariant tori. Linearizing NLS about q yields a variable coefficient (in x and t) linear equation that governs the linear stability of $q(x, t)$. Quadratic products of solutions of the Zakharov-Shabat linear system (2.2) generate a basis of solutions of this linearization [5], [1]. With this basis one can assess the linear stability properties of the solution $\vec{q}$.

First (in the absence of higher order multiple points), the basis splits into two parts, one labeled by simple eigenvalues and one labeled by double points. There is no exponential growth associated with that part of the basis associated to the simple eigenvalues, nor to that part associated to real double points. The only possible exponential instabilities are labeled by complex double points. These are at most finite in number. Typically, for each complex double point there is one exponentially growing and one exponentially decaying linearized solution — although examples do exist of situations [8], with complex double points, for which there is no instability.

In summary, all instabilities are associated with complex multiple points, of which any solution has at most a finite number. For the integrable setting, this analysis, based upon the spectral transform, generalizes the classical analysis of the "modulational instability", which is fundamental in the theory of nonlinear waves.

2.3. Monitoring with the Spectrum of $\hat{L}$. This spectral theory of $\hat{L}$ is also useful for flows that are perturbations of the NLS equation. For example, one can use the spectrum of $\hat{L}$ to monitor a chaotic time series for the perturbed flow. Consider a chaotic trajectory $q(x, t)$ for the perturbed NLS equation, represented numerically. At each time t, one can construct numerically the spectrum of $\hat{L}[q(t)]$. By monitoring this spectrum as a function of t, one can measure the number and degree of excitations of the "nonlinear normal modes" of the wave, as well as the appearance of complex multiple points in the spectrum. This procedure is described in detail in [20].

The result of this numerical monitoring of the spectrum establishes, as much as any numerical experiment can establish, that chaotic behavior for the pde can occur in a region of function space of quite low dimension, but one that contains q's with one or more complex double points. Thus, the type of instability behind the chaotic behavior is identified through these spectral measurements.

These measurements have been performed for both damped-driven and conservative perturbations. Different phenomena and behavior occur in the two cases. For detailed descriptions, see [20], [22], [9].

3. Global Representations of Whiskered Tori

Returning to the integrable setting, one can use Bäcklund transformations to exponentiate the linearized instabilities to obtain global solutions of the NLS equation. Fix a periodic solution $q(x,t)$ of NLS that is quasiperiodic in t, for which the linear operator $\hat{L}$ has a complex double point ν of geometric multiplicity 2 *that is associated with an NLS instability.* We denote two linearly independent solutions of the Zakharov-Shabat linear system at $\lambda = \nu$ by $(\vec{\phi}^{+}, \vec{\phi}^{-})$. Thus, a general solution of the linear system at $(\vec{q}, \nu)$ is given by

$$\vec{\phi}(x,t;\nu;c_{+},c_{-}) = c_{+}\vec{\phi}^{+} + c_{-}\vec{\phi}^{-}. \tag{3.1}$$

We use $\vec{\phi}$ to define a transformation matrix [24] G by

$$G = G(\lambda;\nu;\vec{\phi}) \equiv N \left(\begin{array}{cc} \lambda - \nu & 0 \\ 0 & \lambda - \bar{\nu} \end{array} \right) N^{-1}, \tag{3.2}$$

where

$$N \equiv \left[\begin{array}{cc} \phi_{1} & -\bar{\phi}_{2} \\ \phi_{2} & \bar{\phi}_{1} \end{array} \right]. \tag{3.3}$$

Then we define Q and $\vec{\Psi}$ by

$$Q(x,t) \equiv q(x,t) + 2(\nu - \bar{\nu}) \frac{\phi_{1}\bar{\phi}_{2}}{\phi_{1}\bar{\phi}_{1} + \phi_{2}\bar{\phi}_{2}} \tag{3.4}$$

and $\quad \vec{\Psi}(x,t;\lambda) \equiv G(\lambda;\nu;\vec{\phi})\,\vec{\psi}(x,t;\lambda), \tag{3.5}$

where $\vec{\psi}$ solves the linear system (3.1) at $(\vec{q}, \nu)$. Formulas (3.4) and (3.5) are Bäcklund transformations for the potential and eigenfunctions, respectively. We have the following:

THEOREM 3.1 *Let $q(x,t)$ denote a periodic solution of NLS that is linearly unstable with an exponential instability associated to a complex double point ν in $\sigma\left(\hat{L}(\vec{q})\right)$. Let the complex double point ν have geometric multiplicity 2, with eigenbasis $(\vec{\phi}^{+}, \vec{\phi}^{-})$ for linear system (3.1), and define $Q(x,t)$ and $\vec{\Psi}(x,t;\lambda)$ by (3.4) and (3.5). Then*

(i) $Q(x,t)$ is a solution of NLS, with spatial period 1;

(ii) $\sigma(\hat{L}(\vec{Q})) = \sigma(\hat{L}(\vec{q}))$;

(iii) $Q(x,t)$ is homoclinic to $q(x,t)$ in the sense that $Q(x,t) \longrightarrow q_{\theta_\pm}(x,t)$ exponentially as $\exp(-\sigma_\nu|t|)$ as $t \longrightarrow \pm\infty$. Here $q_{\theta_\pm}$ is a "torus translate" of q, σ_ν is the nonvanishing growth rate associated to the complex double point ν, and explicit formulas can be developed for this growth rate and for the translation parameters $\theta_\pm$.

(iv) $\vec{\Psi}(x,t;\lambda)$ solves the linear system (3.1) at $(\vec{Q},\lambda)$.

This theorem is quite general, constructing homoclinic solutions from a wide class of starting solutions $\vec{q}(x,t)$. Its proof is one of direct verification, following the sine-Gordon model [6]. In references [19] and [20], several qualitative features of these homoclinic orbits are emphasized: (i) $Q(x,t)$ is homoclinic to a torus that itself possesses rather complicated spatial and temporal structure, and is not just a fixed point. (ii) Nevertheless, the homoclinic orbit typically has still more complicated spatial structure than its "target torus". (iii) When there are several complex double points, each with nonvanishing growth rate, one can iterate the Bäcklund transformations to generate more complicated homoclinic manifolds. (iv) The number of complex double points with nonvanishing growth rates counts the dimension of the unstable manifold of the critical torus in that two unstable directions are coordinatized by the complex ratio c_+/c_-. Under even symmetry only one real dimension satisfies the constraint of evenness, as will be clearly illustrated in the following example. (v) These Bäcklund formulas provide coordinates for the stable and unstable manifolds of the critical tori; thus, they provide explicit representations of the critical level sets that consist of "whiskered tori". (vi) These "whiskered tori" may be constructed for any of the soliton equations that possess instabilities. Thus, they provide a large collection of explicit homoclinic structures in the setting of integrable nonlinear waves.

3.1. An Example: The Spatially Uniform Plane Wave.

As a concrete example, consider the spatially uniform plane wave:

$$q = c\exp\left[-i(2c^2t + \gamma)\right].$$

For this x-independent q, the homoclinic orbits can be explicitly computed. A single Bäcklund transformation at one purely imaginary double point yields $Q = Q_H(x,t;c,\gamma;k=\pi,c_+/c_-)$:

$$Q_H = \left[\frac{\cos 2p - \sin p \ \text{sech}\,\tau\cos(2kx + \phi) - i\sin 2p\tanh\tau}{1 + \sin p \ \text{sech}\,\tau\cos(2kx + \phi)}\right]ce^{-i(2c^2t+\gamma)} \quad (3.6)$$

$$\rightarrow e^{\mp 2ip}ce^{-i(2c^2t+\gamma)} \qquad \text{as } \rho \rightarrow \mp\infty,$$

where $c_+/c_- \equiv \exp(\rho + i\beta)$ and p is defined by $k + \nu = c\exp(ip)$, $\tau \equiv \sigma t - \rho$, and $\phi \equiv p - (\beta + \pi/2)$.

In this example the target is always the plane wave; hence, it is always a circle of dimension one, and in this example we are really constructing only whiskered circles. On the other hand, in this example the dimension of the whiskers need not be one, but is determined by the number of purely imaginary double points, which in turn is controlled by the amplitude c of the plane wave target and by the spatial period. (The dimension of the whiskers increases linearly with the spatial period.) When there are several complex double points, Bäcklund transformations must be iterated to produce complete representations. Although these iterated formulas are quite complicated, their parameterizations admit rather direct qualitative interpretations [22].

Returning to the example with one double point, it will be important to impose the constraint of *even about* $x = 0$. To do so, one fixes the phase of the complex transformation parameter c_+/c_- to be one of two values

$$\phi = 0, \pi \qquad\qquad \text{(evenness)}.$$

Each choice fixes one whisker. Although the target q is independent of x, each of these whiskers has x dependence through the $\cos(2kx)$. One whisker has exactly this dependence and can be interpreted as a spatial excitation located near $x = 0$, whereas the second whisker has the dependence $\cos(2k(x - \pi/2k))$, which we interpret as spatial structure located near $x = 1/2$ (since here $k = \pi$).

3.2. Natural Symbols The simple example just described is actually quite important for our numerical experiments on chaotic behavior. First, our numerical experiments were performed under *even* periodic boundary conditions. Second, the spectral transform measurements of the chaotic signal have shown that the observed chaotic behavior takes place near the plane wave in function space. Moreover, that complex double point, which we used to construct the orbit Q_H, frequently appears in the measurements of the chaotic signal. Thus, based on the spatial structures of the two whiskers under even symmetry, one anticipates that "center-edge" behavior may be characteristic of the chaotic signal, and this was indeed observed in the numerical experiments. The spatial profiles of the chaotic signal consisted of one localized spatial structure (solitary wave), which jumped between "center" and "edge" following what appears to be an irregular pattern of jumps. In this simple case, the "center-edge" locations of the spatially localized excitations provide natural candidates for a symbol dynamics.

4. Persistence Results

Given the above type of information, we were ready to begin a dynamical systems style of analysis on the perturbed pde. That study is currently in progress and is rather technical. Here, limitations on space only permit a very short intuitive description. Detailed mathematical results may be found in the references.

We begin with the perturbed NLS equation,

$$-2iq_t + q_{xx} + (qq^* - 1)q = i\epsilon \left[\alpha q - \hat{\beta}q + \Gamma \right] , \qquad (4.1)$$

in H^1, under even periodic boundary conditions. (The operator $\hat{\beta}$ is bounded and dissipative, and we work in the region of function space with one complex double point.)

The plane of constants (in x) is an invariant plane (Π) for both the unperturbed and perturbed dynamics. On this plane the perturbed problem has three fixed points, two of which emerge from the unit circle S of fixed points for the unperturbed problem. One of these fixed points, p_ϵ, is a sink on the plane; the other, q_ϵ, is a saddle on the plane. Both live in a "resonance band" on the plane, in which the motion is very slow. Both inherit from the unperturbed integrable pde its one-dimensional, rapid, unstable behavior off the plane Π. Thus, typical motion has at least two time scales that impose a singular perturbation aspect upon the analysis.

Next, one [17] establishes the existence of two codimension 1 manifolds, W_ϵ^{cs} and W_ϵ^{cu}, both of which are invariant under the perturbed dynamics. These two manifolds are persistent deformations of the center-stable $W^{cs}(S)$ and center-unstable $W^{cu}(S)$ manifolds of the circle S of fixed points in the unperturbed, integrable case. Their intersection $W \equiv W_\epsilon^{cs} \bigcap W_\epsilon^{cu}$ is a codimension 2 invariant manifold that is *normally hyperbolic*, and which is the persistent image of an unperturbed center manifold $W^c(S)$ of the circle S.

In addition to the existence of these invariant manifolds, one needs a representation of them that is useful for two time scale singular perturbation calculations. For this, we extend some finite-dimensional representations of Fenichel [7] into the infinite-dimensional pde framework. Specifically, we [17] represent W_ϵ^{cs} and W_ϵ^{cu} as fiber bundles, with one-dimensional fibers over the base W. Intuitively, motion on the base W is slow, and the fibers factor out fast one-dimensional expansion and contraction by identifying each point on a given fiber with its base point on the slow manifold. These fibers are at least twice differentiable with respect to the perturbation parameter ϵ, which makes the representations useful for perturbation calculations.

To this point the constructions are local, in a neighborhood of an invariant circle on the plane Π. These local constructions are possible because we have a complete understanding of motion on the plane Π, and control near Π. Next comes the global part of the argument, for which control is obtained through the global representations of the homoclinic orbits for the unperturbed, integrable pde.

First, one uses a "Melnikov argument" to establish that the unstable manifold of the fixed point q_ϵ intersects the manifold W_ϵ^{cs}, $W_\epsilon^u(q_\epsilon) \bigcap W_\epsilon^{cs} \neq W_c^u(q_\epsilon) \cap \pi$. Additional arguments are then used to describe the fate in forward time of orbits in this intersection.

In order to carry out this program, we first developed the arguments for some carefully chosen finite-dimensional models [2], [20], [12], [13], [21], [16],[11]. Detailed results for the pde may be found in [17] and [10], with some preliminary, but key, ideas in the thesis [14].

In this manner, homoclinic and heteroclinic orbits may be constructed that are singular perturbations of the integrable orbits. Some of these orbits have complex "center-edge" patterns [11], [10], which establish the presence of "long, chaotic transients". At this time, we have not established a symbol dynamics in the pde case. However, this work certainly provides a rich example illustrating that methods from dynamical systems theory can provide a guide for global analysis of certain nonlinear wave pde's.

References

[1] R. F. Bikbaev and S. B. Kuksin, Algebra and Analysis, No. 3, to appear 1992.

[2] A. R. Bishop, R. Flesch, M. G. Forest, D. W. McLaughlin, and E. A. Overman II, *Correlations between chaos in a perturbed sine-Gordon equation and a truncated model system*, SIAM J. Math. Anal., 21:1511–1536, 1990.

[3] A. R. Bishop, M. G. Forest, D. W. McLaughlin, and E. A. Overman II, *A quasi-periodic route to chaos in a near-integrable pde*, Phys. D, 23:293–328, 1986.

[4] A. R. Bishop, M. G. Forest, D. W. McLaughlin, and E. A. Overman II, *A quasiperi-odic route to chaos in a near-integrable PDE: Homoclinic crossings*, Phys. Lett. A, 127:335–340, 1988.

[5] N. M. Ercolani, M. G. Forest, and D. W. McLaughlin, *Geometry of the modulational instability I, II*, unpublished manuscript, 1987.

[6] N. M. Ercolani, M. G. Forest, and D. W. McLaughlin, *Geometry of the modulational instability, Part III: Homoclinic orbits for the periodic sine-Gordon equation*, Phys. D, 43:349–384, 1990.

[7] N. Fenichel, *Geometric singular perturbation theory for ordinary differential equations*, J. Differential Equations, 31:53–98, 1979.

[8] R. Flesch, M. G. Forest, and A. Sinha, *Numerical inverse spectral transform for the periodic sine-Gordon equation theta function solutions and their linearized stability*, Phys. D, 48:169–231, 1991.

[9] M. G. Forest, C. Geddy, and A. Sinha, *Chaotic transport and integrable instabilities*, Phys. D, 67:347–386, 1993.

[10] G. Haller, *Orbits homoclinic to resonances: The Hamiltonian Pde case*, in preparation, 1994.

[11] G. Haller and S. Wiggins, *Orbits homoclinic to resonances: The Hamiltonian case*, Phys. D, 66:298–346, 1993.

[12] G. Kovacic, *Dissipative dynamics of orbits homoclinic to a resonance band*, Phys. Lett. A, 167:143–150, 1992.

[13] G. Kovacic and S. Wiggins, *Orbits homoclinic to resonances, with an application to chaos in a model of the forced and damped sine-Gordon equation*, Phys. D, 57:185–225, 1992.

[14] Y. Li, *Chaotic behavior in PDE's*, Ph.D. thesis, Princeton University, 1993.

[15] Y. Li and D. W. McLaughlin, *Morse and Melnikov functions for NLS Pdes*, Comm. Math Phys., 162:175–214, 1994.

[16] Y. Li and D. W. McLaughlin, *Homoclinic orbits for a perturbed discrete NLS equation I*, in preparation, 1994.

[17] Y. Li, D. W. McLaughlin, J. Shatah, and S. Wiggins, *Persistent homoclinic orbits for a perturbed NLS equation*, in preparation, 1994.

[18] W. Magnus and W. Winkler, Hill's Equation, Interscience-Wiley (New York), 1966.

[19] D. W. McLaughlin, *Whiskered Tori for the NLS Equation*, in: Important Developments in Soliton Theory (A. S. Fokas and V. E. Zakharov, Springer Series in Nonlinear Dynamics, 537–558, 1993.

[20] D. W. McLaughlin and E. A. Overman, *Whiskered tori for integrable pdes and chaotic behavior in near integrable pdes*, Surveys in Appl. Math. 1, 1994.

[21] D. W. McLaughlin, E. A. Overman, S. Wiggins, and C. Xiong, *Homoclinic orbits in a four dimensional model of a perturbed NLS equation*, Dynamics Reports, accepted, pending revision, 1994.

[22] D. W. McLaughlin and C. M. Schober, *Chaotic and homoclinic behavior for numerical discretizations of the nonlinear Schrödinger equation*, Phys. D, 57:447–465, 1992.

[23] J. Poschel and E. Trubowitz, Inverse Spectral Theory, Academic Press (New York), 1987.

[24] D. H. Sattinger and V. D. Zurkowski, *Gauge theory of backlund transformations*, Phys. D, 26:225–250, 1987.

[25] V. E. Zakharov and A. B. Shabat, *Exact theory of two-dimensional self-focusing and one-dimensional self-modulation of waves in nonlinear media*, Soviet Phys. JETP, 34(1):62–69, 1972.

Formulas for Finding Coefficients from Nodes/Nodal Lines

Joyce R. McLaughlin

Rensselaer Polytechnic Institute
Troy, NY 12181, USA

1. Introduction

We ask the question: What can be determined about a vibrating system from the positions of nodal lines? The question is answered for a rectangular membrane. We give a formula for finding a potential from the nodal line positions. A uniqueness result is presented. Analogous one-dimensional results are presented; three-dimensional results are announced. A case study showing the effect of structural damping on measurements is described.

Two experiments can be performed both to motivate the choice of nodal position data for the inverse problem and to motivate the mathematical difficulties in the two-dimensional case. One experiment is a vibrating beam, the other a vibrating plate. The vibrating beam is driven at one end and free at the other. This experiment is performed twice, once for a homogeneous beam and once for a beam with mass added in a small subregion of the beam. The changes in natural frequencies, which are lowered, and changes in the nodal positions, which move toward the added mass, can be measured. For the second experiment, a plate driven at the center and free on the edges is excited at several different natural frequencies. Sand is distributed on the plate. At each frequency the sand accumulates along the corresponding nodal lines. This experiment is similar to the Chladni experiments, see [17]. One observes that the connected domains defined by the nodal lines are frequently long, curved strips with occasional smaller enclosed domains.

These experiments, especially the plate experiment, are demonstration experiments. One can ask then: Can the nodal lines or nodal points be measured accurately? One method to do this is to direct a laser at the vibrating surface. The Doppler shift in the backscatter is measured. The lines, or points, where the Doppler shift is minimized are the nodal lines.

The mathematical results in higher dimensions build on previous one-dimensional results. These results, obtained by McLaughlin and Hald [11], [6], [7], [8], present formulas, uniqueness results, and numerical calculations. In the second section we briefly present a few of these results. The model we choose is the longitudinal motion of a beam in the case where both the elasticity coefficient p and the density ρ may have discontinuities. Two formulas are given. One gives a piecewise constant approximation to the elasticity coefficient when the density is constant. This formula converges to the elasticity coefficient at every point of continuity. The other formula gives a piecewise constant approximation to the density when the elasticity coefficient is constant. This formula converges to the

Proceedings of the International Congress
of Mathematicians, Zürich, Switzerland 1994
© Birkhäuser Verlag, Basel, Switzerland 1995

density at every point of continuity. To establish our results a bound on the square of the eigenvalues is established for the case where both p and ρ are of bounded variation. This work is presented in the first section.

For the two-dimensional problem, see [9] and [12], we consider a rectangular membrane. There is a force on the membrane that depends linearly on the displacement. The amplitude of the force is unknown. We solve the inverse problem: find the amplitude q from nodal line positions. We show that at a dense set of points q can be approximated by the difference of two eigenvalues. To establish this result there are two difficulties. One is that in order to establish perturbation results for the eigenvalues and eigenfunctions, we must solve a *small divisor* problem. This problem is solved by establishing criteria for eigenvalues to be well separated, thus making it possible to bound terms that contain the small divisors. Almost all eigenvalues satisfy the criteria and perturbation results are established for only those eigenvalues. The second difficulty is that even though the nodal domains can be long, thin, curved domains, we must cut these domains to define small approximate nodal domains. The methods for solving each of these problems, as well as the presentation of the formula for q, are given in the third section. The three-dimensional results are discussed briefly in the third Section as well.

We discuss briefly the effect of structural damping on nodal point measurement in the last section.

1.1. The One-Dimensional Bounded Variation Problem.

We consider the mathematical model for the longitudinal vibrations of a beam with fixed ends. The elasticity coefficient $p > 0$ and density $\rho > 0$ are of bounded variation.

$$
\begin{aligned}
(p\, u_x)_x + \lambda \rho u &= 0, & 0 < x < L, \quad p, \rho \in BV[0, L], & \qquad (1) \\
u(0) = u(L) &= 0.
\end{aligned}
$$

This problem has a set of eigenvalues satisfying $0 < (\omega_1)^2 < (\omega_2)^2 < \dots$. The nth eigenfunction has exactly $n - 1$ nodes, which we label in increasing order, x_j^n, $j = 1, \dots, n - 1$.

To establish our piecewise constant approximations to p and ρ we require an estimate for $\omega_n(p, \rho)$. Letting $V(f)$ represent the total variation of a function f of bounded variation, we have shown, see [8],

THEOREM 1 *Let $p > 0$, $\rho > 0$ satisfy p, $\rho \in BV[0, L]$. Then the eigenvalues for* (1), $\{[\omega_n(p, \rho)]^2\}_{n=1}^{\infty}$, *obey the bound*

$$
\left| \omega_n(p, \rho) \int_0^L \sqrt{\frac{p}{\rho}}\, dx \;-\; n\pi \right| \le \frac{1}{4} V(\ln p\rho).
$$

Our goal is to find piecewise constant approximations to either p or ρ from measurements of the eigenvalues and nodal positions. To do this, let $x_0^n = 0$ and $x_n^n = L$ and define

$$
\rho_n = \left[\frac{\pi}{\omega_n(x_j^n - x_{j-1}^n)} \right]^2 \qquad \text{when} \qquad x_{j-1}^n \le x < x_j^n, \qquad j = 1, \dots, n
$$

together with $\rho_n(L) = \rho_n(x_{n-1}^n)$. Define

$$p_n = \left[\frac{\omega_n\,(x_j^n - x_{j-1}^n)}{\pi}\right]^2 \qquad \text{when} \qquad x_{j-1}^n \le x < x_j^n, \quad j = 1, \dots, n$$

together with $p_n(L) = p_n(x_{n-1}^n)$. Then we have shown in [8],

THEOREM 2 *Let* $p = 1$, $\rho > 0$, *and* $\rho \in BV[0, L]$ *in* (1). *Then* ρ_n *converges pointwise to* ρ *at every point of continuity.*

THEOREM 3 *Let* $\rho = 1$, $p > 0$, *and* $p \in BV[0, L]$ *in* (1). *Then* p_n *converges pointwise to* p *at every point of continuity.*

Results of numerical experiments are presented in [8].

2. The Two-Dimensional Smooth Potential Problem

Here we consider the mathematical model for a rectangular vibrating membrane fixed on the boundary; the rectangle $R = [0, \pi/a] \times [0, \pi]$ with a^2 chosen to be irrational. Letting $q \in C_0^\infty(R)$, the mathematical model for the eigenvalue problem is then

$$\begin{aligned} -\Delta u + q\,u &= \lambda u, & x \in R, && (2) \\ u &= 0, & x \in \partial R. \end{aligned}$$

The object is to find q from the nodal line positions of the eigenfunctions for this problem.

Before stating our main results, we establish notation for the $q = 0$ problem. Letting $\alpha = (an, m)$ we define the lattice L to be

$$L = \{\alpha = (an, m)| \; n, m = 1, 2, 3, \dots\}.$$

Then the eigenvalue and normalized eigenfunction pairs for the $q = 0$ problem can be naturally indexed by $\alpha \in L$ as

$$\lambda_{\alpha 0} = |\alpha|^2 = a^2 n^2 + m^2, \qquad u_{\alpha 0} = \frac{2\sqrt{a}}{\pi}\sin anx \sin my.$$

Our theory requires that almost all of the eigenvalues $\{\lambda_{\alpha 0}\}_{\alpha \in L}$ be well separated. To be well separated the eigenvalues must satisfy five conditions. We describe three of these conditions in words here. The first condition is that there is a small interval about $\lambda_{\alpha 0} = |\alpha|^2$ that contains no other eigenvalues of the $q = 0$ problem. The length of the interval decreases slowly as $|\alpha| \to \infty$. The second condition is that for lattice points β near α the corresponding eigenvalues $\lambda_{\beta 0} = |\beta|^2$ are a large distance from $\lambda_{\alpha 0}$. This distance $\lambda_{\beta 0} - \lambda_{\alpha 0}$ increases rapidly as $|\alpha| \to \infty$. The third condition is that the number of oscillations of $u_{\alpha 0}$ in the x and y directions are comparable. All five conditions are stated explicitly in the Appendix at the end of this paper.

We do not show our main results for all irrational a^2 but only for those values that are poorly approximated by rational numbers. Our condition is almost the same as that given by Moser [13]. That is, fixing $A_0 > 1$ and $0 < \delta < \frac{1}{2}$, we require $a \in V$ where

$$V = \left\{ a \quad | \quad 1 < a < A_0 \quad \text{and there exists} \quad K > 0 \quad \text{so that for all} \right.$$

$$\left. \text{integers} \quad p, q > 0, \quad |a^2 - \frac{p}{q}| > \frac{K}{q^{2+\delta}} \right\}.$$

We note that meas $V = A_0 - 1$. Further we can identify specific $a \in V$. By a theorem of Roth [14], if a^2 is irrational and algebraic then $a \in V$.

We can now state our main theorems. We do not give our sharpest possible results here but give a presentation that makes our ideas more accessible to the reader. Letting $\fint q = \frac{2\sqrt{a}}{\pi} \int_R q$ be the integral average of q we can show, see [9], the uniqueness theorem,

THEOREM 4 *Let $a \in V$ and $q \in C_0^\infty(R)$. The translated potential $q - \fint q$ in (2) is uniquely determined by a subset of nodal lines of the eigenfunctions.*

The uniqueness theorem above is a corollary of the following theorem, which gives a formula for approximating q.

THEOREM 5 *Let $a \in V$. Let $q \in C_0^\infty(R)$. There exists an infinite set $L(a) \backslash M(a)$ and a dense set of $x' \in R$ with corresponding subsets $\Omega_\alpha' \subset R$ defined by the nodal lines, and indexed by $\alpha \in L(a) \backslash M(a)$, so that the potential in (2) satisfies*

$$\left| q(x') - \fint q - [\lambda_{\alpha 0} - \lambda_{1,0}(\Omega_\alpha')] \right| < \frac{1}{2} |\alpha|^{-7/4}.$$

Here $\lambda_{1,0}(\Omega_\alpha')$ is the smallest eigenvalue for

$$-\Delta u = \lambda u, \; x \in \Omega_\alpha', \qquad u = 0, \; x \in \partial\Omega_\alpha'.$$

REMARK: We show that the set $L(a) \backslash M(a)$ has density one in $L(a)$. This means

$$\lim_{r \to \infty} \frac{\#\{\alpha \in L(a) \backslash M(a) \; | \; |\alpha| < r\}}{\{\alpha \in L(a) \; | \; |\alpha| < r\}} = 1.$$

Further, for α to be in $L(a) \backslash M(a)$ it is necessary, though not sufficient, for the three conditions described above to be satisfied. Finally, for any $N > 0$ we can choose our dense set $\{x'\}$ so that the formula for approximating q satisfies the above bound for some $|\alpha| > N$.

To establish Theorem 5, and hence Theorem 4, we proved the following perturbation result whose demonstration requires the solution of a "small divisor" problem, see [9].

THEOREM 6 *Let $a \in V$. Let $q \in C_0^\infty(R)$. Then there exists a set $L(a) \backslash M(a)$ of density one such that for $\alpha \in L(a) \backslash M(a)$ there is a unique eigenvalue, normalized eigenfunction pair $\lambda_{\alpha q}, u_{\alpha q}$ of (2) satisfying*

$$\left| \lambda_{\alpha q} - \fint q - \lambda_{\alpha 0} \right| \leq |\alpha|^{-15/8},$$

$$\left\| u_{\alpha q} - u_{\alpha 0} - \sum_{\beta \neq \alpha} \frac{(q u_{\alpha 0}, u_{\beta 0})}{\lambda_{\alpha 0} - \lambda_{\beta 0}} \right\|_\infty \leq \sqrt{a}\, |\alpha|^{-15/8}.$$

REMARK: Our proof of this result was influenced by the perturbation results in [1], [2], [3]. There L^2 bounds for the eigenfunctions were obtained. We require L^∞ bounds for the eigenfunctions for two reasons. One is so that we can get good estimates for the positions of the nodal lines for $u_{\alpha q}$. The second reason is to obtain a sharp estimate for $u_{\alpha q}$ near the points of intersection of the nodal lines of $u_{\alpha 0}$.

Having established the perturbation result we encounter one more difficulty: the nodal domains for $u_{\alpha q}$ are, in general, not small, see [15],[16]. The following figure gives a typical case.

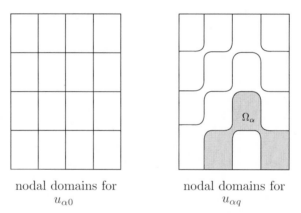

nodal domains for   nodal domains for
$u_{\alpha 0}$ $u_{\alpha q}$

To illustrate our result we consider the shaded region, which we call Ω_α. We cut the nodal domain Ω_α with the straight diagonal lines, D_1, D_2, illustrated below.

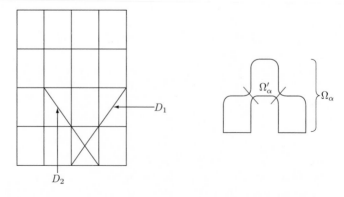

The upper cut domain of Ω_α is called Ω'_α; see the figure on the right above. We establish a bound for $u_{\alpha q}$ near $\Omega_\alpha \cap D_i$, $i = 1, 2$, and hence show that there exists an $x' \in \Omega'_\alpha$ satisfying

$$\left| q(x') - \fint q - [\lambda_{\alpha 0} - \lambda_{1,0}(\Omega'_\alpha)] \right| \le \frac{1}{2|\alpha|^{7/4}}.$$

The choice of the particular cut domain Ω'_α was rather arbitrary. Each nodal domain of $u_{\alpha q}$, for $\alpha \in L(a) \backslash M(a)$ can be cut in similar ways to obtain approximate nodal domains where a similar approximation to $q - \fint q$ can be obtained. Because of the conditions satisfied by $L(a) \backslash M(a)$ we can establish Theorem 6 by selecting all the approximate domains for $u_{\alpha q}$ for any infinite sequence of $\alpha \in L(a) \backslash M(a)$.

The above results are for the two-dimensional inverse nodal problem. The three-dimensional inverse nodal problem has also been considered. In that case the nodal sets are surfaces instead of lines. There are substantially more eigenvalues on the real line. Nonetheless, results analogous to those of the two-dimensional case have been obtained in [10]. That is, conditions have been established so that the eigenvalues are well separated; perturbation results have been established for a set of density one eigenvalues and eigenfunctions; a formula for an approximation to the potential has been obtained. Bounds for the difference between the potential and its approximation have been established. A uniqueness theorem has been proved.

3. A Related Result

We briefly discuss one other result. This is a specific case study to examine the effect of structural damping on nodal position measurements. It is important to show that structural damping has a small effect on nodal position measurement. The reason for this is that the mathematical models used to develop the the formulas for material parameters do not contain terms that model damping. Our study was not for elastic systems for which second order mathematical models are used but rather for stiff vibrating systems where structural damping may have a larger effect. Hence we chose the Euler-Bernoulli beam, which models transverse vibration; we also chose Kelvin-Voigt damping.

For this specific case, the beam is free at each end and driven with an oscillating force at the center. The natural frequencies are defined as those frequencies where the response of the beam is maximized. When the beam is driven at a natural frequency, the nodal positions of the damped, driven beam are defined as the positions where the amplitude of the displacement is minimized. We compare those positions with the nodal positions of the mode shapes of the undamped beam and establish a bound on the difference. The bound shows that the difference is small when the beam is driven at low frequencies. This result is contained in [4], [5].

Acknowledgments. Partial support for this research came from ONR, Grant no. N00014-91J-1166 and from NSF, grant no. VPW-8902967.

Appendix

In order to establish the perturbation results, we require that $\alpha \in L(a)$ satisfy five conditions. We give three of these conditions first. The description is somewhat technical. We require that α is in each of the following three sets:

$$L(a)\backslash M_{10}(a) = \left\{ \alpha \in L \mid \forall \beta \in L \quad \text{with} \quad \beta \neq \alpha, \quad |\,|\alpha|^2 - |\beta|^2| > \frac{4}{|\alpha|^{\frac{1}{4}}} \right\};$$

let $C_1, C_2 \geq 1$ and label
$L(a)\backslash M_{11}(a) =$

$$\left\{ \alpha \in L \mid \forall \beta \in L \text{ with } \beta \neq \alpha, \, |\alpha - \beta| < C_1 |\alpha|^{\frac{1}{20}}, \, |\,|\alpha|^2 - |\beta|^2| > C_2 |\alpha|^{\frac{15}{16}} \right\};$$

let $\zeta = 5^{\frac{1}{2}} 2^{-\frac{1}{8}}$ and label

$$L(a)\backslash M_{12}(a) = \left\{ \alpha \in L \mid m > (\zeta C_1)^2 (an)^{\frac{1}{2}} \quad \text{and} \quad an > (\zeta C_1)^2 m^{\frac{1}{2}} \right\}.$$

In words, $L(a)\backslash M_{10}(a)$ contains those $\alpha \in L(a)$ for which the corresponding eigenvalue $|\alpha|^2$ has no other eigenvalue $|\beta|^2$ in a small interval about $|\alpha|^2$. Note that the size of this interval decreases as $|\alpha|^2$ increases. For the $\alpha \in L(a)\backslash M_{11}(a)$ the distance between $|\alpha|^2$ and $|\beta|^2$ is large when the corresponding $u_{\alpha 0}$, $u_{\beta 0}$ have nearly the same oscillations in both the x and y directions. Note that, in this case, the distance between $|\alpha|^2$ and $|\beta|^2$ increases as $|\alpha|^2$ increases. Finally, the elements in $L(a)\backslash M_{12}(a)$ are such that an and m are always comparable.

The three subsets, $M_{10}(a)$, $M_{11}(a)$, $M_{12}(a)$ are combined with two others. We define

$$M_1(a) = M_{10} \cup M_{11} \cup M_{12} \cup \{\alpha \in L \mid |\alpha| < 20\} \cup \left\{ \alpha \in L \mid |\alpha| < \left(\frac{4}{C_1}\right)^{20} \right\}.$$

We show that $L(a)\backslash M_1(a)$ has density zero in $L(a)$. The full set of density one, $L(a)\backslash M(a)$, for which we establish perturbation results, is contained in $L(a)\backslash M_1(a)$. The elements of $M(a)\backslash M_1(a)$ are determined without specific description of their lattice properties.

References

[1] J. Feldman, H. Knörrer, and E. Trubowitz, *The perturbatively stable spectrum of a periodic Schrödinger operator*, Invent. Math., 100 (1990), pp. 250–300.

[2] L. Friedlander, *On certain spectral properties of very weak nonselfadjoint perturbations of selfadjoint operators*, Trans. Moscow Math. Soc. (1982), pp. 185–218.

[3] L. Friedlander, *On the spectrum of the periodic problem for the Schrödinger operator*, Comm. Partial Differential Equations, Vol. 15 (1990), pp. 1631–1647.

[4] B. Geist and J. R. McLaughlin, *The effect of structural damping on the nodes of a uniform, pinned, Euler-Bernoulli beam: A specific case study*, Rensselaer Polytechnic Institute Mathematics Technical Report Number 199 (1991).

[5] B. Geist and J. R. McLaughlin, *The effect of structural damping on nodes for the Euler-Bernoulli beam: A specific case study*, Appl. Math. Lett., 7 (1994), pp. 51–55.

[6] O. H. Hald and J. R. McLaughlin, *Inverse problems using nodal position data: Uniqueness results, algorithms and bounds*, Special Program on Inverse Problems, Proc. Centre Math. Anal., Austral. Nat. Univ., 17 (1988), pp. 32–59.

[7] O. H. Hald and J. R. McLaughlin, *Solutions of inverse nodal problems*, Inverse Problems, 5 (1989), pp. 307–347.

[8] O. H. Hald and J. R. McLaughlin, *Inverse nodal problems: Recovery of BV coefficients from nodes*, to appear.

[9] O. H. Hald and J. R. McLaughlin, *Inverse nodal problems: Finding the potential from nodal lines*, accepted for publication in Mem. Amer. Math. Soc.

[10] Yu. Karpeshina and J. R. McLaughlin, *Inverse nodal problems: Finding the potential from nodal surfaces*, to appear.

[11] J. R. McLaughlin, *Inverse spectral theory using nodal points as data — Uniqueness result*, J. Differential Equations 73 (1988), pp. 354–363.

[12] J. R. McLaughlin and O. H. Hald, *A formula for finding the potential from nodal lines*, Bull. Amer. Math. Soc., 32 (1995), pp. 241–247.

[13] J. Moser, *Lectures on Hamiltonian systems*, Mem. Amer. Math. Soc. 81, 1968.

[14] K. F. Roth, *Rational approximations to algebraic numbers*, Mathematika, 2 (1955), pp. 1–20, p. 168.

[15] K. Uhlenbeck, *Eigenfunctions of Laplace operators*, Bull. Amer. Math. Soc., 78 (1972), pp. 1073–1076.

[16] K. Uhlenbeck, *Generic properties of eigenfunctions*, Amer. J. Math., 98 (1976) pp. 1059–1078.

[17] M.D. Waller, Chladni Figures: A Study in Symmetry, Bell, London, 1961.

Backward Stochastic Differential Equations and Applications

Etienne Pardoux

Institut Universitaire de France
and
Labo. d'Analyse, Topologie, Probabilités
Centre de Mathématiques et d'Informatique
Université de Provence
Rue Joliot Curie, 13453 Marseille Cedex 13
France

Introduction

A new type of stochastic differential equation, called the backward stochastic differentil equation (BSDE), where the value of the solution is prescribed at the final (rather than the initial) point of the time interval, but the solution is nevertheless required to be at each time a function of the past of the underlying Brownian motion, has been introduced recently, independently by Peng and the author in [16], and by Duffie and Epstein in [7]. This class of equations is a natural nonlinear extension of linear equations that appear both as the equation for the adjoint process in the maximum principle for optimal stochastic control (see [2]), and as a basic model for asset pricing in financial mathematics. It was soon after discovered (see [22], [17]) that those BSDEs provide probabilistic formulas for solutions of certain semilinear partial differential equations (PDEs), which generalize the well-known Feynmann-Kac formula for second order linear PDEs. This provides a new additional tool for analyzing solutions of certain PDEs, for instance reaction-diffusion equations.

1 Backward stochastic differential equations

Let $\{B_t;\ 0 \leq t \leq T\}$ denote a d-dimensional Brownian motion defined on a probability space $(\Omega, \mathcal{F}, P)$. For $0 \leq t \leq T$, we denote by $\mathcal{F}_t$ the σ-algebra $\sigma\{B_s;\ 0 \leq s \leq t\}$, augmented with the P-null sets of $\mathcal{F}$. We are given two objects:
- a *final condition* $\xi \in L^2(\Omega, \mathcal{F}_T, P; \mathbb{R}^k)$;
- a *coefficient* f, which is a mapping from $\Omega \times [0, T] \times \mathbb{R}^k \times \mathbb{R}^{k \times d}$ into $\mathbb{R}^k$, and is such that there exists $K > 0$ with:

$$f(\cdot, y, z) \text{ is progressively measurable,}^1 \ \forall y \in \mathbb{R}^k, z \in \mathbb{R}^{k \times d}; \qquad \text{(i)}$$

---

1)   That is for each t, the restriction to $[0, t] \times \Omega$ of the mapping $(s, \omega) \rightarrow f(s, \omega, y, , z)$ is $\mathcal{B}(0, t) \otimes \mathcal{F}_t$ measurable.

Proceedings of the International Congress
of Mathematicians, Zürich, Switzerland 1994
© Birkhäuser Verlag, Basel, Switzerland 1995

$$E \int_0^T |f(t,0,0)|^2 \, dt < \infty; \qquad \text{(ii)}$$

$$|f(t,y,z) - f(t,y',z')| \leq K(|y-y'| + |z-z'|), \ \forall t, y, y', z, z'. \qquad \text{(iii)}$$

A solution of the BSDE(ξ, f) is a pair $\{(Y_t, Z_t), 0 \leq t \leq T\}$ of *progressively measurable processes* with values in $\mathbb{R}^k \times \mathbb{R}^{k \times d}$ such that

$$E \int_0^T |Z_t|^2 \, dt < \infty \qquad \text{(j)}$$

$$Y_t = \xi + \int_t^T f(s, Y_s, Z_s) \, ds - \int_t^T Z_s \, dB_s, \ 0 \leq t \leq T. \qquad \text{(jj)}$$

We have the following.

THEOREM 1.1. [16] *Under the above conditions, in particular* (i), (ii), *and* (iii), *the BSDE(ξ, f) has a unique solution* $(Y_t, Z_t), 0 \leq t \leq T$.

REMARK 1.2. *The constraint that the solution be progressively measurable (i.e. adapted to the past of B at each time t) is in a sense rather unnatural for the solution of a backward equation. This is the reason why we need to have the freedom of choosing Z independently of Y.*

REMARK 1.3. *What makes the solution random is the randomness of ξ and f. The stochastic integral is there to make Y progressively measurable, which in fact means "reducing the randomness of Y". In particular, if for some stopping time $\tau \leq T$, ξ and f are $\mathcal{F}_\tau$ measurable, then on the interval $[\tau, T]$, $Z = 0$ and Y is given by the solution of the ODE*

$$dY_t/dt = -f(t, Y_t, 0), \ Y_T = \xi.$$

REMARK 1.4. *Using Itô calculus, it follows from the square integrability of ξ, (ii), (iii), (j), and (jj) that if (Y, Z) solves the BSDE(ξ, f), $E[\sup_{0 \leq t \leq T} |Y_t|^2] < \infty$. It is easy to check that a solution of the BSDE($\xi, 0$) is given as follows. $Y_t = E[\xi / \mathcal{F}_t]$, and Z is given by Itô's representation theorem of functionals of Brownian motion, which says that $\xi = E[\xi] + \int_0^T Z_t \, dB_t$, for a certain progressively measurable process Z satisfying (j). Note that if we require only $\int_0^T |Z_t|^2 \, dt < \infty$, then the uniqueness of Z is not guaranteed.*

2 Applications

Before giving some indication for the proof of Theorem 1.1, let us motivate that notion by presenting several applications. Later we shall present our main application, which is to semilinear PDEs.

2.1 Application in financial mathematics. Consider a typical model for continuous time asset pricing. Let V_t denote the total wealth of an agent at time t, which he can invest in $n+1$ different assets, one nonrisky asset, whose price per unit P_t^0 is governed by the linear ordinary differential equation (ODE) $dP_t^0/dt = P_t^0 r_t$, and n risky assets, where the price process for one share of the ith stock is governed by the linear stochastic differential equation (SDE) $dP_t^i = P_t^i[\mu_t^i\, dt + \sum_{j=1}^n \sigma_t^{ij}\, dB_t^j]$. The asset pricing problem is as follows. Given a contingent claim ξ, which is an $\mathcal{F}_T$-measurable random variable that we suppose to be square integrable, find an initial wealth V_0 and a portfolio $(\Pi_t^i,\ 0 \le t \le T,\ 1 \le i \le n)$ such that the wealth at time T is exactly ξ. Hence, we need to solve the following linear BSDE ($\mathbf{1}$ denotes the d-dimensional vector whose coordinates are all equal to 1):

$$V_t = \xi - \int_t^T r_s[V_s - \Pi_s^*\mathbf{1}]\, ds - \int_0^T \Pi_s^*[\mu_s\, ds + \sigma_s\, dB_s].$$

This linear BSDE is a very classical model in financial mathematics. It is in particular the starting point of the celebrated Black-Scholes formula for option pricing. No general theory is necessary to study such a linear equation. However, there is at least one unreasonable assumption in our model: $V_t - \Pi_t^*\mathbf{1}$ represents an amount of money that is deposited in the bank whenever it is positive, but it represents an amount of money that is borrowed from the bank if it is negative. As the interest rate for borrowing is in fact bigger than the bond rate, we should rather write the above equation as a nonlinear BSDE, with some interest rate process $R_t > r_t$

$$V_t = \xi - \int_t^T r_s[V_s - \Pi_s^*\mathbf{1}]^+\, ds + \int_t^T R_s[V_s - \Pi_s^*\mathbf{1}]^-\, ds - \int_0^T \Pi_s^*[\mu_s\, ds + \sigma_s\, dB_s].$$

Note that this last BSDE is of the type considered in Section 1, in the particular case $k = 1$. There are several other reasons for using nonlinear BSDEs as models in financial mathematics, including taking into account technology constraints, as well as the notion of recursive utility (see [7], [8], [9], [10], [11] and the bibliographies therein).

2.2 Application in stochastic control (see [11], [22]). Suppose now that $k = 1$, and the coefficient f of our BSDE is concave in the variables y and z. We define the following "polar" process:

$$F(t, \beta, \gamma) := \sup_{y \in \mathbb{R},\, z \in \mathbb{R}^d} [f(t, y, z) - \beta y - \gamma \cdot z].$$

It follows from a measurable selection theorem that to each progressively measurable process (Y_t, Z_t), one can associate a progressively measurable pair (β_t^*, γ_t^*) such that

$$F(t, \beta_t^*, \gamma_t^*) = f(t, Y_t, Z_t) - \beta_t^* Y_t - \gamma_t^* \cdot Z_t,\ 0 \le t \le T.$$

Let $\mathcal{A}$ denote the set of progressively measurable "control" processes (β_t, γ_t) that satisfy $E \int_0^T F(t, \beta_t, \gamma_t)^2\, dt < \infty$. Consider for each $t \ge 0$ the scalar forward linear SDE

$$\Gamma_{t,s}^{\beta,\gamma} = 1 + \int_t^s \Gamma_{t,r}^{\beta,\gamma}[\beta_r\, dr + \gamma_r\, dB_r],\ s \ge t.$$

We then have the following.

THEOREM 2.1. [11] *Let (Y_t, Z_t) be the unique solution of the BSDE(ξ, f). Then for each $0 \leq t \leq T$, Y_t is the value function of a stochastic control problem, in the sense that*

$$Y_t = \sup_{(\beta,\gamma) \in \mathcal{A}} E\left[\int_t^T \Gamma_{t,s}^{\beta,\gamma} F(s, \beta_s, \gamma_s) \, ds + \Gamma_{t,T}^{\beta,\gamma} \xi \Big| \mathcal{F}_t\right].$$

2.3 Application in stochastic geometry (see [6]). One can show that the construction of a gamma-martingale (which is a notion of martingale adapted to processes with values in a manifold equipped with a connection Γ) with prescribed final value ξ can be achieved by solving a backward SDE where the coefficient f takes the form

$$f_i(y, z) = \sum_{j,k,q} \Gamma_{j,k}^i(y) z_{j,q} z_{k,q}.$$

One can assume that Γ is bounded and Lipschitz, however f here is not Lipschitz in z, hence Theorem 1.1 does not apply directly. However, combining BSDE and gamma-martingale techniques, one can show existence and uniqueness of a solution in this case.

3 Proof of Theorem 1.1

We now indicate a proof of our basic Theorem 1.1. The notation is as in Section 1. Let $\mathcal{B}^2$ denote the set of pairs $\{(Y_t, Z_t), 0 \leq t \leq T\}$ of *progressively measurable processes* with values in $\mathbb{R}^k \times \mathbb{R}^{k \times d}$ satisfying (j) and $E \int_0^T |Y_t|^2 \, dt < \infty$.

We define a mapping Φ from $\mathcal{B}^2$ into itself as follows. Given $(U, V) \in \mathcal{B}^2$, let $(Y, Z) = \Phi(U, V)$ be defined by: $Y_t = E[\xi + \int_t^T f(s, U_s.V_s) \, ds | \mathcal{F}_t]$, where Z is the process given by Itô's martingale representation theorem applied to the martingale $M_t = E[\xi + \int_0^T f(s, U_s.V_s) \, ds | \mathcal{F}_t]$. We then have that

$$Y_t = \xi + \int_t^T f(s, U_s, V_s) \, ds - \int_t^T Z_s \, dB_s, \quad 0 \leq t \leq T.$$

Define the following norm on $\mathcal{B}^2$, for $r > 0$:

$$\|(Y, Z)\|_r = \left(E \int_0^T e^{rt} \left[|Y_t|^2 + |Z_t|^2\right] dt\right)^{1/2}.$$

One can show by Itô calculus that, if r is large enough, the mapping Φ is a strict contraction on $\mathcal{B}^2$ equipped with the norm $\|\cdot\|_r$, hence it has a unique fixed point, which means that the BSDE(ξ, f) has a unique solution.

REMARK 3.1. *Note that one has a similar result to Theorem 1.1, if one assumes $\xi \in L^p(\Omega, \mathcal{F}, P; \mathbb{R}^k)$ for some $p > 1$, instead of $p = 2$. There is apparently no general theory for BSDEs with locally Lipschitz coefficient f. However, in addition to the result of [6], one can find some results in that direction in [19].*

One important tool for analyzing BSDEs is the following comparison theorem.

THEOREM 3.2. *Suppose $k = 1$, and let (ξ, f) and (ξ', f') be two pairs of data that satisfy the assumptions of Section 1. Suppose in addition that*

$$\xi \leq \xi' \ a.s. \quad and \quad f(t, y, z) \leq f'(t, y, z) \ \forall (t, y, z) \in [0, T] \times \mathbb{R} \times \mathbb{R}^d \ and \ a.s.,$$

then $Y_t \leq Y_t'$, $0 \leq t \leq T$, a.s.

4 BSDEs and viscosity solutions of second order semilinear PDEs

We now describe the relation between BSDEs and systems of second order semilinear PDEs. It turns out that solutions of BSDEs are naturally related with viscosity solutions of PDEs. This approach allows us to minimize the regularity requirements on the coefficients, while we will have to restrict ourselves to the case where the ith component of f depends on the ith line of the matrix Z only.

Before introducing the system of PDEs, we need to put the BSDE in a Markovian framework, i.e. to let ξ and f be functions of the Brownian motion B through a Markov-diffusion process, solution of a forward SDE driven by B.

Let $b : \mathbb{R}^d \to \mathbb{R}^d$ and $\sigma : \mathbb{R}^d \to \mathbb{R}^{d \times d}$ be Lipschitz functions, and for any $(t, x) \in [0, T] \times \mathbb{R}^d$, let $\{X_s^{t,x}; \ t \leq s \leq T\}$ denote the solution of the forward SDE

$$X_s^{t,x} = x + \int_t^s b(X_r^{t,x}) \, dr + \int_t^s \sigma(X_r^{t,x}) \, dB_r, \tag{4.1}$$

and consider the backward SDE

$$Y_s^{t,x} = g(X_T^{t,x}) + \int_s^T f(r, X_r^{t,x}, Y_r^{t,x}, Z_r^{t,x}) \, dr - \int_s^T Z_r^{t,x} \, dB_r, \ t \leq s \leq T, \tag{4.2}$$

where g and f map respectively $\mathbb{R}^d$ and $[0, T] \times \mathbb{R}^d \times \mathbb{R}^k \times \mathbb{R}^{k \times d}$ into $\mathbb{R}^k$, g and $f(t, \cdot, y, z)$ are jointly continuous and there exist constants K, p such that for all $t \in [0, T]$, $x \in \mathbb{R}^d$, $y, y' \in \mathbb{R}^k$, $z, z' \in \mathbb{R}^{k \times d}$:

$$|g(x)| \leq K(1 + |x|^p), \quad |f(t, x, y, z)| \leq K(1 + |x|^p + |y| + |z|),$$

$$|f(t, x, y, z) - f(t, x, y', z')| \leq K(|y - y'| + |z - z'|).$$

We now associate to (4.1) and (4.2) the following system of parabolic second order semilinear PDEs:

$$\begin{cases} \dfrac{\partial u}{\partial t}(t, x) + Lu(t, x) + f(t, x, u(t, x), \nabla u \sigma(t, x)) = 0, \ (t, x) \in [0, T] \times \mathbb{R}^d, \\ \hspace{3cm} u(T, x) = g(x), \ x \in \mathbb{R}^d, \end{cases} \tag{4.3}$$

where

$$Lu = \begin{pmatrix} Lu_1 \\ \vdots \\ Lu_k \end{pmatrix}, \quad L = \frac{1}{2}(\sigma\sigma^*)_{i,j} \frac{\partial^2}{\partial x_i \partial x_j} + b_i \frac{\partial}{\partial x_i}$$

(with the convention of summation over repeated indices).

Let us first recall the notion of viscosity solutions of the system of PDEs (4.3) (see [5], [14]).

DEFINITION 4.1. $u \in C([0,T] \times \mathbb{R}^d; \mathbb{R}^k)$ is called a viscosity sub-solution of (4.3) whenever $u_i(T,x) \leq g_i(x)$, $1 \leq i \leq k$, $x \in \mathbb{R}^d$, and for each $1 \leq i \leq k$, $(t,x) \in (0,T) \times \mathbb{R}^d$, $\varphi \in C^{1,2}((0,T) \times \mathbb{R}^d)$ such that (t,x) is a local maximum of $u_i - \varphi$,

$$-\frac{\partial \varphi}{\partial t}(t,x) - L\varphi(t,x) - f_i(t,x,u(t,x),\nabla \varphi \sigma(t,x)) \leq 0.$$

$u \in C([0,T] \times \mathbb{R}^d; \mathbb{R}^k)$ is called a viscosity super-solution of (4.3) whenever $u_i(T,x) \geq g_i(x)$, $1 \leq i \leq k$, $x \in \mathbb{R}^d$, and for each $1 \leq i \leq k$, $(t,x) \in [0,T] \times \mathbb{R}^d$, $\varphi \in C^{1,2}((0,T) \times \mathbb{R}^d)$ such that (t,x) is a local minimum of $u_i - \varphi$,

$$-\frac{\partial \varphi}{\partial t}(t,x) - L\varphi(t,x) - f_i(t,x,u(t,x),\nabla \varphi \sigma(t,x)) \geq 0.$$

$u \in C([0,T] \times \mathbb{R}^d; \mathbb{R}^k)$ is called a viscosity solution of (4.3) if it is both a viscosity sub- and super-solution of (4.3).

We can now state the main result of this section. ($C_p([0,T] \times \mathbb{R}^d; \mathbb{R}^k)$ denotes the set of continuous functions from $[0,T] \times \mathbb{R}^d$ into $\mathbb{R}^k$, which grow at most polynomialy at infinity).

THEOREM 4.2. $u(t,x) := Y_t^{t,x}$ belongs to $C_p([0,T] \times \mathbb{R}^d; \mathbb{R}^k)$ and is the unique viscosity solution of (4.3).

Proof. Uniqueness is proved by methods from viscosity solutions, see [5]. In order to prove that $\{Y_t^{t,x}\}$ is a sub-solution we assume that $1 \leq i \leq k$, $(t,x) \in [0,T] \times \mathbb{R}^d$, $\varphi \in C^{1,2}((0,T) \times \mathbb{R}^d)$ are such that $u_i(t,x) = \varphi(t,x)$ and $u_i(s,y) \leq \varphi(s,y)$, $(s,y) \in [0,T] \times \mathbb{R}^d$. We first note that from uniqueness of the solution of the BSDE, $Y_{t+h}^{t,x} = u(t+h, X_{t+h}^{t,x})$. Hence, deleting the superscripts t,x for notational simplicity,

$$u_i(t,x) = u_i(t+h, X_{t+h}) + \int_t^{t+h} f_i(X_s, Y_s, Z_s^i)\, ds - \int_t^{t+h} Z_s^i\, dB_s.$$

Let $(\bar{Y}_s, \bar{Z}_s, t \leq s \leq t+h)$ be the solution of the BSDE

$$\bar{Y}_s = \varphi(t+h, X_{t+h}) + \int_s^{t+h} f_i(X_r, (\bar{Y}_r, \tilde{Y}_r^i), \bar{Z}_r)\, dr - \int_s^{t+h} \bar{Z}_r\, dB_r,$$

where $(\bar{Y}_r, \tilde{Y}_r^i)$ denotes the vector whose ith component equals $\bar{Y}_r$, and the others equal the corresponding components of the vector Y_r. From the comparison theorem, $u_i(t,x) \leq \bar{Y}_t$. Applying Itô's formula to $\varphi(s,X_s)$ and defining $\hat{Y}_s = \bar{Y}_s - \varphi(s,X_s)$, $\hat{Z}_s = \bar{Z}_s - \nabla \varphi \sigma(s,X_s)$, we have that

$$0 \leq \hat{Y}_t = E \int_t^{t+h} \Big[\big(\frac{\partial \varphi}{\partial s} + L\varphi \big)(s, X_s)$$

$$+ f_i(X_s, (\hat{Y}_s + \varphi(s,X_s), \tilde{Y}_s^i), \hat{Z}_s + \nabla \varphi \sigma(s,X_s)) \Big]\, ds.$$

It remains essentially to divide by h and let $h \to 0$.

REMARK 4.3. *Suppose that $k = 1$ and $f(t, x, y, z) = c(t, x) y$. Then by the variation of constants formula,*

$$Y_t^{t,x} = g(X_T^{t,x}) \exp[\int_t^T c(s, X_s^{t,x}) ds] - \int_t^T \exp[\int_t^s c(r, X_r^{t,x}) dr](Z_s^{t,x}, dB_s)$$

$$= EY_t^{t,x}$$

$$= E\left(g(X_T^{t,x}) \exp[\int_t^T c(s, X_s^{t,x} ds] \right),$$

which is the well-known Feynman-Kac formula.

Let us now indicate some results on reaction-diffusion equations that have been obtained recently, with the help of the above stochastic representation, following [23]. Consider a reaction-diffusion equation of the type

$$\frac{\partial u}{\partial t}(t, x) = \frac{\partial^2 u}{\partial x^2}(t, x) + f(u(t, x)), \ t \geq 0, \ x \in \mathbb{R},$$

where for instance f is of the "KPP type", $f(r) = r(1 - r)$. Suppose the initial condition is of the form $u_0(x) = \mathbf{1}_{\mathbb{R}_-}(x)$. The function $u^\varepsilon(t, x) = u(t/\varepsilon, x/\varepsilon)$ behaves for small ε as $v(\frac{x - \alpha t}{\varepsilon})$, where v decreases from 1 to 0 on $\mathbb{R}$. Moreover α, the speed of the front, can be computed in terms of the derivative of f at 0^+. For proving this type of result, Freidlin [13] uses the Feynman-Kac formula (which is implicit here, because of the nonlinearity of f) and results from the theory of large deviations. Using BSDEs, one can write an explicit probabilistic formula for the solution of the PDE, even when the nonlinear term f depends on the first derivative of u, and obtain asymptotic results on u^ε in that more general set-up.

5 Extensions

We now indicate several results that have been obtained recently and generalize the results presented above.

5.1 BSDEs with respect to Brownian motion and Poisson random measure. It is possible to solve BSDEs with the Brownian motion replaced by a general martingale, see [3]. One can also consider a diffusion process with jumps X, solution of a forward SDE driven by both a Brownian motion and a Poisson random measure, and consider a backward SDE whose final condition and coefficient are functions of X. This provides a stochastic formula for a semilinear integro-partial differential equation [1], [4]. One can also use these results for giving a probabilistic formula for the solution of a system of parabolic PDEs, where the second order PDE operator is different from one line to the other, i.e. with the notation of the previous section, $Lu = (L_1 u_1, \dots, L_k u_k)'$, see [20].

5.2 Coupled forward-backward SDEs [15]. In the last section, the diffusion process X was perturbing the coefficients of the BSDE for (Y, Z). Suppose now that we can solve a pair of forward-backward SDEs, where (Y, Z) appear in the coefficients of the forward SDE for X, and X appears in the coefficients of the backward SDE for (Y, Z). Then $u(t, x) := Y_t^{t,x}$ would solve a general type of quasilinear PDE. However, it seems so far that in this case the pair of SDEs can be solved directly only in very restricted cases, and that results on the corresponding PDE are necessary to solve the pair of SDEs under rather general assumptions.

5.3 Equations with boundary conditions. So far we have considered only PDEs in the whole space $\mathbb{R}^d$. It is possible to consider PDEs with Dirichlet or Neumann boundary conditions at the boundary of a domain $D \subset \mathbb{R}^d$, provided one replaces the diffusion X by a diffusion either stopped at the boundary of D, or reflected at the same boundary.

5.4 Infinite-time horizon BSDEs, and elliptic PDEs. If one replaces the final time T by $+\infty$, or by a stopping time τ, it is possible to give probabilistic formulas for quasilinear elliptic PDEs. Of course, more restrictive assumptions on the coefficient f are then required. For a result on infinite-time horizon BSDEs, see [10].

5.5 Reflected BSDEs. The following reflected BSDE has been studied in [12].

$$Y_t = \xi + \int_t^T f(t, X_t, Y_t, Z_t)\, dt + K_T - K_t - \int_t^T Z_s\, dB_s,\ 0 \leq t \leq T$$

$$Y_t \geq S_t,\ 0 \leq t \leq T;\ \int_0^T (Y_t - S_t)\, dK_t = 0,$$

where ξ and f are as in Section 1, and $\{S_t,\ 0 \leq t \leq T\}$ is a continuous process satisfying $E(\sup_t |S_t|^2) < \infty$ and $S_T \leq \xi$ a.s. The solution is a triple (Y, Z, K) of progressively measurable processes, where K is continuous and increasing. Note that, unlike the case of reflected forward SDEs, whenever S is a semimartingale, the increasing process K is absolutely continuous.

In the case where the data (ξ, f, S) is a given function of a diffusion X, we get the probabilistic interpretation of an obstacle problem for a quasilinear PDE.

5.6 Backward doubly stochastic differential equations and stochastic PDEs [18]. If we introduce another independent Brownian motion $\{W_t\}$, the equation

$$Y_t = \xi + \int_t^T f(s, Y_s, Z_s)\, ds + \int_t^T g(s, Y_s, Z_s)\, dW_s - \int_t^T Z_s\, dB_s$$

has a unique solution $\{(Y_t, Z_t);\ 0 \leq t \leq T\}$ which is adapted at each time t to the sup of the past of B and the future increments of W, provided both f and g are Lipschitz. When put in a Markovian framework, we obtain a formula for a system of quasilinear SPDEs driven by $\{W_t\}$.

References

[1] G. Barles, R. Buckdahn, and E. Pardoux, *BSDEs and integro-partial differential equations*, preprint.

[2] J. M. Bismut, Théorie probabiliste du contrôle des diffusions, Mem. Amer. Math. Soc. **176** (1973).

[3] R. Buckdahn, *Backward stochastic differential equations driven by a martingale*, LATP prépublication 93-5.

[4] R. Buckdahn and E. Pardoux, *BSDEs with jumps and associated integro-partial differential equations*, Proc. Conf. Metz 1994, to appear.

[5] M. Crandall, H. Ishii, and P. L. Lions, *User's guide to viscosity solutions of second order partial differential equations*, Bull. Amer. Math. Soc. (New Ser.) **27** (1992), 1–67.

[6] R. Darling, *Constructing Gamma-martingales with prescribed limit, using backward SDE*, LATP prépublication 94-2.

[7] D. Duffie and L. Epstein, *Stochastic differential utility*, Econometrica **60** (1992), 353–394.

[8] D. Duffie and L. Epstein, *Asset pricing with stochastic differential utility*, Rev. Financial Stud. **5** (1992), 411–436.

[9] D. Duffie and P. L. Lions, *PDE solutions of stochastic differential utility*, J. Math. Econom. **21** (1992).

[10] D. Duffie, J. Ma, and J. Yong, *Black's consol rate conjecture*, IMA preprint series # 1164 (1993).

[11] N. El Karoui, S. Peng, and M. C. Quenez, *Backward stochastic differential equations in finance*, preprint.

[12] N. El Karoui, C. Kapoudjian, E. Pardoux, S. Peng, and M. C. Quenez, *Reflecting BSDE and associated obstacle problem for PDEs*, preprint.

[13] M. Freidlin, *Semi-linear PDEs and limit theorems for large deviations*, in: Ecole d'été de probabilités de St Flour 1990, Lecture Notes in Mathematics **1527** (1992), 1–109, Springer-Verlag, Berlin and New York.

[14] H. Ishii and S. Koike, *Viscosity solutions for monotone systems of second order elliptic PDEs*, Comm. Partial Differential Equations **16** (1991), 1095–1128.

[15] J. Ma, P. Protter, and J. Yong, *Solving forward-backward stochastic differential equations explicitly — A four-step scheme*, preprint.

[16] E. Pardoux and S. Peng, *Adapted solutions of backward stochastic equations*, Systems Control Lett. **14** (1990), 55–61.

[17] E. Pardoux and S. Peng, *Backward stochastic differential equations and quasilinear parabolic partial differential equations*, in: Stochastic Partial Differential Equations and Their Applications (B. L. Rozovskii and R. S. Sowers, eds.), LNCIS **176** (1992), 200–217, Springer-Verlag, Berlin and New York.

[18] E. Pardoux and S. Peng, *Backward doubly stochastic differential equations and systems of quasilinear SPDEs*, Probab. Theory Related Fields **98** (1994), 209–227.

[19] E. Pardoux and S. Peng, Some backward SDEs with non-Lipschitz coefficients, Proc. Conf. Metz 1994, to appear.

[20] E. Pardoux, F. Pradeilles, and Z. Rao, *BSDEs and viscosity solutions of systems of second order parabolic PDEs*, preprint.

[21] S. Peng, *Probabilistic interpretation for systems of quasilinear parabolic partial differential equations*, Stochastics and Stochastics Reports **37** (1991), 61–74.

[22] S. Peng, *Backward stochastic differential equations and applications to optimal control*, Appl. Math. Optim. **27** (1993), 125–144.

[23] F. Pradeilles, *Equations différentielles stochastiques rétrogrades et équations de réaction-diffusion*, in preparation.

Max-Plus Algebra and Applications to System Theory and Optimal Control

MAX-PLUS WORKING GROUP,* PRESENTED BY JEAN-PIERRE QUADRAT

INRIA-Rocquencourt, B.P. 105
F-78153 Le Chesnay Cedex, France.

In the modeling of human activities, in contrast to natural phenomena, quite frequently only the operations max (or min) and + are needed. A typical example is the performance evaluation of synchronized processes such as those encountered in manufacturing (dynamic systems made up of storage and queuing networks). Another typical example is the computation of a path of maximum weight in a graph and more generally of the optimal control of dynamical systems. We give examples of such situations. The max-plus algebra is a mathematical framework well suited to handle such situations. We present results on (i) linear algebra, (ii) system theory, and (iii) duality between probability and optimization based on this algebra.

1. Max-Plus Linear Algebra

DEFINITION 1   *1. An* abelian monoid $\mathcal{K}$ *is a set endowed with one operation* $\oplus$, *which is associative, commutative, and has a* zero *element* ε.

2. *A* semiring *is an abelian monoid endowed with a second operation* $\otimes$, *which is associative and distributive with respect to* $\oplus$, *which has an* identity *element denoted* e, *with* ε absorbing *(that is* $\varepsilon \otimes a = a \otimes \varepsilon = \varepsilon$*)*.

3. *A* dioid *is a semiring that is* idempotent *(that is* $a \oplus a = a$, $\forall a \in \mathcal{K}$*)*.

4. *A* semifield *is a semiring having its second operation invertible on* $\mathcal{K}_\star = \mathcal{K} \setminus \{\varepsilon\}$.

5. *A semifield that is also a dioid is called an* idempotent semifield.

6. *We will say that these structures are* commutative *when the product is also commutative*.

7. *We call* $\mathbb{R}_{\max}$ *(resp.* $\mathbb{R}_{\min}$*) the set* $\mathbb{R} \cup \{-\infty\}$ *(resp.* $\mathbb{R} \cup \{+\infty\}$*) endowed with the two operations* $\oplus = \max$ *(resp* $\oplus = \min$*) and* $\otimes = +$.

---

*Currently consisting of M. Akian, G. Cohen, S. Gaubert, J.-P. Quadrat, and M. Viot.

Proceedings of the International Congress
of Mathematicians, Zürich, Switzerland 1994
© Birkhäuser Verlag, Basel, Switzerland 1995

8. *We call $\mathbb{R}_{\max}^{n\times n}$ and analogously $\mathbb{R}_{\min}^{n\times n}$ the set of $(n \times n)$ matrices with entries belonging to $\mathbb{R}_{\max}$ endowed with $\oplus$ denoting the* max *entry by entry and $\otimes$ defined by*

$$[AB]_{ij} \overset{\mathrm{def}}{=} [A \otimes B]_{ij} \overset{\mathrm{def}}{=} \max_k [A_{ik} + B_{kj}] = \oplus_k A_{ik} \otimes B_{kj} .$$

9. *We call $\mathcal{S}_{\max}$ (resp. $\mathcal{I}_{\max}$) the set of functions (resp. increasing functions), from $\mathbb{R}$ into $\mathbb{R}_{\max}$ endowed with $\oplus$ denoting the pointwise maximum and $\otimes$ the* sup-convolution *defined by*

$$[f \otimes g](x) \overset{\mathrm{def}}{=} [f \,\square\, g](x) \overset{\mathrm{def}}{=} \sup_t [f(x - t) + g(t)] .$$

   Analogously we define $\mathcal{S}_{\min}$ (resp. $\mathcal{I}_{\min}$). The set $\mathcal{I}_{\min}^d$ is the restriction of $\mathcal{I}_{\min}$ to piecewise constant increasing functions with jumps at positive integer abscissas.

10. *We call $\mathcal{C}_x$ (resp. $\mathcal{C}_v$) the set of lower (resp. upper) semicontinuous and proper (never equal to $-\infty$ (resp. ∞)) convex (concave) functions endowed with the $\oplus$ operator denoting the pointwise maximum (minimum) and the $\otimes$ operator denoting the pointwise sum.*

11. *We call $\mathcal{C}_0$ the set of lower semicontinuous and proper strictly convex functions having 0 as infimum endowed with the $\otimes$ operator denoting the inf-convolution of two functions.*

Clearly the algebraic structures $\mathbb{R}_{\max}$ *and* $\mathbb{R}_{\min}$ *are idempotent commutative semi-fields,* $\mathbb{R}_{\max}^{n\times n}$, $\mathbb{R}_{\min}^{n\times n}$, $\mathcal{S}_{\max}$, $\mathcal{S}_{\min}$, $\mathcal{I}_{\max}$, $\mathcal{I}_{\min}$, $\mathcal{I}_{\min}^d$, $\mathcal{C}_x$, *and* $\mathcal{C}_v$ *are dioids, and* $\mathcal{C}_0$ *is a commutative monoid. We will call all these vectorial structures based on* $\mathbb{R}_{\max}$ *or* $\mathbb{R}_{\min}$ *max-plus algebras. Working with these structures shows that idempotency is as useful as the existence of a symmetric element in the simplification of formulas and therefore that these structures are very effective to make algebraic computations.*

APPLICATION 2   *1. These mathematical structures introduce a linear algebra point of view to dynamic programming problems.*

   Given C in $\mathbb{R}_{\min}^{n\times n}$ we call precedence graph $\mathcal{G}(C)$ *the graph having (i) n nodes, and (ii) oriented arcs (i, j) of weight C_{ji} if $C_{ji} \neq \varepsilon$ in the matrix C.*

   The min-plus linear dynamical system

$$X^{m+1} = C \otimes X^m, \ X_j^0 = e, \ for \ j = i, \ X_j^0 = \varepsilon \ elsewhere, \tag{1}$$

   is a dynamic programming equation. The number X_j^m is equal to the least weight of all paths from i to j (the weight of a path is the sum of the weights of its arcs) of length m (composed of m arcs).

   The minimal average weight by arc of paths having their lengths going to infinity is obtained by computing the λ solution of the spectral problem

$$\lambda \otimes X = C \otimes X .$$

The computation of the minimal weight of paths from i to a region described by $d \in \mathbb{R}^n_{\min}$ ($d_j = e$ if j belongs to the region, $d_j = \varepsilon$ elsewhere) is equal to the X_i solution of

$$X = C \otimes X \oplus d .$$

2. *The evaluation of some systems where synchronization between tasks appears (as in event graphs a subset of Petri nets) can be modeled linearly in $\mathbb{R}_{\max}$ or dually in $\mathbb{R}_{\min}$ by*

$$X^{m+1} = F \otimes X^m \oplus G \otimes U^m, \; Y^{m+1} = H \otimes X^{m+1} . \tag{2}$$

In $\mathbb{R}_{\max}$, the number X_i^m has the interpretation of the earliest date that the mth occurrence of the event i (for example the starting time of a task on a machine in manufacturing) has happened. The max operator models the fact that tasks can be performed as soon as all the preconditions are fulfilled. The vector U models the timing of the input preconditions. The vector Y denotes the timing of the outputs of the system.

In $\mathbb{R}_{\min}$ the number X_i^m has the interpretation of the maximum number of events of kind i that can occur before the date m. We can pass from (F, G, H) over $\mathbb{R}_{\max}$ to the one over $\mathbb{R}_{\min}$ by interchanging the role of the delays and the coefficients (see [5] for more details).

3. *Clearly infinite dimensional and/or continuous time versions of equation exist (1). For $c, \psi \in \mathcal{C}_0$ the problem*

$$v_x^m = \min_u \left[\sum_{i=m}^{N-1} c(u^i) + \psi(x^N) \mid x^m = x \right], \; x^{i+1} = x^i - u^i ,$$

may be called dynamic programming with independent instantaneous costs (c depends only on u and not on x). Clearly v satisfies the linear recurrence in $\mathcal{C}_0$

$$v^m = c \, \square \, v^{m+1}, \; v^N = \psi .$$

To solve some of these applications we have to solve max-plus linear equations in $\mathbb{R}^{n \times n}_{\max}$ or $\mathbb{R}^{n \times n}_{\min}$. The general one can be written $A \otimes X \oplus b = C \otimes X \oplus d$. In this section we use three points of view (contraction, residuation, combinatorial) to study this kind of equations.

1.1. Spectral Equations, Contraction and Residuation.   As in conventional algebra all the linear iterations are not contractions. We can characterize the contractions using the max-plus spectral theory. To simplify the discussion we give a simplified result under restrictive hypotheses on the connectivity of the associated incidence graph. The general result will be found for example in [5].

THEOREM 3   *1. If the graph $\mathcal{G}(C)$ associated with the matrix C has only a strongly connected component there exists a unique λ solution of $\lambda \otimes X =$*

$C \otimes X$. *It has the graph interpretation*

$$\lambda = \max_{\zeta} \frac{|\zeta|_w}{|\zeta|_l} \, ,$$

where $|\zeta|_w$ denotes the weight of the circuit ζ and $|\zeta|_l$ its length.

2. *We denote C_λ the matrix defined by $C_\lambda \stackrel{\mathrm{def}}{=} \lambda^{-1} \otimes C$, $C^* \stackrel{\mathrm{def}}{=} E \oplus C \oplus C^2 \oplus \cdots \oplus C^{n-1}$ where E denotes the identity matrix and $C^+ \stackrel{\mathrm{def}}{=} CC^*$. A column i of $[C_\lambda]^+$ such that $[C_\lambda]_{ii}^+ = e$ is an eigenvector. In C_λ^+ there exists at least one such column.*

3. *There exists c such that for m large enough we have*

$$C^{m+c} = \lambda^c C^m.$$

If $\mathcal{G}(C)$ has more than one strongly connected component, C may have more than one eigenvalue. The largest one is called the *spectral radius* of the matrix C and is denoted by $\rho(C)$.

THEOREM 4 *The equation $\mu X = CX \oplus d$ has a least solution $X = [C_\mu]^* d_\mu$ when $\rho(C) \leq \mu$. The solution is unique when $\rho(C) < \mu$.*

The equation $Ax = d$ does not always have a solution but its greatest subsolution can be computed explicitly

$$x = A \backslash d \stackrel{\mathrm{def}}{=} \max\{x \mid Ax \leq d\} = \min_{j}(d_j - a_j) \, .$$

This computation, well known in residuation theory, defines a new binary operator $\backslash$ which can be seen as the dual operator of $\otimes$. The $\backslash$ is distributive with respect to $\wedge$ (defined as the min operator in the $\mathbb{R}_{\max}^{n \times n}$ context). With these two operators dual linear equations may be written.

COROLLARY 5 *The equation $\mu \backslash X = (C \backslash X) \wedge d$ has a solution as soon as $\mu \geq \rho(C)$. The largest X solution of this equation is*

$$X = [C_\mu]^* \backslash \mu d = \mu d \wedge (C_\mu \backslash \mu d) \wedge (C_\mu \backslash C_\mu \backslash \mu d) \wedge \cdots \, .$$

APPLICATION 6 *In the event graphs framework described previously this kind of equation appears when we compute the the latest date at which an event must occur if we want respect due times coded in d (see [5] for more details).*

1.2. Symmetrization of the Max-Plus Algebra. Because every idempotent group is reduced to the zero element it is not possible to symmetrize the max operation. Nevertheless we can adapt the idea of the construction of $\mathbb{Z}$ from $\mathbb{N}$ to build an extension of $\mathbb{R}_{\max}$ such that the general linear scalar equation has always a solution.

Let us consider the set of pairs $\mathbb{R}^2_{\max}$ endowed with the natural idempotent semiring structure

$$(x', x'') \oplus (y', y'') = (x' \oplus y', x'' \oplus y'') \,,$$

$$(x', x'') \otimes (y', y'') = (x'y' \oplus x''y'', x'y'' \oplus x''y') \,,$$

with $(\varepsilon, \varepsilon)$ as the zero element and (e, ε) as the identity element and $\ominus(x', x'') \overset{\text{def}}{=} (x'', x')$.

DEFINITION 7 *Let $x = (x', x'')$ and $y = (y', y'')$. We say that x balances y (which is denoted $x \nabla y$) if $x' \oplus y'' = x'' \oplus y'$.*

It is fundamental to notice that ∇ is *not* transitive and thus is not a congruence. However, we can introduce the congruence $\mathcal{R}$ on $\mathbb{R}^2_{\max}$ closely related to the balance relation:

$$(x', x'')\mathcal{R}(y', y'') \Leftrightarrow \begin{cases} x' \oplus y'' = x'' \oplus y' & \text{if } x' \neq x'', y' \neq y'' \,, \\ (x', x'') = (y', y'') & \text{otherwise.} \end{cases}$$

We denote $\mathbb{S} \overset{\text{def}}{=} \mathbb{R}^2_{\max}/\mathcal{R}$.

We distinguish three kinds of equivalence classes:

$\{(t, x'') \mid x'' < t\}$,	called positive elements,	represented by t;
$\{(x', t) \mid x' < t\}$,	called negative elements,	represented by $\ominus t$;
$\{(t, t)\}$,	called balanced elements,	represented by $t^\bullet$.

The set of positive (resp. negative, resp. balanced) elements is denoted $\mathbb{S}^\oplus$ (resp. $\mathbb{S}^\ominus$, resp. $\mathbb{S}^\bullet$). This yields the decomposition

$$\mathbb{S} = \mathbb{S}^\oplus \cup \mathbb{S}^\ominus \cup \mathbb{S}^\bullet \,.$$

We also denote $\mathbb{S}^\vee \overset{\text{def}}{=} \mathbb{S}^\oplus \cup \mathbb{S}^\ominus$ and $\mathbb{S}^\vee_\star = \mathbb{S}^\vee \setminus \{\varepsilon\}$.

If $x \nabla y$ and $x, y \in \mathbb{S}^\vee$, we have $x = y$. We call this result the *reduction of balances*.

We now consider a solution X, in $\mathbb{R}^n_{\max}$, of the equation $AX \oplus b = CX \oplus d$; then the definition of the balance relation implies that $(A \ominus C)X \oplus (b \ominus d) \nabla \varepsilon$. Conversely, assuming that X is a positive solution of $AX \oplus b \nabla CX \oplus d$, with $AX \oplus b$ and $CX \oplus d \in \mathbb{S}^\oplus$, using the reduction of balances we obtain that X is a solution of $AX \oplus b = CX \oplus d$.

THEOREM 8 ((CRAMER'S RULE)) *Let $A \in \mathbb{S}^{n \times n}$, $b \in \mathbb{S}^n$, $|A|$ be the determinant of the matrix A (defined by replacing $+$ by $\oplus$, $-$ by $\ominus$, and $\times$ by $\otimes$ in the conventional definition) and A_i be the matrix obtained from A by replacing the ith column by b. Then if $|A| \in \mathbb{S}^\vee_\star$ and $|A_i| \in \mathbb{S}^\vee$, $\forall i = 1, \cdots, n$, then there exists a unique solution of $AX \nabla b$, belonging to $(\mathbb{S}^\vee)^n$, that satisfies*

$$X_i = |A_i|/|A| \,.$$

2. Min-Plus Linear System Theory

System theory is concerned with the input (u)-output (y) relation of a dynamical system ($\mathcal{S}$) denoted $y = S(u)$ and by the improvement of this input-output relation (based on some engineering criterion) by altering the system through a feedback control law $u = F(y, v)$. Then the new input (v)-output (y) relation is defined implicitly by $y = S(F(y, v))$. Not surprisingly, system theory is well developed in the particular case of linear shift-invariant systems. Analogously, a min-plus version of this theory can also be developed. The typical application is the performance evaluation of systems that can be described in terms of event graphs.

2.1. Inf-convolution and Shift-Invariant Max-Plus Linear Systems.

DEFINITION 9   *1. A signal u is a mapping from $\mathbb{R}$ into $\mathbb{R}_{\min}$. The signals set, denoted $\mathcal{Y}$, is endowed with two operations, namely the pointwise minimum of signals denoted $\oplus$, and the addition of a constant to a signal denoted $\otimes$, which plays the role of the external product of a signal by a scalar.*

2. A system is an operator $S : \mathcal{Y} \to \mathcal{Y}, u \mapsto y$. We call u (respectively y) the input (respectively output) of the system. We say that the system is min-plus linear when the corresponding operator is linear.

3. The set of linear systems is endowed with two internal and one external operations, namely

(i) parallel composition $S = S_1 \oplus S_2$ defined by pointwise minimum of output signals corresponding to the same input;

(ii) series composition $S = S_1 \otimes S_2$, or more briefly, $S_1 S_2$ defined by the composition of operators;

(iii) amplification $T = a \otimes S$, $a \in \mathbb{R}_{\min}$ defined by $T(k) = a \otimes S(k)$.

4. The improved input (v)-output (y) relation of a system S by a linear feedback $u = F(y) \oplus G(v)$ is obtained by solving the equation $y = S(F(y)) \oplus S(G(v))$ in y.

5. A linear system is called shift invariant when it commutes with the shift operators on signals ($u(.) \mapsto u(. + k)$).

THEOREM 10   *1. For a shift-invariant continuous[1] min-plus linear system S there exists $h : \mathbb{R} \mapsto \mathbb{R}_{\min}$ called the impulse response such that*

$$y = h \otimes u \overset{\text{def}}{=} h \,\square\, u \,.$$

2. The set of impulse responses endowed with the pointwise minimum and the inf-convolution is the dioid $\mathcal{S}_{\min}$.

---

[1] Linear also for infinite linear combinations.

3. *If f (resp. g) denotes the impulse response of the system SF (resp. SG), the impulse response h of a system S altered by the linear feedback $u = F(y) \oplus G(v)$ is a solution of*

$$h = f \otimes h \oplus g \, .$$

2.2. Fenchel Transform. The Fourier and Laplace transforms are important tools in automatic control and signal processing because the exponentials diagonalize all the convolution operators simultaneously and consequently the convolutions are converted into multiplications by the Fourier transform. Analogous tools exist in the framework of the min-plus algebra.

DEFINITION 11 *Let $c \in \mathcal{C}_x$. Its Fenchel transform is the function in $\mathcal{C}_x$ defined by $\hat{c}(\theta) = [\mathcal{F}(c)](\theta) \overset{\text{def}}{=} \sup_x [\theta x - c(x)]$.*

For example, setting $l_a(x) = ax$ we have $[\mathcal{F}(l_a)](\theta) = \chi_a(\theta)$ with

$$\chi_a(\theta) = \begin{cases} +\infty & \text{for } \theta \neq a, \\ 0 & \text{for } \theta = a. \end{cases}$$

THEOREM 12 *For $f, g \in \mathcal{C}_x$ we have (i) $\mathcal{F}(f) \in \mathcal{C}_x$, (ii) $\mathcal{F}$ is an involution that is $\mathcal{F}(\mathcal{F}(f)) = f$, (iii) $\mathcal{F}(f \,\square\, g) = \mathcal{F}(f) + \mathcal{F}(g)$, and (iv) $\mathcal{F}(f + g) = \mathcal{F}(f) \,\square\, \mathcal{F}(g)$.*

THEOREM 13 *The response to a conventional affine input (min-plus exponential) is a conventional affine output with the same slope. If $y = h \,\square\, u$ and $u = l_a$ we have*

$$y = l_a / [\mathcal{F}(h)](a).$$

Unfortunately, the class of min-plus linear combinations of affine functions is only the set of concave functions, which is not sufficient to describe all the interesting inputs of min-plus linear systems.

2.3. Rational Systems. A general impulse response is too complicated to be used in practice as it involves an infinite number of operations to be defined.

DEFINITION 14   1. *An impulse response $h \in \mathcal{I}^d_{\min}$ is rational if it can be computed with a finite number of $\oplus$, $\otimes$, and $*$[2] operations, from the functions $a \otimes e$ ($a \in \mathbb{R}_{\min}$) and $\chi_1 \otimes e$ where*

$$e(t) \overset{\text{def}}{=} \begin{cases} e & \text{for } t \leq 0, \\ \varepsilon & \text{for } t > 0. \end{cases}$$

2. *It is called realizable if there exists (F, G, H) such that $h^m = FG^m H$. Then there exists X such that*

$$X^{m+1} = F \otimes X^m \oplus G \otimes U^m, \quad Y^m = H \otimes X^m \, .$$

The vector X is called the state *of the realization.*

---

[2]For an impulse response h we define the operator $*$ by $h^* \overset{\text{def}}{=} e \oplus h \oplus h^2 \cdots$

3. *The system is called* ultimately periodic *if* $h^{m+c} = c \times \lambda + h^m$, *for m large enough.*

4. *The number λ is called* the ultimate slope *of h.*

THEOREM 15 *For SISO systems having an impulse response in $\mathcal{I}^d_{\min}$ the three notions of rationality, ultimate periodicity, and realizability are equivalent.*

This theorem is a min-plus version of the Kleene-Schutzenberger theorem. The realization of an impulse response with a vectorial state X of minimal dimension is an open problem in the discrete time case.

2.4. Feedback Stabilization.   Feedback can be used to stabilize a system without slowing down its throughput (the ultimate slope of its impulse response).

DEFINITION 16   1. *A realization of a rational system is* internally stable *if all the ultimate slopes of the impulse responses from any input to any state are the same.*

2. *A realization is* structurally controllable *if every state can be reached by a path from at least one input.*

3. *A realization is* structurally observable *if from every state there exists a path to at least one output.*

THEOREM 17 *Any structurally controllable and observable realization can be made internally stable by a dynamic output feedback without changing the ultimate slope of the impulse response of the system.*

3.   Bellman Processes

The functions stable by inf-convolution are known. They are the dynamic programming counterparts of the stable distributions of probability calculus. They are the following functions:

$$\mathcal{M}^p_{m,\sigma}(x) = \frac{1}{p}(|x - m|/\sigma)^p, \text{ with } \mathcal{M}^p_{m,0}(x) = \chi_m(x), \quad p \geq 1, \, m \in \mathbb{R}, \, \sigma \in \mathbb{R}^+ .$$

We have $\mathcal{M}^p_{m,\sigma} \,\square\, \mathcal{M}^p_{\bar{m},\bar{\sigma}} = \mathcal{M}^p_{m+\bar{m},[\sigma^{p'}+\bar{\sigma}^{p'}]^{1/p'}}$ with $1/p + 1/p' = 1$.

3.1.   Cramer Transform.   The Cramer transform ($\mathcal{C} \overset{\text{def}}{=} \mathcal{F} \circ \log \circ \mathcal{L}$, where $\mathcal{L}$ denotes the Laplace transform) maps probability measures to convex functions and transforms convolutions into inf-convolutions:

$$\mathcal{C}(f * g) = \mathcal{C}(f) \,\square\, \mathcal{C}(g).$$

Therefore it converts the problem of adding independent random variables into a dynamic programming problem with independent costs. In Table 1 we give some properties of the Cramer transform. For a systematic study of the Cramer transform see Azencott [4].

Table 1: Properties of the Cramer transform.

$\mathcal{M}$	$\log(\mathcal{L}(\mathcal{M})) = \mathcal{F}(\mathcal{C}(\mathcal{M}))$	$\mathcal{C}(\mathcal{M})$
μ	$\hat{c}(\theta) = \log \int e^{\theta x}\, d\mu(x)$	$c(x) = \sup_\theta(\theta x - \hat{c}(\theta))$
$\mu \geq 0$	$\hat{c}$ convex l.s.c.	c convex l.s.c.
$m_0 \overset{\text{def}}{=} \int d\mu = 1$	$\hat{c}(0) = 0$	$\inf_x c(x) = 0$
$m_0 = 1,\ m \overset{\text{def}}{=} \int x\, d\mu$	$\hat{c}'(0) = m$	$c(m) = 0$
$m_0 = 1,\ m_2 \overset{\text{def}}{=} \int x^2 d\mu$	$\hat{c}''(0) = \sigma^2 \overset{\text{def}}{=} m_2 - m^2$	$c''(m) = 1/\sigma^2$
$m_0 = 1$ $\hat{c} = |\sigma\theta|^{p'}/p' + o(|\theta|^{p'})$	$\hat{c}^{(p')}(0^+) = \Gamma(p')\sigma^{p'}$	$c^{(p)}(0^+) = \Gamma(p)/\sigma^p$
$\frac{1}{\sigma\sqrt{2\pi}} e^{-\frac{1}{2}(x-m)^2/\sigma^2}$	$m\theta + \frac{1}{2}(\sigma\theta)^2$	$\mathcal{M}_{m,\sigma}^2$
stable distrib. Feller [10]	$m\theta + \frac{1}{p'}|\sigma\theta|^{p'}$	$\mathcal{M}_{m,\sigma}^p$ with $p > 1,\ 1/p + 1/p' = 1$

3.2. Decision Space, Decision Variables. These remarks suggest the existence of a formalism anologous to probability calculus adapted to optimization. We start by defining cost measures, which can be viewed as the normalized idempotent measures of Maslov [13].

DEFINITION 18   1. We call a decision space *the triplet $(U, \mathcal{U}, \mathbb{K})$ where U is a topological space, $\mathcal{U}$ is the set of the open subsets of U, and $\mathbb{K}$ is a map from $\mathcal{U}$ into $\overline{\mathbb{R}}^{+3}$ such that (i) $\mathbb{K}(U) = 0$, (ii) $\mathbb{K}(\emptyset) = +\infty$, and (iii) $\mathbb{K}\left(\bigcup_n A_n\right) = \inf_n \mathbb{K}(A_n)$ for any $A_n \in \mathcal{U}$.*

2. *The map $\mathbb{K}$ is called a* cost measure.

3. *A map $c : u \in U \mapsto c(u) \in \overline{\mathbb{R}}^+$ such that $\mathbb{K}(A) = \inf_{u \in A} c(u)$, $\forall A \in \mathcal{U}$, is called a* cost density *of the cost measure $\mathbb{K}$.*

4. *The* conditional cost excess *to take the best decision in A knowing that it must be taken in B is*

$$\mathbb{K}(A|B) \overset{\text{def}}{=} \mathbb{K}(A \cap B) - \mathbb{K}(B)\,.$$

THEOREM 19 *Given a l.s.c. positive real valued function c such that $\inf_u c(u) = 0$, the expression $\mathbb{K}(A) = \inf_{u \in A} c(u)$ for all $A \in \mathcal{U}$ defines a cost measure. Conversely any cost measure defined on the open subsets of a Polish space admits a unique minimal extension $\mathbb{K}_*$ to $\mathcal{P}(U)$ (the set of the parts of U) having a density c^4 that is a l.s.c. function on U satisfying $\inf_u c(u) = 0$.*

This precise result is proved in Akian [1].

---

[3] $\overline{\mathbb{R}}^+ \overset{\text{def}}{=} \mathbb{R}^+ \cup \{+\infty\}$.

[4] We extend the previous definition to a general subset of U.

By analogy with random variables we define decision variables and related notions.

DEFINITION 20   *1. A decision variable X on $(U, \mathcal{U}, \mathbb{K})$ is a mapping from U into E a topological space. It induces $\mathbb{K}_X$ a cost measure on $(E, \mathcal{B})$ ($\mathcal{B}$ denotes the set of open sets of E) defined by $\mathbb{K}_X(A) = \mathbb{K}_*(X^{-1}(A))$, $\forall A \in \mathcal{B}$. The cost measure $\mathbb{K}_X$ has a l.s.c. density denoted c_X.*

2. When $E = \mathbb{R}$ (resp. $\mathbb{R}^n$, resp. $\mathbb{R}_{\min}$) with the topology induced by the absolute value (resp. the euclidian distance, resp. $d(x,y) = |e^{-x} - e^{-y}|$) then X is called a real (resp. vectorial, resp. cost) decision variable.

3. Two decision variables X and Y are said to be independent when

$$c_{X,Y}(x, y) = c_X(x) + c_Y(y).$$

4. The optimum of a real decision variable is defined by $\mathbb{O}(X) \stackrel{\text{def}}{=} \arg\min_x c_X(x)$ when the minimum exists. When a decision variable X satisfies $\mathbb{O}(X) = 0$, we say that it is centered.

5. When the optimum of a real decision variable X is unique and when near the optimum, we have

$$c_X(x) = \frac{1}{p} \left| \frac{x - \mathbb{O}(X)}{\sigma} \right|^p + o(|x - \mathbb{O}(X)|^p),$$

we define the sensitivity of order p of $\mathbb{K}$ by $\sigma^p(X) \stackrel{\text{def}}{=} \sigma$. When a decision variable satisfies $\sigma^p(X) = 1$, we say that it is of order p and normalized.

6. The numbers

$$|X|_p \stackrel{\text{def}}{=} \inf \left\{ \sigma \mid c_X(x) \geq \frac{1}{p} |(x - \mathbb{O}(X))/\sigma|^p \right\} \quad and \quad \|X\|_p \stackrel{\text{def}}{=} |X|_p + |\mathbb{O}(X)|$$

define respectively a seminorm and a norm on the set of decision variables having a unique optimum such that $\|X\|_p$ is finite. The corresponding set of decision variables is called $\mathbb{D}^p$. The space $\mathbb{D}^p$ is a conventional vector space and $\mathbb{O}$ is a linear operator on $\mathbb{D}^p$.

7. The characteristic function of a real decision variable is $\mathbb{F}(X) \stackrel{\text{def}}{=} \mathcal{F}(c_X)$ (clearly $\mathbb{F}$ characterizes only decision variables with cost in $\mathcal{C}_x$).

The role of the Laplace or Fourier transform in probability calculus is played by the Fenchel transform in decision calculus.

THEOREM 21 *If the cost density of a decision variable is convex, admits a unique minimum, and is of order p, we have:*[5]

$$\mathbb{F}(X)'(0) = \mathbb{O}(X), \quad [\mathbb{F}(X - \mathbb{O}(X))]^{(p')}(0) = \Gamma(p')[\sigma^p(X)]^{p'}, \quad with \ 1/p + 1/p' = 1.$$

---

[5]Γ denotes the classical gamma function.

THEOREM 22 *For two independent decision variables X and Y of order p and $k \in \mathbb{R}$ we have*

$$c_{X+Y} = c_X \,\square\, c_Y, \quad \mathbb{F}(X+Y) = \mathbb{F}(X) + \mathbb{F}(Y), \quad [\mathbb{F}(kX)](\theta) = [\mathbb{F}(X)](k\theta),$$

$$\mathbb{O}(X+Y) = \mathbb{O}(X) + \mathbb{O}(Y), \quad \mathbb{O}(kX) = k\mathbb{O}(X), \quad \sigma^p(kX) = |k|\sigma^p(X),$$

$$[\sigma^p(X+Y)]^{p'} = [\sigma^p(X)]^{p'} + [\sigma^p(Y)]^{p'}, \quad (|X+Y|_p)^{p'} \le (|X|_p)^{p'} + (|Y|_p)^{p'}.$$

3.3.   Limit Theorems for Decision Variables.   We now study the behavior of normalized sums of real decision variables. They correspond to asymptotic theorems (when the number of steps goes to infinity) for dynamic programming. We have first to define convergence of sequences of decision variables. We have defined counterparts of each of the four classical kinds of convergences used in probability in previous papers (see [3]). Let us recall the definition of the two most important ones.

DEFINITION 23 *For the decision variable sequence $\{X^m, m \in \mathbb{N}\}$ we say that*

1. *X^m weakly converges towards X, denoted $X^m \xrightarrow{w} X$, if for all f in $\mathcal{C}_b(E)$ (where $\mathcal{C}_b(E)$ denotes the set of uniformly continuous and lower bounded functions on E into $\mathbb{R}_{\min}$), $\lim_m \mathbb{M}[f(X^m)] = \mathbb{M}[f(X)]$, with $\mathbb{M}(f(X)) \overset{\mathrm{def}}{=} \inf_x(f(x) + c_X(x))$.*

2. *$X^m \in \mathbb{D}^p$ converges in p-sensitivity towards $X \in \mathbb{D}^p$, denoted $X^m \xrightarrow{\mathbb{D}^p} X$, if $\lim_m \|X^m - X\|_p = 0$.*

THEOREM 24 *Convergence in sensitivity implies convergence and the converse is false.*

The proof is given in Akian [2].

We have the analogue of the law of large numbers and the central limit theorem.

THEOREM 25 ((LARGE NUMBERS AND CENTRAL LIMIT)) *Given a sequence $\{X^m, m \in \mathbb{N}\}$ of independent identically costed (i.i.c.) real decision variables belonging to $\mathbb{D}^p$, $p \ge 1$, we have*

$$\lim_{N \to \infty} \frac{1}{N} \sum_{m=0}^{N-1} X^m = \mathbb{O}(X^0),$$

where the limit is taken in the sense of p-sensitivity convergence.

Moreover if $\{X^m, m \in \mathbb{N}\}$ is centered and of order p we have

$$^6\mathrm{weak}^* \lim_N \frac{1}{N^{1/p'}} \sum_{m=0}^{N-1} X^m = X, \quad \text{with } 1/p + 1/p' = 1,$$

where X is a decision variable with cost equal to $\mathcal{M}^p_{0,\sigma^p(X^0)}$.

---

[6] The weak* convergence corresponds to the restriction of test functions to the conventional linear ones in the definition of the weak convergence.

The analogues of Markov chains, continuous time Markov processes, and Brownian and diffusion processes have also been given in [3].

References

[1] Akian, M., Idempotent integration and cost measures, INRIA Report (1994), to appear.

[2] Akian, M., Theory of cost measures: convergence of decision variables, INRIA Report (1994), to appear.

[3] Akian, M.; Quadrat, J.-P., and Viot, M., Bellman Processes, Proceedings of the 11th International Conference on Analysis and Optimization of Systems, LNCIS, Springer-Verlag, Berlin and New York (June 1994).

[4] Azencott, R.; Guivarc'h, Y.; and Gundy, R. F.,École d'été de Saint Flour 8, Lecture Notes in Math., Springer-Verlag, Berlin and New York (1978).

[5] Baccelli, F.; Cohen, G.; Olsder, G. J.; and Quadrat, J. P., Synchronization and Linearity: An Algebra for Discrete Event Systems, John Wiley & Sons, New York (1992).

[6] Bellalouna, F., *Processus de décision min-markovien*, thesis dissertation, University of Paris-Dauphine (1992).

[7] Bellman, R., and Karush, W., *Mathematical programming and the maximum transform*, SIAM J. Appl. Math. **10** (1962).

[8] Cuninghame-Green, R., Minimax Algebra, Lecture Notes on Economics and Mathematical Systems, **166**, Springer-Verlag (1979).

[9] Del Moral, P., *Résolution particulaire des problèmes d'estimation et d'optimisation non-linéaires*, thèse Toulouse, France (Juin 1994).

[10] Feller, W.; An Introduction to Probability Theory and its Applications, John Wiley & Sons, New York (1966).

[11] Freidlin, M. I., and Wentzell, A. D., Random Perturbations of Dynamical Systems, Springer-Verlag, Berlin and New York (1979).

[12] Gaubert, S., *Théorie des systèmes linéaires dans les dioides*, thèse École des Mines de Paris, (Juillet 1992).

[13] Maslov, V., Méthodes Opératorielles, Éditions MIR, Moscou (1987).

[14] Maslov, V., and Samborski, S. N., Idempotent Analysis, Adv. in Sov. Math. **13** Amer. Math. Soc., Providence, RI (1992).

[15] Quadrat, J.-P., *Théorèmes asymptotiques en programmation dynamique*, Note CRAS Paris **311** (1990) 745–748.

[16] Rockafellar, R. T., Convex Analysis, Princeton University Press, Princeton, NJ (1970).

[17] Varadhan, S. R. S., Large deviations and applications, CBMS-NSF Regional Conference Series in Applied Mathematics, **46**, SIAM Philadelphia, PA, (1984).

[18] Whittle, P., Risk Sensitive Optimal Control, John Wiley & Sons, New York (1990).

Statistical Mechanics and Hydrodynamical Turbulence

RAOUL ROBERT

CNRS, Laboratoire d'Analyse Numérique, Université Lyon 1
43, Bd du 11 novembre 1918
F-69622 Villeurbanne Cedex, France

1 Introduction

An important progress made in the study of fluid turbulence in the last decades is a clearer understanding of the differences between the two- and three-dimensional cases. The most striking feature of 2D turbulence is the emergence of a large-scale organization of the flow (leading to structures usually called coherent structures), while the energy is conserved. By contrast, in the 3D case energy dissipation actually occurs, and the main experimental observations are the power-law energy spectrum and intermittency effects (the dissipation of energy does not seem to occur homogeneously in space).

Because, like a gas of molecules, slightly viscous turbulent flows have a large number of degrees of freedom, we expect to explain these properties by statistical mechanics arguments. The introduction of statistical mechanics ideas and methods in hydrodynamics has a long story, beginning with Onsager's 1949 pioneering paper. But great difficulties occurred on the way, and they seemed to have raised (for a time) some doubts as to the applicability of statistical mechanics to this field.

These difficulties are of a very different nature in two or three dimensions. In 2D, we have known for a long time of the existence of solutions (for all time) for Euler equations for an inviscid incompressible fluid flow. Indeed, nice estimates come from the existence of an infinite family of constants of the motion, associated to the law of vorticity conservation along the trajectories of the fluid particles. Until recently, this family of invariants was thought to be a serious technical obstacle to a relevant statistical mechanics approach. By contrast, this family of invariants disappears in the 3D case where, in spite of considerable efforts (see for example Majda [14]), our understanding of the dynamics remains poor (What about the existence of weak solutions for all time?, What about the dissipation of energy?). This is obviously a considerable obstacle to the application of statistical mechanics (What is the phase space?).

The importance of the subject was recalled by Chorin's lectures [6], which renewed the interest in Onsager's ideas, and some works were devoted to the statistical mechanics of point- vortices systems (see [16] for references).

In recent works [16], [20], [22], we showed that the main difficulties raised by the 2D case can be overcome by using two devices. The first is to work in

Proceedings of the International Congress
of Mathematicians, Zürich, Switzerland 1994
© Birkhäuser Verlag, Basel, Switzerland 1995

a relevant extended phase space (the space of Young measures) on which the constants of the motion of the system set natural constraints, and the second is to use modern large deviation theory (such as Baldi's large deviation theorem) to perform a thermodynamic limit in this framework. This approach works for a large class of dynamical systems (which can be described as the convection of a scalar density by an incompressible velocity field).

2 Statistical equilibrium states for a class of infinite-dimensional dynamical systems

2.1 A class of dynamical systems

A large class of evolution equations, coming from the modelling of various physical phenomena displaying complex turbulent behavior, can be described as the convection of a scalar density by an incompressible velocity field. More precisely, they are of the form:

$$\text{(I)} \quad \left\{ \begin{array}{l} q_t + \operatorname{div}(q\mathbf{u}) = 0, \\ \mathbf{u} = L(q), \operatorname{div}(\mathbf{u}) = 0, \end{array} \right\}$$

where $q(t, \mathbf{x})$ is some scalar density function defined on $\mathbb{R} \times \Omega$ (Ω is a bounded connected smooth domain of $\mathbb{R}^d$), $\mathbf{u}(t, \mathbf{x})$ is an incompressible velocity field taking its values in $\mathbb{R}^d$, which can be recovered from q by solving a P.D.E. system. Thus L denotes a (not necessarily linear) integro-differential operator. Let us give some well-known examples of such systems.

(1) The simplest example of (I) is the linear transport equation, where $\mathbf{u}(\mathbf{x})$ is a given incompressible velocity field on Ω.

(2) 2D incompressible Euler equations in the usual velocity-vorticity formulation are clearly of the form (I). Take for q the vorticity $q = \operatorname{curl} \mathbf{u}$, then $\mathbf{u}$ is given by:

$$\left\{ \begin{array}{l} \operatorname{curl} \mathbf{u} = q, \\ \operatorname{div} \mathbf{u} = 0, \\ \mathbf{u} \cdot \mathbf{n} = 0 \text{ on } \partial\Omega. \end{array} \right\}$$

This is a particular case of the quasi-geostrophic model used in geophysical fluid dynamics [17].

(3) Collisionless kinetic equations such as the Boltzmann-Poisson equation of stellar dynamics and Vlasov-Maxwell equations of plasmas can also be written in the form (I).

The first step in our program is to define a flow associated to (I) on the phase space $L^\infty(\Omega)$ (for reasons which will appear later, this is a good phase space because it contains small-scale oscillating step functions). Unfortunately, to our knowledge, there is no general existence-uniqueness result for the Cauchy problem for systems like (I). Examples (1) and (2) are well known, but for kinetic equations, although some existence results are available [1], [10] it seems that the uniqueness problem is not yet solved.

To proceed further, we shall assume that the system (I) defines a flow $\Phi_t :$ $L^\infty(\Omega) \longrightarrow L^\infty(\Omega)$, which satisfies some (rather technical, see [16] for details) continuity property. This means that for any given initial datum $q_0(\mathbf{x})$, the solution of the Cauchy problem for (I) is $q(t, \mathbf{x}) = (\Phi_t q_0)(\mathbf{x})$. This hypothesis on the system is satisfied in Example (1) if $\mathbf{u}$ is a C^1 velocity field on $\overline{\Omega}$ that is tangent to the boundary. It is also satisfied in Example (2): this is the classical Youdovitch's theorem for a perfect fluid [31].

Constants of the motion. For systems of the form (I), there is a family of constants of the motion that will play a crucial role. These are the functionals:

$$C_f(q) = \int_\Omega f(q(\mathbf{x})) \, d\mathbf{x},$$

for any given continuous function f on $\mathbb{R}$. Let us define the distribution measure of q, π_q by $\langle \pi_q, f \rangle = C_f(q)$. Then π_q is conserved by the flow.

According to each particular case, we will also have to take into account the classical constants of the motion of the system, such as energy and angular momentum. For example, in the case of Euler equations, the energy $E(q) = \frac{1}{2} \int_\Omega \mathbf{u}^2 \, d\mathbf{x}$ is also conserved. Integrating by parts, it will be convenient to write $E(q) = \frac{1}{2} \int_\Omega \Psi q \, d\mathbf{x}$, where Ψ is the stream function of $\mathbf{u} : -\Delta \Psi = q$, $\Psi = 0$ on $\partial \Omega$.

2.2 Long-time dynamics and Young measures

Let us consider a system of the form (I), and an initial datum q_0. It is well known that, in general, as time evolves, $\Phi_t q_0$ becomes a very intricate oscillating function. Let us denote $r = \| q_0 \|_{L^\infty(\Omega)}$. Because the measure π_q is conserved, $\Phi_t q_0$ will remain, for all time, in the ball $L_r^\infty = \{ q : \| q \|_\infty \leq r \}$. Extracting a subsequence (if necessary), we may suppose that, as time goes to infinity, $\Phi_t q_0$ converges weakly (for the weak-star topology $\sigma(L^\infty, L^1)$) towards some function q^*:

$$\Phi_t q_0 \overset{w}{\longrightarrow} q^*.$$

We can easily see that $C_f(\Phi_t q_0)$ does not converge towards $C_f(q^*)$ if f is nonlinear, whereas some other invariants can converge, as is the case for the energy in Euler equations. So, much information (given by the constants of the motion) is lost in this limit process. Thus the weak space $L^\infty(\Omega)$ is not a good one to describe the long-time limits of our system. Fortunately, the relevant space to do this is well known. The need to describe in some macroscopic way the small-scale oscillations of functions was understood long ago by Young [32]. To solve problems from the calculus of variations, Young introduced a natural generalization of the notion of function: at each point $\mathbf{x}$ in Ω, we no longer associate a well-determined real value, but only some probability distribution on $\mathbb{R}$ (such a mapping is called a Young measure on $\Omega \times \mathbb{R}$). More precisely, a Young measure ν on $\Omega \times \mathbb{R}$ is a measurable mapping $\mathbf{x} \to \nu_\mathbf{x}$ from Ω to the set $M_1(\mathbb{R})$ of the Borel probability measures on $\mathbb{R}$, endowed with the narrow topology (weak topology associated to the continuous bounded functions).

Clearly, ν defines a positive Borel measure on $\Omega \times \mathbb{R}$ (which we will also denote by ν) by:

$$\langle \nu, f \rangle = \int_\Omega \langle \nu_\mathbf{x}, f(\mathbf{x}, \cdot) \rangle \, d\mathbf{x},$$

for every real function $f(\mathbf{x}, z)$, continuous and compactly supported on $\Omega \times \mathbb{R}$.

To any measurable real function g on Ω, we associate the Young measure $\delta_g : \mathbf{x} \to \delta_{g(\mathbf{x})}$, Dirac mass at $g(\mathbf{x})$. We shall denote by $\mathcal{M}$ the convex set of Young measures on $\Omega \times \mathbb{R}$, and we recall some useful properties.

– $\mathcal{M}$ is closed in the space of all bounded Borel measures on $\Omega \times \mathbb{R}$ (with the narrow topology). In the sequel, $\mathcal{M}$ will be endowed with the narrow topology. If we replace $\mathbb{R}$ by the compact interval $[-r, r]$, the space $\mathcal{M}_r$ of Young measures on $\Omega \times [-r, r]$ is compact.
– $\{\delta_g | g : \Omega \to [-r, r] \text{ measurable}\}$ is a dense subset of $\mathcal{M}_r$.

We can now identify the long-time limits of the system as Young measures. Indeed, $\mathcal{M}_r$ is a suitable compactification of L_r^∞ because the narrow convergence (when t goes to infinity) of $\delta_{\Phi_t q}$ towards some Young measure ν preserves the information given by the constants of the motion; that is, for all functions $f(z)$:

$$\int_\Omega f(\Phi_t q(\mathbf{x})) \, d\mathbf{x} \to \int_\Omega \langle \nu_\mathbf{x}, f \rangle \, d\mathbf{x},$$

but the left-hand side is constant and equal to $\langle \pi_q, f \rangle$, so that:

$$\int_\Omega \nu_\mathbf{x} \, d\mathbf{x} = \pi_q.$$

The same kinds of arguments apply to the other invariants. For example, in the case of Euler equations, because $\Phi_t q$ converges weakly towards $\bar{\nu}(\mathbf{x}) = \int z \, d\nu_\mathbf{x}(z)$, we have, for the energy, $E(\Phi_t q) \to E(\bar{\nu})$, which is the energy of the Young measure ν.

Thus we see that the constants of the motion of (I) set constraints on the possible long-time limits.

2.3 Approximate Liouville measures, thermodynamic limit, and large deviation theory

A natural way to define equilibrium states is to construct invariant Gibbs measures on the phase space. But we do not know how to construct such measures on the natural phase space $L^\infty(\Omega)$ for systems like (I). In the case of Euler equations some work has been devoted to the study of Gibbs measures with formal densities given by the enstrophy ($\int q^2 d\mathbf{x}$) and the energy [3], and also to Gibbs measures associated to the law of vorticity conservation along the trajectories of the fluid particles [4]. Unfortunately all these measures are supported by "large" functional spaces so that not only the mean energy and enstrophy of these states are infinite but the phase space $L^\infty(\Omega)$ is of null measure. So, it is only at a formal level that this makes sense. Moreover this approach fails to give any prediction on the long-time dynamics corresponding to a given initial vorticity function.

The most common approach to overcome these difficulties is to use a convenient finite dimensional approximation of the system, possessing an invariant Liouville measure. Then one can consider the canonical measures associated to the constants of the motion and try to perform a thermodynamic limit in the space of generalized functions when the number of degrees of freedom goes to infinity. For example, for Euler equations one can consider the N Fourier-mode approximation or the point-vortex approximation. Two difficulties arise in this approach. The first is to choose a relevant scaling to perform the limit; an interesting comment on this point can be found in [18], see also [4]. The second is even more fundamental: generally, the approximate system will have fewer constants of the motion than the continuous one (I), so that the long-time dynamics of that system may be very different from that of the continuous one. For more comments and references on these attempts see for example [16], [17], [18], [20], [22].

Our program. We first define a family $\Pi_{\mathcal{O}}$ of approximate Liouville measures for Φ_t, and then we take the thermodynamic limit of these measures with the conditioning given by all the constants of the motion of (I).

The measures $\Pi_{\mathcal{O}}$. We shall say that a partition $\mathcal{O} = \{\Omega^i | i = 1, \ldots, n(\mathcal{O})\}$ of Ω is an equipartition if the subsets Ω^i are measurable and $|\Omega^i| = |\Omega^j|$ for all i, j ($|\Omega^i|$ denotes the Lebesgue measure of Ω^i). $d(\mathcal{O}) = \sup_i \sup_{\mathbf{x}, \mathbf{x}' \in \Omega^i} |\mathbf{x} - \mathbf{x}'|$ is the diameter of $\mathcal{O}$.

Given any equipartition $\mathcal{O}$, we denote $E_{\mathcal{O}}$ the subset of the functions of L_r^∞ that are constant on the sets Ω^i of $\mathcal{O}$. On $E_{\mathcal{O}}$ we define the probability distribution $\Pi_{\mathcal{O}} = \overset{n}{\otimes} \pi_0$ (where π_0 is any probability distribution on $[-r, r]$, in what follows we shall take $\pi_0 = \frac{1}{|\Omega|} \pi_{q_0}$). Of course, $\Pi_{\mathcal{O}}$ cannot be exactly conserved by Φ_t ($E_{\mathcal{O}}$ is not!), but it is conserved in some approximate sense [16]:

For all t, and any Borel subset B of L_r^∞, we have $\Pi_{\mathcal{O}}(\Phi_t(B)) \sim \Pi_{\mathcal{O}}(B)$, when $d(\mathcal{O}) \to 0$.

REMARK. A closely related issue is the construction of an approximation (on $E_{\mathcal{O}}$) of the flow Φ_t for which $\Pi_{\mathcal{O}}$ is exactly conserved. Although not entirely solved, this question is investigated in [16].

The thermodynamic limit. As noticed above, it is convenient to work in the space $\mathcal{M}_r$. The mapping $f \to \delta_f$ from $(E_{\mathcal{O}}, \Pi_{\mathcal{O}})$ into $\mathcal{M}_r$ defines a random variable (which we denote $\delta_{\mathcal{O}}$) on $\mathcal{M}_r$.

Now we have to take the thermodynamic limit (when $d(\mathcal{O}) \to 0$) of the random Young measures $\delta_{\mathcal{O}}$ conditioned by the "microcanonical" constraints:

$$\int_{\Omega} \nu_{\mathbf{x}} \, d\mathbf{x} = \pi_{q_0},$$

and other constraints such as $E(\bar{\nu}) = E(q_0)$, in the case of Euler equations.

The key ingredient to perform the limit is to prove that the random Young measure $\delta_{\mathcal{O}}$ has a large deviation property with constants $n(\mathcal{O})/|\Omega|$ and rate function $I_\pi(\nu)$.

This means (roughly speaking, see [16] for precise statements) that for any Borel subset $\mathcal{B}$ of $\mathcal{M}_r$, we have:

$$\mathrm{Prob}(\delta_{\mathcal{O}} \in \mathcal{B}) \sim \exp\left(-\frac{n(\mathcal{O})}{|\Omega|} \inf_{\nu \in \mathcal{B}} I_\pi(\nu) \right) \quad \text{(when } d(\mathcal{O}) \to 0\text{)},$$

where $\pi = d\mathbf{x} \otimes \pi_0$, and $I_\pi(\nu)$ is the classical Kullback information functional defined on $\mathcal{M}$ by:

$$I_\pi(\nu) = \int_{\Omega \times [-r,r]} \log \frac{d\nu}{d\pi} d\nu, \text{ if } \nu \text{ is absolutely continuous with respect to } \pi,$$

$I_\pi(\nu) = +\infty$ otherwise.

Modern large deviation theory (see for example [30]) is a powerful tool in proving such estimates. In [16] we apply Baldi's theorem [2], which is an elegant result giving general conditions under which a family of probability measures on a locally convex topological vector space has the large deviation property.

A straightforward consequence of this large deviation property is that the random Young measure $\delta_\mathcal{O}$, conditioned by the constraints $\int_\Omega \nu_\mathbf{x} \, d\mathbf{x} = \pi_{q_0}$, and (eventually) $E(\bar\nu) = E(q_0)$, is exponentially concentrated about the set $\mathcal{E}^*$ of the solutions of the variational problem

(V.P.) $I_\pi(\nu^*) = \inf\{I_\pi(\nu) | \nu \in \mathcal{E}\}$,

where $\mathcal{E}$ is the closed set of the Young measures satisfying

(∗) $\int_\Omega \nu_\mathbf{x} \, d\mathbf{x} = \pi_{q_0}$,

(∗∗)   other constraints (energy...).

2.4 The Gibbs states (or mean field) equation

Because $\mathcal{M}_r$ is compact and I_π is a lower semi-continuous functional, the set $\mathcal{E}^*$ of the solutions of (V.P.) is always nonempty. Using Lagrange multipliers, we get the equation satisfied by the critical point of (V.P.): the Gibbs states equation.

To fix the ideas, let us use the computations in the particular case of Euler equations. Let us suppose that ν^* is a critical point of (V.P.), then we show [20] that ν^* can be written:

$$\nu^* = \rho^*(\mathbf{x}, z)\pi, \text{ with } \rho^*(\mathbf{x}, z) = \frac{\exp(-\alpha(z) - \beta z \Psi^*(\mathbf{x}))}{Z(\Psi^*(\mathbf{x}))}, \text{ where :}$$

- β is the Lagrange multiplier of the energy,
- $\alpha(z)$ is a continuous function associated with the (infinite-dimensional) constraint (∗),
- Ψ^* is the stream function associated with the mean vorticity $\int z\rho^*(\mathbf{x}, z)\, d\pi_0(z)$,
- $Z(\Psi) = \int \exp(-\alpha(z) - \beta z \Psi)d\pi_0(z)$.

Thus Ψ^* must satisfy the following Gibbs states' equation:

(G.S.E.) $\left\{ \begin{array}{c} -\Delta\Psi = -\dfrac{1}{\beta}\dfrac{d}{d\Psi} \log Z, \\[2mm] \Psi = 0 \text{ on } \partial\Omega. \end{array} \right\}$

This nonlinear elliptic equation always has solutions. The solution is unique when β is greater than some negative value β_c, but when $-\beta$ is sufficiently large, bifurcations to multiple solutions generally occur [26].

3 Miscellaneous comments

In our opinion, the key in understanding the main features of 2D turbulence is the existence of statistical equilibria. Indeed, once we know the entropy functional associated with turbulent motion, we can study the relaxation process towards the equilibrium, and this yields naturally new evolution equations modelling the effect of small scales on the large ones in turbulent flows [21], [23].

The relevance of this equilibrium theory was tested, in the case of Euler equations, by experiments [9] and numerical simulations using Navier-Stokes equations at large Reynolds number [21], [26]. From these works the tentative conclusion emerges that the entropy functional given by the theory is accurate, even if the complex dynamics of the system can limit the complete relaxation towards the global equilibrium.

In [17], [27], [28] we show that this equilibrium theory provides a natural explanation of the spectacular organization of the atmospheric flow on the Jovian planets (Jupiter, Saturn, Uranus, and Neptune). Indeed, they are gaseous fast-rotating planets and their atmospheric dynamics satisfy an equation of the form (I). A detailed study of the equilibrium states displays the main common features of these flows such as latitudinal banding and the existence of large permanent vortices (among which the most famous is the great red spot of Jupiter).

The idea of such equilibrium states was previously investigated by Lynden-Bell [13] in the context of stellar dynamics. The same equation of Gibbs states was derived independently by Miller [18], from a more physical point of view. See also the closely related work by Shnirelman [24].

In contrast with the 2D case, the energy dissipation is not negligible in 3D turbulence, and a clear understanding of the mechanisms producing the main observed statistical properties of turbulence (such as power law energy spectrum and intermittency effects) is still missing. The existence of weak solutions of 3D Euler equations, which dissipate the energy, is certainly a key issue (as was suggested by Onsager, see Eyink [11] and Constantin [8]). This naturally leads to the study of the formation and statistics of shocks. The statistics of shocks was studied in the simpler (but enlightening) one-dimensional case of inviscid Burgers equation by Sinai [25], and recent progress was made with the elegant and decisive contribution by Carraro and Duchon [5].

Last but not least, we will evoke the pioneering and intuitive work of Chorin on the statistical mechanics of 3D turbulence. Chorin [7] exploits an analogy between vortex tubes and self-avoiding walks and applies methods from the statistical mechanics of polymers. It appears that power law energy spectrum and intermittency effects can be recovered by this method.

Acknowledgments

It is a pleasure for me to thank here J. Duchon and J. Sommeria whose thoughts on turbulence have greatly influenced me, C. Staquet, C. Rosier, and T. Dumont who performed enlightening numerical simulations with ability and patience, and J. Michel whose thesis is a gifted contribution to the subject.

References

[1] A. Arsenev, *Global existence of a weak solution of system of equations*, U.S.S.R. Comput. Math. and Math. Phys. 15, 131–143 (1975).

[2] P. Baldi, *Large deviations and stochastic homogenization*, Ann. Mat. Pura Appl. (4) 151, 161–177 (1988).

[3] G. Benfatto, P. Pico, and M. Pulvirenti, J. Statist. Phys. 46, 729 (1987).

[4] C. Boldrighini and S. Frigio, *Equilibrium states for a plane incompressible perfect fluid*, Comm. Math. Phys. 72, 55–76 (1980).

[5] L. Carraro and J. Duchon, *Solutions statistiques intrinsèques de l'équation de Burgers et processus de Levy*, C. R. Acad. Sci. Paris, to appear.

[6] A. Chorin, *Lectures on turbulence theory*, Publish or Perish, 1975.

[7] A. Chorin, *Statistical mechanics and vortex motion*, Lectures in Appl. Math. 28, Amer. Math. Soc. 85 (1991).

[8] P. Constantin, Comm. Math. Phys., to appear.

[9] M. A. Denoix, J.Sommeria, and A. Thess, *Two-dimensional turbulence: The prediction of coherent structures by statistical mechanics*, to appear.

[10] R. Di Perna and P. L. Lions, *Global weak solutions of Vlasov-Maxwell system*, Comm. Pure Appl. Math. 6, 729–757 (1989).

[11] G. Eyink, *Energy dissipation without viscosity in ideal hydrodynamics*, to appear.

[12] A. P. Ingersol, *Atmospheric dynamics of the outer planets*, Science, 248, p. 308 (1990).

[13] D. Lynden-Bell, *Statistical mechanics of violent relaxation in stellar systems*, Mon. Not. R. Astr. Soc. 136, 101–121 (1967).

[14] A. Majda, *The interaction of nonlinear analysis and modern applied mathematics*, Proc. Internat. Congress Math., Kyoto, 1990.

[15] J. Michel, Thesis, Université Lyon 1, 1993.

[16] J. Michel and R. Robert, *Large deviations for Young measures and statistical mechanics of infinite dimensional dynamical systems with conservation law*, Comm. Math. Phys. 159, 195–215 (1994).

[17] J. Michel and R. Robert, *Statistical mechanical theory of the great red spot of Jupiter*, J. Statist. Phys., to appear 1994.

[18] J. Miller, P. B. Weichman, and M. C. Cross, *Statistical mechanics, Euler equations, and Jupiter's red spot*, Phys. Rev. A 45, 2328–2359 (1992).

[19] L. Onsager, *Statistical hydrodynamics*, Nuovo Cimento Supl. 6, 279 (1949)

[20] R. Robert, *A maximum entropy principle for two-dimensional Euler equations*, J. Statist. Phys. 65, 3/4, 531–553 (1991).

[21] R. Robert and C. Rosier, to appear.

[22] R. Robert and J. Sommeria, *Statistical equilibrium states for two-dimensional flows*, J. Fluid Mech. 229, 291–310 (1991).

[23] R. Robert and J. Sommeria, *Relaxation towards a statistical equilibrium state in two-dimensional perfect fluid dynamics*, Phys. Rev. Lett. 69, 2276–2279 (1992).

[24] A. I. Shnirelman, *Lattice theory and flows of ideal incompressible fluid*, Russian J. Math. Phys. vol 1, no. 1, 105–114 (1993).

[25] Y. Sinai, *The statistics of shocks in the solutions of inviscid Burgers equation*, Comm. Math. Phys. 148, no. 3, 601–621 (1992).

[26] J. Sommeria, C. Staquet, and R. Robert, *Final equilibrium state of a two-dimensional shear layer*, J. Fluid Mech. 233, 661–689 (1991).

[27] J. Sommeria, C. Nore, T. Dumont, and R. Robert, *Théorie statistique de la tache rouge de Jupiter*, C. R. Acad. Sci. Paris, t. 312, Série II, 999–1005 (1991).

[28] J. Sommeria, R. Robert, and T. Dumont, to appear.

[29] B. Turkington and R. Jordan, *Turbulent relaxation of a magnetofluid: a statistical equilibrium model*, Proceedings, International conference on advances in geometric analysis and continuum mechanics, Stanford University, August 1993.

[30] S. R. S. Varadhan, *Large deviations and applications*, Ecole d'été de probabilités de Saint-Flour XV–XVII, 1985–1987.

[31] V. I. Youdovitch, *Non-stationary flow of an incompressible liquid*, Zh. Vychisl. Mat. i Mat. Fiz. 3, 1032–1066 (1963).

[32] L. C. Young, *Generalized surfaces in the calculus of variations*, Ann. Math. t. 43, 84–103 (1942).

Spaces of Observables in Nonlinear Control

EDUARDO D. SONTAG

Department of Mathematics, Rutgers University
New Brunswick, NJ 08903, USA

Engineering design and optimization techniques for control typically rely upon the theory of irreducible finite-dimensional representations of linear shift-invariant integral operators. A representation of $\mathcal{F} : [\mathcal{L}_{\infty,\mathrm{loc}}(0, \infty)]^m \to [C_0(0, \infty)]^p$ is specified by a triple of linear maps $A : \mathbb{R}^n \to \mathbb{R}^n$, $B : \mathbb{R}^m \to \mathbb{R}^n$, and $C : \mathbb{R}^n \to \mathbb{R}^p$ so that, for each "input" ω, $\mathcal{F}(\omega)(t) = C\xi(t)$, where the state ξ is the solution of the initial value problem $\xi'(t) - A\xi(t) = B\omega(t)$, $\xi(0) = 0$.

For such state-space realizations to exist, it is an elementary and well-known fact that the following equivalent properties must hold, if $\mathcal{F}(\omega)(t) = \int_0^t K(t-\tau)\omega(\tau)\,d\tau$ and the entries of the $(p \times m)$-matrix kernel $K(t)$ are analytic and of exponential order $|K_{ij}(t)| < \alpha e^{ct}$: rationality of the Laplace transform matrix $\mathcal{K}(s) = \int_0^\infty K(t)e^{-st}dt$; existence of some nontrivial algebraic-differential equation $\mathcal{E}(\omega(t), \omega'(t), \ldots \omega^{(s)}(t); \eta(t), \eta'(t), \ldots, \eta^{(r)}(t)) = 0$ relating inputs and outputs $\eta = \mathcal{F}(\omega)$; and finiteness of the rank of the block Hankel matrix $H = (H_{ij})_{i,j=0}^\infty$ that is defined, in terms of the Taylor expansion of K, by the $(p \times m)$-submatrix entries $\left(\partial^{i+j} K / \partial t^{i+j}\right)(0)$.

Irreducible representations are exactly those of minimal dimension, which equals the rank ϱ of H, and they have desirable control-theoretic properties. Most significant are the facts that the elementary observables $x \mapsto Ce^{tA}x + C \int_0^t e^{(t-\tau)A} B\omega(s)\,ds$ separate points, and that states can be asymptotically steered to the equilibrium $x_0 = 0$ by means of linear feedback laws $\omega(t) = Fx(t)$ that render $\mathrm{Re}\,\lambda < 0$ for all eigenvalues λ of $A + BF$.

The study of representability and the analysis of qualitative properties of minimal realizations have their roots in the nineteenth century, in particular in the work of Lord Kelvin regarding the use of integrators for solving differential equations, Kronecker's contributions to linear algebra (to a great extent motivated by essentially these questions), and Hurwitz' and Routh's stability criteria. The theory, which is at the core of modern multivariable linear control, achieved full development mainly during the 1960s. Standard textbooks (e.g. [21]) cover this material, which forms the basis of widely used computer-aided design packages. Much effort has been directed since the early 1970s towards extensions to nonlinear operators, including the characterization of representability by means of explicit numerical invariants generalizing ϱ, the equivalence to high-order differential constraints, and the synthesis of steering control laws. A still-developing but fairly detailed body of knowledge is by now available, covering both global algebraic and local analytic aspects.

Proceedings of the International Congress
of Mathematicians, Zürich, Switzerland 1994
© Birkhäuser Verlag, Basel, Switzerland 1995

This paper will focus on a narrow but fundamental and unifying subtopic, namely the role played by observables, which are the functions on states induced by experiments. I will start with a brief introduction to control systems and the questions to be studied, followed by an outline of results.

1. Introduction

To *control* something means to influence its behavior so as to achieve a desired goal. Sophisticated regulation mechanisms are ubiquitous in nature as well as in modern technology, where they appear in a wide range of industrial and consumer applications, such as anti-lock brakes, fly-by-wire high-performance aircraft, automation robots, or precision controllers for CD players. Control theory postulates mathematical models of control systems and deals with the basic principles underlying their analysis and design.

The basic paradigm is that of a *(controlled) system* Σ, specified by a right action

$$\mathcal{X} \times \Omega \to \mathcal{X} : \quad (x, \omega) \mapsto x \cdot \omega$$

of a monoid Ω, whose elements are called *inputs* or *controls*, on a set $\mathcal{X}$, the *state space*, together with a map, the *output function*,

$$h : \mathcal{X} \to \mathcal{Y}$$

into a set $\mathcal{Y}$, of *output* or *measurement* values. (Partial actions are also of interest, particularly in the context of the differential systems discussed below, but at this abstract level they can be subsumed merely by adjoining to $\mathcal{X}$ an "undefined" element, invariant under all ω, as well as an extra element to $\mathcal{Y}$.) Typically the elements of Ω are functions of a discrete or continuous time variable, and one interprets the action $x \cdot \omega$ as defining a forced dynamical system with phase space $\mathcal{X}$. The function h expresses constraints on the information readily available about states. Often the control objective is to find appropriate functions ω that force the new state $x \cdot \omega$ to have some particular desired characteristic, such as being close to a certain target set or optimizing a cost criterion, using only information about the initial state x inferred from outputs.

Different algebraic, topological, and/or analytic structures are then superimposed on this basic setup in order to model specific applications and to develop nontrivial results. For instance, as with classical (noncontrolled) dynamical systems, one manner in which actions often arise is through the integration of ordinary differential equations. Let $\mathcal{X}$ be a (second countable) differentiable manifold, with tangent bundle projection $\pi : T\mathcal{X} \to \mathcal{X}$, and let $\mathcal{U}$ be a separable locally compact metric space (of *input values*). A *continuous-time differential system* is specified by a continuous mapping $f : \mathcal{X} \times \mathcal{U} \to T\mathcal{X}$ such that $\pi(f(x, u)) = x$ for each $(x, u) \in \mathcal{X} \times \mathcal{U}$, of class C^1 on $\mathcal{X}$ and with f_x continuous on $\mathcal{X} \times \mathcal{U}$, together with a continuous $h : \mathcal{X} \to \mathcal{Y}$ into another metric space. For each $T > 0$, let $\mathcal{L}_\infty^{\mathcal{U}}[0, T] =$ measurable and essentially compact maps from $[0, T]$ into $\mathcal{U}$. For each $\omega \in \mathcal{L}_\infty^{\mathcal{U}}[0, T]$ and $x \in \mathcal{X}$, there is a well-posed initial value problem on $[0, T]$

$$\xi'(t) = f(\xi(t), \omega(t)), \quad \xi(0) = x. \tag{1}$$

Solutions exist at least for small $t > 0$; let $x \cdot \omega$ be the value $\xi(T)$, if defined, of this solution. The *concatenation* of $\omega \in \mathcal{L}_\infty^\mathcal{U}[0, T]$ and $\nu \in \mathcal{L}_\infty^\mathcal{U}[0, S]$ is the element $\omega \sharp \nu \in \mathcal{L}_\infty^\mathcal{U}[0, T + S]$, which is almost everywhere equal to $\omega(t)$ on $[0, T]$ and to $\nu(t - T)$ on $[T, T + S]$. Let $\Omega_\mathcal{U}$ be the disjoint union of the sets $\mathcal{L}_\infty^\mathcal{U}[0, T]$, over all $T \geq 0$, including for $T = 0$ the zero-length input $\diamond$; this is a monoid under $\sharp$, with identity $\diamond$. A controlled system as above results. Often $\mathcal{Y}$ is a Euclidean space and the components of $h(x)$ designate coordinates of the state x that can be instantaneously measured. As a concrete example, the dynamic and kinematic equations of a rigid body subject to torques and translational forces give rise to a differential system evolving on $\mathcal{X} =$ tangent bundle of the Euclidean group. The input values are in $\mathcal{U} = \mathbb{R}^m$ if there are m independent external torques and forces acting on the system. An appropriate measurement function $h : TE(3) \to \mathbb{R}^3$ is included in the system specification if one can only directly measure the body's angular momentum, but not its $SO(3)$ orientation component or its translational coordinates.

In order to attain a desired control objective, it is usually necessary to determine the current state x of the system. This motivates the *state estimation*, or in its stochastic formulation, the Kalman filtering problem: find x on the basis of experiments consisting of applying a test input and measuring the ensuing response. That is to say, one needs to reconstruct x from the values $h^\omega(x)$ of the *observables*

$$h^\omega : \mathcal{X} \to \mathcal{Y} : x \mapsto h(x \cdot \omega).$$

A necessary condition for state estimation is that $\{h^\omega, \omega \in \Omega\}$ separate points; algorithmic and well-posedness requirements lead in turn to several refinements of this condition.

In many practical situations it is impossible to derive flow models like differential equations from physical principles. Sometimes the system to be controlled is only known implicitly, through its external behavior, but no dynamical model (action, output map) is available. The only data is the response of the system to the various possible inputs ω, when starting from some initial or "relaxed" state x_0. Mathematically, one is given a mapping $F : \Omega \to \mathcal{Y}$ rather than a system Σ in the form defined above. Thus, a preliminary step in control design requires the solution of the *realization* problem: passing from an external or *input-output* (I/O) description to a well-formulated internal or *state-space* dynamical model. This is the inverse problem of representing the given F in the form

$$F(\omega) = h^\omega(x_0) = h(x_0 \cdot \omega)$$

for some system Σ and initial state x_0. Typically, moreover, one wants to find a Σ that satisfies additional constraints — the state space has a topological structure, its dynamics arise as the flow of a differential equation, etc. — so as to permit the eventual application of numerical optimization techniques in order to solve control problems.

For state estimation and realization questions, the observables h^ω, together with their infinitesimal versions for differential systems, obviously play a central role. It is perhaps surprising that their study is also extremely useful when dealing with many other control issues, due in part to the dualities between "input to state" and "state to output" maps, and between $\mathcal{X}$ and functions on $\mathcal{X}$. Studying the duality between observables and states is particularly fruitful in control theory, perhaps more so than in physics.

Since the mid-1970s, various algebraic structures associated to observables have been introduced[*] and shown to be fundamental ingredients in providing new insights into realization, observation, and other control-theory problems.

In this paper, I give a brief and selective account of basic concepts and recent developments in the program of study that deals with the systematic use of spaces of observables. Considered are questions such as: Given an I/O mapping, how does one classify its possible state-space representations? With what algebraic, topological, and/or analytic structures are state spaces naturally endowed? How does one characterize those operators which admit representations in terms of finite systems of first order ordinary differential equations? How do algebraic-differential constraints on input/output data relate to such representability? What are implications of finite dimensionality, finite generation, and finite transcendence degree of linear spaces, algebras, and fields of observables, respectively, upon the classification of internal models? and Which input functions are rich enough to permit all information about systems and states to be deduced from their associated observables? Several answers are outlined, along with applications to the numerical solution of path planning problems for nonholonomic mechanical systems.

The results reported here represent the contributions of many researchers; as far as my own work in this area is concerned, it has benefited greatly from discussions with and the insight of many colleagues, including especially Jean-Michel Coron, Michel Fliess, Bronek Jakubczyk, Héctor Sussmann, and Yuan Wang.

In the interest of preserving clarity of exposition, the formulations in this talk are not the most general possible. For instance, inputs and outputs are often taken to lie in Euclidean spaces; although this covers the most interesting cases for applications, many aspects can be developed in far more generality, and this is indeed what is done in many of the references. Undefined concepts and terminology from control theory are as in [21].

2. Global Algebraic Aspects

The most fundamental level on which to formulate the construction of observables is as follows. Let $\mathcal{X}$ be a set endowed with an action by a monoid (semigroup with identity) Ω, and let $\mathcal{H} \subseteq \mathbb{R}^{\mathcal{X}}$ be a collection of real-valued functions on $\mathcal{X}$. For each $\omega \in \Omega$ and $\ell \in \mathbb{R}^{\mathcal{X}}$, let $\omega \cdot \ell := \ell^{\omega}$, where $\ell^{\omega}(x) := \ell(x \cdot \omega)$. This induces a left action of Ω on $\mathbb{R}^{\mathcal{X}}$, and seeing the latter as an algebra with pointwise operations, each map $\ell \mapsto \ell^{\omega}$ is a homomorphism. The *observation space* $\mathcal{O}(\mathcal{X}, \mathcal{H})$ and *observation algebra* $\mathcal{A}(\mathcal{X}, \mathcal{H})$ are the smallest Ω-invariant $\mathbb{R}$-linear subspace and subalgebra of $\mathbb{R}^{\mathcal{X}}$, respectively, that contain $\mathcal{H}$. Their generating elements ℓ^{ω}, $\ell \in \mathcal{H}$, are the *elementary (global) observables*.

---

[*]As with so many other notions central to control, observation spaces and algebras were first systematically studied by Rudolf Kalman, now at the E.T.H., so this topic is particularly appropriate for a Zürich ICM.

An *algebraic controlled system* is given by an action $\mathcal{X} \times \Omega \to \mathcal{X}$ and output map $h : \mathcal{X} \to \mathcal{Y}$ for which each of the sets $\mathcal{X}$ and $\mathcal{Y}$ is endowed with the structure of a real affine scheme, Ω acts by morphisms, and h is a morphism. The state space $\mathcal{X}$ comes equipped with both the Zariski topology and the strong topology obtained by requiring that all elements of the algebra of real-valued regular functions $A(\mathcal{X})$ be continuous. Such systems can be viewed as "generalized polynomial systems," because in the particular case when the algebras of functions $A(\mathcal{X})$ and $A(\mathcal{Y})$ are finitely generated, the schemes $\mathcal{X}$ and $\mathcal{Y}$ are algebraic sets and h as well as each of the maps $x \mapsto x \cdot \omega$ are expressed by vector polynomial functions. (All affine schemes X are here assumed to be reduced over $\mathbb{R}$, meaning reduced and real points are dense. Identifying X with the set of its real points, X can be seen as the set $\mathrm{Spec}_{\mathbb{R}}(A)$ of all homomorphisms $A \to \mathbb{R}$, for some $\mathbb{R}$-algebra $A = A(X)$ that is reduced over $\mathbb{R}$.) Several basic results for algebraic controlled systems, some of which are summarized next, were developed in [19]. (This reference dealt specifically with discrete time systems, but the results hold in more generality.)

For such a system $\Sigma = (\mathcal{X}, h)$, let $\mathcal{H}$ be the set of coordinates $\{\varphi \circ h, \varphi \in A(\mathcal{Y})\}$ of h. The system Σ is said to be *algebraically observable* (ao) if $A^{\Sigma} := A(\mathcal{X}, \mathcal{H}) = A(\mathcal{X})$. This condition is stronger than merely stipulating that observables must separate points; it corresponds to the requirement that states must be recoverable from input/output experiments by means of purely algebraic operations. In the case of finitely generated algebras, it means precisely that each coordinate of the state must be expressible as a polynomial combination of the results of a finite number of experiments. With respect to a fixed *initial state* $x_0 \in \mathcal{X}$, the action is *algebraically reachable* (ar) if $x_0 \cdot \Omega$ is Zariski-dense in $\mathcal{X}$. This property is in general weaker than complete reachability — i.e. transitivity of the action, $x_0 \cdot \Omega = \mathcal{X}$ — and corresponds to the nonexistence of nontrivial algebraic invariants of the orbit $x_0 \cdot \Omega$. An initialized system $\Sigma = (\mathcal{X}, h, x_0)$ is *algebraically irreducible* or *canonical* if it is both ao and ar.

An *(I/O) response* is any map $F : \Omega \to \mathcal{Y}$. A representation or *realization* of F is a $\Sigma = (\mathcal{X}, h, x_0)$ so that $F(\omega) = h(x_0 \cdot \omega)$ for all ω. Initialized algebraic systems form a category under the natural notion of morphism $T : \Sigma^1 \to \Sigma^2$, namely a scheme morphism $T : \mathcal{X}^1 \to \mathcal{X}^2$, with $T(x_0^1) = x_0^2$, such that $h^1(x) = h^2(T(x))$ and $T(x \cdot \omega) = T(x) \cdot \omega$ for all x and ω. Isomorphisms can be interpreted as "changes of coordinates" in the state space. Two isomorphic systems always give rise to the same response.

THEOREM [23], [19]. *For any response F there exists a canonical realization given by an initialized algebraic controlled system. Any two canonical realizations of the same response are necessarily isomorphic.*

A proof of the existence part of this result is quite simple and provides a starting point for the study of algebraic realizations, so it is worth sketching. Considering Ω acting on itself on the right, $A(\Omega, \{\varphi \circ F, \varphi \in A(\mathcal{Y})\})$ is the *observation algebra* $\mathcal{A}^F$ of F. As Ω acts by homomorphisms on $\mathcal{A}^F$, duality provides an algebraic action of Ω on $\mathcal{X}_F := \mathrm{Spec}_{\mathbb{R}}(\mathcal{A}^F)$. On the other hand, the map F induces a homomorphism $A(\mathcal{Y}) \to \mathcal{A}^F$ via $\varphi \mapsto \varphi \circ F$, which in turn by duality provides an output morphism $h : \mathcal{X}_F \to \mathcal{Y}$. The construction is completed defining $x_0 \in \mathcal{X}_F$ by $x_0(\psi) := \psi(\diamond)$ (evaluation at the identity). An important feature of this constructive proof is that finiteness conditions on spaces of observables, which can be in principle verified directly from input/output data, are immediately reflected upon corresponding finiteness properties of canonical realizations.

2.1. Finiteness Conditions.

For simplicity, assume from now on that the output value space is Euclidean, $\mathcal{Y} = \mathbb{R}^p$ for some integer p (the number of "output channels").

For Ω acting on itself and F_i the ith coordinate of F, $\mathcal{O}(\Omega, \{F_1, \ldots, F_p\})$ is the *observation space* $\mathcal{O}^F \subseteq \mathcal{A}^F$. Finite dimensionality of $\mathcal{O}^F$ as a real vector space translates into realizability by state-affine systems, for which $\mathcal{X}$ is Euclidean and transitions $x \mapsto x \cdot \omega$ and output h are given by affine maps [14], [19], [6]. This is analogous to Hochschild-Mostow "representative" functions on Lie groups [11], those whose translates span a finite-dimensional space, but here translates are being taken with respect to a semigroup action.

Finite generation of $\mathcal{A}^F$ as an algebra over $\mathbb{R}$ corresponds to canonical realizability by systems evolving on algebraic varieties, and tools from algebraic geometry lead to stronger results. As one illustration of such results, take two realizations of the same response whose state spaces are nonsingular varieties. Assume further that both systems are reachable and observable in the sense that the algebra $A(\mathcal{X})$ separates complex points (this is considerably weaker than algebraic observability). An argument based on Zariski's main theorem shows that the two systems must then be isomorphic ([19], Section 26).

When $\mathcal{A}^F$ is an integral domain, one may introduce its field of fractions $\mathcal{K}^F$, the *observation field* of F. Natural finiteness conditions are then finite generation of $\mathcal{K}^F$ or finite transcendence degree as a field extension of $\mathbb{R}$. For classes of discrete-time responses, these two turn out to be equivalent. They characterize realizability in terms of systems with dynamics definable by rational difference equations, or alternatively by piecewise regular functions on a stratification into quasi-affine varieties of dimension at most tr.deg $\mathcal{K}^F$ ([19], Section 27). This dimension can be explicitly computed from F, and finiteness is equivalent to existence of algebraic difference equations relating inputs and outputs.

Most often encountered in engineering are *linear* responses. These are well understood (e.g. [21]), but it is worth recalling the basic facts in order to appreciate the context of the above and later results. In continuous time, fix a positive integer m, the "number of input channels," and let $\mathcal{U} = \mathbb{R}^m$ and $\Omega = \Omega_{\mathcal{U}}$. A *linear response* is one defined by a convolution operator $F(\omega) = \int_0^T K(T-s)\omega(s)\,ds$, for each $\omega \in \mathcal{L}_\infty^{\mathcal{U}}[0, T]$, where K is analytic (or more generally $K \in (\mathcal{L}_{1,\mathrm{loc}}[0, \infty))^{p \times m}$). The natural finite-dimensional realizations in this case are linear differential systems, which have Euclidean state spaces $\mathcal{X} = \mathbb{R}^n$, linear $f(x, u) = Ax + Bu$ in their differential equation descriptions (1), initial state 0, and linear output map $h(x) = Cx$. Linear isomorphisms, basis changes in the state space, lead to an action $(A, B, C) \mapsto (T^{-1}AT, T^{-1}B, CT)$ of $GL(n)$ on these representations.

THEOREM. *Let F be a linear response. The following are equivalent: (1) the vector space $\mathcal{O}^F$ is finite dimensional; (2) the algebra $\mathcal{A}^F$ is finitely generated; (3) tr.deg $\mathcal{K}^F < \infty$; and (4) there exists a linear realization. Moreover, there is always in that case a canonical linear realization, whose dimension $\varrho^F = \dim \mathcal{O}^F =$*

$tr.deg \mathcal{K}^F$ *is minimal among all possible linear realizations, and any two such re-alizations are in the same $GL(n)$-orbit.*

One then studies the family of all responses with fixed $\varrho^F = n$, or equivalently the quotient space of the open subset of canonical triples $\mathcal{M}_{n,m,p} \subseteq \mathbb{R}^{n(n+m+p)}$ under $GL(n)$, seen as a smooth action of a Lie group on a manifold. This action is free, and the quotient has a structure of differentiable manifold for which the map $\mathcal{M}_{n,m,p} \rightarrow \mathcal{M}_{n,m,p}/GL(n)$ is a smooth submersion and in fact defines a prin-cipal fibre bundle ([21], Section 5.6). Moreover, $\mathcal{M}_{n,m,p}/GL(n)$ is a nonsingular algebraic variety, and the moduli problem, fundamental for the understanding of parameterization problems for identification applications, is solved [27]. The op-erator A in a minimal linear realization is the infinitesimal generator of a shift operator in the observation space. However, it may also be viewed as a derivation on a space of jets of observables. This alternative characterization, which holds in the linear case, motivates the study of *infinitesimal* observation vector spaces and algebras. These objects can be defined for "analytic" classes of nonlinear re-sponses, which arise when viewing $\int_0^T K(T-s)\omega(s)$ as the first term in a higher order functional Taylor (nonlinear Volterra) expansion. In continuous time, such spaces are computationally far more useful than their global versions, as they do not involve integration of differential equations. A convenient way to introduce this approach is by means of generating series.

3. Local Analytic Responses

I continue to assume that $\mathcal{U} = \mathbb{R}^m$ and $\mathcal{Y} = \mathbb{R}^p$ for some positive integers m and p. Analytic responses are defined in terms of power series in finitely many noncommuting variables, so these need to be reviewed first.

3.1. Generating Series.
Let $\Theta = \{\theta_0, \ldots, \theta_m\}$ be a set of symbols, $L\langle\Theta\rangle$ the free real Lie algebra on the set Θ, $\mathbb{R}\langle\Theta\rangle$ its enveloping algebra, and $\mathbb{R}\langle\langle\Theta\rangle\rangle$ the completion of $\mathbb{R}\langle\Theta\rangle$ with respect to the maximal ideal (Θ). Thus, $\mathbb{R}\langle\langle\Theta\rangle\rangle$ is the set of formal power series

$$c = \sum_{\alpha \in \Theta^*} \langle c | \alpha \rangle \, \alpha \, .$$

In these terms, the associative noncommutative $\mathbb{R}$-algebra structure in $\mathbb{R}\langle\langle\Theta\rangle\rangle$ is that whose product extends concatenation in Θ^*, $\mathbb{R}\langle\Theta\rangle$ is the polynomial sub-algebra consisting of series with finitely many nonvanishing coefficients, the free associative $\mathbb{R}$-algebra on Θ, and $L\langle\Theta\rangle$ is the Lie subalgebra generated by Θ. Finally, the Lie algebra of Lie series $L\langle\langle\Theta\rangle\rangle$ consists of those $c \in \mathbb{R}\langle\langle\Theta\rangle\rangle$ whose homogeneous components are in $L\langle\Theta\rangle$, and the set of exponential Lie series $G\langle\langle\Theta\rangle\rangle = \exp(L\langle\langle\Theta\rangle\rangle)$ forms a multiplicative subgroup of $\mathbb{R}\langle\langle\Theta\rangle\rangle$ (Campbell-Hausdorff formula). There is a linear duality between $\mathbb{R}\langle\langle\Theta\rangle\rangle$ and $\mathbb{R}\langle\Theta\rangle$:

$$\langle c | \lambda \rangle = \sum_{\alpha \in \Theta^*} \langle c | \alpha \rangle \langle \lambda | \alpha \rangle \, . \tag{2}$$

There is also a commutative associative product on $\mathbb{R}\langle\langle\Theta\rangle\rangle$, the *shuffle product*, with the empty word as unit and with $\beta\theta_i \sqcup\!\sqcup \alpha\theta_j = ((\beta\theta_i)\sqcup\!\sqcup\alpha)\theta_j + (\beta\sqcup\!\sqcup(\alpha\theta_j))\theta_i$ for all $\beta, \alpha \in \Theta^*$. For each $d \in \mathbb{R}\langle\langle\Theta\rangle\rangle$ and $\lambda \in \mathbb{R}\langle\Theta\rangle$, $\lambda^{-1}d \in \mathbb{R}\langle\langle\Theta\rangle\rangle$ is the adjoint defined by $\langle\lambda^{-1}d|\alpha\rangle := \langle d|\lambda\alpha\rangle$ ($d\lambda^{-1}$ is defined analogously). The action $d \mapsto \lambda^{-1}d$ makes $\mathbb{R}\langle\langle\Theta\rangle\rangle$ into a right module over $\mathbb{R}\langle\Theta\rangle$. Restricting λ to $L\langle\Theta\rangle$ defines an action by derivations of $L\langle\Theta\rangle$ on $\mathbb{R}\langle\langle\Theta\rangle\rangle$ seen as a shuffle product algebra (Friedrichs' criterion amounts to the converse: $d \mapsto \lambda^{-1}d$ being a derivation implies $\lambda \in L\langle\Theta\rangle$; see the excellent exposition [16]). The series $c \in \mathbb{R}\langle\langle\Theta\rangle\rangle$ is *convergent* if there is a positive (radius of convergence) ρ and a K so that $|\langle c|\alpha\rangle| \leq K\underline{\alpha}!\rho^{\underline{\alpha}}$ for each $\alpha \in \Theta^*$, $\underline{\alpha} = $ length of α. The set of convergent series is invariant under $d \mapsto \lambda^{-1}d$.

3.2. Chen-Fliess Embedding of Inputs.

For each $\omega \in \mathcal{L}_\infty^\mathcal{U}[0,T] = (\mathcal{L}_\infty[0,T])^m$ and $S_0 \in \mathbb{R}\langle\langle\Theta\rangle\rangle$, consider the initial value problem

$$S'(t) = \left(\theta_0 + \sum_{i=1}^m \omega_i(t)\theta_i\right) S(t), \quad S(0) = S_0 \tag{3}$$

seen as a differential equation over $\mathbb{R}\langle\langle\Theta\rangle\rangle$ (derivative taken coefficientwise). There is a unique solution S_{ω,S_0} defined on $[0,T]$, with absolutely continuous coefficients, that can be characterized as a fixed point of the corresponding integral equation; successive approximations give rise to the Peano-Baker formula, which exhibits the solution in terms of iterated integrals. In particular, $S_{\omega,1}(T)$ defines the *Chen-Fliess* series $\mathrm{CF}[\omega]$ of ω [2], [5], [26]. By uniqueness of solutions of (3), the mapping $\omega \mapsto \mathrm{CF}[\omega]$ is an (anti-) homomorphism from $\Omega = \Omega_\mathcal{U}$ into the multiplicative structure of $\mathbb{R}\langle\langle\Theta\rangle\rangle$, and as moments of ω are among the coefficients of $\mathrm{CF}[\omega]$, the map is 1-1. It can be proved that the elements in the image lie in $G\langle\langle\Theta\rangle\rangle$, so one has a natural group embedding of Ω into $G\langle\langle\Theta\rangle\rangle$. Furthermore, if the components of $\omega \in \mathcal{L}_\infty^\mathcal{U}[0,T]$ have magnitude less than 1 then $|\langle \mathrm{CF}[\omega]|\alpha\rangle| \leq T^{\underline{\alpha}}/\underline{\alpha}!$ for each $\alpha \in \Theta^*$. The pairing (2) extends to such a series $\lambda = \mathrm{CF}[\omega]$, provided that c is convergent with radius satisfying $T\rho(m+1) < 1$; the series defining $\langle c|\mathrm{CF}[\omega]\rangle$ then converges absolutely, uniformly on the restrictions $\omega|_t$ of ω to initial subintervals $[0,t]$, $t < T$.

3.3. Germs of Responses and Systems.

Let $\Omega_0 \subseteq \Omega$ contain for some $T > 0$ a neighborhood of $0 \in \mathcal{L}_\infty^\mathcal{U}[0,T]$ and be closed under restrictions to initial subintervals. A map $F : \Omega_0 \to \mathcal{Y}$ defined on some such Ω_0, with coordinates $F_i(\omega) := \langle c_i|\mathrm{CF}[\omega]\rangle$ for some vector $c = (c_1, \ldots, c_p)$ of convergent generating series c, is a *local analytic response*. (Various more global definitions of analytic response can be given; see e.g. [12].) From now on, I will identify any two F's that coincide on the intersection of their domains and "response" will thus mean a germ of local analytic response. With this convention, responses are in 1-1 correspondence with p-vectors of convergent generating series. Furthermore, to make the presentation simpler, and because most interesting cases for applications are encompassed by this subclass, *system* will mean a differential system Σ analytic and affine in controls: $\mathcal{X}$ is an analytic manifold, h and f are analytic, and $f(x,u)$ is affine in u. That is, in (1) one has $\xi' = g_0(\xi) + \sum_{j=1}^m \omega_j g_j(\xi)$, for some $m+1$ analytic vector fields g_j. Fixing an initial state $x_0 \in \mathcal{X}$, the complete I/O behavior $F^{\Sigma,x_0}(\omega) = h(x_0 \cdot \omega)$ is by analytic continuation uniquely determined by its restriction to small times and controls, the response characterized by the Fliess generating series

$$\langle c_i | \theta_{j_1} \dots \theta_{j_k} \rangle := (g_{j_k} \dots g_{j_1} h_i)(x_0),$$

where h_i is the ith coordinate of h (cf. [5], which generalized Gröbner's "Lie series" [8] for autonomous systems). A *realization* of a response will mean a local realization in this sense: specifying a manifold, initial state, h, and vector fields that represent the germ.

3.4. Infinitesimal Observables of the First Kind, Realizability.

For any given system Σ, the *infinitesimal observables of the first kind*, which summarize information contained in jets of global observables [14], [10], [6, 21], are the functions

$$g_{j_1} \dots g_{j_k} h_i, \quad (j_1, \dots, j_k) \in \{0, \dots, m\}^k, \ k \ge 0, \ i = 1, \dots, p.$$

The observation space (resp., algebra) of the first kind $\mathcal{O}^{\Sigma}$ (resp., $\mathcal{A}^{\Sigma}$) is defined as the linear span (resp., algebra under pointwise products) of all these functions. The field of fractions $\mathcal{K}^{\Sigma}$ is well-defined if the manifold is connected. Starting instead with a response F, with series $\tilde{c} = (c_1, \dots, c_p) \in (\mathbb{R}\langle\!\langle\Theta\rangle\!\rangle)^p$, there is an *infinitesimal observable of the first kind*, $\alpha^{-1} c_i$, for each $\alpha \in \Theta^*$ and $i = 1, \dots, p$. (These elements correspond to certain derivatives of F that can be defined when using piecewise constant controls ω.) Taking the smallest $\mathbb{R}$-linear subspace of $\mathbb{R}\langle\!\langle\Theta\rangle\!\rangle$, shuffle subalgebra, and quotient field, containing all elements $\alpha^{-1} c_i$, there result the observation linear space $\mathcal{O}^F$, algebra $\mathcal{A}^F$, and field of observables $\mathcal{K}^F$ of the first kind associated to F (or $\tilde{c}$). When (Σ, x_0) realizes F, Σ is accessible (see below), and $\mathcal{X}$ is connected, $\mathcal{O}^F \simeq \mathcal{O}^{\Sigma}$, $\mathcal{A}^F \simeq \mathcal{A}^{\Sigma}$, $\mathcal{K}^F \simeq \mathcal{K}^{\Sigma}$.

The system Σ is *accessible at* x_0 if the reachable set from x_0 has nonempty interior; equivalently (Chow's theorem) the accessibility rank condition (ARC) holds: $\mathcal{L}^{\Sigma}(x_0) = T_{x_0}\mathcal{X}$, where $\mathcal{L}^{\Sigma}$ is the (accessibility) Lie algebra of vector fields generated by $\{g_i, i = 0, \dots, m\}$. It is *locally observable at* x_0 if observables corresponding to small-time controls separate points near x_0; equivalently, the observability rank condition (ORC) holds: $d\mathcal{O}^{\Sigma}(x_0) = T^*_{x_0}\mathcal{X}$. The system Σ is *analytically canonical at* x_0 (from now on, just "canonical") if it is both accessible and locally observable at x_0. Canonical realizations of any response are unique up to local diffeomorphisms, and a global result also holds (cf. Sussmann [24], as well as Fliess [5], which related to Singer and Sternberg's work on local equivalence of pseudogroups induced by isomorphic Lie algebras [17]).

In complete analogy to the global algebraic case discussed earlier, algebraic finiteness conditions on infinitesimal observables associated to F reflect differential realizability properties [1], [31]. Finite generation of $\mathcal{A}^F$ relates to canonical realizations describable by polynomial differential equations, and $\mathrm{tr.deg}\,\mathcal{K} < \infty$ to rational realizability.

Of far wider applicability is an elegant general condition for realizability due to Fliess, which can also be expressed in terms of $\mathcal{A}^F$. Assume that (Σ, x_0) is any realization of F. The set $\{\ell \mapsto X\ell(x_0), X \in \mathcal{L}^{\Sigma}\}$ of linear maps $\mathcal{A}^{\Sigma} \to \mathbb{R}$ identifies with the subquotient $T^{\Sigma}_{x_0} = \mathcal{L}^{\Sigma}(x_0)/\mathcal{L}^{\Sigma}(x_0) \bigcap (d\mathcal{O}^{\Sigma}(x_0))^0$ of $T_{x_0}\mathcal{X}$. The crucial insight is that an intrinsic definition of $T^{\Sigma}_{x_0}$, independent of the particular realization, is possible. The elements of $L\langle\Theta\rangle$ act as derivations on the (shuffle) ring of observables $\mathcal{A}^F$ and hence can be thought of as formal vector fields. Vectors should be obtained by evaluations of these vector fields at a point playing the role of x_0. As $\mathcal{A}^F$ is an algebra of functions on Chen-Fliess series, a candidate for such an evaluation is $\langle\cdot|1\rangle$. (In fact, one could also think of the group extension of Chen-Fliess series as the state space for a formal, accessible, but not observable, realization.) Thus, it is natural to define T_0^F as the vector space of linear operators $d \mapsto \langle\lambda^{-1}d|1\rangle$ on $\mathcal{A}^F$. The dimension of T_0^F, which is isomorphic to $\mathrm{span}\{\widetilde{c}\lambda^{-1}, \lambda \in L\langle\Theta\rangle\} \subseteq (\mathbb{R}\langle\langle\Theta\rangle\rangle)^p$, is the *Lie rank* ϱ^F; it can be computed algebraically from the coefficients of the series $\widetilde{c}$ and generalizes the Hankel rank from the linear case. It is easy to see that $T_0^F \simeq T_{x_0}^\Sigma$ for all possible realizations Σ; moreover, $T_{x_0}^\Sigma = T_{x_0}\mathcal{X}$ if and only if Σ is canonical. Thus $\varrho^F \leq \dim\mathcal{X}$, with equality in the canonical case.

A result ensuring existence of Σ provided $\varrho^F < \infty$ was stated by Fliess, motivated by formal groups work of Guillemin, Sternberg, and Singer [17], [9] in connection with Cartan's fundamental theorems, and various alternative proofs have been given:

THEOREM [5], [15], [12]. *F is realizable if and only if $\varrho^F < \infty$; ϱ^F is then the dimension of the canonical realization, and is the minimal possible dimension of any realization.*

3.5. Infinitesimal Observables of the Second Kind, I/O Equations.

For each smooth control ω and each $k \geq 0$, let $\delta(\omega, k) := S_{\omega,1}^{(k)}(0) \in \mathbb{R}\langle\Theta\rangle$. This is the kth derivative, evaluated at $t = 0$, of the solution defining the Chen-Fliess series of ω. Given an F, with generating series $\widetilde{c}$, and any $i = 1, \ldots, p$ and k and ω, there is an *infinitesimal observable of the second kind* $\delta(\omega, k)^{-1}c_i$. These elements span a linear space, a shuffle algebra, and a field $\mathcal{O}_\star^F$, $\mathcal{A}_\star^F$, and $\mathcal{K}_\star^F$. They characterize jets of outputs, because $\langle c_i|\delta(\omega, k)\rangle$ is the ith coordinate of $(d^k/dt^k)(F(\omega|_t))|_{t=0}$. The following fundamental equalities, valid for any F, are central to further results, and can be proved by establishing that the elements $\delta(\omega, k)$ form a set of generators for the algebra $\mathbb{R}\langle\Theta\rangle$.

THEOREM [28]. $\mathcal{O}^F = \mathcal{O}_\star^F$, $\mathcal{A}^F = \mathcal{A}_\star^F$, and $\mathcal{K}^F = \mathcal{K}_\star^F$.

Often in applications, one is given a differential equation directly linking inputs and outputs. Let $\mathcal{E} : \mathcal{U}^k \times \mathbb{R}^{k+1} \to \mathbb{R}$ be analytic, nontrivial on the last variable. A response F with $\mathcal{Y} = \mathbb{R}$ *satisfies the input/output equation*

$$\mathcal{E}(\omega(t), \omega'(t), \ldots \omega^{(k-1)}(t); \eta(t), \eta'(t), \ldots, \eta^{(k)}(t)) = 0$$

of order k if this equality holds for all pairs of functions $(\omega, \eta(t) = F(\omega|_t))$ with ω smooth of sufficiently small magnitude and all small t. (For $\mathcal{Y} = \mathbb{R}^p$, $p > 1$, an equation would be imposed on each output coordinate, but for simplicity I only consider the special case $p = 1$.) The *differential rank* $\varrho_\star^F$ of F is the smallest k (possibly $+\infty$) so that F satisfies an i/o equation of order k.

For linear responses, it is well-known — an immediate consequence of the theory of linear recurrences — that realizability is equivalent to the existence of "autoregressive moving average" representations, that is, I/O equations with $\mathcal{E}$ linear (or in harmonic analysis terms, rationality of transfer functions). In general, an I/O equation establishes constraints on jets. For instance, if $\mathcal{E}$ is a polynomial function, the above equation says that $\eta^{(k)}$ is algebraic over a field generated by lower derivatives of inputs and outputs (the precise formulation is in terms of differential algebra). Appropriate finiteness conditions on $\mathcal{O}_\star^F$, $\mathcal{A}_\star^F$, and $\mathcal{K}_\star^F$ are equivalent to the existence of equations with $\mathcal{E}$ linear or polynomial on y, and to corresponding special forms of realizations [31]. For the general analytic case one has:

LEMMA [30]. *If F satisfies an I/O equation ($\varrho_\star^F < +\infty$) then it is realizable ($\varrho < +\infty$).*

Observation fields play a central role in the proof. Sketch: $\varrho_\star^F < +\infty$ implies that $\mathcal{K}_\star^F$ is a meromorphically finitely generated extension of $\mathbb{R}$, and by the previous fundamental equalities the same holds for $\mathcal{K}^F$. Using coordinates for the latter, one obtains a formal realization. However, this realization may have singularities at its initial state x_0, so it may not define a true analytic system. On the other hand, in this realization, the responses corresponding to nonsingular states near x_0 give rise to generating series with finite, in fact uniformly bounded, ϱ. Using lower semicontinuity of Lie rank gives that $\varrho^F < +\infty$.

3.6. Universal Inputs, Orders of Equations. For each fixed smooth control ω and response F, one may consider the linear span $\mathcal{O}_\omega^F$ of the elements $\delta(\omega, k)^{-1} c_i$, $i = 1, \ldots, p$, $k \geq 0$. In general this is a proper subspace of $\mathcal{O}_\star^F$ (the sum of $\mathcal{O}_\omega^F$ over all ω). However, for appropriate ω, the projections on the constant term, being either 0 or $\mathbb{R}$, may produce a pointwise equality. For any $O \subseteq \mathbb{R}\langle\langle\Theta\rangle\rangle$, let $\langle O|1\rangle = \{\langle l|1\rangle, l \in O\}$. A smooth control ω is *universal for the family of responses* $\mathcal{F}$ if $\langle\mathcal{O}_\omega^F|1\rangle = \langle\mathcal{O}_\star^F|1\rangle$ for each $F \in \mathcal{F}$. Given a system Σ, consider the family of responses $\mathcal{F}_\Sigma$ that is obtained by including, for each fixed state x_0, the response F^{Σ,x_0}, and for each x_0 and each vector $v \in T_{x_0}\mathcal{X}$ also the response $dF^{\Sigma,x_0,v}$ defined by the series with $\langle c_i|\theta_{j_1} \ldots \theta_{j_k}\rangle := (dg_{j_k} \ldots g_{j_1} h_i)(x_0)v$.

THEOREM. *For each Σ, there exist analytic controls universal for $\mathcal{F}_\Sigma$. Moreover, the set of smooth controls universal for $\mathcal{F}_\Sigma$ is generic.*

Here, a set Ω_0 of smooth controls is *generic* if the set $\{(\omega(0), \omega'(0), \ldots), \omega \in \Omega_0\}$ of jets contains a countable intersection of open dense subsets of $\prod_{i=0}^\infty \mathbb{R}^m$ (product topology). This transversality result can be traced to a sequence of papers including [7], [18], [25], [29], [4], [32] by Grasselli and Isidori, Sussmann, Wang, Coron, and the author. (A somewhat more general result ensures existence of controls universal uniformly on all $\mathcal{F}_\Sigma$. Also, an alternative theorem can be stated in terms of genericity in a Whitney topology, for controls of fixed length.)

It is an immediate consequence of the theorem that, for each observable system Σ, there exists some analytic control with the property that the output function when using this particular control uniquely determines the internal state ("universal inputs for observability"). Also, given any two initialized systems, there is an analytic control that distinguishes them. Another application is as follows. Here $p = 1$.

THEOREM [29], [32]. *For each F, $\varrho \leq \varrho_\star^F$. If there is a canonical realization of F in terms of rational vector fields, equality holds.*

This generalizes the classical linear case, where $\varrho = \varrho_\star^F$. By the Lemma, $\varrho_\star^F < +\infty$ implies $\varrho < +\infty$. The critical step in the proof is to show that if Σ is a canonical realization of dimension n, but there is an equation of order $d < n$, then for each ω there would exist x and v so that $\langle \mathcal{O}_\omega^G | 1 \rangle$ is a proper subspace of $\langle \mathcal{O}_\star^G | 1 \rangle$, for $G = dF^{\Sigma, x_0, v}$. The additional property in the rational case follows by straightforward elimination theory.

Yet another application is, by duality, to controllability problems. Assume that $\alpha_x : \omega \mapsto x \cdot \omega$ is defined on a $\mathcal{L}_\infty^\mathcal{U}[0, T]$-neighborhood of ω_0. The control ω_0 is *nonsingular for* x if α_x is a submersion at ω_0. (Fréchet derivative $\alpha_x'[\omega_0]$ is onto, or equivalently the variational equation along the ensuing trajectory is controllable as a time-varying linear system.) The control ω is *nonsingular for the system* Σ if for each state x there is a restriction of ω to some initial subinterval $[0, t]$ that is nonsingular for x. (It follows that, if $x \cdot \omega$ is defined, ω itself must be nonsingular for x.) Numerical techniques based on linearizations rely upon such controls, so it is of interest to study their existence.

The system Σ is *strongly accessible* from a state x if there is some $T > 0$ so that the reachable set from x in time exactly T has nonempty interior; equivalently, Σ satisfies the strong accessibility rank condition (SARC) at x: $\mathcal{L}_0^\Sigma(x) = T_x \mathcal{X}$, where $\mathcal{L}_0^\Sigma$ is the smallest Lie ideal of $\mathcal{L}^\Sigma$ containing $\{g_i, i = 1, \ldots, m\}$. For each x there is some ω that is nonsingular for x if and only if the SARC holds at all x. One implication is immediate from the implicit function theorem, and the converse follows by a standard argument involving Brouwer's fixed point theorem, which allows restricting to countable families of controls, hence permitting application of Sard's lemma [20]. A stronger result holds:

COROLLARY. *If the SARC holds from each state then there is an analytic control nonsingular for Σ. Moreover, the set of smooth controls nonsingular for Σ is generic.*

This result can be found, in this form, in [22]; a weaker form in which controls are multiplied by a scalar function of x was given in [3], for general smooth, not necessarily analytic, systems with $g_0 = 0$, and stronger results are now available as well [4]. The basic observation needed in the proof is that nonsingularity can be expressed as the nonvanishing of the output of an extended system Σ_e obtained from Σ by adjoining a variational equation and a matrix equation that computes the controllability Gramian. The problem is reduced to finding inputs universal for $\mathcal{F}_{\Sigma_e}$.

I end with an illustration, closely connected with the results in [3], of how this corollary can be used to numerically approach certain control problems.

4. An Application: Steering Nonholonomic Systems

It is often of interest to explicitly compute motions for mechanical systems, especially those subject to constraints such as the nonslippage of rolling wheels. Specified are a (for the present purposes, analytic) manifold $\mathcal{X}$, the *configuration space*, and a constant-rank codistribution D on $\mathcal{X}$, which describes the *kinematic constraints*. The objective is to find, for each pair of states x_0 and x_f, a curve tangential to the kernel of D, whose initial point is x_0 and final point equals, or is sufficiently near, the target $x_{\rm f}$. Assuming that $\ker D$ can be globally spanned by independent (analytic) vector fields $g_1, \ldots, g_m$, one may introduce a system as in Section 3 ($g_0 = 0$), and for this system the problem becomes one of finding a control ω so that $x_0 \cdot \omega$ is equal to or close to $x_{\rm f}$. In this case complete controllability, that is, solvability of the exact problem for all pairs $(x_0, x_{\rm f})$, is equivalent to the SARC (or the ARC, as $g_0 = 0$) holding globally. Many sophisticated synthesis procedures have been proposed, most based on a nontrivial analysis of the structure of $\mathcal{L}_0^\Sigma$, and a rich literature exists (e.g. [13] and references there). When the structure $\mathcal{L}_0^\Sigma$ is too complicated for a detailed analysis, a numerical technique as follows could in principle be used.

For simplicity of exposition, I'll assume that $\mathcal{X} = \mathbb{R}^n$ is Euclidean and $x_{\rm f} = 0$. Multiplying the vector fields g_i by a suitable scalar function, one may assume that the system is complete. Thus, controls defined on a fixed interval, say, $[0, 1/2]$, provide well-defined trajectories, and by the results previously stated, smooth ones are generically nonsingular.

For any one such control ω, one may consider the antisymmetric extension $\widetilde{\omega}$ of ω to $[0,1]$ having $\widetilde{\omega}(1-t) = -\omega(t)$ for $t \in [0,1)$. This defines a measurable control that is again nonsingular for the system, but now in addition $x \cdot \widetilde{\omega} = x$ for each state. Thus, $x \cdot (\widetilde{\omega} + v) = x + \alpha_x'[\widetilde{\omega}](v) + o(v)$. By nonsingularity, there is some v so that $\alpha_x'[\widetilde{\omega}](v) = -x$. One choice for v is the pseudo-inverse $v = N(x) = -(\alpha_x'[\widetilde{\omega}])^\#(x)$. Thus, $x \cdot (\widetilde{\omega} + hv) = (1-h)x + o(h)$ for such v and small h. With the alternative choice of the adjoint $v = N(x) = -(\alpha_x'[\widetilde{\omega}])^*(x)$, there results $x \cdot (\widetilde{\omega} + hv) = (I - hQ)x + o(h)$, where Q is positive definite and self-adjoint. In either case a contraction results for small h. Moreover, the following result holds for both of these choices of operator N (which correspond respectively to Newton and steepest descent algorithms, and can be explicitly computed in terms of variational equations), as well as for a larger class defined in abstract terms; it concerns the convergence of the iteration $F_h(x) := x \cdot (\widetilde{\omega} + hN(x))$.

THEOREM [22]. *Let $B_1 \subseteq B_2$ be any two balls in $\mathbb{R}^n$ centered at 0. Then, for generic ω, and for each $h > 0$ small enough, there is some integer N so that $F_h^N(B_2) \subseteq B_1$.*

References

[1] Z. Bartosiewicz, *Minimal polynomial realizations*, Math. Control Signals Systems **1** (1988), 227–231.

[2] K. T. Chen, *Iterated path integrals*, Bull. Amer. Math. Soc. **83** (1977), 831–879.

[3] J.-M. Coron, *Global asymptotic stabilization for controllable systems without drift*, Math. Control Signals Systems **5** (1992), 295–312.

[4] ———, *Linearized control systems and applications to smooth stabilization*, SIAM J. Control Optim. **32** (1994), 358–386.

[5] M. Fliess, *Réalisation locale des systèmes non linéaires, algèbres de Lie filtrées transitives et séries génératrices non commutatives*, Invent. Math. **71** (1983), 521–537.

[6] M. Fliess and I. Kupka, *A finiteness criterion for nonlinear input-output differential systems*, SIAM J. Control Optim. **21** (1983), 721–728.

[7] O.M. Grasselli and A. Isidori, *An existence theorem for observers of bilinear systems*, IEEE Trans. Automat. Control **26** (1981), 1299–1301.

[8] W. Gröbner, Die Lie-Reihen und ihre Anwendungen, VEB Deutscher Verlag der Wissenschaften, Berlin, 1967.

[9] V. W. Guillemin and S. Sternberg, An algebraic model of transitive differential geometry, Bull. Amer. Math. Soc. **70** (1964), 16–47.

[10] R. Hermann and A. J. Krener, Nonlinear controllability and observability, IEEE Trans. Automat. Control **22** (1977), 728–740.

[11] G. Hochschild and G. D. Mostow, Representations and representative functions of Lie groups, Ann. of Math. (2) **66** (1957), 495–542.

[12] B. Jakubczyk, Realization theory for nonlinear systems; three approaches, in Algebraic and Geometric Methods in Nonlinear Control Theory (M. Fliess and M. Hazewinkel, eds.), Reidel, Dordrecht, 1986, pp. 3–31.

[13] Z.X. Li and J.F. Canny, eds., Nonholonomic Motion Planning, KAP, Boston, 1993.

[14] J. T. Lo, Global bilinearization of systems with controls appearing linearly, SIAM J. Control **13** (1975), 879–885.

[15] C. Reutenauer, The local realization of generating series of finite Lie rank, in Algebraic and Geometric Methods in Nonlinear Control Theory (M. Fliess and M. Hazewinkel, eds.), Reidel, Dordrecht, 1986, pp. 33–44.

[16] ———, Free Lie algebras, Oxford University Press, New York, 1993.

[17] I. M. Singer and S. Sternberg, The infinite groups of Lie and Cartan, Israel J. Anal. Math **15** (1965), 1–114.

[18] E. D. Sontag, On the observability of polynomial systems, SIAM J. Control Optim. **17** (1979), 139–151.

[19] ———, Polynomial Response Maps, Springer-Verlag, Berlin, 1979.

[20] ———, Finite dimensional open-loop control generators for nonlinear systems, Internat. J. Control **47** (1988), 537–556.

[21] ———, Mathematical Control Theory, Deterministic Finite Dimensional Systems, Springer, Berlin and New York, 1990.

[22] ———, Control of systems without drift via generic loops, IEEE Trans. Automat. Control **40** (1995), 1210–1219.

[23] E. D. Sontag and Y. Rouchaleau, On discrete-time polynomial systems, J. Nonlinear Anal. **1** (1976), 55–64.

[24] H. J. Sussmann, Existence and uniqueness of minimal realizations of nonlinear systems, Math. Systems Theory **10** (1977), 263–284.

[25] ———, Single-input observability of continuous-time systems, Math. Systems Theory **12** (1979), 371–393.

[26] ———, Lie brackets and local controllability, A sufficient condition for scalar-input systems, SIAM J. Control Optim. **21** (1983), 686–713.

[27] A. Tannenbaum, Invariance and System Theory, Algebraic and Geometric Aspects, Springer, Berlin and New York, 1980.

[28] Y. Wang and E. D. Sontag, On two definitions of observation spaces, Systems Control Lett. **13** (1989), 279–289.

[29] ———, I/O equations for nonlinear systems and observation spaces, Proc. IEEE Conf. Decision and Control, Brighton, UK, Dec. 1991, IEEE Publications, 1991, pp. 720–725.

[30] ———, Generating series and nonlinear systems, analytic aspects, local realizability, and i/o representations, Forum Mathematicum **4** (1992), 299–322.

[31] ———, Algebraic differential equations and rational control systems, SIAM J. Control Optim. **30** (1992), 1126–1149.

[32] ———, Orders of input/output differential equations and state space dimensions, SIAM J. Control Optim.**33** (1995), 1102–1127.

Changes in the Teaching of Undergraduate Mathematics: The Role of Technology

DEBORAH HUGHES-HALLETT

Harvard University
Cambridge, MA 02138, USA

1. What is the Impetus for Change?

The past decade has seen significant changes in the teaching of mathematics in the United States. There are new curriculum guidelines for primary and secondary school mathematics written by the National Council of Teachers of Mathematics. There is a flurry of activity around the teaching of calculus, both at the university and at the secondary school level. There has been a tremendous upsurge in the number of presentations on educational issues at professional meetings.

These changes have come about for reasons that are frequently as relevant outside the United States as inside. One is the concern over access. Do enough students take mathematics? Do enough students pass mathematics? Do the right students pass? Do we orient the courses to future mathematicians at the expense of future engineers? Do we educate men at the expense of women? Students from one race or culture at the expense of others? To mathematicians accustomed to teaching the best course possible and then seeing who rises to the top, these questions might seem unusual or out of place. But in a world in which mathematics can be a ticket to economic success, mathematics education can have a political component.

The second impetus for change in the United States is dissatisfaction with students' performance. In many countries, the sentiment that "students aren't what they used to be" has led to a cascade of complaints blaming students' preparation on the level below. University professors blame high school teachers; high school teachers blame elementary school teachers, who in turn blame the home environment.

The past ten years have seen a gradual acknowledgment of the fact that progress will involve working with students as they are, rather than as we might wish they were. This has led to new approaches to involving students. An outstanding example is *Calculus&Mathematica* [1] at the University of Illinois, which is written in a medium in which students are very much at home (computer generated visual images) and uses language that is obviously theirs and not their professors'. The fact that formal definitions are postponed until students have developed an intuitive understanding of concepts from graphs has enabled a wider range of students to understand calculus.

Proceedings of the International Congress
of Mathematicians, Zürich, Switzerland 1994
© Birkhäuser Verlag, Basel, Switzerland 1995

A third force for change, and the one responsible for getting many mathematicians involved, is the advance in technology. The fact that calculators and computers can now easily compute definite integrals, sketch graphs, solve equations, and find high powers of of matrices is having an impact on what we teach. Within a year, inexpensive calculators will be able to do all the algebraic manipulations that have been the backbone of high school mathematics for decades. What does that mean for our courses? Should we stop teaching students to do these procedures? Should we not allow students to use technology in the classroom? How much should technology drive the curriculum?

These are extraordinarily difficult questions. It is my hope that the research community will heed the recent call to "work, not fight" and will cooperate actively with the education community to find answers that are mathematically sound as well as pedagogically robust. We must have answers that accurately reflect mathematics *and* that make sense to real students and real teachers.

2.   The Central Problem

Before embarking on any change, one should first establish goals. Discussions in the Consortium based at Harvard have convinced us that our primary concern should be to affect students' views of mathematics.

Consider, for example, the calculus student who objected to being asked to suggest formulas for functions whose graphs were given. His reaction was that he did his graphs "in the other order." He meant that if he were given the formulas, he could draw the graphs, but not the other way round.

Another student objected to the following problem on the grounds that it was "too vague:"

A particle starts at the origin and moves along the graph of $y = x^2/2$ at a speed of 10 units/second.

(a) Write down the integral that shows how far the particle has travelled when it reaches the point where $x = a$.

(b) Estimate the x-coordinate of the point the particle has reached after it has been travelling 2 seconds.

Further investigation made it clear that the student thought the problem was vague because it didn't contain the words "arc length" and therefore he didn't know what to do.

Such examples are unfortunately not unique. All over the world students come to mathematics asking, as one of mine did recently, "When do we get to the part where we just have to do it?" For these students, mathematics is a set of procedures to be followed, rather than ideas to be understood. Such procedures are indeed important, but learning algorithms alone is not learning mathematics. Unfortunately, many students currently do well in their mathematics courses without learning to think at all.

Consequently, the central problem facing teachers of mathematics is, as it always has been, getting students to think mathematically: to understand and use concepts, and to be ready to do problems for which no template has been given.

Technology may be useful in addressing this problem, but technology is not the main issue.

3.   Multiple Representations: The Role of Technology

To focus a course on the central ideas of mathematics, observe that most concepts are better communicated in pictures or words, or sometimes numbers, rather than in symbols. For example, the idea of a derivative can be thought of as a rate of change (words), or the slope of a curve at a point (pictures), or as the limit of the difference quotients (numbers). Thus, focusing on concepts requires verbal, graphical, and numerical presentations in addition to a symbolic one.

It is this renewed emphasis on graphical and numerical representations that makes technology important. Using technology, graphs and numerical work, previously hard for students, become easy to include in any course. Thus, although the technology does not drive change, it does make the change possible.

As an example, consider two problems that use graphing technology to focus on concepts. These problems are from the materials written by the Calculus Consortium based at Harvard [2].

> *Example 1:* Determine the windows (domains and ranges) that make the graphs of $y = 3^x$ and $y = x^4$ look like those in Figure 1.

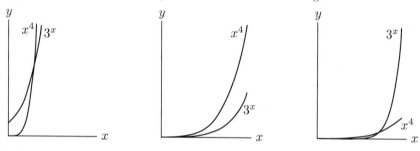

Figure 1

You might think that such a problem is easy to do by trial and error and that therefore the only skill being tested is calculator button pushing. Watching students work on this problem, however, shows that it requires them to consider the qualitative behavior of power and exponential functions: which function is greater where, and so on. Thus it is, in fact, a problem about the effects of scale on power and exponential functions.

> *Example 2:* The number of hours, H, of daylight in Madrid is approximated by a sine function of t, the number of days since the start of the year. Figure 2 shows a one-month portion of the graph of H.
>
> (a) Comment on the shape of the graph. Why does it look like a straight line?
>
> (b) What month does the graph show? How do you know?
>
> (c) What is the approximate slope of the line? What does this slope mean in practical terms?

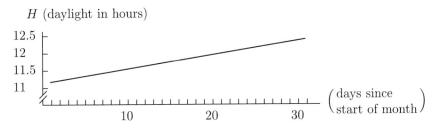

Figure 2

With a calculator that can "zoom in," the idea of local linearity that lies behind this problem becomes vivid to students in a way hand-drawn pictures and words can never achieve.

Two aspects of this problem that strike students as unusual are that they need information from outside mathematics to solve it (that Madrid is in the northern hemisphere, for example) and that they are expected to explain what the slope means in words. It is this insistence on explaining graphical ideas in words that helps students learn to think mathematically.

4.   The Relationship Between Intuition and Rigor

There is a remarkable difference between the way in which professional mathematicians work and the way in which they teach. In research, experimentation and examples come first, followed by conjecture, more examples, and finally proof. In teaching, theorem and proof often come first and examples afterward. We should consider the fact that students may learn mathematics better if they are taught in a way modelled on the way in which mathematics is actually done.

A question of great importance for mathematics is how to teach students to think rigorously. Past experience shows that teaching computational algorithms is not enough. We have many students who learn how to do computations without knowing what they mean. (For example, the student described previously knew how to compute arc length, but did not recognize it as giving the distance travelled by a particle moving along the curve.) Precise arguments are impossible without a clear picture of the meaning of the symbols and computations involved. Thus, a necessary prerequisite for writing a sound mathematical argument is a clear intuitive understanding of concepts. We need to change many of our undergraduate courses to foster such understanding.

For many students, however, it is not practical to go from the introduction of an idea to a precise formulation in one course. Some time is needed for the intuitive ideas to gel before the formal definitions can be used with fluency. Unfortunately, in the past we have often skipped over the development of geometric and intuitive understanding and attempted to get students to argue formally from the start. This is a shortcut that seldom works.

Students will progress further mathematically if they first obtain a clear visual understanding of the concepts. The next stage is to learn to explain the concepts in everyday language. The third stage, becoming adept at using formal definitions and precise language, depends upon the ideas being already well understood graphically and verbally. To enable students to be successful at the third stage, it is crucial that enough time be spent at the first two stages.

References

[1] Bill Davis, Horacio Porta, and Jerry Uhl, Calculus & Mathematica, Addison-Wesley, Reading, MA, 1994.

[2] Deborah Hughes-Hallett, Andrew Gleason, et al., Calculus, John Wiley & Sons, New York, 1994.

Issues for the Popularization of Mathematics

JOEL SCHNEIDER*

Children's Television Workshop, One Lincoln Plaza
New York, NY 10023, USA

The popularization of mathematics is a new topic for these congresses. Although it has always been one of the main congress functions, popularizarion here is set at a high level and for a very special audience, one for which mathematics is already a congenial friend. But mathematics is not congenial to most people, even though it may be important to them. This leads us to think of popularizing mathematics to general audiences, young and old, learned or not, indeed any group we can reach. The very etymology of the word "popular" impels us to adopt the widest possible scope in our efforts.

My aim here is to popularize the popularization of mathematics. I will identify, discuss briefly, and illustrate some issues concerning its practice. Most of my illustrations will come from *Square One TV*, a television series about mathematics for children.

Background

The popularization of mathematics has a longer history than one might realize. In Reid's biography we read that David Hilbert gave popular lectures in 1921 for students returning to the university after the war and continued the series through the 1920s. [8, p. 154]. Lucas' "Towers of Hanoi" game is an example with a much broader impact [cf. 4]. One of my favorite examples is *The Ladies' Diary*, published from 1704–1841 by the Company of Stationers (London). It advertised itself as "Containing new improvements in arts and sciences and many entertaining particulars...." The "particulars" included mathematics problems and letters about them [cf. 2, pp. 155 and 455].

Even though popularization has a long tradition, practitioners have only begun to think about it systematically in the last few years. The stimulus for this was an invitation by the International Commission on Mathematics Instruction (ICMI) to participate in a study seminar at the University of Leeds (UK) in 1989. In announcing the seminar, Howson, Kahane, and Pollak [5] described the need, a framework, some principles, and methods of popularization.

Some ask, "Why do it at all?" For a long time, the mathematical community has maintained itself as a sort of priesthood. We've been rather passive with regard to recruits, erected a difficult series of trials for anyone who would join

*)   joels@columbia.edu

Proceedings of the International Congress
of Mathematicians, Zürich, Switzerland 1994
© Birkhäuser Verlag, Basel, Switzerland 1995

us, and even militated against participation by members of some groups. However, the world has changed so that more people need comfort and facility with mathematics to get on with their lives. Local, regional, and global political and economic realities generate increasingly complex demands, many involving some mathematics. We might expect mathematicians, if only through self-interest, to take some responsibility for the ways in which the general public sees and uses mathematics. A society comfortable with mathematics is more likely to tolerate and support those who want to work with it.

By its duration and intensity, school experience dominates all other influences on people's attitudes toward mathematics. As most people have a bad experience with mathematics in school, we need to improve the school situation in order to hope for significant change in the popular view of mathematics. All mathematicians should take note of the vigorous reform movement for mathematics education that is at work in many places. Meanwhile, many people are beyond the reach of school. For them and for the support of today's students, we must create opportunities to learn about and to practice mathematics outside school. Programs to popularize mathematics among a broad audience serve this purpose. Before proposing several issues for our consideration, I will describe some exemplary programs.

Principles and Examples

In reviewing and comparing existing programs of popularization, several principles emerge that seem to guide them. We go to people where they are — watching television, reading a newspaper or shopping for clothing — rather than expecting them to come to us. A program must be attractive to draw participants, as participation is voluntary. The primary attraction may not be mathematics, but rather something else such as music, humor, or physical activity. Without willing participants, without an audience, there is no possibility of success, no matter how worthwhile the mathematics. What we hope is that the satisfaction in the experience will include pleasure in the mathematics and encourage a favorable attitude and a readiness to consider more. The effects of any one experience are often slight and diffuse, but popular activities are repeatable. One can revisit a museum, watch a film again, follow a television series. The effects accumulate and interact. A discussion of these principles appears in [5].

All of this is very much in the spirit of the Leeds conference. The conference proceedings [6] are worth reviewing because they describe a variety of projects in several formats and venues: lectures, competitions, games, exhibitions, magazines and newspapers, and broadcast media — radio and television. Another source of examples is the ICMI report in the proceedings of the Seventh International Congress on Mathematics Education [7]. Following are a few examples from the Leeds' papers by way of illustration.

Shannon describes his experience in talking to a Rotary Club unit in Sydney (Australia). He alerts us to an important audience, available to each of us, namely the local business and professional groups in our home communities. It's worth noting that they often represent the local political and economic power. This is a project that can be taken up by any mathematician.

Competitions can be great fun and rewarding for those who enjoy them. Burjan and Vrba describe an extensive national system of competitions. We usually think of the International Mathematics Olympiads and the national competitions that produce the teams, but for our goal of affecting a broad audience, general competitions are more important. The programs of Gilles Cohen (La Fédération Française des Jeux Mathématiques et Logiques) [cf. 7] in France and of George Lenchner (Mathematics Olympiads for Elementary Schools) in the USA both show that there is a large general audience for competitions even at the elementary level.

Games give many people their earliest experiences with probability, strategy, and patterns. They are an especially effective format for popularization in that they readily involve parents and children. DeGuzman discusses games in terms of popularization.

Exhibitions are increasingly successful means for attracting attention and interest. Brown and Porter describe the problems that arise in constructing an effective exhibition and their experience in creating "Mathematics and Knots". This interesting exhibit was also included in the PopMaths Roadshow at the Leeds conference [cf. 6, Foreword].

Mathematics and newspapers and other varieties of print are natural vehicles for popularization. Many popular science magazines feature problems or a column on a mathematical topic. Barbeau, Emmer, and Larsen all discuss writing about mathematics. Of course, each of them is a mathematician writing about mathematics. Steen addresses the difficulties in promoting articles on mathematics in the newspapers, where writers and editors will usually not be friendly to mathematics. It is worth noting that the Zürich newspapers ran at least five articles [3] on mathematics during the congress. Of course, one would rather not have to go to the trouble of convening an international conference to read about mathematics in the newspapers.

The broadcast media, both radio and television, are powerful tools for delivering information and for shaping public opinion. They are prominent elements of popular culture. Power and prominence make them attractive to us, too. Four of the Leeds papers deal with these media. Barbeau writes about discussing mathematics on a radio interview program; Emmer about making films relating art and mathematics; and Hoyles about *Fun and Games*, a televised mathematical game show. *Fun and Games* is an important project because of its success as a program for an adult audience broadcast in a prime viewing time. Esty and Schneider describe *Square One TV*, a television series for children, broadcast in the USA.

Square One TV is a daily series broadcast in the USA from 1987–1994 late in the afternoon. The primary audience is 8-to-12-year-old children viewing at home, not in school. Each of the 230 half-hour shows comprises 6–12 independent segments drawn from a pool of 1100. The segments are humorous parodies of television broadcasting conventions: dramas, musicals, game shows, commercials, and so on. In keeping with the principles of popularization mentioned above, we tried to produce a series that would compete for viewers among the great variety of entertaining alternatives available at the same time on commercial television. The primary audience varies greatly in age, taste, and social as well as mathematical sophistication. In response, we also varied the shows in style and format and in level and type of mathematics. Our Leeds paper describes the project in detail.

Issues

In the course of producing *Square One TV*, we learned a lot about dealing with mathematics in a popular medium, as have all of the other practitioners. In reviewing a large number of programs, one notices, in addition to guiding principles, at least nine issues that should concern us:

1.   What is the relation to school mathematics?
2.   What is the influence of the setting or venue?
3.   What is the nature of cooperation with partners?
4.   What is the interplay with the wider culture?
5.   What is the relation to the problem of women's participation in mathematics at all levels?
6.   What is the relation to the problem of cultural minorities' participation in mathematics at all levels?
7.   What is the relationship with popularization of science?
8.   How do we define and assess the impact of projects?
9.   How do we promote a healthy flow of information and encourage collaboration among practitioners?

I propose that we begin to look at these issues to improve our practice and to increase our effectiveness. I will draw on my experience with Square One to illustrate some of them.

School. Surely any effort to popularize mathematics should support school reform. Successful programs might and, often, do migrate into the schools in some form. Popular lectures may be repeated to new audiences, especially by using film or video recording. Schools are a natural place to prepare for competitions, which in turn have the potential to influence curricula. Many teachers use games in their teaching. School trips to visit exhibitions are common. Magazines and newspapers are a standard feature of many classrooms. We can also have a migration from the medium of open-circuit broadcast television, such as in the case of *Square One TV*.

Even though we were producing *Square One TV* for an audience at home, we were alert to the possibility that the shows might be useful in school. In fact, some teachers used the shows at the very beginning. With this encouragement, we are producing a version of the series specifically designed for classroom use. The derivative series, *Square One TV Math Talk*, comprises twenty new 15 minute shows. Each of the new shows will deal with a single topic (e.g., bilateral symmetry). We will provide a book that describes ways for teachers to use the shows in classrooms. An instructional television network will broadcast the shows for teachers to videotape, and we will also offer them on video cassette for purchase.

Each *Math Talk* show features two animated cartoon characters as hosts, Maria Lopez and her partner, Buster, who is a parrot. They respond to telephone calls from people asking questions about mathematics and they illustrate their responses with video from the *Square One TV* library. For example, in the show dealing with bilateral symmetry, Maria and Buster illustrate the concept with a segment of "Mathman" (a parody of the once-popular video game "Pacman"). In this segment, the character must identify geometric shapes with a line of symmetry. Maria and Buster continue with a segment of "General Mathpital" (a parody of "General Hospital", a popular soap opera). In it, surgeons must operate on an asymmetric shape so as to reassemble the pieces into one that is bilaterally symmetric. In doing so, the surgeons discuss the concept and explore several solutions to the problem. The show concludes with a portion of one of the nine *Square One TV* game shows. Of course, it features a task for the contestants that involves symmetry.

Context. In producing *Square One TV*, we placed our material deeply in the context of our medium. Soap operas and game shows are perennial favorites among radio and television audiences. Video games are more recent innovations, but just as popular. The conventions and constraints of the medium governed many of our decisions and certainly impinged on the mathematics. For example, in producing *Square One TV*, we had a persistent conflict between our wanting to give a viewer time to think about a problem and the producers' sense that the best show is a fast-paced show.

Partners. In using a popular medium, we have not been able to act alone or to act solely from the standpoint of mathematics. Television production involves a large number of people — producers, directors, writers, carpenters, and many more. The goal for each of these talented people was to produce an attractive piece of television that would be a good addition to the *Square One TV* library — lively and repeatable. They were not also expected to attend to the additional goal that it convey some worthwhile information about mathematics. That is, mathematics was not the concern for most of the group. In fact, they typified our audience in terms of attitudes toward mathematics. This was a useful check on those of us who were responsible for the mathematics.

Because the primary goal was to attract an audience, our partners controlled the overall content, style, and tone of the product as well as the packaging. This is a valid model, whether the medium is a magazine or a newspaper, radio or television. Of course, we expect that all sides engage in friendly and respectful negotiation. Related questions for us include how to find and encourage partners, how to turn their attention in our direction, and how to create opportunities for them and for us. It is important to realize that they are entrepreneurs. They have a business to run, even if they run it in the public interest. They naturally want to continue their work and to do so they must not only generate and satisfy an audience, but also justify their decisions to the sources of their money. A decision to carry out a project on mathematics, instead of some other worthwhile subject, may have more to do with the availability of money than with their interest in one subject or another.

Culture. Although the culture in which we work is always important, music videos raise the issue most directly. The music video is a very popular television format.

We produced more than fifty of them and often had the cooperation of well-known popular singers. Associating celebrities with a product is a long-standing commercial practice and is useful even if the product is mathematics. We produced music in several genres: blues, heavy metal, rock and roll, country and western, rap, and others. No one style appeals to everyone. We broadcast in the USA, which has a complex web of subcultures. What does a black, urban rap lyric mean to viewers in rural Iowa or suburban Phoenix? *Square One TV* was also licensed for broadcast in more than twenty other countries. What does *Square One TV* mean to a viewer in Bermuda, Indonesia, or Zimbabwe? It is obvious that culture plays a role in any of the means of popularization that we have discussed. This is a complex issue and worthy of careful thought. Bishop's book [1] is a good starting point in addition to the growing literature of the field of ethnomathematics [cf. 7].

A new project based on *Square One TV* speaks to this issue. *Risky Numbers* is the working title of a half-hour game show based on the mathematical game shows that are a part of *Square One TV*. We are negotiating with producers in several countries for them to produce their versions of this show. Although the new format stems from *Square One TV*, these new productions will be rooted in their local cultures. The prospect is that we will have several variants of *Risky Numbers* for comparison and contrast within a few years. *Szalone Liczby*, the Polish version of *Risky Numbers*, premiered in January 1995. An Indonesian version will appear in January 1996.

Other Issues. The related issues of women and mathematics and of cultural minorities and mathematics are very much part of the politics of the mathematics community. Should programs to popularize mathematics promote the interests of special groups? Can they do so? In the case of *Square One TV*, we deliberately cast our actors with these issues in mind. Popularization of science is better established than popularization of mathematics in popular culture. Science centers and science museums exist in many places, but mathematics centers and museums are rare. How can popularization of mathematics cooperate with popularization of science? How do we avoid losing the mathematics in the science? In the case of *Square One TV*, our nonrealistic, comedy-variety format allowed us to effectively highlight the mathematics and we made no particular attempt to expose aspects of science. We need to develop techniques to assess programs for their impact. We need to understand what we mean by impact or value. The *Square One TV* project had a large research and evaluation component, which generated a mass of reports [cf. our Leeds paper in 6]. The ninth issue transcends the others. We need to develop convenient means for effective communication among practitioners. Given the increasing availability of electronic tools, we do not need to wait for international conferences and publication of their proceedings to learn from others' experiences. I call on the community to invest some of their energies in this direction.

Conclusion

The popularization of mathematics has a long tradition, but relatively recent impetus for systematic scrutiny. There are several issues to consider as we develop and improve our practice. Although it is often incorrectly identified with "informal mathematics education", popularization also has a valid function for professional mathematicians and for formal mathematics education. Even so, we need to concentrate on developing effective programs to help a broad, general population to develop a fruitful appreciation and facility for mathematics.

Note on Video

The audience for this lecture from which this paper stems viewed three selections from *Square One TV*. First was a portion of the *Square One TV Math Talk* show on symmetry, described above. Second was "Rule of Thumb", a music video featuring Kid 'N Play. It deals with measurement by estimation. Third was a mock commercial for geometry with the tag line: "Geometry — another division of mathematics. It's more than just arithmetic."

References

[1]   A. Bishop, Mathematical Enculturation, Kluwer, Dordrecht, 1991.

[2]   L. E. Dickson, History of the Theory of Numbers, Chelsea Publishing Co., New York, 1966.

[3]   Five newspaper articles covering the congress:

   – *Die Mathematiker kommen*, p. 45 in Neue Zürcher Zeitung, 28. Juli 1994.

   – *Mathematiker geehrt*, p. 1 in Tagesanzeiger, 4. August 1994.

   – T. Müller, *Die Mathematiker sind in der Stadt*, p. 17 in Tagesanzeiger, 4. August 1994.

   – B. Eckmann, *Mathematik: Fragen und Antworten*, p. 49 in Neue Zürcher Zeitung, 3. August 1994.

   – *Bundesrätin Dreifuss mahnt zur Verantwortung*, p. 41 in Neue Zürcher Zeitung, 3. August 1994.

[4]   A. M. Hinz, *The Tower of Hanoi*, L'Enseignement Mathématique **35** (1989), 289–321.

[5]   A. G. Howson, J.-P. Kahane, and H. Pollak, *The popularization of mathematics*, L'Enseig. Math. **34** (1988). [This is the discussion paper for the Leeds Conference.]

[6]   A. G. Howson and J.-P. Kahane (eds.), The Popularization of Mathematics, ICMI Study Series, Cambridge University Press, Cambridge (UK), 1990.

   – G. Howson and J.-P. Kahane, *A study overview*;

   – *Mathematics in different cultures* (report of the working group);

   – E. J. Barbeau, *Mathematics for the public*;

   – V. Burjan and A. Vrba, *The role of mathematical competitions in the popularization of mathematics in Czechoslovakia*;

   – M. de Guzman, *Games and mathematics*;

   – M. Emmer, *Mathematics and the media*;

   – E. Esty and J. Schneider, *Square One TV: A venture in the popularization of mathematics*;

   – G. Hatch and C. Shiu, *Frogs and candles — Tales from a mathematics workshop*;

   – C. Hoyles, *Mathematics in prime-time television: The story of fun and games*;

   – G. Knight, *Cultural alienation and mathematics*;

   – M. Larsen, *Solving the problem of popularizing mathematics through problems*;

   – B. Mortimer and J. Poland, *Popularizing mathematics at the undergraduate level*;

- T. Nemetz, *The Popularization of mathematics in Hungary*;
- T. Shannon, *Sowing mathematical seeds in the local community*;
- L. Steen, *Mathematical news that's fit to print*;
- C. Zeeman, *Christmas lectures and mathematics masterclasses*;
- D.Z. Zhang, H.K. Kiu and S. Yu, *Some aspects of the popularization of mathematics in China.*

[7]   H. Pollak, *ICMI Study 2, The popularization of mathematics*, in Proceedings of the 7th International Congress on Mathematical Education (C. Gaulin et al., eds.), Les Presses de L'Université Laval, Sainte-Foy, 1994.

[8]   C. Reid, Hilbert, Springer-Verlag, New York, 1970.

Some other items of interest:

[9]   B. Cipra, What's Happening in the Mathematical Sciences, v. 1–2 (P. Zorn, ed.), AMS, Providence (RI), 1993, 1994. (Recent advances in mathematics for a sophisticated audience.)

[10] V. Crane et al., Informal Science Learning, Research Communications Ltd., Dedham (MA), 1994. (Papers on a research basis for informal science and mathematics education.)

[11] A. Joseph, F. Mignot, F. Murat, B. Prum, and R. Rentschler (eds.), First European Congress of Mathematics, Round Tables, Birkhäuser, Basel, Boston, and Berlin, 1994. (Includes papers on Mathematics and Society with a section on Mathematics and the General Public.)

[12] J.-P. Kahane, *The Popularization of Mathematics*, in Proceedings of the UCSMP International Conference on Mathematics Education, Developments in School Mathematics Education Around the World, v. 3 (I. Wirzup and R. Streip, eds.), National Council of Teachers of Mathematics, Reston (VA), 1992.

[13] The Popularization of Mathematics, ICMI Secretariat, Southampton, 1989. (The working papers for the Leeds conference.)

Number Theory as a Core Mathematical Discipline

JOHN STILLWELL

Monash University
Clayton, Victoria 3168, Australia

1. Introduction

In recent years there has been much discussion of the role of calculus in mathematics education. Calculus is the de facto qualification for entry to higher mathematics at most institutions, but is it still the best? Compare its role with that of Euclidian geometry, which was the entry qualification until last century. By the end of the nineteenth century it was clear that Euclid was no longer a sufficiently broad basis for geometry, let alone the rest of mathematics. Analysis took over because it was then the supreme mathematical discipline, and calculus was its entry point.

Since that time, however, analysis has burst its nineteenth-century bounds, by taking on large amounts of algebra and topology. Calculus gives no inkling of these developments. Moreover the traditional drill in differentiation and integration is becoming an embarrassment now that computer packages can do most of the questions on a typical calculus exam. It could even be argued that calculus is today inferior to Euclid as a qualification for higher mathematics (particularly when Euclid's theory of integers and real numbers is included).

What is to be done? My suggestion is that mathematics, *from kindergarten onwards*, should be built around a core that is

- Interesting at all levels
- Capable of unlimited development
- Strongly connected to all parts of mathematics.

My paper attempts to show that number theory meets these requirements, and that it is natural to build modern mathematics around such a core.

Before doing so I would like to relate an experience that shows the need for a core reaching down to elementary levels. First year calculus students at Monash this year found themselves unable to do a question from a previous year's exam: integrate $\frac{1}{2-\sin\theta}$. Why? This year, for the first time, we stopped teaching the substitution $\sin\theta = \frac{2t}{1+t^2}$. The reason given was that students can now use a computer package to do such integrals, so they no longer need to know tricky substitutions. Fair enough, but the tragedy is that now they may *never know* that $\sin\theta = \frac{2t}{1+t^2}$, because the formula has already been dropped from school trigonometry. If they don't know that, what can they know? In my opinion, they cannot know basic algebra and geometry, so an understanding of calculus is out of the question.

Proceedings of the International Congress
of Mathematicians, Zürich, Switzerland 1994
© Birkhäuser Verlag, Basel, Switzerland 1995

However, it is not good enough to restore traditional algebra and geometry to high schools, even if that were possible. As already indicated, we need more than a foundation for traditional calculus. We need to rebuild from a more fundamental level, to support all the higher disciplines in a unified and efficient manner. What follows is a sketch of such a development. It is taken far enough, I hope, to show that number theory is a natural starting point for all the disciplines, even calculus.

2.   Why number theory?

Most people would agree that mathematics begins with numbers. We all began with counting and arithmetic, and in fact this is the *only* mathematics remembered by most adults. By the end of primary school, most children have learned some important concepts of number theory: division with remainder, gcd, and lcm. They may also have heard of prime numbers. It would be easy, in years 7, 8, and 9, to build on this knowledge by introducing the Euclidean algorithm for gcd and using it to prove unique prime factorization in **Z**. By the end of secondary school, students could be fluent in basic number theory and its applications, say as far as Fermat's little theorem and public key cryptography.

Unfortunately, around year 7, this train of thought is broken with the arrival of algebra and other branches of mathematics seen as more "grown up" than arithmetic. If the student ever takes up number theory again it will be after a hiatus of 5 or 6 years, and even simple things like division with remainder will have to be relearned. What a waste! The continuation of number theory through secondary school need not be in competition with algebra, geometry, trigonometry, etc., but could support and unify them.

What follows are some suggestions for making fruitful links between these disciplines (and also with calculus), using number theory as a source. They are by no means confined to school mathematics. Indeed, until the dream of school number theory is achieved, they will probably be more relevant to university mathematics.

Experience at Monash has shown that first year students can pick up basic number theory, from the Euclidean algorithm to Fermat's little theorem, in a minicourse of about eight lectures. The main steps are as follows:

- For any nonzero $a, b \in \mathbf{Z}$, there are $q, r \in \mathbf{Z}$ ("quotient" and "remainder") such that
$$a = qb + r \quad \text{with} \quad 0 \leq |r| < |b|$$

- The Euclidean algorithm on a, b (repeatedly dividing the larger number by the smaller and keeping the remainder) gives $\gcd(a, b)$.

- It follows from the Euclidean Algorithm that there are $m, n \in \mathbf{Z}$ such that $\gcd(a, b) = ma + nb$

- As $\gcd(p, a) = 1 = mp + na$ for a prime p with $p \nmid a$, it follows that
$$p | ab \Rightarrow p | a \quad \text{or} \quad p | b$$

- Unique prime factorization

- a has a multiplicative inverse mod $k \Leftrightarrow \gcd(a, k) = 1$
- Fermat's little theorem: for p prime, $\gcd(a, p) = 1$,

$$a^{p-1} \equiv 1 \pmod{p}.$$

3. Algebra

With the support of the above number theory, high school algebra could be much more than solution of equations. All of the following are easy consequences of division with remainder.

- Divisibility theory of polynomials (patterned on $\mathbf{Z}$, with degree in place of absolute value and irreducibles in place of primes)
- The finite fields $\mathbf{Z}/p\mathbf{Z}$ and the analogous algebraic number fields $\mathbf{Q}[x]/p(x)\,\mathbf{Q}[x]$
- nth degree polynomial has $\leq n$ roots over a field
- Some properties of abelian groups, e.g.
  $\gcd(\operatorname{order}(a), \operatorname{order}(b)) = 1 \Rightarrow \operatorname{order}(a + b) = \operatorname{lcm}(\operatorname{order}(a), \operatorname{order}(b))$.

Incidentally, this is not surprising, because most basic commutative algebra is derived from Gauss's *Disquisitiones Arithmeticae* via Dirichlet and Dedekind. Algebra was intended to serve number theory by simplifying the proofs of classic theorems, such as the following.

- Existence of primitive roots mod p
  Proof. $(\mathbf{Z}/p\mathbf{Z})^\times$ not cyclic
  $\Rightarrow$ an element x of maximal order $n < p - 1$
  $\Rightarrow$ all $p - 1$ elements of $(\mathbf{Z}/p\mathbf{Z})^\times$ satisfy the nth degree equation $x^n = 1$.

- -1 is a square mod $p \Leftrightarrow p = 4n + 1$
  Proof. Use the fact the squares are the even powers of a primitive root.

- $p = 4n + 1$ is a sum of two squares
  Proof. $-1 \equiv m^2 \pmod{p}$
  $\Rightarrow kp = m^2 + 1 = (m + i)(m - i)$
  $\Rightarrow p | (m + i)(m - i)$ but $p \nmid m \pm i$
  $\Rightarrow p$ is not a Gaussian prime (see next section)
  $\Rightarrow p = (a + ib)(a - ib) = a^2 + b^2$.

Another wonderful constellation of results comes from forming the product of elements in an abelian group in two ways:

★ Fermat's little theorem $a^{p-1} \equiv 1 \pmod{p}$
  Proof. Form the product of $\{1, 2, \ldots, p - 1\}$, which equals the set $\{a, 2a, \ldots (p - 1)a\}$ mod p.

★ Wilson's theorem $(p - 1)! \equiv -1 \pmod{p}$
  Proof. $(p - 1)! = 1 \times 2 \times \cdots \times (p - 2) \times (p - 1)$
  $\equiv -2 \times \cdots \times (p - 2)$ as $p - 1 \equiv -1 \pmod{p}$
  $\equiv -1$, pairing inverses mod p.

★ Euler's criterion $q^{\frac{p-1}{2}} \equiv \left(\frac{q}{p}\right) \bmod p$

Proof. $\left(\frac{a}{p}\right) = -1 \Rightarrow a \not\equiv m^2 \pmod p$

$\Rightarrow a\times$ inverse of $m \not\equiv m \pmod p$

$\Rightarrow \left(\frac{a}{p}\right) = -1 \equiv 1 \times 2 \times \cdots \times (p-1) \equiv a^{\frac{p-1}{2}} \pmod p$

pairing each m with $a\times$ (inverse of m).

$\left(\frac{a}{p}\right) = 1 \Rightarrow a \equiv m^2 \pmod p$ for some m

$\Rightarrow a^{\frac{p-1}{2}} \equiv m^{p-1} \equiv 1 = \left(\frac{a}{p}\right) \pmod p.$

★ Quadratic reciprocity $\left(\frac{q}{p}\right)\left(\frac{p}{q}\right) = (-1)^{\frac{p-1}{2} \cdot \frac{q-1}{2}}$

Proof. Form the product of the members of $(\mathbf{Z}/pq\mathbf{Z})^\times / \{1, -1\}$, then rewrite it using the Chinese remainder theorem: $(\mathbf{Z}/pq\mathbf{Z})^\times \cong (\mathbf{Z}/p\mathbf{Z})^\times \times (\mathbf{Z}/q\mathbf{Z})^\times$.

4. Geometry

In some ways geometry is remote from number theory, being intuitive, visual, and noncomputational. One would expect geometry to be at best a complement to number theory, if not a competitor. However, the two fields interact at such a fundamental level that it is unwise to separate them. Geometry can throw light on numerical facts that seem at first to be none of its business. The classic example is the *parametrization of Pythagorean triples*, the positive integer triples (a, b, c) such that

$$a^2 + b^2 = c^2.$$

The number theoretic approach to the solution of this equation begins with the removal of any common divisor from a, b, c and an analysis of even and odd squares to reduce to the case of a even, b odd, c odd. Then the equation is rewritten

$$\left(\frac{a}{2}\right)^2 = \frac{c+b}{2} \cdot \frac{c-b}{2}$$

and the theory of divisibility (unique prime factorization or an equivalent) is used to conclude that the integer factors $\frac{c+b}{2}$, $\frac{c-b}{2}$ of the integer square $\left(\frac{a}{2}\right)^2$ are themselves squares u^2, v^2, whence

$$\begin{aligned} a &= 2uv \\ b &= u^2 - v^2 \\ c &= u^2 + v^2. \end{aligned}$$

Thus, one gets a parametrization of primitive Pythagorean triples (those with no common divisor) in terms of two integer variables u, v.

This is algebraically equivalent to a parametrization of the *ratios* $y = \frac{a}{c}$, $x = \frac{b}{c}$ by a single rational variable $t = \frac{u}{v}$, namely

$$y = \frac{a}{c} = \frac{2t}{1+t^2}, \quad x = \frac{b}{c} = \frac{1-t^2}{1+t^2}.$$

The surprising thing is that t has a *geometric meaning*. It is the slope of the line $y = t(x + 1)$ through the trivial rational point $(-1, 0)$ on the circle $x^2 + y^2 = 1$ and the general rational point $y = \frac{2t}{1+t^2}$, $x = \frac{1-t^2}{1+t^2}$, as one finds by solving the simultaneous equations $y = t(x + 1)$ and $x^2 + y^2 = 1$.

Indeed, it follows from the Pythagorean theorem that primitive Pythagorean triples (a, b, c) correspond to points on the unit circle with nonzero rational coefficients $(\frac{a}{c}, \frac{b}{c})$. These points in turn correspond to lines through $(-1, 0)$ with rational slope. Hence the (geometrically motivated) process of solving the equations $y = t(x + 1)$ and $x^2 + y^2 = 1$ is fully equivalent to the arithmetic method of finding the primitive Pythagorean triples. I find this remarkable, because the geometric process avoids the theory of divisibility of integers.

The process seems to have been discovered by Diophantus, though he did not mention its geometric interpretation. In Book II, Problem 8, of his *Arithmetica* he found two nonzero rationals x, y whose squares add to 16, essentially by intersecting the circle $x^2 + y^2 = 16$ with the line $y = 2x - 4$ through the "obvious" rational point $x = 0$, $y = -4$ on the circle.

The intimate relation between geometry and divisibility shows up again in the theory of the Gaussian integers $\mathbf{Z}[i] = \{a + ib : a, b \in \mathbf{Z}\}$. In $\mathbf{Z}[i]$, geometry is actually the basis for the divisibility theory – for any nonzero Gaussian integers α, β it gives a remainder ρ smaller than β in absolute value when α is divided by β. This can be seen as follows. First notice that the Gaussian integer multiples $\mu\beta$ of β form a lattice of squares in the complex plane: the typical square is the one with corners 0, β, $i\beta$, $(1 + i)\beta$. Now, as the distance $|\alpha - \mu\beta|$ from any point α to the nearest lattice point $\mu\beta$ is less than the side length $|\beta|$ of a square, we have: *for any nonzero $\alpha, \beta \in \mathbf{Z}[i]$ there are $\mu, \rho \in \mathbf{Z}[i]$ ("quotient" and "remainder") with*

$$\alpha = \mu\beta + \rho \quad \text{and} \quad 0 \leq |\rho| < |\beta|.$$

This fact is the key to the divisibility theory of $\mathbf{Z}[i]$, giving a Euclidean algorithm and unique prime factorization as in $\mathbf{Z}$.

These two examples show, I think, how number theory and geometry are intimately related at the foundational level. Of course, we are well aware how strong the relationship is at higher levels — the mere mention of "elliptic curves" should suffice. I do not wish even to define elliptic curves here, but I cannot avoid mentioning an elementary problem that is actually connected with them. This is the question: Are there positive integers x, y, z such that $x^4 + y^4 = z^2$? Fermat showed that there are not, by an ingenious argument using little more than the parametrization of Pythagorean triples. Here is the gist of it.

Suppose that there are positive integers x, y, z such that $x^4 + y^4 = z^2$, or in other words, $(x^2)^2 + (y^2)^2 = z^2$. This says that x^2, y^2, z is a Pythagorean triple, which we can take to be primitive, hence there are integers u, v such that

$$x^2 = 2uv, \quad y^2 = u^2 - v^2, \quad z = u^2 + v^2.$$

The middle equation says that v, y, u is also a Pythagorean triple, and it is also primitive, hence there are integers s, t such that

$$v = 2st, \quad y = s^2 - t^2, \quad u = s^2 + t^2.$$

This gives

$$x^2 = 2uv = 4st(s^2 + t^2),$$

so the integers s, t and $s^2 + t^2$ have product equal to the square $(x/2)^2$. They are also relatively prime, so each is itself a square, say

$$s = x_1^2, \quad t = y_1^2, \quad s^2 + t^2 = z_1^2,$$

and hence

$$x_1^4 + y_1^4 = z_1^2.$$

Thus, we have found another sum of two fourth powers equal to a square, and by retracing the argument we find that the new square z_1^2 is *smaller* than the old, z^2, but still nonzero. By repeating the process we can therefore obtain an infinite descending sequence of positive integers, which is a contradiction.

Fermat's argument shows *a fortiori* that there are no positive integers a, b, c such that $a^4 + b^4 = c^4$ — the first instance of "Fermat's last theorem". It is worth mentioning that Fermat found both these theorems as spinoffs of a theorem about right-angled triangles: *the area of a right-angled triangle whose sides are rational numbers cannot be a rational square.* (Just which numbers *do* occur as areas of rational right-angled triangles turns out to be a deep question about elliptic curves. For further information see Weil [1984] and Koblitz [1985].)

5. Trigonometry

While on the subject of right-angled triangles, let us consider the multiplication of complex numbers from the geometric viewpoint. If

$$z_1 = a_1 + ib_1, \quad z_2 = a_2 + ib_2$$

are complex numbers, we know that

$$z_1 z_2 = (a_1 a_2 - b_1 b_2) + i(a_1 b_2 + a_2 b_1).$$

This formula is equivalent to a pair of geometric facts, namely

$$|z_1 z_2| = |z_1||z_2|,$$

and

$$\arg(z_1 z_2) = \arg(z_1) + \arg(z_2),$$

where $|a + ib| = \sqrt{a^2 + b^2}$ is the distance of $a + ib$ from O, and $\arg(a + ib) = \tan^{-1}(b/a)$ is the angle it subtends at O. Both of these facts were known from number theory long before the invention of complex numbers.

The multiplicative property of $|\ |$ was known to Diophantus in the form of the identity

$$(a_1 a_2 - b_1 b_2)^2 + (a_1 b_2 + a_2 b_1)^2 = (a_1{}^2 + b_1{}^2)(a_2{}^2 + b_2{}^2),$$

as can be seen, at least implicitly, from his *Arithmetica*, Book III, Problem 19. Here Diophantus remarks

> 65 is naturally divided into two squares in two ways, namely into $7^2 + 4^2$ and $8^2 + 1^2$, which is due to the fact that 65 is the product of 13 and 5, each of which is the sum of two squares.

This undoubtedly means that he was aware of the even more general identity

$$(a_1 a_2 \mp b_1 b_2)^2 + (a_1 b_2 \pm a_2 b_1)^2 = (a_1{}^2 + b_1{}^2)(a_2{}^2 + b_2{}^2).$$

The additive property of argument was discovered by Viète in his *Genesis triangulorum* of around 1590. Given two right-angled triangles, with sides a_1, b_1 and a_2, b_2 (and hence hypotenuses $\sqrt{a_1^2 + b_1^2}$, $\sqrt{a_2^2 + b_2^2}$ respectively), he forms the triangle with sides $(a_1 a_2 - b_1 b_2)$, $(a_1 b_2 + a_2 b_1)$. The hypotenuse of the third triangle is therefore the product of the hypotenuses of the first two, by Diophantus' identity, but Viète also observes that its angle (between the side $a_1 a_2 - b_1 b_2$ and the hypotenuse) is the sum of the corresponding angles of the first two. In effect, his construction completely encodes the multiplication of $a_1 + ib_1$ and $a_2 + ib_2$, *and* reveals its geometric properties. As we know, the additive property of argument under multiplication of complex numbers gives an easy proof of the addition theorems for sine and cosine (de Moivre's theorem). Viète already observed that these theorems follow from his construction.

The trigonometric formulae $\sin\theta = \frac{2t}{1+t^2}$ and $\cos\theta = \frac{1-t^2}{1+t^2}$, where $t = \tan\frac{\theta}{2}$, also relate to number theory. They follow immediately from the fact used to parametrize Pythagorean triples in Section 4 — that the line of slope t through $(-1,0)$ meets the unit circle at the point $\left(\frac{1-t^2}{1+t^2}, \frac{2t}{1+t^2}\right)$.

Once again, we find a mathematical topic where number theory is relevant from the lowest levels. As elsewhere, its relevance only gets stronger at higher levels. Thanks to Gauss, for example, we know that number theory has much to gain from the theory of cyclotomic fields. The name "cyclotomic" ("circle-dividing") says it all. It concerns the number $\zeta_n = \cos\frac{2\pi}{n} + i\sin\frac{2\pi}{n}$ whose powers divide the unit circle into n equal arcs. The algebraic properties of ζ_n are the key to deciding whether the regular n-gon is constructible by straightedge and compass, but they are also magically effective in pure number theory. In particular, Gauss's favorite theorem — quadratic reciprocity — can be proved via properties of ζ_n.

6. Calculus

Calculus, as it is usually taught, is mostly algebra. It is true that calculus has an extra ingredient, the limit concept, but serious discussion of limits is usually postponed until an analysis course. It is also true that the motivating problems of calculus — tangents, areas, speed — come from geometry and physics, but they are quickly boiled down to algebraic manipulation. In fact, we choose problems for their algebraic tractability, almost unconsciously, without thinking how tractability is decided, or how it might be explained to students.

A typical problem is to "rationalize" $\sqrt{1 - x^2}$ by substituting a suitable function for x. One such substitution is $x = \sin\theta$, but knowing that $\sin\theta = \frac{2t}{1+t^2}$ we can achieve the same result by the rational substitution $x = \frac{2t}{1+t^2}$, obtaining $\sqrt{1 - x^2} = \frac{1-t^2}{1+t^2}$. This rationalization enables us to integrate any function that is rational except for occurrences of $\sqrt{1 - x^2}$. It is similar for functions rational except for $\sqrt{a + bx + cx^2}$. The main problem is to rationalize the square root, and the necessary manipulation can be done by the methods of Diophantus. Thus, the tractability of such problems can be traced back to the parametrization of Pythagorean triples.

After these few successes, integration becomes frustrating. Square roots of higher degree polynomials cannot in general be rationalized, and calculus cannot tell us why. Generations of students have hit the wall of elliptic integrals and have received no explanation of the nature of the obstacle. We simply tell them that such integrals can't be done.

There is no need to give up so easily! With a little number theory, it is quite easy to explain why $\sqrt{1 - x^4}$, for example, cannot be rationalized. Suppose, on the contrary, that for some rational function $x = x(t)$, $1 - x(t)^4$ is the square $y(t)^2$ of a rational function. Then we have an equation

$$1 - x(t)^4 = y(t)^2,$$

where $x(t)$, $y(t)$ are quotients of polynomials. Multiplying through by a common denominator gives an equation between polynomials $X(t)$, $Y(t)$, $Z(t)$:

$$Z(t)^4 - X(t)^4 = Y(t)^2.$$

This looks like the equation proved impossible by Fermat, and indeed this one can be proved impossible by a similar argument — if X, Y, Z are integers. Of course, they aren't, but they *are* polynomials, and we know that polynomials behave a lot like integers. Fermat's argument can be carried over if we assume that $X(t)$, $Y(t)$, $Z(t)$ are polynomials not all of zero degree, in which a contradiction is gained by showing that any solution implies a solution of lower total degree.

Thus, number theory *can* throw light on calculus, but is it reasonable to expect calculus students to have the necessary background knowledge? To answer this question I would like to go back to the work of Leibniz and the Bernoulli brothers Jacob and Johann. These three created most of calculus as we know it today, in particular the techniques for expressing basic integrals in "closed form". (Newton, who of course is also a giant of calculus, was not very interested in such solutions, and was content to express integrals as infinite series.) They are known for their contributions to calculus and its applications, but certainly *not* for contributions to number theory. Nevertheless, Leibniz [1702] wrote:

> I . . . remember having suggested (what could seem strange to some) that the progress of our integral calculus depended in good part upon the development of that type of arithmetic which, so far as we know, Diophantus has been the first to treat systematically.

Even more surprising, Jacob Bernoulli [1704] tried to explain why $\sqrt{1 - x^4}$ cannot be rationalized, appealing to Fermat's theorem essentially as I have done above (though he confined his argument to integers and did not extend it to polynomials). On another occasion [1696], when he had to rationalize $\sqrt{2x - x^2}$, he directly credited Diophantus with the substitution that does it. Incidentally, he used the latter substitution to transform the integral expressing arc length of a circle into the form $\int \frac{dt}{1+t^2}$, thus obtaining the canonical derivation of the infinite series for π:

$$\frac{\pi}{4} = \int_0^1 \frac{dt}{1+t^2} = \int_0^1 (1 - t^2 + t^4 - t^6 + \cdots)\,dt = 1 - \frac{1}{3} + \frac{1}{5} - \frac{1}{7} + \cdots$$

To my mind, this shows that number theory was historically decisive in the development of calculus. The fact that number theory is no longer considered relevant to calculus only shows how much we have forgotten, and how out of touch with the rest of mathematics calculus has become.

7. Conclusions

In this paper I have tried to show that number theory is the best basis for a mathematical education because it supports, or at least throws light on, all other mathematical disciplines. In particular, calculus rests on algebra and ultimately on number theory, and all understanding of calculus is lost when this support is withdrawn. It is no doubt impractical to suggest that number theory should immediately *replace* calculus as the entry qualification for higher mathematics. Such a revolutionary act would not be tolerated by most mathematics departments, let alone the science and engineering schools that rely on us for service teaching. However, I believe that number theory should be given a larger role in secondary schools and universities, with an emphasis on its connections with other parts of mathematics.

I have stressed the connections of number theory with traditional disciplines — algebra, geometry, trigonometry, calculus — because these connections seem to be the most neglected and forgotten. If anything, it is easier to make a case for number theory as a support for newer disciplines such as computer science, where the importance of primes, factorization, and number fields are well known. However, it should not be necessary to appeal to current fashion. Numbers have always been important in mathematics, they always will be, and our teaching of mathematics should always reflect that fact. Hilbert used to say that we do not really understand a piece of mathematics until we can explain it to the first person we meet in the street. In most cases, that means understanding mathematics in terms of numbers.

References

Bernoulli, Jacob
[1696] *Positionum de seriebus infinitis pars tertia*, Werke, **4**, 85–106.
[1704] *Positionum de seriebus infinitis ... pars quinta*, Werke, **4**, 127–147.

Koblitz, N.
[1985] Introduction to Elliptic Curves and Modular Forms, Springer-Verlag, Berlin New York.

Leibniz, G. W.
[1702] *Specimum novum analyseos pro scientia infiniti circa summas et quadraturas*, Mathematische Schriften, **5**, 350–361.

Weil, A.
[1984] Number Theory. An Approach Through History, Birkhäuser, Basel and Boston, MA.

Mathematics in Medieval Islamic Spain

JAN P. HOGENDIJK

Department of Mathematics, University of Utrecht,
P.O. Box 80.010, 3508 TA Utrecht, Netherlands

1. Introduction

From the seventh to the eleventh century, a large part of present-day Spain and Portugal belonged to the Islamic world. I will use the term "Islamic Spain" to indicate the part of the Iberian peninsula that was under Muslim rule. The term Islamic Spain is not strictly correct, because Spain did not exist in the early Middle Ages, but the important medieval scientific centers (Córdoba, Zaragoza, Toledo) are all in present-day Spain. Until recently the general view has been that Islamic Spain was important in the history of mathematics only because of its role in the transmission of mathematics from Arabic to Latin. In the last 15 years this view has changed as a result of the investigation of unpublished manuscript sources. We now know that there were creative mathematicians in Islamic Spain in the eleventh century. In this paper I will try to give you an impression of their work. No previous acquaintance with the history of mathematics will be assumed, and I will begin with some general remarks on the historical context.

2. The historical context

Before 300 b.c. the ancient Greeks developed geometry as a deductive system. Greek mathematics flourished until the third century a.d. and then declined. The Romans were not interested in theoretical mathematics, and the tradition remained dormant until it was revived by the Muslims. Around a.d. 800, the caliphs made Baghdad into the scientific capital of the world. They had Arabic translations made of many Greek texts on mathematics and astronomy (including Euclid's *Elements*, the works of Archimedes, Apollonius, etc.), and also of Sanskrit works from India. This is the beginning of Arabic science, i.e. science written in Arabic. When we use a term such as "Arabic mathematics", we should bear in mind that large contributions were made by non-Arabs, notably the Iranians.

Islamic Spain is far away from Baghdad, and it took a while for science to reach the area. In the tenth century there was much interest in learning in Islamic Spain, and there was a library with more than 400,000 books in the capital, Córdoba. At that time mathematics and astronomy were studied on a rudimentary level, necessary for practical applications, such as astrology and timekeeping. Around the year 1000 the interest in theoretical mathematics and astronomy deepened. The eleventh century is the golden age of Islamic Spanish science. After the

Proceedings of the International Congress
of Mathematicians, Zürich, Switzerland 1994
© Birkhäuser Verlag, Basel, Switzerland 1995

eleventh century, the size of Islamic Spain was reduced and there was a decline in the scientific activity.

The Christians reconquered a large part of present-day Spain in the eleventh and twelfth centuries: Toledo was captured in 1085, Zaragoza in 1118. The end of Islamic domination of these areas did not mean the complete end of the scientific tradition, and in the twelfth century, many Arabic manuscripts from Spain were translated into Latin. Not many people realize how important this event was for the history of mathematics. In the early Middle Ages mathematics was practically nonexistent in western Europe; a few facts from elementary mathematics were mentioned in Latin texts and encyclopedias, but hardly anybody knew how to prove a theorem. By means of the twelfth century Latin translations, the Christians came in contact with mathematics as a deductive science. The creative period of Islamic Spanish mathematics occurred just before the takeover by the Christians of most of Spain and the translation movement.

3. Sources and the state of research

Some texts by eleventh-century mathematicians and astronomers have come down to us in the original Arabic. These are not the manuscripts written by the authors themselves, but are always later copies, made in the Islamic world or in medieval Christian Spain, in places where Arabic astronomy was studied (such as the court of Alfonso the Wise in the thirteenth century). A few texts survive only in a medieval Latin or Hebrew translation. One can also find traces of eleventh-century mathematics in texts by later authors (from Islamic or Christian Spain or North Africa). Even so, the evidence we have is very incomplete, and the history of mathematics in the eleventh century has to be patched together from various bits and pieces of information. Not all relevant sources have been studied, and many manuscripts still await edition and translation. There are various researchers all over the world who are working on these materials. The most important center for the study of Islamic Spanish science is the Department of Arabic Philology in the University of Barcelona under the direction of Julio Samsó. The center publishes a series of editions of sources, and Samsó has recently published the first reliable survey of science in Islamic Spain, including the results of research up until 1992 [5].

4. General remarks

In the history of mathematics in Islamic Spain, a distinction can be made between arithmetic and algebra on the one hand, and geometry and trigonometry on the other hand. In arithmetic and algebra, Islamic Spain seems to have lagged behind the East. The mathematicians of the eleventh century worked with Eastern texts that had become obsolete in the East, such as the arithmetic and the algebra of al-Khwārizmī (ca. 830). Recent advances in the East were unknown in Islamic Spain. For geometry and trigonometry the situation was different. The Islamic Spanish mathematicians were working on the same level as their Eastern Islamic colleagues, and they were aware of many recent developments in the East.

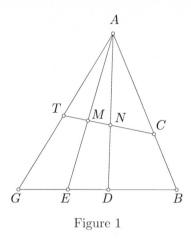

Figure 1

I will not be giving you a list of all mathematicians with their known works and contributions to geometry and trigonometry, because I believe such lists are boring for the nonspecialist.[1] Instead, I will discuss two concrete examples in some detail.

5. Al-Mu'taman and his "Book of Perfection"

My first example is al-Mu'taman ibn Hūd, the king of the kingdom of Zaragoza (in northeastern Spain), who died in 1085. The kingdom of Zaragoza was one of the so-called petty kingdoms into which Islamic Spain had disintegrated in the eleventh century. Al-Mu'taman was also a mathematician, who wrote a very long mathematical work, entitled *The Book of Perfection* (in Arabic: Istikmāl) [1]. For a long time this work was believed to be lost, but numerous fragments have recently turned up in four anonymous Arabic manuscripts [2]. The work is in a very poor state of preservation because the most important manuscript (in Copenhagen) was damaged and many leaves are missing. As a result, there are 11 gaps in the text we have. [2] In the *Book of Perfection* al-Mu'taman presents the essentials of mathematics, philosophically arranged, with all the proofs. The work resembles, to some extent, the *Éléments de Mathématique* of N. Bourbaki. Al-Mu'taman adapted most of the material in his *Book of Perfection* from existing works by Euclid, Archimedes, Apollonius, and other mathematicians from antiquity or the eastern Islamic world. However, some theorems are "new", in the sense that they are not found anywhere else in the ancient and medieval literature we know. The following two examples still play a role in modern geometry:

---

[1] Thus, I will not be talking about al-Zarqāllu's new variant of the *astrolabe*, which has received a great deal of attention in the literature. Like the standard astrolabe, this variant is based on stereographic projection, but the pole of projection is shifted from the north pole to the vernal point.

[2] See note added in proof.

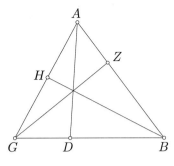

Figure 2

(1) (Figure 1) In "proposition 16 of section 3 of species 1 of species 3" al-Mu'taman considers a triangle ABG with points D and E on its base, and a transversal $TMNC$ that intersects AG at T, AE at M, AD at N, and AB at C.[3]

He proves $(TC : CN) \cdot (NM : MT) = (GB : BD) \cdot (DE : EG)$. In modern terms this is the oldest extant statement of the perspective invariance of cross-ratios in a rather general situation. (There is an ancient Greek proof, in the *Mathematical Collection* of Pappus of Alexandria, for the special case where B and C coincide.)

(2) The so-called theorem of Ceva (Figure 2) is stated and proved in the margin of the *Book of Perfection*, as "proposition 18 of section 3 of species 1 of species 3".[4] Consider a triangle ABG with points D on side BG, H on side AG, and Z on side AB. Then lines AD, BH, and GZ intersect at one point if and only if $(BZ : ZA) \cdot (AH : HG) = (BD : DG)$. (This is proved by two applications of the theorem of Menelaus.) The theorem has been named after Giovanni Ceva, who stated it in 1678, but it should now perhaps be renamed the "theorem of al-Mu'taman."

In the extant parts of the *Book of Perfection*, al-Mu'taman never distinguishes between his own contributions and theorems that he adapted from other sources. Cross-ratios occur very rarely in Arabic mathematics, but we know that they were used extensively in Euclid's lost work on *Porisms*, which was transmitted to Arabic in some form. I therefore believe that al-Mu'taman took his cross-ratio theorem from an Arabic translation of a lost Greek work (if this is true, we get new information on Greek mathematics from an eleventh-century Islamic Spanish source). I do not know whether the so-called theorem of Ceva was a contribution by al-Mu'taman or not.

---

[3]In ancient Greek and medieval mathematics, there was no real number concept. The line segment is a basic concept; a line segment does not have a length, but it is a (positive) length. Line segments can be compared, and there existed a theory of ratios between line segments. Ratios could be ordered, and a ratio between two line segments could be compared to a ratio between two integers. All straight lines were bounded; a straight line could be produced indefinitely, but an infinite straight line did not exist.

[4]Because the theorem has a proposition number, it belonged to the original text, and was first left out by an oversight.

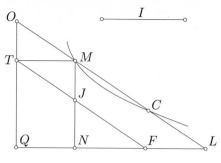

Figure 3

There are only a few rather complicated propositions that we can attribute to al-Mu'taman with some certainty. I will now discuss one example, namely his simplification of a construction by Ibn al-Haytham (a geometer who worked in Egypt in the early eleventh century), which is based on an ancient Greek construction. The following explanation will take some time but it will be worthwhile because it gives an idea of al-Mu'taman's ability as a mathematician. I will try to give the general ideas and avoid the details of the proofs, which involve proportions and similar triangles. My notations are those of Ibn al-Haytham [4, pp. 315–318].

The ancient Greek construction is as follows (Figure 3). We are given a rectangle $TQNM$ and a straight segment I. We want to construct a straight segment TJF that intersects MN at J and QN extended at F in such a way that $FJ = I$. Solution: Draw a hyperbola through M with asymptotes TQ and QN. (In antiquity and the Middle Ages, a hyperbola was always a single-branch hyperbola.) Draw a circle with center M and radius I. Let them intersect at C. Draw TF // MC. This is the desired line.

The proof is easy: Extend MC to meet QT at O and QN at L. By parallelograms, $TJ = OM$ and $TF = ML$, so $FJ = ML - OM$. By a property of the hyperbola: $OM=CL$. Hence $FJ = MC = I$. I note that the problem cannot be solved by ruler and compass.

This construction was used in rather a confusing way by the Egyptian geometer Ibn al-Haytham around 1040 in his work on *Optics*, in a series of a preliminaries for the study of reflection in circular mirrors.

Ibn al-Haytham considers a circle with a given diameter BG and on this circle a given point A (Figure 4). A straight segment EK is also given. He wants to construct a straight line through A that intersects the circle at H and the diameter at D in such a way that $DH = EK$. Point D is assumed to be outside the circle.

The basic idea is as follows. First suppose that DH is arbitrary. Ibn al-Haytham draws line GZ parallel to BA, meeting DH at Z. He proves $AZ : DG = BG : DH$ (I omit the details).

Therefore $DH=EK$ if and only if $AZ : DG = BG : EK$. Unfortunately, we do not know the length of DG, but we know the angles at G: $\angle DGZ = \angle GBA$, and $\angle ZGA = \angle GAB = 90°$.

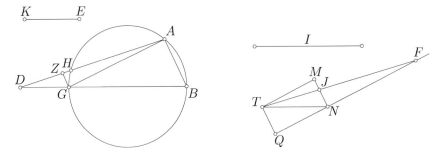

Figure 4

Ibn al-Haytham now uses one of his favorite techniques, which consists of constructing an auxiliary figure similar to the original one. Choose a segment NT of arbitrary known length, and make rectangle $TMNQ$ such that $\angle TNM = \angle DGZ$.

Choose I such that $I : NT = BG : EK$, and find (by the Greek construction) TJF such that $FJ = I$. Then construct D such that $\angle GAD = \angle NFT$. Then $GAZD$ is similar to $NFJT$. Thus $AZ : DG = FJ : NT = I : NT = BG : EK$. Because $AZ : DG = BG : DH$, we have $DH = EK$ as required. Note that the known segment NT in the auxiliary figure corresponds to the unknown segment GD in the original figure.

Al-Mu'taman simplified and generalized the construction of Ibn al-Haytham, on the basis of the following three ideas:

(1) Because in Figure 4 NT is an arbitrary segment, we can identify it with a known segment in the original figure, such as BG. If this is done, the auxiliary figure and the original figure coincide in a nice way, and the construction is as follows (Figure 5). Complete rectangle $GABQ$, draw the hyperbola with center A and asymptotes GQ and QB. Choose I such that $I : BG = BG : EK$ and find C as a point of intersection of the hyperbola and the circle with center A and radius I. Extend CA to intersect circle ABG again at H and line BG at D. Only a small number of proportions and similar triangles[5] are needed to prove $AC : BG = BG : DH$. Because $AC=I$ and $I : BG = BG : EK$ we have $DH=EK$ as required.

(2) Thus far we have been discussing the case where point D is outside the circle. Al-Mu'taman saw that one can solve the similar problem for D inside the circle, by finding the points at which the other branch of the hyperbola intersects the circle with center A and radius I, and by exactly the same reasoning (broken line in Figure 5). Al-Mu'taman gives a general proof for the two different cases. Ibn al-Haytham has a new and essentially different solution for D inside the circle.

(3) Al-Mu'taman observed that $GABQ$ can be an arbitrary parallelogram instead of a rectangle. Thus, it is not necessary to assume that BG is a diameter;

---

[5]Details: as A, B, G, H are on the same circle, $DA{\cdot}DH = DB{\cdot}DG$, so $DH : DG = DB : DA =$ (by similar triangles) $BG : AZ$. Hence $BG : DH = AZ : DG =$ (by similar triangles) $JF : BG$. $JF = AC$ is proved just as $JF = MC$ in Figure 3. The text will appear in [6].

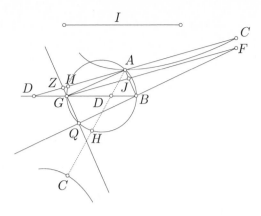

Figure 5

A, B, and G can be three arbitrary different points on a circle. In general, point Q is not on the circle.

Summarizing: al-Mu'taman solved a more general problem than Ibn al-Haytham in a much shorter way. Ibn al-Haytham's solution of the two cases is ten times as long as al-Mu'taman's general solution. This shows that al-Mu'taman was able to improve on Ibn al-Haytham, who was one of the most important geometers of the Islamic tradition.

Al-Mu'taman must be the author of the general solution (Figure 5) for the following reasons. The solution occurs in a series of geometrical constructions that are clearly adapted from the *Optics*. Ibn al-Haytham wrote this work around the year 1030 in Egypt at an advanced age, less than 50 years before al-Mu'taman wrote his *Book of Perfection*. In the Eastern part of the Islamic world, nobody seems to have seriously studied Ibn al-Haytham's *Optics* in the eleventh century. We have a description of Islamic Spanish science made around 1065 by the biographer Ṣāʿid al-Andalusī, in which we can read that there are only three people interested in "natural philosophy", namely Al-Mu'taman and two others (whose names are mentioned). The two other people are not known to have written any mathematical works, and their mathematical reputations among their contemporaries are nowhere near that of al-Mu'taman. This leaves al-Mu'taman as the only plausible author.

His authorship is confirmed by the fact that there are many more solutions of a similar nature in the *Book of Perfection*. An example is the famous problem of Apollonius (to construct by ruler and compass a circle tangent to three given circles). Al-Mu'taman presents a solution that was also inspired by the same Ibn al-Haytham, but that is also much shorter and much clearer than Ibn al-Haytham's confused original. Another example is al-Mu'taman's simplification of a quadrature of the parabola by Ibrāhīm ibn Sinān, a geometer from tenth century Baghdad. Al-Mu'taman was not the only eleventh-century Islamic Spanish geometer working on complicated subjects. Between 1087 and 1096, the geometer Ibn Sayyid of Valencia developed a theory of higher-order curves, and he used these to divide an angle into an arbitrary number of parts and to construct an arbitrary number of geometric means between two given lines.[6] His work is lost; all we have is rather a vague description of it by the philosopher Ibn Bājja. [1].

These examples show that geometry was studied on a much higher level in Islamic Spain than had been thought 15 years ago.

6. Ibn Mu^{c}ādh and the astrological "aspects"

One cannot get an adequate view of mathematics in Islamic Spain without looking at the applications in astronomy and astrology. To give you some of the flavor, I have chosen a nontrivial problem from astrology, which was solved by the eleventh-century mathematician Ibn Mu^{c}ādh. Astrology is nowadays considered to be a pseudo-science, but it was very important in the history of mathematics in the Middle Ages, because it was one of the main fields where mathematics was applied in a nontrivial way. I first recall a few astronomical preliminaries.

In ancient and medieval planetary theory, the positions of the planets were represented on the celestial sphere, in coordinates called "celestial longitude" and "celestial latitude". The basic circle of reference is the ecliptic; that is, the apparent orbit of the sun around the earth against the background of the fixed stars. The zero point on the ecliptic is the vernal point, that is the position of the sun at the beginning of spring, and the celestial longitude of a planet is the arc between the vernal point and the perpendicular projection of the planet on the ecliptic. This arc is measured in the direction of the motion of the sun. I will ignore the celestial latitude, that is the arc measuring the deviation of the planet from the ecliptic. In the Islamic Middle Ages, there was a satisfactory theory for the prediction of the celestial longitudes of the sun, moon, and planets on the ecliptic at any place and at any moment of time, and there were many handbooks explaining the necessary computations and containing the necessary tables. The fact that astronomy was geocentric, not heliocentric, did not affect the exactness of these predictions.

I will now explain the astrological concept of "aspect". The basic assumption of medieval astrology is that the sun, moon, and planets have an influence on events on earth, and that the positions of these celestial bodies can be used to predict the future with some probability. Such predictions were very complicated, and one of the many things that the astrologer had to take into account was certain special configurations between two planets, the so-called aspects. The idea is as follows. The astrologers believed that from its position on the ecliptic each planet emitted seven "visual rays" to other points of the ecliptic. (According to the ancient theory of vision, a human being sees by emitting visual rays from the eye, and not because light enters his eye.) If a second planet is sufficiently close to the endpoint of such a visual ray, it is "seen" by the first planet ("looked at", in Latin: adspectus), and then the planets are said to have an aspect.

There were two theories for the computation of these aspects. According to the simplest theory, the seven rays are emitted to points in the ecliptic at angular

---

[6]n geometric means between two given line segments a and b are n straight segments $x_1 \ldots x_n$ such that $a : x_1 = x_1 : x_2 = \cdots = x_n : b$.

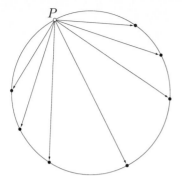

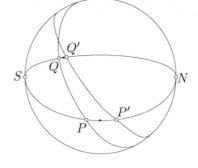

Figure 6   Figure 7

distances of 60°, 90°, 120°, and 180° in both directions (Figure 6). Thus, two planets have an aspect if the difference in their celestial longitude is 60°, etc. This does not lead to interesting mathematical problems.

For reasons unknown to me, several medieval Islamic astrologers believed in a more complicated theory, which I will call the "equatorial" theory of aspects. The idea is as follows. We first project the planet P, which is on the ecliptic, on the intersection P' of the celestial equator[7] with the great semicircle through P and the north point N and the south point S of the horizon. From P' we measure the angles of 60°, 90°, 120°, and 180° on the celestial equator in both directions (in Figure 7 we have $P'Q'=60°$). For each of the seven points Q' that we find this way, we apply the inverse projection: we draw the great semicircle $NQ'S$ and we call Q its intersection with the ecliptic. Then according to the "equatorial" theory the planet at P emits its seven visual rays to the seven points Q.

If $P'Q' = 60°$, 90°, or 120°, arc PQ has a variable magnitude, which depends not only on the celestial longitude of P, but also, oddly enough, on the geographical latitude of the astrologer and on the (local) time of day. (This leads to the rather surprising conclusion that the planet emits its visual rays differently for different observers on earth. This does not seem to have worried the astrologers.)

The "equatorial" theory of aspects was quite popular, in spite of (or maybe because of) its difficulty. In the ninth century, al-Khwārizmī computed tables for the computation of the aspects according to the equatorial theory for Baghdad, using a rather crude approximation. This al-Khwārizmī is the famous mathematician who wrote on algebra and whose name has been corrupted in the word *algorithm*.

These tables were recomputed for Córdoba, around the year 1000, by the leading Islamic Spanish mathematician of that time, Maslama ibn Aḥmad al-Majrīṭī. (Al-Majrīṭī is Arabic for: of Madrid.) Maslama used a different geographical latitude and a more sophisticated (but still approximate) mathematical method.

I will now discuss the exact computation of the aspects according to the "equatorial" theory by Ibn Muʿādh al-Jayyānī [3]. He was one of the best math-

---

[7]This is the intersection of the celestial sphere and the plane through the observer perpendicular to the world axis, i.e. the line through the celestial north and south pole.

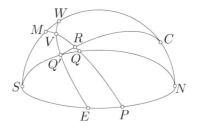

Figure 8

ematicians of the eleventh century, and also a qāḍī, that is an Islamic judge. He lived in Jaén, in southern Spain.

For simplicity of explanation, I will assume that the planet P is on the eastern horizon and that the vernal point is close to the meridian. The general case is not essentially more difficult. I will skip some computations that were standard in Ibn Muʿādh's time, and I will use modern algebraical notation for the sake of clarity. My source is an unpublished Arabic text by Ibn Muʿādh on the theory of "aspects", which survives in a unique Arabic manuscript, written in thirteenth century (Christian) Spain.

Figure 8 displays the quadrant of the celestial sphere above the horizon and East of the meridian. All arcs in this figure are arcs of great circles, i.e. intersections of the sphere with planes through its center. Points E, S, and N are the East, South, and North points of the horizon. Because point P is on the eastern horizon, its "projection" P' coincides with the East point E. Point C is the celestial north pole, $NCWMS$ is the meridian, which intersects the ecliptic at M and the celestial equator WVE at W. Point V is the vernal point. Then $\angle PVE = \epsilon$, the obliquity of the ecliptic, a known angle for which the value $23°35'$ was often used in Arabic astronomy; $\angle VES = 90°-\phi$; here ϕ is the geographical latitude, which is also assumed to be known (we assume that the astrologer is in the Northern Hemisphere, between the equator and the Arctic Circle). Note that arc $CN = \phi$. We also assume that we know arc VP, that is the celestial longitude of P.

To compute the so-called right sextile aspect of P, we mark off $EQ'=60°$ on the celestial equator, and we draw semicircle $SQ'N$ to intersect the ecliptic at Q. We want to compute arc VQ.

Ibn Muʿādh draws arc CQ' to meet the ecliptic at point R. Because C is the celestial north pole and Q' is on the equator, $\angle CQ'V$ is a right angle.

Arc VE, the so-called "rising time of point P" can be found by a standard computation, or can be looked up in a "table for rising times" for the particular geographical latitude.

On the celestial equator, we know arc $EQ'=60°$, $EW=90°$, so from EV we find $Q'V$ and VW. Using these arcs we can compute or look up the following arcs: VR, VM (from a right ascension table; these are the longitudes that belong to right ascensions VQ', VW) and $Q'R$, MW (from a declination table). We have arc $WS = 90°-\phi$, hence we find arc $MS =$ arc $WS-$ arc WM. The following

quantities are now known: $a = $ arc MS, $\Lambda = $ arc RM, $\delta = $ arc $Q'R$; we also know
arc $CS = 180° - \phi$.

The next step in the argument is crucial.

I set $x = $ arc QR. Ibn Muʿādh uses the spherical theorem of Menelaus, to
the effect that

$$\frac{\sin MQ}{\sin QR} = \frac{\sin MS}{\sin SC} \cdot \frac{\sin CQ'}{\sin Q'R}.$$

Menelaus was a Greek astronomer who lived in Rome around a.d. 70, and
whose work on geometry of the sphere is lost in Greek but extant in Arabic.
Menelaus used a different trigonometric function, the "chord", which the Arabs
replaced by the sine, a function of Indian origin. The medieval sine is a constant
factor times the modern sine. Because we are dealing with proportions between
sines, the constant factor can be ignored.

Thus, we get

$$\frac{\sin(\Lambda + x)}{\sin x} = \frac{\sin a}{\sin \phi} \cdot \frac{\sin 90°}{\sin \delta}.$$

The right-hand side of this equation is a known quantity c. Thus, Ibn Muʿādh has
to solve x from

$$\frac{\sin(\Lambda + x)}{\sin x} = c, \tag{1}$$

for known Λ, c.

In an earlier work [7], Ibn Muʿādh had shown how equation (1) can be reduced
to

$$\tan(\frac{\Lambda}{2} + x) = \frac{c+1}{c-1} \cdot \tan\frac{\Lambda}{2}. \tag{2}$$

There Ibn Muʿādh had also tabulated the tangent[8] function for every degree up
to 89°.[9] By means of this table, x can easily be computed. We have arc $QV = $ arc
$RV + x$.

Ibn Muʿādh reduced (1) to (2) by means of the following geometrical reasoning (Figure 9, [7, Fig. 4]). On a circle of reference we represent the known arc Λ
as arc AG, bisected at B, and the unknown arc x as arc GL. Let D be the center,
extend lines DL and AG to meet at point K, and drop perpendiculars GN, BZ,
and AM onto DK. Let BD intersect AG at O.

Now $\tan(\frac{\Lambda}{2} + x) = BZ : ZD$. By similar triangles, $BZ : ZD = KO : OD =$
$(KO : OG) \cdot (OG : OD)$. We have $AK : KG = AM : GN = \sin(\Lambda + x) : \sin x = c$.
Thus, $KO : OG = (KG + KA) : AG = (c+1) : (c-1)$. Finally, $OG : OD = \tan\frac{\Lambda}{2}$.

I have discussed the solution of the problem for a particular configuration.
The procedure is general; in some cases one gets a minus sign instead of a plus
sign in equation (1). Ibn Muʿādh also discusses this situation in [7].

It is not certain whether Ibn Muʿādh ever computed tables for the aspects,
to replace the tables of his famous predecessors al-Khwārizmī and al-Majrīṭī. We
know, however, that Ibn Muʿādh computed tables of the so-called astrological

---

[8]Ibn Muʿādh did not have a special name for the tangent; he called it the quotient of the sine
and the cosine.

[9]Also for every quarter of a degree between 89° and 89°45′.

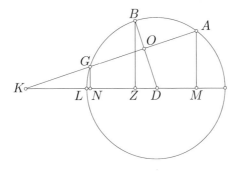

Figure 9

houses on the basis of the same computation. The astrologers divided the celestial sphere into 12 "houses" somewhat like an orange into 12 parts, and according to Ibn Mu^{c}ādh this should be done in the same "equatorial" way as in the theory of aspects. The arcs of the equator between the meridian and the horizon had to be divided into three equal parts of $30°$, and through the division points Q' one had to draw semicircles $NQ'S$, etc., which were the boundaries of the houses. Ibn Mu^{c}ādh computed tables for the celestial longitude of the arcs VQ as a function of the celestial longitude of point P on the Eastern horizon, so that the astrologers could use his system of houses.

It is clear that astrology created nontrivial work for the mathematicians. Not all astrologers needed to understand the computations, but there was always a certain demand for people who understood the technicalities. The applications in astrology and similar ones in astronomy were much more advanced than the trivial applications of mathematics in land measurement, administration, and commerce.

The purpose of my two examples (of al-Mu'taman and Ibn Mu^{c}ādh) has been to convey some idea of the work of the eleventh-century Islamic Spanish mathematicians in geometry and trigonometry. There were no revolutions in the mathematics of this period, comparable to those in antiquity and the seventeenth century. However, the geometrical methods of antiquity and the Islamic East were handled independently and creatively. In these areas the Western European mathematicians did not catch up with their Islamic Spanish predecessors until the Renaissance.[10] Therefore Islamic Spain has a unique position in the history of mathematics in medieval Europe.

Note added in proof: A complete manuscript of a 13-th century recension of the Istikmāl has turned up in 1995.

---

[10]The Renaissance scholar Regiomontanus (1436–1476) was influenced by Ibn Mu^{c}ādh; they both used the same system of astrological "houses".

References

[1] A. Djebbar, *Deux mathématiciens peu connus de l'Espagne du XIe siècle: al-Mu'taman et Ibn Sayyid*, M. Folkerts et al. (eds.), Vestigia Mathematica, Studies in Medieval and Early Modern Mathematics in Honour of H.L.L. Busard, Amsterdam 1993, pp. 79–91.

[2] J. P. Hogendijk, *The geometrical parts of the* Istikmāl *of Yūsuf al-Mu'taman ibn Hūd (11th century), an analytical table of contents*, Arch. Internat. Hist. Sci. **41** (1991), 207–281.

[3] J. P. Hogendijk, *Applied mathematics in 11th century Islamic Spain: Ibn Mu^{c}ādh's computation of the astrological houses and rays*, to appear in Centaurus.

[4] A. I. Sabra, *Ibn al-Haytham's lemmas for solving "Alhazen's problem"*, Arch. Hist. Exact Sci. **26** (1982), 299–324.

[5] J. Samsó, Las Ciencias de los Antiguos en al-Andalus, Madrid, 1992.

[6] J. Samsó (ed.), Proceedings of symposium 49, the transmission of scientific ideas in the field of the exact sciences between Eastern and Western Islam in the middle ages (Zaragoza 1993), Barcelona, 1995.

[7] M. Villuendas, La trigonometría europea en el siglo XI. Estudio de la obra de Ibn Mu^{c}ād, El-Kitāb maŷhūlāt, Barcelona, 1979.

Mathematics in National Contexts (1875–1900): An International Overview

KAREN HUNGER PARSHALL

Departments of Mathematics and History
University of Virginia
Charlottesville, VA 22903-3199, USA

Within the international mathematical community, the last three decades have witnessed a striking number of centennial celebrations. To name just a few, the London Mathematical Society (LMS) entered a new century in 1965 with the Société mathématique de France (SMF) following in 1972, the *American Journal of Mathematics* and the Circolo matematico di Palermo (CMP) saw their centenaries in 1978 and 1984, respectively, and the American Mathematical Society (AMS) passed its century mark in 1988 preceding the Deutsche Mathematiker-Vereinigung (DMV) by two years.[1] These milestones suggest, at the very least, that the mathematical endeavor developed in important ways in diverse national settings during the closing quarter of the nineteenth century.

In recent years, numerous historians of mathematics have labored to document, explain, and interpret this development within distinct national settings.[2] Their work has detailed the seemingly idiosyncratic causes behind the emergence of mathematical communities in individual European countries as well as in the Americas. Depending on the national venue, those causes have included, for example, educational reforms brought on, at least to some extent, by political unification in the case of Italy, by the loss of the Franco-Prussian War in the case of France, and by the philanthropy of individuals made wealthy in an age of expansion and industrialization in the United States. As this brief list indicates, however, although the underlying reasons for it may have differed rather dramatically from country to country, educational reform nevertheless represented a sort of international common denominator in the formation of these national mathematical constituencies in the period, roughly speaking, from 1875 to 1900. A natural question would then seem to be: Are there other such common denominators?

In this paper, I show that this question has a resoundingly affirmative answer by examining some of the recent historical literature on the national mathematical

---

[1]For publications honoring the centennials of the LMS, AMS, and DMV, see [6], [8], and [11], respectively. Gispert detailed the history of the SMF from 1872 to 1914 in her book [12], and Brigaglia and Masotto chronicled the early history of the CMP in [5].

[2]Owing to space limitations, I am able to indicate only a small fraction of this work in the bibliography accompanying this paper. I have tried, however, to select references that also provide bibliographies useful for further research. In this regard, see especially [13].

Proceedings of the International Congress
of Mathematicians, Zürich, Switzerland 1994
© Birkhäuser Verlag, Basel, Switzerland 1995

scenes during this quarter century in Germany, France, Italy, Spain, Russia, Great Britain, and the United States. I set the stage for the analysis by sketching some of the well-known, nineteenth-century contours of the situation in Germany. In the twenty-five-year period under discussion, it is undeniable that, as Bottazzini has so cogently put it, "Germany slowly achieved European hegemony in mathematics" [4, p. 283],[3] and that hegemony extended to the Americas as well. Indeed, in each of the countries considered here, educators in general and mathematicians in particular reacted to, adopted, or adapted — whether directly or indirectly — specific aspects of what they viewed as the German model. I uncover a remarkable sameness amidst the apparent diversity of individual national mathematical histories by selectively tracing the naturalization in many soils of predominantly German institutional and mathematical values. In so doing, I seek to demonstrate the fruitfulness of an international — in addition to a more strictly national — approach in reaching a deeper understanding of a key era in the history of Western mathematics. Through the complex process of professionalization, the twenty-five years from 1875 to 1900 witnessed the formation not only of many of the institutions and values so characteristic of mathematics today but also of an international mathematical endeavor.[4] In short, I hope to provide at least some indication of answers to broader questions such as: How did mathematics become international? and to suggest potentially fertile lines for future historical inquiry. With these goals in mind, let us now turn to the German scene.

A Glimpse at the Contours of the German Context

The opening decade of the nineteenth century was one of great political reorganization in the German states owing to the effects of the Napoleonic Wars. In Prussia, for example, the years from 1806 and the Prussian defeat at the battle of Jena to 1810 and the founding of the University of Berlin witnessed a series of fundamental political, socio-economic, and educational reforms. The latter, spearheaded by Wilhelm von Humboldt, came quickly to dominate not only the educational system in Prussia but those in the other predominantly Protestant German states as well.[5] Von Humboldt's vision of higher education stressed the importance of pure research over the utilitarian concerns perceived as dominant within the post-Revolutionary educational system in France. This emphasis on research accompanied and complemented a strong insistence on academic freedom that developed into the ideals of *Lehr- und Lernfreiheit*, that is, the freedom to teach and to learn without political or religious interference. Such educational reforms aimed not only to support the faculty's search for new knowledge but also to train independent-minded, creative, and original thinkers within an atmosphere of disinterested, scholarly pursuit.

---

[3]All translations presented here of quotations originally in languages other than English are my own.

[4]Because the concept of professionalization in mathematics has been defined and discussed in many places, among them, [19], [12], and [16], I do not redefine it here.

[5]This panoply of issues has also been thoroughly examined. See, for example, the references provided in [16, pp. 24–26] and [19]. The situation, however, was different in the predominantly Catholic, southern German states [20].

These latter aims came to characterize the Prussian system as teaching *and* research increasingly defined the university professor's mission. In the specific case of mathematics, this new research ethic ultimately brought with it greater specialization in the field, as mathematicians and mathematicians-to-be tended to focus their studies in an effort to make their own personal contributions. At the same time, the emphasis on disinterested — as opposed to more applications-oriented — research resulted in the evolution of a fundamentally purist approach to the discipline.[6] Perhaps nowhere were these interrelated aspects of the development of mathematics in Germany more in evidence early on than at the University of Berlin under Dirichlet, Kummer, Weierstrass, and Kronecker [3] and later at the formerly Hanoverian university in Göttingen.

At Göttingen following his assumption of a chair there in 1886, Felix Klein, in fact, set the standards for late nineteenth-century mathematical teaching and activism. As a professor, he brilliantly employed the seminar, an institution that had evolved in the German context as the principal vehicle for the active training of young researchers, to animate a thriving mathematical community in Göttingen [16, pp. 189–234, 239–254].[7] Moreover, as a mathematical activist, he lobbied energetically and successfully for mathematics with Prussian ministerial officials, edited the *Mathematische Annalen*, supported and participated in the activities of the Deutsche Mathematiker-Vereinigung [18], and generally served as an advocate for the field in his efforts to stimulate further the German mathematical community as a whole. These institutions — the graduate seminar, the specialized journal, the specialized society — together with the twin values of research and teaching largely defined the profession and, in subtler ways, the discipline of mathematics as it had developed in Germany by the end of the nineteenth century. These same institutions and values informed the emergence of mathematical research communities in a number of other countries as well and thereby served to build a common foundation for the subsequent internationalization of the field.

The Reverberations in Mathematics of Educational Reform

It was not accidental that this brief sketch of the context of German mathematical developments in the nineteenth century opened with a discussion of educational reform. Changes in higher education and in its overall objectives naturally spurred changes at the level of the individual disciplines. Educational reform tended to affect mathematics even more directly, because one of the key features distinguishing the mathematical endeavor of the nineteenth century from that of the preceding hundred-year period was its venue, namely, the university as opposed to an Academy of Sciences, a royal court, or elsewhere [19, p. 111]. Its effects were not always positive relative to the development of *research-level* mathematics in

---

[6]The processes underlying these developments as well as the interrelations between pure and applied mathematics are, however, much more complicated than these statements might suggest. See, for example, [19] and [20].

[7]Obviously, Berlin and Göttingen did not support the only active graduate programs in mathematics in late nineteenth-century Germany. Paul Gordan and Max Noether at Erlangen, Sophus Lie at Leipzig, and others throughout Germany contributed to the overall profile of higher education in mathematics. For more on the Leipzig seminar, in particular, consult [2].

a given national setting, though, as a comparison of the situations in France and Spain underscores.[8]

In a France awakened from complacency by its loss of the Franco-Prussian War in 1870–1871, Gaston Darboux had already had cause to remark that "we need to mend our [system of] higher education. The Germans get the better of us there as elsewhere. I think that if that continues, the Italians will surpass us before too long" [12, p. 19]. In fact, spurred largely by the military defeat and its implication that the so-called *grandes écoles* were perhaps not grand enough to prepare the French adequately for times of crisis, leaders of the newly formed Third Republic sought to strengthen their political position, at least in part, by fostering an intellectual élite associated not with the *grandes écoles* as had been the case with previous regimes but rather with the *facultés* in each of France's administrative regions. In order to fit the latter for this purpose, the French politicians and educational reformers consciously refashioned them along the lines of the German model [21, pp. 302–303].

In a series of major reforms that took place between 1876 and 1900, the French established new chairs (principally in the provinces) and a new type of position (the salaried post of *maître de conférence*), loosened the old disciplinary boundaries through the creation of chairs in various subdisciplines (as exemplified by Camille Jordan's chair not in mathematics but in higher algebra *per se*), and adopted research productivity as a criterion for determining salary. All of these changes contributed to the rise of a more specialized, research-oriented mathematical profession in France on a par by 1900 with that in place in Germany [12, pp. 59–63].[9]

Still, such reform did not necessarily have a positive impact relative to the development of mathematics at the research level. In 1857, Spain adopted a centralized educational system modeled on the one put in place in France under Napoleon around the turn of the nineteenth century, yet under Madrid's firm control, further change came only slowly. In mathematics, that control translated into the dominance in the advanced curriculum of the projective geometry that Karl von Staudt had developed around mid-century and that Madrid's Eduardo Torroja y Caballé embraced beginning in the 1870s. Although Torroja did advocate *doing* mathematics in his courses at Madrid, he clung doggedly to an area that, over the closing decades of the nineteenth century, grew increasingly distant from, for instance, the more purist Riemannian frontiers of geometrical research [13, p. 1508]. In so doing, Torroja and his adherents in Madrid obstructed the efforts of others in Spain, like García de Galdeano [14, pp. 112–114], to encourage the sort of mathematics being done elsewhere in Europe and particularly in Germany, France, and Italy [1, pp. 162–163].

As the examples of France and Spain illustrate, widespread educational reforms in the last quarter of the nineteenth century affected the development of

---

[8]Space limitations do not allow for the inclusion of the cases of Italy and England, which were presented in the version of this paper delivered at ICM94.

[9]Although the circumstances surrounding their "emergent periods" were certainly quite different, France and the United States were influenced by many of the same external factors in the years from 1875 to 1900. Compare [12] and [16].

mathematics in countries throughout Europe (the United States could be cited here as an example as well). The creation of new academic chairs and institutions, the addition of new grades of instructors, the direct emulation of the German ideals linking teaching, research, and the production of future researchers — these aspects of reform complemented one another in those countries where mathematics at the research level came to define the professional standard. The absence of one or more of them, however, tended to thwart that sort of development.

The Production of Future Researchers

In turn-of-the-century France, Émile Picard summarized well the key role educational reforms had played in the professionalization of high-level mathematics. "Beyond their mission of making the sciences known and understood," he wrote, "the institutions of higher education ... have another [mission], just as noble as all the others, that of advancing science and of continually initiating new generations of researchers to the methods of invention and of discovery" [12, p. 60]. As he clearly stressed, a sense of the importance of the training of future researchers represented one crucial byproduct of these German-inspired reforms. Thus, educators and mathematicians in other countries who looked toward Germany and France for their inspiration and guidance in the final quarter of the nineteenth century tended to conceive of this "noble mission" as an integral part of their endeavor. The United States and Russia provide just two of the possible examples we could examine of this sort of influence.

The years between 1875 and 1900 represented a period of growth and financial prosperity in the United States that had important repercussions in higher education. As great fortunes were made on the railroads, the telegraphs, and industrial expansion in general, individuals like Johns Hopkins and John D. Rockefeller endowed universities through their private philanthropy. The presidents of these new schools, well aware of the educational scenes abroad and especially in Germany, France, and Great Britain, crafted their new institutional philosophies informed by the examples of those foreign systems. In particular, many of them adopted the production of research and of future researchers as explicit missions for their faculties and schools [16, pp. 261–294].

At the University of Chicago, for example, a university financed by Rockefeller and opened in 1892, the American, Eliakim Hastings Moore, and the two Germans, Oskar Bolza and Heinrich Maschke, implemented a training program in mathematics rivaling that of many of their German competitors [16, p. 367]. This comes as no surprise in light of the facts that Bolza and Maschke had learned their trade from Felix Klein and that Moore had spent a year abroad studying mathematics in Göttingen and Berlin. In addition to the regular lecture courses they offered in the established areas of late nineteenth-century mathematics — invariant theory, the theory of substitutions, elliptic function theory, among others — the Chicago mathematicians also incorporated the seminar into their overall pedagogical approach. As especially Bolza and Maschke knew from firsthand experience, the seminar served as a fertile seedbed for the germination of new mathematical ideas along more specialized lines. The educational atmosphere fostered by this faculty produced in short order a number of first-rate mathematicians, notably

Leonard Dickson, Oswald Veblen, Robert L. Moore, and George D. Birkhoff [16, pp. 372–393]. To get a sense of the fruitfulness of the training process these members of the next generation of American mathematicians underwent, let us briefly consider the case of the algebraist, Dickson.

An 1896 Chicago Ph.D. and student of E. H. Moore, Dickson quite naturally pursued a topic in his dissertation reflective of the research interests of his adviser. In an 1893 paper, Moore had examined a number of specific questions in the theory of finite simple groups. In particular, he had presented a codification of the known simple groups of order 600 or less, which had led him to the discovery and explicit proof of the simplicity of what he called a two-parameter family of groups of order $\frac{p^n(p^{2n}-1)}{2}$ for p a prime and $(p,n) \neq (2,1),(3,1)$. (Today, these groups are denoted $PSL_2(p^n)$.) In so doing, he had also explored the nature of the finite fields of order p^n upon which his new groups depended and had proven that every abstract finite field $F(s)$ is, in fact, a Galois field $GF[p^n]$ where $s = p^n$. (For Moore and his contemporaries, the Galois field $GF[p^n]$ was the set of p^n equivalence classes of $\mathbb{Z}_p[X]/(f(x))$ for an indeterminate X and an irreducible monic polynomial $f(X) \in \mathbb{Z}[X]$ of degree n over the prime field $\mathbb{Z}_p = \mathbb{Z}/p\mathbb{Z}$.) The realization that any arbitrary finite field actually had the structure of a Galois field allowed Moore to apply that well-known theory in his analysis of his finite simple groups [16, pp. 375–379].

Moore encouraged Dickson to pursue other algebraic questions involving finite fields $F = GF[p^n]$. For a polynomial $\phi(X)$ of degree $k \leq p^n$ with coefficients in F, Dickson looked, in his dissertation's first part, at the associated mapping $\phi : F \to F$, $\xi \mapsto \phi(\xi)$ and explicitly determined all of the bijective mappings ϕ for $k < 7$, obtaining partial results for $k = 7, 11$. In the second and final part, he shifted to an analysis of the general linear group $GL_m(F)$, consciously extending Camille Jordan's earlier results for the special case of $F = GF[p]$. In studying the composition series of these groups, Dickson proved one of the main results in his thesis, namely, if Z is the center of $SL_m(F)$, then $SL_m(F)/Z$ is simple provided $(m,n,p) \neq (2,1,2),\ (2,1,3)$. Thus, he extended Moore's ideas of 1893 to three-parameter systems of simple groups [16, pp. 379–381].

Dickson continued his work on the theory of linear groups for over a decade before moving on to other lines of algebraic research, most notably in the theory of algebras. As a faculty member at Chicago from 1900 until his retirement in 1939, Dickson perpetuated, through his ongoing work, the style of algebraic research he had learned at the feet of his adviser and then colleague, Moore. His training of almost seventy graduate students, among them his own successor at Chicago, A. Adrian Albert, further solidified that research tradition throughout the United States.[10]

---

[10]Della Dumbaugh Fenster has recently completed a dissertation, entitled "Leonard Eugene Dickson and His Work in the Theory of Algebras" (University of Virginia, 1994), that analyzes Dickson's work in the theory of algebras within the broader context of the consolidation and growth of this algebraic research at Chicago. Compare [16, pp. 427–431] for an analytical framework of the history of mathematics in America in terms of periodization. The years from roughly 1900 to 1930 are characterized there as a period of "consolidation and growth."

Although the broader cultural and political circumstances in Moscow could perhaps not have been more different than those in the Chicago of the late nineteenth century, Moscow University, like the University of Chicago, supported a mathematics program under an activist attuned to contemporaneous mathematical developments in both Germany and France. Nikolai Bugaev, who had studied in both Berlin and Paris for two-and-a-half years beginning in 1863, returned to Moscow to influence a corps of colleagues and students through his broader conception of mathematics. For Bugaev, mathematics involved communication which he fostered through his vigorous support of the Moscow Mathematical Society and of its journal *Matematicheskii Sbornik*, founded in 1864 and 1866, respectively. It also hinged on its university setting, which he worked to strengthen and enhance at Moscow through his efforts first as secretary and then as dean of its faculty of physics and mathematics. Most importantly, it depended on training students capable of contributing to its further development. To the latter end, Bugaev taught a wide range of courses in, for example, number theory, the theory of elliptic functions, the calculus of variations, and the theory of analytic functions, which aimed to introduce his students to these subjects at the research level. He also fostered and contributed to a philosophical atmosphere in which mathematics was interpreted essentially as a theory of functions and where the theory of discontinuous functions played a key role. This conception not only proved conducive to the acceptance of Georg Cantor's novel set-theoretic ideas but also served as the foundation of the Moscow school of function theory, spearheaded in the early decades of the twentieth century by Bugaev's student, D. F. Egorov [7], and perpetuated by Egorov's disciple, N. N. Luzin. This school, which also included such influential twentieth-century mathematicians as P. S. Aleksandrov, A. Ya. Khinchin, and D. E. Menshov, made seminal contributions to the advancement of measure theory and the general theory of functions of a real variable [17].

The cases of both Moscow University and the University of Chicago drive home the obvious point that the success of the mathematical endeavor in a given national context depends crucially on the process of training talented students in areas rich in interesting, open questions. At its core, mathematics undeniably involves proving theorems, and these students not only learned how to carry out that creative process successfully but also embraced the belief that they should pass on their insights to a subsequent generation. As they had been trained, so should they train — this philosophy came to characterize the mathematical mission internationally in the latter quarter of the nineteenth century. Moreover, in concert with the other factors examined above, it encouraged the formation of self-sustaining mathematical communities, that is, interacting groups of people linked by common interests.

The Establishment of Lines of Communication

The formation of a community, however, also turns upon the ability of its members to communicate effectively. Our time period, one in which telegraphy, railroad systems, steamships, and the printed word linked nations internally and with each other, witnessed the widespread creation of at least two sorts of communications vehicles dependent on this new level of mobility: the mathematical society and the specialized mathematical journal.

Although the Moscow Mathematical Society predated it, the London Mathematical Society, which first met under that name in January of 1865, served as a model for mathematical organizers throughout Europe and in the United States. Not only did it bring together mathematicians in and around London and eventually throughout England for the presentation and discussion of mathematical results, but it also published from the outset the *Proceedings of the LMS* for the further dissemination of original research [6, pp. 577–581].

The example of the English mathematicians informed, at least partly, initiatives taken in Palermo for the promotion of research-level mathematics.[11] Giovanni Battista Guccia, a student of both Brioschi and Cremona, established the Circolo matematico di Palermo in 1884 inspired by the examples of both the French Association for the Advancement of Science and the LMS. His objectives for the new society, however, were in some sense more outward looking than those of either of the societies that served as his model. Guccia sought to create an organization that united mathematicians internationally at the same time that it spurred advanced work in Italy by drawing it into the wider international arena [5, pp. 51–75]. He worked to achieve these goals by actively soliciting foreign members as well as by publishing, beginning in 1885, the Circolo's *Rendiconti*. Guccia recognized the distinct advantages of an international as opposed to a strictly national posture for the overall vitality of mathematics at the research level.

As this brief discussion of the establishment of national mathematical societies highlights, mathematicians during the closing quarter of the nineteenth century recognized the importance of communication both in person and in print for the advancement of their discipline. Their organization of societies further reflected their growing sense of mathematics as a profession, while their creation of new publication outlets reinforced the standards adopted for that profession.[12] By the late nineteenth century, to be a mathematician meant the same thing internationally, namely, to produce and to share the results of original research with like-minded members of an extended community of mathematical scholars both at home and abroad.[13]

---

[11] Michel Chasles also looked to the example of the LMS in calling for the establishment of an analogous society in France in a report of 1870. The institutional void that he sensed was ultimately filled in 1872 following the formation of the Société mathématique de France [12, pp. 14–17]. The LMS directly influenced the founders of the American Mathematical Society as well [16, pp. 267–268].

[12] Here, we could clearly also cite many examples of journals founded during this time period that were independent of the mathematical societies formed: the *Mathematische Annalen* founded by Alfred Clebsch in Germany in 1868, Gaston Darboux's *Bulletin des sciences mathématiques* started in 1870, and the *American Journal of Mathematics* begun by James Joseph Sylvester in the United States in 1878, to name just three of the earlier periodicals.

[13] For a quantitative sense of the depth of the American mathematical research community, see, for example, [9] and [10]. [12] provides an analysis of the broader French mathematical constituency, and [15] gives some indication of the situation in Spain.

An International Overview: Some Common Denominators

From this comparative inquiry the following composite of the parameters of the mathematical endeavor during the last quarter of the nineteenth century now emerges. First and foremost, the establishment internationally of a mathematical *profession* during this time period largely — although not exclusively — hinged on changes in higher education in the various national settings. Although these changes came about often through very different sequences of events — political, financial, philosophical, pedagogical — and so for very different reasons in different national contexts, they nevertheless tended to provide more and more conducive settings within higher education for the study and pursuit of research-level mathematics. In particular, the adoption of nationally tailored versions of the twin German principles of *Lehr- und Lernfreiheit* in various countries brought with it a redefinition of the role of the professor of mathematics that increasingly encompassed the dual activities of teaching *and* the production of original research. Education at a graduate level thus developed in order to train students capable of realizing these two goals.

As the discovery of new mathematical results came to set the standard of entry into the evolving profession, the discipline defined itself in more specialized terms. Universities split their chairs of mathematics *and* physics or of mathematics *and* astronomy and even created chairs in specific mathematical subdisciplines. This specialization resulted both in the sharpening definition of mathematical areas and in an increase in the number of positions available in the field. This latter aspect of the evolution of a profession was also influenced by the establishment of new grades of instructors under the professor (*Dozenten*, *maîtres de conférences*, assistant and associate professors, etc.). As individuals sought out this graduate training, as they assumed these new positions, as they adopted these values of teaching and research, they banded together in national or broadly based mathematical societies and shared their new research through specialized journals targeted at an appreciative and understanding audience. The individual nationalization of mathematics was thus well underway by the end of the nineteenth century, and because the model emulated was largely the same in the various national contexts, this implies that the internationalization of the field was likewise in process. Perhaps no one piece of evidence supports this latter conclusion better than the fact that Zürich hosted the first International Congress of Mathematicians in 1897.

By taking an international perspective on the development of mathematics over the period from roughly 1875 to 1900, this analysis has thus uncovered a number of factors common to particular national settings, which strictly nationally oriented studies tend perhaps to obscure. In so doing, it has hopefully also shed some light on the complexity of the process of the internationalization of mathematics and has suggested at least implicitly some fruitful lines for further historical research.

References

[1]   Elena Ausejo and Ana Millán, *The Spanish Mathematical Society and its periodi-
      cals in the first third of the 20th century*, in Messengers of Mathematics: European
      Mathematical Journals (1800–1946), (eds.) Elena Ausejo and Mariano Hormignón,
      Madrid: Siglo XXI de España Editores, 1993, pp. 159–187.

[2]   Herbert Beckert and Horst Schumann (eds.), 100 Jahre Mathematisches Seminar der
      Karl Marx-Universität Leipzig, Berlin: VEB Deutscher Verlag der Wissenschaften,
      1981.

[3]   Kurt-R. Biermann, Die Mathematik und ihre Dozenten an der Berliner Universität,
      1810–1933, Berlin: Akademie-Verlag, 1988.

[4]   Umberto Bottazzini, *Il diciannovesimo secolo in Italia*, in Mathematica: Un profilo
      storico, by Dirk Struik, Bologna: Il Mulino, 1981, pp. 249–312.

[5]   Aldo Brigaglia and Guido Masotto, Il Circolo matematico di Palermo, Bari: Edizioni
      Dedalo, 1982.

[6]   E. F. Collingwood, *A century of the London Mathematical Society*, J. London Math.
      Soc. **41** (1966), 577–594.

[7]   Sergei Demidov, *N. V. Bougaiev et la création de l'école de Moscou de la théorie
      des fonctions d'une variable réelle*, in Mathemata: Festschrift für Helmuth Gericke,
      (eds.) Menso Folkerts and Uta Lindgren, Stuttgart: Franz Steiner Verlag Wiesbaden
      GMBH, 1985, pp. 651–673.

[8]   Peter Duren *et al.* (eds.), A century of mathematics in America, History of Mathe-
      matics, vols. 1–3, Providence: Amer. Math. Soc., 1988–1989.

[9]   Della Dumbaugh Fenster and Karen Hunger Parshall, *A profile of the American
      mathematical research community: 1891–1906*, in The History of Modern Mathe-
      matics, (eds.) Eberhard Knobloch and David E. Rowe, vol. 3, Boston: Academic
      Press, 1994, pp. 179–227.

[10] _____, *Women in the American mathematical research community: 1891–1906*, in
      The History of Modern Mathematics, (eds.) Eberhard Knobloch and David E. Rowe,
      vol. 3, Boston: Academic Press, 1994, pp. 229–261.

[11] Gerd Fischer, Friedrich Hirzebruch, Winfried Scharlau, and Willi Törnig (eds.), Ein
      Jahrhundert Mathematik 1890–1990: Festschrift zum Jubiläum der DMV, Doku-
      mente zur Geschichte der Mathematik, vol. 6, Braunschweig/Wiesbaden: Friedr.
      Vieweg & Sohn, 1990.

[12] Hélène Gispert, La France mathématique: La Société mathématique de France
      (1872–1914), Paris: Société française d'histoire des sciences et des techniques &
      Société mathématique de France, 1991.

[13] Ivor Grattan-Guinness (ed.), Companion Encyclopedia of the History and Philos-
      ophy of the Mathematical Sciences, 2 vols., London/New York: Routledge, 1994,
      2:1427–1539 on higher education and institutions.

[14] Mariano Hormigón, *García de Galdeano and el progreso matematico*, in Messengers
      of Mathematics: European Mathematical Journals (1800–1946), (eds.) Elena Ausejo
      and Mariano Hormignón, Madrid: Siglo XXI de España Editores, 1993, pp. 95–115.

[15] Eduardo Ortiz, *El rol de las revistas matemáticas intermedias en el establecimiento
      de contactos entre las comunidades de Francia y España hacia fines del siglo XIX*, in
      Contra los titanes de la rutina/Contre les titans de la routine, (eds.) Santiago Garma,
      Dominique Flament, and Victor Navarro, Madrid: Comunidad de Madrid/C.S.I.C.,
      1994, pp. 367–382.

[16] Karen Hunger Parshall and David E. Rowe, The emergence of the American mathematical research community (1876–1900): J. J. Sylvester, Felix Klein, E. H. Moore, History of Mathematics, vol. 8, Providence: Amer. Math. Soc. and London: London Math. Soc., 1994.

[17] Esther Phillips, *Nicolai Nicolaevich Luzin and the Moscow school of the theory of functions*, Historia Math. **5** (1978), 275–305.

[18] Norbert Schappacher, *Fachverband — Institut — Staat*, in [11], pp. 1–82.

[19] Gert Schubring, *The conception of pure mathematics as an instrument in the professionalization of mathematics*, in Social History of Nineteenth Century Mathematics, (eds.) Herbert Mehrtens, Henk Bos, and Ivo Schneider, Boston/Basel/Stuttgart: Birkhäuser, 1981, pp. 111–134.

[20] _____, *Pure and applied mathematics in divergent institutional settings in Germany: The Role and Impact of Felix Klein*, in The History of Modern Mathematics, (eds.) David E. Rowe and John McCleary, 2 vols., Boston: Academic Press, 1989, 2:171–220.

[21] Terry Shinn, *The French science faculty system, 1808–1914: Institutional changes and research potential in mathematics and the physical sciences*, Historical Studies in the Physical Sciences **10** (1979), 271–332.

Hermann Weyl's "Purely Infinitesimal Geometry"

ERHARD SCHOLZ

Fachbereich 7: Mathematik, Bergische Universität
Gauß-Straße 20, D-42097 Wuppertal, Germany

Weyl's view of the continuum

The years 1916 and 1921/22 delimit a phase of Weyl's work during which he made
his most radical contributions to the foundations of mathematics as well as of
highly innovative contributions to differential geometry and classical field theory.
All of this work had a strongly speculative background. Although Weyl lived and
worked in Zürich from 1913 onward, he continued to be a German citizen and had
been drafted into the German army in May 1915. He was discharged from the
service after an intervention by the Swiss government in August 1916.[1] The deep
crisis of European culture in the years surrounding World War I was deeply felt
in a very personal way by Weyl, although his reaction had much in common with
the way this crisis was felt by other German academic intellectuals, particularly
in the cultural sciences.[2]

Weyl's conception of the continuum during that time shifted from the semi-
constructivist position in [W3] to a clear adherence to Brouwer's intuitionism and
became more cautious with some openness to Hilbert's position from the middle
of the 1920s. A common feature during the shifts was that he wanted to link the
mathematical concept of continuum to a "directly experienced continuity" [W5, p.
527] so that the latter could never be identified with an extensionally closed realm
or even object of thought. Weyl prized Brouwer's theory as it offered a means of
representing the continuum as a "medium of free becoming", an expression carrying rather clear Fichtean connotations.[3] And, in fact, he chose deliberately to link
the continuum with a concept structure elaborated in the dialectical philosophy of
the early nineteenth century, a notion whose logical features may appear surprising
for a twentieth-century mathematician.

A general concept, *the whole* (here the continuum), has to be presupposed
in order to give meaning to an individual determination, the particular, or the
part (here the point).[4] On the other hand, *the whole*, the general concept (the
continuum) is constituted in a process of common generation by the particulars,
(the parts).

---

[1] [S, 61ff.]; [FS, 20ff.]

[2] A broad cultural picture of the self-understanding of cultural scientists in Germany has been
given by Ringer [R].

[3] Weyl had already admitted the Fichtean inspiration of his thoughts in the foundations of
mathematics in his book *Das Kontinuum* [W3, p. 2].

[4] So far this conceptual figure was also shared by Riemann in his inaugural lecture of 1854.

Proceedings of the International Congress
of Mathematicians, Zürich, Switzerland 1994
© Birkhäuser Verlag, Basel, Switzerland 1995

This conceptual figure came close to those procedures that had been called "impredicative" by Russell and Poincaré and that had been blamed as being responsible for contradictions of the type of Russell's antinomy. One example in real analysis, debated by Weyl himself, was the definition of a real number a as $a := \sup(A)$ for a bounded set $A \in \mathbf{R}$, as in this definition reference to the totality of the reals is made. So, for Weyl, it was clear that an attempt to translate too directly what he regarded to be the "essence of the continuum" into mathematical form might easily lead to logical difficulties.[5] Weyl saw and discussed two conceptual strategies to avoid the dilemma:

Strategy I: Design of an "atomistic" theory of the continuum (either contructive or axiomatic).

This approach might be logically consistent but would, in Weyl's view, never lead to sufficiently rich expressions of the intuitive idea of the continuum, as Weyl immediately remarked in a discussion of his own "atomistic" theory in [W3, p. 83]; [W5, p. 527]. A set-theoretic approach with a topological characterization of continuity appeared even less convincing to him. Even in 1925, when his most radical phase in the foundational debate had already ended, Weyl made quite clear that such an approach was not convincing to him because a set-theoretic approach:

> ... contradicts the essence of the continuum, which by its very nature cannot be battered into a set of separated elements. Not the relationship of an element to a set, but that of a part to the whole should serve as the basis for an analysis of the continuum. [W6, p. 5]

But what, then, might be understood as the "relationship of a part to the whole" in the case of the continuum, independent from set theory? Weyl never gave a final answer to that obviously very difficult and perhaps unanswerable question, but he did outline how he wanted to approach it.

Strategy II: Development of mathematical theories that symbolically explore the "relationship between the part and the whole" for the case of the continuum.

Fundamental for this strategy was an idea expressed by Weyl in his 1925 article for the Lobachevsky prize as follows:

> ... a manifold is continuous if the points are joined together in such a way that it is impossible to single out a point just for itself, but always only together with a vaguely delimited surrounding halo (Hof), with a neighbourhood. [W6, p. 2]

As examples, Weyl referred to the characterization of a function element in complex analysis and the characterization of neighborhood systems of a point as a limit structure in topology. His "infinitesimal geometry" of 1918 was designed as another, and he hoped for a time, far-reaching, contribution to this strategy. This work should therefore be read with this context in mind.

---

[5]Actually Weyl discussed this type of conceptual problem inherent in the classical approach to analysis from his point of view as the necessity to distinguish between "extensionally definite" and (intensionally) well-determined properties in his letter to Hölder, published as [W4].

The design of a "purely infinitesimal geometry"

From the perspective of Weyl's view of the continuum around 1918, differential geometry may be considered as one line of access to the question of how to link the infinitesimal "halos" of the point to the structure of the *whole*, the manifold. From this point of view, the differential geometric structures should be defined such that only relations in each infinitesimal neighborhood are immediately meaningful. Relations between quantities in different neighborhoods (of finite distance) ought to be considered meaningful only by mediation of the *whole*, more technically expressed by an integration process over paths joining the two points at the centers of the neighborhoods. From the point of view of building the continuum from its *smallest parts*, so Weyl claimed over and over again in the years following 1918, Riemann's differential geometry did not appear completely convincing. In Riemannian geometry the relationship between lengths of vectors ξ and η is well defined, independent of the points p and q of the manifold to which they are attached ($\xi \in T_pM, \eta \in T_qM, p \neq q$). From Weyl's continuum-based view of differential geometry such a comparison appeared unmotivated, and he stipulated instead that a

> ... truly infinitesimal geometry (wahrhafte Nahegeometrie) ... should know a transfer principle for length measurements between infinitely close points only. [W2, p. 30]

In this formulation Weyl alluded to Levi-Civita's transfer principle of direction in a Riemannian manifold embedded in a sufficiently high-dimensional Euclidean space, locally given by

$$\xi'^i = \xi^i - \Gamma^i{}_{jk}\xi^j dx^k$$

with the dx^i to be interpreted as the coordinate representation of a displacement vector between two infinitesimally close points so that the direction vector ξ^i has been transferred to ξ'^i.

Weyl immediately recognized that Levi-Civita's concept of parallel displacement wonderfully suited his nascent ideas about how to build differential geometry strictly on the basis of infinitesimal neighborhoods. He discussed it in this light during his lecture on general relativity in the summer of 1917 at Zürich, not yet knowing how to proceed similarly for the measurement and comparison of lengths.[6]

This was the motivation behind Weyl's effort to separate logically the concept of parallel displacement from metrics and to introduce what he called an *affine connection* Γ on a (differentiable) manifold as, speaking in later terminology, a linear torsion-free connection. Guided by the example of affine connections Weyl also proceeded to build up the metric in a manifold from a "purely infinitesimal view". The result was his introduction of a generalized Riemannian metric, a *Weylian metric* on a differentiable manifold M, which is given by:

---

[6]Weyl described in retrospect (1946) a discussion he had during his 1917 lecture course with one of his students, Willy Scherrer, that triggered these doubts. This story and its importance for Weyl's infinitesimal geometry has been described convincingly by Sigurdsson [S, p. 154].

1. a *conformal structure* on M, i.e. a class of (semi-) Riemannian metrics $[g]$ in local coordinates given by $g_{ij}(x)$ or $\tilde{g}_{ij}(x) = \lambda(x)g_{ij}(x)$, with multiplication by $\lambda(x) > 0$ (real valued) representing what Weyl considered to be a *gauge transformation* of the representative of $[g]$, and

2. a *length connection* on M, i.e. a class of differential 1-forms φ in local coordinates represented by $\varphi_i dx_i$, $\tilde{\varphi}_i dx_i = \varphi_i dx_i - d \log \lambda$ (representing the *gauge transformation* of the representative of φ).

In modern terms, Weyl introduced a connection ω in the line bundle $\pi : L \longrightarrow M$ associated to the $GL(n, \mathbf{R})$ principal bundle of frames in the tangent bundle TM or — assuming orientability — in the principal bundle of positive 1-frames in L with group $G = (\mathbf{R}_+, \cdot)$. Here ω assumes values in the Lie algebra $\mathcal{G} = \mathbf{R}$ with trivial multiplication ($[x, y] = 0$) and can, after local trivialization, be identified with a 1-form ω. Change of trivialization is here given by a pointwise multiplication by $\lambda(x)$ with values in $\mathbf{R}_+$. That leads to a gauge transformation

$$\tilde{\omega} = \lambda^{-1}d\lambda + \lambda^{-1}\omega\lambda = \lambda^{-1}d\lambda + \omega = d \log \lambda + \omega \,,$$

which coincides up to sign with Weyl's formula. Weyl, of course, derived the gauge transformation as a compatibility condition for length transfer expressed in different representatives of the conformal metric.

The length transfer by a connection form φ on M was introduced by Weyl in close analogy to the direction transfer in M by an affine connection. Specification of length calibration in some point $p \in M$ may be represented as the determination of a positive real value $l(p)$. Without further specifications this calibration is meaningful only as a measuring device at the point p in M. In order to compare lengths of vectors attached to different points such a measure has to be transported from one point to another. For two infinitesimally close points p, p' with displacement vector α (for a modern reader to be understood as $\alpha \in T_pM$) transport of length measure was introduced by Weyl as fulfilling the condition

$$l(p') = l(p) - \varphi(\alpha)l(p) = l(p)(1 - \varphi(\alpha)) \;;$$

i.e. length has to recalibrated in an infinitesimal transfer along the displacement vector α by the factor $(1 - \varphi(\alpha))$. For two finitely distant points $p, q \in M$, recalibration is given by a factor of the form

$$1 - \int_I \varphi(\dot{c}(t)) \, dt$$

integrated along a path parameterized by I. So length comparison between finitely distant points becomes in general path dependent, just like comparison of direction for affine connections.

In consequence, Weyl introduced *length curvature* as giving the length difference $l - l'$ for transfer along the periphery of an infinitely small parallelogram with sides α, β. This turned out to be a 2-form[7]

$$l - l' = f(\alpha, \beta) = -f(\beta, \alpha) \text{ with } f = d\varphi.$$

---

[7]For modern gauge theorists this is no surprise, as with hindsight one notes that Weyl worked with an abelian gauge group, therefore curvature should be $R = d\omega + \frac{1}{2}[\omega, \omega] = d\omega$.

Weyl explored his new geometry and derived, among others, the following basic properties.

THEOREM 1 *A Weylian metric is (semi-)Riemannian*[8] *if and only if $f \equiv 0$.*

THEOREM 2 *In a Weylian manifold there exists exactly one compatible affine connection Γ.*[9]

So there exists a curvature tensor $F^i{}_{jkl}$ of Γ that is moreover, as Weyl remarked, gauge invariant. This appeared quite remarkable for Weyl as he considered gauge invariance an important criterion for physical relevance of quantities in his new geometrical structure, in addition to general covariance.

The derivation of the length curvature from the length connection is structurally identical with the relationship between the relativistic electromagnetic field f_{ij} and its 4-potential φ_i. Thus, obviously, the *first set of Maxwell equations* hold

$$df = dd\varphi = 0, \quad \frac{\partial f_{jk}}{\partial x^i} + \frac{\partial f_{ki}}{\partial x^j} + \frac{\partial f_{ij}}{\partial x^k} = 0.$$

Guided by Mie's relativistic formulation of Maxwell's theory Weyl considered the vector densities $\tilde{f}^{ij} := \sqrt{-\det(g_{ij})} f^{ij}$ and the the source equation for them as an expression of the *second set of Maxwell's equations* in his theoretical framework

$$\frac{\partial \tilde{f}^{ij}}{\partial x^j} = \rho^i.$$

For readers of Weyl's papers of about 1918 such an approach may have appeared as a nice but still rather formal analogy. But for Weyl it appeared to be much more; in fact he appeared to be carried away by this analogy and took it as an indicator of a more or less obvious and forceful semantical bridge between mathematics and physics. So, for example, in his first paper on the subject in *Mathematische Zeitschrift*, addressed to a mathematical audience, Weyl barely made any distinction between mathematical concepts and physical interpretation [W1]. He seemed to be completely sure that his analysis had led him to a

> ... world metric from which not only gravitation but also the electromagnetic effects result, which therefore gives account of all physical processes, as one may assume with good reason. [W1, p. 2]

For the reader the "good reasons" for such a realistic semantical turn in the argument could not be at all so clear as they seemed to be for Weyl. Taken at face value they were not much more than analogies between the curvature equations and Mie's theory of the electromagnetic field. That was, of course, sufficient reason to investigate whether it was possible to extend this connection to a more elaborate theoretical link. For Weyl, however, there appeared to be not so much a question

---

[8] A Weylian metric reduces to the semi-Riemannian case, if $\varphi \equiv 0$ after proper choice of gauge, so that the calibration can be integrated independently of the path.

[9] Compatibility of an affine connection Γ with a Weylian metric means conservation of angles by parallel transport according to Γ and length transported as would be done by φ.

but rather already an answer. Surprising as such a semantical jump may appear for later readers, we want to take a glance at Weyl's "good reasons" for accepting the connection between his new geometry and physics on the spot. For that we have to take Weyl's philosophical background, in particular his Fichtean studies, into account.

Weyl's Fichtean background

In the period between 1916 and 1922 Weyl was strongly fascinated by Fichte's work because in it he found "metaphysical idealism in its most unreserved and forceful expression". With that position Weyl took up a motif that was shared by a group of intellectuals in Germany at the beginning of the twentieth century who felt the rise of positivism as an indicator of a cultural crisis and looked for strong idealist counterforces in classical German philosophy. For Weyl, Husserl's phenomenology, which he had started to study even earlier, shifted into the background during these years. It continued to be represented in his personal universe of discourse, however, by his wife Helene, who had been a student of Husserl's and was now an expert in phenomenology. Her views thus served Weyl as a sort of critical counterbalance to Fichte's highly speculative version of idealism. In our context Fichte's derivation of the concepts of space, time, and matter, in addition to those features of his philosophy that made Fichte "a constructivist of the purest water", as Weyl later said [W8, p. 637], are of particular concern.

The constructivist aspect in Fichte's philosophy of knowledge is most clearly expressed in the latter's essay on "transcendental logic" [F2]. Here Fichte discussed the relationship between classical logic and *transcendental logic*, which Fichte used synonymously with the terms *philosophical* or *dialectical* logic. Reading Fichte today and from the point of view of the history of mathematics, we find striking similarities between Fichte's opposition between *classical* and *transcendental logic* and the differences between Hilbert and Weyl with respect to formalist versus intuitionist mathematics.[10] Fichte criticized classical logic because it took concepts as something given and analyzed their relationships by formal means only. Philosophical (*transcendental*) logic would, on the contrary, generate the concepts in a sort of reconstruction ("Nachconstruction") always trying to go back to some presupposed final origin ("das Ursprüngliche") [F2, p. 122]. In this context Fichte gave a very general explication of what he thought *construction* should achieve:

> Construction is the instruction to invent the concept by the power of imagination (Einbildungskraft) such that evidence is gained. [F2, p. 188]

So, for Fichte, *construction* is not reducible to technical means of concept generation that are specified once and for all, although such technical means are obviously not excluded. Transcendental logic or dialectics should even give a "lawful method"

---

[10]Even the rather well-known chess metaphor used by Weyl to describe the nature of Hilbert's formalist proof theory [W5, p. 535ff.] had been used in a completely analogous way by Fichte to characterize the procedures of classical logic from the viewpoint of transcendental logic [F2, p. 387].

for such a construction, i.e. characterize the conditions of reproducibility for a generation process of the mind. For a reader like Weyl, the correlation with Brouwers' intuitive view of mathematical concepts could not be closer. Obviously it might be quite instructive to reread Weyl's "constructivist" approach to mathematics in light of this Fichtean background, but that is not the goal of this paper.

Fichte exemplified his general views by considering the construction of the concepts of space, time, force, and matter; notions that appear over and over again in his *Wissenschaftslehre*. The basic figure of Fichte's generative dialectics of knowledge was the opposition of a so-called *self* (Ich) to a *non-self* (Nicht-Ich), which lies at the bottom of the Fichtean generative myth of all the concepts of the world. At the beginning and at the end of this generation process, the self is considered as "absolute", i.e. not opposed to any non-self, and was declared by Fichte as "absolutely free". During the transitional steps the self becomes limited and its freedom restricted by the opposition to the non-self. All the derivations in Fichte's *Wissenschaftslehre* deal with different steps toward an assimilation of certain features of the non-self, limiting the activities of the self. Part of these derivations is the generation of the concept of space, which Fichte described as follows:

— Take separate "products of the non-self". They are posed by the imagination (by the "self") into separate points X, Y[11] and endowed with separate "spheres of action (Sphären der Wirksamkeit)" z_X, z_Y, mutually excluding each other.

— Yet, on the other hand, between the mutually excluding "spheres of activity" there is, according to Fichte, "necessarily continuity". In this sense a unity between any such "spheres of activity" is posed by the imagination.

— The unity is produced in a medium of "continuous free activity" of the mind. This medium of free activity *is space*, symbolized as O by Fichte. [F1, p. 194–196]

Thus, in Fichte's derivation, Kant's transcendental space concept was taken up (*space* as unifying principle of the mind for determinations of localized activities) and radicalized (medium of "continuous free activity"). Thus Fichte's notion of space was devoid of any specific a priori structure like, for example, Euclidean geometry. With respect to mathematical structure his approach could be considered as generalizing Kant's, and Fichte went on from here to the construction of the concept of matter, considering the "spheres of action" as *intensities* irreducible to, but united in *extension O*. He considered the "intensities" of the "spheres of action" as being "forces" that fill the space with their product [F1, p. 201]. Matter, finally, was to him nothing but the external manifestation of the activities of forces.

The last motif, matter as nothing but the manifestation of forces [F2, p. 356f.], was rather widespread in German natural philosophy around the turn of the eighteenth to the nineteenth century. It had been expressed by Kant, radicalized by Fichte, turned into a more subtle, qualitatively differentiated version of productivity by Schelling, and finally assimilated by Hegel into his system. For Fichte, as for the other post-Kantians, there was no empty space as an a priori form in which the senses could locate their empirical information. Space was the connecting activity between different "spheres of action" reconstructed by the self. The latter figured for Fichte as a sort of semantical meeting point for quite different

---

[11]Symbolism X, Y used by Fichte.

features. They constituted the "infinitely small parts of space" and the building forces of matter. They represented the "interior forces of the non-self acting in absolute freedom" [12] but had to be reconstructed by the self as "spheres of action" of the productive imagination [F1, p. 201f.].[13]

When Weyl tried, beginning around 1918, to conceive the mathematical continuum as the uniting medium of the individual determinations that cannot be "battered" into its single elements without loss of meaning, he was taking up an old topic in the philosophy of mathematics, which can be traced back to Aristotle. But in Weyl's rhetoric (the continuum as "medium of free becoming") the ties to Fichte's conception of continuity as the medium of free activity of the self were apparent to those who had read the latter. Fichte's "spheres of activity" bound inseparably to the points stood in close affinity to Weyl's goal of constituting the concept of mathematical continuum by some structural link binding the points inseparably to the general concept. But of course such a reference was not enough in itself to become mathematically productive. That the conceptual figures elaborated in nuce in an adaptation of Fichtean topics could take hold for several years so strongly on Weyl's mind and actually bear mathematical fruit depended on another development that had taken place in mathematical physics and that fit nicely into his philosophical scheme.

Mie – Hilbert – Weyl

In 1912 the Greifswald physicist Gustav Mie had formulated a research strategy to explain the basic phenomena of matter on a purely electromagnetic basis, in particular the existence, mass, and stability of the electron [M1]. This program continued pre-relativistic attempts by Lorentz, Kaufmann, and others to develop a purely electromagnetic theory of the electron and its mass. Mie started in 1912 from the established concept of special relativity and in particular its energy-mass equivalence. He considered Lorentz-invariant scalar density functions L on Minkowski space-time depending on the values of a 4-potential φ and its first derivatives $D\varphi$ (Mie, of course, expressed these in coordinates $\varphi_i, \frac{\partial \varphi_i}{\partial x^k}, i, k = 1, \ldots 4$). This led him to a variational criterion for potentials

$$\delta \int L(\varphi, D\varphi)\, dx = 0$$

with respect to infinitesimal variations of φ. Starting from the observation that the Maxwell vacuum equations result from the particular case

$$L(\varphi, D\varphi) = \frac{1}{2} F_{ik} F^{ik} \ , \quad F_{ik} = \frac{\partial \varphi_i}{\partial x^k}$$

---

[12] "Absolute" freedom of the non-self to be understood as unlimited by the activities of the self.

[13] The concept of space was thus generated by Fichte in a figure very close to the one used by him (and the other dialectical philosophers of the early nineteenth century) to characterize the relationship between the individual and society (the state, in Fichte's language). Such a social metaphor for the general figure of generation of the space concept was taken up by Weyl in his [W7, p. 46f.].

he proposed to look for modifications of L leading to nonlinear terms in the field equation and solutions with space-like local energy concentrations ("energy knots"), which, as Mie hoped, might turn out to be stable, i.e. nondispersive in time.

Mie thus tried to bring the old dream of a dynamistic matter explanation, which had been philosophically expressed by Kant, to physico-mathematical maturity, although the technical framework in which Mie located his approach reduced the general concept of force to the electromagnetic setting.[14] This reduction of the dynamistic matter explanation to purely electromagnetic actions had also enjoyed a long tradition among natural scientists going back deep into the nineteenth century. As a conjectural approach it had already been expressed by Weber, Riemann's teacher in physics. It was actively pursued by Zöllner, and to a certain degree by Riemann. Around the turn of the century, Kaufmann, Wiechert, and again to a certain degree Lorentz, had taken this approach up, placing it in the setting of Maxwell's theory of electrodynamics.[15]

The daring mathematical design of Mie's program, combining special relativity with variational viewpoints, and with the Kant-Weber-Riemann connection looming in the natural philosophical background, may have contributed to its enthusiastic reception by Hilbert. As is well known, Hilbert's research for a generally covariant formulation of relativity in 1915 was built on and modified Mie's program [EG], [V, p. 54ff.]. Hilbert extended Mie's variational principle to a Lagrangian $L(g, Dg, D^2g, \varphi, D\varphi)$ depending on a varying Lorentz metric g (expressed, of course, in coordinates as $g_{ij}(x)$), as well as its derivatives up to the second order, and a 1-form φ with its first derivatives, where the Lagrangian had to fulfill the condition of invariance under general coordinate change. So Hilbert's Lagrangian L lived on a Lorentz manifold, and for Hilbert the underlying invariance postulate had a rather direct semantical justification. Combining Riemann's and Minkowski's views, physically meaningful functions are defined on space-time events and thus on a physical manifold, which ought to be infinitesimally Minkowskian rather than Euclidean. Read today, that may seem a rather apparent step. That it was not so at the time, however, is shown by the long journey Einstein had to take between 1912 to 1915 before he finally convinced himself that (semi-)Riemannian geometry was more than a formal game and could be endowed with strong physical meaning.[16] Hilbert's goal, moreover, in that respect also differed from Einstein's, as it was not primarily directed towards gravitation theory.[17] Hilbert was much more ambitious; following Mie he wanted to outline the basis for a universal theory of matter. He was quite frank in expressing this ambitious goal, and, even more, he actually believed for some time that he had come close to it. He thus stated his conviction that by means of his generalized version of Mie's field theory of matter

---

[14]Mie was apparently completely aware of the neo-Kantian connotations of his research program. An explicit formulation is to be found in [M2]; and there is no reason to doubt that he was already aware of this connection in 1912.

[15]See [McC] for more details.

[16]Compare [Sta] and [Rea].

[17]Nevertheless, Hilbert derived a version of field equations that could be considered as equivalent to Einstein's if read from the point of view of theory of gravitation [H1].

... the most intimate hidden processes in the interior of the atom can (...) be clarified.

Taking such strong hopes into account, expressed by Hilbert, it may appear less surprising that the much more "emphatic" thinker[18] Weyl was nearly overwhelmed by the prospect of a close link between his Fichtean-inspired researches on the conceptual structures of the continuum and the Mie-Hilbert theory of matter. Surely he must have felt strong and surprising semantical resonances between the two different worlds of discourse in which he lived in 1918, the mathematical/physical one centered at Göttingen and the philosophical/cultural circles he moved in at that time in his local Zürich environment.[19] These resonances must have appeared even more seductive for Weyl, as with regard to foundational questions proper the two realms of discourse brought him into strong opposition with his former teacher Hilbert, stronger than he would have liked personally. Weyl could now hope for a basic agreement between him and Hilbert, at least in questions of mathematical philosophy of nature as well as a place for himself in a renewed Göttingen universe of mathematics.

Instead, however, he had to live through disillusionment in several respects before he could continue to explore his basic ideas in a direction that found acceptance at Göttingen. With respect to the theory of matter the reaction of physicists was split. For a short time he found some strong support from Sommerfeld, Pauli and Eddington, whereas Einstein was more than skeptical from the outset. But even the early supporters turned away from Weyl's theory after the first possibilities of directly tapping its explicative power for physical phenomena had to be weakened.[20] Even more disillusioning for Weyl may have been Hilbert's rather strong rejection of Weyl's theory in the first years after its publication. Thus, in a lecture course given in the winter semester 1919/1920 at Göttingen, Hilbert criticized it as a kind

... of Hegelian physics trying to predetermine the whole world process which would not transcend the limited content of a finite thought, if one takes that view seriously. [H3, p. 100]

This flat rejection cannot be understood from a purely mathematical point of view,[21] and in fact it was explicitly formulated by Hilbert as a difference in the mathematical philosophy of nature. Of course, it was easy for Hilbert to identify a strong idealist reductionism in Weyl's approach,[22] as Weyl quite frankly expressed his Fichte-inspired drive to give direct physical meaning to the concepts of "infinitesimal geometry". But, in effect and read more carefully, Hilbert's strong rejection of Weyl's theory argued (perhaps unknowingly) against temptations he, Hilbert, in fact shared with Weyl. Hilbert himself was a strong proponent

---

[18]The characterization of Weyl's style of thought as "emphatic" goes back to E. Brieskorn in a comparison with the thought style of Hausdorff.

[19]To give names: F. Medicus and Helene Weyl.

[20][Str], [S], [V].

[21]Compare Rowe's remark in [H3, p. xif].

[22]Although Hilbert slightly misrepresented Weyl's philosophical motivations ("Hegelian physics") if examined more closely.

of mathematical reductionism in the study of nature (and probably a more rigid one than Weyl), and he was himself much more prone to shortcut solutions when it came to the dialectics of finiteness and infiniteness in mathematical theories, as was shown by his involvement in the debates and research on the foundations of mathematics. Hilbert's brillant former student, Weyl, now (around 1918) deviated from central tenets governing the Göttingen mathematical discourse. He was thus ideally suited as an external target for a projection of basic problems inherent in these rules, problems which were only vaguely intuited by Hilbert and never accepted by him as inherent to his own approach.

Only after Weyl had partially withdrawn his radical opposition to Hilbert's views with regard to foundational questions and outlined how a compatibility between both approaches could be established on a pragmatic level [W5] did Hilbert offer a positive reevaluation of Weyl's contributions to differential geometry and the Mie-Hilbert theory of matter [H2]. On the other hand, Weyl's own revision of his matter concept starting in 1922/23 and his later contribution to Dirac's electron theory at the end of the 1920s transformed his early gauge-theoretic ideas in a way that made them much more accessible and capable of a piecemeal assimilation by physicists. This complicated development of change, transfer, and transformation from Weyl's early gauge geometry to the gauge theories of the 1960s is a different story, although it is linked to the young Weyl's reformulation of Riemann's geometry following his dreams of a philosophically more acceptable concept of the continuum.[23]

References

[EG]   Earman, John, and Glymour, Clark, *Einstein and Hilbert: Two months in the history of general relativity*, Arch. Hist. Exact Sci. **1** (1978), 291–308.

[F1]   Fichte, Johann Gottlieb, Grundriss des Eigenthümlichen der Wissenschaftslehre in Rücksicht auf das theoretische Vermögen. Jena – Leipzig 1802. Gesamtausgabe **I.3**. Hrsg. Reinhard Lauth, Hans Jacob. Stuttgart – Bad Cannstadt, 142–208.

[F2]   Fichte, Johann Gottlieb, *Ueber das Verhältnis der Logik zur Philosophie oder transcendentale Logik*. Nachgelassene Werke **1**. Hrsg. I. H. Fichte. Berlin 1834 (Nachdruck Berlin: de Gruyter 1971), 103–400.

[FS]   Frei, Günther, and Stammbach, Urs, Hermann Weyl und die Mathematik an der ETH Zürich, 1913–1930. Basel usw.: Birkhäuser 1992.

[H1]   Hilbert, David, *Die Grundlagen der Physik. Erste Mitteilung*, Nachrichten Gesellschaft der Wissenschaften Göttingen (1915), 395–407.

[H2]   Hilbert, David, *Referat über die geometrischen Schriften und Abhandlungen Hermann Weyls, erstattet der Physiko-Mathematischen Gesellschaft an der Universität Kasan*, Bulletin de la Société Phys. Math. de Kazan (3) **2** (1927), 66–70.

[H3]   Hilbert, David, *Natur und mathematisches Erkennen*, Vorlesung 1919–1920 Göttingen, nach der Ausarbeitung von P. Bernays. Hrsg. D. Rowe. Basel etc.: Birkhäuser 1992.

[McC] McCormach, Russell, *H. A. Lorentz and the electromagnetic view of nature*, Isis **61** (1970), 459–497.

---

[23]I owe my thanks to D. Rowe who did his best to make this paper more readable.

[M1]   Mie, Gustav, *Grundlagen einer Theorie der Materie. 3 Teile*, Ann. Physik **37** (1912), 511–534, **39** (1912), 1–40, **40** (1913), 1–66.

[M2]   Mie, Gustav, *Aus meinem Leben*, Zeitwende **19** (1948), 733–743.

[Rea]   Renn, Jürgen, e.a. *Das Züricher Notizbuch*, in G. Castagnetti, P. Damerow, W. Heinrich, J. Renn, T. Sauer. Wissenschaft zwischen Grundlagenkrise und Politik: Einstein in Berlin. Arbeitsbericht der Arbeitsstelle Albert Einstein. MPI für Bildungsforschung Berlin 1994, 24–57.

[R]   Ringer, Fritz, Die Gelehrten. Der Niedergang der deutschen Mandarine, 1890–1933, München: DTV 1987.

[S]   Sigurdsson, Skuli, *Hermann Weyl, Mathematics and physics, 1900–1927*, dissertation, Harvard University, Cambridge, MA, 1991.

[Sta]   Stachel, John, *Einstein's search for general covariance, 1912–1915*, in Howard, Don; Stachel, John, Einstein and the History of General Relativity, Basel etc.: Birkhäuser 1989, 63–100.

[Str]   Straumann, Norbert, *Zum Ursprung der Eichtheorien bei Hermann Weyl*, Physikalische Blätter **43** (11) (1987), 414–421.

[V]   Vizgin, Vladimir P., Unified Field Theories in the First Third of the 20th Century, Translated from the Russian by J. B. Barbour, Basel etc.: Birkhäuser 1994.

[W1]   Weyl, Hermann, *Reine Infinitesimalgeometrie*, Math. Z. **2** (1918), 384–411. [W9] **2**, 1–28 [30].

[W2]   Weyl, Hermann, *Gravitation und Elektrizität*, Sitzungsberichte Akademie der Wissenschaften Berlin, 465–480. [W9] **2** (1918), 29–42 [31].

[W3]   Weyl, Hermann, Das Kontinuum. Kritische Untersuchungen über die Grundlagen der Analysis, Leipzig 1918. Neudruck New York 1960.

[W4]   Weyl, Hermann, *Der circulus vitiosus in der heutigen Begründung der Analysis*, Jber. DMV **28** (1919), 85–92. [W9] **2**, 43–50 [32].

[W5]   Weyl, Hermann, *Die heutige Erkenntnislage in der Mathematik*, Symposion **1** (1925), 1–32. [W9] **2**, 511–542 [67].

[W6]   Weyl, Hermann, Riemanns geometrische Ideen, ihre Auswirkungen und ihre Verknüpfung mit der Gruppentheorie, Ed. K. Chandrasekharan, Berlin etc.: Springer 1988.

[W7]   Weyl, Hermann, Mathematische Analyse des Raumproblems, Vorlesungen gehalten in Barcelona und Madrid. Berlin etc.: Springer 1923, Neudruck Darmstadt (Wissenschaftliche Buchgesellschaft) 1963.

[W8]   Weyl, Hermann, *Erkenntnis und Besinnung*, Studia Philosophica 1954. [W9] **4**, 631–649. [166]

[W9]   Weyl, Hermann, Gesammelte Abhandlungen, 4 Bände, Hrsg. K. Chandrasekharan, Berlin etc.: Springer 1968.

Author Index